PE Environmental
Review

Michael R. Lindeburg, PE

PPI2PASS.COM
A KAPLAN COMPANY

Report Errors for This Book

PPI is grateful to every reader who notifies us of a possible error. Your feedback allows us to improve the quality and accuracy of our products. Report errata at **ppi2pass.com**.

PE ENVIRONMENTAL REVIEW

Current release of this edition: 4

Release History

date	edition number	revision number	update
Jun 2019	1	2	Minor corrections. Minor cover updates.
Jul 2020	1	3	Minor corrections.
Feb 2021	1	4	Minor corrections.

PPI
ppi2pass.com

ISBN: 978-1-59126-575-7

Topics

Water

Air

Solid & Hazard. Waste

Site Assessment & Remediation

Environmental Health & Safety

Assoc. Eng. Principles

Support Material

Table of Contents

Topic VI: Associated Engineering Principles

Topic VII: Support Material

Index

Appendix
Table of Contents

Preface

The purpose of this book, *PE Environmental Review* (or *Review*), is to prepare you for the Principles and Practice of Engineering (PE) exam for environmental engineering. The PE exam is developed, administered, and scored by the National Council of Examiners for Engineering and Surveying (NCEES). State and other local licensing boards use the PE exam as a way of uniformly assessing the preparation and competency of engineers practicing within their jurisdictions.

The NCEES PE environmental exam is a computer-based test. You may not bring your own reference books, notes, calculator, or scratch paper with you. During the test, your only reference for important formulas and data will be an electronic copy of the *NCEES PE Environmental Reference Handbook* (or *Handbook*). You will also use an on-screen calculator and write your calculations on a reusable notepad that is supplied to you.

To help you prepare for this challenge, the content in *Review* has been written with two main goals in mind: strengthening your mastery of exam knowledge areas, and familiarizing you with the contents and structure of the *Handbook*. The content in *Review* is organized into the same knowledge areas as the exam is, and equations and other data are annotated with their locations in the *Handbook*.

PE Environmental Review is just one of a collection of PPI study tools designed to get you ready for your exam day. The companion book *PE Environmental Practice* offers 500 exam-like problems organized into chapters that correspond to the chapters of this book. In the online PPI Learning Hub, you will find electronic versions of these books, diagnostic exams, full-length practice exams, and a quiz generator with hundreds more practice problems. Use the PPI Learning Hub to integrate these tools with a custom study plan that takes into account your exam date, how much time you have for study, and the subject areas you're already strong in and those you need more review in. To learn how, visit **ppi2pass.com**.

Acknowledgments

A team of subject matter experts was indispensable in helping me to prepare this edition: R. Wane Schneiter, PhD, PE; Sabeen Cochinwala, MS, PE, QSD/P, PMP; Michael Goldrich, MS, PE; and Julia Kolberg, BS, PE. Anil Acharya checked the calculations. My deepest thanks go to them.

Their technical expertise was further refined with the help of the PPI product management and editorial teams. I could not have completed this work without the help of Steve Shea, product manager; Megan Synnestvedt, senior product manager; Scott Marley, lead and senior copy editor; Tyler Hayes, senior copy editor; Scott Rutherford, copy editor; Kim Wimpsett, freelance copy editor and proofreader; Richard Iriye, copy editor and typesetter; Bradley Burch, production editor; Tom Bergstrom, technical illustrator and cover designer; Kelly Winquist and Kim Burton-Weisman, freelance proofreaders; Ellen Nordman, publishing systems specialist; Sam Webster, product data operations manager; Cathy Schrott, editorial operations manager; and Grace Wong, director of editorial operations.

Despite our best efforts, if you see errors as you make your way through the book, please bring such discoveries to PPI's attention through **ppi2pass.com/errata**. Valid submitted errors will be incorporated into future printings of this book.

Best wishes on your examination experience.

Michael R. Lindeburg, PE

Introduction

1. ASSESS, REVIEW, PRACTICE

PE Environmental Review (or *Review*) is written for one purpose, and one purpose only: to get you ready for the NCEES PE Environmental exam. Use it along with the other PPI PE Environmental study tools to assess, review, and practice until you pass your exam.

Assess

PE Environmental Review is designed to be one of a collection of study tools available at PPI Learning Hub (**ppi2pass.com**). To pinpoint the subject areas where you need more study, use the diagnostic exams on the PPI Learning Hub. How you perform on these diagnostic exams will tell you which topics you need to spend more time on and which you can review more lightly.

Review

PE Environmental Review and its companion book *PE Environmental Practice* (or *Practice*), by R. Wane Schneiter, PhD, PE, offer a complete review for the PE Environmental exam. The topics and chapters in both books correspond neatly to the PE Environmental exam specifications. In turn, the chapters in *Review* pair exactly with the chapters in *Practice*, making it easy to find practice problems on a subject area you have just studied. If you don't fully understand the solution to a problem in *Practice*, just turn to the corresponding chapter in *Review* to review the subject area.

Practice

PE Environmental Review and *PE Environmental Practice* are designed to be used together. *Practice* contains exam-like problems which are organized into the same knowledge areas as *Review*.

The PPI Learning Hub also features a quiz generator that contains hundreds of practice problems, letting you create quizzes on any subject area or combination of subject areas.

It's a very good idea to take a practice exam within a few weeks of your actual exam. This lets you concentrate on sharpening your test-taking skills without the distraction of rusty recall. The PPI Learning Hub offers online practice exams that simulate the exam-day experience with the same number of problems, a similar level of difficulty, and the same time limits and break periods as the actual exam.

2. ABOUT THE *NCEES PE ENVIRONMENTAL HANDBOOK*

When you take the computer-based exam, the only reference material you will have access to is the NCEES *PE Environmental Reference Handbook* (or *Handbook*). The *Handbook* contains approximately 200 pages of formulas, tables, charts, and other information for use on the exam.

Because you are allowed no other reference books, it is crucial that you study for the exam with the *Handbook* always at hand. You need to become proficient with the *Handbook*'s equations, even when they are arranged differently or use different variables or units from the forms you are used to. It is important to get used to the *Handbook*'s terminology, because these are the search terms you must use to locate things quickly. And you need to become familiar with how the *Handbook* is organized, so that you don't waste valuable exam time scrolling through the wrong section of the PDF in search of the information you need.

This book, then, is written to follow the *Handbook* as closely as possible. Equations are given in the same form as in the *Handbook* and with the same variables and units. Explanations and examples of how to use these equations use the same terminology.

Each time a *Handbook* equation is referenced or information is taken from a *Handbook* table or chart, the heading of the *Handbook* section is given in blue. For example,

> If necessary, h_f can be converted to an actual pressure drop in lbf/ft^2 or Pa by multiplying by the fluid density. [Energy Equation (Bernoulli)]
>
> $$\Delta p = h_f \times \rho g \qquad \text{[SI]} \qquad 2.54(a)$$
>
> $$\Delta p = h_f \times \rho\left(\frac{g}{g_c}\right) = h_f \gamma \qquad \text{[U.S.]} \qquad 2.54(b)$$
>
> The Hagen-Poiseuille equation can also be expressed for Q in terms of the pressure drop, ΔP_f.
>
> Pressure Drop for Laminar Flow
>
> $$Q = \frac{\pi R^4 \Delta P_f}{8\mu L} = \frac{\pi D^4 \Delta P_f}{128\mu L} \qquad 2.55$$

These phrases in blue are the search terms you can use on exam day to find information in the *Handbook* quickly. In a few cases, where the same heading (such as "Nomenclature") appears several times in the *Handbook*, a heading and a subheading are both given, separated by a colon—for example, Site Assessment and Remediation: Nomenclature. To find the information quickly, search on the first heading (before the colon), then scroll down until you see the second heading (after the colon).

It's important to become familiar with the *Handbook*'s usage, but this is made a little more difficult by the *Handbook*'s own inconsistency. In various parts of the *Handbook*, for example, the variable used for specific gravity is SG, $S.G.$, or S. The universal gas constant may be either $\bar{R}$ or R; the latter is also used for the specific gas constant. Volume may be either v or V; velocity may be either of these as well as an unitalicized v.

As much as possible, then, this book follows the usage of the *Handbook* section being discussed; but in areas where equations and other information are brought together from multiple parts of the *Handbook*, some small inconsistencies in usage could not be avoided.

3. HOW TO USE THIS BOOK

Learn to Use the *NCEES PE Environmental Handbook*

Download a PDF of the *NCEES PE Environmental Reference Handbook* from the NCEES website. As you solve the problems in *Review* and *Practice*, use the *Handbook* as your reference. Although you could print out the *Handbook* and use it that way, it will be better for your preparations if you use it in PDF form on your laptop or computer. This is how you will be referring to it and searching in it during the actual exam.

It is critical that you get to know what the *Handbook* includes and how to find what you need efficiently. Even if you know how to find the equations and data you need more quickly in other references, take the time to search for them in the *Handbook*. Get to know the terms and section titles used in the *Handbook* so that you can use them as your search terms during the exam.

A step-by-step solution is provided for each problem in *Review* and *Practice*. In these solutions, wherever an equation, a figure, or a table is used from the *Handbook*, the section heading from the *Handbook* is given in blue. Getting to know the content in the *Handbook* will save you valuable time on the exam.

When you take a practice exam at the PPI Learning Hub, you'll also be getting practice in using the *Handbook* under simulated exam conditions. During your practice exam, problems are displayed on one side of your computer screen and the searchable *Handbook* is displayed on the other, as in the actual exam.

Access the PPI Learning Hub

Although *PE Environmental Review* and *PE Environmental Practice* can be used on their own, they are designed to work with the PPI Learning Hub. In the PPI Learning Hub, you can study from online versions of both books side by side, displaying a chapter of *Review* on one side of your screen and the corresponding chapter of *Practice* on the other. You can test your understanding of what you've just studied by solving a practice problem on the same subject and, if you don't solve the problem correctly, you can read the step-by-step solution or return to *Review* to review concepts related to the solution.

At the PPI Learning Hub, you can also access

- a personal study plan, keyed to your exam date, to help keep you on track

- diagnostic exams to help you identify the subject areas where you are strong and where you need more review

- a quiz generator containing hundreds of additional exam-like problems that cover all knowledge areas on the PE environmental exam

- NCEES-like, computer-based practice exams to familiarize you with the exam day experience and let you hone your time management and test-taking skills

- electronic versions of *Review* and *Practice*

For more about the PPI Learning Hub, visit PPI's website at **ppi2pass.com**.

Be Thorough

Really do the work.

Time and again, customers ask us for the easiest way to pass the exam. The short answer is this: Pass it the first time you take it. Put the time in. Take advantage of the problems provided, and practice, practice, practice! Take the practice exams and time yourself so you will feel comfortable during the exam. When you are prepared you will know it. Yes, the reports in the PPI Learning Hub will agree with your conclusion but, most importantly, if you have followed the PPI study plan and done the work, it is more likely than not that you will pass the exam.

Some people think they can read a problem statement, think about it for 10 seconds, read the solution, and then say, "Yes, that's what I was thinking of, and that's what I would have done." Sadly, these people find out too late that the human brain makes many more mistakes when under time pressure, and that there are many ways to get messed up in solving a problem even if you understand the concepts. It may be in the use of your calculator, like using *log* instead of *ln* or forgetting to set the angle to radians instead of degrees. It may be rusty math, like forgetting exactly how to factor a polynomial. Maybe you can't find the conversion factor you need, or you don't remember what joules per kilogram is in SI base units.

For real exam preparation, you'll have to spend some time with a stubby pencil. You have to make these mistakes during your exam prep so that you do not make them during the actual exam. So do the problems—all of them. Do not look at the solutions until you have sweated a little.

EVERYTHING YOU EVER WANTED TO KNOW ABOUT THE PE EXAM

What Is the Format of the Exam?

The NCEES PE Environmental exam is a computer-based (CBT) test that contains 80 multiple-choice problems given over two consecutive sessions (sections, parts, etc.). The problems require a variety of approaches and methodologies, and you must answer all problems in each session correctly to receive full credit. There are no optional problems.

The exam is nine hours long, and includes a tutorial and an optional scheduled break. The actual time you will have to complete the exam problems is eight hours.

Exams are given year round at a Pearson VUE test center. The exam is closed book, with an electronic reference, the *NCEES PE Environmental Reference Handbook.*

What Subjects Are on the Exam?

NCEES has published a description of subjects on the examination. Irrespective of the published examination structure, the exact number of problems that will appear in each subject area cannot be predicted reliably.

There is no guarantee that any single subject will occur in any quantity. One of the reasons for this is that some of the problems span several disciplines. You might consider a pump selection problem to come from the subject of fluids, while someone else might categorize it as sanitary engineering.

Table 1 lists the subject areas and topics on the current exam specifications and the number of questions on the exam for each area. Most examinees find the list to be formidable in appearance. NCEES adds,

> The exam is developed with problems that will require a variety of approaches and methodologies including design, analysis, application, and operations. The knowledge areas specified as examples of kinds of knowledge are not exclusive or exhaustive categories.

What Is the Typical Problem Format?

Almost all of the problems are stand-alone—that is, they are completely independent. Problem types include traditional multiple-choice problems, as well as alternative item types (AITs). AITs include, but are not limited to

- multiple correct, which allows you to select multiple answers
- point and click, which requires you to click on a part of a graphic to answer
- drag and drop, with requires you to click on and drag items to match, sort, rank, or label
- fill in the blank, which provides a space for you to enter a response to the problem

Although AITs are a recent addition to the PE Environmental exam and may take some getting used to, they are not inherently difficult to master. For your reference, additional AIT resources are available on the PPI Learning Hub (**ppi2pass.com**).

Traditional multiple-choice problems will have four answer options, labeled A, B, C, and D. If the four answer options are numerical, they will be displayed in increasing value. One of the answer options is correct (or will be "most nearly correct," as described in a later section). The remaining answer options are incorrect and may consist of one or more "logical distractors," the term used by NCEES to designate incorrect options that look plausibly correct.

NCEES intends the problems to be unrelated. Questions are independent or start with new given data. A mistake on one of the problems shouldn't cause you to get a subsequent problem wrong. However, considerable time may be required to repeat previous calculations with a new set of given data.

The NCEES Nondisclosure Agreement

At the beginning of your CBT experience, a nondisclosure agreement will appear on the screen. In order to begin the exam, you must accept the agreement within two minutes. If you do not accept within two minutes, your CBT experience will end, and you will forfeit your appointment and exam fees. The nondisclosure agreement, as stated in the NCEES Examinee Guide, is as follows.

> This exam is confidential and secure, owned and copyrighted by NCEES and protected by the laws of the United States and elsewhere. It is made available to you, the examinee, solely for valid assessment and licensing purposes. In order to take this exam, you must agree not to disclose, publish, reproduce, or transmit this exam, in whole or in part, in any form or by any means, oral or written,

Table 1 *PE Environmental Exam Specifications*

I. Water: 21 to 35 questions

A. principles (hydraulics and fluid mechanics; chemistry; biology and microbiology; fate and transport; sampling and measurement methods; hydrology and hydrogeology; codes, standards, regulations, and guidelines): 3 to 5 questions

B. wastewater (sources of pollution; minimization and prevention; treatment technologies and management; collection systems; residuals (sludge) management; water reuse): 7 to 11 questions

C. stormwater (sources of pollution; treatment technologies and management; collection systems): 2 to 4 questions

D. potable water (source water quality; treatment technologies and management; distribution systems; residuals management—solid, liquid, and gas): 7 to 11 questions

E. water resources (sources of pollution; watershed management and planning; source supply and protection): 2 to 4 questions

II. Air: 14 to 22 questions

A. principles (sampling and measurement methods; codes, standards, regulations, and guidelines; chemistry; fate and transport; atmospheric science and meteorology): 7 to 11 questions

B. pollution control (sources of pollution; characterization, calculations, and inventory of emissions; treatment and control technologies; pollution minimization and prevention): 7 to 11 questions

III. Solid and Hazardous Waste: 11 to 18 questions

A. principles (chemistry; fate and transport; codes, standards, regulations, and guidelines; risk assessment; sampling and measurement methods; minimization, reduction, and recycling; mass and energy balance; hydrology, hydrogeology, and geology): 5 to 8 questions

B. municipal and industrial solid waste (storage, collection, and transportation systems; treatment and disposal technologies and management): 4 to 6 questions

C. hazardous, medical, and radioactive waste (storage, collection, and transportation systems; treatment and disposal technologies and management): 2 to 4 questions

IV. Site Assessment and Remediation: 12 to 19 questions

A. principles (codes, standards, regulations, and guidelines; chemistry and biology; hydrology and hydrogeology; sampling and measurement methods): 5 to 8 questions

B. applications (site assessment and characterization; risk assessment; fate and transport; remediation alternative identification; remediation technologies and management): 7 to 11 questions

V. Environmental Health and Safety: 7 to 11 questions

A. principles (health and safety; security, emergency plans, and incident response procedures; codes, standards, regulations, and guidelines): 3 to 5 questions

B. applications (industrial hygiene; exposure assessments; indoor air quality): 4 to 6 questions

VI. Associated Engineering Principles: 5 to 9 questions

A. principles (statistics; sustainability): 2 to 4 questions

B. applications (engineering economics, project management, mass and energy balance; data management): 3 to 5 questions

electronic or mechanical, for any purpose, without the prior express written permission of NCEES. This includes agreeing not to post or disclose any test questions or answers from this exam, in whole or in part, on any websites, online forums, or chat rooms, or in any other electronic transmissions, at any time.

Your Exam Is Unique

The exam that you take will not be the exam taken by the person sitting next to you. Differences between exams go beyond mere sequencing differences. NCEES says that the CBT system will randomly select different, but equivalent, problems from its database for each examinee using a linear-on-the-fly (LOFT) algorithm. Each examinee will have a unique exam of equivalent difficulty. That translates into each examinee having a slightly different minimum passing score.

There is no way to determine exactly how NCEES ensures that each examinee is given an equivalent exam. All that can be said is that looking at your neighbor's monitor would be a waste of time.

The Exam Interface

The on-screen exam interface contains only minimal navigational tools. On-screen navigation is limited to selecting an answer, advancing to the next problem, going back to the previous problem, and flagging the current problem for later review. The interface also includes a timer, the current problem number (e.g., 45 of 110), a pop-up scientific calculator, and access to an on-screen version of the *Handbook*. During the exam, you can advance sequentially through the problems, but you cannot jump to any specific problem, whether or not it has been flagged. After you have completed the last problem in a session, however, the navigation capabilities change, and you are permitted to review problems in any sequence and navigate to flagged problems.

The *Handbook* Interface

Examinees are provided with a 24-inch computer monitor that will simultaneously display both the exam problems and a searchable PDF of the *Handbook*. The PDF's table of contents consists of live links. The search function is capable of finding anything in the *Handbook*, down to and including individual variables. However, the search function finds only precise search terms (e.g., "Dupuit Formula" will not locate "Dupuit's Formula").

What Does "Most Nearly" Really Mean?

One of the more disquieting aspects of these problems is that the available answer choices for problems requiring math are seldom exact. Numerical answer choices generally have only two or three significant digits. Exam problems ask, "Which answer choice is most nearly the correct value?" or they instruct you to complete the sentence, "The value is approximately... " A lot of self-

confidence is required to move on to the next problem when you don't find an exact match for the answer you calculated, and you have had to split the difference because no available answer choice is really close.

NCEES describes it like this:

> Many of the problems on NCEES exams require calculations to arrive at a numerical answer. Depending on the method of calculation used, it is very possible that examinees working correctly will arrive at a range of answers. The phrase "most nearly" is used to accommodate answers that have been derived correctly but which may be slightly different from the correct answer choice given on the exam. You should use good engineering judgment when selecting your choice of answer. For example, if the problem asks you to calculate an electrical current or determine the load on a beam, you should literally select the answer option that is most nearly what you calculated, regardless of whether it is more or less than your calculated value. However, if the problem asks you to select a fuse or circuit breaker to protect against a calculated current or to size a beam to carry a load, you should select an answer option that will safely carry the current or load. Typically, this requires selecting a value that is closest to but larger than the current or load.

The difference is significant. Suppose you were asked to calculate "most nearly" the volumetric pure airflow required to dilute a contaminated air stream to an acceptable concentration. Suppose, also, that you calculated 823 cfm. If the answer choices were (A) 600 cfm, (B) 800 cfm, (C) 1000 cfm, and (D) 1200 cfm, you would go with answer choice (B), because it is most nearly what you calculated. If, however, you were asked to select a fan or duct with the same rated capacities, you would have to go with choice (C). Got it?

How Much Mathematics Is Needed for the Exam?

There are no pure mathematics problems (algebra, geometry, trigonometry, etc.) on the exam. However, you will need to apply your knowledge of these subjects to the exam problems.

Generally, only simple algebra, trigonometry, and geometry are needed on the PE exam. You will need to use the trigonometric, logarithm, square root, exponentiation, and similar buttons on your calculator. There is no need to use any other method for these functions.

Except for simple quadratic equations, you will probably not need to find the roots of higher-order equations. Occasionally, it will be convenient to use the equation-solving capability of an advanced calculator. However, other solution methods will always exist. For second-order (quadratic) equations, the exam does not care if you find roots by factoring, completing the square, using the quadratic equation, or using your calculator's root finder.

There is essentially no use of calculus on the exam. Rarely, you may need to take a simple derivative to find a maximum or minimum of some simple algebraic function. Even rarer is the need to integrate to find an average.

There is essentially no need to solve differential equations. Problems involving radioactive decay and fluid mixing have appeared from time to time. However, these applications can usually be handled without having to solve differential equations.

Basic statistical analysis of observed data may be necessary. Statistical calculations are generally limited to finding means, medians, standard deviations, variances, and confidence limits. The only population distribution you need to be familiar with is the normal curve. Probability, reliability, hypothesis testing, and statistical quality control are not explicit exam subjects.

The PE exam is concerned with numerical answers, not with proofs or derivations. You will not be asked to prove or derive formulas, use deductive reasoning, or validate theorems, corollaries, or lemmas.

Occasionally, a calculation may require an iterative solution method. Generally, there is no need to complete more than two iterations. You will not need to obtain an "exact" answer, nor will you generally need to use complex numerical methods. Inasmuch as first assumptions can significantly affect the rate of convergence, problems requiring trial-and-error solutions are unlikely.

How About Engineering Economics?

For most of the early years of engineering licensing, problems on engineering economics appeared frequently on the examinations. This is no longer the case. However, the degree of engineering economics knowledge may have decreased somewhat, the basic economic concepts (e.g., time value of money, present worth, nonannual compounding, comparison of alternatives, etc.) are still valid test subjects.

If engineering economics is incorporated into other questions, its "disguise" may be totally transparent. For example, you might need to compare the economics of buying and operating two blowers for remediation of a hydrocarbon spill—blowers whose annual costs must be calculated from airflow rates and heads.

What About Professionalism and Ethics?

For decades, NCEES has considered adding professionalism and ethics problems to the PE exam. However, these subjects are not parts of the test outline, and there has yet to be an ethics problem on the exam.

Is the Exam Tricky?

Other than providing superfluous data, the PE exam is not a "tricky exam." It does not overtly try to get you to fail. Examinees manage to fail on a regular basis with perfectly straightforward problems. The exam questions are difficult in their own right. NCEES does not need to provide misleading or conflicting statements. However, you will find that commonly made mistakes are represented in the available answer choices. Thus, the alternative answers (known as *distractors*) will be logical.

Problems are generally practical, dealing with common and plausible situations that you might experience in your job. You will not be asked to analyze trickling sludge digesters on Mars, to design an air filter manufactured from bamboo, or to develop PELs from a statistical analysis of meal worm deaths.

Does NCEES Write Exam Problems Around This Book?

Only NCEES knows what NCEES uses to write its exam problems. However, it is irrelevant, because this book is not intended to (1) be everything you need to pass the exam, (2) expose exam secrets or exam questions, or (3) help you pass when you don't deserve to pass. NCEES knows about this book, but worrying about NCEES writing exam problems based on information that is or is not in this book means you are placing too much dependency on this book. This book, for example, will provide instruction in certain principles. Expecting that you will not need to learn anything else is unrealistic. This book presents many facts, definitions, and numerical values. Expecting that you will not need to know other facts, definitions, and numerical values is unrealistic. What NCEES uses to write exam problems won't have any effect on what you need to do to prepare for the exam.

Does the PE Exam Use SI Units?

The PE exam in environmental engineering requires working in both customary U.S. units (also known as "English units," "inch-pound units," and "British units") and a variety of metric systems, including SI. Problems use the units that correspond to commonly accepted industry standards. And, in this regard, the standard is a mixture of units. Some problems, such as those covering fluid and air-handling, use units of pounds, feet, seconds, gallons, and degrees Fahrenheit. Metric units are used in most subjects involving mixtures and chemical concentrations as well as water supply and wastewater concentration (mg/L) problems.

The exam does not differentiate between lbf and lbm (pounds-force and pounds-mass) as is done in this book. However, the environmental exam does not cover many subjects that involve units of force. Therefore, the distinction probably does not have to be made very often.

Why Does NCEES Reuse Some Problems?

NCEES reuses some of the more reliable problems from each exam. The percentage of repeat problems isn't high—no more than 25% of the exam. NCEES repeats problems in order to equate the performance of one group of examinees with the performance of an earlier group. The repeated problems are known as *equaters*, and together, they are known as the *equating subtest*.

Occasionally, a new problem appears on the exam that very few of the examinees do well on. Usually, the reason for this is that the subject is too obscure or the question is too difficult. Also, there have been cases where a low percentage of the examinees gets the answer correct because the problem was inadvertently stated in a poor or confusing manner. Problems that everyone gets correct are also considered defective.

NCEES tracks the usage and "success" of each of the exam problems. "Rogue" problems are not repeated without modification. This is one of the reasons historical analysis of problem types shouldn't be used as the basis of your review.

Does NCEES Use the Exam to Pre-Test Future Problems?

NCEES does not use the PE exam to "pre-test" or qualify future problems. (It does use this procedure on the FE exam, however.) All of the problems you work will contribute toward your final score.

Are the Example Problems in This Book Representative of the Exam?

The example problems in this book are intended to be instructional and informative. They were written to illustrate how their respective concepts can be implemented. Example problems given with the *Handbook* content (i.e., example problems with blue labels) are intended to be solved using the *Handbook*, and serve as examples of how you might need to use the *Handbook* during the exam.

What Reference Material Is Permitted in the Exam?

The PE examination is a closed-book exam. Your only reference for the exam is the *Handbook*. You will be provided an electronic copy of the *Handbook* that you can access during your exam.

What About Calculators?

You may bring one NCEES-approved calculator, without its cover, into the testing room. NCEES maintains a list of approved calculators on its website. It is a good idea to bring a spare calculator and extra batteries just in case, but these must be left outside the testing room along with the cover from your first calculator. Most testing centers provide small lockers for storing personal items not permitted in the testing room.

Ideally, your spare calculator should be the same model as your first one. If it isn't, spend plenty of time solving practice problems with it. You should be as comfortable and familiar with it as you are with your first calculator.

Are Cell Phones Permitted?

You may not possess or use a walkie-talkie, cell phone, smartphone, or other communications or text-messaging device during the exam, regardless of whether it is on. You won't be frisked upon entrance to the exam, but should a proctor discover that you are in possession of a communication device, you should expect to be politely excluded from the remainder of the examination.

How Is the Exam Graded and Scored?

The maximum number of points you can earn on the PE Environmental exam is 80. The minimum number of points for passing (referred to by NCEES as the *cut score*) varies from exam to exam. The cut score is determined through a rational procedure, without the benefit of knowing examinees' performance on the exam. That is, the exam is not graded on a curve. The cut score is selected on the basis of what you are expected to know, not on the basis of passing a certain percentage of engineers.

Each of the problems is worth one point. Grading is straightforward, since a computer grades your score sheet. You either get the problem right or you don't. However, if you mark two or more answers for the same problem, no credit is given for the problem.

Your score may or may not be revealed to you, depending on your state's procedure. Even if the score is reported to you, it may have been scaled or normalized to 100%. It may be difficult to determine whether the reported score is out of 80 or is out of 100.

How You Should Guess

There is no deduction for incorrect answers, so guessing is encouraged. NCEES produces defensible licensing exams, so there is no pattern to the placement of correct responses. Since the quantitative responses are sequenced according to increasing values, the placement of a correct answer among other numerical distractors is a function of the distractors, not of some statistical normalizing routine. Therefore, it is not important whether you randomly guess all "A," "B," "C," or "D" when you get into guessing mode during the last minute or two of the exam.

The proper way to guess is as an engineer. You should use your knowledge of the subject to eliminate illogical answer choices. Illogical answer choices are those that violate good engineering principles, that are outside normal operating ranges, or that require extraordinary assumptions. Of course, this requires you to have some basic understanding of the subject in the first place. Otherwise, it's back to random guessing. That's the reason that the minimum passing score is higher than 25%.

You won't get any points using the "test-taking skills" that helped you in college—the skills that helped with tests prepared by amateurs. You won't be able to eliminate any [verb] answer choices from "Which [noun]..." problems. You won't find problems with options of the "more than 50" and "less than 50" variety. You won't find one answer choice among the four that has a different number of significant digits, or has a verb in a different tense, or has some singular/plural discrepancy with the stem. The distractors will always match the stem, and they will be logical.

How Is the Cut Score Established?

The raw cut score may be established by NCEES before or after the exam is administered. Final adjustments may be made following the exam date.

NCEES uses a process known as the modified *Angoff procedure* to establish the cut score. This procedure starts with a small group (the cut score panel) of professional engineers and educators selected by NCEES. Each individual in the group reviews each problem and makes an estimate of its difficulty. Specifically, each individual estimates the number of minimally qualified engineers out of a hundred examinees who should know the correct answer to the problem. (This is equivalent to predicting the percentage of minimally qualified engineers who will answer correctly.)

Next, the panel assembles, and the estimates for each problem are openly compared and discussed. Eventually, a consensus value is obtained for each. When the panel has established a consensus value for every problem, the values are summed and divided by 100 to establish the cut score.

Various minor adjustments can be made to account for examinee population (as characterized by the average performance on any equater problems) and any flawed problems. Rarely, security breaches result in compromised problems or examinations. How equater questions, examination flaws, and security issues affect examinee performance is not released by NCEES to the public.

Cheating and Exam Subversion

There aren't very many ways to cheat on a computer-based test. The proctors are well trained in spotting the few ways that do exist. It goes without saying that you should not talk to other examinees in the room, nor should you pass notes back and forth. You should not take notes on the contents of the exam. You shouldn't use your cell phone. The number of people who are released to use the restroom may be limited to prevent discussions.

NCEES regularly reuses good problems that have appeared on previous exams. Therefore, examination integrity is a serious issue with NCEES, which goes to great lengths to make sure nobody copies the problems.

The proctors are concerned about exam subversion, which generally means any activity that might invalidate the examination or the examination process. The most common form of exam subversion involves trying to copy exam problems for future use.

NCEES has become increasingly unforgiving about loss of its intellectual property. NCEES routinely prosecutes violators and seeks financial redress for loss of its examination problems, as well as invalidating any engineering license you may have earned by taking one of its examinations while engaging in prohibited activities. Your state board may impose additional restrictions on your right to retake any examination if you are convicted of such activities. In addition to tracking down the sources of any examination problem compilations that it becomes aware of, NCEES is also aggressive in pursuing and prosecuting examinees who disclose the contents of the exam in internet forum and "chat" environments. Your constitutional rights to free speech and expression will not protect you from civil prosecution for violating the nondisclosure agreement that NCEES requires you to sign before taking the examination. If you wish to participate in a dialog about a particular exam subject, you must do so in such a manner that does not violate the essence of your nondisclosure agreement. This requires decoupling your discussion from the examination and reframing the problem to avoid any examination particulars.

HOW TO PREPARE FOR AND PASS THE PE EXAM

What Should You Study?

The exam covers many diverse subjects. Strictly speaking, you don't have to study every subject on the exam in order to pass. However, the more subjects you study, the greater your chances of passing. You should decide early in the preparation process which subjects you are going to study.

If you're not sure where you should be concentrating your studying efforts, use the diagnostic exams on the PPI Learning Hub as a tool to help assess your strengths and weaknesses. Your performance on these exams will tell you which topics you need to devote more time to learning and which you can review more lightly.

Do You Need a Classroom Prep Course?

Approximately 60% of first-time PE examinees take a prep course of some form. Live classroom, audio, video, and internet courses of various types are available for some or all of the exam topics. Live courses and instructor-moderated internet courses provide several significant advantages over self-directed study, some of which may apply to you.

- A course structures and paces your review. It keeps you going forward without getting bogged down in one subject.

- A course focuses you on a limited amount of material. Without a course, you may not know which subjects to study.

- A course provides you with the problems you need to solve. You won't have to spend time looking for them.

- The course instructor can answer your problems when you are stuck.

How Long Should You Study?

We've all heard stories of the person who didn't crack a book until the week before the exam and still passed it with flying colors. Yes, these people really exist. But I'm not one of them, and you probably aren't either. In fact, after having taught thousands of engineers in my own classes, I'm convinced that these people are as rare as the ones who have taken the exam five times and still can't pass it.

A thorough review takes approximately 300 hours. Most of this time is spent solving problems. Some of it may be spent in class; some is spent at home. Some examinees spread this time over a year. Others try to cram it all into two months. Most classroom prep courses last for three or four months. The best time to start studying will depend on how much time you can spend per week.

Should You Look for Old Exams?

The traditional approach to preparing for standardized tests includes working sample tests. However, NCEES does not release old tests or problems after they are used. Therefore, there are no official problems or tests available from legitimate sources. NCEES has published a booklet of sample problems and solutions to illustrate the format of the exam. However, these questions have been compiled from various previous exams, and the resulting publication is not a true "old exam." Furthermore, NCEES sometimes constructs its sample problems books from problems that have been pulled from active use for various reasons, including poor performance. Such marginal problems, while accurately reflecting the format of the examination, are not always representative of actual exam subjects.

What Should You Memorize?

You get lucky here, because it isn't necessary to actually memorize anything. The *NCEES Handbook* will be available to you for the duration of the exam, so you can look up most procedures, formulae, or pieces of information you might need. You can speed up your problem-solving response time significantly if you don't have to look up the conversion from Btu/sec to horsepower, the definition of the sine of an angle, and the chemical formula for carbon dioxide, but you don't even have to memorize these kinds of things. As you work practice problems in *PE Environmental Practice*, you will automatically memorize the things that you come across more than a few times.

Do You Need a Study Plan?

It is important that you develop and adhere to a review outline and schedule. Once you have decided which subjects you are going to study, you can allocate the available time to those subjects in a manner that makes sense to you. If you are not taking a classroom prep course (where the order of preparation is determined by the lectures), you should make an outline of subjects for self-study to use for scheduling your preparation.

If you want a head start in making a plan, you'll have access to an interactive, adjustable, and personalized study plan at the PPI Learning Hub, **ppi2pass.com**. You can input the days you personally have available to study for the test (even exempting days that you know you will not be able to study, like vacations or holidays) and it will produce a plan you can follow as you prepare. Log on to your PPI account to access your custom study plan.

How You Can Make Your Review Realistic

In the exam, you must be able to quickly recall solution procedures, formulas, and important data. You must remain sharp for eight hours or more. When you played a sport back in school, your coach tried to put you in game-related situations. Preparing for the PE exam isn't much different from preparing for a big game. Some part of your preparation should be realistic and representative of the examination environment.

Learning to use your time wisely is one of the most important lessons you can learn during your review. You will undoubtedly encounter problems that end up taking much longer than you expected. In some instances, you will cause your own delays by spending too much time looking through the *Handbook* for things you need. Other times, the problems will just entail too much work. Learn to recognize these situations so that you can make an intelligent decision about skipping such problems in the exam.

How to Solve Multiple-Choice and AIT Problems

When you begin each session of the exam, observe the following suggestions.

- Do not spend an inordinate amount of time on any single problem. If you have not answered a problem in a reasonable amount of time, make a note of it and move on.

- Set your wristwatch alarm for five minutes before the end of each four-hour session, and use that remaining time to guess at all of the remaining problems. Odds are that you will be successful with about 25% of your guesses, and these points will more than make up for the few points that you might earn by working during the last five minutes.

- Make mental notes about any problems for which you cannot find a correct response, which appears to have two correct responses, or which you believe have some technical flaw. Errors in the exam are rare, but they do occur. Such errors are almost always discovered during the scoring process and discounted from the examination, so it is not necessary to tell your proctor, but be sure to mark the one best answer before moving on.

Solve Problems Carefully

Many points are lost to carelessness. Keep the following items in mind when you are solving the problems. Hopefully, these suggestions will be automatic in the exam.

[] Did you recheck your mathematical equations?

[] Do the units cancel out in your calculations?

[] Did you convert between radius and diameter?

[] Did you convert between feet and inches?

[] Did you convert from gage to absolute pressures?

[] Did you convert between kPa and Pa?

[] Did you use the universal gas constant that corresponds to the set of units used in the calculation?

[] Did you recheck all data obtained from other sources, tables, and figures? (In finding the friction factor, did you enter the Moody diagram at the correct Reynolds number?)

What to Do a Few Days Before the Exam

There are a few things that you should do a week or so before the examination. You should make arrangements for childcare and transportation. Since the examination does not always start or end at the designated time, make sure that your childcare and transportation arrangements are flexible.

Check PPI's website for last-minute updates and errata to this book.

If you haven't already done so, read the "Advice from Past Examinees" section of PPI's website.

Make sure your spare calculator is in good working order. Make sure you have fresh spare batteries for both of your calculators.

If it is convenient, visit the exam location in order to find the building, parking areas, examination room, and restrooms. If it is not convenient, you may find driving directions and/or site maps on the web.

What to Do the Day Before the Exam

Take the day before the examination off from work to relax. Do not cram the last night. A good night's sleep is the best way to start the examination. If you live a considerable distance from the examination site, consider getting a hotel room in which to spend the night.

Calculate your wake-up time and set the alarms on two bedroom clocks. Select and lay out your clothing items. (Dress in layers.) Select and lay out your breakfast items.

Make sure you have gas in your car and money in your wallet.

What to Do the Day of the Exam

Turn off the quarterly and hourly alerts on your wristwatch. Leave your cell phone in the car. If you must bring it, you'll need to leave it in the locker and set it to silent or off.

You should arrive at least 30 minutes before the examination starts. This will allow time for finding a convenient parking place, getting to the examination room, and calming down.

Should You Talk to Other Examinees After the Exam?

The jury is out on this problem. People react quite differently to the examination experience. Some people are energized. Most are exhausted. Some people need to unwind by talking with other examinees, describing every detail of their experience, and dissecting every examination problem. Others need lots of quiet space, and prefer just to get into a hot tub to soak and sulk. Most engineers, apparently, are in this latter category.

Since everyone who took the exam has seen it, you will not be violating your "oath of silence" if you talk about the details with other examinees immediately after the exam. It's difficult not to ask how someone else approached a problem that had you completely stumped. However, keep in mind that it is very disquieting to think you answered a problem correctly, only to have someone tell you where you went wrong.

To ensure you do not violate the nondisclosure agreement you signed before taking the exam, make sure you do not discuss any exam particulars with people who have not also taken the exam.

After the Exam

Here's what I suggest you do as soon as you get home, before you collapse.

[ ] Thank your partner and children for helping you during your preparation.

[ ] Take any paperwork you received on exam day out of your pocket, purse, or wallet. Put this inside your copy of *PE Environmental Review*.

[ ] Reflect on any statements regarding exam secrecy to which you signed your agreement in the exam.

[ ] If you participated in a PPI Prep Course, log on one last time to thank the instructors. (Prep courses remain open for a week after the exam.)

[ ] Call your employer and tell him/her that you need to take a mental health day off.

A few days later, when you can face the world again, do the following.

[ ] Make notes about anything you would do differently if you had to take the exam over again.

[ ] Consolidate all of your application paperwork, correspondence to/from your state, and any paperwork that you received on exam day.

[ ] If you took a live prep course, call or email the instructor (or write a note) to say "Thanks."

[ ] Return any books you borrowed.

[ ] Write thank-you notes to all of the people who wrote letters of recommendation or reference for you.

Finally

By the time that you've "undone" all of your preparations, you might have thought of a few things that could help future examinees. If you have any sage comments about how to prepare, any suggestions about what to do in or bring to the exam, any comments on how to improve this book, or any funny anecdotes about any aspect of your experience, I hope you will share these with me. By this time, you'll be the "expert," and I'll be your biggest fan.

And Then, There's the Wait...

Waiting for the exam results is its own form of mental torture.

Yes, I know the exam is 100% online, and grading should be almost instantaneous. But you are going to wait, nevertheless. Exam results are typically available 7–10 days after you take the exam. You will receive an email notification from NCEES with instructions to view your results in your MyNCEES account. You will also receive information specific to your licensing board about how you should proceed depending on your performance.

And When You Pass...

[ ] Celebrate.

[ ] Notify the people who wrote letters of recommendation or reference for you.

[ ] Read "FAQs About What Happens After You Pass the Exam" on PPI's website.

[ ] Ask your employer for a raise.

[ ] Tell the folks at PPI (who have been rootin' for you all along) the good news.

IF YOU ARE AN INSTRUCTOR

If you are teaching a prep course for the PE examination without the benefit of recent, first-hand exam experience, you can use the material in this book as a guide to prepare your lectures. You should emphasize the subjects in each chapter and avoid subjects that have been omitted. You can feel confident that subjects omitted from this book have rarely, if ever, appeared on the PE exam.

I have always tried to overprepare my students. For that reason, the non-NCEES example problems are often more difficult and varied than actual examination problems. Also, you will appreciate the fact that it is more efficient to cover several procedural steps in one example problem than to ask simple "one-liners" or definition problems. That is the reason the example problems are often longer and harder than actual exam problems. In your course, plan to provide both the longer, more comprehensive example problems, as well as the shorter, NCEES-exam like example problems.

There are many practice problems for each major examination subject in this book's companion, *PE Environmental Practice*. All problems are assigned in my prep courses. To do all the problems and answer all of the problems requires approximately 15 to 20 hours of preparation per week for approximately 14 weeks.

"Capacity assignment" is the goal in my prep courses. If you assign 20 hours of homework and a student is able to put in only 10 hours of preparation that week, that student will have worked to his or her capacity. After the PE examination, that student will honestly say that he or she could not have prepared any more than he or she did in your course. For that reason, you have to assign homework on the basis of what is required to become proficient in the subjects of your lecture. You must resist assigning only the homework that you think can be completed in an arbitrary number of hours.

Homework assignments in my courses are not individually graded. Instead, students are permitted to make use of existing solutions to learn procedures and techniques to the problems in their homework set, such as those in the accompanying *Practice*, which contains solutions to all practice problems. However, each student must turn in a completed set of problems for credit each week. Though I don't correct the homework problems, I address special needs or problems written on the assignments.

I believe that students should start preparing for the PE exam at least six months before the examination date. However, most wait until three or four months before getting serious. Because of that, I have found that a 14-week format works well for a PE prep course. It's a little rushed, but the course is over before everyone gets bored with my jokes. Each week there is a three-hour meeting, which includes lecture and a short break. However, I don't think you can cover the full breadth of material in much less time or in many fewer weeks.

I have tried to order the subjects in a logical, progressive manner, keeping my eye on "playing the high-probability subjects." I cover the subjects that everyone can learn (e.g., fluids) early in the course. I leave the subjects that only daily practitioners should attempt to the end.

Lecture coverage of some examination subjects is necessarily brief; other subjects are not covered at all. These omissions are intentional; they are not the result of scheduling omissions. Why? First, time is not on our side in a prep course. Second, some subjects rarely contribute to the examination. Third, some subjects are not well received by the students. For example, I have found that very few people try to become proficient in process control theory if they don't already work in that area.

All the skipped chapters and any related practice problems are presented as floating assignments to be made up in the students' "free time."

I strongly believe in exposing my students to a realistic sample examination, but I no longer administer an in-class mock exam. Since the prep course usually ends only a few days before the real PE examination, I hesitate to make students sit for several hours in the late evening to take a "final exam." Rather, I assign a take-home sample exam at the first meeting of the prep course.

Even better, point students to PPI's online practice exam on the PPI Learning Hub. This exam simulates the exam experience so students can familiarize themselves with the format and type of content that they'll encounter in the PE Environmental exam.

If the practice test is to be used as an indication of preparedness, caution your students not to even look at the sample exam prior to taking it. Looking at the sample exam, or otherwise using it to direct their review, will produce unwarranted specialization in subjects contained in the sample exam.

There are many ways to organize a PE prep course depending on your available time, budget, intended audience, facilities, and enthusiasm. However, all good course formats have the same result: the students struggle with the workload during the course, and then they breeze through the examination after the course.

Topic I: Water

Chapter

Fluid Properties and Statics

Content in blue refers to the *NCEES Handbook*.

1. CHARACTERISTICS OF A FLUID

Liquids and gases can both be categorized as fluids, although this chapter is primarily concerned with incompressible liquids. There are certain characteristics shared by all fluids, and these characteristics can be used, if necessary, to distinguish between liquids and gases.[1]

- *compressibility:* Liquids are only slightly compressible and are assumed to be incompressible for most purposes. Gases are highly compressible.

- *shear resistance:* Liquids and gases cannot support shear, and they deform continuously to minimize applied shear forces.

- *shape and volume:* As a consequence of their inability to support shear forces, liquids and gases take on the shapes of their containers. Only a liquid will form a free surface. Liquids have fixed volumes, regardless of their container volumes, and these volumes are not significantly affected by temperature and pressure. Unlike liquids, gases take on the volumes of their containers. If allowed to do so, the density of a gas will change as temperature and pressure are varied.

- *resistance to motion:* Due to viscosity, liquids resist instantaneous changes in velocity, but the resistance stops when liquid motion stops. Gases have very low viscosities.

- *molecular spacing:* Molecules in liquids are relatively close together and are held together with strong forces of attraction. Liquid molecules have low kinetic energy. The distance each liquid molecule travels between collisions is small. In gases, the molecules are relatively far apart, and the attractive forces are weak. Kinetic energy of the molecules is high. Gas molecules travel larger distances between collisions.

- *pressure:* The pressure at a point in a fluid is the same in all directions. Pressure exerted by a fluid on a solid surface (e.g., container wall) is always normal to that surface.

[1]The differences between liquids and gases become smaller as temperature and pressure are increased. Gas and liquid properties become the same at the critical temperature and pressure.

2. TYPES OF FLUIDS

For computational convenience, fluids are generally divided into two categories: ideal fluids and real fluids. (See Fig. 1.1.) *Ideal fluids* are assumed to have no viscosity (and therefore, no resistance to shear), be incompressible, and have uniform velocity distributions when flowing. In an ideal fluid, there is no friction between moving layers of fluid, and there are no eddy currents or turbulence.

Real fluids exhibit finite viscosities and nonuniform velocity distributions, are compressible, and experience friction and turbulence in flow. Real fluids are further divided into *Newtonian fluids* and *non-Newtonian fluids*, depending on their viscous behavior. The differences between Newtonian and the various types of non-Newtonian fluids are described in Sec. 1.9.

For convenience, most fluid problems assume real fluids with Newtonian characteristics. This is an appropriate assumption for water, air, gases, steam, and other simple fluids (alcohol, gasoline, acid solutions, etc.). However, slurries, pastes, gels, suspensions, and polymer/electrolyte solutions may not behave according to simple fluid relationships.

Figure 1.1 *Types of Fluids*

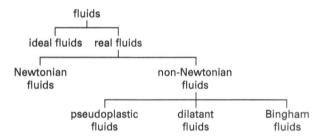

3. FLUID PRESSURE AND VACUUM

In the English system, fluid pressure is measured in pounds per square inch (lbf/in² or psi) and pounds per square foot (lbf/ft² or psf), although tons (2000 pounds) per square foot (tsf) are occasionally used. In SI units, pressure is measured in pascals (Pa). Because a pascal is very small, kilopascals (kPa) and megapascals (MPa) are usually used. Other units of pressure include bars, millibars, atmospheres, inches and feet of water, torrs, and millimeters, centimeters, and inches of mercury. (See Fig. 1.2.)

Fluid pressures are measured with respect to two pressure references: zero pressure and atmospheric pressure. Pressures measured with respect to a true zero pressure reference are known as *absolute pressures*. Pressures measured with respect to atmospheric pressure are known as *gage pressures*.[2] Most pressure gauges read the

Figure 1.2 *Relative Sizes of Pressure Units*

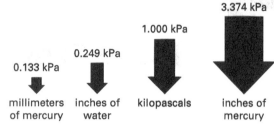

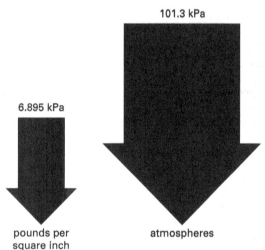

excess of the test pressure over atmospheric pressure (i.e., the gage pressure). To distinguish between these two pressure measurements, the letters "a" and "g" are traditionally added to the unit symbols in the English unit system (e.g., 14.7 psia and 4015 psfg). For SI units, the actual words "gage" and "absolute" can be added to the measurement (e.g., 25.1 kPa absolute). Alternatively, the pressure is assumed to be absolute unless the "g" is used (e.g., 15 kPag or 98 barg).

Absolute and gage pressures are related by Eq. 1.1. Atmospheric pressure in Eq. 1.1 is the actual atmospheric pressure existing when the gage measurement is taken. It is not standard atmospheric pressure, unless that pressure is implicitly or explicitly applicable. Also, since a barometer measures atmospheric pressure, *barometric pressure* is synonymous with atmospheric pressure. Table 1.1 lists standard atmospheric pressure in various units.

Pressure Field in a Static Liquid

$$\text{Absolute pressure} = \text{atmospheric pressure} + \text{gauge pressure reading} \qquad \textit{1.1}$$

---

[2]The spelling *gage* persists even though pressures are measured with *gauges*. In some countries, the term *meter pressure* is used instead of gage pressure.

Table 1.1 Standard Atmospheric Pressure

1.000 atm	(atmosphere)
14.696 psia	(pounds per square inch absolute)
2116.2 psfa	(pounds per square foot absolute)
407.1 in wg	(inches of water, inches water gage)
33.93 ft wg	(feet of water, feet water gage)
29.921 in Hg	(inches of mercury)
760.0 mm Hg	(millimeters of mercury)
760.0 torr	
1.013 bars	
1013 millibars	
1.013×10^5 Pa	(pascals)
101.3 kPa	(kilopascals)

A *vacuum* measurement is implicitly a pressure below atmospheric (i.e., a negative gage pressure). It must be assumed that any measured quantity given as a vacuum is a quantity to be subtracted from the atmospheric pressure. For example, when a condenser is operating with a vacuum of 4.0 in Hg (4 in of mercury), the absolute pressure is 29.92 in Hg − 4.0 in Hg = 25.92 in Hg. Vacuums are generally stated as positive numbers.

Pressure Field in a Static Liquid

$$\text{Absolute pressure}$$
$$= \text{atmospheric pressure} \qquad 1.2$$
$$\quad -\text{vacuum gauge pressure reading}$$

A difference in two pressures may be reported with units of *psid* (i.e., a *differential* in psi).

4. DENSITY

The *density*, ρ, of a fluid is its mass per unit volume.[3] In SI units, density is measured in kg/m³. In a consistent English system, density would be measured in slugs/ft³, even though fluid density is exclusively reported in lbm/ft³.

The density of a fluid in a liquid form is usually given, known in advance, or easily obtained from tables in any one of a number of sources. (See Table 1.2.) Most English fluid data are reported on a per pound basis, and the data included in this book follow that tradition. To convert pounds to slugs, divide by g_c.

$$\rho_{\text{slugs}} = \frac{\rho_{\text{lbm}}}{g_c} \qquad 1.3$$

The density of an ideal gas can be found from the specific gas constant and the ideal gas law.

$$\rho = \frac{P}{RT} \qquad 1.4$$

Table 1.2 Approximate Densities of Common Fluids

fluid	lbm/ft³	kg/m³
air (STP)	0.0807	1.29
air (70°F, 1 atm)	0.075	1.20
alcohol	49.3	790
ammonia	38	602
gasoline	44.9	720
glycerin	78.8	1260
mercury	848	13 600
water	62.4	1000

(Multiply lbm/ft³ by 16.01 to obtain kg/m³.)

Example 1.1

The density of water is typically taken as 62.4 lbm/ft³ for engineering problems where greater accuracy is not required. What is the value in (a) slugs/ft³ and (b) kg/m³?

Solution

(a) Equation 1.3 can be used to calculate the slug-density of water.

$$\rho = \frac{\rho_{\text{lbm}}}{g_c} = \frac{62.4 \; \dfrac{\text{lbm}}{\text{ft}^3}}{32.2 \; \dfrac{\text{lbm-ft}}{\text{lbf-sec}^2}} = 1.94 \; \text{lbf-sec}^2/\text{ft-ft}^3$$

$$= 1.94 \; \text{slugs/ft}^3$$

(b) The conversion between lbm/ft³ and kg/m³ is approximately 16.0, derived as follows.

$$\rho = \left(62.4 \; \frac{\text{lbm}}{\text{ft}^3}\right) \left(\frac{35.31 \; \dfrac{\text{ft}^3}{\text{m}^3}}{2.205 \; \dfrac{\text{lbm}}{\text{kg}}}\right)$$

$$= \left(62.4 \; \frac{\text{lbm}}{\text{ft}^3}\right) \left(16.01 \; \frac{\text{kg-ft}^3}{\text{m}^3\text{-lbm}}\right)$$

$$= 999 \; \text{kg/m}^3$$

---

[3]Mass is an absolute property of a substance. Weight is not absolute, since it depends on the local gravity. The equations using γ that result (such as Bernoulli's equation) cannot be used with SI data, since the equations are not consistent. Thus, engineers end up with two different equations for the same thing.

In SI problems, it is common to take the density of water as 1000 kg/m^3.

5. SPECIFIC VOLUME

Specific volume, v, is the volume occupied by a unit mass of fluid.[4] Since specific volume is the reciprocal of density, typical units will be ft^3/lbm, ft^3/lbmol, or m^3/kg.[5]

State Functions (properties)
$$v = V/m \qquad \textbf{1.5}$$

6. SPECIFIC GRAVITY

Specific gravity, SG, is a dimensionless ratio of a fluid's density to some standard reference density.[6] For liquids and solids, the reference is the density of pure water. There is some variation in this reference density, however, since the temperature at which the water density is evaluated is not standardized. Temperatures of 39.2°F (4°C), 60°F (16.5°C), and 70°F (21.1°C) have been used.[7]

Fortunately, the density of water is the same to three significant digits over the normal ambient temperature range: 62.4 lbm/ft^3 or 1000 kg/m^3. However, to be precise, the temperature of both the fluid and water should be specified (e.g., "... the specific gravity of the 20°C fluid is 1.05 referred to 4°C water ...").

$$SG_{\text{liquid}} = \frac{\rho_{\text{liquid}}}{\rho_{\text{water}}} \qquad \textbf{1.6}$$

Since the SI density of water is very nearly 1.000 g/cm^3 (1000 kg/m^3), the numerical values of density in g/cm^3 and specific gravity are the same. Such is not the case with English units.

The standard reference used to calculate the specific gravity of gases is the density of air. Since the density of a gas depends on temperature and pressure, both must be specified for the gas and air (i.e., two temperatures and two pressures must be specified). While STP (standard temperature and pressure) conditions are commonly specified, they are not universal.[8] Table 1.3 lists several common sets of standard conditions.

$$SG_{\text{gas}} = \frac{\rho_{\text{gas}}}{\rho_{\text{air}}} \qquad \textbf{1.7}$$

Table 1.3 *Commonly Quoted Values of Standard Temperature and Pressure*

system	temperature	pressure
SI	273.15K	101.325 kPa
scientific	0.0°C	760 mm Hg
U.S. engineering	32°F	14.696 psia
natural gas industry (U.S.)	60°F	14.65, 14.73, or 15.025 psia
natural gas industry (Canada)	60°F	14.696 psia

If it is known or implied that the temperature and pressure of the air and gas are the same, the specific gravity of the gas will be equal to the ratio of molecular weights and the inverse ratio of specific gas constants. The density of air evaluated at STP is listed in Table 1.2. At 70°F (21.1°C) and 1.0 atm, the density is approximately 0.075 lbm/ft^3 (1.20 kg/m^3).

$$SG_{\text{gas}} = \frac{M_{\text{gas}}}{M_{\text{air}}} = \frac{M_{\text{gas}}}{29.0} = \frac{R_{\text{air}}}{R_{\text{gas}}}$$
$$= \frac{53.3 \; \dfrac{\text{ft-lbf}}{\text{lbm-°R}}}{R_{\text{gas}}} \qquad \textbf{1.8}$$

Specific gravities of petroleum liquids and aqueous solutions (of acid, antifreeze, salts, etc.) can be determined by use of a *hydrometer*. (See Fig. 1.3.) In its simplest form, a hydrometer is constructed as a graduated scale weighted at one end so it will float vertically. The height at which the hydrometer floats depends on the density of the fluid, and the graduated scale can be calibrated directly in specific gravity.[9]

There are two standardized hydrometer scales (i.e., methods for calibrating the hydrometer stem).[10] Both state specific gravity in degrees, although temperature is not being measured. The *American Petroleum Institute* (API) scale (°API) may be used with all liquids, not only with oils or other hydrocarbons. For the specific gravity value, a standard reference temperature of 60°F (15.6°C) is implied for both the liquid and the water.

$$°\text{API} = \frac{141.5}{SG} - 131.5 \qquad \textbf{1.9}$$

---

[4]Care must be taken to distinguish between the symbol upsilon, υ, used for specific volume, and italic "vee," v, used for velocity in many engineering textbooks.

[5]Units of ft^3/slug are also possible, but this combination of units is almost never encountered.

[6]The symbols S.G., sp.gr., S, and G are also used. In fact, petroleum engineers in the United States use γ, a symbol that civil engineers use for specific weight. There is no standard engineering symbol for specific gravity.

[7]Density of liquids is sufficiently independent of pressure to make consideration of pressure in specific gravity calculations unnecessary.

[8]The abbreviation "SC" (standard conditions) is interchangeable with "STP."

[9]This is a direct result of the buoyancy principle of Archimedes.

[10]In addition to °Be and °API mentioned in this chapter, the *Twaddell scale* (°Tw) is used in chemical processing, the *Brix* and *Balling scales* are used in the sugar industry, and the *Salometer scale* is used to measure salt (NaCl and CaCl$_2$) solutions.

Figure 1.3 Hydrometer

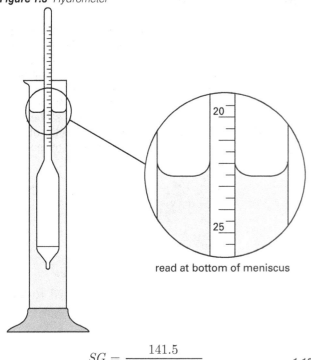

read at bottom of meniscus

$$SG = \frac{141.5}{°API + 131.5} \qquad 1.10$$

The *Baumé scale* (°Be) is used in the wine, honey, and acid industries. It is somewhat confusing because there are actually two Baumé scales—one for liquids heavier than water and another for liquids lighter than water. (There is also a discontinuity in the scales at $SG = 1.00$.) As with the API scale, the specific gravity value assumes 60°F (15.6°C) is the standard temperature for both scales.

$$SG = \frac{140.0}{130.0 + °Be} \qquad [SG < 1.00] \qquad 1.11$$

$$SG = \frac{145.0}{145.0 - °Be} \qquad [SG > 1.00] \qquad 1.12$$

Example 1.2

Determine the specific gravity of carbon dioxide gas (molecular weight = 44) at 66°C (150°F) and 138 kPa (20 psia) using STP air as a reference. The specific gas constant of carbon dioxide is 35.1 ft-lbf/lbm-°R.

SI Solution

Since the specific gas constant for carbon dioxide was not given in SI units, it must be calculated from the universal gas constant.

$$R = \frac{R^*}{M} = \frac{8314.47 \; \dfrac{J}{kmol \cdot K}}{44 \; \dfrac{kg}{kmol}} = 189 \; J/kg \cdot K$$

$$\rho = \frac{P}{RT} = \frac{1.38 \times 10^5 \; Pa}{\left(189 \; \dfrac{J}{kg \cdot K}\right)(66°C + 273°)}$$

$$= 2.15 \; kg/m^3$$

From Table 1.2, the density of STP air is 1.29 kg/m^3. From Eq. 1.7, the specific gravity of carbon dioxide at the conditions given is

$$SG = \frac{\rho_{gas}}{\rho_{air}} = \frac{2.15 \; \dfrac{kg}{m^3}}{1.29 \; \dfrac{kg}{m^3}} = 1.67$$

Customary U.S. Solution

Since the conditions of the carbon dioxide and air are different, Eq. 1.8 cannot be used. Therefore, it is necessary to calculate the density of the carbon dioxide from Eq. 1.4. Absolute temperature (degrees Rankine) must be used. The density is

$$\rho = \frac{P}{RT} = \frac{\left(20 \; \dfrac{lbf}{in^2}\right)\left(12 \; \dfrac{in}{ft}\right)^2}{\left(35.1 \; \dfrac{ft\text{-}lbf}{lbm\text{-}°R}\right)(150°F + 460°)}$$

$$= 0.135 \; lbm/ft^3$$

From Table 1.2, the density of STP air is 0.0807 lbm/ft^3. From Eq. 1.7, the specific gravity of carbon dioxide at the conditions given is

$$SG = \frac{\rho_{gas}}{\rho_{air}} = \frac{0.135 \; \dfrac{lbm}{ft^3}}{0.0807 \; \dfrac{lbm}{ft^3}} = 1.67$$

Water

7. SPECIFIC WEIGHT

Specific weight (unit weight), γ, is the weight of fluid per unit volume. The use of specific weight is most often encountered in civil engineering projects in the United States, where it is commonly called "density." The usual units of specific weight are lbf/ft^3.[11] Specific weight is not an absolute property of a fluid, since it depends not only on the fluid, but on the local gravitational field as well.

Field Equation

$$\gamma = \rho g \qquad \text{[SI]} \quad 1.13(a)$$

Unit and Conversion Factors

$$SW = \rho g / g_c \qquad \text{[U.S.]} \quad 1.13(b)$$

If the gravitational acceleration is 32.2 ft/sec^2, as it is almost everywhere on earth, the specific weight in lbf/ft^3 will be numerically equal to the density in lbm/ft^3. This concept is demonstrated in Ex. 1.3.

Example 1.3

What is the sea level ($g = 32.2 \text{ ft/sec}^2$) specific weight (in lbf/ft^3) of liquids with densities of (a) 1.95 slug/ft^3 and (b) 58.3 lbm/ft^3?

Solution

(a) Equation 1.13(a) can be used with any consistent set of units, including densities involving slugs.

Field Equation

$$\gamma = \rho g$$
$$= \left(1.95 \; \frac{\text{slug}}{\text{ft}^3}\right)\left(32.2 \; \frac{\text{ft}}{\text{sec}^2}\right)$$
$$= \left(1.95 \; \frac{\text{lbf-sec}^2}{\text{ft-ft}^3}\right)\left(32.2 \; \frac{\text{ft}}{\text{sec}^2}\right)$$
$$= 62.8 \; \text{lbf/ft}^3$$

(b) Use Eq. 1.13(b).

Unit and Conversion Factors

$$SW = \rho g / g_c$$
$$= \frac{\left(58.3 \; \dfrac{\text{lbm}}{\text{ft}^3}\right)\left(32.2 \; \dfrac{\text{ft}}{\text{sec}^2}\right)}{32.2 \; \dfrac{\text{lbm-ft}}{\text{lbf-sec}^2}}$$
$$= 58.3 \; \text{lbf/ft}^3$$

8. MOLE FRACTION

Mole fraction is an important parameter in many practical engineering problems, particularly in chemistry and chemical engineering. The composition of a fluid consisting of two or more distinctly different substances, A, B, C, and so on, can be described by the mole fractions, x_A, x_B, x_C, and so on, of each substance. (There are also other methods of specifying the composition.) The mole fraction of component i is the number of moles of that component, N_i, divided by the total number of moles in the combined fluid mixture.

Ideal Gas Mixtures

$$x_i = N_i / N \qquad 1.14$$

$$N = \sum N_i \qquad 1.15$$

$$\sum x_i = 1 \qquad 1.16$$

Equation 1.14 and Eq. 1.15 can be combined to form Eq. 1.17.

$$x_i = \frac{N_i}{\sum N_i} \qquad 1.17$$

Mole fraction is a number between 0 and 1. *Mole percent* is the mole fraction multiplied by 100%, expressed as a percentage.

Mole fractions can be converted to mass fractions using Eq. 1.18. Multiply each component's mole fraction by its molecular weight to get its mass, then divide by the total mass of all components to get the component's mass fraction.

$$y = \frac{xM}{\sum (x_i M_i)} \qquad 1.18$$

As with mole fractions, a mass fraction is a number between 0 and 1.

9. VISCOSITY

The *viscosity* of a fluid is a measure of that fluid's resistance to flow when acted upon by an external force such as a pressure differential or gravity. Some fluids, such as heavy oils, jellies, and syrups, are very viscous. Other fluids, such as water, lighter hydrocarbons, and gases, are not as viscous.

The more viscous the fluid, the more time will be required for the fluid to leak out of a container. *Saybolt seconds universal* (SSU) and *Saybolt seconds FUROL* (SSF) are scales of such viscosity measurement based on the smaller and larger orifices, respectively. Seconds can

---

[11]Notice that the units are lbf/ft^3, not lbm/ft^3. Pound-mass (lbm) is a mass unit, not a weight (force) unit.

be converted (empirically) to viscosity in other units. The following relations are approximate conversions between SSU and stokes.

- For SSU < 100 sec,

$$\nu_{\text{stokes}} = 0.00226(\text{SSU}) - \frac{1.95}{\text{SSU}} \qquad \textit{1.19}$$

- For SSU > 100 sec,

$$\nu_{\text{stokes}} = 0.00220(\text{SSU}) - \frac{1.35}{\text{SSU}} \qquad \textit{1.20}$$

Most common liquids will flow more easily when their temperatures are raised. However, the behavior of a fluid when temperature, pressure, or stress is varied will depend on the type of fluid. The different types of fluids can be determined with a *sliding plate viscometer test*.[12]

Consider two plates of area A separated by a fluid with thickness y_0, as shown in Fig. 1.4. The bottom plate is fixed, and the top plate is kept in motion at a constant velocity, v_0, by a force, F.

Figure 1.4 *Sliding Plate Viscometer*

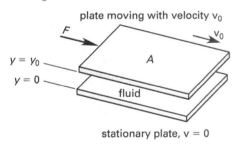

Experiments with water and most common fluids have shown that the force, F, required to maintain the velocity, v_0, is proportional to the velocity and the area and is inversely proportional to the separation of the plates. That is,

$$\frac{F}{A} \propto \frac{dv}{dy} \qquad \textit{1.21}$$

The constant of proportionality needed to make Eq. 1.21 an equality is the *absolute viscosity*, μ, also known as the *coefficient of viscosity*.[13] The reciprocal of absolute viscosity, $1/\mu$, is known as the *fluidity*.

$$\frac{F}{A} = \mu \frac{dv}{dy} \qquad \textit{1.22}$$

F/A is the *fluid shear stress*, τ. The quantity dv/dy (v_0/y_0) is known by various names, including *rate of strain*, *shear rate*, *velocity gradient*, and *rate of shear formation*. Equation 1.22 is known as *Newton's law of viscosity*, from which *Newtonian fluids* get their name. Sometimes Eq. 1.23 is written with a minus sign to compare viscous behavior with other behavior. However, the direction of positive shear stress is arbitrary. Equation 1.23 is simply the equation of a straight line.

Stress, Pressure, and Viscosity

$$\tau = \mu(dv/dy) \qquad \textit{1.23}$$

Not all fluids are Newtonian (although most common fluids are), and Eq. 1.23 is not universally applicable. Figure 1.5 (known as a *rheogram*) illustrates how differences in fluid shear stress behavior (at constant temperature and pressure) can be used to define Bingham, pseudoplastic, and dilatant fluids, as well as Newtonian fluids.

Figure 1.5 *Shear Stress Behavior for Different Types of Fluids*

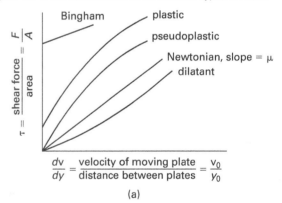

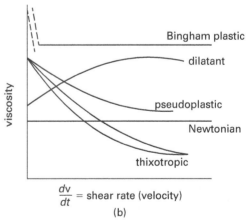

---

[12]This test is conceptually simple but is not always practical, since the liquid leaks out between the plates. In research work with liquids, it is common to determine viscosity with a *concentric cylinder viscometer*, also known as a *cup-and-bob viscometer*. Viscosities of perfect gases can be predicted by the kinetic theory of gases. Viscosity can also be measured by a *Saybolt viscometer*, which is essentially a container that allows a given quantity of fluid to leak out through one of two different-sized orifices.

[13]Another name for absolute viscosity is *dynamic viscosity*. The name *absolute viscosity* is preferred, if for no other reason than to avoid confusion with *kinematic viscosity*.

Gases, water, alcohol, and benzene are examples of Newtonian fluids. In fact, all liquids with a simple chemical formula are Newtonian. Also, most solutions of simple compounds, such as sugar and salt, are Newtonian. For a more viscous fluid, the straight line will be closer to the τ axis (i.e., the slope will be higher). (See Fig. 1.5.) For low-viscosity fluids, the straight line will be closer to the dv/dy axis (i.e., the slope will be lower).

Pseudoplastic fluids (muds, motor oils, polymer solutions, natural gums, and most slurries) exhibit viscosities that decrease with an increasing velocity gradient. Such fluids present no serious pumping problems.

Plastic materials, such as tomato catsup, behave similarly to pseudoplastic fluids once movement begins; that is, their viscosities decrease with agitation. However, a finite force must be applied before any fluid movement occurs.

Bingham fluids (Bingham plastics), typified by toothpaste, jellies, bread dough, and some slurries, are capable of indefinitely resisting a small shear stress but move easily when the stress becomes large—that is, Bingham fluids exhibit Newtonian fluid behavior when the applied shear stress increases above the yield stress.

Dilatant fluids are rare but include clay slurries, various starches, some paints, milk chocolate with nuts, and other candy compounds. They exhibit viscosities that increase with increasing agitation (i.e., with increasing velocity gradients), but they return rapidly to their normal viscosity after the agitation ceases. Pump selection is critical for dilatant fluids because these fluids can become almost solid if the shear rate is high enough.

Viscosity can also change with time (all other conditions being constant). If viscosity decreases with time during agitation, the fluid is said to be a *thixotropic fluid*. If viscosity increases (usually up to a finite value) with time during agitation, the fluid is a *rheopectic fluid*. Viscosity does not change in time-independent fluids. *Colloidal materials*, such as gelatinous compounds, lotions, shampoos, and low-temperature solutions of soaps in water and oil, behave like *thixotropic liquids*—their viscosities decrease as the agitation continues. However, viscosity does not return to its original state after the agitation ceases.

Molecular cohesion is the dominating cause of viscosity in liquids. As the temperature of a liquid increases, these cohesive forces decrease, resulting in a decrease in viscosity.

In gases, the dominant cause of viscosity is random collisions between gas molecules. This molecular agitation increases with increases in temperature. Therefore, viscosity in gases increases with temperature.

Although viscosity of liquids increases slightly with pressure, the increase is insignificant over moderate pressure ranges. Therefore, the absolute viscosity of both gases and liquids is usually considered to be essentially independent of pressure.[14]

The units of absolute viscosity, as derived from Eq. 1.23, are lbf-sec/ft^2. Such units are actually used in the English engineering system.[15] Absolute viscosity is measured in pascal-seconds (Pa·s) in SI units. Another common unit used throughout the world is the *poise* (abbreviated P), equal to a dyne·s/cm^2. These dimensions are the same primary dimensions as in the English system, $F\theta/L^2$ or $M/L\theta$, and are functionally the same as a g/cm·s. Since the poise is a large unit, the *centipoise* (abbreviated cP) scale is generally used. The viscosity of pure water at room temperature is approximately 1 cP.

Example 1.4

A liquid ($\mu = 5.2 \times 10^{-5}$ lbf-sec/ft^2) is flowing in a rectangular duct. The equation of the symmetrical facial velocity distribution (in ft/sec) is approximately v = $3y^{0.7}$ ft/sec, where y is measured in inches from the wall. (a) What is the velocity gradient at $y = 3.0$ in from the duct wall? (b) What is the shear stress in the fluid at that point?

Solution

(a) The velocity is not a linear function of y, so dv/dy must be calculated as a derivative.

$$\frac{dv}{dy} = \frac{d}{dy} 3y^{0.7}$$
$$= (3)(0.7y^{-0.3})$$
$$= 2.1y^{-0.3}$$

At $y = 3$ in,

$$\frac{dv}{dy} = (2.1)(3)^{-0.3} = 1.51 \text{ ft/sec-in}$$

(b) From Eq. 1.23, the shear stress is

Stress, Pressure, and Viscosity

$$\tau = \mu(dv/dy)$$
$$= \left(5.2 \times 10^{-5} \frac{\text{lbf-sec}}{\text{ft}^2}\right)\left(1.51 \frac{\text{ft}}{\text{sec-in}}\right)\left(12 \frac{\text{in}}{\text{ft}}\right)$$
$$= 9.42 \times 10^{-4} \text{ lbf/ft}^2$$

---

[14]This is not true for kinematic viscosity, however.
[15]Units of lbm/ft-sec are also used for absolute viscosity in the English system. These units are obtained by multiplying lbf-sec/ft^2 units by g_c.

10. KINEMATIC VISCOSITY

Another quantity with the name *viscosity* is the ratio of absolute viscosity to mass density. This combination of variables, known as *kinematic viscosity*, v, appears sufficiently often in fluids and other problems as to warrant its own symbol and name. Typical units are ft^2/sec and cm^2/s (the *stoke*, St). It is also common to give kinematic viscosity in *centistokes*, cSt. The SI units of kinematic viscosity are m^2/s.

Stress, Pressure, and Viscosity

$$v = \mu/\rho \qquad \text{[SI]} \qquad 1.24(a)$$

$$v = \frac{\mu g_c}{\rho} = \frac{\mu g}{\gamma} \qquad \text{[U.S.]} \qquad 1.24(b)$$

It is essential that consistent units be used with Eq. 1.24. The following sets of units are consistent.

$$ft^2/sec = \frac{lbf\text{-}sec/ft^2}{slugs/ft^3}$$

$$m^2/s = \frac{Pa\cdot s}{kg/m^3}$$

$$St \text{ (stoke)} = \frac{P \text{ (poise)}}{g/cm^3}$$

$$cSt \text{ (centistokes)} = \frac{cP \text{ (centipoise)}}{g/cm^3}$$

Unlike absolute viscosity, kinematic viscosity is greatly dependent on both temperature and pressure, since these variables affect the density of the fluid. Referring to Eq. 1.24, even if absolute viscosity is independent of temperature or pressure, the change in density will change the kinematic viscosity.

11. VISCOSITY CONVERSIONS

The most common units of absolute and kinematic viscosity are listed in Table 1.4. Table 1.5 contains conversions between the various viscosity units.

Table 1.4 *Common Viscosity Units*

	absolute, μ	kinematic, ν
English	lbf-sec/ft^2 (slug/ft-sec)	ft^2/sec
conventional metric	dyne·s/cm^2 (poise)	cm^2/s (stoke)
SI	Pa·s (N·s/m^2)	m^2/s

Table 1.5 *Viscosity Conversions**

multiply	by	to obtain
absolute viscosity, μ		
dyne·s/cm^2	0.10	Pa·s
lbf-sec/ft^2	478.8	P
lbf-sec/ft^2	47,880	cP
lbf-sec/ft^2	47.88	Pa·s
slug/ft-sec	47.88	Pa·s
lbm/ft-sec	1.488	Pa·s
cP	1.0197×10^{-4}	kgf·s/m^2
cP	2.0885×10^{-5}	lbf-sec/ft^2
cP	0.001	Pa·s
Pa·s	0.020885	lbf-sec/ft^2
Pa·s	1000	cP
reyn	144	lbf-sec/ft^2
reyn	1.0	lbf-sec/in^2
kinematic viscosity, ν		
ft^2/sec	92,903	cSt
ft^2/sec	0.092903	m^2/s
m^2/s	10.7639	ft^2/sec
m^2/s	1×10^6	cSt
cSt	1×10^{-6}	m^2/s
cSt	1.0764×10^{-5}	ft^2/sec
absolute viscosity to kinematic viscosity		
cP	$1/\rho$ in g/cm^3	cSt
cP	$6.7195 \times 10^{-4}/\rho$ in lbm/ft^3	ft^2/sec
lbf-sec/ft^2	$32.174/\rho$ in lbm/ft^3	ft^2/sec
kgf·s/m^2	$9.807/\rho$ in kg/m^3	m^2/s
Pa·s	$1000/\rho$ in g/cm^3	cSt
kinematic viscosity to absolute viscosity		
cSt	ρ in g/cm^3	cP
cSt	$0.001 \times \rho$ in g/cm^3	Pa·s
cSt	$1.6 \times 10^{-5} \times \rho$ in lbm/ft^3	Pa·s
m^2/s	$0.10197 \times \rho$ in kg/m^3	kgf·s/m^2
m^2/s	$1000 \times \rho$ in g/cm^3	Pa·s
m^2/s	ρ in kg/m^3	Pa·s
ft^2/sec	$0.031081 \times \rho$ in lbm/ft^3	lbf-sec/ft^2
ft^2/sec	$1488.2 \times \rho$ in lbm/ft^3	cP

*cP: centipoise; cSt: centistoke; kgf: kilogram-force; P: poise

Example 1.5

Water at 60°F has a specific gravity of 0.999 and a kinematic viscosity of 1.12 cSt. What is the absolute viscosity in lbf-sec/ft²?

Solution

The density of a liquid expressed in g/cm³ is numerically equal to its specific gravity.

$$\rho = 0.999 \text{ g/cm}^3$$

The centistoke (cSt) is a measure of kinematic viscosity. Kinematic viscosity is converted first to the absolute viscosity units of centipoise. From Table 1.5,

$$\mu_{cP} = v_{cSt}\rho_{g/cm^3}$$

$$= (1.12 \text{ cSt})\left(0.999 \ \frac{g}{cm^3}\right)$$

$$= 1.119 \text{ cP}$$

Next, centipoise is converted to lbf-sec/ft².

$$\mu_{lbf\text{-}sec/ft^2} = \mu_{cP}(2.0885 \times 10^{-5})$$

$$= (1.119 \text{ cP})(2.0885 \times 10^{-5})$$

$$= 2.34 \times 10^{-5} \text{ lbf-sec/ft}^2$$

12. VISCOSITY GRADE

The ISO *viscosity grade* (VG) as specified in ISO 3448 is commonly used to classify oils. (See Table 1.6.) Viscosity at 104°F (40°C), the approximate temperature of machinery, in centistokes (same as mm²/s), is used as the index. Each subsequent viscosity grade within the classification has approximately a 50% higher viscosity, whereas the minimum and maximum values of each grade range ±10% from the midpoint.

13. VISCOSITY INDEX

Viscosity index (VI) is a measure of a fluid's viscosity sensitivity to changes in temperature. It has traditionally been applied to crude and refined oils through use of a 100-point scale.[16] The viscosity is measured at two temperatures: 100°F and 210°F (38°C and 99°C). These viscosities are converted into a viscosity index in accordance with standard ASTM D2270.

Table 1.6 *ISO Viscosity Grade*

ISO 3448 viscosity grade	kinematic viscosity at 40°C (cSt)		
	minimum	midpoint	maximum
ISO VG 2	1.98	2.2	2.42
ISO VG 3	2.88	3.2	3.52
ISO VG 5	4.14	4.6	5.06
ISO VG 7	6.12	6.8	7.48
ISO VG 10	9.0	10	11.0
ISO VG 15	13.5	15	16.5
ISO VG 22	19.8	22	24.2
ISO VG 32	28.8	32	35.2
ISO VG 46	41.4	46	50.6
ISO VG 68	61.2	68	74.8
ISO VG 100	90	100	110
ISO VG 150	135	150	165
ISO VG 220	198	220	242
ISO VG 320	288	320	352
ISO VG 460	414	460	506
ISO VG 680	612	680	748
ISO VG 1000	900	1000	1100
ISO VG 1500	1350	1500	1650

14. VAPOR PRESSURE

Molecular activity in a liquid will allow some of the molecules to escape the liquid surface. Strictly speaking, a small portion of the liquid vaporizes. Molecules of the vapor also condense back into the liquid. The vaporization and condensation at constant temperature are equilibrium processes. The equilibrium pressure exerted by these free molecules is known as the *vapor pressure* or *saturation pressure*. (Vapor pressure does not include the pressure of other substances in the mixture.) Typical values of vapor pressure are given in Table 1.7.

Some liquids, such as propane, butane, ammonia, and Freon, have significant vapor pressures at normal temperatures. Liquids near their boiling points or that vaporize easily are said to be *volatile liquids*.[17] Other liquids, such as mercury, have insignificant vapor pressures at normal temperatures. Liquids with low vapor pressures are used in accurate barometers.

---

[16]Use of the *viscosity index* has been adopted by other parts of the chemical process industry (CPI), including in the manufacture of solvents, polymers, and other synthetics. The 100-point scale may be exceeded (on both ends) for these uses. Refer to standard ASTM D2270 for calculating extreme values of the viscosity index.

[17]Because a liquid that vaporizes easily has an aroma, the term *aromatic liquid* is also occasionally used.

Table 1.7 Typical Vapor Pressures

fluid	lbf/ft², 68°F	kPa, 20°C
mercury	0.00362	0.000173
turpentine	1.115	0.0534
water	48.9	2.34
ethyl alcohol	122.4	5.86
ether	1231	58.9
butane	4550	218
Freon-12	12,200	584
propane	17,900	855
ammonia	18,550	888

(Multiply lbf/ft² by 0.04788 to obtain kPa.)

The tendency toward vaporization is dependent on the temperature of the liquid. *Boiling* occurs when the liquid temperature is increased to the point that the vapor pressure is equal to the local ambient pressure. Therefore, a liquid's boiling temperature depends on the local ambient pressure as well as on the liquid's tendency to vaporize.

Vapor pressure is usually considered to be a nonlinear function of temperature only. It is possible to derive correlations between vapor pressure and temperature, and such correlations usually involve a logarithmic transformation of vapor pressure.[18] Vapor pressure can also be graphed against temperature in a (logarithmic) *Cox chart* when values are needed over larger temperature extremes. Although there is also some variation with external pressure, the external pressure effect is negligible under normal conditions.

15. OSMOTIC PRESSURE

Osmosis is a special case of diffusion in which molecules of the *solvent* move under pressure from one fluid to another (i.e., from the *solvent* to the *solution*) in one direction only, usually through a *semipermeable membrane*.[19] Osmosis continues until sufficient solvent has passed through the membrane to make the activity (or solvent pressure) of the solution equal to that of the solvent.[20] The pressure at equilibrium is known as the *osmotic pressure*, π.

Figure 1.6 illustrates an *osmotic pressure apparatus*. The fluid column can be interpreted as the result of an osmotic pressure that has developed through diffusion into the solution. The fluid column will continue to increase in height until equilibrium is reached. Alternatively, the fluid column can be adjusted so that the solution pressure just equals the osmotic pressure that would develop otherwise, in order to prevent the flow of

solvent. For the arrangement in Fig. 1.6, the osmotic pressure can be calculated from the difference in fluid level heights, h.

$$\pi = \rho g h \qquad \text{[SI]} \qquad 1.25(a)$$

$$\pi = \frac{\rho g h}{g_c} \qquad \text{[U.S.]} \qquad 1.25(b)$$

Figure 1.6 Osmotic Pressure Apparatus

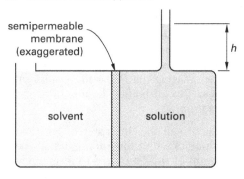

In dilute solutions, osmotic pressure follows the ideal gas law. The solute acts like a gas in exerting pressure against the membrane. The solvent exerts no pressure since it can pass through. In Eq. 1.26, M is the molarity (concentration). The value of the *universal gas constant*, R^*, depends on the units used. Common values include 1545.35 ft-lbf/lbmol-°R, 8314.47 J/kmol·K, and 0.08206 atm·L/mol·K. Consistent units must be used.

$$\pi = MR^*T \qquad 1.26$$

Example 1.6

An aqueous solution is in isopiestic equilibrium with a 0.1 molarity sucrose solution at 22°C. What is the osmotic pressure?

Solution

Use Eq. 1.26. [Temperature Conversions]

$M = 0.1$ mol/L of solution

$T = 22°C + 273.15° = 295.15$K

$\pi = MR^*T = \left(0.1 \ \frac{\text{mol}}{\text{L}}\right)\left(0.08206 \ \frac{\text{atm·L}}{\text{mol·K}}\right)(295.15\text{K})$

$= 2.42$ atm

---

[18]The *Clausius-Clapeyron equation* and *Antoine equation* are two such logarithmic correlations of vapor pressure with temperature.
[19]A semipermeable membrane will be impermeable to the solute but permeable for the solvent.
[20]Two solutions in equilibrium (i.e., whose activities are equal) are said to be in *isopiestic equilibrium*.

Water

16. SURFACE TENSION

The membrane or "skin" that seems to form on the free surface of a liquid is due to the intermolecular cohesive forces and is known as *surface tension*, σ. Surface tension is the reason that a needle is able to float and insects are able to walk on water. Surface tension also causes bubbles and droplets to take on a spherical shape, since any other shape would have more surface area per unit volume.

Data on the surface tension of liquids is important in determining the performance of heat-, mass-, and momentum-transfer equipment, including heat transfer devices.[21] Surface tension data is needed to calculate the nucleate boiling point (i.e., the initiation of boiling) of liquids in a pool (using the *Rohsenow equation*) and the maximum heat flux of boiling liquids in a pool (using the *Zuber equation*).

Surface tension can be interpreted as the tension between two points a unit distance apart on the surface or as the amount of work required to form a new unit of surface area in an apparatus similar to that shown in Fig. 1.7. Typical units of surface tension are lbf/ft (ft-lbf/ft^2), dyne/cm, and N/m.

Figure 1.7 *Wire Frame for Stretching a Film*

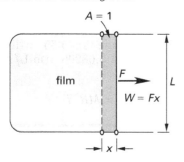

The apparatus shown in Fig. 1.7 consists of a wire frame with a sliding side that has been dipped in a liquid to form a film. Surface tension is determined by measuring the force necessary to keep the sliding side stationary against the surface tension pull of the film.[22] (The film does not act like a spring, since the force, F, does not increase as the film is stretched.) Since the film has two surfaces (i.e., two surface tensions), the surface tension is

$$\sigma = \frac{F}{2L} \qquad 1.27$$

Alternatively, surface tension can also be determined by measuring the force required to pull a wire ring out of the liquid, as shown in Fig. 1.8.[23] Since the ring's inner and outer sides are in contact with the liquid, the wetted perimeter is twice the circumference. The surface tension is

$$\sigma = \frac{F}{4\pi r} \qquad 1.28$$

Figure 1.8 *Du Nouy Ring Surface Tension Apparatus*

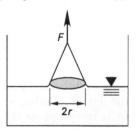

Surface tension depends slightly on the gas in contact with the free surface. Surface tension values are usually quoted for air contact. Typical values of surface tension are listed in Table 1.8.

Table 1.8 *Approximate Values of Surface Tension (air contact)*

fluid	lbf/ft, 68°F	N/m, 20°C
n-octane	0.00149	0.0217
ethyl alcohol	0.00156	0.0227
acetone	0.00162	0.0236
kerosene	0.00178	0.0260
carbon tetrachloride	0.00185	0.0270
turpentine	0.00186	0.0271
toluene	0.00195	0.0285
benzene	0.00198	0.0289
olive oil	0.0023	0.034
glycerin	0.00432	0.0631
water	0.00499	0.0728
mercury	0.0356	0.519

(Multiply lbf/ft by 14.59 to obtain N/m.)

(Multiply dyne/cm by 0.001 to obtain N/m.)

At temperatures below freezing, the substance will be a solid, so surface tension is a moot point. As the temperature of a liquid is raised, the surface tension decreases because the cohesive forces decrease. Surface tension is zero at a substance's critical temperature. If a

---

[21]Surface tension plays a role in processes involving dispersion, emulsion, flocculation, foaming, and solubilization. Surface tension data are particularly important in determining the performance of equipment in the chemical process industry (CPI), such as distillation columns, packed towers, wetted-wall columns, strippers, and phase-separation equipment.
[22]The force includes the weight of the sliding side wire if the frame is oriented vertically, with gravity acting on the sliding side wire to stretch the film.
[23]This apparatus is known as a *Du Nouy torsion balance*. The ring is made of platinum with a diameter of 4.00 cm.

substance's critical temperature is known, the *Othmer correlation*, Eq. 1.29, can be used to determine the surface tension at one temperature from the surface tension at another temperature.

$$\sigma_2 = \sigma_1 \left(\frac{T_c - T_2}{T_c - T_1} \right)^{11/9} \qquad 1.29$$

Surface tension is the reason that the pressure on the inside of bubbles and droplets is greater than on the outside. Equation 1.30 gives the relationship between the surface tension in a hollow bubble surrounded by a gas and the difference between the inside and outside pressures. For a spherical droplet or a bubble in a liquid, where in both cases there is only one surface in tension, the surface tension is twice as large. (r is the radius of the bubble or droplet.)

$$\sigma_{\text{bubble}} = \frac{r(P_{\text{inside}} - P_{\text{outside}})}{4} \qquad 1.30$$

$$\sigma_{\text{droplet}} = \frac{r(P_{\text{inside}} - P_{\text{outside}})}{2} \qquad 1.31$$

17. CAPILLARY ACTION

Capillary action (*capillarity*) is the name given to the behavior of a liquid in a thin-bore tube. Capillary action is caused by surface tension between the liquid and a vertical solid surface.[24] In the case of liquid water in a glass tube, the adhesive forces between the liquid molecules and the surface are greater than (i.e., dominate) the cohesive forces between the water molecules themselves.[25] The adhesive forces cause the water to attach itself to and climb a solid vertical surface. It can be said that the water "reaches up and tries to wet as much of the interior surface as it can." In so doing, the water rises above the general water surface level. The surface is *hydrophilic* (*lyophilic*). This is illustrated in Fig. 1.9.

Figure 1.9 also illustrates that the same surface tension forces that keep a droplet spherical are at work on the surface of the liquid in the tube. The curved liquid surface, known as the *meniscus*, can be considered to be an incomplete droplet. If the inside diameter of the tube is less than approximately 0.1 in (2.5 mm), the meniscus is essentially hemispherical, and $r_{\text{meniscus}} = r_{\text{tube}}$.

For a few other liquids, such as mercury, the molecules have a strong affinity for each other (i.e., the cohesive forces dominate). The liquid avoids contact with the tube surface. The surface is *hydrophobic* (*lyophobic*). In such liquids, the meniscus in the tube will be below the general surface level.

Figure 1.9 *Capillarity of Liquids*

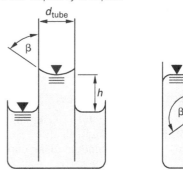

(a) adhesive force dominates   (b) cohesive force dominates

The *angle of contact*, β, is an indication of whether adhesive or cohesive forces dominate. For contact angles less than 90°, adhesive forces dominate. For contact angles greater than 90°, cohesive forces dominate.

Equation 1.32 can be used to predict the capillary rise in a small-bore tube. Surface tension and contact angles can be obtained from Table 1.8 and Table 1.9, respectively.

Table 1.9 *Contact Angles, B*

materials	angle
mercury–glass	140°
water–paraffin	107°
water–silver	90°
silicone oil–glass	20°
kerosene–glass	26°
glycerin–glass	19°
water–glass	0°
ethyl alcohol–glass	0°

$$h = \frac{4\sigma \cos \beta}{\rho d_{\text{tube}} g} \qquad \text{[SI]} \qquad 1.32(a)$$

$$h = \frac{4\sigma \cos \beta}{\rho d_{\text{tube}}} \times \frac{g_c}{g} \qquad \text{[U.S.]} \qquad 1.32(b)$$

$$\sigma = \frac{h \rho d_{\text{tube}} g}{4 \cos \beta} \qquad \text{[SI]} \qquad 1.33(a)$$

$$\sigma = \frac{h \rho d_{\text{tube}}}{4 \cos \beta} \times \frac{g}{g_c} \qquad \text{[U.S.]} \qquad 1.33(b)$$

$$r_{\text{meniscus}} = \frac{d_{\text{tube}}}{2 \cos \beta} \qquad 1.34$$

---

[24]In fact, observing the rise of liquid in a capillary tube is another method of determining the surface tension of a liquid.
[25]*Adhesion* is the attractive force between molecules of different substances. *Cohesion* is the attractive force between molecules of the same substance.

If it is assumed that the meniscus is hemispherical, then $r_{\text{meniscus}} = r_{\text{tube}}$, $\beta = 0°$, and $\cos\beta = 1.0$, and the previous equations can be simplified. (Such an assumption can only be made when the diameter of the capillary tube is less than 0.1 in.)

Example 1.7

Ethyl alcohol's density is 49 lbm/ft^3 (790 kg/m^3). To what height will 68°F (20°C) ethyl alcohol rise in a 0.005 in (0.127 mm) internal diameter glass capillary tube?

SI Solution

From Table 1.8 and Table 1.9, respectively, the surface tension and contact angle are

$$\sigma = 0.0227 \text{ N/m}$$
$$\beta = 0°$$

From Eq. 1.32, the height is

$$h = \frac{4\sigma\cos\beta}{\rho d_{\text{tube}}g} = \frac{(4)\left(0.0227\ \dfrac{\text{N}}{\text{m}}\right)(1.0)\left(1000\ \dfrac{\text{mm}}{\text{m}}\right)}{\left(790\ \dfrac{\text{kg}}{\text{m}^3}\right)(0.127\ \text{mm})\left(9.81\ \dfrac{\text{m}}{\text{s}^2}\right)}$$

$$= 0.0923 \text{ m}$$

Customary U.S. Solution

From Table 1.8 and Table 1.9, respectively, the surface tension and contact angle are

$$\sigma = 0.00156 \text{ lbf/ft}$$
$$\beta = 0°$$

From Eq. 1.32, the height is

$$h = \frac{4\sigma\cos\beta}{\rho d_{\text{tube}}} \times \frac{g_c}{g}$$

$$= \frac{(4)\left(0.00156\ \dfrac{\text{lbf}}{\text{ft}}\right)(1.0)\left(32.2\ \dfrac{\text{lbm-ft}}{\text{lbf-sec}^2}\right)\left(12\ \dfrac{\text{in}}{\text{ft}}\right)}{\left(49\ \dfrac{\text{lbm}}{\text{ft}^3}\right)(0.005\ \text{in})\left(32.2\ \dfrac{\text{ft}}{\text{sec}^2}\right)}$$

$$= 0.306 \text{ ft}$$

18. COMPRESSIBILITY[26]

Compressibility (also known as the *coefficient of compressibility*), β, is the fractional change in the volume of a fluid per unit change in pressure in a constant-temperature process.[27] Typical units are in^2/lbf, ft^2/lbf, $1/\text{atm}$, and $1/\text{kPa}$. (See Table 1.10.) It is the reciprocal of the bulk modulus, a quantity that is more commonly tabulated than compressibility. Equation 1.35 is written with a negative sign to show that volume decreases as pressure increases.

$$\beta = \frac{-\dfrac{\Delta V}{V_0}}{\Delta P} = \frac{1}{E} \qquad \text{1.35}$$

Compressibility can also be written in terms of partial derivatives.

$$\beta = \left(\frac{-1}{V_0}\right)\left(\frac{\partial V}{\partial P}\right)_T = \left(\frac{1}{\rho_0}\right)\left(\frac{\partial \rho}{\partial P}\right)_T \qquad \text{1.36}$$

Compressibility changes only slightly with temperature. The small compressibility of liquids is typically considered to be insignificant, giving rise to the common understanding that liquids are incompressible.

The density of a compressible fluid depends on the fluid's pressure. For small changes in pressure, Eq. 1.37 can be used to calculate the density at one pressure from the density at another pressure.

$$\rho_2 \approx \rho_1\big(1 + \beta(P_2 - P_1)\big) \qquad \text{1.37}$$

Gases, of course, are easily compressed. The compressibility of an ideal gas depends on its pressure, P, its ratio of specific heats, k, and the nature of the process.[28] Depending on the process, the compressibility may be known as *isothermal compressibility* or *(adiabatic) isentropic compressibility*. Of course, compressibility is zero for constant-volume processes and is infinite (or undefined) for constant-pressure processes.

$$\beta_T = \frac{1}{P} \quad \text{[isothermal ideal gas processes]} \qquad \text{1.38}$$

$$\beta_s = \frac{1}{kP} \quad \text{[adiabatic ideal gas processes]} \qquad \text{1.39}$$

---

[26]Compressibility should not be confused with the *thermal coefficient of expansion*, $(1/V_0)(\partial V/\partial T)_p$, which is the fractional change in volume per unit temperature change in a constant-pressure process (with units of 1/°F or 1/°C), or the dimensionless *compressibility factor*, Z, which is used with the ideal gas law.
[27]Other symbols used for compressibility are c, C, and K.
[28]For air, $k = 1.4$.

Table 1.10 *Approximate Compressibilities of Common Liquids at 1 atm*

liquid	temperature	β (in^2/lbf)	β (1/atm)
mercury	32°F	0.027×10^{-5}	0.39×10^{-5}
glycerin	60°F	0.16×10^{-5}	2.4×10^{-5}
water	60°F	0.33×10^{-5}	4.9×10^{-5}
ethyl alcohol	32°F	0.68×10^{-5}	10×10^{-5}
chloroform	32°F	0.68×10^{-5}	10×10^{-5}
gasoline	60°F	1.0×10^{-5}	15×10^{-5}
hydrogen	20K	11×10^{-5}	160×10^{-5}
helium	2.1K	48×10^{-5}	700×10^{-5}

(Multiply 1/psi by 14.696 to obtain 1/atm.)

(Multiply in^2/lbf by 0.145 to obtain 1/kPa.)

Example 1.8

Water at 68°F (20°C) and 1 atm has a density of 62.3 lbm/ft^3 (997 kg/m^3). What is the new density if the pressure is isothermally increased from 14.7 lbf/in^2 to 400 lbf/in^2 (100 kPa to 2760 kPa)? The bulk modulus has a constant value of 320,000 lbf/in^2 (2.2×10^6 kPa).

SI Solution

Compressibility is the reciprocal of the bulk modulus.

$$\beta = \frac{1}{E} = \frac{1}{2.2 \times 10^6 \text{ kPa}} = 4.55 \times 10^{-7} \text{ 1/kPa}$$

From Eq. 1.37,

$$\rho_2 = \rho_1\big(1 + \beta(P_2 - P_1)\big)$$

$$= \left(997 \ \frac{\text{kg}}{\text{m}^3}\right)\left(\begin{array}{l} 1 + \left(4.55 \times 10^{-7} \ \dfrac{1}{\text{kPa}}\right) \\ \quad \times (2760 \text{ kPa} - 100 \text{ kPa}) \end{array}\right)$$

$$= 998.2 \text{ kg/m}^3$$

Customary U.S. Solution

Compressibility is the reciprocal of the bulk modulus.

$$\beta = \frac{1}{E} = \frac{1}{320,000 \ \dfrac{\text{lbf}}{\text{in}^2}} = 0.3125 \times 10^{-5} \text{ in}^2/\text{lbf}$$

From Eq. 1.37,

$$\rho_2 = \rho_1(1 + \beta(P_2 - P_1))$$

$$= \left(62.3 \ \frac{\text{lbm}}{\text{ft}^3}\right)\left(\begin{array}{l} 1 + \left(0.3125 \times 10^{-5} \ \dfrac{\text{in}^2}{\text{lbf}}\right) \\ \quad \times \left(400 \ \dfrac{\text{lbf}}{\text{in}^2} - 14.7 \ \dfrac{\text{lbf}}{\text{in}^2}\right) \end{array}\right)$$

$$= 62.38 \text{ lbm/ft}^3$$

19. BULK MODULUS

The *bulk modulus*, E, of a fluid is analogous to the modulus of elasticity of a solid.[29] Typical units are lbf/in^2, atm, and kPa. The term ΔP in Eq. 1.40 represents an increase in stress. The term $\Delta V/V_0$ is a *volumetric strain*. Analogous to Hooke's law describing elastic formation, the *bulk modulus* of a fluid (liquid or gas) is given by Eq. 1.40.

$$E = \frac{\text{stress}}{\text{strain}} = \frac{-\Delta P}{\dfrac{\Delta V}{V_0}} \qquad \text{[SI]} \quad 1.40(a)$$

$$E = -V_0\left(\frac{\partial P}{\partial V}\right)_T \qquad \text{[U.S.]} \quad 1.40(b)$$

The term *secant bulk modulus* is associated with Eq. 1.40(a) (the average slope), while the terms *tangent bulk modulus* and *point bulk modulus* are associated with Eq. 1.40(b) (the instantaneous slope).

The bulk modulus is the reciprocal of compressibility.

$$E = \frac{1}{\beta} \qquad 1.41$$

The bulk modulus changes only slightly with temperature. The bulk modulus of water is usually taken as 300,000 lbf/in^2 (2.1×10^6 kPa) unless greater accuracy is required, in which case Table 1.11 can be used.

Table 1.11 *Approximate Bulk Modulus of Water*

pressure (lbf/in^2)	32°F	68°F	120°F	200°F	300°F
	(thousands of lbf/in^2)				
15	292	320	332	308	–
1500	300	330	340	319	218
4500	317	348	362	338	271
15,000	380	410	420	405	350

(Multiply lbf/in^2 by 6.8948 to obtain kPa.)

Reprinted with permission from Victor L. Streeter, *Handbook of Fluid Dynamics*, © 1961, by McGraw-Hill Book Company.

---

[29]To distinguish it from the modulus of elasticity, the bulk modulus is represented by the symbol B when dealing with solids.

20. PROPERTIES OF MIXTURES OF NONREACTING LIQUIDS

There are very few convenient ways of predicting the properties of nonreacting, nonvolatile organic and aqueous solutions (acids, brines, alcohol mixtures, coolants, etc.) from the individual properties of the components.

Volumes of two nonreacting organic liquids (e.g., acetone and chloroform) in a mixture are essentially additive. The volume change upon mixing will seldom be more than a few tenths of a percent. The volume change in aqueous solutions is often slightly greater, but is still limited to a few percent (e.g., 3% for some solutions of methanol and water). Therefore, the specific gravity (density, specific weight, etc.) can be considered to be a volumetric weighting of the individual specific gravities.

A rough estimate of the absolute viscosity of a mixture of two or more liquids having different viscosities (at the same temperature) can be found from the mole fractions, x_i, of the components.

$$\mu_{\text{mixture}} \approx \frac{1}{\sum_i \dfrac{x_i}{\mu_i}} = \frac{\sum_i \dfrac{m_i}{M_i}}{\sum_i \dfrac{m_i}{M_i \mu_i}} \qquad 1.42$$

A three-step procedure for calculating a more reliable estimate of the viscosity of a mixture of two or more liquids starts by using the *Refutas equation* to calculate a linearized *viscosity blending index*, VBI (or, *viscosity blending number*, VBN), for each component.

$$\text{VBI}_i = 10.975 + 14.534 \times \ln\big(\ln(\nu_{i,\text{cSt}} + 0.8)\big) \qquad 1.43$$

The second step calculates the average VBI of the mixture from the gravimetrically weighted component VBIs.

$$\text{VBI}_{\text{mixture}} = \sum_i G_i \times \text{VBI}_i \qquad 1.44$$

The final step extracts the mixture viscosity from the mixture VBI by inverting the Refutas equation.

$$\nu_{\text{mixture,cSt}} = \exp\left(\exp\left(\frac{\text{VBI}_{\text{mixture}} - 10.975}{14.534}\right)\right) - 0.8 \qquad 1.45$$

Components of mixtures of hydrocarbons may be referred to as pseudocomponents.[30] *Raoult's law* can be used to calculate the vapor pressure above a liquid mixture from the vapor pressures of the liquid components.

$$P_{v,\text{mixture}} = \sum_i x_i P_{v,i} \qquad 1.46$$

21. PRESSURE-MEASURING DEVICES

There are many devices for measuring and indicating fluid pressure. Some devices measure gage pressure; others measure absolute pressure. The effects of nonstandard atmospheric pressure and nonstandard gravitational acceleration must be determined, particularly for devices relying on columns of liquid to indicate pressure. Table 1.12 lists the common types of devices and the ranges of pressure appropriate for each.

Table 1.12 *Common Pressure-Measuring Devices*

device	approximate range (in atm)
water manometer	0–0.1
mercury barometer	0–1
mercury manometer	0.001–1
metallic diaphragm	0.01–200
transducer	0.001–15,000
Bourdon pressure gauge	1–3000
Bourdon vacuum gauge	0.1–1

The *Bourdon pressure gauge* is the most common pressure-indicating device. (See Fig. 1.10.) This mechanical device consists of a C-shaped or helical hollow tube that tends to straighten out (i.e., unwind) when the tube is subjected to an internal pressure. The gauge is referred to as a *C-Bourdon gauge* because of the shape of the hollow tube. The degree to which the coiled tube unwinds depends on the difference between the internal and external pressures. A Bourdon gauge directly indicates *gage pressure*. Extreme accuracy is generally not a characteristic of Bourdon gauges.

Figure 1.10 *C-Bourdon Pressure Gauge*

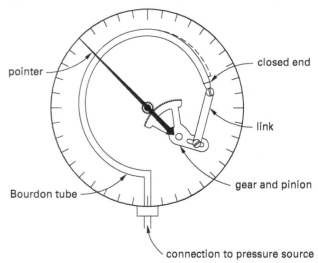

---

[30]A *pseudocomponent* represents a mixture of components and has the thermodynamic behavior of a mixture, rather than that of a single component. The term can refer to the performance of a blend of known components, or to the performance of a component of a known mixture.

In non-SI installations, gauges are always calibrated in psi (or psid). Vacuum pressure is usually calibrated in inches of mercury. SI gauges are marked in kilopascals (kPa) or bars, although kg/cm² may be used in older gauges. Negative numbers are used to indicate vacuum. The gauge dial will be clearly marked if other units are indicated.

The *barometer* is a common device for measuring the absolute pressure of the atmosphere.[31] It is constructed by filling a long tube open at one end with mercury (or alcohol, or some other liquid) and inverting the tube so that the open end is below the level of a mercury-filled container. If the vapor pressure of the mercury in the tube is neglected, the fluid column will be supported only by the atmospheric pressure transmitted through the container fluid at the lower, open end.

Strain gauges, *diaphragm gauges*, *quartz-crystal trans-ducers*, and other devices using the *piezoelectric effect* are also used to measure stress and pressure, particularly when pressure fluctuates quickly (e.g., as in a rocket combustion chamber). With these devices, calibration is required to interpret pressure from voltage generation or changes in resistance, capacitance, or inductance. These devices are generally unaffected by atmospheric pressure or gravitational acceleration.

Manometers (*U-tube manometers*) can also be used to indicate small pressure differences, and for this purpose they provide great accuracy. (Manometers are not suitable for measuring pressures much larger than 10 lbf/in² (70 kPa), however.) A difference in manometer fluid surface heights is converted into a pressure difference. If one end of a manometer is open to the atmosphere, the manometer indicates gage pressure. It is theoretically possible, but impractical, to have a manometer indicate absolute pressure, since one end of the manometer would have to be exposed to a perfect vacuum.

A *static pressure tube* (*piezometer tube*) is a variation of the manometer. (See Fig. 1.11.) It is a simple method of determining the static pressure in a pipe or other vessel, regardless of fluid motion in the pipe. A vertical transparent tube is connected to a hole in the pipe wall.[32] (No part of the tube projects into the pipe.) The static pressure will force the contents of the pipe up into the tube. The height of the contents will be an indication of gage pressure in the pipe.

The device used to measure the pressure should not be confused with the method used to obtain exposure to the pressure. For example, a static pressure *tap* in a pipe

is merely a hole in the pipe wall. A Bourdon gauge, manometer, or transducer can then be used with the tap to indicate pressure.

Tap holes are generally ⅛–¼ in (3–6 mm) in diameter, drilled at right angles to the wall, and smooth and flush with the pipe wall. No part of the gauge or connection projects into the pipe. The tap holes should be at least 5 to 10 pipe diameters downstream from any source of turbulence (e.g., a bend, fitting, or valve).

Figure 1.11 *Static Pressure Tube*

22. MANOMETERS

Figure 1.12 illustrates a simple U-tube manometer used to measure the difference in pressure between two vessels. When both ends of the manometer are connected to pressure sources, the name *differential manometer* is used. If one end of the manometer is open to the atmosphere, the name *open manometer* is used.[33] The open manometer implicitly measures gage pressures.

Figure 1.12 *Simple U-Tube Manometer*

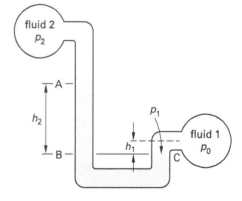

---

[31]A barometer can be used to measure the pressure inside any vessel. However, the barometer must be completely enclosed in the vessel, which may not be possible. Also, it is difficult to read a barometer enclosed within a tank.

[32]Where greater accuracy is required, multiple holes may be drilled around the circumference of the pipe and connected through a manifold (*piezometer ring*) to the pressure-measuring device.

[33]If one of the manometer legs is inclined, the term *inclined manometer* or *draft gauge* is used. Although only the vertical distance between the manometer fluid surfaces should be used to calculate the pressure difference, with small pressure differences it may be more accurate to read the inclined distance (which is larger than the vertical distance) and compute the vertical distance from the angle of inclination.

Since the pressure at point B in Fig. 1.12 is the same as at point C, the pressure differential produces the vertical fluid column of height h. In Eq. 1.48, A is the cross-sectional area of the tube.

$$F_{net} = F_C - F_A = \text{weight of fluid column AB} \quad \textit{1.47}$$

$$(P_2 - P_1)A = \rho g h A \quad \textit{1.48}$$

In Eq. 1.49, the area cancels from both sides of the equation. In the absence of any capillary action, then, the inside diameter of the manometer tube is irrelevant. The difference in pressure between the two vessels is

Pressure Field in a Static Liquid

$$P_2 - P_1 = -\gamma(z_2 - z_1) = -\gamma h = -\rho g h \quad \textit{1.49}$$

Equation 1.49 assumes that the manometer fluid height is small, or that only low-density gases fill the tubes above the manometer fluid. If a high-density fluid (such as water) is present above the measuring fluid, or if the columns h_1 or h_2 are very long, corrections will be necessary. (See Fig. 1.13.)

Figure 1.13 *Manometer Requiring Corrections*

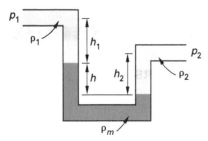

Fluid column h_2 "sits on top" of the manometer fluid, forcing the manometer fluid to the left. This increase must be subtracted out. Similarly, the column h_1 restricts the movement of the manometer fluid. The observed measurement must be increased to correct for this restriction.

$$P_2 - P_1 = g(\rho_m h + \rho_1 h_1 - \rho_2 h_2) \quad \text{[SI]} \quad \textit{1.50(a)}$$

$$\begin{aligned} p_2 - p_1 &= (\rho_m h + \rho_1 h_1 - \rho_2 h_2) \times \frac{g}{g_c} \quad \text{[U.S.]} \quad \textit{1.50(b)} \\ &= \gamma_m h + \gamma_1 h_1 - \gamma_2 h_2 \end{aligned}$$

When a manometer is used to measure the pressure difference across an orifice or other fitting where the same liquid exists in both manometer sides (shown in Fig. 1.14), it is not necessary to correct the manometer reading for all of the liquid present above the manometer fluid. This is because parts of the correction for both sides of the manometer are the same. Therefore, the distance y in Fig. 1.14 is an irrelevant distance.

Figure 1.14 *Irrelevant Distance, y*

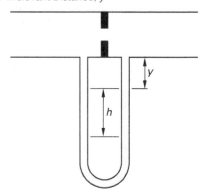

Manometer tubes are generally large enough in diameter to avoid significant capillary effects. Corrections for capillarity are seldom necessary.

Example 1.9

The pressure at the bottom of a water tank ($\rho = 62.4$ lbm/ft^3 (998 kg/m^3)) is measured with a mercury manometer located below the tank bottom, as shown. (The density of mercury is 848 lbm/ft^3 (13 575 kg/m^3).) What is the gage pressure at the bottom of the water tank?

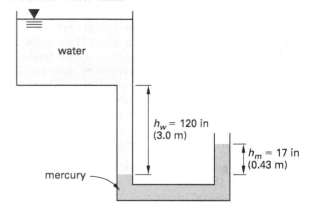

SI Solution

From Eq. 1.50(a),

$$\begin{aligned} \Delta P &= g(\rho_m h_m - \rho_w h_w) \\ &= \left(9.81 \; \frac{m}{s^2}\right) \left(\begin{array}{l} \left(13\,575 \; \dfrac{kg}{m^3}\right)(0.43 \text{ m}) \\[6pt] -\left(998 \; \dfrac{kg}{m^3}\right)(3.0 \text{ m}) \end{array} \right) \\ &= 27\,892 \text{ Pa} \quad (27.9 \text{ kPa gage}) \end{aligned}$$

Customary U.S. Solution

From Eq. 1.50(b),

$$\Delta P = (\rho_m h_m - \rho_w h_w) \times \frac{g}{g_c}$$

$$= \frac{\left(848 \ \frac{\text{lbm}}{\text{ft}^3}\right)(17 \ \text{in}) - \left(62.4 \ \frac{\text{lbm}}{\text{ft}^3}\right)(120 \ \text{in})}{\left(12 \ \frac{\text{in}}{\text{ft}}\right)^3}$$

$$\times \left(\frac{32.2 \ \frac{\text{ft}}{\text{sec}^2}}{32.2 \ \frac{\text{lbm-ft}}{\text{lbf-sec}^2}}\right)$$

$$= 4.01 \ \text{lbf/in}^2 \quad (4.01 \ \text{psig})$$

23. HYDROSTATIC PRESSURE

Hydrostatic pressure is the pressure a fluid exerts on an immersed object or container walls.[34] Pressure is equal to the force per unit area of surface.

$$P = \frac{F}{A} \qquad\qquad 1.51$$

Hydrostatic pressure in a stationary, incompressible fluid behaves according to the following characteristics.

- Pressure is a function of vertical depth (and density) only. The pressure will be the same at two points with identical depths.

- Pressure varies linearly with (vertical and inclined) depth.

- Pressure is independent of an object's area and size and the weight (mass) of water above the object. Figure 1.15 illustrates the *hydrostatic paradox*. The pressures at depth h are the same in all four columns because pressure depends on depth, not volume.

Figure 1.15 *Hydrostatic Paradox*

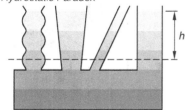

- Pressure at a point has the same magnitude in all directions (*Pascal's law*). Therefore, pressure is a scalar quantity.

- Pressure is always normal to a surface, regardless of the surface's shape or orientation. (This is a result of the fluid's inability to support shear stress.)

- The resultant of the pressure distribution acts through the *center of pressure*.

- The center of pressure rarely coincides with the average depth.

24. FLUID HEIGHT EQUIVALENT TO PRESSURE

Pressure varies linearly with depth. The relationship between pressure and depth (i.e., the *hydrostatic head*) for an incompressible fluid is given by Eq. 1.52.

$$P = \rho g h \qquad \text{[SI]} \quad 1.52(a)$$

$$P = \frac{\rho g h}{g_c} = \gamma h \qquad \text{[U.S.]} \quad 1.52(b)$$

Since ρ and g are constants, Eq. 1.52 shows that P and h are linearly related. Knowing one determines the other.[35] For example, the height of a fluid column needed to produce a pressure is

$$h = \frac{P}{\rho g} \qquad \text{[SI]} \quad 1.53(a)$$

$$h = \frac{P g_c}{\rho g} = \frac{P}{\gamma} \qquad \text{[U.S.]} \quad 1.53(b)$$

Table 1.13 lists six common fluid height equivalents.[36]

Table 1.13 *Approximate Fluid Height Equivalents at 68°F (20°C)*

liquid	height equivalents	
water	0.0361 psi/in	27.70 in/psi
water	62.4 psf/ft	0.01603 ft/psf
water	9.81 kPa/m	0.1019 m/kPa
water	0.4329 psi/ft	2.31 ft/psi
mercury	0.491 psi/in	2.036 in/psi
mercury	133.3 kPa/m	0.00750 m/kPa

---

[34]The term *hydrostatic* is used with all fluids, not only with water.
[35]In fact, pressure and height of a fluid column can be used interchangeably. The height of a fluid column is known as *head*. For example: "The fan developed a static head of 3 in of water," or "The pressure head at the base of the water tank was 8 m." When the term "head" is used, it is essential to specify the fluid.
[36]Of course, these values are recognized to be the approximate specific weights of the liquids.

A barometer is a device that measures atmospheric pressure by the height of a fluid column. If the vapor pressure of the barometer liquid is neglected, the atmospheric pressure will be given by Eq. 1.54.

$$P_a = \rho g h \qquad \text{[SI]} \qquad \textit{1.54(a)}$$

$$P_a = \frac{\rho g h}{g_c} = \gamma h \qquad \text{[U.S.]} \qquad \textit{1.54(b)}$$

If the vapor pressure of the barometer liquid is significant (as it would be with alcohol or water), the vapor pressure effectively reduces the height of the fluid column, as Eq. 1.55 illustrates.

$$P_a - P_v = \rho g h \qquad \text{[SI]} \qquad \textit{1.55(a)}$$

$$P_a - P_v = \frac{\rho g h}{g_c} = \gamma h \qquad \text{[U.S.]} \qquad \textit{1.55(b)}$$

Example 1.10

A vacuum pump is used to drain a flooded mine shaft of 68°F (20°C) water.[37] The vapor pressure of water at this temperature is 0.34 lbf/in² (2.34 kPa). The pump is incapable of lifting the water higher than 400 in (10.16 m). What is the atmospheric pressure?

SI Solution

From Eq. 1.55,

$$P_a = P_v + \rho g h$$

$$= 2.34 \text{ kPa} + \frac{\left(998 \ \dfrac{\text{kg}}{\text{m}^3}\right)\left(9.81 \ \dfrac{\text{m}}{\text{s}^2}\right)(10.16 \text{ m})}{1000 \ \dfrac{\text{Pa}}{\text{kPa}}}$$

$$= 101.8 \text{ kPa}$$

(Alternate SI solution, using Table 1.13)

$$P_a = P_v + \rho g h = 2.34 \text{ kPa} + \left(9.81 \ \frac{\text{kPa}}{\text{m}}\right)(10.16 \text{ m})$$

$$= 102 \text{ kPa}$$

Customary U.S. Solution

From Table 1.13, the height equivalent of water is approximately 0.0361 psi/in. The unit psi/in is the same

as lbf/in³, the units of γ. From Eq. 1.55, the atmospheric pressure is

$$P_a = P_v + \rho g h = \rho_v + \gamma h$$

$$= 0.34 \ \frac{\text{lbf}}{\text{in}^2} + \left(0.0361 \ \frac{\text{lbf}}{\text{in}^3}\right)(400 \text{ in})$$

$$= 14.78 \text{ lbf/in}^2 \quad (14.78 \text{ psia})$$

25. MULTIFLUID BAROMETERS

It is theoretically possible to fill a barometer tube with several different immiscible fluids.[38] Upon inversion, the fluids will separate, leaving the most dense fluid at the bottom and the least dense fluid at the top. All of the fluids will contribute, by superposition, to the balance between the external atmospheric pressure and the weight of the fluid column.

$$P_a - P_v = g \sum \rho_i h_i \qquad \text{[SI]} \qquad \textit{1.56(a)}$$

$$P_a - P_v = \frac{g}{g_c} \sum \rho_i h_i = \sum \gamma_i h_i \qquad \text{[U.S.]} \qquad \textit{1.56(b)}$$

The pressure at any intermediate point within the fluid column is found by starting at a location where the pressure is known, and then adding or subtracting $\rho g h$ terms to get to the point where the pressure is needed. Usually, the known pressure will be the atmospheric pressure located in the barometer barrel at the level (elevation) of the fluid outside of the barometer.

Example 1.11

Neglecting vapor pressure, what is the pressure of the air at point E in the container shown? The atmospheric pressure is 1.0 atm (14.7 lbf/in², 101 300 Pa).

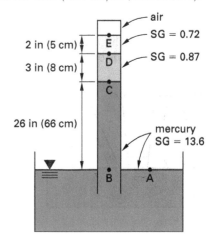

SI Solution

$$P_E = P_a - g\rho_w \sum (SG_i) h_i$$

$$= 101\,300 \text{ Pa} - \left(9.81 \frac{\text{m}}{\text{s}^2}\right) \left(1000 \frac{\text{kg}}{\text{m}^3}\right)$$

$$\times \left(\begin{array}{c} (0.66 \text{ m})(13.6) + (0.08 \text{ m})(0.87) \\ + (0.05 \text{ m})(0.72) \end{array} \right)$$

$$= 12\,210 \text{ Pa} \quad (12.2 \text{ kPa})$$

Customary U.S. Solution

The pressure at point B is the same as the pressure at point A, which is 1.0 atm. The density of mercury is 13.6×0.0361 lbm/in³ $= 0.491$ lbm/in³, and the specific weight is 0.491 lbf/in³. Therefore, the pressure at point C is

$$P_C = 14.7 \frac{\text{lbf}}{\text{in}^2} - (26 \text{ in}) \left(0.491 \frac{\frac{\text{lbf}}{\text{in}^2}}{\text{in}} \right)$$

$$= 1.93 \text{ lbf/in}^2 \quad (1.93 \text{ psia})$$

Similarly, the pressure at point E (and anywhere within the captive air space) is

$$P_E = 14.7 \frac{\text{lbf}}{\text{in}^2} - (26 \text{ in}) \left(0.491 \frac{\text{lbf}}{\text{in}^3} \right)$$

$$- (3 \text{ in})(0.87) \left(0.0361 \frac{\text{lbf}}{\text{in}^3} \right)$$

$$- (2 \text{ in})(0.72) \left(0.0361 \frac{\text{lbf}}{\text{in}^3} \right)$$

$$= 1.79 \text{ lbf/in}^2 \quad (1.79 \text{ psia})$$

26. PRESSURE ON A HORIZONTAL PLANE SURFACE

The pressure on a horizontal plane surface is uniform over the surface because the depth of the fluid is uniform. (See Fig. 1.16.) The resultant of the pressure distribution acts through the center of pressure of the surface, which corresponds to the centroid of the surface.

Figure 1.16 Hydrostatic Pressure on a Horizontal Plane Surface

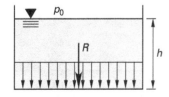

The uniform pressure at depth h is given by Eq. 1.57.[39] [Pressure Field in a Static Liquid]

$$P = \rho g h \qquad \text{[SI]} \qquad 1.57(a)$$

$$P = \frac{\rho g h}{g_c} = \gamma h \qquad \text{[U.S.]} \qquad 1.57(b)$$

The total vertical force on the horizontal plane of area A is given by Eq. 1.58.

$$R = PA \qquad 1.58$$

It is tempting, but not always correct, to calculate the vertical force on a submerged surface as the weight of the fluid above it. Such an approach works only when there is no change in the cross-sectional area of the fluid above the surface. This is a direct result of the *hydrostatic paradox*. (See Sec. 1.23.) Figure 1.17 illustrates two containers with the same pressure distribution (force) on their bottom surfaces.

Figure 1.17 Two Containers with the Same Pressure Distribution

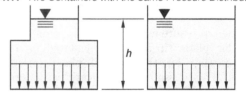

27. PRESSURE ON A RECTANGULAR VERTICAL PLANE SURFACE

The pressure on a vertical rectangular plane surface increases linearly with depth. The pressure distribution will be triangular, as in Fig. 1.18(a), if the plane surface extends to the surface; otherwise, the distribution will be trapezoidal, as in Fig. 1.18(b).

Figure 1.18 Hydrostatic Pressure on a Vertical Plane Surface

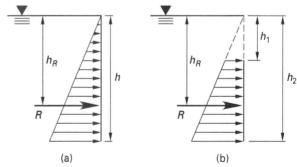

(a)   (b)

The resultant force is calculated from the *average pressure*.

$$\overline{P} = \tfrac{1}{2}(P_1 + P_2) \qquad 1.59$$

---

[39]The phrase *pressure at a depth* is universally understood to mean the *gage pressure*, as given by Eq. 1.57.

$$\overline{P} = \tfrac{1}{2}\rho g(h_1 + h_2) \qquad \text{[SI]} \quad 1.60\text{(a)}$$

$$\overline{P} = \frac{\tfrac{1}{2}\rho g(h_1 + h_2)}{g_c} = \tfrac{1}{2}\gamma(h_1 + h_2) \qquad \text{[U.S.]} \quad 1.60\text{(b)}$$

$$R = \overline{P} A \qquad 1.61$$

Although the resultant is calculated from the average depth, it does not act at the average depth. The resultant of the pressure distribution passes through the centroid of the pressure distribution. For the triangular distribution of Fig. 1.18(a), the resultant is located at a depth of $h_R = \tfrac{2}{3}h$. For the more general case of Fig. 1.18(b), the resultant is located from Eq. 1.62.

$$h_R = \tfrac{2}{3}\left(h_1 + h_2 - \frac{h_1 h_2}{h_1 + h_2}\right) \qquad 1.62$$

28. PRESSURE ON A RECTANGULAR INCLINED PLANE SURFACE

The average pressure and resultant force on an inclined rectangular plane surface are calculated in a similar fashion as that for the vertical plane surface. (See Fig. 1.19.) The pressure varies linearly with depth. The resultant is calculated from the average pressure, which, in turn, depends on the average depth.

Figure 1.19 *Hydrostatic Pressure on an Inclined Rectangular Plane Surface*

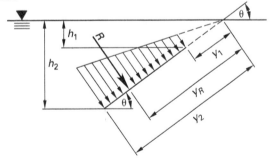

The average pressure and resultant are found using Eq. 1.63 through Eq. 1.65.

$$\overline{P} = \tfrac{1}{2}(P_1 + P_2) \qquad 1.63$$

$$\overline{P} = \tfrac{1}{2}\rho g(h_1 + h_2)$$
$$= \tfrac{1}{2}\rho g(h_3 + h_4)\sin\theta \qquad \text{[SI]} \quad 1.64\text{(a)}$$

$$\overline{P} = \frac{\tfrac{1}{2}\rho g(h_1 + h_2)}{g_c} = \frac{\tfrac{1}{2}\rho g(h_3 + h_4)\sin\theta}{g_c} \qquad \text{[U.S.]} \quad 1.64\text{(b)}$$
$$= \tfrac{1}{2}\gamma(h_1 + h_2) = \tfrac{1}{2}\gamma(h_3 + h_4)\sin\theta$$

$$R = \overline{P} A \qquad 1.65$$

As with the vertical plane surface, the resultant acts at the centroid of the pressure distribution, not at the average depth. Equation 1.62 is rewritten in terms of inclined depths.[40]

$$h_R = \left(\frac{\tfrac{2}{3}}{\sin\theta}\right)\left(h_1 + h_2 - \frac{h_1 h_2}{h_1 + h_2}\right)$$
$$= \tfrac{2}{3}\left(h_3 + h_4 - \frac{h_3 h_4}{h_3 + h_4}\right) \qquad 1.66$$

Example 1.12

The tank shown is filled with water ($\rho = 62.4$ lbm/ft^3 (1000 kg/m^3)). (a) What is the total resultant force on a 1 ft (1 m) width of the inclined portion of the wall?[41] (b) At what depth (vertical distance) is the resultant force on the inclined portion of the wall located?

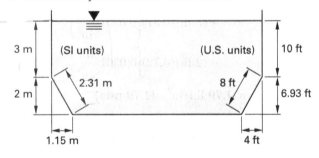

SI Solution

(a) The depth of the tank bottom is

$$h_2 = 3\text{ m} + 2\text{ m} = 5\text{ m}$$

---

[40]h_R is an inclined distance. If a vertical distance is wanted, it must usually be calculated from h_R and sin θ. Equation 1.66 can be derived simply by dividing Eq. 1.62 by sin θ.
[41]Since the width of the tank (the distance into and out of the illustration) is unknown, it is common to calculate the pressure or force per unit width of tank wall. This is the same as calculating the pressure on a 1 ft (1 m) wide section of wall.

From Eq. 1.64, the average gage pressure on the inclined section is

$$\overline{P} = \frac{1}{2}\rho g(h_1 + h_2)$$

$$= \left(\frac{1}{2}\right)\left(1000 \ \frac{\text{kg}}{\text{m}^3}\right)\left(9.81 \ \frac{\text{m}}{\text{s}^2}\right)(3 \text{ m} + 5 \text{ m})$$

$$= 39\,240 \text{ Pa} \quad \text{(gage)}$$

The total resultant force on a 1 m section of the inclined portion of the wall is

$$R = \overline{P}A$$

$$= (39\,240 \text{ Pa})(2.31 \text{ m})(1 \text{ m})$$

$$= 90\,644 \text{ N} \quad (90.6 \text{ kN})$$

(b) θ must be known to determine h_R.

$$\theta = \arctan\frac{2 \text{ m}}{1.15 \text{ m}}$$

$$= 60°$$

From Eq. 1.66, the location of the resultant can be calculated once h_3 and h_4 are known.

$$h_3 = \frac{3 \text{ m}}{\sin 60°}$$

$$= 3.464 \text{ m}$$

$$h_4 = \frac{5 \text{ m}}{\sin 60°}$$

$$= 5.774 \text{ m}$$

$$h_R = \left(\frac{2}{3}\right)\left(3.464 \text{ m} + 5.774 \text{ m} - \frac{(3.464 \text{ m})(5.774 \text{ m})}{3.464 \text{ m} + 5.774 \text{ m}}\right)$$

$$= 4.715 \text{ m} \quad \text{[inclined]}$$

The vertical depth at which the resultant force on the inclined portion of the wall acts is

$$h = h_R \sin\theta$$

$$= (4.715 \text{ m})\sin 60°$$

$$= 4.08 \text{ m} \quad \text{[vertical]}$$

Customary U.S. Solution

(a) The depth of the tank bottom is

$$h_2 = 10 \text{ ft} + 6.93 \text{ ft}$$

$$= 16.93 \text{ ft}$$

From Eq. 1.64, the average gage pressure on the inclined section is

$$\overline{P} = \frac{\frac{1}{2}\rho g(h_1 + h_2)}{g_c}$$

$$= \frac{\left(\frac{1}{2}\right)\left(62.4 \ \frac{\text{lbm}}{\text{ft}^3}\right)\left(32.2 \ \frac{\text{ft}}{\text{sec}^2}\right)}{32.2 \ \frac{\text{lbm-ft}}{\text{lbf-sec}^2}}$$

$$\times (10 \text{ ft} + 16.93 \text{ ft})$$

$$= 840.2 \text{ lbf/ft}^2 \quad \text{(gage)}$$

The total resultant force on a 1 ft section of the inclined portion of the wall is

$$R = \overline{P}A = \left(840.2 \ \frac{\text{lbf}}{\text{ft}^2}\right)(8 \text{ ft})(1 \text{ ft}) = 6722 \text{ lbf}$$

(b) θ must be known to determine h_R.

$$\theta = \arctan\frac{6.93 \text{ ft}}{4 \text{ ft}} = 60°$$

From Eq. 1.66, the location of the resultant can be calculated once h_3 and h_4 are known.

$$h_3 = \frac{10 \text{ ft}}{\sin 60°} = 11.55 \text{ ft}$$

$$h_4 = \frac{16.93 \text{ ft}}{\sin 60°} = 19.55 \text{ ft}$$

From Eq. 1.66,

$$h_R = \left(\frac{2}{3}\right)\left(11.55 \text{ ft} + 19.55 \text{ ft} - \frac{(11.55 \text{ ft})(19.55 \text{ ft})}{11.55 \text{ ft} + 19.55 \text{ ft}}\right)$$

$$= 15.89 \text{ ft} \quad \text{[inclined]}$$

The vertical depth at which the resultant force on the inclined portion of the wall acts is

$$h = h_R \sin\theta = (15.89 \text{ ft})\sin 60°$$

$$= 13.76 \text{ ft} \quad \text{[vertical]}$$

29. PRESSURE ON A GENERAL PLANE SURFACE

Figure 1.20 illustrates a nonrectangular plane surface that may or may not extend to the liquid surface and that may or may not be inclined, as was shown in Fig. 1.19. As with other regular surfaces, the resultant force depends on the average pressure and acts through the *center of pressure* (CP). The average pressure is calculated from the depth of the surface's centroid (center of gravity, CG).

$$\overline{P} = \rho g h_c \sin\theta \qquad \text{[SI]} \quad 1.67(a)$$

$$\overline{P} = \frac{\rho g h_c \sin\theta}{g_c} = \gamma h_c \sin\theta \qquad \text{[U.S.]} \quad 1.67(b)$$

$$R = \overline{P} A \qquad 1.68$$

Figure 1.20 *General Plane Surface*

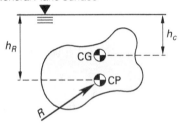

The resultant force acts at depth h_R normal to the plane surface. I_c in Eq. 1.69 is the centroidal area moment of inertia about an axis parallel to the surface. Both h_c and h_R are measured parallel to the plane surface. That is, if the plane surface is inclined, h_c and h_R are inclined distances.

$$h_R = h_c + \frac{I_c}{A h_c} \qquad 1.69$$

Example 1.13

The top edge of a vertical circular observation window is located 4.0 ft (1.25 m) below the surface of the water. The window is 1.0 ft (0.3 m) in diameter. The water's density is 62.4 lbm/ft³ (1000 kg/m³). Neglect the salinity of the water. (a) What is the resultant force on the window? (b) At what depth does the resultant force act?

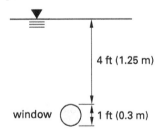

SI Solution

(a) The radius of the window is

$$r = \frac{0.3 \text{ m}}{2} = 0.15 \text{ m}$$

The depth at which the centroid of the circular window is located is

$$h_c = 1.25 \text{ m} + 0.15 \text{ m} = 1.4 \text{ m}$$

The area of the circular window is

$$A = \pi r^2 = \pi(0.15 \text{ m})^2 = 0.0707 \text{ m}^2$$

The average pressure is

$$\overline{P} = \rho g h_c = \left(1000 \ \frac{\text{kg}}{\text{m}^3}\right)\left(9.81 \ \frac{\text{m}}{\text{s}^2}\right)(1.4 \text{ m})$$
$$= 13\,734 \text{ Pa} \quad \text{(gage)}$$

The resultant is calculated from Eq. 1.65.

$$R = \overline{P} A = (13\,734 \text{ Pa})(0.0707 \text{ m}^2) = 971 \text{ N}$$

(b) The centroidal area moment of inertia of a circle is

$$I_c = \frac{\pi}{4} r^4 = \left(\frac{\pi}{4}\right)(0.15 \text{ m})^4 = 3.976 \times 10^{-4} \text{ m}^4$$

From Eq. 1.69, the depth at which the resultant force acts is

$$h_R = h_c + \frac{I_c}{A h_c} = 1.4 \text{ m} + \frac{3.976 \times 10^{-4} \text{ m}^4}{(0.0707 \text{ m}^2)(1.4 \text{ m})}$$
$$= 1.404 \text{ m}$$

Customary U.S. Solution

(a) The radius of the window is

$$r = \frac{1 \text{ ft}}{2} = 0.5 \text{ ft}$$

The depth at which the centroid of the circular window is located is

$$h_c = 4.0 \text{ ft} + 0.5 \text{ ft} = 4.5 \text{ ft}$$

The area of the circular window is

$$A = \pi r^2 = \pi(0.5 \text{ ft})^2 = 0.7854 \text{ ft}^2$$

The average pressure is

$$\overline{P} = \gamma h_c = \left(62.4 \ \frac{\text{lbf}}{\text{ft}^3}\right)(4.5 \text{ ft})$$
$$= 280.8 \text{ lbf/ft}^2 \quad \text{(psfg)}$$

The resultant is calculated from Eq. 1.65.

$$R = \overline{P}A = \left(280.8 \ \frac{\text{lbf}}{\text{ft}^2}\right)(0.7854 \ \text{ft}^2)$$
$$= 220.5 \ \text{lbf}$$

(b) The centroidal area moment of inertia of a circle is

$$I_c = \frac{\pi}{4}r^4 = \left(\frac{\pi}{4}\right)(0.5 \ \text{ft})^4$$
$$= 0.049 \ \text{ft}^4$$

From Eq. 1.69, the depth at which the resultant force acts is

$$h_R = h_c + \frac{I_c}{Ah_c}$$
$$= 4.5 \ \text{ft} + \frac{0.049 \ \text{ft}^4}{(0.7854 \ \text{ft}^2)(4.5 \ \text{ft})}$$
$$= 4.514 \ \text{ft}$$

30. SPECIAL CASES: VERTICAL SURFACES

Several simple wall shapes and configurations recur frequently. Figure 1.21 indicates the depths, h_R, of their hydrostatic pressure resultants (*centers of pressure*). In all cases, the surfaces are vertical and extend to the liquid's surface.

Figure 1.21 Centers of Pressure for Common Configurations

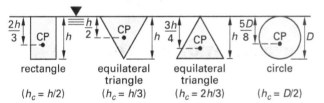

rectangle	equilateral triangle	equilateral triangle	circle
($h_c = h/2$)	($h_c = h/3$)	($h_c = 2h/3$)	($h_c = D/2$)

31. FORCES ON CURVED AND COMPOUND SURFACES

Figure 1.22 illustrates a curved surface cross section, BA. The resultant force acting on such a curved surface is not difficult to determine, although the x- and y-components of the resultant usually must be calculated first. The magnitude and direction of the resultant are found by conventional methods.

$$R = \sqrt{R_x^2 + R_y^2} \qquad 1.70$$

$$\theta = \arctan \frac{R_y}{R_x} \qquad 1.71$$

The horizontal component of the resultant hydrostatic force is found in the same manner as for a vertical plane surface.

Figure 1.22 Pressure Distributions on a Curved Surface

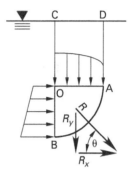

The fact that the surface is curved does not affect the calculation of the horizontal force. In Fig. 1.22, the horizontal pressure distribution on curved surface BA is the same as the horizontal pressure distribution on imaginary projected surface BO.

The vertical component of force on the curved surface is most easily calculated as the weight of the liquid above it.[42] In Fig. 1.22, the vertical component of force on the curved surface BA is the weight of liquid within the area ABCD, with a vertical line of action passing through the centroid of the area ABCD.

Figure 1.23 illustrates a curved surface with no liquid above it. However, it is not difficult to show that the resultant force acting upward on the curved surface HG is equal in magnitude (and opposite in direction) to the force that would be acting downward due to the missing area EFGH. Such an imaginary area used to calculate hydrostatic pressure is known as an *equivalent area*.

Figure 1.23 Equivalent Area

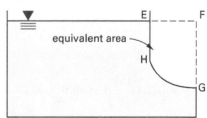

---

[42]Calculating the vertical force component in this manner is not in conflict with the hydrostatic paradox as long as the cross-sectional area of liquid above the curved surface does not decrease between the curved surface and the liquid's free surface. If there is a change in the cross-sectional area, the vertical component of force is equal to the weight of fluid in an unchanged cross-sectional area (i.e., the equivalent area).

Example 1.14

What is the total resultant force on a 1 ft section of the entire wall in Ex. 1.12?

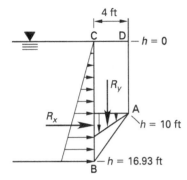

Solution

The average depth is

$$\bar{h} = \left(\frac{1}{2}\right)(0 + 16.93 \text{ ft}) = 8.465 \text{ ft}$$

The average pressure and horizontal component of the resultant on a 1 ft section of wall are

$$\bar{P} = \gamma \bar{h} = \left(62.4 \ \frac{\text{lbf}}{\text{ft}^3}\right)(8.465 \text{ ft})$$

$$= 528.2 \text{ lbf/ft}^2 \quad (528.2 \text{ psfg})$$

$$R_x = \bar{P}A = \left(528.2 \ \frac{\text{lbf}}{\text{ft}^2}\right)(16.93 \text{ ft})(1 \text{ ft})$$

$$= 8942 \text{ lbf}$$

The volume of a 1 ft section of area ABCD is

$$V_{\text{ABCD}} = (1 \text{ ft})\left((4 \text{ ft})(10 \text{ ft}) + \left(\frac{1}{2}\right)(4 \text{ ft})(6.93 \text{ ft})\right)$$

$$= 53.86 \text{ ft}^3$$

The vertical component is

$$R_y = \gamma V = \left(62.4 \ \frac{\text{lbf}}{\text{ft}^3}\right)(53.86 \text{ ft}^3) = 3361 \text{ lbf}$$

The total resultant force is

$$R = \sqrt{(8942 \text{ lbf})^2 + (3361 \text{ lbf})^2}$$

$$= 9553 \text{ lbf}$$

32. TORQUE ON A GATE

When an openable gate or door is submerged in such a manner as to have unequal depths of liquid on its two sides, or when there is no liquid present on one side, the hydrostatic pressure will act to either open or close the door. If the gate does not move, this pressure is resisted, usually by a latching mechanism on the gate itself.[43] The magnitude of the resisting latch force can be determined from the *hydrostatic torque* (*hydrostatic moment*) acting on the gate. (See Fig. 1.24.) The moment is almost always taken with respect to the gate hinges.

Figure 1.24 *Torque on a Hinge (Gate)*

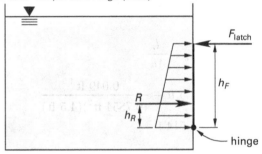

The applied moment is calculated as the product of the resultant force on the gate and the distance from the hinge to the resultant on the gate. This applied moment is balanced by the resisting moment, calculated as the latch force times the separation of the latch and hinge.

$$M_{\text{applied}} = M_{\text{resisting}} \qquad 1.72$$

$$R y_R = F_{\text{latch}} y_F \qquad 1.73$$

33. HYDROSTATIC FORCES ON A DAM

The concepts presented in the preceding sections are applicable to dams. That is, the horizontal force on the dam face can be found as in Ex. 1.14, regardless of inclination or curvature of the dam face. The vertical force on the dam face is calculated as the weight of the water above the dam face. Of course, the vertical force is zero if the dam face is vertical.

Figure 1.25 illustrates a typical dam, defining its *heel*, *toe*, and *crest*. x_{CG}, the horizontal distance from the toe to the dam's center of gravity, is not shown.

There are several stability considerations for gravity dams.[44] Most notably, the dam must not tip over or slide away due to the hydrostatic pressure. Furthermore, the pressure distribution within the soil under the dam is not uniform, and soil loading must not be excessive.

---

[43]Any contribution to resisting force from stiff hinges or other sources of friction is typically neglected.
[44]A *gravity dam* is one that is held in place and orientation by its own mass (weight) and the friction between its base and the ground.

Figure 1.25 Dam

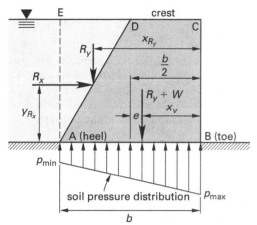

The *overturning moment* is a measure of the horizontal pressure's tendency to tip the dam over, pivoting it about the toe of the dam (point B in Fig. 1.25). (Usually, moments are calculated with respect to the pivot point.) The overturning moment is calculated as the product of the horizontal component of hydrostatic pressure (i.e., the x-component of the resultant) and the vertical distance between the toe and the line of action of the force.

$$M_{\text{overturning}} = R_x y_{R_x} \qquad 1.74$$

In most configurations, the overturning is resisted jointly by moments from the dam's own weight, W, and the vertical component of the resultant, R_y (i.e., the weight of area EAD in Fig. 1.25).[45]

$$M_{\text{resisting}} = R_y x_{R_y} + W x_{\text{CG}} \qquad 1.75$$

The *factor of safety against overturning* is

$$(\text{FS})_{\text{overturning}} = \frac{M_{\text{resisting}}}{M_{\text{overturning}}} \qquad 1.76$$

In addition to causing the dam to tip over, the horizontal component of hydrostatic force will also cause the dam to tend to slide along the ground. This tendency is resisted by the frictional force between the dam bottom and soil. The frictional force, F_f, is calculated as the product of the *normal force*, N, and the *coefficient of static friction*, μ.

$$F_f = \mu_{\text{static}} N = \mu_{\text{static}}(W + R_y) \qquad 1.77$$

The *factor of safety against sliding* is

$$(\text{FS})_{\text{sliding}} = \frac{F_f}{R_x} \qquad 1.78$$

The soil pressure distribution beneath the dam is usually assumed to vary linearly from a minimum to a maximum value. (The minimum value must be greater than zero, since soil cannot be in a state of tension. The maximum pressure should not exceed the allowable soil pressure.) Equation 1.79 predicts the minimum and maximum soil pressures.

$$P_{\text{max}}, P_{\text{min}} = \left(\frac{R_y + W}{b}\right)\left(1 \pm \frac{6e}{b}\right) \quad \text{[per unit width]} \quad 1.79$$

The *eccentricity*, e, in Eq. 1.79 is the distance between the mid-length of the dam and the location, x_v, given by Eq. 1.80. (The eccentricity must be less than $b/6$ for the entire base to be in compression.) Distances x_v and x_{CG} are different.

$$e = \frac{b}{2} - x_v \qquad 1.80$$

$$x_v = \frac{M_{\text{resisting}} - M_{\text{overturning}}}{R_y + W} \qquad 1.81$$

34. PRESSURE DUE TO SEVERAL IMMISCIBLE LIQUIDS

Figure 1.26 illustrates the nonuniform pressure distribution due to two immiscible liquids (e.g., oil on top and water below).

Figure 1.26 Pressure Distribution from Two Immiscible Liquids

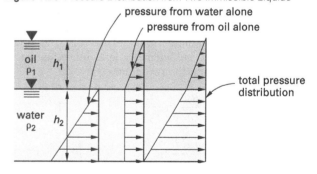

The pressure due to the upper liquid (oil), once calculated, serves as a *surcharge* to the liquid below (water). The pressure at the tank bottom is given by Eq. 1.82. (The principle can be extended to three or more immiscible liquids as well.)

$$P_{\text{bottom}} = \rho_1 g h_1 + \rho_2 g h_2 \qquad \text{[SI]} \quad 1.82(a)$$

$$P_{\text{bottom}} = \frac{\rho_1 g h_1}{g_c} + \frac{\rho_2 g h_2}{g_c} \qquad \text{[U.S.]} \quad 1.82(b)$$
$$= \gamma_1 h_1 + \gamma_2 h_2$$

---

[45]The density of concrete or masonry with steel reinforcing is usually taken to be approximately 150 lbm/ft³ (2400 kg/m³).

35. PRESSURE FROM COMPRESSIBLE FLUIDS

Fluid density, thus far, has been assumed to be independent of pressure. In reality, even "incompressible" liquids are slightly compressible. Sometimes, the effect of this compressibility cannot be neglected.

The familiar $P = \rho g h$ equation is a special case of Eq. 1.83. (It is assumed that $h_2 > h_1$. The minus sign in Eq. 1.83 indicates that pressure decreases as elevation (height) increases.)

$$\int_{P_1}^{P_2} \frac{dP}{\rho g} = -(h_2 - h_1) \qquad \text{[SI]} \qquad 1.83(a)$$

$$\int_{P_1}^{P_2} \frac{g_c dP}{\rho g} = -(h_2 - h_1) \qquad \text{[U.S.]} \qquad 1.83(b)$$

If the fluid is a perfect gas, and if compression is an isothermal (i.e., constant temperature) process, then the relationship between pressure and density is given by Eq. 1.84. The isothermal assumption is appropriate, for example, for the earth's *stratosphere* (i.e., above 35,000 ft (11 000 m)), where the temperature is assumed to be constant at approximately $-67°F$ ($-55°C$).

$$P v = \frac{P}{\rho} = R T = \text{constant} \qquad 1.84$$

In the isothermal case, Eq. 1.84 can be rewritten as Eq. 1.85. (For air, $R = 53.35$ ft-lbf/lbm-°R (287.03 J/kg·K).) Of course, the temperature, T, must be in degrees absolute (i.e., in °R or K). Equation 1.85 is known as the *barometric height relationship*[46] because knowledge of atmospheric temperature and the pressures at two points is sufficient to determine the elevation difference between the two points.

$$h_2 - h_1 = \frac{R T}{g} \ln \frac{P_1}{P_2} \qquad \text{[SI]} \qquad 1.85(a)$$

$$h_2 - h_1 = \frac{g_c R T}{g} \ln \frac{P_1}{P_2} \qquad \text{[U.S.]} \qquad 1.85(b)$$

The pressure at an elevation (height) h_2 in a layer of perfect gas that has been isothermally compressed is given by Eq. 1.86.

$$P_2 = P_1 e^{g(h_1 - h_2)/RT} \qquad \text{[SI]} \qquad 1.86(a)$$

$$P_2 = P_1 e^{g(h_1 - h_2)/g_c RT} \qquad \text{[U.S.]} \qquad 1.86(b)$$

If the fluid is a perfect gas, and if compression is an *adiabatic process*, the relationship between pressure and density is given by Eq. 1.87,[47] where k is the *ratio of specific heats*, a property of the gas. ($k = 1.4$ for air, hydrogen, oxygen, and carbon monoxide, among others.)

$$P v^k = P \left(\frac{1}{\rho} \right)^k = \text{constant} \qquad 1.87$$

The following three equations apply to adiabatic compression of an ideal gas.

$$h_2 - h_1 = \left(\frac{k}{k-1} \right) \left(\frac{R T_1}{g} \right) \left[1 - \left(\frac{P_2}{P_1} \right)^{(k-1)/k} \right] \qquad \text{[SI]} \qquad 1.88(a)$$

$$h_2 - h_1 = \left(\frac{k}{k-1} \right) \left(\frac{g_c}{g} \right) R T_1 \left[1 - \left(\frac{P_2}{P_1} \right)^{(k-1)/k} \right] \qquad \text{[U.S.]} \qquad 1.88(b)$$

$$P_2 = P_1 \left[1 - \left(\frac{k-1}{k} \right) \left(\frac{g}{R T_1} \right)(h_2 - h_1) \right]^{k/(k-1)} \qquad \text{[SI]} \qquad 1.89(a)$$

$$P_2 = P_1 \left[1 - \left(\frac{k-1}{k} \right) \left(\frac{g}{g_c} \right) \left(\frac{h_2 - h_1}{R T_1} \right) \right]^{k/(k-1)} \qquad \text{[U.S.]} \qquad 1.89(b)$$

$$T_2 = T_1 \left[1 - \left(\frac{k-1}{k} \right) \left(\frac{g}{R T_1} \right)(h_2 - h_1) \right] \qquad \text{[SI]} \qquad 1.90(a)$$

$$T_2 = T_1 \left[1 - \left(\frac{k-1}{k} \right) \left(\frac{g}{g_c} \right) \left(\frac{h_2 - h_1}{R T_1} \right) \right] \qquad \text{[U.S.]} \qquad 1.90(b)$$

The three adiabatic compression equations can be used for the more general *polytropic compression* case simply by substituting the *polytropic exponent*, n, for k.[48] Unlike the ratio of specific heats, the polytropic exponent is a function of the process, not of the gas. The polytropic compression assumption is appropriate for the earth's *troposphere*.[49] Assuming a linear decrease in temperature along with an altitude of $-0.00356°F/ft$ ($-0.00649°C/m$), the polytropic exponent is $n = 1.235$.

---

[46]This is equivalent to the work done in an isothermal compression process. The elevation (height) difference, $h_2 - h_1$ (with units of feet), can be interpreted as the work done per unit mass during compression (with units of ft-lbf/lbm).
[47]There is no heat or energy transfer to or from the ideal gas in an adiabatic process. However, this is not the same as an isothermal process.
[48]Actually, polytropic compression is the general process. Isothermal compression is a special case ($n = 1$) of the polytropic process, as is adiabatic compression ($n = k$).
[49]The *troposphere* is the part of the earth's atmosphere we live in and where most atmospheric disturbances occur. The *stratosphere*, starting at approximately 35,000 ft (11 000 m), is cold, clear, dry, and still. Between the troposphere and the stratosphere is the *tropopause*, a transition layer that contains most of the atmosphere's dust and moisture. Temperature actually increases with altitude in the stratosphere and decreases with altitude in the troposphere, but is constant in the tropopause.

Example 1.15

The air pressure and temperature at sea level are 1.0 standard atmosphere and 68°F (20°C), respectively. Assume polytropic compression with $n = 1.235$. What is the pressure at an altitude of 5000 ft (1525 m)?

SI Solution

The absolute temperature of the air is 20°C + 273.15° = 293.15K. [Temperature Conversions]

From Eq. 1.89 (substituting $k = n = 1.235$ for polytropic compression), the pressure at 1525 m altitude is

$$P_2 = (1.0 \text{ atm})$$

$$\times \left(1 - \left(\frac{1.235 - 1}{1.235} \right) \left(9.81 \; \frac{\text{m}}{\text{s}^2} \right) \right)^{1.235/(1.235-1)}$$
$$\times \left(\frac{1525 \text{ m}}{\left(287.03 \; \frac{\text{J}}{\text{kg·K}} \right) (293.15\text{K})} \right)$$

$$= 0.834 \text{ atm}$$

Customary U.S. Solution

The absolute temperature of the air is 68°F + 459.69° = 527.69°R. [Temperature Conversions]

From Eq. 1.89 (substituting $k = n = 1.235$ for polytropic compression), the pressure at an altitude of 5000 ft is

$$P_2 = (1.0 \text{ atm})$$

$$\times \left(1 - \left(\frac{1.235 - 1}{1.235} \right) \left(\frac{32.2 \; \frac{\text{ft}}{\text{sec}^2}}{32.2 \; \frac{\text{lbm-ft}}{\text{lbf-sec}^2}} \right) \right)^{1.235/(1.235-1)}$$
$$\times \left(\frac{5000 \text{ ft}}{\left(53.35 \; \frac{\text{ft-lbf}}{\text{lbm-°R}} \right) (527.69°\text{R})} \right)$$

$$= 0.835 \text{ atm}$$

36. EXTERNALLY PRESSURIZED LIQUIDS

If the gas above a liquid in a closed tank is pressurized to a gage pressure of P_t, this pressure will add to the hydrostatic pressure anywhere in the fluid. The pressure at the tank bottom illustrated in Fig. 1.27 is given by Eq. 1.91.

$$P_{\text{bottom}} = P_t + \rho g h \qquad \text{[SI]} \quad 1.91(a)$$

$$P_{\text{bottom}} = P_t + \frac{\rho g h}{g_c} = P_t + \gamma h \qquad \text{[U.S.]} \quad 1.91(b)$$

Figure 1.27 Externally Pressurized Liquid

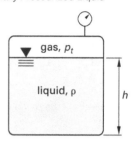

37. HYDRAULIC RAM

A *hydraulic ram* (*hydraulic jack, hydraulic press, fluid press,* etc.) is illustrated in Figure 1.28. This is a force-multiplying device. A force, F_p, is applied to the *plunger,* and a useful force, F_r, appears at the *ram.* Even though the pressure in the hydraulic fluid is the same on the ram and plunger, the forces on them will be proportional to their respective cross-sectional areas.

Figure 1.28 Hydraulic Ram

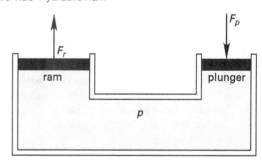

Since the pressure, P, is the same on both the plunger and ram, Eq. 1.92 can be solved for it.

$$F = PA = P\pi \left(\frac{d^2}{4} \right) \qquad 1.92$$

$$P_p = P_r \qquad 1.93$$

$$\frac{F_p}{A_p} = \frac{F_r}{A_r} \qquad 1.94$$

$$\frac{F_p}{d_p^2} = \frac{F_r}{d_r^2} \qquad 1.95$$

A small, manually actuated hydraulic ram will have a lever handle to increase the *mechanical advantage* of the ram from 1.0 to M, as illustrated in Figure 1.29. In most cases, the pivot and connection mechanism will not be frictionless, and some of the applied force will be used to overcome the friction. This friction loss is accounted for by a *lever efficiency* or *lever effectiveness*, η.

$$M = \frac{L_1}{L_2} \qquad \text{1.96}$$

$$F_p = \eta M F_l \qquad \text{1.97}$$

$$\frac{\eta M F_l}{A_p} = \frac{F_r}{A_r} \qquad \text{1.98}$$

Figure 1.29 *Hydraulic Ram with Mechanical Advantage*

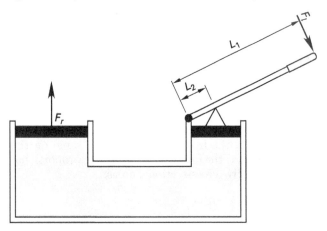

38. BUOYANCY

The *buoyant force* is an upward force that acts on all objects that are partially or completely submerged in a fluid. The fluid can be a liquid, as in the case of a ship floating at sea, or the fluid can be a gas, as in a balloon floating in the atmosphere.

There is a buoyant force on all submerged objects, not just those that are stationary or ascending. There will be, for example, a buoyant force on a rock sitting at the bottom of a pond. There will also be a buoyant force on a rock sitting exposed on the ground, since the rock is "submerged" in air. For partially submerged objects floating in liquids, such as icebergs, a buoyant force due to displaced air also exists, although it may be insignificant.

Buoyant force always acts to counteract an object's weight (i.e., buoyancy acts against gravity). The magnitude of the buoyant force is predicted from *Archimedes' principle* (*the buoyancy theorem*): The buoyant force on a submerged object is equal to the weight of the displaced fluid.[50] An equivalent statement of Archimedes' principle is: A floating object displaces liquid equal in weight to its own weight.

$$F_{\text{buoyant}} = \rho g V_{\text{displaced}} \qquad \text{[SI]} \qquad \text{1.99(a)}$$

$$F_{\text{buoyant}} = \frac{\rho g V_{\text{displaced}}}{g_c} = \gamma V_{\text{displaced}} \qquad \text{[U.S.]} \qquad \text{1.99(b)}$$

In the case of stationary (i.e., not moving vertically) objects, the buoyant force and object weight are in equilibrium. If the forces are not in equilibrium, the object will rise or fall until equilibrium is reached. That is, the object will sink until its remaining weight is supported by the bottom, or it will rise until the weight of displaced liquid is reduced by breaking the surface.[51]

The specific gravity, SG, of an object submerged in water can be determined from its dry and submerged weights. Neglecting the buoyancy of any surrounding gases,

$$SG = \frac{W_{\text{dry}}}{W_{\text{dry}} - W_{\text{submerged}}} \qquad \text{1.100}$$

Figure 1.30 illustrates an object floating partially exposed in a liquid. Neglecting the insignificant buoyant force from the displaced air (or other gas), the fractions, x, of volume exposed and submerged are easily determined.

$$x_{\text{submerged}} = \frac{\rho_{\text{object}}}{\rho_{\text{liquid}}} = \frac{(SG)_{\text{object}}}{(SG)_{\text{liquid}}} \qquad \text{1.101}$$

$$x_{\text{exposed}} = 1 - x_{\text{submerged}} \qquad \text{1.102}$$

Figure 1.30 *Partially Submerged Object*

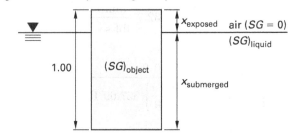

---

[50]The volume term in Eq. 1.99 is the total volume of the object only in the case of complete submergence.
[51]An object can also stop rising or falling due to a change in the fluid's density. The buoyant force will increase with increasing depth in the ocean due to an increase in density at great depths. The buoyant force will decrease with increasing altitude in the atmosphere due to a decrease in density at great heights.

Figure 1.31 illustrates a somewhat more complicated situation—that of an object floating at the interface between two liquids of different densities. The fractions of immersion in each liquid are given by the following equations.

$$x_1 = \frac{(SG)_2 - (SG)_{\text{object}}}{(SG)_2 - (SG)_1} \qquad 1.103$$

$$x_2 = 1 - x_1 = \frac{(SG)_{\text{object}} - (SG)_1}{(SG)_2 - (SG)_1} \qquad 1.104$$

Figure 1.31 *Object Floating in Two Liquids*

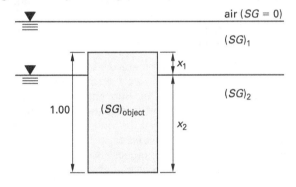

A more general case of a floating object is shown in Fig. 1.32. Situations of this type are easily evaluated by equating the object's weight with the sum of the buoyant forces.

In the case of Fig. 1.32 (with two liquids), the following relationships apply, where x_0 is the fraction, if any, extending into the air above.

$$(SG)_{\text{object}} = x_1(SG)_1 + x_2(SG)_2 \qquad 1.105$$

$$(SG)_{\text{object}} = (1 - x_0 - x_2)(SG)_1 + x_2(SG)_2 \qquad 1.106$$

$$(SG)_{\text{object}} = x_1(SG)_1 + (1 - x_0 - x_1)(SG)_2 \qquad 1.107$$

Figure 1.32 *General Two-Liquid Buoyancy Problem*

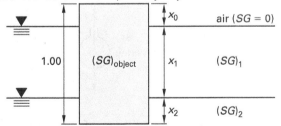

Example 1.16

An empty polyethylene telemetry balloon and payload have a mass of 500 lbm (225 kg). The balloon is filled with helium when the atmospheric conditions are 60°F (15.6°C) and 14.8 psia (102 kPa). The specific gas constant of helium is 2079 J/kg·K (386.3 ft-lbf/lbm-°R). What minimum volume of helium is required for lift-off from a sea-level platform?

SI Solution

The gas densities are

$$\rho_{\text{air}} = \frac{P}{RT} = \frac{1.02 \times 10^5 \text{ Pa}}{\left(287.03 \dfrac{\text{J}}{\text{kg·K}}\right)(15.6°\text{C} + 273°)}$$

$$= 1.231 \text{ kg/m}^3$$

$$\rho_{\text{helium}} = \frac{1.02 \times 10^5 \text{ Pa}}{\left(2079 \dfrac{\text{J}}{\text{kg·K}}\right)(288.6\text{K})} = 0.17 \text{ kg/m}^3$$

The total mass of the balloon, payload, and helium is

$$m = 225 \text{ kg} + \left(0.17 \dfrac{\text{kg}}{\text{m}^3}\right)V_{\text{He}}$$

The buoyant force is the weight of the displaced air. Neglecting the payload volume, the displaced air volume is the same as the helium volume.

$$m_b = \left(1.231 \dfrac{\text{kg}}{\text{m}^3}\right)V_{\text{He}}$$

At lift-off, the weight of the balloon is just equal to the buoyant force. Equate the two and solve for the helium volume.

$$225 \text{ kg} + \left(0.17 \dfrac{\text{kg}}{\text{m}^3}\right)V_{\text{He}} = \left(1.231 \dfrac{\text{kg}}{\text{m}^3}\right)V_{\text{He}}$$

$$V_{\text{He}} = 212.1 \text{ m}^3$$

Customary U.S. Solution

The gas densities are

$$\rho_{\text{air}} = \frac{P}{RT} = \frac{\left(14.8 \dfrac{\text{lbf}}{\text{in}^2}\right)\left(12 \dfrac{\text{in}}{\text{ft}}\right)^2}{\left(53.35 \dfrac{\text{ft-lbf}}{\text{lbm-°R}}\right)(60°\text{F} + 460°)}$$

$$= 0.07682 \text{ lbm/ft}^3$$

$$\gamma_{\text{air}} = \rho \times \frac{g}{g_c} = 0.07682 \text{ lbf/ft}^3$$

$$\rho_{\text{helium}} = \frac{\left(14.8 \ \dfrac{\text{lbf}}{\text{in}^2}\right)\left(12 \ \dfrac{\text{in}}{\text{ft}}\right)^2}{\left(386.3 \ \dfrac{\text{ft-lbf}}{\text{lbm-}^\circ\text{R}}\right)(520^\circ\text{R})}$$

$$= 0.01061 \ \text{lbm/ft}^3$$

$$\gamma_{\text{helium}} = 0.01061 \ \text{lbf/ft}^3$$

The total weight of the balloon, payload, and helium is

$$W = 500 \ \text{lbf} + \left(0.01061 \ \frac{\text{lbf}}{\text{ft}^3}\right) V_{\text{He}}$$

The buoyant force is the weight of the displaced air. Neglecting the payload volume, the displaced air volume is the same as the helium volume.

$$F_b = \left(0.07682 \ \frac{\text{lbf}}{\text{ft}^3}\right) V_{\text{He}}$$

At lift-off, the weight of the balloon is just equal to the buoyant force. Equate the two and solve for the helium volume.

$$W = F_b$$

$$500 \ \text{lbf} + \left(0.01061 \ \frac{\text{lbf}}{\text{ft}^3}\right) V_{\text{He}} = \left(0.07682 \ \frac{\text{lbf}}{\text{ft}^3}\right) V_{\text{He}}$$

$$V_{\text{He}} = 7552 \ \text{ft}^3$$

39. BUOYANCY OF SUBMERGED PIPELINES

Whenever possible, submerged pipelines for river crossings should be completely buried at a level below river scour. This will reduce or eliminate loads and movement due to flutter, scour and fill, drag, collisions, and buoyancy. Submerged pipelines should cross at right angles to the river. For maximum flexibility and ductility, pipelines should be made of thick-walled mild steel.

Submerged pipelines should be weighted to achieve a minimum of 20% negative buoyancy (i.e., an average density of 1.2 times the environment, approximately 72 lbm/ft³ or 1200 kg/m³). Metal or concrete clamps can be used for this purpose, as well as concrete coatings. Thick steel clamps have the advantage of a smaller lateral exposed area (resulting in less drag from river flow), while brittle concrete coatings are sensitive to pipeline flutter and temperature fluctuations.

Due to the critical nature of many pipelines and the difficulty in accessing submerged portions for repair, it is common to provide a parallel auxiliary line. The auxiliary and main lines are provided with crossover and mainline valves, respectively, on high ground at both sides of the river to permit either or both lines to be used.

40. INTACT STABILITY: STABILITY OF FLOATING OBJECTS

A stationary object is said to be in *static equilibrium*. However, an object in static equilibrium is not necessarily stable. For example, a coin balanced on edge is in static equilibrium, but it will not return to the balanced position if it is disturbed. An object is said to be *stable* (i.e., in *stable equilibrium*) if it tends to return to the equilibrium position when slightly displaced.

Stability of floating and submerged objects is known as *intact stability*.[52] There are two forces acting on a stationary floating object: the buoyant force and the object's weight. The buoyant force acts upward through the centroid of the displaced volume. This centroid is known as the *center of buoyancy*. The gravitational force on the object (i.e., the object's weight) acts downward through the object's center of gravity.

For a totally submerged object (as in the balloon and submarine shown in Figure 1.33) to be stable, the center of buoyancy must be above the center of gravity. The object will be stable because a righting moment will be created if the object tips over, since the center of buoyancy will move outward from the center of gravity.

Figure 1.33 Stability of a Submerged Object

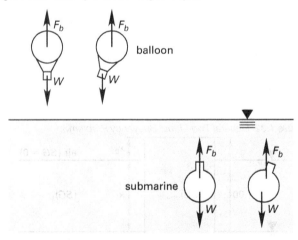

The stability criterion is different for partially submerged objects (e.g., surface ships). If the vessel shown in Figure 1.34 heels (i.e., lists or rolls), the location of the center of gravity of the object does not change.[53]

---

[52]The subject of intact stability, being a part of naval architecture curriculum, is not covered extensively in most fluids books. However, it is covered extensively in basic ship design and naval architecture books.
[53]The verbs *roll, list,* and *heel* are synonymous.

However, the center of buoyancy shifts to the centroid of the new submerged section 123. The centers of buoyancy and gravity are no longer in line.

Figure 1.34 Stability of a Partially Submerged Floating Object

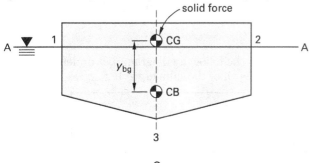

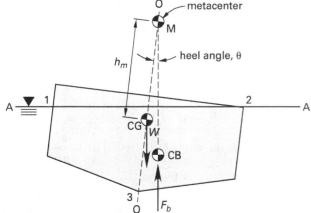

This righting couple exists when the extension of the buoyant force, F_b, intersects line O–O above the center of gravity at M, the *metacenter*. For partially submerged objects to be stable, the metacenter must be above the center of gravity. If M lies below the center of gravity, an overturning couple will exist. The distance between the center of gravity and the metacenter is called the *metacentric height*, and it is reasonably constant for heel angles less than 10°. Also, for angles less than 10°, the center of buoyancy follows a locus for which the metacenter is the instantaneous center.

The metacentric height is one of the most important and basic parameters in ship design. It determines the ship's ability to remain upright as well as the ship's roll and pitch characteristics.

"Acceptable" minimum values of the metacentric height have been established from experience, and these depend on the ship type and class. For example, many submarines are required to have a metacentric height of 1 ft (0.3 m) when surfaced. This will increase to approximately 3.5 ft (1.2 m) for some of the largest surface ships. If an acceptable metacentric height is not achieved initially, the center of gravity must be lowered or the keel depth increased. The beam width can also be increased slightly to increase the waterplane moment of inertia.

For a surface vessel rolling through an angle less than approximately 10°, the distance between the vertical center of gravity and the metacenter can be found from Eq. 1.108. Variable I is the centroidal area moment of inertia of the original waterline (free surface) cross section about a longitudinal (fore and aft) waterline axis; V is the displaced volume.

If the distance, y_{bg}, separating the centers of buoyancy and gravity is known, Eq. 1.108 can be solved for the metacentric height. y_{bg} is positive when the center of gravity is above the center of buoyancy. This is the normal case. Otherwise, y_{bg} is negative.

$$y_{bg} + h_m = \frac{I}{V} \qquad \text{1.108}$$

The *righting moment* (also known as the *restoring moment*) is the stabilizing moment exerted when the ship rolls. Values of the righting moment are typically specified with units of foot-tons (MN·m).

$$M_{\text{righting}} = h_m \gamma_w V_{\text{displaced}} \sin \theta \qquad \text{1.109}$$

The transverse (roll) and longitudinal (pitch) *periods* also depend on the metacentric height. The roll characteristics are found from the differential equation formed by equating the righting moment to the product of the ship's transverse mass moment of inertia and the angular acceleration. Larger metacentric heights result in lower roll periods. If k is the radius of gyration about the roll axis, the roll period is

$$T_{\text{roll}} = \frac{2\pi k}{\sqrt{gh_m}} \qquad \text{1.110}$$

The roll and pitch periods must be adjusted for the appropriate level of crew and passenger comfort. A "beamy" ice-breaking ship will have a metacentric height much larger than normally required for intact stability, resulting in a short, nauseating roll period. The designer of a passenger ship, however, would have to decrease the intact stability (i.e., decrease the metacentric height) in order to achieve an acceptable ride characteristic. This requires a moderate metacentric height that is less than approximately 6% of the beam length.

Example 1.17

A 600,000 lbm (280 000 kg) rectangular barge has external dimensions of 24 ft width, 98 ft length, and 12 ft height (7 m × 30 m × 3.6 m). It floats in seawater ($\gamma_w = 64.0$ lbf/ft³; $\rho_w = 1024$ kg/m³). The center of gravity is 7.8 ft (2.4 m) from the top of the barge as loaded. Find (a) the location of the center of buoyancy when the barge is floating on an even keel, and (b) the approximate location of the metacenter when the barge experiences a 5° heel.

SI Solution

(a) Refer to the following diagram. Let dimension y represent the depth of the submerged barge.

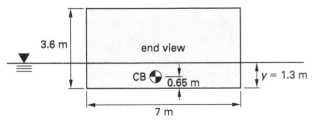

From Archimedes' principle, the buoyant force equals the weight of the barge. This, in turn, equals the weight of the displaced seawater.

$$F_b = W = V\rho_w g$$

$$(280\,000 \text{ kg})\left(9.81 \ \frac{\text{m}}{\text{s}^2}\right) = y\big((7 \text{ m})(30 \text{ m})\big)\left(1024 \ \frac{\text{kg}}{\text{m}^3}\right)$$

$$\times \left(9.81 \ \frac{\text{m}}{\text{s}^2}\right)$$

$$y = 1.3 \text{ m}$$

The center of buoyancy is located at the centroid of the submerged cross section. When floating on an even keel, the submerged cross section is rectangular with a height of 1.3 m. The height of the center of buoyancy above the keel is

$$\frac{1.3 \text{ m}}{2} = 0.65 \text{ m}$$

(b) While the location of the new center of buoyancy can be determined, the location of the metacenter does not change significantly for small angles of heel. For approximate calculations, the angle of heel is not significant.

The area moment of inertia of the longitudinal waterline cross section is

$$I = \frac{Lw^3}{12} = \frac{(30 \text{ m})(7 \text{ m})^3}{12} = 858 \text{ m}^4$$

The submerged volume is

$$V = (1.3 \text{ m})(7 \text{ m})(30 \text{ m}) = 273 \text{ m}^3$$

The distance between the center of gravity and the center of buoyancy is

$$y_{\text{bg}} = 3.6 \text{ m} - 2.4 \text{ m} - 0.65 \text{ m} = 0.55 \text{ m}$$

The metacentric height measured above the center of gravity is

$$h_m = \frac{I}{V} - y_{\text{bg}} = \frac{858 \text{ m}^4}{273 \text{ m}^3} - 0.55 \text{ m} = 2.6 \text{ m}$$

Customary U.S. Solution

(a) Refer to the following diagram. Let dimension y represent the depth of the submerged barge.

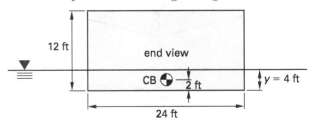

From Archimedes' principle, the buoyant force equals the weight of the barge. This, in turn, equals the weight of the displaced seawater.

$$F_b = W = V\gamma_w$$

$$600,000 \text{ lbf} = y\big((24 \text{ ft})(98 \text{ ft})\big)\left(64 \ \frac{\text{lbf}}{\text{ft}^3}\right)$$

$$y = 4 \text{ ft}$$

The center of buoyancy is located at the centroid of the submerged cross section. When floating on an even keel, the submerged cross section is rectangular with a height of 4 ft. The height of the center of buoyancy above the keel is

$$\frac{4 \text{ ft}}{2} = 2 \text{ ft}$$

(b) While the location of the new center of buoyancy can be determined, the location of the metacenter does not change significantly for small angles of heel. Therefore, for approximate calculations, the angle of heel is not significant.

The area moment of inertia of the longitudinal waterline cross section is

$$I = \frac{Lw^3}{12} = \frac{(98 \text{ ft})(24 \text{ ft})^3}{12} = 112,900 \text{ ft}^4$$

The submerged volume is

$$V = (4 \text{ ft})(24 \text{ ft})(98 \text{ ft}) = 9408 \text{ ft}^3$$

The distance between the center of gravity and the center of buoyancy is

$$y_{\text{bg}} = 12 \text{ ft} - 7.8 \text{ ft} - 2 \text{ ft} = 2.2 \text{ ft}$$

The metacentric height measured above the center of gravity is

$$h_m = \frac{I}{V} - y_{bg} = \frac{112{,}900 \text{ ft}^4}{9408 \text{ ft}^3} - 2.2 \text{ ft}$$
$$= 9.8 \text{ ft}$$

41. FLUID MASSES UNDER EXTERNAL ACCELERATION

If a fluid mass is subjected to an external acceleration (moved sideways, rotated, etc.), an additional force will be introduced. This force will change the equilibrium position of the fluid surface as well as the hydrostatic pressure distribution.

Figure 1.35 illustrates a liquid mass subjected to constant accelerations in the vertical and/or horizontal directions. (a_y is negative if the acceleration is downward.) The surface is inclined at the angle predicted by Eq. 1.111. The planes of equal hydrostatic pressure beneath the surface are also inclined at the same angle.[54]

$$\theta = \arctan \frac{a_x}{a_y + g} \qquad \text{1.111}$$

$$P = \rho(g + a_y)h \qquad \text{[SI]} \quad \text{1.112(a)}$$

$$P = \frac{\rho(g + a_y)h}{g_c} = \gamma h\left(1 + \frac{a_y}{g}\right) \qquad \text{[U.S.]} \quad \text{1.112(b)}$$

Figure 1.35 *Fluid Mass Under Constant Linear Acceleration*

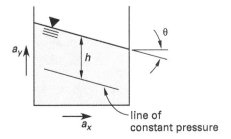

Figure 1.36 illustrates a fluid mass rotating about a vertical axis at constant angular velocity, ω, in rad/sec.[55] The resulting surface is parabolic in shape. The elevation of the fluid surface at point A at distance r from the axis of rotation is given by Eq. 1.114. The

distance h in Fig. 1.36 is measured from the lowest fluid elevation during rotation. h is not measured from the original elevation of the stationary fluid.

$$\theta = \arctan \frac{\omega^2 r}{g} \qquad \text{1.113}$$

$$h = \frac{(\omega r)^2}{2g} = \frac{\text{v}^2}{2g} \qquad \text{1.114}$$

Figure 1.36 *Rotating Fluid Mass*

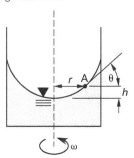

42. NOMENCLATURE

a	acceleration	ft/sec^2	m/s^2
A	area	ft^2	m^2
e	eccentricity	ft	m
E	bulk modulus	lbf/ft^2	Pa
F	force	lbf	N
FS	factor of safety	–	–
g	gravitational acceleration, 32.2 (9.81)	ft/sec^2	m/s^2
g_c	gravitational constant, 32.2	lbm-ft/lbf-sec^2	n.a.
G	gravimetric (mass) fraction	–	–
h	height	ft	m
I	moment of inertia	ft^4	m^4
k	radius of gyration	ft	m
k	ratio of specific heats	–	–
L	length	ft	m
m	mass	lbm	kg
M	mechanical advantage	–	–
M	molar concentration	lbmol/ft^3	kmol/m^3
M	moment	ft-lbf	N·m
M	molecular weight	lbm/lbmol	kg/kmol
n	number of moles	–	–
n	polytropic exponent	–	–
N	normal force	lbf	N
N	number of moles	–	–
p, P	pressure	lbf/ft^2	Pa

---

[54]Once the orientation of the surface is known, the pressure distribution can be determined without considering the acceleration. The hydrostatic pressure at a point depends only on the height of the liquid above that point. The acceleration affects that height but does not change the $P = \rho gh$ relationship.

[55]Even though the rotational speed is not increasing, the fluid mass experiences a constant *centripetal acceleration* radially outward from the axis of rotation.

r	radius	ft	m
R	resultant force	lbf	N
R	specific gas constant	ft-lbf/lbm-°R	J/kg·K
R^*	universal gas constant, 1545.35 (8314.47)	ft-lbf/lbmol-°R	J/kmol·K
SG	specific gravity	–	–
T	absolute temperature	°R	K
v	kinematic viscosity	ft²/sec	m²/s
v	specific volume	ft³/lbm	m³/kg
v	velocity	ft/sec	m/s
V	volume	ft³	m³
w	width	ft	m
W	weight	lbf	n.a.
x	distance	ft	m
x	fraction	–	–
x	mole fraction	–	–
y	distance	ft	m
y	mass fraction	–	–
Z	compressibility factor	–	–

Symbols

β	compressibility	ft²/lbf	Pa⁻¹
β	contact angle	deg	deg
γ	specific weight	lbf/ft³	n.a.
η	efficiency	–	–
θ	angle	deg	deg
μ	absolute viscosity	lbf-sec/ft²	Pa·s
π	osmotic pressure	lbf/ft²	Pa
σ	surface tension	lbf/ft	N/m
τ	shear stress	lbf/ft²	Pa
ω	angular velocity	rad/sec	rad/s

Subscripts

0	zero velocity (wall face), constant velocity
a	atmospheric
b	buoyant
bg	between CB and CG
c	centroidal
c	critical
f	frictional
m	manometer fluid, mercury, or metacentric
p	plunger
r	ram
R	resultant
s	constant entropy
t	tank
T	constant temperature
w	water
v	vapor or vertical

2 Fluid Flow, Dynamics, Hydraulics

Content in blue refers to the *NCEES Handbook*.

1. INTRODUCTION TO FLUID ENERGY UNITS

Several important fluids and thermodynamics equations, such as Bernoulli's equation and the steady-flow energy equation, are special applications of the *conservation of energy* concept. However, it is not always obvious how some formulations of these equations can be termed "energy." For example, elevation, z, with units of feet, is often called *gravitational energy*.

Fluid energy is expressed per unit mass, as indicated by the name, *specific energy*. Units of fluid specific energy are commonly ft-lbf/lbm and J/kg.[1]

2. KINETIC ENERGY

Energy is needed to accelerate a stationary body. Therefore, a moving mass of fluid possesses more energy than an identical, stationary mass. The energy difference is the *kinetic energy* of the fluid.[2] Equation 2.1 gives the kinetic energy corresponding to a fluid flow with an average velocity of v.

---

[1]The ft-lbf/lbm unit may be thought of as just feet, although pounds of force and pounds of mass do not really cancel out. The combination of variables $E \times (g_c/g)$ is required in order for the units to resolve into feet.

[2]The terms *velocity energy* and *dynamic energy* are used less often.

Water

Unit and Conversion Factors

$$KE = mv^2/2g_c \qquad \qquad 2.1$$

Kinetic energy is often evaluated per unit mass, in which case the term *specific kinetic energy* may be used (though in practice the word *specific* is often omitted, being understood). The units of specific kinetic energy in the SI system are m²/s², which is equivalent to J/kg. The units in the customary U.S. system are ft-lbf/lbm, which are usually taken as equivalent to feet.

$$KE = \frac{mv^2}{2} \qquad \text{[SI]} \quad 2.2(a)$$

$$E_v = \frac{mv^2}{2g_c} \qquad \text{[U.S.]} \quad 2.2(b)$$

3. POTENTIAL ENERGY

Work is performed in elevating a body. Therefore, a mass of fluid at a high elevation will have more energy than an identical mass of fluid at a lower elevation. The energy difference is the *potential energy* of the fluid.[3]

Unit and Conversion Factors

$$PE = mgh/g_c \qquad \qquad 2.3$$

Like kinetic energy, potential energy is often expressed per unit mass. Equation 2.4 gives the specific potential energy of fluid at an elevation h.[4]

$$PE = mgh \qquad \text{[SI]} \quad 2.4(a)$$

$$E_z = \frac{gh}{g_c} \qquad \text{[U.S.]} \quad 2.4(b)$$

The units of potential energy in the SI system are again m²/s² or J/kg. The units in the customary U.S. system are ft-lbf/lbm or feet.

h is the elevation of the fluid. The reference point (i.e., zero elevation point) is entirely arbitrary and can be chosen for convenience. This is because potential energy always appears in a difference equation (i.e., ΔPE), and the reference point cancels out.

4. PRESSURE ENERGY

Work is performed and energy is added when a substance is compressed. Therefore, a mass of fluid at a high pressure will have more energy than an identical mass of fluid at a lower pressure. The energy difference is the *pressure energy* (also called *pressure head*) of the fluid.[5] Pressure energy is usually found in equations along with kinetic and potential energies and is expressed as energy per unit mass or in feet. Equation 2.5 gives the pressure energy of fluid at pressure P. In Eq. 2.5, w is the specific weight of the fluid.

$$E_P = \frac{P}{w} \qquad \qquad 2.5$$

5. BERNOULLI EQUATION

The *Bernoulli equation* is an ideal energy conservation equation based on several reasonable assumptions. The equation assumes the following.

- The fluid is incompressible.

- There is no fluid friction.

- Changes in thermal energy are negligible.[6]

The Bernoulli equation states that the *total energy* of a fluid flowing without friction losses in a pipe is constant.[7] The total energy possessed by the fluid is the sum of its pressure, kinetic, and potential energies. Drawing on Eq. 2.2, Eq. 2.4, and Eq. 2.5, the Bernoulli equation can be written as

$$E_t = E_z + E_P + E_v \qquad \qquad 2.6$$

The Bernoulli equation can be expressed in terms of total head, H.

Energy Equation (Bernoulli)

$$H = Z_a + \frac{P_a}{w} + \frac{V_a^2}{2g} \qquad \qquad 2.7$$

Equation 2.7 is valid for both laminar and turbulent flows. Since the original research by Bernoulli assumed laminar flow, when Eq. 2.7 is used for turbulent flow, it may be referred to as the "steady-flow energy equation" instead of the "Bernoulli equation." It can also be used for gases and vapors if the incompressibility assumption is valid.

---

[3]The term *gravitational energy* is also used.
[4]Since $g = g_c$ (numerically), it is common for many engineers in the United States to treat potential energy and elevation as the same thing.
[5]The terms *static energy* and *flow energy* are also used. The name *flow energy* results from the need to push (pressurize) a fluid to get it to flow through a pipe. However, flow energy and kinetic energy are not the same.
[6]In thermodynamics, the fluid flow is said to be *adiabatic*.
[7]Strictly speaking, this is the *total specific energy*, since the energy is per unit mass. However, the word "specific," being understood, is seldom used. Of course, "the total energy of the system" means something else and requires knowing the fluid mass in the system.

Total head, H, and *total pressure*, P, can be calculated from total energy.

$$H = \frac{E}{g} \qquad \text{[SI]} \qquad 2.8(a)$$

$$H = E \times \frac{g_c}{g} \qquad \text{[U.S.]} \qquad 2.8(b)$$

$$P = \rho g H \qquad \text{[SI]} \qquad 2.9(a)$$

$$P = \rho H \times \frac{g}{g_c} \qquad \text{[U.S.]} \qquad 2.9(b)$$

Example 2.1

A pipe draws water from the bottom of a reservoir and discharges it freely at point C, 100 ft (30 m) below the surface. The flow is frictionless. (a) What is the total specific energy at an elevation 50 ft (15 m) below the water surface (i.e., point B)? (b) What is the velocity at point C?

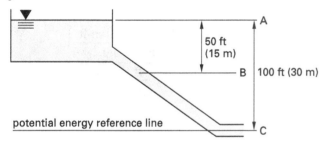

SI Solution

(a) At point A, the velocity and gage pressure are both zero. Therefore, the total energy consists only of potential energy. Choose point C as the reference ($z = 0$) elevation.

$$P_A = z_A g = (30 \text{ m})\left(9.81 \ \frac{\text{m}}{\text{s}^2}\right)$$
$$= 294.3 \text{ m}^2/\text{s}^2 \quad (\text{J/kg})$$

At point B, the fluid is moving and possesses kinetic energy. The fluid is also under hydrostatic pressure and possesses pressure energy. These energy forms have come at the expense of potential energy. (This is a direct result of the Bernoulli equation.) Also, the flow is frictionless. Therefore, there is no net change in the total energy between points A and B.

$$E_B = E_A$$
$$= 294.3 \text{ m}^2/\text{s}^2 \quad (\text{J/kg})$$

(b) At point C, the gage pressure and pressure energy are again zero, since the discharge is at atmospheric pressure. The potential energy is zero, since $z = 0$. The total energy of the system has been converted to kinetic energy. From Eq. 2.6,

$$E_t = 294.3 \ \frac{\text{m}^2}{\text{s}^2}$$
$$= 0 + \frac{\text{v}^2}{2} + 0$$
$$\text{v} = 24.3 \text{ m/s}$$

Customary U.S. Solution

(a)

$$E_A = \frac{z_A g}{g_c}$$
$$= \frac{(100 \text{ ft})\left(32.2 \ \dfrac{\text{ft}}{\text{sec}^2}\right)}{32.2 \ \dfrac{\text{lbm-ft}}{\text{lbf-sec}^2}}$$
$$= 100 \text{ ft-lbf/lbm}$$
$$E_t = E_B$$
$$= E_A$$
$$= 100 \text{ ft-lbf/lbm}$$

(b)

$$E_t = E_C$$
$$= 100 \ \frac{\text{ft-lbf}}{\text{lbm}}$$
$$= 0 + \frac{\text{v}^2}{2g_c} + 0$$
$$\text{v}^2 = 2g_c E_t$$
$$= (2)\left(32.2 \ \frac{\text{lbm-ft}}{\text{lbf-sec}^2}\right)\left(100 \ \frac{\text{ft-lbf}}{\text{lbm}}\right)$$
$$= 6440 \text{ ft}^2/\text{sec}^2$$
$$\text{v} = \sqrt{6440 \ \frac{\text{ft}^2}{\text{sec}^2}}$$
$$= 80.2 \text{ ft/sec}$$

Example 2.2

Water (62.4 lbm/ft³; 1000 kg/m³) is pumped up a hillside into a reservoir. The pump discharges water with a velocity of 6 ft/sec (2 m/s) and a pressure of 150 psig (1000 kPa). Disregarding friction, what is the maximum elevation (above the centerline of the pump's discharge) of the reservoir's water surface?

SI Solution

At the centerline of the pump's discharge, the potential energy is zero. The atmospheric pressure at the pump inlet is the same as (and counteracts) the atmospheric pressure at the reservoir surface, so gage pressures may be used. The pressure and velocity energies are

$$E_P = \frac{P}{\rho} = \frac{(1000 \text{ kPa})\left(1000 \frac{\text{Pa}}{\text{kPa}}\right)}{1000 \frac{\text{kg}}{\text{m}^3}}$$

$$= 1000 \text{ J/kg}$$

$$E_\text{v} = \frac{\text{v}^2}{2} = \frac{\left(2 \frac{\text{m}}{\text{s}}\right)^2}{2} = 2 \text{ J/kg}$$

The total energy at the pump's discharge is

$$E_{t,1} = E_P + E_\text{v} = 1000 \frac{\text{J}}{\text{kg}} + 2 \frac{\text{J}}{\text{kg}} = 1002 \text{ J/kg}$$

Since the flow is frictionless, the same energy is possessed by the water at the reservoir's surface. Since the velocity and gage pressure at the surface are zero, all of the available energy has been converted to potential energy.

$$E_{t,2} = E_{t,1}$$

$$z_2 g = 1002 \text{ J/kg}$$

$$z_2 = \frac{E_{t,2}}{g} = \frac{1002 \frac{\text{J}}{\text{kg}}}{9.81 \frac{\text{m}}{\text{s}^2}} = 102.1 \text{ m}$$

The volumetric flow rate of the water is not relevant since the water velocity was known. Similarly, the pipe size is not needed.

Customary U.S. Solution

The atmospheric pressure at the pump inlet is the same as (and counteracts) the atmospheric pressure at the reservoir surface, so gage pressures may be used.

$$E_P = \frac{P}{\rho} = \frac{\left(150 \frac{\text{lbf}}{\text{in}^2}\right)\left(12 \frac{\text{in}}{\text{ft}}\right)^2}{62.4 \frac{\text{lbm}}{\text{ft}^3}}$$

$$= 346.15 \text{ ft-lbf/lbm}$$

$$E_\text{v} = \frac{\text{v}^2}{2g_c}$$

$$= \frac{\left(6 \frac{\text{ft}}{\text{sec}}\right)^2}{(2)\left(32.2 \frac{\text{lbm-ft}}{\text{lbf-sec}^2}\right)}$$

$$= 0.56 \text{ ft-lbf/lbm}$$

$$E_{t,1} = E_P + E_\text{v} = 346.15 \frac{\text{ft-lbf}}{\text{lbm}} + 0.56 \frac{\text{ft-lbf}}{\text{lbm}}$$

$$= 346.71 \text{ ft-lbf/lbm}$$

$$E_{t,2} = E_{t,1}$$

$$\frac{z_2 g}{g_c} = 346.71 \text{ ft-lbf/lbm}$$

$$z_2 = \frac{E_{t,2} g_c}{g}$$

$$= \frac{\left(346.71 \frac{\text{ft-lbf}}{\text{lbm}}\right)\left(32.2 \frac{\text{lbm-ft}}{\text{lbf-sec}^2}\right)}{32.2 \frac{\text{ft}}{\text{sec}^2}}$$

$$= 346.71 \text{ ft}$$

6. FIELD EQUATION

When flow is one-dimensional, friction losses are negligible, and no energy is gained through a pump or lost through a turbine, then from the principle of conservation of energy, the total energy will be equal at every section in the system. Applying the energy equation to two sections 1 and 2 in such a system gives Eq. 2.10, the *field equation*.

Field Equation

$$\frac{P_2}{\gamma} + \frac{\text{v}_2^2}{2g} + z_2 = \frac{P_1}{\gamma} + \frac{\text{v}_1^2}{2g} + z_1 \qquad \textbf{2.10}$$

Multiplying both sides of Eq. 2.10 by gravitational acceleration, *g*, gives an alternative form of the field equation.

Field Equation

$$\frac{P_2}{\rho} + \frac{\text{v}_2^2}{2} + z_2 g = \frac{P_1}{\rho} + \frac{\text{v}_1^2}{2} + z_1 g \qquad \textbf{2.11}$$

P_1 and P_2 are the pressures, v_1 and v_2 are the average fluid velocities and z_1 and z_2 are the elevations at sections 1 and 2, respectively.

7. IMPACT ENERGY

Impact energy, E_i (also known as *stagnation energy* and *total energy*), is the sum of the kinetic and pressure energy terms.[8] Equation 2.13 is applicable to liquids and gases flowing with velocities less than approximately Mach 0.3.

$$E_i = E_P + E_v \qquad 2.12$$

$$E_i = \frac{P}{\rho} + \frac{v^2}{2} \qquad \text{[SI]} \quad 2.13(a)$$

$$E_i = \frac{P}{\rho} + \frac{v^2}{2g_c} \qquad \text{[U.S.]} \quad 2.13(b)$$

Impact head, h_i, is calculated from the impact energy in a manner analogous to Eq. 2.8. Impact head represents the height the liquid will rise in a piezometer-pitot tube when the liquid has been brought to rest (i.e., stagnated) in an adiabatic manner. Such a case is illustrated in Fig. 2.1. If a gas or high-velocity, high-pressure liquid is flowing, it will be necessary to use a mercury manometer or pressure gauge to measure stagnation head.

Figure 2.1 *Pitot Tube-Piezometer Apparatus*

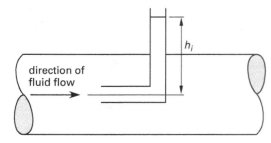

8. PITOT TUBE

A *pitot tube* (also known as an *impact tube* or *stagnation tube*) is simply a hollow tube that is placed longitudinally in the direction of fluid flow, allowing the flow to enter one end at the fluid's *velocity of approach*. (See Fig. 2.2.) It is used to measure velocity of flow and finds uses in both subsonic and supersonic applications.

Figure 2.2 *Pitot Tube*

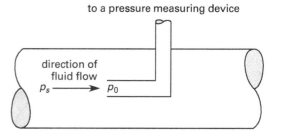

When the fluid enters the pitot tube, it is forced to come to a stop (at the *stagnation point*), and the velocity energy is transformed into pressure energy. If the fluid is a low-velocity gas, the stagnation is assumed to occur without compression heating of the gas. If there is no friction (the common assumption), the process is said to be adiabatic.

Bernoulli's equation can be used to predict the static pressure at the stagnation point. Since the velocity of the fluid within the pitot tube is zero, the upstream velocity can be calculated if the static and stagnation pressures are known.

$$\frac{P_1}{\rho} + \frac{v_1^2}{2} = \frac{P_2}{\rho} \qquad 2.14$$

Pitot Tube
$$v = \sqrt{(2/\rho)(P_0 - P_s)} = \sqrt{2g(P_0 - P_s)/\gamma} \qquad 2.15$$

In reality, both friction and heating occur, and the fluid may be compressible. These errors are taken care of by a correction factor known as the *impact factor*, C_i, which is applied to the derived velocity. C_i is usually very close to 1.00 (e.g., 0.99 or 0.995).

$$v_{actual} = C_i v_{indicated} \qquad 2.16$$

Since accurate measurements of fluid velocity are dependent on one-dimensional fluid flow, it is essential that any obstructions or pipe bends be more than 10 pipe diameters upstream from the pitot tube.

Example 2.3

The static pressure of air (0.075 lbf/ft³; 1.20 kg/m³) flowing in a pipe is measured by a precision gauge to be 10.00 psig (68.95 kPa). A pitot tube-manometer indicates 20.6 in (0.523 m) of mercury. The density

---

[8]It is confusing to label Eq. 2.12 *total* when the gravitational energy term has been omitted. However, the reference point for gravitational energy is arbitrary, and in this application the reference coincides with the centerline of the fluid flow. In truth, the effective pressure developed in a fluid which has been brought to rest adiabatically does not depend on the elevation or altitude of the fluid. This situation is seldom ambiguous. The application will determine which definition of total head or total energy is intended.

Water

of mercury is 0.491 lbm/in³ (13 600 kg/m³). Losses are insignificant. What is the velocity of the air in the pipe?

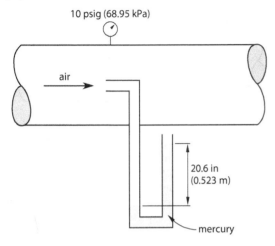

10 psig (68.95 kPa)

air

20.6 in
(0.523 m)

mercury

SI Solution

The impact pressure is

$$P_0 = \rho g h$$

$$= \frac{\left(13\,600\ \frac{\text{kg}}{\text{m}^3}\right)\left(9.81\ \frac{\text{m}}{\text{s}^2}\right)(0.523\ \text{m})}{1000\ \frac{\text{Pa}}{\text{kPa}}}$$

$$= 69.78\ \text{kPa}$$

From Eq. 2.15, the velocity is

Pitot Tube

$$v = \sqrt{(2/\rho)(P_0 - P_s)}$$

$$= \sqrt{\frac{\left(\dfrac{2}{1.20\ \dfrac{\text{kg}}{\text{m}^3}}\right)}{\times \left(\begin{array}{c}(69.78\ \text{kPa} - 68.95\ \text{kPa})\\ \times \left(1000\ \dfrac{\text{Pa}}{\text{kPa}}\right)\end{array}\right)}}$$

$$= 37.2\ \text{m/s}$$

Customary U.S. Solution

The impact pressure is

$$P_0 = \frac{\rho g h}{g_c} = \frac{\left(0.491\ \dfrac{\text{lbm}}{\text{in}^3}\right)\left(32.2\ \dfrac{\text{ft}}{\text{sec}^2}\right)(20.6\ \text{in})}{32.2\ \dfrac{\text{lbm-ft}}{\text{lbf-sec}^2}}$$

$$= 10.11\ \text{lbf/in}^2 \quad (10.11\ \text{psig})$$

From Eq. 2.15, the velocity is

Pitot Tube

$$v = \sqrt{2g(P_0 - P_s)/\gamma}$$

$$= \sqrt{\frac{(2)\left(32.2\ \dfrac{\text{ft}}{\text{sec}^2}\right)\times\left(\begin{array}{c}\left(10.11\ \dfrac{\text{lbf}}{\text{in}^2} - 10.00\ \dfrac{\text{lbf}}{\text{in}^2}\right)\\ \times \left(12\ \dfrac{\text{in}}{\text{ft}}\right)^2\end{array}\right)}{0.075\ \dfrac{\text{lbf}}{\text{ft}^3}}}$$

$$= 116.6\ \text{ft/s}$$

9. HYDRAULIC RADIUS

The *hydraulic radius* is defined as the area in flow divided by the *wetted perimeter*. (The hydraulic radius is not the same as the radius of a pipe.) The area in flow is the cross-sectional area of the fluid flowing. When a fluid is flowing under pressure in a pipe (i.e., *pressure flow* in a *pressure conduit*), the area in flow will be the internal area of the pipe. However, the fluid may not completely fill the pipe and may flow simply because of a sloped surface (i.e., *gravity flow* or *open channel flow*).

The wetted perimeter is the length of the line representing the interface between the fluid and the pipe or channel. It does not include the *free surface* length (i.e., the interface between fluid and atmosphere). [Geometric Elements of Channel Sections]

Flow in Noncircular Conduits

$$R_H = \frac{\text{cross-sectional area}}{\text{wetted perimeter}}$$

$$= \frac{A}{P} \hspace{2cm} 2.17$$

Consider a circular pipe flowing completely full. The area in flow is πr^2. The wetted perimeter is the entire circumference, $2\pi r$. The hydraulic radius is

Flow in Noncircular Conduits

$$R_{H,\text{pipe}} = \frac{\pi r^2}{2\pi r} = \frac{r}{2} = \frac{D_H}{4} \hspace{1cm} 2.18$$

The hydraulic radius of a pipe flowing half full is also $r/2$, since the flow area and wetted perimeter are both halved. However, it is time-consuming to calculate the hydraulic radius for pipe flow at any intermediate depth, due to the difficulty in evaluating the flow area and wetted perimeter.

Example 2.4

A pipe (internal diameter = 6) carries water with a depth of 2 flowing under the influence of gravity. Calculate the hydraulic radius analytically.

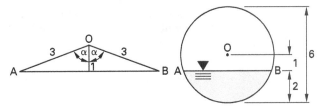

Solution

The equations for a circular segment must be used. The radius is $6/2 = 3$.

Points A, O, and B are used to find the central angle of the circular segment.

$$\phi = 2\alpha = 2\arccos\frac{1}{3} = (2)(70.53°)$$
$$= 141.06°$$

ϕ must be expressed in radians.

$$\phi = 2\pi\left(\frac{141.06°}{360°}\right) = 2.46 \text{ rad}$$

The area of the circular segment (i.e., the area in flow) is

Mensuration of Areas and Volumes: Circular Segment

$$A = [r^2(\phi - \sin\phi)]/2$$
$$= \frac{(3)^2\left(2.46 \text{ rad} - \sin(2.46 \text{ rad})\right)}{2}$$
$$= 8.235$$

The arc length (i.e., the wetted perimeter) is

$$P = r\phi = (3)(2.46 \text{ rad}) = 7.38$$

The hydraulic radius is

$$R_H = \frac{A}{P} = \frac{8.235}{7.38} = 1.12$$

10. HYDRAULIC DIAMETER

Many fluid, thermodynamic, and heat transfer processes are dependent on the physical length of an object. This controlling variable is generally known as the *characteristic dimension*. The characteristic dimension in evaluating fluid flow is the *hydraulic diameter* (also known as the *equivalent hydraulic diameter*).[9] The hydraulic diameter for a full-flowing pipe is simply its inside diameter. The hydraulic diameters of other cross sections in flow are given in Table 2.1. If the hydraulic radius is known, it can be used to calculate the hydraulic diameter.

$$D_H = 4R_H \qquad 2.19$$

Table 2.1 *Hydraulic Diameters for Common Conduit Shapes*

conduit cross section	D_H
flowing full	
circle	D
annulus (outer diameter D_o, inner diameter D_i)	$D_o - D_i$
square (side L)	L
rectangle (sides L_1 and L_2)	$\dfrac{2L_1L_2}{L_1 + L_2}$
flowing partially full	
half-filled circle (diameter D)	D
rectangle (h deep, L wide)	$\dfrac{4hL}{L + 2h}$
wide, shallow stream (h deep)	$4h$
triangle, vertex down (h deep, L broad, s side)	$\dfrac{hL}{s}$
trapezoid (h deep, a wide at top, b wide at bottom, s side)	$\dfrac{2h(a + b)}{b + 2s}$

---

[9]The engineering community is very inconsistent, but the three terms *hydraulic depth*, *hydraulic diameter*, and *equivalent diameter* do not have the same meanings. Hydraulic depth (flow area divided by exposed surface width) is a characteristic length used in Froude number and other open channel flow calculations. Hydraulic diameter (four times the area in flow divided by the wetted surface) is a characteristic length used in Reynolds number and friction loss calculations. Equivalent diameter $(1.3(ab)^{0.625}/(a + b)^{0.25})$ is the diameter of a round duct or pipe that will have the same friction loss per unit length as a noncircular duct. Unfortunately, these terms are often used interchangeably.

Example 2.5

Determine the hydraulic diameter and hydraulic radius for the open trapezoidal channel shown.

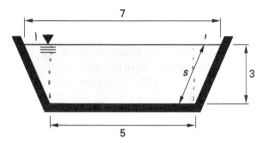

Solution

The batter of the inclined walls is $(7 - 5)/2$ walls $= 1$.

$$s = \sqrt{(3)^2 + (1)^2} = 3.16$$

Using Table 2.1,

$$D_H = \frac{2h(a + b)}{b + 2s} = \frac{(2)(3)(7 + 5)}{5 + (2)(3.16)} = 6.36$$

From Eq. 2.19,

Flow in Noncircular Conduits

$$R_H = \frac{D_H}{4} = \frac{6.36}{4} = 1.59$$

11. REYNOLDS NUMBER

The *Reynolds number*, Re, is a dimensionless number interpreted as the ratio of inertial forces to viscous forces in the fluid.[10]

$$\mathrm{Re} = \frac{\text{inertial forces}}{\text{viscous forces}} \qquad 2.20$$

The inertial forces are proportional to the flow diameter, velocity, and fluid density. (Increasing these variables will increase the momentum of the fluid in flow.) The viscous force is represented by the fluid's absolute viscosity, μ. Thus, the Reynolds number is calculated as

$$\mathrm{Re} = \frac{\mathrm{v}\rho D_H}{\mu} \qquad \text{[SI]} \quad 2.21(a)$$

$$\mathrm{Re} = \frac{\mathrm{v}\rho D_H}{g_c\mu} \qquad \text{[U.S.]} \quad 2.21(b)$$

Since μ/ρ is defined as the *kinematic viscosity*, ν, Eq. 2.21 can be simplified.

Reynolds Number (Newtonian Fluid)
$$\mathrm{Re} = \mathrm{v}D\rho/\mu = \mathrm{v}D/\nu \qquad 2.22$$

Occasionally, the *mass flow rate per unit area*, $G = \rho\mathrm{v}$, will be known. This variable expresses the quantity of fluid flowing in kg/m^2·s or lbm/ft^2-sec.

$$G = \frac{\dot{m}}{A} \qquad 2.23$$

$$\mathrm{Re} = \frac{D_H G}{\mu} \qquad \text{[SI]} \quad 2.24(a)$$

$$\mathrm{Re} = \frac{D_H G}{g_c\mu} \qquad \text{[U.S.]} \quad 2.24(b)$$

12. LAMINAR FLOW

Laminar flow gets its name from the word *laminae* (layers). If all of the fluid particles move in paths parallel to the overall flow direction (i.e., in layers), the flow is said to be *laminar*. (The terms *viscous flow* and *streamline flow* are also used.) This occurs in pipeline flow when the Reynolds number is less than (approximately) 2100. Laminar flow is typical when the flow channel is small, the velocity is low, and the fluid is viscous. Viscous forces are dominant in laminar flow.

In laminar flow, a stream of dye inserted in the flow will continue from the source in a continuous, unbroken line with very little mixing of the dye and surrounding liquid. The fluid particle paths coincide with imaginary *streamlines*. (Streamlines and velocity vectors are always tangent to each other.) A "bundle" of these streamlines (i.e., a *streamtube*) constitutes a complete fluid flow.

13. TURBULENT FLOW

A fluid is said to be in *turbulent flow* if the Reynolds number is greater than (approximately) 4000. Turbulent flow is characterized by a three-dimensional movement of the fluid particles superimposed on the overall direction of motion. A stream of dye injected into a turbulent flow will quickly disperse and uniformly mix with the surrounding flow. Inertial forces dominate in turbulent flow. At very high Reynolds numbers, the flow is said to be *fully turbulent*.

---

[10]Engineering authors are not in agreement about the symbol for the Reynolds number. In addition to Re (used in this book), engineers commonly use **Re**, R, $\Re$, $\mathrm{N_{Re}}$, and $\mathrm{N_R}$.

14. CRITICAL FLOW

The flow is said to be in a *critical zone* or *transition region* when the Reynolds number is between 2100 and 4000. These numbers are known as the lower and upper *critical Reynolds numbers* for fluid flow, respectively. (Critical Reynolds numbers for other processes are different.) It is difficult to design for the transition region, since fluid behavior is not consistent and few processes operate in the critical zone. In the event a critical zone design is required, the conservative assumption of turbulent flow will result in the greatest value of friction loss.

15. FLUID VELOCITY DISTRIBUTION IN PIPES

With laminar flow, the viscosity makes some fluid particles adhere to the pipe wall. The closer a particle is to the pipe wall, the greater the tendency will be for the fluid to adhere to the pipe wall. The following statements characterize laminar flow.

- The velocity distribution is parabolic.

- The velocity is zero at the pipe wall.

- The velocity is maximum at the center and equal to twice the average velocity.

$$v_{ave} = \frac{Q}{A} = \frac{v_{max}}{2} \quad \text{[laminar]} \qquad 2.25$$

With turbulent flow, there is generally no distinction made between the velocities of particles near the pipe wall and particles at the pipe centerline.[11] All of the fluid particles are assumed to have the same velocity. This velocity is known as the *average* or *bulk velocity*. It can be calculated from the volume flowing.

$$v_{ave} = \frac{Q}{A} \quad \text{[turbulent]} \qquad 2.26$$

Laminar and turbulent velocity distributions are shown in Fig. 2.3. In actuality, no flow is completely turbulent, and there is a difference between the *centerline velocity* and the average velocity. The error decreases as the Reynolds number increases. The ratio v_{ave}/v_{max} starts at approximately 0.75 for Re = 4000 and increases to approximately 0.86 at Re = 10^6. Most problems ignore the difference between v_{ave} and v_{max}, but care should be taken when a centerline measurement (as from a pitot tube) is used to evaluate the average velocity.

Figure 2.3 Laminar and Turbulent Velocity Distributions

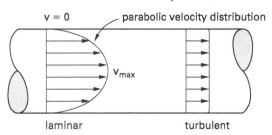

The ratio of the average velocity to maximum velocity is known as the *pipe coefficient* or *pipe factor*. Considering all the other coefficients used in pipe flow, these names are somewhat vague and ambiguous. Therefore, they are not in widespread use.

For turbulent flow (Re $\approx 10^5$) in a smooth, circular pipe of radius r_o, the velocity at a radial distance r from the centerline is given by the $\frac{1}{7}$-*power law*.

$$v_r = v_{max} \left(\frac{r_o - r}{r_o} \right)^{1/7} \quad \text{[turbulent flow]} \qquad 2.27$$

The fluid's *velocity profile*, given by Eq. 2.27, is valid in smooth pipes up to a Reynolds number of approximately 100,000. Above that, up to a Reynolds number of approximately 400,000, an exponent of $\frac{1}{8}$ fits experimental data better. For rough pipes, the exponent is larger (e.g., $\frac{1}{5}$).

Equation 2.27 can be integrated to determine the average velocity.

$$v_{ave} = \left(\frac{49}{60} \right) v_{max}$$
$$= 0.817 v_{max} \quad \text{[turbulent flow]} \qquad 2.28$$

When the flow is laminar, the velocity profile within a pipe will be parabolic and of the form of Eq. 2.29. (Equation 2.27 is for turbulent flow and does not describe a parabolic velocity profile.) The velocity at a radial distance r from the centerline in a pipe of radius r_o is

$$v_r = v_{max} \left(\frac{r_o^2 - r^2}{r_o^2} \right) \quad \text{[laminar flow]} \qquad 2.29$$

When the velocity profile is parabolic, the flow rate and pressure drop can easily be determined. The average velocity is half of the maximum velocity given in the velocity profile equation.

$$v_{ave} = \frac{1}{2} v_{max} \quad \text{[laminar flow]} \qquad 2.30$$

---

[11]This disregards the *boundary layer*, a thin layer near the pipe wall, where the velocity goes from zero to v_{ave}.

The average velocity is used to determine the flow quantity and friction loss. The friction loss is determined by traditional means.

Continuity Equation

$$Q = A\text{v}_{\text{ave}} \qquad 2.31$$

The kinetic energy of laminar flow can be found by integrating the velocity profile equation, resulting in Eq. 2.32.

$$E_{\text{v}} = \text{v}_{\text{ave}}^2 \quad \text{[laminar flow]} \qquad \text{[SI]} \qquad 2.32(a)$$

$$E_{\text{v}} = \frac{\text{v}_{\text{ave}}^2}{g_c} \quad \text{[laminar flow]} \qquad \text{[U.S.]} \qquad 2.32(b)$$

16. ENERGY GRADE LINE

The *energy grade line* (EGL) is a graph of the total energy (total specific energy) along a length of pipe.[12] In a frictionless pipe without pumps or turbines, the total specific energy is constant, and the EGL will be horizontal. (This is a restatement of the Bernoulli equation.)

$$\text{elevation of EGL} = h_p + h_{\text{v}} + h_z \qquad 2.33$$

The *hydraulic grade line* (HGL) is the graph of the sum of the pressure and gravitational heads, plotted as a position along the pipeline. Since the pressure head can increase at the expense of the velocity head, the HGL can increase in elevation if the flow area is increased.

$$\text{elevation of HGL} = h_p + h_z \qquad 2.34$$

The difference between the EGL and the HGL is the velocity head, h_{v}, of the fluid.

$$h_{\text{v}} = \text{elevation of EGL} - \text{elevation of HGL} \qquad 2.35$$

The following rules apply to these grade lines in a frictionless environment, in a pipe flowing full (i.e., under pressure), without pumps or turbines. (See Fig. 2.4.)

- The EGL is always horizontal.

- The HGL is always equal to or below the EGL.

- For still (v = 0) fluid at a free surface, EGL = HGL (i.e., the EGL coincides with the fluid surface in a reservoir).

- If flow velocity is constant (i.e., flow in a constant-area pipe), the HGL will be horizontal and parallel to the EGL, regardless of pipe orientation or elevation.

---

[12]The term *energy line* (EL) is also used.

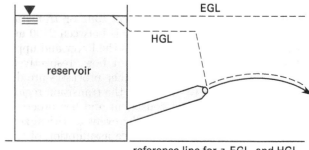

Figure 2.4 *Energy and Hydraulic Grade Lines Without Friction*

- When the flow area decreases, the HGL decreases.

- When the flow area increases, the HGL increases.

- In a free jet (i.e., a stream of water from a hose), the HGL coincides with the jet elevation, following a parabolic path.

17. ENERGY AND HYDRAULIC GRADE LINES WITH FRICTION

The *energy grade line* (EGL, also known as *total energy line*) is a graph of the total energy versus position in a pipeline. Since a pitot tube measures total (stagnation) energy, EGL will always coincide with the elevation of a pitot-piezometer fluid column. When friction is present, the EGL will always slope down, in the direction of flow. Figure 2.5 illustrates the EGL for a complex pipe network. The difference between $\text{EGL}_{\text{frictionless}}$ and $\text{EGL}_{\text{with friction}}$ is the energy loss due to friction.

The EGL line in Fig. 2.5 is discontinuous at point 2, since the friction in pipe section B–C cannot be portrayed without disturbing the spatial correlation of points in the figure. Since the friction loss is proportional to v^2, the slope is steeper when the fluid velocity increases (i.e., when the pipe decreases in flow area), as it does in section D–E. Disregarding air friction, the EGL becomes horizontal at point 6 when the fluid becomes a free jet.

The *hydraulic grade line* (HGL) is a graph of the sum of pressure and potential energies versus position in the pipeline. (That is, the EGL and HGL differ by the kinetic energy.) The HGL will always coincide with the height of the fluid column in a static piezometer tube. The reference point for elevation is arbitrary, and the pressure energy is usually referenced to atmospheric pressure. Therefore, the pressure energy, E_p, for a free jet will be zero, and the HGL will consist only of the potential energy, as shown in section G–H.

The easiest way to draw the energy and hydraulic grade lines is to start with the EGL. The EGL can be drawn simply by recognizing that the rate of divergence from

Figure 2.5 *Energy and Hydraulic Grade Lines*

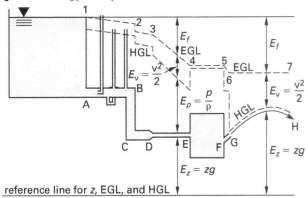

reference line for *z*, EGL, and HGL

the horizontal $\text{EGL}_{\text{frictionless}}$ line is proportional to v^2. Then, since EGL and HGL differ by the velocity head, the HGL can be drawn parallel to the EGL when the pipe diameter is constant. The larger the pipe diameter, the closer the two lines will be.

The EGL for a pump will increase in elevation by E_A across the pump. (The actual energy "path" taken by the fluid is unknown, and a dotted line is used to indicate a lack of knowledge about what really happens in the pump.) The placement of the HGL for a pump will depend on whether the pump increases the fluid velocity and elevation, as well as the fluid pressure. In most cases, only the pressure will be increased. Figure 2.6 illustrates the HGL for the case of a pressure increase only.

Figure 2.6 *EGL and HGL for a Pump*

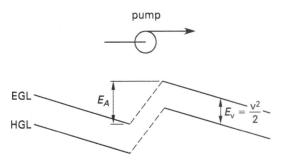

The EGL and HGL for minor losses (fittings, contractions, expansions, etc.) are shown in Fig. 2.7.

Figure 2.7 *EGL and HGL for Minor Losses*

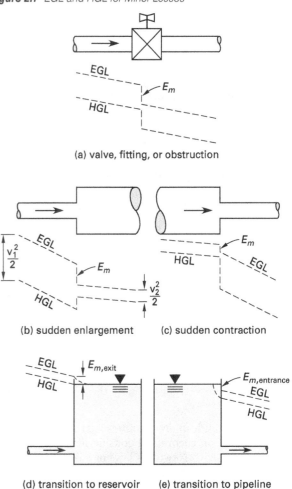

(a) valve, fitting, or obstruction

(b) sudden enlargement   (c) sudden contraction

(d) transition to reservoir   (e) transition to pipeline

18. SPECIFIC ENERGY

Specific energy is a term that is used primarily with open channel flow. It is the total energy with respect to the channel bottom, consisting of pressure and velocity energy contributions only.

$$E_{\text{specific}} = E_P + E_v \qquad 2.36$$

Since the channel bottom is chosen as the reference elevation ($z = 0$) for gravitational energy, there is no contribution by gravitational energy to specific energy.

$$E_{\text{specific}} = \frac{P}{\rho} + \frac{v^2}{2} \qquad \text{[SI]} \quad 2.37(a)$$

$$E_{\text{specific}} = \frac{P}{\rho} + \frac{v^2}{2g_c} \qquad \text{[U.S.]} \quad 2.37(b)$$

However, since P is the hydrostatic pressure at the channel bottom due to a fluid depth, the P/ρ term can be interpreted as the depth of the fluid, y.

Open-Channel Flow and Pipe Flow of Water

$$E = \alpha \frac{V^2}{2g} + y \qquad 2.38$$

Specific energy is constant when the flow depth and width are constant (i.e., *uniform flow*). A change in channel width will cause a change in flow depth, and since width is not part of the equation for specific energy, there will be a corresponding change in specific energy. There are other ways that specific energy can decrease, also.[13]

19. HYDRAULICS AND HYDRODYNAMICS

This chapter covers fluid moving through pipes, measurements with venturis and orifices, and other motion-related topics such as model theory, lift and drag, and pumps. In a strict interpretation, any fluid-related phenomenon that is not hydro*statics* should be hydro*dynamics*. However, tradition has separated the study of moving fluids into the fields of hydraulics and hydrodynamics.

In a general sense, *hydraulics* is the study of the practical laws of fluid flow and resistance in pipes and open channels. Hydraulic formulas are often developed from experimentation, empirical factors, and curve fitting, without an attempt to justify why the fluid behaves the way it does.

On the other hand, *hydrodynamics* is the study of fluid behavior based on theoretical considerations. Hydrodynamicists start with Newton's laws of motion and try to develop models of fluid behavior. Models developed in this manner are complicated greatly by the inclusion of viscous friction and compressibility. Therefore, hydrodynamic models assume a perfect fluid with constant density and zero viscosity. The conclusions reached by hydrodynamicists can differ greatly from those reached by hydraulicians.[14]

20. CONSERVATION OF MASS

Fluid mass is always conserved in fluid systems, regardless of the pipeline complexity, orientation of the flow, and fluid. This single concept is often sufficient to solve simple fluid problems.

$$\dot{m}_1 = \dot{m}_2 \qquad 2.39$$

When applied to fluid flow, the conservation of mass law is known as the *continuity equation*.

$$\rho_1 A_1 v_1 = \rho_2 A_2 v_2 \qquad 2.40$$

If the fluid is incompressible, then $\rho_1 = \rho_2$.

Continuity Equation

$$A_1 v_1 = A_2 v_2 \qquad 2.41$$

Since the volumetric flow is the product of the cross-sectional area for flow and the average flow velocity,

$$Q_1 = Q_2 \qquad 2.42$$

Calculation of flow rates is often complicated by the interdependence between flow rate and friction loss. Each affects the other, so many pipe flow problems must be solved iteratively. Usually, a reasonable friction factor is assumed and used to calculate an initial flow rate. The flow rate establishes the flow velocity, from which a revised friction factor can be determined.

21. TYPICAL VELOCITIES IN PIPES

Fluid friction in pipes is kept at acceptable levels by maintaining reasonable fluid velocities. Table 2.2 lists typical maximum fluid velocities. Higher velocities may be observed in practice, but only with a corresponding increase in friction and pumping power.

22. HEAD LOSS DUE TO FRICTION

The original Bernoulli equation was based on an assumption of frictionless flow. In actual practice, friction occurs during fluid flow. This friction "robs" the fluid of energy, E, so that the fluid at the end of a pipe section has less energy than it does at the beginning.[15]

$$E_1 > E_2 \qquad 2.43$$

Most formulas for calculating friction loss use the symbol h_f to represent the *head loss due to friction*.[16] This loss is added into the original Bernoulli equation to restore the equality. Of course, the units of h_f must be the same as the units for the other terms in the Bernoulli equation. (See Eq. 2.52.) If the Bernoulli equation is written in terms of energy, the units will be ft-lbf/lbm or J/kg.

$$E_1 = E_2 + E_f \qquad 2.44$$

Consider the constant-diameter, horizontal pipe in Fig. 2.8. An incompressible fluid is flowing at a steady rate. Since the elevation of the pipe, z, does not change, the potential energy is constant. Since the pipe has a constant area, the kinetic energy (velocity) is constant. Therefore, the friction energy loss must show up as a

---

[13]Specific energy changes dramatically in a *hydraulic jump* or *hydraulic drop*.
[14]Perhaps the most disparate conclusion is *D'Alembert's paradox*. In 1744, D'Alembert derived theoretical results "proving" that there is no resistance to bodies moving through an ideal (nonviscous) fluid.
[15]The friction generates minute amounts of heat. The heat is lost to the surroundings.
[16]Other names and symbols for this friction loss are *friction head loss* (h_L), *lost work* (LW), *friction heating* ($\mathscr{F}$), *skin friction loss* (F_f), and *pressure drop due to friction* (ΔP_f). All terms and symbols essentially mean the same thing, although the units may be different.

Table 2.2 Typical Full-Pipe Bulk Fluid Velocities

fluid and application	velocity	
	ft/sec	m/s
water: city service	2–10	0.6–2.1
3 in diameter	4	1.2
6 in diameter	5	1.5
12 in diameter	9	2.7
water: boiler feed	8–15	2.4–4.5
water: pump suction	4	1.2
water: pump discharge	4–8.5	1.2–2.5
water, sewage: partially filled sewer	2.5 (min)	0.75 (min)
brine, water: chillers and coolers	6–8 typ (3–10)	1.8–2.4 typ (0.9–3)
air: compressor suction	75–200	23–60
air: compressor discharge	100–250	30–75
air: HVAC forced air	15–25	5–8
natural gas: overland pipeline	< 150 (60 typ)	< 45 (18 typ)
steam, saturated: heating	65–100	20–30
steam, saturated: miscellaneous	100–200	30–60
50–100 psia	< 150	< 45
150–400 psia	< 130	< 39
400–600 psia	< 100	< 30
steam, superheated: turbine feed	160–250	50–75
hydraulic fluid: fluid power	7–15	2.1–4.6
liquid sodium ($T > 525°C$): heat transfer	10 typ (0.3–40)	3 typ (0.1–12)
ammonia: compressor suction	85 (max)	25 (max)
ammonia: compressor discharge	100 (max)	30 (max)
oil, crude: overland pipeline	4–12	1.2–3.6
oil, lubrication: pump suction	< 2	< 0.6
oil, lubrication: pump discharge	3–7	0.9–2.1

(Multiply ft/sec by 0.3048 to obtain m/s.)

decrease in pressure energy. Since the fluid is incompressible, this can only occur if the pressure, P, decreases in the direction of flow.

Figure 2.8 Pressure Drop in a Pipe

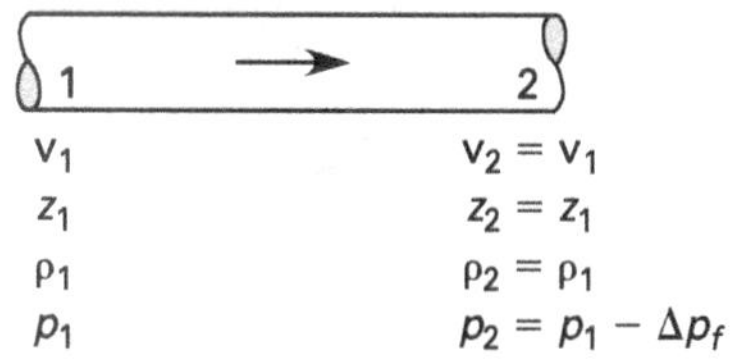

v_1	$v_2 = v_1$
z_1	$z_2 = z_1$
ρ_1	$\rho_2 = \rho_1$
p_1	$p_2 = p_1 - \Delta p_f$

23. RELATIVE ROUGHNESS

It is intuitive that pipes with rough inside surfaces will experience greater friction losses than smooth pipes.[17] *Specific roughness*, ϵ, is a parameter that measures the average size of imperfections inside the pipe. Table 2.3 lists values of ϵ for common pipe materials.

Table 2.3 Values of Specific Roughness, ϵ, for Common Pipe Materials

material	ϵ	
	ft	m
plastic (PVC, ABS)	0.000005	1.5×10^{-6}
copper and brass	0.000005	1.5×10^{-6}
steel	0.0002	6.0×10^{-5}
plain cast iron	0.0008	2.4×10^{-4}
concrete	0.004	1.2×10^{-3}

(Multiply ft by 0.3048 to obtain m.)

However, an imperfection the size of a sand grain will have much more effect in a small-diameter hydraulic line than in a large-diameter sewer. Therefore, the *relative roughness*, ϵ/D, is a better indicator of pipe roughness. Both ϵ and D have units of length (e.g., feet or meters), and the relative roughness is dimensionless.

24. FRICTION FACTOR

The *Darcy friction factor*, f, is one of the parameters used to calculate friction loss.[18] The friction factor is not constant but decreases as the Reynolds number (fluid velocity) increases, up to a certain point known as *fully turbulent flow* (or *rough-pipe flow*). Once the flow is fully turbulent, the friction factor remains constant and depends only on the relative roughness and not on the Reynolds number. (See Fig. 2.9.) For very smooth pipes, fully turbulent flow is achieved only at very high Reynolds numbers.

---

[17]Surprisingly, this intuitive statement is valid only for turbulent flow. The roughness does not (ideally) affect the friction loss for laminar flow.

Figure 2.9 *Friction Factor as a Function of Reynolds Number*

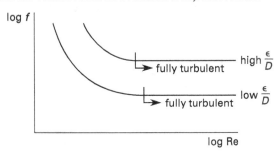

Table 2.4 *Friction Factors for Laminar Flow in Various Cross Sections**

tube geometry	D_H (full)	c/d or θ	friction factor, f
circle	D	–	$64.00/\mathrm{Re}$
rectangle	$\dfrac{2cd}{c+d}$	1 2 3 4 6 8 ∞	$56.92/\mathrm{Re}$ $62.20/\mathrm{Re}$ $68.36/\mathrm{Re}$ $72.92/\mathrm{Re}$ $78.80/\mathrm{Re}$ $82.32/\mathrm{Re}$ $96.00/\mathrm{Re}$
ellipse	$\dfrac{cd}{\sqrt{\frac{1}{2}(c^2+d^2)}}$	1 2 4 8 16	$64.00/\mathrm{Re}$ $67.28/\mathrm{Re}$ $72.96/\mathrm{Re}$ $76.60/\mathrm{Re}$ $78.16/\mathrm{Re}$
isosceles triangle	$\dfrac{d\sin\theta}{2\left(1+\sin\frac{\theta}{2}\right)}$	$10°$ $30°$ $60°$ $90°$ $120°$	$50.80/\mathrm{Re}$ $52.28/\mathrm{Re}$ $53.32/\mathrm{Re}$ $52.60/\mathrm{Re}$ $50.96/\mathrm{Re}$

*$\mathrm{Re} = \mathrm{v_{bulk}}D_H/\nu$, and $D_H = 4A/P$.

The friction factor is not dependent on the material of the pipe but is affected by the roughness. For example, for a given Reynolds number, the friction factor will be the same for any smooth pipe material (glass, plastic, smooth brass and copper, etc.).

The friction factor is determined from the relative roughness, ϵ/D, and the Reynolds number, Re, by various methods. These methods include explicit and implicit equations, the Moody diagram, and tables. The values obtained are based on experimentation, primarily the work of J. Nikuradse in the early 1930s.

When a moving fluid initially encounters a parallel surface (as when a moving gas encounters a flat plate or when a fluid first enters the mouth of a pipe), the flow will generally not be turbulent, even for very rough surfaces. The flow will be laminar for a certain *critical distance* before becoming turbulent.

Friction Factors for Laminar Flow

The easiest method of obtaining the friction factor for laminar flow (Re < 2100) is to calculate it. Equation 2.45 illustrates that roughness is not a factor in determining the frictional loss in ideal laminar flow.

$$f = \frac{64}{\mathrm{Re}} \quad \text{[circular pipe]} \qquad 2.45$$

Table 2.4 gives friction factors for laminar flow in various cross sections.

Friction Factors for Turbulent Flow: by Formula

One of the earliest attempts to predict the friction factor for turbulent flow in smooth pipes resulted in the *Blasius equation* (claimed "valid" for 3000 < Re < 100,000).

$$f = \frac{0.316}{\mathrm{Re}^{0.25}} \qquad 2.46$$

The *Nikuradse equation* can also be used to determine the friction factor for smooth pipes (i.e., when $\epsilon/D = 0$). Unfortunately, this equation is implicit in f and must be solved iteratively.

$$\frac{1}{\sqrt{f}} = 2.0\log_{10}(\mathrm{Re}\sqrt{f}) - 0.80 \qquad 2.47$$

The *Karman-Nikuradse equation* predicts the fully turbulent friction factor (i.e., when Re is very large).

$$\frac{1}{\sqrt{f}} = 1.74 - 2\log_{10}\frac{2\epsilon}{D} \qquad 2.48$$

The most widely known method of calculating the friction factor for any pipe roughness and Reynolds number is another implicit formula, the *Colebrook equation*.

---

[18]There are actually two friction factors: the Darcy friction factor and the *Fanning friction factor*, f_{Fanning}, also known as the *skin friction coefficient* and *wall shear stress factor*. Both factors are in widespread use, sharing the same symbol, f. Civil and (most) mechanical engineers use the Darcy friction factor. The Fanning friction factor is encountered more often by chemical engineers. One can be derived from the other: $f_{\text{Darcy}} = 4f_{\text{Fanning}}$.

Most other equations are variations of this equation. (Notice that the relative roughness, ϵ/D, is used to calculate f.)

$$\frac{1}{\sqrt{f}} = -2\log_{10}\left(\frac{\dfrac{\epsilon}{D}}{3.7} + \frac{2.51}{\mathrm{Re}\sqrt{f}}\right) \qquad 2.49$$

A suitable approximation would appear to be the *Swamee-Jain equation*, which claims to have less than 1% error (as measured against the Colebrook equation) for relative roughnesses between 0.000001 and 0.01, and for Reynolds numbers between 5000 and 100,000,000.[19] Even with a 1% error, this equation produces more accurate results than can be read from the Moody friction factor chart.

$$f = \frac{0.25}{\left(\log_{10}\left(\dfrac{\dfrac{\epsilon}{D}}{3.7} + \dfrac{5.74}{\mathrm{Re}^{0.9}}\right)\right)^2} \qquad 2.50$$

Friction Factors for Turbulent Flow: by Moody Chart

The *Moody friction factor chart*, shown in Fig. 2.10, presents the friction factor graphically as a function of Reynolds number and relative roughness. There are different lines for selected discrete values of relative roughness. Due to the complexity of this graph, it is easy to mislocate the Reynolds number or use the wrong curve. Nevertheless, the Moody chart remains the most common method of obtaining the friction factor. [Moody (Stanton) Diagram]

Friction Factors for Turbulent Flow: by Table

Appendix 2.B (based on the Colebrook equation), or a similar table, will usually be the most convenient method of obtaining friction factors for turbulent flow.

Example 2.6

Determine the friction factor for a Reynolds number of Re = 400,000 and a relative roughness of $\epsilon/D = 0.004$ using (a) the Moody diagram, (b) App. 2.B, and (c) the Swamee-Jain approximation. (d) Check the table value of f with the Colebrook equation.

Solution

(a) From Fig. 2.10, the friction factor is approximately 0.028.

(b) Appendix 2.B lists the friction factor as 0.0287.

(c) From Eq. 2.50,

$$f = \frac{0.25}{\left(\log_{10}\left(\dfrac{\dfrac{\epsilon}{D}}{3.7} + \dfrac{5.74}{\mathrm{Re}^{0.9}}\right)\right)^2}$$

$$= \frac{0.25}{\left(\log_{10}\left(\dfrac{0.004}{3.7} + \dfrac{5.74}{(400,000)^{0.9}}\right)\right)^2}$$

$$= 0.0288$$

(d) From Eq. 2.49,

$$\frac{1}{\sqrt{f}} = -2\log_{10}\left(\frac{\dfrac{\epsilon}{D}}{3.7} + \frac{2.51}{\mathrm{Re}\sqrt{f}}\right)$$

$$\frac{1}{\sqrt{0.0287}} = -2\log_{10}\left(\frac{0.004}{3.7} + \frac{2.51}{400,000\sqrt{0.0287}}\right)$$

$$5.903 = 5.903$$

25. ENERGY LOSS DUE TO FRICTION: LAMINAR FLOW

Two methods are available for calculating the frictional energy loss for fluids experiencing laminar flow. The most common is the *Darcy equation* (which is also known as the *Weisbach equation* or the *Darcy-Weisbach equation*), which can be used for both laminar and turbulent flow.[20] One of the advantages of using the Darcy equation is that the assumption of laminar flow does not need to be confirmed if f is known.

Head Loss Due to Flow

$$h_f = f\frac{L}{D}\frac{\mathrm{v}^2}{2g} \qquad 2.51$$

$$E_f = h_f g = \frac{fL\mathrm{v}^2}{2D} \qquad \text{[SI]} \qquad 2.52(a)$$

$$E_f = h_f \times \frac{g}{g_c} = \frac{fL\mathrm{v}^2}{2Dg_c} \qquad \text{[U.S.]} \qquad 2.52(b)$$

---

[19]American Society of Civil Engineers. *Journal of Hydraulic Engineering* 102 (May 1976): 657. This is not the only explicit approximation to the Colebrook equation in existence.
[20]The difference is that the friction factor can be derived by hydrodynamics: $f = 64/\mathrm{Re}$. For turbulent flow, f is empirical.

Figure 2.10 *Moody Friction Factor Chart*

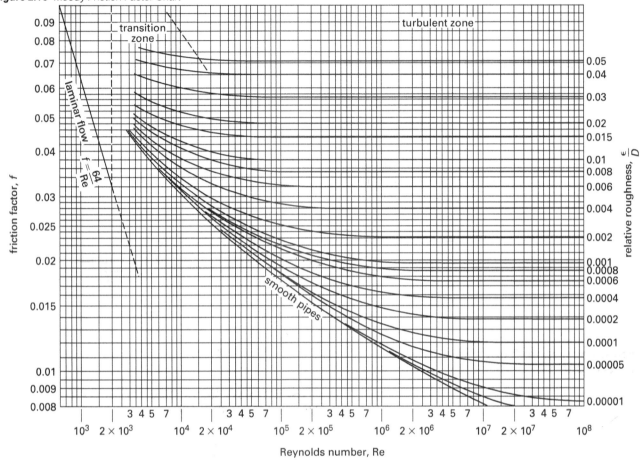

Reprinted with permission from L. F. Moody, "Friction Factor for Pipe Flow," *ASME Transactions*, Vol. 66, published by the American Society of Mechanical Engineers, copyright © 1944.

If the flow is truly laminar and the fluid is flowing in a circular pipe, then the *Hagen-Poiseuille equation* can be used.

$$E_f = \frac{32\mu v L}{D^2 \rho} = \frac{128\mu Q L}{\pi D^4 \rho} \qquad 2.53$$

If necessary, h_f can be converted to an actual pressure drop in lbf/ft^2 or Pa by multiplying by the fluid density. [Energy Equation (Bernoulli)]

$$\Delta p = h_f \times \rho g \qquad \text{[SI]} \quad 2.54(a)$$

$$\Delta p = h_f \times \rho \left(\frac{g}{g_c}\right) = h_f \gamma \qquad \text{[U.S.]} \quad 2.54(b)$$

The Hagen-Poiseuille equation can also be expressed for Q in terms of the pressure drop, ΔP_f.

Pressure Drop for Laminar Flow

$$Q = \frac{\pi R^4 \Delta P_f}{8\mu L} = \frac{\pi D^4 \Delta P_f}{128\mu L} \qquad 2.55$$

Values of the Darcy friction factor, f, are often quoted for new, clean pipe. The friction head losses and pumping power requirements calculated from these values are minimum values. Depending on the nature of the service, scale and impurity buildup within pipes may decrease the pipe diameters over time. Since the frictional loss is proportional to the fifth power of the diameter, such diameter decreases can produce dramatic increases in the friction loss.

$$\frac{h_{f,\text{scaled}}}{h_{f,\text{new}}} = \left(\frac{D_{\text{new}}}{D_{\text{scaled}}}\right)^5 \qquad 2.56$$

Equation 2.56 accounts for only the decrease in diameter. Any increase in roughness (i.e., friction factor) will produce a proportional increase in friction loss.

Because the "new, clean" condition is transitory in most applications, an uprating factor of 10–30% is often applied to either the friction factor, f, or the head loss, h_f. Of course, even larger increases should be considered when extreme fouling is expected.

Another approach eliminates the need to estimate the scaled pipe diameter. This simplistic approach multiplies the initial friction loss by a factor based on the age of the pipe. For example, for schedule-40 pipe between 4 in and 10 in (10 cm and 25 cm) in diameter, the multipliers of 1.4, 2.2, and 5.0 have been proposed for pipe ages of 5, 10, and 20 years, respectively. For larger pipes, the corresponding multipliers are 1.3, 1.6, and 2.0. Obviously, use of these values should be based on a clear understanding of the method's limitations.

26. ENERGY LOSS DUE TO FRICTION: TURBULENT FLOW

The *Darcy equation* is used almost exclusively to calculate the head loss due to friction for turbulent flow.

Head Loss Due to Flow

$$h_f = f \frac{L}{D} \frac{v^2}{2g} \qquad \textbf{2.57}$$

The head loss can be converted to pressure drop.

$$\Delta p = h_f \times \rho g \qquad \text{[SI]} \quad \textbf{2.58(a)}$$

$$\Delta p = h_f \times \rho \left(\frac{g}{g_c} \right) = h_f \gamma \qquad \text{[U.S.]} \quad \textbf{2.58(b)}$$

The Hagen-Poiseuille equation can also be expressed for Q in terms of the pressure drop, ΔP_f.

Pressure Drop for Laminar Flow

$$Q = \frac{\pi R^4 \Delta P_f}{8 \mu L} = \frac{\pi D^4 \Delta P_f}{128 \mu L} \qquad \textbf{2.59}$$

In problems where the pipe size is unknown, it will be impossible to obtain an accurate initial value of the friction factor, f (since f depends on velocity). In such problems, an iterative solution will be necessary.

Civil engineers commonly use the *Hazen-Williams equation* to calculate head loss. This method requires knowing the Hazen-Williams *roughness coefficient*, C, values of which are widely tabulated.[21] (See a table of specific roughness and Hazen-Williams constants for various pipe materials.) The advantage of using this equation is that C does not depend on the Reynolds number. The Hazen-Williams equation is empirical and is not dimensionally homogeneous. The length, L, and diameter, D, of the pipe must both be in feet, and the flow, Q, is in cubic feet per second. It is taken as a matter of faith that the units of h_f are feet.

Circular Pipe Head Loss Equation (Head Loss Expressed in Feet)

$$h_f = \frac{4.73L}{C^{1.852} D^{4.87}} Q^{1.852} \qquad \textbf{2.60}$$

The Hazen-Williams equation should be used only for turbulent flow. It gives good results for liquids that have kinematic viscosities around 1.2×10^{-5} ft²/sec (1.1×10^{-6} m²/s), which corresponds to the viscosity of 60°F (16°C) water. At extremely high and low temperatures, the Hazen-Williams equation can be 20% or more in error for water.

Example 2.7

50°F water is pumped through 1000 ft of 4 in, schedule-40 welded steel pipe at the rate of 300 gpm (0.67 cfs). What friction loss (in ft-lbf/lbm) is predicted by the Darcy equation?

Solution

The fluid viscosity, pipe dimensions, and other parameters can be found from the appendices.

From a table of water properties at atmospheric pressure, $\nu = 1.41 \times 10^{-5}$ ft²/sec.

From a table of specific roughness and Hazen-Williams constants for various pipe materials, $\epsilon = 0.0002$ ft.

From App. 2.A,

$$D = 0.3355 \text{ ft}$$
$$A = 0.0884 \text{ ft}^2$$

The flow quantity is converted from gallons per minute to cubic feet per second.

$$Q = \frac{300 \dfrac{\text{gal}}{\text{min}}}{\left(7.4805 \dfrac{\text{gal}}{\text{ft}^3}\right)\left(60 \dfrac{\text{sec}}{\text{min}}\right)} = 0.6684 \text{ ft}^3/\text{sec}$$

The velocity is

$$v = \frac{Q}{A} = \frac{0.6684 \dfrac{\text{ft}^3}{\text{sec}}}{0.0884 \text{ ft}^2} = 7.56 \text{ ft/sec}$$

---

[21]An approximate value of $C = 140$ is often chosen for initial calculations for new water pipe. $C = 100$ is more appropriate for water pipe that has been in service for some time. For sludge, C values are 20–40% lower than the equivalent water pipe values.

The Reynolds number is

$$\text{Re} = \frac{D\text{v}}{\nu} = \frac{(0.3355 \text{ ft})\left(7.56 \dfrac{\text{ft}}{\text{sec}}\right)}{1.41 \times 10^{-5} \dfrac{\text{ft}^2}{\text{sec}}}$$

$$= 1.8 \times 10^5$$

The relative roughness is

$$\frac{\epsilon}{D} = \frac{0.0002 \text{ ft}}{0.3355 \text{ ft}} = 0.0006$$

From the friction factor table, App. 2.B (or the Moody friction factor chart), $f = 0.0195$. Equation 2.52(b) is used to calculate the friction loss.

$$E_f = h_f \times \frac{g}{g_c} = \frac{fL\text{v}^2}{2Dg_c}$$

$$= \frac{(0.0195)(1000 \text{ ft})\left(7.56 \dfrac{\text{ft}}{\text{sec}}\right)^2}{(2)(0.3355 \text{ ft})\left(32.2 \dfrac{\text{lbm-ft}}{\text{lbf-sec}^2}\right)}$$

$$= 51.6 \text{ ft-lbf/lbm}$$

Example 2.8

Calculate the head loss due to friction for the pipe in Ex. 2.7 using the Hazen-Williams formula. Assume $C = 100$.

Solution

From Eq. 2.60,

Circular Pipe Head Loss Equation (Head Loss Expressed in Feet)

$$h_f = \frac{4.73L}{C^{1.852}D^{1.87}} Q^{1.852}$$

$$= \left(\frac{(4.73)(1000 \text{ ft})}{(100)^{1.852}(0.3355)^{4.87}}\right)\left(0.67 \frac{\text{ft}^3}{\text{sec}}\right)^{1.852}$$

$$= 90.9 \text{ ft}$$

27. FRICTION LOSS FOR WATER FLOW IN STEEL PIPES

Since water's specific volume is essentially constant within the normal temperature range, tables and charts can be used to determine water velocity. Friction loss and velocity for water flowing through steel pipe (as well as for other liquids and other pipe materials) in table and chart form are widely available. Tables and charts almost always give the friction loss per 100 ft or 10 m of pipe. The pressure drop is proportional to the length, so the value read can be scaled for other pipe lengths. Flow velocity is independent of pipe length.

These tables and charts are unable to compensate for the effects of fluid temperature and different pipe roughness. Unfortunately, the assumptions made in developing the tables and charts are seldom listed. Another disadvantage is that the values can be read to only a few significant figures. Friction loss data should be considered accurate to only $\pm 20\%$. Alternatively, a 20% safety margin should be established in choosing pumps and motors.

28. FRICTION LOSS IN NONCIRCULAR DUCTS

The frictional energy loss by a fluid flowing in a rectangular, annular, or other noncircular duct can be calculated from the Darcy equation by using the *hydraulic diameter*, D_H, in place of the diameter, D.[22] The friction factor, f, is determined in any of the conventional manners.

29. EFFECT OF VISCOSITY ON HEAD LOSS

Friction loss in a pipe is affected by the fluid viscosity. For both laminar and turbulent flow, viscosity is considered when the Reynolds number is calculated. When viscosities substantially increase without a corresponding decrease in flow rate, two things usually happen: (a) the friction loss greatly increases, and (b) the flow becomes laminar.

It is sometimes necessary to estimate head loss for a new fluid viscosity based on head loss at an old fluid viscosity. The estimation procedure used depends on the flow regimes for the new and old fluids.

For laminar flow, the friction factor is directly proportional to the viscosity. If the flow is laminar for both fluids, the ratio of new-to-old head losses will be equal to the ratio of new-to-old viscosities. Therefore, if a flow is already known to be laminar at one viscosity and the fluid viscosity increases, a simple ratio will define the new friction loss.

If both flows are fully turbulent, the friction factor will not change. If flow is fully turbulent and the viscosity decreases, the Reynolds number will increase. Theoretically, this will have no effect on the friction loss.

---

[22]Although it is used for both, this approach is better suited for turbulent flow than for laminar flow. Also, the accuracy of this method decreases as the flow area becomes more noncircular. The friction drop in long, narrow slit passageways is poorly predicted, for example. However, there is no other convenient method of predicting friction drop. Experimentation should be used with a particular flow geometry if extreme accuracy is required.

There are no analytical ways of estimating the change in friction loss when the flow regime changes between laminar and turbulent or between semiturbulent and fully turbulent. Various graphical methods are used, particularly by the pump industry, for calculating power requirements.

30. FRICTION LOSS WITH SLURRIES AND NON-NEWTONIAN FLUIDS

A *slurry* is a mixture of a liquid (usually water) and a solid (e.g., coal, paper pulp, foodstuff). The liquid is generally used as the transport mechanism (i.e., the *carrier*) for the solid.

Friction loss calculations for slurries vary in sophistication depending on what information is available. In many cases, only the slurry's specific gravity is known. In that case, use is made of the fact that friction loss can be reasonably predicted by multiplying the friction loss based on the pure carrier (e.g., water) by the specific gravity of the slurry.

Another approach is possible if the density and viscosity in the operating range are known. The traditional Darcy equation and Reynolds number can be used for thin slurries as long as the flow velocity is high enough to keep solids from settling. (See Eq. 2.57.) (Settling is more of a concern for laminar flow. With turbulent flow, the direction of velocity components fluctuates, assisting the solids to remain in suspension.)

The most analytical approach to slurries or other non-Newtonian fluids requires laboratory-derived rheological data. *Non-Newtonian viscosity* (η, in Pa·s) is fitted to data of the shear rate (dv/dy, in s^{-1}) according to two common models: the power-law model and the Bingham-plastic model. These two models are applicable to both laminar and turbulent flow, although each has its advantages and disadvantages.

The *power-law model* has two empirical constants, m and n, that must be determined.

$$\eta = m \left(\frac{dv}{dy} \right)^{n-1} \qquad 2.61$$

The *Bingham-plastic model* also requires finding two empirical constants: the *yield* (or *critical*) *stress*, τ_0 (in units of Pa) below which the fluid is immobile, and the *Bingham-plastic limiting viscosity*, μ_∞ (in units of Pa·s).

$$\eta = \frac{\tau_0}{\dfrac{dv}{dy}} + \mu_\infty \qquad 2.62$$

Once m and n (or τ_0 and μ_∞) have been determined, the friction factor is determined from one of various models (e.g., Buckingham-Reiner, Dodge-Metzner, Metzner-Reed, Hanks-Ricks, Darby, or Hanks-Dadia). Specialized texts and articles cover these models in greater detail. The friction loss is calculated from the traditional Darcy equation.

31. MINOR LOSSES

In addition to the frictional energy lost due to viscous effects, friction losses also result from fittings in the line, changes in direction, and changes in flow area. These losses are known as *minor losses* or *local losses*, since they are usually much smaller in magnitude than the pipe wall frictional loss.[23] Two methods are used to calculate minor losses: equivalent lengths and loss coefficients.

With the *method of equivalent lengths*, each fitting or other flow variation is assumed to produce friction equal to the pipe wall friction from an *equivalent length* of pipe. For example, a 2 in globe valve may produce the same amount of friction as 54 ft (its equivalent length) of 2 in pipe. The equivalent lengths for all minor losses are added to the pipe length term, L, in the Darcy equation. The method of equivalent lengths can be used with all liquids, but it is usually limited to turbulent flow by the unavailability of laminar equivalent lengths, which are significantly larger than turbulent equivalent lengths.

$$L_t = L + \sum L_e \qquad 2.63$$

Equivalent lengths are simple to use, but the method depends on having a table of equivalent length values. The actual value for a fitting will depend on the fitting manufacturer, as well as the fitting material (e.g., brass, cast iron, or steel) and the method of attachment (e.g., weld, thread, or flange).[24] Because of these many variations, it may be necessary to use a "generic table" of equivalent lengths during the initial design stages.

An alternative method of calculating the minor loss for a fitting is to use the *method of resistance coefficients*. Each fitting has a *resistance coefficient* (also called *loss coefficient*), C, associated with it, which, when multiplied by the kinetic energy, gives the loss. (See Table 2.5.)

Minor Losses in Pipe Fittings, Contractions, and Expansions

$$h_{f,\text{fitting}} = C \frac{v^2}{2g} \qquad [v^2/2g = 1 \text{ velocity head}] \qquad 2.64$$

---

[23]Example and practice problems often include the instruction to "ignore minor losses." In some industries, valves are considered to be "components," not fittings. In such cases, instructions to "ignore minor losses in fittings" would be ambiguous, since minor losses in valves would be included in the calculations. However, this interpretation is rare in examples and practice problems.

[24]In the language of pipe fittings, a *threaded fitting* is known as a *screwed fitting*, even though no screws are used.

Table 2.5 Typical Resistance Coefficients, C^a

device	C
angle valve	5
bend, close return	2.2
butterfly valve,^{b} 2–8 in	$45f_t$
butterfly valve, 10–14 in	$35f_t$
butterfly valve, 16–24 in	$25f_t$
check valve, swing, fully open	2.3
corrugated bends	1.3–1.6 times value for smooth bend
standard 90° elbow	0.9
long radius 90° elbow	0.6
45° elbow	0.42
gate valve, fully open	0.19
gate valve, $\frac{1}{4}$ closed	1.15
gate valve, $\frac{1}{2}$ closed	5.6
gate valve, $\frac{3}{4}$ closed	24
globe valve	10
meter disk or wobble	3.4–10
meter, rotary (star or cog-wheel piston)	10
meter, reciprocating piston	15
meter, turbine wheel (double flow)	5–7.5
tee, standard	1.8

^{a}The actual loss coefficient will usually depend on the size of the valve. Average values are given.
^{b}Loss coefficients for butterfly valves are calculated from the friction factors for the pipes with complete turbulent flow.

Therefore, a resistance coefficient is the minor loss expressed in fractions (or multiples) of the velocity head.

$$h_m = C h_v \qquad \text{2.65}$$

Equation 2.66 gives a generally accepted nominal value for head loss in a well-streamlined, gradual contraction.

Minor Losses in Pipe Fittings, Contractions, and Expansions

$$h_{f,\text{fitting}} = 0.04 v^2 / 2g \qquad \text{2.66}$$

The resistance coefficient for any minor loss can be calculated if the equivalent length is known. However, there is no advantage to using one method over the other, other than convention and for consistency in calculations.

$$C = \frac{f L_e}{D} \qquad \text{2.67}$$

Exact friction resistance coefficients for bends, fittings, and valves are unique to each manufacturer. Furthermore, except for contractions, enlargements, exits, and entrances, the coefficients decrease fairly significantly

(according to the fourth power of the diameter ratio) with increases in valve size. Therefore, a single C value is seldom applicable to an entire family of valves. Nevertheless, generic tables and charts have been developed. These compilations can be used for initial estimates as long as the general nature of the data is recognized.

Resistance coefficients for specific fittings and valves must be known in order to be used. They cannot be derived theoretically. However, the loss coefficients for certain changes in flow area can be calculated from the following equations.[25]

- sudden enlargements (D_1 is the smaller of the two diameters)

$$C = \left[1 - \left(\frac{D_1}{D_2} \right)^2 \right]^2 \qquad \text{2.68}$$

- sudden contractions (D_1 is the smaller of the two diameters)

$$C = \frac{1}{2} \left[1 - \left(\frac{D_1}{D_2} \right)^2 \right] \qquad \text{2.69}$$

- pipe exit (projecting exit, sharp-edged, or rounded) [Minor Losses in Pipe Fittings, Contractions, and Expansions]

$$C = 1.0 \qquad \text{2.70}$$

- pipe entrance

reentrant: $C = 0.78$
sharp-edged: $C = 0.50$
rounded:

bend radius $\dfrac{D}$	C
0.02	0.28
0.04	0.24
0.06	0.15
0.10	0.09
0.15	0.04

- tapered diameter changes

$$\beta = \frac{\text{small diameter}}{\text{large diameter}} = \frac{D_1}{D_2}$$

ϕ = wall-to-horizontal angle

For enlargement, $\phi \leq 22°$:

$$C = 2.6 \sin \phi (1 - \beta^2)^2 \qquad \text{2.71}$$

For enlargement, $\phi > 22°$:

$$C = (1 - \beta^2)^2 \qquad \text{2.72}$$

[25]No attempt is made to imply great accuracy with these equations. Correlation between actual and theoretical losses is fair.

For contraction, $\phi \leq 22°$:

$$C = 0.8 \sin \phi (1 - \beta^2) \qquad \text{2.73}$$

For contraction, $\phi > 22°$:

$$C = 0.5 \sqrt{\sin \phi} \, (1 - \beta^2) \qquad \text{2.74}$$

Example 2.9

A pipeline contains one gate valve, five regular 90° elbows, one tee (flow through the run), and 228 ft of straight pipe. All fittings are 1 in screwed steel pipe. Disregard entrance and exit losses. Determine the total equivalent length of the piping system shown.

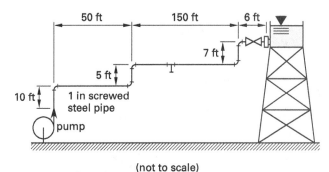

(not to scale)

Solution

From a table of equivalent lengths of straight pipe for various fittings, the individual and total equivalent lengths are

1	gate valve	1×0.84 ft	=	0.84 ft
5	regular elbows	5×5.2 ft	=	26.00 ft
1	tee run	1×3.2 ft	=	3.20 ft
	straight pipe		=	228.00 ft
		total L_t	=	258.04 ft

32. VALVE FLOW COEFFICIENTS

Valve flow capacities depend on the geometry of the inside of the valve. The *flow coefficient*, C_v, for a valve (particularly a control valve) relates the flow quantity (in gallons per minute) of a fluid with specific gravity to the pressure drop (in pounds per square inch). (The flow coefficient for a valve is not the same as the coefficient of flow for an orifice or venturi meter.) As Eq. 2.75 shows, the flow coefficient is not dimensionally homogeneous.

Metricated countries use a similar concept with a different symbol, K_v, (not the same as the loss coefficient, K) to distinguish the valve flow coefficient from customary U.S. units. K_v is defined[26] as the flow rate in cubic meters per hour of water at a temperature of 16°C with a pressure drop across the valve of 1 bar. To further distinguish it from its U.S. counterpart, K_v may also be referred to as a *flow factor*. C_v and K_v are linearly related by Eq. 2.76.

$$Q_{\text{m}^3/\text{h}} = K_v \sqrt{\frac{\Delta P_{\text{bars}}}{SG}} \qquad \text{[SI]} \quad \text{2.75(a)}$$

$$Q_{\text{gpm}} = C_v \sqrt{\frac{\Delta P_{\text{psi}}}{SG}} \qquad \text{[U.S.]} \quad \text{2.75(b)}$$

$$K_v = 0.86 \, C_v \qquad \text{2.76}$$

When selecting a control valve for a particular application, the value of C_v is first calculated. Depending on the application and installation, C_v may be further modified by dividing by *piping geometry* and *Reynolds number factors*. (These additional procedures are often specified by the valve manufacturer.) Then, a valve with the required value of C_v is selected.

Although the flow coefficient concept is generally limited to control valves, its use can be extended to all fittings and valves. The relationship between C_v and the loss coefficient, K, is

$$C_v = \frac{29.9 d_{\text{in}}^2}{\sqrt{K}} \quad \text{[U.S.]} \qquad \text{2.77}$$

33. SHEAR STRESS IN CIRCULAR PIPES

Shear stress in fluid always acts to oppose the motion of the fluid. (That is the reason the term *pipe friction* is used.) Shear stress for a fluid in laminar flow can be calculated from the basic definition of absolute viscosity.

Stress, Pressure, and Viscosity

$$\tau_l = \mu (d\text{v}/dy) \qquad \text{[SI]} \quad \text{2.78(a)}$$

Unit and Conversion Factors

$$\tau = (\mu/g_c)(d\text{v}/dy) \qquad \text{[U.S.]} \quad \text{2.78(b)}$$

---

[26]Several definitions of both C_v and K_v are in use. A definition of C_v based on Imperial gallons is used in Great Britain. Definitions of K_v based on pressure drops in kilograms-force and volumes in liters per minute are in use. Other differences in the definition include the applicable temperature, which may be given as 5–30°C or 5–40°C instead of 16°C.

In the case of the flow in a circular pipe, dr can be substituted for dy in the expression for *shear rate* (*velocity gradient*), dv/dy.

$$\tau = \mu \frac{d v}{d r} \qquad 2.79$$

Equation 2.80 calculates the shear stress between fluid layers a distance r from the pipe centerline in terms of the pressure drop across a length L of the pipe.[27] Equation 2.80 is valid for both laminar and turbulent flows.

$$\tau = \frac{(p_1 - p_2)r}{2L} \quad \left[r \le \frac{D}{2}\right] \qquad 2.80$$

The quantity $(p_1 - p_2)$ can be calculated from the Darcy equation. (See Eq. 2.57.) If v is the average flow velocity, the shear stress at the wall (where $r = D/2$) is

Momentum Transfer

$$\tau_w = -\frac{f \rho V^2}{8} \qquad 2.81$$

Equation 2.80 can be rearranged to give the relationship between the pressure gradient along the flow path and the shear stress at the wall.

$$\frac{dp}{dL} = \frac{4 \tau_{\text{wall}}}{D} \qquad 2.82$$

Equation 2.81 can be combined with the Hagen-Poiseuille equation (given in Eq. 2.53) if the flow is laminar. (v in Eq. 2.83 is the average velocity of fluid flow.)

$$\tau = \frac{16 \mu v r}{D^2} \quad \left[\text{laminar}; r \le \frac{D}{2}\right] \qquad 2.83$$

At the pipe wall, $r = D/2$, and the shear stress is maximum. (See Fig. 2.11.)

$$\tau_{\text{wall}} = \frac{8 \mu v}{D} \qquad 2.84$$

Figure 2.11 *Shear Stress Distribution in a Circular Pipe*

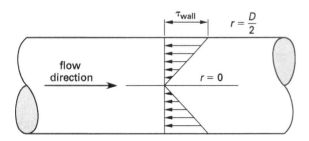

34. EXTENDED BERNOULLI EQUATION

The original Bernoulli equation assumes frictionless flow and does not consider the effects of pumps and turbines. When friction is present and when there are minor losses such as fittings and other energy-related devices in a pipeline, the energy balance is affected. The *extended Bernoulli equation* takes these additional factors into account. [Minor Losses in Pipe Fittings, Contractions, and Expansions]

$$(E_P + E_v + E_z)_1 + E_A$$
$$= (E_P + E_v + E_z)_2 + E_E + E_f + E_m \qquad 2.85$$

$$\frac{P_1}{\rho} + \frac{v_1^2}{2} + z_1 g + E_A$$
$$= \frac{P_2}{\rho} + \frac{v_2^2}{2} + z_2 g + E_E + E_f + E_m \qquad \text{[SI]} \quad 2.86(a)$$

$$\frac{P_1}{\rho} + \frac{v_1^2}{2g_c} + \frac{z_1 g}{g_c} + E_A$$
$$= \frac{P_2}{\rho} + \frac{v_2^2}{2g_c} + \frac{z_2 g}{g_c} + E_E + E_f + E_m \qquad \text{[U.S.]} \quad 2.86(b)$$

As defined, E_A, E_E, and E_f are all positive terms. None of the terms in Eq. 2.85 and Eq. 2.86 are negative.

The concepts of sources and sinks can be used to decide whether the friction, pump, and turbine terms appear on the left or right side of the Bernoulli equation. An *energy source* puts energy into the system. The incoming fluid and a pump contribute energy to the system. An *energy sink* removes energy from the system. The leaving fluid, friction, and a turbine remove energy from the system. In an energy balance, all energy must be accounted for, and the energy sources just equal the energy sinks.

$$\sum E_{\text{sources}} = \sum E_{\text{sinks}} \qquad 2.87$$

Therefore, the energy added by a pump always appears on the entrance side of the Bernoulli equation. Similarly, the frictional energy loss always appears on the discharge side.

35. DISCHARGE FROM TANKS

The velocity of a jet issuing from an orifice in a tank can be determined by comparing the total energies at the free fluid surface and the jet itself. (See Fig. 2.12.) At the fluid surface, $P_1 = 0$ (atmospheric) and $v_1 = 0$. (v_1 is known as the *velocity of approach*.) The only energy the fluid has is potential energy. At the jet, $P_2 = 0$. All of the potential energy difference $(z_1 - z_2)$ has been converted to kinetic energy. The theoretical velocity of the

---

[27]In highly turbulent flow, shear stress is not caused by viscous effects but rather by momentum effects. Equation 2.80 is derived from a shell momentum balance. Such an analysis requires the concept of *momentum flux*. In a circular pipe with laminar flow, momentum flux is maximum at the pipe wall, zero at the flow centerline, and varies linearly in between.

jet can be derived from the Bernoulli equation. Equation 2.88 is known as the equation for *Torricelli's speed of efflux.*

$$v_t = \sqrt{2gh} \qquad 2.88$$

$$h = z_1 - z_2 \qquad 2.89$$

Figure 2.12 *Discharge from a Tank*

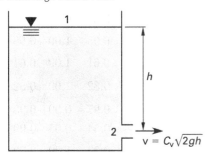

The actual jet velocity is affected by the orifice geometry, as shown by Eq. 2.90. The *coefficient of velocity*, C_v, is an empirical factor that accounts for the friction and turbulence at the orifice. Typical values of C_v are given in Table 2.6.

$$v = C_v\sqrt{2gh} \qquad 2.90$$

$$C_v = \frac{\text{actual velocity}}{\text{theoretical velocity}} = \frac{v}{v_t} \qquad 2.91$$

The specific energy loss due to turbulence and friction at the orifice is calculated as a multiple of the jet's kinetic energy.

$$E_f = \left(\frac{1}{C_v^2} - 1\right)\frac{v^2}{2} = (1 - C_v^2)gh \qquad \text{[SI]} \quad 2.92(a)$$

$$E_f = \left(\frac{1}{C_v^2} - 1\right)\frac{v^2}{2g_c} = (1 - C_v^2)h \times \frac{g}{g_c} \qquad \text{[U.S.]} \quad 2.92(b)$$

The total head producing discharge (*effective head*) is the difference in elevations that would produce the same velocity from a frictionless orifice.

$$h_{\text{effective}} = C_v^2 h \qquad 2.93$$

The orifice guides quiescent water from the tank into the jet geometry. Unless the orifice is very smooth and the transition is gradual, momentum effects will continue to cause the jet to contract after it has passed through. The velocity calculated from Eq. 2.90 is usually assumed to be the velocity at the *vena contracta*, the section of smallest cross-sectional area. (See Fig. 2.13.)

Figure 2.13 *Vena Contracta of a Fluid Jet*

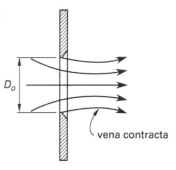

For a thin plate or sharp-edged orifice, the vena contracta is often assumed to be located approximately one half an orifice diameter past the orifice, although the actual distance can vary from $0.3D_o$ to $0.8D_o$. The area of the vena contracta can be calculated from the orifice area and the *coefficient of contraction*, C_c. For water flowing with a high Reynolds number through a small sharp-edged orifice, the contracted area is approximately 61–63% of the orifice area. [Orifices]

$$A_2 = C_c A_0 \qquad 2.94$$

$$C_c = \frac{A_2}{A_0} \qquad 2.95$$

The theoretical discharge rate from a tank is $Q_t = A\sqrt{2gh}$. However, this relationship needs to be corrected for friction and contraction by multiplying by C_v and C_c. The *discharge coefficient*, C, is the product of C_v and C_c. [Orifice Discharging Freely into Atmosphere]

$$Q = C_c v A_0 = C v_t A_0 = C A_0 \sqrt{2gh} \qquad 2.96$$

$$C = C_v C_c \frac{Q}{Q_t} \qquad 2.97$$

36. DISCHARGE FROM PRESSURIZED TANKS

If the gas or vapor above the liquid in a tank is at gage pressure p, and the discharge is to atmospheric pressure, the head causing discharge will be

$$h = z_1 - z_2 + \frac{P}{\rho g} \qquad \text{[SI]} \quad 2.98(a)$$

$$h = z_1 - z_2 + \frac{P}{\rho} \times \frac{g_c}{g} = z_1 - z_2 + \frac{P}{\gamma} \qquad \text{[U.S.]} \quad 2.98(b)$$

The discharge velocity can be calculated from Eq. 2.90 using the increased discharge head. (See Fig. 2.14.)

Water

Table 2.6 *Approximate Orifice Coefficients, C, for Turbulent Water*

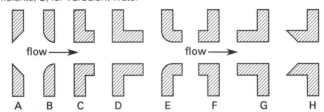

illustration	description	C	C_c	C_v
A	sharp-edged	0.62	0.63	0.98
B	round-edged	0.98	1.00	0.98
C	short tube* (fluid separates from walls)	0.61	1.00	0.61
D	sharp tube (no separation)	0.82	1.00	0.82
E	short tube with rounded entrance	0.97	0.99	0.98
F	reentrant tube, length less than one-half of pipe diameter	0.54	0.55	0.99
G	reentrant tube, length two to three pipe diameters	0.72	1.00	0.72
H	Borda	0.51	0.52	0.98
(none)	smooth, well-tapered nozzle	0.98	0.99	0.99

*A short tube has a length less than approximately three pipe diameters.

Figure 2.14 *Discharge from a Pressurized Tank*

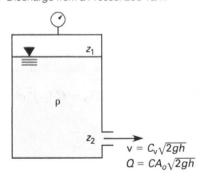

37. COORDINATES OF A FLUID STREAM

Fluid discharged from an orifice in a tank gets its initial velocity from the conversion of potential energy. After discharge, no additional energy conversion occurs, and all subsequent velocity changes are due to external forces. (See Fig. 2.15.)

In the absence of air friction (drag), there are no decelerating or accelerating forces in the x-direction on the fluid stream. The x-component of velocity is constant. Projectile motion equations can be used to predict the path of the fluid stream.

$$v_x = v \quad \text{[horizontal discharge]} \qquad 2.99$$

$$x = vt = v\sqrt{\frac{2y}{g}} = 2C_v\sqrt{hy} \qquad 2.100$$

Figure 2.15 *Coordinates of a Fluid Stream*

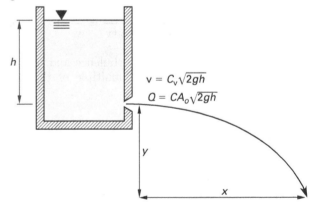

After discharge, the fluid stream is acted upon by a constant gravitational acceleration. The y-component of velocity is zero at discharge but increases linearly with time.

$$v_y = gt \qquad 2.101$$

$$y = \frac{gt^2}{2} = \frac{gx^2}{2v^2} = \frac{x^2}{4hC_v^2} \qquad 2.102$$

Water

38. DISCHARGE FROM A SUBMERGED ORIFICE

Equation 2.103 gives the rate of discharge for a submerged orifice under steady-flow conditions.

Submerged Orifice (Operating Under Steady-Flow Conditions)

$$Q = A_2 v_2 = C_c C_v A \sqrt{2g(h_1 - h_2)}$$
$$= CA\sqrt{2g(h_1 - h_2)} \qquad \text{2.103}$$

The orifice is submerged on both sides. h_1 is the distance from the center of the orifice to the fluid surface on the entrance side of the orifice, and h_2 is the same distance on the exit side of the orifice. A is the area of the orifice. v_2 and A_2 are the velocity and area of flow just after exiting the orifice.

39. DISCHARGE FROM LARGE ORIFICES

When an orifice diameter is large compared with the discharge head, the jet velocity at the top edge of the orifice will be less than the velocity at the bottom edge. Since the velocity is related to the square root of the head, the distance used to calculate the effective jet velocity should be measured from the fluid surface to a point above the centerline of the orifice.

This correction is generally neglected, however, since it is small for heads of more than twice the orifice diameter. Furthermore, if an orifice is intended to work regularly with small heads, the orifice should be calibrated in place. The discrepancy can then be absorbed into the discharge coefficient, C.

40. TIME TO EMPTY A TANK

If the fluid in an open or vented tank is not replenished at the rate of discharge, the static head forcing discharge through the orifice will decrease with time. If the tank has a varying cross section, A_t, Eq. 2.104 specifies the basic relationship between the change in elevation and elapsed time. (The negative sign indicates that z decreases as t increases.)

$$Q\,dt = -A_t\,dz \qquad \text{2.104}$$

If A_t can be expressed as a function of h, Eq. 2.105 can be used to determine the time to lower the fluid elevation from z_1 to z_2.

$$t = \int_{z_1}^{z_2} \frac{-A_t\,dz}{CA_0\sqrt{2gz}} \qquad \text{2.105}$$

For a tank with a constant cross-sectional area, A_t, the time required to lower the fluid elevation is

$$t = \frac{2A_t(\sqrt{z_1} - \sqrt{z_2})}{CA_0\sqrt{2g}} \qquad \text{2.106}$$

If a tank is replenished at a rate of $\dot{V}_{in}$, Eq. 2.107 can be used to calculate the discharge time. If the tank is replenished at a rate greater than the discharge rate, t in Eq. 2.107 will represent the time to raise the fluid level from z_1 to z_2.

$$t = \int_{z_1}^{z_2} \frac{A_t\,dz}{CA_0\sqrt{2gz} - Q_{in}} \qquad \text{2.107}$$

If the tank is not open or vented but is pressurized, the elevation terms, z_1 and z_2, in Eq. 2.106 must be replaced by the total head terms, h_1 and h_2, that include the effects of pressurization.

Example 2.10

A vertical, cylindrical tank 15 ft in diameter discharges 150°F water ($\rho = 61.20$ lbm/ft³) through a sharp-edged, 1 in diameter orifice ($C = 0.62$) in the tank bottom. The original water depth is 12 ft. The tank is continually pressurized to 50 psig. How long does it take, in seconds, to empty the tank?

Solution

The area of the orifice is

$$A_0 = \frac{\pi D^2}{4} = \frac{\pi(1\text{ in})^2}{(4)\left(12\,\dfrac{\text{in}}{\text{ft}}\right)^2} = 0.00545\text{ ft}^2$$

The tank area is constant with respect to depth.

$$A_t = \frac{\pi D^2}{4} = \frac{\pi(15\text{ ft})^2}{4} = 176.7\text{ ft}^2$$

The total initial head includes the effect of the pressurization. Use Eq. 2.98.

$$h_1 = z_1 - z_2 + \frac{P}{\rho} \times \frac{g_c}{g}$$

$$= 12\text{ ft} + \frac{\left(50\,\dfrac{\text{lbf}}{\text{in}^2}\right)\left(12\,\dfrac{\text{in}}{\text{ft}}\right)^2}{61.2\,\dfrac{\text{lbm}}{\text{ft}^3}} \times \frac{32.2\,\dfrac{\text{lbm-ft}}{\text{lbf-sec}^2}}{32.2\,\dfrac{\text{ft}}{\text{sec}^2}}$$

$$= 12\text{ ft} + 117.6\text{ ft}$$
$$= 129.6\text{ ft}$$

When the fluid has reached the level of the orifice, the fluid potential head will be zero, but the pressurization will remain.

$$h_2 = 117.6 \text{ ft}$$

The time needed to empty the tank is given by Eq. 2.106.

$$t = \frac{2A_t(\sqrt{z_1} - \sqrt{z_2})}{CA_0\sqrt{2g}}$$

$$= \frac{(2)(176.7 \text{ ft}^2)(\sqrt{129.6 \text{ ft}} - \sqrt{117.6 \text{ ft}})}{(0.62)(0.00545 \text{ ft}^2)\sqrt{(2)\left(32.2 \ \dfrac{\text{ft}}{\text{sec}^2}\right)}}$$

$$= 7036 \text{ sec}$$

41. JET PROPULSION

A basic application of the impulse-momentum principle is the analysis of jet propulsion. Air enters a jet engine and is mixed with fuel. The air and fuel mixture is compressed and ignited, and the exhaust products leave the engine at a greater velocity than was possessed by the original air. The change in momentum of the air produces a force on the engine. (See Fig. 2.16.)

Figure 2.16 *Jet Engine*

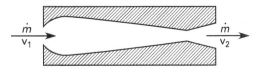

The governing equation for a jet engine is Eq. 2.108. The mass of the jet fuel is small compared with the air mass, and the fuel mass is commonly disregarded.

$$F_x = \dot{m}(v_2 - v_1) \qquad \qquad 2.108$$

$$F_x = Q_2\rho_2 v_2 - Q_1\rho_1 v_1 \qquad \text{[SI]} \quad 2.109(a)$$

$$F_x = \frac{Q_2\rho_2 v_2 - Q_1\rho_1 v_1}{g_c} \qquad \text{[U.S.]} \quad 2.109(b)$$

42. OPEN JET ON A VERTICAL FLAT PLATE

Figure 2.17 illustrates an open jet on a vertical flat plate. The fluid approaches the plate with no vertical component of velocity; it leaves the plate with no horizontal component of velocity. (This is another way of saying there is no splash-back.) Therefore, all of the

velocity in the *x*-direction is canceled. (The minus sign in Eq. 2.110 indicates that the force is opposite the initial velocity direction.)

$$\Delta v = -v \qquad \qquad 2.110$$

$$F_x = -\dot{m}v \qquad \qquad \text{[SI]} \quad 2.111(a)$$

$$F_x = \frac{-\dot{m}v}{g_c} \qquad \qquad \text{[U.S.]} \quad 2.111(b)$$

Figure 2.17 *Jet on a Vertical Plate*

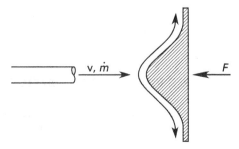

Since the flow is divided, half going up and half going down, the net velocity change in the *y*-direction is zero. There is no force in the *y*-direction on the fluid.

43. OPEN JET ON A HORIZONTAL FLAT PLATE

If a jet of fluid is directed upward, its velocity will decrease due to the effect of gravity. The force exerted on the fluid by the plate will depend on the fluid velocity at the plate surface, v_y, not the original jet velocity, v_o. All of this velocity is canceled. Since the flow divides evenly in both horizontal directions ($\Delta v_x = 0$), there is no force component in the *x*-direction. (See Fig. 2.18.)

$$v_y = \sqrt{v_o^2 - 2gh} \qquad \qquad 2.112$$

$$\Delta v_y = -\sqrt{v_o^2 - 2gh} \qquad \qquad 2.113$$

$$F_y = -\dot{m}\sqrt{v_o^2 - 2gh} \qquad \text{[SI]} \quad 2.114(a)$$

$$F_y = \frac{-\dot{m}\sqrt{v_o^2 - 2gh}}{g_c} \qquad \text{[U.S.]} \quad 2.114(b)$$

44. OPEN JET ON AN INCLINED PLATE

An open jet will be diverted both up and down (but not laterally) by a stationary, inclined plate, as shown in Fig. 2.19. In the absence of friction, the velocity in each diverted flow will be v, the same as in the approaching jet. The fractions f_1 and f_2 of the jet that are diverted up and down can be found from Eq. 2.115 through Eq. 2.118.

Figure 2.18 *Open Jet on a Horizontal Plate*

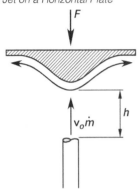

$$f_1 = \frac{1 + \cos \theta}{2}$$

2.115

$$f_2 = \frac{1 - \cos \theta}{2}$$ 2.116

$$f_1 - f_2 = \cos \theta$$ 2.117

$$f_1 + f_2 = 1.0$$ 2.118

Figure 2.19 *Open Jet on an Inclined Plate*

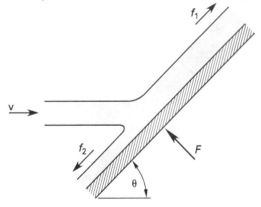

If the flow along the plate is frictionless, there will be no force component parallel to the plate. The force perpendicular to the plate is given by Eq. 2.119.

$$F = \dot{m} v \sin \theta$$ [SI] *2.119(a)*

$$F = \frac{\dot{m} v \sin \theta}{g_c}$$ [U.S.] *2.119(b)*

45. OPEN JET ON A SINGLE STATIONARY BLADE

Figure 2.20 illustrates a fluid jet being turned through an angle θ by a stationary blade (also called a *vane*). It is common to assume that $|v_2| = |v_1|$, although this will not be strictly true if friction between the blade and fluid is considered. Since the fluid is both decelerated (in the x-direction) and accelerated (in the y-direction), there will be two components of force on the fluid.

$$\Delta v_x = v_2 \cos \theta - v_1$$ 2.120

$$\Delta v_y = v_2 \sin \theta$$ 2.121

$$F_x = \dot{m}(v_2 \cos \theta - v_1)$$ [SI] *2.122(a)*

$$F_x = \frac{\dot{m}(v_2 \cos \theta - v_1)}{g_c}$$ [U.S.] *2.122(b)*

$$F_y = \dot{m} v_2 \sin \theta$$ [SI] *2.123(a)*

$$F_y = \frac{\dot{m} v_2 \sin \theta}{g_c}$$ [U.S.] *2.123(b)*

$$F = \sqrt{F_x^2 + F_y^2}$$ 2.124

Figure 2.20 *Open Jet on a Stationary Blade*

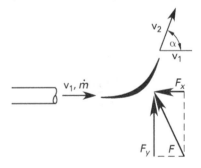

46. OPEN JET ON A SINGLE MOVING BLADE

If a blade is moving away at velocity v_b from the source of the fluid jet, only the *relative velocity difference* between the jet and blade produces a momentum change. Furthermore, not all of the fluid jet overtakes the moving blade. The equations used for the single stationary blade can be used by substituting $(v - v_b)$ for v and by using the effective mass flow rate, $\dot{m}_{eff}$. (See Fig. 2.21.)

$$\Delta v_x = (v - v_b)(\cos \theta - 1)$$ 2.125

$$\Delta v_y = (v - v_b)\sin \theta$$ 2.126

$$\dot{m}_{eff} = \left(\frac{v - v_b}{v}\right)\dot{m}$$ 2.127

$$F_x = \dot{m}_{\text{eff}}(\text{v} - \text{v}_b)(\cos\theta - 1) \qquad \text{[SI]} \qquad 2.128(a)$$

$$F_x = \frac{\dot{m}_{\text{eff}}(\text{v} - \text{v}_b)(\cos\theta - 1)}{g_c} \qquad \text{[U.S.]} \qquad 2.128(b)$$

$$F_y = \dot{m}_{\text{eff}}(\text{v} - \text{v}_b)\sin\theta \qquad \text{[SI]} \qquad 2.129(a)$$

$$F_y = \frac{\dot{m}_{\text{eff}}(\text{v} - \text{v}_b)\sin\theta}{g_c} \qquad \text{[U.S.]} \qquad 2.129(b)$$

Figure 2.21 *Open Jet on a Moving Blade*

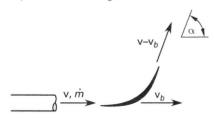

47. OPEN JET ON A MULTIPLE-BLADED WHEEL

An *impulse turbine* consists of a series of blades (buckets or vanes) mounted around a wheel. (See Fig. 2.22.) The tangential velocity of the blades is approximately parallel to the jet. The effective mass flow rate, $\dot{m}_{\text{eff}}$, used in calculating the reaction force is the full discharge rate, since when one blade moves away from the jet, other blades will have moved into position. All of the fluid discharged is captured by the blades. Equation 2.128 and Eq. 2.129 are applicable if the total flow rate is used. The tangential blade velocity is

$$\text{v}_b = \frac{2\pi r n_{\text{rpm}}}{60} = \omega r \qquad 2.130$$

Figure 2.22 *Impulse Turbine*

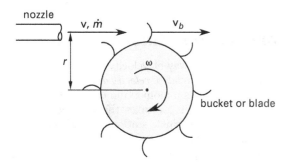

48. IMPULSE TURBINE POWER

The total power potential of a fluid jet can be calculated from the kinetic energy of the jet and the mass flow rate.[28] (This neglects the pressure energy, which is small by comparison.)

$$P_{\text{jet}} = \frac{\dot{m}\text{v}^2}{2} \qquad \text{[SI]} \qquad 2.131(a)$$

$$P_{\text{jet}} = \frac{\dot{m}\text{v}^2}{2g_c} \qquad \text{[U.S.]} \qquad 2.131(b)$$

The power transferred from a fluid jet to the blades of a turbine is calculated from the x-component of force on the blades. The y-component of force does no work.

$$P = F_x \text{v}_b \qquad 2.132$$

$$P = \dot{m}\text{v}_b(\text{v} - \text{v}_b)(1 - \cos\theta) \qquad \text{[SI]} \qquad 2.133(a)$$

$$P = \frac{\dot{m}\text{v}_b(\text{v} - \text{v}_b)(1 - \cos\theta)}{g_c} \qquad \text{[U.S.]} \qquad 2.133(b)$$

The maximum theoretical blade velocity is the velocity of the jet: $\text{v}_b = \text{v}$. This is known as the *runaway speed* and can only be achieved when the turbine is unloaded. If Eq. 2.133 is maximized with respect to v_b, the maximum power will occur when the blade is traveling at half of the jet velocity: $\text{v}_b = \text{v}/2$. The power (force) is also affected by the deflection angle of the blade. Power is maximized when $\theta = 180°$. Figure 2.23 illustrates the relationship between power and the variables θ and v_b.

Figure 2.23 *Turbine Power*

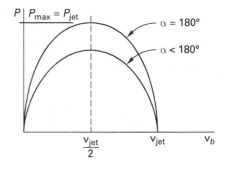

Putting $\theta = 180°$ and $\text{v}_b = \text{v}/2$ into Eq. 2.133 results in $P_{\text{max}} = \dot{m}\text{v}^2/2$, which is the same as P_{jet} in Eq. 2.131. If the machine is 100% efficient, 100% of the jet power can be transferred to the machine.

---

[28]The full jet discharge is used in this section. If only a single blade is involved, the effective mass flow rate, $\dot{m}_{\text{eff}}$, must be used.

49. CONFINED STREAMS IN PIPE BENDS

Momentum can also be changed by pressure forces. Such is the case when fluid enters a pipe fitting or bend. (See Fig. 2.24.) Since the fluid is confined, the forces due to static pressure must be included in the analysis. (The effects of gravity and friction are neglected.)

$$F_x = P_2 A_2 \cos\theta - P_1 A_1 + \dot{m}(v_2 \cos\theta - v_1) \quad \text{[SI]} \quad \textbf{2.134(a)}$$

$$F_x = P_2 A_2 \cos\theta - P_1 A_1$$
$$+ \frac{\dot{m}(v_2 \cos\theta - v_1)}{g_c} \quad \text{[U.S.]} \quad \textbf{2.134(b)}$$

$$F_y = (P_2 A_2 + \dot{m} v_2)\sin\theta \quad \text{[SI]} \quad \textbf{2.135(a)}$$

$$F_y = \left(P_2 A_2 + \frac{\dot{m} v_2}{g_c}\right)\sin\theta \quad \text{[U.S.]} \quad \textbf{2.135(b)}$$

Figure 2.24 *Pipe Bend*

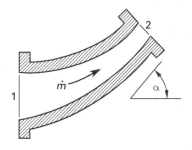

Example 2.11

60°F water ($\rho = 62.4$ lbm/ft^3) at 40 psig enters a 12 in × 8 in reducing elbow at 8 ft/sec and is turned through an angle of 30°. Water leaves 26 in higher in elevation. (a) What is the resultant force exerted on the water by the elbow? (b) What other forces should be considered in the design of supports for the fitting?

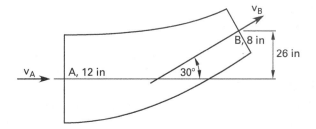

Solution

(a) The velocity and pressure at point B are both needed. The velocity is easily calculated from the continuity equation.

$$A_A = \frac{\pi D_A^2}{4} = \frac{\pi\left(\dfrac{12 \text{ in}}{12 \, \dfrac{\text{in}}{\text{ft}}}\right)^2}{4}$$
$$= 0.7854 \text{ ft}^2$$

$$A_B = \frac{\pi D_B^2}{4} = \frac{\pi\left(\dfrac{8 \text{ in}}{12 \, \dfrac{\text{in}}{\text{ft}}}\right)^2}{4}$$
$$= 0.3491 \text{ ft}^2$$

$$v_B = \frac{v_A A_A}{A_B}$$
$$= \left(8 \, \frac{\text{ft}}{\text{sec}}\right)\left(\frac{0.7854 \text{ ft}^2}{0.3491 \text{ ft}^2}\right)$$
$$= 18 \text{ ft/sec}$$

$$P_A = \left(40 \, \frac{\text{lbf}}{\text{in}^2}\right)\left(12 \, \frac{\text{in}}{\text{ft}}\right)^2$$
$$= 5760 \text{ lbf/ft}^2$$

The Bernoulli equation is used to calculate P_B. (Gage pressures are used for this calculation. Absolute pressures could also be used, but the addition of p_{atm}/ρ to both sides of the Bernoulli equation would not affect P_B.)

$$\frac{P_A}{\rho} + \frac{v_A^2}{2g_c} = \frac{P_B}{\rho} + \frac{v_B^2}{2g_c} + \frac{zg}{g_c}$$

$$\frac{5760 \, \dfrac{\text{lbf}}{\text{ft}^2}}{62.4 \, \dfrac{\text{lbm}}{\text{ft}^3}} + \frac{\left(8 \, \dfrac{\text{ft}}{\text{sec}}\right)^2}{(2)\left(32.2 \, \dfrac{\text{lbm-ft}}{\text{lbf-sec}^2}\right)}$$

$$= \frac{P_B}{62.4 \, \dfrac{\text{lbm}}{\text{ft}^3}} + \frac{\left(18 \, \dfrac{\text{ft}}{\text{sec}}\right)^2}{(2)\left(32.2 \, \dfrac{\text{lbm-ft}}{\text{lbf-sec}^2}\right)}$$

$$+ \frac{26 \text{ in}}{12 \, \dfrac{\text{in}}{\text{ft}}} \times \frac{32.2 \, \dfrac{\text{ft}}{\text{sec}^2}}{32.2 \, \dfrac{\text{lbm-ft}}{\text{lbf-sec}^2}}$$

$$P_B = 5373 \text{ lbf/ft}^2$$

The mass flow rate is

$$\dot{m} = Q\rho = \text{v}A\rho = \left(8\ \frac{\text{ft}}{\text{sec}}\right)(0.7854\ \text{ft}^2)\left(62.4\ \frac{\text{lbm}}{\text{ft}^3}\right)$$
$$= 392.1\ \text{lbm/sec}$$

From Eq. 2.134,

$$F_x = P_2 A_2 \cos\theta - P_1 A_1 + \frac{\dot{m}(\text{v}_2\cos\theta - \text{v}_1)}{g_c}$$

$$= \left(5373\ \frac{\text{lbf}}{\text{ft}^2}\right)(0.3491\ \text{ft}^2)\cos 30°$$

$$\quad - \left(5760\ \frac{\text{lbf}}{\text{ft}^2}\right)(0.7854\ \text{ft}^2)$$

$$\quad + \frac{\left(392.1\ \dfrac{\text{lbm}}{\text{sec}}\right)\left[\left(18\ \dfrac{\text{ft}}{\text{sec}}\right)(\cos 30°) - 8\ \dfrac{\text{ft}}{\text{sec}}\right]}{32.2\ \dfrac{\text{lbm-ft}}{\text{lbf-sec}^2}}$$

$$= -2807\ \text{lbf}$$

From Eq. 2.135,

$$F_y = \left(P_2 A_2 + \frac{\dot{m}\text{v}_2}{g_c}\right)\sin\theta$$

$$= \left[\begin{array}{c}\left(5373\ \dfrac{\text{lbf}}{\text{ft}^2}\right)(0.3491\ \text{ft}^2) \\[6pt] + \dfrac{\left(392.1\ \dfrac{\text{lbm}}{\text{sec}}\right)\left(18\ \dfrac{\text{ft}}{\text{sec}}\right)}{32.2\ \dfrac{\text{lbm-ft}}{\text{lbf-sec}^2}}\end{array}\right]\sin 30°$$

$$= 1047\ \text{lbf}$$

The resultant force on the water is

$$R = \sqrt{F_x^2 + F_y^2} = \sqrt{(-2807\ \text{lbf})^2 + (1047\ \text{lbf})^2}$$
$$= 2996\ \text{lbf}$$

(b) In addition to counteracting the resultant force, R, the support should be designed to carry the weight of the elbow and the water in it. Also, the support must carry a part of the pipe and water weight tributary to the elbow.

50. DRAG

Drag is a frictional force that acts parallel but opposite to the direction of motion. The total drag force is made up of *skin friction* and *pressure drag* (also known as *form drag*). These components, in turn, can be subdivided and categorized into *wake drag*, *induced drag*, and *profile drag*. (See Fig. 2.25.)

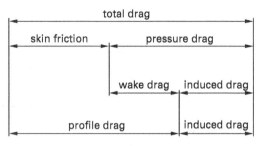

Figure 2.25 Components of Total Drag

Most aeronautical engineering books contain descriptions of these drag terms. However, the difference between the situations where either skin friction drag or pressure drag predominates is illustrated in Fig. 2.26.

Figure 2.26 Extreme Cases of Pressure Drag and Skin Friction

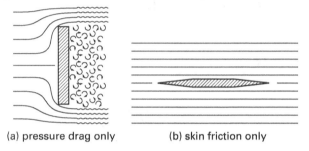

(a) pressure drag only   (b) skin friction only

Total drag is most easily calculated from the dimensionless *drag coefficient*, C_D. The drag coefficient depends only on the Reynolds number.

Drag Coefficient for Spheres, Disks, and Cylinders

$$C_D = \frac{2F_D}{\rho\text{v}^2 A} \qquad \textit{2.136}$$

In most cases, the area, A, in Eq. 2.136 is the projected area (i.e., the *frontal area*) normal to the stream. This is appropriate for spheres, cylinders, and automobiles. In a few cases (e.g., for airfoils and flat plates) as determined by how C_D was derived, the area is a projection of the object onto a plane parallel to the stream.

Typical drag coefficients for production cars vary from approximately 0.25 to approximately 0.70, with most modern cars being nearer the lower end. By comparison, other low-speed drag coefficients are approximately 0.05 (aircraft wing), 0.10 (sphere in turbulent flow), and 1.2 (flat plate). (See Table 2.7.)

Aero horsepower is a term used by automobile manufacturers to designate the power required to move a car horizontally at 50 mi/hr (80.5 km/h) against the drag force. Aero horsepower varies from approximately 7 hp (5.2 kW) for a streamlined subcompact car (approximate drag coefficient 0.35) to approximately 100 hp (75 kW) for a box-shaped truck (approximate drag coefficient 0.9).

Table 2.7 *Typical Ranges of Vehicle Drag Coefficients*

vehicle	low	medium	high
experimental race	0.17	0.21	0.23
sports	0.27	0.31	0.38
performance	0.32	0.34	0.38
'60s muscle car	0.38	0.44	0.50
sedan	0.34	0.39	0.50
motorcycle	0.50	0.90	1.00
truck	0.60	0.90	1.00
tractor-trailer	0.60	0.77	1.20

Used with permission from *Unit Operations*, by George Granger Brown, et al., John Wiley & Sons, Inc., © 1950.

In aerodynamic studies performed in *wind tunnels*, accuracy in lift and drag measurement is specified in *drag counts*. One drag count is equal to a C_D (or, C_L) of 0.0001. Drag counts are commonly used when describing the effect of some change to a fuselage or wing geometry. For example, an increase in C_D from 0.0450 to 0.0467 would be reported at an increase of 17 counts. Some of the best wind tunnels can measure C_D down to 1 count; most can only get to about 5 counts. Outside of the wind tunnel, an accuracy of 1 count is the target in computational fluid dynamics.

51. DRAG ON SPHERES AND DISKS

The drag coefficient varies linearly with the Reynolds number for laminar flow around a sphere or disk. In laminar flow, the drag is almost entirely due to skin friction. For Reynolds numbers below approximately 0.4, experiments have shown that the drag coefficient can be calculated from Eq. 2.137.[29] In calculating the Reynolds number, the sphere or disk diameter should be used as the characteristic dimension.

Drag Coefficient for Spheres, Disks, and Cylinders

$$C_D = \frac{24}{\mathrm{Re}} \qquad 2.137$$

Substituting this value of C_D into Eq. 2.136 results in *Stokes' law* (see Eq. 2.138), which is applicable to laminar slow motion (ascent or descent) of spherical particles and bubbles through a fluid. Stokes' law is based on the assumptions that (a) flow is laminar, (b) Newton's law of viscosity is valid, and (c) all higher-order velocity terms (v^2, etc.) are negligible.

$$F_D = 6\pi\mu vR = 3\pi\mu vD \quad \text{[laminar]} \qquad 2.138$$

The drag coefficients for disks and spheres operating outside the region covered by Stokes' law have been determined experimentally. In the turbulent region, pressure drag is predominant. Figure 2.27 can be used to obtain approximate values for C_D.

Figure 2.27 shows that there is a dramatic drop in the drag coefficient around $\mathrm{Re} = 10^5$. The explanation for this is that the point of separation of the boundary layer shifts, decreasing the width of the wake. (See Fig. 2.28.) Since the drag force is primarily pressure drag at higher Reynolds numbers, a reduction in the wake reduces the pressure drag. Therefore, anything that can be done to a sphere (scuffing or wetting a baseball, dimpling a golf ball, etc.) to induce a smaller wake will reduce the drag. There can be no shift in the boundary layer separation point for a thin disk, since the disk has no depth in the direction of flow. Therefore, the drag coefficient for a thin disk remains the same at all turbulent Reynolds numbers.

Figure 2.27 *Drag Coefficients for Spheres and Circular Flat Disks*

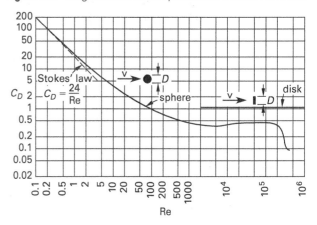

Binder, *Fluid Mechanics*, 5th, © 1973. Printed and electronically reproduced by permission of Pearson Education, Inc., New York, New York.

Figure 2.28 *Turbulent Flow Around a Sphere at Various Reynolds Numbers*

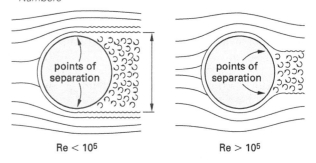

[29]Some sources report that the region in which Stokes' law applies extends to Re = 1.0.

Water

52. TERMINAL VELOCITY

The velocity of an object falling through a fluid will continue to increase until the drag force equals the net downward force (i.e., the weight less the buoyant force). The maximum velocity attained is known as the *terminal velocity* (*settling velocity*).

$$F_D = mg - F_b \quad \begin{bmatrix} \text{at terminal} \\ \text{velocity} \end{bmatrix} \quad \text{[SI]} \quad 2.139(a)$$

$$F_D = \frac{mg}{g_c} - F_b \quad \begin{bmatrix} \text{at terminal} \\ \text{velocity} \end{bmatrix} \quad \text{[U.S.]} \quad 2.139(b)$$

If the drag coefficient is known, the terminal velocity can be calculated from Eq. 2.140. For small, heavy objects falling in air, the buoyant force (represented by the ρ_f term) can be neglected.[30]

General Spherical

$$v_t = \sqrt{\frac{4g(\rho_p - \rho_f)d}{3C_D \rho_f}} \qquad 2.140$$

If the spherical particle is very small, Stokes' law may apply. In that case, the terminal velocity can be calculated from Eq. 2.141.

Stokes' Law

$$v_t = \frac{g(\rho_p - \rho_f)d^2}{18\mu} = \frac{g\rho_f(S.G. - 1)d^2}{18\mu} \qquad 2.141$$

53. INTRODUCTION TO PUMPS AND TURBINES

A *pump* adds energy to the fluid flowing through it. (See Fig. 2.29.) The amount of energy that a pump puts into the fluid stream can be determined by the difference between the total energy on either side of the pump. The specific energy added (a positive number) on a per-unit mass basis (i.e., ft-lbf/lbm or J/kg) is given by Eq. 2.142. In most situations, a pump will add primarily pressure energy.

$$E_A = E_{t,2} - E_{t,1} \qquad 2.142$$

The *head added* by a pump, H_A, is

$$H_A = \frac{H_A}{g} \qquad \text{[SI]} \quad 2.143(a)$$

$$H_A = \frac{E_A g_c}{g} \qquad \text{[U.S.]} \quad 2.143(b)$$

Figure 2.29 *Pump and Turbine Representation*

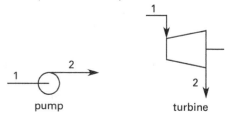

The specific energy added by a pump can also be calculated from the input power if the mass flow rate is known. The input power to the pump will be the output power of the electric motor or engine driving the pump.

$$E_A = \frac{\left(1000 \; \frac{\text{W}}{\text{kW}}\right) P_{\text{kW,input}} \eta_{\text{pump}}}{\dot{m}} \qquad \text{[SI]} \quad 2.144(a)$$

$$E_A = \frac{\left(550 \; \frac{\text{ft-lbf}}{\text{sec-hp}}\right) P_{\text{hp,input}} \eta_{\text{pump}}}{\dot{m}} \qquad \text{[U.S.]} \quad 2.144(b)$$

The fluid power, $\dot{W}_{\text{fluid}}$ (also known as *hydraulic power*), is the amount of power actually entering the fluid.

$$\dot{W}_{\text{fluid}} = \dot{W} \eta_{\text{pump}} \qquad 2.145$$

In Eq. 2.145, $\dot{W}$ is the brake pump power, the power delivered to the pump.

A *turbine* extracts energy from the fluid flowing through it. As with a pump, the energy extraction can be obtained by determining the total energy on both sides of the turbine and taking the difference. The energy extracted (a positive number) on a per-unit mass basis is given by Eq. 2.146.

$$E_E = E_{t,1} - E_{t,2} \qquad 2.146$$

54. TYPES OF PUMPS

Pumps can be classified according to how energy is transferred to the fluid: intermittently or continuously.

The most common types of *positive displacement pumps* (*PD pumps*) are *reciprocating action pumps* (which use pistons, plungers, diaphragms, or bellows) and *rotary action pumps* (which use vanes, screws, lobes, or progressing cavities). Such pumps discharge a fixed volume for each stroke or revolution. Energy is added intermittently to the fluid.

Kinetic pumps transform fluid kinetic energy into fluid static pressure energy. The pump imparts the kinetic energy; the pump mechanism or housing is constructed in a manner that causes the

---

[30]A skydiver's terminal velocity can vary from approximately 120 mi/hr (54 m/s) in horizontal configuration to 200 mi/hr (90 m/s) in vertical configuration.

transformation. *Jet pumps* and *ejector pumps* fall into the kinetic pump category, but centrifugal pumps are the primary examples.

In the operation of a *centrifugal pump*, liquid flowing into the *suction side* (the *inlet*) is captured by the *impeller* and thrown to the outside of the pump casing. Within the casing, the velocity imparted to the fluid by the impeller is converted into pressure energy. The fluid leaves the pump through the *discharge line* (the *exit*). It is a characteristic of most centrifugal pumps that the fluid is turned approximately 90° from the original flow direction. (See Table 2.8.)

Table 2.8 *Generalized Characteristics of Positive Displacement and Kinetic Pumps*

characteristic	positive displacement pumps	kinetic pumps
flow rate	low	high
pressure rise per stage	high	low
constant quantity over operating range	flow rate	pressure rise
self-priming	yes	no
discharge stream	pulsing	steady
works with high viscosity fluids	yes	no

55. TERMINOLOGY OF HYDRAULIC MACHINES AND PIPE NETWORKS

A pump has an inlet (designated the *suction*) and an outlet (designated the *discharge*). The subscripts *s* and *d* refer to the inlet and outlet of the pump, not of the pipeline.

All of the terms that are discussed in this section are *head* terms and, as such, have units of length. When working with hydraulic machines, it is common to hear such phrases as "a pressure head of 50 feet" and "a static discharge head of 15 meters." The term *head* is often substituted for pressure or pressure drop. Any head term (*pressure head, atmospheric head, vapor pressure head,* etc.) can be calculated from pressure by using Eq. 2.147.[31]

$$H = \frac{P}{g\rho} \qquad \text{[SI]} \qquad 2.147(a)$$

$$H = \frac{P}{\rho} \times \frac{g_c}{g} = \frac{P}{\gamma} \qquad \text{[U.S.]} \qquad 2.147(b)$$

Some of the terms used in the description of pipe networks appear to be similar (e.g., suction head and total suction head). The following general rules will help to clarify the meanings.

Rule 1: The word *suction* or *discharge* limits the quantity to the suction line or discharge line, respectively. The absence of either word implies that both the suction and discharge lines are included. Example: discharge head.

Rule 2: The word *static* means that static head only is included (not velocity head, friction head, etc.). Example: static suction head.

Rule 3: The word *total* means that static head, velocity head, and friction head are all included. (Total does not mean the combination of suction and discharge.) Example: total suction head.

The following terms are commonly encountered.

- *friction head,* h_f: The head required to overcome resistance to flow in the pipes, fittings, valves, entrances, and exits.

Head Loss Due to Flow

$$h_f = f \frac{L}{D} \frac{v^2}{2g} \qquad 2.148$$

- *velocity head,* H_v: The specific kinetic energy of the fluid. Also known as *dynamic head.*[32]

$$H_v = \frac{v^2}{2g} \qquad 2.149$$

- *static suction head,* $H_{z(s)}$: The vertical distance above the centerline of the pump inlet to the free level of the fluid source. If the free level of the fluid is below the pump inlet, $H_{z(s)}$ will be negative and is known as *static suction lift.* (See Fig. 2.30.)

- *static discharge head,* $H_{z(d)}$: The vertical distance above the centerline of the pump inlet to the point of free discharge or surface level of the discharge tank. (See Fig. 2.31.)

---

[31]Equation 2.147 can be used to define *pressure head, atmospheric head,* and *vapor pressure head,* whose meanings and derivations should be obvious.
[32]The term *dynamic* is not as consistently applied as are the terms described in the rules. In particular, it is not clear whether dynamic head includes friction head.

Figure 2.30 Static Suction Lift

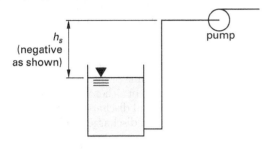

Figure 2.31 Static Discharge Head

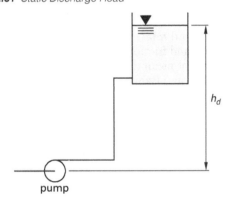

The ambiguous term *effective head* is not commonly used when discussing hydraulic machines, but when used, the term most closely means *net head* (i.e., starting head less losses). Consider a hydroelectric turbine that is fed by water with a static head of H. After frictional and other losses, the net head available to the turbine will be less than H. The turbine output will coincide with an ideal turbine being acted upon by the net or effective head. Similarly, the actual increase in pressure across a pump will be the effective head added (i.e., the head net of internal losses and geometric effects).

Example 2.12

Write the symbolic equations for the following terms: (a) the total suction head, (b) the total discharge head, and (c) the total head added.

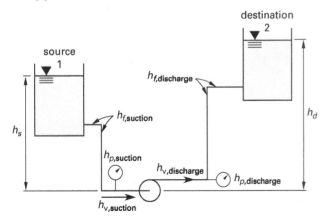

Solution

(a) The *total suction head* at the pump inlet is the sum of static (pressure) head and velocity head at the pump suction.

$$H_{t(s)} = H_{P(s)} + H_{v(s)}$$

Total suction head can also be calculated from the conditions existing at the source (1), in which case suction line friction would also be considered. (With an open reservoir, $H_{p(1)}$ will be zero if gage pressures are used, and $H_{v(1)}$ will be zero if the source is large.)

$$H_{t(s)} = H_{P(1)} + H_{z(s)} + H_{v(1)} - h_{f(s)}$$

(b) The *total discharge head* at the pump outlet is the sum of the static (pressure) and velocity heads at the pump outlet. Friction head is zero since the fluid has not yet traveled through any length of pipe when it is discharged.

$$H_{t(d)} = H_{P(d)} + H_{v(d)}$$

The total discharge head can also be evaluated at the destination (2) if the friction head, $h_{f(d)}$, between the discharge and the destination is known. (With an open reservoir, $H_{p(2)}$ will be zero if gage pressures are used, and $H_{v(2)}$ will be zero if the destination is large.)

$$H_{t(d)} = H_{P(2)} + H_{z(d)} + H_{v(2)} + h_{f(d)}$$

(c) The *total head added* by the pump is the total discharge head less the total suction head.

$$H_t = H_A = H_{t(d)} - H_{t(s)}$$

Assuming suction from and discharge to reservoirs exposed to the atmosphere and assuming negligible reservoir velocities,

$$H_t = H_A \approx H_{z(d)} - H_{z(s)} + h_{f(d)} + h_{f(s)}$$

56. PUMPING POWER

The energy (head) added by a pump can be determined from the difference in total energy on either side of the pump. Writing the Bernoulli equation for the discharge and suction conditions produces Eq. 2.150, an equation for the *total dynamic head*, often abbreviated TDH.

$$H_A = H_t = H_{t(d)} - H_{t(s)} \qquad 2.150$$

$$H_A = \frac{P_d - P_s}{\rho g} + \frac{v_d^2 - v_s^2}{2g} + z_d - z_s \qquad [\text{SI}] \quad 2.151(a)$$

$$H_A = \frac{(P_d - P_s)g_c}{\rho g} + \frac{v_d^2 - v_s^2}{2g} + z_d - z_s \quad [\text{U.S.}] \quad 2.151(b)$$

In most applications, the change in velocity and potential heads is either zero or small in comparison to the increase in pressure head. Equation 2.151 then reduces to Eq. 2.152.

$$H_A = \frac{P_d - P_s}{\rho g} \qquad [\text{SI}] \quad 2.152(a)$$

$$H_A = \frac{P_d - P_s}{\rho} \times \frac{g_c}{g} \qquad [\text{U.S.}] \quad 2.152(b)$$

It is important to recognize that the variables in Eq. 2.151 and Eq. 2.152 refer to the conditions at the pump's immediate inlet and discharge, not to the distant ends of the suction and discharge lines. However, the total dynamic head added by a pump can be calculated in another way. For example, for a pump raising water from one open reservoir to another, the total dynamic head would consider the total elevation rise, the velocity head (often negligible), and the friction losses in the suction and discharge lines.

The head added by the pump can also be calculated from the impeller and fluid speeds. Equation 2.153 is useful for radial- and mixed-flow pumps for which the incoming fluid has little or no rotational velocity component (i.e., up to a specific speed of approximately 2000 U.S. or 40 SI). In Eq. 2.153, v_{impeller} is the tangential impeller velocity at the radius being considered, and v_{fluid} is the average tangential velocity imparted to the fluid by the impeller. The impeller efficiency, η_{impeller}, is typically 0.85–0.95. This is much higher than the total pump efficiency because it does not include mechanical and fluid friction losses. (See Sec. 2.57.)

$$H_A = \frac{\eta_{\text{impeller}} v_{\text{impeller}} v_{\text{fluid}}}{g} \qquad 2.153$$

Head added can be thought of as the energy added per unit mass. The total pumping power depends on the head added, H_A, and the mass flow rate. For example, the product $\dot{m} H_A$ has the units of foot-pounds per second (in customary U.S. units), which can be easily converted to horsepower. Pump output power is known as *hydraulic power* or *water power*. Hydraulic power is the net power actually transferred to the fluid.

Horsepower is a common unit of power, which results in the terms *hydraulic horsepower* and *water horsepower*, $\dot{W}_{\text{fluid}}$, being used to designate the power that is transferred into the fluid. Various relationships for finding the hydraulic horsepower are given in Table 2.9.

The unit of power in SI units is the watt (kilowatt). Table 2.10 can be used to determine *hydraulic kilowatts*, WkW.

Table 2.9 *Hydraulic Horsepower Equations*

	Q (gal/min)	$\dot{m}$ (lbm/sec)	Q (ft³/sec)
H_A in feet	$\dfrac{H_A Q(SG)}{3956}$	$\dfrac{H_A \dot{m}}{550} \times \dfrac{g}{g_c}$	$\dfrac{H_A Q(SG)}{8.814}$
ΔP in psi	$\dfrac{\Delta P Q}{1714}$	$\dfrac{\Delta P \dot{m}}{(238.3)(SG)} \times \dfrac{g}{g_c}$	$\dfrac{\Delta P Q}{3.819}$
ΔP in psf[b]	$\dfrac{\Delta P Q}{2.468 \times 10^5}$	$\dfrac{\Delta P \dot{m}}{(34,320)(SG)} \times \dfrac{g}{g_c}$	$\dfrac{\Delta P Q}{550}$
W in $\dfrac{\text{ft-lbf}}{\text{lbm}}$	$\dfrac{W Q(SG)}{3956}$	$\dfrac{W \dot{m}}{550}$	$\dfrac{W Q(SG)}{8.814}$

(Multiply horsepower by 0.7457 to obtain kilowatts.)
[a]based on $\rho_{\text{water}} = 62.4 \text{ lbm/ft}^3$ and $g = 32.2 \text{ ft/sec}^2$
[b]Velocity head changes must be included in ΔP.

Example 2.13

A pump adds 550 ft of pressure head to 100 lbm/sec of water ($\rho = 62.4 \text{ lbm/ft}^3$ or 1000 kg/m³). (a) Complete the following table of performance data. (b) What is the hydraulic power in horsepower and kilowatts?

item	customary U.S.	SI
$\dot{m}$	100 lbm/sec	__ kg/s
h	550 ft	__ m
ΔP	__ lbf/ft²	__ kPa
Q	__ ft³/sec	__ m³/s
W	__ ft-lbf/lbm	__ J/kg
$\dot{W}$	__ hp	__ kW

Solution

(a) Work initially with the customary U.S. data.

$$\Delta P = \rho h \times \frac{g}{g_c} = \left(62.4 \, \frac{\text{lbm}}{\text{ft}^3}\right)(550 \text{ ft}) \times \frac{g}{g_c}$$
$$= 34,320 \text{ lbf/ft}^2$$

$$Q = \frac{\dot{m}}{\rho} = \frac{100 \dfrac{\text{lbm}}{\text{sec}}}{62.4 \dfrac{\text{lbm}}{\text{ft}^3}} = 1.603 \ \text{ft}^3/\text{sec}$$

$$W = h \times \frac{g}{g_c} = 550 \ \text{ft} \times \frac{g}{g_c}$$
$$= 550 \ \text{ft-lbf/lbm}$$

Convert to SI units.

$$\dot{m} = \frac{100 \dfrac{\text{lbm}}{\text{sec}}}{2.201 \dfrac{\text{lbm}}{\text{kg}}} = 45.43 \ \text{kg/s}$$

$$h = \frac{550 \ \text{ft}}{3.281 \dfrac{\text{ft}}{\text{m}}} = 167.6 \ \text{m}$$

$$\Delta P = \left(34{,}320 \ \frac{\text{lbf}}{\text{ft}^2}\right)\left(\frac{1}{\left(12 \dfrac{\text{in}}{\text{ft}}\right)^2}\right)\left(6.895 \ \frac{\text{kPa}}{\dfrac{\text{lbf}}{\text{in}^2}}\right)$$
$$= 1643 \ \text{kPa}$$

$$Q = \left(1.603 \ \frac{\text{ft}^3}{\text{sec}}\right)\left(0.0283 \ \frac{\text{m}^3}{\text{ft}^3}\right) = 0.0454 \ \text{m}^3/\text{s}$$

$$W = \left(550 \ \frac{\text{ft-lbf}}{\text{lbm}}\right)\left(1.356 \ \frac{\text{J}}{\text{ft-lbf}}\right)\left(2.201 \ \frac{\text{lbm}}{\text{kg}}\right)$$
$$= 1642 \ \text{J/kg}$$

(b) From Table 2.9, the hydraulic horsepower is

$$\dot{W}_{\text{fluid}} = \frac{H_A \dot{m}}{550} \times \frac{g}{g_c}$$
$$= \frac{(550 \ \text{ft})\left(100 \ \dfrac{\text{lbm}}{\text{sec}}\right)}{550 \ \dfrac{\text{ft-lbf}}{\text{hp-sec}}} \times \frac{g}{g_c}$$
$$= 100 \ \text{hp}$$

From Table 2.10, the power is

$$\text{WkW} = \frac{\Delta P \dot{m}}{(1000)(SG)} = \frac{(1643 \ \text{kPa})\left(45.43 \ \dfrac{\text{kg}}{\text{s}}\right)}{\left(1000 \ \dfrac{\text{W}}{\text{kW}}\right)(1.0)}$$
$$= 74.6 \ \text{kW}$$

Table 2.10 *Hydraulic Kilowatt Equations*[a]

	Q (L/s)	$\dot{m}$ (kg/s)	Q (m³/s)
H_A in meters	$\dfrac{9.81 H_A Q(SG)}{1000}$	$\dfrac{9.81 H_A \dot{m}}{1000}$	$9.81 H_A Q(SG)$
ΔP in kPa[b]	$\dfrac{\Delta P Q}{1000}$	$\dfrac{\Delta P \dot{m}}{1000(SG)}$	$\Delta P Q$
W in $\dfrac{\text{J}}{\text{kg}}$[b]	$\dfrac{W Q(SG)}{1000}$	$\dfrac{W \dot{m}}{1000}$	$W Q(SG)$

(Multiply kilowatts by 1.341 to obtain horsepower.)
[a]based on $\rho_{\text{water}} = 1000 \ \text{kg/m}^3$ and $g = 9.81 \ \text{m/s}^2$
[b]Velocity head changes must be included in ΔP.

57. PUMPING EFFICIENCY

Fluid power (or *hydraulic power*) is the net energy actually transferred to the fluid per unit time.

Equation 2.154 is the equation for fluid power.

Centrifugal Pump Characteristics

$$\dot{W}_{\text{fluid}} = \rho g H Q \qquad \textbf{2.154}$$

The input power delivered by the motor to the pump is known as the *brake pump power* (also called *brake horsepower*). Due to frictional losses between the fluid and the pump and mechanical losses in the pump itself, the brake pump power will be greater than the hydraulic power. The ratio of hydraulic power to brake pump power is the pump efficiency, η_{pump}.

$$\eta_{\text{pump}} = \frac{\dot{W}_{\text{fluid}}}{\dot{W}} = \frac{\rho g H Q}{\dot{W}} \qquad \textbf{2.155}$$

The equation for brake pump power, then, is

Centrifugal Pump Characteristics

$$\dot{W} = \frac{\rho g H Q}{\eta_{\text{pump}}} \qquad \textbf{2.156}$$

Figure 2.32 gives typical pump efficiencies as a function of the pump's specific speed.

The difference between the brake pump power and the fluid power is known as the *friction power*.

$$\dot{W}_{\text{friction}} = \dot{W} - \dot{W}_{\text{fluid}} \qquad \textbf{2.157}$$

Pumping efficiency is not constant for any specific pump; rather, it depends on the operating point and the speed of the pump. A pump's characteristic efficiency curves will be published by its manufacturer.

Figure 2.32 *Average Pump Efficiency Versus Specific Speed*

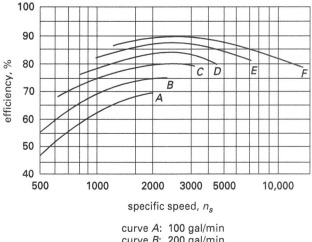

curve A: 100 gal/min
curve B: 200 gal/min
curve C: 500 gal/min
curve D: 1000 gal/min
curve E: 3000 gal/min
curve F: 10,000 gal/min

With pump characteristic curves given by the manufacturer, the efficiency is not determined from the intersection of the system curve and the efficiency curve. Rather, the efficiency is a function of only the flow rate. Therefore, the operating efficiency is read from the efficiency curve directly above or below the operating point. (See Fig. 2.33.)

Figure 2.33 *Typical Centrifugal Pump Efficiency Curves*

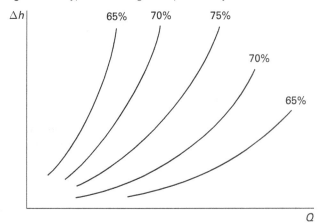

Efficiency curves published by a manufacturer will not include such losses in the suction elbow, discharge diffuser, couplings, bearing frame, seals, or pillow blocks. Up to 15% of the motor horsepower may be lost to these factors. Upon request, the manufacturer may provide the pump's installed *wire-to-water efficiency* (i.e., the fraction of the electrical power drawn that is converted to hydraulic power).

The *purchased power* is the power delivered to the motor. Due to mechanical losses in the motor, not all the purchased power will be converted to brake pump power. The ratio of brake pump power to purchased power is the motor efficiency, η_{motor}.

Centrifugal Pump Characteristics

$$\dot{W}_{\text{purchased}} = \frac{\dot{W}}{\eta_{\text{motor}}} \qquad 2.158$$

Combining Eq. 2.156 and Eq. 2.158 gives the *pump power equation*.

Pump Power Equation

$$\dot{W} = Q\gamma h/\eta = Q\rho g h/\eta_t \qquad 2.159$$

In Eq. 2.160, η_t is the total efficiency of the pump and motor.

Pump Power Equation

$$\eta_t = \eta_{\text{pump}} \times \eta_{\text{motor}} \qquad 2.160$$

58. REACTOR HYDRAULICS

Reactors may be classified on the basis of *ideal flow* versus *nonideal flow*. Ideal flow has two forms: *complete mix* and *plug flow*. In both cases, the flows that enter and exit the reactor are continuous.

In an ideal *complete-mix reactor*, the contents of the reactor are fully and immediately mixed so that all particles are uniformly distributed throughout the reactor at all times and the effluent is of consistent composition. In an ideal *plug-flow reactor*, there is no mixing at all, so particles pass through and exit the reactor in the same order as they enter. Complete mix and plug flow are the two extremes of all possible flow regimes.

In practice, both complete mixing and a complete lack of mixing are difficult to achieve. Most reactors exhibit nonideal flow, also called *arbitrary flow*. Nonideal flow is any flow regime existing on the continuum between complete mix and plug flow.

The characteristics and efficiency of a reactor can be assessed by monitoring effluent concentrations of the target contaminants or tracers at uniform time intervals. By observing the characteristics curves, general conclusions can be reached regarding the degree to which complete-mix or plug-flow conditions dominate.

Water

The *hydraulic residence time* (also called *hydraulic detention time* or *hydraulic retention time*), θ, is the amount of time that fluid spends in the reactor. Hydraulic residence time is calculated with Eq. 2.161.

Stokes' Law

$$\theta = V/Q \qquad 2.161$$

V is the reactor volume. If the actual volume of the reactor is used for V, then V/Q represents the *theoretical residence time*, the residence time of the tank when 100% hydraulically efficient.

The *mean residence time*, or *average residence time*, t, is the average time the fluid actually remains in the reactor. The mean residence time is defined as the centroid of the relative concentration-time plot (C/C_0 against t/θ). For ideal flow (either complete mix or plug flow), $t = \theta$. For arbitrary flow,

$$t = \frac{\sum C_i t_i \Delta t}{\sum C_i \Delta t} \qquad 2.162$$

Δt is the sampling interval.

The closer t is to θ, the more efficient the reactor is. The *hydraulic efficiency* of a reactor, η, is calculated with Eq. 2.163.

$$\eta = \frac{t}{\theta} \times 100\% \qquad 2.163$$

The *effective volume* of the reactor, V_e, is that part of the reactor actually being used. V_e is related to t and θ by Eq. 2.164 and Eq. 2.165.

$$t = \frac{V_e}{Q} \qquad 2.164$$

$$V_e = V\left(\frac{t}{\theta}\right) = V\left(\frac{\eta}{100\%}\right) \qquad 2.165$$

Lower efficiencies indicate the existence of dead spaces and short-circuiting.

Example 2.14

Determine the mean residence time and the hydraulic efficiency for the reactor represented by the following data. The theoretical residence time for the reactor is 23 min.

elapsed time (min)	concentration (mg/L)	$C_i t_i \Delta_t$ (min²·mg/L)	$C_i \Delta_t$ (min·mg/L)
0	0.00	0.0	0.00
2	0.00	0.0	0.00
4	0.02	0.32	0.08
6	0.6	7.2	1.2
8	1.0	1.6	2.0
10	1.5	30	3.0
14	1.7	95	6.8
18	1.6	115	6.4
22	1.3	114	5.2
26	0.9	94	3.6
30	0.6	72	2.4
34	0.4	54	1.6
38	0.08	12	0.32
42	0.01	1.7	0.04
sum		597	33

Solution

Use Eq. 2.162 to find the mean residence time.

$$t = \frac{\sum C_i t_i \Delta t}{\sum C_i \Delta t} = \frac{597 \text{ min}^2 \dfrac{\text{mg}}{\text{L}}}{33 \text{ min} \dfrac{\text{mg}}{\text{L}}}$$

$$= 18 \text{ min}$$

Use Eq. 2.163 to find the hydraulic efficiency.

$$\eta = \frac{t}{\theta} \times 100\% = \frac{18 \text{ min}}{23 \text{ min}} \times 100\%$$

$$= 78\%$$

59. FLOW MODELS

Several different flow regimes are possible, depending on the reactor configuration and the needs of the unit process. The general flow models encountered include

- batch reactor
- plug-flow reactor (PFR)
- completely mixed or continuous-flow stirred tank reactor (CSTR)
- two or more CSTRs in series

Among these various reactor types, fixed bed (or packed bed), moving bed, and fluidized-bed reactors are also employed.

Batch Reactors

A batch reactor is filled at the beginning of the reaction and emptied at the end. Flow does not enter or leave the reactor during the treatment process. The residence time is determined by the reaction rate; it is not a function of the flow rate. Any change in the concentration of target contaminants is a function of decay or accumulation within the reactor. Batch reactors are most commonly, but not always, employed for small or intermittent flows.

The contents of a batch reactor are usually completely mixed (this is not the same as continuously mixed). Such reactors may be referred to as *completely mixed batch* (CMB) reactors.

For a first-order reaction, a batch reactor can be modeled by Eq. 2.166.

First-Order Irreversible Reaction Kinetics
$$\ln(C_A/C_{A0}) = -kt \qquad \textit{2.166}$$

The negative reaction rate constant, $-k$, indicates that the concentration is being degraded as the reaction proceeds. The reactor in Ex. 2.14 is a first-order batch reactor.

Plug-Flow Reactors

In a plug-flow reactor (PFR), a constant relative position is maintained among the components in the reactor as material moves from the inlet to the outlet. The mass balance includes input, outputs, and changes internal to the reactor. However, because the relative position of the materials in the reactor remains constant, the plug-flow reactor is modeled using the same equation for a first-order reaction as is used for batch reactors. For comparable reaction and residence times, batch and plug-flow reactors will perform comparably. Plug-flow reactors usually have a rectangular or baffled configuration to reduce mixing.

Continuous-Flow Stirred Tank Reactors

In a continuous-flow stirred tank reactor (CSTR), the concentration of a component is uniform at every location in the reactor. It is assumed that the inputs are nearly instantaneously mixed as they enter the reactor, and, therefore, the concentration anywhere in the reactor is the same as the concentration of the output. To perform effectively, CSTRs must be properly mixed and are usually configured with cubic dimensions to promote proper mixing. For a first-order reaction, a CSTR is modeled by Eq. 2.167.

$$\frac{C_0}{C} = 1 + kt \qquad \textit{2.167}$$

Continuous-Flow Stirred Tank Reactors in Series

CSTRs tend to have longer hydraulic residence times than PFRs to accomplish the same level of treatment. For this reason, it is sometimes desirable to use several CSTRs of the same volume in series instead of a single large reactor. The effective result will be to create a series of CSTRs with the same residence time as a PFR of similar size. This effect is observed as the number of reactors in series approaches about 10.

Equation 2.168 is a generalized equation for CSTRs in series under first-order conditions.

CSTR in Series
$$\frac{C_0}{C_N} = (1 + kt)^N \qquad \textit{2.168}$$

In Eq. 2.168, N is the number of reactors in series, C_N is the effluent concentration of the Nth reactor, and t is the residence time for each reactor in the series.

Example 2.15

Determine the residence time for 90% degradation of a contaminant using a reaction rate constant of 0.45 h^{-1} for (a) a batch reactor, (b) a PFR, (c) a single CSTR, and (d) a series of nine CSTRs.

Solution

For 90% conversion, $C/C_0 = 0.1$ and $C_0/C = 10$.

(a) For a batch reactor, use Eq. 2.166 and solve for the residence time.

First-Order Irreversible Reaction Kinetics
$$\ln(C_A/C_{A0}) = -kt$$

$$t = \frac{\ln \dfrac{C_A}{C_{A0}}}{-k} = \frac{\ln 0.1}{-0.45 \dfrac{1}{h}}$$

$$= 5.12 \text{ h}$$

(b) For a PFR, the calculation is the same as for a batch reactor, so $t = 5.12$ h.

(c) For a CSTR, use Eq. 2.167 and solve for the residence time.

$$\frac{C_0}{C} = 1 + kt$$

$$t = \frac{\dfrac{C_0}{C} - 1}{k} = \frac{10 - 1}{0.45 \dfrac{1}{h}}$$

$$= 20 \text{ h}$$

(d) For a series of nine CSTRs, use Eq. 2.168 and solve for the residence time for a single reactor.

CSTR in Series

$$\frac{C_0}{C_N} = (1 + kt)^N$$

$$t = \frac{\sqrt[N]{\dfrac{C_0}{C_N}} - 1}{k} = \frac{\sqrt[9]{10} - 1}{0.45 \ \dfrac{1}{\text{h}}}$$

$$= 0.648 \text{ h}$$

For nine reactors in series, the total residence time is

$$t_{\text{total}} = (9)(0.648 \text{ h}) = 5.83 \text{ h}$$

60. BED REACTORS

A reactor may contain some sort of porous or reactive solid medium occupying a portion of the reactor volume. This medium is present either to control hydraulic characteristics or to effect a chemical, biological, or physical reaction.

Bed reactors can be categorized by the behavior of their media. In a *packed-bed reactor*, the porosity of the medium is not influenced by flow. The solid materials remain in place as the fluid passes through the voids in the material. A completely filled packed-bed reactor typically exhibits plug-flow characteristics. The bed medium may be completely immersed in water, or the contaminant may be passed through the medium in the form of a cascade, mist, or vapor.

In an *expanded-bed reactor* or *fluidized-bed reactor*, the fluid travels upward through a solid, granular medium (such as sand or gravel) at a sufficiently high velocity to cause the granules to move. In this way, the porosity of the medium is influenced by flow, with porosity increasing as flow increases. In an expanded-bed reactor, the fluid velocity is enough to cause the bed to expand in volume. In a fluidized-bed reactor, the fluid velocity is high enough to cause the granules to become suspended so that the medium behaves as though it were a fluid. To create the fluidized or expanded bed conditions, the flow in these reactors is from bottom to top.

In a *moving-bed reactor*, the medium moves countercurrent to the liquid flow and is continuously replenished or recirculated. The medium is usually reactive and requires reactivation or regeneration before reuse.

61. NOMENCLATURE

a	length	ft	m
a	speed of sound	ft/sec	m/s
b	length	ft	m
A	area	ft^2	m^2
C	coefficient	–	–
C	concentration	mg/L	mg/L
C	Hazen-Williams roughness coefficient	–	–
C_i	impact factor	–	–
d	distance	ft	m
D	diameter	ft	m
D_H	hydraulic diameter	ft	m
E	energy	ft-lbf/lbm	J/kg
f	friction factor	–	–
F	force	lbf	N
g	gravitational acceleration, 32.2 (9.81)	ft/sec^2	m/s^2
g_c	gravitational constant, 32.2	lbm-ft/lbf-sec^2	n.a.
G	mass flow rate per unit area	lbm/ft^2-sec	kg/m^2·s
h	head or height	ft	m
H	total head	ft	m
k	ratio of specific heats	–	–
K	coefficient	–	–
KE	kinetic energy	ft-lbf	J
K_v	valve flow coefficent (metric)	–	n.a.
L	length	ft	m
m	mass	lbm	kg
$\dot{m}$	mass flow rate	lbm/sec	kg/s
n	flow rate exponent	–	–
n_s	specific speed	rpm	rpm
p	pressure	lbf/ft^2	Pa
P	momentum	lbm-ft/sec	kg·m/s
P	power	ft-lbf/sec	W
P	pressure, total pressure	lbf/ft^2	Pa
P	wetted perimeter	ft	m
PE	potential energy	ft-lbf	J
Q	volumetric flow rate	ft^3/sec, gal/min	m^3/s
r	distance or radius	ft	m
R	resultant force	lbf	N
R	universal gas constant, 1545.35 (8314.47)	ft-lbf/lbmol-°R	J/kmol·K
Re	Reynolds number	–	–
SG	specific gravity	–	–
t	time	sec	s
u	x-component of velocity	ft/sec	m/s
v, v	velocity	ft/sec	m/s
v	y-component of velocity	ft/sec	m/s
V	volume	ft^3	m^3
V_a	velocity	ft/sec	m/s
w	specific weight	lbf/ft^3	n.a.
W	work	ft-lbf	J
x	x-coordinate of position	ft	m
y	y-coordinate of position	ft	m
z	elevation	ft	m
Z_a	elevation	ft	m

Water

Symbols

β	diameter ratio	–	–
γ	specific weight	lbf/ft^3	N/m^3
ϵ	specific roughness	ft	m
η	efficiency	–	–
η	non-Newtonian viscosity	lbf-sec/ft^2	Pa·s
θ	angle	deg	deg
θ	hydraulic resistance time	sec	s
μ	absolute viscosity	lbf-sec/ft^2	Pa·s
ν	kinematic viscosity	ft^2/sec	m^2/s
ρ	density	lbm/ft^3	kg/m^3
τ	shear stress	lbf/ft^2	Pa
τ	residence time	sec	s
υ	specific volume	ft^3/lbm	m^3/kg
ϕ	angle	deg	deg
ω	angular velocity	rad/sec	rad/s

Subscripts

0	critical (yield)or initial
a	assumed
A	added (by pump)
b	blade or buoyant
c	contraction
d	discharge
D	drag
e	effiective or equivalent
E	extracted (by turbine)
f	flow or friction
h, H	hydraulic
i	impact, inner, or inside
L	lift
N	N^{th} reactor
o	orifice, outer, or outside
p	pressure or prototype
P	pressure or pump
r	radial distance or ratio
s	static
t	tank, theoretical, or total
υ	valve
v	velocity
z	potential

3 Chemistry

Content in blue refers to the *NCEES Handbook*.

1. ATOMIC STRUCTURE

An *element* is a substance that cannot be decomposed into simpler substances during ordinary chemical reactions.[1] An *atom* is the smallest subdivision of an element that can take part in a chemical reaction. A *molecule* is the smallest subdivision of an element or compound that can exist in a natural state.

The atomic nucleus consists of neutrons and protons, known as *nucleons*. The masses of neutrons and protons are essentially the same—one *atomic mass unit*, amu. One amu is exactly $\frac{1}{12}$ of the mass of an atom of carbon-12, approximately equal to 1.66×10^{-27} kg.[2] The *relative atomic weight* or *atomic weight*, A, of an atom is approximately equal to the number of protons and neutrons in the nucleus.[3] The *atomic number*, Z, of an atom is equal to the number of protons in the nucleus.

The atomic number and atomic weight of an element E are written in symbolic form as $_Z\mathrm{E}^A$, E_Z^A, or $_Z^A\mathrm{E}$. For example, carbon is the sixth element; radioactive carbon has an atomic mass of 14. Therefore, the symbol for carbon-14 is C_6^{14}. Since the atomic number is superfluous if the chemical symbol is given, the atomic number can be omitted (e.g., C^{14}).

2. ISOTOPES

Although an element can have only a single atomic number, atoms of that element can have different atomic weights. Many elements possess *isotopes*. The nuclei of isotopes differ from one another only in the number of neutrons. Isotopes behave the same way chemically.[4] Therefore, isotope separation must be done physically (e.g., by centrifugation or gaseous diffusion) rather than chemically.

Hydrogen has three isotopes. H_1^1 is *normal hydrogen* with a single proton nucleus. H_1^2 is known as *deuterium* (*heavy hydrogen*), with a nucleus of a proton and neutron. (This nucleus is known as a *deuteron*.) Finally, H_1^3 (*tritium*) has two neutrons in the nucleus. While normal

---

[1]Atoms of an element can be decomposed into subatomic particles in nuclear reactions.
[2]Until 1961, the atomic mass unit was defined as $\frac{1}{16}$ of the mass of one atom of oxygen-16.
[3]The term *weight* is used even though all chemical calculations involve mass. The atomic weight of an atom includes the mass of the electrons. Published *chemical atomic weights* of elements are averages of all the atomic weights of stable isotopes, taking into consideration the relative abundances of the isotopes.
[4]There are slight differences, known as *isotope effects*, in the chemical behavior of isotopes. These effects usually influence only the rate of reaction, not the kind of reaction.

hydrogen and deuterium are stable, tritium is radioactive. Many elements have more than one stable isotope. Tin, for example, has 10.

The *relative abundance*, x_i, of an isotope, i, is equal to the fraction of that isotope in a naturally occurring sample of the element. The *chemical atomic weight* is the weighted average of the isotope weights.

$$A_{\text{average}} = x_1 A_1 + x_2 A_2 + \cdots \qquad 3.1$$

3. PERIODIC TABLE

The *periodic table* is organized around the *periodic law:* The properties of the elements depend on the atomic structure and vary with the atomic number in a systematic way. Elements are arranged in order of increasing atomic numbers from left to right. Adjacent elements in horizontal rows differ decidedly in both physical and chemical properties. However, elements in the same column have similar properties. Graduations in properties, both physical and chemical, are most pronounced in the *periods* (i.e., the horizontal rows). [Periodic Table of Elements]

The vertical columns are known as *groups*, numbered in Roman numerals. Elements in a group are called *cogeners*. Each vertical group except 0 and VIII has A and B subgroups (*families*). The elements of a family resemble each other more than they resemble elements in the other family of the same group. Graduations in properties are definite but less pronounced in vertical families. The trend in any family is toward more *metallic properties* as the atomic weight increases.

Metals (elements at the left end of the periodic chart) have low electron affinities and electronegativities, are reducing agents, form positive ions, and have positive oxidation numbers. They have high electrical conductivities, luster, generally high melting points, ductility, and malleability.

Nonmetals (elements at the right end of the periodic chart) have high electron affinities and electronegativities, are oxidizing agents, form negative ions, and have negative oxidation numbers. They are poor electrical conductors, have little or no luster, and form brittle solids. Of the common nonmetals, fluorine has the highest electronic affinity and electronegativity, with oxygen having the next highest values.

The *metalloids* (e.g., boron, silicon, germanium, arsenic, antimony, tellurium, and polonium) have characteristics of both metals and nonmetals. Electrically, they are semiconductors.

The electron-attracting power of an atom is called its *electronegativity*. Metals have low electronegativities. Group VIIA elements (fluorine, chlorine, etc.) are most strongly electronegative. The alkali metals (Group IA) are the most weakly electronegative. Generally, the most *electronegative elements* are those at the right ends of the periods. Elements with low electronegativities are found at the beginning (i.e., left end) of the periods. Electronegativity decreases as you go down a group.

Elements in the periodic table are often categorized into the following groups.

- *actinides:* same as actinons
- *actinons:* elements 90–103[5]
- *alkali metals:* group IA
- *alkaline earth metals:* group IIA
- *halogens:* group VIIA
- *heavy metals:* metals near the center of the chart
- *inner transition elements:* same as transition metals
- *lanthanides:* same as lanthanons
- *lanthanons:* elements 58–71[6]
- *light metals:* elements in the first two groups
- *metals:* everything except the nonmetals
- *metalloids:* elements along the dark line in the chart separating metals and nonmetals
- *noble gases:* group 0
- *nonmetals:* elements 2, 5–10, 14–18, 33–36, 52–54, 85, and 86
- *rare earths:* same as lanthanons
- *transition elements:* same as transition metals
- *transition metals:* all B families and group VIII B[7]

4. OXIDATION NUMBER

The *oxidation number (oxidation state)* is an electrical charge assigned by a set of prescribed rules. It is actually the charge assuming all bonding is ionic. The sum of the oxidation numbers equals the net charge. For monoatomic ions, the oxidation number is equal to the charge. The oxidation numbers of some common atoms and radicals are given in Table 3.1.

---

[5]The *actinons* resemble element 89, *actinium*. Therefore, element 89 is sometimes included as an actinon.
[6]The *lanthanons* resemble element 57, *lanthanum*. Therefore, element 57 is sometimes included as a lanthanon.
[7]The *transition metals* are elements whose electrons occupy the d sublevel. They can have various oxidation numbers, including +2, +3, +4, +6, and +7.

Table 3.1 Oxidation Numbers of Selected Atoms and Charge Numbers of Radicals

name	symbol	oxidation or charge number
acetate	$C_2H_3O_2$	−1
aluminum	Al	+3
ammonium	NH_4	+1
barium	Ba	+2
borate	BO_3	−3
boron	B	+3
bromine	Br	−1
calcium	Ca	+2
carbon	C	+4, −4
carbonate	CO_3	−2
chlorate	ClO_3	−1
chlorine	Cl	−1
chlorite	ClO_2	−1
chromate	CrO_4	−2
chromium	Cr	+2, +3, +6
copper	Cu	+1, +2
cyanide	CN	−1
dichromate	Cr_2O_7	−2
fluorine	F	−1
gold	Au	+1, +3
hydrogen	H	+1 (−1 in hydrides)
hydroxide	OH	−1
hypochlorite	ClO	−1
iron	Fe	+2, +3
lead	Pb	+2, +4
lithium	Li	+1
magnesium	Mg	+2
mercury	Hg	+1, +2
nickel	Ni	+2, +3
nitrate	NO_3	−1
nitrite	NO_2	−1
nitrogen	N	−3, +1, +2, +3, +4, +5
oxygen	O	−2 (−1 in peroxides)
perchlorate	ClO_4	−1
permanganate	MnO_4	−1
phosphate	PO_4	−3
phosphorus	P	−3, +3, +5
potassium	K	+1
silicon	Si	+4, −4
silver	Ag	+1
sodium	Na	+1
sulfate	SO_4	−2
sulfite	SO_3	−2
sulfur	S	−2, +4, +6
tin	Sn	+2, +4
zinc	Zn	+2

In covalent compounds, all of the bonding electrons are assigned to the ion with the greater electronegativity. For example, nonmetals are more electronegative than metals. Carbon is more electronegative than hydrogen.

For atoms in a free-state molecule, the oxidation number is zero. Hydrogen gas is a diatomic molecule, H_2. Therefore, the oxidation number of the hydrogen molecule, H_2, is zero. The same is true for the atoms in O_2, N_2, Cl_2, and so on. Also, the sum of all the oxidation numbers of atoms in a neutral molecule is zero.

Fluorine is the most electronegative element, and it has an oxidation number of −1. Oxygen is second only to fluorine in electronegativity. Usually, the oxidation number of oxygen is −2, except in peroxides, where it is −1, and when combined with fluorine, where it is +2. Hydrogen is usually +1, except in hydrides, where it is −1.

For a charged *radical* (a group of atoms that combine as a single unit), the net oxidation number is equal to the charge on the radical, known as the *charge number*.

Example 3.1

What are the oxidation numbers of all the elements in the chlorate (ClO_3^{-1}) and permanganate (MnO_4^{-1}) ions?

Solution

For the chlorate ion, the oxygen is more electronegative than the chlorine. (Only fluorine is more electronegative than oxygen.) Therefore, the oxidation number of oxygen is −2. In order for the net oxidation number to be −1, the chlorine must have an oxidation number of +5.

For the permanganate ion, the oxygen is more electronegative than the manganese. Therefore, the oxidation number of oxygen is −2. For the net oxidation number to be −1, the manganese must have an oxidation number of +7.

5. FORMATION OF COMPOUNDS

Compounds form according to the *law of definite (constant) proportions:* A pure compound is always composed of the same elements combined in a definite proportion by mass. For example, common table salt is always NaCl. It is not sometimes NaCl and other times Na_2Cl or $NaCl_3$ (which do not exist, in any case).

Furthermore, compounds form according to the *law of (simple) multiple proportions:* When two elements combine to form more than one compound, the masses of one element that combine with the same mass of the other are in the ratios of small integers.

In order to evaluate whether a compound formula is valid, it is necessary to know the *oxidation numbers* of the interacting atoms. Although some atoms have more than one possible oxidation number, most do not.

The sum of the oxidation numbers must be zero if a neutral compound is to form. For example, H_2O is a valid compound since the two hydrogen atoms have a total positive oxidation number of $2 \times 1 = +2$. The oxygen ion has an oxidation number of -2. These oxidation numbers sum to zero.

On the other hand, $NaCO_3$ is not a valid compound formula. The sodium (Na) ion has an oxidation number of $+1$. However, the carbonate radical has a charge number of -2. The correct sodium carbonate molecule is Na_2CO_3.

6. NAMING COMPOUNDS

Combinations of elements are known as *compounds*. *Binary compounds* contain two elements; *ternary (tertiary) compounds* contain three elements. A *chemical formula* is a representation of the relative numbers of each element in the compound. For example, the formula $CaCl_2$ shows that there are one calcium atom and two chlorine atoms in one molecule of calcium chloride. [Common Names and Molecular Formulas of Some Industrial (Inorganic and Organic) Chemicals]

Generally, the numbers of atoms are reduced to their lowest terms. However, there are exceptions. For example, acetylene is C_2H_2, and hydrogen peroxide is H_2O_2.

For binary compounds with a metallic element, the positive metallic element is listed first. The chemical name ends in the suffix "-ide." For example, NaCl is sodium chloride. If the metal has two oxidation states, the suffix "-ous" is used for the lower state, and "-ic" is used for the higher state. Alternatively, the element name can be used with the oxidation number written in Roman numerals. For example,

$FeCl_2$: ferrous chloride, or iron (II) chloride

$FeCl_3$: ferric chloride, or iron (III) chloride

For binary compounds formed between two nonmetals, the more positive element is listed first. The number of atoms of each element is specified by the prefixes "di-" (2), "tri-" (3), "tetra-" (4), and "penta-" (5), and so on. For example,

N_2O_5: dinitrogen pentoxide

Binary acids start with the prefix "hydro-," list the name of the nonmetallic element, and end with the suffix "-ic." For example,

HCl: hydrochloric acid

Ternary compounds generally consist of an element and a radical. The positive part is listed first in the formula. *Ternary acids* (also known as *oxyacids*) usually contain hydrogen, a nonmetal, and oxygen, and can be grouped into families with different numbers of oxygen atoms. The most common acid in a family (i.e., the root acid) has the name of the nonmetal and the suffix "-ic." The acid with one more oxygen atom than the root is given the prefix "per-" and the suffix "-ic." The acid containing one less oxygen atom than the root is given the ending "-ous." The acid containing two less oxygen atoms than the root is given the prefix "hypo-" and the suffix "-ous."

For example,

HClO: hypochlorous acid

$HClO_2$:   chlorous acid

$HClO_3$:   chloric acid (the root)

$HClO_4$:   perchloric acid

7. MOLES AND AVOGADRO'S LAW

The *mole* is a measure of the quantity of an element or compound. Specifically, a mole of an element will have a mass equal to the element's atomic (or molecular) weight. The three main types of moles are based on mass being measured in grams, kilograms, and pounds. Obviously, a gram-based mole of carbon (12.0 grams) is not the same quantity as a pound-based mole of carbon (12.0 pounds). Although "mol" is understood in SI countries to mean a gram-mole, the term *mole* is ambiguous, and the units mol (gmol), kmol (kgmol), or lbmol must be specified.[8]

One gram-mole of any substance has the same number of particles (atoms, molecules, ions, electrons, etc.), 6.022×10^{23}, *Avogadro's number*, N_A. A pound-mole contains approximately 454 times the number of particles in a gram-mole.

Avogadro's law (hypothesis) holds that equal volumes of all gases at the same temperature and pressure contain equal numbers of gas molecules. Specifically, at standard scientific conditions (1.0 atm and 0°C), one gram-mole of any gas contains 6.022×10^{23} molecules and occupies 22.4 L. A pound-mole occupies 454 times that volume, 359 ft^3.

"Molar" is used as an adjective when describing properties of a mole. For example, a *molar volume* is the volume of a mole.

Example 3.2

How many electrons are in 0.01 g of gold? ($A = 196.97$; $Z = 79$.)

---

[8]There are also variations on the presentation of these units, such as g mol, gmole, g-mole, kmole, kg-mol, lb-mole, pound-mole, and p-mole. In most cases, the intent is clear.

Solution

The number of gram-moles of gold present is

$$n = \frac{m}{M} = \frac{0.01 \text{ g}}{196.97 \frac{\text{g}}{\text{mol}}} = 5.077 \times 10^{-5} \text{ mol}$$

The number of gold nuclei is

$$N = nN_A = (5.077 \times 10^{-5} \text{ mol})\left(6.022 \times 10^{23} \frac{\text{nuclei}}{\text{mol}}\right)$$

$$= 3.057 \times 10^{19} \text{ nuclei}$$

Since the atomic number is 79, there are 79 protons and 79 electrons in each gold atom. The number of electrons is

$$N_{\text{electrons}} = (3.057 \times 10^{19})(79) = 2.42 \times 10^{21}$$

8. FORMULA AND MOLECULAR WEIGHTS

The *formula weight*, FW, of a molecule (compound) is the sum of the atomic weights of all elements in the molecule. The *molecular weight*, M, is generally the same as the formula weight. The units of molecular weight are g/mol, kg/kmol, or lbm/lbmol. However, units are sometimes omitted because weights are relative. For example,

$$\text{CaCO}_3: \text{FW} = M = 40.1 + 12 + 3 \times 16 = 100.1$$

An *ultimate analysis* (which determines how much of each element is present in a compound) will not necessarily determine the molecular formula. It will determine only the formula weight based on the relative proportions of each element. Therefore, except for hydrated molecules and other linked structures, the molecular weight will be an integer multiple of the formula weight.

For example, an ultimate analysis of hydrogen peroxide (H_2O_2) will show that the compound has one oxygen atom for each hydrogen atom. In this case, the formula would be assumed to be HO and the formula weight would be approximately 17, although the actual molecular weight is 34.

For *hydrated molecules* (e.g., $\text{FeSO}_4 \cdot 7\text{H}_2\text{O}$), the mass of the *water of hydration* (also known as the *water of crystallization*) is included in the formula and in the molecular weight.

9. EQUIVALENT WEIGHT

The *equivalent weight*, EW (i.e., an *equivalent*), is the amount of substance (in grams) that supplies one gram-mole (i.e., 6.022×10^{23}) of reacting units. For acid-base reactions, an acid equivalent supplies one gram-mole of H$^+$ ions. A base equivalent supplies one gram-mole of OH$^-$ ions. In oxidation-reduction reactions, an equivalent of a substance gains or loses a gram-mole of electrons. Similarly, in electrolysis reactions an equivalent weight is the weight of substance that either receives or donates one gram-mole of electrons at an electrode.

The equivalent weight can be calculated as the molecular weight divided by the change in oxidation number experienced by a compound in a chemical reaction. A compound can have several equivalent weights.

$$\text{EW} = \frac{M}{\Delta \text{ oxidation number}} \qquad 3.2$$

Example 3.3

What are the equivalent weights of the following compounds?

(a) Al in the reaction

$$\text{Al}^{3+} + 3\text{e}^- \rightarrow \text{Al}$$

(b) H_2SO_4 in the reaction

$$\text{H}_2\text{SO}_4 + \text{H}_2\text{O} \rightarrow 2\text{H}^+ + \text{SO}_4^{2-} + \text{H}_2\text{O}$$

(c) NaOH in the reaction

$$\text{NaOH} + \text{H}_2\text{O} \rightarrow \text{Na}^+ + \text{OH}^- + \text{H}_2\text{O}$$

Solution

(a) The atomic weight of aluminum is approximately 27. Since the change in the oxidation number is 3, the equivalent weight is $27/3 = 9$.

(b) The molecular weight of sulfuric acid is approximately 98. Since the acid changes from a neutral molecule to ions with two charges each, the equivalent weight is $98/2 = 49$.

(c) Sodium hydroxide has a molecular weight of approximately 40. The originally neutral molecule goes to a singly charged state. Therefore, the equivalent weight is $40/1 = 40$.

10. GRAVIMETRIC FRACTION

The *gravimetric fraction*, x_i, of an element i in a compound is the fraction by weight m of that element in the compound. The gravimetric fraction is found from an *ultimate analysis* (also known as a *gravimetric analysis*) of the compound.

$$x_i = \frac{m_i}{m_1 + m_2 + \cdots + m_i + \cdots + m_n} = \frac{m_i}{m_t} \qquad 3.3$$

The *percentage composition* is the gravimetric fraction converted to percentage.

$$\% \text{ composition} = x_i \times 100\% \qquad 3.4$$

If the gravimetric fractions are known for all elements in a compound, the *combining weights* of each element can be calculated. (The term *weight* is used even though mass is the traditional unit of measurement.)

$$m_i = x_i m_t \qquad 3.5$$

11. EMPIRICAL FORMULA DEVELOPMENT

It is relatively simple to determine the *empirical formula* of a compound from the atomic and combining weights of elements in the compound. The empirical formula gives the relative number of atoms (i.e., the formula weight is calculated from the empirical formula).

step 1: Divide the gravimetric fractions (or percentage compositions) by the atomic weight of each respective element.

step 2: Determine the smallest ratio from step 1.

step 3: Divide all of the ratios from step 1 by the smallest ratio.

step 4: Write the chemical formula using the results from step 3 as the numbers of atoms. Multiply through as required to obtain all integer numbers of atoms.

Example 3.4

A clear liquid is analyzed, and the following gravimetric percentage compositions are recorded: carbon, 37.5%; hydrogen, 12.5%; oxygen, 50%. What is the chemical formula for the liquid?

Solution

step 1: Divide the percentage compositions by the atomic weights.

$$C: \frac{37.5}{12} = 3.125$$

$$H: \frac{12.5}{1} = 12.5$$

$$O: \frac{50}{16} = 3.125$$

step 2: The smallest ratio is 3.125.

step 3: Divide all ratios by 3.125.

$$C: \frac{3.125}{3.125} = 1$$

$$H: \frac{12.5}{3.125} = 4$$

$$O: \frac{3.125}{3.125} = 1$$

step 4: The empirical formula is CH_4O.

If it had been known that the liquid behaved as though it contained a hydroxyl (OH) radical, the formula would have been written as CH_3OH. This is recognized as methyl alcohol.

12. CHEMICAL REACTIONS

During chemical reactions, bonds between atoms are broken and new bonds are usually formed. The starting substances are known as *reactants*; the ending substances are known as *products*. In a chemical reaction, reactants are either converted to simpler products or synthesized into more complex compounds. There are four common types of reactions.

- *direct combination* (or *synthesis*): This is the simplest type of reaction where two elements or compounds combine directly to form a compound.

$$2H_2 + O_2 \rightarrow 2H_2O$$
$$SO_2 + H_2O \rightarrow H_2SO_3$$

- *decomposition* (or *analysis*): Bonds within a compound are disrupted by heat or other energy to produce simpler compounds or elements.

$$2HgO \rightarrow 2Hg + O_2$$
$$H_2CO_3 \rightarrow H_2O + CO_2$$

- *single displacement* (or *replacement*[9]): This type of reaction has one element and one compound as reactants.

$$2Na + 2H_2O \rightarrow 2NaOH + H_2$$
$$2KI + Cl_2 \rightarrow 2KCl + I_2$$

- *double displacement* (or *replacement*): These are reactions with two compounds as reactants and two compounds as products.

$$AgNO_3 + NaCl \rightarrow AgCl + NaNO_3$$
$$H_2SO_4 + ZnS \rightarrow H_2S + ZnSO_4$$

---

[9]Another name for replacement is *metathesis*.

13. BALANCING CHEMICAL EQUATIONS

The coefficients in front of element and compound symbols in chemical reaction equations are the numbers of molecules or moles taking part in the reaction. (For gaseous reactants and products, the coefficients also represent the numbers of volumes. This is a direct result of Avogadro's hypothesis that equal numbers of molecules in the gas phase occupy equal volumes under the same conditions.)[10]

Since atoms cannot be changed in a normal chemical reaction (i.e., mass is conserved), the numbers of each element must match on both sides of the equation. When the numbers of each element match, the equation is said to be "balanced." The total atomic weights on both sides of the equation will be equal when the equation is balanced.

Balancing simple chemical equations is largely a matter of deductive trial and error. More complex reactions require use of oxidation numbers.

Example 3.5

Balance the following reaction equation.

$$Al + H_2SO_4 \rightarrow Al_2(SO_4)_3 + H_2$$

Solution

As written, the reaction is not balanced. For example, there is one aluminum on the left, but there are two on the right. The starting element in the balancing procedure is chosen somewhat arbitrarily. [Common Chemicals in Water and Wastewater Processing]

step 1: Since there are two aluminums on the right, multiply Al by 2.

$$2Al + H_2SO_4 \rightarrow Al_2(SO_4)_3 + H_2$$

step 2: Since there are three sulfate radicals (SO_4) on the right, multiply H_2SO_4 by 3.

$$2Al + 3H_2SO_4 \rightarrow Al_2(SO_4)_3 + H_2$$

step 3: Now there are six hydrogens on the left, so multiply H_2 by 3 to balance the equation.

$$2Al + 3H_2SO_4 \rightarrow Al_2(SO_4)_3 + 3H_2$$

14. STOICHIOMETRIC REACTIONS

Stoichiometry is the study of the proportions in which elements and compounds react and are formed. A *stoichiometric reaction* (also known as a *perfect reaction* or

an *ideal reaction*) is one in which just the right amounts of reactants are present. After the reaction stops, there are no unused reactants.

Stoichiometric problems are known as *weight and proportion problems* because their solutions use simple ratios to determine the masses of reactants required to produce given masses of products, or vice versa. The procedure for solving these problems is essentially the same regardless of the reaction.

step 1: Write and balance the chemical equation.

step 2: Determine the atomic (molecular) weight of each element (compound) in the equation.

step 3: Multiply the atomic (molecular) weights by their respective coefficients and write the products under the formulas.

step 4: Write the given mass data under the weights determined in step 3.

step 5: Fill in the missing information by calculating simple ratios.

Example 3.6

Caustic soda (NaOH) is made from sodium carbonate (Na_2CO_3) and slaked lime ($Ca(OH)_2$) according to the given reaction. How many kilograms of caustic soda can be made from 2000 kg of sodium carbonate?

Solution

	Na_2CO_3	$+ Ca(OH)_2$	$\rightarrow 2NaOH$	$+ CaCO_3$
molecular weights	106	74	2×40	100
given data	2000 kg		m kg	

The simple ratio used is

$$\frac{NaOH}{Na_2CO_3} = \frac{80}{106} = \frac{m}{2000 \text{ kg}}$$

Solving for the unknown mass, $m = 1509$ kg.

15. NONSTOICHIOMETRIC REACTIONS

In many cases, it is not realistic to assume a stoichiometric reaction because an excess of one or more reactants is necessary to assure that all of the remaining reactants take part in the reaction. Combustion is an example where the stoichiometric assumption is, more often than not, invalid. Excess air is generally needed to ensure that all of the fuel is burned.

---

[10]When water is part of the reaction, the interpretation that the coefficients are volumes is valid only if the reaction takes place at a high enough temperature to vaporize the water.

With nonstoichiometric reactions, the reactant that is used up first is called the *limiting reactant*. The amount of product will be dependent on (limited by) the limiting reactant.

The *theoretical yield* or *ideal yield* of a product is the maximum mass of product per unit mass of limiting reactant that can be obtained from a given reaction if the reaction goes to completion. The *percentage yield* is a measure of the efficiency of the actual reaction.

$$\text{percentage yield} = \frac{\text{actual yield} \times 100\%}{\text{theoretical yield}} \quad 3.6$$

16. SOLUTIONS OF GASES IN LIQUIDS

When a liquid is exposed to a gas, a small amount of the gas will dissolve in the liquid. Diffusion alone is sufficient for this to occur; bubbling or collecting the gas over a liquid is not necessary. Given enough time, at equilibrium, the concentration of the gas will reach a maximum known as the *saturation concentration*.

Due to the large amount of liquid compared to the small amount of dissolved gas, a liquid exposed to multiple gases will eventually become saturated by all of the gases; the presence of one gas does not affect the solubility of another gas.

The characteristics of a solution of one or more gases in a liquid is predicted by *Henry's law*. In one formulation specifically applicable to liquids exposed to mixtures of gases, Henry's law states that, at equilibrium, the partial pressure, P_i, of a gas in a mixture will be proportional to the gas mole fraction, x_i, of that dissolved gas in solution.

In Eq. 3.7, Henry's law constant, h, has units of pressure, typically reported in the literature in atmospheres (same as atm/mole fraction). (See Table 3.2.) P is total pressure, and y_i is the mole fraction of component i in the gas mixture.

Henry's Law at Constant Temperature
$$P_i = Py_i = hx_i \quad 3.7$$

Since, for mixtures of ideal gases, the mole fraction, x_i, volumetric fraction, B_i, and partial pressure fraction, P_i/P, all have the same numerical values, these measures can all be integrated into Henry's law.

Table 3.2 *Approximate Values of Henry's Law Constant (solutions of gases in water)*

gas	Henry's law constant, h (atm) (Multiply all values by 10^3.)	
	20°C	30°C
CO	53.6	62.0
CO_2	1.42	1.86
H_2S	48.3	60.9
N_2	80.4	92.4
NO	26.4	31.0
O_2	40.1	47.5
SO_2	0.014	0.016

Adapted from *Scrubber Systems Operational Review* (APTI Course SI:412C), Second Edition, *Lesson 11: Design Review of Absorbers Used for Gaseous Pollutants*, 1998, North Carolina State University for the U.S. Environmental Protection Agency.

Henry's law is stated in several incompatible formulations, and the corresponding equations and Henry's law constants are compatible only with their own formulations. Also, a variety of variables are used for Henry's law constant, including H, k_H, K_H, and so on. The context and units of the Henry's law constant must be used to determine Henry's law.

equation	typical units of Henry's law constant	Henry's law statement (at equilibrium)
$P_i = hx_i$	h: pressure (atm)	Partial pressure is proportional to mole fraction.
$P_i = k_{h(p/C),i}C_i$	$k_{H(p/C)}$: pressure divided by concentration (atm·L/mol; atm·L/mg)	Partial pressure is proportional to concentration.*
$P_i = \dfrac{C_i}{k_{h(C/p),i}}$	$k_{H(C/p)}$: concentration divided by pressure (mol/atm·L; mg/atm·L)	
$C_{i,\text{gas}} = \alpha_i C_{i,\text{liquid}}$	α: dimensionless ($L_{\text{gas}}/L_{\text{liquid}}$)	Concentration in the gas mixture is proportional to concentration in the liquid solution.

*The statements of Henry's law are the same. However, the values of Henry's law constants are inverses.

The dimensionless form of the Henry's law constant is also known as the *absorption coefficient*, *coefficient of absorption*, and *solubility coefficient*. It represents the volume of a gas at a specific temperature and pressure that can be dissolved in a unit volume of liquid.

Typical units are L/L (dimensionless). Approximate values for gases in water at 1 atm and 20°C are: CO, 0.023; CO_2, 0.88; He, 0.009; H_2, 0.017; H_2S, 2.62; N_2, 0.015; NH_3, 710; O_2, 0.028.

The amount of gas dissolved in a liquid varies with the temperature of the liquid and the concentration of dissolved salts in the liquid. Generally, the solubility of gases in liquids decreases with increasing temperature.

Tables are available that list the saturation values of dissolved oxygen in water at various temperatures and for various amounts of chloride ion (also referred to as *salinity*). [Dissolved-Oxygen Concentration in Water]

Example 3.7

At 20°C and 1 atm, 1 L of water can absorb 0.043 g of oxygen and 0.017 g of nitrogen. Atmospheric air is 20.9% oxygen by volume, and the remainder is assumed to be nitrogen. What masses of oxygen and nitrogen will be absorbed by 1 L of water exposed to 20°C air at 1 atm?

Solution

Since partial pressure is volumetrically weighted,

$$m_{\text{oxygen}} = (0.209)\left(0.043 \ \frac{\text{g}}{\text{L}}\right)$$
$$= 0.009 \ \text{g/L}$$
$$m_{\text{nitrogen}} = (1.000 - 0.209)\left(0.017 \ \frac{\text{g}}{\text{L}}\right)$$
$$= 0.0134 \ \text{g/L}$$

Example 3.8

At an elevation of 4000 ft, the barometric pressure is 660 mm Hg. What is the dissolved oxygen concentration of 18°C water with a 800 mg/L chloride concentration at that elevation?

Solution

From a table of values, calculate oxygen's saturation concentration for 18°C water corrected for a 800 mg/L chloride concentration. [Dissolved-Oxygen Concentration in Water]

$$C_s = 9.45 \ \frac{\text{mg}}{\text{L}} - \left(\frac{800 \ \frac{\text{mg}}{\text{L}}}{100 \ \frac{\text{mg}}{\text{L}}}\right)\left(0.0083 \ \frac{\text{mg}}{\text{L}}\right)$$
$$= 9.384 \ \text{mg/L}$$

Correct for the barometric pressure.

$$C_s' = C_s\left(\frac{P - p}{760 \ \text{mm} - p}\right)$$
$$= \left(9.384 \ \frac{\text{mg}}{\text{L}}\right)\left(\frac{660 \ \text{mm} - 15.49 \ \text{mm}}{760 \ \text{mm} - 15.49 \ \text{mm}}\right)$$
$$= 8.12 \ \text{mg/L}$$

17. PROPERTIES OF SOLUTIONS

There are very few convenient ways of predicting the properties of nonreacting, nonvolatile organic and aqueous solutions (acids, brines, alcohol mixtures, coolants, etc.) from the individual properties of the components.

Volumes of two nonreacting organic liquids (e.g., acetone and chloroform) in a mixture are essentially additive. The volume change upon mixing will seldom be more than a few tenths of a percent. The volume change in aqueous solutions is often slightly greater, but is still limited to a few percent (e.g., 3% for some solutions of methanol and water). Therefore, the specific gravity (density, specific weight, etc.) can be considered to be a volumetric weighting of the individual specific gravities.

Most other fluid properties of aqueous solutions, such as viscosity, compressibility, surface tension, and vapor pressure, must be measured.

18. SOLUTIONS OF SOLIDS IN LIQUIDS

When a solid is added to a liquid, the solid is known as the *solute*, and the liquid is known as the *solvent*.[11] If the dispersion of the solute throughout the solvent is at the molecular level, the mixture is known as a *solution*. If the solute particles are larger than molecules, the mixture is known as a *suspension*.[12]

In some solutions, the solvent and solute molecules bond loosely together. This loose bonding is known as *solvation*. If water is the solvent, the bonding process is also known as *aquation* or *hydration*.

The solubility of most solids in liquid solvents usually increases with increasing temperature. Pressure has very little effect on the solubility of solids in liquids.

When the solvent has absorbed as much solute as it can, it is a *saturated solution*.[13] Adding more solute to an already saturated solution will cause the excess solute to settle to the bottom of the container, a process known as *precipitation*. Other changes (in temperature, concentration, etc.) can be made to cause precipitation from saturated and unsaturated solutions. Precipitation in a chemical reaction is indicated by a downward arrow

---

[11]The term *solvent* is often associated with volatile liquids, but the term is more general than that. (A *volatile liquid* evaporates rapidly and readily at normal temperatures.) Water is the solvent in aqueous solutions.
[12]An *emulsion* is not a mixture of a solid in a liquid. It is a mixture of two immiscible liquids.
[13]Under certain circumstances, a *supersaturated solution* can exist for a limited amount of time.

(i.e., "↓"). For example, the precipitation of silver chloride from an aqueous solution of silver nitrate ($AgNO_3$) and potassium chloride (KCl) would be written as

$$AgNO_3(aq) + KCl(aq) \rightarrow AgCl(s){\downarrow} + KNO_3(aq)$$

19. UNITS OF CONCENTRATION

Several units of concentration are commonly used to express solution strengths.

F— *formality:* The number of gram formula weights (i.e., molecular weights in grams) per liter of solution.

m— *molality:* The number of gram-moles of solute per 1000 grams of solvent. A "molal" solution contains 1 gram-mole per 1000 grams of solvent.

M— *molarity:* The number of gram-moles of solute per liter of solution. A "molar" (i.e., 1 M) solution contains 1 gram-mole per liter of solution. Molarity is related to normality as shown in Eq. 3.8.

$$N = M \times \Delta \,\text{oxidation number} \qquad 3.8$$

N— *normality:* The number of gram equivalent weights of solute per liter of solution. A solution is "normal" (i.e., 1 N) if there is exactly one gram equivalent weight per liter of solution.

x— *mole fraction:* The number of moles of solute divided by the number of moles of solvent and all solutes.

meq/L— *milligram equivalent weights of solute per liter of solution:* calculated by multiplying normality by 1000 or dividing concentration in mg/L by equivalent weight.

mg/L— *milligrams per liter:* The number of milligrams of solute per liter of solution. Same as ppm for solutions of water.

ppm— *parts per million:* The number of pounds (or grams) of solute per million pounds (or grams) of solution. Same as mg/L for solutions of water.

ppb— *parts per billion:* The number of pounds (or grams) of solute per billion (10^9) pounds (or grams) of solution. Same as $\mu g/L$ for solutions of water.

For compounds whose molecules do not dissociate in solution (e.g., table sugar), there is no difference between molarity and formality. There is a difference, however, for compounds that dissociate into ions (e.g., table salt). Consider a solution derived from 1 gmol of magnesium nitrate $Mg(NO_3)_2$ in enough water to bring the volume to 1 L. The formality is 1 F (i.e., the solution is 1 formal). However,

3 mol of ions will be produced: 1 mol of Mg^{2+} ions and 2 moles of NO_3^- ions. Therefore, molarity is 1 M for the magnesium ion and 2 M for the nitrate ion.

The use of formality avoids the ambiguity in specifying concentrations for ionic solutions. Also, the use of formality avoids the problem of determining a molecular weight when there are no discernible molecules (e.g., as in a crystalline solid such as NaCl). Unfortunately, the distinction between molarity and formality is not always made, and molarity may be used as if it were formality.

Example 3.9

A solution is made by dissolving 0.353 g of $Al_2(SO_4)_3$ in 730 g of water. If ionization is 100%, what is the concentration expressed as normality, molarity, and mg/L?

Solution

The molecular weight of $Al_2(SO_4)_3$ is

$$M = (2)\left(26.98 \,\frac{\text{g}}{\text{mol}}\right) + (3)\left(32.06 \,\frac{\text{g}}{\text{mol}}\right)$$
$$+ (4)(3)\left(16 \,\frac{\text{g}}{\text{mol}}\right)$$
$$= 342.14 \,\text{g/mol}$$

Either the aluminum or sulfate ion can be used to determine the net charge transfer (i.e., the oxidation number). Since each aluminum ion has a charge of 3, and since there are two aluminum ions in the molecule, the oxidation number is $(3)(2) = 6$.

The equivalent weight is

$$\text{EW} = \frac{M}{\text{oxidation number}} = \frac{342.14 \,\frac{\text{g}}{\text{mol}}}{6}$$
$$= 57.02 \,\text{g/mol}$$

The number of gram equivalent weights used is

$$\frac{0.353 \,\text{g}}{57.02 \,\frac{\text{g}}{\text{mol}}} = 6.19 \times 10^{-3} \,\text{GEW}$$

The volume of solution (same as the solvent volume if the small amount of solute is neglected) is 0.73 L.

The normality is

$$N = \frac{6.19 \times 10^{-3} \,\text{GEW}}{0.73 \,\text{L}} = 8.48 \times 10^{-3}$$

The number of moles of solute used is

$$\frac{0.353 \text{ g}}{342.14 \frac{\text{g}}{\text{mol}}} = 1.03 \times 10^{-3} \text{ mol}$$

The molarity is

$$M = \frac{1.03 \times 10^{-3} \text{ mol}}{0.73 \text{ L}} = 1.41 \times 10^{-3} \text{ mol/L}$$

The concentration is

$$C = \frac{m}{V} = \frac{(0.353 \text{ g})\left(1000 \frac{\text{mg}}{\text{g}}\right)}{0.73 \text{ L}} = 483.6 \text{ mg/L}$$

20. pH AND pOH

A standard measure of the strength of an acid or base is the number of hydrogen or hydroxide ions in a liter of solution. Since these are very small numbers, a logarithmic scale is used.

Acids, Bases, and pH

$$\text{pH} = -\log_{10}[\text{H}^+] = \log_{10}\frac{1}{[\text{H}^+]} \qquad 3.9$$

$$\text{pOH} = -\log_{10}[\text{OH}^-] = \log_{10}\frac{1}{[\text{OH}^-]} \qquad 3.10$$

The quantities $[\text{H}^+]$ and $[\text{OH}^-]$ in square brackets are the *ionic concentrations* in moles of ions per liter. The number of moles can be calculated from Avogadro's law by dividing the actual number of ions per liter by 6.022×10^{23}. Alternatively, for a partially ionized compound in a solution of known molarity, M, the ionic concentration is

$$[\text{ion}] = XM \qquad 3.11$$

A *neutral solution* has a pH of 7. Solutions with a pH below 7 are acidic; the smaller the pH, the more acidic the solution. Solutions with a pH above 7 are basic.

The relationship between pH and pOH is

$$\text{pH} + \text{pOH} = 14 \qquad 3.12$$

Example 3.10

A 4.2% ionized 0.01M ammonia solution is prepared from ammonium hydroxide (NH_4OH). Calculate the pH, pOH, and concentrations of $[\text{H}^+]$ and $[\text{OH}^-]$.

Solution

From Eq. 3.11,

$$[\text{OH}^-] = XM = (0.042)(0.01)$$
$$= 4.2 \times 10^{-4} \text{ mol/L}$$

From Eq. 3.10,

$$\text{pOH} = -\log[\text{OH}^-] = -\log(4.2 \times 10^{-4})$$
$$= 3.38$$

From Eq. 3.12,

$$\text{pH} = 14 - \text{pOH} = 14 - 3.38$$
$$= 10.62$$

The $[\text{H}^+]$ ionic concentration can be extracted from the definition of pH.

$$[\text{H}^+] = 10^{-\text{pH}} = 10^{-10.62}$$
$$= 2.4 \times 10^{-11} \text{ mol/L}$$

21. BUFFERS

A *buffer solution* resists changes in acidity and maintains a relatively constant pH when a small amount of an acid or base is added to it. Buffers are usually combinations of weak acids and their salts. A buffer is most effective when the acid and salt concentrations are equal.

22. NEUTRALIZATION

Acids and bases neutralize each other to form water.

$$\text{H}^+ + \text{OH}^- \rightarrow \text{H}_2\text{O}$$

Assuming 100% ionization of the solute, the volumes, V, required for complete neutralization can be calculated from the normalities, N, or the molarities, M.

$$V_b N_b = V_a N_a \qquad 3.13$$

$$V_b M_b \Delta_{b,\text{charge}} = V_a M_a \Delta_{a,\text{charge}} \qquad 3.14$$

23. REVERSIBLE REACTIONS

Reversible reactions are capable of going in either direction and do so to varying degrees (depending on the concentrations and temperature) simultaneously. These reactions are characterized by the simultaneous presence of all reactants and all products. For example, the chemical equation for the exothermic formation of ammonia from nitrogen and hydrogen is

$$N_2 + 3H_2 \rightleftharpoons 2NH_3 \quad (\Delta H = -92.4 \text{ kJ})$$

At *chemical equilibrium*, reactants and products are both present. Concentrations of the reactants and products do not change after equilibrium is reached.

24. LE CHATELIER'S PRINCIPLE

Le Châtelier's principle predicts the direction in which a reversible reaction initially at equilibrium will go when some condition (e.g., temperature, pressure, concentration) is "stressed" (i.e., changed). The principle says that when an equilibrium state is stressed by a change, a new equilibrium is formed that reduces that stress.

Consider the formation of ammonia from nitrogen and hydrogen. (See Sec. 3.23.) When the reaction proceeds in the forward direction, energy in the form of heat is released and the temperature increases. If the reaction proceeds in the reverse direction, heat is absorbed and the temperature decreases. If the system is stressed by increasing the temperature, the reaction will proceed in the reverse direction because that direction absorbs heat and reduces the temperature.

For reactions that involve gases, the reaction equation coefficients can be interpreted as volumes. In the nitrogen-hydrogen reaction, four volumes combine to form two volumes. If the equilibrium system is stressed by increasing the pressure, then the forward reaction will occur because this direction reduces the volume and pressure.[14]

If the concentration of any participating substance is increased, the reaction proceeds in a direction away from the substance with the increase in concentration. (For example, an increase in the concentration of the reactants shifts the equilibrium to the right, increasing the amount of products formed.)

The *common ion effect* is a special case of Le Châtelier's principle. If a salt containing a common ion is added to a solution of a weak acid, almost all of the salt will dissociate, adding large quantities of the common ion to the solution. Ionization of the acid will be greatly suppressed, a consequence of the need to have an unchanged equilibrium constant.

25. IRREVERSIBLE REACTION KINETICS

The rate at which a compound is formed or used up in an irreversible (one-way) reaction is known as the *rate of reaction*, *speed of reaction*, *reaction velocity*, and so on. The rate, r_A, is the change in concentration per unit time, usually measured in mol/L·s.

Site Assessment and Remediation: Nomenclature

$$-r_A = -\frac{1}{V}\frac{dN_A}{dt} \qquad \textit{3.15}$$

When V is constant,

Site Assessment and Remediation: Nomenclature

$$-r_A = \frac{-dC_A}{dt} \qquad \textit{3.16}$$

According to the *law of mass action*, the rate of reaction varies with the concentrations of the reactants and products. Specifically, the rate is proportional to the molar concentrations (i.e., the molarities). The rate of the formation or conversion of substance A is represented in various forms, such as r_A, dA/dt, and $d[A]/dt$, where the variable A or [A] can represent either the mass or the concentration of substance A. Substance A can be either a pure element or a compound.

The rate of reaction is generally not affected by pressure, but it does depend on five other factors.

- *type of substances in the reaction:* Some substances are more reactive than others.

- *exposed surface area:* The rate of reaction is proportional to the amount of contact between the reactants.

- *concentrations:* The rate of reaction increases with increases in concentration.

- *temperature:* The rate of reaction approximately doubles with every 10°C increase in temperature.

- *catalysts:* If a catalyst is present, the rate of reaction increases. However, the equilibrium point is not changed. (A catalyst is a substance that increases the reaction rate without being consumed in the reaction.)

26. ORDER OF THE REACTION

The *order of the reaction* is the total number of reacting molecules in or before the slowest step in the process.[15] The order must be determined experimentally. However, for an irreversible elementary reaction, the order is usually assumed from the stoichiometric reaction equation as the

---

[14]The exception to this rule is the addition of an inert or nonparticipating gas to a gaseous equilibrium system. Although there is an increase in total pressure, the position of the equilibrium is not affected.
[15]This definition is valid for elementary reactions. For complex reactions, the order is an empirical number that need not be an integer.

Table 3.3 Reaction Rates and Half-Life Equations

reaction	order	rate equation	integrated forms
$A \to B$	zero	$\dfrac{dC_A}{dt} = -k_0$	Zero-Order Irreversible Reaction Kinetics $C_A = C_{A0} - kt$ $t_{1/2} = \dfrac{C_{A0}}{2k_0}$
$A \to B$	first	$\dfrac{dC_A}{dt} = -kC_A$	First-Order Irreversible Reaction Kinetics $\ln\left(C_A/C_{A0}\right) = -kt$ $t_{1/2} = \dfrac{1}{-k}\ln 2$
$A + A \to P$	second, type I	$\dfrac{dC_A}{dt} = -k_2[A]^2$	Second-Order Irreversible Reaction Kinetics $1/C_A - 1/C_{A0} = kt$ $t_{1/2} = \dfrac{1}{k_2 C_{A0}}$
$aA + bB \to P$	second, type II	$\dfrac{dC_A}{dt} = -k_2 C_A[B]$	$\ln\dfrac{C_{A0} - [B]}{[B]_0 - \left(\dfrac{b}{a}\right)[X]} = \ln\dfrac{C_A}{[B]}$ $= \left(\dfrac{bC_{A0} - a[B]_0}{a}\right)k_2 t + \ln\dfrac{C_{A0}}{[B]_0}$ $t_{1/2} = \left(\dfrac{a}{k_2\left(bC_{A0} - a[B]_0\right)}\right)\ln\left(\dfrac{a[B]_0}{2a[B]_0 - bC_{A0}}\right)$

sum of the combining coefficients for the reactants.[16,17] For example, for the reaction $m\,A + n\,B \to p\,C$, the overall order of the forward reaction is assumed to be $m + n$.

Many reactions (e.g., dissolving metals in acid or the evaporation of condensed materials) have *zero-order reaction rates*. These reactions do not depend on the concentrations or temperature at all, but rather, are affected by other factors such as the availability of reactive surfaces or the absorption of radiation. The formation (conversion) rate of a compound in a zero-order reaction is constant. That is, $dA/dt = -k_0$. k_0 is known as the *reaction rate constant*. (The subscript "0" refers to the zero-order.) Since the concentration (amount) of the substance decreases with time, dA/dt is negative. Since the negative sign is explicit in rate equations, the reaction rate constant is generally reported as a positive number.

Once a reaction rate equation is known, it can be integrated to obtain an expression for the concentration (mass) of the substance at various times. The time for half of the substance to be formed (or converted) is the *half-life*, $t_{1/2}$. Table 3.3 contains reaction rate equations and half-life equations for various types of low-order reactions.

Example 3.11

Nitrogen pentoxide decomposes according to the following first-order reaction.

$$N_2O_5 \to 2NO_2 + \tfrac{1}{2}O_2$$

At a particular temperature, the decomposition of nitrogen pentoxide is 85% complete after 11 min. What is the reaction rate constant?

---

[16] The overall order of the reaction is the sum of the orders with respect to the individual reactants. For example, in the reaction $2NO + O_2 \to 2NO_2$, the reaction is second order with respect to NO, first order with respect to O_2, and third order overall.

[17] In practice, the order of the reaction must be known, given, or determined experimentally. It is not always equal to the sum of the combining coefficients for the reactants. For example, in the reaction $H_2 + I_2 \to 2HI$, the overall order of the reaction is indeed 2, as expected. However, in the reaction $H_2 + Br_2 \to 2HBr$, the overall order is found experimentally to be 3/2, even though the two reactions have the same stoichiometry, and despite the similarities of iodine and bromine.

Solution

The reaction is given as first order. Use the integrated reaction rate equation from Table 3.3. Since the decomposition reaction is 85% complete, the surviving fraction is 15% (0.15).

First-Order Irreversible Reaction Kinetics

$$\ln(C_A / C_{A0}) = -kt$$

$$\ln(0.15) = -k(11 \text{ min})$$

$$-k = -0.172 \text{ 1/min} \quad (0.172 \text{ 1/min})$$

(The rate constant is reported as a positive number.)

27. REVERSIBLE REACTION KINETICS

Consider the following reversible reaction.

Site Assessment and Remediation: Nomenclature

$$aA + bB \rightleftharpoons cC + dD \qquad 3.17$$

In Eq. 3.17 and Eq. 3.18, the *reaction rate constants* are $k_{forward}$ and $k_{reverse}$. The order of the forward reaction is $a + b$; the order of the reverse reaction is $c + d$.

$$v_{forward} = k_{forward}[A]^a [B]^b \qquad [SI] \quad 3.18(a)$$

$$v_{reverse} = k_{reverse}[C]^c [D]^d \qquad [U.S.] \quad 3.18(b)$$

At equilibrium, the forward and reverse speeds of reaction are equal.

$$v_{forward} = v_{reverse}|_{equilibrium} \qquad 3.19$$

28. EQUILIBRIUM CONSTANT

For reversible reactions, the *equilibrium constant, K*, is proportional to the ratio of the reverse rate of reaction to the forward rate of reaction.[18] Except for catalysis, the equilibrium constant depends on the same factors affecting the reaction rate. For the complex reversible reaction given by Eq. 3.17, the equilibrium constant is given by the *law of mass action*.

$$K_a = \frac{(\hat{a}_C^c)(\hat{a}_D^d)}{(\hat{a}_A^a)(\hat{a}_B^b)} = \prod_i (\hat{a}_i)^{r_i} \qquad 3.20$$

If any of the reactants or products are in pure solid or pure liquid phases, their concentrations are omitted from the calculation of the equilibrium constant. For example, in weak aqueous solutions, the concentration of water, H_2O, is very large and essentially constant; therefore, that concentration is omitted.

For gaseous reactants and products, the concentrations (i.e., the numbers of atoms) will be proportional to the partial pressures. Therefore, an equilibrium constant can be calculated directly from the partial pressures and is given the symbol K_p. For example, for the formation of ammonia gas from nitrogen and hydrogen, the equilibrium constant is [Chemical Reaction Equilibrium: For Mixtures of Ideal Gases]

$$K_p = \frac{[p_{NH_3}]^2}{[p_{N_2}][p_{H_2}]^3} \qquad 3.21$$

K and K_p are not numerically the same, but they are related by Eq. 3.22. Δn is the number of moles of products minus the number of moles of reactants.

$$K_p = K(R^* T)^{\Delta n} \qquad 3.22$$

Example 3.12

A particularly weak solution of acetic acid ($HC_2H_3O_2$) in water has the ionic concentrations (in mol/L) given. What is the equilibrium constant?

$$HC_2H_3O_2 + H_2O \rightleftharpoons H_3O^+ + C_2H_3O_2^-$$
$$[HC_2H_3O_2] = 0.09866$$
$$[H_2O] = 55.5555$$
$$[H_3O^+] = 0.00134$$
$$[C_2H_3O_2^-] = 0.00134$$

Solution

The concentration of the water molecules is not included in the calculation of the equilibrium or ionization constant. Therefore, the equilibrium constant is [Chemical Reaction Equilibrium]

$$K = K_a = \frac{[H_3O^+][C_2H_3O_2^-]}{[HC_2H_3O_2]} = \frac{(0.00134)(0.00134)}{0.09866}$$
$$= 1.82 \times 10^{-5}$$

29. IONIZATION CONSTANT

The equilibrium constant for a weak solution is essentially constant and is known as the *ionization constant* (also known as a *dissociation constant*). (See Table 3.4.) For weak acids, the symbol K_a and name *acid constant*

---

[18]The symbols K_c (in molarity units) and K_{eq} are occasionally used for the equilibrium constant.

Table 3.4 *Approximate Ionization Constants of Common Water Supply Chemicals*

substance	0°C	5°C	10°C	15°C	20°C	25°C
$Ca(OH)_2$						3.74×10^{-3}
$HClO$	2.0×10^{-8}	2.3×10^{-8}	2.6×10^{-8}	3.0×10^{-8}	3.3×10^{-8}	3.7×10^{-8}
$HC_2H_3O_2$	1.67×10^{-5}	1.70×10^{-5}	1.73×10^{-5}	1.75×10^{-5}	1.75×10^{-5}	1.75×10^{-5}
$HBrO$					$\approx 2 \times 10^{-9}$	
$H_2CO_3\ (K_1)$	2.6×10^{-7}	3.04×10^{-7}	3.44×10^{-7}	3.81×10^{-7}	4.16×10^{-7}	4.45×10^{-7}
$HClO_2$					$\approx 1.1 \times 10^{-2}$	
NH_3	1.37×10^{-5}	1.48×10^{-5}	1.57×10^{-5}	1.65×10^{-5}	1.71×10^{-5}	1.77×10^{-5}
NH_4OH						1.79×10^{-5}
water*	14.9435	14.7338	14.5346	14.3463	14.1669	13.9965

*$-\log_{10}K$ given

are used. For weak bases, the symbol K_b and the name *base constant* are used. For example, for the ionization of hydrocyanic acid,

$$HCN \rightleftharpoons H^+ + CN^-$$

$$K_a = \frac{[H^+][CN^-]}{[HCN]}$$

Pure water is itself a very weak electrolyte and ionizes only slightly.

$$2H_2O \rightleftharpoons H_3O^+ + OH^- \qquad \textbf{\textit{3.23}}$$

At equilibrium, the ionic concentrations are equal.

$$[H_3O^+] = 10^{-7}$$

$$[OH^-] = 10^{-7}$$

From Eq. 3.20, the ionization constant (*ion product*) for pure water is

$$K_w = K_{a,water} = [H_3O^+][OH^-] = (10^{-7})(10^{-7})$$

$$= 10^{-14} \qquad \textbf{\textit{3.24}}$$

If the molarity, M, and *fraction of ionization*, X, are known, the ionization constant can be calculated from Eq. 3.25.

$$K_{ionization} = \frac{MX^2}{1-X} \quad [K_a \text{ or } K_b] \qquad \textbf{\textit{3.25}}$$

The reciprocal of the ionization constant is the *stability constant* (*overall stability constant*), also known as the *formation constant*. Stability constants are used to describe complex ions that dissociate readily.

Example 3.13

A 0.1 molar (0.1 M) acetic acid solution is 1.34% ionized. Find the (a) hydrogen ion concentration, (b) acetate ion concentration, (c) un-ionized acid concentration, and (d) ionization constant.

Solution

(a) From Eq. 3.11, the hydrogen (hydronium) ion concentration is

$$[H_3O^+] = XM = (0.0134)(0.1)$$

$$= 0.00134 \text{ mol/L}$$

(b) Since every hydronium ion has a corresponding acetate ion, the acetate and hydronium ion concentrations are the same.

$$[C_2H_3O_2^-] = [H_3O^+] = 0.00134 \text{ mol/L}$$

(c) The concentration of un-ionized acid can be derived from Eq. 3.11.

$$[HC_2H_3O_2] = (1-X)M = (1-0.0134)(0.1)$$

$$= 0.09866 \text{ mol/L}$$

(d) The ionization constant is calculated from Eq. 3.25.

$$K_a = \frac{MX^2}{1-X} = \frac{(0.1)(0.0134)^2}{1-0.0134}$$

$$= 1.82 \times 10^{-5}$$

Example 3.14

The ionization constant for acetic acid is 1.82×10^{-5}. What is the hydrogen ion concentration for a 0.2 M solution?

Solution

From Eq. 3.25,

$$K_a = \frac{MX^2}{1-X}$$

$$1.82 \times 10^{-5} = \frac{0.2X^2}{1-X}$$

Since acetic acid is a weak acid, X is known to be small. Therefore, the computational effort can be reduced by assuming that $1 - X \approx 1$.

$$1.82 \times 10^{-5} = 0.2X^2$$
$$X = 9.54 \times 10^{-3}$$

From Eq. 3.11, the concentration of the hydrogen ion is

$$[H_3O^+] = XM = (9.54 \times 10^{-3})(0.2)$$
$$= 1.9 \times 10^{-3} \text{ mol/L}$$

Example 3.15

The ionization constant for acetic acid ($HC_2H_3O_2$) is 1.82×10^{-5}. What is the hydrogen ion concentration of a solution with 0.1 mol of 80% ionized ammonium acetate ($NH_4C_2H_3O_2$) in one liter of 0.1 M acetic acid?

Solution

The acetate ion ($C_2H_3O_2^-$) is a common ion, since it is supplied by both the acetic acid and the ammonium acetate. Both sources contribute to the ionic concentration. However, the ammonium acetate's contribution dominates. Since the acid dissociates into an equal number of hydrogen and acetate ions,

$$[C_2H_3O_2^-]_{\text{total}} = [C_2H_3O_2^-]_{\text{acid}}$$
$$+ [C_2H_3O_2^-]_{\text{ammonium acetate}}$$
$$= [H_3O^+] + (0.8)(0.1)$$
$$\approx (0.8)(0.1) = 0.08$$

As a result of the common ion effect and Le Châtelier's law, the acid's dissociation is essentially suppressed by the addition of the ammonium acetate. The concentration of un-ionized acid is

$$[HC_2H_3O_2] = 0.1 - [H_3O^+]$$
$$\approx 0.1$$

The ionization constant is unaffected by the number of sources of the acetate ion.

$$K_a = \frac{[H_3O^+][C_2H_3O_2^-]}{[HC_2H_3O_2]}$$

$$1.82 \times 10^{-5} = \frac{[H_3O^+](0.08)}{0.1}$$

$$[H_3O^+] = 2.3 \times 10^{-5} \text{ mol/L}$$

30. IONIZATION CONSTANTS FOR POLYPROTIC ACIDS

A *polyprotic acid* has as many ionization constants as it has acidic hydrogen atoms. For oxyacids (see Sec. 3.6), each successive ionization constant is approximately 10^5 times smaller than the preceding one. For example, phosphoric acid (H_3PO_4) has three ionization constants.

$$K_1 = 7.1 \times 10^{-3} \quad (H_3PO_4)$$
$$K_2 = 6.3 \times 10^{-8} \quad (H_2PO_4^-)$$
$$K_3 = 4.4 \times 10^{-13} \quad (HPO_4^{2-})$$

31. SOLUBILITY PRODUCT

When an ionic solid is dissolved in a solvent, it dissociates. For example, consider the ionization of silver chloride in water.

$$AgCl(s) \rightleftharpoons Ag^+(aq) + Cl^-(aq)$$

If the equilibrium constant is calculated, the terms for pure solids and liquids (in this case, [AgCl] and [H_2O]) are omitted. Therefore, the *solubility product*, K_{SP}, consists only of the ionic concentrations. As with the general case of ionization constants, the solubility product for slightly soluble solutes is essentially constant at a standard value.

Chemistry Definitions

$$K_{SP} = [A^+]^m [B^-]^n$$

3.26

When the product of terms exceeds the standard value of the solubility product, solute will precipitate out until the product of the remaining ion concentrations attains the standard value. If the product is less than the standard value, the solution is not saturated.

The solubility products of nonhydrolyzing compounds are relatively easy to calculate. (Example 3.16 demonstrates a method.) Such is the case for chromates (CrO_4^{2+}), halides (F^-, Cl^-, Br^-, I^-), sulfates (SO_4^{2+}), and iodates (IO_3^-). However, compounds that hydrolyze (i.e., combine with water molecules) must be treated differently. The method used in Ex. 3.16 cannot be used for hydrolyzing compounds.

Example 3.16

At a particular temperature, it takes 0.038 grams of lead sulfate ($PbSO_4$, molecular weight = 303.25) per liter of water to prepare a saturated solution. What is the solubility product of lead sulfate if all of the lead sulfate ionizes?

Solution

Sulfates are not one of the hydrolyzing ions. Therefore, the solubility product can be calculated from the concentrations.

Since one liter of water has a mass of 1 kg, the number of moles of lead sulfate dissolved per saturated liter of solution is

$$n = \frac{m}{M} = \frac{0.038 \text{ g}}{303.25 \dfrac{\text{g}}{\text{mol}}} = 1.25 \times 10^{-4} \text{ mol}$$

Lead sulfate ionizes according to the following reaction.

$$PbSO_4(s) \rightleftharpoons Pb^{2+}(aq) + SO_4^{2-}(aq) \quad [\text{in water}]$$

Since all of the lead sulfate ionizes, the number of moles of each ion is the same as the number of moles of lead sulfate. Therefore,

Chemistry Definitions

$$\begin{aligned} K_{SP} &= [A^+]^m[B^-]^n \\ &= [Pb^{2+}][SO_4^{2-}] \\ &= (1.25 \times 10^{-4})(1.25 \times 10^{-4}) \\ &= 1.56 \times 10^{-8} \end{aligned}$$

The equilibrium constant applies to equations at equilibrium, but does not alone provide information about the direction of the reaction. Comparing the equilibrium constant to the *reaction quotient*, Q, will reveal whether the reaction is at equilibrium, proceeding from left to right, or proceeding from right to left.

$$Q = \frac{[C]^c[D]^d}{[A]^a[B]^b} \qquad \textit{3.27}$$

- If Q is equal to K, the reaction is at equilibrium.

- If Q is less than K, the reaction will proceed from left to right.

- If Q is greater than K, the reaction will proceed from right to left.

Example 3.17

For the following reaction, Ca^{2+} is present at 20 mg/L and F^- is present at 0.1 mg/L.

$$CaF_2 \rightarrow Ca^{2+} + 2F^-$$

The solubility product is 1.53×10^{-10}. Will CaF_2 dissolve or precipitate?

Solution

Because the reaction is a solubility reaction, assume $[CaF_2] = 1$ mol/L. Divide each element's concentration by its molar mass to find its ionic concentration.

$$\begin{aligned} [Ca^{2+}] &= \frac{20 \dfrac{\text{mg}}{\text{L}}}{\left(40 \dfrac{\text{g}}{\text{mol}}\right)\left(1000 \dfrac{\text{mg}}{\text{g}}\right)} \\ &= 5.00 \times 10^{-4} \text{ mol/L} \end{aligned}$$

$$\begin{aligned} [F^-] &= \frac{0.1 \dfrac{\text{mg}}{\text{L}}}{\left(19 \dfrac{\text{g}}{\text{mol}}\right)\left(1000 \dfrac{\text{mg}}{\text{g}}\right)} \\ &= 5.26 \times 10^{-6} \text{ mol/L} \end{aligned}$$

Use Eq. 3.27 to find the reaction quotient.

$$\begin{aligned} Q &= \frac{[Ca^{2+}][F^-]^2}{[CaF_2]} \\ &= \frac{\left(5.00 \times 10^{-4} \dfrac{\text{mol}}{\text{L}}\right)\left(5.26 \times 10^{-6} \dfrac{\text{mol}}{\text{L}}\right)^2}{1 \dfrac{\text{mol}}{\text{L}}} \\ &= 1.38 \times 10^{-14} \text{ mol/L} \end{aligned}$$

The solubility product, K_{SP}, is an equilibrium constant applied to the special circumstance of solubility. Compare Q with K_{SP}.

$$Q = 1.38 \times 10^{-14} < K_{SP} = 1.53 \times 10^{-10}$$

Q is smaller than K_{SP}, so the reaction proceeds from left to right (as written) and CaF_2 will dissolve.

32. ENTHALPY OF FORMATION

Enthalpy, H, is the useful energy that a substance possesses by virtue of its temperature, pressure, and phase.[19] The *enthalpy of formation (heat of formation),* ΔH_f, of a compound is the energy absorbed during the formation of 1 gmol of the compound from the elements in their free, standard states.[20] The enthalpy of formation is assigned a value of zero for elements in their free states at 25°C and 1 atm. This is the so-called *standard state* for enthalpies of formation.

Table 3.5 contains enthalpies of formation for some common elements and compounds. The enthalpy of formation depends on the temperature and phase of the compound. A standard temperature of 25°C is used in most tables of enthalpies of formation.[21] Compounds are solid (*s*) unless indicated to be gaseous (*g*) or liquid (*l*). Some aqueous (*aq*) values are also encountered.

33. ENTHALPY OF REACTION

The *enthalpy of reaction (heat of reaction),* ΔH_r, is the energy absorbed during a chemical reaction under constant volume conditions. It is found by summing the enthalpies of formation of all products and subtracting the sum of enthalpies of formation of all reactants. This is essentially a restatement of the energy conservation principle and is known as *Hess' law of energy summation.*

$$(\Delta H_r^\circ) = \sum_{products} v_i (\Delta H_f^\circ)_i - \sum_{reactants} v_i (\Delta H_f^\circ)_i \quad \text{3.28}$$

Heats of Reaction

Reactions that give off energy (i.e., have negative enthalpies of reaction) are known as *exothermic reactions.* Many (but not all) exothermic reactions begin spontaneously. On the other hand, *endothermic reactions* absorb energy and require thermal or electrical energy to begin.

Example 3.18

Using enthalpies of formation, calculate the heat of stoichiometric combustion (standardized to 25°C) of gaseous methane (CH_4) and oxygen.

Solution

The balanced chemical equation for the stoichiometric combustion of methane is

$$CH_4 + 2O_2 \rightarrow 2H_2O + CO_2$$

The enthalpy of formation of oxygen gas (its free-state configuration) is zero. Using enthalpies of formation from Table 3.5 in Eq. 3.28, the enthalpy of reaction per mole of methane is

Heats of Reaction

$$(\Delta H_r^\circ) = \sum_{products} v_i (\Delta H_f^\circ)_i - \sum_{reactants} v_i (\Delta H_f^\circ)_i$$

$$= (2)\left(-57.80 \ \frac{kcal}{mol}\right) + \left(-94.05 \ \frac{kcal}{mol}\right)$$

$$\quad - \left(-17.90 \ \frac{kcal}{mol}\right) - (2)(0)$$

$$= -191.75 \ kcal/mol \ CH_4 \quad [exothermic]$$

Using the footnote of Table 3.5, this value can be converted to Btu/lbm. The molecular weight of methane is

$$MW_{CH_4} = 12 + (4)(1) = 16$$

$$\text{higher heating value} = \frac{\left(191.75 \ \frac{kcal}{mol}\right)\left(1800 \ \frac{Btu\text{-}mol}{lbm\text{-}kcal}\right)}{16}$$

$$= 21{,}572 \ Btu/lbm$$

34. TESTS OF WASTEWATER CHARACTERISTICS

The most common wastewater analyses used to determine characteristics of a municipal wastewater determine biochemical oxygen demand (BOD) and suspended solids (SS). BOD and flow data are basic requirements for the operation of biological treatment units. The concentration of suspended solids relative to BOD indicates the degree that organic matter is removable by primary settling. Additionally, temperature, pH, chemical oxygen demand (COD), dissolved and suspended solids, alkalinity, color, grease, and the quantity of heavy metals are necessary to characterize municipal wastewater characteristics.

35. BIOCHEMICAL OXYGEN DEMAND

When oxidizing organic material in water, biological organisms also remove oxygen from the water. This is typically considered to occur through oxidation of organic material to CO_2 and H_2O by microorganisms at the molecular level and is referred to as *biochemical oxygen demand* (BOD). Therefore, oxygen use is an indication of the organic waste content.

---

[19]The older term *heat* is rarely encountered today.
[20]The symbol *H* is used to denote molar enthalpies. The symbol *h* is used for specific enthalpies (i.e., enthalpy per kilogram or per pound).
[21]It is possible to correct the enthalpies of formation to account for other reaction temperatures.

Table 3.5 Standard Enthalpies of Formation (at 25°C)

element/compound	ΔH_f (kcal/mol)
Al (s)	0.00
Al_2O_3 (s)	−399.09
C (graphite)	0.00
C (diamond)	0.45
C (g)	171.70
CO (g)	−26.42
CO_2 (g)	−94.05
CH_4 (g)	−17.90
C_2H_2 (g)	54.19
C_2H_4 (g)	12.50
C_2H_6 (g)	−20.24
CCl_4 (g)	−25.5
$CHCl_4$ (g)	−24
CH_2Cl_2 (g)	−21
CH_3Cl (g)	−19.6
CS_2 (g)	27.55
COS (g)	−32.80
$(CH_3)_2S$ (g)	−8.98
CH_3OH (g)	−48.08
C_2H_5OH (g)	−56.63
$(CH_3)_2O$ (g)	−44.3
C_3H_6 (g)	9.0
C_6H_{12} (g)	−29.98
C_6H_{10} (g)	−1.39
C_6H_6 (g)	19.82
Fe (s)	0.00
Fe (g)	99.5
Fe_2O_3 (s)	−196.8
Fe_3O_4 (s)	−267.8
H_2 (g)	0.00
H_2O (g)	−57.80
H_2O (l)	−68.32
H_2O_2 (g)	−31.83
H_2S (g)	−4.82
N_2 (g)	0.00
NO (g)	21.60
NO_2 (g)	8.09
NO_3 (g)	13
NH_3 (g)	−11.04
O_2 (g)	0.00
O_3 (g)	34.0
S (g)	0.00
SO_2 (g)	−70.96
SO_3 (g)	−94.45

(Multiply kcal/mol by 4.184 to obtain kJ/mol.)
(Multiply kcal/mol by 1800/MW to obtain Btu/lbm.)

Typical values of BOD for various industrial wastewaters are given in Table 3.6. A BOD of 100 mg/L is considered to be a weak wastewater; a BOD of 200–250 mg/L is considered to be a medium strength wastewater; and a BOD above 300 mg/L is considered to be a strong wastewater.

Table 3.6 Typical BOD and COD of Industrial Wastewaters

industry/ type of waste	BOD	COD
canning		
corn	19.5 lbm/ton corn	
tomatoes	8.4 lbm/ton tomatoes	
dairy milk		
processing	1150 lbm/ton raw milk	1900 mg/L
	1000 mg/L	
beer brewing	1.2 lbm/barrel beer	
commercial		
laundry	1250 lbm/1000 lbm dry	2400 mg/L
	700 mg/L	
slaughterhouse	7.7 lbm/animal	2100 mg/L
(meat packing)	1400 mg/L	
papermill	121 lbm/ton pulp	
synthetic textile	1500 mg/L	3300 mg/L
chlorophenolic		
manufacturing	4300 mg/L	5400 mg/L
milk bottling	230 mg/L	420 mg/L
cheese production	3200 mg/L	5600 mg/L
candy production	1600 mg/L	3000 mg/L

In the past, the BOD has been determined in a lab using a traditional standardized BOD testing procedure. Since this procedure requires five days of incubating, inline measurement systems have been developed to provide essentially instantaneous and continuous BOD information at wastewater plants.

The standardized BOD test consists of adding a measured amount of wastewater (which supplies the organic material) to a measured amount of dilution water (which reduces toxicity and supplies dissolved oxygen) and then incubating the mixture at a specific temperature. The standard procedure calls for a five-day incubation period at 68°F (20°C), though other temperatures are used. The measured BOD is designated as BOD_5 or BOD-5. The BOD of a biologically active sample at the end of the incubation period is given by Eq. 3.29.

BOD Test Solution and Seeding Procedures

$$\text{BOD, mg/L} = \frac{D_1 - D_2}{P} \qquad \textbf{3.29}$$

In Eq. 3.29, P is the fraction of the sample volume to the total diluted volume.

$$\left[P = \frac{V_{sample}}{V_{total}} = \frac{V_{sample}}{V_{sample} + V_{dilution}} \right] \qquad \textbf{3.30}$$

If more than one identical sample is prepared, the increase in BOD over time can be determined, rather than just the final BOD. Figure 3.1 illustrates a typical plot of BOD versus time. The BOD at any time t is known as the *BOD exertion*.

Figure 3.1 *BOD Exertion*

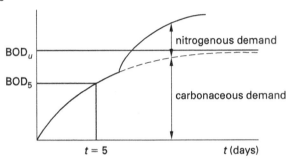

BOD exertion can be found from Eq. 3.31. The ultimate BOD, L, is the total oxygen used by *carbonaceous bacteria* if the test is run for a long period of time, usually taken as 20 days. (The symbols BOD_u, BOD_L, and BOD_{ult} are also widely used to represent the ultimate BOD.)

BOD Exertion
$$y_t = L(1 - e^{-kt}) \qquad 3.31$$

Rearranging to solve for L, ultimate BOD,

$$L = \frac{y_t}{1 - e^{-kt}} \qquad 3.32$$

The ultimate BOD, L, usually cannot be found from long-term studies due to the effect of nitrogen-consuming bacteria in the sample. However, if k is 0.1 day^{-1}, the ultimate BOD can be approximated from Eq. 3.33 derived from Eq. 3.31.

$$L \approx 1.463 BOD_5 \qquad 3.33$$

The approximate variation in first-stage BOD (either BOD_5 or L) municipal wastewater with temperature is given by Eq. 3.34.[22]

$$BOD_{T°C} = BOD_{20°C}(0.02 T_{°C} + 0.6) \qquad 3.34$$

The deviation from the expected exponential growth curve in Fig. 3.1 is due to nitrification and is considered to be *nitrogenous demand*. *Nitrification* is the use of oxygen by autotrophic bacteria. Autotrophic bacteria oxidize ammonia to nitrites and nitrates.

The number of autotrophic bacteria in a wastewater sample is initially small. Generally, 6 to 10 days are required for the autotrophic population to become sufficiently large to affect a BOD test. Therefore, the standard BOD test is terminated at five days, before the autotrophic contribution to BOD becomes significant.

Until recently, BOD has not played a significant role in the daily operations decisions of treatment plants because of the length of time it took to obtain BOD data. Other measurements, such as COD, TOC, turbidity, suspended solids, respiration rates, or flow have been used.

Example 3.19

Ten 5 mL samples of wastewater are placed in 300 mL BOD bottles and diluted to full volume. Half of the bottles are titrated immediately, and the average initial concentration of dissolved oxygen is 7.9 mg/L. The remaining bottles are incubated for five days, after which the average dissolved oxygen is determined to be 4.5 mg/L. The deoxygenation rate constant is known to be 0.13 day^{-1} (base-10). What are the (a) standard BOD and (b) ultimate carbonaceous BOD?

Solution

(a) Use Eq. 3.29.

BOD Test Solution and Seeding Procedures

$$BOD, \text{mg/L} = \frac{D_1 - D_2}{P}$$

$$= \frac{D_1 - D_2}{\dfrac{V_{sample}}{V_{total}}}$$

$$= \frac{7.9\ \dfrac{mg}{L} - 4.5\ \dfrac{mg}{L}}{\dfrac{5\ mL}{300\ mL}}$$

$$= 204\ mg/L$$

(b) Use Eq. 3.31, and solve for the ultimate BOD.

BOD Exertion

$$y_t = L(1 - e^{-kt})$$

$$L = \frac{y_t}{1 - e^{-kt}}$$

$$= \frac{204\ \dfrac{mg}{L}}{1 - e^{-(0.13\ day^{-1})(5\ days)}}$$

$$= 426.82\ \ mg/l$$

---

[22]As reported in *Water-Resources Engineering*, 4th ed., Ray K. Linsley, Joseph B. Franzini, David Freyberg, and George Tchobanoglous, McGraw-Hill Book Co., 1992, New York.

36. SEEDED BOD

Industrial wastewater may lack sufficient microorganisms to metabolize the organic matter. In this situation, the standard BOD test will not accurately determine the amount of organic matter unless seed organisms are added.

The BOD in seeded samples is found by measuring dissolved oxygen in the seeded sample after 15 min (DO_i) and after five days (DO_f), as well as by measuring the dissolved oxygen of the seed material itself after 15 min (DO_i^*) and after five days (DO_f^*). In Eq. 3.35, x is the ratio of the volume of seed added to the sample to the volume of seed used to find DO.

$$\text{BOD} = \frac{DO_i - DO_f - x(DO_i^* - DO_f^*)}{\dfrac{V_{\text{sample}}}{V_{\text{sample}} + V_{\text{dilution}}}} \qquad 3.35$$

37. CHEMICAL OXYGEN DEMAND

Chemical oxygen demand (COD) is a measure of oxygen removed by both biological organisms feeding on organic material, as well as oxidizable inorganic compounds (unlike biochemical oxygen demand, which is a measure of oxygen removed only by biological organisms). Therefore, COD is a good measure of total effluent strength (organic and inorganic contaminants).

COD testing is required where there is industrial chemical pollution. In such environments, the organisms necessary to metabolize organic compounds may not exist. Furthermore, the toxicity of the wastewater may make the standard BOD test impossible to carry out. Standard COD test results are usually available in a matter of hours.

If toxicity is low, BOD and COD test results can be correlated. The BOD/COD ratio typically varies from 0.4–0.8. This is a wide range but, for any given treatment plant and waste type, the correlation is essentially constant. The correlation can, however, vary along the treatment path.

38. RELATIVE STABILITY

The *relative stability* test is easier to perform than the BOD test, although it is less accurate. The relative stability of an effluent is defined as the percentage of initial BOD that has been satisfied. The test consists of taking a sample of effluent and adding a small amount of methylene blue dye. The mixture is then incubated, usually at 20°C. When all oxygen has been removed from the water, anaerobic bacteria start to remove the dye. The amount of time it takes for the color to start degrading is known as the *stabilization time* or *decoloration time*. The relative stability can be found from the stabilization time by using Table 3.7.

Table 3.7 *Relative Stability[a]*

stabilization time (day)	relative stability[b] (%)
½	11
1	21
1½	30
2	37
2½	44
3	50
4	60
5	68
6	75
7	80
8	84
9	87
10	90
11	92
12	94
13	95
14	96
16	97
18	98
20	99

[a]incubation at 20°C
[b]calculated as $(100\%)(1 - 0.794^t)$

39. NOMENCLATURE

A	atomic weight	lbm/lbmol	kg/kmol
B	volumetric fraction	–	–
BOD	biochemical oxygen demand	mg/L	mg/L
C	concentration	n.a.	mg/L
COD	chemical oxygen demand	mg/L	mg/L
EW	equivalent weight	lbm/lbmol	kg/kmol
F	formality	n.a.	FW/L
FW	formula weight	lbm/lbmol	kg/kmol
g	gaseous	n.a.	n.a.
GEW	gram equivalent weight	mol	mol
H	enthalpy	Btu/lbmol	kcal/mol
H	Henry's law constant	atm	atm
i	isotope or element	–	–
k	reaction rate constant	1/min	1/min
K	equilibrium constant	–	–
l	liquid	n.a.	n.a.
L	ultimate BOD	mg/L	mg/L
m	mass	lbm	kg
m	molality	n.a.	mol/1000 g
M	molarity	n.a.	mol/L
M, MW	molecular weight	lbm/lbmol	g/mol
N	normality	n.a.	GEW/L
n	number of moles	–	–
N	normality	n.a.	GEW/L
N	number of subatomic particles	–	–
N_A	Avogadro's number	n.a.	1/mol
p, P	pressure	lbf/ft²	Pa
r	rate of reaction	mol/L·sec	mol/L·sec
Q	reaction quotient	–	–
R^*	universal gas constant, 1545.35 (8314.47)	ft-lbf/lbmol-°R	J/kmol·K
s	solid	n.a.	n.a.
t	time	min or sec	min or s
T	temperature	°R	K
v	rate of reaction	n.a.	mol/L·s
V	volume	ft³	m³
x	gravimetric fraction	–	–
x	mole fraction	–	–
x	relative abundance	–	–
X	fraction ionized	–	–
y	mole fraction	–	–
y_t	amount of BOD exerted at time t	mg/L	mg/L
Z	atomic number	–	–

Symbols

α	gas/liquid volume ratio	–	–

Subscripts

0	zero-order
a	acid
A	Avogadro
b	base
eq	equilibrium
f	formation
H	Henry's law constant
L	ultimate
p	partial pressure
r	reaction
s	saturation
sp, SP	solubility product
t	time, total
u, ult	ultimate

 **Biology and Microbiology**

1. MICROORGANISMS

Microorganisms occur in untreated water and represent a potential human health risk if they enter the water supply. Microorganisms include viruses, bacteria, fungi, algae, protozoa, worms, rotifers, and crustaceans.

Microorganisms are organized into three broad groups based upon their structural and functional differences. The groups are called *kingdoms*. The three kingdoms are animals (rotifers and crustaceans), plants (mosses and ferns), and *Protista* (bacteria, algae, fungi, and protozoa). Bacteria and protozoa of the kingdom *Protista* constitute the major groups of microorganisms in the biological systems that are used in secondary treatment of wastewater.

2. PATHOGENS

Many infectious diseases in humans or animals are caused by organisms categorized as *pathogens*. Pathogens are found in fecal wastes that are transmitted and transferred through the handling of wastewater. Pathogens will proliferate in areas where sanitary disposal of feces is not adequately practiced and where contamination of water supply from infected individuals is not

properly controlled. The wastes may also be improperly discharged into surface waters, making the water *nonpotable* (unfit for drinking). Certain shellfish can become toxic when they concentrate pathogenic organisms in their tissues, increasing the toxic levels much higher than the levels in the surrounding waters.

Organisms that are considered to be pathogens include bacteria, protozoa, viruses, and helminths (worms). Table 4.1 lists potential waterborne diseases, the causative organisms, and the typical infection sources.

Not all microorganisms are considered pathogens. Some microorganisms are exploited for their usefulness in wastewater processing. Most wastewater engineering (and an increasing portion of environmental engineering) involves designing processes and operating facilities that utilize microorganisms to destroy organic and inorganic substances.

3. MICROBE CATEGORIZATION

Carbon is the basic building block for cell synthesis, and it is prevalent in large quantities in wastewater. Wastewater treatment converts carbon into microorganisms that are subsequently removed from the water by settling. Therefore, the growth of organisms that use organic material as energy is encouraged.

If a microorganism uses organic material as a carbon supply, it is *heterotrophic*. *Autotrophs* require only carbon dioxide to supply their carbon needs. Organisms that rely only on the sun for energy are called *phototrophs*. *Chemotrophs* extract energy from organic or inorganic oxidation/reduction (redox) reactions. *Organotrophs* use organic materials, while *lithotrophs* oxidize inorganic compounds.

Most microorganisms in wastewater treatment processes are bacteria. Conditions in the treatment plant are readjusted so that chemoheterotrophs predominate.

Each species of bacteria reproduces most efficiently within a limited range of temperatures. Four temperature ranges are used to classify bacteria. Those bacteria that grow best below 68°F (20°C) are called *psychrophiles* (i.e., are *psychrophilic*, also known as *cryophilic*). *Mesophiles* (i.e., *mesophilic bacteria*) grow best at temperatures in a range starting around 68°F (20°C) and ending around 113°F (45°C). Between 113°F (45°C) and

Table 4.1 *Potential Pathogens*

name of organism	major disease	source
Bacteria		
Salmonella typhi	typhoid fever	human feces
Salmonella paratyphi	paratyphoid fever	human feces
other *Salmonella*	salmonellosis	human/animal feces
Shigella	bacillary dysentery	human feces
Vibrio cholerae	cholera	human feces
Enteropathogenic coli	gastroenteritis	human feces
Yersinia enterocolitica	gastroenteritis	human/animal feces
Campylobacter jejuni	gastroenteritis	human/animal feces
Legionella pneumophila	acute respiratory illness	thermally enriched waters
Mycobacterium	tuberculosis	human respiratory exudates
other *Mycobacteria*	pulmonary illness	soil and water
opportunistic bacteria	variable	natural waters
Enteric Viruses/Enteroviruses		
Polioviruses	poliomyelitis	human feces
Coxsackieviruses A	aseptic meningitis	human feces
Coxsackieviruses B	aseptic meningitis	human feces
Echoviruses	aseptic meningitis	human feces
other *Enteroviruses*	encephalitis	human feces
Reoviruses	upper respiratory and gastrointestinal illness	human/animal feces
Rotaviruses	gastroenteritis	human feces
Adenoviruses	upper respiratory and gastrointestinal illness	human feces
Hepatitis A virus	infectious hepatitis	human feces
Norwalk and related gastrointestinal viruses	gastroenteritis	human feces
Fungi		
Aspergillus	ear, sinus, lung, and skin infections	airborne spores
Candida	yeast infections	various
Protozoa		
Acanthamoeba castellani	amoebic meningoencephalitis	soil and water
Balantidium coli	balantidosis (dysentery)	human feces
Cryptosporidium[*]	cryptosporidiosis	human/animal feces
Entamoeba histolytica	amoebic dysentery	human feces
Giardia lamblia	giardiasis (gastroenteritis)	human/animal feces
Naegleria fowleri	amoebic meningoencephalitis	soil and water
Algae (blue-green)		
Anabaena flos-aquae	gastroenteritis (possible)	natural waters
Microcystis aeruginosa	gastroenteritis (possible)	natural waters
Alphanizomenon flos-aquae	gastroenteritis (possible)	natural waters
Schizothrix calciola	gastroenteritis (possible)	natural waters
Helminths (intestinal parasites/worms)		
Ascaris lumbricoides (roundworm)	digestive disturbances	ingested worm eggs
E. vericularis (pinworm)	any part of the body	ingested worm eggs
Hookworm	pneumonia, anemia	ingested worm eggs
Threadworm	abdominal pain, nausea, weight loss	ingested worm eggs
T. trichiuro (whipworm)	trichinosis	ingested worm eggs
Tapeworm	digestive disturbances	ingested worm eggs

[*]Disinfectants have little effect on *Cryptosporidia*. Most large systems now use filtration, the most effective treatment to date against *Cryptosporidia*.

140°F (60°C), the *thermophiles* (*thermophilic bacteria*) grow best. Above 140°F (60°C), *stenothermophiles* grow best.

Because most reactions proceed slowly at these temperatures, cells use *enzymes* to speed up the reactions and control the rate of growth. Enzymes are proteins, ranging from simple structures to complex conjugates, and are specialized for the reactions they catalyze.

The temperature ranges are qualitative and somewhat subjective. The growth range of facultative thermophiles extends from the thermophilic range into the mesophilic range. Bacteria will grow in a wide range of temperatures and will survive at a very large range of temperatures. *E. coli*, for example, is classified as a mesophile. It grows best at temperatures between 68°F (20°C) and 122°F (50°C) but can continue to reproduce at temperatures down to 32°F (0°C).

Nonphotosynthetic bacteria are classified into two heterotrophic and autotrophic groups depending on their sources of nutrients and energy. *Heterotrophs* use organic matter as both an energy source and a carbon source for synthesis. Heterotrophs are further subdivided into groups depending on their behavior toward free oxygen: aerobes, anaerobes, and facultative bacteria. *Obligate aerobes* require free dissolved oxygen while they decompose organic matter to gain energy for growth and reproduction. *Obligate anaerobes* oxidize organics in the complete absence of dissolved oxygen by using the oxygen bound in other compounds, such as nitrate and sulfate. *Facultative bacteria* comprise a group that uses free dissolved oxygen when available but that can also behave anaerobically in the absence of free dissolved oxygen (i.e., *anoxic conditions*). Under anoxic conditions, a group of facultative anaerobes, called *denitrifiers*, utilizes nitrites and nitrates instead of oxygen. Nitrate nitrogen is converted to nitrogen gas in the absence of oxygen. This process is called *anoxic denitrification*.

Autotrophic bacteria (*autotrophs*) oxidize inorganic compounds for energy, use free oxygen, and use carbon dioxide as a carbon source. Significant members of this group are the *Leptothrix* and *Crenothrix* families of *iron bacteria*. These have the ability to oxidize soluble ferrous iron into insoluble ferric iron. Because soluble iron is often found in well waters and iron pipe, these bacteria deserve some attention. They thrive in water pipes where dissolved iron is available as an energy source and bicarbonates are available as a carbon source. As the colonies die and decompose, they release foul tastes and odors and have the potential to cause staining of porcelain or fabrics.

4. VIRUSES

Viruses are parasitic organisms that pass through filters that retain bacteria, can only be seen with an electron microscope, and grow and reproduce only inside living cells, but they can survive outside the host. They are not cells, but particles composed of a protein sheath surrounding a nucleic-acid core. Most viruses of interest in water supply range in size from 10 nm to 25 nm.

The *viron particles* invade living cells, and the viral genetic material redirects cell activities toward production of new viral particles. A large number of viruses are released to infect other cells when the infected cell dies. Viruses are host-specific, attacking only one type of organism.

There are more than 100 types of human enteric viruses. Those of interest in drinking water are *Hepatitis A*, *Norwalk*-type viruses, *Rotaviruses*, *Adenoviruses*, *Enteroviruses*, and *Reoviruses*.

5. BACTERIA

Bacteria are microscopic organisms (*microorganisms*) having round, rodlike, spiral or filamentous single-celled or noncellular bodies. Bacteria are *prokaryotes* (i.e., they lack nucleii structures). They are often aggregated into colonies. Bacteria use soluble food and reproduce through binary fission. Most bacteria are not pathogenic to humans, but they do play a significant role in the decomposition of organic material and can have an impact on the aesthetic quality of water.

6. FUNGI

Fungi are aerobic, multicellular, nonphotosynthetic, heterotrophic, eukaryotic protists. Most fungi are saprophytes that degrade dead organic matter. Fungi grow in low-moisture areas, and they are tolerant of low-pH environments. Fungi release carbon dioxide and nitrogen during the breakdown of organic material.

Fungi are obligate aerobes that reproduce by a variety of methods including fission, budding, and spore formation. They form normal cell material with one-half the nitrogen required by bacteria. In nitrogen-deficient wastewater, they may replace bacteria as the dominant species.

7. ALGAE

Algae are autotrophic, photosynthetic organisms (*photoautotrophs*) and may be either unicellular or multicellular. They take on the color of the pigment that is the catalyst for photosynthesis. In addition to chlorophyll (green), different algae have different pigments, such as carotenes (orange), phycocyanin (blue), phycoerythrin (red), fucoxanthin (brown), and xanthophylls (yellow).

Algae derive carbon from carbon dioxide and bicarbonates in water. The energy required for cell synthesis is obtained through photosynthesis. Algae utilize oxygen for respiration in the absence of light. Algae and bacteria have a symbiotic relationship in aquatic systems, with the algae producing oxygen used by the bacterial population.

In the presence of sunlight, the photosynthetic production of oxygen is greater than the amount used in respiration. At night algae use up oxygen in respiration. If the daylight hours exceed the night hours by a reasonable amount, there is a net production of oxygen. Excessive algal growth (*algal blooms*) can result in supersaturated oxygen conditions in the daytime and anaerobic conditions at night.

Some algae cause tastes and odors in natural water. While they are not generally considered pathogenic to humans, algae do cause turbidity, and turbidity provides a residence for microorganisms that are pathogenic.

8. PROTOZOA

Protozoa are single-celled animals that reproduce by *binary fission* (dividing in two). Most are aerobic chemoheterotrophs (i.e., *facultative heterotrophs*). Protozoa have complex digestive systems and use solid organic matter as food, including algae and bacteria. Therefore, they are desirable in wastewater effluent because they act as polishers in consuming the bacteria.

Flagellated protozoa are the smallest protozoans. The *flagella* (long hairlike strands) provide motility by a whiplike action. Amoebas move and take in food through the action of a mobile protoplasm. Free-swimming protozoa have *cilia*, small hairlike features, used for propulsion and gathering in organic matter.

9. WORMS AND ROTIFERS

Rotifers are aerobic, multicellular chemoheterotrophs. The rotifer derives its name from the apparent rotating motion of two sets of cilia on its head. The cilia provide mobility and a mechanism for catching food. Rotifers consume bacteria and small particles of organic matter.

Many *worms* are aquatic parasites. *Flatworms* of the class *Trematoda* are known as *flukes*, and the *Cestoda* are tapeworms. *Nemotodes* of public health concern are *Trichinella*, which causes trichinosis; *Necator*, which causes pneumonia; *Ascaris*, which is the common roundworm that takes up residence in the human intestine (ascariasis); and *Filaria*, which causes filariasis.

10. MOLLUSKS

Mollusks, such as mussels and clams, are characterized by a shell structure. They are aerobic chemoheterotrophs that feed on bacteria and algae. They are a source of food for fish and are not found in wastewater

treatment systems to any extent, except in underloaded lagoons. Their presence is indicative of a high level of dissolved oxygen and a very low level of organic matter.

Macrofouling is a term referring to infestation of water inlets and outlets by clams and mussels. *Zebra mussels*, accidentally introduced into the United States in 1986, are particularly troublesome for several reasons. First, young mussels are microscopic, easily passing through intake screens. Second, they attach to anything, even other mussels, producing thick colonies. Third, adult mussels quickly sense some biocides, most notably those that are halogen-based (including chlorine), quickly closing and remaining closed for days or weeks.

The use of biocides in the control of zebra mussels is controversial. Chlorination is a successful treatment that is recommended with some caution since it results in increased toxicity, affecting other species and THM production. Nevertheless, an ongoing biocide program aimed at pre-adult mussels, combined with slippery polymer-based and copper-nickel alloy surface coatings, is probably the best approach at prevention. Once a pipe is colonized, mechanical removal by scraping or water blasting is the only practical option.

11. INDICATOR ORGANISMS

The techniques for comprehensive bacteriological examination for pathogens are complex and time-consuming. Isolating and identifying specific pathogenic microorganisms is a difficult and lengthy task. Many of these organisms require sophisticated tests that take several days to produce results. Because of these difficulties, and also because the number of pathogens relative to other microorganisms in water can be very small, *indicator organisms* are used as a measure of the quality of the water. The primary function of an indicator organism is to provide evidence of recent fecal contamination from warm-blooded animals.

The characteristics of a good indicator organism are: (a) The indicator is always present when the pathogenic organism of concern is present. It is absent in clean, uncontaminated water. (b) The indicator is present in fecal material in large numbers. (c) The indicator responds to natural environmental conditions and to treatment processes in a manner similar to the pathogens of interest. (d) The indicator is easy to isolate, identify, and enumerate. (e) The ratio of indicator to pathogen should be high. (f) The indicator and pathogen should come from the same source (e.g., gastrointestinal tract).

While there are several microorganisms that meet these criteria, *total coliform* and *fecal coliform* are the indicators generally used. *Total coliform* refers to the group of aerobic and facultatively anaerobic, gram-negative, non-spore-forming, rod-shaped bacteria that ferment lactose with gas formation within 48 hr at 95°F (35°C). This encompasses a variety of organisms, mostly of intestinal origin, including *Escherichia coli* (*E. coli*), the most numerous facultative bacterium in the feces of warm-blooded animals. Unfortunately, this group also includes *Enterobacter*, *Klebsiella*, and *Citrobacter*, which are present in wastewater but can be derived from other environmental sources such as soil and plant materials.

Fecal coliforms are a subgroup of the total coliforms that come from the intestines of warm-blooded animals. They are measured by running the standard total coliform fermentation test at an elevated temperature of 112°F (44.5°C), providing a means to distinguish false positives in the total coliform test.

Results of fermentation tests are reported as a *most probable number* (MPN) *index*. This is an index of the number of coliform bacteria that, more probably than any other number, would give the results shown by the laboratory examination; it is not an actual enumeration.

12. METABOLISM/METABOLIC PROCESSES

Metabolism is a term given to describe all chemical activities performed by a cell. The cell uses *adenosine triphosphate* (ATP) as the principle energy currency in all processes. Those processes that allow the bacterium to synthesize new cells from the energy stored within its body are said to be *anabolic*. All biochemical processes in which cells convert substrate into useful energy and waste products are said to be *catabolic*.

13. DECOMPOSITION OF WASTE

Decomposition of waste involves oxidation/reduction reactions and is classified as aerobic or anaerobic. The type of electron acceptor available for catabolism determines the type of decomposition used by a mixed culture of microorganisms. Each type of decomposition has its own peculiar characteristics that affect its use in waste treatment.

14. AEROBIC DECOMPOSITION

Molecular oxygen (O_2) must be present as the terminal electron acceptor in order for decomposition to proceed by aerobic oxidation. As in natural water bodies, the dissolved oxygen content is measured. When oxygen is present, it is the only terminal electron acceptor used. Hence, the chemical end products of decomposition are primarily carbon dioxide, water, and new cell material as demonstrated by Table 4.2. Odoriferous, gaseous end products are kept to a minimum. In healthy natural water systems, aerobic decomposition is the principal means of self-purification.

A wider spectrum of organic material can be oxidized by aerobic decomposition than by any other type of decomposition. Because of the large amount of energy released in aerobic oxidation, most aerobic organisms are capable of high growth rates. Consequently, there is a relatively large production of new cells in comparison with the other oxidation systems. This means that more biological sludge is generated in aerobic oxidation than in the other oxidation systems.

Aerobic decomposition is the preferred method for large quantities of dilute ($BOD_5 < 500$ mg/L) wastewater because decomposition is rapid and efficient and has a low odor potential. For high-strength wastewater ($BOD_5 > 1000$ mg/L), aerobic decomposition is not suitable because of the difficulty in supplying enough oxygen and because of the large amount of biological sludge that is produced.

15. ANOXIC DECOMPOSITION

Some microorganisms will use nitrates in the absence of oxygen needed to oxidize carbon. The end products from such *denitrification* are nitrogen gas, carbon dioxide, water, and new cell material. The amount of energy made available to the cell during denitrification is about the same as that made available during aerobic decomposition. The production of cells, although not as high as in aerobic decomposition, is relatively high.

Denitrification is of importance in wastewater treatment when nitrogen must be removed. In such cases, a special treatment step is added to the conventional process for removal of carbonaceous material. One other important aspect of denitrification is in final clarification of the treated wastewater. If the final clarifier becomes anoxic, the formation of nitrogen gas will cause large masses of sludge to float to the surface and escape from the treatment plant into the receiving water. Thus, it is necessary to ensure that anoxic conditions do not develop in the final clarifier.

16. ANAEROBIC DECOMPOSITION

In order to achieve anaerobic decomposition, molecular oxygen and nitrate must not be present as terminal electron acceptors. Sulfate, carbon dioxide, and organic compounds that can be reduced serve as terminal electron acceptors. The reduction of sulfate results in the production of hydrogen sulfide (H_2S) and a group of equally odoriferous organic sulfur compounds called *mercaptans*.

The anaerobic decomposition of organic matter, also known as *fermentation*, generally is considered to be a two-step process. In the first step, complex organic compounds are fermented to low molecular weight *fatty acids* (*volatile acids*). In the second step, the organic acids are converted to methane. Carbon dioxide serves as the electron acceptor.

Anaerobic decomposition yields carbon dioxide, methane, and water as the major end products. Additional end products include ammonia, hydrogen sulfide, and

Table 4.2 *Waste Decomposition End Products*

substrates	representative end products		
	aerobic decomposition	anoxic decomposition	anaerobic decomposition
proteins and other organic nitrogen compounds	amino acids ammonia → nitrites→nitrates alcohols organic acids $\Big\} \to CO_2 + H_2O$	amino acids nitrates → nitrites → N_2 alcohols organic acids $\Big\} \to CO_2 + H_2O$	amino acids ammonia hydrogen sulfide methane carbon dioxide alcohols organic acids
carbohydrates	alcohols fatty acids $\Big\} \to CO_2 + H_2O$	alcohols fatty acids $\Big\} \to CO_2 + H_2O$	carbon dioxide alcohols fatty acids
fats and related substances	fatty acids + glycerol alcohols lower fatty acids $\Big\} \to CO_2 + H_2O$	fatty acids + glycerol alcohols lower fatty acids $\Big\} \to CO_2 + H_2O$	fatty acids + glycerol carbon dioxide alcohols lower fatty acids

mercaptans. As a consequence of these last three compounds, anaerobic decomposition is characterized by a malodorous stench.

Because only small amounts of energy are released during anaerobic oxidation, the amount of cell production is low. Thus, sludge production is correspondingly low. Wastewater treatment based on anaerobic decomposition is used to stabilize sludges produced during aerobic and anoxic decomposition.

Direct anaerobic decomposition of wastewater generally is not feasible for dilute waste. The optimum growth temperature for the anaerobic bacteria is at the upper end of the mesophilic range. Thus, to get reasonable biodegradation, the temperature of the culture must first be elevated. For dilute wastewater, this is not practical. For concentrated wastes ($BOD_5 > 1000$ mg/L), anaerobic digestion is quite appropriate.

17. MICROBIAL GROWTH

For the large numbers and mixed cultures of microorganisms found in waste treatment systems, it is more convenient to measure biomass than numbers of organisms. This is accomplished by measuring suspended or volatile suspended solids. An expression that depicts the rate of conversion of food into biomass for wastewater treatment was developed by Monod.

Equation 4.1 calculates the rate of growth, r_g, of a bacterial culture as a function of the rate of concentration of limiting food in solution. The concentration, X, of

microorganisms, with dimensions of mass per unit volume (e.g., mg/L) is given by Eq. 4.1. k is a specific growth rate factor with dimensions of time^{-1}.

$$r_g = \frac{dX}{dt} = kX \qquad 4.1$$

When one essential nutrient (referred to as a *substrate*) is present in a limited amount, the specific growth rate will increase up to a maximum value, as shown in Fig. 4.1 and as given by *Monod's equation*, Eq. 4.2. k_o is the *maximum specific growth rate coefficient* with dimensions of time^{-1}. S is the concentration of the growth-limiting nutrient with dimensions of mass/unit volume. K_s is the *half-velocity coefficient*, the nutrient concentration at one half of the maximum growth rate, with dimensions of mass per unit volume.

$$k = k_o \left(\frac{S}{K_s + S} \right) \qquad 4.2$$

Combining Eq. 4.1 and Eq. 4.2, the rate of growth is

$$r_g = \frac{k_o X S}{K_s + S} \qquad 4.3$$

Figure 4.1 *Bacterial Growth Rate with Limited Nutrient*

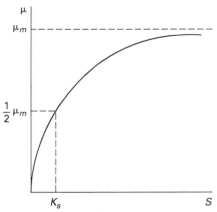

The rate of substrate (nutrient) utilization is easily calculated if the *maximum yield coefficient*, Y, defined as the ratio of the mass of cells formed to the mass of nutrient consumed (with dimensions of mass/mass) is known.

$$r_{su} = \frac{dS}{dt}$$
$$= \frac{-k_o X S}{Y(K_s + S)} \qquad 4.4$$

The maximum rate of substrate utilization per unit mass of microorganisms is

$$k = \frac{k_o}{Y} \qquad 4.5$$

Combining Eq. 4.4 and Eq. 4.5,

$$\frac{dS}{dt} = \frac{-k X S}{K_s + S} \qquad 4.6$$

The Monod equation assumes that all microorganisms are the same age. The actual distribution of ages and other factors (death and predation of cells) decreases the rate of cell growth. The decrease is referred to as *endogenous decay* and is accounted for with an *endogenous decay coefficient*, k_d, with dimensions of time^{-1}. The rate of endogenous decay is

$$r_d = -k_d X \qquad 4.7$$

The *net rate of cell growth* and *net specific growth rate* are

$$r_g' = \frac{k_o X S}{K_s + S} - k_d X$$
$$= -Y r_{su} - k_d X \qquad 4.8$$

$$k' = k_o \left(\frac{S}{K_s + S} \right) - k_d \qquad 4.9$$

18. DISSOLVED OXYGEN IN WASTEWATER

A body of water exposed to the atmosphere will normally become saturated with oxygen. If the dissolved oxygen content, DO, of water is less than the saturated values, there will be good reason to believe that the water is organically polluted.

The difference between the saturated and actual dissolved oxygen concentration is known as the *oxygen deficit*, D.

Stream Modeling
$$DO = DO_{sat} - D \qquad 4.10$$

19. REOXYGENATION

The oxygen deficit can be reduced (i.e., the dissolved oxygen concentration can be increased) only by aerating the water. This occurs naturally in free-running water (e.g., in rivers). The rate of reaeration is

$$r_r = k_2(DO_{sat} - DO) \qquad 4.11$$

k_2 is the *reoxygenation (reaeration) rate constant*, which depends on the type of flow and temperature and with common units of day^{-1}. (k_2 may be written as K_r and K_R by some authorities.) Typical values of k_2 (base-10) are: small stagnant ponds, 0.05–0.10; sluggish streams, 0.10–0.15; large lakes, 0.10–0.15; large streams, 0.15–0.30; swiftly flowing rivers and streams, 0.3–0.5; whitewater and waterfalls, 0.5 and above.

An exponential model is used to predict the oxygen deficit, D, as a function of time during a pure reoxygenation process. Reoxygenation constants are given for use with a base-10 and natural logarithmic base.

$$D_t = D_0 10^{-k_2 t} \text{ [base-10]} \qquad 4.12$$

$$D_t = D_0 e^{-k_2' t} \text{ [base-}e\text{]} \qquad 4.13$$

The base-e constant k_2' may be written as $k_{2,\text{base-}e}$ by some authorities. The constants k_2 and k_2' are different but related.

$$k_2' = 2.303 k_2 \qquad 4.14$$

k_2' can be approximated from the *O'Connor and Dobbins formula* for moderate to deep natural streams with low to moderate velocities. Both v and d are average values.

$$k_{2,20°C}' \approx \frac{3.93\sqrt{v_{m/s}}}{d_m^{1.5}}$$

$$\begin{bmatrix} 0.3 \text{ m } < d < 9.14 \text{ m} \\ 0.15 \text{ m/s } < v < 0.49 \text{ m/s} \end{bmatrix}$$

[SI]   *4.15(a)*

$$k_{2,68°F}' \approx \frac{12.9\sqrt{v_{ft/sec}}}{d_{ft}^{1.5}}$$

$$\begin{bmatrix} 1 \text{ ft} < d < 30 \text{ ft} \\ 0.5 \text{ ft/sec } < v < 1.6 \text{ ft} \end{bmatrix}$$

[U.S.]   *4.15(b)*

An empirical formula that has been proposed for faster moving water is the *Churchill formula*.

$$k_{2,20°C}' \approx \frac{5.049 v_{m/s}^{0.969}}{d_m^{1.67}}$$

$$\begin{bmatrix} 0.61 < d < 3.35 \text{ m} \\ 0.55 \text{ m/s } < v < 1.52 \text{ m/s} \end{bmatrix}$$

[SI]   *4.16(a)*

$$k_{2,68°F}' \approx \frac{11.61 v_{ft/sec}^{0.969}}{d_{ft}^{1.67}}$$

$$\begin{bmatrix} 2 \text{ ft} < d < 11 \text{ ft} \\ 1.8 \text{ ft/sec } < v < 5 \text{ ft/sec} \end{bmatrix}$$

[U.S.]   *4.16(b)*

The variation in k_2' with temperature is given approximately by Eq. 4.17. The temperature constant $\theta = 1.024$ has been reported as 1.016 by some authorities.

Kinetic Temperature Corrections

$$k_T = k_{20}(\theta)^{T-20}$$

4.17

Equation 4.17 is actually a special case of Eq. 4.18, which gives the relationship between values of k_2' between any two temperatures.

$$k_{T_1} = k_{T_2}\theta^{T_1-T_2}$$

4.18

20. DEOXYGENATION

The rate of deoxygenation at time t is

$$r_{d,t} = -K_d DO$$

4.19

The *deoxygenation rate constant*, K_d, for treatment plant effluent is approximately 0.05–0.10 day^{-1} and is typically taken as 0.1 day^{-1} (base-10 values). (K_d may be represented as K and K_1 by some authorities.) For

highly polluted shallow streams, K_d can be as high as 0.25 day^{-1}. For raw sewage, it is approximately 0.15–0.30 day^{-1}. K_d' is the base-e version of K_d.

$$K_d' = 2.303 K_d$$

4.20

K_d for other temperatures can be found from Eq. 4.21. The *temperature variation constant*, θ_d, is often quoted in literature as 1.047. However, this value should not be used with temperatures below 68°F (20°C). Additional research suggests that θ_d varies from 1.135 for temperatures between 39°F and 68°F (4°C and 20°C) up to 1.056 for temperatures between 68°F and 86°F (20°C and 30°C).

$$K_{d,T}' = K_{d,20°C}'\theta_d^{T-20°C}$$

4.21

Equation 4.21 is actually a special case of Eq. 4.22, which gives the relationship between values of K_d' between any two temperatures.

$$K_{d,T_1}' = K_{d,T_2}'\theta_d^{T_1-T_2}$$

4.22

21. NOMENCLATURE

d	depth of flow	ft	m
D	oxygen deficit	mg/L	mg/L
DO	dissolved oxygen	mg/L	mg/L
k	specific growth rate factor	time^{-1}	time^{-1}
k	maximum rate of substrate utilization per unit mass	L/day-mg	L/d·mg
k_o	maximum specific growth rate coefficient	time^{-1}	time^{-1}
k_2	reoxgenation rate constant	day^{-1}	d^{-1}
k_d	endogenous decay coefficient	time^{-1}	time^{-1}
K	rate constant	day^{-1}	d^{-1}
K_d	deoxygenation rate constant	day^{-1}	d^{-1}
K_s	half-velocity coefficient	–	–
r	rate of change	time^{-1}	time^{-1}
S	concentration of growth-limiting nutrient	mg/L	mg/L
t	time	sec	s
T	temperature	°F	°C
v	velocity	ft/sec	m/s
X	microorganism concentration	mg/L	mg/L
Y	maximum yield coefficient	–	–

Symbols

θ	temperature constant	–	–
θ_d	temperature variation constant	–	–

Subscripts

0	initial
d	decay or deoxygenation
g	growth
r	reaeration, reoxygenation or river
sat	saturated
su	substrate utilization
t	time
T, T	temperature

5

Fate and Transport

1. ASSIMILATIVE CAPACITY

Assimilative capacity (or *absorptive capacity*) is the capacity of the environment to absorb discharges of waste. As long as the assimilative capacity of the environment is not reached, discharges will not lead to pollution. When a discharge causes the assimilative capacity to be exceeded, however, pollution results.

Assimilative capacity is considered relative to stock pollutants and fund pollutants. A *stock pollutant* is a substance or material for which the assimilative capacity is very small. Stock pollutants are toxic, and they accumulate in the environment over time with little or very slow degradation; essentially any discharge of a stock pollutant will result in an unacceptable negative impact. Examples of substances that are usually considered stock pollutants are dioxin (2,3,7,8-tetrachlorodibenzodioxin), which has a maximum contaminant level (MCL) of 3×10^{-8} μg/L, and selenium, which has an MCL of 50 μg/L.

A *fund pollutant* is a substance or material for which the assimilative capacity is relatively large. Fund pollutants produce little accumulation in the environment over time. Examples of fund pollutants are human and animal waste. For municipal wastewater, common discharge limits to surface waters are 20 mg/L biochemical oxygen demand (BOD) and 20 mg/L total suspended solids (TSS); this is about 10^{12} times greater than the MCL for dioxins.

Because they do not accumulate in the environment, fund pollutants are managed in two ways: by allowing natural degradation to convert the pollutant to an innocuous form and by creating conditions for enhanced degradation.

A common example involving both is the repeated withdrawal and discharge of water from a river that passes by several cities on its way to the ocean. A city takes water from the river, treats it to satisfy health concerns for human use, and distributes the water to homes and businesses, where it is contaminated with fund pollutants. This contaminated water is subsequently discharged to the sewers.

The sewer waters are collected for treatment at a wastewater treatment plant, then discharged back into the river. The water discharged to the river still contains some of the fund pollutants, and the last increment of treatment occurs in the river itself. This process repeats itself numerous times as the river passes other cities downstream. It is the nonaccumulative characteristic of the wastewater that allows this repeated use.

When wastewater is discharged to a receiving water such as a river or lake, the intent is to stay within the capacity of the water to assimilate the discharge. Assimilative capacity can be exceeded by organic or nutrient loading. Where a receiving water has a low assimilative capacity, the discharge criteria applied by the regulatory authority will be more stringent.

2. SELF-PURIFICATION

The *self-purification* of streams and rivers is a process closely related to assimilative capacity. Self-purification occurs when untreated or partially treated wastewater is discharged into a receiving water such as a stream or river. After the wastewater has mixed with the river water, the river experiences both deoxygenation (from the biological activity of the waste) and reaeration (from the agitation of the flow). If the body is large and is adequately oxygenated, the sewage's BOD may be satisfied without putrefaction.

When two streams merge, the characteristics of the streams are blended in the combined flow. If values of the characteristics of interest are known for the upstream flows, the blended concentration can be found as a weighted average. Equation 5.1 can be used to calculate the final temperature, dissolved oxygen, BOD, or suspended solids content immediately after two flows are mixed. (The subscript a is also used in literature to represent the condition immediately after mixing.)

$$C_f = \frac{C_1 Q_1 + C_2 Q_2}{Q_1 + Q_2} \qquad 5.1$$

As the discharged water flows away from the discharge point, the changing dissolved oxygen (DO) concentration will influence the ecosystem of the receiving water. This influence can be defined by the

Figure 5.1 *Self-Purification Zones*

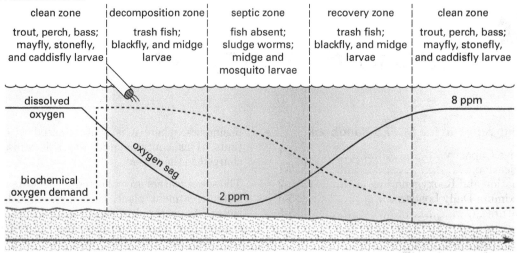

five stream self-purification zones shown in Fig. 5.1. The five zones are defined by unique physical, chemical, and biological characteristics and roughly coincide with the oxygen sag curve.

Example 5.1

6 MGD of wastewater with a dissolved oxygen concentration of 0.9 mg/L is discharged into a 50°F river flowing at 40 ft³/sec. The river is initially saturated with oxygen. What is the oxygen content of the river immediately after mixing?

Solution

Both flow rates must have the same units. Convert MGD to ft³/sec.

$$Q_w = \frac{(6 \text{ MGD})\left(10^6 \dfrac{\text{gal}}{\text{MG}}\right)}{\left(7.48 \dfrac{\text{gal}}{\text{ft}^3}\right)\left(24 \dfrac{\text{hr}}{\text{day}}\right)\left(60 \dfrac{\text{min}}{\text{hr}}\right)\left(60 \dfrac{\text{sec}}{\text{min}}\right)}$$

$$= 9.28 \text{ ft}^3/\text{sec}$$

The saturated oxygen content at 50°F (10°C) is 11.3 mg/L. [Dissolved-Oxygen Concentration in Water]

From Eq. 5.1,

$$C_f = \frac{C_1 Q_1 + C_2 Q_2}{Q_1 + Q_2}$$

$$= \frac{\left(0.9 \dfrac{\text{mg}}{\text{L}}\right)\left(9.28 \dfrac{\text{ft}^3}{\text{sec}}\right) + \left(11.3 \dfrac{\text{mg}}{\text{L}}\right)\left(40 \dfrac{\text{ft}^3}{\text{sec}}\right)}{9.28 \dfrac{\text{ft}^3}{\text{sec}} + 40 \dfrac{\text{ft}^3}{\text{sec}}}$$

$$= 9.34 \text{ mg/L}$$

3. DEOXYGENATION AND REOXYGENATION

The oxygen deficit is the difference between actual and saturated oxygen concentrations, expressed in mg/L. Since reoxygenation and deoxygenation of a polluted river occur simultaneously, the oxygen deficit will increase if the reoxygenation rate is less than the deoxygenation rate. If the oxygen content goes to zero, anaerobic decomposition and putrefaction will occur.

The minimum dissolved oxygen concentration that will protect aquatic life in the river is the *dissolved oxygen standard* for the river. 4–6 mg/L is the generally accepted range of dissolved oxygen required to support fish populations. 5 mg/L is adequate as is evidenced in high-altitude trout lakes. However, 6 mg/L is preferable, particularly for large fish populations.

The oxygen deficit at time t is given by the *Streeter-Phelps equation*. In Eq. 5.2, L_0 is the ultimate carbonaceous BOD of the river immediately after mixing, and D_0 is the initial dissolved oxygen deficit.

Stream Modeling

$$D = \frac{k_1 L_0}{k_2 - k_1}\left(e^{-k_1 t} - e^{-k_2 t}\right) + D_0 e^{-k_2 t} \qquad 5.2$$

A graph of dissolved oxygen versus time (or distance downstream) is known as a *dissolved oxygen sag curve* and is shown in Fig. 5.2. The location at which the lowest dissolved oxygen concentration occurs is the *critical point*. The location of the critical point is found from the river velocity and flow time to that point.

$$x_c = v t_c \qquad 5.3$$

Figure 5.2 *Oxygen Sag Curve*

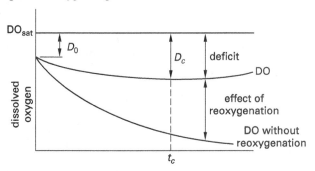

The time to the critical point is given by Eq. 5.4. The ratio k_2/k_1 in Eq. 5.4 is known as the *self-purification constant*.

Stream Modeling

$$t_c = \frac{1}{k_2 - k_1} \ln\left[\frac{k_2}{k_1}\left(1 - D_0\frac{(k_2 - k_1)}{k_1 L_0}\right)\right] \qquad 5.4$$

The critical oxygen deficit can be found by substituting t_c from Eq. 5.4 into Eq. 5.2. However, it is more expedient to use Eq. 5.5.

$$D_c = \left(\frac{k_1 L_0}{k_2}\right)e^{-k_1 t_c} \qquad 5.5$$

Example 5.2

A treatment plant effluent has the following characteristics.

quantity	$15 \text{ ft}^3/\text{sec}$
$BOD_{5,20°C}$	45 mg/L
DO	2.9 mg/L
temperature	$24°C$
$K_{d,20°C}$	0.23 day^{-1} (base e)
K_r	0.519 day^{-1} (base e)
temperature variation constant for K_d, θ_d	1.047
temperature variation constant for K_r, θ_r	1.016

The outfall is located in a river carrying $120 \text{ ft}^3/\text{sec}$ of water with the following characteristics.

quantity	$120 \text{ ft}^3/\text{sec}$
velocity	0.55 ft/sec
average depth	4 ft
$BOD_{5,20°C}$	4 mg/L
DO	8.3 mg/L
temperature	$16°C$

When the treatment plant effluent and river are mixed, the mixture temperature persists for a long distance downstream.

Determine the BOD_5, DO, and temperature immediately after mixing.

Solution

(a) To find the river conditions immediately after mixing, use Eq. 5.1 three times.

$$C_f = \frac{C_1 Q_1 + C_2 Q_2}{Q_1 + Q_2}$$

$$BOD_{5,20°C} = \frac{\left(15 \frac{\text{ft}^3}{\text{sec}}\right)\left(45 \frac{\text{mg}}{\text{L}}\right) + \left(120 \frac{\text{ft}^3}{\text{sec}}\right)\left(4 \frac{\text{mg}}{\text{L}}\right)}{15 \frac{\text{ft}^3}{\text{sec}} + 120 \frac{\text{ft}^3}{\text{sec}}}$$

$$= 8.56 \text{ mg/L}$$

$$DO = \frac{\left(15 \frac{\text{ft}^3}{\text{sec}}\right)\left(2.9 \frac{\text{mg}}{\text{L}}\right) + \left(120 \frac{\text{ft}^3}{\text{sec}}\right)\left(8.3 \frac{\text{mg}}{\text{L}}\right)}{15 \frac{\text{ft}^3}{\text{sec}} + 120 \frac{\text{ft}^3}{\text{sec}}}$$

$$= 7.7 \text{ mg/L}$$

$$T = \frac{\left(15 \frac{\text{ft}^3}{\text{sec}}\right)(24°C) + \left(120 \frac{\text{ft}^3}{\text{sec}}\right)(16°C)}{15 \frac{\text{ft}^3}{\text{sec}} + 120 \frac{\text{ft}^3}{\text{sec}}}$$

$$= 16.89°C$$

Calculate the approximate rate constants. Although the temperature is low, θ_d is specified in the problem statement.

Kinetic Temperature Corrections

$$K_{d,T} = K_{d,20°C}(1.047)^{T-20°C}$$

$$K_{d,16.89°C} = (0.23 \text{ day}^{-1})(1.047)^{16.89°C-20°C}$$

$$= 0.199 \text{ day}^{-1}$$

Using the given value of θ,

Kinetic Temperature Corrections

$$K_{r,T} = K_{r,20°C}(1.016)^{T-20°C}$$

$$K_{r,16.89°C} = (0.519 \text{ day}^{-1})(1.016)^{16.89°C-20°C}$$

$$= 0.494 \text{ day}^{-1}$$

To estimate BOD_u, use the equation for BOD exertion. Represent y_t by BOD_5 and L by BOD_u, then rearrange to solve for BOD_u.

BOD Exertion

$$y_t = L(1 - e^{-kt})$$

$$BOD_{5,20°C} = BOD_u(1 - e^{-K_d t})$$

$$BOD_u = \frac{BOD_{5,20°C}}{1 - e^{-K_d t}} = \frac{8.56 \ \frac{mg}{L}}{1 - e^{(-0.23 \text{ day}^{-1})(5 \text{ days})}}$$

$$= 12.52 \text{ mg/L}$$

Calculate the original oxygen deficit. By interpolation from a table of dissolved-oxygen concentration as a function of temperature and salinity, the dissolved oxygen concentration at 16.89°C is approximately 9.67 mg/L.

Dissolved Oxygen Concentration in Water

$$D_0 = 9.67 \ \frac{mg}{L} - 7.7 \ \frac{mg}{L}$$

$$= 1.97 \text{ mg/L}$$

Calculate t_c from Eq. 5.4.

$$t_c = \left(\frac{1}{K_r - K_d}\right)$$

$$\times \ln\left(\left(\frac{K_d BOD_u - K_r D_0 + K_d D_0}{K_d BOD_u}\right)\left(\frac{K_r}{K_d}\right)\right)$$

$$= \left(\frac{1}{0.494 \text{ day}^{-1} - 0.199 \text{ day}^{-1}}\right)$$

$$\times \ln\left(\begin{array}{c} \dfrac{\left(\begin{array}{c}(0.199 \text{ day}^{-1})\left(11.74 \ \frac{mg}{L}\right) \\ -(0.494 \text{ day}^{-1})\left(1.97 \ \frac{mg}{L}\right) \\ +(0.199 \text{ day}^{-1})\left(1.97 \ \frac{mg}{L}\right)\end{array}\right)}{(0.199 \text{ day}^{-1})\left(11.74 \ \frac{mg}{L}\right)} \\ \times \left(\dfrac{0.494 \text{ day}^{-1}}{0.199 \text{ day}^{-1}}\right)\end{array}\right)$$

$$= 2.16 \text{ days}$$

The distance downstream is

$$x_c = v t_c$$

$$= \frac{\left(0.55 \ \frac{ft}{sec}\right)(2.16 \text{ days})\left(86,400 \ \frac{sec}{day}\right)}{5280 \ \frac{ft}{mi}}$$

$$= 19.4 \text{ mi}$$

(b) The critical oxygen deficit is found from Eq. 5.5.

$$D_c = \left(\frac{k_1 L_0}{k_2}\right) e^{-k_1 t_c}$$

$$= \left(\frac{(0.199 \text{ day}^{-1})\left(12.52 \ \frac{mg}{L}\right)}{0.494 \text{ day}^{-1}}\right)$$

$$\times \left(e^{(-0.199 \text{ day}^{-1})(2.16 \text{ days})}\right)$$

$$= 3.28 \text{ mg/L}$$

The saturated oxygen content at the critical point with a water temperature of 16.89°C is 9.67 mg/mL. The actual oxygen content is

$$DO = DO_{sat} - D_c = 9.67 \ \frac{mg}{L} - 3.28 \ \frac{mg}{L}$$

$$= 6.39 \text{ mg/L}$$

Since this is greater than 4–6 mg/L, fish life is supported.

4. TOTAL MAXIMUM DAILY LOADS

In some cases, receiving waters may be impaired under Sec. 303(d) of the Clean Water Act. These are waters that do not meet defined water quality standards; consequently, the jurisdictional authority is required to develop *total maximum daily loads* (TMDLs) for these waters. The TMDL establishes the maximum allowed loading to the receiving water. The TMDL is calculated using Eq. 5.6.

$$\text{TMDL} = \sum \text{WLA} + \sum \text{LA} + \text{MOS} \qquad 5.6$$

5. NOMENCLATURE

BOD	biochemical oxygen demand	mg/L	mg/L
C	concentration	mg/L	mg/L
C	river condition	–	–
D	oxygen deficit	mg/L	mg/L
DO	dissolved oxygen	mg/L	mg/L
k_1	deoxygenation rate constant	base e (days^{-1})	base e (d^{-1})
k_2	reaeration rate constant	base e (days^{-1})	base e (d^{-1})
K	rate constant	day^{-1}	d^{-1}
L	carbonaceous BOD	mg/L	mg/L
LA	load allocation for nonpoint sources	g/day	g/d
MCL	maximum contaminant level	mμg/L	
MOS	margin of safety	g/day	g/d
Q	flow rate	gal/day	L/d
t	time	sec	s
T	temperature	°F	°C
TMDL	total maximum daily load	g/day	g/d
TSS	total suspended solids	mg/L	mg/L
v	velocity	ft/sec	m/s
WLA	waste load allocation	g/day	g/d
x	distance	ft	m
X	river condition	–	–
y_t	BOD exerted at time t	–	mg/L

Symbols

θ	temperature	°F	°C

Subscripts

0	initial, ultimate
c	critical
d	decay or deoxygenation
f	final
r	reoxygenation or river
sat	saturated
t	time
T	temperature
u	ultimate

 **Sampling and Measurement Methods**

Content in blue refers to the *NCEES Handbook*.

1. WASTEWATER COMPOSITING

Proper sampling techniques are essential for accurate evaluation of wastewater flows. Samples should be well mixed, representative of the wastewater flow stream in composition, taken at regular time intervals, and stored properly until analysis can be performed. *Compositing* is the sampling procedure that accomplishes these goals.

Samples are collected (i.e., "grabbed," hence the name *grab sample*) at regular time intervals (e.g., every hour on the hour), stored in a refrigerator or ice chest, and then integrated to formulate the desired combination for a particular test. Flow rates are measured at each sampling to determine the wastewater flow pattern.

The total volume of the composite sample desired depends on the kinds and number of laboratory tests to be performed.

Sampling and measurement methods are part of any program to assess, manage, treat, collect, distribute, or protect water as a resource or commodity. These methods are the same regardless of the end goal. Whether for the treatment of potable water or wastewater, the same sampling and measurement methods are used to define the types and levels of treatment required. The effects of stormwater runoff and the effects of weather from both urban and rural landscapes are evaluated using the same methods.

Table 6.1 lists basic parameters used in evaluating water quality that are of common interest to environmental engineers.

The collection and analysis of water samples for these parameters follow protocol defined by *Standard Methods for the Examination of Water and Wastewater*, a joint publication of the American Public Health Association (APHA), the American Water Works Association (AWWA), and the Water Environment Federation (WEF). The U.S. Environmental Protection Agency also publishes protocol for the analysis of chemical, physical, and biological components of water and wastewater samples required by the Clean Water Act (CWA) and the Safe Drinking Water Act (SDWA).

Table 6.1 Common Water Quality Parameters

category	parameter
solids	total solids (TS)
	total suspended solids (TSS)
	total dissolved solids (TDS)
	volatile suspended solids (VSS)
oxygen	biochemical oxygen demand (BOD)
	chemical oxygen demand (COD)
	theoretical oxygen demand (ThOD)
ionic materials	alkalinity
	total hardness
	carbonate hardness
	noncarbonate hardness
	specific conductance
nutrients	total Kjeldahl nitrogen (TKN)
	organic nitrogen
	ammonia nitrogen (NH_3)
	nitrates (NO_3^-) and nitrites (NO_2^-)
	total phosphorus
	phosphates (PO_4^{3-})
pathogens	viruses
	bacteria
	parasites
chemicals	metals
	trace organic chemicals

Sampling and measurement methods related to specific areas are discussed in more detail in Chap. 4 (biology and microbiology), Chap. 9 (minimization and prevention of pollution), and Chap. 16 (potable water).

7 Hydrology and Hydrogeology

1. HYDROLOGIC CYCLE

The *hydrologic cycle* is the full "life cycle" of water. The cycle begins with *precipitation*, which encompasses all of the hydrometeoric forms, including rain, snow, sleet, and hail from a storm. Precipitation can (a) fall on vegetation and structures and evaporate back into the atmosphere, (b) be absorbed into the ground and either make its way to the water table or be absorbed by plants after which it evapotranspires back into the atmosphere, or (c) travel as surface water to a depression, watershed, or creek from which it either evaporates back into the atmosphere, infiltrates into the ground water system, or flows off in streams and rivers to an ocean or lakes. The cycle is completed when lake and ocean water evaporates into the atmosphere.

The *water balance equation* (*water budget equation*) is the application of conservation to the hydrologic cycle.

$$
\begin{aligned}
\text{total} &= \text{net change in surface water removed}\\
\text{precipitation} &\\
&+ \text{net change in ground water removed}\\
&+ \text{evapotranspiration}\\
&+ \text{interception evaporization} \qquad \textit{7.1}\\
&+ \text{net increase in surface water storage}\\
&+ \text{net increase in ground water storage}\\
P &= Q + E + \Delta S
\end{aligned}
$$

The total amount of water that is intercepted (and subsequently evaporates) and absorbed into ground water before runoff begins is known as the *initial abstraction*. Even after runoff begins, the soil continues to absorb some infiltrated water. Initial abstraction and infiltration do not contribute to surface runoff. Equation 7.1 can be restated as Eq. 7.2.

$$
\begin{aligned}
\text{total}\atop\text{precipitation} &= \text{initial abstraction} + \text{infiltration}\\
&\quad + \text{surface runoff}
\end{aligned} \qquad \textit{7.2}
$$

2. STORM CHARACTERISTICS

Storm rainfall characteristics include the duration, total volume, intensity, and areal distribution of a storm. Storms are also characterized by their recurrence intervals. (See Sec. 7.6.)

The duration of storms is measured in hours and days. The volume of rainfall is simply the total quantity of precipitation dropping on the watershed. Average rainfall intensity is the volume divided by the duration of the storm. Average rainfall can be considered to be generated by an equivalent theoretical storm that drops the same volume of water uniformly and constantly over the entire watershed area.

A *storm hyetograph* is the instantaneous rainfall intensity measured as a function of time, as shown in Fig. 7.1(a). Hyetographs are usually bar graphs showing constant rainfall intensities over short periods of time. Hyetograph data can be reformulated as a *cumulative rainfall curve*, as shown in Fig. 7.1(b). Cumulative rainfall curves are also known as *rainfall mass curves*.

3. PRECIPITATION DATA

Precipitation data on rainfall can be collected in a number of ways, but use of an open precipitation rain gauge is quite common. This type of gauge measures only the volume of rain collected between readings, usually 24 hr.

The *average precipitation* over a specific area can be found from station data in several ways.

Method 1: If the stations are uniformly distributed over a flat site, their precipitations can be averaged. This also requires that the individual precipitation records not vary too much from the mean.

Method 2: The *Thiessen method* calculates the average weighting station measurements by the area of the assumed watershed for each station. These assumed

Figure 7.1 *Storm Hyetograph and Cumulative Rainfall Curves*

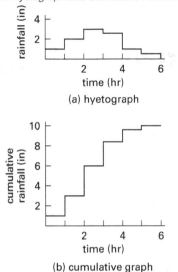

(a) hyetograph

(b) cumulative graph

watershed areas are found by drawing dotted lines between all stations and bisecting these dotted lines with solid lines (which are extended outward until they connect with other solid lines). The solid lines will form a polygon whose area is the assumed watershed area.

Method 3: The *isohyetal method* requires plotting lines of constant precipitation (*isohyets*) and weighting the isohyet values by the areas enclosed by the lines. This method is the most accurate of the three methods, as long as there are enough observations to permit drawing of isohyets. Station data are used to draw isohyets, but they are not used in the calculation of average rainfall.

4. ESTIMATING UNKNOWN PRECIPITATION

If a precipitation measurement at a location is unknown, it may still be possible to estimate the value by one of the following procedures.

Method 1: Choose three stations close to and evenly spaced around the location with missing data. If the normal annual precipitations at the three sites do not vary more than 10% from the missing station's normal annual precipitation, the rainfall can be estimated as the arithmetic mean of the three neighboring stations' precipitations for the period in question.

Method 2: If the precipitation difference between locations is more than 10%, the *normal-ratio method* can be used. In Eq. 7.3, P_x is the precipitation at the missing station; N_x is the long-term normal precipitation at the missing station; P_A, P_B, and P_C are the precipitations at known stations; and N_A, N_B, and N_C are the long-term normal precipitations at the known stations.

$$P_x = \frac{1}{3}\left(\left(\frac{N_x}{N_A}\right)P_A + \left(\frac{N_x}{N_B}\right)P_B + \left(\frac{N_x}{N_C}\right)P_C\right) \qquad 7.3$$

Method 3: Use data from stations in the four nearest quadrants (north, south, east, and west of the unknown station) and weight the data with the inverse squares of the distance between the stations. In Eq. 7.4, P_x, P_A, P_B, P_C, and P_D are defined as in Method 2, and $d_{A\text{-}x}$, $d_{B\text{-}x}$, $d_{C\text{-}x}$, and $d_{D\text{-}x}$ are the distances between stations A and x, B and x, and so on, respectively.

$$P_x = \frac{\dfrac{P_A}{d_{A\text{-}x}^2} + \dfrac{P_B}{d_{B\text{-}x}^2} + \dfrac{P_C}{d_{C\text{-}x}^2} + \dfrac{P_D}{d_{D\text{-}x}^2}}{\dfrac{1}{d_{A\text{-}x}^2} + \dfrac{1}{d_{B\text{-}x}^2} + \dfrac{1}{d_{C\text{-}x}^2} + \dfrac{1}{d_{D\text{-}x}^2}} \qquad 7.4$$

5. TIME OF CONCENTRATION

Time of concentration, t, is defined as the time of travel from the hydraulically most remote (timewise) point in the watershed to the watershed outlet or other design point. For points (e.g., manholes) along storm drains being fed from a watershed, time of concentration is taken as the largest combination of overland flow time (sheet flow), swale or ditch flow (shallow concentrated flow), and storm drain, culvert, or channel time. It is unusual for time of concentration to be less than 0.1 hr (6 min) when using the Natural Resources Conservation Service (NRCS, previously known as the Soil Conservation Service (SCS)) method, or less than 10 min when using the rational method.

$$t = t_{\text{sheet}} + t_{\text{shallow}} + t_{\text{channel}} \qquad 7.5$$

The NRCS specifies using the *Manning kinematic equation* (Overton and Meadows 1976 formulation) for calculating *sheet flow* or *overland flow* travel time over distances less than 300 ft (100 m). In Eq. 7.6, C is the Manning roughness coefficient for sheet or overland flow, as given in Table 7.1. i is the rainfall intensity in in/hr, and S is the slope of the hydraulic grade line in ft/ft.

Time of Concentration

$$t_i = C(L/Si^2)^{1/3} \qquad 7.6$$

Table 7.1 *Manning Roughness Coefficient for Sheet or Overland Flow*

surface	C
smooth surfaces (concrete, asphalt, gravel, or bare soil)	0.011
fallow (no residue cover)	0.05
cultivated soils	
residue cover $\leq 20\%$	0.06
residue cover $> 20\%$	0.17
grasses	
short prairie grass	0.15
dense grass[a]	0.24
Bermuda grass	0.41
range, natural	0.13
woods[b]	
light underbrush	0.40
dense underbrush	0.80

[a]This includes species such as weeping lovegrass, bluegrass, buffalo grass, blue grama grass, and native grass mixtures.
[b]When selecting a value of n, consider the cover to a height of about 0.1 ft (3 cm). This is the only part of the plant that will obstruct sheet flow.
Reprinted from *Urban Hydrology for Small Watersheds*, Technical Release TR-55, United States Department of Agriculture, Natural Resources Conservation Service, Table 3-1, after Engman (1986).

After about 300 ft (100 m), the flow usually becomes a shallow concentrated flow (swale, ditch flow). Travel time is calculated as L/v. Velocity can be found from the Manning equation if the flow geometry is well defined, but must be determined from other correlations, such as those specified by the NRCS in Eq. 7.7 and Eq. 7.8.

$$v_{\text{shallow,ft/sec}} = 16.1345\sqrt{S_{\text{decimal}}} \quad \text{[unpaved]} \qquad 7.7$$

$$v_{\text{shallow,ft/sec}} = 20.3282\sqrt{S_{\text{decimal}}} \quad \text{[paved]} \qquad 7.8$$

Storm drain (channel) time is found by dividing the storm drain length by the actual or an assumed channel velocity. Storm drain velocity is found from either the Manning or the Hazen-Williams equation. Since size and velocity are related, an iterative trial-and-error solution is generally required.[1]

There are a variety of methods available for estimating time of concentration. Early methods include Kirpich (1940), California Culverts Practice (1942), Hathaway (1945), and Izzard (1946). More recent methods include those from the Federal Aviation Administration, or FAA (1970), the kinematic wave formulas of Morgali (1965) and Aron (1973), the NRCS lag equation (1975), and NRCS average velocity charts (1975). Estimates of time of concentration from these methods can vary by

as much as 100%. These differences carry over into estimates of peak flow, hence the need to carefully determine the validity of any method used.

The distance L_o in the various equations that follow is the longest distance to the collection point, as shown in Fig. 7.2.

Figure 7.2 *Overland Flow Distances*

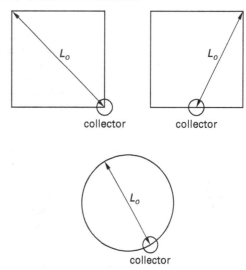

For irregularly shaped drainage areas, it may be necessary to evaluate several alternative overland flow distances. For example, Fig. 7.3 shows a drainage area with a long tongue. Although the tongue area contributes little to the drainage area, it does lengthen the overland flow time. Depending on the intensity-duration-frequency curve, the longer overland flow time (resulting in a lower rainfall intensity) may offset the increase in area due to the tongue. Therefore, two runoffs need to be compared, one ignoring and the other including the tongue.

Figure 7.3 *Irregular Drainage Area*

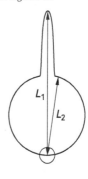

---

[1]If the pipe or channel size is known, the velocity can be found from $Q = Av$. If the pipe size is not known, the area will have to be estimated. In that case, one might as well estimate velocity instead. 5 ft/sec (1.5 m/s) is a reasonable flow velocity for open channel flow. The minimum velocity for a self-cleansing pipe is 2 ft/sec (0.6 m/s).

The *FAA formula*, Eq. 7.9, was developed from airfield drainage data collected by the Army Corps of Engineers. However, it has been widely used for urbanized areas. C is the rational method runoff coefficient. The slope S, in Eq. 7.9 is in percent. L_o is in ft.

$$t_{\min} = \frac{(1.8)(1.1 - C)\sqrt{L_{o,\text{ft}}}}{S_{\text{percent}}^{1/3}} \qquad 7.9$$

The *kinematic wave formula* is particularly accurate for uniform planar homogenous areas (e.g., paved areas such as parking lots and streets). It requires iteration, since both intensity and time of concentration are unknown. In Eq. 7.10, L_o is in ft, n is the *Manning overland roughness coefficient (retardance roughness coefficient)*, I is the intensity in in/hr, and S_{decimal} is the slope in ft/ft. Recommended values of n for this application are different than for open channel flow and are: smooth impervious surfaces, 0.011–0.014; smooth bare-packed soil, free of stones, 0.05; poor grass, moderately bare surface, 0.10; pasture or average grass cover, 0.20; and dense grass or forest, 0.40. Equation 7.10 is solved iteratively since the intensity depends on the time to concentration.

$$t_{\min} = \frac{0.94 L_{o,\text{ft}}^{0.6} n^{0.6}}{I_{\text{in/hr}}^{0.4} S_{\text{decimal,ft/ft}}^{0.3}} \qquad 7.10$$

The NRCS *lag equation* was developed from observations of agricultural watersheds where overland flow paths are poorly defined and channel flow is absent. However, it has been adapted to small urban watersheds under 2000 ac. The equation performs reasonably well for areas that are completely paved, as well. Correction factors are used to account for channel improvement and impervious areas. L_o is in feet, CN is the NRCS runoff curve number, S_{in} is the potential maximum retention in the watershed after runoff begins in inches, and S_{percent} is the average slope in percent. The factor 1.67 converts the watershed lag time to the time of concentration. Since the formula overestimates time for mixed areas, different adjustment factors have been proposed. The NRCS lag equation performs poorly when channel flow is a significant part of the time of concentration.

$$t_{\min} = 1.67 t_{\text{watershed lag time, min}}$$

$$= \frac{(1.67)\left(60\,\dfrac{\min}{\text{hr}}\right) L_{o,\text{ft}}^{0.8} (S_{\text{in}} + 1)^{0.7}}{1900\sqrt{S_{\text{percent}}}}$$

$$\qquad 7.11$$

$$= \frac{(1.67)\left(60\,\dfrac{\min}{\text{hr}}\right) L_{o,\text{ft}}^{0.8}\left(\dfrac{1000}{\text{CN}} - 9\right)^{0.7}}{1900\sqrt{S_{\text{percent}}}}$$

If the velocity of runoff water is known, the time of concentration can be easily determined. The NRCS has published charts of average velocity as functions of watercourse slope and surface cover. (See Fig. 7.4.) The charts are best suited for flow paths of at least several hundred feet (at least 70 m). The time of concentration is easily determined from these charts as

$$t = \frac{\sum L_{o,\text{ft}}}{\text{v}\left(60\,\dfrac{\text{sec}}{\min}\right)} \qquad 7.12$$

Figure 7.4 *NRCS Average Velocity Chart for Overland Flow Travel Time*

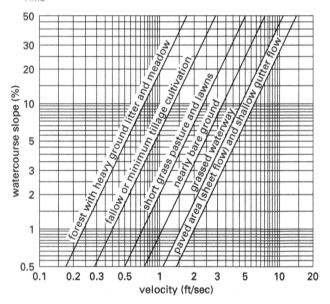

Reprinted from NRCS TR-55-1975. TR-55-1986 contains a similar graph for shallow-concentrated flow over paved and unpaved surfaces, but it does not contain this particular graph.

6. RAINFALL INTENSITY

Effective design of a surface feature depends on its geographical location and required degree of protection. Once the location of the feature is known, the *design storm* (or *design flood*) must be determined based on some probability of recurrence.

Rainfall intensity is the amount of precipitation per hour. The instantaneous intensity changes throughout the storm. However, it may be averaged over short time intervals or over the entire storm duration. Average intensity will be low for most storms, but it can be high for some. These high-intensity storms can be expected infrequently, say, every 20, 50, or 100 years. The average number of years between storms of a given intensity is known as the *frequency of occurrence (recurrence interval, return interval,* or *storm frequency)*.

In general, the design storm may be specified by its recurrence interval (e.g., "100-year storm"), its annual probability of occurrence (e.g., "1% storm"), or a nickname (e.g.,"century storm"). A 1% storm is a storm that would be exceeded in severity only once every hundred years on the average.

The average intensity of a storm over a time period t can be calculated from Eq. 7.13 (and similar correlations). In the United States, it is understood that the units of intensity calculated using Eq. 7.13 will be in in/hr.

$$I = \frac{K' F^a}{(t+b)^c} \qquad 7.13$$

K', F, a, b, and c are constants that depend on the conditions, recurrence interval, and location of a storm. For many reasons, these constants may be unavailable. The *Steel formula* is a simplification of Eq. 7.13. Steel formula rainfall regions are shown in Fig. 7.5.

$$I = \frac{K}{t+b} \qquad 7.14$$

Figure 7.5 *Steel Formula Rainfall Regions*

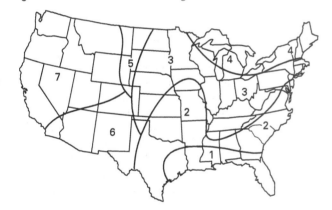

Values of the constants K and b in Eq. 7.14 are not difficult to obtain once the intensity-duration-frequency curve is established. Although a logarithmic transformation could be used to convert the data to straight-line form, an easier method exists. This method starts by taking the reciprocal of Eq. 7.14 and converting the equation to a straight line.

$$\begin{aligned} \frac{1}{I} &= \frac{t+b}{K} \\ &= \frac{t}{K} + \frac{b}{K} \qquad 7.15 \\ &= C_1 t + C_2 \end{aligned}$$

Once C_1 and C_2 have been found, K and b can be calculated.

$$K = \frac{1}{C_1} \qquad 7.16$$

$$b = \frac{C_2}{C_1} \qquad 7.17$$

Published values of K and b can be obtained from compilations, but these values are suitable only for very rough estimates. Table 7.2 is typical of some of this general data.

The total rainfall can be calculated from the average intensity and duration.

$$P = It \qquad 7.18$$

Table 7.2 *Steel Formula Coefficients (for intensities of in/hr)*

frequency in years	coefficients	region 1	2	3	4	5	6	7
2	K	206	140	106	70	70	68	32
	b	30	21	17	13	16	14	11
4	K	247	190	131	97	81	75	48
	b	29	25	19	16	13	12	12
10	K	300	230	170	111	111	122	60
	b	36	29	23	16	17	23	13
25	K	327	260	230	170	130	155	67
	b	33	32	30	27	17	26	10
50	K	315	350	250	187	187	160	65
	b	28	38	27	24	25	21	8
100	K	367	375	290	220	240	210	77
	b	33	36	31	28	29	26	10

(Multiply in/hr by 2.54 to obtain cm/h.)

Rainfall data can be compiled into *intensity-duration-frequency curves* (*IDF curves*) similar to those shown in Fig. 7.6.

Figure 7.6 *Typical Intensity-Duration-Frequency Curves*

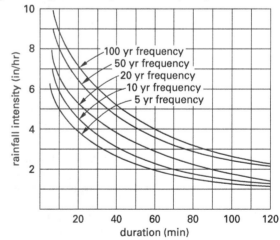

Example 7.1

A storm has an intensity given by

$$I = \frac{100}{t + 10}$$

15 min are required for runoff from the farthest corner of a 5 ac watershed to reach a discharge culvert. What is the design intensity?

Solution

The intensity is

$$I = \frac{100}{t + 10} = \frac{100 \; \frac{\text{in-min}}{\text{hr}}}{15 \; \text{min} + 10 \; \text{min}}$$
$$= 4 \; \text{in/hr}$$

7. FLOODS

A *flood* occurs when more water arrives than can be drained away. When a watercourse (i.e., a creek or river) is too small to contain the flow, the water overflows the banks.

The flooding may be categorized as nuisance, damaging, or devastating. *Nuisance floods* result in inconveniences such as wet feet, tire spray, and soggy lawns. *Damaging floods* soak flooring, carpeting, and first-floor furniture. *Devastating floods* wash buildings, vehicles, and live-stock downstream.

Although rain causes flooding, large storms do not always cause floods. The size of a flood depends not only on the amount of rainfall, but also on the conditions within the watershed before and during the storm. Run-off will occur only when the rain falls on a very wet watershed that is unable to absorb additional water, or when a very large amount of rain falls on a dry water-shed faster than it can be absorbed.

Specific terms are sometimes used to designate the degree of protection required. For example, the *probable maximum flood* (PMF) is a hypothetical flood that can be expected to occur as a result of the most severe com-bination of critical meteorologic and hydrologic condi-tions possible within a region.

Designing for the *probable maximum precipitation* (PMP) or probable maximum flood is very conservative and usually uneconomical since the recurrence interval for these events exceeds 100 years and may even approach 1000 years. Designing for 100-year floods and floods with even lower recurrence intervals is more com-mon. (100-year floods are not necessarily caused by 100-year storms.)

The *design flood* or *design basis flood* (DBF) depends on the site. It is the flood that is adopted as the basis for design of a particular project. The DBF is usually deter-mined from economic considerations, or it is specified as part of the contract document.

The *standard flood* or *standard project flood* (SPF) is a flood that can be selected from the most severe combi-nations of meteorological and hydrological conditions reasonably characteristic of the region, excluding extremely rare combinations of events. SPF volumes are commonly 40–60% of the PMF volumes.

The probability that a flooding event in any given year will equal a design basis flood with a *recurrence interval frequency* (*return interval*) of T years is

$$P(X \geq x_T \text{ in one year}) = \frac{1}{T} \qquad \textit{7.19}$$

The probability of a T event occurring in n years is

Storm Return Period

$$P\left(\begin{array}{c} X \geq x_T \text{ at least once} \\ \text{in } n \text{ years} \end{array}\right) = 1 - \left(1 - \frac{1}{T}\right)^n \qquad \textit{7.20}$$

Planning for a 1% flood has proven to be a good compro-mise between not doing enough and spending too much. Although the 1% flood is a common choice for the design basis flood, shorter recurrence intervals are often used, par-ticularly in low-value areas such as cropland. For example, a 5-year value can be used in residential areas, a 10-year value in business sections, and a 15-year value for high-value districts where flooding will result in more extensive damage. The ultimate choice of recurrence interval, how-ever, must be made on the basis of economic considerations and trade-offs.

Example 7.2

A wastewater treatment plant has been designed to be in use for 40 years. What is the probability that a 1% flood will occur within the useful lifetime of the plant?

Solution

Use Eq. 7.20.

$$P(X \geq x_T \text{ at least once in } n \text{ years}) = 1 - \left(1 - \frac{1}{T}\right)^n$$

$$P(X \geq x_{100} \text{ at least once in 40 years}) = 1 - \left(1 - \frac{1}{100}\right)^{40}$$
$$= 0.33 \quad (33\%)$$

8. TOTAL SURFACE RUNOFF FROM STREAM HYDROGRAPH

After a rain, runoff and groundwater increases stream flow. A plot of the stream discharge versus time is known as a *hydrograph*. Hydrograph periods may be very short (e.g., hours) or very long (e.g., days, weeks, or months). A typical hydrograph is shown in Fig. 7.7. The *time base* is the length of time that the stream flow exceeds the original *base flow*. The flow rate increases on the *rising limb* (*concentration curve*) and decreases on the *falling limb* (*recession curve*).

Figure 7.7 *Stream Hydrograph*

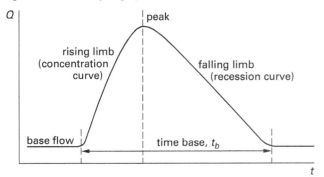

9. HYDROGRAPH SEPARATION

The stream discharge consists of both surface runoff and subsurface groundwater flows. A procedure known as *hydrograph separation* or *hydrograph analysis* is used to separate runoff (*surface flow, net flow,* or *overland flow*) and groundwater (*subsurface flow, base flow*).[2]

There are several methods of separating groundwater from runoff. Most of the methods are somewhat arbitrary. Three methods that are easily carried out manually are presented here.

Method 1: In the *straight-line method*, a horizontal line is drawn from the start of the rising limb to the falling limb. All of the flow under the horizontal line is considered base flow. This assumption is not theoretically accurate, but the error can be small. This method is illustrated in Fig. 7.8.

Figure 7.8 *Straight-Line Method*

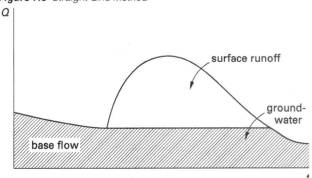

Method 2: In the *fixed-base method*, shown in Fig. 7.9, the base flow existing before the storm is projected graphically down to a point directly under the peak of the hydrograph. Then, a straight line is used to connect the projection to the falling limb. The duration of the recession limb is determined by inspection, or it can be calculated from correlations with the drainage area.

Figure 7.9 *Fixed-Base Method*

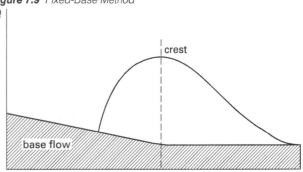

Method 3: The *variable-slope method*, as shown in Fig. 7.10, recognizes that the shape of the base flow curve before the storm will probably match the shape of the base flow curve after the storm. The groundwater curve after the storm is projected back under the hydrograph to a point under the inflection point of the falling limb. The separation line under the rising limb is drawn arbitrarily.

Figure 7.10 *Variable-Slope Method*

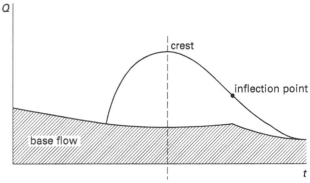

Once the base flow is separated out, the hydrograph of surface runoff will have the approximate appearance of Fig. 7.11.

10. UNIT HYDROGRAPH

Once the overland flow hydrograph for a watershed has been developed, the total runoff (i.e., "excess rainfall") volume, V, from the storm can be found as the area under the curve. Although this can be found by

---

[2]The total rain dropped by a storm is the *gross rain*. The rain that actually appears as immediate runoff can be called *surface runoff, overland flow, surface flow,* and *net rain*. The water that is absorbed by the soil and that does not contribute to the surface runoff can be called *base flow, groundwater, infiltration,* and *dry weather flow*.

Water

Figure 7.11 *Overland Flow Hydrograph*

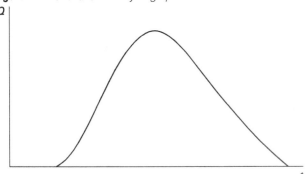

integration, planimetry, or computer methods, it is often sufficiently accurate to approximate the hydrograph with a histogram and to sum the areas of the rectangles. (See Fig. 7.12.)

Figure 7.12 *Hydrograph Histogram*

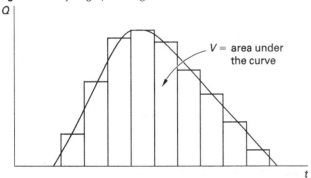

Since the area of the watershed is known, the average depth, P_{ave}, of the excess precipitation can be calculated. (Consistent units must be used in Eq. 7.21.)

$$V = A_d P_{ave,excess} \qquad 7.21$$

A *unit hydrograph* is developed by dividing every point on the overland flow hydrograph by the average excess precipitation, $P_{ave,excess}$. This is a hydrograph of a storm dropping 1 in (1 cm) of excess precipitation (runoff) evenly on the entire watershed. Units of the unit hydrograph are in/in (cm/cm). Figure 7.13 shows how a unit hydrograph compares to its surface runoff hydrograph.

Once a unit hydrograph has been developed from historical data of a particular storm volume, it can be used for other storm volumes. Such application is based on several assumptions: (a) all storms in the watershed have the same duration, (b) the time base is constant for all storms, (c) the shape of the rainfall curve is the same for all storms, and (d) only the total amount of rainfall varies from storm to storm.

The hydrograph of a storm producing more or less than 1 in (1 cm) of rain is found by multiplying all ordinates on the unit hydrograph by the total precipitation of the storm.

Figure 7.13 *Unit Hydrograph*

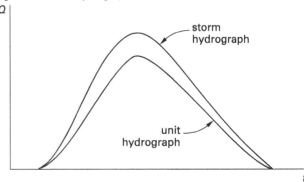

A unit hydrograph can be used to predict the runoff for storms that have durations somewhat different than the storms used to develop the unit hydrograph. Generally, storm durations differing by up to $\pm 25\%$ are considered to be equivalent.

Example 7.3

After a 2 hr storm, a station downstream from a 45 mi^2 (115 km^2) drainage watershed records a peak discharge of 9400 ft^3/sec (250 m^3/s) and a total runoff of 3300 ac-ft (4×10^6 m^3). (a) What is the unit hydrograph peak discharge? (b) What would be the peak runoff and design flood volume if a 2 hr storm dropped 2.5 in (6 cm) net precipitation?

SI Solution

(a) Use Eq. 7.21 to find the average precipitation for the drainage watershed.

$$P_{ave} = \frac{V}{A_d}$$

$$= \frac{4 \times 10^6 \text{ m}^3}{(115 \text{ km}^2)\left(1000 \ \dfrac{\text{m}}{\text{km}}\right)^2}$$

$$= 0.0348 \text{ m} \quad (3.5 \text{ cm})$$

The ordinates for the unit hydrograph are found by dividing every point on the 3.5 cm hydrograph by 3.5 cm. Therefore, the peak discharge for the unit hydrograph is

$$q_{p,unit} = \frac{250 \ \dfrac{\text{m}^3}{\text{s}}}{3.5 \text{ cm}}$$

$$= 71.4 \text{ m}^3/\text{s·cm}$$

(b) The hydrograph for a storm that is producing more than 1 cm of rain is found by multiplying the ordinates of the unit hydrograph by the total precipitation. For a 6 cm storm, the peak discharge is

$$q_p = \left(71.4 \ \frac{\text{m}^3}{\text{s}\cdot\text{cm}}\right)(6 \text{ cm})$$

$$= 428 \text{ m}^3/\text{s}$$

To find the design flood volume, first use Eq. 7.21 to find the unit hydrograph total volume.

$$V = A_d P$$

$$= \frac{(115 \text{ km}^2)\left(1000 \ \frac{\text{m}}{\text{km}}\right)^2 \left(1 \ \frac{\text{cm}}{\text{cm}}\right)}{100 \ \frac{\text{cm}}{\text{m}}}$$

$$= 1.15 \times 10^6 \text{ m}^3/\text{cm}$$

The design flood volume for a 6 cm storm is

$$V = (6 \text{ cm})\left(1.15 \times 10^6 \ \frac{\text{m}^3}{\text{cm}}\right)$$

$$= 6.9 \times 10^6 \text{ m}^3$$

Customary U.S. Solution

(a) Use Eq. 7.21 to find the average precipitation for the drainage watershed.

$$P_{\text{ave}} = \frac{V}{A_d}$$

$$= \frac{(3300 \text{ ac-ft})\left(43{,}560 \ \frac{\text{ft}^2}{\text{ac}}\right)\left(12 \ \frac{\text{in}}{\text{ft}}\right)}{(45 \text{ mi}^2)\left(5280 \ \frac{\text{ft}}{\text{mi}}\right)^2}$$

$$= 1.375 \text{ in}$$

The ordinates for the unit hydrograph are found by dividing every point on the 1.375 in hydrograph by 1.375 in. Therefore, the peak discharge for the unit hydrograph is

$$q_{p,\text{unit}} = \frac{9400 \ \frac{\text{ft}^3}{\text{sec}}}{1.375 \text{ in}}$$

$$= 6836 \text{ ft}^3/\text{sec-in}$$

(b) The hydrograph for a storm that is producing more than 1 in of rain is found by multiplying the ordinates of the unit hydrograph by the total precipitation. For a 2.5 in storm, the peak discharge is

$$q_p = \left(6836 \ \frac{\text{ft}^3}{\text{sec-in}}\right)(2.5 \text{ in})$$

$$= 17{,}090 \text{ ft}^3/\text{sec}$$

To find the design flood volume, first use Eq. 7.21 to find the unit hydrograph total volume.

$$V = A_d P$$

$$= (45 \text{ mi}^2)\left(\frac{640 \ \frac{\text{ac}}{\text{mi}^2}}{12 \ \frac{\text{in}}{\text{ft}}}\right)\left(1 \ \frac{\text{in}}{\text{in}}\right)$$

$$= 2400 \text{ ac-ft/in}$$

The design flood volume for a 2.5 in storm is

$$V = (2.5 \text{ in})\left(2400 \ \frac{\text{ac-ft}}{\text{in}}\right) = 6000 \text{ ac-ft}$$

Example 7.4

A 6 hr storm rains on a 25 mi² (65 km²) drainage watershed. Records from a stream gaging station draining the watershed are shown. (a) Construct the unit hydrograph for the 6 hr storm. (b) Find the runoff rate at $t = 15$ hr from a two-storm system if the first storm drops 2 in (5 cm) starting at $t = 0$ and the second storm drops 5 in (12 cm) starting at $t = 12$ hr.

t (hr)	Q (ft³/sec)	Q (m³/s)
0	0	0
3	400	10
6	1300	35
9	2500	70
12	1700	50
15	1200	35
18	800	20
21	600	15
24	400	10
27	300	10
30	200	5
33	100	3
36	0	0
totals	9500	263

SI Solution

(a) Plot the stream gaging data.

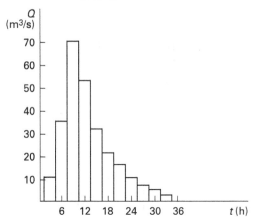

The total runoff is the total area under the curve, found by multiplying the average runoff by the rainfall interval. This is equivalent to summing the areas of each rectangle of the histogram. Each of the histogram bars is 3 h "wide."

$$V = \left(263 \ \frac{m^3}{s}\right)(3 \ h)\left(3600 \ \frac{s}{h}\right)$$
$$= 2.84 \times 10^6 \ m^3$$

The watershed drainage area is

$$A_d = (65 \ km^2)\left(1000 \ \frac{m}{km}\right)^2$$
$$= 65 \times 10^6 \ m^2$$

The average precipitation is calculated from Eq. 7.21.

$$P = \frac{V}{A_d}$$
$$= \frac{(2.84 \times 10^6 \ m^3)\left(100 \ \frac{cm}{m}\right)}{65 \times 10^6 \ m^2}$$
$$= 4.37 \ cm$$

The unit hydrograph has the same shape as the actual hydrograph with all ordinates reduced by a factor of 4.37.

(b) To find the flow at 15 h, add the contributions from each storm. For the 5 cm storm, the contribution is the 15 h runoff multiplied and divided by its scaling factors; for the 12 cm storm, the contribution is the 15 h − 12 h = 3 h runoff multiplied by its scaling factors.

$$Q = \frac{(5 \ cm)\left(35 \ \frac{m^3}{s}\right)}{4.37 \ cm} + \frac{(12 \ cm)\left(10 \ \frac{m^3}{s}\right)}{4.37 \ cm}$$
$$= 67.5 \ m^3/s$$

Customary U.S. Solution

(a) Plot the stream gaging data.

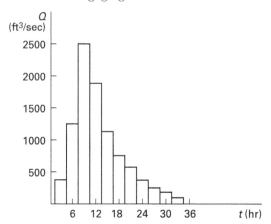

The total runoff is the area under the curve, found by multiplying the average runoff by the rainfall interval. This is equivalent to summing the areas of each rectangle of the histogram. Each of the histogram bars is 3 hr "wide."

$$V = \left(9500 \ \frac{ft^3}{sec}\right)(3 \ hr)\left(3600 \ \frac{sec}{hr}\right)$$
$$= 1.026 \times 10^8 \ ft^3$$

The watershed drainage area is

$$A_d = (25 \ mi^2)\left(5280 \ \frac{ft}{mi}\right)^2$$
$$= 6.97 \times 10^8 \ ft^2$$

The average precipitation is calculated from Eq. 7.21.

$$P = \frac{V}{A_d}$$
$$= \frac{(1.026 \times 10^8 \ ft^3)\left(12 \ \frac{in}{ft}\right)}{6.97 \times 10^8 \ ft^2}$$
$$= 1.766 \ in$$

The unit hydrograph has the same shape as the actual hydrograph with all ordinates reduced by a factor of 1.766.

(b) To find the flow at $t = 15$ hr, add the contributions from each storm. For the 2 in storm, the contribution is the 15 hr runoff multiplied and divided by its scaling factors; for the 5 in storm, the contribution is the 15 hr − 12 hr = 3 hr runoff multiplied by its scaling factors.

$$Q = \frac{(2 \text{ in})\left(1200 \dfrac{\text{ft}^3}{\text{sec}}\right)}{1.766 \text{ in}} + \frac{(5 \text{ in})\left(400 \dfrac{\text{ft}^3}{\text{sec}}\right)}{1.766 \text{ in}}$$

$$= 2492 \text{ ft}^3/\text{sec}$$

11. NRCS SYNTHETIC UNIT HYDROGRAPH

If a watershed is unmonitored such that no historical records are available to produce a unit hydrograph, the *synthetic hydrograph* can still be reasonably approximated. The process of developing a synthetic hydrograph is known as *hydrograph synthesis*.

Pioneering work was done in 1938 by Snyder, who based his analysis on Appalachian highland watersheds with areas between 10 mi^2 and 10,000 mi^2. Snyder's work has been largely replaced by more sophisticated analyses, including the NRCS methods.

The NRCS developed a synthetic unit hydrograph based on the *curve number*, CN. The method was originally intended for use with rural watersheds up to 2000 ac, but it appears to be applicable for urban conditions up to 4000–5000 ac.

In order to draw the NRCS synthetic unit hydrograph, it is necessary to calculate the time to peak flow, T_p, and the peak discharge, q_p. Provisions for calculating both of these parameters are included in the method.

$$T_p = 0.5 t_R + t_1 \qquad 7.22$$

t_R in Eq. 7.22 is the storm duration (i.e., of the rainfall). t_1 in Eq. 7.22 is the *lag time* (i.e., the time from the centroid of the rainfall distribution to the peak discharge). Lag time can be determined from correlations with geographical region and drainage area or calculated from Eq. 7.23. Although Eq. 7.23 was developed for natural watersheds, limited studies of urban watersheds indicate that it does not change significantly for urbanized watersheds. S in Eq. 7.23 is the soil water storage capacity in inches, computed as a function of the curve number with Eq. 7.24.

$$t_{1,\text{hr}} = \frac{L_{o,\text{ft}}^{0.8}(S+1)^{0.7}}{1900\sqrt{S_{\text{percent}}}} \qquad 7.23$$

$$S = \frac{1.000}{CN} - 10 \qquad 7.24$$

The peak runoff is calculated as

$$q_p = \frac{0.756 A_{d,\text{ac}}}{t_p} \qquad 7.25$$

$$q_p = \frac{484 A_{d,\text{mi}^2}}{t_p} \qquad 7.26$$

$$q_p = \frac{484 \, A Q}{T_p} \qquad 7.27$$

q_p and T_p only contribute one point to the construction of the unit hydrograph. To construct the remainder, Table 7.3 must be used. Using time as the independent variable, selections of time (different from T_p) are arbitrarily made, and the ratio t/T_p is calculated. The curve is then used to obtain the ratio of Q/q_p.

12. NRCS SYNTHETIC UNIT TRIANGULAR HYDROGRAPH

The NRCS unit triangular hydrograph is shown in Fig. 7.14. It is found from the peak runoff, the time to peak, and the duration of runoff. Peak runoff, q_p, is found from Eq. 7.25 or Eq. 7.26. Time to peak and duration are correlated with the time of concentration. These are generalizations that apply to specific storm and watershed types.

Figure 7.14 NRCS Synthetic Unit Triangular Hydrograph

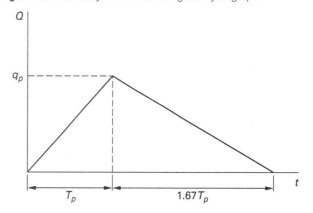

Equation 7.28 can be used to estimate the time to peak.

$$T_p = 0.5 t_R + t_1 \qquad 7.28$$

$$t_1 \approx 0.6 t \qquad 7.29$$

Table 7.3 *NRCS Dimensionless Unit Hydrograph and Mass Curve Ratios*

time ratios (t/T_p)	discharge ratios (Q/q_p)	cumulative mass curve fraction
0.0	0.000	0.000
0.1	0.030	0.001
0.2	0.100	0.006
0.3	0.190	0.012
0.4	0.310	0.035
0.5	0.470	0.065
0.6	0.660	0.107
0.7	0.820	0.163
0.8	0.930	0.228
0.9	0.990	0.300
1.0	1.000	0.375
1.1	0.990	0.450
1.2	0.930	0.522
1.3	0.860	0.589
1.4	0.780	0.650
1.5	0.680	0.700
1.6	0.560	0.751
1.7	0.460	0.790
1.8	0.390	0.822
1.9	0.330	0.849
2.0	0.280	0.871
2.2	0.207	0.908
2.4	0.147	0.934
2.6	0.107	0.953
2.8	0.077	0.967
3.0	0.055	0.977
3.2	0.040	0.984
3.4	0.029	0.989
3.6	0.021	0.993
3.8	0.015	0.995
4.0	0.011	0.997
4.5	0.005	0.999
5.0	0.000	1.000

Source: National Engineering Handbook, Part 630, Hydrology, NRCS, 1972.

Alternatively, the time to peak has been roughly correlated to the time of concentration.

$$T_p = 0.67t \qquad 7.30$$

The total duration of the unit hydrograph is the sum of time to peak and length of recession limb, assumed to be $1.67\,T_p$.

$$t_b = T_p + t_{\text{recession}} = T_p + 1.67\,T_p = 2.67\,T_p \qquad 7.31$$

13. ESPEY SYNTHETIC UNIT HYDROGRAPH

The *Espey method* calculates the time to peak (T_p, in min), peak discharge (q_p, in ft^3/sec), total hydrograph base (t_b, in min), and the hydrograph widths at 50% and 75% of the peak discharge rates (W_{50} and W_{75}, in min). These values depend on the watershed area (A, in mi^2), main channel flow path length (L, in ft), slope (S, in ft/ft), roughness, and percent imperviousness (Imp). ϕ is a dimensionless watershed conveyance factor ($0.6 < \phi < 1.3$) that depends on the percent imperviousness and weighted main channel Manning roughness coefficient, n. ϕ is found graphically from Fig. 7.15.

$$T_p = 3.1 L^{0.23} S_{\text{decimal}}^{-0.25} (\text{Imp}_{\text{percent}})^{-0.18} \phi^{1.57} \qquad 7.32$$

$$q_p = (31.62 \times 10^3) A^{0.96} T_p^{-1.07} \qquad 7.33$$

$$t_b = (125.89 \times 10^3) A q_p^{-0.95} \qquad 7.34$$

$$W_{50} = (16.22 \times 10^3) A^{0.93} q_p^{-0.92} \qquad 7.35$$

$$W_{75} = (3.24 \times 10^3) A^{0.79} q_p^{-0.78} \qquad 7.36$$

To use this method, the geometric slope, S, used in Eq. 7.32 is specifically calculated from Eq. 7.37. H is the difference in elevation of points A and B. Point A is the channel bottom a distance $0.2L$ downstream from the upstream watershed boundary. Point B is the channel bottom at the downstream watershed boundary.

$$S_{\text{decimal}} = \frac{H}{0.8L} \qquad 7.37$$

The unit hydrograph is drawn by manually "fitting" a smooth curve over the seven computed points. The widths of W_{50} and W_{75} are allocated in a 1:2 ratio to the rising and falling hydrograph limbs, respectively. After the curve is drawn, it is adjusted to be a unit hydrograph. The resulting curve is sometimes referred to as an *Espey 10-minute unit hydrograph*. (See Fig. 7.16.)

14. HYDROGRAPH SYNTHESIS

If a storm's duration is not the same as, or close to, the hydrograph base length, the unit hydrograph cannot be used to predict runoff. For example, the runoff from a six hour storm cannot be predicted from a unit hydrograph derived from a two hour storm. However, the technique of hydrograph synthesis can be used to construct the hydrograph of the longer storm from the unit hydrograph of a shorter storm.

Figure 7.15 *Espey Watershed Conveyance Factor*

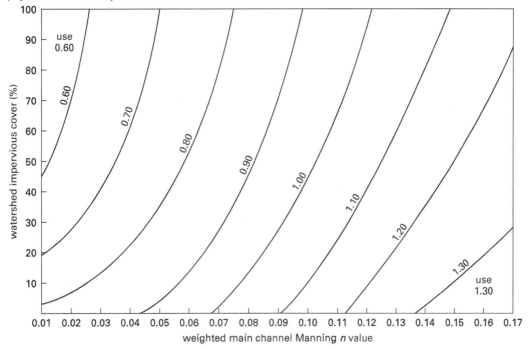

Reprinted from *Recommended Hydrologic Procedures for Computing Urban Runoff from Small Watersheds in Pennsylvania*, Commonwealth of Pennsylvania, Fig. 6-3, 1982, Department of Environmental Resources.

A. Lagging Storm Method

If a unit hydrograph for a storm of duration t_R is available, the *lagging storm method* can be used to construct the hydrograph of a storm whose duration is a whole multiple of t_R. (See Fig. 7.17.) For example, a six-hour storm hydrograph can be constructed from a two-hour unit hydrograph.

Let the whole multiple number be n. To construct the longer hydrograph, draw n unit hydrographs, each separated by time t_R. Then add the ordinates to obtain a hydrograph for an nt_R duration storm. Since the total rainfall from this new hydrograph is n inches (having been constructed from n unit hydrographs), the curve will have to be reduced (i.e., divided) by n everywhere to produce a unit hydrograph.

B. S-Curve Method

The *S-curve method* can be used to construct hydrographs from unit hydrographs with longer or shorter durations, even when the storm durations are not multiples. This method begins by adding the ordinates of many unit hydrographs, each lagging the other by time t_R, the duration of the storm that produced the unit hydrograph. After a sufficient number of lagging unit hydrographs have been added together, the accumulation will level off and remain constant. At that point, the lagging can be stopped. The resulting accumulation is known as an S-curve. (See Fig. 7.18.)

If two S-curves are drawn, one lagging the other by time t_R', the area between the two curves represents a hydrograph area for a storm of duration t_R'. (See Fig. 7.19.) The differences between the two curves can be plotted and scaled to a unit hydrograph by multiplying by the ratio of t_R/t_R'.

Figure 7.16 *Espey Synthetic Hydrograph*

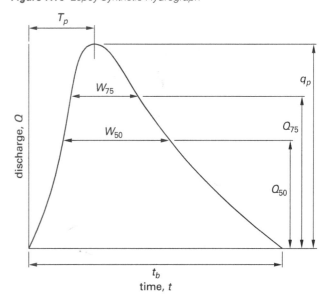

Figure 7.17 *Lagging Storm Method*

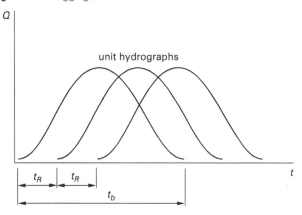

Figure 7.19 *Using S-Curves to Construct a t' Hydrograph*

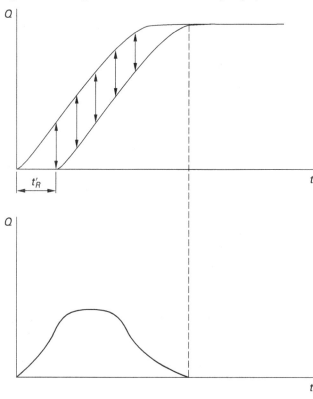

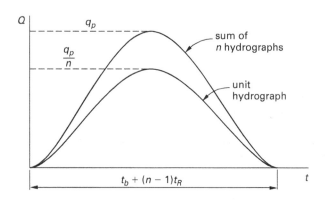

Figure 7.18 *Constructing the S-Curve*

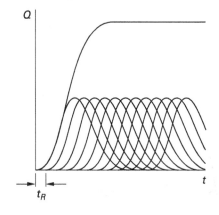

15. NOMENCLATURE

a	storm constant	–	–
A	area	ft^2	m^2
A_d	drainage area	ac	km^2
b	storm constant	min	min
c	storm constant	–	–
C	constant	–	–
C	Manning roughness coefficient	–	–
C	rational runoff coefficient	–	–
CN	NRCS runoff curve number	–	–
d	distance between stations	mi	km
E	evaporation	in/day	cm/d
F	storm constant	–	–
F	frequency of occurrence	1/yr	1/yr
H	elevation difference	ft	m
i, I	rainfall intensity	in/hr	cm/h
K	coefficient	–	–
K, K'	storm constant	in-min/hr	cm·min/h
K_p	pan coefficient	–	–
L	length	ft	m
L_o	longest distance to collection point	ft	m
L	distance of overland flow	ft	m
M	order number	–	–
n	Manning roughness coefficient	–	–
N	normal long-term precipitation	in	cm
P	precipitation	in	cm
q	runoff	ft^3/mi^2-in	m^3/km^2·cm
q_p	peak discharge	ft^3/sec	m^2/s
Q	flow rate	ft^3/sec	m^3/s
Q	runoff	in	cm
S	storage capacity	in	cm
$S, S_{decimal}$	slope	ft/ft	m/m
$S_{percent}$	slope	%	%
t, T	time	min	min
t	time of concentration	min	min
t_b	total hydrographic base	min	min
t_i	time of overland flow	min	min
t_1	lag time	min	min
t_R	rainstorm duration	hr	h
T_p	time from start of storm to peak runoff	hr	h
v	flow velocity	ft/sec	m/s
V	volume	ft^3	m^3
W	width of unit hydrograph	min	min
x_T	recurrence frequency interval of T yr	yr	yr

Symbols

ϕ	watershed conveyance factor	–	–

Subscripts

ave	average
b	base, total duration
c	concentration
d	drainage
n	number or period
o	overland
p	peak, pond, or pan
R	rainfall (storm) or reservoir
t	time
x	missing station

8 Codes, Standards, Regulations, Guidelines

In the United States, the two primary laws governing water resources are the Clean Water Act (CWA) of 1972 and the Safe Drinking Water Act (SDWA) of 1974, along with their associated regulations. In addition to these regulations, *Recommended Standards for Water Works* and *Recommended Standards for Wastewater Facilities* (known together as the *Ten States' Standards*), among other guidelines, provide design direction for treating water and wastewater.

1. CLEAN WATER ACT

The Clean Water Act was originally called the Water Pollution Control Act (WPCA). It primarily addresses discharges to waters of the United States by imposing effluent limitations through a pretreatment and permit program. The Clean Water Act is administered by the U.S. Environmental Protection Agency (EPA) and by individual states.

Discharge Criteria

Discharge criteria establish pretreatment standards for categorical discharges from industrial facilities to wastewater treatment plants (WWTPs) or publicly owned treatment works (POTWs). These regulations are found in the *Code of Federal Regulations*, Title 40, Parts 400–471 (40 CFR 400–471). Discharger categories are defined in the regulations for specific industrial categories (e.g., Electroplating and Metal Finishing Pretreatment Standards) and apply to nondomestic waste generators that discharge to a POTW.

Secondary treatment standards are applicable to discharges to receiving waters from POTWs, according to 40 CFR 133. These include treatment standards for specific conventional pollutants such as biochemical oxygen demand (BOD), total suspended solids (TSS), and pH, as well as for specific toxic substances. Secondary treatment standards are minimum, technology-based requirements for POTWs. Secondary treatment standards also regulate treatment plants that receive flows from combined sewers, treatment plants that received industrial wastes, waste stabilization ponds and lagoons, and treatment plants receiving dilute wastewater for combined and separate sewers.

The National Pollutant Discharge Elimination System (NPDES) regulates discharges to waters of the United States. The NPDES program is administered by individual states. For example, in Virginia, the NPDES permit is a Virginia Pollutant Discharge Elimination System (VPDES) permit. These permits consider site-specific conditions in establishing discharge criteria defined by the permit. The regulations are found in 40 CFR 133.

Wetlands

Section 404 of the CWA protects wetlands by prohibiting the discharge of any dredged or fill materials into waters of the United States, including wetlands, unless authorized by permit through the U.S. Army Corps of Engineers (USACE). The USACE may issue two kinds of permits, a general permit and an individual permit. *General permits* are applied to a wide range of activities and require compliance with all permit conditions. General permits are typically issued when environmental impacts from fill activities are expected to be minimal, such that a detailed individual review of each activity is not necessary. *Individual permits* address specific activities with the potential for greater negative environmental impact not addressed in a general permit. Individual permits receive close scrutiny and require deeper analysis than general permits. Individual permit review includes a published public notice describing the proposed action.

Priority Pollutants

The 1977 amendments to the CWA included a list of 65 priority pollutants (specific chemicals and classes of chemicals) to be used for defining toxic substances. The original list has been expanded to the current list of 129 priority pollutants. *Priority pollutants* are those chemicals with relatively high toxicity and high production volume. The priority pollutants list is currently outdated, having not been revised since 1981. The list contains some chemicals whose manufacture in the United Stated is prohibited. Because the list is outdated, EPA also regulates other toxic, conventional, and nonconventional pollutants considered to be important for the protection of surface waters. The priority pollutants list is in 40 CFR 423, App. A.

Environmental Justice

Established by executive order in 1994, the EPA manages a program of environmental justice (EJ) for impacts related to CWA activities. The EPA defines *environmental justice* as "the fair treatment and meaningful involvement of all people regardless of race, color, national origin, or income with respect to the development, implementation, and enforcement of environmental laws, regulations, and policies." The goal of EJ is to ensure that all populations bear an equal burden in sharing the potentially negative environmental impacts resulting from activities or policies of government, industrial, and commercial interests. The EPA Office of Environmental Justice accomplishes its goals by

- providing an online screening tool for stakeholder use to map and screen data describing environmental and demographic indictors; the tool, EJSCREEN, allows stakeholders to monitor EPA decisions regarding environmental impacts on affected communities

- publishing documents to provide consistent guidance for policy makers as they consider agency actions and consequent EJ impacts

- providing training and workshops for interested stakeholders

- establishing legal policy to guide the EPA in protecting human health and the environment in minority and low-income communities

2. SAFE DRINKING WATER ACT

The Safe Drinking Water Act and its implementing regulations are the current applicable law intended to protect public health by mandating minimum drinking water quality. The SDWA's regulations are administered by the EPA and by states through parallel programs. The SDWA has mandated national primary and secondary drinking water standards applicable to drinking water sources including contaminated surface and groundwater.

National Primary Drinking Water Standards

The national primary drinking water standards, published in 40 CFR 141 and 142, were established to protect public health.

These standards set limits on the amount of various substances in drinking water. Every public water supply serving at least 15 service connections or 25 or more people must ensure that its water meets these minimum standards.

Accordingly, the EPA has established the National Primary Drinking Water Standards and the National Secondary Drinking Water Standards. The primary standards establish *maximum contaminant levels* (MCL) and *maximum contaminant level goals* (MCLG) for materials that are known or suspected health hazards. The MCL is the enforceable level that the water supplier must not exceed, while the MCLG is an unenforceable health goal equal to the maximum level of a contaminant that is not expected to cause any adverse health effects over a lifetime of exposure.

National Secondary Drinking Water Standards

The SDWA identifies national secondary drinking water standards. Unlike the primary standards, which deal with hazards to public health, the secondary standards provide helpful guidelines regarding the taste, odor, color, and other aesthetic aspects of drinking water. [Secondary Drinking Water Standards]

3. TEN STATES' STANDARDS

In addition to the CWA and SDWA regulations, other public and private entities publish guidelines for the treatment of water and wastewater. Most prominent among these are *Recommended Standards for Water Works* and *Recommended Standards for Wastewater Facilities* (known together as *Ten States' Standards*), prepared by the Great Lakes–Upper Mississippi River Board of State and Provincial Public Health and Environmental Managers. The *Ten States' Standards* provides design guidance for water, wastewater, and individual sewage systems.

Wastewater Pollution, Minimization, Prevention

1. WASTEWATER SOURCES

Sanitary wastewater refers to liquid and waterborne wastes from domestic, commercial, and industrial sources. Sanitary wastewater is generally collected through a sanitary sewer system and treated in a centralized facility such as a publicly owned treatment works (POTW) or other wastewater treatment plant. Use of a central facility provides the advantage of economies of scale and reduces the regulatory effort needed to monitor and permit the discharge.

Domestic wastewater is wastewater that has been discharged from residences. This category is generally taken to include wastewater discharged from institutions and commercial buildings such as schools, offices, and hotels, which is usually similar in composition; wastewater from manufacturing facilities, however, is excluded.

Domestic wastewater is generally measured in gallons per capita per day, abbreviated gpcd (liters per capita per day, abbreviated Lpcd). The volume of domestic wastewater varies from 50 gpcd to 250 gpcd (190–950 Lpcd), depending on sewer uses. A more common range for domestic wastewater flow is 100–120 gpcd (380–450 Lpcd), which assumes that residential dwellings have major water-using appliances, such as dishwashers and washing machines, or use potable water for landscape irrigation.

Industrial wastewater is also generally discharged to a central sewer system. Uncontaminated streams, such as of cooling water, can often be discharged into sewers directly. However, most industrial wastewater must undergo some level of pretreatment at the facility that has generated it before it can be discharged to the sewer. The manufacturing plant may also be required to equalize flow by holding the wastewater in a basin before discharging it. Where industrial activities produce wastewater that is unique or that requires special treatment, the generating facility is often required to pretreat the wastewater to be compatible with the central POTW processes and to reduce the loading to the POTW. In the *joint processing* of wastewater, the municipality accepts responsibility for final treatment and disposal.

To minimize the impact of industrial wastewater on the sewage treatment plant, consideration is given to modifications in industrial processes, segregation of wastes, flow equalization, and waste strength reduction. Modern industrial and manufacturing processes require segregation of separate waste streams for individual pretreatment, controlled mixing, and/or separate disposal. Process changes, equipment modifications, by-product recovery, and in-plant wastewater reuse can result in cost savings for both water supply and wastewater treatment.

Toxic waste streams are not generally accepted into the municipal treatment plant. Toxic substances require appropriate pretreatment before being disposed of by other means.

2. MUNICIPAL WASTEWATER

Municipal wastewater is the general name given to the liquid collected in sanitary sewers and routed to municipal sewage treatment plants. Many older cities have *combined sewer systems* where storm water and sanitary wastewaters are collected in the same lines. The combined flows are conveyed to the treatment plant for processing during dry weather. During wet weather, when the combined flow exceeds the plant's treatment capacity, the excess flow often bypasses the plant and is discharged directly into the watercourse.

3. WASTEWATER QUANTITY

Approximately 70–80% of a community's domestic and industrial water supply returns as wastewater. This water is discharged into the sewer systems, which may or may not also function as storm drains. Therefore, the nature of the return system must be known before sizing can occur.

Infiltration, due to cracks and poor joints in old or broken lines, can increase the sewer flow significantly. Infiltration per mile (kilometer) per in (mm) of pipe diameter is limited by some municipal codes to 500 gpd/mi-in (46 Lpd/km·mm). Modern piping materials and joints easily reduce the infiltration to 200 gpd/in-mi (18 Lpd/km·mm) and below. Infiltration can also be roughly estimated as 3–5% of the peak hourly domestic rate or as 10% of the average rate.

Inflow is another contributor to the flow in sewers. Inflow is water discharged into a sewer system from such sources as roof down spouts, yard and area drains, parking area catch basins, curb inlets, and holes in manhole covers.

Sanitary sewer sizing is commonly based on an assumed average of 100–125 gpcd (380–474 Lpcd). There will be variations in the flow over time, although the variations are not as pronounced as they are for water supply. Hourly variations are the most significant. The flow rate pattern is essentially the same from day to day. Weekend flow patterns are not significantly different from weekday flow patterns. Seasonal variation depends on the location, local industries, and infiltration.

Table 9.1 lists typical *peaking factors* (i.e., peak multipliers) for treatment plant influent volume. Due to storage in ponds, clarifiers, and sedimentation basins, these multipliers are not applicable throughout all processes in the treatment plant.

Table 9.1 *Typical Variations in Wastewater Flows (based on average annual daily flow)*

flow description	typical time	peaking factor
daily average	–	1.0
daily peak	10–12 a.m.	2.25
daily minimum	4–5 a.m.	0.4
seasonal average	May, June	1.0
seasonal peak	late summer	1.25
seasonal minimum	late winter	0.9

Recommended Standards for Sewage Works (*Ten States' Standards*, abbreviated TSS) specifies that new sanitary sewer systems should be designed to have an average flow of 100 gpcd (380 Lpcd or 0.38 m³/d), which includes an allowance for normal infiltration [TSS Sec. 33.94]. However, the sewer pipe must be sized to carry the peak flow as a gravity flow. In the absence of any studies or other justifiable methods, the ratio of peak hourly flow to average flow should be estimated from the following relationship, in which P is the population served in thousands of people at a particular point in the network.

$$\frac{Q_{\text{peak}}}{Q_{\text{ave}}} = \frac{18 + \sqrt{P}}{4 + \sqrt{P}} \qquad 9.1$$

The peaking factor can also be estimated by using *Harmon's peaking factor* equation.

$$\frac{Q_{\text{peak}}}{Q_{\text{ave}}} = \frac{1 + 14}{4 + \sqrt{P}} \geq 2.5 \qquad 9.2$$

Collectors (i.e., *collector sewers*, *trunks*, or *mains*) are pipes that collect wastewater from individual sources and carry it to interceptors. (See Fig. 9.1.) Collectors must be designed to handle the maximum hourly flow, including domestic and infiltration, as well as additional discharge from industrial plants nearby. Peak flows of 400 gpcd (1500 Lpcd) for laterals and submains flowing full and 250 gpcd (950 Lpcd) for main, trunk, and outfall sewers can be assumed for design purposes, making the peaking factors approximately 4.0 and 2.5 for submains and mains, respectively. Both of these generalizations include generous allowances for infiltration. The lower flow rates take into consideration the averaging effect of larger contributing populations and the damping (storage) effect of larger distances from the source.

Interceptors are major sewer lines receiving wastewater from collector sewers and carrying it to a treatment plant or to another interceptor.

Figure 9.1 *Classification of Sewer Lines*

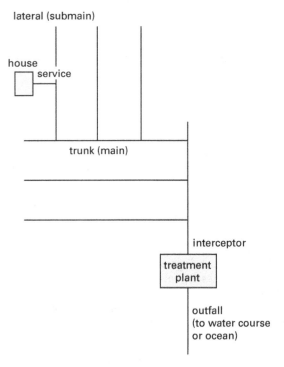

4. TREATMENT PLANT LOADING

Ideally, the quantity and organic strength of wastewater should be based on actual measurements taken throughout the year in order to account for variations that result from seasonal, climatic changes and other factors.

The *hydraulic loading* for over 75% of the sewage treatment plants in the United States is 1 MGD or less. Plants with flows less than 1 MGD are categorized as *minors*. Treatment plants handling 1 MGD or more, or serving equivalent service populations of 10,000, are categorized as *majors*.

The *organic loading* of treatment units is expressed in terms of pounds (kilograms) of biochemical oxygen demand (BOD) per day or pounds (kilograms) of solids per day. In communities where a substantial portion of household kitchen wastes is discharged to the sewer system through garbage disposals, the average organic matter contributed by each person each day to domestic wastewater is approximately 0.24 lbm (110 g) of suspended solids and approximately 0.17–0.20 lbm (77–90 g) of BOD.

Since the average BOD of domestic waste is typically taken as 0.2 pounds per capita day (90 g per capita day), the *population equivalent* of any wastewater source, including industrial sources, can be calculated.

$$P_{e,1000s} = \frac{\text{BOD}_{mg/L} Q_{ML/d}}{90 \dfrac{g}{\text{person·d}}} \quad \text{[SI]} \quad \textit{9.3(a)}$$

$$P_{e,1000s} = \frac{\text{BOD}_{mg/L} Q_{gal/day} \times \left(8.345 \dfrac{\text{lbm-L}}{\text{MG-mg}}\right)}{\left(10^6 \dfrac{\text{gal}}{\text{MG}}\right)(1000 \text{ persons}) \times \left(0.20 \dfrac{\text{lbm}}{\text{person-day}}\right)} \quad \text{[U.S.]} \quad \textit{9.3(b)}$$

5. WASTEWATER CHARACTERISTICS

Domestic wastewater may be characterized by the concentrations it contains of organic and inorganic solid matter, dissolved organic compounds and dissolved organic matter, nutrients, and other inorganic substances such as metals and salts. Although domestic wastewater from various sources share common characteristics, the exact concentrations of the constituents may vary substantially. Table 9.2 lists typical values for strong and weak domestic wastewater.

Solids

Solids in wastewater are categorized in the same manner as in water supplies. *Total solids* consist of *suspended* and *dissolved solids*. Generally, total solids constitute only a small amount of the incoming flow—less than 1/10% by mass. Therefore, wastewater fluid transport properties are

essentially those of water. Figure 9.2 illustrates the solids categorization, along with typical percentages. Each category can be further divided into organic and inorganic groups.

The amount of *volatile solids* can be used as a measure of the organic pollutants capable of affecting the oxygen content. As in water supply testing, volatile solids are measured by igniting filtered solids and measuring the decrease in mass.

Table 9.2 *Strong and Weak Domestic Wastewater**

constituent	strong	weak
solids, total	1200	350
dissolved, total	850	250
fixed	525	145
volatile	325	105
suspended, total	350	100
fixed	75	30
volatile	275	70
settleable solids (mL/L)	20	5
biochemical oxygen demand, five-day, 20°C	300	100
total organic carbon	300	100
chemical oxygen demand	1000	250
nitrogen (total as N)	85	20
organic	35	8
free ammonia	50	12
nitrites	0	0
nitrates	0	0
phosphorus (total as P)	20	6
organic	5	2
inorganic	15	4
chlorides	100	30
alkalinity (as $CaCO_3$)	200	50
grease	150	50

*All concentrations are in mg/L unless otherwise noted.

Figure 9.2 *Wastewater Solids**

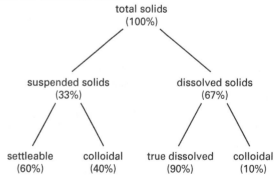

*typical percentages given

Refractory solids (*refractory organics*) are solids that are difficult to remove by common wastewater treatment processes.

Organic Compounds

Biodegradable organic matter in municipal wastewater is classified into three major categories: proteins, carbohydrates, and greases (fats).

Proteins are long strings of amino acids containing carbon, hydrogen, oxygen, nitrogen, and phosphorus.

Carbohydrates consist of sugar units containing the elements of carbon, hydrogen, and oxygen. They are identified by the presence of a saccharide ring. The rings range from simple monosaccharides to polysaccharides (long chain sugars categorized as either readily degradable starches found in potatoes, rice, corn, and other edible plants, or as cellulose found in wood and similar plant tissues).

Greases are a variety of biochemical substances that have the common property of being soluble to varying degrees in organic solvents (acetone, ether, ethanol, and hexane) while being only sparingly soluble in water. The low solubility of grease in water causes problems in pipes and tanks where it accumulates, reduces contact areas during various filtering processes, and produces a sludge that is difficult to dispose of.

Of the organic matter in wastewater, 60–80% is readily available for biodegradation. Degradation of greases by microorganisms occurs at a very slow rate. A simple fat is a triglyceride composed of a glycerol unit with short-chain or long-chain fatty acids attached.

The majority of carbohydrates, fats, and proteins in wastewater are in the form of large molecules that cannot penetrate the cell membrane of microorganisms. Bacteria, in order to metabolize high molecular-weight substances, must be capable of breaking down the large molecules into diffusible fractions for assimilation into the cell.

Several organic compounds, such as cellulose, long-chain saturated hydrocarbons, and other complex compounds, although available as a bacterial substrate, are considered nonbiodegradable because of the time and environmental limitations of biological wastewater treatment systems.

Volatile organic chemicals (VOCs) are released in large quantities in industrial, commercial, agricultural, and household activities. The adverse health effects of VOCs include cancer and chronic effects on the liver, kidney, and nervous system. *Synthetic organic chemicals* (SOCs) and *inorganic chemicals* (IOCs) are used in agricultural and industrial processes as pesticides, insecticides, and herbicides.

Organic petroleum derivatives, detergents, pesticides, and other synthetic, organic compounds are particularly resistant to biodegradation in wastewater treatment plants; some are toxic and inhibit the activity of microorganisms in biological treatment processes.

Metals

Metals are classified by four categories: *Dissolved metals* are those constituents of an unacidified sample that pass through a 0.45 μm membrane filter. *Suspended metals* are those constituents of an unacidified sample retained by a 0.45 μm membrane filter. *Total metals* are the concentration determined on an unfiltered sample after vigorous digestion or the sum of both dissolved and suspended fractions. *Acid-extractable metals* remain in solution after treatment of an unfiltered sample with hot dilute mineral acid.

The metal concentration in the wastewater is calculated from Eq. 9.4. The sample size is selected so that the product of the sample size (in mL) and the metal concentration (C, in mg/L) is approximately 1000.

$$C = \frac{C_{\text{digestate}} V_{\text{digested solution}}}{V_{\text{sample}}} \qquad 9.4$$

Some metals are biologically essential; others are toxic and adversely affect wastewater treatment systems and receiving waters.

6. WASTEWATER MINIMIZATION

Inflow contributes hydraulic loading to wastewater treatment plants through direct and unintended connections to sewers. Inflow is generally associated with sources such as foundation drains, eave troughs, and sump pumps. Some sources may produce a steady, continuous flow, such as those from foundation drains fed by springs; others may produce a delayed and intermittent flow, such as those connected to storm events.

Infiltration contributes hydraulic loading to wastewater treatment plants through cracks in sewer pipes, loose pipe joints, and other defects that allow groundwater to infiltrate into the sewer. Infiltration may also include runoff that enters the sewer through poorly fitted manhole covers. Infiltration can be continuous or intermittent, depending on the proximity of sewer lines to the normal groundwater table and on influences from storm events.

Taken together, inflow and infiltration are called *I and I* (I/I). I/I presents problems for wastewater treatment because it increases the hydraulic load to the POTW. I/I can be controlled by local ordinances that prohibit inflow connections and by maintaining the sewer system.

To reduce hydraulic loading to POTWs, some building codes require the use of low-flow, water-conserving fixtures in home and commercial construction. Incentives for installing low-flow fixtures in existing buildings can also contribute substantially to decreasing hydraulic loading.

Although it is not often feasible to separate existing combined stormwater and sanitary sewers, it is feasible to segregate them in new construction. Especially in locales receiving regular rainfall, separate stormwater and sanitary sewers are effective in minimizing wastewater volume.

Wastewater flow to a POTW is not uniform but fluctuates according to diurnal and seasonal cycles. As the flow rate changes, so does the loading. Seasonal influences are evident as people's habits for bathing, food preparation, laundry, and other activities change from season to season. Other seasonal influences can occur with changes in I/I from wet-weather to dry-weather flows, institutional activities such as K-12 schools and resident student populations at colleges and universities, and tourism influences on commercial activities such as hotels and restaurants. These variations impede some efforts to minimize wastewater flows, beyond application of low-flow fixtures.

7. WASTEWATER REUSE

Wastewater reuse in the United States has been limited primarily to irrigation of areas where human contact with the water or the irrigated vegetation is controlled. Although there are few technical constraints to reusing nearly all wastewater, people are uncomfortable with a direct connection from the sewer, through treatment, to the tap. These reservations apply to food processors that, understandably, will not challenge consumer anxiety over wastewater reuse in their food supply. However, there does not seem to be much hesitation by the population to reuse wastewater once a discharge has been mixed with the water in a river and traveled a few miles downstream.

8. NOMENCLATURE

BOD	biochemical oxygen demand	mg/L	mg/L
C	concentration	mg/L	mg/L
P	population	–	–
Q	flow quantity	gal/day	L/d
V	volume	mL	mL

Subscripts

ave	average
e	equivalent

10 Wastewater Treatment and Management

Content in blue refers to the *NCEES Handbook*.

1. INDUSTRIAL WASTEWATER TREATMENT

The *National Pollution Discharge Elimination System* (NPDES) places strict controls on the discharge of industrial wastewaters into municipal sewers. Any industrial wastes that would harm subsequent municipal treatment facilities or that would upset subsequent biological processes need to be pretreated. Manufacturing plants may also be required to equalize wastewaters by holding them in basins for stabilization prior to their discharge to the sewer. Table 10.1 lists typical limitations on industrial wastewaters.

Table 10.1 *Typical Industrial Wastewater Effluent Limitations*

characteristic	concentration[a] (mg/L)
COD	300–2000
BOD	100–300
oil and grease or TPH[b]	15–55
total suspended solids	15–45
pH	6.0–9.0
temperature	less than 40°C
color	2 color units
NH_3/NO_3	1.0–10
phosphates	0.2
heavy metals	0.1–5.0
surfactants	0.5–1.0 (total)
sulfides	0.01–0.1
phenol	0.1–1.0
toxic organics	1.0 total
cyanide	0.1

[a]maximum permitted at discharge
[b]total petroleum hydrocarbons

2. CESSPOOLS

A *cesspool* is a covered pit into which domestic (i.e., household) sewage is discharged. Cesspools for temporary storage can be constructed as watertight enclosures. However, most are *leaching cesspools* that allow seepage of liquid into the soil. Cesspools are rarely used today. They are acceptable for disposal for only very small volumes (e.g., from a few families).

3. SEPTIC TANKS

A *septic tank* is a simply constructed tank that holds domestic sewage while sedimentation and digestion occur. (See Fig. 10.1.) Typical residence (or detention) times are 8–24 hr. Only 30–50% of the suspended solids are digested in a septic tank. The remaining solids settle, eventually clogging the tank, and must be removed. Semi-clarified effluent percolates into the surrounding soil through lateral lines placed at the *flow line* that lead into an underground leach field. In the past, clay drainage tiles were used. The terms "tile field" and "tile bed" are still encountered even though perforated plastic pipe is now widely used.

Figure 10.1 *Septic Tank*

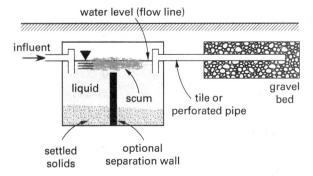

Most septic tanks are built for use by one to three families. Larger communal tanks can be constructed for small groups. (They should be designed to hold 12–24 hr of flow plus stored sludge.) A general rule is to allow at least 30 gal (0.1 m³) of storage per person served by the tank. Typical design parameters of domestic septic tanks are given in Table 10.2.

Proper design of the percolation field is the key to successful operation. Soils studies are essential to ensure adequate absorption into the soil.

4. DISPOSAL OF SEPTAGE

Septage is the water and solid material pumped periodically from septic tanks, cesspools, or privies. In the United States, disposal of septage is controlled by federal sludge disposal regulations, and most local and state governments have their own regulations as well. While surface and land spreading were acceptable

Table 10.2 *Typical Characteristics of Domestic Septic Tanks*

minimum capacity below flow line	300–500 gal (1.1–1.9 m³), plus 30 gal (0.1 m³) for each person served over 5
plan aspect ratio	1:2
minimum depth below flow line	3–4 ft (0.9–1.2 m)
minimum freeboard above flow line	1 ft (0.3 m)
tank burial depth	1–2 ft (0.3–0.6 m)
drainage field tile length	30 ft (9 m) per person
maximum drainage field tile length	60 ft (18 m)
minimum tile depth	1.5–2.5 ft (0.45–0.75 m)
lateral line spacing	6 ft (1.8 m)
gravel bed	0.33 ft (0.1 m) above lateral, 1–3 ft (0.3–1 m) below
soil layer below tile bed	10 ft (3 m)

(Multiply gal by 0.003785 to obtain m³.)
(Multiply ft by 0.3048 to obtain m.)

disposal methods in the past, septage haulers are now turning to municipal treatment plants for disposal. (Table 10.3 compares septage characteristics to those of municipal sewage.) Since septage is many times more concentrated than sewage, it can cause *shock loads* in preliminary treatment processes. Typical problems are plugged screens and aerator inlets, reduced efficiency in grit chambers and aeration basins, and increased odors. The impact on subsequent processes is less pronounced due to the effect of dilution.

5. WASTEWATER TREATMENT PLANTS

For traditional *wastewater treatment plants* (WWTP), *preliminary treatment* of the wastewater stream is essentially a mechanical process intended to remove large objects, rags, and wood. Heavy solids and excessive oils and grease are also eliminated. Damage to pumps and other equipment would occur without preliminary treatment.

Odor control through chlorination or ozonation, freshening of septic waste by aeration, and flow equalization in holding basins can also be loosely categorized as preliminary processes.

After preliminary treatment, there are three "levels" of wastewater treatment: primary, secondary, and tertiary. *Primary treatment* is a mechanical (settling) process used to remove oil and most (i.e., approximately 50%)

Table 10.3 Comparison of Typical Septage and Municipal Sewage[a]

characteristic	septage	sewage[b]
BOD	7000	220
COD	15,000	500
total solids	40,000	720
total volatile solids	25,000	365
total suspended solids	15,000	220
volatile suspended solids	10,000	165
TKN	700	40
NH_3 as N	150	25
alkalinity	1000	100
grease	8000	100
pH	6.0	n.a.

[a]all values except pH in mg/L
[b]medium strength

of the settleable solids. With domestic wastewater, a 25–35% reduction in BOD is also achieved, but BOD reduction is not the goal of primary treatment.

In the United States, secondary treatment is mandatory for all publicly owned wastewater treatment plants. *Secondary treatment* involves biological treatment in trickling filters, rotating contactors, biological beds, and activated sludge processes. Processing typically reduces the suspended solids and BOD content by more than 85%, volatile solids by 50%, total nitrogen by about 25%, and phosphorus by 20%.

Tertiary treatment (also known as *advanced wastewater treatment*, AWT) is targeted at specific pollutants or wastewater characteristics that have passed through previous processes in concentrations that are not allowed in the discharge. *Suspended solids* are removed by microstrainers or polishing filter beds. *Phosphorus* is removed by chemical precipitation. Aluminum and iron coagulants, as well as lime, are effective in removing phosphates. *Ammonia* can be removed by air stripping, biological denitrification, breakpoint chlorination, anion exchange, and algae ponds. Ions from *inorganic salts* can be removed by electrodialysis and reverse osmosis. The so-called *trace organics* or *refractory substances*, which are *dissolved organic solids* that are resistant to biological processes, can be removed by filtering through carbon or ozonation.

6. WASTEWATER PLANT SITING CONSIDERATIONS

Wastewater plants should be located as far as possible from inhabited areas. A minimum distance of 1000 ft (300 m) for uncovered plants and lagoons is desired. Uncovered plants should be located downwind when a definite wind direction prevails. Soil conditions need to

be evaluated, as does the proximity of the water table. Elevation in relationship to the need for sewage pumping (and for dikes around the site) is relevant.

The plant must be protected against flooding. 100-year storms are often chosen as the design flood when designing dikes and similar facilities. Distance to the outfall and possible effluent pumping need to be considered.

Table 10.4 lists the approximate acreage for sizing wastewater treatment plants. Estimates of population expansion should provide for future capacity.

Table 10.4 Treatment Plant Acreage Requirements

type of treatment	surface area required (ac/MGD)
physical-chemical plants	1.5
activated sludge plants	2
trickling filter plants	3
aerated lagoons	16
stabilization basins	20

7. PUMPS USED IN WASTEWATER PLANTS

Wastewater treatment plants should be gravity-fed wherever possible. The influent of most plants is pumped to the starting elevation, and wastewater flows through subsequent processes by gravity thereafter. However, there are still many instances when pumping is required. Table 10.5 lists pump types by application.

At least two identical pumps should be present at every location, each capable of handling the entire peak flow. Three or more pumps are suggested for flows greater than 1 MGD, and peak flow should be handled when one of the pumps is being serviced.

8. FLOW EQUALIZATION

Equalization tanks or ponds are used to smooth out variations in flow that would otherwise overload wastewater processes. Graphical or tabular techniques similar to those used in reservoir sizing can be used to size equalization ponds. (See Fig. 10.2.) In practice, up to 25% excess capacity is added as a safety factor.

In a pure flow equalization process, there is no settling. Mechanical aerators provide the turbulence necessary to keep the solids in suspension while providing oxygen to prevent putrefaction. For typical municipal wastewater, air is provided at the rate of 1.25–2.0 ft^3/min-1000 gal (0.01–0.015 m^3/min·m^3). Power requirements are approximately 0.02–0.04 hp/1000 gal (4–8 W/m^3).

Table 10.5 Pumps Used in Wastewater Plants

flow	flow rate (gal/min (L/min))	pump type
raw sewage	< 50 (190)	pneumatic ejector
	50–200 (190–760)	submersible or end-suction, nonclog centrifugal
	> 200 (760)	end-suction, nonclog centrifugal
settled sewage	< 500 (1900)	end-suction, nonclog centrifugal
	> 500 (1900)	vertical axial or mixed-flow centrifugal
sludge, primary, thickened, or digested	–	plunger
sludge, secondary	–	end-suction, nonclog centrifugal
scum	–	plunger or recessed impeller
grit	–	recessed impeller, centrifugal, pneumatic ejector, or conveyor rake

(Multiply gal/min by 3.785 to obtain L/min.)

Figure 10.2 Equalization Volume: Mass Diagram Method

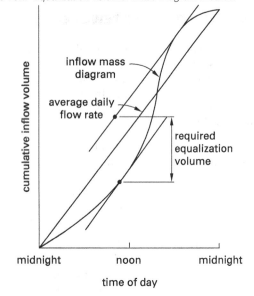

9. STABILIZATION PONDS

The term *stabilization pond* (*oxidation pond* or *stabilization lagoon*) refers to a pond used to treat organic waste by biological and physical processes. (See Table 10.6 for typical characteristics.) Aquatic plants, weeds, algae, and microorganisms stabilize the organic matter. The algae give off oxygen that is used by microorganisms to digest the organic matter. The microorganisms give off carbon dioxide, ammonia, and phosphates that the algae use. Even in modern times, such ponds may be necessary in remote areas (e.g., national parks and campgrounds). To keep toxic substances from leaching into the ground, ponds should not be used without strictly enforced industrial pretreatment requirements.

There are several types of stabilization ponds. *Aerobic ponds* are shallow ponds, less than 4 ft (1.2 m) in depth, where dissolved oxygen is maintained throughout the entire depth, mainly by action of photosynthesis. *Facultative ponds* have an anaerobic lower zone, a facultative middle zone, and an aerobic upper zone. The upper zone is maintained in an aerobic condition by photosynthesis and, in some cases, mechanical aeration at the surface. *Anaerobic ponds* are so deep and receive such a high organic loading that anaerobic conditions prevail throughout the entire pond depth. *Maturation ponds* (*tertiary ponds* or *polishing ponds*) are used for polishing effluent from secondary biological processes. Dissolved oxygen is furnished through photosynthesis and mechanical aeration. *Aerated lagoons* are oxygenated through the action of surface or diffused air aeration. They are often used with activated sludge processes. (See also Sec. 10.11.)

10. FACULTATIVE PONDS

Facultative ponds are the most common pond type selected for small communities. Approximately 25% of the municipal wastewater treatment plants in this country use ponds, and about 90% of these are located in communities of 5000 people or fewer. Long retention times and large volumes easily handle large fluctuations in wastewater flow and strength with no significant effect on effluent quality. Also, capital, operating, and maintenance costs are less than for other biological systems that provide equivalent treatment.

In a facultative pond, raw wastewater enters at the center of the pond. Suspended solids contained in the wastewater settle to the pond bottom where an anaerobic layer develops. A facultative zone develops just

Table 10.6 Typical Characteristics of Non-Aerated Stabilization Ponds

characteristic	aerobic		facultative	facultative with surface agitation	anaerobic
	algae-growth maximizing (high rate)	oxygen-transfer maximizing (low rate)			
size (cell)	0.5–2.5 ac (0.25–1 ha)	<10 ac (<4 ha)	2.5–10 ac (1–4 ha)	2.5–10 ac (1–4 ha)	0.5–2.5 ac (0.25–1 ha)
depth	1.0–1.5 ft (0.3–0.45 m)	3.0–5.0 ft (1–1.5 m)	3.0–7.0 ft (1–2 m)	3.0–8.5 ft (1–2.5 m)	8.0–15 ft (2.5–5 m)
BOD_5 loading	75–150 lbm/ac-day (80–160 kg/ha·d)	25–100 lbm/ac-day (40–120 kg/ha·d)	12–70 lbm/ac-day (15–80 kg/ha·d)	45–175 lbm/ac-day (50–200 kg/ha·d)	175–450 lbm/ac-day (200–500 kg/ha·d)
BOD_5 conversion	80–90%	80–90%	80–90%	80–90%	50–85%
residence time	4–6 days	10–40 days	7–30 days	7–20 days	20–50 days
temperature	40–85°F (5–30°C)	32–85°F (0–30°C)	32–120°F (0–50°C)	32–120°F (0–50°C)	40–120°F (5–50°C)
algal concentration	100–250 mg/L	40–100 mg/L	20–80 mg/L	5–20 mg/L	0–5 mg/L
suspended solids in effluent	150–300 mg/L	80–140 mg/L	40–100 mg/L	40–60 mg/L	80–160 mg/L
cell arrangement	series	parallel or series	parallel or series	parallel or series	series
minimum dike width			8 ft (2.5 m)		
maximum dike wall slope			1:3 (vertical:horizontal)		
minimum dike wall slope			1:4 (vertical:horizontal)		
minimum freeboard			3 ft (1 m)		

(Multiply lbm/ac-day by 1.12 to obtain kg/ha·d.)
(Multiply ft by 0.3048 to obtain m.)
(Multiply ac by 0.4 to obtain ha.)
(Multiply mg/L by 1.0 to obtain g/m³.)

above the anaerobic zone. Molecular oxygen is not available in the region at all times. Generally, the zone is aerobic during the daylight hours and anaerobic during the hours of darkness.

An aerobic zone with molecular oxygen present at all times exists above the facultative zone. Some oxygen is supplied from diffusion across the pond surface, but the majority is supplied through algal photosynthesis.

General guidelines are used to design facultative ponds. Ponds may be round, square, or rectangular. Usually, there are three cells, piped to permit operation in series or in parallel. Two of the three cells should be identical, each capable of handling half of the peak design flow. The third cell should have a minimum volume of one-third of the peak design flow.

11. AERATED LAGOONS

An *aerated lagoon* is a stabilization pond that is mechanically aerated. Such lagoons are typically deeper and have shorter residence times than nonaerated

ponds. In warm climates and with floating aerators, one acre can support several hundred pounds (a hundred kilograms) of BOD per day.

The basis for the design of aerated lagoons is typically the organic loading and/or residence time. Other factors that must be considered in the design process are solids removal requirements, oxygen requirements, temperature effects, and energy for mixing.

Equation 10.1 is the design equation for an aerated lagoon.

BOD_5 for Mixed Lagoons in Series

$$\frac{S}{S^0} = \frac{1}{1 + k_p \theta}$$

(10.1)

In Eq. 10.1, S^0 is the effluent total BOD_5 concentration, and S is the influent total BOD_5 concentration. k_p is the *kinetic constant*, also called the first-order BOD *removal rate constant*. The kinetic constant is a measure of the rate at which BOD is reduced in the lagoon.

Typical values of k_p (as specified by TSS Sec. 93.33) are 0.12 day^{-1} at 68°F (20°C) and 0.06 day^{-1} at 34°F (1°C).

θ is the hydraulic residence time.

Activated Sludge

$$\theta = V/Q \qquad 10.2$$

In designing an aerated lagoon, the time a contaminated waste stream must be detained will reduce the contaminant BOD before the stream is discharged into additional treatment or its final receiving location. Equation 10.2 calculates the hydraulic residence time by taking the constant volume of a tank and dividing by the flow rate through the tank. Both the volume and flow must be constant to maintain a constant residence time.

Equation 10.1 can be rearranged for use in determining the needed residence time from the kinetic constant for the lagoon, k_p, and the desired BOD removal fraction, η, which is equal to $1 - S/S_0$.

$$\theta = \frac{1 - \dfrac{S}{S_0}}{k_p\left(\dfrac{S}{S_0}\right)} = \frac{\eta}{k_p(1 - \eta)} \qquad 10.3$$

Aeration should maintain a minimum oxygen content of 2 mg/L at all times. Design depth is larger than for nonaerated lagoons—10–15 ft (3–4.5 m). Common design characteristics of mechanically aerated lagoons are shown in Table 10.7. Other characteristics are similar to those listed in Table 10.6.

Table 10.7 *Typical Characteristics of Aerated Lagoons*

aspect ratio	less than 3:1
depth	10–15 ft (3.0–4.5 m)
hydraulic residence time	4–10 days
BOD loading	20–400 lbm/day-ac;
	200 lbm/day-ac typical
	(22–440 kg/ha·d;
	220 kg/ha·d typical)
operating temperature	0–38°C (21°C optimum)
typical effluent BOD	20–70 mg/L
oxygen required	0.7–1.4 times BOD removed (mass basis)

(Multiply ft by 0.3048 to obtain m.)
(Multiply lbm/day-ac by 1.12 to obtain kg/ha·d.)

12. RACKS AND SCREENS

Trash racks or *coarse screens* with openings 2 in (51 mm) or larger should precede pumps to prevent clogging. *Medium screens* ($\frac{1}{2}$–$1\frac{1}{2}$ in (13–38 mm) openings) and *fine*

screens ($\frac{1}{16}$–$\frac{1}{8}$ in (1.6–3 mm)) are also used to relieve the load on grit chambers and sedimentation basins. Fine screens are rare except when used with selected industrial waste processing plants. Screens in all but the smallest municipal plants are cleaned by automatic scraping arms. A minimum of two screen units is advisable.

Screen capacities and head losses are specified by the manufacturer. Although the flow velocity must be sufficient to maintain sediment in suspension, the approach velocity should be limited to 3 ft/sec (0.9 m/s) to prevent debris from being forced through the screen.

13. GRIT CHAMBERS

Abrasive *grit* can erode pumps, clog pipes, and accumulate in excessive volumes. In a *grit chamber* (also known as a *grit clarifier* or *detritus tank*), the wastewater is slowed, allowing the grit to settle out but allowing the organic matter to continue through. (See Table 10.8 for typical characteristics.) Grit can be manually or mechanically removed with buckets or screw conveyors. A minimum of two units is needed.

Horizontal flow grit chambers are designed to keep the flow velocity as close to 1 ft/sec (0.3 m/s) as possible. If an analytical design based on settling velocity is required, the *scouring velocity* should not be exceeded. Scouring is the dislodging of particles that have already settled. Scouring velocity is not the same as settling velocity.

Scouring will be prevented if the horizontal velocity is kept below that predicted by Eq. 10.4, the *Camp formula*. SG_p is the specific gravity of the particle, typically taken as 2.65 for sand. d_p is the particle diameter. k is a dimensionless constant with typical values of 0.04 for sand and 0.06 or more for sticky, interlocking matter. The dimensionless Darcy friction factor, f, is approximately 0.02–0.03. Any consistent set of units can be used with Eq. 10.4.

$$v = \sqrt{8k\left(\frac{gd_p}{f}\right)(SG_p - 1)} \qquad 10.4$$

14. AERATED GRIT CHAMBERS

An *aerated grit chamber* is a bottom-hoppered tank, as shown in Fig. 10.3, with a short residence time. Diffused aeration from one side of the tank rolls the water and keeps the organics in suspension while the grit drops into the hopper. The water spirals or rolls through the tank. Influent enters through the side, and degritted wastewater leaves over the outlet weir. A minimum of two units is needed.

Solids are removed by pump, screw conveyer, bucket elevator, or gravity flow. However, the grit will have a significant organic content. A grit washer or cyclone

Table 10.8 Typical Characteristics of Grit Chambers

grit size	0.008 in (0.2 mm) and larger
grit specific gravity	2.65
grit arrival/removal rate	0.5–5 ft^3/MG (4–40 × 10^{-6} m^3/m^3)
depth of chamber	4–10 ft (1.2–3 m)
length	40–100 ft (12–30 m)
width	varies (not critical)
residence time	90–180 sec
horizontal velocity	0.75–1.25 ft/sec (0.23–0.38 m/s)

(Multiply in by 25.4 to obtain mm.)
(Multiply ft^3/MG by 7.48 × 10^{-6} to obtain m^3/m^3.)
(Multiply ft by 0.3048 to obtain m.)
(Multiply ft/sec by 0.3048 to obtain m/s.)

Figure 10.3 Aerated Grit Chamber

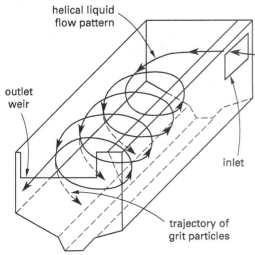

separator can be used to clean the grit. Common design characteristics of aerated grit chambers are shown in Table 10.9.

15. SKIMMING TANKS

If the sewage has more than 50 mg/L of floating grease or oil, a basin 8–10 ft (2.4–3 m) providing 5–15 min of residence time will allow the grease to rise as *floatables* (*scum*) to the surface. An aerating device will help coagulate and float grease to the surface. 40–80 psig (280–550 kPa) air should be provided at the approximate rate of 0.01–0.1 ft^3/gal (0.07–0.7 m^3/m^3) of influent. A small fraction (e.g., 30%) of the influent may be recycled in some cases. Grease rise rates are typically 0.2–1 ft/min (0.06–0.5 m/min), depending on degree of dispersal, amount of colloidal fines, and aeration.

Table 10.9 Typical Characteristics of Aerated Grit Chambers

residence time	2–5 min at peak flow; 3 min typical
air supply shallow tanks	1.5–5 cfm/ft length; 3 typical (0.13–0.45 m^3/min·m; 0.27 typical)
deep tanks	3–8 cfm/ft length; 5 typical (0.27–0.7 m^3/min·m; 0.45 typical)
grit and scum quantity	0.5–25 ft^3/MG; 2 typical (4–190 × 10^{-6} m^3/m^3; 15 typical)
length:width ratio	2.5:1–5:1 (3:1–4:1 typical)
depth	6–15 ft (1.8–4.5 m)
length	20–60 ft (6–18 m)
width	7–20 ft (2.1–6 m)

(Multiply cfm/ft by 0.0929 to obtain m^3/min·m.)
(Multiply ft^3/MG by 7.48 × 10^{-6} to obtain m^3/m^3.)
(Multiply ft by 0.3048 to obtain m.)

Scum is mechanically removed by skimming troughs. Scum may not generally be disposed of by landfilling. It is processed with other solid wastes in anaerobic digesters. *Scum grinding* may be needed to reduce the scum to a small enough size for thorough digestion. In some cases, scum may be incinerated, although this practice may be affected by air-quality regulations.

16. SHREDDERS

Shredders (also called *comminutors*) cut waste solids to approximately ¼ in (6 mm) in size, reducing the amount of screenings that must be disposed of. Shreddings stay with the flow for later settling.

17. PLAIN SEDIMENTATION BASINS/ CLARIFIERS

Plain sedimentation basins/clarifiers (i.e., basins in which no chemicals are added to encourage clarification) are similar in concept and design to those used to treat water supplies. Typical design characteristics for wastewater treatment sedimentation basins are listed in Table 10.10. Since the bottom slopes slightly, the depth varies with location. Therefore, the *side water depth* is usually quoted. The *surface loading* (*surface loading rate*, *overflow rate*, or *settling rate*), v_0, along with sludge storage volume, is the primary design parameter.

$$v_0 = \frac{Q}{A} \qquad 10.5$$

Water

Table 10.10 *Typical Characteristics of Clarifiers**

BOD reduction	20–40% (25–35% typical)
total suspended solids reduction	35–65%
bacteria reduction	50–60%
organic content of settled solids	50–75%
specific gravity of settled solids	1.2 or less
minimum settling velocity	4 ft/hr (1.2 m/hr) typical
plan shape	rectangular (or circular)
basin depth (side water depth)	6–15 ft; 10–12 ft typical (1.8–4.5 m; 3–3.6 m typical)
basin width	10–50 ft (3–15 m)
plan aspect ratio	3:1 to 5:1
basin diameter (circular only)	50–150 ft; 100 ft typical (15–45 m; 30 m typical)
minimum freeboard	1.5 ft (0.45 m)
minimum hopper wall angle	60°
hydraulic residence time	1.5–2.5 hr
flow-through velocity	18 ft/hr (1.5 mm/s)
minimum flow-through time	30% of residence time
weir loading	10,000–20,000 gal/day-ft (125–250 m³/d·m)
surface loading	400–2000 gal/day-ft²; 800–1200 gal/day-ft² typical (16–80 m³/d·m²; 32–50 m³/d·m² typical)
bottom slope to hopper	8%
inlet	baffled to prevent turbulence
scum removal	mechanical or manual

(Multiply ft/hr by 0.3048 to obtain m/h.)
(Multiply ft by 0.3048 to obtain m.)
(Multiply ft/hr by 0.0847 to obtain mm/s.)
(Multiply gal/day-ft by 0.0124 to obtain m³/d·m.)
(Multiply gal/day-ft² by 0.0407 to obtain m³/d·m².)
*See Sec. 10.26 and Sec. 10.27 for intermediate and final clarifiers, respectively.

The *hydraulic residence* (*detention time* or *retention period*) is

$$V/Q = \theta \qquad \text{\small Clarifier} \atop \text{\small \textbf{10.6}}$$

The *weir loading* (*weir overflow rate*) is

$$WOR = Q/\text{Weir Length} \qquad \text{\small Clarifier} \atop \text{\small \textbf{10.7}}$$

18. CHEMICAL SEDIMENTATION BASINS/ CLARIFIERS

Chemical flocculation (*clarification* or *coagulation*) operations in chemical sedimentation basins are similar to those encountered in the treatment of water supplies except that the coagulant doses are greater. Chemical precipitation may be used when plain sedimentation is insufficient, or occasionally when the stream into which the outfall discharges is running low, or when there is a large increase in sewage flow. As with water treatment, the five coagulants used most often are (a) aluminum sulfate, $Al_2(SO_4)_3$; (b) ferric chloride, $FeCl_3$; (c) ferric sulfate, $Fe_2(SO_4)_3$; (d) ferrous sulfate, $FeSO_4$; and (e) chlorinated copperas. Lime and sulfuric acid may be used to adjust the pH for proper coagulation.

Example 10.1

A primary clarifier receives 1.4 MGD of domestic waste. The clarifier has a peripheral weir, is 50 ft in diameter, and is filled to a depth of 7 ft. Determine if the clarifier is operating within typical performance ranges.

Solution

The perimeter length is

$$L = \pi D = \pi(50 \text{ ft})$$
$$= 157 \text{ ft}$$

The surface area is

$$A = \frac{\pi}{4}D^2 = \left(\frac{\pi}{4}\right)(50 \text{ ft})^2$$
$$= 1963 \text{ ft}^2$$

The volume is

$$V = AZ = (1963 \text{ ft}^2)(7 \text{ ft})$$
$$= 13{,}741 \text{ ft}^3$$

According to Table 10.10, surface loading is typically within the range of 400–2000 gal/day-ft^2.

$$v_0 = \frac{Q}{A} = \frac{1.4 \times 10^6 \; \dfrac{\text{gal}}{\text{day}}}{1963 \; \text{ft}^2}$$

$$= 713 \; \text{gal/day-ft}^2 \quad [\text{OK}]$$

The hydraulic residence time is typically within the range of 1.5–2.5 hr.

Clarifier

$$\theta = V/Q$$

$$= \frac{(13{,}741 \; \text{ft}^3) \left(7.48 \; \dfrac{\text{gal}}{\text{ft}^3} \right)}{\left(\dfrac{1.4 \times 10^6 \; \dfrac{\text{gal}}{\text{day}}}{24 \; \dfrac{\text{hr}}{\text{day}}} \right)}$$

$$= 1.76 \; \text{hr} \quad [\text{OK}]$$

The weir loading is typically within the range of 10,000–20,000 gal/day-ft.

$$\frac{Q}{L} = \frac{1.4 \times 10^6 \; \dfrac{\text{gal}}{\text{day}}}{157 \; \text{ft}}$$

$$= 8917 \; \text{gal/day-ft} \quad [\text{low, but probably OK}]$$

Weir loading is lower than the typical performance range, though not greatly.

19. TRICKLING FILTERS

Trickling filters (also known as *biological beds* and *fixed media filters*) consist of beds of rounded river rocks with approximate diameters of 2–5 in (50–125 mm), wooden slats, or modern synthetic media. Wastewater from primary sedimentation processing is sprayed intermittently over the bed. The biological and microbial slime growth attached to the bed purifies the wastewater as it trickles down. The water is introduced into the filter by rotating arms that move by virtue of spray reaction (reaction-type) or motors (motor-type). The clarified water is collected by an underdrain system.

The distribution rate is sometimes given by an *SK rating*, where SK is the water depth in millimeters deposited per pass of the distributor. Though there is strong evidence that rotational speeds of 1–2 rev/hr (high SK) produce significant operational improvement, traditional distribution arms revolve at 1–5 rev/min (low SK).

On the average, one acre of *low-rate filter* (also referred to as *standard-rate filter*) is needed for each 20,000 people served. Trickling filters can remove 70–90% of the suspended solids, 65–85% of the BOD, and 70–95% of the bacteria. Although low-rate filters have rocks to a depth of 6 ft (1.8 m), most of the reduction occurs in the first few feet of bed, and organisms in the lower part of the bed may be in a near-starvation condition.

Due to the low concentration of carbonaceous material in the water near the bottom of the filter, nitrogenous bacteria produce a highly nitrified effluent from low-rate filters. With low-rate filters, the bed will periodically slough off (unload) parts of its slime coating. Therefore, sedimentation after filtering is necessary. *Filter flies* are a major problem with low-rate filters, since fly larvae are provided with an undisturbed environment in which to breed.

Since there are limits to the heights of trickling filters, longer contact times can be achieved by returning some of the collected filter water back to the filter. This is known as *recirculation* or *recycling*. Recirculation is also used to keep the filter medium from drying out and to smooth out fluctuations in the hydraulic loading.

High-rate filters are used in most facilities. The higher hydraulic loading flushes the bed and inhibits excess biological growth. High-rate stone filters may be only 3–6 ft (0.9–1.8 m) deep. The high rate is possible because much of the filter discharge is recirculated. With the high flow rates, fly larvae are washed out, minimizing the filter fly problem. Since the biofilm is less thick and provided with carbon-based nutrients at a high rate, the effluent is nitrified only when the filter experiences low loading.

Super high-rate filters (*oxidation towers*) using synthetic media may be up to 40 ft (12 m) tall. High-rate and super high-rate trickling filters may be used as *roughing filters*, receiving wastewater at high hydraulic or organic loading and providing intermediate treatment or the first step of a multistage biological treatment process.

A BOD balance at the mixing point of a filter with recirculation results in Eq. 10.8, in which S_i is the BOD applied to the filter by the diluted influent, S_o is the BOD of the effluent from the primary settling tank (or the plant influent BOD if there is no primary sedimentation), and S_e is the BOD of the trickling filter effluent. (See Fig. 10.4.)

$$S_o + RS_e = (1 + R)S_i \qquad 10.8$$

$$S_i = \frac{S_o + RS_e}{1 + R} \qquad 10.9$$

Water

Figure 10.4 *Trickling Filter Process*

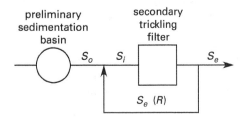

BOD is reduced significantly in a trickling filter. Standard-rate filters produce an 80–85% reduction, and high-rate filters remove 65–80% of BOD, less because of reduced contact area and time. The *removal fraction* (*removal efficiency*) of a single-stage trickling filter is

$$\eta = \frac{S_{\text{removed}}}{S_o} = \frac{S_o - S_e}{S_o} \qquad \textit{10.10}$$

By definition, the *recirculation ratio*, R, is zero for standard-rate filters but can be as high as 4:1 for high-rate filters.

$$R = \frac{Q_R}{Q_w} \qquad \textit{10.11}$$

Filters may be classified as high-rate based on their hydraulic loading, organic loading, or both. The *hydraulic loading* of a trickling filter is the total water flow divided by the plan area. Typical values of hydraulic loading are 25–100 gal/day-ft^2 (1–4 m^3/d·m^2) for standard filters and 250–1000 gal/day-ft^2 (10–40 m^3/d·m^2) or higher for high-rate filters.

$$L_H = \frac{Q_w + Q_R}{A} = \frac{Q_w(1 + R)}{A} \qquad \textit{10.12}$$

The *BOD loading* (*organic loading* or *surface loading*) is calculated without considering recirculated flow. BOD loading for the filter/clarifier combination is essentially the BOD of the incoming wastewater divided by the filter volume. BOD loading is usually given in lbm per 1000 ft^3 per day (hence the 1000 term in Eq. 10.13.) Typical values are 5–25 lbm/1000 ft^3-day (0.08–0.4 kg/m^3·day) for low-rate filters and 25–110 lbm/1000 ft^3-day (0.4–1.8 kg/m^3·d) for high-rate filters.

$$L_{\text{BOD,kg/m}^3\text{·d}} = \frac{Q_{w,\text{m}^3/\text{d}} S_{\text{mg/L}}}{\left(1000 \ \dfrac{\text{mg·m}^3}{\text{kg·L}}\right) V_{\text{m}^3}} \qquad \text{[SI]} \quad \textit{10.13(a)}$$

$L_{\text{BOD,lbm/1000 ft}^3\text{-day}}$

$$= \frac{Q_{w,\text{MGD}} S_{\text{mg/L}} \left(8.345 \ \dfrac{\text{lbm-L}}{\text{MG-mg}}\right) \times \left(1000 \ \dfrac{\text{ft}^3}{1000 \ \text{ft}^3}\right)}{V_{\text{ft}^3}} \qquad \text{[U.S.]} \quad \textit{10.13(b)}$$

The *specific surface area* of the filter is the total surface area of the exposed filter medium divided by the total volume of the filter.

Example 10.2

A single-stage trickling filter plant processes 1.4 MGD of raw domestic waste with a BOD of 170 mg/L. The trickling filter is 90 ft in diameter and has river rock media to a depth of 7 ft. The recirculation rate is 50%, and the filter is classified as high-rate. Water passes through clarification operations both before and after the trickling filter operation. The effluent leaves with a BOD of 45 mg/L. Determine if the trickling filter has been sized properly.

Solution

The filter area is

$$A = \frac{\pi}{4} D^2 = \left(\frac{\pi}{4}\right)(90 \ \text{ft})^2 = 6362 \ \text{ft}^2$$

The rock volume is

$$V = AZ = (6362 \ \text{ft}^2)(7 \ \text{ft}) = 44{,}534 \ \text{ft}^3$$

The hydraulic load for a high-rate filter should be 250–1000 gal/day-ft^2.

$$L_H = \frac{Q_w(1 + R)}{A}$$

$$= \frac{\left(1.4 \times 10^6 \ \dfrac{\text{gal}}{\text{day}}\right)(1 + 0.5)}{6362 \ \text{ft}^2}$$

$$= 330 \ \text{gal/day-ft}^2 \quad \text{[OK]}$$

From Table 10.10, the primary clarification process will remove approximately 30% of the BOD. The remaining BOD is

$$S_o = (1 - 0.3)\left(170 \ \frac{mg}{L}\right)$$

$$= 119 \ mg/L$$

The BOD loading should be approximately 25–110 lbm/1000 ft^3-day.

$$L_{\text{BOD}} = \frac{Q_{w,\text{MGD}} S_{\text{mg/L}}\left(8.345 \ \dfrac{\text{lbm-L}}{\text{MG-mg}}\right)(1000)}{V_{\text{ft}^3}}$$

$$= \frac{\left(1.4 \ \dfrac{\text{MG}}{\text{day}}\right)\left(119 \ \dfrac{\text{mg}}{\text{L}}\right)}{44{,}534 \ \text{ft}^3}$$
$$\times \left(8.345 \ \dfrac{\text{lbm-L}}{\text{MG-mg}}\right)(1000)$$

$$= 31.2 \ \text{lbm/day-1000 ft}^3 \quad [\text{OK}]$$

20. TWO-STAGE TRICKLING FILTERS

If a higher BOD or solids removal fraction is needed, then two filters can be connected in series with an optional intermediate settling tank to form a *two-stage trickling filter* system. The efficiency of the second-stage filter is considerably less than that of the first-stage filter because much of the biological food has been removed from the flow.

21. NATIONAL RESEARCH COUNCIL EQUATION

In 1946, the National Research Council (NRC) studied sewage treatment facilities at military installations. The wastewater at these facilities was stronger than typical municipal wastewater. Not surprisingly, the NRC concluded that the organic loading had a greater effect on removal efficiency than did the hydraulic loading.

If it is assumed that the biological layer and hydraulic loading are uniform, the water is at 20°C, and the filter is single-stage rock followed by a settling tank, then the *NRC equation*, Eq. 10.14, can be used to calculate the BOD removal fraction of the single-stage filter/clarifier combination. Inasmuch as installations with high BOD and low hydraulic loads were used as the basis of the studies, BOD removal efficiencies in typical municipal facilities are higher than predicted by Eq. 10.14. The BOD loading to the filter, W, excludes recirculation returned directly from the filter outlet, which is accounted for in the value of the recirculation factor, F.

National Research Council (NRC) Trickling Filter Performance

$$E_1 = \frac{100}{1 + 0.0561\sqrt{\dfrac{W}{VF}}} \qquad 10.14$$

In Eq. 10.14, E_1 is the BOD removal efficiency expressed as a percentage (not as a decimal fraction). It is also important to recognize the BOD loading, W, in Eq. 10.14, has units of lbm/day, not lbm/day-1000 ft^3, and that the volume of the filter media, V, is expressed in thousands of cubic feet (1000 ft^3). The constant 0.0085 is often encountered in place of 0.0561 in the literature, as in Eq. 10.15. However, this value is for use with filter media volumes, V, expressed in ac-ft, not in 1000 ft^3.

$$E_1 = \frac{100}{1 + 0.0561\sqrt{\dfrac{W}{V_{\text{ac-ft}}F}}} \qquad 10.15$$

There are a number of ways to recirculate water from the output of the trickling filters back to the filter. Water can be brought back to a wet well, to the primary settling tank, to the filter itself, or to a combination of the three. Variations in performance are not significant as long as sludge is not recirculated. Equation 10.14 and Eq. 10.15 can be used with any of the recirculation schemes.

F is the *effective number of passes* of the organic material through a filter. R is the ratio of filter discharge returned to the inlet to the raw influent.

National Research Council (NRC) Trickling Filter Performance

$$F = \frac{1 + R}{(1 + R/10)^2} \qquad 10.16$$

It is time-consuming to extract the organic loading, W, from Eq. 10.14 given E_1 and R. Figure 10.5 can be used for this purpose.

Based on the NRC model, the removal fraction for a two-stage filter with an intermediate clarifier is

$$E_2 = \frac{1}{1 + \left(\dfrac{0.0561}{1 - E_1}\right)\sqrt{\dfrac{W}{F}}} \qquad 10.17$$

In recent years, the NRC equations have fallen from favor for several reasons. These reasons include applicability only to rock media, inability to correct adequately for temperature variations, inapplicability to industrial waste, and empirical basis.

Water

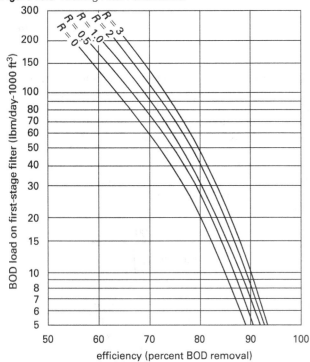

Figure 10.5 *Trickling Filter Performance**

*NRC model, single-stage

22. VELZ EQUATION

Equation 10.18, known as the *Velz equation*, was the first semi-theoretical analysis of trickling filter performance versus depth of media. It is useful in predicting the BOD removal fraction for any generic trickling filter, including those with synthetic media. The original Velz used the term "total removable fraction of BOD," though this is understood to mean the maximum fraction of removable BOD removed, generally S_i.

Values of the *Velz decay rate*, k, an empirical rate constant, are highly dependent on the installation and operating characteristics, particularly the hydraulic loading, which is not included in the formula. Subsequent to Velz' work, the industry has developed a large database of applicable values for numerous application scenarios.

$$\frac{S_e}{S_i} = e^{-kZ} \quad \left[\text{alternatively } \frac{S_e}{S_i} = 10^{-kZ}\right] \qquad 10.18$$

The variation in temperature was assumed to be

Kinetic Temperature Corrections
$$k_{\text{T}} = k_{20}(\theta)^{\text{T}-20} \qquad 10.19$$

The calculated decay rate, k_{T}, is for a specific temperature, T, and a specific treatment process, represented by the constant θ. The Velz decay rate for 20°C, k_{20}, is given, and then this value must be corrected for both the actual temperature and the treatment process being

used. When trickling filters are used, the value of θ is 1.072. For activated sludge, θ equals 1.136 when T is above 20°C and 1.056 when T is below 20°C.

23. MODERN FORMULATIONS

Various researchers have built upon the *Velz equation* and developed correlations of the form of Eq. 10.20 for various situations and types of filters. (In Eq. 10.20, S_a is the *specific surface area* (colonization surface area per unit volume), not the BOD.) Each researcher used different nomenclature and assumptions. For example, the inclusion of the effects of dilution by recirculation is far from universal. The specific form of the equation, values of constants and exponents, temperature coefficients, limitations, and assumptions are needed before such a correlation can be reliably used.

$$\frac{S_e}{S_i} = \exp\left\{-kZS_a^m\left(\frac{A}{Q}\right)^n\right\} \qquad 10.20$$

As an example of the difficulty in finding one model that predicts all trickling filter performance, consider the BOD removal efficiency as predicted by the *Schulze correlation* (1960) for rock media and the *Germain correlation* (1965) for synthetic media. Though both correlations have the same form, both measured filter depths in feet, and both expressed hydraulic loadings in gal/min-ft^2, the Schulze correlation included recirculation while the Germain correlation did not. (See Eq. 10.21.) The treatability constant, k, was 0.51–0.76 day^{-1} for the Schulze correlation and 0.088 day^{-1} for the Germain correlation. The *media factor* exponent, n, was 0.67 for the Schulze correlation (rock media) and 0.5 for the Germain correlation (plastic media).

$$\frac{S_e}{S_i} = \exp\left\{\frac{-kZ}{L_H^n}\right\} \qquad 10.21$$

24. ROTATING BIOLOGICAL CONTACTORS

Rotating biological contactors, RBCs (also known as *rotating biological reactors*), consist of large-diameter plastic disks, partially immersed in wastewater, on which biofilm is allowed to grow. (See Fig. 10.6.) The disks are mounted on shafts that turn slowly. The rotation progressively wets the disks, alternately exposing the biofilm to organic material in the wastewater and to oxygen in the air. The biofilm population, since it is well oxygenated, efficiently removes organic solids from the wastewater. RBCs are primarily used for carbonaceous BOD removal, although they can also be used for nitrification or a combination of both.

Figure 10.6 *Rotating Biological Contactor*

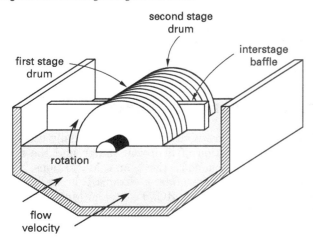

The primary design criterion is hydraulic loading, not organic (BOD) loading. For a specific hydraulic loading, the BOD removal efficiency will be essentially constant, regardless of variations in BOD. Other design criteria are listed in Table 10.11.

RBC operation is more efficient when several stages are used. Recirculation is not common with RBC processes. The process can be placed in series or in parallel with existing trickling filter or activated sludge processes.

Table 10.11 *Typical Characteristics of Rotating Biological Contactors*

number of stages	2–4
disk diameter	10–12 ft (3–3.6 m)
immersion, percentage of area	40%
hydraulic loading secondary treatment	2–4 gal/day-ft^2 (0.08–0.16 m^3/d·m^2)
tertiary treatment (nitrification)	0.75–2 gal/day-ft^2 (0.03–0.08 m^3/d·m^2)
optimum peripheral rotational speed	1 ft/sec (0.3 m/s)
tank volume	0.12 gal/ft^2 (0.0049 m^3/m^2) of biomass area
operating temperature	13–32°C
BOD removal fraction	70–80%

25. SAND FILTERS

For small populations, a *slow sand filter* (*intermittent sand filter*) can be used. Because of the lower flow rate, the filter area per person is higher than in the case of a trickling filter. Roughly one acre is needed for each 1000 people. The filter is constructed as a sand bed 2–3 ft (0.6–0.9 m) deep over a 6–12 in (150–300 mm) gravel bed. The filter is alternately exposed to water from a settling tank and to air (hence the term intermittent).

Straining and aerobic decomposition clean the water. Application rates are usually 2–2.5 gal/day-ft^2 (0.08–0.1 m^3/d·m^2). Up to 95% of the BOD can be satisfied in an intermittent sand filter. The filter is cleaned by removing the top layer of clogged sand.

If the water is applied continuously as a final process following secondary treatment, the filter is known as a *polishing filter* or *rapid sand filter*. The water rate of a polishing filter is typically 2–8 gal/min-ft^2 (0.08–0.32 m^3/min·m^2) but may be as high as 10 gal/min-ft^2 (0.4 m^3/min·m^2). Although the designs are similar to those used in water supply treatment, coarser media are used since the turbidity requirements are less stringent. Backwashing is required more frequently and is more aggressive than with water supply treatment.

26. INTERMEDIATE CLARIFIERS

Sedimentation tanks located between trickling filter stages or between a filter and subsequent aeration are known as *intermediate clarifiers* or *secondary clarifiers*. Typical characteristics of intermediate clarifiers used with trickling filter processes are: maximum overflow rate, 1500 gal/day-ft^2 (61 m^3/d·m^2); water depth, 10–13 ft (3–4 m); and maximum weir loading, 10,000 gal/day-ft (125 m^3/d·m) for plants processing 1 MGD or less and 15,000 gal/day-ft (185 m^3/d·m) for plants processing over 1 MGD.

27. FINAL CLARIFIERS

Sedimentation following secondary treatment occurs in *final clarifiers*. The purpose of final clarification is to collect sloughed-off material from trickling filter processes or to collect sludge and return it for activated sludge processes, not to reduce BOD. The depth is approximately 10–12 ft (3.0–3.7 m), the average overflow rate is 500–600 gal/day-ft^2 (20–24 m^3/d·m^2), and the maximum overflow rate is approximately 1100 gal/day-ft^2 (45 m^3/d·m^2). The maximum weir loading is the same as for intermediate clarifiers, but the lower rates are preferred. For settling following extended aeration, the overflow rate and loading should be reduced approximately 50%.

28. PHOSPHORUS REMOVAL: PRECIPITATION

Phosphorus concentrations of 5–15 mg/L (as P) are experienced in untreated wastewater, most of which originates from synthetic detergents and human waste. Approximately 10% of the total phosphorus is insoluble and can be removed in primary settling. The amount that is removed by absorption in conventional biological processes is small. The remaining phosphorus is soluble and must be removed by converting it into an insoluble precipitate.

Soluble phosphorus is removed by precipitation and settling. Aluminum sulfate, ferric chloride ($FeCl_3$), and lime may be used depending on the nature of the phosphorus radical. Aluminum sulfate is more desirable since lime reacts with hardness and forms large quantities of additional precipitates. (Hardness removal is not as important as it is in water supply treatment.) However, the process requires about 10 lbm (10 kg) of aluminum sulfate for each pound (kilogram) of phosphorus removed. The process also produces a chemical sludge that is difficult to dewater, handle, and dispose of.

$$Al_2(SO_4)_3 + 2PO_4 \rightleftharpoons 2AlPO_4 + 3SO_4 \qquad 10.22$$

$$FeCl_3 + PO_4 \rightleftharpoons FePO_4 + 3Cl \qquad 10.23$$

Due to the many other possible reactions the compounds can participate in, the dosage should be determined from testing. The stoichiometric chemical reactions describe how the phosphorus is removed, but they do not accurately predict the quantities of coagulants needed.

29. AMMONIA REMOVAL: AIR STRIPPING

Ammonia may be removed by either biological processing or air stripping. In the biological *nitrification and denitrification process*, ammonia is first aerobically converted to nitrite and then to nitrate (nitrification) by bacteria. Then, the nitrates are converted to nitrogen gas, which escapes (denitrification).

In the *air-stripping* (*ammonia-stripping*) method, lime is added to water to increase its pH to about 10. This causes the ammonium ions, NH_4^+, to change to dissolved ammonia gas, NH_3. The water is passed through a packed tower into which air is blown at high rates. The air strips the ammonia gas out of the water. Recarbonation follows to remove the excess lime.

30. CHLORINE DEMAND AND DOSE

Chlorination destroys bacteria, hydrogen sulfide, and other compounds and substances. The *chlorine demand* (*chlorine dose*) must be determined by careful monitoring of coliform counts and free residuals since there are several ways that chlorine can be used up without producing significant disinfection. Only after uncombined (free) chlorine starts showing up is it assumed that all chemical reactions and disinfection are complete.

Chlorine demand is the amount of chlorine (or its chloramine or hypochlorite equivalent) required to leave the desired residual (usually 0.5 mg/L) 15 min after mixing. Fifteen minutes is the recommended mixing and holding time prior to discharge, since during this time nearly all pathogenic bacteria in the water will have been killed.

Typical doses for wastewater effluent depend on the application point and are widely variable, though doses rarely exceed 30 mg/L. For example, chlorine may be applied at 5–25 mg/L prior to primary sedimentation, 2–6 mg/L after sand filtration, and 3–15 mg/L after trickle filtration.

31. BREAKPOINT CHLORINATION

Because of their reactivities, chlorine is initially used up in the neutralization of hydrogen sulfide and the rare ferrous and manganous (Fe^{2+} and Mn^{2+}) ions. For example, hydrogen sulfide is oxidized according to Eq. 10.24. The resulting HCl, $FeCl_2$, and $MnCl_2$ ions do not contribute to disinfection. They are known as *unavailable combined residuals*.

$$H_2S + 4H_2O + 4Cl_2 \rightarrow H_2SO_4 + 8HCl \qquad 10.24$$

Ammonia nitrogen combines with chlorine to form the family of *chloramines*. Depending on the water pH, monochloramines (NH_2Cl), dichloramines ($NHCl_2$), or trichloramines (nitrogen trichloride, NCl_3) may form. Chloramines have long-term disinfection capabilities and are therefore known as *available combined residuals*. Equation 10.25 is a typical chloramine formation reaction.

$$NH_4^+ + HOCl \rightleftharpoons NH_2Cl + H_2O + H^+ \qquad 10.25$$

Continued addition of chlorine after chloramines begin forming changes the pH and allows chloramine destruction to begin. Chloramines are converted to nitrogen gas (N_2) and nitrous oxide (N_2O). Equation 10.26 is a typical chloramine destruction reaction.

$$2NH_2Cl + HOCl \rightleftharpoons N_2 + 3HCl + H_2O \qquad 10.26$$

The destruction of chloramines continues, with the repeated application of chlorine, until no ammonia remains in the water. The point at which all ammonia has been removed is known as the *breakpoint*.

In the *breakpoint chlorination* method, additional chlorine is added after the breakpoint in order to obtain free chlorine residuals. The free residuals have a high disinfection capacity. Typical free residuals are free chlorine (Cl_2), hypochlorous acid (HOCl), and hypochlorite ions. Equation 10.27 and Eq. 10.28 illustrate the formation of these free residuals.

$$Cl_2 + H_2O \rightarrow HCl + HOCl \qquad 10.27$$

$$HOCl \rightarrow H^+ + ClO^- \qquad 10.28$$

There are several undesirable characteristics of breakpoint chlorination. First, it may not be economical to use breakpoint chlorination unless the ammonia nitrogen has been reduced. Second, free chlorine residuals produce trihalomethanes. Third, if free residuals are prohibited to prevent trihalomethanes, the water may need to be dechlorinated using sulfur dioxide gas or sodium bisulfate. (Where small concentrations of free residuals are permitted, dechlorination may be needed only during the dry months. During winter storm months, the chlorine residuals may be adequately diluted with rainwater.)

32. CHLORINATION

Chlorination to disinfect and deodorize is one of the final steps prior to discharge. Vacuum-type feeders are used predominantly with chlorine gas. Chlorine under vacuum is combined with wastewater to produce a chlorine solution. A *flow-pacing chlorinator* will reduce the chlorine solution feed rate when the wastewater flow decreases (e.g., at night).

The size of the *contact tank* varies, depending on economics and other factors. An average design residence time is 30 min at average flow, some of which can occur in the plant outfall after the contact basin. Contact tanks are baffled to prevent short-circuiting that would otherwise reduce chlorination time and effectiveness.

Alternatives to disinfection by chlorine include sodium hypochlorite, ozone, ultraviolet light, bromine (as bromine chloride), chlorine dioxide, and hydrogen peroxide. All alternatives have one or more disadvantages when compared to chlorine gas.

33. DECHLORINATION

Toxicity, by-products, and strict limits on *total residual oxidants* (TROs) now make dechlorination mandatory at many installations. Sulfur dioxide (SO_2) and sodium thiosulfite ($Na_2S_2SO_3$) are the primary compounds used as dechlorinators today. Other compounds seeing limited use are sodium metabisulfate ($Na_2S_2O_5$), sodium bisulfate ($NaHSO_3$), sodium sulfite (Na_2SO_3), hydrogen peroxide (H_2O_2), and granular activated carbon.

Though reaeration was at one time thought to replace oxygen in water depleted by sulfur dioxide, this is now considered to be unnecessary unless required to meet effluent discharge requirements.

34. EFFLUENT DISPOSAL

Organic material and bacteria present in wastewater are generally not removed in their entireties, though the removal efficiency is high (e.g., better than 95% for some processes). Therefore, effluent must be discharged to large bodies of water where the remaining contaminants can be substantially diluted. Discharge to flowing surface water and oceans is the most desirable. Discharge to lakes and reservoirs should be avoided.

In some areas, *combined sewer overflow* (CSO) outfalls still channel wastewater into waterways during heavy storms when treatment plants are overworked. Such pollution can be prevented by the installation of large retention basins (*diversion chambers*), miles of tunnels, reservoirs to capture overflows for later controlled release to treatment plants, screening devices to separate solids from wastewater, swirl concentrators to capture solids for treatment while permitting the clearer portion to be chlorinated and released to waterways, and more innovative vortex solids separators.

35. ACTIVATED SLUDGE PROCESS

The *activated sludge process* is a secondary biological wastewater treatment technique in which a mixture of wastewater and sludge solids is aerated. (See Fig. 10.7.) The sludge mixture produced during this oxidation process contains an extremely high concentration of aerobic bacteria, most of which are near starvation. This condition makes the sludge an ideal medium for the destruction of any organic material in the mixture. Since the bacteria are voraciously active, the sludge is called *activated sludge*.

The well-aerated mixture of wastewater and sludge, known as *mixed liquor*, flows from the aeration tank to a secondary clarifier where the sludge solids settle out. Most of the settled sludge solids are returned to the aeration tank in order to maintain the high population of bacteria needed for rapid breakdown of the organic material. However, because more sludge is produced than is needed, some of the return sludge is diverted ("wasted") for subsequent treatment and disposal. This wasted sludge is referred to as *waste activated sludge*, WAS. The volume of sludge returned to the aeration basin is typically 20–30% of the wastewater flow. The liquid fraction removed from the secondary clarifier weir is chlorinated and discharged.

Though diffused aeration and mechanical aeration are the most common methods of oxygenating the mixed liquor, various methods of staging the aeration are used, each having its own characteristic ranges of operating parameters. (See Table 10.12 for typical characteristics of activated sludge plants, Table 10.13 for additional information, and Table 10.14 for *Ten States' Standards*.)

In a traditional activated sludge plant using conventional aeration, the wastewater is typically aerated for 6–8 hours in long, rectangular aeration basins. Sufficient air, about eight volumes for each volume of wastewater treated, is provided to keep the sludge in suspension. The air is injected near the bottom of the aeration tank through a system of diffusers.

36. AERATION STAGING METHODS

Small wastewater quantities can be treated with *extended aeration*. (See Fig. 10.8 for available aeration methods.) This method uses mechanical floating or fixed subsurface aerators to oxygenate the mixed liquor for 24–36 hours in a large lagoon. There is no primary clarification, and there is generally no sludge wasting

Water

Figure 10.7 *Typical Activated Sludge Plant*

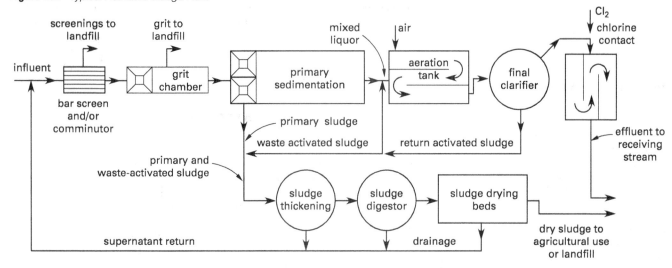

Table 10.12 *Typical Characteristics of Conventional Activated Sludge Plants*

BOD reduction	90–95%
effluent BOD	5–30 mg/L
effluent COD	15–90 mg/L
effluent suspended solids	5–30 mg/L
F:M	0.2–0.5
aeration chamber volume	5000 ft³ (140 m³) max.
aeration chamber depth	10–15 ft (3–4.5 m)
aeration chamber width	20 ft (6 m)
length:width ratio	5:1 or greater
aeration air rate	0.5–2 ft³ air/gal (3.6–14 m³/m³) raw wastewater
minimum dissolved oxygen	1 mg/L
MLSS	1000–4000 mg/L
sludge volume index	50–150
settling basin depth	15 ft (4.5 m)
settling basin residence time	4–8 hr
basin overflow rate	400–2000 gal/day-ft²; 1000 gal/day-ft² typical (16–81 m³/d·m²; 40 m³/d·m² typical)
settling basin weir loading	10,000 gal/day-ft (130 m³/d·m)
fractional sludge recycle	0.20–0.30
frequency of sludge recycle	hourly

(Multiply ft³ by 0.0283 to obtain m³.)
(Multiply ft by 0.3048 to obtain m.)
(Multiply ft³/gal by 7.48 to obtain m³/m³.)
(Multiply gal/day-ft² by 0.0407 to obtain m³/d·m².)
(Multiply gal/day-ft by 0.0124 to obtain m³/d·m.)

process. Sludge is allowed to accumulate at the bottom of the lagoon for several months. Then, the system is shut down and the lagoon is pumped out. Sedimentation basins are sized very small, with low overflow rates of 200–600 gal/day-ft² (8.1–24 m³/d·m²) and long retention times.

An *oxidation ditch* is a type of extended aeration system, configured in a continuous long and narrow oval. The basin contents circulate continuously to maintain mixing and aeration. Mixing occurs by brush aerators or by a combination of low-head pumping and diffused aeration.

In *conventional aeration*, the influent is taken from a primary clarifier and then aerated. The amount of aeration is usually decreased as the wastewater travels along the aeration path since the BOD also decreases along the route. This is known as *tapered aeration*.

With *step-flow aeration*, aeration is constant along the length of the aeration path, but influent is introduced at various points along the path.

With *complete-mix aeration*, wastewater is added uniformly and the mixed liquor is removed uniformly over the length of the tank. The mixture is essentially uniform in composition throughout.

Units for the *contact stabilization* (*biosorption*) process are typically factory-built and brought to the site for installation, although permanent facilities can be built for the same process. Prebuilt units are compact but not as economical or efficient as larger plants running on the same process. The aeration tank is called a *contact tank*. The *stabilization tank* takes the sludge from the clarifier and aerates it. With this process, colloidal solids are absorbed in the activated sludge during the 30–90 min of aeration in the contact tank. Then, the sludge is removed by clarification and the return sludge is aerated for 3–6 more hours in the stabilization tank. Less time and space is required for this process because the sludge

Table 10.13 Representative Operating Conditions for Aeration[a]

type of aeration	plant flow rate (MGD)	mean cell residence time, θ_c (days)	oxygen required (lbm/lbm BOD removed)	waste sludge (lbm/lbm BOD removed)	total plant BOD load (lbm/day)	aerator BOD load, L_{BOD} (lbm/day-1000 ft³)	F:M (lbm/lbm-day)	MLSS (mg/L)	R (%)	η_{BOD} (%)
conventional	0–0.5	7.5	0.8–1.1	0.4–0.6	0–1000	30	0.2–0.5	1500–3000	30	90–95
	0.5–1.5	7.5–6.0			1000–3000	30–40				
	1.5 up	6.0			3000 up	40				
contact stabilization	0–0.5	3.0[b]	0.8–1.1	0.4–0.6	0–1000	30	0.2–0.5	1000–3000[b]	100	85–90
	0.5–1.5	3.0–2.0[b]	0.4–0.6		1000–3000	30–50				
	1.5 up	2.0–1.5[b]	0.4–0.6		3000 up	50				
extended	0–0.5	24	1.4–1.6	0.15–0.3	all	10.0	0.05–0.1	3000–6000	100	85–95
	0.5–1.5	20				12.5				
	1.5 up	16				15.0				
high rate	0–0.5	4.0	0.7–0.9	0.5–0.7	2000 up	100	1.0 or less	4000–10,000	100	80–85
	0.5–1.5	3.0								
	1.5 up	2.0								
step aeration	0–0.5	7.5			0–1000	30	0.2–0.5	2000–3500	50	85–95
	0.5–1.5	7.5–5.0			1000–3000	30–50				
	1.5 up	5.0			3000 up	50				
high purity oxygen		3.0–1.0				100–200	0.6–1.5	6000–8000	50	90–95
oxidation ditch		36–12				5–30				

(Multiply MGD by 3785.4 to obtain m³/d.)

(Multiply lbm/day by 0.4536 to obtain kg/d.)

(Multiply lbm/day-1000 ft³ by 0.016 to obtain kg/d·m³.)

[a]compiled from a variety of sources

[b]in contact unit only

stabilization is done while the sludge is still concentrated. This method is very efficient in handling colloidal wastes.

The *high-rate aeration* method uses mechanical mixing along with aeration to decrease the aeration period and increase the BOD load per unit volume.

The *high purity oxygen aeration* method requires the use of bottled or manufactured oxygen that is introduced into closed/covered aerating tanks. Mechanical mixers are needed to take full advantage of the oxygen supply because little excess oxygen is provided. Retention times in aeration basins are longer than for other aerobic systems, producing higher concentrations of MLSS and better stabilization of sludge that is ultimately wasted. This method is applicable to high-strength sewage and industrial wastewater.

Sequencing batch reactors (SBRs) operate in a fill-and-drain sequence. The reactor is filled with influent, contents are aerated, and sludge is allowed to settle. Effluent is drawn from the basin along with some of the settled sludge, and new effluent is added to repeat the process. The "return sludge" is the sludge that remains in the basin.

37. FINAL CLARIFIERS

Table 10.15 lists typical operating characteristics for clarifiers in activated sludge processes. Sludge should be removed rapidly from the entire bottom of the clarifier.

Table 10.14 Ten States' Standards for Activated Sludge Processes

process	aeration tank organic loading—lbm BOD_5/day per 1000 ft³ (kg/d·m³)	F:M lbm BOD_5/day per lbm MLVSS	MLSS[a] (mg/L)
conventional			
step aeration	40 (0.64)	0.2–0.5	1000–3000
complete mix			
contact stabilization	50[b] (0.80)	0.2–0.6	1000–3000
extended aeration			
oxidation ditch	15 (0.24)	0.05–0.1	3000–5000[c]

(Multiply lbm/day-1000 ft³ by 0.016 to obtain kg/d·m³.)
(Multiply lbm/day-lbm by 1.00 to obtain kg/d·kg.)

[a]MLSS values are dependent upon the surface area provided for sedimentation and the rate of sludge return as well as the aeration process.
[b]Total aeration capacity; includes both contact and reaeration capacities. Normally the contact zone equals 30–35% of the total aeration capacity.
[c]Reprinted from *Recommended Standards for Sewage Works*, 2004, Sec. 92.31, Great Lakes-Upper Mississippi River Board of State Sanitary Engineers, published by Health Education Service, Albany, NY.

Figure 10.8 Methods of Aeration

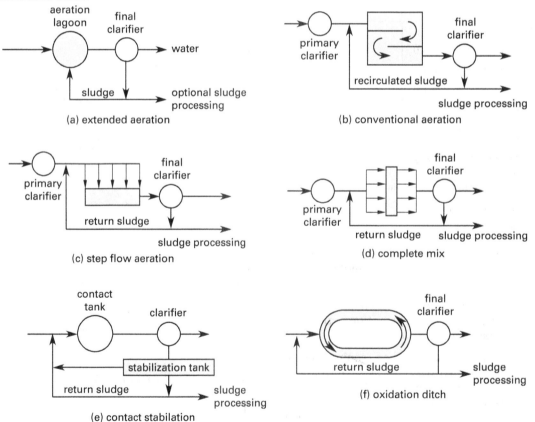

Table 10.15 Characteristics of Final Clarifiers for Activated Sludge Processes

type of aeration	design flow (MGD)	minimum residence time (hr)	maximum overflow rate (gpd/ft^2)
conventional, high rate and step	< 0.5	3.0	600
	0.5–1.5	2.5	700
	> 1.5	2.0	800
contact stabilization	< 0.5	3.6	500
	0.5–1.5	3.0	600
	>1.5	2.5	700
extended aeration	< 0.05	4.0	300
	0.05–0.15	3.6	300
	>0.15	3.0	600

(Multiply MGD by 0.0438 to obtain m^3/s.)
(Multiply gal/day-ft^2 by 0.0407 m^3/d·m^2.)

38. SLUDGE PARAMETERS

The bacteria and other suspended material in the mixed liquor is known as *mixed liquor suspended solids* (MLSS) and is measured in mg/L. Suspended solids are further divided into fixed solids and volatile solids. *Fixed solids* (also referred to as *nonvolatile solids*) are those inert solids that are left behind after being fired in a furnace. *Volatile solids* are essentially those carbonaceous solids that are consumed in the furnace. The volatile solids are considered to be the measure of solids capable of being digested. Approximately 60–75% of sludge solids are volatile. The volatile material in the mixed liquor is known as the *mixed liquor volatile suspended solids* (MLVSS).

The organic material in the incoming wastewater constitutes "food" for the activated organisms. The food arrival rate is given by Eq. 10.29. S_o is usually taken as the incoming BOD$_5$, although COD is used in rare situations.

$$F = Q_o S_o \qquad 10.29$$

The mass of microorganisms, M, is determined from the *volatile suspended solids concentration*, X, in the aeration tank.

$$M = V_a X_A \qquad 10.30$$

The *food-to-microorganism ratio*, (F:M) is displayed in Eq. 10.31. For conventional aeration, typical values are 0.20–0.50 lbm/lbm-day (0.20–0.50 kg/kg·d), though values between 0.05 and 1.0 have been reported. θ in Eq. 10.31 is the *hydraulic residence time*.

Activated Sludge

$$(\text{F:M}) = Q_0 S_0 / (Vol \ X_A) \qquad 10.31$$

The liquid fraction of the wastewater passes through an activated sludge process in a matter of hours. However, the sludge solids are recycled continuously and have an average stay much longer in duration. There are two measures of *sludge age*: the *mean cell residence time* (also known as the *age of the suspended solids* and *solids residence time*), essentially the age of the microorganisms, and *age of the BOD*, essentially the age of the food. For conventional aeration, typical values of the mean cell residence time, θ_c, are 6 to 15 days for high-quality effluent and sludge.

Activated Sludge

$$\theta_c = \text{Solids residence time} = \frac{V(X_A)}{Q_w X_w + Q_e X_e} \qquad 10.32$$

The *BOD sludge age* (*age of the BOD*), θ_{BOD}, is the reciprocal of the food-to-microorganism ratio.

$$\theta_{\text{BOD}} = \frac{1}{\text{F:M}} = \frac{(Vol \ X_A)}{Q_0 S_0} \qquad 10.33$$

The *sludge volume index*, SVI, is a measure of the sludge's settleability. SVI can be used to determine the tendency toward *sludge bulking*. (See Sec. 10.50.) SVI is determined by taking 1 L of mixed liquor and measuring the volume of settled solids after 30 minutes. SVI is the volume in mL occupied by 1 g of settled volatile and nonvolatile suspended solids.

Activated Sludge

$$\text{SVI} = \frac{\text{Sludge volume after settling (mL/L)} \times 1{,}000}{\text{MLSS (mg/L)}} \qquad 10.34$$

The concentration of total suspended solids, TSS, in the recirculated sludge can be found from Eq. 10.35. This equation is useful in calculating the solids concentration for any sludge that is wasted from the return sludge line. (Suspended sludge solids wasted include both fixed and volatile portions.) In Eq. 10.35, TSS is in mg/L, and SVI is in mL/g.

$$\text{TSS} = \frac{\left(1000 \ \frac{\text{mg}}{\text{g}}\right)\left(1000 \ \frac{\text{mL}}{\text{L}}\right)}{\text{SVI}} \qquad 10.35$$

39. SOLUBLE BOD ESCAPING TREATMENT

S is a variable used to indicate the growth-limiting substrate in solution. Specifically, S (without a subscript) is defined as the soluble BOD$_5$ escaping treatment (i.e., the effluent BOD$_5$ leaving the activated sludge process).

This may need to be calculated from the effluent suspended solids and the fraction, G, of the suspended solids that is ultimately biodegradable.

$$
\begin{aligned}
\text{BOD}_e &= \text{BOD}_{\text{escaping treatment}} \\
&\quad + \text{BOD}_{\text{effluent suspended solids}} \\
&= S + S_e \\
&= S + 1.42fGX_e
\end{aligned} \qquad 10.36
$$

Care must be taken to distinguish between the standard five-day BOD and the ultimate BOD. The ratio of these two parameters, typically approximately 70%, is

$$
f = \frac{\text{BOD}_5}{\text{BOD}_u} \qquad 10.37
$$

In some cases, the terms in the product $1.42fG$ in Eq. 10.36 are combined, and S_e is known directly as a percentage of X_e. However, this format is less common.

40. PROCESS EFFICIENCY

The *treatment efficiency* (*BOD removal efficiency, removal fraction*, etc.), typically 90–95% for conventional aeration, is given by Eq. 10.38. S_0 is the BOD as received from the primary settling tank. If the overall plant efficiency is wanted, the effluent S_e is taken as the total BOD_5 of the effluent. Depending on convention, efficiency can also be based on only the soluble BOD escaping treatment (i.e., S from Eq. 10.36).

$$
\eta_{\text{BOD}} = \frac{S_0 - S}{S_0} \qquad 10.38
$$

41. NITROGEN CONTROL

Nitrogen exists in wastewater in four forms: organic nitrogen, ammonia nitrogen (NH_3), nitrate nitrogen (NO_3^-), and nitrite nitrogen (NO_2^-). The total nitrogen is defined as the sum of these four forms.

Another common measure is *total Kjeldahl nitrogen* (TKN), which is the sum of the total organic nitrogen and the total ammonia nitrogen. TKN can be used with ammonia nitrogen to estimate organic nitrogen concentration.

Nitrogen removal occurs by a two-step process. The first step is *nitrification*, the aerobic biological conversion of ammonia in the wastewater to nitrate and nitrite. The second step is *denitrification*, the anaerobic conversion of nitrite and nitrate to nitrogen gas. The nitrogen gas, which is inert, is released to the atmosphere. These two steps, one aerobic and one anaerobic, require different environments within the treatment system. Each

conversion process can employ either attached growth or suspended growth treatment processes; suspended growth processes are more commonly used.

The treatment processes employed for nitrogen removal are typically the same as those employed for BOD removal, with kinetic coefficients evaluated for the specific purpose of nitrogen control.

42. NITRIFICATION

During the nitrification process, ammonia is converted first to nitrite and then to nitrate according to the following overall reaction.

$$
\text{NH}_4^+ + 2\text{O}_2 \rightarrow \text{NO}_3^- + \text{H}_2\text{O} + 2\text{H}^+ \qquad 10.39
$$

For this reaction to go to completion, 4.57 g of oxygen is needed for each 1.0 g of ammonia. This is known as the *nitrogenous biochemical oxygen demand* (NBOD). Consequently, the problem with nitrogen in the wastewater is the oxygen demand exerted as the water is discharged to the environment.

Nitrification is influenced by the *nitrifier fraction*, the fraction of organisms comprising the mixed liquor volatile suspended solids (MLVSS) that are capable of nitrifying. Typical values for the nitrifier fraction range from 0.03 to 0.05.

Kinetic coefficients are corrected for influences of temperature, pH, and DO relevant to nitrification by Eq. 10.40.

$$
\begin{aligned}
\mu_m' &= \mu_m e^{0.098(T-15)} \left(\frac{DO}{K_o + DO} \right) \\
&\quad \times \left(1 - 0.883(7.2 - \text{pH}) \right)
\end{aligned} \qquad 10.40
$$

In Eq. 10.40, μ_m is the maximum growth rate, and μ_m' is the corrected maximum growth rate. The minimum mean cell residence time, θ_c^m, can be defined by

$$
\frac{1}{\theta_c^m} = \frac{\mu_m S_0}{K_s + S_0} - k_d \qquad 10.41
$$

If the corrected value, μ_m', is substituted for μ_m and the influent ammonia concentration, N_0, is substituted for S_0, then θ_c^m can be determined by Eq. 10.42.

$$
\frac{1}{\theta_c^m} = \frac{\mu_m' N_0}{K_s + N_0} - k_d \qquad 10.42
$$

By applying a safety factor, the design mean cell residence time, θ_c^d, can be defined by

$$
\theta_c^d = (\text{SF})\theta_c^m \qquad 10.43
$$

When θ_c^d has been defined, it can be used with the standard complete-mix or plug-flow equation to find the residence time, θ, for nitrification. θ_c^d is calculated for both nitrification and BOD removal, and the larger resulting value is applied to the design of the bioreactor. Values of θ_c and θ are usually higher for nitrification than for BOD removal.

For BOD removal, the biomass concentration is given by Eq. 10.44, where θ_c is the design mean cell residence time (θ_c^d in Eq. 10.43).

Activated Sludge

$$X_A = \frac{\theta_c Y (S_0 - S_e)}{\theta(1 + k_d \theta_c)} \qquad \text{10.44}$$

Equation 10.44 can be rearranged to form Eq. 10.45, which can be used to find θ_c.

$$\frac{1}{\theta_c} = \frac{Y(S_0 - S_e)}{\theta X_A} - k_d \qquad \text{10.45}$$

Substituting N for S and X_N for X_A into Eq. 10.44 gives

$$X_N = \frac{\theta_c Y (N_0 - N_e)}{\theta(1 + k_d \theta_c)} \qquad \text{10.46}$$

Equation 10.46 can be rearranged to form Eq. 10.47, which can be used to find θ_c for nitrification.

$$\frac{1}{\theta_c} = \frac{Y(N_0 - N_e)}{\theta X_N} - k_d \qquad \text{10.47}$$

Example 10.3

Size a complete-mix activated sludge unit process to treat 10 000 m³/d of the wastewater defined by the following parameters.

influent BOD from primary clarifier	185 mg/L
influent ammonia nitrogen from the primary clarifier	40 mg/L as N
effluent total BOD	20 mg/L
effluent ammonia nitrogen	1 mg/L as N
mixed liquor volatile suspended solids	1900 mg/L
pH	6.9
temperature	18°C
dissolved oxygen	7.2 mg/L
yield coefficient	0.2
kinetic constant	0.05 d⁻¹ at 20°C
maximum growth rate	0.5 d⁻¹ at 20°C
half-velocity constant	0.8 mg/L
safety factor	3.5

Solution

Use Eq. 10.40 to calculate the corrected maximum growth rate.

$$\mu_m' = \mu_m e^{0.098(T-15)} \left(\frac{DO}{K_o + DO} \right)$$
$$\times \left(1 - 0.883(7.2 - \text{pH}) \right)$$
$$= \left(0.5 \frac{1}{\text{d}} \right) e^{(0.098)(18°\text{C} - 15°\text{C})}$$
$$\times \left(\frac{7.2 \frac{\text{mg}}{\text{L}}}{1.3 \frac{\text{mg}}{\text{L}} + 7.2 \frac{\text{mg}}{\text{L}}} \right)$$
$$\times \left(1 - (0.883)(7.2 - 6.9) \right)$$
$$= 0.43 \ \text{d}^{-1}$$

Use Eq. 10.42 to calculate the mean cell residence time for nitrification.

$$\frac{1}{\theta_{c,\text{nitr}}^m} = \frac{\mu_m' N_0}{K_s + N_0} - k_d$$
$$= \frac{\left(0.43 \frac{1}{\text{d}} \right) \left(40 \frac{\text{mg}}{\text{L}} \right)}{0.8 \frac{\text{mg}}{\text{L}} + 40 \frac{\text{mg}}{\text{L}}} - 0.05 \frac{1}{\text{d}}$$
$$= 0.37 \ \text{d}^{-1}$$
$$\theta_{c,\text{nitr}}^m = \frac{1}{0.37 \frac{1}{\text{d}}} = 2.7 \ \text{d}$$

Use Eq. 10.41 to calculate the mean cell residence time for BOD removal.

$$\frac{1}{\theta_{c,\text{BOD}}^m} = \frac{\mu_m S_0}{K_s + S_0} - k_d$$
$$= \frac{\left(0.5 \frac{1}{\text{d}} \right) \left(185 \frac{\text{mg}}{\text{L}} \right)}{0.8 \frac{\text{mg}}{\text{L}} + 185 \frac{\text{mg}}{\text{L}}} - 0.05 \frac{1}{\text{d}}$$
$$= 0.45 \ \text{d}^{-1}$$
$$\theta_{c,\text{BOD}}^m = \frac{1}{0.45 \frac{1}{\text{d}}} = 2.2 \ \text{d}$$

Nitrification controls the design with $\theta_c^m = 2.7$ d. Use Eq. 10.43 to calculate the design mean cell residence time.

$$\theta_c^d = (\text{SF})\theta_c^m = (3.5)(2.7 \ \text{d})$$
$$= 9.45 \ \text{d} \quad [\text{say } 10 \ \text{d}]$$

Assume a typical nitrifier fraction of 0.04. This means that about 4% of the MLVSS is capable of converting ammonia to nitrite and nitrate. Confirm that nitrification controls design by calculating the hydraulic residence times for nitrification and BOD removal. For nitrification, use Eq. 10.46 and solve for θ.

$$X_N = 0.04 X_A = (0.04)\left(1900 \ \frac{mg}{L}\right)$$

$$= 77 \ mg/L$$

$$X_N = \frac{\theta_c Y(N_0 - N_e)}{\theta(1 + k_d \theta_c)}$$

$$\theta = \frac{\theta_c Y(N_0 - N_e)}{X_N(1 + k_d \theta_c)}$$

$$= \frac{(10 \ d)(0.2)\left(40 \ \frac{mg}{L} - 1 \ \frac{mg}{L}\right)}{\left(77 \ \frac{mg}{L}\right)\left(1 + \left(0.05 \ \frac{1}{d}\right)(10 \ d)\right)}$$

$$= 0.675 \ d$$

For BOD removal, use Eq. 10.44 and solve for θ.

Activated Sludge

$$X_A = \frac{\theta_c Y(S_0 - S_e)}{\theta(1 + k_d \theta_c)}$$

$$\theta = \frac{\theta_c Y(S_0 - S_e)}{X_A(1 + k_d \theta_c)}$$

$$= \frac{(10 \ d)(0.2)\left(185 \ \frac{mg}{L} - 20 \ \frac{mg}{L}\right)}{\left(1900 \ \frac{mg}{L}\right)\left(1 + \left(0.05 \ \frac{1}{d}\right)(10 \ d)\right)}$$

$$= 0.116 \ d$$

This is confirmation that nitrification controls design.

43. DENITRIFICATION

After nitrification, denitrification may occur in one of two ways.

- in a separate reactor using an external carbon source
- in a reactor combined with wastewater source BOD removal

Where the influent BOD has been essentially removed during nitrification, it is necessary to provide a substrate for the growth of the denitrifying organisms. Methanol is typically used as the external carbon source for this purpose.

Where an external carbon source is added, careful design and operation are needed to prevent adding BOD to the effluent. The design is similar to that of an ordinary activated sludge system, with an additional step required to remove nitrogen gas entrained in the biomass and in the treated wastewater.

Typical values of parameters used with the equations for denitrification are

$$\mu_m = 0.3 \ d^{-1}$$
$$K_N = 0.1 \ mg/L$$
$$Y = 0.8$$
$$k_d = 0.04 \ d^{-1}$$

Equations in the following form represent a correction to the kinetic coefficients as described for nitrification.

$$\mu'_D = \mu_{mD}(0.0025 \ T^2)\left(\frac{C_m}{K_m + C_m}\right)\left(\frac{N_0}{K_N + N_0}\right)$$

The steps for designing suspended growth denitrification processes are

1. Determine the required methanol concentration.

Methanol Requirement for Biologically Treated Wastewater

$$C_m = 2.47 N_o + 1.53 N_1 + 0.87 D_o \qquad 10.48$$

2. Calculate θ_c^d for denitrification.

$$\frac{1}{\theta_c^m} = \mu'_D - k_d$$

$$\theta_c^d = SF \times \theta_c^m$$

3. Calculate the *substrate utilization rate*, U.

$$\frac{1}{\theta_c^d} = YU - k_d$$

4. Calculate the effluent nitrogen concentration, N_e.

$$U = \frac{kN_e}{K_N + N_e}$$

5. Determine the hydraulic retention time.

$$\theta = \frac{N_o - N_e}{UX_A}$$

Example 10.4

Determine the volume of the bioreactor for an activated sludge denitrification process described by the following.

flow rate	$15\,000$ m³/d
water temperature	$18°C$
initial nitrate concentration	30 mg/L
initial nitrite concentration	6 mg/L
initial DO concentration	1.8 mg/L
MLVSS	3000 mg/L
half-velocity constant for methanol	20 mg/L
SF	2

The methanol concentration is

Methanol Requirement for Biologically Treated Wastewater

$$C_m = 2.47 N_o + 1.53 N_1 + 0.87 D_o$$

$$= (2.47)\left(30\ \frac{\text{mg}}{\text{L}}\right) + (1.53)\left(6\ \frac{\text{mg}}{\text{L}}\right) + (0.87)\left(1.8\ \frac{\text{mg}}{\text{L}}\right)$$

$$= 85\ \text{mg/L}$$

Using typical values, calculate the corrected maximum growth rate for denitrification

$$\mu_{mD} = 0.3\ \text{d}^{-1}$$
$$k_d = 0.04\ \text{d}^{-1}$$
$$Y = 0.8$$
$$K_N = 0.1\ \text{mg/L}$$

$$N = 30\ \frac{\text{mg}}{\text{L}} + 6\ \frac{\text{mg}}{\text{L}} = 36\ \text{mg/L}$$

$$\mu'_D = \mu_{mD}(0.0025\,T^2)\left(\frac{C_m}{K_m + C_m}\right)\left(\frac{N}{K_N + N}\right)$$

$$= \left(0.3\ \frac{1}{\text{d}}\right)(0.0025)(18°C)^2$$

$$\times \left(\frac{85\ \frac{\text{mg}}{\text{L}}}{20\ \frac{\text{mg}}{\text{L}} + 85\ \frac{\text{mg}}{\text{L}}}\right)\left(\frac{36\ \frac{\text{mg}}{\text{L}}}{0.1\ \frac{\text{mg}}{\text{L}} + 36\ \frac{\text{mg}}{\text{L}}}\right)$$

$$= 0.196\ \text{d}^{-1}$$

The minimum mean cell residence time for denitrification is

$$\frac{1}{\theta_c^m} = \mu'_D - k_d = 0.196\ \frac{1}{\text{d}} - 0.04\ \frac{1}{\text{d}}$$

$$= 0.156\ \text{d}^{-1}$$

$$\theta_c^m = \frac{1}{0.156\ \frac{1}{\text{d}}} = 6.4\ \text{d}$$

The design mean cell residence time is

$$\theta_c^d = (\text{SF})\theta_c^m = (2)(6.4\ \text{d}) = 12.8\ \text{d}$$

Calculate the substrate utilization rate.

$$\frac{1}{\theta_c^d} = YU - k_d$$

$$U = \frac{\dfrac{1}{\theta_c^d} + k_d}{Y} = \frac{\dfrac{1}{12.8\ \text{d}} + 0.04\ \dfrac{1}{\text{d}}}{0.8}$$

$$= 0.15\ \text{d}^{-1}$$

Calculate the effluent nitrogen concentration.

$$k = \frac{\mu'_D}{Y} = \frac{0.196\ \dfrac{1}{\text{d}}}{0.8} = 0.245\ \text{d}^{-1}$$

$$U = \frac{kN_e}{K_N + N_e}$$

$$N_e = \frac{K_N U}{k - U} = \frac{\left(0.1\ \dfrac{\text{mg}}{\text{L}}\right)\left(0.15\ \dfrac{1}{\text{d}}\right)}{0.245\ \dfrac{1}{\text{d}} - 0.15\ \dfrac{1}{\text{d}}}$$

$$= 0.16\ \text{mg/L}$$

The hydraulic retention time is

$$\theta = \frac{N_o - N_e}{UX_A} = \frac{36\ \dfrac{\text{mg}}{\text{L}} - 0.16\ \dfrac{\text{mg}}{\text{L}}}{\left(\dfrac{0.15\ \dfrac{1}{\text{d}}}{24\ \dfrac{\text{h}}{\text{d}}}\right)\left(3000\ \dfrac{\text{mg}}{\text{L}}\right)}$$

$$= 1.9\ \text{h}$$

The aeration tank volume is

$$V = Q\theta = \left(15{,}000\ \frac{\text{m}^3}{\text{d}}\right)(0.08\ \text{d})$$

$$= 1200\ \text{m}^3$$

44. COMBINED NITRIFICATION AND DENITRIFICATION

In a separate-stage denitrification system, it is expensive to add an external carbon source such as methanol for substrate. For this reason, processes have been developed to accomplish both nitrification and denitrification using influent BOD and biomass as the carbon source. This can be accomplished by providing alternating aerobic and anaerobic (anoxic) zones in a single bioreactor or by using a sequence of separate tanks. The objective is to provide some level of nitrification followed by some level of denitrification, incrementally removing ammonia and nitrate as the process sequences between the aerobic and anoxic zones.

45. ATMOSPHERIC AIR

Atmospheric air is a mixture of oxygen, nitrogen, and small amounts of carbon dioxide, water vapor, argon, and other inert gases. If all constituents except oxygen are grouped with the nitrogen, the air composition is as given in Table 10.16. It is necessary to supply by weight $1/0.2315 = 4.32$ masses of air to obtain one mass of oxygen. The average molecular weight of air is 28.9.

Table 10.16 *Composition of Air*[a]

component	fraction by mass	fraction by volume
oxygen	0.2315	0.209
nitrogen	0.7685	0.791
ratio of nitrogen to oxygen	3.320	3.773[b]
ratio of air to oxygen	4.320	4.773

[a]Inert gases and CO_2 are included in N_2.
[b]This value is also reported by various sources as 3.76, 3.78, and 3.784.

46. AERATION TANKS

The *aeration period* is the same as the *hydraulic residence time* and is calculated without considering recirculation.

$$\theta = \frac{V}{Q_0} \qquad 10.49$$

The organic (BOD) *volumetric loading* rate (with units of kg $BOD_5/d \cdot m^3$) for the aeration tank is given by Eq. 10.50. In the United States, volumetric loading is often specified in lbm/day-1000 ft³.

$$L_{BOD} = \frac{S_0 Q_0}{V\left(1000 \ \dfrac{mg \cdot m^3}{kg \cdot L}\right)} \qquad [SI] \quad 10.50(a)$$

$$L_{BOD} = \frac{S_{0,mg/L} Q_{0,MGD} \times \left(8.345 \ \dfrac{lbm \text{-} L}{MG \text{-} mg}\right)(1000)}{V_{ft^3}} \qquad [U.S.] \quad 10.50(b)$$

The *rate of oxygen transfer* from the air to the mixed liquor during aeration is given by Eq. 10.51. K_t is a macroscopic transfer coefficient that depends on the equipment and characteristics of the mixed liquor and that has typical dimensions of time^{-1}. The dissolved oxygen deficit, D in mg/L, is given by Eq. 10.52. β is the mixed liquor's *oxygen saturation coefficient*, approximately 0.8 to 0.9. β corrects for the fact that the mixed liquor does not absorb as much oxygen as pure water does. The minimum dissolved oxygen content is approximately 0.5 mg/L, below which the processing would be limited by oxygen. However, a lower limit of 2 mg/L is typically specified.

$$\dot{m}_{oxygen} = K_t D \qquad 10.51$$

$$D = \beta DO_{saturated\,water} - DO_{mixed\,liquor} \qquad 10.52$$

The *oxygen demand* is given by Eq. 10.53. The factor 1.42 is the theoretical gravimetric ratio of oxygen required for carbonaceous organic material based on an ideal stoichiometric reaction. Equation 10.53 neglects nitrogenous demand, which can also be significant.

$$\dot{m}_{oxygen,kg/d} = \frac{Q_{0,m^3/d}(S_0 - S_e)_{mg/L}}{f\left(1000 \ \dfrac{mg \cdot m^3}{kg \cdot L}\right)} - 1.42 P_{x,kg/d} \qquad [SI] \quad 10.53(a)$$

$$\dot{m}_{oxygen,lbm/day} = \frac{Q_{0,MGD}(S_0 - S_e)_{mg/L} \times \left(8.345 \ \dfrac{lbm \text{-} L}{MG \text{-} mg}\right)}{f} - 1.42 P_{x,lbm/day} \qquad [U.S.] \quad 10.53(b)$$

Air is approximately 23.2% oxygen by mass. Considering a *transfer efficiency* of $\eta_{transfer}$ (typically less than 10%), the air requirement is calculated from the oxygen demand.

$$\dot{m}_{\text{air}} = \frac{\dot{m}_{\text{oxygen}}}{0.232\eta_{\text{transfer}}} \qquad 10.54$$

The volume of air required is given by Eq. 10.55. The density of air is approximately 0.075 lbm/ft^3 (1.2 kg/m^3). Typical volumes for conventional aeration are 500–900 $\text{ft}^3/\text{lbm BOD}_5$ (30–$55 \text{ m}^3/\text{kg BOD}_5$) for F:M ratios greater than 0.3, and 1200–1800 $\text{ft}^3/\text{lbm BOD}_5$ (75–$115 \text{ m}^3/\text{kg BOD}_5$) for F:M ratios less than 0.3 day^{-1}. Air flows in SCFM (*standard cubic feet per minute*) are based on a temperature of 70°F (21°C) and pressure of 14.7 psia (101 kPa).

$$\dot{V}_{\text{air}} = \frac{\dot{m}_{\text{air}}}{\rho_{\text{air}}} \quad \text{[PFR and CSTR]} \qquad 10.55$$

Aeration equipment should be designed with an excess-capacity safety factor of 1.5–2.0. *Ten States' Standards* requires 200% of calculated capacity for air diffusion systems, which is assumed to be 1500 $\text{ft}^3/\text{lbm BOD}_5$ ($94 \text{ m}^3/\text{kg BOD}_5$) [TSS Sec. 92.332].

The air requirement per unit volume (m^3/m^3 or ft^3/MG, calculated using consistent units) is V_{air}/Q_0. The air requirement in cubic meters per kilogram of soluble BOD_5 removed is

$$V_{\text{air,m}^3/\text{kg BOD}} = \frac{\dot{V}_{\text{air,m}^3/\text{d}}\left(1000 \, \dfrac{\text{mg}\cdot\text{m}^3}{\text{kg}\cdot\text{L}}\right)}{Q_{0,\text{m}^3/\text{d}}(S_0 - S_e)_{\text{mg/L}}} \quad \text{[SI]} \qquad 10.56(a)$$

$$V_{\text{air,ft}^3/\text{lbm BOD}} = \frac{\dot{V}_{\text{air,ft}^3/\text{day}}}{Q_{0,\text{MGD}}(S_0 - S_e)_{\text{mg/L}}}$$
$$\times \left(8.345 \, \frac{\text{lbm-L}}{\text{MG-mg}}\right) \quad \text{[U.S.]} \qquad 10.56(b)$$

Example 10.5

Mixed liquor ($\beta = 0.9$, $DO = 3 \text{ mg/L}$) is aerated at 20°C. The transfer coefficient, K_t, is 2.7 h^{-1}. What is the rate of oxygen transfer?

Solution

At 20°C, the saturated oxygen content of pure water is 9.07 mg/L. The oxygen deficit is found from Eq. 10.52.

$$D = \beta DO_{\text{saturated water}} - DO_{\text{mixed liquor}}$$
$$= (0.9)\left(9.07 \, \frac{\text{mg}}{\text{L}}\right) - 3 \, \frac{\text{mg}}{\text{L}}$$
$$= 5.16 \text{ mg/L}$$

The rate of oxygen transfer is found from Eq. 10.51.

$$\dot{m}_{\text{oxygen}} = K_t D = (2.7 \text{ h}^{-1})\left(5.16 \, \frac{\text{mg}}{\text{L}}\right)$$
$$= 13.93 \text{ mg/L·h}$$

47. AERATION POWER AND COST

The work required to compress a unit mass of air from atmospheric pressure to the discharge pressure depends on the nature of the compression. Various assumptions can be made, but generally an *isentropic compression* is assumed. Inefficiencies (thermodynamic and mechanical) are combined into a single *compressor efficiency*, η_c, which is essentially an *isentropic efficiency* for the compression process.

The ideal power, P, to compress a volumetric flow rate, $\dot{V}$, or mass flow rate, $\dot{m}$, of air from pressure p_1 to pressure p_2 in an isentropic steady-flow compression process is given by Eq. 10.57. k is the ratio of specific heats, which has an approximate value of 1.4. c_p is the specific heat at constant pressure, which has a value of 0.24 Btu/lbm-°R (1005 J/kg·K). R_{air} is the specific gas constant, 53.3 ft-lbf/lbm-°R (287 J/kg·K). In Eq. 10.57, temperatures must be absolute. Volumetric and mass flow rates are per second.

$$P_{\text{ideal,kW}} = -\left(\frac{kP_1\dot{V}_1}{(k-1)\left(1000 \, \dfrac{\text{W}}{\text{kW}}\right)}\right)$$
$$\times \left[1 - \left(\frac{P_2}{P_1}\right)^{(k-1)/k}\right]$$
$$= -\left(\frac{k\dot{m}R_{\text{air}}T_1}{(k-1)\left(1000 \, \dfrac{\text{W}}{\text{kW}}\right)}\right) \quad \text{[SI]} \qquad 10.57(a)$$
$$\times \left[1 - \left(\frac{P_2}{P_1}\right)^{(k-1)/k}\right]$$
$$= -\left(\frac{c_p\dot{m}T_1}{1000 \, \dfrac{\text{W}}{\text{kW}}}\right)\left[1 - \left(\frac{P_2}{P_1}\right)^{(k-1)/1}\right]$$

$$P_{\text{ideal,hp}} = -\left[\frac{kP_1\dot{V}_1}{(k-1)\left(550\ \dfrac{\text{ft-lbf}}{\text{hp-sec}}\right)}\right]$$
$$\times\left[1-\left(\frac{P_2}{P_1}\right)^{(k-1)/k}\right]$$
$$= -\left[\frac{k\dot{m}R_{\text{air}}T_1}{(k-1)\left(550\ \dfrac{\text{ft-lbf}}{\text{hp-sec}}\right)}\right]\quad\text{[U.S.]}\quad \textit{10.57(b)}$$
$$\times\left[1-\left(\frac{P_2}{P_1}\right)^{(k-1)/k}\right]$$
$$= -\left(\frac{c_p J\dot{m}T_1}{550\ \dfrac{\text{ft-lbf}}{\text{hp-sec}}}\right)\left[1-\left(\frac{P_2}{P_1}\right)^{(k-1)/k}\right]$$

The actual compression power is

$$P_{\text{actual}} = \frac{P_{\text{ideal}}}{\eta_c} \qquad \textit{10.58}$$

The cost of running the compressor for a duration, t, is

$$\text{total cost} = C_{\text{kW-hr}}P_{\text{actual}}t \qquad \textit{10.59}$$

For calculating the power requirement, P_W, of a blower, an alternative to Eq. 10.57 is Eq. 10.60.

Blowers

$$P_W = \frac{WRT_1}{Cne}\left[\left(\frac{P_2}{P_1}\right)^{0.283}-1\right] \qquad \textit{10.60}$$

In Eq. 10.60, the conversion constant C is equal to 29.7 for SI units and 550 ft-lbf/sec-hp for U.S. units. The constant n is equal to $(k-1)/k$ and has a value of 0.283 for air. The blower efficiency, e, is usually between 0.7 and 0.9. The calculated power requirement, P_W, can be used to estimate the compression of a mass flow, W, from an initial pressure, P_1, to a final pressure, P_2.

48. RECYCLE RATIO AND RECIRCULATION RATE

For any given wastewater and treatment facility, the influent flow rate, BOD, and tank volume cannot be changed. Of all the variables appearing in Eq. 10.31, only the suspended solids can be controlled. Therefore, the primary process control variable is the amount of organic material in the wastewater. This is controlled by the sludge return rate. Equation 10.61

gives the sludge *recycle ratio*, R, which is typically 0.20–0.30. (Some authorities use the symbol α to represent the recycle ratio.)

$$R = \frac{Q_R}{Q_0} \qquad \textit{10.61}$$

The actual *recirculation rate* can be found by writing a suspended solids mass balance around the inlet to the reactor and solving for the recirculation rate, Q_r, noting $X_0 = 0$.

$$X_0Q_0 + X_rQ_R = X_A(Q_0+Q_R) \qquad \textit{10.62}$$

$$\frac{Q_R}{Q_0+Q_R} = \frac{X_A}{X_r} \qquad \textit{10.63}$$

The theoretical required return rate for the current MLSS content can be calculated from the *sludge volume index* (SVI; see Sec. 10.38). Figure 10.9 shows the theoretical relationship between the SVI test and the return rate. Equation 10.64 assumes that the settling tank responds identically to the graduated 1000 mL cylinder used in the SVI test. Equation 10.63 and Eq. 10.64 are analogous.

$$\frac{Q_R}{Q_0+Q_R} = \frac{V_{\text{settled,mL/L}}}{1000\ \dfrac{\text{mL}}{\text{L}}} \qquad \textit{10.64}$$

$$R = \frac{V_{\text{settled,mL/L}}}{1000\ \dfrac{\text{mL}}{\text{L}} - V_{\text{settled,mL/L}}}$$
$$= \frac{1}{\dfrac{10^6}{(\text{SVI}_{\text{mL/g}})(\text{MLSS}_{\text{mg/L}})}-1} \qquad \textit{10.65}$$

Equation 10.65 is based on settling and represents a theoretical value, but it does not necessarily correspond to the actual recirculation rate. Greater-than-necessary recirculation returns clarified liquid unnecessarily, and less recirculation leaves settled solids in the clarifier.

Figure 10.9 *Theoretical Relationship Between SVI Test and Return Rate*

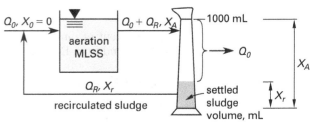

Example 10.6

Two 1 L samples of mixed liquor are taken from an aerating lagoon. After settling for 30 min in a graduated cylinder, 250 mL of solids have settled out in the first sample. The total suspended solids concentration in the second sample is found to be 2300 mg/L. (a) What is the sludge volume index? (b) What is the theoretical required sludge recycle rate?

Solution

(a) Use Eq. 10.34.

Activated Sludge

$$
\text{SVI} = \frac{\left(1000\ \dfrac{\text{mg}}{\text{g}}\right) V_{\text{settled, mL/L}}}{\text{MLSS}_{\text{mg/L}}}
$$

$$
= \frac{\left(1000\ \dfrac{\text{mg}}{\text{g}}\right)\left(250\ \dfrac{\text{mL}}{\text{L}}\right)}{2300\ \dfrac{\text{mg}}{\text{L}}}
$$

$$
= 109\ \text{mL/g}
$$

(b) The theoretical required recycle rate is given by Eq. 10.65.

$$
R = \frac{V_{\text{settled, mL/L}}}{1000\ \dfrac{\text{mL}}{\text{L}} - V_{\text{settled, mL/L}}}
$$

$$
= \frac{250\ \dfrac{\text{mL}}{\text{L}}}{1000\ \dfrac{\text{mL}}{\text{L}} - 250\ \dfrac{\text{mL}}{\text{L}}}
$$

$$
= 0.33
$$

Example 10.7

An activated sludge plant receives 4.0 MGD of wastewater with a BOD of 200 mg/L. The primary clarifier removes 30% of the BOD. The aeration-settle-waste cycle takes 6 hr. The oxygen transfer efficiency is 6%, and one pound of oxygen is required for each pound of BOD oxidized. The food-to-microorganism ratio based on MLSS is 0.33. The SVI is 100. 30% of the MLSS is wasted each day from the aeration tank. The MLVSS/MLSS ratio is 0.7. The surface settling rate of the secondary clarifier is 800 gal/day-ft². The overflow rate in the secondary clarifier is 800 gal/day-ft². The final effluent has a BOD of 10 mg/L.

Calculate the (a) secondary BOD removal fraction, (b) aeration tank volume, (c) MLSS, (d) total suspended solids content in the recirculated sludge, (e) theoretical recirculation rate, (f) mass of MLSS wasted each day, (g) clarifier surface area, and (h) volumetric air requirements.

Solution

(a) Wastewater leaves primary settling with a BOD of $(1 - 0.3)(200\ \text{mg/L}) = 140\ \text{mg/L}$. The efficiency of the activated sludge processing is

$$
\eta_{\text{BOD}} = \frac{140\ \dfrac{\text{mg}}{\text{L}} - 10\ \dfrac{\text{mg}}{\text{L}}}{140\ \dfrac{\text{mg}}{\text{L}}} = 0.929\ \ (92.9\%)
$$

(b) The aeration tank volume is

$$
V = Q_0 \theta = \frac{(4\ \text{MGD})\left(10^6\ \dfrac{\text{gal}}{\text{MG}}\right)(6\ \text{hr})}{\left(24\ \dfrac{\text{hr}}{\text{day}}\right)\left(7.48\ \dfrac{\text{gal}}{\text{ft}^3}\right)}
$$

$$
= 1.34 \times 10^5\ \text{ft}^3
$$

(c) The total MLSS (X_A) can be determined from the food-to-microorganism ratio given. Use Eq. 10.31.

Activated Sludge

$$
(\text{F:M}) = Q_0 S_0 / (Vol\ X_A)
$$

$$
= \frac{\left(140\ \dfrac{\text{mg}}{\text{L}}\right)(4\ \text{MGD})\left(10^6\ \dfrac{\text{gal}}{\text{MG}}\right)}{\left(0.33\ \dfrac{\text{lbm}}{\text{lbm-day}}\right)(1.34 \times 10^5\ \text{ft}^3)\left(7.48\ \dfrac{\text{gal}}{\text{ft}^3}\right)}
$$

$$
= 1693\ \text{mg/L}
$$

(d) The suspended solids content in the recirculated sludge is given by Eq. 10.35.

$$
\text{TSS}_{\text{mg/L}} = \frac{\left(1000\ \dfrac{\text{mg}}{\text{g}}\right)\left(1000\ \dfrac{\text{mL}}{\text{L}}\right)}{\text{SVI}_{\text{mL/g}}}
$$

$$
= \frac{\left(1000\ \dfrac{\text{mg}}{\text{g}}\right)\left(1000\ \dfrac{\text{mL}}{\text{L}}\right)}{100\ \dfrac{\text{mL}}{\text{g}}}
$$

$$
= 10{,}000\ \text{mg/L}
$$

(e) The recirculation ratio can be calculated from the SVI test. However, the settled volume must first be calculated.

Use Eq. 10.34.

$$V_{\text{settled, mL/L}} = \frac{\text{MLSS}_{\text{mg/L}}\text{SVI}_{\text{mL/g}}}{1000 \, \dfrac{\text{mg}}{\text{g}}}$$

$$= \frac{\left(1693 \, \dfrac{\text{mg}}{\text{L}}\right)\left(100 \, \dfrac{\text{mL}}{\text{g}}\right)}{1000 \, \dfrac{\text{mg}}{\text{g}}}$$

$$= 169.3 \, \text{mL/L}$$

Use Eq. 10.64.

$$R = \frac{V_{\text{settled, mL/L}}}{1000 \, \dfrac{\text{mL}}{\text{L}} - V_{\text{settled, mL/L}}}$$

$$= \frac{169.3 \, \dfrac{\text{mL}}{\text{L}}}{1000 \, \dfrac{\text{mL}}{\text{L}} - 169.3 \, \dfrac{\text{mL}}{\text{L}}}$$

$$= 0.204$$

The recirculation rate is given by Eq. 10.61.

$$Q_R = RQ_0 = (0.204)(4 \, \text{MGD})$$

$$= 0.816 \, \text{MGD}$$

(f) The sludge wasted is

$$(0.30)\left(4 \, \text{MGD}\right)\left(1693 \, \dfrac{\text{mg}}{\text{L}}\right)$$

$$\times \left(8.345 \, \dfrac{\text{lbm-L}}{\text{mg-MG}}\right)$$

$$= 16,954 \, \text{lbm/day}$$

(g) The recirculated flow is removed by the bottom drain, so it does not influence the surface area. The clarifier surface area is

$$A = \frac{Q_0}{v_0}$$

$$= \frac{(4 \, \text{MGD})\left(10^6 \, \dfrac{\text{gal}}{\text{MG}}\right)}{800 \, \dfrac{\text{gal}}{\text{day-ft}^2}}$$

$$= 5000 \, \text{ft}^2$$

(h) From part (a), the fraction of BOD reduction in the activated sludge process aeration (exclusive of the wasted sludge) is 0.929.

Since one pound of oxygen is required for each pound of BOD removed, the required oxygen mass is

$$m_{\text{oxygen}} = (0.929)(4 \, \text{MGD})\left(140 \, \dfrac{\text{mg}}{\text{L}}\right)$$

$$\times \left(8.345 \, \dfrac{\text{lbm-L}}{\text{mg-MG}}\right)$$

$$= 4341 \, \text{lbm/day}$$

The density of air is approximately $0.075 \, \text{lbm/ft}^3$, and air is 20.9% oxygen by volume. Considering the efficiency of the oxygenation process, the air required is

$$\dot{V}_{\text{air}} = \frac{4341 \, \dfrac{\text{lbm}}{\text{day}}}{(0.209)(0.06)\left(0.075 \, \dfrac{\text{lbm}}{\text{ft}^3}\right)}$$

$$= 4.62 \times 10^6 \, \text{ft}^3/\text{day}$$

49. SLUDGE WASTING

The activated sludge process is a biological process. Influent brings in BOD ("biomass" or organic food), and the return activated sludge (RAS) brings in microorganisms (bacteria) to digest the BOD. The food-to-microorganism ratio (F:M) is balanced by adjusting the aeration and how much sludge is recirculated and how much is wasted. As the microorganisms grow and are mixed with air, they clump together (flocculate) and are readily settled as sludge in the secondary clarifier. More sludge is produced in the secondary clarifier than is needed, and without wasting, the sludge would continue to build up.

Sludge wasting is the main control of the effluent quality and microorganism population size. The wasting rate affects the mixed liquor suspended solids (MLSS) concentration and the mean cell residence time. Excess sludge may be wasted either from the sludge return line or directly from the aeration tank as mixed liquor. Wasting from the aeration tank is preferred as the sludge concentration is fairly steady in that case. The wasted volume is normally 40% to 60% of the wastewater flow. When wasting is properly adjusted, the MLSS level will remain steady. With an excessive F:M ratio, sludge production will increase, and eventually, sludge will appear in the clarifier effluent.

50. OPERATIONAL DIFFICULTIES

Sludge bulking refers to a condition in which the sludge does not settle out. Since the solids do not settle, they leave the sedimentation tank and cause problems in subsequent processes. The sludge volume index can often (but not always) be used as a measure of settling characteristics. If the SVI is less than 100, the settling process is probably operating satisfactorily. If SVI is greater than 150, the sludge is bulking. Remedies include addition of lime, chlorination, additional aeration, and a reduction in MLSS.

Some sludge may float back to the surface after settling, a condition known as *rising sludge*. Rising sludge occurs when nitrogen gas is produced from denitrification of the nitrates and nitrites. Remedies include increasing the return sludge rate and decreasing the mean cell residence time. Increasing the speed of the sludge scraper mechanism to dislodge nitrogen bubbles may also help.

It is generally held that after 30 min of settling, the sludge should settle to between 20% and 70% of its original volume. If it occupies more than 70%, there are too many solids in the aeration basin, and more sludge should be wasted. If the sludge occupies less than 20%, less sludge should be wasted.

Sludge washout (*solids washout*) can occur during a period of peak flow or even with excess recirculation. Washout is the loss of solids from the sludge blanket in the settling tank. This often happens with insufficient wasting, such that solids build up (to more than 70% in a settling test), hinder settling in the clarifier, and cannot be contained.

51. NOMENCLATURE

A	area	ft^2	m^2
BOD	biochemical oxygen demand	mg/L	mg/L
c_p	specific heat	Btu/lbm-°F	J/kg·°C
C	concentration	mg/L	mg/L
C	conversion constant	–	–
C_m	methanol concentration	mg/L	mg/L
d_p	particle diameter	ft	m
D	diameter	ft	m
D	dissolved oxygen deficit	mg/L	mg/L
D_0, D_o	initial dissolved oxygen concentration	mg/L	mg/L
DO	dissolved oxygen concentration	mg/L	mg/L
e	blower efficiency	–	–
E_1	BOD removal efficiency	%	%
f	Darcy friction factor	–	–
F	effective number of passes	–	–
F	food arrival rate	mg/day	mg/d
F:M	food to microorganism ratio	lbm/lbm-day	kg/kg·d
g	acceleration of gravity, 32.2 (9.81)	ft/sec^2	m/s^2
G	fraction of solids that are biodegradable	–	–
k	Camp formula constant	–	–
k	maximum rate of substrate utilization per unit mass	L/day-mg	L/d·mg
k	ratio of specific heats	–	–
k	removal rate constant	day^{-1}	d^{-1}
k	treatability constant	–	–
k	Velz constant	1/ft	1/m
k_d	endogenous decay rate constant	day^{-1}	d^{-1}
k_d	microbial death ratio	day^{-1}	d^{-1}
k_p	kinetic constant	day^{-1}	d^{-1}
k_t, k_T	decay rate	day^{-1}	d^{-1}
K_N	half-velocity constant for nitrogen	mg/L	mg/L
K_t	macroscopic transfer coefficient	$time^{-1}$	$time^{-1}$
L	length	ft	m
L_{BOD}	BOD loading rate	lbm/1000 ft^3-day	kg BOD_5/ d·m^3
L_H	hydraulic loading	gal/day-ft^2	m^3/d·m^2
m	exponent	–	–
$\dot{m}$	mass flow rate	lbm/day	kg/d
M	mass of microorganisms	lbm	kg
MLSS	total mixed liquor suspended solids	mg/L	mg/L
n	$(k-1)/k$ constant	–	–
n	rotational speed	rev/min	rev/min
n	exponent	–	–
N	nitrate concentration	mg/L	mg/L
N_0	initial nitrate concentration	mg/L	mg/L
N_1	initial nitrite-nitrogen concentration	mg/L	mg/L
N_e	effluent nitrogen concentration	mg/L	mg/L
N_o	initial nitrate-nitrogen concentration	mg/L	mg/L
P	power	hp	kW
P_1	absolute inlet pressure	lbf/in^2	kPa
P_2	absolute outlet pressure	lbf/in^2	kPa
P_W	blower power requirement	hp	hp

Q	air flow quantity	ft^3/min	L/s
Q	flow rate	ft^3/min	L/s
Q	water flow quantity	gal/day	m^3/d
Q_r	recirculation rate	gal/day	m^3/d
R	filter discharge ratio	–	–
R	recycle ratio	–	–
R	recirculation ratio	–	–
R_{air}	specific gas constant	ft-lbf/lbm-°R	J/kg·K
S	soluble BOD$_5$ escaping treatment or influent total BOD$_5$ concentration	mg/L	mg/L
S	concentration of growth-limiting nutrient	mg/L	mg/L
S_0	BOD from settling tank	mg/L	mg/L
S_e	trickling filter effluent BOD	mg/L	mg/L
S_i	diluted influent BOD	mg/L	mg/L
S_a	specific surface area	ft^2/ft^3	m^2/m^3
S_o	effluent BOD concentration or incoming BOD$_5$	mg/L	mg/L
S^0	effluent BOD concentration	mg/L	mg/L
SF	safety factor	–	–
SG	specific gravity	–	–
SVI	sludge volume index	mL/g	mL/g
t	time	days	d
T, T	temperature	°F	°C
T_1	absolute inlet temperature	°R	°R
TSS	total suspended solids	mg/L	mg/L
U	substrate utilization rate	day^{-1}	d^{-1}
v	velocity	ft/sec	m/s
V	volume	ft^3	m^3
$\dot{V}$	volumetric flow rate	ft^3/day	m^3/d
w	empirical weighting factor	–	–
W	BOD loading rate	lbm/day	n.a.
W	compression of mass flow	lbf/in^3	Pa
WOR	weir overflow rate	gal/day-ft^2	m^3/d·m^2
X	microorganism concentration	mg/L	mg/L
X	mixed liquor volatile suspended solids	mg/L	mg/L
Y	yield coefficient	–	–
Z	depth (of filter or basin)	ft	m

Symbols

β	oxygen saturation coefficient	–	–
η	BOD removal fraction	–	–
η	efficiency, transfer efficiency	–	–
θ	residence time	day	d
θ	temperature constant, treatment process	–	–
θ_{BOD}	BOD sludge age	day	d
θ_c^d	design mean cell residence time	day	d
θ_c^m	minimum mean cell residence time	day	d
μ_m	maximum specific growth rate	day^{-1}	d^{-1}
μ'_m	corrected maximum growth rate	day^{-1}	d^{-1}
ρ	density	lbm/ft^3	kg/m^3

Subscripts

0	initial
5	5-day
a	area
c	cell
d	detention
e	effluent
H	hydraulic
i	in (influent)
m	methanol
o	original (entry)
p	particle
r, R	recirculation
T	at temperature T
u	ultimate
w	wastewater

11 Wastewater Collection Systems

1. RECIPROCATING POSITIVE DISPLACEMENT PUMPS

Reciprocating positive displacement (PD) pumps can be used with all fluids, and are useful with viscous fluids and slurries (up to about 8000 Saybolt seconds universal (SSU), when the fluid is sensitive to shear, and when a high discharge pressure is required.[1] By entrapping a volume of fluid in the cylinder, reciprocating pumps provide a fixed-displacement volume per cycle. They are self-priming and inherently leak-free. Within the pressure limits of the line and pressure relief valve and the current capacity of the motor circuit, reciprocating pumps can provide an infinite discharge pressure.[2]

There are three main types of reciprocating pumps: power, direct-acting, and diaphragm. A *power pump* is a *cylinder-operated pump*. It can be single-acting or double-acting. A *single-acting pump* discharges liquid (or takes suction) only on one side of the piston, and there is only one transfer operation per crankshaft revolution. A *double-acting pump* discharges from both sides, and there are two transfers per revolution of the crank.

Traditional reciprocating pumps with pistons and rods can be either single-acting or double-acting and are suitable up to approximately 2000 psi (14 MPa). *Plunger pumps* are only single-acting and are suitable up to approximately 10,000 psi (70 MPa).

Simplex pumps have one cylinder, *duplex pumps* have two cylinders, *triplex pumps* have three cylinders, and so forth. *Direct-acting pumps* (sometimes referred to as *steam pumps*) are always double-acting. They use steam, unburned fuel gas, or compressed air as a motive fluid.

PD pumps are limited by both their NPSHR characteristics, acceleration head, and (for rotary pumps) slip.[3] Because the flow is unsteady, a certain amount of energy, the *acceleration head*, H_{ac}, is required to accelerate the fluid flow each stroke or cycle. If the acceleration head is too large, the NPSHR requirements may not be attainable. Acceleration head can be reduced by increasing the pipe diameter, shortening the suction piping, decreasing the pump speed, or placing a *pulsation damper* (*stabilizer*) in the suction line.[4]

Generally, friction losses with pulsating flows are calculated based on the maximum velocity attained by the fluid. Since this is difficult to determine, the maximum

---

[1]For viscosities of SSU greater than 240, multiply SSU viscosity by 0.216 to get viscosity in centistokes.
[2]For this reason, a relief valve should be included in every installation of positive displacement pumps. Rotary pumps typically have integral relief valves, but external relief valves are often installed to provide easier adjusting, cleaning, and inspection.
[3]Manufacturers of PD pumps prefer the term *net positive inlet pressure* (NPIP) to NPSH. NPIPA corresponds to NPSHA; NPIPR corresponds to NPSHR. Pressure and head are related by $P = \gamma H$.
[4]Pulsation dampers are not needed with rotary-action PD pumps, as the discharge is essentially constant.

velocity can be approximated by multiplying the average velocity (calculated from the rated capacity) by the factors in Table 11.1.

Table 11.1 Typical v_{max}/v_{ave} Velocity Ratios[a,b]

pump type	single-acting	double-acting
simplex	3.2	2.0
duplex	1.6	1.3
triplex	1.1	1.1
quadriplex	1.1	1.1
quintuplex and up	1.05	1.05

[a]Without stabilization. With properly sized stabilizers, use 1.05–1.1 for all cases.
[b]Multiply the values by 1.3 for metering pumps where lost fluid motion is relied on for capacity control.

When the suction line is "short," the acceleration head can be calculated from the length of the suction line, the average velocity in the line, and the rotational speed.[5]

In Eq. 11.1, C and K are dimensionless factors. K represents the relative compressibility of the liquid. (Typical values are 1.4 for hot water; 1.5 for amine, glycol, and cold water; and 2.5 for hot oil.) Values of C are given in Table 11.2.

$$H_{ac} = \frac{C}{K}\left(\frac{L_{suction}v_{ave}N}{g}\right)$$ *11.1*

Table 11.2 Typical Acceleration Head C-Values[*]

pump type	single-acting	double-acting
simplex	0.4	0.2
duplex	0.2	0.115
triplex	0.066	0.066
quadriplex	0.040	0.040
quintuplex and up	0.028	0.028

[*]Typical values for common connecting rod lengths and crank radii.

2. ROTARY PUMPS

Rotary pumps are positive displacement (PD) pumps that move fluid by means of screws, progressing cavities, gears, lobes, or vanes turning within a fixed casing (the *stator*). Rotary pumps are useful for high viscosities (up to 4×10^6 SSU for screw pumps). The rotation creates a cavity of fixed volume near the pump input; atmospheric or external pressure forces the fluid into that cavity. Near the outlet, the cavity is collapsed, forcing the fluid out. Figure 11.1 illustrates the external circumferential piston rotary pump.

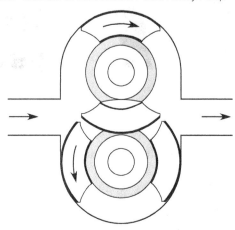

Figure 11.1 External Circumferential Piston Rotary Pump

Discharge from rotary pumps is relatively smooth. Acceleration head is negligible. Pulsation dampers and suction stabilizers are not required.

Slip in rotary pumps is the amount (sometimes expressed as a percentage) of each rotational fluid volume that "leaks" back to the suction line on each revolution. Slip reduces pump capacity. It is a function of clearance, differential pressure, and viscosity. Slip is proportional to the third power of the clearance between the rotating element and the casing. Slip decreases with increases in viscosity; it increases linearly with increases in differential pressure. Slip is not affected by rotational speed. The *volumetric efficiency* is defined in terms of volumetric flow rate, Q, by Eq. 11.2. Figure 11.2 illustrates the relationship between flow rate, speed, slip, and differential pressure.

$$\eta_v = \frac{Q_{actual}}{Q_{ideal}} = \frac{Q_{ideal} - Q_{slip}}{Q_{ideal}}$$ *11.2*

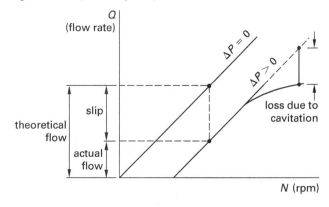

Figure 11.2 Slip in Rotary Pumps

Except for screw pumps, rotary pumps are generally not used for handling abrasive fluids or materials with suspended solids.

---

[5]With a properly designed pulsation damper, the effective length of the suction line is reduced to approximately 10 pipe diameters.

3. DIAPHRAGM PUMPS

Hydraulically operated *diaphragm pumps* have a diaphragm that completely separates the pumped fluid from the rest of the pump. A reciprocating plunger pressurizes and moves a hydraulic fluid that, in turn, flexes the diaphragm. Single-ball check valves in the suction and discharge lines determine the direction of flow during both phases of the diaphragm action.

Metering is a common application of diaphragm pumps. Diaphragm metering pumps have no packing and are essentially leakproof. This makes them ideal when fugitive emissions are undesirable. Diaphragm pumps are suitable for pumping a wide range of materials, from liquefied gases to coal slurries, though the upper viscosity limit is approximately 3500 SSU. Within the limits of their reactivities, hazardous and reactive materials can also be handled.

Diaphragm pumps are limited by capacity, suction pressure, and discharge pressure and temperature. Because of their construction and size, most diaphragm pumps are limited to discharge pressures of 5000 psi (35 MPa) or less, and most high-capacity pumps are limited to 2000 psi (14 MPa). Suction pressures are similarly limited to 5000 psi (35 MPa). A pressure range of 3–9 psi (20–60 kPa) is often quoted as the minimum liquid-side pressure for metering applications.

The discharge is inherently pulsating, and dampers or stabilizers are often used. (The acceleration head term is required when calculating NPSHR.) The discharge can be smoothed out somewhat by using two or three (i.e., duplex or triplex) plungers.

Diaphragms are commonly manufactured from stainless steel (type 316) and polytetrafluorethylene (PTFE) or other elastomers. PTFE diaphragms are suitable in the range of −50°F to 300°F (−45°C to 150°C) while metal diaphragms (and some ketone resin diaphragms) are used up to approximately 400°F (200°C) with life expectancy being reduced at higher temperatures. Although most diaphragm pumps usually operate below 200 spm (strokes per minute), diaphragm life will be improved by limiting the maximum speed to 100 spm.

4. CENTRIFUGAL PUMPS

Centrifugal pumps and their impellers can be classified according to the way energy is imparted to the fluid. Each category of pump is suitable for a different application and (specific) speed range. (See Table 11.6 later in this chapter.) Figure 11.3 illustrates a typical centrifugal pump and its schematic symbol.

Figure 11.3 Centrifugal Pump and Symbol

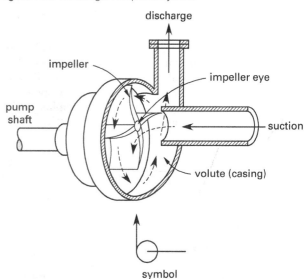

Radial-flow impellers impart energy primarily by centrifugal force. Fluid enters the impeller at the hub and flows radially to the outside of the casing. Radial-flow pumps are suitable for adding high pressure at low fluid flow rates. *Axial-flow impellers* impart energy to the fluid by acting as compressors. Fluid enters and exits along the axis of rotation. Axial-flow pumps are suitable for adding low pressures at high fluid flow rates.[6]

Radial-flow pumps can be designed for either single- or double-suction operation. In a *single-suction pump*, fluid enters from only one side of the impeller. In a *double-suction pump*, fluid enters from both sides of the impeller.[7] (That is, the impeller is two-sided.) Operation is similar to having two single-suction pumps in parallel. (See Fig. 11.4.)

Figure 11.4 Radial- and Axial-Flow Impellers

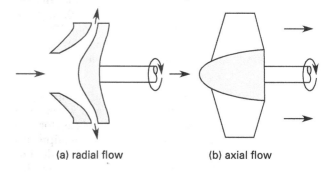

(a) radial flow   (b) axial flow

A *multiple-stage pump* consists of two or more impellers within a single casing. The discharge of one stage feeds the input of the next stage, and operation is similar to

---

[6]There is a third category of centrifugal pumps known as *mixed-flow pumps*. Mixed-flow pumps have operational characteristics between those of radial flow and axial flow pumps.

[7]The double-suction pump can handle a greater fluid flow rate than a single-suction pump with the same specific speed. Also, the double-suction pump will have a lower NPSHR.

having several pumps in series. In this manner, higher heads are achieved than would be possible with a single impeller.

The *pitch* (*circular blade pitch*) is the impeller's circumference divided by the number of impeller vanes. The impeller *tip speed* is calculated from the impeller diameter and rotational speed. The impeller "tip speed" is actually the tangential velocity at the periphery. Tip speed is typically somewhat less than 1000 ft/sec (300 m/s).

$$v_{\text{tip}} = \frac{\pi DN}{60\ \frac{\text{sec}}{\text{min}}} = \frac{D\omega}{2} \qquad \textit{11.3}$$

5. SEWAGE PUMPS

The primary consideration in choosing a pump for sewage and large solids is resistance to clogging. Centrifugal pumps should always be the single-suction type with nonclog, open impellers. (Double-suction pumps are prone to clogging because rags catch and wrap around the shaft extending through the impeller eye.) Clogging can be further minimized by limiting the number of impeller blades to two or three, providing for large passageways, and using a bar screen ahead of the pump.

Though made of heavy construction, nonclog pumps are constructed for ease of cleaning and repair. Horizontal pumps usually have a split casing, half of which can be removed for maintenance. A hand-sized cleanout opening may also be built into the casing. A sewage pump should normally be used with a grit chamber for prolonged bearing life.

The solids-handling capacity of a pump may be specified in terms of the largest sphere that can pass through it without clogging, usually about 80% of the inlet diameter. For example, a wastewater pump with a 6 in (150 mm) inlet should be able to pass a 4 in (100 mm) sphere. The pump must be capable of handling spheres with diameters slightly larger than the bar screen spacing.

Figure 11.5 shows a simplified wastewater pump installation. Not shown are instrumentation and water level measurement devices, baffles, lighting, drains for the dry well, electrical power, pump lubrication equipment, and access ports. (Totally submerged pumps do not require dry wells. However, such pumps without dry wells are more difficult to access, service, and repair.)

The multiplicity and redundancy of pumping equipment is not apparent from Fig. 11.5. The number of pumps used in a wastewater installation largely depends on the expected demand, pump capacity, and design criteria for backup operation. It is good practice to install pumps in sets of two, with a third backup pump being available for each set of pumps that performs the same

Figure 11.5 Typical Wastewater Pump Installation (greatly simplified)

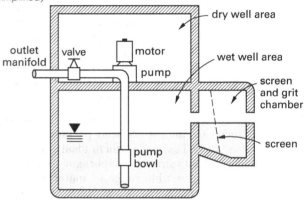

function. The number of pumps and their capacities should be able to handle the peak flow when one pump in the set is out of service.

6. SLUDGE PUMPS AND GRAVITY FLOW

Centrifugal and reciprocating pumps are extensively used for pumping sludge. Progressive cavity screw impeller pumps are also used.

As further described in Sec. 11.22, the pumping power is proportional to the specific gravity. Accordingly, pumping power for dilute and well-digested sludges is typically only 10–25% higher than for water. However, most sludges are non-Newtonian fluids, often flow in a laminar mode, and have characteristics that may change with the season. Also, sludge characteristics change greatly during the pumping cycle; engineering judgment and rules of thumb are often important in choosing sludge pumps. For example, a general rule is to choose sludge pumps capable of developing at least 50–100% excess head.

One method of determining the required pumping power is to multiply the power required for pumping pure water by a service factor. Historical data is the best method of selecting this factor. Choice of initial values is a matter of judgment. Guidelines are listed in Table 11.3.

Table 11.3 Typical Pumping Power Multiplicative Factors

solids concentration	digested sludge	untreated, primary, and concentrated sludge
0%	1.0	1.0
2%	1.2	1.4
4%	1.3	2.5
6%	1.7	4.1
8%	2.2	7.0
10%	3.0	10.0

Derived from *Wastewater Engineering: Treatment, Disposal, Reuse*, 3rd ed., by Metcalf & Eddy, et al., © 1991, with permission from The McGraw-Hill Companies, Inc.

Generally, sludge will thin out during a pumping cycle. The most dense sludge components will be pumped first, with more watery sludge appearing at the end of the pumping cycle. With a constant power input, the reduction in load at the end of pumping cycles may cause centrifugal pumps to operate far from the desired operating point and to experience overload failures. The operating point should be evaluated with high-, medium-, and low-density sludges.

To avoid cavitation, sludge pumps should always be under a positive suction head of at least 4 ft (1.2 m), and suction lifts should be avoided. The minimum diameters of suction and discharge lines for pumped sludge are typically 6 in (150 mm) and 4 in (100 mm), respectively.

Not all sludge is moved by pump action. Some installations rely on gravity flow to move sludge. The minimum diameter of sludge gravity transfer lines is typically 8 in (200 mm), and the recommended minimum slope is 3%.

To avoid clogging due to settling, sludge velocity should be above the transition from laminar to turbulent flow, known as the *critical velocity*. The critical velocity for most sludges is approximately 3.5 ft/sec (1.1 m/s). Velocity ranges of 5–8 ft/sec (1.5–2.4 m/s) are commonly quoted and are adequate.

7. STANDARD MOTOR SIZES AND SPEEDS

An effort should be made to specify standard motor sizes when selecting the source of pumping power. Table 11.4 lists NEMA (National Electrical Manufacturers Association) standard motor sizes by horsepower *nameplate rating.*[8] Other motor sizes may also be available by special order.

Table 11.4 *NEMA Standard Motor Sizes (brake horsepower)*

$\frac{1}{8}$,*	$\frac{1}{6}$,*	$\frac{1}{4}$,*	$\frac{1}{3}$,*	$\frac{1}{2}$,*	$\frac{3}{4}$,*		
1,	1.5,	2,	3,	5,	7.5	10,	15
20,	25,	30,	40,	50,	60,	75,	100
125,	150,	200,	250,	300,	350,	400,	450
500,	600,	700,	800,	900,	1000,	1250,	1500
1750,	2000,	2250,	2500,	2750,	3000,	3500,	4000
4500,	5000,	6000,	7000,	8000			

*fractional horsepower series

For industrial-grade motors, the rated (nameplate) power is the power at the output shaft that the motor can produce continuously while operating at the nameplate ambient conditions and without exceeding a particular temperature rise dependent on its wiring type. However, if a particular motor is housed in a larger, open frame that has better cooling, it will be capable of producing more power than the nameplate rating. NEMA defines the *service factor*, SF, as the amount of continual overload capacity designed into a motor without reducing its useful life (provided voltage, frequency, and ambient temperature remain normal). Common motor service factors are 1.0, 1.15, and 1.25. When a motor develops more than the nameplate power, efficiency, power factor, and operating temperature will be affected adversely.

$$\text{SF} = \frac{\dot{W}_{\text{max continuous}}}{\dot{W}_{\text{rated}}} \qquad 11.4$$

Larger horsepower motors are usually three-phase induction motors. The *synchronous speed*, n in rpm, of such motors is the speed of the rotating field, which depends on the number of poles per stator phase and the frequency, f. The number of poles must be an even number. The frequency is typically 60 Hz, as in the United States, or 50 Hz, as in European countries. Table 11.5 lists common synchronous speeds.

$$N = \frac{120f}{\text{no. of poles}} \qquad 11.5$$

Table 11.5 *Common Synchronous Speeds*

number of poles	N (rpm)	
	60 Hz	50 Hz
2	3600	3000
4	1800	1500
6	1200	1000
8	900	750
10	720	600
12	600	500
14	514	428
18	400	333
24	300	250
48	150	125

Induction motors do not run at their synchronous speeds when loaded. Rather, they run at slightly less than synchronous speed. The deviation is known as the *slip*. Slip is typically around 4% and is seldom greater than 10% at full load for motors in the 1–75 hp range.

$$\text{slip (in rpm)}$$
$$= \text{synchronous speed} - \text{actual speed} \qquad 11.6$$

---

[8]The nameplate rating gets its name from the information stamped on the motor's identification plate. Besides the horsepower rating, other nameplate data used to classify the motor are the service class, voltage, full-load current, speed, number of phases, frequency of the current, and maximum ambient temperature (or, for older motors, the motor temperature rise).

slip (in percent)

$$= \frac{\text{synchronous speed} - \text{actual speed}}{\text{synchronous speed}} \qquad \textit{11.7}$$
$$\times 100\%$$

Induction motors may also be specified in terms of their kVA (kilovolt-amp) ratings. The kVA rating is not the same as the power in kilowatts, although one can be derived from the other if the motor's *power factor* is known. Such power factors typically range from 0.8 to 0.9, depending on the installation and motor size.

$$\text{kVA rating} = \frac{\text{motor power in kW}}{\text{power factor}} \qquad \textit{11.8}$$

Example 11.1

A pump driven by an electrical motor moves 25 gal/min of water from reservoir A to reservoir B, lifting the water a total of 245 ft. The efficiencies of the pump and motor are 64% and 84%, respectively. Electricity costs $0.08/kW-hr. Neglect velocity head, friction, and minor losses. (a) What minimum size motor is required? (b) How much does it cost to operate the pump for 6 hr?

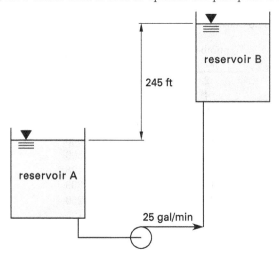

Solution

(a) The head added is 245 ft. Incorporating the pump efficiency, find the motor power required. [Unit and Conversion Factors].

$$\dot{W} = \frac{\rho g Q H_A}{\eta_{\text{pump}}} = \frac{(245 \text{ ft})\left(25 \ \frac{\text{gal}}{\text{min}}\right)\left(8.34 \ \frac{\text{lbm}}{\text{gal}}\right)(1.0)}{\left(33{,}000 \ \frac{\text{ft-lbf}}{\text{min-hp}}\right)(0.64)}$$
$$= 2.42 \text{ hp}$$

From Table 11.4, select a 3 hp motor.

(b) The developed power, not the motor's rated power, is used. From Eq. 11.43 and Eq. 11.44,

$$\text{cost} = (\text{cost per kW-hr})\left(\frac{\dot{W} t}{\eta_m}\right)$$
$$= \left(0.08 \ \frac{\$}{\text{kW-hr}}\right)\left(0.7457 \ \frac{\text{kW}}{\text{hp}}\right)\left(\frac{(2.42 \text{ hp})(6 \text{ hr})}{0.84}\right)$$
$$= \$1.03$$

8. PUMP SHAFT LOADING

The torque on a pump or motor shaft can be calculated from the brake power and rotational speed. [Fans, Pumps, and Compressors]

$$T_{\text{in-lbf}} = \frac{63{,}025 \, \dot{W}}{N} \qquad \textit{11.9}$$

$$T_{\text{ft-lbf}} = \frac{5252 \, \dot{W}}{N} \qquad \textit{11.10}$$

$$T_{\text{N·m}} = \frac{9549 \, \dot{W}}{N} \qquad \textit{11.11}$$

The actual (developed) torque can be calculated from the change in momentum of the fluid flow. For radial impellers, the fluid enters through the eye and is turned 90°. The direction change is related to the increase in momentum and the shaft torque. When fluid is introduced axially through the eye of the impeller, the tangential velocity at the inlet (eye), $v_{t(s)}$, is zero.[9]

$$T_{\text{actual}} = \dot{m}\left(v_{t(d)} r_{\text{impeller}} - v_{t(s)} r_{\text{eye}}\right) \qquad \text{[SI]} \quad \textit{11.12(a)}$$

$$T_{\text{actual}} = \frac{\dot{m}}{g_c}\left(v_{t(d)} r_{\text{impeller}} - v_{t(s)} r_{\text{eye}}\right) \qquad \text{[U.S.]} \quad \textit{11.12(b)}$$

Centrifugal pumps can be driven directly from a motor, or a speed changer can be used. Rotary pumps generally require a speed reduction. *Gear motors* have integral speed reducers. V-belt drives are widely used because of their initial low cost, although timing belts and chains can be used in some applications.

When a belt or chain is used, the pump's and motor's maximum overhung loads must be checked. This is particularly important for high-power, low-speed applications (such as rotary pumps). *Overhung load* is the side load (force) put on shafts and bearings. The overhung load is calculated from Eq. 11.13. The empirical factor K is 1.0 for chain drives, 1.25 for timing belts, and 1.5 for V-belts.

$$\text{overhung load} = \frac{2KT}{D_{\text{sheave}}} \qquad \textit{11.13}$$

---

[9]The tangential component of fluid velocity is sometimes referred to as the *velocity of whirl*.

If a direct drive cannot be used and the overhung load is excessive, the installation can be modified to incorporate a jack shaft or outboard bearing.

Example 11.2

A centrifugal pump delivers 275 lbm/sec (125 kg/s) of water while turning at 850 rpm. The impeller has straight radial vanes and an outside diameter of 10 in (25.4 cm). Water enters the impeller through the eye. The driving motor delivers 30 hp (22 kW). What are the (a) theoretical torque, and (b) pump efficiency?

SI Solution

(a) From Eq. 11.3, the impeller's tangential velocity is

$$v_t = \frac{\pi DN}{60 \frac{s}{min}} = \frac{\pi(25.4 \text{ cm})\left(850 \frac{rev}{min}\right)}{\left(60 \frac{s}{min}\right)\left(100 \frac{cm}{m}\right)}$$

$$= 11.3 \text{ m/s}$$

Since water enters axially, the incoming water has no tangential component. From Eq. 11.12, the developed torque is

$$T = \dot{m} v_{t(d)} r_{impeller} = \frac{\left(125 \frac{kg}{s}\right)\left(11.3 \frac{m}{s}\right)\left(\frac{25.4 \text{ cm}}{2}\right)}{100 \frac{cm}{m}}$$

$$= 179.4 \text{ N·m}$$

(b) From Eq. 11.11, the developed power is

$$\dot{W} = \frac{NT_{N·m}}{9549} = \frac{\left(850 \frac{rev}{min}\right)(179.4 \text{ N·m})}{9549 \frac{N·m}{kW·min}}$$

$$= 15.97 \text{ kW}$$

The pump efficiency is

$$\eta_p = \frac{\dot{W}_{developed}}{\dot{W}_{input}}$$

$$= \frac{15.97 \text{ kW}}{22 \text{ kW}}$$

$$= 0.726 \quad (72.6\%)$$

Customary U.S. Solution

(a) From Eq. 11.3, the impeller's tangential velocity is

$$v_t = \frac{\pi DN}{60 \frac{sec}{min}} = \frac{\pi(10 \text{ in})\left(850 \frac{rev}{min}\right)}{\left(60 \frac{sec}{min}\right)\left(12 \frac{in}{ft}\right)}$$

$$= 37.09 \text{ ft/sec}$$

Since water enters axially, the incoming water has no tangential component. From Eq. 11.12, the developed torque is

$$T = \frac{\dot{m} v_{t(d)} r_{impeller}}{g_c} = \frac{\left(275 \frac{lbm}{sec}\right)\left(37.08 \frac{ft}{sec}\right)\left(\frac{10 \text{ in}}{2}\right)}{\left(32.2 \frac{lbm\text{-}ft}{lbf\text{-}sec^2}\right)\left(12 \frac{in}{ft}\right)}$$

$$= 131.9 \text{ ft-lbf}$$

(b) From Eq. 11.10, the developed power is

$$\dot{W}_{hp} = \frac{T_{ft\text{-}lbf}N}{5252} = \frac{(131.9 \text{ ft-lbf})\left(850 \frac{rev}{min}\right)}{5252 \frac{ft\text{-}lbf}{hp\text{-}min}}$$

$$= 21.35 \text{ hp}$$

The pump efficiency is

$$\eta_p = \frac{\dot{W}_{developed}}{\dot{W}_{input}} = \frac{21.35 \text{ hp}}{30 \text{ hp}}$$

$$= 0.712 \quad (71.2\%)$$

9. SPECIFIC SPEED

The capacity and efficiency of a centrifugal pump are partially governed by the impeller design. For a desired flow rate and added head, there will be one optimum impeller design. The quantitative index used to optimize the impeller design is known as *specific speed*, N_s, also known as *impeller specific speed*. Table 11.6 lists the impeller designs that are appropriate for different specific speeds.[10]

Highest heads per stage are developed at low specific speeds. However, for best efficiency, specific speed should be greater than 650 (13 in SI units). If the specific speed for a given set of conditions drops below 650 (13), a multiple-stage pump should be selected.[11]

---

[10]Specific speed is useful for more than just selecting an impeller type. Maximum suction lift, pump efficiency, and net positive suction head required ·(NPSHR) can be correlated with specific speed.
[11]*Partial emission, forced vortex centrifugal pumps* allow operation down to specific speeds of 150 (3 in SI). Such pumps have been used for low-flow, high-head applications, such as high-pressure petrochemical cracking processes.

Table 11.6 *Specific Speed Versus Impeller Design*

impeller type	approximate range of specific speed (rpm)	
	customary U.S. units	SI units
radial vane	500 to 1000	10 to 20
Francis (mixed) vane	2000 to 3000	40 to 60
mixed flow	4000 to 7000	80 to 140
axial flow	9000 and above	180 and above

(Divide customary U.S. specific speed by 51.64 to obtain SI specific speed.)

Specific speed is a function of a pump's capacity, head, and rotational speed at peak efficiency, as shown in Eq. 11.14. For a given pump and impeller configuration, the specific speed remains essentially constant over a range of flow rates and heads. (Q or $\dot{V}$ in Eq. 11.14 is half of the full flow rate for double-suction pumps.)

$$N_s = \frac{N\sqrt{\dot{V}}}{H_A^{0.75}} \qquad \text{[SI]} \quad \textit{11.14(a)}$$

$$N_s = \frac{N\sqrt{Q}}{H_A^{0.75}} \qquad \text{[U.S.]} \quad \textit{11.14(b)}$$

A common definition of specific speed is the speed (in rpm) at which a *homologous pump* would have to turn in order to deliver 1 gal/min at 1 ft total added head.[12] This definition is implicit to Eq. 11.14 but is not very useful otherwise. While specific speed is not dimensionless, the units are meaningless. Specific speed may be assigned units of rpm, but most often it is expressed simply as a pure number.

The numerical range of acceptable performance for each impeller type is redefined when SI units are used. The SI specific speed is obtained by dividing the customary U.S. specific speed by 51.64.

Specific speed can be used to determine the type of impeller needed. Once a pump is selected, its specific speed and Eq. 11.14 can be used to determine other operational parameters (e.g., maximum rotational speed). Specific speed can be used with Fig. 11.6 to obtain an approximate pump efficiency.

Example 11.3

A centrifugal pump powered by a direct-drive induction motor is needed to discharge 150 gal/min against a 300 ft total head when turning at the fully loaded speed of 3500 rpm. What type of pump should be selected?

Figure 11.6 *Average Pump Efficiency Versus Specific Speed*

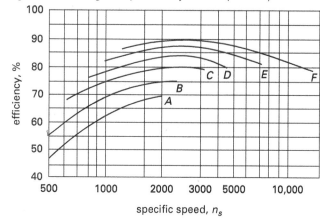

curve *A*: 100 gal/min
curve *B*: 200 gal/min
curve *C*: 500 gal/min
curve *D*: 1000 gal/min
curve *E*: 3000 gal/min
curve *F*: 10,000 gal/min

Solution

From Eq. 11.14, the specific speed is

$$N_s = \frac{N\sqrt{Q}}{H_A^{0.75}}$$

$$= \frac{\left(3500 \; \frac{\text{rev}}{\text{min}}\right)\sqrt{150 \; \frac{\text{gal}}{\text{min}}}}{(300 \text{ ft})^{0.75}}$$

$$= 595$$

From Table 11.6, the pump should be a radial vane type. However, pumps achieve their highest efficiencies when specific speed exceeds 650. (See Fig. 11.6.) To increase the specific speed, the rotational speed can be increased, or the total added head can be decreased. Since the pump is direct-driven and 3600 rpm is the maximum speed for induction motors, the total added head should be divided evenly between two stages, or two pumps should be used in series. (See Table 11.5.)

In a two-stage system, the specific speed would be

$$N_s = \frac{\left(3500 \; \frac{\text{rev}}{\text{min}}\right)\sqrt{150 \; \frac{\text{gal}}{\text{min}}}}{\left(\frac{300 \text{ ft}}{2}\right)^{0.75}}$$

$$= 1000$$

This is satisfactory for a radial vane pump.

---

[12]*Homologous pumps* are geometrically similar. This means that each pump is a scaled up or down version of the others. Such pumps are said to belong to a *homologous family*.

Example 11.4

An induction motor turning at 1200 rpm is to be selected to drive a single-stage, single-suction centrifugal water pump through a direct drive. The total dynamic head added by the pump is 26 ft. The flow rate is 900 gal/min. What is the approximate pump efficiency?

Solution

The specific speed is

$$N_s = \frac{N\sqrt{Q}}{H_A^{0.75}} = \frac{\left(1200 \ \frac{\text{rev}}{\text{min}}\right)\sqrt{900 \ \frac{\text{gal}}{\text{min}}}}{(26 \ \text{ft})^{0.75}}$$
$$= 3127$$

From Fig. 11.6, the pump efficiency will be approximately 82%.

From Table 11.4, select a 7.5 hp or larger motor.

Example 11.5

A single-stage pump driven by a 3600 rpm motor is currently delivering 150 gal/min. The total dynamic head is 430 ft. What would be the approximate increase in efficiency per stage if the single-stage pump is replaced by a double-stage pump?

Solution

The specific speed is

$$N_s = \frac{N\sqrt{Q}}{H_A^{0.75}} = \frac{\left(3600 \ \frac{\text{rev}}{\text{min}}\right)\sqrt{150 \ \frac{\text{gal}}{\text{min}}}}{(430 \ \text{ft})^{0.75}}$$
$$= 467$$

From Fig. 11.6, the approximate efficiency is 45%.

In a two-stage pump, each stage adds half of the head. The specific speed per stage would be

$$N_s = \frac{N\sqrt{Q}}{H_A^{0.75}} = \frac{\left(3600 \ \frac{\text{rev}}{\text{min}}\right)\sqrt{150 \ \frac{\text{gal}}{\text{min}}}}{\left(\dfrac{430 \ \text{ft}}{2}\right)^{0.75}}$$
$$= 785$$

From Fig. 11.6, the efficiency for this configuration is approximately 60%.

The increase in stage efficiency is 60% − 45% = 15%. Whether or not the cost of multistaging is worthwhile in this low-volume application would have to be determined.

10. CAVITATION

Cavitation is a spontaneous vaporization of the fluid inside the pump, resulting in a degradation of pump performance. Wherever the fluid pressure is less than the vapor pressure, small pockets of vapor will form. These pockets usually form only within the pump itself, although cavitation slightly upstream within the suction line is also possible. As the vapor pockets reach the surface of the impeller, the local high fluid pressure collapses them. Noise, vibration, impeller pitting, and structural damage to the pump casing are manifestations of cavitation.

Cavitation can be caused by any of the following conditions.

- discharge head far below the pump head at peak efficiency

- high suction lift or low suction head

- excessive pump speed

- high liquid temperature (i.e., high vapor pressure)

11. NET POSITIVE SUCTION HEAD

The occurrence of cavitation is predictable. Cavitation will occur when the net pressure in the fluid drops below the vapor pressure. This criterion is commonly stated in terms of head: Cavitation occurs when the available head is less than the required head for satisfactory operation. (See Eq. 11.20.)

The minimum fluid energy required at the pump inlet for satisfactory operation (i.e., the required head) is known as the *net positive suction head required*, NPSHR.[13] NPSHR is a function of the pump and will be given by the pump manufacturer as part of the pump performance data.[14] NPSHR is dependent on the flow rate. However, if NPSHR is known for one flow rate, it can be determined for another flow rate from Eq. 11.20.

$$\frac{\text{NPSHR}_2}{\text{NPSHR}_1} = \left(\frac{Q_2}{Q_1}\right)^2 \qquad \text{11.15}$$

---

[13]If NPSHR (a head term) is multiplied by the fluid specific weight, it is known as the *net inlet pressure required*, NIPR. Similarly, NPSHA can be converted to NIPA.

[14]It is also possible to calculate NPSHR from other information, such as suction specific speed. However, this still depends on information provided by the manufacturer.

Net positive suction head available, NPSHA, is the actual total fluid energy at the pump inlet. There are two different methods for calculating NPSHA, both of which are correct and will yield identical answers. Equation 11.16 is based on the conditions at the fluid surface at the top of an open fluid source (e.g., tank or reservoir). There is a potential energy term but no kinetic energy term. Equation 11.17 is based on the conditions at the immediate entrance (suction, sub-script s) to the pump. At that point, some of the potential head has been converted to velocity head. Frictional losses are implicitly part of the reduced pressure head, as is the atmospheric pressure head. Since the pressure head, $H_{P(s)}$, is absolute, it includes the atmospheric pressure head, and the effect of higher altitudes is explicit in Eq. 11.16 and implicit in Eq. 11.17. If the source was pressurized instead of being open to the atmosphere, the pressure head would replace H_{atm} in Eq. 11.16 but would be implicit in $H_{P(s)}$ in Eq. 11.17.

$$\text{NPSHA} = H_{\text{atm}} + H_{z(s)} - H_{f(s)} - H_{\text{vp}} \qquad \textit{11.16}$$

$$\text{NPSHA} = H_{P(s)} + H_{v(s)} - H_{\text{vp}} \qquad \textit{11.17}$$

The net positive suction head available (NPSHA) for most positive displacement pumps includes a term for acceleration head.[15]

$$\text{NPSHA} = H_{\text{atm}} + H_{z(s)} - H_{f(s)} - H_{\text{vp}} - H_{\text{ac}} \quad \textit{11.18}$$

$$\text{NPSHA} = H_{P(s)} + H_{v(s)} - H_{\text{vp}} - H_{\text{ac}} \qquad \text{[SI]} \quad \textit{11.19(a)}$$

Centrifugal Pump Characteristics

$$NPSH_A = \frac{P_{\text{atm}}}{\rho g} \pm H_s - H_f - \frac{V^2}{2g} - \frac{P_{\text{vapor}}}{\rho g} \quad \text{[U.S.]} \quad \textit{11.19(b)}$$

If NPSHA is less than NPSHR, the fluid will cavitate. The criterion for cavitation is given by Eq. 11.20. (In practice, it is desirable to have a safety margin.)

$$\text{NPSHA} < \text{NPSHR} \quad \begin{bmatrix} \text{criterion for} \\ \text{cavitation} \end{bmatrix} \qquad \textit{11.20}$$

Example 11.6

2.0 ft^3/sec (56 L/s) of 60°F (16°C) water is pumped from an elevated feed tank to an open reservoir through 6 in (15.2 cm), schedule-40 steel pipe, as shown. The friction loss for the piping and fittings in the suction line is 2.6 ft (0.9 m). The friction loss for the piping and fittings in the discharge line is 13 ft (4.3 m). The atmospheric pressure is 14.7 psia (101 kPa). What is the NPSHA?

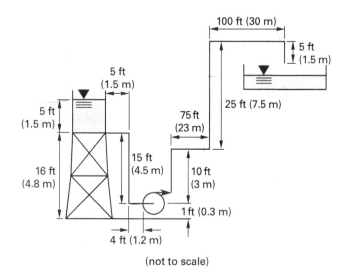

(not to scale)

SI Solution

The density of water is approximately 1000 kg/m^3. The atmospheric head is

$$H_{\text{atm}} = \frac{P}{\rho g}$$

$$= \frac{(101 \text{ kPa})\left(1000 \ \dfrac{\text{Pa}}{\text{kPa}}\right)}{\left(1000 \ \dfrac{\text{kg}}{\text{m}^3}\right)\left(9.81 \ \dfrac{\text{m}}{\text{s}^2}\right)}$$

$$= 10.3 \text{ m}$$

For 16°C water, the vapor pressure is approximately 0.01818 bars. The vapor pressure head is

$$H_{\text{vp}} = \frac{P}{\rho g}$$

$$= \frac{(0.01818 \text{ bar})\left(1 \times 10^5 \ \dfrac{\text{Pa}}{\text{bar}}\right)}{\left(1000 \ \dfrac{\text{kg}}{\text{m}^3}\right)\left(9.81 \ \dfrac{\text{m}}{\text{s}^2}\right)}$$

$$= 0.2 \text{ m}$$

From Eq. 11.16, the NPSHA is

$$\begin{aligned} \text{NPSHA} &= H_{\text{atm}} + H_{z(s)} - H_{f(s)} - H_{\text{vp}} \\ &= 10.3 \text{ m} + 6 \text{ m} + 4.8 \text{ m} - 0.3 \text{ m} \\ &\quad - 0.9 \text{ m} - 0.2 \text{ m} \\ &= 19.7 \text{ m} \end{aligned}$$

---

[15]The friction loss and the acceleration terms are both maximum values, but they do not occur in phase. Combining them is conservative.

Customary U.S. Solution

The specific weight of water is approximately 62.4 lbf/ft^3. The atmospheric head is

$$H_{\text{atm}} = \frac{P}{\gamma} = \frac{\left(14.7 \ \frac{\text{lbf}}{\text{in}^2}\right)\left(12 \ \frac{\text{in}}{\text{ft}}\right)^2}{62.4 \ \frac{\text{lbf}}{\text{ft}^3}}$$

$$= 33.9 \text{ ft}$$

For 60°F water, the vapor pressure head is 0.59 ft. Use 0.6 ft.

From Eq. 11.16, the NPSHA is

$$\text{NPSHA} = H_{\text{atm}} + H_{z(s)} - H_{f(s)} - H_{\text{vp}}$$
$$= 33.9 \text{ ft} + 20 \text{ ft} + 16 \text{ ft} - 1 \text{ ft} - 2.6 \text{ ft} - 0.6 \text{ ft}$$
$$= 65.7 \text{ ft}$$

12. PREVENTING CAVITATION

Cavitation is prevented by increasing NPSHA or decreasing NPSHR. NPSHA can be increased by

- increasing the height of the fluid source

- lowering the pump

- reducing friction and minor losses by shortening the suction line or using a larger pipe size

- reducing the temperature of the fluid at the pump entrance

- pressurizing the fluid supply tank

- reducing the flow rate or velocity (i.e., reducing the pump speed)

NPSHR can be reduced by

- placing a throttling valve or restriction in the discharge line[16]

- using an oversized pump

- using a double-suction pump

- using an impeller with a larger eye

- using an inducer

High NPSHR applications, such as boiler feed pumps needing 150–250 ft (50–80 m), should use one or more booster pumps in front of each high-NPSHR pump. Such booster pumps are typically single-stage, double-suction pumps running at low speed. Their NPSHR can be 25 ft (8 m) or less.

Throttling the input line to a pump and venting or evacuating the receiving tank both increase cavitation. Throttling the input line increases the friction head and decreases NPSHA. Evacuating the receiving tank increases the flow rate, increasing NPSHR while simultaneously increasing the friction head and reducing NPSHA.

13. CAVITATION COEFFICIENT

The *cavitation coefficient* (or *cavitation number*), σ, is a dimensionless number that can be used in modeling and extrapolating experimental results. The actual cavitation coefficient is compared with the *critical cavitation number* obtained experimentally. If the actual cavitation number is less than the critical cavitation number, cavitation will occur. Absolute pressure must be used for the fluid pressure term, P.

$$\sigma < \sigma_{\text{cr}} \quad \text{[criterion for cavitation]} \qquad \textit{11.21}$$

$$\sigma = \frac{2(P - P_{\text{vp}})}{\rho v^2} = \frac{\text{NPSHA}}{H_A} \qquad \text{[SI]} \quad \textit{11.22(a)}$$

$$\sigma = \frac{2g_c(P - P_{\text{vp}})}{\rho v^2} = \frac{\text{NPSHA}}{H_A} \qquad \text{[U.S.]} \quad \textit{11.22(b)}$$

The two forms of Eq. 11.22 yield slightly different results. The first form is essentially the ratio of the net pressure available for collapsing a vapor bubble to the velocity pressure creating the vapor. It is useful in model experiments. The second form is applicable to tests of production model pumps.

14. SUCTION SPECIFIC SPEED

The formula for *suction specific speed*, N_{ss}, can be derived by substituting NPSHR for total head in the expression for specific speed. Q and $\dot{V}$ are halved for double-suction pumps.

$$N_{\text{ss}} = \frac{N\sqrt{\dot{V}}}{(\text{NPSHR in m})^{0.75}} \qquad \text{[SI]} \quad \textit{11.23(a)}$$

$$N_{\text{ss}} = \frac{N\sqrt{Q}}{(\text{NPSHR in ft})^{0.75}} \qquad \text{[U.S.]} \quad \textit{11.23(b)}$$

Suction specific speed is an index of the suction characteristics of the impeller. Ideally, it should be approximately 8500 (165 in SI) for both single- and double-suction pumps. This assumes the pump is operating at or near its point of optimum efficiency.

---

[16]This will increase the total head, h_A, added by the pump, thereby reducing the pump's output and driving the pump's operating point into a region of lower NPSHR.

Suction specific speed can be used to determine the maximum recommended operating speed by substituting 8500 (165 in SI) for N_{ss} in Eq. 11.23 and solving for N.

If the suction specific speed is known, it can be used to determine the NPSHR. If the pump is known to be operating at or near its optimum efficiency, an approximate NPSHR value can be found by substituting 8500 (165 in SI) for N_{ss} in Eq. 11.23 and solving for NPSHR.

Suction specific speed available, SA, is obtained when NPSHA is substituted for total head in the expression for specific speed. The suction specific speed available must be less than the suction specific speed required to prevent cavitation.[17]

15. PUMP PERFORMANCE CURVES

For a given impeller diameter and constant speed, the head added will decrease as the flow rate increases. This is shown graphically on the *pump performance curve* (*pump curve*) supplied by the pump manufacturer. Other operating characteristics (e.g., power requirement, NPSHR, and efficiency) also vary with flow rate, and these are usually plotted on a common graph, as shown in Fig. 11.7.[18] Manufacturers' pump curves show performance over a limited number of calibration speeds. If an operating point is outside the range of published curves, the affinity laws can be used to estimate the speed at which the pump gives the required performance. [Centrifugal Pump Characteristics]

Figure 11.7 *Pump Performance Curves*

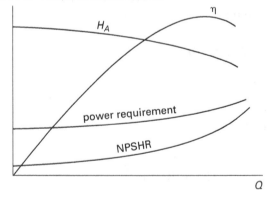

On the pump curve, the *shutoff point* (also known as *churn*) corresponds to a closed discharge valve (i.e., zero flow); the *rated point* is where the pump operates with rated 100% of capacity and head; the *overload point* corresponds to 65% of the rated head.

Figure 11.7 is for a pump with a fixed impeller diameter and rotational speed. The characteristics of a pump operated over a range of speeds or for different impeller diameters are illustrated in Fig. 11.8.

Figure 11.8 *Centrifugal Pump Characteristics Curves*

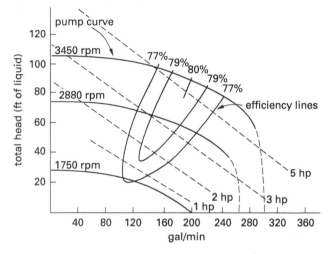

(a) variable speed

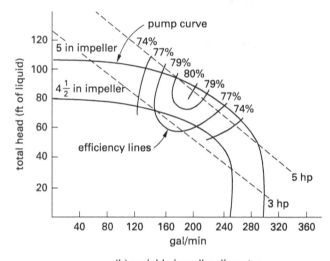

(b) variable impeller diameter

16. SYSTEM CURVES

A *system curve* (or *system performance curve*) is a plot of the static and friction energy losses experienced by the fluid for different flow rates. Unlike the pump curve, which depends only on the pump, the system curve depends only on the configuration of the suction and discharge lines. (The following equations assume equal pressures at the fluid source and destination surfaces, which is the case for pumping from one atmospheric reservoir to another. The velocity head is insignificant and is disregarded.)

$$H_A = H_z + H_f \qquad 11.24$$

$$H_z = H_{z(d)} - H_{z(s)} \qquad 11.25$$

$$H_f = H_{f(s)} + H_{f(d)} \qquad 11.26$$

---

[17]Since speed and flow rate are constants, this is another way of saying NPSHA must equal or exceed NPSHR.
[18]The term *pump curve* is commonly used to designate the H_A versus Q characteristics, whereas *pump characteristics curve* refers to all of the pump data.

The relationship of the different head variables to each other is clearly defined in the following equation.

Energy Equation (Bernoulli)

$$H = Z_a + \frac{P_a}{w} + \frac{V_a^2}{2g}$$

11.27

If the fluid reservoirs are large, or if the fluid reservoir levels are continually replenished, the net static suction head $(H_{z(d)} - H_{z(s)})$ will be constant for all flow rates. The friction loss, H_f, varies with V^2 (and, therefore, with Q^2) in the Darcy friction formula. This makes it easy to find friction losses for other flow rates (subscript 2) once one friction loss (subscript 1) is known.[19]

$$\frac{H_{f,1}}{H_{f,2}} = \left(\frac{Q_1}{Q_2}\right)^2$$

11.28

Figure 11.9 illustrates a system curve following Eq. 11.24 with a positive added head (i.e., a fluid source below the fluid destination). The system curve is shifted upward, intercepting the vertical axis at some positive value of H_A.

Figure 11.9 *System Curve*

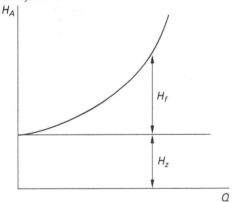

17. OPERATING POINT

The intersection of the pump curve and the system curve determines the *operating point*, as shown in Fig. 11.10. The operating point defines the system head and system flow rate.

When selecting a pump, the system curve is plotted on manufacturers' pump curves for different speeds and/or impeller diameters (i.e., Fig. 11.8). There will be several possible operating points corresponding to the various pump curves shown. Generally, the design operating point should be close to the highest pump efficiency. This, in turn, will determine speed and impeller diameter.

Figure 11.10 *Extreme Operating Points*

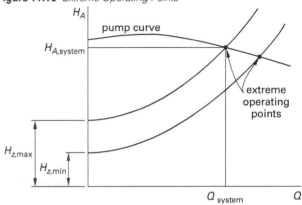

In some systems, the static head varies as the source reservoir is drained or as the destination reservoir fills. The system head is then defined by a pair of matching system friction curves intersecting the pump curve. The two intersection points are called the *extreme operating points*—the maximum and minimum capacity requirements.

After a pump is installed, it may be desired to change the operating point. This can be done without replacing the pump by placing a throttling valve in the discharge line. The operating point can then be moved along the pump curve by partially opening or closing the valve, as is illustrated in Fig. 11.11. (A throttling valve should never be placed in the suction line since that would reduce NPSHA.)

Figure 11.11 *Throttling the Discharge*

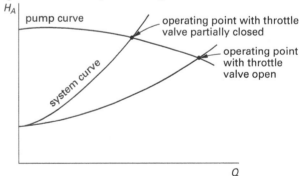

18. PUMPS IN PARALLEL

Parallel operation is obtained by having two pumps discharging into a common header. This type of connection is advantageous when the system demand varies greatly or when high reliability is required. A single pump providing total flow would have to operate far from its optimum efficiency at one point or another. With two

---

[19]Equation 11.28 implicitly assumes that the friction factor, f, is constant. This may be true over a limited range of flow rates, but it is not true over large ranges. Nevertheless, Eq. 11.28 is often used to quickly construct preliminary versions of the system curve.

pumps in parallel, one can be shut down during low demand. This allows the remaining pump to operate close to its optimum efficiency point.

Figure 11.12 illustrates that parallel operation increases the capacity of the system while maintaining the same total head.

Figure 11.12 *Pumps Operating in Parallel*

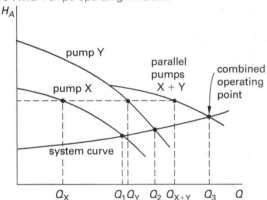

The performance curve for a set of pumps in parallel can be plotted by adding the capacities of the two pumps at various heads. A second pump will operate only when its discharge head is greater than the discharge head of the pump already running. Capacity does not increase at heads above the maximum head of the smaller pump.

When the parallel performance curve is plotted with the system head curve, the operating point is the intersection of the system curve with the X + Y curve. With pump X operating alone, the capacity is given by Q_1. When pump Y is added, the capacity increases to Q_3 with a slight increase in total head.

19. PUMPS IN SERIES

Series operation is achieved by having one pump discharge into the suction of the next. This arrangement is used primarily to increase the discharge head, although a small increase in capacity also results. (See Fig. 11.13.)

The performance curve for a set of pumps in series can be plotted by adding the heads of the two pumps at various capacities.

20. AFFINITY LAWS

Most parameters (impeller diameter, speed, and flow rate) determining a specific pump's performance can be modified. If the impeller diameter is held constant and

Figure 11.13 *Pumps Operating in Series*

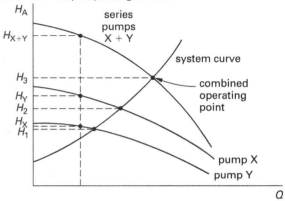

the speed is varied, and with an unchanged system curve, the following ratios are maintained with no change in efficiency.[20] [Fans, Pumps, and Compressors]

$$\frac{Q_2}{Q_1} = \frac{N_2}{N_1} \qquad 11.29$$

$$\frac{H_2}{H_1} = \left(\frac{N_2}{N_1}\right)^2 = \left(\frac{Q_2}{Q_1}\right)^2 \qquad 11.30$$

$$\frac{\dot{W}_2}{\dot{W}_1} = \left(\frac{N_2}{N_1}\right)^3 = \left(\frac{Q_2}{Q_1}\right)^3 \qquad 11.31$$

If the speed is held constant and the impeller size is reduced (i.e., the impeller is trimmed), while keeping the pump body, volute, shaft diameter, and suction and discharge openings the same, the following ratios may be used.[21]

$$\frac{Q_2}{Q_1} = \frac{D_2}{D_1} \qquad 11.32$$

$$\frac{H_2}{H_1} = \left(\frac{D_2}{D_1}\right)^2 \qquad 11.33$$

$$\frac{P_2}{P_1} = \left(\frac{D_2}{D_1}\right)^3 \qquad 11.34$$

The affinity laws are based on the assumption that the efficiency stays the same. In reality, larger pumps are somewhat more efficient than smaller pumps, and extrapolations to greatly different sizes should be avoided. Equation 11.35 can be used to estimate the

---

[20]See Sec. 11.21 if the entire pump is scaled to a different size.
[21]One might ask, "How is it possible to change a pump's impeller diameter?" In practice, a different impeller may be available from the manufacturer, but more often the impeller is taken out and shaved down on a lathe. Equation 11.32, Eq. 11.33, and Eq. 11.34 are limited in use to radial flow machines, and with reduced accuracy, to mixed-flow impellers. Changing the impeller diameter significantly impacts other design relationships, and the accuracy of performance prediction decreases if the diameter is changed much more than 20%.

efficiency of a differently sized pump. The dimensionless exponent, n, varies from 0 to approximately 0.26, with 0.2 being a typical value.

$$\frac{1 - \eta_{\text{smaller}}}{1 - \eta_{\text{larger}}} = \left(\frac{D_{\text{larger}}}{D_{\text{smaller}}}\right)^n \qquad 11.35$$

Example 11.7

A pump operating at 1770 rpm delivers 500 gal/min against a total head of 200 ft. It is desired to have the pump deliver a total head of 375 ft. At what speed should this pump be operated to achieve this new head at the same efficiency?

Solution

From Eq. 11.30,

$$N_2 = N_1\sqrt{\frac{H_2}{H_1}} = \left(1770 \ \frac{\text{rev}}{\text{min}}\right)\sqrt{\frac{375 \ \text{ft}}{200 \ \text{ft}}}$$

$$= 2424 \ \text{rpm}$$

Example 11.8

A pump is required to pump 500 gal/min against a total dynamic head of 425 ft. The hydraulic system has no static head change. Only the 1750 rpm performance curve is known for the pump. At what speed must the pump be turned to achieve the desired performance with no change in efficiency or impeller size?

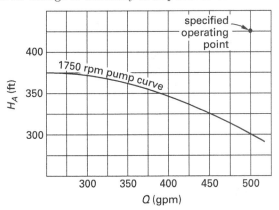

Solution

A flow of 500 gal/min with a head of 425 ft does not correspond to any point on the 1750 rpm curve.

From Eq. 11.30, the quantity H/Q^2 is constant.

$$\frac{H}{Q^2} = \frac{425 \ \text{ft}}{\left(500 \ \dfrac{\text{gal}}{\text{min}}\right)^2} = 1.7 \times 10^{-3} \ \text{ft-min}^2/\text{gal}^2$$

In order to use the affinity laws, the operating point on the 1750 rpm curve must be determined. Random values of Q are chosen and the corresponding values of H are determined such that the ratio H/Q^2 is unchanged.

Q	H
475	383
450	344
425	307
400	272

These points are plotted as the system curve. The intersection of the system and 1750 rpm pump curve at 440 gal/min defines the operating point at 1750 rpm. From Eq. 11.29, the required pump speed is

$$N_2 = \frac{N_1 Q_2}{Q_1} = \frac{\left(1750 \ \dfrac{\text{rev}}{\text{min}}\right)\left(500 \ \dfrac{\text{gal}}{\text{min}}\right)}{440 \ \dfrac{\text{gal}}{\text{min}}}$$

$$= 1989 \ \text{rpm}$$

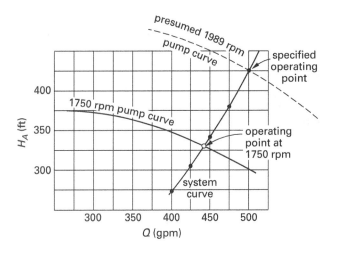

21. PUMP SIMILARITY

The performance of one pump can be used to predict the performance of a *dynamically similar* (*homologous*) pump. This can be done by using Eq. 11.36 through Eq. 11.41. [Fans, Pumps, and Compressors]

$$\left(\frac{H}{N^2 D^2}\right)_1 = \left(\frac{H}{N^2 D^2}\right)_2 \qquad 11.36$$

$$\left(\frac{Q}{N D^3}\right)_1 = \left(\frac{Q}{N D^3}\right)_2 \qquad 11.37$$

$$\left(\frac{P}{\rho N^2 D^2}\right)_1 = \left(\frac{P}{\rho N^2 D^2}\right)_2 \qquad 11.38$$

$$\left(\frac{Q}{ND^3}\right)_1 = \left(\frac{Q}{ND^3}\right)_2 \qquad \textit{11.39}$$

$$\left(\frac{\dot{W}}{\rho N^3 D^5}\right)_1 = \left(\frac{\dot{W}}{\rho N^3 D^5}\right)_2 \qquad \textit{11.40}$$

$$\frac{N_1\sqrt{Q_1}}{H_1^{0.75}} = \frac{N_2\sqrt{Q_2}}{H_2^{0.75}} \qquad \textit{11.41}$$

These *similarity laws* (also known as *scaling laws*) assume that both pumps

- operate in the turbulent region
- have the same pump efficiency
- operate at the same percentage of wide-open flow

Similar pumps also will have the same specific speed and cavitation number.

As with the affinity laws, these relationships assume that the efficiencies of the larger and smaller pumps are the same. In reality, larger pumps will be more efficient than smaller pumps. Therefore, extrapolations to much larger or much smaller sizes should be avoided.

Example 11.9

A 6 in pump operating at 1770 rpm discharges 1500 gal/min of cold water (SG = 1.0) against an 80 ft head at 85% efficiency. A homologous 8 in pump operating at 1170 rpm is being considered as a replacement. (a) What total head and capacity can be expected from the new pump? (b) For a hydraulic horsepower of 36.92 hp, what would be the new motor horsepower requirement?

Solution

(a) From Eq. 11.36,

$$H_2 = \left(\frac{D_2 N_2}{D_1 N_1}\right)^2 H_1 = \left(\frac{(8 \text{ in})\left(1170 \frac{\text{rev}}{\text{min}}\right)}{(6 \text{ in})\left(1770 \frac{\text{rev}}{\text{min}}\right)}\right)^2 (80 \text{ ft})$$

$$= 62.14 \text{ ft}$$

From Eq. 11.39,

$$Q_2 = \left(\frac{N_2 D_2^3}{N_1 D_1^3}\right)Q_1 = \left(\frac{\left(1170 \frac{\text{rev}}{\text{min}}\right)(8 \text{ in})^3}{\left(1770 \frac{\text{rev}}{\text{min}}\right)(6 \text{ in})^3}\right)\left(1500 \frac{\text{gal}}{\text{min}}\right)$$

$$= 2350 \text{ gal/min}$$

From Eq. 11.35,

$$\eta_{\text{larger}} = 1 - \frac{1 - \eta_{\text{smaller}}}{\left(\dfrac{D_{\text{larger}}}{D_{\text{smaller}}}\right)^n} = 1 - \frac{1 - 0.85}{\left(\dfrac{8 \text{ in}}{6 \text{ in}}\right)^{0.2}}$$

$$= 0.858$$

$$\dot{W}_2 = \frac{\dot{W}_{\text{fluid},2}}{\eta_p} = \frac{36.92 \text{ hp}}{0.858}$$

$$= 43.0 \text{ hp}$$

22. PUMPING LIQUIDS OTHER THAN COLD WATER

Many liquid pump parameters are determined from tests with cold, clear water at 85°F (29°C). The following guidelines can be used when pumping water at other temperatures or when pumping other liquids.

- Head developed is independent of the liquid's specific gravity. Pump performance curves from tests with water can be used with other Newtonian fluids (e.g., gasoline, alcohol, and aqueous solutions) having similar viscosities.

- The hydraulic horsepower depends on the specific gravity of the liquid. If the pump characteristic curve is used to find the operating point, multiply the horsepower reading by the specific gravity.

- Efficiency is not affected by changes in temperature that cause only the specific gravity to change.

- Efficiency is nominally affected by changes in temperature that cause the viscosity to change. Equation 11.42 is an approximate relationship suggested by the Hydraulics Institute when extrapolating the efficiency (in decimal form) from cold water to hot water. n is an experimental exponent established by the pump manufacturer, generally in the range of 0.05–0.1.

$$\eta_{\text{hot}} = 1 - (1 - \eta_{\text{cold}})\left(\frac{\nu_{\text{hot}}}{\nu_{\text{cold}}}\right)^n \qquad \textit{11.42}$$

- NPSHA depends significantly on liquid temperature.

- NPSHR is not significantly affected by variations in the liquid temperature.

- When hydrocarbons are pumped, the NPSHR determined from cold water can usually be reduced. This reduction is apparently due to the slow vapor release of complex organic liquids. If the hydrocarbon's vapor pressure at the pumping temperature is known, Fig. 11.14 will give the percentage of the cold-water NPSHR.

Water

- Pumping many fluids requires expertise that goes far beyond simply extrapolating parameters in proportion to the fluid's specific gravity. Such special cases include pumping liquids containing abrasives, liquids that solidify, highly corrosive liquids, liquids with vapor or gas, highly viscous fluids, paper stock, and hazardous fluids.

- Head, flow rate, and efficiency are all reduced when pumping highly viscous non-Newtonian fluids. No exact method exists for determining the reduction factors, other than actual tests of an installation using both fluids. Some sources have published charts of correction factors based on tests over limited viscosity and size ranges.[22]

Figure 11.14 *Hydrocarbon NPSHR Correction Factor*

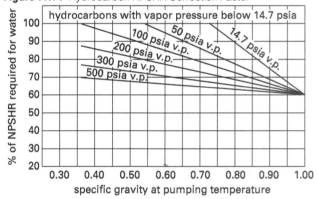

Example 11.10

A centrifugal pump has an NIPR (NPSHR) of 12 psi based on cold water. 10°F liquid isobutane has a specific gravity of 0.60 and a vapor pressure of 15 psia. What NPSHR should be used with 10°F liquid isobutane?

Solution

From Fig. 11.14, the intersection of a specific gravity of 0.60 and 15 psia is above the horizontal 100% line. The full NIPR of 12 psi should be used.

23. COST OF ELECTRICITY

The power utilization of pump motors is usually measured in kilowatts. The kilowatt usage represents the rate that energy is transferred by the pump motor. The total amount of work, W, done by the pump motor is found by multiplying the rate of energy usage (i.e., the delivered power, P), by the length of time the pump is in operation.

$$W = Pt \qquad 11.43$$

Although the units horsepower-hours are occasionally encountered, it is more common to measure electrical work in *kilowatt-hours* (kW-hr). Accordingly, the cost of electrical energy is stated per kW-hr (e.g., $0.10 per kW-hr).

$$\text{cost} = \frac{W_{\text{kW-hr}}(\text{cost per kW-hr})}{\eta_m} \qquad 11.44$$

24. SEWER PIPE MATERIALS

In the past, sewer pipes were constructed from clay, concrete, asbestos-cement, steel, and cast iron. Due to cost, modern large-diameter sewer lines are now almost always concrete, and new small-diameter lines are generally plastic.

Concrete pipe is used for gravity sewers and pressure mains. Circular pipe is used in most applications, although special shapes (e.g., arch, egg, elliptical) can be used. Concrete pipe in diameters up to 24 in (610 mm) is available in standard 3 ft and 4 ft (0.9 m and 1.2 m) lengths and is usually not reinforced. *Reinforced concrete pipe* (RCP) in diameters ranging from 12 in to 144 in (305–3660 mm) is available in lengths from 4 ft to 12 ft (1.2–3.6 m).

Concrete pipe is used for large diameter (16 in (406 mm) or larger) trunk and interceptor sewers. In some geographical regions, concrete pipe is used for smaller domestic sewers. However, concrete domestic lines should be selected only where stale or septic sewage is not anticipated. They should not be used with corrosive wastes or in corrosive soils.

Cast-iron pipe is particularly suited to installations where scour, high velocity waste, and high external loads are anticipated. It can be used for domestic connections, although it is more expensive than plastic. Special linings, coatings, wrappings, or encasements are required for corrosive wastes and soils.

Polyvinyl chloride (PVC) and acrylonitrile-butadiene-styrene (ABS) are two *plastic pipe* compositions that can be used for normal domestic sewage and industrial wastewater lines. They have excellent resistance to corrosive soils. However, special attention and care must be given to trench loading and pipe bedding.

ABS plastic can also be combined with concrete reinforcement for collector lines for use with corrosive domestic sewage and industrial waste. Such pipe is known as *truss pipe* due to its construction. The plastic is extruded with inner and outer web-connected pipe walls. The annular voids and the inner and outer walls are filled with lightweight concrete.

For pressure lines (i.e., *force mains*), welded steel pipe with an epoxy liner and cement-lined and coated-steel pipe are also used.

---

[22]A chart published by the Hydraulics Institute is widely distributed.

Vitrified clay pipe is resistant to acids, alkalies, hydrogen sulfide (septic sewage), erosion, and scour. Two strengths of clay pipe are available. The standard strength is suitable for pipes less than 12 in (300 mm) in diameter, D, for any depth of cover if the "$4D/3 + 8$ in trench width" ($4D/3 + 200$ mm) rule is observed. Double-strength pipe is recommended for large pipe that is deeply trenched. Clay is seldom used for diameters greater than 36 in (910 mm).

Asbestos-cement pipe has been used in the past for both gravity and pressure sewers carrying non-septic and non-corrosive waste through non-corrosive soils. The lightweight and longer laying lengths are inherent advantages of asbestos-cement pipe. However, such pipes have fallen from favor due to the asbestos content.

25. GRAVITY AND FORCE COLLECTION SYSTEMS

Wastewater collection systems are made up of a network of discharge and flow lines, drains, inlets, valve works, and connections for transporting domestic and industrial wastewater flows to treatment facilities.

Flow through gradually sloping *gravity sewers* is the most desirable means of moving sewage since it does not require pumping energy. In some instances, though, it may be necessary to use pressurized *force mains* to carry sewage uphill or over long, flat distances.

Alternative sewers are used in some remote housing developments for domestic sewage where neither the conventional sanitary sewer nor the septic tank/leach field is acceptable. Alternative sewers use pumps to force raw or communited sewage through small-diameter plastic lines to a more distant communal treatment or collection system.

There are four categories of alternative sewer systems. In the *G-P system*, a grinder-pump pushes chopped-up but untreated wastewater through a small-diameter plastic pipe. The *STEP system* (septic-tank-effluent pumping) uses a septic tank at each house, but there is no leach field. The clear overflow goes to a pump that pushes the effluent through a small-diameter plastic sewer line operating under low pressure. The *vacuum sewer system* uses a vacuum valve at each house that periodically charges a slug of wastewater into a vacuum sewer line. The vacuum is created by a central pumping station. The *small-diameter gravity sewer* (SDG) also uses a small-diameter plastic pipe but relies on gravity instead of pumps. The pipe carries the effluent from a septic tank at each house. Costs in all four systems can be reduced greatly by having two or more houses share septic tanks, pumps, and vacuum valves.

26. SEWER VELOCITIES

The minimum design velocity actually depends on the particulate matter size. However, 2 ft/sec (0.6 m/s) is commonly quoted as the minimum *self-cleansing velocity*, although 1.5 ft/sec (0.45 m/s) may be acceptable if the line is occasionally flushed out by peak flows [TSS Sec. 33.42]. Table 11.7 gives minimum flow velocities based on fluid type.

Table 11.8 lists the approximate minimum slope needed to achieve a 2 ft/sec (0.6 m/s) flow. Slopes slightly less than those listed may be permitted (with justification) in lines where design average flow provides a depth of flow greater than 30% of the pipe diameter. Velocity greater than 10–15 ft/sec (3–4.5 m/s) requires special provisions to protect the pipe and manholes against erosion and displacement by shock hydraulic loadings.

Table 11.7 Minimum Flow Velocities

fluid	minimum velocity to keep particles in suspension (ft/sec)	(m/s)	minimum resuspension velocity (ft/sec)	(m/s)
raw sewage	2.5	(0.75)	3.5	(1.1)
grit tank effluent	2	(0.6)	2.5	(0.75)
primary settling tank effluent	1.5	(0.45)	2	(0.6)
mixed liquor	1.5	(0.45)	2	(0.6)
trickling filter effluent	1.5	(0.45)	2	(0.6)
secondary settling tank effluent	0.5	(0.15)		(0.3)

(Multiply ft/sec by 0.3 to obtain m/s.)

27. SEWER SIZING

The Manning equation is traditionally used to size gravity sewers. Depth of flow at the design flow rate is usually less than 70–80% of the pipe diameter. (See Table 11.7.) A pipe should be sized to be able to carry the peak flow (including normal infiltration) at a depth of approximately 70% of its diameter.

In general, sewers in the collection system (including laterals, interceptors, trunks, and mains) should have diameters of at least 8 in (203 mm). Building service connections can be as small as 4 in (101 mm).

28. STREET INLETS

Street inlets are required at all low points where ponding could occur, and they should be placed no more than 600 ft (180 m) apart; a limit of 300 ft (90 m) is preferred. A common practice is to install three inlets in a sag vertical curve—one at the lowest point and one on each side with an elevation of 0.2 ft (60 mm) above the center inlet.

Table 11.8 Minimum Slopes and Capacities for Sewers[a]

sewer diameter		minimum change in elevation[b] (ft/100 ft or m/100 m)	full flow discharge at minimum slope[c,d] (ft³/sec)
(in)	(mm)		
8	(200)	0.40	0.771
9	(230)	0.33	0.996
10	(250)	0.28	1.17
12	(300)	0.22	1.61
14	(360)	0.17	2.23
15	(380)	0.15	2.52
16	(410)	0.14	2.90
18	(460)	0.12	3.67
21	(530)	0.10	5.05
24	(610)	0.08	6.45
27	(690)	0.067	8.08
30	(760)	0.058	9.96
36	(910)	0.046	14.4

(Multiply in by 25.4 to obtain mm.)
(Multiply ft/100 ft by 1 to obtain m/100 m.)
(Multiply ft³/sec by 448.8 to obtain gal/min.)
(Multiply ft³/sec by 28.32 to obtain L/s.)
(Multiply gal/min by 0.0631 to obtain L/s.)

[a]to achieve a velocity of 2 ft/sec (0.6 m/s) when flowing full
[b]as specified in Sec. 24.31 of *Recommended Standards for Sewage Works* (RSSW), also known as "*Ten States' Standards*" (TSS), published by the Health Education Service, Inc. [TSS Sec. 33.41]
[c]$n = 0.013$ assumed
[d]For any diameter in inches and $n = 0.013$, calculate the full flow as $Q = 0.0472 D_{in}^{8/3} \sqrt{S}$

Depth of gutter flow is found using Manning's equation. Inlet capacities of street inlets have traditionally been calculated from semi-empirical formulas.

Grate-type street inlets (known as *gutter inlets*) less than 0.4 ft (120 mm) deep have approximate capacities given by Eq. 11.45. Equation 11.45 should also be used for combined curb-grate inlets. All dimensions are in feet. For gutter inlets, the bars should ideally be parallel to the flow and be at least 1.5 ft (450 mm) long. Gutter inlets are more efficient than curb inlets, but clogging is a problem. Depressing the grating level below the street level increases the capacity.

$$Q_{ft^3/sec} = 3.0(\text{grate perimeter length})$$
$$\times (\text{inlet flow depth})^{3/2} \qquad 11.45$$

The capacity of a *curb inlet* (where there is an opening in the vertical plane of the gutter rather than in the horizontal plane) is given by Eq. 11.46. All dimensions are in feet. Not all curb inlets have inlet depressions. A typical curb inlet depression is 0.4 ft (130 mm).

$$Q_{ft^3/sec} = 0.7(\text{curb opening length})$$
$$\times (\text{inlet flow depth} \qquad 11.46$$
$$+ \text{curb inlet depression})^{3/2}$$

29. MANHOLES

Manholes along sewer lines should be provided at sewer line intersections and at changes in elevation, direction, size, diameter, and slope. If a sewer line is too small for a person to enter, manholes should be placed every 400 ft (120 m) to allow for cleaning. Recommended maximum spacings are 400 ft (120 m) for pipes with diameters less than 18 in (460 mm), 500 ft (150 m) for 18–48 in (460–1220 mm) pipes, and 600–700 ft (180–210 m) for larger pipes.

30. SULFIDE ATTACK

Wastewater flowing through sewers often turns septic and releases *hydrogen sulfide* gas, H_2S. The common *Thiobacillus* sulfur bacterium converts the hydrogen sulfide to sulfuric acid. The acid attacks the crowns of concrete pipes that are not flowing full.

$$2H_2S + O_2 \rightarrow 2S + 2H_2O \qquad 11.47$$

$$2S + 3O_2 + 2H_2O \rightarrow 2H_2SO_4 \qquad 11.48$$

In warm climates, hydrogen sulfide can be generated in sanitary sewers placed on flat grades. Hydrogen sulfide also occurs in sewers supporting food processing industries. Sulfide generation is aggravated by the use of home garbage disposals and, to a lesser extent, by recycling. Interest in nutrition and fresh foods has also added to food wastes.

Sulfide attack in partially full sewers can be prevented by maintaining the flow rate above 5 ft/sec (1.5 m/s), raising the pH above 10.4, using biocides, precipitating the sulfides, or a combination thereof. Not all methods may be possible or economical, however. In sewers, the least expensive method generally involves intermittent (e.g., biweekly during the critical months) *shock treatments* of sodium hydroxide and ferrous chloride. The sodium hydroxide is a sterilization treatment that works by raising the pH, while the ferrous chloride precipitates the sulfides.

Sulfide attack is generally not an issue in force mains because sewage is continually moving and always makes contact with all of the pipe wall.

31. NOMENCLATURE

C	pump factor	–	–
D	diameter	ft	m
f	Darcy friction factor	–	–
f	frequency	Hz	Hz
g	gravitational acceleration, 32.2 (9.81)	ft/sec^2	m/s^2
g_c	gravitational constant, 32.2	lbm-ft/ lbf-sec^2	n.a.
h, H	height or head	ft	m
K	empirical factor or relative compressibility factor	–	–
L	length	ft	m
$\dot{m}$	mass flow rate	lbm/sec	kg/s
n	dimensionless exponent	–	–
n, N	rotational speed	rev/min	rev/ min
N_{ss}	suction specific speed	rev/min	rev/ min
NPSHA, i$NPSH_A$	net positive suction head available	ft	m
NPSHR	net positive suction head required	ft	m
p, P_a	pressure	lbf/ft^2	Pa
Q	flow quantity	gal/day	L/d
Q	volumetric flow rate	gal/min	L/s
r	radius	ft	m
SF	service factor	–	–
SG	specific gravity	–	–
t	time	sec	s
T	temperature	°F	°C
T	torque	ft-lbf	N·m
v	velocity	ft/sec	m/s
V, V_a	fluid velocity	ft/sec	m/s
V	volume	mL	mL
$\dot{V}$	volumetric flow rate	ft^3/sec	m^3/s
W	work	ft-lbf	kW·h
$\dot{W}$	power	ft-lbf/sec	W
w	specific weight	lbf/ft^3	n.a.
z, Z_a	elevation	ft	m

Symbols

η	efficiency	–	–
γ	specific weight	lbm/ft^3	n.a.
ν	kinematic viscosity	ft^2/sec	m^2/s
ρ	density	lbm/ft^3	kg/m^3
σ	cavitation number	–	–
ω	angular velocity	rad/sec	rad/s

Subscripts

ac	acceleration
atm	atmospheric
A	added (by pump)
cr	critical
f	final
f	friction
m	motor
p	pressure or pump
s	specific or suction
ss	suction specific
t	at time t
t	tangential or total
T	at temperature T
v	volumetric
vp	vapor pressure
z	potential

12 Wastewater Residuals Management

Content in blue refers to the *NCEES Handbook*.

1. SLUDGE

Sludge is the mixture of water, organic and inorganic solids, and treatment chemicals that accumulates in settling tanks. The term is also used to refer to the dried residue (screenings, grit, filter cake, and drying bed scraping) from separation and drying processes, although the term *biosolids* is becoming more common in this regard. (The term *residuals* is also used, though this term more commonly refers to sludge from water treatment plants.)

2. QUANTITIES OF SLUDGE

The procedure for determining the volume and weight of sludge from activated sludge processing is essentially the same as for sludge from sedimentation and any other processes. Table 12.1 gives characteristics of sludge. The primary variables other than weight and volume are unit weight and/or specific gravity.

The weight of wet sludge is

$$W_{\text{sludge}} = V\rho_{\text{sludge}} = VS\rho_{\text{water}} \qquad 12.1$$

In Eq. 12.1, S is the specific gravity of wet sludge.

Raw sludge is approximately 95–99% water, and the specific gravity of sludge is only slightly greater than 1.0. The actual specific gravity of sludge can be calculated from the solids content, s, and the specific gravity of the sludge solids, which is approximately

Table 12.1 Characteristics of Sludge

origin of sludge	fraction solids, s	dry weight (lbf/day-person)
primary settling tank	0.06–0.08	0.12
trickling filter	0.04–0.06	0.04
mixed primary settling and trickling filter	0.05	0.16
conventional activated sludge	0.005–0.01	0.07
mixed primary and conventional activated sludge	0.02–0.03	0.19
high-rate activated sludge	0.025–0.05	0.06
mixed primary settling and high-rate activated sludge	0.05	0.18
extended aeration activated sludge	0.02	0.02
filter backwashing water	0.01–0.1	n.a.
softening sludge	0.03–0.15	n.a.

(Multiply lbf/day-person by 0.45 to obtain kg/d·person.)

2.5. (If the sludge specific gravity is already known, Eq. 12.2 can be solved for the specific gravity of the solids.)

$$
\begin{aligned}
\frac{1}{S} &= \frac{1-s}{1} + \frac{s}{S_{\text{solids}}} \\
&= \frac{1 - s_{\text{fixed}} - s_{\text{volatile}}}{1} + \frac{s_{\text{fixed}}}{S_{\text{fixed solids}}} \\
&\quad + \frac{s_{\text{volatile}}}{S_{\text{volatile solids}}}
\end{aligned}
\qquad 12.2
$$

In Eq. 12.2, s is the percentage of the sludge that is solids.

If the weight of the dry solids and the concentration of solids are both known, the volume of waste sludge can be found with Eq. 12.3.

Specific Gravity for a Solids Slurry

$$V = \frac{W_s}{(s/100)\gamma S} = \frac{W_s}{[(100-p)/100]\gamma S} \qquad 12.3$$

In Eq. 12.3, p is the percentage of the sludge that is water. Because the specific gravity of sludge is close to 1.0, the volume of sludge can also be estimated from just the weight of the dry solids.

$$V = \frac{W_s}{s\rho_{sludge}} \approx \frac{W_s}{s\rho_{water}} \qquad 12.4$$

The dried mass of sludge solids from primary settling basins is easily determined from the decrease in solids. The decrease in suspended solids, ΔSS, due to primary settling is approximately 50% of the total incoming suspended solids.

$$m_{dried,kg/d} = \frac{(\Delta SS)_{mg/L}\, Q_{0,m^3/d}}{1000\ \dfrac{mg \cdot m^3}{kg \cdot L}} \qquad [SI] \quad 12.5(a)$$

$$m_{dried,lbm/day} = (\Delta SS)_{mg/L}\, Q_{0,MGD}$$
$$\times \left(8.345\ \frac{lbm\text{-}L}{MG\text{-}mg}\right) \qquad [U.S.] \quad 12.5(b)$$

The dried mass of solids for biological filters and secondary aeration (e.g., activated sludge) can be estimated on a macroscopic basis as a fraction of the change in BOD. In Eq. 12.6, K is the *cell yield*, also known as the *removal efficiency*, the fraction of the total influent BOD that ultimately appears as excess (i.e., settled) biological solids. The cell yield can be estimated from the food-to-microorganism ratio from Fig. 12.1. (The cell yield, K, and the *yield coefficient*, also known as the *sludge yield* or *biomass yield*, Y, are not the same, because K is based on total incoming BOD and Y is based on consumed BOD. The term "cell yield" is often used for both Y and K.)

$$m_{dried,kg/d} = \frac{KS_{0,mg/L}\, Q_{0,m^3/d}}{1000\ \dfrac{mg \cdot m^3}{kg \cdot L}}$$
$$= \frac{Y(S_0 - S)_{mg/L}\, Q_{0,m^3/d}}{1000\ \dfrac{mg \cdot m^3}{kg \cdot L}} \qquad [SI] \quad 12.6(a)$$

$$m_{dried,lbm/day} = KS_{0,mg/L}\, Q_{0,MGD}$$
$$\times \left(8.345\ \frac{lbm\text{-}L}{MG\text{-}mg}\right)$$
$$= Y(S_0 - S)_{mg/L}\, Q_{0,MGD} \qquad [U.S.] \quad 12.6(b)$$
$$\times \left(8.345\ \frac{lbm\text{-}L}{MG\text{-}mg}\right)$$

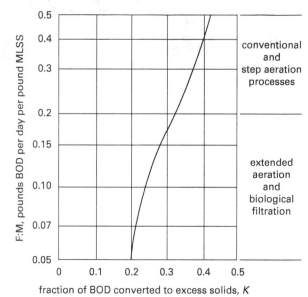

Figure 12.1 *Approximate Food: Microorganism Ratio Versus Cell Yield*, K*

*calculated assuming an effluent BOD of approximately 30 mg/L

Hammer, Mark J., *Water and Wastewater Technology*, 3rd, © 1996. Printed and electronically reproduced by permission of Pearson Education, Inc., New York, New York.

Example 12.1

A trickling filter plant processes 4 MGD of domestic wastewater with 190 mg/L BOD and 230 mg/L suspended solids. The primary sedimentation tank removes 50% of the suspended solids and 30% of the BOD. The sludge yield for the trickling filter is 0.25. What is the total sludge volume from the primary sedimentation tank and trickling filter if the combined solids content of the sludge is 5%?

Solution

The mass of solids removed in primary settling is given by Eq. 12.5.

$$m_{dried} = (\Delta SS)_{mg/L}\, Q_{0,MGD}\left(8.345\ \frac{lbm\text{-}L}{MG\text{-}mg}\right)$$
$$= (0.5)\left(230\ \frac{mg}{L}\right)(4\ MGD)\left(8.345\ \frac{lbm\text{-}L}{MG\text{-}mg}\right)$$
$$= 3839\ lbm/day$$

The mass of solids removed from the trickling filter's clarifier is calculated from Eq. 12.6.

$$m_{\text{dried}} = K\Delta S_{\text{mg/L}} Q_{0,\text{MGD}}\left(8.345 \ \frac{\text{lbm-L}}{\text{MG-mg}}\right)$$

$$= (0.25)(1 - 0.3)\left(190 \ \frac{\text{mg}}{\text{L}}\right)(4 \ \text{MGD})$$

$$\times\left(8.345 \ \frac{\text{lbm-L}}{\text{MG-mg}}\right)$$

$$= 1110 \ \text{lbm/day}$$

The volume is calculated from Eq. 12.4.

$$V \approx \frac{W_s}{s\rho_{\text{water}}}$$

$$= \frac{3839 \ \dfrac{\text{lbm}}{\text{day}} + 1110 \ \dfrac{\text{lbm}}{\text{day}}}{(0.05)\left(62.4 \ \dfrac{\text{lbm}}{\text{ft}^3}\right)}$$

$$= 1586 \ \text{ft}^3/\text{day}$$

3. SLUDGE THICKENING

Waste-activated sludge (WAS) has a typical solids content of 0.5–1.0%. *Thickening* of sludge is used to reduce the volume of sludge prior to digestion or dewatering. Thickening is accomplished by decreasing the liquid fraction $(1 - s)$, thus increasing the solids fraction, s. Equation 12.3 shows that the volume of wet sludge is inversely proportional to its solids content.

For dewatering, thickening to at least 4% solids (i.e., 96% moisture) is required; for digestion, thickening to at least 5% solids is required. Depending on the nature of the sludge, polymers may be used with all of the thickening methods. Other chemicals (e.g., lime for stabilization through pH control and potassium permanganate to react with sulfides) can also be used. However, because of cost and increased disposal problems, the trend is toward the use of fewer inorganic chemicals with new thickening or dewatering applications unless special conditions apply.

Gravity thickening occurs in circular sedimentation tanks similar to primary and secondary clarifiers. The settling process is categorized into four zones: the *clarification zone*, containing relatively clear *supernatant*; the *hindered settling zone*, where the solids move downward at essentially a constant rate; the *transition zone*, characterized by a decrease in solids settling rate; and the *compression zone*, where motion is essentially zero.

In *batch gravity thickening*, the tank is filled with thin sludge and allowed to stand. Supernatant is decanted, and the tank is topped off with more thin sludge. The operation is repeated continually or a number of times before the underflow sludge is removed. A heavy-duty (deep-truss) scraper mechanism pushes the settled solids into a hopper in the tank bottom, and the clarified effluent is removed in a peripheral weir. A doubling of solids content is usually possible with gravity thickening. Table 12.2 contains typical characteristics of gravity thickening tanks.

With *dissolved air flotation* (DAF) *thickening* (DAFT), fine air bubbles are released into the sludge as it enters the DAF tank. The solids particles adhere to the air bubbles and float to the surface where they are skimmed away as scum. The scum has a solids content of approximately 4%. Up to 85% of the total solids may be recovered in this manner, and chemical flocculants (e.g., polymers) can increase the recovery to 97–99%.

Table 12.2 *Typical Characteristics of Gravity Thickening Tanks*

shape	circular
minimum number of tanks*	2
overflow rate	600–800 gal/day-ft² (24–33 m³/d·m²)
maximum dry solids loading	
primary sludge	22 lbm/day-ft² (107 kg/d·m²)
primary and trickling filter sludge	15 lbm/day-ft² (73 kg/d·m²)
primary and modified aeration activated sludge	12 lbm/day-ft² (59 kg/d·m²)
primary and conventional aeration activated sludge	8 lbm/day-ft² (39 mg/d·m²)
waste-activated sludge	4 lbm/day-ft² (20 kg/d·m²)
minimum detention time	6 hr
minimum sidewater depth	10 ft (3 m)
minimum freeboard	1.5 ft (0.45 m)

(Multiply gal/day-ft² by 0.0407 to obtain m³/d·m².)
(Multiply lbm/day-ft² by 4.882 to obtain kg/d·m².)
(Multiply ft by 0.3048 to obtain m.)

*unless alternative methods of thickening are available

With *gravity belt thickening* (GBT), sludge is spread over a drainage belt. Multiple stationary plows in the path of the moving belt may be used to split, turn, and recombine the sludge so that water does not pool on top of the sludge. After thickening, the sludge cake is removed from the belt by a doctor blade.

Gravity thickening is usually best for sludges from primary and secondary settling tanks, while dissolved air flotation and centrifugal thickening are better suited for activated sludge. The current trend is toward using gravity thickening for primary sludge and flotation thickening for activated sludge, then blending the thickened sludge for further processing.

Table 12.3 *Typical Thickening Solids Fraction and Capture Efficiencies*

operation	solids capture range (%)	solids content range (%)
gravity thickeners		
primary sludge only	85–92	4–10
primary and waste-activated	80–90	2–6
flotation thickeners		
with chemicals	90–98	3–6
without chemicals	80–95	3–6
centrifuge thickeners		
with chemicals	90–98	4–8
without chemicals	80–90	3–6
vacuum filtration with chemicals	90–98	15–30
belt filter press with chemicals	85–98	15–30
filter press with chemicals	90–98	20–50
centrifuge dewatering		
with chemicals	85–98	10–35
without chemicals	55–90	10–30

Not all of the solids in sludge are captured in thickening processes. Some sludge solids escape with the liquid portion. Table 12.3 lists typical solids capture fractions.

4. SLUDGE STABILIZATION

Raw sludge is too bulky, odorous, and putrescible to be dewatered easily or disposed of by land spreading. Such sludge can be "stabilized" through chemical addition or by digestion prior to dewatering and landfilling. (Sludge that is incinerated does not need to be stabilized.) Stabilization converts the sludge to a stable, inert form that is essentially free from odors and pathogens.

Sludge can be stabilized chemically by adding lime dust to raise the pH. A rule of thumb is that adding enough lime to raise the pH to 12 for at least 2 hr will inhibit or kill bacteria and pathogens in the sludge. Kiln and cement dust is also used in this manner.

Alternatively, ammonia compounds can be added to the sludge prior to land spreading. This has the additional advantage of producing a sludge with high plant nutrient value.

5. AEROBIC DIGESTION

Aerobic digestion occurs in an open holding tank digester and is preferable for stabilized primary and combined primary-secondary sludges. Up to 70% (typically 40–50%) of the volatile solids can be removed in an aerobic digester. Mechanical aerators are used in a manner similar to aerated lagoons. Construction details of aerobic digesters are similar to those of aerated lagoons.

6. ANAEROBIC DIGESTION

Anaerobic digestion occurs in the absence of oxygen. Anaerobic digestion is more complex and more easily upset than aerobic digestion. However, it has a lower operating cost. Table 12.4 gives typical characteristics of aerobic digesters.

Three types of anaerobic bacteria are involved. The first type converts organic compounds into simple fatty or amino acids. The second group of *acid-formed bacteria* converts these compounds into simple organic acids, such as acetic acid. The third group of *acid-splitting bacteria* converts the organic acids to methane, carbon dioxide, and some hydrogen sulfide. The third phase takes the longest time and sets the rate and loadings. The pH should be 6.7–7.8, and temperature should be in the mesophilic range of 85–100°F (29–38°C) for the methane-producing bacteria to be effective. Sufficient alkalinity must be added to buffer acid production.

A simple single-stage sludge digestion tank consists of an inlet pipe, outlet pipes for removing both the digested sludge and the clear supernatant liquid, a dome, an outlet pipe for collecting and removing the digester gas, and a series of heating coils for circulating hot water. (See Fig. 12.2.)

In a single-stage, floating-cover digester, sludge is brought into the tank at the top. The contents of the digester stratify into four layers: scum on top, clear supernatant, a layer of actively digesting sludge, and a bottom layer of concentrated sludge. Some of the contents may be withdrawn, heated, and returned in order to maintain a proper digestion temperature.

Figure 12.2 *Simple Anaerobic Digester*

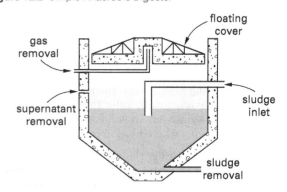

Table 12.4 *Typical Characteristics of Aerobic Digesters*

minimum number of units	2
size	2–3 ft^3/person $(0.05$–0.08 m^3/person)
aeration period	
activated sludge only	10–15 days
activated sludge and primary settling sludge	12–18 days
mixture of primary sludge and activated or trickling filter sludge	15–20 days
volatile solids loading	0.1–0.3 lbm/day-ft^3 $(1.6$–$4.8 \text{ kg/d·m}^3)$
volatile solids reduction	40–50%
sludge age	
primary sludge	25–30 days
activated sludge	15–20 days
oxygen required (primary sludge)	1.6–1.9 lbm O_2/lbm BOD removed $(1.6$–1.9 kg O_2/kg BOD removed)
minimum dissolved oxygen level	1–2 mg/L
mixing energy required by mechanical aerators	0.75–$1.5 \text{ hp}/1000 \text{ ft}^3$ $(0.02$–$0.04 \text{ kW/m}^3)$
maximum depth	15 ft (4.5 m)

(Multiply ft^3 by 0.0283 to obtain m^3.)
(Multiply lbm/day-ft^3 by 16.0 to obtain kg/d·m^3.)
(Multiply lbm/lbm by 1.0 to obtain kg/kg.)
(Multiply $\text{hp}/1000 \text{ ft}^3$ by 0.0263 to obtain kW/m^3.)
(Multiply ft by 0.3048 to obtain m.)

Supernatant is removed along the periphery of the digester and returned to the input of the processing plant. Digested sludge is removed from the bottom and dewatered prior to disposal. Digester gas is removed from the gas dome. Heat from burning the methane can be used to warm the sludge that is withdrawn or to warm raw sludge prior to entry.

A reasonable loading and well-mixed digester can produce a well-stabilized sludge in 15 days. With poorer mixing, 30–60 days may be required.

A single-stage digester performs the functions of digestion, gravity thickening, and storage in one tank. In a two-stage process, two digesters in series are used. Heating and mechanical mixing occur in the first digester. Since the sludge is continually mixed, it will not settle. Settling and further digestion occur in the unheated second tank.

Germany began experimenting with large spherical and egg-shaped digesters in the 1950s. These and similar designs are experiencing popularity in the United States. Volumes are on the order of 100,000–400,000 ft^3 (3000–12 000 m^3), and digesters can be constructed from either prestressed concrete or steel. Compared with conventional cylindrical digesters, the egg-shaped units minimize problems associated with the accumulation of solids in the lower corners and collection of foam and scum on the upper surfaces. The shape also promotes better mixing that enhances digestion. However, variations in the shape greatly affect grit collection and required mixing power. Mixing can be accomplished with uncombined gas mixing, mechanical mixing using an impeller and draft-tube, and pumped recirculation. Typical volatile suspended solids loading rates are 0.062–0.17 lbm VSS/ft^3-day (1–2.8 kg VSS/m^3·d) with a 15–25 day retention time. Operational problems include grit accumulation, gas leakage, and foaming.

7. METHANE PRODUCTION

Wastes that contain biologically degradable organics and are free of substances that are toxic to microorganisms (e.g., acids) will be biodegraded. Methane gas is a product of anaerobic degradation. Under anaerobic conditions, degradation of organic wastes yields methane. Inorganic wastes are not candidates for biological treatment and do not produce methane.

The theoretical volume of methane produced from sludge digestion is 5.61 ft^3 (0.35 m^3) per pound (per kilogram) of BOD converted. Wasted sludge does not contribute to methane production, and the theoretical amount is reduced by various inefficiencies. Therefore, the actual volume of methane produced is predicted by Eq. 12.7. The total volume of digester gas is larger than the methane volume alone since methane constitutes about $\frac{2}{3}$ of the total volume. E is the efficiency of waste utilization, typically 0.6–0.9. S_0 is the ultimate BOD of the influent. ES_0 can be replaced with $S_0 - S$, as the two terms are equivalent.

$$V_{methane,m^3/d} = \left(0.35\ \frac{m^3}{kg}\right)$$
$$\times \left(\frac{ES_{0,mg/L}Q_{m^3/d}}{1000\ \frac{g}{kg}} - 1.42P_{x,kg/d}\right) \quad [SI] \quad 12.7(a)$$

$$V_{methane,ft^3/day} = \left(5.61\ \frac{ft^3}{lbm}\right)$$
$$\times \left(\frac{\begin{array}{c}ES_{0,mg/L}\\ \times Q_{MGD}\\ \times \left(8.345\ \frac{lbm\text{-}L}{MG\text{-}mg}\right)\end{array}}{} - 1.42P_{x,lbm/day}\right) \quad [U.S.] \quad 12.7(b)$$

The heating value of digester gas listed in Table 12.5 is approximately 600 Btu/ft³ (22 MJ/m³). This value is appropriate for digester gas with the composition given: 65% methane and 35% carbon dioxide by volume. Carbon dioxide is not combustible and does not contribute to the heating effect. The actual heating value will depend on the volumetric fraction of methane. The temperature and pressure also affect the gas volume. At 60°F (15.6°C) and 1 atm, the pure methane has a lower heating value, LHV, of 913 Btu/ft³ (34.0 MJ/m³). The higher heating value of 1013 Btu/ft³ (37.7 MJ/m³) should not normally be used, as the water vapor produced during combustion is not allowed to condense out.

The total heating energy is

$$\dot{Q} = (LHV)V_{fuel} \quad 12.8$$

8. HEAT TRANSFER AND LOSS

In an adiabatic reactor, energy is required only to heat the influent sludge to the reactor temperature. In a non-adiabatic reactor, energy will also be required to replace energy lost to the environment through heat transfer from the exposed surfaces.

The energy required to increase the temperature of sludge from T_1 to T_2 is given by Eq. 12.9. The specific heat, c_p, of sludge is assumed to be the same as for water, 1.0 Btu/lbm-°F (4190 J/kg·°C).

State Functions (properties)

$$\dot{Q} = m_{sludge}c_p(T_2 - T_1)$$
$$= V_{sludge}\rho_{sludge}c_p(T_2 - T_1) \quad 12.9$$

Table 12.5 *Typical Characteristics of Single-Stage Heated Anaerobic Digesters*

minimum number of units	2
size	
trickling filter sludge	3–4 ft³/person (0.08–0.11 m³/person)
primary and secondary sludge	6 ft³ (0.16 m³/person)
volatile solids loading	0.13–0.2 lbm/day-ft³ (2.2–3.3 kg/d·m³)
temperature	85–100°F; 95–98°F optimum (29–38°C; 35–36.7°C optimum)
pH	6.7–7.8; 6.9–7.2 optimum
gas production	7–10 ft³/lbm volatile solids (0.42–0.60 m³/kg volatile solids)
	0.5–1.0 ft³/day-person (0.014–0.027 m³/d·person)
gas composition	65% methane; 35% carbon dioxide
gas heating value	500–700; 600 Btu/ft³ typical (19–26 MJ/m³; 22 MJ/m³)
sludge final moisture content	90–95%
sludge detention time	30–90 days (conventional) 15–25 days (high rate)
depth	20–45 ft (6–14 m)
minimum freeboard	2 ft (0.6 m)

(Multiply ft³ by 0.0283 to obtain m³.)
(Multiply lbm/day-ft³ by 16.0 to obtain kg/d·m³.)
(Multiply ft³/lbm by 0.0624 to obtain m³/kg.)
(Multiply Btu/ft³ by 0.0372 to obtain MJ/m³.)
(Multiply ft by 0.3048 to obtain m.)

Equation 12.9 shows that the relationship between heat capacity and temperature for heat transfer is analogous to that for fluid flow, as shown in Eq. 12.10. ($\dot{Q}$ is used for heat flow to avoid confusion with liquid flow rate, Q.)

State Functions (properties)

$$Q = mc\Delta T \quad 12.10$$

The energy lost from a heated digester can be calculated in several ways. If the surface temperature is known, the convective heat lost from a heated surface to the surrounding air depends on the convective outside *film coefficient, h*. The film coefficient depends on the surface temperature, air temperature, orientation, and prevailing wind. Heat losses from digester sides, top, and other surfaces must be calculated separately and combined.

Heat Transfer: Convection

$$\dot{Q} = hA(T_w - T_\infty) \qquad 12.11$$

If the temperature of the digester contents is known, it can be used with the *overall coefficient of heat transfer, U* (a combination of inside and outside film coefficients), as well as the conductive resistance of the digester walls to calculate the heat loss.

$$\dot{Q} = UA(T_{\text{contents}} - T_\infty) \qquad 12.12$$

If the temperatures of the inner and outer surfaces of the digester walls are known, the energy loss can be calculated as a conductive loss. k is the average thermal conductivity of the digester wall, and L is the wall thickness.

Heat Transfer: Conduction Through a Plane Wall

$$\dot{Q} = \frac{-kA(T_2 - T_1)}{L} \qquad 12.13$$

9. SLUDGE DEWATERING

Once sludge has been thickened, it can be digested or dewatered prior to disposal. At water contents of 75%, sludge can be handled with shovel and spade, and is known as *sludge cake*. A 50% moisture content represents the general lower limit for conventional drying methods. Several methods of dewatering are available, including vacuum filtration, pressure filtration, centrifugation, sand and gravel drying beds, and lagooning.

Generally, vacuum filtration and centrifugation are used with undigested sludge. A common form of vacuum filtration occurs in a *vacuum drum filter*. A hollow drum covered with filter cloth is partially immersed in a vat of sludge. The drum turns at about 1 rpm. Suction is applied from within the drum to attract sludge to the filter surface and to extract moisture. The sludge cake is dewatered to about 75–80% moisture content (solids content of 20–25%) and is then scraped off the belt by a blade or delaminated by a sharp bend in the belt material. This degree of dewatering is sufficient for sanitary landfill. A higher solids content of 30–33% is needed, however, for direct incineration unless external fuel is provided. Chemical flocculants (e.g., polymers or ferric chloride, with or without lime) may be used to collect finer particles on the filter drum. Performance is

measured in terms of the amount of dry solids removed per unit area. A typical performance is 3.5 lbm/hr-ft^2 (17 kg/h·m^2).

Belt filter presses accept a continuous feed of conditioned sludge and use a combination of gravity drainage and mechanically applied pressure. Sludge is applied to a continuous-loop woven belt (cotton, wool, nylon, polyester, woven stainless steel, etc.) where the sludge begins to drain by gravity. As the belt continues on, the sludge is compressed under the action of one or more rollers. Belt presses typically have belt widths of 80–140 in (2.0–3.5 m) and are loaded at the rate of 60–450 lbm/ft-hr (90–680 kg/m·h). Moisture content of the cake is typically 70–80%.

Pressure filtration is usually chosen only for sludges with poor dewaterability (such as waste-activated sludge) and where it is desired to dewater to a solids content higher than 30%. *Recessed plate filter presses* operate by forcing sludge at pressures of up to 120 psi (830 kPa) into cavities of porous filter cloth. The filtrate passes through the cloth, leaving the filter cake behind. When the filter becomes filled with cake, it is taken offline and opened. The filter cake is either manually or automatically removed.

Filter press sizing is based on a continuity of mass. In Eq. 12.14, s is the solids content as a decimal fraction in the filter cake.

$$\begin{aligned} V_{\text{press}}\rho_{\text{filter cake}}s_{\text{filter cake}} \\ = V_{\text{sludge,per cycle}}s_{\text{sludge}}\rho_{\text{sludge}} \qquad 12.14 \\ = V_{\text{sludge,per cycle}}s_{\text{sludge}}\rho_{\text{water}}S_{\text{sludge}} \end{aligned}$$

Centrifuges can operate continuously and reduce water content to about 70%, but the effluent has a high percentage of suspended solids. The types of centrifuges used include solid bowl (both cocurrent and countercurrent varieties) and imperforate basket decanter centrifuges.

With *solid bowl centrifuges*, sludge is continuously fed into a rotating bowl where it is separated into a dilute stream (the *centrate*) and a dense cake consisting of approximately 70–80% water. Since the centrate contains fine, low-density solids, it is returned to the wastewater treatment system. Equation 12.15 gives the number of gravities, G, acting on a rotating particle. Rotational speed, n, is typically 1500–2000 rpm. A polymer may be added to increase the capture efficiency (same as reducing the solids content of the centrate) and, accordingly, increase the feed rate.

$$G = \frac{\omega^2 r}{g} = \left(\frac{2\pi n}{60}\right)^2 \left(\frac{r}{g}\right) \qquad 12.15$$

Sand drying beds are preferable when digested sludge is to be disposed of in a landfill since sand beds produce sludge cake with a low moisture content. Drying beds are enclosed by low concrete walls. A broken stone/

gravel base that is 8–20 in (20–50 cm) thick is covered by 4–8 in (10–20 cm) of sand. Drying occurs primarily through drainage, not evaporation. Water that seeps through the sand is removed by a system of drainage pipes. Typically, sludge is added to a depth of about 8–12 in (20–30 cm) and allowed to dry for 10–30 days.

There is little theory-based design, and drying beds are sized by rules of thumb. The drying area required is approximately 1–1.5 ft^2 (0.09–1.4 m^2) per person for primary digested sludge; 1.25–1.75 ft^2 (0.11–0.16 m^2) per person for primary and trickling filter digested sludge; 1.75–2.5 ft^2 (0.16–0.23 m^2) per person for primary and waste-activated digested sludge; and 2–2.5 ft^2 (0.18–0.23 m^2) per person for primary and chemically precipitated digested sludge. This area can be reduced by approximately 50% if transparent covers are placed over the beds, creating a "greenhouse effect" and preventing the spread of odors.

Other types of drying beds include paved beds (both drainage and decanting types), vacuum-assisted beds, and artificial media (e.g., stainless steel wire and high-density polyurethane panels) beds.

Sludge dewatering by freezing and thawing in exposed beds is effective during some months in parts of the country.

Example 12.2

A belt filter press handles an average flow of 19,000 gal/day of thickened sludge containing 2.5% solids by weight. Operation is 8 hr/day and 5 days/wk. The belt filter press loading is specified as 500 lbm/m-hr (lbm/hr per meter of belt width). The dewatered sludge is to have 25% solids by weight, and the suspended solids concentration in the filtrate is 900 mg/L. The belt is continuously washed on its return path at a rate of 24 gal/m-min. The specific gravities of the sludge feed, dewatered cake, and filtrate are 1.02, 1.07, and 1.01, respectively. (a) What size belt is required? (b) What is the volume of filtrate rejected? (c) What is the solids capture fraction?

Solution

(a) The daily masses of wet and dry influent sludge are

$$m_{\text{wet}} = V\rho_{\text{sludge}}$$
$$= V\rho_{\text{water}}S_{\text{sludge}}$$
$$= \left(19{,}000 \ \frac{\text{gal}}{\text{day}}\right)\left(\frac{\left(62.4 \ \frac{\text{lbm}}{\text{ft}^3}\right)(1.02)}{7.48 \ \frac{\text{gal}}{\text{ft}^3}}\right)$$
$$= 1.617 \times 10^5 \ \text{lbm/day}$$

$$m_{\text{dry}} = (\text{fraction solids})\, m_{\text{wet}}$$
$$= (0.025)\left(1.617 \times 10^5 \ \frac{\text{lbm}}{\text{day}}\right)$$
$$= 4042 \ \text{lbm/day}$$

Since the influent is received continuously but the belt filter press operates only 5 days a week for 8 hours per day, the average hourly processing rate is

$$\frac{\left(4042 \ \frac{\text{lbm}}{\text{day}}\right)\left(\frac{7 \ \text{days}}{5 \ \text{days}}\right)}{8 \ \frac{\text{hr}}{\text{day}}} = 707 \ \text{lbm/hr}$$

The belt size (width) required is

$$w = \frac{707 \ \frac{\text{lbm}}{\text{hr}}}{500 \ \frac{\text{lbm}}{\text{m-hr}}}$$
$$= 1.41 \ \text{m}$$

Use one 1.5 m belt filter press. A second standby unit may also be specified.

(b) The mass of solids received each day is

$$\left(707 \ \frac{\text{lbm}}{\text{hr}}\right)\left(8 \ \frac{\text{hr}}{\text{day}}\right) = 5656 \ \text{lbm/day}$$

The fractional solids in the filtrate is

$$s_{\text{filtrate}} = \frac{900 \ \frac{\text{mg}}{\text{L}}}{\left(1000 \ \frac{\text{mg}}{\text{g}}\right)\left(1000 \ \frac{\text{g}}{\text{L}}\right)} = 0.0009$$

The solids balance is

influent solids = sludge cake solids + filtrate solids

$$5656 \ \frac{\text{lbm}}{\text{day}} = \left(\frac{Q_{\text{cake, gal/day}}}{7.48 \ \frac{\text{gal}}{\text{ft}^3}}\right)\left(62.4 \ \frac{\text{lbm}}{\text{ft}^3}\right)(1.07)(0.25)$$
$$+ \left(\frac{Q_{\text{filtrate, gal/day}}}{7.48 \ \frac{\text{gal}}{\text{ft}^3}}\right)\left(62.4 \ \frac{\text{lbm}}{\text{ft}^3}\right)$$
$$\times (1.01)(0.0009)$$

The total liquid flow rate is the sum of the sludge flow rate and the washwater flow rate.

$$\dot{Q}_{total} = \dot{Q}_{cake} + \dot{Q}_{filtrate}$$
$$= \left(19{,}000 \ \frac{\text{gal}}{\text{day}}\right)\left(\frac{7 \ \text{days}}{5 \ \text{days}}\right)$$
$$+ \left(24 \ \frac{\text{gal}}{\text{m·min}}\right)(1.5 \ \text{m})\left(60 \ \frac{\text{min}}{\text{hr}}\right)\left(8 \ \frac{\text{hr}}{\text{day}}\right)$$
$$= 43{,}880 \ \text{gal/day}$$

Solving the solids and liquid flow rate equations simultaneously,

$$\dot{Q}_{cake} = 2396 \ \text{gal/day}$$

$$\dot{Q}_{filtrate} = 41{,}484 \ \text{gal/day}$$

(c) The solids in the filtrate are

$$\dot{m}_{filtrate} = \left(\frac{41{,}484 \ \dfrac{\text{gal}}{\text{day}}}{7.48 \ \dfrac{\text{gal}}{\text{ft}^3}}\right)\left(62.4 \ \frac{\text{lbm}}{\text{ft}^3}\right)(1.01)(0.0009)$$
$$= 315 \ \text{lbm/day}$$

The solids capture fraction is

$$\frac{\text{influent solids} - \text{filtrate solids}}{\text{influent solids}} = \frac{5656 \ \dfrac{\text{lbm}}{\text{day}} - 315 \ \dfrac{\text{lbm}}{\text{day}}}{5656 \ \dfrac{\text{lbm}}{\text{day}}}$$
$$= 0.944$$

10. SLUDGE DISPOSAL

Liquid sludge and sludge cake disposal options are limited to landfills, incineration, nonagricultural land application (e.g., strip-mine reclamation, agricultural land application, commercial applications (known as *distribution and marketing* and *beneficial reuse*) such as composting, special monofills, and surface impoundments), and various state-of-the-art methods. Different environmental contaminants are monitored with each disposal method, making the options difficult to evaluate.

If a satisfactory site is available, sludge, sludge cake, screenings, and scum can be discarded in municipal *landfills*. Approximately 40% of the sludge produced in the United States is disposed of in this manner. This has traditionally been the cheapest disposal method, even when transportation costs are included.

The sludge composition and its potential effects on landfill leachate quality must be considered. Sludge with minute concentrations of monitored substances (e.g., heavy metals such as cadmium, chromium, copper, lead, nickel, and zinc) may be classified as hazardous waste, requiring disposal in hazardous waste landfills.

Sludge-only *monofills* and other surface impoundments may be selected in place of municipal landfills. Monofills are typically open pits protected by liners, slurry cutoff walls, and levees (as protection against floods). Monofills may include a pug mill to combine sludge with soil from the site.

Approximately 25% of the sludge produced in the United States is disposed of in a *beneficial reuse program* such as surface spreading (i.e., "land application") and/or composting. Isolated cropland, pasture, and forest land (i.e., "silviculture") are perfect for surface spreading. However, sludge applied to the surface may need to be harrowed into the soil soon after spreading to prevent water pollution from sludge runoff and possible exposure to or spread of bacteria. Alternatively, the sludge can be injected 12–18 in (300–460 mm) below grade directly from distribution equipment. The actual application rate depends on sludge strength (primarily nitrogen content), the condition of the receiving soil, the crop or plants using the applied nutrients, and the spreading technology.

By itself, filter cake is not highly useful as a fertilizer since its nutrient content is low. However, it is a good filler and soil conditioner for more nutrient-rich fertilizers. Other than as filler, filter cake has few practical applications.

Where sludges have low metals contents, *composting* (in static piles or vessels) can be used. *Compost* is a dry, odor-free soil conditioner with applications in turf grass, landscaping, land reclamation, and other horticultural industries. These markets can help to defray the costs of the program. Control of odors, cold weather operations, and concern about airborne pathogens are problems with *static pile composting*. Expensive air-quality scrubbers add to the cost of processing. With *in-vessel composting*, dewatered raw sludge is mixed with a bulking material, placed in an enclosed tank, and mixed with air in the tank.

Incineration is the most expensive treatment, though it results in almost total destruction of volatile solids, is odor free, and is independent of the weather. Approximately 20% of sludge is disposed of in this manner. A solids content of 30–33% is needed for direct incineration without the need to provide external fuel (i.e., so that the sludge burns itself). Air quality controls, equipment, and the regulatory permitting needed to meet strict standards may make incineration prohibitively expensive in some areas. The incineration ash produced represents a disposal problem of its own.

State-of-the-art options include using sludge solids in bricks, tiles, and other building materials; below-grade injection; subsurface injection wells; multiple-effect (Carver-Greenfield) evaporation; combined incineration with solid waste in "mass burn" units; and fluidized-bed gasification. The first significant installation attempting to use sludge solids as powdered fuel for power generation was the 25 MW Hyperion Energy Recovery System at the Hyperion treatment plant in Los Angeles. This combined-cycle cogeneration plant unsuccessfully attempted to operate using the Carver-Greenfield process and a fluidized-bed gasifier to convert sludge solids to fuel.

Ocean dumping, once a popular option for coastal areas, has been regulated out of existence and is no longer an option for new or existing facilities.

11. NOMENCLATURE

A	area	ft^2	m^2
c_p	specific heat	Btu/lbm-°F	J/kg·°C
E	efficiency of waste utilization	–	–
g	acceleration of gravity, 32.2 (9.81)	ft/sec^2	m/s^2
G	number of gravities	–	–
h	convection heat transfer or film coefficient	Btu/hr-ft²-°F	W/m²·K
k	thermal conductivity	Btu-ft/hr-ft²-°F	W/m·K
K	cell yield constant (removal efficiency)	lbm/lbm	mg/mg
L	thickness	ft	m
LHV	lower heating value	Btu/ft^3	kJ/m^3
m	mass	lbm	kg
$\dot{m}$	mass flow rate	lbm/day	kg/d
n	rotational speed	rev/min	rev/min
p	percentage of water sludge	%	%
P_x	mass of sludge wasted	lbm/day	kg/d
Q	flow rate	ft^3/day	m^3/d
$\dot{Q}$	heat flow rate	Btu/sec	kJ/s
r	distance from center of rotation	ft	m
s	percentage of solid sludge	%	%
S	specific gravity	–	–
S_0	ultimate BOD	mg/L	mg/L
SS	suspended solids	mg/L	mg/L
T	absolute temperature	°R	K
U	overall coefficient of heat transfer	Btu/hr-ft²-°F	W/m²·K
V	volume	ft^3	m^3
w	width	ft	m
W	weight	lbf	N
Y	yield coefficient	lbm/lbm	mg/mg

Symbols

γ	unit weight of water, 62.4 (1000)	lbf/ft^3	kg/m^3
ρ	density	lbm/ft^3	kg/m^3
ω	rotational speed	rad/sec	rad/s

Subscripts

0	initial
s	solid
w	wall
∞	bulk fluid

13 Wastewater Reuse

Content in blue refers to the *NCEES Handbook*.

1. WASTEWATER RECLAMATION

Treated wastewater can be used for irrigation, fire-fighting, road maintenance, or flushing. Public access to areas where reclaimed water is disposed of through spray-irrigation (e.g., in sod farms, fodder crops, and pasture lands) should be restricted. When the reclaimed water is to be used for irrigation of public areas, the water quality standards regarding suspended solids, fecal coliforms, and viruses should be strictly controlled.

Although it has been demonstrated that wastewater can be made potable at great expense, such practice has not gained widespread favor.

14 Stormwater Pollution, Treatment, Management

1. SOURCES OF WATER POLLUTION

Sources of water pollution, whether to surface waters or to groundwater, fall into three categories.

- point source discharges

- nonpoint source discharges

- one-time releases from underground tanks, oil tankers and platforms, pipelines, and other similar events

Point source discharges take on a variety of forms, but each is traceable to an identifiable end of pipe. Nonpoint source discharges are much more difficult to identify and, consequently, much more difficult to regulate and control.

For example, nonpoint source pollution occurs when rain washes pesticides and fertilizers from croplands into surface waters. This is a significant problem with drainage to the Chesapeake Bay and the Gulf of Mexico. Nonpoint source pollution is difficult to assess and control because of its nonspecific source.

Nonpoint source discharge control strategies require improved management practices to reduce the potential for pollution to occur. Overfertilization, improper pesticide application and residue disposal, and employing salt as a road deicer all contribute to nonpoint source pollution but can be controlled by strict management practices.

Pollutants that occur in water contain many materials common to both point sources and nonpoint sources. However, most point source pollutants occur at higher concentrations than do nonpoint source pollutants. Typical pollutants that may be generated from point and nonpoint sources are listed in Table 14.1, along with typical values for strong and weak domestic sewage. These pollutants are precursors to wastewater

discharges from publicly owned treatment works (POTWs) to receiving waters and represent substantial potential impacts on receiving water quality. Organic matter associated with oxygen demand and nitrogen and phosphorus are particular targets for wastewater treatment because of their significant potential negative impacts on receiving water quality.

Table 14.1 *Strong and Weak Domestic Sewages[*]*

constituent	strong	weak
solids, total	1200	350
dissolved, total	850	250
fixed	525	145
volatile	325	105
suspended, total	350	100
fixed	75	30
volatile	275	70
settleable solids (mL/L)	20	5
biochemical oxygen demand, five-day, 20°C	300	100
total organic carbon	300	100
chemical oxygen demand	1000	250
nitrogen (total as N)	85	20
organic	35	8
free ammonia	50	12
nitrites	0	0
nitrates	0	0
phosphorus (total as P)	20	6
organic	5	2
inorganic	15	4
chlorides	100	30
alkalinity (as $CaCO_3$)	200	50
grease	150	50

[*]All concentrations are in mg/L unless otherwise noted.

In addition to those listed in Table 14.1 and their subsequent occurrence in waters discharged from POTWs, other pollutants of concern from POTW discharges are pharmaceuticals and endocrine disruptors. Pharmaceuticals are associated with products purchased by humans for their personal care or that of their pets and other animals. Chemicals from these products enter the environment through improper disposal and as metabolic by-products in urine and feces. Pharmaceuticals are not all readily degraded or captured in conventional wastewater treatment and may pass through a POTW to a receiving water unaltered and little diminished in concentration.

Endocrine disruptors are chemical compounds that, as pollutants, alter the normal function of the endocrine system and interfere with hormone production. Endocrine disruptors may occur naturally, but most are chemical byproducts of pesticides and other synthetic organic chemicals. Endocrine disruptors such as phthalates, a group of compounds commonly used in plastics, may also enter the environment through leaching and as the plastics degrade. Endocrine disruptors can be found in both point and nonpoint pollutant sources.

Organic matter and nitrogen combine to act as pollutants by exerting an oxygen demand. Nitrogen adds to the oxygen demand through the process of nitrification, which is the biological aerobic conversion of ammonia (the ammonium ion NH_4^+) to nitrate (NO_3^-). Nitrification occurs according to the reaction

$$NH_4^+ + 2O_2 \rightleftharpoons NO_3^- + H_2O + 2H^+$$

The formation of nitrate as a product in the reaction creates the oxygen demand, which is called *nitrogenous biochemical oxygen demand* (NBOD). The NBOD is calculated with Eq. 14.1.

$$
\begin{aligned}
\text{NBOD} &= \frac{\text{mass of oxygen used}}{\text{mass of nitrogen oxydized}} \\
&= \frac{(2 \text{ mol})\left(\dfrac{2 \times 16 \text{ g O}_2}{\text{mol}}\right)}{(1 \text{ mol})\left(\dfrac{14 \text{ g N}}{\text{mol}}\right)} \\
&= \frac{4.57 \text{ g O}_2}{\text{gN}}
\end{aligned}
\qquad 14.1
$$

Pathogens may enter the water through incomplete disinfection of POTW effluent. A potentially more significant source of pollution, however, is from agricultural activities involving livestock. The grazing and corralling of livestock near a waterway provides a nonpoint source of pollution as rainfall runoff carries both pathogens and nutrients to the receiving water. Solids, both biological and inorganic, can also be carried by runoff, especially where erosion is occurring under intense rainfall and sparse groundcover.

Other agricultural and landscaping activities contribute more pollutants in the forms of pesticides, herbicides, and fertilizers. These can be significant sources of pollution in both rural and urban settings, especially where overapplication occurs. Impervious surfaces such as roofs, roads, driveways, and parking lots are another widespread source of pollution because the salts, oils, dust, dirt, and other materials that are deposited on such surfaces remain there and accumulate until they are washed away by runoff.

Although surface waters often receive the most attention, groundwater resources are also at risk from pesticides, herbicides, and other substances that can percolate through the soil into an underlying aquifer. In rural areas particularly, septic tank effluent may impact groundwater quality. Releases from underground tanks —which are used to store everything from salad oil to organic solvents—are also potentially significant contributors to groundwater contamination.

2. MINIMIZATION

The most significant opportunity to minimize negative impacts to waterways from pollutants is through control of nonpoint source discharges. Regulations and their enforcement have been effective in reducing polluting discharges from POTWs, industrial wastewater treatment plants, and other identifiable end-of-pipe sources. It has proved much more difficult, however, to control nonpoint sources discharges from agricultural sources and from urban landscapes.

Pollution from these sources has been best controlled through implementation of best management practices and innovative design elements. For example, farmers can help reduce agricultural runoff by creating buffer zones around fertilized fields, employing irrigation techniques that limit overwatering, timing chemical applications during periods where runoff potential is small, and applying pesticides, herbicides, and fertilizers using methods and at application rates that reduce overuse and dispersion.

Urban pollution sources can be reduced by employing landscaping techniques that retain runoff, using pervious pavements for some roadways and for parking lots, providing retention basins to allow solids and other nonsoluble materials to settle, enforcing programs to detect and eliminate illicit discharges to storm drains, and implementing engineered control measures and best management practices.

When storm sewers are segregated from sanitary sewers, stormwater is typically discharged directly to a receiving water such as a river or ocean. The collected stormwater may be held in a retention basin, but in such a case the retention basin is usually used to reduce the stormwater flow rate from developed land, not to provide active treatment.

When storm sewers are combined with sanitary sewers, the combined stormwater and sanitary wastewater is conveyed to a publically owned treatment works (POTW) for treatment. The stormwater has the effect of diluting the sanitary flow and increasing the hydraulic loading to the POTW. However, other than this diluting effect, the combined stormwater and sanitary wastewater are subjected to the same kinds of treatment that would be applied to undiluted sanitary wastewater.

3. STABILIZATION PONDS

The term *stabilization pond* (*oxidation pond* or *stabilization lagoon*) refers to a pond used to treat organic waste by biological and physical processes. Aquatic plants, weeds, algae, and microorganisms stabilize the organic matter. The algae give off oxygen that is used by microorganisms to digest the organic matter. The microorganisms give off carbon dioxide, ammonia, and phosphates that the algae use.

There are several types of stabilization ponds, any of which may be appropriate for treating stormwater. *Aerobic ponds* are shallow ponds, less than 4 ft (1.2 m) in depth, where dissolved oxygen is maintained throughout the entire depth, mainly by action of photosynthesis. *Facultative ponds* have an anaerobic lower zone, a facultative middle zone, and an aerobic upper zone. The upper zone is maintained in an aerobic condition by photosynthesis and, in some cases, mechanical aeration at the surface. *Maturation ponds* (*tertiary ponds* or *polishing ponds*) are used for polishing effluent from secondary biological processes. Dissolved oxygen is furnished through photosynthesis and mechanical aeration. *Aerated lagoons* are oxygenated through the action of surface or diffused air aeration. They are often used with activated sludge processes.

4. PLAIN SEDIMENTATION BASINS/ CLARIFIERS

Plain sedimentation basins/clarifiers (i.e., basins in which no chemicals are added to encourage clarification) used to treat stormwater are similar in concept and design to those used to treat water supplies and wastewater. Typical design characteristics for stormwater treatment sedimentation basins are listed in Table 14.2. Since the bottom slopes slightly, the depth varies with location. Therefore, the *side water depth* is usually quoted. The *surface loading* (*surface loading rate*, *overflow rate*, or *settling rate*), v_o, along with sludge storage volume, is the primary design parameter.

$$v_o = Q/A_{surface} \qquad \text{Clarifier 14.2}$$

The *detention time* (*hydraulic residence time* or *retention period*) is

$$\theta = V/Q \qquad \text{Clarifier 14.3}$$

The *weir overflow* (*weir overflow rate*, WOR) is

$$WOR = Q/\text{Weir Length} \qquad \text{Clarifier 14.4}$$

Table 14.2 *Typical Characteristics of Clarifiers**

BOD reduction	20–40% (25–35% typical)
total suspended solids reduction	35–65%
bacteria reduction	50–60%
specific gravity of settled solids	1.2 or less
minimum settling velocity	4 ft/hr (1.2 m/hr) typical
plan shape	rectangular (or circular)
basin depth (side water depth)	6–15 ft; 10–12 ft typical (1.8–4.5 m; 3–3.6 m typical)
basin width	10–50 ft (3–15 m)
plan aspect ratio	3:1 to 5:1
basin diameter (circular only)	50–150 ft; 100 ft typical (15–45 m; 30 m typical)
minimum freeboard	1.5 ft (0.45 m)
detention time	1.5–2.5 hr
flow-through velocity	18 ft/hr (1.5 mm/s)
weir loading	10,000–20,000 gal/day-ft (125–250 m³/d·m)
surface loading	400–2000 gal/day-ft²; 800–1200 gal/day-ft² typical (16–80 m³/d·m²; 32–50 m³/d·m² typical)

5. CHEMICAL SEDIMENTATION BASINS/ CLARIFIERS

Chemical flocculation (*clarification* or *coagulation*) operations in chemical sedimentation basins used in stormwater treatment are similar to those encountered in the treatment of water and wastewater except that the coagulant doses are greater. The five coagulants used most often are (a) aluminum sulfate, $Al_2(SO_4)_3$; (b) ferric chloride, $FeCl_3$; (c) ferric sulfate, $Fe_2(SO_4)_3$; (d) ferrous sulfate, $FeSO_4$; and (e) chlorinated copperas. Lime and sulfuric acid may be used to adjust the pH for proper coagulation.

6. ACTIVATED SLUDGE PROCESS

The *activated sludge process* is a secondary biological stormwater treatment technique in which a mixture of stormwater and sludge solids is aerated. The activated sludge process can also be applied to treat stormwater where elevated concentrations of organic matter are present. The sludge mixture produced during this oxidation process contains an extremely high concentration of aerobic bacteria, most of which are near starvation. This condition makes the sludge an ideal medium for the destruction of any organic material in the mixture. Since the bacteria are voraciously active, the sludge is called *activated sludge*.

The well-aerated mixture of stormwater and sludge, known as *mixed liquor*, flows from the aeration tank to a secondary clarifier where the sludge solids settle out. Most of the settled sludge solids are returned to the aeration tank in order to maintain the high population of bacteria needed for rapid breakdown of the organic material. However, because more sludge is produced than is needed, some of the return sludge is diverted ("wasted") for subsequent treatment and disposal. This wasted sludge is referred to as *waste activated sludge*, WAS. The volume of sludge returned to the aeration basin is typically 20–30% of the stormwater flow. The liquid fraction removed from the secondary clarifier weir is chlorinated and discharged.

Though diffused aeration and mechanical aeration are the most common methods of oxygenating the mixed liquor, various methods of staging the aeration are used, each having its own characteristic ranges of operating parameters.

In a traditional activated sludge plant using conventional aeration, the stormwater is typically aerated for 6–8 hours in long, rectangular aeration basins. Sufficient air, about eight volumes for each volume of stormwater treated, is provided to keep the sludge in suspension. The air is injected near the bottom of the aeration tank through a system of diffusers.

7. FINAL CLARIFIERS

Table 14.3 lists typical operating characteristics for clarifiers in activated sludge processes. Sludge should be removed rapidly from the entire bottom of the clarifier.

Table 14.3 *Characteristics of Final Clarifiers for Activated Sludge Processes*

type of aeration	design flow (MGD)	minimum detention time (hr)	maximum overflow rate (gpd/ft^2)
conventional, high rate and step	< 0.5	3.0	600
	0.5–1.5	2.5	700
	> 1.5	2.0	800
contact stabilization	< 0.5	3.6	500
	0.5–1.5	3.0	600
	>1.5	2.5	700
extended aeration	< 0.05	4.0	300
	0.05–0.15	3.6	300
	>0.15	3.0	600

(Multiply MGD by 0.0438 to obtain m^3/s.)
(Multiply gal/day-ft^2 by 0.0407 m^3/d·m^2.)

8. NITROGEN

Nitrogen occurs in water in organic, ammonia, nitrate, nitrite, and dissolved gaseous forms. Nitrogen in stormwater runoff results from application of nitrogen fertilizers to landscaping and crops. Bacterial decomposition and the hydrolysis of urea produces ammonia, NH_3. Ammonia in water forms the ammonium ion NH_4^+, also known as *ammonia nitrogen*. Ammonia nitrogen must be removed from the waste effluent stream due to potential exertion of oxygen demand. The total of organic and ammonia nitrogen is known as *total Kjeldahl nitrogen*, or TKN. *Total nitrogen*, TN, includes TKN plus inorganic nitrates and nitrites.

Nitrification (the oxidation of ammonia nitrogen) occurs as follows.

$$NH_4^+ + 2O_2 \rightarrow NO_3^- + H_2O + 2H^+ \qquad 14.5$$

Denitrification, the reduction of nitrate nitrogen to nitrogen gas by facultative heterotrophic bacteria, occurs as follows.

$$2NO_3^- + \text{organic matter} \rightarrow N_2 + CO_2 + H_2O \quad 14.6$$

Ammonia nitrogen can be removed chemically from water without first converting it to nitrate form by raising the pH level. This converts the ammonium ion back into ammonia, which can then be stripped from the water by passing large quantities of air through the water. Lime is added to provide the hydroxide for the reaction. The *ammonia stripping* reaction (a physical-chemical process) is

$$NH_4^+ + OH^- \rightarrow NH_3 + H_2O \qquad 14.7$$

Ammonia stripping has no effect on nitrate, which is commonly removed in an activated sludge process with a short cell detention time.

9. NOMENCLATURE

A	area	ft^2	m^2
NBOD	nitrogenous biochemical oxygen demand	g/g	g/g
Q	flow rate	ft^3/day	m^3/d
v_o	surface loading	ft/day	m/d
V	volume	ft^3	m^3
WOR	weir overflow rate	gal/day-ft^2	L/d·m^2

Symbols

θ	hydraulic residence time	day	d

15 Stormwater Collection Systems

Content in blue refers to the *NCEES Handbook*.

1. INTRODUCTION

An *open channel* is a fluid passageway that allows part of the fluid to be exposed to the atmosphere. This type of channel includes natural waterways, canals, culverts, flumes, and pipes flowing under the influence of gravity (as opposed to pressure conduits, which always flow full). A *reach* is a straight section of open channel with uniform shape, depth, slope, and flow quantity.

There are difficulties in evaluating open channel flow. The unlimited geometric cross sections and variations in roughness have contributed to a relatively small number of scientific observations upon which to estimate the required coefficients and exponents. Therefore, the analysis of open channel flow is more empirical and less exact than that of pressure conduit flow. This lack of precision, however, is more than offset by the percentage error in runoff calculations that generally precedes the channel calculations.

Flow can be categorized on the basis of the channel material, such as concrete or metal pipe or earth material. Except for a short discussion of erodible canals in Sec. 15.36, this chapter assumes the channel is nonerodible.

2. TYPES OF FLOW

Flow in open channels is almost always turbulent; laminar flow will occur only in very shallow channels or at very low fluid velocities. However, within the turbulent category are many somewhat confusing categories of flow. Flow can be a function of time and location. If the flow quantity (volume per unit of time across an area in flow) is invariant, it is said to be *steady flow*. (Flow that varies with time, such as stream flow during a storm, known as *varied flow*, is not covered in this chapter.) If the flow cross section does not depend on the location along the channel, it is said to be *uniform flow*. Steady flow can also be *nonuniform*, as in the case of a river with a varying cross section or on a steep slope. Furthermore, uniform channel construction does not ensure uniform flow, as will be seen in the case of hydraulic jumps.

Table 15.1 summarizes some of the more common categories and names of steady open channel flow. All of the subcategories are based on variations in depth and flow area with respect to location along the channel.

Water

Table 15.1 Categories of Steady Open Channel Flow

subcritical flow (tranquil flow)
  uniform flow
    normal flow
  nonuniform flow
    accelerating flow
    decelerating flow (decelerated flow)
critical flow
supercritical flow (rapid flow, shooting flow)
  uniform flow
    normal flow
  nonuniform flow
    accelerating flow
    decelerating flow

3. MINIMUM VELOCITIES

The minimum permissible velocity in a sewer or other nonerodible channel is the lowest that prevents sedimentation and plant growth. Velocity ranges of 2 ft/sec to 3 ft/sec (0.6 m/s to 0.9 m/s) keep all but the heaviest silts in suspension. 2.5 ft/sec (0.75 m/s) is considered the minimum to prevent plant growth.

4. VELOCITY DISTRIBUTION

Due to the adhesion between the wetted surface of the channel and the water, the velocity will not be uniform across the area in flow. The velocity term used in this chapter is the *mean velocity*. The mean velocity, when multiplied by the flow area, gives the flow quantity.

Continuity Equation
$$Q = Av \qquad \text{15.1}$$

The location of the mean velocity depends on the distribution of velocities in the waterway, which is generally quite complex. The procedure for measuring the velocity of a channel (called *stream gauging*) involves measuring the average channel velocity at multiple locations and depths across the channel width. These subaverage velocities are averaged to give a grand average (mean) flow velocity.

Stream gauging is performed by measuring the flow velocity and the channel cross section. The stream width is divided into 20 to 30 equal segments, and flow velocity is measured at these intervals and at standard depths. Velocity is typically measured using a Price

current meter or propeller-type current meter. Where the water depth is adequate, the velocity readings are taken at 0.2 and 0.8 of the segment depth, and the readings are averaged for the segment velocity. For shallower depths, such as those near stream banks, a single reading is taken at 0.6 of the segment depth.

Once the segment width and average depth are obtained, the segment area is calculated. The segment area is multiplied by the segment average velocity to give the segment flow. The segment flows are then summed to find the total stream discharge. For streams with higher flow velocities, the meter depth will be more difficult to determine, since the meter will move downstream off the vertical. A downstream deflection of 12° will produce an error of about 2%. Figure 15.1 illustrates.

Figure 15.1 Stream Gauging

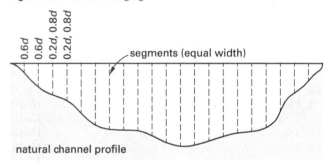

5. PARAMETERS USED IN OPEN CHANNEL FLOW

The *hydraulic radius* is the ratio of the area in flow to the wetted perimeter.[1]

Flow in Noncircular Conduits
$$R_H = \frac{\text{cross-sectional area}}{\text{wetted perimeter}} \qquad \text{15.2}$$

For a circular channel flowing either full or half-full, the hydraulic radius is one-fourth of the *hydraulic diameter*, $D/4$. The hydraulic radii of other channel shapes are easily calculated from the basic definition. For very wide channels such as rivers, the hydraulic radius is approximately equal to the depth. [Geometric Elements of Channel Sections]

The *hydraulic depth* is the ratio of the area in flow to the width of the channel at the fluid surface.[2]

Geometric Elements of Channel Sections
$$D = \frac{A}{T} \qquad \text{15.3}$$

---

[1]The hydraulic radius is also referred to as the *hydraulic mean depth*. However, this name is easily confused with "mean depth" and "hydraulic depth," both of which have different meanings. Therefore, the term "hydraulic mean depth" is not used in this chapter.
[2]For a rectangular channel, $D = d$.

The uniform flow *section factor* represents a frequently occurring variable group. The section factor is often evaluated against depth of flow when working with discharge from irregular cross sections.

$$Z = AR^{2/3} \quad \text{[general uniform flow]} \qquad 15.4$$

$$Z = A\sqrt{\frac{A}{T}} \quad \text{[critical flow only]} \qquad 15.5$$

The *slope*, S, is the gradient of the energy line. In general, the slope can be calculated from the Bernoulli equation as the energy loss per unit length of channel. For small slopes typical of almost all natural waterways, the channel length and horizontal run are essentially identical.

$$S = dE/dL \qquad 15.6$$

If the flow is uniform, the slope of the energy line will parallel the water surface and channel bottom, and the *energy gradient* will equal the *geometric slope*, S_0.

$$S_0 = \frac{\Delta z}{L} \approx S \quad \text{[uniform flow]} \qquad 15.7$$

Any open channel performance equation can be written using the geometric slope, S_0, instead of the hydraulic slope, S, but only under the condition of uniform flow.

In most problems, the geometric slope is a function of the terrain and is known. However, it may be necessary to calculate the slope that results in some other specific parameter. The slope that produces flow at some normal depth, b, is called the *normal slope*. The slope that produces flow at some critical depth, b_c, is called the *critical slope*. Both are determined by solving the Manning equation for slope.

6. GOVERNING EQUATIONS FOR UNIFORM FLOW

Since water is incompressible, the continuity equation is

Continuity Equation
$$A_1 v_1 = A_2 v_2 \qquad 15.8$$

The most common equation used to calculate the flow velocity in open channels is the 1768 *Chezy equation*.[3]

$$v = C\sqrt{RS} \qquad 15.9$$

Various methods for evaluating the *Chezy coefficient*, C, or "Chezy's C," have been proposed.[4] If the channel is small and very smooth, Chezy's own formula can be used. The friction factor, f, is dependent on the Reynolds number and can be found in the usual manner from the Moody diagram.

$$C = \sqrt{\frac{8g}{f}} \qquad 15.10$$

If the channel is large and the flow is fully turbulent, the friction loss will not depend so much on the Reynolds number as on the channel roughness. The 1888 *Manning formula* is frequently used to evaluate the constant C.[5] The value of C depends only on the channel roughness and geometry.

$$C = \left(\frac{1}{n}\right)R^{1/6} \qquad \text{[SI]} \quad 15.11(a)$$

$$C = \left(\frac{1.486}{n}\right)R^{1/6} \qquad \text{[U.S.]} \quad 15.11(b)$$

n is the *Manning roughness coefficient (Manning constant)*. Judgment is needed in selecting values since tabulated values often differ by as much as 30%. More important to recognize for sewer work is the layer of slime that often coats the sewer walls. Since the slime characteristics can change with location in the sewer, there can be variations in Manning's roughness coefficient along the sewer length. Typical values of Manning's roughness coefficient are tabulated in various codebooks and standards. [Manning's Values of Roughness Coefficient, n]

Combining Eq. 15.9 and Eq. 15.11 produces the *Manning equation*, also known as the *Chezy-Manning equation*. For SI units, K is equal to 1.0; for customary U.S. units, K is equal to 1.486.

Manning's Equation
$$v = (K/n)R_H^{2/3}S^{1/2} \qquad 15.12$$

All of the coefficients and constants in the Manning equation may be combined into the *conveyance*, K.

$$Q = vA = (K/n)AR_H^{2/3}S^{1/2} = KS^{1/2} \qquad 15.13$$

---

[3]Pronounced "Shay'-zee." This equation does not appear to be dimensionally consistent. However, the coefficient C is not a pure number. Rather, it has units of length$^{1/2}$/time (i.e., acceleration$^{1/2}$).
[4]Other methods of evaluating C include the *Kutter equation* (also known as the *G.K. formula*) and the *Bazin formula*. These methods are interesting from a historical viewpoint, but both have been replaced by the Manning equation.
[5]This equation was originally proposed in 1868 by Gaukler and again in 1881 by Hagen, both working independently. For some reason, the Frenchman Flamant attributed the equation to an Irishman, R. Manning. In Europe and many other places, the Manning equation may be known as the *Strickler equation*.

Example 15.1

A rectangular channel on a 0.002 slope is constructed of finished concrete. The channel is 8 ft (2.4 m) wide. Water flows at a depth of 5 ft (1.5 m). What is the flow rate?

SI Solution

The hydraulic radius is

$$R = \frac{A}{P} = \frac{(2.4 \text{ m})(1.5 \text{ m})}{1.5 \text{ m} + 2.4 \text{ m} + 1.5 \text{ m}}$$
$$= 0.667 \text{ m}$$

The roughness coefficient for the finished concrete is 0.012. [Manning's Values of Roughness Coefficient, n]

The Manning coefficient is determined from Eq. 15.13.

$$Q = vA = (K/n)AR_H^{2/3}S^{1/2}$$
$$= \left(\frac{1}{0.012}\right)(3.6 \text{ m}^2)(0.667 \text{ m})^{2/3}(0.002)^{1/2}$$
$$= 10.24 \text{ m}^3/\text{s}$$

Customary U.S. Solution

The hydraulic radius is

$$R = \frac{A}{P} = \frac{(8 \text{ ft})(5 \text{ ft})}{5 \text{ ft} + 8 \text{ ft} + 5 \text{ ft}}$$
$$= 2.22 \text{ ft}$$

The roughness coefficient for finished concrete is 0.012. [Manning's Values of Roughness Coefficient, n]

The Manning coefficient is determined from Eq. 15.13.

$$Q = vA = (K/n)AR_H^{2/3}S^{1/2}$$
$$= \left(\frac{1.49}{0.012}\right)(40 \text{ ft}^2)(2.22 \text{ ft})^{2/3}(0.002)^{1/2}$$
$$= 377.9 \text{ ft}^3/\text{sec}$$

7. VARIATIONS IN THE MANNING CONSTANT

The value of n also depends on the depth of flow, leading to a value (n_{full}) specifically intended for use with full flow. (It is seldom clear from tabulations whether the values are for full flow or general use.)

For most calculations, however, n is assumed to be constant. The accuracy of other parameters used in open-flow calculations often does not warrant considering the variation of n with depth, and the choice to use a constant or varying n-value is left to the engineer.

If it is desired to acknowledge variations in n with respect to depth, it is expedient to use tables or graphs of hydraulic elements prepared for that purpose. Table 15.2 lists such hydraulic elements under the assumption that n varies. A hydraulic elements graph for circular sewers can be used for both varying and constant n.

Table 15.2 *Circular Channel Ratios (varying n)*

$\dfrac{d}{D}$	$\dfrac{Q}{Q_{full}}$	$\dfrac{v}{v_{full}}$
0.1	0.02	0.31
0.2	0.07	0.48
0.3	0.14	0.61
0.4	0.26	0.71
0.5	0.41	0.80
0.6	0.56	0.88
0.7	0.72	0.95
0.8	0.87	1.01
0.9	0.99	1.04
0.95	1.02	1.03
1.00	1.00	1.00

Example 15.2

2.5 ft³/sec (0.07 m³/s) of water flows in a 20 in (0.5 m) diameter sewer line ($n = 0.015$, $S = 0.001$). The Manning coefficient, n, varies with depth. Flow is uniform and steady. What are the velocity and depth?

SI Solution

The hydraulic radius is

$$R = \frac{D}{4}$$
$$= \frac{0.5 \text{ m}}{4}$$
$$= 0.125 \text{ m}$$

From Eq. 15.12,

Manning's Equation

$$v = (K/n)R_H^{2/3}S^{1/2}$$
$$= \left(\frac{1}{0.015}\right)(0.125 \text{ m})^{2/3}\sqrt{0.001}$$
$$= 0.53 \text{ m/s}$$

If the pipe was flowing full, it would carry Q_{full}.

$$Q_{\text{full}} = v_{\text{full}} A$$
$$= \left(0.53 \ \frac{m}{s}\right)\left(\frac{\pi}{4}\right)(0.5 \ m)^2$$
$$= 0.10 \ m^3/s$$

$$\frac{Q}{Q_{\text{full}}} = \frac{0.07 \ \frac{m^3}{s}}{0.10 \ \frac{m^3}{s}} = 0.7$$

From a hydraulic elements graph for circular sewers, $d/D = 0.68$, and $v/v_{\text{full}} = 0.94$.

$$v = (0.94)\left(0.53 \ \frac{m}{s}\right) = 0.50 \ m/s$$
$$d = (0.68)(0.5 \ m) = 0.34 \ m$$

Customary U.S. Solution

The hydraulic radius is

$$R = \frac{D}{4} = \frac{\dfrac{20 \ in}{12 \ \frac{in}{ft}}}{4} = 0.417 \ ft$$

From Eq. 15.12,

Manning's Equation

$$v = (K/n)R_H^{2/3}S^{1/2}$$
$$= \left(\frac{1.486}{0.015}\right)(0.417 \ ft)^{2/3}\sqrt{0.001}$$
$$= 1.75 \ ft/sec$$

If the pipe was flowing full, it would carry Q_{full}.

$$Q_{\text{full}} = v_{\text{full}} A$$
$$= \left(1.75 \ \frac{ft}{sec}\right)\left(\frac{\pi}{4}\right)\left(\frac{20 \ in}{12 \ \frac{in}{ft}}\right)^2$$
$$= 3.82 \ ft^3/sec$$

$$\frac{Q}{Q_{\text{full}}} = \frac{2.5 \ \frac{ft^3}{sec}}{3.82 \ \frac{ft^3}{sec}} = 0.65$$

From a hydraulic elements graph for circular sewers, $d/D = 0.66$, and $v/v_{\text{full}} = 0.92$.

$$v = (0.92)\left(1.75 \ \frac{ft}{sec}\right) = 1.61 \ ft/sec$$
$$d = (0.66)(20 \ in) = 13.2 \ in$$

8. HAZEN-WILLIAMS VELOCITY

The empirical Hazen-Williams open channel velocity equation was developed in the early 1920s. It is still occasionally used in the United States for sizing gravity sewers. It is applicable to water flows at reasonably high Reynolds numbers and is based on sound dimensional analysis. However, the constants and exponents were developed experimentally.

The equation uses the Hazen-Williams coefficient, C, to characterize the roughness of the channel. Since the equation is used only for water within "normal" ambient conditions, the effects of temperature, pressure, and viscosity are disregarded. The primary advantage of this approach is that the coefficient, C, depends only on the roughness, not on the fluid characteristics. This is also the method's main disadvantage, since professional judgment is required in choosing the value of C. For SI units, $k_1 = 0.849$; for customary U.S. units, $k_1 = 1.318$.

Hazen-Williams Equation

$$V = k_1 C R_H^{0.63} S^{0.54} \qquad \textbf{\textit{15.14}}$$

9. NORMAL DEPTH

When the depth of flow is constant along the length of the channel (i.e., the depth is neither increasing nor decreasing), the flow is said to be *uniform*. The depth of flow in that case is known as the *normal depth*, y. If the normal depth is known, it can be compared with the actual depth of flow to determine if the flow is uniform.[6]

The difficulty with which the normal depth is calculated depends on the cross section of the channel. If the width is very large compared to the depth, the flow cross section will essentially be rectangular and the Manning equation can be used. (Equation 15.15 assumes that the hydraulic radius equals the normal depth.)

$$y = \left(\frac{nQ}{w\sqrt{S}}\right)^{3/5} \quad [w \gg d_n] \qquad \text{[SI]} \quad \textbf{\textit{15.15(a)}}$$

$$y = 0.788\left(\frac{nQ}{w\sqrt{S}}\right)^{3/5} \quad [w \gg d_n] \quad \text{[U.S.]} \quad \textbf{\textit{15.15(b)}}$$

---

[6]Normal depth is a term that applies only to uniform flow. The two alternate depths that can occur in nonuniform flow are not normal depths.

Normal depth in circular channels can be calculated directly only under limited conditions. If the circular channel is flowing full, the normal depth is the inside pipe diameter.

$$D = y = 1.548 \left(\frac{nQ}{\sqrt{S}} \right)^{3/8} \quad \text{[full]} \qquad \text{[SI]} \quad \textit{15.16(a)}$$

$$D = y = 1.335 \left(\frac{nQ}{\sqrt{S}} \right)^{3/8} \quad \text{[full]} \qquad \text{[U.S.]} \quad \textit{15.16(b)}$$

If a circular channel is flowing half full, the normal depth is half of the inside pipe diameter.

$$D = 2y = 2.008 \left(\frac{nQ}{\sqrt{S}} \right)^{3/8} \quad \text{[half full]} \qquad \text{[SI]} \quad \textit{15.17(a)}$$

$$D = 2y = 1.731 \left(\frac{nQ}{\sqrt{S}} \right)^{3/8} \quad \text{[half full]} \qquad \text{[U.S.]} \quad \textit{15.17(b)}$$

For other cases of uniform flow (trapezoidal, triangular, etc.), it is more difficult to determine normal depth. Various researchers have prepared tables and figures to assist in the calculations. For example, Table 15.2 can be used for circular channels flowing other than full or half full.

In the absence of tables or figures, trial-and-error solutions are required. The appropriate expressions for the flow area and hydraulic radius are used in the Manning equation. Trial values are used in conjunction with graphical techniques, linear interpolation, or extrapolation to determine the normal depth. The Manning equation is solved for flow rate with various assumed values of y. The calculated value is compared to the actual known flow quantity, and the normal depth is approached iteratively.

For a rectangular channel whose width is small compared to the depth, the hydraulic radius and area in flow are

Geometric Elements of Channel Sections

$$R = \frac{by}{b + 2y} \qquad \textit{15.18}$$

$$A = by \qquad \textit{15.19}$$

$$Q = \left(\frac{1}{n} \right) by \left(\frac{by}{b + 2y} \right)^{2/3} \sqrt{S} \quad \text{[rectangular]} \qquad \text{[SI]} \quad \textit{15.20(a)}$$

$$Q = \left(\frac{1.49}{n} \right) by \left(\frac{by}{b + 2y} \right)^{2/3} \sqrt{S} \quad \text{[rectangular]} \quad \text{[U.S.]} \quad \textit{15.20(b)}$$

For a trapezoidal channel with exposed surface width T, base width b, normal depth of flow y, and side slope 1:z (vertical:horizontal) the hydraulic radius and area in flow are

Geometric Elements of Channel Sections

$$A = (b + zy)y \qquad \textit{15.21}$$

$$R = \frac{(b + zy)y}{b + 2y\sqrt{1 + z^2}} \qquad \textit{15.22}$$

For a symmetrical triangular channel with exposed surface width T, side slope 1:z (vertical:horizontal), and normal depth of flow y, the hydraulic radius and area in flow are

Geometric Elements of Channel Sections

$$R = \frac{zy}{2\sqrt{1 + z^2}} \quad \text{[triangular]} \qquad \textit{15.23}$$

$$A = zy^2 \quad \text{[triangular]} \qquad \textit{15.24}$$

10. ENERGY AND FRICTION RELATIONSHIPS

Bernoulli's equation is an expression for the conservation of energy along a fluid streamline. The Bernoulli equation can also be written for two points along the bottom of an open channel.

Minor Losses in Pipe Fittings, Contractions, and Expansions

$$\frac{P_1}{\rho g} + z_1 + \frac{v_1^2}{2g} = \frac{P_2}{\rho g} + z_2 + \frac{v_2^2}{2g} + h_f + h_{f,\text{fitting}} \quad \textit{15.25}$$

$$\frac{P_1}{\gamma} + z_1 + \frac{v_1^2}{2g} + = \frac{P_2}{\gamma} + z_2 + \frac{v_2^2}{2g} + h_f + h_{f,\text{fitting}} \quad \textit{15.26}$$

However, $P/\rho g = P/\gamma = d$.

$$d_1 + \frac{v_1^2}{2g} + z_1 = d_2 + \frac{v_2^2}{2g} + z_2 + h_f \qquad \textit{15.27}$$

And, since $d_1 = d_2$ and $v_1 = v_2$ for uniform flow at the bottom of a channel,

$$h_f = z_1 - z_2 \qquad \textit{15.28}$$

$$S_0 = \frac{z_1 - z_2}{L} \qquad \textit{15.29}$$

The channel slope, S_0, and the hydraulic energy gradient, S, are numerically the same for uniform flow. Therefore, the total friction loss along a channel is

$$h_f = LS \qquad \textit{15.30}$$

Combining Eq. 15.30 with the Manning equation results in a method for calculating friction loss. (See Eq. 15.12.)

$$h_f = \frac{Ln^2 v^2}{R^{4/3}} \qquad \text{[SI]} \quad 15.31(a)$$

$$h_f = \frac{Ln^2 v^2}{2.208 R^{4/3}} \qquad \text{[U.S.]} \quad 15.31(b)$$

Example 15.3

The velocities upstream and downstream, v_1 and v_2, of a 12 ft (4.0 m) wide sluice gate are both unknown. The upstream and downstream depths are 6 ft (2.0 m) and 2 ft (0.6 m), respectively. Flow is uniform and steady. What is the downstream velocity, v_2?

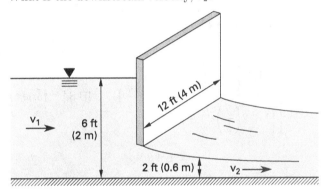

SI Solution

Since the channel bottom is essentially level on either side of the gate, $z_1 = z_2$. Bernoulli's equation reduces to

$$d_1 + \frac{v_1^2}{2g} = d_2 + \frac{v_2^2}{2g}$$

$$2 \text{ m} + \frac{v_1^2}{2g} = 0.6 \text{ m} + \frac{v_2^2}{2g}$$

v_1 and v_2 are related by continuity.

$$Q_1 = Q_2$$
$$A_1 v_1 = A_2 v_2$$
$$(2 \text{ m})(4 \text{ m}) v_1 = (0.6 \text{ m})(4 \text{ m}) v_2$$
$$v_1 = 0.3 v_2$$

Substituting the expression for v_1 into the Bernoulli equation gives

$$2 \text{ m} + \frac{(0.3 v_2)^2}{(2)\left(9.81 \ \dfrac{\text{m}}{\text{s}^2}\right)} = 0.6 \text{ m} + \frac{v_2^2}{(2)\left(9.81 \ \dfrac{\text{m}}{\text{s}^2}\right)}$$

$$2 \text{ m} + 0.004587 v_2^2 = 0.6 \text{ m} + 0.050968 v_2^2$$

$$v_2 = 5.5 \text{ m/s}$$

Customary U.S. Solution

Since the channel bottom is essentially level on either side of the gate, $z_1 = z_2$. Bernoulli's equation reduces to

$$d_1 + \frac{v_1^2}{2g} = d_2 + \frac{v_2^2}{2g}$$

$$6 \text{ ft} + \frac{v_1^2}{2g} = 2 \text{ ft} + \frac{v_2^2}{2g}$$

v_1 and v_2 are related by continuity.

$$Q_1 = Q_2$$
$$A_1 v_1 = A_2 v_2$$
$$(6 \text{ ft})(12 \text{ ft}) v_1 = (2 \text{ ft})(12 \text{ ft}) v_2$$
$$v_1 = \frac{v_2}{3}$$

Substituting the expression for v_1 into the Bernoulli equation gives

$$6 \text{ ft} + \frac{v_2^2}{(3)^2 (2)\left(32.2 \ \dfrac{\text{ft}}{\text{sec}^2}\right)} = 2 \text{ ft} + \frac{v_2^2}{(2)\left(32.2 \ \dfrac{\text{ft}}{\text{sec}^2}\right)}$$

$$6 \text{ ft} + 0.00173 v_2^2 = 2 \text{ ft} + 0.0155 v_2^2$$

$$v_2 = 17.0 \text{ ft/sec}$$

Example 15.4

In Ex. 15.1, the open channel experiencing normal flow had the following characteristics: $S = 0.002$, $n = 0.012$, $v = 9.447$ ft/sec (2.9 m/s), and $R = 2.22$ ft (0.68 m). What is the energy loss per 1000 ft (100 m)?

SI Solution

There are two methods for finding the energy loss. From Eq. 15.30,

$$h_f = LS = (100 \text{ m})(0.002)$$
$$= 0.2 \text{ m}$$

From the Darcy-Weisbach equation,

Head Loss Due to Flow

$$h_f = f \frac{L}{D} \frac{\mathrm{v}^2}{2g}$$

$$= (0.006) \left(\frac{100 \ \mathrm{m}}{(2)(0.68 \ \mathrm{m})} \right) \left(\frac{\left(2.9 \ \dfrac{\mathrm{m}}{\mathrm{s}} \right)^2}{(2) \left(9.8 \ \dfrac{\mathrm{m}}{\mathrm{s}^2} \right)} \right)$$

$$= 0.2 \ \mathrm{m}$$

Customary U.S. Solution

There are two methods for finding the energy loss. From Eq. 15.30,

$$h_f = LS = (1000 \ \mathrm{ft})(0.002)$$
$$= 2 \ \mathrm{ft}$$

From the Darcy-Weisbach equation,

Head Loss Due to Flow

$$h_f = f \frac{L}{D} \frac{\mathrm{v}^2}{2g}$$

$$= (0.006) \left(\frac{1000 \ \mathrm{ft}}{(2)(2.22 \ \mathrm{ft})} \right) \left(\frac{\left(9.447 \ \dfrac{\mathrm{ft}}{\mathrm{sec}} \right)^2}{(2) \left(32.2 \ \dfrac{\mathrm{ft}}{\mathrm{sec}} \right)} \right)$$

$$= 2 \ \mathrm{ft}$$

11. SIZING TRAPEZOIDAL AND RECTANGULAR CHANNELS

Trapezoidal and rectangular cross sections are commonly used for artificial surface channels. The flow through a trapezoidal channel is easily determined from the Manning equation when the cross section is known. However, when the cross section or uniform depth is unknown, a trial-and-error solution is required. (See Fig. 15.2.)

Figure 15.2 *Trapezoidal Cross Section*

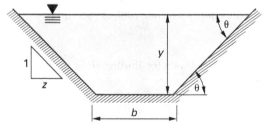

For such problems involving rectangular and trapezoidal channels, it is common to calculate and plot the *conveyance*, K (or alternatively, the product Kn), against depth. For trapezoidal sections, it is particularly convenient to write the uniform flow, Q, in terms of a modified conveyance, K'. b is the base width of the channel, d is the depth of flow, and m is the cotangent of the side slope angle. m and the ratio d/b are treated as independent variables.

$$Q = \frac{K' b^{8/3} \sqrt{S_0}}{n} \qquad \text{15.32}$$

$$K' = \left(\frac{\left(1 + z \left(\dfrac{y}{b} \right) \right)^{5/3}}{\left(1 + 2 \left(\dfrac{d}{b} \right) \sqrt{1 + z^2} \right)^{2/3}} \right) \left(\frac{y}{b} \right)^{5/3} \qquad \text{[SI]} \quad \text{15.33(a)}$$

$$K' = \left(\frac{1.49 \left(1 + z \left(\dfrac{y}{b} \right) \right)^{5/3}}{\left(1 + 2 \left(\dfrac{y}{b} \right) \sqrt{1 + z^2} \right)^{2/3}} \right) \left(\frac{y}{b} \right)^{5/3} \qquad \text{[U.S.]} \quad \text{15.33(b)}$$

$$z = \cot \theta \qquad \text{15.34}$$

For any fixed value of z, enough values of K' are calculated over a reasonable range of the y/b ratio ($0.05 < y/b < 0.5$) to define a curve. Given specific values of Q, n, S_0, and b, the value of K' can be calculated from the expression for Q. The graph is used to determine the ratio y/b, giving the depth of uniform flow, y, since b is known.

When the ratio of d/b is very small (less than 0.02), it is satisfactory to consider the trapezoidal channel as a wide rectangular channel with area $A = by$.

12. MOST EFFICIENT CROSS SECTION

The most efficient open channel cross section will maximize the flow for a given Manning coefficient, slope, and flow area. Accordingly, the Manning equation requires that the hydraulic radius be maximum. For a given flow area, the wetted perimeter will be minimum.

Semicircular cross sections have the smallest wetted perimeter; therefore, the cross section with the highest efficiency is the semicircle. Although such a shape can be constructed with concrete, it cannot be used with earth channels.

The most efficient cross section is also generally assumed to minimize construction cost. This is true only in the most simplified cases, however, since the labor and material costs of excavation and formwork must be considered. Rectangular and trapezoidal channels are much easier to form than semicircular channels.

So in this sense the "least efficient" (i.e., most expensive) cross section (i.e., semicircular) is also the "most efficient." (See Fig. 15.3.)

Figure 15.3 Circles Inscribed in Efficient Channels

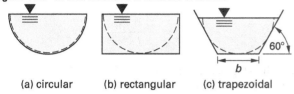

(a) circular   (b) rectangular   (c) trapezoidal

The most efficient rectangle is one having depth equal to one-half of the width (i.e., is one-half of a square).

$$y = \frac{b}{2} \quad \text{[most efficient rectangle]} \qquad 15.35$$

$$A = yb = \frac{b^2}{2} = 2y^2 \qquad 15.36$$

$$P = y + b + y = 2b = 4y \qquad 15.37$$

$$R = \frac{b}{4} = \frac{y}{2} \qquad 15.38$$

The most efficient trapezoidal channel is always one in which the flow depth is twice the hydraulic radius. If the side slope is adjustable, the sides of the most efficient trapezoid should be inclined at 60° from the horizontal. Since the surface width will be equal to twice the sloping side length, the most efficient trapezoidal channel will be half of a regular hexagon (i.e., three adjacent equilateral triangles of side length $2y/\sqrt{3}$). If the side slope is any other angle, only the $y = 2R$ criterion is applicable.

$$y = 2R \quad \text{[most efficient trapezoid]} \qquad 15.39$$

$$b = \frac{2y}{\sqrt{3}} \qquad 15.40$$

$$A = \sqrt{3}\, y^2 \qquad 15.41$$

$$P = 3b = 2\sqrt{3}\, d \quad \text{[most efficient trapezoid]} \qquad 15.42$$

$$R = \frac{d}{2} \qquad 15.43$$

A semicircle with its center at the middle of the water surface can always be inscribed in a cross section with maximum efficiency.

Example 15.5

A rubble masonry open channel is being designed to carry 500 ft³/sec (14 m³/s) of water on a 0.0001 slope. Using $n = 0.017$, find the most efficient dimensions for a rectangular channel.

SI Solution

Let the depth and width be y and b, respectively. For an efficient rectangle, $y = b/2$.

$$A = yw = \left(\frac{b}{2}\right)b = \frac{b^2}{2}$$

$$P = y + b + y = \frac{b}{2} + b + \frac{b}{2} = 2b$$

$$R = \frac{A}{P} = \frac{\dfrac{b^2}{2}}{2b} = \frac{b}{4}$$

Using Eq. 15.13,

$$Q = \left(\frac{1}{n}\right)AR^{2/3}\sqrt{S}$$

$$14\ \frac{\text{m}^3}{\text{s}} = \left(\frac{1}{0.017}\right)\left(\frac{b^2}{2}\right)\left(\frac{b}{4}\right)^{2/3}\sqrt{0.0001}$$

$$14\ \frac{\text{m}^3}{\text{s}} = 0.1167 b^{8/3}$$

$$b = 6.02\ \text{m}$$

$$d = \frac{b}{2} = \frac{6.02\ \text{m}}{2} = 3.01\ \text{m}$$

Customary U.S. Solution

Let the depth and width be d and w, respectively. For an efficient rectangle, $d = w/2$.

$$A = yb = \left(\frac{b}{2}\right)b = \frac{b^2}{2}$$

$$P = y + b + y = \frac{b}{2} + b + \frac{b}{2} = 2b$$

$$R = \frac{A}{P} = \frac{\dfrac{b^2}{2}}{2b} = \frac{b}{4}$$

Using Eq. 15.13,

$$Q = \left(\frac{1.49}{n}\right)AR^{2/3}\sqrt{S}$$

$$500\ \frac{\text{ft}^3}{\text{sec}} = \left(\frac{1.49}{0.017}\right)\left(\frac{b^2}{2}\right)\left(\frac{b}{4}\right)^{2/3}\sqrt{0.0001}$$

$$500\ \frac{\text{ft}^3}{\text{sec}} = 0.1739 w^{8/3}$$

$$b = 19.82\ \text{ft}$$

$$y = \frac{b}{2} = \frac{19.82\ \text{ft}}{2} = 9.91\ \text{ft}$$

13. ANALYSIS OF NATURAL WATERCOURSES

Natural watercourses do not have uniform paths or cross sections. This complicates their analysis considerably. Frequently, analyzing the flow from a river is a matter of making the most logical assumptions. Many evaluations can be solved with a reasonable amount of error.

As was seen in Eq. 15.31, the friction loss (and hence the hydraulic gradient) depends on the square of the roughness coefficient. Therefore, an attempt must be made to evaluate the roughness constant as accurately as possible. If the channel consists of a river with flood plains (as in Fig. 15.4), it should be treated as parallel channels. The flow from each subdivision can be calculated independently and the separate values added to obtain the total flow. (The common interface between adjacent subdivisions is not included in the wetted perimeter.) Alternatively, a composite value of the roughness coefficient, n_c, can be approximated from the *Horton-Einstein equation* using the individual values of n and the corresponding wetted perimeters. Equation 15.44 assumes the flow velocities and lengths are the same in all cross sections.

$$n_c = \left(\frac{\sum P_i n_i^{3/2}}{\sum P_i} \right)^{2/3} \hspace{2cm} 15.44$$

Figure 15.4 *River with Flood Plain*

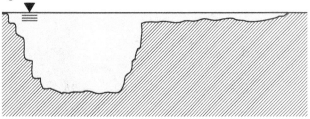

If the channel is divided (as in Fig. 15.5) by an island into two channels, some combination of flows will usually be known. For example, if the total flow, Q, is known, Q_1 and Q_2 may be unknown. If the slope is known, Q_1 and Q_2 may be known. Iterative trial-and-error solutions are often required.

Figure 15.5 *Divided Channel*

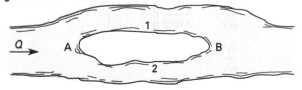

Since the elevation drop $(z_A - z_B)$ between points A and B is the same regardless of flow path,

$$S_1 = \frac{z_A - z_B}{L_1} \hspace{2cm} 15.45$$

$$S_2 = \frac{z_A - z_B}{L_2} \hspace{2cm} 15.46$$

Once the slopes are known, initial estimates Q_1 and Q_2 can be calculated from Eq. 15.13. The sum of Q_1 and Q_2 will probably not be the same as the given flow quantity, Q. In that case, Q should be prorated according to the ratios of Q_1 and Q_2 to $Q_1 + Q_2$.

If the lengths L_1 and L_2 are the same or almost so, the Manning equation may be solved for the slope by writing Eq. 15.47.

$$\begin{aligned} Q &= Q_1 + Q_2 \\ &= \left(\left(\frac{A_1}{n_1} \right) R_1^{2/3} + \left(\frac{A_2}{n_2} \right) R_2^{2/3} \right) \sqrt{S} \end{aligned} \hspace{0.5cm} \text{[SI]} \hspace{0.5cm} 15.47(a)$$

$$\begin{aligned} Q &= Q_1 + Q_2 \\ &= 1.49 \left(\left(\frac{A_1}{n_1} \right) R_1^{2/3} + \left(\frac{A_2}{n_2} \right) R_2^{2/3} \right) \sqrt{S} \end{aligned} \hspace{0.3cm} \text{[U.S.]} \hspace{0.3cm} 15.47(b)$$

Equation 15.47 yields only a rough estimate of the flow quantity, as the geometry and roughness of a natural channel changes considerably along its course.

14. FLOW MEASUREMENT WITH WEIRS

A *weir* is an obstruction in an open channel over which flow occurs. Although a dam spillway is a specific type of weir, most weirs are intended specifically for flow measurement.

Measurement weirs consist of a vertical flat plate with sharp edges. Because of their construction, they are called *sharp-crested weirs*. Sharp-crested weirs are most frequently rectangular, consisting of a straight, horizontal crest. However, weirs may also have trapezoidal and triangular openings.

For any given width of weir opening (referred to as the *weir length*), the discharge will be a function of the head over the weir. The head (or sometimes surface elevation) can be determined by a standard *staff gauge* mounted adjacent to the weir.

The full channel flow usually goes over the weir. However, it is also possible to divert a small portion of the total flow through a measurement channel. The full channel flow rate can be extrapolated from a knowledge of the split fractions.

If a rectangular weir is constructed with an opening width less than the channel width, the falling liquid sheet (called the *nappe*) decreases in width as it falls.

Because of this *contraction* of the nappe, these weirs are known as *contracted weirs*, although it is the nappe that is actually contracted, not the weir. If the opening of the weir extends the full channel width, the weir is known as a *suppressed weir*, since the contractions are suppressed. (See Fig. 15.6.)

Figure 15.6 *Contracted and Suppressed Weirs*

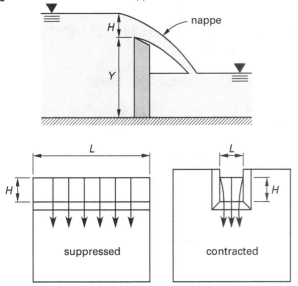

The derivation of an expression for the quantity flowing over a weir is dependent on many simplifying assumptions. The basic weir equation (see Eq. 15.48 or Eq. 15.49) is, therefore, an approximate result requiring correction by experimental coefficients.

If it is assumed that the contractions are suppressed, upstream velocity is uniform, flow is laminar over the crest, nappe pressure is zero, the nappe is fully ventilated, and viscosity, turbulence, and surface tension effects are negligible, then the following equation may be derived from the Bernoulli equation.

$$Q = \frac{2}{3} L \sqrt{2g} \left[\left(H + \frac{v^2}{2g} \right)^{3/2} - \left(\frac{v^2}{2g} \right)^{3/2} \right] \quad \text{15.48}$$

If the velocity of approach, v_1, is negligible, then

$$Q = \frac{2}{3} L \sqrt{2g}\, H^{3/2} \quad \text{15.49}$$

Equation 15.49 must be corrected for all of the assumptions made, primarily for a nonuniform velocity distribution. This is done by introducing an empirical discharge coefficient, C. Equation 15.50 is known as the *Francis weir equation*.

Weir Formulas
$$Q = CLH^{3/2} \quad \text{15.50}$$

For a free discharge contracted weir, Eq. 15.51 is used.

Weir Formulas
$$Q = C(L - 0.2H)H^{3/2} \quad \text{15.51}$$

Many investigations have been done to evaluate C_1 analytically. Perhaps the most widely known is the coefficient formula developed by *Rehbock*.[7]

$$C = \frac{2}{3} \sqrt{2g} \left(0.6035 + 0.0813 \left(\frac{H}{Y} \right) + \frac{0.000295}{Y} \right)$$
$$\times \left(1 + \frac{0.00361}{H} \right)^{3/2} \quad \text{15.52}$$

$$C = \frac{2}{3} \sqrt{2g} \left(0.602 + 0.083 \frac{H}{Y} \right) \quad \text{15.53}$$

When $H/Y < 0.2$, C approaches 0.61–0.62. In most cases, a value in this range is adequate. Other constants (i.e., $\frac{2}{3}$ and $\sqrt{2g}$) can be taken out of Eq. 15.50. In that case,

$$Q \approx 1.84 LH^{3/2} \quad \text{[SI]} \quad \text{15.54(a)}$$

$$Q \approx 3.33 LH^{3/2} \quad \text{[U.S.]} \quad \text{15.54(b)}$$

If the contractions are not suppressed (i.e., one or both sides do not extend to the channel sides), then the actual width, L, should be replaced with the *effective width*. In Eq. 15.55, N is 1 if one side is contracted and N is 2 if there are two end contractions.

$$L_{\text{effective}} = L_{\text{actual}} - 0.1NH \quad \text{15.55}$$

A *submerged rectangular weir* requires a more complex analysis because of the difficulty in measuring H and because the discharge depends on both the upstream and downstream depths. (See Fig. 15.7.) The following equation, however, may be used with little difficulty.

$$Q_{\text{submerged}} = Q_{\text{free flow}} \left[1 - \left(\frac{H_{\text{downstream}}}{H_{\text{upstream}}} \right)^{3/2} \right]^{0.385} \quad \text{15.56}$$

Equation 15.56 is used by first finding the flow rate, Q, from Eq. 15.50 and then correcting it with the bracketed quantity.

---

[7]There is much variation in how different investigators calculate the discharge coefficient, C_1. For ratios of H/b less than 5, $C_1 = 0.622$ gives a reasonable value. With the questionable accuracy of some of the other variables used in open channel flow problems, the pursuit of greater accuracy is of dubious value.

Figure 15.7 *Submerged Weir*

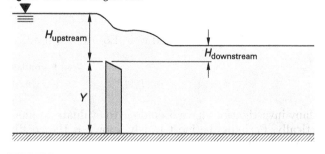

Example 15.6

The crest of a sharp-crested, rectangular weir with two contractions is 2.5 ft (1.0 m) high above the channel bottom. The crest is 4 ft (1.6 m) long. A 4 in (100 mm) head exists over the weir. What is the velocity of approach?

SI Solution

$$H = \frac{100 \text{ mm}}{1000 \ \frac{\text{mm}}{\text{m}}} = 0.1 \text{ m}$$

The number of contractions, N, is 2. From Eq. 15.55, the effective width is

$$L_{\text{effective}} = L_{\text{actual}} - 0.1NH = 1.6 \text{ m} - (0.1)(2)(0.1 \text{ m})$$
$$= 1.58 \text{ m}$$

From Eq. 15.53, the Rehbock coefficient is

$$C = \frac{2}{3}\sqrt{2g}\left(0.602 + 0.083\frac{H}{Y}\right)$$
$$= 0.602 + (0.083)\left(\frac{0.1}{1}\right)$$
$$= 0.61$$

From Eq. 15.50, the flow is

Weir Formulas

$$Q = CLH^{3/2}$$
$$= \left(\frac{2}{3}\right)(0.61)(1.58 \text{ m})\sqrt{(2)\left(9.81 \ \frac{\text{m}}{\text{s}^2}\right)}(0.10 \text{ m})^{3/2}$$
$$= 0.090 \text{ m}^3/\text{s}$$

$$v = \frac{Q}{A} = \frac{0.090 \ \frac{\text{m}^3}{\text{s}}}{(1.6 \text{ m})(1.0 \text{ m} + 0.1 \text{ m})}$$
$$= 0.051 \text{ m/s}$$

Customary U.S. Solution

$$H = \frac{4 \text{ in}}{12 \ \frac{\text{in}}{\text{ft}}} = 0.333 \text{ ft}$$

The number of contractions, N, is 2. From Eq. 15.55, the effective width is

$$b_{\text{effective}} = b_{\text{actual}} - 0.1NH = 4 \text{ ft} - (0.1)(2)(0.333 \text{ ft})$$
$$= 3.93 \text{ ft}$$

From Eq. 15.52, the Rehbock coefficient is

$$C = \frac{2}{3}\sqrt{2g}\left(0.6035 + 0.0813\left(\frac{H}{Y}\right) + \frac{0.000295}{Y}\right)$$
$$\times \left(1 + \frac{0.00361}{H}\right)^{3/2}$$
$$= \left(0.6035 + (0.0813)\left(\frac{0.333 \text{ ft}}{2.5 \text{ ft}}\right) + \frac{0.000295}{2.5 \text{ ft}}\right)$$
$$\times \left(1 + \frac{0.00361}{0.333 \text{ ft}}\right)^{3/2}$$
$$= 0.624$$

From Eq. 15.50, the flow is

Weir Formulas

$$Q = CLH^{3/2}$$
$$= \left(\frac{2}{3}\right)(0.624)(3.93 \text{ ft})\sqrt{(2)\left(32.2 \ \frac{\text{ft}}{\text{sec}^2}\right)}(0.333 \text{ ft})^{3/2}$$
$$= 2.52 \text{ ft}^3/\text{sec}$$

$$v = \frac{Q}{A} = \frac{2.52 \ \frac{\text{ft}^3}{\text{sec}}}{(4 \text{ ft})(2.5 \text{ ft} + 0.333 \text{ ft})}$$
$$= 0.222 \text{ ft/sec}$$

15. TRIANGULAR WEIRS

Triangular weirs (V-notch weirs) should be used when small flow rates are to be measured. The flow coefficient over a triangular weir depends on the notch angle, θ, but generally varies from 0.58 to 0.61. For a 90° weir, $C = 1.40$ in SI units and 2.54 for customary U.S. units. (See Fig. 15.8.)

$$Q = C\left(\frac{8}{15}\tan\frac{\theta}{2}\right)\sqrt{2g}\,H^{5/2} \qquad \textbf{15.57}$$

Weir Formulas

$$Q = CH^{5/2} \qquad \textbf{15.58}$$

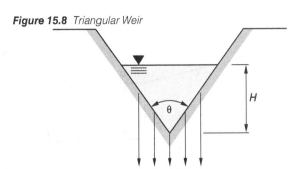

Figure 15.8 *Triangular Weir*

16. TRAPEZOIDAL WEIRS

A *trapezoidal weir* is essentially a rectangular weir with a triangular weir on either side. (See Fig. 15.9.) If the angle of the sides from the vertical is approximately 14° (i.e., 4 vertical and 1 horizontal), the weir is known as a *Cipoletti weir*. The discharge from the triangular ends of a Cipoletti weir approximately make up for the contractions that would reduce the flow over a rectangular weir. Therefore, no correction is theoretically necessary. This is not completely accurate, and for this reason, Cipoletti weirs are not used where great accuracy is required. The discharge is

$$Q = \tfrac{2}{3} C_d b \sqrt{2g}\, H^{3/2} \qquad \text{15.59}$$

The average value of the discharge coefficient is 0.63. The discharge from a Cipoletti weir is found by using Eq. 15.60.

$$Q = 1.86 b H^{3/2} \qquad \text{[SI]} \quad \text{15.60(a)}$$

$$Q = 3.367 b H^{3/2} \qquad \text{[U.S.]} \quad \text{15.60(b)}$$

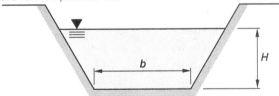

Figure 15.9 *Trapezoidal Weir*

17. BROAD-CRESTED WEIRS AND SPILLWAYS

Most weirs used for flow measurement are sharp-crested. However, the flow over spillways, broad-crested weirs, and similar features can be calculated using Eq. 15.50 even though flow measurement is not the primary function of the feature. (A weir is broad-crested if the weir thickness is greater than half of the head, H.)

A *dam's spillway (overflow spillway)* is designed for a capacity based on the dam's inflow hydrograph, turbine capacity, and storage capacity. Spillways frequently have a cross section known as an *ogee*, which closely approximates the underside of a nappe from a sharp-crested weir. This cross section minimizes the cavitation that is likely to occur if the water surface breaks contact with the spillway due to upstream heads that are higher than designed for.[8]

Discharge from an overflow spillway is derived in the same manner as for a weir. Equation 15.61 can be used for broad-crested weirs ($C_1 = 0.5$–0.57) and ogee spillways ($C_1 = 0.60$–0.75).

$$Q = \tfrac{2}{3} C_1 b \sqrt{2g}\, H^{3/2} \qquad \text{15.61}$$

The *Horton equation* for broad-crested weirs combines all of the coefficients into a spillway (weir) coefficient and adds the velocity of approach to the upstream head. (See Eq. 15.62.) The *Horton coefficient*, C_s, is specific to the Horton equation. (C_s and C_1 differ by a factor of about 5 and cannot easily be mistaken for each other.)

$$Q = C_s b \left(H + \frac{\mathrm{v}^2}{2g} \right)^{3/2} \qquad \text{15.62}$$

If the velocity of approach is insignificant, the discharge is found using Eq. 15.63.

$$Q = C_s b H^{3/2} \qquad \text{15.63}$$

C_s is a *spillway coefficient*, which varies from about 3.3 ft$^{1/2}$/sec to 3.98 ft$^{1/2}$/sec (1.8 m$^{1/2}$/s to 2.2 m$^{1/2}$/s) for ogee spillways. 3.97 ft$^{1/2}$/sec (2.2 m$^{1/2}$/s) is frequently used for first approximations. For broad-crested weirs, C_s varies between 2.63 ft$^{1/2}$/sec and 3.33 ft$^{1/2}$/sec (1.45 m$^{1/2}$/s and 1.84 m$^{1/2}$/s). (Use 3.33 ft$^{1/2}$/sec (1.84 m$^{1/2}$/sec) for initial estimates.) C_s increases as the upstream design head above the spillway top, H, increases, and the larger values apply to the higher heads.

Broad-crested weirs and spillways can be calibrated to obtain greater accuracy in predicting flow rates.

Scour protection is usually needed at the toe of a spillway to protect the area exposed to a hydraulic jump. This protection usually takes the form of an extended horizontal or sloping apron. Other measures, however, are needed if the tailwater exhibits large variations in depth.

---

[8]Cavitation and separation will not normally occur as long as the actual head, H, is less than twice the design value. The shape of the ogee spillway will be a function of the design head.

18. PROPORTIONAL WEIRS

The *proportional weir (Sutro weir)* is used in water level control because it demonstrates a linear relationship between Q and H. Figure 15.10 illustrates a proportional weir whose sides are hyperbolic in shape.

$$Q = C_d K \left(\frac{\pi}{2}\right) \sqrt{2g}\, H \qquad 15.64$$

$$K = 2x \sqrt{y} \qquad 15.65$$

Figure 15.10 *Proportional Weir*

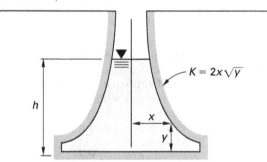

19. FLOW MEASUREMENT WITH PARSHALL FLUMES

The Parshall flume is widely used for measuring open channel wastewater flows. It performs well when head losses must be kept to a minimum and when there are high amounts of suspended solids. (See Fig. 15.11.)

The Parshall flume is constructed with a converging upstream section, a throat, and a diverging downstream section. The walls of the flume are vertical, but the floor of the throat section drops. The length, width, and height of the flume are essentially predefined by the anticipated flow rate.[9]

The throat geometry in a Parshall flume has been designed to force the occurrence of critical flow (described in Sec. 15.25) at that point. Following the critical section is a short length of supercritical flow followed by a hydraulic jump. (See Sec. 15.33.) This design eliminates any dead water region where debris and silt can accumulate (as are common with flat-topped weirs).

The discharge relationship for a wide Parshall flume is given for submergence ratios of H_b/H_a up to 0.7. Above 0.7, the true discharge is less than predicted by Eq. 15.66. Values of K are given in Table 15.3, although using a value of 4.0 is accurate for most purposes.

Parshall (Venturi) Flume

$$Q = 4WH_a^{1.522\,W^{0.026}} \qquad 15.66$$

Figure 15.11 *Parshall Flume*

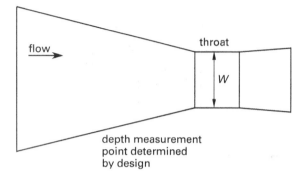

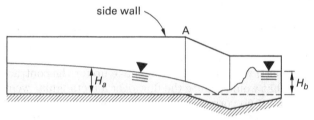

Above a certain tailwater height, the Parshall flume no longer operates in the *free-flow mode*. Rather, it operates in a *submerged mode*. A very high tailwater reduces the flow rate through the flume. Equation 15.66 predicts the flow rate with reasonable accuracy, however, even for 50–80% submergence (calculated as H_b/H_a). For large submergence, the tailwater height must be known and a different analysis method must be used.

Table 15.3 *Parshall Flume K-Values*

b (ft (m))	K
0.25 (0.076)	3.97
0.50 (0.15)	4.12
0.75 (0.229)	4.09
1.0 (0.305)	4.00
1.5 (0.46)	4.00
2.0 (0.61)	4.00
3.0 (0.92)	4.00
4.0 (1.22)	4.00

(Multiply ft by 0.3048 to obtain m.)

20. UNIFORM AND NONUNIFORM STEADY FLOW

Steady flow is constant-volume flow. However, the flow may be uniform or nonuniform (varied) in depth. There may be significant variations over long and short distances without any change in the flow rate.

---

[9]This chapter does not attempt to design the Parshall flume, only to predict flow rates through its use.

Figure 15.12 illustrates the three definitions of "slope" existing for open channel flow. These three slopes are the slope of the channel bottom, the slope of the water surface, and the slope of the energy gradient line.

Figure 15.12 *Slopes Used in Open Channel Flow*

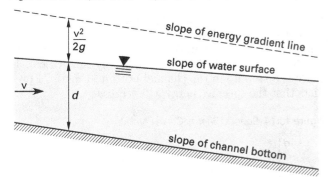

Under conditions of uniform flow, all of these three slopes are equal since the flow quantity and flow depth are constant along the length of flow.[10] With nonuniform flow, however, the flow velocity and depth vary along the length of channel and the three slopes are not necessarily equal.

If water is introduced down a path with a steep slope (as after flowing over a spillway), the effect of gravity will cause the velocity to increase. As the velocity increases, the depth decreases in accordance with the continuity of flow equation. The downward velocity is opposed by friction. Because the gravitational force is constant but friction varies with the square of velocity, these two forces eventually become equal. When equal, the velocity stops increasing, the depth stops decreasing, and the flow becomes uniform. Until they become equal, however, the flow is nonuniform (varied).

21. SPECIFIC ENERGY

The total head possessed by a fluid is given by the Bernoulli equation.

Energy Equation (Bernoulli)

$$H = Z_a + \frac{P_a}{w} + \frac{V_a^2}{2g}$$ 15.67

Specific energy is defined as the total head with respect to the channel bottom. In this case, $Z = 0$ and $P/W = d$.

$$H = d + \frac{V_a^2}{2g}$$ 15.68

Equation 15.68 is not meant to imply that the potential energy is an unimportant factor in open channel flow problems. The concept of specific energy is used for convenience only, and it should be clear that the Bernoulli equation is still the valid energy conservation equation.

In uniform flow, total head also decreases due to the frictional effects, but specific energy is constant. In nonuniform flow, total head also decreases, but specific energy may increase or decrease.

Since $v = Q/A$, Eq. 15.68 can be written as

$$H = d + \frac{Q^2}{2gA^2}$$ 15.69

For a rectangular channel, the velocity can be written in terms of the width and flow depth.

$$v = \frac{Q}{A} = \frac{Q}{wd}$$ 15.70

The specific energy equation for a rectangular channel is given by Eq. 15.71 and shown in Fig. 15.13.

$$H = d + \frac{Q^2}{2g(wd)^2}$$ 15.71

Specific energy can be used to differentiate between flow regimes. Figure 15.13 illustrates how specific energy is affected by depth and, accordingly, how specific energy relates to critical depth (described in Sec. 15.25) and the Froude number (described in Sec. 15.27).

Figure 15.13 *Specific Energy Diagram*

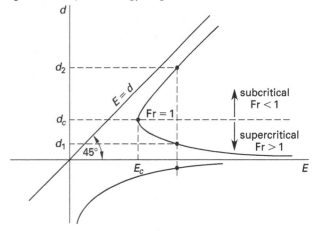

22. SPECIFIC FORCE

The *specific force* of a general channel section is the total force per unit weight acting on the water. Equivalently, specific force is the total force that a submerged

---

[10]As a simplification, this chapter deals only with channels of constant width. If the width is varied, changes in flow depth may not coincide with changes in flow quantity.

object would experience. In Eq. 15.72, y is the distance from the free surface to the centroid of the flowing area cross section, A.

Open-Channel Flow and Pipe Flow of Water

$$E = \alpha \frac{V^2}{2g} + y = \frac{\alpha Q^2}{2gA^2} + y \qquad \textbf{15.72}$$

Graphs of specific force and specific energy are similar in appearance and predict equivalent results for the critical and alternate depths. (See Sec. 15.24.)

23. CHANNEL TRANSITIONS

Sudden changes in channel width or bottom elevation are known as *channel transitions*. (Contractions in width are not covered in this chapter.) For sudden vertical steps in channel bottom, the Bernoulli equation, written in terms of the specific energy, is used to predict the flow behavior (i.e., the depth).

$$E_1 + Z_1 = E_2 + Z_2 \qquad \textbf{15.73}$$

$$E_1 - E_2 = Z_2 - Z_1 \qquad \textbf{15.74}$$

The maximum possible change in bottom elevation without affecting the energy equality occurs when the depth of flow over the step is equal to the critical depth (d_2 equals d_c). (See Sec. 15.25.)

24. ALTERNATE DEPTHS

Since the area depends on the depth, fixing the channel shape and slope and assuming a depth will determine the flow rate, Q, as well as the specific energy. Since Eq. 15.71 is a cubic equation, there are three values of depth of flow, d, that will satisfy it. One of them is negative, as Fig. 15.13 shows. Since depth cannot be negative, that value can be discarded. The two remaining values are known as *alternate depths*.

For a given flow rate, the two alternate depths have the same energy. One represents a high velocity with low depth; the other represents a low velocity with high depth. The former is called *supercritical (rapid) flow*; the latter is called *subcritical (tranquil) flow*.

If the flow depth at supercritical flow is designated y_1 and that at subcritical flow is designated y_2, the relationships between the two depths is given by Eq. 15.75.

Specific Energy Diagram

$$y_2 = \frac{y_1}{2} \left(\sqrt{1 + 8Fr_1^2} - 1 \right) \qquad \textbf{15.75}$$

The Bernoulli equation cannot predict which of the two alternate depths will occur for any given flow quantity. The concept of *accessibility* is required to evaluate the two depths. Specifically, the upper and lower limbs of the energy curve are not accessible from each other unless there is a local restriction in the flow.

Energy curves can be drawn for different flow quantities, as shown in Fig. 15.14 for flow quantities Q_A and Q_B. Suppose that flow is initially at point 1. Since the flow is on the upper limb, the flow is initially subcritical. If there is a step up in the channel bottom, Eq. 15.74 predicts that the specific energy will decrease.

Figure 15.14 *Specific Energy Curve Families*

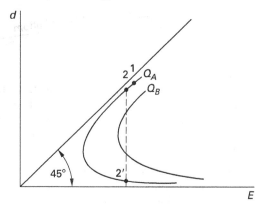

However, the flow cannot arrive at point 2' without the flow quantity changing (i.e., going through a specific energy curve for a different flow quantity). Therefore, point 2' is not accessible from point 1 without going through point 2 first.[11]

If the flow is well up on the top limb of the specific energy curve (as it is in Ex. 15.7), the water level will drop only slightly. Since the upper limb is asymptotic to a 45° diagonal line, any change in specific energy will result in almost the same change in depth.[12] Therefore, the surface level will remain almost the same.

$$\Delta d \approx \Delta E \quad \text{[fully subcritical]} \qquad \textbf{15.76}$$

However, if the initial point on the limb is close to the critical point (i.e., the nose of the curve), then a small change in the specific energy (such as might be caused by a small variation in the channel floor) will cause a large change in depth. That is why severe turbulence commonly occurs near points of critical flow.

---

[11]Actually, specific energy curves are typically plotted for flow per unit width, $q = Q/w$. If that is the case, a jump from one limb to the other could take place if the width were allowed to change as well as depth.
[12]A rise in the channel bottom does not always produce a drop in the water surface. Only if the flow is initially subcritical will the water surface drop upon encountering a step. The water surface will rise if the flow is initially supercritical.

Example 15.7

4 ft/sec (1.2 m/s) of water flow in a 7 ft (2.1 m) wide, 6 ft (1.8 m) deep open channel. The flow encounters a 1.0 ft (0.3 m) step in the channel bottom. What is the depth of flow above the step?

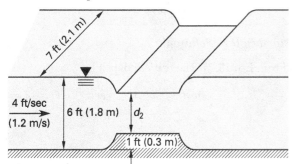

SI Solution

The initial specific energy is found from Eq. 15.68.

$$H = d + \frac{V_a^2}{2g}$$

$$= 1.8 \text{ m} + \frac{\left(1.2 \dfrac{\text{m}}{\text{s}}\right)^2}{(2)\left(9.81 \dfrac{\text{m}}{\text{s}^2}\right)}$$

$$= 1.87 \text{ m}$$

From Eq. 15.74, the specific energy above the step is

$$E_2 = E_1 + Z_1 - Z_2$$
$$= 1.87 \text{ m} + 0 - 0.3 \text{ m}$$
$$= 1.57 \text{ m}$$

The quantity flowing is

$$Q = Av$$
$$= (2.1 \text{ m})(1.8 \text{ m})\left(1.2 \dfrac{\text{m}}{\text{s}}\right)$$
$$= 4.54 \text{ m}^3/\text{s}$$

Substituting Q into Eq. 15.71 gives

$$H = d + \frac{Q^2}{2g(wd)^2}$$

$$1.57 \text{ m} = d_2 + \frac{\left(4.54 \dfrac{\text{m}^3}{\text{s}}\right)^2}{(2)\left(9.81 \dfrac{\text{m}}{\text{s}^2}\right)(2.1 \text{ m})^2 d_2^2}$$

By trial and error or a calculator's equation solver, the alternate depths are $d_2 = 0.46$ m, 1.46 m.

Since the 0.46 m depth is not accessible from the initial depth of 1.8 m, the depth over the step is 1.5 m. The drop in the water level is

$$1.8 \text{ m} - (1.46 \text{ m} + 0.3 \text{ m}) = 0.04 \text{ m}$$

Customary U.S. Solution

The initial specific energy is found from Eq. 15.68.

$$H = d + \frac{V_a^2}{2g}$$

$$= 6 \text{ ft} + \frac{\left(4 \dfrac{\text{ft}}{\text{sec}}\right)^2}{(2)\left(32.2 \dfrac{\text{ft}}{\text{sec}^2}\right)}$$

$$= 6.25 \text{ ft}$$

From Eq. 15.74, the specific energy over the step is

$$E_2 = E_1 + Z_1 - Z_2$$
$$= 6.25 \text{ ft} + 0 - 1 \text{ ft}$$
$$= 5.25 \text{ ft}$$

The quantity flowing is

$$Q = Av$$
$$= (7 \text{ ft})(6 \text{ ft})\left(4 \dfrac{\text{ft}}{\text{sec}}\right)$$
$$= 168 \text{ ft}^3/\text{sec}$$

Substituting Q into Eq. 15.71 gives

$$H = d + \frac{Q^2}{2g(wd)^2}$$

$$5.25 \text{ ft} = d_2 + \frac{\left(168 \dfrac{\text{ft}^3}{\text{sec}}\right)^2}{(2)\left(32.2 \dfrac{\text{ft}}{\text{sec}^2}\right)(7 \text{ ft})^2 d_2^2}$$

By trial and error or a calculator's equation solver, the alternate depths are $d_2 = 1.6$ ft, 4.9 ft.

Since the 1.6 ft depth is not accessible from the initial depth of 6 ft, the depth over the step is 4.9 ft. The drop in the water level is

$$6 \text{ ft} - (4.9 \text{ ft} + 1 \text{ ft}) = 0.1 \text{ ft}$$

Water

25. CRITICAL FLOW AND CRITICAL DEPTH IN RECTANGULAR CHANNELS

There is one depth, known as the *critical depth*, that minimizes the energy of flow. (The depth is not minimized, however.) The critical depth for a given flow depends on the shape of the channel.

For a rectangular channel, if Eq. 15.71 is differentiated with respect to depth in order to minimize the specific energy, Eq. 15.77 results.

Open-Channel Flow and Pipe Flow of Water

$$y_c = \left(\frac{q^2}{g} \right)^{1/3} \qquad \textbf{15.77}$$

Geometrical and analytical methods can be used to correlate the critical depth and the minimum specific energy.

$$y_c = \frac{2E}{\alpha + 2} \qquad \textbf{15.78}$$

For a rectangular channel, $q = y_c w v_c$. Substituting this into Eq. 15.77 produces an equation for the *critical velocity*.

$$v_c = \sqrt{g y_c} \qquad \textbf{15.79}$$

The expression for critical velocity also coincides with the expression for the velocity of a low-amplitude *surface wave (surge wave)* moving in a liquid of depth d_c. Since surface disturbances are transmitted as ripples upstream (and downstream) at velocity v_c, it is apparent that a surge wave will be stationary in a channel moving at the critical velocity. Such motionless waves are known as *standing waves*.

If the flow velocity is less than the surge wave velocity (for the actual depth), then a ripple can make its way upstream. If the flow velocity exceeds the surge wave velocity, the ripple will be swept downstream.

Example 15.8

500 ft³/sec (14 m³/s) of water flows in a 20 ft (6m) wide rectangular channel. What are the (a) critical depth and (b) critical velocity?

SI Solution

(a) From Eq. 15.77, the critical depth is

Open-Channel Flow and Pipe Flow of Water

$$y_c = \left(\frac{q^2}{g} \right)^{1/3} = \left(\frac{\left[\dfrac{14 \ \frac{m^3}{s}}{6 \ m} \right]^2}{9.81 \ \frac{m}{s^2}} \right)^{1/3} = 0.822 \ m$$

(b) From Eq. 15.79, the critical velocity is

$$v_c = \sqrt{g y_c}$$
$$= \sqrt{\left(9.81 \ \frac{m}{s^2} \right)(0.822 \ m)}$$
$$= 2.84 \ m/s$$

Customary U.S. Solution

(a) From Eq. 15.77, the critical depth is

Open-Channel Flow and Pipe Flow of Water

$$y_c = \left(\frac{q^2}{g} \right)^{1/3}$$
$$= \left(\frac{\left[\dfrac{500 \ \frac{ft^3}{sec}}{20 \ ft} \right]^2}{32.2 \ \frac{ft}{sec^2}} \right)^{1/3}$$
$$= 2.687 \ ft$$

(b) From Eq. 15.79, the critical velocity is

$$v_c = \sqrt{g y_c} = \sqrt{\left(32.2 \ \frac{ft}{sec^2} \right)(2.687 \ ft)}$$
$$= 9.30 \ ft/sec$$

26. CRITICAL FLOW AND CRITICAL DEPTH IN NONRECTANGULAR CHANNELS

For nonrectangular shapes (including trapezoidal channels), the critical depth can be found by trial and error from the following equation in which T is the surface width. To use Eq. 15.80, assume trial values of the critical depth, use them to calculate dependent quantities in the equation, and then verify the equality.

Open-Channel Flow and Pipe Flow of Water

$$\frac{Q^2}{g} = \frac{A^3}{T} \qquad [\text{nonrectangular}] \qquad \textbf{15.80}$$

Equation 15.80 is particularly difficult to use with circular channels. Appendix 15.A is a convenient method of determining critical depth in circular channels.

27. FROUDE NUMBER

The dimensionless *Froude number*, Fr, is a convenient index of the flow regime. It can be used to determine whether the flow is subcritical or supercritical. y_h is the *hydraulic depth*, also referred to as the *characteristic length*, *characteristic (length) scale*, hydraulic depth, mean hydraulic depth, and others, depending on the channel configuration. d is the depth corresponding to velocity v. For circular channels flowing half full, $y_h = \pi D/8$. For a rectangular channel, $y_h = d$. For trapezoidal and semicircular channels, and in general, y_h is the area in flow divided by the top width, T.

Open-Channel Flow and Pipe Flow of Water

$$\text{Fr} = \frac{\text{v}}{\sqrt{gy_h}} \qquad 15.81$$

When the Froude number is less than one, the flow is subcritical (i.e., the depth of flow is greater than the critical depth) and the velocity is less than the critical velocity.

For convenience, the Froude number can be written in terms of the flow rate per average unit width.

$$\text{Fr} = \frac{\dfrac{Q}{b}}{\sqrt{gd^3}} \quad \text{[rectangular]} \qquad 15.82$$

$$\text{Fr} = \frac{\dfrac{Q}{b_{\text{ave}}}}{\sqrt{g\left(\dfrac{A}{b_{\text{ave}}}\right)^3}} \quad \text{[nonrectangular]} \qquad 15.83$$

When the Froude number is greater than one, the flow is supercritical. The depth is less than critical depth, and the flow velocity is greater than the critical velocity.

When the Froude number is equal to one, the flow is critical.[13]

The Froude number has another form. Dimensional analysis determines it to be v^2/gL, a form that is also used in analyzing similarity of models. Whether the derived form or the square root form is used can sometimes be determined by observing the form of the intended application. If the Froude number is squared (as it is in Eq. 15.84), then the square root form is probably intended. For open channel flow, the Froude number is always the square root of the derived form.

28. PREDICTING OPEN CHANNEL FLOW BEHAVIOR

Upon encountering a variation in the channel bottom, the behavior of an open channel flow is dependent on whether the flow is initially subcritical or supercritical. Open channel flow is governed by Eq. 15.84 in which the Froude number is the primary independent variable.

$$\frac{dd}{dx}(1 - \text{Fr}^2) + \frac{dz}{dx} = 0 \qquad 15.84$$

The quantity dd/dx is the slope of the surface (i.e., it is the derivative of the depth with respect to the channel length). The quantity dz/dx is the slope of the channel bottom.

For an upward step, $dz/dx > 0$. If the flow is initially subcritical (i.e., $\text{Fr} < 1$), then Eq. 15.84 requires that $dd/dx < 0$, a drop in depth.

This logic can be repeated for other combinations of the terms. Table 15.4 lists the various behaviors of open channel flow surface levels based on Eq. 15.84.

Table 15.4 *Surface Level Change Behavior*

initial flow	step up	step down
subcritical	surface drops	surface rises
supercritical	surface rises	surface drops

If $dz/dx = 0$ (i.e., a horizontal slope), then either the depth must be constant or the Froude number must be unity. The former case is obvious. The latter case predicts critical flow. Such critical flow actually occurs where the slope is horizontal over broad-crested weirs and at the top of a rounded spillway. Since broad-crested weirs and spillways produce critical flow, they represent a class of controls on flow.

29. OCCURRENCES OF CRITICAL FLOW

The critical depth not only minimizes the energy of flow, but also maximizes the quantity flowing for a given cross section and slope. Critical flow is generally quite turbulent because of the large changes in energy that occur with small changes in elevation and depth. Critical depth flow is often characterized by successive water surface undulations over a very short stretch of channel. (See Fig. 15.15.)

For any given discharge and cross section, there is a unique slope that will produce and maintain flow at critical depth. Once y_c is known, this critical slope can be found from the Manning equation. In all of the instances of critical depth, Eq. 15.79 can be used to calculate the actual velocity.

---

[13]The similarity of the Froude number to the Mach number used to classify gas flows is more than coincidental. Both bodies of knowledge employ parallel concepts.

Figure 15.15 *Occurrence of Critical Depth*

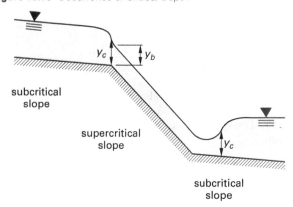

Critical depth occurs at free outfall from a channel of mild slope. The occurrence is at the point of curvature inversion, which is just upstream from the brink. (See Fig. 15.16.) For mild slopes, the *brink depth* is approximately

$$y_b = 0.715 y_c \qquad \text{15.85}$$

Figure 15.16 *Free Outfall*

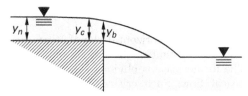

Critical flow can occur across a broad-crested weir, as shown in Fig. 15.17.[14] With no obstruction to hold the water, it falls from the normal depth to the critical depth, but it can fall no more than that because there is no source to increase the specific energy (to increase the velocity). This is not a contradiction of the previous free outfall case where the brink depth is less than the critical depth. The flow curvatures in free outfall are a result of the constant gravitational acceleration.

Figure 15.17 *Broad-Crested Weir*

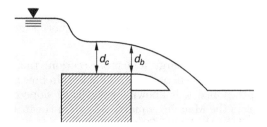

Critical depth can also occur when a channel bottom has been raised sufficiently to choke the flow. A raised channel bottom is essentially a broad-crested weir. (See Fig. 15.18.)

Figure 15.18 *Raised Channel Bottom with Choked Flow*

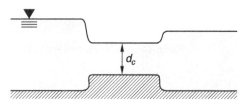

Example 15.9

At a particular point in an open rectangular channel ($n = 0.013$, $S = 0.002$, and $w = 10$ ft (3 m)), the flow is 250 ft^3/sec (7 m^3/s) and the depth is 4.2 ft (1.3 m).

(a) Is the flow tranquil, critical, or rapid?

(b) What is the normal depth?

(c) If the channel ends in a free outfall, what is the brink depth?

SI Solution

(a) From Eq. 15.77, the critical depth is

Open-Channel Flow and Pipe Flow of Water

$$y_c = \left(\frac{q^2}{g} \right)^{1/3}$$

$$= \left(\frac{\left(7 \dfrac{\text{m}^3}{\text{s}} \right)^2}{(3 \text{ m})^2 \left(9.81 \dfrac{\text{m}}{\text{s}^2} \right)} \right)^{1/3}$$

$$= 0.82 \text{ m}$$

Since the actual depth exceeds the critical depth, the flow is tranquil.

(b) From Eq. 15.13,

$$Q = vA = (K/n)AR_H^{2/3}S^{1/2} = KS^{1/2}$$

$$R = \frac{A}{P} = \frac{y_n(3 \text{ m})}{2y_n + 3 \text{ m}}$$

---

[14]Figure 15.17 is an example of a *hydraulic drop*, the opposite of a hydraulic jump. A hydraulic drop can be recognized by the sudden decrease in depth over a short length of channel.

Substitute the expression for R into Eq. 15.13 and solve for y_n.

$$7 \; \frac{m^3}{s} = \left(\frac{1}{0.013}\right) y_n(3 \; m) \left(\frac{y_n(3 \; m)}{2d_n + 3 \; m}\right)^{2/3} \sqrt{0.002}$$

By trial and error or a calculator's equation solver, $y_n = 0.97$ m. Since the actual and normal depths are different, the flow is nonuniform.

(c) From Eq. 15.85, the brink depth is

$$y_b = 0.715y_c = (0.715)(0.82 \; m)$$
$$= 0.59 \; m$$

Customary U.S. Solution

(a) From Eq. 15.77, the critical depth is

Open-Channel Flow and Pipe Flow of Water

$$y_c = \left(\frac{q^2}{g}\right)^{1/3}$$

$$= \left(\frac{\left(250 \; \frac{ft^3}{sec}\right)^2}{\left(32.2 \; \frac{ft}{sec^2}\right)(10 \; ft)^2}\right)^{1/3}$$

$$= 2.69 \; ft$$

Since the actual depth exceeds the critical depth, the flow is tranquil.

(b) From Eq. 15.13,

$$Q = vA = (K/n)AR_H^{2/3}S^{1/2} = KS^{1/2}$$

$$R = \frac{A}{P} = \frac{y_n(10 \; ft)}{2y_n + 10 \; ft}$$

Substitute the expression for R into Eq. 15.13 and solve for y_n.

$$250 \; \frac{ft^3}{sec} = \left(\frac{1.49}{0.013}\right) y_n(10 \; ft) \left(\frac{d_n(10 \; ft)}{2d_n + 10 \; ft}\right)^{2/3} \sqrt{0.002}$$

By trial and error or a calculator's equation solver, $y_n = 3.1$ ft. Since the actual and normal depths are different, the flow is nonuniform.

(c) From Eq. 15.85, the brink depth is

$$y_b = 0.715y_c$$
$$= (0.715)(2.69 \; ft)$$
$$= 1.92 \; ft$$

30. CONTROLS ON FLOW

In general, any feature that affects depth and discharge rates is known as a *control on flow*. Controls may consist of constructed control structures (weirs, gates, sluices, etc.), forced flow through critical depth (as in a free outfall), sudden changes of slope (which forces a hydraulic jump or hydraulic drop to the new normal depth), or free flow between reservoirs of different surface elevations. A downstream control may also be an upstream control, as Fig. 15.19 shows.

If flow is subcritical, then a disturbance downstream will be able to affect the upstream conditions. Since the flow velocity is less than the critical velocity, a ripple will be able to propagate upstream to signal a change in the downstream conditions. Any object downstream that affects the flow rate, velocity, or depth upstream is known as a *downstream control*.

If a flow is supercritical, then a downstream obstruction will have no effect upstream, since disturbances cannot propagate upstream faster than the flow velocity. The only effect on supercritical flow is from an upstream obstruction. Such an obstruction is said to be an *upstream control*.

Figure 15.19 Control on Flow

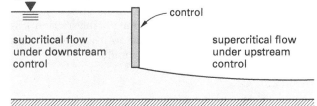

subcritical flow under downstream control

control

supercritical flow under upstream control

31. FLOW CHOKING

A channel feature that causes critical flow to occur is known as a *choke*, and the corresponding flow past the feature and downstream is known as *choked flow*.

In the case of vertical transitions (i.e., upward or downward steps in the channel bottom), choked flow will occur when the step size is equal to the difference between the upstream specific energy and the critical flow energy.

$$\Delta z = E_1 - E_c \quad \text{[choked flow]} \qquad \textit{15.86}$$

In the case of a rectangular channel, combining Eq. 15.68 and Eq. 15.79 the maximum variation in channel bottom will be

$$\Delta z = E_1 - \left(y + \frac{V^2}{2g} \right) \qquad 15.87$$

$$= E_1 - \frac{3}{2} y$$

The flow downstream from a choke point can be subcritical or supercritical, depending on the downstream conditions. If there is a downstream control, such as a sluice gate, the flow downstream will be subcritical. If there is additional gravitational acceleration (as with flow down the side of a dam spillway), then the flow will be supercritical.

32. VARIED FLOW

Accelerated flow occurs in any channel where the actual slope exceeds the friction loss per foot.

$$S_0 > \frac{h_f}{L} \qquad 15.88$$

Decelerated flow occurs when the actual slope is less than the unit friction loss.

$$S_0 < \frac{h_f}{L} \qquad 15.89$$

In sections AB and CD of Fig. 15.20, the slopes are less than the energy gradient, so the flows are decelerated. In section BC, the slope is greater than the energy gradient, so the velocity increases (i.e., the flow is accelerated). If section BC were long enough, the friction loss would eventually become equal to the accelerating energy and the flow would become uniform.

Figure 15.20 *Varied Flow*

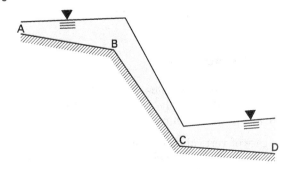

The distance between points 1 and 2 with two known depths in accelerated or decelerated flow can be determined from the average velocity. Equation 15.90 and Eq. 15.91 assume that the friction losses are the same for varied flow as for uniform flow. These two equations

are equivalent to Manning's equation, Eq. 15.12, but expressed in terms of S with values of K taken as 1.0 and 1.49.

$$S_{ave} = \left(\frac{n v_{ave}}{R_{ave}^{2/3}} \right)^2 \qquad \text{[SI]} \quad 15.90(a)$$

$$S_{ave} = \left(\frac{n v_{ave}}{1.49 R_{ave}^{2/3}} \right)^2 \qquad \text{[U.S.]} \quad 15.90(b)$$

$$v_{ave} = \frac{1}{2}(v_1 + v_2) \qquad 15.91$$

S is the slope of the energy gradient from Eq. 15.90, not the channel slope S_0. The usual method of finding the *depth profile* is to start at a point in the channel where d_2 and v_2 are known. Then, assume a depth d_1, find v_1 and S, and solve for L. Repeat as needed.

$$L = \frac{\left(d_1 + \frac{v_1^2}{2g} \right) - \left(d_2 + \frac{v_2^2}{2g} \right)}{S - S_0} \qquad 15.92$$

$$= \frac{E_1 - E_2}{S - S_0}$$

In Eq. 15.92, d_1 is always the smaller of the two depths.

Example 15.10

How far from the point described in Ex. 15.9 will the depth be 4 ft (1.2 m)?

SI Solution

The difference between 1.3 m and 1.2 m is small, so a one-step calculation will probably be sufficient.

$$d_1 = 1.2 \text{ m}$$

$$v_1 = \frac{Q}{A} = \frac{7 \frac{m^3}{s}}{(1.2 \text{ m})(3 \text{ m})}$$

$$= 1.94 \text{ m/s}$$

$$E_1 = d_1 + \frac{v_1^2}{2g} = 1.2 \text{ m} + \frac{\left(1.94 \frac{m}{s} \right)^2}{(2)\left(9.81 \frac{m}{s^2} \right)}$$

$$= 1.39 \text{ m}$$

$$R_1 = \frac{A_1}{P_1} = \frac{(1.2 \text{ m})(3 \text{ m})}{1.2 \text{ m} + 3 \text{ m} + 1.2 \text{ m}}$$
$$= 0.67 \text{ m}$$
$$d_2 = 1.3 \text{ m}$$

$$v_2 = \frac{7 \dfrac{\text{m}^3}{\text{s}}}{(1.3 \text{ m})(3 \text{ m})}$$
$$= 1.79 \text{ m/s}$$

$$E_2 = d_2 + \frac{v_2^2}{2g} = 1.3 \text{ m} + \frac{\left(1.79 \dfrac{\text{m}}{\text{s}}\right)^2}{(2)\left(9.81 \dfrac{\text{m}}{\text{s}^2}\right)}$$
$$= 1.46 \text{ m}$$

$$R_2 = \frac{A_2}{P_2} = \frac{(1.3 \text{ m})(3 \text{ m})}{1.3 \text{ m} + 3 \text{ m} + 1.3 \text{ m}}$$
$$= 0.70 \text{ m}$$

$$v_{\text{ave}} = \tfrac{1}{2}(v_1 + v_2) = \left(\frac{1}{2}\right)\left(1.94 \frac{\text{m}}{\text{s}} + 1.79 \frac{\text{m}}{\text{s}}\right)$$
$$= 1.865 \text{ m/s}$$

$$R_{\text{ave}} = \tfrac{1}{2}(R_1 + R_2) = \left(\frac{1}{2}\right)(0.70 \text{ m} + 0.67 \text{ m})$$
$$= 0.685 \text{ m}$$

From Eq. 15.90,

$$S = \left(\frac{n\,v_{\text{ave}}}{R_{\text{ave}}^{2/3}}\right)^2$$
$$= \left(\frac{(0.013)\left(1.865 \dfrac{\text{m}}{\text{s}}\right)}{(0.685 \text{ m})^{2/3}}\right)^2$$
$$= 0.000973$$

From Eq. 15.92,

$$L = \frac{E_1 - E_2}{S - S_0}$$
$$= \frac{1.39 \text{ m} - 1.46 \text{ m}}{0.000973 - 0.002}$$
$$= 68.2 \text{ m}$$

Customary U.S. Solution

The difference between 4 ft and 4.2 ft is small, so a one-step calculation will probably be sufficient.

$$d_1 = 4 \text{ ft}$$

$$v_1 = \frac{Q}{A} = \frac{250 \dfrac{\text{ft}^3}{\text{sec}}}{(4 \text{ ft})(10 \text{ ft})}$$
$$= 6.25 \text{ ft/sec}$$

$$E_1 = d_1 + \frac{v_1^2}{2g}$$
$$= 4 \text{ ft} + \frac{\left(6.25 \dfrac{\text{ft}}{\text{sec}}\right)^2}{(2)\left(32.2 \dfrac{\text{ft}}{\text{sec}^2}\right)}$$
$$= 4.607 \text{ ft}$$

$$R_1 = \frac{A_1}{P_1} = \frac{(4 \text{ ft})(10 \text{ ft})}{4 \text{ ft} + 10 \text{ ft} + 4 \text{ ft}}$$
$$= 2.22 \text{ ft}$$
$$d_2 = 4.2 \text{ ft}$$
$$= 2.25 \text{ ft}$$

$$v_2 = \frac{250 \dfrac{\text{ft}^3}{\text{sec}}}{(4.2 \text{ ft})(10 \text{ ft})}$$
$$= 5.95 \text{ ft/sec}$$

$$E_2 = d_2 + \frac{v_2^2}{2g} = 4.2 \text{ ft} + \frac{\left(5.95 \dfrac{\text{ft}}{\text{sec}}\right)^2}{(2)\left(32.2 \dfrac{\text{ft}}{\text{sec}^2}\right)}$$
$$= 4.75 \text{ ft}$$

$$R_2 = \frac{A_2}{P_2} = \frac{(4.2 \text{ ft})(10 \text{ ft})}{4.2 \text{ ft} + 10 \text{ ft} + 4.2 \text{ ft}}$$
$$= 2.28 \text{ ft}$$

$$v_{\text{ave}} = \tfrac{1}{2}(v_1 + v_2)$$
$$= \left(\frac{1}{2}\right)\left(6.25 \frac{\text{ft}}{\text{sec}} + 5.95 \frac{\text{ft}}{\text{sec}}\right)$$
$$= 6.1 \text{ ft/sec}$$

$$R_{\text{ave}} = \tfrac{1}{2}(R_1 + R_2)$$
$$= \left(\frac{1}{2}\right)(2.22 \text{ ft} + 2.28 \text{ ft})$$
$$= 2.25 \text{ ft}$$

From Eq. 15.90,

$$S = \left(\frac{n \, \mathrm{v}_{\mathrm{ave}}}{1.49 R_{\mathrm{ave}}^{2/3}} \right)^2$$

$$= \left(\frac{(0.013)\left(6.1 \, \dfrac{\mathrm{ft}}{\mathrm{sec}} \right)}{(1.49)(2.25 \, \mathrm{ft})^{2/3}} \right)^2$$

$$= 0.000961$$

From Eq. 15.92,

$$L = \frac{E_1 - E_2}{S - S_0} = \frac{4.607 \, \mathrm{ft} - 4.75 \, \mathrm{ft}}{0.000961 - 0.002}$$

$$= 138 \, \mathrm{ft}$$

33. HYDRAULIC JUMP

If water is introduced at high (supercritical) velocity to a section of slow-moving (subcritical) flow (as shown in Fig. 15.21), the velocity will be reduced rapidly over a short length of channel. The abrupt rise in the water surface is known as a *hydraulic jump*. The increase in depth is always from below the critical depth to above the critical depth.[15] The depths on either side of the hydraulic jump are known as *conjugate depths*. The conjugate depths and the relationship between them are as follows.

$$d_1 = -\frac{1}{2}d_2 + \sqrt{\frac{2\mathrm{v}_2^2 d_2}{g} + \frac{d_2^2}{4}} \quad \begin{bmatrix} \text{rectangular} \\ \text{channels} \end{bmatrix} \quad 15.93$$

$$d_2 = -\frac{1}{2}d_1 + \sqrt{\frac{2\mathrm{v}_1^2 d_1}{g} + \frac{d_1^2}{4}} \quad \begin{bmatrix} \text{rectangular} \\ \text{channels} \end{bmatrix} \quad 15.94$$

$$\frac{d_2}{d_1} = \frac{1}{2}\left(\sqrt{1 + 8(\mathrm{Fr}_1)^2} - 1 \right) \quad \begin{bmatrix} \text{rectangular} \\ \text{channels} \end{bmatrix} \quad 15.95(a)$$

$$\frac{d_1}{d_2} = \frac{1}{2}\left(\sqrt{1 + 8(\mathrm{Fr}_2)^2} - 1 \right) \quad \begin{bmatrix} \text{rectangular} \\ \text{channels} \end{bmatrix} \quad 15.95(b)$$

If the depths d_1 and d_2 are known, then the upstream velocity can be found from Eq. 15.96.

$$\mathrm{v}_1^2 = \left(\frac{gd_2}{2d_1} \right)(d_1 + d_2) \quad \begin{bmatrix} \text{rectangular} \\ \text{channels} \end{bmatrix} \quad 15.96$$

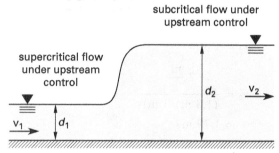

Figure 15.21 *Conjugate Depths*

Conjugate depths are not the same as alternate depths. Alternate depths are derived from the conservation of energy equation (i.e., a variation of the Bernoulli equation). Conjugate depths (calculated in Eq. 15.93 through Eq. 15.95) are derived from a conservation of momentum equation. Conjugate depths are calculated only when there has been an abrupt energy loss such as occurs in a hydraulic jump or drop.

Hydraulic jumps have practical applications in the design of stilling basins. Stilling basins are designed to intentionally reduce energy of flow through hydraulic jumps. In the case of a concrete apron at the bottom of a dam spillway, the apron friction is usually low, and the water velocity will decrease only gradually. However, supercritical velocities can be reduced to much slower velocities by having the flow cross a series of baffles on the channel bottom.

The specific energy lost in the jump is the energy lost per pound of water flowing.

$$\Delta E = \left(d_1 + \frac{\mathrm{v}_1^2}{2g} \right) - \left(d_2 + \frac{\mathrm{v}_2^2}{2g} \right) \approx \frac{(d_2 - d_1)^3}{4d_1 d_2} \quad 15.97$$

Evaluation of hydraulic jumps in stilling basins starts by determining the depth at the toe. The depth of flow at the toe of a spillway is found from an energy balance. Neglecting friction, the total energy at the toe equals the total upstream energy before the spillway. The total upstream energy before the spillway is

$$E_{\mathrm{upstream}} = E_{\mathrm{toe}} \quad 15.98$$

$$y_{\mathrm{crest}} + H + \frac{\mathrm{v}^2}{2g} = d_{\mathrm{toe}} + \frac{\mathrm{v}_{\mathrm{toe}}^2}{2g} \quad 15.99$$

The upstream velocity, v, is the velocity before the spillway (which is essentially zero), not the velocity over the brink. If the brink depth is known, the velocity over the brink can be used with the continuity equation to calculate the upstream velocity, but the velocity over the brink should not be used with H to determine total energy since $\mathrm{v}_b \neq \mathrm{v}$. (See Fig. 15.22.)

---

[15]This provides a way of determining if a hydraulic jump can occur in a channel. If the original depth is above the critical depth, the flow is already subcritical. Therefore, a hydraulic jump cannot form. Only a hydraulic drop could occur.

Figure 15.22 *Total Energy Upstream of a Spillway*

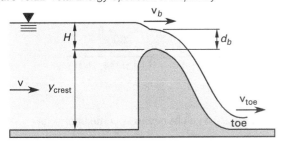

If water is drained quickly from the apron so that the tailwater depth is small or zero, no hydraulic jump will form. This is because the tailwater depth is already less than the critical depth.

A hydraulic jump will form along the apron at the bottom of the spillway when the actual tailwater depth equals the conjugate depth d_2 corresponding to the depth at the toe. That is, the jump is located at the toe when $d_2 = d_{tailwater}$, where d_2 and $d_1 = d_{toe}$ are conjugate depths. The tailwater and toe depths are implicitly the conjugate depths. This is shown in Fig. 15.23(a) and is the proper condition for energy dissipation in a stilling basin.

When the actual tailwater depth is less than the conjugate depth d_2 corresponding to d_{toe}, but still greater than the critical depth, flow will continue along the apron until the depth increases to conjugate depth d_1 corresponding to the actual tailwater depth. This is shown in Fig. 15.23(b). A hydraulic jump will form at that point to increase the depth to the tailwater depth. (Another way of saying this is that the hydraulic jump moves downstream from the toe.) This is an undesirable condition, since the location of the jump is often inadequately protected from scour.

If the tailwater depth is greater than the conjugate depth corresponding to the depth at the toe, as in Fig. 15.23(c), the hydraulic jump may occur up on the spillway, or it may be completely submerged (i.e., it will not occur at all).

Example 15.11

A hydraulic jump is produced at a point in a 10 ft (3 m) wide channel where the depth is 1 ft (0.3 m). The flow rate is 200 ft³/sec (5.7 m³/s). (a) What is the depth after the jump? (b) What is the total power dissipated?

SI Solution

(a) From Eq. 15.70,

$$v_1 = \frac{Q}{A} = \frac{5.7 \ \frac{m^3}{s}}{(3 \ m)(0.3 \ m)}$$
$$= 6.33 \ m/s$$

Figure 15.23 *Hydraulic Jump to Reach Tailwater Level*

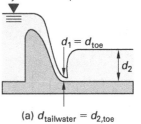

(a) $d_{tailwater} = d_{2,toe}$

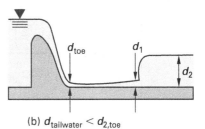

(b) $d_{tailwater} < d_{2,toe}$

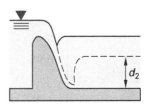

(c) $d_{tailwater} > d_{2,toe}$

From Eq. 15.94,

$$d_2 = -\frac{1}{2}d_1 + \sqrt{\frac{2v_1^2 d_1}{g} + \frac{d_1^2}{4}}$$

$$= -\left(\frac{1}{2}\right)(0.3 \ m)$$

$$+ \sqrt{\frac{(2)\left(6.33 \ \frac{m}{s}\right)^2 (0.3 \ m)}{9.81 \ \frac{m}{s^2}} + \frac{(0.3 \ m)^2}{4}}$$

$$= 1.42 \ m$$

(b) The mass flow rate is

$$\dot{m} = \left(5.7 \ \frac{m^3}{s}\right)\left(1000 \ \frac{kg}{m^3}\right)$$
$$= 5700 \ kg/s$$

The velocity after the jump is

$$v_2 = \frac{Q}{A_2} = \frac{5.7 \ \frac{m^3}{s}}{(3 \ m)(1.42 \ m)}$$
$$= 1.33 \ m/s$$

From Eq. 15.97, the change in specific energy is

$$\Delta E = \left(d_1 + \frac{v_1^2}{2g}\right) - \left(d_2 + \frac{v_2^2}{2g}\right)$$

$$= \left(0.3 \text{ m} + \frac{\left(6.33 \frac{\text{m}}{\text{s}}\right)^2}{(2)\left(9.81 \frac{\text{m}}{\text{s}^2}\right)}\right)$$

$$- \left(1.423 \text{ m} + \frac{\left(1.33 \frac{\text{m}}{\text{s}}\right)^2}{(2)\left(9.81 \frac{\text{m}}{\text{s}^2}\right)}\right)$$

$$= 0.83 \text{ m}$$

The total power dissipated is

$$P = \dot{m}g\Delta E$$

$$= \left(5700 \frac{\text{kg}}{\text{s}}\right)\left(9.81 \frac{\text{m}}{\text{s}^2}\right)\left(\frac{0.83 \text{ m}}{1000 \frac{\text{W}}{\text{kW}}}\right)$$

$$= 46.4 \text{ kW}$$

Customary U.S. Solution

(a) From Eq. 15.70,

$$v_1 = \frac{Q}{A} = \frac{200 \frac{\text{ft}^3}{\text{sec}}}{(10 \text{ ft})(1 \text{ ft})}$$

$$= 20 \text{ ft/sec}$$

From Eq. 15.94,

$$d_2 = -\frac{1}{2}d_1 + \sqrt{\frac{2v_1^2 d_1}{g} + \frac{d_1^2}{4}}$$

$$= -\left(\frac{1}{2}\right)(1 \text{ ft}) + \sqrt{\frac{(2)\left(20 \frac{\text{ft}}{\text{sec}}\right)^2(1 \text{ ft})}{32.2 \frac{\text{ft}}{\text{sec}^2}} + \frac{(1 \text{ ft})^2}{4}}$$

$$= 4.51 \text{ ft}$$

(b) The mass flow rate is

$$\dot{m} = \left(200 \frac{\text{ft}^3}{\text{sec}}\right)\left(62.4 \frac{\text{lbm}}{\text{ft}^3}\right)$$

$$= 12{,}480 \text{ lbm/sec}$$

The velocity after the jump is

$$v_2 = \frac{Q}{A_2} = \frac{200 \frac{\text{ft}^3}{\text{sec}}}{(10 \text{ ft})(4.51 \text{ ft})}$$

$$= 4.43 \text{ ft/sec}$$

From Eq. 15.97, the change in specific energy is

$$\Delta E = \left(d_1 + \frac{v_1^2}{2g}\right) - \left(d_2 + \frac{v_2^2}{2g}\right)$$

$$= \left(1 \text{ ft} + \frac{\left(20 \frac{\text{ft}}{\text{sec}}\right)^2}{(2)\left(32.2 \frac{\text{ft}}{\text{sec}^2}\right)}\right)$$

$$- \left(4.51 \text{ ft} + \frac{\left(4.43 \frac{\text{ft}}{\text{sec}}\right)^2}{(2)\left(32.2 \frac{\text{ft}}{\text{sec}^2}\right)}\right)$$

$$= 2.4 \text{ ft}$$

The total power dissipated is

$$P = \frac{\dot{m}g\Delta E}{g_c}$$

$$= \frac{\left(12{,}480 \frac{\text{lbm}}{\text{sec}}\right)\left(32.2 \frac{\text{ft}}{\text{sec}^2}\right)(2.4 \text{ ft})}{\left(32.2 \frac{\text{lbm-ft}}{\text{lbf-sec}^2}\right)\left(550 \frac{\text{ft-lbf}}{\text{hp-sec}}\right)}$$

$$= 54.5 \text{ hp}$$

34. LENGTH OF HYDRAULIC JUMP

For practical stilling basin design, it is helpful to have an estimate of the length of the hydraulic jump. Lengths of hydraulic jumps are difficult to measure because of the difficulty in defining the endpoints of the jumps. However, the length of the jump, L, varies within the limits of $5 < L/d_2 < 6.5$, in which d_2 is the conjugate depth after the jump. Where greater accuracy is warranted, Table 15.5 can be used. This table correlates the length of the jump to the upstream Froude number.

Table 15.5 *Approximate Lengths of Hydraulic Jumps*

Fr_1	L/d_2
3	5.25
4	5.8
5	6.0
6	6.1
7	6.15
8	6.15

35. HYDRAULIC DROP

A *hydraulic drop* is the reverse of a hydraulic jump. If water is introduced at low (subcritical) velocity to a section of fast-moving (supercritical) flow, the velocity will be increased rapidly over a short length of channel. The abrupt drop in the water surface is known as a hydraulic drop. The decrease in depth is always from above the critical depth to below the critical depth.

Water flowing over a spillway and down a long, steep chute typically experiences a hydraulic drop, with critical depth occurring just before the brink. This is illustrated in Fig. 15.17 and Fig. 15.22.

The depths on either side of the hydraulic drop are the *conjugate depths*, which are determined from Eq. 15.93 and Eq. 15.94. The equations for calculating specific energy and power changes are the same for hydraulic jumps and drops.

36. ERODIBLE CHANNELS

Given an appropriate value of the Manning coefficient, the analysis of channels constructed of erodible materials is similar to that for concrete or pipe channels.

However, for design problems, maximum velocities and permissible side slopes must also be considered. The present state of knowledge is not sufficiently sophisticated to allow for precise designs. The usual uniform flow equations are insufficient because the stability of erodible channels is dependent on the properties of the channel material rather than on the hydraulics of flow. Two methods of design exist: (a) the tractive force method and (b) the simpler maximum permissible velocity method. Maximum velocities that should be used with erodible channels are given in Table 15.6.

The sides of the channel should not have a slope exceeding the natural angle of repose for the material used. Although there are other factors that determine the maximum permissible side slope, Table 15.7 lists some guidelines.

Table 15.6 *Suggested Maximum Velocities*

soil type or lining (earth; no vegetation)	maximum permissible velocities (ft/sec)		
	clear water	water carrying fine silts	water carrying sand and gravel
fine sand (noncolloidal)	1.5	2.5	1.5
sandy loam (noncolloidal)	1.7	2.5	2.0
silt loam (noncolloidal)	2.0	3.0	2.0
ordinary firm loam	2.5	3.5	2.2
volcanic ash	2.5	3.5	2.0
fine gravel	2.5	5.0	3.7
stiff clay (very colloidal)	3.7	5.0	3.0
graded, loam to cobbles (noncolloidal)	3.7	5.0	5.0
graded, silt to cobbles (colloidal)	4.0	5.5	5.0
alluvial silts (noncolloidal)	2.0	3.5	2.0
alluvial silts (colloidal)	3.7	5.0	3.0
coarse gravel (noncolloidal)	4.0	6.0	6.5
cobbles and shingles	5.0	5.5	6.5
shales and hard pans	6.0	6.0	5.0

(Multiply ft/sec by 0.3048 to obtain m/s.)

Source: Special Committee on Irrigation Research, ASCE, 1926.

Table 15.7 *Recommended Side Slopes*

type of channel	side slope (horizontal:vertical)
firm rock	vertical to $\frac{1}{4}$:1
concrete-lined stiff clay	$\frac{1}{2}$:1
fissured rock	$\frac{1}{2}$:1
firm earth with stone lining	1:1
firm earth, large channels	1:1
firm earth, small channels	$1\frac{1}{2}$:1
loose, sandy earth	2:1
sandy, porous loam	3:1

37. CULVERTS[16]

A *culvert* is a pipe that carries water under or through some feature (usually a road or highway) that would otherwise block the flow of water. For example, highways are often built at right angles to ravines draining hillsides and other watersheds. Culverts under the highway keep the construction fill from blocking the natural runoff.

Culverts are classified according to which of their ends controls the discharge capacity: inlet control or outlet control. If water can flow through and out of the culvert faster than it can enter, the culvert is under *inlet control*. If water can flow into the culvert faster than it can flow through and out, the culvert is under *outlet control*. Culverts under inlet control will always flow partially full. Culverts under outlet control can flow either partially full or full.

The culvert length is one of the most important factors in determining whether the culvert flows full. A culvert may be known as "hydraulically long" if it runs full and "hydraulically short" if it does not.[17]

All culvert design theory is closely dependent on energy conservation. However, due to the numerous variables involved, no single formula or procedure can be used to design a culvert. Culvert design is often an empirical, trial-and-error process. Figure 15.24 illustrates some of the important variables that affect culvert performance.

Figure 15.24 *Flow Profiles in Culvert Design*

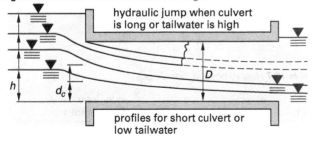

A culvert can operate with its entrance partially or totally submerged. Similarly, the exit can be partially or totally submerged, or it can have free outfall. The upstream head, h, is the water surface level above the lowest part of the culvert barrel, known as the *invert*.[18]

In Fig. 15.24, the three lowermost surface level profiles are of the type that would be produced with inlet control. Such a situation can occur if the culvert is short and the slope is steep. Flow at the entrance is critical as the water falls over the brink. Since critical flow occurs,

the flow is choked and the inlet controls the flow rate. Downstream variations cannot be transmitted past the critical section.

If the tailwater covers the culvert exit completely (i.e., a submerged exit), the culvert will be full at that point, even though the inlet control forces the culvert to be only partially full at the inlet. The transition from partially full to totally full occurs in a hydraulic jump, the location of which depends on the flow resistance and water levels. If the flow resistance is very high, or if the headwater and tailwater levels are high enough, the jump will occur close to or at the entrance.

If the flow in a culvert is full for its entire length, then the flow is under outlet control. The discharge will be a function of the differences in tailwater and headwater levels, as well as the flow resistance along the barrel length.

38. DETERMINING TYPE OF CULVERT FLOW

For convenience, culvert flow is classified into six different types on the basis of the type of control, the steepness of the barrel, the relative tailwater and headwater heights, and in some cases, the relationship between critical depth and culvert size. These parameters are quantified through the use of the ratios in Table 15.8.[19] The six types are illustrated in Fig. 15.25. Identification of the type of flow beyond the guidelines in Table 15.8 requires a trial-and-error procedure.

In the following cases, several variables appear repeatedly. C_d is the discharge coefficient, a function of the barrel inlet geometry. Orifice data can be used to approximate the discharge coefficient when specific information is unavailable. v_1 is the average velocity of the water approaching the culvert entrance and is often insignificant. The velocity-head coefficient, α, also called the *Coriolis coefficient*, accounts for a nonuniform distribution of velocities over the channel section. However, it represents only a second-order correction and is normally neglected (i.e., assumed equal to 1.0). d_c is the critical depth, which may not correspond to the actual depth of flow. (It must be calculated from the flow conditions.) h_f is the friction loss in the identified section. For culverts flowing full, the friction loss can be found in the usual manner developed for pipe flow: from the Darcy formula and the Moody friction factor chart. For partial flow, the Manning equation and its variations (e.g., Eq. 15.31) can also be used. The Manning equation is particularly useful since it eliminates the need for

---

[16]The methods of culvert flow analysis in this chapter are based on Bodhaine, G.L., 1968, *Measurement of Peak Discharge by Indirect Methods*, U.S. Geological Survey, *Techniques of Water Resources Investigations*, book 3, chapter A3.
[17]Proper design of culvert entrances can reduce the importance of length on culvert filling.
[18]The highest part of the culvert barrel is known as the *soffit* or *crown*.
[19]The six cases presented here do not exhaust the various possibilities for entrance and exit control. Culvert design is complicated by this multiplicity of possible flows. Since only the easiest problems can be immediately categorized as one of the six cases, each situation needs to be carefully evaluated.

Table 15.8 *Culvert Flow Classification Parameters*

flow type	$\dfrac{h_1 - z}{D}$	$\dfrac{h_4}{h_c}$	$\dfrac{h_4}{D}$	culvert slope	barrel flow	location of control	kind of control
1	< 1.5	< 1.0	≤ 1.0	steep	partial	inlet	critical depth
2	< 1.5	< 1.0	≤ 1.0	mild	partial	outlet	critical depth
3	< 1.5	> 1.0	≤ 1.0	mild	partial	outlet	backwater
4	> 1.0		≥ 1.0	any	full	outlet	backwater
5	≥ 1.5		≤ 1.0	any	partial	inlet	entrance geometry
6	≥ 1.5		≤ 1.0	any	full	outlet	entrance and barrel geometry

trial-and-error solutions. The friction head loss between sections 2 and 3, for example, can be calculated from Eq. 15.100.

$$h_{f,2\text{-}3} = \frac{LQ^2}{K_2 K_3} \qquad \textit{15.100}$$

$$K = \left(\frac{1}{n}\right) R^{2/3} A \qquad \text{[SI]} \quad \textit{15.101(a)}$$

$$K = \left(\frac{1.49}{n}\right) R^{2/3} A \qquad \text{[U.S.]} \quad \textit{15.101(b)}$$

The total hydraulic head available, H (equal to $h_1 - h_4$ in Fig. 15.25), is divided between the velocity head in the culvert, the entrance loss from Table 15.9 (if considered), and the friction.

$$H = \frac{\mathrm{v}^2}{2g} + k_e\left(\frac{\mathrm{v}^2}{2g}\right) + \frac{\mathrm{v}^2 n^2 L}{R^{4/3}} \qquad \text{[SI]} \quad \textit{15.102(a)}$$

$$H = \frac{\mathrm{v}^2}{2g} + k_e\left(\frac{\mathrm{v}^2}{2g}\right) + \frac{\mathrm{v}^2 n^2 L}{2.21 R^{4/3}} \qquad \text{[U.S.]} \quad \textit{15.102(b)}$$

Equation 15.102 can be solved directly for the velocity. Equation 15.103 is valid for culverts of any shape.

$$\mathrm{v} = \sqrt{\frac{H}{\dfrac{1 + k_e}{2g} + \dfrac{n^2 L}{R^{4/3}}}} \qquad \text{[SI]} \quad \textit{15.103(a)}$$

$$\mathrm{v} = \sqrt{\frac{H}{\dfrac{1 + k_e}{2g} + \dfrac{n^2 L}{2.21 R^{4/3}}}} \qquad \text{[U.S.]} \quad \textit{15.103(b)}$$

Table 15.9 *Minor Entrance Loss Coefficients*

k_e	condition of entrance
0.08	smooth, tapered
0.10	flush concrete groove
0.10	flush concrete bell
0.15	projecting concrete groove
0.15	projecting concrete bell
0.50	flush, square-edged
0.90	projecting, square-edged

For all types of flow, the discharge is found from Eq. 15.104. The product of C_c and C_v in this equation is equivalent to the discharge coefficient, C.

Submerged Orifice (Operating Under Steady-Flow Conditions)

$$\begin{aligned} Q &= C_c C_v A \sqrt{2g(h_1 - h_2)} \\ &= CA\sqrt{2g(h_1 - h_2)} \end{aligned} \qquad \textit{15.104}$$

A. Type-1 Flow

Water passes through the critical depth near the culvert entrance, and the culvert flows partially full. The slope of the culvert barrel is greater than the critical slope, and the tailwater elevation is less than the elevation of the water surface at the control section. The area, A, used in the discharge equation is not the culvert area since the culvert does not flow full. Instead, use the area in flow at the critical section.

B. Type-2 Flow

As in type-1 flow, flow passes through the critical depth at the culvert outlet, and the barrel flows partially full. The slope of the culvert is less than critical, and the tailwater elevation does not exceed the elevation of the water surface at the control section. The area, A, used in the discharge equation is not the culvert area since the culvert does not flow full. Instead, use the area in flow at the critical section.

Figure 15.25 *Culvert Flow Classifications*

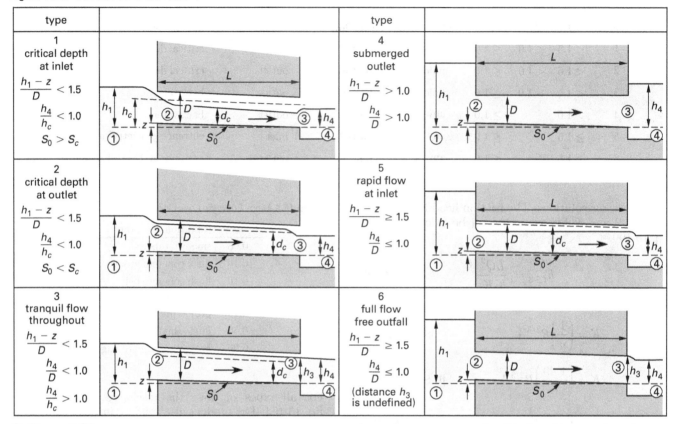

C. Type-3 Flow

When backwater is the controlling factor in culvert flow, the critical depth cannot occur. The upstream water-surface elevation for a given discharge is a function of the height of the tailwater. For type-3 flow, flow is subcritical for the entire length of the culvert, with the flow being partial. The outlet is not submerged, but the tailwater elevation does exceed the elevation of critical depth at the terminal section. The area, A, used in the discharge equation is not the culvert area since the culvert does not flow full. Instead, use the area in flow at numbered section 3 (i.e., the exit).

D. Type-4 Flow

As in type-3 flow, the backwater elevation is the controlling factor in this case. Critical depth cannot occur, and the upstream water surface elevation for a given discharge is a function of the tailwater elevation. Discharge is independent of barrel slope. The culvert is submerged at both the headwater and the tailwater. No differentiation between low head and high head is made for this case. If the velocity head at section 1 (the entrance), the entrance friction loss, and the exit friction loss are neglected, the discharge can be calculated. The area, A, is the culvert area. The complicated term in the

denominator corrects for friction. For rough estimates and for culverts less than 50 ft (15 m) long, the friction loss can be ignored.

E. Type-5 Flow

Partially full flow under a high head is classified as type-5 flow. The flow pattern is similar to the flow downstream from a sluice gate, with rapid flow near the entrance. Usually, type-5 flow requires a relatively square entrance that causes contraction of the flow area to less than the culvert area. In addition, the barrel length, roughness, and bed slope must be sufficient to keep the velocity high throughout the culvert.

It is difficult to distinguish in advance between type-5 and type-6 flow. Within a range of the important parameters, either flow can occur.[20] The area, A, is the culvert area.

F. Type-6 Flow

Type-6 flow, like type-5 flow, is considered a high-head flow. The culvert is full under pressure with free outfall. The discharge is easily calculated from Eq. 15.105.[21] A_o is the culvert area.

$$Q = A_o \mathrm{v} \qquad 15.105$$

---

[20]If the water surface ever touches the top of the culvert, the passage of air in the culvert will be prevented and the culvert will flow full everywhere. This is type-6 flow.
[21]Equation 15.105 does not include the discharge coefficient. Velocity, v, when calculated from Eq. 15.103, is implicitly the velocity in the barrel.

Example 15.12

Size a square culvert with an entrance fluid level 5 ft above the barrel top and a free exit to operate with the following characteristics.

$$\text{slope} = 0.01$$
$$\text{length} = 250 \text{ ft}$$
$$\text{capacity} = 45 \text{ ft}^3/\text{sec}$$
$$n = 0.013$$

Solution

Since the h_1 dimension is measured from the culvert invert, it is difficult to classify the type of flow at this point. However, either type 5 or type 6 is likely since the head is high.

step 1: Assume a trial culvert size. Select a square opening with 1 ft sides.

step 2: Calculate the flow assuming case 5 (entrance control). The entrance will act like an orifice.

$$A = (1 \text{ ft})(1 \text{ ft}) = 1 \text{ ft}^2$$
$$H = h_1 - h_2$$
$$= \big(5 \text{ ft} + 1 \text{ ft} + (0.01)(250 \text{ ft})\big)$$
$$\quad -(0.01)(250 \text{ ft})$$
$$= 6 \text{ ft}$$

C_d is approximately 0.62 for square-edged openings with separation from the wall. From Eq. 15.105,

Submerged Orifice (Operating Under Steady-Flow Conditions)

$$Q = C_c C_v A \sqrt{2g(h_1 - h_2)} = CA\sqrt{2g(h_1 - h_2)}$$

$$= (0.62)(1 \text{ ft}^2)\sqrt{(2)\Big(32.2 \ \frac{\text{ft}}{\text{sec}^2}\Big)(6 \text{ ft})}$$

$$= 12.2 \text{ ft}^3/\text{sec}$$

Since this size has insufficient capacity, try a larger culvert. Choose a square opening with 2 ft sides.

Submerged Orifice (Operating Under Steady-Flow Conditions)

$$H = h_1 - h_2 = 5 \text{ ft} + 2 \text{ ft} = 7 \text{ ft}$$

$$Q = C_c C_v A \sqrt{2g(h_1 - h_2)} = CA\sqrt{2g(h_1 - h_2)}$$

$$= (0.62)(4 \text{ ft}^2)\sqrt{(2)\Big(32.2 \ \frac{\text{ft}}{\text{sec}^2}\Big)(7 \text{ ft})}$$

$$= 52.7 \text{ ft}^3/\text{sec}$$

step 3: Begin checking the entrance control assumption by calculating the maximum hydraulic radius. The upper surface of the culvert is not wetted because the flow is entrance controlled. The hydraulic radius is maximum at the entrance.

$$R = \frac{A_o}{P}$$
$$= \frac{4 \text{ ft}^2}{2 \text{ ft} + 2 \text{ ft} + 2 \text{ ft}}$$
$$= 0.667 \text{ ft}$$

step 4: Calculate the velocity using the Manning equation for open channel flow. Since the hydraulic radius is maximum, the velocity will also be maximum.

$$v = \Big(\frac{1.49}{n}\Big) R^{2/3}\sqrt{S}$$
$$= \Big(\frac{1.49}{0.013}\Big)(0.667 \text{ ft})^{2/3}\sqrt{0.01}$$
$$= 8.75 \text{ ft}/\text{sec}$$

step 5: Calculate the normal depth, y_n.

$$y_n = \frac{Q}{vw} = \frac{45 \ \dfrac{\text{ft}^3}{\text{sec}}}{\Big(8.75 \ \dfrac{\text{ft}}{\text{sec}}\Big)(2 \text{ ft})}$$
$$= 2.57 \text{ ft}$$

Since the normal depth is greater than the culvert size, the culvert will flow full under pressure. (It was not necessary to calculate the critical depth since the flow is implicitly subcritical.) The entrance control assumption was, therefore, not valid for this size culvert.[22] At this point, two things can be done: A larger culvert can be chosen if entrance control is desired, or the solution can continue by checking to see if the culvert has the required capacity as a pressure conduit.

---

[22]If the normal depth had been less than the barrel diameter, it would still be necessary to determine the critical depth of flow. If the normal depth was less than the critical depth, the entrance control assumption would have been valid.

step 6: Check the capacity as a pressure conduit. H is the total available head.

$$H = h_1 - h_3$$
$$= \left(5 \text{ ft} + 2 \text{ ft} + (0.01)(250 \text{ ft})\right) - 2 \text{ ft}$$
$$= 7.5 \text{ ft}$$

step 7: Since the pipe is flowing full, the hydraulic radius is

$$R = \frac{A}{P} = \frac{4 \text{ ft}^2}{8 \text{ ft}} = 0.5 \text{ ft}$$

step 8: Equation 15.103 can be used to calculate the flow velocity. Since the culvert has a square-edged entrance, a loss coefficient of $k_e = 0.5$ is used. However, this does not greatly affect the velocity.

$$v = \sqrt{\frac{H}{\dfrac{1+k_e}{2g} + \dfrac{n^2 L}{2.21 R^{4/3}}}}$$

$$= \sqrt{\frac{7.5 \text{ ft}}{\dfrac{1+0.5}{(2)\left(32.2 \dfrac{\text{ft}}{\text{sec}}\right)} + \dfrac{(0.013)^2(250 \text{ ft})}{(2.21)(0.5 \text{ ft})^{4/3}}}}$$

$$= 10.24 \text{ ft/sec}$$

step 9: Check the capacity.

$$Q = vA_o = \left(10.24 \frac{\text{ft}}{\text{sec}}\right)(4 \text{ ft}^2)$$
$$= 40.96 \text{ ft}^3/\text{sec}$$

The culvert size is not acceptable since its discharge under the maximum head does not have a capacity of 45 ft^3/sec.

step 10: Repeat from step 2, trying a larger-size culvert. With a 2.5 ft side, the following values are obtained.

$$A = (2.5 \text{ ft})(2.5 \text{ ft}) = 6.25 \text{ ft}^2$$
$$H = 5 \text{ ft} + 2.5 \text{ ft} = 7.5 \text{ ft}$$
$$Q = (0.62)(6.25 \text{ ft}^2)$$
$$\times \sqrt{(2)\left(32.2 \frac{\text{ft}}{\text{sec}^2}\right)(7.5 \text{ ft})}$$
$$= 85.2 \text{ ft}^3/\text{sec}$$
$$R = \frac{6.25 \text{ ft}^2}{7.5 \text{ ft}} = 0.833 \text{ ft}$$

$$v = \left(\frac{1.49}{0.013}\right)(0.833 \text{ ft})^{2/3}\sqrt{0.01}$$
$$= 10.15 \text{ ft/sec}$$

$$y_n = \frac{Q}{vw} = \frac{45 \dfrac{\text{ft}^3}{\text{sec}}}{\left(10.15 \dfrac{\text{ft}}{\text{sec}}\right)(2.5 \text{ ft})}$$

$$= 1.77 \text{ ft}$$

step 11: Calculate the critical depth. For rectangular channels, Eq. 15.77 can be used.

Open-Channel Flow and Pipe Flow of Water

$$y_r = \left(\frac{q^2}{g}\right)^{1/3}$$

$$= \left(\frac{\left(45 \dfrac{\text{ft}^3}{\text{sec}}\right)^2}{\left(32.2 \dfrac{\text{ft}}{\text{sec}^2}\right)(2.5 \text{ ft})^2}\right)^{1/3}$$

$$= 2.16 \text{ ft}$$

Since the normal depth is less than the critical depth, the flow is supercritical. The entrance control assumption was correct for the culvert. The culvert has sufficient capacity to carry 45 ft^3/sec.

39. CULVERT DESIGN

Designing a culvert is somewhat easier than culvert analysis because of common restrictions placed on designers and the flexibility to change almost everything else. For example, culverts may be required to (a) never be more than 50% full (deep), (b) always be under inlet control, or (c) always operate with some minimum head (above the centerline or crown). In the absence of any specific guidelines, a culvert may be designed using the following procedure.

step 1: Determine the required flow rate.

step 2: Determine all water surface elevations, lengths, and other geometric characteristics.

step 3: Determine the material to be used for the culvert and its roughness.

step 4: Assume type 1 flow (inlet control).

step 5: Select a trial diameter.

step 6: Assume a reasonable slope.

step 7: Position the culvert entrance such that the ratio of headwater depth (inlet to water surface) to culvert diameter is 1:2 to 1:2.5.

step 8: Calculate the flow. Repeat step 5 through step 7 until the capacity is adequate.

step 9: Determine the location of the outlet. Check for outlet control. If the culvert is outlet controlled, repeat step 5 through step 7 using a different flow model.

step 10: Calculate the discharge velocity. Specify rip-rap, concrete, or other protection to prevent erosion at the outlet.

40. PRESSURE CULVERTS

A *culvert* is a water path (usually a large diameter pipe) used to channel water around or through an obstructing feature. (See Fig. 15.26.) Commonly, a culvert is a concrete pipe or corrugated metal pipe (CMP). In most instances, a culvert is used to restore a natural water path obstructed by a manufactured feature. For example, when a road is built across (perpendicular to) a natural drainage, a culvert is used to channel water under the road.

Figure 15.26 *Simple Pipe Culvert*

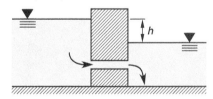

Because culverts often operate only partially full and with low heads, Torricelli's equation does not apply. Therefore, most culvert designs are empirical. However, if the entrance and exit of a culvert are both submerged, the culvert will flow full, and the discharge will be independent of the barrel slope. Equation 15.106 can be used to calculate the discharge.

Orifice Discharging Freely into Atmosphere

$$Q = CA_0\sqrt{2gh} \qquad \text{15.106}$$

If the culvert is long (more than 60 ft or 20 m), or if the entrance is not gradual, the available energy will be divided between friction and velocity heads. The effective head used in Eq. 15.106 should be

$$h_{\text{effective}} = h - h_{f,\text{barrel}} - h_{m,\text{entrance}} \qquad \text{15.107}$$

The friction loss in the barrel can be found in the usual manner, from either the Darcy equation or the Hazen-Williams equation. The entrance loss is calculated using the standard method of loss coefficients. Representative values of the loss coefficient, K, are given in Table 15.10. Since the fluid velocity is not initially known but is needed to find the friction factor, a trial-and-error solution will be necessary.

Table 15.10 *Representative Loss Coefficients, K, for Culvert Entrances*

entrance	K
smooth and gradual transition	0.08
flush vee or bell shape	0.10
projecting vee or bell shape	0.15
flush, square-edged	0.50
projecting, square-edged	0.90

41. SIPHONS

A *siphon* is a bent or curved tube that carries fluid from a fluid surface at a high elevation to another fluid surface at a lower elevation. Normally, it would not seem difficult to have a fluid flow to a lower elevation. However, the fluid seems to flow "uphill" in a portion of a siphon. Figure 15.27 illustrates a siphon.

Figure 15.27 *Siphon*

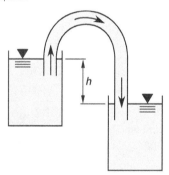

Starting a siphon requires the tube to be completely filled with liquid. Then, since the fluid weight is greater in the longer arm than in the shorter arm, the fluid in the longer arm "falls" out of the siphon, "pulling" more liquid into the shorter arm and over the bend.

Operation of a siphon is essentially independent of atmospheric pressure. The theoretical discharge is the same as predicted by the Torricelli equation. A correction for discharge is necessary, but little data is available on typical values of C. Therefore, siphons should be tested and calibrated in place. The discharge can be found using Eq. 15.106

42. NOMENCLATURE

A	area	ft^2	m^2
b	bottom width	ft	m
b	depth of flow	ft	m
C	Chezy coefficient	ft$^{1/3}$/sec	m$^{1/3}$/s
C	coefficient	–	–
C	Williams coefficient	ft$^{1/2}$/sec	m$^{1/2}$/s
C_s	Horton coefficient	–	–
d	depth	in	m
d	diameter	in	m
D	diameter	ft	m
D	hydraulic depth	ft	m
E	specific energy	ft	m
f	friction factor	–	–
Fr	Froude number	–	–
g	gravitational acceleration, 32.2 (9.81)	ft/sec^2	m/s^2
g_c	gravitational constant, 32.2	lbm-ft/lbf-sec^2	n.a.
h	head	ft	m
H	head	ft	m
k	minor loss coefficient	–	–
K	conveyance	ft^3/sec	m^3/s
K'	modified conveyance	–	–
L	channel length	ft	m
L	width	ft	m
$\dot{m}$	mass flow rate	lbm/sec	kg/s
n	Manning roughness coefficient	–	–
P	power	ft-lbf/sec	W
P	wetted perimeter	ft	m
q	unit discharge	ft/sec	m/s
Q	flow rate	gal/min	n.a.
R	radius	ft	m
R	hydraulic radius	ft	m
S	hydraulic slope	–	–
S	slope of energy line (energy gradient)	–	–
S_0	channel slope or geometric slope	–	–
T	width at surface or top width	ft	m
v	velocity	ft/sec	m/s
w	width	ft	m
y	normal depth	ft	m
Y	weir height	ft	m
z	height above datum	ft	m
Z	section factor	various	various

Symbols

α	kinetic energy correction factor	–	–
α	velocity-head coefficient	–	–
γ	specific weight	lbf/ft^3	n.a.
ρ	density	lbm/ft^3	kg/m^3
θ	angle	radian	radian

Subscripts

2	conjugate
b	brink
c	critical or composite
d	discharge
e	entrance or equivalent
f	friction
H	hydraulic
n	normal
o	channel or culvert barrel
s	spillway
t	total
w	weir

16 Potable Water Source Quality

1. WATER SUPPLY CHEMISTRY

Most municipal water supply composition data are not given in units of molarity, normality, molality, and so on. Rather, the most common measure of solution strength is the *CaCO₃ equivalent*. With this method, substances are reported in milligrams per liter (mg/L, same as parts per million, ppm) "as $CaCO_3$," even when $CaCO_3$ is unrelated to the substance or reaction that produced the substance.

Actual gravimetric amounts of a substance can be converted to amounts as $CaCO_3$ by use of the conversion factors in App. 16.A. These factors are easily derived from stoichiometric principles.

The reason for converting all substance quantities to amounts as $CaCO_3$ is to simplify calculations for non-technical personnel. Equal $CaCO_3$ amounts constitute stoichiometric reaction quantities. For example,

100 mg/L as $CaCO_3$ of sodium ion (Na^+) will react with 100 mg/L as $CaCO_3$ of chloride ion (Cl^-) to produce 100 mg/L as $CaCO_3$ of salt (NaCl), even though the gravimetric quantities differ and $CaCO_3$ is not part of the reaction.

Example 16.1

Lime is added to water to remove carbon dioxide gas.

$$CO_2 + Ca(OH)_2 \rightarrow CaCO_3 \downarrow + H_2O$$

If water contains 5 mg/L of CO_2, how much lime is required for its removal?

Solution

From App. 16.A, the factor that converts CO_2 as substance to CO_2 as $CaCO_3$ is 2.27.

$$CO_2 \text{ as } CaCO_3 \text{ equivalent} = (2.27)\left(5 \ \frac{mg}{L}\right)$$
$$= 11.35 \text{ mg/L as } CaCO_3$$

Therefore, the $CaCO_3$ equivalent of lime required will also be 11.35 mg/L.

From App. 16.A again, the factor that converts lime as $CaCO_3$ to lime as substance is (1/1.35).

$$Ca(OH)_2 \text{ substance} = \frac{11.35 \ \dfrac{mg}{L}}{1.35}$$
$$= 8.41 \text{ mg/L as substance}$$

This problem could also have been solved stoichiometrically.

2. CATIONS AND ANIONS IN NEUTRAL SOLUTIONS

Equivalency concepts provide a useful check on the accuracy of water analyses. For the water to be electrically neutral, the sum of anion equivalents must equal the sum of cation equivalents.

Concentrations of dissolved compounds in water are usually expressed in mg/L, not equivalents. However, anionic and cationic substances can be converted to their equivalent concentrations in milliequivalents per liter (meq/L) by dividing their concentrations in mg/L by their equivalent weights. Appendix 16.A is useful for this purpose.

Since water is an excellent solvent, it will contain the ions of the inorganic compounds to which it is exposed. A chemical analysis listing these ions does not explicitly determine the compounds from which the ions originated. Several graphical methods, such as bar graphs and Piper (Hill) trilinear diagrams, can be used for this purpose. The bar graph, also known as a *milliequivalent per liter bar chart*, is constructed by listing the cations (positive ions) in the sequence of calcium (Ca^{2+}), magnesium (Mg^{2+}), sodium (Na^+), and potassium (K^+), and pairing them with the anions (negative ions) in the sequence of carbonate (CO_3^{2-}), bicarbonate (HCO_3^-), sulfate (SO_4^{2-}), and chloride (Cl^-). A bar chart, such as the one shown in Ex. 16.2, can be used to deduce the hypothetical combinations of positive and negative ions that would have resulted in the given water analysis.

Example 16.2

A water analysis reveals the following ionic components in solution: Ca^{2+}, 29.0 mg/L; Mg^{2+}, 16.4 mg/L; Na^+, 23.0 mg/L; K^+, 17.5 mg/L; HCO_3^-, 171 mg/L; SO_4^{2-}, 36.0 mg/L; Cl^-, 24.0 mg/L. Verify that the analysis is reasonably accurate. Draw the milliequivalent per liter bar chart.

Solution

Use the equivalent weights in App. 16.A to complete the following table.

compound	concentration (mg/L)	equivalent weight (g/mol)	equivalent (meq/L)
cations			
Ca^{2+}	29.0	20.0	1.45
Mg^{2+}	16.4	12.2	1.34
Na^+	23.0	23.0	1.00
K^+	17.5	39.1	0.44
		total	4.23
anions			
HCO_3^-	171.0	61.0	2.80
SO_4^{2-}	36.0	48.0	0.75
Cl^-	24.0	35.5	0.68
		total	4.23

The sums of the cation equivalents and anion equivalents are equal. The analysis is presumed to be reasonably accurate.

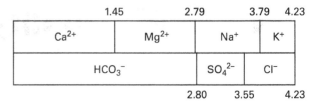

3. ACIDITY

Acidity is a measure of acids in solution. Acidity in surface water is caused by formation of carbonic acid (H_2CO_3) from carbon dioxide in the air. (See Eq. 16.1.) Carbonic acid is aggressive and must be neutralized to eliminate a cause of water pipe corrosion. If the pH of water is greater than 4.5, carbonic acid ionizes to form bicarbonate. (See Eq. 16.2.) If the pH is greater than 8.3, carbonate ions form. (See Eq. 16.3.)

$$CO_2 + H_2O \rightarrow H_2CO_3 \qquad \textit{16.1}$$

$$H_2CO_3 + H_2O \rightarrow HCO_3^- + H_3O^+ \text{ (pH > 4.5)} \qquad \textit{16.2}$$

$$HCO_3^- + H_2O \rightarrow CO_3^{2-} + H_3O^+ \text{ (pH > 8.3)} \qquad \textit{16.3}$$

Measurement of acidity is done by titration with a standard basic measuring solution. Acidity, A, in water is typically given in terms of the $CaCO_3$ equivalent that would neutralize the acid. The constant 50 000 used in Eq. 16.4 is the product of the equivalent weight of $CaCO_3$ (50 g) and 1000 mg/g.

$$A_{\text{mg/L as } CaCO_3} = \frac{V_{\text{titrant,mL}} N_{\text{titrant}} (50\,000)}{V_{\text{sample,mL}}} \qquad \textit{16.4}$$

The standard titration procedure for determining acidity measures the amount of titrant needed to raise the pH to 3.7 plus the amount needed to raise the pH to 8.3. The total of these two is used in Eq. 16.4. Most water samples, unless grossly polluted with industrial wastes, will exist at a pH greater than 3.7, so the titration will be a one-step process.

4. ALKALINITY

Alkalinity is a measure of the ability of a water to neutralize acids (i.e., to absorb hydrogen ions without significant pH change). The principal alkaline ions are OH^-, CO_3^{2-}, and HCO_3^-. Other radicals, such as NO_3^-, also contribute to alkalinity, but their presence is rare. The measure of alkalinity is the sum of concentrations of each of the substances measured as equivalent $CaCO_3$.

The standard titration method for determining alkalinity measures the amount of acidic titrant needed to lower the pH to 8.3 plus the amount needed to lower the pH to 4.5. Therefore, alkalinity, M, is the sum of all titratable base concentrations down to a pH of 4.5. The constant 50 000 in Eq. 16.5 is the product of the equivalent weight of $CaCO_3$ (50 g) and 1000 mg/g.

$$M_{mg/L\,as\,CaCO_3} = \frac{V_{titrant,mL}N_{titrant}(50\,000)}{V_{sample,mL}} \quad \text{16.5}$$

Example 16.3

Water from a city well is analyzed and found to contain 20 mg/L as substance of HCO_3^- and 40 mg/L as substance of CO_3^{2-}. What is the alkalinity of the water expressed as $CaCO_3$?

Solution

The equivalent weight of HCO_3^- is 61 g/mol, the equivalent weight of CO_3^{2-} is 30 g/mol, and the equivalent weight of $CaCO_3$ is 50 g/mol. [Periodic Table of Elements]

$$M_{mg/L\,of\,CaCO_3} = \left(20\,\frac{mg}{L}\right)\left(\frac{50\,\frac{g}{mol}}{61\,\frac{g}{mol}}\right)$$

$$+\left(40\,\frac{mg}{L}\right)\left(\frac{50\,\frac{g}{mol}}{30\,\frac{g}{mol}}\right)$$

$$= 83.1 \text{ mg/L as } CaCO_3$$

5. ACIDITY AND ALKALINITY IN MUNICIPAL WATER SUPPLIES

Acidity is a measure of acids in solutions. Acidity in surface water (e.g., lakes and streams) is caused by formation of *carbonic acid* (H_2CO_3) from carbon dioxide in the air.[1] Acidity in water is typically given in terms of the $CaCO_3$ equivalent.

$$CO_2 + H_2O \rightarrow H_2CO_3 \quad \text{16.6}$$

$$H_2CO_3 + H_2O \rightarrow HCO_3^- + H_3O^+ \quad [pH > 4.5] \quad \text{16.7}$$

$$HCO_3^- + H_2O \rightarrow CO_3^{2-} + H_3O^+ \quad [pH > 8.3] \quad \text{16.8}$$

Alkalinity is a measure of the amount of negative (basic) ions in the water. Specifically, OH^-, CO_3^{2-}, and HCO_3^- all contribute to alkalinity.[2] The measure of alkalinity is the sum of concentrations of each of the substances measured as $CaCO_3$.

Alkalinity and acidity of a titrated sample are determined from color changes in indicators added to the titrant.

Example 16.4

Water from a city well is analyzed and is found to contain 20 mg/L as substance of HCO_3^- and 40 mg/L as substance of CO_3^{2-}. What is the alkalinity of this water as $CaCO_3$?

Solution

From App. 16.A, the factors converting HCO_3^- and CO_3^{2-} ions to $CaCO_3$ equivalents are 0.82 and 1.67, respectively.

$$\text{alkalinity} = (0.82)\left(20\,\frac{mg}{L}\right) + (1.67)\left(40\,\frac{mg}{L}\right)$$

$$= 83.2 \text{ mg/L as } CaCO_3$$

6. INDICATOR SOLUTIONS

End points for acidity and alkalinity titrations are determined by color changes in indicator dyes that are pH sensitive. Several commonly used indicators are listed in Table 16.1.

Table 16.1 Indicator Solutions Commonly Used in Water Chemistry

indicator	titration	end point pH	color change
bromophenol blue	acidity	3.7	yellow to blue
phenolphthalein	acidity	8.3	colorless to red-violet
phenolphthalein	alkalinity	8.3	red-violet to colorless
mixed bromocresol/ green-methyl red	alkalinity	4.5	grayish to orange-red

Depending on pH, alkaline samples can contain hydroxide alone, hydroxide and carbonate, carbonate alone, carbonate and bicarbonate, or bicarbonate alone. Samples containing hydroxide or hydroxide and carbonate have a high pH, usually greater than 10. If the titration

---

[1]Carbonic acid is very aggressive and must be neutralized to eliminate the cause of water pipe corrosion. If the pH of water is greater than 4.5, carbonic acid ionizes to form bicarbonate. (See Eq. 16.7.) If the pH is greater than 8.3, carbonate ions form that cause water hardness by combining with calcium. (See Eq. 16.8.)

[2]Other ions, such as NO_3^-, also contribute to alkalinity, but their presence is rare. If detected, they should be included in the calculation of alkalinity.

is complete at the phenolphthalein end point (i.e., the mixed bromocresol green-methyl red indicator does not change color), the alkalinity is hydroxide alone. Samples containing carbonate and bicarbonate alkalinity have a pH greater than 8.3, and the titration to the phenolphthalein end point represents stoichiometrically one-half of the carbonate alkalinity. If the volume of phenolphthalein titrant equals the volume of titrant needed to reach the mixed bromocresol green-methyl red end point, all of the alkalinity is in the form of carbonate.

If abbreviations of P for the measured phenolphthalein alkalinity and M for the total alkalinity are used, the following relationships define the possible alkalinity states of the sample. In sequence, the relationships can be used to determine the actual state of a sample.

(state I)	hydroxide $= P = M$
(state II)	hydroxide $= 2P - M$ and
	carbonate $= 2(M - P)$
(state III)	carbonate $= 2P = M$
(state IV)	carbonate $= 2P$ and
	bicarbonate $= M - 2P$
(state V)	bicarbonate $= M$

Example 16.5

A 100 mL sample is titrated for alkalinity by using 0.02 N sulfuric acid solution. To reach the phenolphthalein end point requires 3.0 mL of the acid solution, and an additional 12.0 mL is added to reach the mixed bromocresol green-methyl red end point. Calculate the (a) phenolphthalein alkalinity and (b) total alkalinity. (c) What are the ionic forms of alkalinity present?

Solution

(a) Use Eq. 16.5.

$$P = \frac{V_{titrant,mL}N_{titrant}(50\,000)}{V_{sample,mL}}$$

$$= \frac{(3.0 \text{ mL})\left(0.02 \frac{gEW}{L}\right)\left(50\,000 \frac{mg}{gEW}\right)}{100 \text{ mL}}$$

$$= 30 \text{ mg/L (as CaCO}_3\text{)}$$

(b) Use Eq. 16.5.

$$M = \frac{V_{titrant,mL}N_{titrant}(50\,000)}{V_{sample,mL}}$$

$$= \frac{(3.0 \text{ mL} + 12.0 \text{ mL})}{\times\left(0.02 \frac{gEW}{L}\right)\left(50\,000 \frac{mg}{gEW}\right)}{100 \text{ mL}}$$

$$= 150 \text{ mg/L (as CaCO}_3\text{)}$$

(c) Test the state relationships in sequence.

(I) hydroxide $= P = M$:

$$30 \text{ mg/L} \neq 150 \text{ mg/L}$$
$$\text{(invalid: } P \neq M\text{)}$$

(II) hydroxide $= 2P - M$:

$$(2)\left(30 \frac{mg}{L}\right) - 150 \frac{mg}{L}$$
$$= -90 \text{ mg/L}$$
$$\text{(invalid: negative value)}$$

(III) carbonate $= 2P = M$:

$$(2)\left(30 \frac{mg}{L}\right) \neq 150 \text{ mg/L}$$
$$\text{(invalid: } 2P \neq M\text{)}$$

(IV) carbonate $= 2P$:

$$(2)\left(30 \frac{mg}{L}\right) = 60 \text{ mg/L}$$

bicarbonate $= M - 2P$

$$150 \frac{mg}{L} - (2)\left(30 \frac{mg}{L}\right) = 90 \text{ mg/L}$$

The process ends at (IV) since a valid answer is obtained. It is not necessary to check relationship (V), as (IV) gives consistent results. The alkalinity is composed of 60 mg/L as CaCO$_3$ of carbonate and 90 mg/L as CaCO$_3$ of bicarbonate.

7. HARDNESS

Hardness in natural water is caused by the presence of polyvalent (but not singly charged) metallic cations. Principal cations causing hardness in water and the major anions associated with them are presented in Table 16.2. Because the most prevalent of these species are the divalent cations of calcium and magnesium, total hardness is typically defined as the sum of the concentration of these two elements and is expressed in terms of milligrams per liter as CaCO$_3$. (Hardness is occasionally expressed in units of *grains per gallon*, where 7000 grains are equal to a pound.)

Carbonate hardness is caused by cations from the dissolution of calcium or magnesium carbonate and bicarbonate in the water. Carbonate hardness is hardness that is chemically equivalent to alkalinity, where most of the alkalinity in natural water is caused by the bicarbonate and carbonate ions.

Noncarbonate hardness is caused by cations from calcium (i.e., calcium hardness) and magnesium (i.e., magnesium hardness) compounds of sulfate, chloride, or silicate that are dissolved in the water. Noncarbonate hardness is equal to the total hardness minus the carbonate hardness.

Table 16.2 *Principal Cations and Anions Indicating Hardness*

cations	anions
Ca^{2+}	HCO_3^-
Mg^{2+}	SO_4^{2-}
Sr^{2+}	Cl^-
Fe^{2+}	NO_3^-
Mn^{2+}	SiO_3^{2-}

Hardness can be classified as shown in Table 16.3. Although high values of hardness do not present a health risk, they have an impact on the aesthetic acceptability of water for domestic use. (Hardness reacts with soap to reduce its cleansing effectiveness and to form scum on the water surface.) Where feasible, carbonate hardness in potable water should be reduced to the 25–40 mg/L range and total hardness reduced to the 50–75 mg/L range.

Table 16.3 *Relationship of Hardness Concentration to Classification*

hardness (mg/L as $CaCO_3$)	classification
0 to 60	soft
61 to 120	moderately hard
121 to 180	hard
181 to 350	very hard
> 350	saline; brackish

Water containing bicarbonate (HCO_3^-) can be heated to precipitate carbonate (CO_3^{2-}) as a *scale*. Water used in steam-producing equipment (e.g., boilers) must be essentially hardness-free to avoid deposit of scale.

Noncarbonate hardness, also called *permanent hardness*, cannot be removed by heating. It can be removed by precipitation softening processes (typically the lime-soda ash process) or by ion exchange processes using resins selective for ions causing hardness.

Hardness is measured in the laboratory by titrating the sample using a standardized solution of ethylenediaminetetraacetic acid (EDTA) and an indicator dye such as Eriochrome Black T. The sample is titrated at a pH of approximately 10 until the dye color changes from red to blue. The standardized solution of EDTA is usually prepared such that 1 mL of EDTA is equivalent to 1 mg/L of hardness, but Eq. 16.9 can be used to determine hardness with any strength EDTA solution.

$$H_{mg/L\,as\,CaCO_3} = \frac{V_{titrant,mL}\times(CaCO_3 \text{ equivalent of EDTA})(1000)}{V_{sample,mL}} \quad \text{16.9}$$

Example 16.6

A water sample contains sodium (Na^+, 15 mg/L), magnesium (Mg^{2+}, 70 mg/L), and calcium (Ca^{2+}, 40 mg/L). What is the hardness?

Solution

Sodium is singly charged, so it does not contribute to hardness. The necessary approximate equivalent weights are found in a periodic table. [Periodic Table of Elements]

Mg: 12.2 g/mol

Ca: 20.0 g/mol

$CaCO_3$: 50 g/mol

The equivalent hardness is

$$H = \left(70\ \frac{mg}{L}\right)\left(\frac{50\ \frac{g}{mol}}{12.2\ \frac{g}{mol}}\right) + \left(40\ \frac{mg}{L}\right)\left(\frac{50\ \frac{g}{mol}}{20\ \frac{g}{mol}}\right)$$

$$= 387\ \text{mg/L (as } CaCO_3)$$

Example 16.7

A 75 mL water sample was titrated using 8.1 mL of an EDTA solution formulated such that 1 mL of EDTA is equivalent to 0.8 mg/L of $CaCO_3$. What is the hardness of the sample, measured in terms of mg/L of $CaCO_3$?

Solution

Use Eq. 16.9.

$$H = \frac{V_{\text{titrant,mL}}(\text{CaCO}_3 \text{ equivalent of EDTA})(1000)}{V_{\text{sample,mL}}}$$

$$= \frac{(8.1 \text{ mL})\left(0.8 \dfrac{\text{gEW}}{\text{L}}\right)\left(1000 \dfrac{\text{mg}}{\text{gEW}}\right)}{75 \text{ mL}}$$

$$= 86.4 \text{ mg/L (as CaCO}_3)$$

8. WATER HARDNESS

Water hardness is caused by multivalent (doubly charged, triply charged, etc., but not singly charged) positive metallic ions such as calcium, magnesium, iron, and manganese. (Iron and manganese are not as common, however.) Hardness reacts with soap to reduce its cleansing effectiveness and to form scum on the water surface and a ring around the bathtub.

Water containing bicarbonate (HCO_3^-) ions can be heated to precipitate carbonate molecules.[3] This hardness is known as *temporary hardness* or *carbonate hardness*.[4]

Remaining hardness due to sulfates, chlorides, and nitrates is known as *permanent hardness* or *noncarbonate hardness* because it cannot be removed by heating. The amount of permanent hardness can be determined numerically by causing precipitation, drying, and then weighing the precipitate.

Total hardness is the sum of temporary and permanent hardnesses, both expressed in mg/L as CaCO_3.

9. HARDNESS AND ALKALINITY

Hardness is caused by multi-positive ions. Alkalinity is caused by negative ions. Both positive and negative ions are present simultaneously. Therefore, an alkaline water can also be hard.

With some assumptions and minimal information about the water composition, it is possible to determine the ions in the water from the hardness and alkalinity. For example, Fe^{2+} is an unlikely ion in most water supplies, and it is often neglected. Figure 16.1 can be used to quickly deduce the compounds in the water from hardness and alkalinity.

If hardness and alkalinity (both as CaCO_3) are the same and there are no monovalent cations, then there are no SO_4^{2-}, Cl^-, or NO_3^- ions present. That is, there is no noncarbonate (permanent) hardness. If hardness is greater than the alkalinity, however, then noncarbonate hardness is present, and the carbonate (temporary) hardness

is equal to the alkalinity. If hardness is less than the alkalinity, then all hardness is carbonate hardness, and the extra HCO_3^- comes from other sources (such as NaHCO_3).

Example 16.8

A water sample analysis results in the following: alkalinity, 220 mg/L; hardness, 180 mg/L; Ca^{2+}, 140 mg/L; OH^-, insignificant. All concentrations are expressed as CaCO_3. (a) What is the noncarbonate hardness? (b) What is the Mg^{2+} content in mg/L as substance?

Solution

(a) Use Fig. 16.1 with $M > H$ (alkalinity greater than hardness).

$$[\text{NaHCO}_3] = M - H = 220 \frac{\text{mg}}{\text{L}} - 180 \frac{\text{mg}}{\text{L}}$$
$$= 40 \text{ mg/L (as CaCO}_3)$$
$$[\text{Mg(HCO}_3)_2] = H - \text{Ca} = 180 \frac{\text{mg}}{\text{L}} - 140 \frac{\text{mg}}{\text{L}}$$
$$= 40 \text{ mg/L (as CaCO}_3)$$
$$[\text{Ca(HCO}_3)_2] = \text{Ca} = 140 \text{ mg/L (as CaCO}_3)$$
$$\text{carbonate hardness} = H = 180 \text{ mg/L (as CaCO}_3)$$
$$\text{noncarbonate hardness} = 0$$

(b) The Mg^{2+} ion content as CaCO_3 is equal to the $\text{Mg(HCO}_3)_2$ content as CaCO_3. Use a periodic table to convert CaCO_3 equivalents to amounts as substance. To convert Mg^{2+} as CaCO_3 to Mg^{2+} as substance, use a factor of 4.1. [Periodic Table of Elements]

$$\text{Mg}^{2+} = \frac{40 \dfrac{\text{mg}}{\text{L}}}{4.1} = 9.8 \text{ mg/L (as substance)}$$

10. COMPARISON OF ALKALINITY AND HARDNESS

Hardness measures the presence of positive, multivalent ions in the water supply. Alkalinity measures the presence of negative (basic) ions such as hydrates, carbonates, and bicarbonates. Since positive and negative ions coexist, an alkaline water can also be hard.

If certain assumptions are made, it is possible to draw conclusions about the water composition from the hardness and alkalinity. For example, if the effects of Fe^{2+} and OH^- are neglected, the following rules apply. (All concentrations are measured as CaCO_3.)

---

[3]Hard water forms scale when heated. This scale, if it forms in pipes, eventually restricts water flow. Even in small quantities, the scale insulates boiler tubes. Therefore, water used in steam-producing equipment must be essentially hardness-free.
[4]The hardness is known as *carbonate* hardness even though it is caused by *bicarbonate* ions, not carbonate ions.

Figure 16.1 *Hardness and Alkalinity*[a,b]

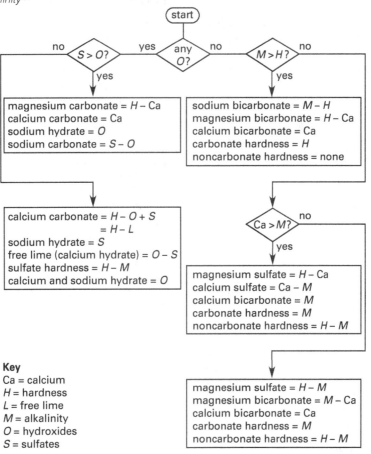

Key
Ca = calcium
H = hardness
L = free lime
M = alkalinity
O = hydroxides
S = sulfates

[a]All concentrations are expressed as $CaCO_3$.
[b]Not for use when other ionic species are present in significant quantities.

- *hardness = alkalinity:* There is no noncarbonate hardness. There are no SO_4^{2-}, Cl^-, or NO_3^- ions present.

- *hardness > alkalinity:* Noncarbonate hardness is present.

- *hardness < alkalinity:* All hardness is carbonate hardness. The extra HCO_3^- comes from other sources (e.g., $NaHCO_3$).

Titration with indicator solutions is used to determine the alkalinity. The *phenolphthalein alkalinity* (or "P reading" in mg/L as $CaCO_3$) measures hydrate alkalinity and half of the carbonate alkalinity. The *methyl orange alkalinity* (or "M reading" in mg/L as $CaCO_3$) measures the total alkalinity (including the phenolphthalein alkalinity). Table 16.4 can be used to interpret these tests.[5]

Table 16.4 *Interpretation of Alkalinity Tests*

case	hydrate as $CaCO_3$	carbonate as $CaCO_3$	bicarbonate as $CaCO_3$
$P = 0$	0	0	M
$0 < P < \dfrac{M}{2}$	0	2P	M − 2P
$P = \dfrac{M}{2}$	0	2P	0
$\dfrac{M}{2} < P < M$	2P − M	2(M − P)	0
$P = M$	M	0	0

---

[5]The titration may be affected by the presence of silica and phosphates, which also contribute to alkalinity. The effect is small, but the titration may not be a completely accurate measure of carbonates and bicarbonates.

11. IRON

Even at low concentrations, iron is objectionable because it stains porcelain bathroom fixtures, causes a brown color in laundered clothing, and can be tasted. Typically, iron is a problem in groundwater pumped from anaerobic aquifers in contact with iron compounds. Soluble ferrous ions can be formed under these conditions, which, when exposed to atmospheric air at the surface or to dissolved oxygen in the water system, are oxidized to the insoluble ferric state, causing the color and staining problems mentioned. Iron determinations are made through colorimetric (i.e., wet titration) analysis.

12. MANGANESE

Manganese ions are similar in formation, effect, and measurement to iron ions.

13. FLUORIDE

Natural fluoride is found in groundwaters as a result of dissolution from geologic formations. Surface waters generally contain much smaller concentrations of fluoride. An absence or low concentration of ingested fluoride causes the formation of tooth enamel less resistant to decay, resulting in a high incidence of dental cavities in children's teeth. Excessive concentration of fluoride causes *fluorosis*, a brownish discoloration of dental enamel. The MCL of 4.0 mg/L established by the EPA is to prevent unsightly fluorosis.

Communities with water supplies deficient in natural fluoride may chemically add fluoride during the treatment process. Since water consumption is influenced by climate, the recommended optimum concentrations listed in Table 16.5 are based on the annual average of the maximum air temperatures based on a minimum of five years of records.

Compounds commonly used as fluoride sources in water treatment are listed in Table 16.6.

Table 16.5 *Recommended Optimum Concentrations of Fluoride in Drinking Water*

average air temperature range (°F (°C))	recommended optimum concentration (mg/L)
53.7 (12.0) and below	1.2
53.8 to 58.3 (12.1–14.6)	1.1
58.4 to 63.8 (14.7–17.6)	1.0
63.9 to 70.6 (17.7–21.4)	0.9
70.7 to 79.2 (21.5–26.2)	0.8
79.3 to 90.5 (26.3–32.5)	0.7

Table 16.6 *Fluoridation Chemicals*

compound	formula	percentage F^- ion (%)
sodium fluoride	NaF	45
sodium silicofluoride	Na_2SiF_6	61
hydrofluosilicic acid	H_2SiF_6	79
ammonium silicofluoride*	$(NH_4)_2SiF_6$	64

*used in conjunction with chlorine disinfection where it is desired to maintain a chloramine residual in the distribution system

Example 16.9

A liquid feeder adds a 4.0% saturated sodium fluoride solution to a water supply, increasing the fluoride concentration from the natural fluoride level of 0.4 mg/L to 1.0 mg/L. The commercial NaF powder used to prepare the NaF solution contains 45% fluoride by weight. (a) How many pounds (kilograms) of NaF are required per million gallons (liters) treated? (b) What volume of 4% NaF solution is used per million gallons (liters)?

SI Solution

$$m_{NaF} = \frac{\left(1.0\ \frac{mg}{L} - 0.4\ \frac{mg}{L}\right)\left(10^6\ \frac{L}{ML}\right)}{(0.45)\left(10^6\ \frac{mg}{kg}\right)}$$

$$= 1.33\ kg/ML \quad (kg\ per\ million\ liters)$$

(b) One liter of water has a mass of one kilogram (10^6 mg). Therefore, a 4% NaF solution contains 40 000 mg NaF per liter. The concentration needs to be increased from 0.4 mg/L to 1.0 mg/L.

$$\left(40\,000\ \frac{mg}{L}\right)(0.45) = 18\,000\ mg/L$$

$$V = \frac{\left(1.0 \times 10^6\ \frac{L}{ML}\right) \times \left(1.0\ \frac{mg}{L} - 0.4\ \frac{mg}{L}\right)}{18\,000\ \frac{mg}{L}}$$

$$= 33.3\ L/ML$$

Customary U.S. Solution

$$m_{NaF} = \left(\frac{1.0\ \frac{mg}{L} - 0.4\ \frac{mg}{L}}{0.45}\right)\left(8.345\ \frac{lbm\text{-}L}{mg\text{-}MG}\right)$$

$$= 11.1\ lbm/MG \quad (lbm\ per\ million\ gallons)$$

(b) A 4% NaF solution contains 40 000 mg NaF per liter. The concentration needs to be increased from 0.4 mg/L to 1.0 mg/L.

$$\left(40\,000 \; \frac{\text{mg}}{\text{L}}\right)(0.45) = 18\,000 \; \text{mg/L}$$

$$V = \frac{\left(1.0 \times 10^6 \; \frac{\text{gal}}{\text{MG}}\right) \times \left(1.0 \; \frac{\text{mg}}{\text{L}} - 0.4 \; \frac{\text{mg}}{\text{L}}\right)}{18\,000 \; \frac{\text{mg}}{\text{L}}}$$

$$= 33.3 \; \text{gal/MG}$$

14. PHOSPHORUS

Phosphate content is more of a concern in wastewater treatment than in supply water, although phosphorus can enter water supplies in large amounts from runoff. Excessive phosphate discharge contributes to aquatic plant (phytoplankton, algae, and macrophytes) growth and subsequent *eutrophication*. (Eutrophication is an "over-fertilization" of receiving waters.)

Phosphorus is of considerable interest in the management of lakes and reservoirs because phosphorus is a nutrient that has a major effect on aquatic plant growth. Algae normally have a phosphorus content of 0.1–1% of dry weight. The molar N:P ratio for ideal algae growth is 16:1.

Phosphorus exists in several forms in aquatic environments. Soluble phosphorus occurs as *orthophosphate*, as condensed *polyphosphates* (from detergents), and as various organic species. Orthophosphates ($H_2PO_4^-$, HPO_4^{2-}, and PO_4^{3-}) and polyphosphates (such as $Na_3(PO_3)_6$) result from the use of synthetic detergents (*syndets*). A sizable fraction of the soluble phosphorus is in the organic form, originating from the decay or excretion of nucleic acids and algal storage products. *Particulate phosphorus* occurs in the organic form as a part of living organisms and detritus as well as in the inorganic form of minerals such as apatite.

Phosphorus is normally measured by colorimetric or digestion methods. The results are reported in terms of mg/L of phosphorus (e.g., "mg/L of P" or "mg/L of total P"), although the tests actually measure the concentration of orthophosphate. The concentration of a particular compound is found by multiplying the concentration as P by the molecular weight of the compound and dividing by the atomic weight of phosphorus (30.97).

$$[X] = \frac{[P](MW_X)}{30.97} \qquad \textit{16.10}$$

A substantial amount of the phosphorus that enters lakes is probably not available to aquatic plants. Bioavailable compounds include orthophosphates, polyphosphates, most soluble organic phosphorus, and a portion of the particulate fraction. Studies have indicated that bioavailable phosphorus generally does not exceed 60% of the total phosphorus.

In aquatic systems, phosphorus does not enter into any redox reactions, nor are any common species volatile. Therefore, lakes retain a significant portion of the entering phosphorus. The main mechanism for retaining phosphorus is simple sedimentation of particles containing the phosphorus. Particulate phosphorus can originate from the watershed (*allochthonous material*) or can be formed within the lake (*autochthonous material*).

A large fraction of the phosphorus that enters a lake is recycled, and much of the recycling occurs at the sediment-water interface. Recycling of phosphorus is linked to iron and manganese recycling. Soluble phosphorus in the water column is removed by adsorption onto iron and manganese hydroxides, which precipitate under aerobic conditions. However, when the *hypolimnion* (i.e., the lower part of the lake that is essentially stagnant) becomes anaerobic, the iron (or manganese) is reduced, freeing up phosphorus. This is consistent with a fairly general observation that phosphorus release rates are nearly an order of magnitude higher under anaerobic conditions than under aerobic conditions.

Factors that control phosphorus recycling rates are not well understood, although it is clear that oxygen status, phosphorus speciation, temperature, and pH are important variables. Phosphorus release from sediments is usually considered to be constant (usually less than 1 mg/m²·d under aerobic conditions).

In addition to direct regeneration from sediments, *macrophytes* (large aquatic plants) often play a significant role in phosphorus recycling. Macrophytes with highly developed root systems derive most of their phosphorus from the sediments. Regeneration to the water column can occur by excretion or through decay, effectively "pumping" phosphorus from the sediments. The internal loading generated by recycling is particularly important in shallow, eutrophic lakes. In several cases where phosphorus inputs have been reduced to control algal blooms, regeneration of phosphorus from phosphorus-rich sediments has slowed the rate of recovery.

15. NITROGEN

Compounds containing nitrogen are not abundant in virgin surface waters. However, nitrogen can reach large concentrations in ground waters that have been contaminated with barnyard runoff or that have percolated through heavily fertilized fields. Sources of surface water contamination include agricultural runoff and discharge from sewage treatment facilities.

Of greatest interest, in order of decreasing oxidation state, are nitrates (NO_3^-), nitrites (NO_2^-), ammonia (NH_3), and organic nitrogen. These three compounds are reported as *total nitrogen*, TN, with units of "mg/L of N" or "mg/L of total N." The concentration of a particular compound is found by multiplying the concentration as N by the molecular weight of the compound and dividing by the atomic weight of nitrogen (14.01).

$$[X] = \frac{[N](MW_X)}{14.01} \qquad 16.11$$

Excessive amounts of nitrate in water can contribute to the illness in infants known as *methemoglobinemia* ("blue baby" syndrome). As with phosphorus, nitrogen stimulates aquatic plant growth.

Un-ionized ammonia is a colorless gas at standard temperature and pressure. A pungent odor is detectable at levels above 50 mg/L. Ammonia is very soluble in water at low pH.

Ammonia levels in zero-salinity surface water increase with increasing pH and temperature (see Table 16.7). At low pH and temperature, ammonia combines with water to produce ammonium (NH_4^+) and hydroxide (OH^-) ions. The ammonium ion is nontoxic to aquatic life and not of great concern. The un-ionized ammonia (NH_3), however, can easily cross cell membranes and have a toxic effect on a wide variety of fish. The EPA has established the criteria for fresh and saltwater fish that depend on temperature, pH, species, and averaging period.

Ammonia is usually measured by a distillation and titration technique or with an ammonia-selective electrode. The results are reported as ammonia nitrogen. Nitrites are measured by a colorimetric method. The results are reported as nitrite nitrogen. Nitrates are measured by ultraviolet spectrophotometry, selective electrode, or reduction methods. The results are reported as nitrate nitrogen. Organic nitrogen is determined by a digestion process that identifies organic and ammonia nitrogen combined. Organic nitrogen is found by subtracting the ammonia nitrogen value from the digestion results.

Table 16.7 *Percentage of Total Ammonia Present in Toxic, Un-Ionized Form*

temp	pH								
(°F (°C))	6.0	6.5	7.0	7.5	8.0	8.5	9.0	9.5	10.0
41 (5)	0.013	0.040	0.12	0.39	1.2	3.8	11	28	56
50 (10)	0.019	0.059	0.19	0.59	1.8	5.6	16	37	65
59 (15)	0.027	0.087	0.27	0.86	2.7	8.0	21	46	73
68 (20)	0.040	0.13	0.40	1.2	3.8	11	28	56	80
77 (25)	0.057	0.18	0.57	1.8	5.4	15	36	64	85
86 (30)	0.080	0.25	0.80	2.5	7.5	20	45	72	89

16. COLOR

Color in water is caused by substances in solution, known as *true color*, and by substances in suspension, mostly organics, known as *apparent* or *organic color*. Iron, copper, manganese, and industrial wastes all can cause color. Color is aesthetically undesirable, and it stains fabrics and porcelain bathroom fixtures.

Water color is determined by comparison with standard platinum/cobalt solutions or by spectrophotometric methods. The standard color scales range from 0 (clear) to 70. Water samples with more intense color can be evaluated using a dilution technique.

17. TURBIDITY

Turbidity is a measure of the light-transmitting properties of water and is comprised of suspended and colloidal material. Turbidity is expressed in *nephelometric turbidity units* (NTU). Viruses and bacteria become attached to these particles, where they can be protected from the bactericidal and viricidal effects of chlorine, ozone, and other disinfecting agents. The organic material included in turbidity has also been identified as a potential precursor to carcinogenic disinfection by-products.

Turbidity in excess of 5 NTU is noticeable by visual observation. Turbidity in a typical clear lake is approximately 25 NTU, and muddy water exceeds 100 NTU. Turbidity is measured using an electronic instrument called a nephelometer, which detects light scattered by the particles when a focused light beam is shown through the sample.

18. SOLIDS

Solids present in a sample of water can be classified in several ways.

- *total solids* (TS): Total solids are the material residue left in the vessel after the evaporation of the sample at 103–105°C. Total solids include total suspended solids and total dissolved solids.

- *total suspended solids* (TSS): The material retained on a standard glass-fiber filter disk is defined as the suspended solids in a sample. The filter is weighed before filtration, dried at 103–105°C, and weighed again. The gain in weight is the amount of suspended solids. Suspended solids can also be categorized into *volatile suspended solids* (VSS) and *fixed suspended solids* (FSS).

- *total dissolved solids* (TDS): These solids are in solution and pass through the pores of the standard glass-fiber filter. Dissolved solids are determined by passing the sample through a filter, collecting the filtrate in a weighed drying dish, and evaporating the liquid at 180°C. The gain in weight represents the dissolved solids. Dissolved solids can be categorized into *volatile dissolved solids* (VDS) and *fixed dissolved solids* (FDS).

- *total volatile solids* (TVS): The residue from one of the previous determinations is ignited to constant weight in an electric muffle furnace at 550°C. The loss in weight during the ignition process represents the volatile solids.

- *total fixed solids* (TFS): The weight of solids that remain after the ignition used to determine volatile solids represents the fixed solids.

- *settleable solids*: The volume (mL/L) of settleable solids is measured by allowing a sample to settle for one hour in a graduated conical container (*Imhoff cone*).

The following relationships exist.

$$TS = TSS + TDS = TVS + TFS \quad 16.12$$

$$TSS = VSS + FSS \quad 16.13$$

$$TDS = VDS + FDS \quad 16.14$$

$$TVS = VSS + VDS \quad 16.15$$

$$TFS = FSS + FDS \quad 16.16$$

Waters with high concentrations of suspended solids are classified as *turbid waters*. Waters with high concentrations of dissolved solids often can be tasted. Therefore, a limit of 500 mg/L has been established in the National Secondary Drinking Water Standards for dissolved solids.

19. CHLORINE AND CHLORAMINES

Chlorine is the most common disinfectant used in water treatment. It is a strong oxidizer that deactivates microorganisms. Its oxidizing capability also makes it useful in removing soluble iron and manganese ions.

Chlorine gas in water forms *hydrochloric* and *hypochlorous acids*. At a pH greater than 9, hypochlorous acid dissociates to hydrogen and hypochlorite ions, as shown in Eq. 16.17.

$$Cl_2 + H_2O \underset{pH<4}{\overset{pH>4}{\rightleftharpoons}} HCl$$

$$+ HOCl \underset{pH<9}{\overset{pH>9}{\rightleftharpoons}} H^+ + OCl^- \quad 16.17$$

Free chlorine, hypochlorous acid, and hypochlorite ions left in water after treatment are known as *free chlorine residuals*. Hypochlorous acid reacts with ammonia (if it is present) to form *chloramines*. Chloramines are known as *combined residuals*. Chloramines are more stable than free residuals, but their disinfecting ability is less.

$$HOCl + NH_3 \rightarrow H_2O + NH_2Cl \text{ (monochloramine)} \quad 16.18$$

$$HOCl + NH_2Cl \rightarrow H_2O + NHCl_2 \text{ (dichloramine)} \quad 16.19$$

$$HOCl + NHCl_2 \rightarrow H_2O + NCl_3 \text{ (trichloramine)} \quad 16.20$$

Free and combined residual chlorine can be determined by color comparison, by titration, and with chlorine-sensitive electrodes. Color comparison is the most common field method.

20. HALOGENATED COMPOUNDS

Halogenated compounds are a subject of concern in the treatment of water due to their potential as carcinogens. The use of chlorine as a disinfectant in water treatment generates these compounds by reacting with organic substances in the water. (For this reason there are circumstances under which chlorine may not be the most appropriate disinfectant.)

The organic substances, called *precursors*, are not in themselves harmful, but the chlorinated end products, known by the term *disinfection by-products* (DBPs) raise serious health concerns. Typical precursors are decay by-products such as humic and fulvic acids. Several of the DBPs contain bromine, which is found in low concentrations in most surface waters and can also occur as an impurity in commercial chlorine gas.

The DBPs can take a variety of forms depending on the precursors present, the concentration of free chlorine, the contact time, the pH, and the temperature. The most common ones are the trihalomethanes (THMs), haloacetic acids (HAAs), dihaloacetonitriles (DHANs), and various trichlorophenol isomers.

21. TRIHALOMETHANES

Only four trihalomethane (THM) compounds are normally found in chlorinated waters.

$CHCl_3$	trichloromethane (chloroform)
$CHBrCl_2$	bromodichloromethane
$CHBr_2Cl$	dibromochloromethane
$CHBr_3$	tribromomethane (bromoform)

Trihalomethanes are regulated by the EPA under the National Primary Drinking Water Standards. Standards for total *haloacetic acids* (referred to as "HAA5" in consideration of the five compounds identified) have also been established.

22. HALOACETIC ACIDS

The haloacetic acids (HAAs) exist in tri-, di-, and mono- forms, abbreviated as THAAs, DHAAs, and MHAAs. All of the haloacetic acids are toxic and are suspected or proven carcinogens.

$C_2HCl_3O_2$	trichloroacetic acid	(TCAA)
$C_2HBrCl_2O_2$	bromodichloroacetic acid	(BDCAA)
$C_2HBr_2ClO_2$	dibromochloroacetic acid	(DBCAA)
$C_2HBr_3O_2$	tribromoacetic acid	(TBAA)
$C_2H_2Cl_2O_2$	dichloroacetic acid	(DCAA)
$C_2H_2BrClO_2$	bromochloroacetic acid	(BCAA)
$C_2H_2Br_2O_2$	dibromoacetic acid	(DBAA)
$C_2H_3ClO_2$	monochloroacetic acid	(MCAA)
$C_2H_3BrO_2$	monobromoacetic acid	(MBAA)

23. DIHALOACETONITRILES

The dihaloacetonitriles (DHANs) are formed when acetonitrile (methyl cyanide: C_2H_3N (structurally H_3CCN)) is exposed to chlorine. All are toxic and suspected carcinogens. Maximum concentration limits have not been established by the EPA for DHANs.

C_2HCl_2N	dichloroacetonitrile	(DCAN)
$C_2HBrClN$	bromochloroacetonitrile	(BCAN)
C_2HBr_2N	dibromoacetonitrile	(DBAN)

24. TRICHLOROPHENOL

Trichlorophenol can exist in six isomeric forms, with varying potential toxicities. As with all halogenated organics, they are potential carcinogens.

2,3,4-trichlorophenol	
2,3,5-trichlorophenol	
2,3,6-trichlorophenol	
2,4,5-trichlorophenol	irritant
2,4,6-trichlorophenol	fungicide, bactericide
3,4,5-trichlorophenol	

25. AVOIDANCE OF DISINFECTION BY-PRODUCTS IN DRINKING WATER

Reduction of DBPs can best be achieved by avoiding their production in the first place. The best strategy dictates using source water with few or no organic precursors.

Often source water choices are limited, necessitating tailoring treatment processes to produce the desired result. This entails removing the precursors prior to the application of chlorine, applying chlorine at certain points in the treatment process that minimize production of DBPs, using disinfectants that do not produce significant DBPs, or a combination of these techniques.

Removal of precursors is achieved by preventing growth of vegetative material (algae, plankton, etc.) in the source water and by collecting source water at various depths to avoid concentrations of precursors. Oxidizers such as potassium permanganate and chlorine dioxide can often reduce the concentration of the precursors without forming the DBPs. Use of activated carbon, pH-adjustment processes or dechlorination can reduce the impact of chlorination.

Chlorine application should be delayed if possible until after the flocculation, coagulation, settling, and filtration processes have been completed. In this manner, turbidity and common precursors will be reduced. If it is necessary to chlorinate early to facilitate treatment processes, chlorination can be followed by dechlorination to reduce contact time. Granular activated carbon has been used to some extent to remove DBPs after they form, but the carbon needs frequent regeneration.

Alternative disinfectants include ozone, chloramines, chlorine dioxide, iodine, bromine, potassium permanganate, hydrogen peroxide, and ultraviolet radiation. Ozone and chloramines, singularly or together, are often used for control of THMs. However, ozone creates other DBPs, including aldehydes, hydrogen peroxide, carboxylic acids, ketones, and phenols. When ozone is used as the primary disinfectant, a secondary disinfectant such as chlorine or chloramine must be used to provide an active residual that can be measured within the distribution system.

26. NOMENCLATURE

A	acidity	mg/L	mg/L
Ca	calcium	mg/L	mg/L
FDS	fixed dissolved solids	lbf	kg
FSS	fixed suspended solids	lbf	kg
H	total hardness	mg/L	mg/L
L	free lime	mg/L	mg/L
m	mass	lbm	kg
M	alkalinity	mg/L	mg/L
MW	molecular weight	lbm/lbmol	kg/kmol
N	normality	gEW/L	gEW/L
O	hydroxides	mg/L	mg/L
P	phenolphthalein alkalinity	mg/L	mg/L
S	sulfate	mg/L	mg/L
TDS	total dissolved solids	lbf	kg
TFS	total fixed solids	lbf	kg
TN	total nitrogen	n.a.	mg/L of N
TS	total solids	lbf	kg
TSS	total suspended solids	lbf	kg
TVS	total volatile solids	lbf	kg
V	volume	ft^3	m^3
VDS	volatile dissolved solids	lbf	kg
VSS	volatile suspended solids	lbf	kg
X	compound	n.a.	n.a.

Subscripts

X	compound

17 Potable Water Treatment and Management

1. WATER TREATMENT PLANT LOCATION

The location chosen for a water treatment plant is influenced by many factors. The most common factors include availability of resources such as (a) local water, (b) power, and (c) sewerage services; economic factors such as (d) land cost and (e) annual taxes; and environmental factors such as (f) traffic and (g) other concerns that would be identified on an environmental impact report.

It is imperative to locate water treatment plants (a) above the flood plain and (b) where a 15–20 ft (4.5–6 m) elevation difference exists. This latter requirement will eliminate the need to pump the water between processes. Traditional treatment requires a total head of approximately 15 ft (4.5 m), whereas advanced processes such as granular activated charcoal and ozonation increase the total head required to approximately 20 ft (6 m).

The average operational life of equipment and facilities is determined by economics. However, it is unlikely that a lifetime of less than 50 years would be economically viable. Equipment lifetimes of 25–30 years are typical.

2. PROCESS INTEGRATION

The processes and sequences used in a water treatment plant depend on the characteristics of the incoming water. However, some sequences are more appropriate than others due to economic and hydraulic considerations. *Conventional filtration*, also referred to as *complete filtration*, is a term used to describe the traditional sequence of adding coagulation chemicals, flash mixing, coagulation-flocculation, sedimentation, and subsequent filtration. Coagulants, chlorine (or an alternative disinfectant), fluoride, and other chemicals are added at various points along the path, as indicated by Fig. 17.1. Conventional filtration is still the best choice when incoming water has high color, turbidity, or other impurities.

Direct filtration refers to a modern sequence of adding coagulation chemicals, flash mixing, minimal flocculation, and subsequent filtration. In direct filtration, the physical chemical reactions of flocculation occur to some extent, but special flocculation and sedimentation facilities are eliminated. This reduces the amount of sludge that has to be treated and disposed of. Direct filtration is applicable when the incoming water is of high initial quality.

Water

Figure 17.1 *Chemical Application Points*

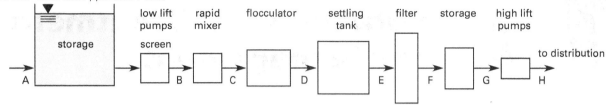

typical flow diagram of water treatment plant

category of chemicals	possible points of application							
	A	B	C	D	E	F	G	H
algicide	X				X			
disinfectant		X	X		X	X	X	X
activated carbon		X	X	X	X			
coagulants		X	X					
coagulation aids		X	X		X			
alkali								
for flocculation			X					
for corrosion control						X		
for softening			X					
acidifier			X			X		
fluoride						X		
cupric-chloramine						X		
dechlorinating agent						X		X

Note: With solids contact reactors, point C is the same as point D.

In-line filtration refers to another modern sequence that starts with adding coagulation chemicals at the filter inlet pipe. Mixing occurs during the turbulent flow toward a filter, which is commonly of the pressure-filter variety. As with direct filtration, flocculation and sedimentation facilities are not used.

Table 17.1 provides guidelines based on incoming water characteristics for choosing processes required to achieve satisfactory quality.

3. PRETREATMENT

Preliminary treatment is a general term that usually includes all processes prior to the first flocculation operation (i.e., pretreatment, screening, presedimentation, microstraining, aeration, and chlorination). Flow measurement is usually considered to be part of the pretreatment sequence.

4. SCREENING

Screens are used to protect pumps and mixing equipment from large objects. The degree of screening required will depend on the nature of solids expected. Screens can be either manually or automatically cleaned.

5. MICROSTRAINING

Microstrainers are effective at removing 50–95% of the algae in incoming water. Microstrainers are constructed from woven stainless steel fabric mounted on a hollow drum that rotates at 4–7 rpm. Flow is usually radially outward through the drum. The accumulated filter cake is removed by backwashing.

6. ALGAE PRETREATMENT

Biological growth in water from impounding reservoirs, lakes, storage reservoirs, and settling basins can be prevented or eliminated with an *algicide* such as copper sulfate. Such growth can produce unwanted taste and odors, clog fine-mesh filters, and contribute to the buildup of slime.

Typical dosages of copper sulfate vary from 0.54 lbm/ac-ft to 5.4 lbm/ac-ft (0.2 mg/L to 2.0 mg/L), with the lower dose usually being adequate for waters that are soft or have alkalinities less than 50 mg/L as $CaCO_3$. (For aqueous solutions, units of mg/L and ppm are equivalent.) Since it is toxic, copper sulfate should not be used without considering and monitoring the effects on aquatic life (e.g., fish).

Table 17.1 *Treatment Methods*

incoming water quality		pretreatment				treatment					special treatments			
constituents	concentration (mg/L)	screening	prechlorination	plain settling	aeration	lime softening	coagulation and sedimentation	rapid sand filtration	slow sand filtration	postchlorination	superchlorination[a] or chlorammoniation	active carbon	special chemical treatment	salt water conversion[b]
coliform monthly average (MPN[e]/100 mL)	0–20									E				
	20–100			O			O	O	O	E				
	100–5000		E				E	E	O	E				
	>5000		E	O[c]			E	E		E	O			
suspended solids	0–100	O							O					
	100–200	O					E	E						
	>200	O		O[d]			E	E						
color, (mg/L)	20–70						O	O			O			
	>70						E	E			O			
tastes and odors	noticeable		O		O					O	O	E		
CaCO₃, (mg/L)	>200					E	E	E					E	
pH	6.5–8.5 (normal)													
iron and manganese, (mg/L)	≤0.3		O	O										
	0.3–1.0				O		E	E	O					
	>1.0		E		E		E	E	O				O	
chloride, (mg/L)	0–250													
	250–500													O
	500+													E
phenolic compounds, (mg/L)	0–0.005						O	O			O	O		
	>0.005						E	E			O	E	O	
toxic chemicals							E	E				E	O	
less critical chemicals							O	O				O	O	

Note: E = essential, O = optional
[a]Superchlorination shall be followed by dechlorination.
[b]As an alternative, dilute with low chloride water.
[c]Double settling shall be provided for coliform exceeding 20,000 MPN/100 mL.
[d]For extremely muddy water, presedimentation by plain settling may be provided.
[e]MPN = most probable number

7. PRECHLORINATION

A prechlorination process was traditionally employed in most water treatment plants. Many plants have now eliminated all or part of the prechlorination due to the formation of THMs. In so doing, benefits such as algae control have been eliminated. Alternative disinfection chemicals (e.g., ozone and potassium permanganate) can be used. Otherwise, coagulant doses must be increased.

8. PRESEDIMENTATION

The purpose of presedimentation is to remove easily settled sand and grit. This can be accomplished by using pure sedimentation basins, sand and grit chambers, and various passive cyclone degritters. *Trash racks* may be integrated into sedimentation basins to remove leaves and other floating debris.

9. FLOW MEASUREMENT

Flow measurement is incorporated into the treatment process whenever the water is conditioned enough to be compatible with the measurement equipment. Flow measurement devices should not be exposed to scour from grit or highly corrosive chemicals (e.g., chlorine).

Flow measurement often takes place in a Parshall flume. Chemicals may be added in the flume to take advantage of the turbulent mixing that occurs at that point. All of the traditional fluid measurement devices are also applicable, including venturi meters, orifice plates, propeller and turbine meters, and modern passive devices such as magnetic and ultrasonic flowmeters.

Water

10. AERATION

Aeration is used to reduce taste- and odor-causing compounds, to lower the concentration of dissolved gases (e.g., hydrogen sulfide), to increase dissolved CO_2 (i.e., recarbonation) or decrease CO_2, to reduce iron and manganese, and to increase dissolved oxygen.

Various types of aerators are used. The best transfer efficiencies are achieved when the air-water contact area is large, the air is changed rapidly, and the aeration period is long. *Force draft air injection* is common. The release depth varies from 10 ft to 25 ft (3 m to 7.5 m). The ideal compression power required to aerate water with a simple air injector depends on the air flow rate, Q, and head (i.e., which must be greater than the release depth), h, at the point where the air is injected. Motor-compression-distribution efficiencies are typically around 75%.

$$P_{kW} = \frac{Q_{L/s} h_m}{100} \qquad \text{[SI]} \quad \textit{17.1(a)}$$

$$P_{hp} = \frac{Q_{cfm} h_{ft}}{528} \qquad \text{[U.S.]} \quad \textit{17.1(b)}$$

Diffused air systems with compressed air at 5–10 psig (35–70 kPa) are the most efficient methods of aerating water. The air injection volume is 0.2–0.3 ft³/gal. However, the equipment required to produce and deliver compressed air is more complex than with simple injectors. The *transfer efficiency* of a diffused air system varies with depth and bubble size. With injection depths of 5–10 ft (1.5–3.0 m), if coarse bubbles are produced, only 4–8% of the available oxygen will be transferred to the water. With medium-sized bubbles, the efficiency can be 6–15%, and it can approach 10–30% with fine bubble systems.

11. SEDIMENTATION PHYSICS

Water containing suspended sediment can be held in a *plain sedimentation tank* (basin) that allows the particles to settle out.[1] Settling velocity and settling time for sediment depends on the water temperature (i.e., viscosity), particle size, and particle specific gravity. (The specific gravity of sand is usually taken as 2.65.) Typical settling velocities are as follows: gravel, 1 m/s; coarse sand, 0.1 m/s; fine sand, 0.01 m/s; and silt, 0.0001 m/s. Bacteria and colloidal particles are generally considered to be nonsettleable during the detention periods available in water treatment facilities.

Settlement time can be calculated from the settling velocity and the depth of the tank. If it is necessary to determine the settling velocity of a particle with diameter D, the following procedure can be used.[2]

step 1: Assume a settling velocity, v_t.

step 2: Calculate the settling Reynolds number, Re.

Reynolds Number (Newtonian Fluid)
$$\mathrm{Re} = v D \rho / \mu = v D / v \qquad \textit{17.2}$$

step 3: (a) If $\mathrm{Re} < 1$, use *Stokes' law*.

Stokes' Law
$$v_t = \frac{g(\rho_p - \rho_f) d^2}{18\mu} = \frac{g\rho_f(S.G. - 1) d^2}{18\mu} \qquad \textit{17.3}$$

(b) If $1 < \mathrm{Re} < 2000$, use Fig. 17.2, which gives theoretical settling velocities in 68°F (20°C) water for spherical particles with specific gravities of 1.05, 1.2, and 2.65. Actual settling velocities will be much less than shown because particles are not actually spherical.

Figure 17.2 *Settling Velocities, 68°F (20°C)*

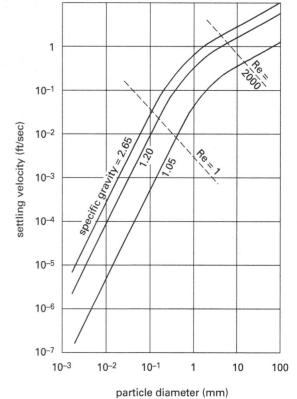

(c) If $\mathrm{Re} > 2000$, use *Newton's first law of motion* and balance the weight of the particle against the buoyant and drag forces. A spherical shape is often assumed in determining the drag coefficient, C_D.

---

[1]The term "plain" refers to the fact that no chemicals are used as coagulants.
[2]This calculation procedure is appropriate for the *Type I settling* that describes sand, sediment, and grit particle performance. It is not appropriate for *Type II settling*, which describes floc and other particles that grow as they settle.

In that case,

- for laminar descent (Re ≤ 1.0), use

$$C_D = 24/\text{Re} \qquad \text{General Spherical} \qquad \textit{17.4}$$

- for transitional descent, use

$$C_D = 24/\text{Re} + 3/(\text{Re}^{1/2}) + 0.34 \qquad \text{General Spherical} \qquad \textit{17.5}$$

- for turbulent descent (Re ≥ 10,000), use

$$C_D = 0.4 \qquad \text{General Spherical} \qquad \textit{17.6}$$

For spherical particles, the terminal settling velocity is

$$v_t = \sqrt{\frac{4g(\rho_p - \rho_f)d}{3C_D\rho_f}} \qquad \text{General Spherical} \qquad \textit{17.7}$$

12. SEDIMENTATION TANKS

Sedimentation tanks are usually concrete, rectangular or circular in plan, and are equipped with scrapers or raking arms to periodically remove accumulated sediment. (See Fig. 17.3.) Steel should be used only for small or temporary installations. Where steel parts are unavoidable, as in the case of some rotor parts, adequate corrosion resistance is necessary.

Figure 17.3 Sedimentation Basin

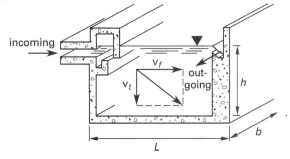

Water flows through the tank at the average *flow-through velocity*, v_f. The flow-through velocity can be found by injecting a colored dye into the tank. It should not exceed 1 ft/min (0.5 cm/sec). The time that water spends in the tank depends on the flow-through velocity and the tank length, L, typically 100–200 ft (30–60 m).

The minimum settling time depends on the tank depth, h, typically 6–15 ft (1.8–4.5 m).

$$t_{\text{settling}} = \frac{h}{v_t} \qquad \textit{17.8}$$

The time that water remains in the basin is known as the *residence time*, θ (detention time, *retention time*, *detention period*, etc.). Typical residence times range from 2 hr to 6 hr, although periods from 1 hr to 12 hr are used depending on the size of particles. All particles will be removed whose t_{settling} is less than θ.

$$\text{Hydraulic residence time} = V/Q = \theta \qquad \text{Stokes' Law} \qquad \textit{17.9}$$

Rectangular basins are preferred. Rectangular basins should be constructed with aspect ratios greater than 3:1, and preferably greater than 4:1. The bottom should be sloped toward the drain at no less than 1%. Multiple inlet ports along the entire inlet wall should be used. If there are fewer than four inlet ports, an inlet baffle should be provided.

Square or circular basins are appropriate only when space is limited. The slope toward the drain should be greater, typically 8%. A baffled center inlet should be provided. For radial-flow basins, the diameter is on the order of 100 ft (30 m).

Basin efficiency can approach 80% for fine sediments. Virtually all of the coarse particles are removed. Theoretically, all particles with settling velocities greater than the *overflow rate*, v^*, also known as the *surface loading* or *critical velocity*, will be removed. In Eq. 17.10, b is the tank width, typically 30–40 ft (9–12 m). At least two basins should be constructed (both normally operating) so that one can be out of service for cleaning during low-volume periods without interrupting plant operation. Therefore, Q_{filter} in Eq. 17.10 should be calculated by dividing the total plant flow by 2 or more. The overflow rate is typically 600–1000 gpd/ft² (24–40 kL/d·m²) for rectangular tanks. For square and circular basins, the range is approximately 500–750 gpd/ft² (20–30 kL/d·m²).

$$v^* = \frac{Q_{\text{filter}}}{A_{\text{surface}}} = \frac{Q_{\text{filter}}}{bL} \qquad \textit{17.10}$$

In some modern designs, sedimentation has been enhanced by the installation of vertically inclined *laminar tubes* or inclined plates (*lamella plates*) at the tank bottom. These passive tubes, which are typically about 2 in (50 mm) in diameter, allow the particles to fall only a short distance in more turbulent water before entering the tube in which the flow is laminar. Incoming solids settle to the lower surface of the tube, slide downward, exit the tube, and settle to the sedimentation basin floor.

Weir loading is the daily flow rate divided by the total effluent weir length. Weir loading is commonly specified as 15,000–20,000 gal/day-ft (190–250 kL/d·m), but certainly less than 50,000 gal/day-ft (630 kL/d·m).

The accumulated sediment is referred to as *sludge*. It is removed either periodically or on a continual basis, when it has reached a concentration of 25 mg/L or is organic. Various methods of removing the sludge are used, including scrapers and pumps. The linear velocity of sludge scrapers should be 15 ft/min (7.5 cm/sec) or higher.

13. SEDIMENTATION REMOVAL EFFICIENCY: UNMIXED BASINS

Sedimentation efficiency, also known as *removal fraction* and *collection efficiency*, refers to the gravimetric fraction (i.e., fraction by weight) of suspended particles that is removed by sedimentation. Since the suspended particles vary in size, shape, and specific gravity, the distribution of particles is characterized by a distribution of settling velocities, v_t. The distribution is determined, almost always, by a *laboratory settling column test*. Turbid water is placed in a 3–8 ft (1–2.4 m) long vertical column and allowed to settle, with the settled mass being measured over time.

The settling test produces multiple discrete cumulative mass fraction settled, $1 - x$, versus settling time, t_s, data pairs. These are transformed to mass fraction remaining, x, versus settling velocity, $v_t = h_{sc}/t_s$ data pairs, where h_{sc} is the height of the settling column. The distribution data can be presented in several ways, including settling test results and cumulative curves of fraction remaining or fraction removed, as shown in Fig. 17.4. The *fraction remaining curve* (shown in Fig. 17.4(c)) plots the mass fraction, x, remaining in suspension and is also known as the *mass fraction curve*. The curve shape is the same as the original settling test results. This curve is usually drawn visually through the settling test data points, although more sophisticated curve fitting methods can be used. The curve may also be drawn simply as linear segments between the data points.

Consider a *quiet sedimentation basin* where water enters on one side and is withdrawn on the other side, without any mixing. Some particles have such a high settling velocity that they will settle no matter where they are introduced into the settling basin. Other particles have such a low settling velocity that they will never settle out during the hydraulic retention time of the basin. In between, some particles will settle to various degrees, depending on their initial depths in the settling basin. This leads to two parameters being used to describe the removal fraction: the removal fraction of particles that are completely removed and the total fraction of all particles removed.[3]

Figure 17.4 Settling Test Data Representations

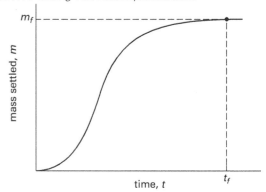

(a) settling test results (mass settled vs. time)

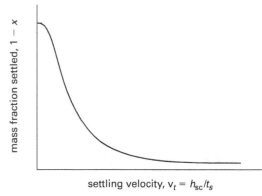

(b) fraction settled vs. settling velocity

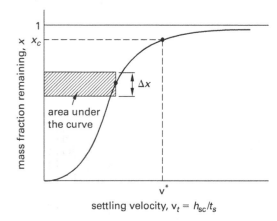

(c) fraction remaining vs. settling velocity

Whether or not a particle settles depends on its settling velocity, the time available to settle, and its initial location. The time available to settle depends on the horizontal component of flow velocity and on the horizontal distance between the settling basin's points of flow entry and removal.

---

[3] That is, the "fraction completely collected" is different from the "total fraction collected."

Due to the continuous outflow, the tank contents experience a horizontal flow-through velocity, v_f, between the entry and removal points. If a particle settles with speed v_t, its vertical fall, y, over the flow path length (generally, the length of the tank), L, is

$$y = v_t t = \frac{v_t L}{v_f} \qquad 17.11$$

The vertical fall distance, y, may be less than or greater than the basin depth, h. If $y \geq h$, the particle will reach the bottom before reaching the end of the basin and will be collected. If $y < h$, the particle may or may not settle, depending on the depth at which it starts. If it starts settling close to the bottom, it will settle; otherwise, it will not fall enough and will escape with the outflow.

The *critical settling velocity*, v_c, is numerically the same as the overflow rate, v^*, and corresponds to the limiting velocity for particles experiencing total removal. Therefore, the fraction of particles experiencing total removal (a collection efficiency of 100%) can be read directly from the mass fraction remaining curve as $1 - x_c$, where x_c is the mass fraction remaining corresponding to $v_t = v_c = v^*$. If the mass fraction curve has not been plotted (or, has been plotted only roughly, as is usually the case), the value of x_c can be determined from the settling test data points by linear interpolation.

The total fraction of all particles removed, f, includes the fraction of particles with settling velocities greater than the critical velocity plus some fraction of the particles with settling velocities less than the critical velocity. For particles with settling velocities less than the critical velocity, the collection efficiency in both rectangular and circular sedimentation basins is

$$\eta = \frac{v_s}{v^*} \quad [v_s \leq v^*] \qquad 17.12$$

The total removal fraction (total collection efficiency) is

$$f = 1 - x_c + \int_{x=0}^{x=x_c} \frac{v_t}{v^*} dx$$
$$\approx 1 - x_c + \frac{1}{v^*} \sum_{x=0}^{x=x_c} v_t \Delta x \qquad 17.13$$

The area beneath the mass fraction curve corresponds to the integral term in Eq. 17.13.[4] Since it will almost never be possible to describe the mass fraction curve mathematically, integrating the area of the curve must be done manually by dividing the area under the curve into rectangles (i.e., replacing the curve with a histogram), as is represented by the second form of Eq. 17.13. The rectangle widths, Δx, may vary and are chosen for convenience. The values of v_t in the second form of Eq. 17.13 correspond to the midpoints of the rectangle widths.

14. SEDIMENTATION REMOVAL EFFICIENCY: MIXED BASINS

Quiet (unmixed) settling basins have the highest removal fractions, but mixing is incorporated into some designs for various reasons. Two types of ideal basins that can be mathematically evaluated are basins with pure transverse mixing and basins with thorough mixing. With transverse mixing, a particle may be moved vertically or side to side in either direction, randomly, but it will not move forward or backward. In other words, the particle is constrained to a transverse plane perpendicular to the overall travel path between entry and discharge points. The collection efficiency as a function of settling velocity for a specific particle is given by Eq. 17.14. While a quiet settling basin has a 100% removal efficiency for a particle with $v_t = v^*$, a basin with transverse mixing has a 63.2% efficiency for the same particle.

$$\eta = 1 - e^{-v_t/v^*} \quad [\text{transverse mixing}] \qquad 17.14$$

For a basin with thorough mixing, particles may move three dimensionally—vertically, horizontally, and longitudinally. The collection efficiency is given by Eq. 17.15. When $v_t = v^*$, the efficiency is 50%.

$$\eta = \frac{v_t}{v_t + v^*} \quad [\text{thorough mixing}] \qquad 17.15$$

As a function of v_t (or as a function of the dimensionless ratio v_t/v^*), the removal efficiencies for transverse- and thoroughly-mixed basins both approach 100% asymptotically. Since large values of v_t/v^* correspond to large basins, any basin removal efficiency can be achieved by making the basin large enough. However, the smallest basin size will be achieved when the basin is unmixed.

15. COAGULANTS

Various chemicals can be added to remove fine solids. There are two main categories of coagulating chemicals: hydrolyzing metal ions (based on either aluminum or iron) and ionic polymers. Since the chemicals work by agglomerating particles in the water to form floc, they are known as *coagulants*. *Floc* is the precipitate that forms when the coagulant allows the colloidal particles to agglomerate.

Common *hydrolyzing metal ion* coagulants are aluminum sulfate ($Al_2(SO_4)_3 \cdot nH_2O$, commonly referred to as "alum"), ferrous sulfate ($FeSO_4 \cdot 7H_2O$, sometimes

---

[4]The variable of integration, x, is the fraction remaining and is plotted vertically on a mass fraction curve. x is not the horizontal axis. The "area under the curve" is approximated by horizontal (not vertical) histogram bars.

referred to as "copperas"), and chlorinated copperas (a mixture of ferrous sulfate and ferric chloride). n is the number of waters of hydration, approximately 14.3.

Alum is, by far, the most common compound used in water treatment.[5] Alum provides the positive charges needed to attract and neutralize negative colloidal particles. It reacts with alkalinity in the water to form gelatinous aluminum hydroxide ($Al(OH)_3$) in the proportion of 1 mg/L alum:$\frac{1}{2}$ mg/L alkalinity. The aluminum hydroxide forms the nucleus for floc agglomeration.

For alum coagulation to be effective, the following requirements must be met: (a) Enough alum must be used to neutralize all of the incoming negative particles. (b) Enough alkalinity must be present to permit as complete as possible a conversion of the aluminum sulfate to aluminum hydroxide. (c) The pH must be maintained within the effective range.

Alum dosage is generally 5–50 mg/L, depending on the turbidity. Alum floc is effective within a wide pH range of 5.5–8.0. However, the hydrolysis of the aluminum ion depends on pH and other factors in complex ways, and stoichiometric relationships rarely tell the entire story. Although a pH of 6–7 is a typical operating range for alum coagulation, depending on the contaminant, the pH can range as high as 9. Alum removal of chromium, for example, is effective within a pH range of 7–9.

Alum reacts with natural alkalinity in the water according to the following reaction.

Coagulation Equations

$$Al_2(SO_4)_3 + 3\,Ca(HCO_3)_2$$
$$\Leftrightarrow 2\,Al(OH)_3 + 3\,CaSO_4 + 6\,CO_2 \qquad \textit{17.16}$$

Since alum is naturally acidic, if the water is not sufficiently alkaline, lime (CaO) and soda ash ("caustic soda," Na_2CO_3) can be added for preliminary pH adjustment.[6] If there is inadequate alkalinity and lime is added, the reaction is

$$Al_2(SO_4)_3 \cdot nH_2O + 3Ca(OH)_2$$
$$\rightarrow 2Al(OH)_3\downarrow + 3CaSO_4 \qquad \textit{17.17}$$
$$+ nH_2O$$

Alum reacts with soda ash according to

Coagulation Equations

$$Al_2(SO_4)_3 + 3\,NaCO_3 + 3\,H_2O$$
$$\Leftrightarrow 2\,Al(OH)_3 + 3\,NaSO_4 + 3\,CO_2 \qquad \textit{17.18}$$

Ferrous sulfate ($FeSO_4 \cdot 7H_2O$) reacts with slaked lime ($Ca(OH)_2$) to flocculate ferric hydroxide ($Fe(OH)_3$). This is an effective method of clarifying turbid water at higher pH.

Ferric sulfate ($Fe_2(SO_4)_3$) reacts with natural alkalinity and lime. It can also be used for color removal at low pH. At high pH, it is useful for iron and manganese removal, as well as a coagulant with precipitation softening.

If alkalinity is high, it may be necessary to use hydrochloric or sulfuric acid for pH control rather than adding lime. The iron salts are effective above a pH of 7, so they may be advantageous in alkaline waters.[7]

Polymers are chains or groups of repeating identical molecules (*mers*) with many available active adsorption sites. Their molecular weights range from several hundred to several million. Polymers can be positively charged (cationic polymers), negatively charged (anionic polymers), or neutral (nonionic polymers). The charge can vary with pH. *Organic polymers* (*polyelectrolytes*), such as starches and polysaccharides, and *synthetic polymers*, such as polyacrylamides, are used.

Polymers are useful in specialized situations, such as when particular metallic ions are to be removed. For example, conventional alum will not remove positive iron ions. However, an anionic polymer will combine with the iron. Polymers are effective in narrow ranges of turbidity and alkalinity. In some cases, it may be necessary to artificially seed the water with clay or alum (to produce floc).

16. FLOCCULATION ADDITIVES

Flocculation additives improve the coagulation efficiency by changing the floc size. Additives include *weighting agents* (e.g., bentonite clays), *adsorbents* (e.g., powdered activated carbon), and *oxidants* (chlorine). Polymers are also used in conjunction with metallic ion coagulants.

17. DOSES OF COAGULANTS AND OTHER COMPOUNDS

The amount of a substance added to a water supply is known as the *dose*. *Purity*, P, and *availability*, G, both affect dose. A substance's purity decreases when it is mixed with another substance. A substance's availability decreases when some of the substance does not dissolve or react chemically.

For example, *lime* is a term that describes various calcium compounds. Lime is typically quarried from naturally occurring limestone ($CaCO_3$). Limestone contains silica and other non-calcium compounds, which reduces the as-delivered lime purity. *Quicklime* (CaO) is

---

[5]Lime is the second most-used compound.
[6]The cost of soda ash used for pH adjustment is about three times that of lime. Also, soda ash leaves sodium ions in solution, an increasingly undesirable contaminant for people with hypertension.
[7]Optimum pH for ferric floc is approximately 7.5. However, iron salts are active over the pH range of 3.8–10.

produced by heating (i.e., burning) $CaCO_3$ in a lime kiln. Accordingly, quicklime may contain some calcium in the form of calcium carbonate rather than calcium oxide. The *total lime* (i.e., total calcium) content includes both calcium compounds, while the *available lime* only includes the CaO form.

Feed rate, F, of a compound with purity P and fractional availability G can be calculated from the *dose equation*, Eq. 17.19. Doses in mg/L and ppm for aqueous solutions are the same. Each 1% by weight of concentration is equivalent to 10 000 mg/L of solution. Doses may be given in terms of grains per gallon. There are 7000 grains per pound.

$$F_{kg/d} = \frac{D_{mg/L}Q_{ML/d}}{PG} \qquad \text{[SI]} \quad \textit{17.19(a)}$$

$$F_{lbm/day} = \frac{D_{mg/L}Q_{MGD}\left(8.345 \ \frac{lbm\text{-}L}{mg\text{-}MG}\right)}{PG} \quad \text{[U.S.]} \quad \textit{17.19(b)}$$

Example 17.1

Incoming water contains 2.5 mg/L as a substance of natural alkalinity (HCO_3^-). The flow rate is 2.5 MGD (9.5 ML/day). (a) What feed rate is required if the alum (purity 48%) dose is 7 mg/L? (b) How much lime (purity 85%) in the form of $Ca(OH)_2$ is required to react completely with the alum?

SI Solution

(a) The dose is specifically in terms of mg/L of alum. Therefore, Eq. 17.19 can be used.

$$F_{kg/d} = \frac{D_{mg/L}Q_{ML/d}}{PG} = \frac{\left(7 \ \frac{mg}{L}\right)\left(9.5 \ \frac{ML}{d}\right)}{(0.48)(1.0)}$$
$$= 138.5 \ kg/d$$

(b) From Eq. 17.16, each alum molecule reacts with six ions of alkalinity.

$$Al_2(SO_4)_3 + 6HCO_3^- \rightarrow$$
$$\text{MW:} \quad 342 \qquad (6)(61)$$
$$\text{mg/L:} \quad X \qquad 2.5$$

By simple proportion, the amount of alum used to counteract the natural alkalinity is

$$X = \frac{\left(2.5 \ \frac{mg}{L}\right)(342)}{(6)(61)} = 2.34 \ mg/L$$

The alum remaining is

$$7 \ \frac{mg}{L} - 2.34 \ \frac{mg}{L} = 4.66 \ mg/L$$

From Eq. 17.17, one alum molecule reacts with three slaked lime molecules.

$$Al_2(SO_4)_3 + 3Ca(OH)_2 \rightarrow$$
$$\text{MW:} \quad 342 \qquad (3)(74)$$
$$\text{mg/L:} \quad 4.66 \qquad X$$

By simple proportion, the amount of lime needed is

$$X = \frac{\left(4.66 \ \frac{mg}{L}\right)(3)(74)}{342} = 3.02 \ mg/L$$

Using Eq. 17.19,

$$F_{kg/d} = \frac{D_{mg/L}Q_{ML/d}}{PG} = \frac{\left(3.02 \ \frac{mg}{L}\right)\left(9.5 \ \frac{ML}{d}\right)}{(0.85)(1.0)}$$
$$= 33.8 \ kg/d$$

Customary U.S. Solution

(a) The dose is specifically in terms of mg/L of alum. Therefore, Eq. 17.19 can be used.

$$F_{lbm/day} = \frac{D_{mg/L}Q_{MGD}\left(8.345 \ \frac{lbm\text{-}L}{mg\text{-}MG}\right)}{PG}$$
$$= \frac{\left(7 \ \frac{mg}{L}\right)(2.5 \ MGD)\left(8.345 \ \frac{lbm\text{-}L}{mg\text{-}MG}\right)}{(0.48)(1.0)}$$
$$= 304.2 \ lbm/day$$

(b) From Eq. 17.16, each alum molecule reacts with six ions of alkalinity.

$$Al_2(SO_4)_3 + 6HCO_3^- \rightarrow$$
$$\text{MW:} \quad 342 \qquad (6)(61)$$
$$\text{mg/L:} \quad X \qquad 2.5$$

By simple proportion, the amount of alum used to counteract the natural alkalinity is

$$X = \frac{\left(2.5 \ \frac{mg}{L}\right)(342)}{(6)(61)}$$
$$= 2.34 \ mg/L$$

The alum remaining is

$$7 \; \frac{\text{mg}}{\text{L}} - 2.34 \; \frac{\text{mg}}{\text{L}} = 4.66 \; \text{mg/L}$$

From Eq. 17.17, one alum molecule reacts with three lime molecules.

$$\text{Al}_2(\text{SO}_4)_3 + 3\text{Ca(OH)}_2 \rightarrow$$

MW:	342	(3)(74)
mg/L:	4.66	X

By simple proportion, the amount of lime needed is

$$\begin{aligned} X &= \frac{\left(4.66 \; \dfrac{\text{mg}}{\text{L}}\right)(3)(74)}{342} \\ &= 3.02 \; \text{mg/L} \end{aligned}$$

Using Eq. 17.19,

$$\begin{aligned} F_{\text{lbm/day}} &= \frac{D_{\text{mg/L}} Q_{\text{MGD}} \left(8.345 \; \dfrac{\text{lbm-L}}{\text{mg-MG}}\right)}{PG} \\ &= \frac{\left(3.02 \; \dfrac{\text{mg}}{\text{L}}\right)(2.5 \; \text{MGD}) \left(8.345 \; \dfrac{\text{lbm-L}}{\text{mg-MG}}\right)}{(0.85)(1.0)} \\ &= 74.1 \; \text{lbm/day} \end{aligned}$$

18. MIXERS AND MIXING KINETICS

Coagulants and other water treatment chemicals are added in *mixers*. If the mixer adds a coagulant for the removal of colloidal sediment, a downstream location (i.e., a tank or basin) with a reduced velocity gradient may be known as a *flocculator*.

There are two basic models: plug flow mixing and complete mixing. The *complete mixing model* is appropriate when the chemical is distributed throughout by impellers or paddles. If the basin volume is small, so that time for mixing is low, the tank is known as a *flash mixer*, *rapid mixer*, or *quick mixer*. The volume of flash mixers is seldom greater than 300 ft³ (8 m³), and flash mixer detention time is usually 30–60 sec. Flash mixing kinetics are described by the complete mixing model.

Flash mixers are usually concrete tanks, square in horizontal cross section, and fitted with vertical shaft impellers. The size of a mixing basin can be determined from various combinations of dimensions that satisfy the volume-flow rate relationship. (Equation 17.20 is a rearrangement of Eq. 17.9.)

$$V = \phi Q \quad\quad\quad 17.20$$

The detention time required for complete mixing in a tank of volume V depends on the *mixing rate constant*, K, and the incoming and outgoing concentrations.

$$t_{\text{complete}} = \frac{V}{Q} = \frac{1}{K}\left(\frac{C_i}{C_o} - 1\right) \quad\quad 17.21$$

The *plug flow mixing* model is appropriate when the water flows through a long narrow chamber, the chemical is added at the entrance, and there is no mechanical agitation. All of the molecules remain in the plug flow mixer for the same amount of time as they flow through. For any mixer, the maximum chemical conversion will occur with plug flow, since all of the molecules have the maximum opportunity to react. The detention time in a plug flow mixer of length L is

$$t_{\text{plug flow}} = \frac{V}{Q} = \frac{L}{\text{v}_f} = \frac{1}{K}\ln\frac{C_i}{C_o} \quad\quad 17.22$$

19. MIXING PHYSICS

The drag force on a paddle is given by the standard fluid drag force equation. For flat plates, the coefficient of drag, C_D, is approximately 1.8.

$$F_D = \frac{C_D A \rho \text{v}_{\text{mixing}}^2}{2} \quad\quad \text{[SI]} \quad 17.23(a)$$

$$\begin{aligned} F_D &= \frac{C_D A \rho \text{v}_{\text{mixing}}^2}{2g_c} \\ &= \frac{C_D A \gamma \text{v}_{\text{mixing}}^2}{2g} \end{aligned} \quad \text{[U.S.]} \quad 17.23(b)$$

The power required is calculated from the drag force and the mixing velocity. The average *mixing velocity*, v_{mixing}, also known as the *relative paddle velocity*, is the difference in paddle and average water velocities. The mixing velocity is approximately 0.7–0.8 times the tip speed. R is the distance from the shaft to the paddle center in ft.

$$\text{v}_{\text{paddle,ft/sec}} = \frac{2\pi R n_{\text{rpm}}}{60 \; \dfrac{\text{sec}}{\text{min}}} \quad\quad 17.24$$

$$\text{v}_{\text{mixing}} = \text{v}_{\text{paddle}} - \text{v}_{\text{water}} \quad\quad 17.25$$

$$P_{kW} = \frac{F_D v_{mixing}}{1000 \frac{W}{kW}}$$

$$= \frac{C_D A \rho v_{mixing}^3}{2\left(1000 \frac{W}{kW}\right)} \qquad \text{[SI]} \quad 17.26(a)$$

$$P_{hp} = \frac{F_D v_{mixing}}{550 \frac{\text{ft-lbf}}{\text{hp-sec}}} = \frac{C_D A \rho v_{mixing}^3}{2g_c\left(550 \frac{\text{ft-lbf}}{\text{hp-sec}}\right)}$$

$$= \frac{C_D A \gamma v_{mixing}^3}{2g\left(550 \frac{\text{ft-lbf}}{\text{hp-sec}}\right)} \qquad \text{[U.S.]} \quad 17.26(b)$$

For slow-moving paddle mixers, the *velocity gradient*, G, varies from 20 sec^{-1} to 75 sec^{-1} for a 15 min to 30 min mixing period. Typical units in Eq. 17.27 are ft-lbf/sec for power (multiply hp by 550 to obtain ft-lbf/sec), lbf-sec/ft^2 for μ, and ft^3 for volume. In the SI system, power in kW is multiplied by 1000 to obtain W, viscosity is in Pa·s, and volume is in m^3. (Multiply viscosity in cP by 0.001 to obtain Pa·s.)

Rapid Mix and Flocculator Design

$$G = \sqrt{\frac{P}{\mu V}} \qquad 17.27$$

Equation 17.27 can also be used for rapid mixers, in which case the mean velocity gradient is much higher: approximately 500–1000 sec^{-1} for 10–30 sec mixing period, or 3000–5000 sec^{-1} for a 0.5–1.0 sec mixing period in an in-line blender configuration.

Equation 17.27 can be rearranged to calculate the power requirement. Power is typically 0.5–1.5 hp/MGD for rapid mixers.

$$P = \mu G^2 V \qquad 17.28$$

The dimensionless product, Gt_d, of the velocity gradient and detention time is known as the *mixing opportunity parameter*. Typical values range from 10^4 to 10^5.

$$Gt_d = \frac{V}{Q}\sqrt{\frac{P}{\mu V}} = \frac{1}{Q}\sqrt{\frac{PV}{\mu}} \qquad 17.29$$

20. IMPELLER CHARACTERISTICS

Mixing equipment uses rotating impellers on rotating shafts. The blades of *radial-flow impellers* (paddle-type impellers, turbine impellers, etc.) are parallel to the drive shaft. *Axial-flow impellers* (propellers, pitched-blade impellers, etc.) have blades inclined with respect to the drive shaft. (See Fig. 17.5.) Axial-flow impellers are better at keeping materials (e.g., water softening chemicals) in suspension.

Figure 17.5 *Typical Axial Flow Mixing Impellers*

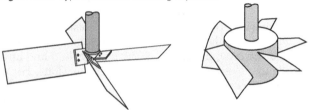

The *mixing Reynolds number* depends on the impeller diameter, D, suspension specific gravity, SG, liquid viscosity, μ, and rotational speed, n, in rev/sec. Flow is in transition for $10 < \text{Re} < 10{,}000$, is laminar below 10, and is turbulent above 10,000.

$$\text{Re} = \frac{D^2 n \rho}{\mu} \qquad \text{[SI]} \quad 17.30(a)$$

$$\text{Re} = \frac{D^2 n \rho}{g_c \mu} \qquad \text{[U.S.]} \quad 17.30(b)$$

The dimensionless *power number*, N_P, of the impeller is defined implicitly by the power required to drive the impeller. For any given level of suspension, Eq. 17.31 through Eq. 17.34 apply to geometrically similar impellers and turbulent flow.

$$P = N_P n^3 D^5 \rho \qquad \text{[SI]} \quad 17.31(a)$$

$$P = \frac{N_P n^3 D^5 \rho}{g_c} \qquad \text{[U.S.]} \quad 17.31(b)$$

$$P = \rho g Q h_v \qquad \text{[SI]} \quad 17.32(a)$$

$$P = \frac{\rho g Q h_v}{g_c} \qquad \text{[U.S.]} \quad 17.32(b)$$

The impeller's dimensionless *flow number*, N_Q, is defined implicitly by the *flow rate equation*.

$$Q = N_Q n D^3 \qquad 17.33$$

The velocity head can be calculated from a combination of the power and flow numbers.

$$h_v = \frac{N_P n^2 D^2}{N_Q g} \qquad 17.34$$

Vibration near the critical speed can be a major problem with modern, high-efficiency, high-speed impellers. Mixing speed should be well below (i.e., less than 80% of) the first critical speed of the shaft. Other important design factors include tip speed and shaft bending moment.

The critical speed of a shaft corresponds to its natural harmonic frequency of vibration when loaded as a flexing beam. If the harmonic frequency, f_c, is in hertz, the shaft's critical rotational speed (in rev/min) will be $60f_c$. The critical speed depends on many factors, including the masses and locations of paddles and other shaft loads, shaft material, length, and end restraints. It is not important whether the shaft is vertical or horizontal, and the damping effects of the mixing fluid and any stabilizing devices are disregarded in calculations. For a lightweight (i.e., massless) mixing impeller that is on a suspended (cantilever) shaft of length L, has a uniform cross section, a mass per unit length of m_L, and is supported by a fixed upper bearing, the first critical speed, f_c, in hertz, will be the same as that of a cantilever beam and can be approximated from Eq. 17.35. Other configurations, including mixed shaft materials, stepped or varying shaft diameters, two or more support bearings, substantial impeller mass, and multiple impellers require different analysis methods.

$$f_c = 0.56 \sqrt{\frac{EI}{L^4 m_L}} \qquad \text{[SI]} \quad 17.35(a)$$

$$f_c = 0.56 \sqrt{\frac{EIg_c}{L^4 m_L}} \qquad \text{[U.S.]} \quad 17.35(b)$$

21. FLOCCULATION

After flash mixing, the floc is allowed to form during a 20–60 min period of gentle mixing. Flocculation is enhanced by the gentle agitation, but floc disintegrates with violent agitation. During this period, the flow-through velocity should be limited to 0.5–1.5 ft/min (0.25–0.75 cm/s). The peripheral speed of mixing paddles should vary approximately from 0.5 ft/sec (0.15 m/s) for fragile, cold-water floc to 3.0 ft/sec (0.9 m/s) for warm-water floc.

Many modern designs make use of *tapered flocculation*, also known as *tapered energy*, a process in which the amount (severity) of flocculation gradually decreases as the treated water progresses through the flocculation basin.

Flocculation is followed by sedimentation for two to eight hours (four hours typical) in a low-velocity portion of the basin. (The flocculation time is determined from settling column data.) A good settling process will remove 90% of the settleable solids. Poor design, usually

resulting in some form of *short-circuiting* of the flow path, will reduce the effective time in which particles have to settle.

22. FLOCCULATOR-CLARIFIERS

A *flocculator-clarifier* combines mixing, flocculation, and sedimentation into a single tank. Such units are called *solid contact units* and *upflow tanks*. They are generally round in construction, with mixing and flocculation taking place near the central hub and sedimentation occurring at the periphery. Flocculator-clarifiers are most suitable when combined with softening, since the precipitated solids help seed the floc.

Typical operational characteristics of flocculator-clarifiers are given in Table 17.2.

Table 17.2 Characteristics of Flocculator-Clarifiers

typical flocculation and mixing time	20–60 min
minimum detention time	1.5–2.0 hr
maximum weir loading	10 gpm/ft (2 L/s·m)
upflow rate	0.8–1.7 gpm/ft^2;
	1.0 gpm/ft^2 typical
	(0.54–1.2 L/s·m^2;
	0.68 L/s·m^2 typical)
maximum sludge formation rate	5% of water flow

(Multiply gpm/ft by 0.207 to obtain L/s·m.)
(Multiply gpm/ft^2 by 0.679 to obtain L/s·m^2.)

23. FILTRATION

Nonsettling floc, algae, suspended precipitates from softening, and metallic ions (iron and manganese) are removed by filtering. *Sand filters* (and in particular, rapid sand filters) are commonly used for this purpose. Sand filters are beds of gravel, sand, and other granulated materials.[8]

Although filter box heights are on the order of 10 ft (3 m) to provide for expansion and freeboard during backwashing, the almost-universal specification used before 1960 for a single-media filter was 24–30 in (610–760 mm) of sand or ground coal. Filters are usually square or nearly square in plan, and they operate with a hydraulic head of 1–8 ft (0.3–2.4 m).

Although most plants have six or more filters, there should be at least three filters so that two can be operational when one is being cleaned. Cleaning of multiple filters should be staged.

---

[8]Most early sand filter beds were designed when a turbidity level of 5 NTU was acceptable. With the U.S. federal MCL at 1 NTU, some states at 0.5 NTU, and planning "on the horizon" for 0.2 NTU, these early conventional filters are clearly inadequate.

Filter operation is enhanced when the top layer of sand is slightly more coarse than the deeper sand. During backwashing, however, the finest sand rises to the top. Various *dual-layer* and *multi-layer* (also known as *dual-media* and *multi-media*) filter designs using layers of coal ("anthrafilt") and (more recently) granular activated carbon, alone or in conjunction with sand, overcome this problem. Since coal has a lower specific gravity than sand, media particle size is larger, as is pore space.

Historically, the *loading rate* (i.e., *flow rate*) for *rapid sand filters* was set at 2–3 gpm/ft² (1.4–2.0 L/s·m²). A range of 4–6 gpm/ft² (2.7–4.1 L/s·m²) is a reasonable minimum rate for dual-media filters. Multi-media filters operate at 5–10 gpm/ft² (3.4–6.8 L/s·m²) and above.[9] The filter loading rate is

$$\text{loading rate} = \frac{Q}{A} \qquad 17.36$$

Filters discharge into a storage reservoir known as a *clearwell*. The *hydraulic head* (the distance between the water surfaces in the filter and clearwell) is usually 9–12 ft (2.7–3.6 m). This allows for a substantial decrease in available head prior to backwashing. Clearwell storage volume is 30–60% of the daily filter output, with a minimum capacity of 12 hr of the maximum daily demand so that demand can be satisfied from the clearwell while the filter is being cleaned or serviced.

24. FILTER BACKWASHING

The most common type of service needed by filters is backwashing, which is needed when the pores between the filter particles clog up. Typically, this occurs after 1 to 3 days of operation, when the head loss reaches 6–8 ft (1.6–2.4 m). There are two parameters that can trigger backwashing: head loss and turbidity. Head loss increases almost linearly with time, while turbidity remains constant for several days before suddenly increasing. The point of sudden increase is known as *breakthrough*.[10] Since head loss is more easily monitored than turbidity, it is desired to have head loss trigger the backwashing cycle.

Backwashing with filtered water pumped back through the filter from the bottom to the top expands the sand layer 30–50%, which dislodges trapped material. Backwashing for 3–5 min at 8–15 gpm/ft² (5.4–10 L/s·m²) is a typical specification.[11] The head loss is reduced to approximately 1 ft (0.3 m) after washing.

Experience has shown that supplementary agitation of the filter media is necessary to prevent "caking" and "mudballs" in almost all installations. Prior to backwashing, the filter material may be expanded by an *air prewash* volume of 1–8 (2–5 typical) times the sand filter volume per minute for 2–10 min (3–5 min typical). Alternatively, turbulence in the filter material may be encouraged during backwashing with an *air wash* or with rotating hydraulic surface jets.

During backwashing, the water in the filter housing will rise at a rate of 1–3 ft/min (0.5–1.5 cm/s). This rise should not exceed the settling velocity of the smallest particle that is to be retained in the filter. The wash water, which is collected in troughs for disposal, constitutes approximately 1–5% of the total processed water. The total water used is approximately 75–100 gal/ft² (3–4 kL/m²). The actual amount of backwash water is

$$V = A_{\text{filter}}(\text{rate of rise})t_{\text{backwash}} \qquad 17.37$$

The temperature of the water used in backwashing is important since the viscosity changes. (The effect of temperature on water density is negligible.) 40°F (4°C) water is more viscous than 70°F (21°C) water. Media particles, therefore, may be expanded to the same extent using lower upflow rates at the lower backwash temperature.

Example 17.2

A filter gallery of seven multimedia sand filters treats 75 700 m³/d of water. Each filter is backwashed once during every 24 h period. Backwashing occurs at 9 m³/m² per hour for 20 min, followed by a conditioning period of 10 min. The filter loading rate is 140 m³/m² per day. What is the net filter production for the plant?

Solution

The flow rate to each filter during periods of no backwash or conditioning, when all filters are in use, is

$$\frac{75\,700\ \dfrac{\text{m}^3}{\text{d}}}{7\ \text{filters}} = 10\,814\ \text{m}^3/\text{filter·d}$$

The flow rate to each filter when one filter is being backwashed or conditioned is

$$\frac{75\,700\ \dfrac{\text{m}^3}{\text{d}}}{6\ \text{filters}} = 12\,617\ \text{m}^3/\text{filter·d}$$

---

[9]Testing has demonstrated that deep-bed, uniformly graded anthracite filters can operate at 10–15 gpm/ft² (7–10 L/s·m²), although this requires high-efficiency preozonation and microflocculation.
[10]Technically, *breakthrough* is the point at which the turbidity rises above the MCL permitted. With low MCLs, this occurs very soon after the beginning of filter performance degradation.
[11]While the maximum may never be used, a maximum backwash rate of 20 gpm/ft² (14 L/s·m²) should be provided for.

The amount of time each day during which when one filter is being backwashed or conditioned is

$$\left(\dfrac{20 \dfrac{\text{min}}{\text{filter·d}} + 10 \dfrac{\text{min}}{\text{filter·d}}}{60 \dfrac{\text{min}}{\text{h}}} \right)$$
$$\times (7 \text{ filters}) = 3.5 \text{ h/d}$$

The amount of time each day when all filters are in use is

$$24 \dfrac{\text{h}}{\text{d}} - 3.5 \dfrac{\text{h}}{\text{d}} = 20.5 \text{ h/d}$$

The gross production per filter per day is

$$\left(10\,814 \dfrac{\text{m}^3}{\text{filter·d}} \right) \left(\dfrac{20.5 \dfrac{\text{h}}{\text{d}}}{24 \dfrac{\text{h}}{\text{d}}} \right)$$
$$+ \left(12\,617 \dfrac{\text{m}^3}{\text{filter·d}} \right) \left(\dfrac{3.5 \dfrac{\text{h}}{\text{d}}}{24 \dfrac{\text{h}}{\text{d}}} \right) = 11\,077 \text{ m}^3/\text{filter·d}$$

The filter bed area is

$$\dfrac{10\,814 \dfrac{\text{m}^3}{\text{filter·d}}}{140 \dfrac{\text{m}^3}{\text{m}^2\text{·d}}} = 77.15 \text{ m}^2/\text{filter}$$

The water lost to backwashing per filter per day is

$$\left(9 \dfrac{\text{m}^3}{\text{m}^2\text{·h}} \right) \left(\dfrac{20 \dfrac{\text{min}}{\text{d}}}{60 \dfrac{\text{min}}{\text{h}}} \right) (77.15 \text{ m}^2/\text{filter}) = 231.5 \text{ m}^3/\text{filter·d}$$

The water lost to conditioning per filter per day is

$$\left(10\,814 \dfrac{\text{m}^3}{\text{filter·d}} \right) \left(\dfrac{10 \dfrac{\text{min}}{\text{d}}}{1440 \dfrac{\text{min}}{\text{d}}} \right) = 75.10 \text{ m}^3/\text{filter·d}$$

The net production per filter per day is

$$11\,077 \dfrac{\text{m}^3}{\text{filter·d}} - 231.5 \dfrac{\text{m}^3}{\text{filter·d}} - 75.10 \dfrac{\text{m}^3}{\text{filter·d}}$$
$$= 10\,770 \text{ m}^3/\text{filter·d}$$

The total net production is

$$\left(10\,770 \dfrac{\text{m}^3}{\text{filter·d}} \right) (7 \text{ filters}) = 75\,390 \text{ m}^3/\text{d}$$

25. OTHER FILTRATION METHODS

Biofilm filtration (*biofilm process*) uses microorganisms to remove selected contaminants (e.g., aromatics and other hydrocarbons). Operation of biofilters is similar to trickling filters used in wastewater processing. Sand filter facilities are relatively easy to modify—sand is replaced with gravel in the 4–14 mm size range, application rates are decreased, and exposure to chlorine from incoming and backwash water is eliminated.

Slow sand filters are primarily of historical interest, though there are some similarities with modern biomethods used to remediate toxic spills. Slow sand filters operate similarly to rapid sand filters except that the exposed surface (loading) area is much larger and the flow rate is much lower (0.05–0.1 gpm/ft^2; 0.03–0.07 L/s·m^2). Slow sand filters are limited to low turbidity applications not requiring chemical treatment and where large rural areas are available to spread out the facilities. Slow sand filters can be operated as biofilm processes if a layer of biological slime is allowed to form on the top of the filter medium.[12] Slow filter cleaning usually involves removing a few inches of sand.

Pressure (*sand*) *filters* for water supply treatment operate similarly to rapid sand filters except that incoming water is typically pressurized 25–75 psig (170–520 kPa gage). Single media filter rates are 2–10 gpm/ft^2, with 4–5 gpm/ft^2 being typical (1.4–14 L/s·m^2 (2.7–3.4 L/s·m^2 typical)), with dual media filters running at 1.5 to 2.0 times these rates. Pressure filters are not used in large installations.

Membrane processes consist of a variety of applications where a semipermeable membrane is used to separate particles or solutes that would ordinarily not be removed by conventional filtration. Of most significance are *reverse osmosis* and *ultrafiltration*, where the opportunity to remove pathogens, dissolved ions, and larger molecular weight organics exists.

Reverse osmosis (RO) and ultrafiltration (UF) are semipermeable membrane processes where pressure is used to reverse the normal process of osmosis and create migration of water through the membrane, leaving contaminants behind.

---

[12]The biological slime is called *schmutzdecke*. The slime forms a physical barrier that traps particles.

In normal osmosis, water moves under osmotic pressure to establish equilibrium in the ionic strength of solutions across a semipermeable membrane. The water moves from the more dilute side of the membrane to the more concentrated side. However, in RO and UF, an external pressure is applied to reverse the flow across the membrane. The water is forced through the membrane from the concentrated side to the dilute side, leaving the contaminants behind in a more concentrated form. (See Fig. 17.6.)

Figure 17.6 *Normal and Reverse Osmosis*

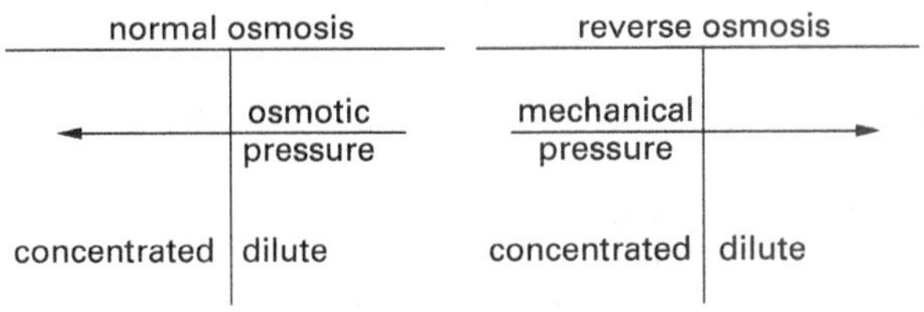

normal osmosis		reverse osmosis	
	osmotic pressure	mechanical pressure	
concentrated	dilute	concentrated	dilute

Operating pressure varies from 15 psi for UF to as much as 1200 psi for RO. Typically, UF pressures range from 15 psi to 75 psi, and RO pressures range from 200 psi to 1200 psi. The type of membrane process selected depends on the size of the targeted contaminant.

Membrane processes are described by

feed water, or raw, untreated water	100% by volume
concentrate, or concentrated brine (the by-product wastewater)	10–40% by volume
permeate, or treated water	60–90% by volume

RO and UF processes can potentially result in a large percentage of the feedwater being lost as concentrate. For this reason, by-product disposal issues are important to the feasibility of these processes.

RO and UF processes usually require pilot testing before final design. Performance is a function of pH and temperature and is influenced by membrane deterioration and membrane scaling or clogging.

The efficiency of the RO process for removal of common contaminants is typically very high. For example,

- 99+% removal for salts (consequently, used for removal of total dissolved solids)

- 100% removal for fruit and vegetable solids (used to concentrate juices)

- 99.9% removal for pathogens

Example 17.3

A reverse osmosis system is needed for the treatment of a drinking water source that is subject to saltwater intrusion. The water and RO system have the following characteristics.

desired freshwater flow rate	$30\,000$ m^3/d
permeate recovery	77%
salt rejection	96%
operating pressure	750 psi
membrane flux rate	0.93 m^3/m^2·d
membrane packing density	800 m^2/m^3
membrane module volume	0.028 m^3
pressure vessel capacity	12 modules
feedwater TDS	540 mg/L

Find the number of pressure vessels needed and the concentration of total dissolved solids (TDS) in the rejected concentrate.

Solution

The feed rate needed to yield the desired flow rate is

$$\frac{30\,000\ \dfrac{\text{m}^3}{\text{d}}}{\dfrac{77\%}{100\%}} = 38\,961\ \frac{\text{m}^3}{\text{d}}$$

The total membrane area is

$$\frac{38\,961\ \dfrac{\text{m}^3}{\text{d}}}{0.93\ \dfrac{\text{m}^3}{\text{m}^2 \cdot \text{d}}} = 41\,894\ \text{m}^2$$

The volume occupied by the membranes is

$$\frac{41\,894\ \text{m}^2}{800\ \dfrac{\text{m}^2}{\text{m}^3}} = 52.4\ \text{m}^3$$

The number of membrane modules needed is

$$\frac{52.4\ \text{m}^3}{0.028\ \dfrac{\text{m}^3}{\text{module}}} = 1871\ \text{modules}$$

The number of pressure vessels needed is

$$\frac{1871 \text{ modules}}{12 \dfrac{\text{modules}}{\text{pressure vessel}}} = 156 \text{ pressure vessels}$$

The TDS concentration in the rejected concentrate is

$$\frac{\left(\dfrac{96\%}{100\%}\right)\left(540 \dfrac{\text{mg}}{\text{L}}\right)}{1 - \dfrac{77\%}{100\%}} = 2254 \text{ mg/L}$$

26. ADSORPTION

Many dissolved organic molecules and inorganic ions can be removed by adsorption processes. *Adsorption* is not a straining process, but occurs when contaminants are trapped on the surface or interior of the adsorption particles.

Granular activated carbon (GAC) is considered to be the best available technology for removal of THMs and synthetic organic chemicals from water.[13] GAC is also useful in removing compounds that contribute to taste, color, and odor. Activated carbon can be used in the form of powder (which must be subsequently removed) or granules. GAC can be integrated into the design of sand filters.[14]

Eventually, the GAC becomes saturated and must be removed and reactivated. In such dual-media filters, the reactivation interval for GAC is approximately one to two years.

27. FLUORIDATION

Fluoridation can occur any time after filtering and can involve solid compounds or liquid solutions. Small utilities may manufacturer their own liquid solution on-site from sodium silicofluoride (Na_2SiF_6, typically 22–30% purity) or sodium fluoride (NaF, typically 90–98% purity) for use with a volumetric metering system. Larger utilities use gravimetric dry feeders with sodium silicofluoride (Na_2SiF_6, typically 98–99% purity) or solution feeders with fluorsilic acid (H_2SiF_6).

Assuming 100% ionization, the application rate, F (in pounds per day), of a compound with fluoride gravimetric fraction, G, and a fractional purity, P, needed to obtain a final concentration, C, of fluoride is

$$F_{\text{kg/d}} = \frac{C_{\text{mg/L}} Q_{\text{L/d}}}{PG} \qquad \text{[SI]} \quad \textit{17.38(a)}$$

$$F_{\text{lbm/day}} = \frac{C_{\text{mg/L}} Q_{\text{MGD}} \left(8.345 \ \dfrac{\text{lbm-L}}{\text{mg-MG}}\right)}{PG} \quad \text{[U.S.]} \quad \textit{17.38(b)}$$

28. IRON AND MANGANESE REMOVAL

Several methods can be used to remove iron and manganese. Most involve aeration with chemical oxidation since manganese is not easily removed by aeration alone. These processes are described in Table 17.3.

29. TASTE AND ODOR CONTROL

A number of different processes affect taste and odor. Some are more effective and appropriate than others. *Microstraining*, using a 35 μm or finer metal cloth, can be used to reduce the number of algae and other organisms in the water, since these are sources of subsequent tastes and odors. Microstraining does not remove dissolved or colloidal organic material, however.

Activated carbon removes more tastes and odors, as well as a wide variety of chemical contaminants.

Aeration can be used when dissolved oxygen is low or when hydrogen sulfide is present. Aeration has little effect on most other tastes and odors.

Chlorination disinfects and reduces odors caused by organic matter and industrial wastes. Normally, the dosage required will be several times greater than those for ordinary disinfection, and the term *superchlorination* is used to describe applying enough chlorine to maintain an excessively large residual. Subsequent dechlorination will be required to remove the excess chlorine. Similar results will be obtained with chlorine dioxide and other disinfection products.

A quantitative odor ranking is the *threshold odor number* (TON), which is determined by adding increasing amounts of odor-free dilution water to a sample until the combined sample is virtually odor free.

$$\text{TON} = \frac{V_{\text{raw sample}} + V_{\text{dilution water}}}{V_{\text{raw sample}}} \qquad \textit{17.39}$$

---

[13]Enhanced coagulation is also a "best available" technology for THM removal that may be economically more attractive.
[14]For example, an 80 in (2032 mm) GAC layer may be placed on top of a 40 in (1016 mm) sand layer.

Table 17.3 Processes for Iron and Manganese Removal

processes	iron and/or manganese removed	pH required	remarks
aeration, settling, and filtration	ferrous bicarbonate	7.5	provide aeration unless incoming water contains adequate dissolved oxygen
	ferrous sulfate	8.0	
	manganous bicarbonate	10.3	
	manganous sulfate	10.0	
aeration, free residual chlorination, settling, and filtration	ferrous bicarbonate manganous bicarbonate	5.0 9.0	provide aeration unless incoming water contains adequate dissolved oxygen
aeration, lime softening, settling, and filtration	ferrous bicarbonate manganous bicarbonate	8.5–9.6	
aeration, coagulation, lime softening, settling, and filtration	colloidal or organic iron colloidal or organic manganese	8.5–9.6 10.0	require lime, and alum or iron coagulant
ion exchange	ferrous bicarbonate	≈ 6.5	water must be devoid of oxygen
	manganous bicarbonate		iron and manganese in raw water not to exceed 2.0 mg/L
			consult manufacturers for type of ion exchange resin to be used

A TON of 3 or less is ideal. Untreated river water usually has a TON between 6 and 24. Treated water normally has a TON between 3 and 6. At TON of 5 and above, customers will begin to notice the taste and odor of their water.

30. PRECIPITATION SOFTENING

Precipitation softening using the *lime-soda ash process* adds lime (CaO), also known as *quicklime*, and soda ash (Na_2CO_3) to remove calcium and magnesium from hard water.[15] Granular quicklime is available with a minimum purity of 90%, and soda ash is available with a 98% purity.

Lime forms *slaked lime* (also known as *hydrated lime*), $Ca(OH)_2$, in an exothermic reaction when added to feed water. The slaked lime is delivered to the water supply as a *milk of lime* suspension.

$$CO + H_2O \rightarrow Ca(OH)_2 + heat \qquad 17.40$$

Slaked lime reacts first with any carbon dioxide dissolved in the water, as in Eq. 17.41. No softening occurs, but the carbon dioxide demand must be satisfied before any reactions involving calcium or magnesium can occur.

Lime-Soda Ash Softening Equations

carbon dioxide removal:

$$CO_2 + Ca(OH)_2 \rightarrow CaCO_3\,(s) + H_2O \qquad 17.41$$

Lime next reacts with any carbonate hardness, precipitating calcium carbonate and magnesium hydroxide, as shown in Eq. 17.42 and Eq. 17.43. The removal of carbonate hardness caused by magnesium (characterized by magnesium bicarbonate in Eq. 17.43) requires two molecules of calcium hydroxide to precipitate calcium carbonate and magnesium hydroxide.

Lime-Soda Ash Softening Equations

calcium carbonate hardness removal:

$$Ca(HCO_3)_2 + Ca(OH)_2 \rightarrow 2CaCO_3(s) + 2H_2O \qquad 17.42$$

Lime-Soda Ash Softening Equations

magnesium carbonate hardness removal:

$$Mg(HCO_3)_2 + 2Ca(OH)_2 \rightarrow$$
$$2CaCO_3(s) + Mg(OH)_2(s) + 2H_2O \qquad 17.43$$

To remove noncarbonate hardness (characterized by sulfate ions in Eq. 17.44 and Eq. 17.45), it is necessary to add soda ash and more lime. The sodium sulfate that remains in solution does not contribute to hardness, for sodium is a single-valent ion (i.e., Na^+).

---

[15]Soda ash does not always need to be used. When lime alone is used, the process may be referred to as *lime softening*.

Lime-Soda Ash Softening Equations

magnesium noncarbonate hardness removal:

$$MgSO_4 + Ca(OH)_2 + Na_2CO_3 \rightarrow$$
$$CaCO_3(s) + Mg(OH)_2(s) + 2Na^+ + SO_4^{2-} \qquad \textit{17.44}$$

Lime-Soda Ash Softening Equations

calcium noncarbonate hardness removal:

$$CaSO_4 + Na_2CO_3 \rightarrow CaCO_3(s) + 2Na^+ + SO_4^{2-} \qquad \textit{17.45}$$

The calcium ion can be effectively reduced by the lime addition shown in the previous equations, raising the pH of the water to approximately 10.3. (Thus, the lime added removes itself.) Precipitation of the magnesium ion, however, requires a higher pH and the presence of excess lime in the amount of approximately 35 mg/L of CaO or 50 mg/L of $Ca(OH)_2$ above the stoichiometric requirements. The practical limits of precipitation softening are 30–40 mg/L of $CaCO_3$ and 10 mg/L of $Mg(OH)_2$, both as $CaCO_3$.

After softening, the water must be recarbonated to lower its pH and to reduce its scale-forming potential. This is accomplished by bubbling carbon dioxide gas through the water.

Lime-Soda Ash Softening Equations

$$Ca^+ + 2OH^- + CO_2 \rightarrow CaCO_3(s) + H_2O \qquad \textit{17.46}$$

The treatment process could be designed such that all of these reactions take place sequentially. That is, first slaked lime would be added (see Eq. 17.42 and Eq. 17.43), then the water would be recarbonated (see Eq. 17.46), then excess lime would be added to raise the pH, followed by soda ash treatment and recarbonation. In fact, essentially such a sequence is used in a *double-stage process:* two chemical application points and two recarbonation points.

A *split process* can be used to reduce the amount of lime that is neutralized in recarbonation. The excess lime needed to raise the pH is added prior to the first flocculator/clarifier stage. The soda ash is added to the first stage effluent, prior to the second flocculator/clarifier stage. Recarbonation, if used, is applied to the effluent of the second stage. A portion of the flow is bypassed (i.e., is not softened) and is later recombined with softened water to obtain the desired hardness.

Example 17.4

Water contains 130 mg/L of calcium bicarbonate $(Ca(HCO_3)_2)$ as $CaCO_3$. How much slaked lime $(Ca(OH)_2)$ is required to remove the hardness?

Solution

The hardness is given as a $CaCO_3$ equivalent. Therefore, the amount of slaked lime required is implicitly the same: 130 mg/L as $CaCO_3$. Convert the quantity to an "as substance" measurement. The conversion factor for $Ca(OH)_2$ is 1.35.

$$Ca(OH)_2: \frac{130 \ \frac{mg}{L}}{1.35} = 96.3 \text{ mg/L as substance}$$

Example 17.5

Water is received with the following characteristics.

total hardness	250 mg/L as $CaCO_3$
alkalinity	150 mg/L as $CaCO_3$
carbon dioxide	5 mg/L as substance

The water is to be treated with precipitation softening and recarbonation. Lime (90% pure) and soda ash (98% pure) are available.

Using 50 mg/L of $Ca(OH)_2$ as substance to raise the pH for magnesium precipitation, what stoichiometric amounts (as substance) of (a) slaked lime, (b) soda ash, and (c) carbon dioxide are required to reduce the hardness of the water to zero?

Solution

(a) First, convert the carbon dioxide concentration to a $CaCO_3$ equivalent. The factor is 2.27.

$$CO_2: \left(5 \ \frac{mg}{L}\right)(2.27) = 11.35 \text{ mg/L as } CaCO_3$$

Since the alkalinity is less than the hardness and no hydroxides or calcium are reported, it is concluded that the carbonate hardness is equal to the alkalinity, and the noncarbonate hardness is equal to the difference in total hardness and alkalinity.

Since the alkalinity is reported as a $CaCO_3$ equivalent, the first-state treatment to remove carbon dioxide and carbonate hardness requires lime in the amount of

$$Ca(OH)_2: \ 11.35 \ \frac{mg}{L} + 150 \ \frac{mg}{L}$$
$$= 161.35 \text{ mg/L as } CaCO_3$$

50 mg/L of $Ca(OH)_2$ are added to raise the pH so that the magnesium (noncarbonate) hardness can be removed. The factor that converts $Ca(OH)_2$ to $CaCO_3$ is 1.35.

$$Ca(OH)_2: \ 161.35 \ \frac{mg}{L} + (1.35)\left(50 \ \frac{mg}{L}\right)$$
$$= 228.85 \ mg/L \ as \ CaCO_3$$

Use the 90% fractional purity to convert to quantity as substance.

$$Ca(OH)_2: \ \frac{228.85 \ \dfrac{mg}{L}}{(1.35)(0.9)} = 188.4 \ mg/L \ as \ substance$$

(b) The noncarbonate hardness is

$$250 \ \frac{mg}{L} - 150 \ \frac{mg}{L} = 100 \ mg/L \ as \ CaCO_3$$

The soda ash requirement is

$$Na_2CO_3: \ \frac{100 \ \dfrac{mg}{L}}{(0.94)(0.98)} = 108.6 \ mg/L \ as \ substance$$

(c) The recarbonation required to lower the pH depends on the amount of excess lime added.

$$CO_2: \ \frac{(1.35)\left(50 \ \dfrac{mg}{L}\right)}{2.27} = 29.74 \ mg/L \ as \ substance$$

31. WATER SOFTENING WITH LIME

Water softening can be accomplished with lime and soda ash to precipitate calcium and magnesium ions from the solution. Lime treatment has the added benefits of disinfection, iron removal, and clarification. Practical limits of *precipitation softening* are 30 mg/L of $CaCO_3$ and 10 mg/L of $Mg(OH)_2$ (as $CaCO_3$) because of intrinsic solubilities. Water treated by this method usually leaves the softening apparatus with a hardness between 50 mg/L and 80 mg/L as $CaCO_3$.

32. ADVANTAGES AND DISADVANTAGES OF PRECIPITATION SOFTENING

Precipitation softening is relatively inexpensive for large quantities of water. Both alkalinity and total solids are reduced. The high pH and lime help disinfect the water.

However, the process produces large quantities of sludge that constitute a disposal problem. The intrinsic solubility of some of the compounds means that complete softening cannot be achieved. Flow rates and chemical feed rates must be closely monitored.

33. WATER SOFTENING BY ION EXCHANGE

In the *ion exchange process* (also known as the *zeolite process* and the *base exchange method*), water is passed through a filter bed of exchange material. Ions in the insoluble exchange material are displaced by ions in the water. The processed water leaves with zero hardness. However, since there is no need for water with zero hardness, some of the water is typically bypassed around the process.

If dissolved solids or sodium concentration of the water are issues, then ion exchange may not be suitable.

There are several types of ion exchange materials. *Green sand (glauconite)* is a natural substance that is mined and treated with manganese dioxide. Green sand is not used commercially.

Synthetic zeolites have long been widely used in water softening and demineralization. Porosity of continuous-phase *gelular resins* is low, and dry contact surface areas of 500 ft²/lbm (0.1 m²/g) or less is common. This makes gel-based zeolites suitable only for small volumes.

Synthetic *macroporous resins (macroreticular resins)* are suitable for use in large-volume water processing systems, when chemical resistance is required, and when specific ions are to be removed. These resins are used in the form of discontinuous, three-dimensional copolymer beads in a rigid-sponge type formation.[16] Each bead is made up of thousands of microspheres of the resin. Porosity is much higher than with gelular resins, and dry contact specific surface areas are approximately 270,000–320,000 ft²/lbm (55–65 m²/g).

There are four primary resin families: strong acid, strong base, weak acid, and weak base. Each family has different resistances to fouling by organic and inorganic chemicals, stabilities, and lifetimes. For example, strong acid resins can last for 20 years. Strong base resins, on the other hand, may have lifetimes of only three years. Each family can be used in either gelular or macroporous forms.

---

[16]The differences in structure between gel polymers and macroporous polymers occur during the polymerization step. Either can be obtained from the same zeolite.

During operation, calcium and magnesium ions in hard water are removed according to the following reaction in which R is the zeolite anion. The resulting sodium compounds are soluble.

$$\begin{Bmatrix} Ca \\ Mg \end{Bmatrix} \begin{bmatrix} (HCO_3)_2 \\ SO_4 \\ Cl_2 \end{bmatrix} + Na_2R$$

$$\rightarrow Na_2 \begin{bmatrix} (HCO_3)_2 \\ SO_4 \\ Cl_2 \end{bmatrix} + \begin{Bmatrix} Ca \\ Mg \end{Bmatrix} R \qquad 17.47$$

The typical saturation capacity ranges of synthetic resins are 1.0–1.5 meq/mL for anion exchange resins and 1.7–1.9 meq/mL for cation exchange resins. However, working capacities are more realistic measures than saturation capacities. The specific working capacity for zeolites is approximately 3–11 kilograins/ft^3 (6.9–25 kg/m^3) of hardness before regeneration.[17] For synthetic resins, the specific working capacity is approximately 10–15 kilograins/ft^3 (23–35 kg/m^3) of hardness.

The volume of water that can be softened per cycle (between regenerations) is

$$V_{water} = \frac{(\text{specific working capacity}) V_{\text{exchange material}}}{\text{hardness}} \qquad 17.48$$

Flow rates are typically 1–6 gpm/ft^3 (2–13 L/s·m^3) of resin volume. The flow rate across the exposed surface will depend on the geometry of the bed, but values of 3–15 gpm/ft^2 (2–10 L/s·m^2) are typical.

Example 17.6

A municipal plant processes water with a total initial hardness of 200 mg/L. The desired discharge hardness is 50 mg/L. If an ion exchange process is used, what is the bypass factor?

Solution

The water passing through the ion exchange unit is reduced to zero hardness. If x is the water fraction bypassed around the zeolite bed,

$$(1 - x)(0) + x\left(200\ \frac{mg}{L}\right) = 50\ \frac{mg}{L}$$
$$x = 0.25$$

34. REGENERATION OF ION EXCHANGE RESINS

Ion exchange material has a finite capacity for ion removal. When the zeolite approaches saturation, it is regenerated (rejuvenated). Standard ion exchange units are regenerated when the alkalinity of their effluent increases to the *set point*. Most units that collect *crud* are operated to a pressure-drop endpoint.[18] The pressure drop through the ion exchange unit is primarily dependent on the amount of crud collected. When the pressure drop reaches a set point, the resin is regenerated.

Regeneration of synthetic ion exchange resins is accomplished by passing a *regenerating solution* over/through the resin. Although regeneration can occur in the ion exchange unit itself, external regeneration is becoming common. This involves removing the bed contents hydraulically, backwashing to separate the components (for mixed beds), regenerating the bed components separately, washing, and then recombining and transferring the bed components back into service. For complete regeneration, a contact time of 30–45 min may be required.

Common regeneration compounds are NaCl (for water hardness removal units), H_2SO_4 (for cation exchange resins), and NaOH (for anion exchange resins). The amount of regeneration solution depends on the resin's degree of saturation. A rule of thumb is to expect to use 5–25 lbm of regeneration compound per cubic foot of resin (80–400 kg/m^3). Alternatively, dosage of the regeneration compound may be specified in terms of hardness removed (e.g., 0.4 lbm of salt per 1000 grains of hardness removed). These rates are applicable to deionization plants for boiler make-up water. For condensate polishing, saturation levels of 10 lbm/ft^3 to 25 lbm/ft^3 (160 kg/m^3 to 400 kg/m^3) are used.

The salt requirement per regeneration cycle is

$$m_{salt} = (\text{specific working capacity}) V_{\text{exchange material}} \times (\text{salt requirement}) \qquad 17.49$$

35. STABILIZATION AND SCALING POTENTIAL

Water that is stable will not gain or lose ions as it passes through the distribution system. Deposits in pipes from dissolved compounds are known as *scale*. Although various types of scale are possible, deposits of calcium carbonate ($CaCO_3$, calcite) are the most prevalent. *Stabilization treatment* is used to eliminate or reduce the potential for scaling in a pipe after the treated water is

---

[17]There are 7000 grains in a pound. 1000 grains of hardness (i.e., $\frac{1}{7}$ of a pound) is known as a *kilograin*. It should not be confused with a kilogram. Other conversions are 1 grain = 64.8 mg, and 1 grain/gal = 17.12 mg/L.
[18]Since water has been filtered before reaching the zeolite bed, *crud* consists primarily of iron-corrosion products ranging from dissolved to particulate matter.

distributed. Various factors influence the stability of water, including temperature, dissolved oxygen, dissolved solids, pH, and alkalinity. Increased temperature greatly increases the likelihood of scaling.

Stability is a general term used to describe a water's tendency to cause scaling. Four indices are commonly used to quantify the stability of water: the Langelier stability index, Ryznar stability index, Puckorius scaling index, and aggressive index. The Langelier and Ryznar indices are widely used in municipal water works, while the Puckorius index is encountered in installations with cooling water (e.g., cooling towers). Each index has its proponents, strengths and weaknesses, and imprecisions. Conclusions drawn from them are not always in agreement.

The common *Langelier stability index* (LSI) (also known as the *Langelier saturation index*) is significantly positive (e.g., greater than 1.5) when a water is supersaturated and will continue to deposit $CaCO_3$ scale in the pipes downstream of the treatment plant. When LSI is significantly negative (e.g., less than -1.5), the water has little scaling potential and may actually dissolve existing scale encountered.

$$LSI = pH - pH_{sat} \qquad 17.50$$

The *Ryznar stability index* (RSI) arranges the same parameters as LSI differently, but avoids the use of negative numbers. Water is considered to be essentially neutral when $6.5 < RSI < 7$. When $RSI > 8$, the water has little scaling potential and will dissolve scale deposits already present. When $RSI < 6.5$, the water has scaling potential.

$$RSI = 2pH_{sat} - pH$$
$$= pH - 2(LSI) \qquad 17.51$$

The *Puckorius scaling index* (PSI) (also known as the *practical scaling index*) does not use the actual pH in its calculation. To account for the buffering effects of other ions, it uses an equilibrium pH. Like RSI, PSI is always positive, and it has the same index interpretation ranges as RSI.

$$PSI = 2pH_{sat} - pH_{eq} \qquad 17.52$$

$$pH_{eq} = 4.54 + 1.465 \log_{10}[M] \qquad 17.53$$

Equation 17.51, Eq. 17.52, and Eq. 17.53 require knowing the pH when the water is saturated with $CaCO_3$. This value depends on the ionic concentrations in a complex manner, and it is calculated or estimated in a variety of ways. The saturation pH can be calculated from Eq. 17.54, where K_{sp} is the solubility product constant for $CaCO_3$, and K_a is the ionization constant for

H_2CO_3.[19,20] M is the alkalinity (often reported as the HCO_3^- concentration) in mg/L as $CaCO_3$. Ca^{2+} is the calcium ion content in mg/L, also as $CaCO_3$.

$$pH_{sat} = (pK_a - pK_{sp}) + pCa + pM$$
$$= -\log_{10}\left(\frac{K_a[Ca^{2+}][M]}{K_{sp}} \right) \qquad 17.54$$

$$M = [HCO_3^-] + 2[CO_3^{2-}] + [OH^-] \qquad 17.55$$

Since the solubility product constants are rarely known and depend in a complex manner on the total ion strength (total dissolved solids) and temperature, if the water chemistry is fully known, the correlation of Eq. 17.56 can be used.[21] Hard water is typically two-thirds calcium hardness and one-third magnesium. However, if only total hardness is known (i.e., calcium hardness is not known) the worst case of calcium hardness equaling total hardness can be assumed. Total dissolved solids (TDS) can be estimated from water conductivity using Table 17.4.

$$pH_{sat} = 9.3 + A + B - C - D \qquad 17.56$$

$$A = \frac{\log_{10}[TDS] - 1}{10} \qquad 17.57$$

$$B = -13.12 \log_{10}(T_{\circ C} + 273°) + 34.55 \qquad 17.58$$

$$C = \log_{10}[Ca^{2+}] - 0.4 \qquad 17.59$$

$$D = \log_{10}[M] \qquad 17.60$$

The *aggressive index* (AI) was originally developed to monitor water in asbestos pipe, but can be encountered in other environments. It is calculated from Eq. 17.61, but disregards the effects of dissolved solids and water temperature. Therefore, it is less accurate than the other indices.

$$AI = pH + \log_{10}[M] + \log_{10}[Ca^{2+}] \qquad 17.61$$

LSI, RSI, PSI, and AI predict tendencies only for calcium carbonate scaling. They do not estimate scaling potential for calcium phosphate, calcium sulfate, silica, or magnesium silicate scale. They are generally only accurate with untreated water. Although water treated with phosphonates and acrylates to dissolve or stabilize calcium carbonate can be evaluated, the scaling potential of water treated with crystal modifiers (e.g., polymaleates, sulfonated styrene/maleic anhydride, and terpolymers) should not be evaluated with these indices.

---

[19]$pK = -\log K$; $pCa = -\log[Ca^{++}]$; $pM = -\log[M]$

[20]Equation 17.54 is an approximation to the stoichiometric reaction equation because it assumes the activities of the H^+, CO_3^-, and HCO_3^- ions present in the water are all 1. This is essentially, but not exactly, true.

[21]As reported in *Industrial Water Treatment, Operation and Maintenance*, UFC 3-240-13FN, Appendix B. Department of Defense (Unified Facilities), 2005.

Table 17.4 *Total Dissolved Solids vs. Conductivity*

conductivity (S/cm)	total dissolved solids (mg/L as $CaCO_3$)
1	0.42
10.6	4.2
21.2	8.5
42.4	17.0
63.7	25.5
84.8	34.0
106.0	42.5
127.3	51.0
148.5	59.5
169.6	68.0
190.8	76.5
212.0	85.0
410.0	170.0
610.0	255.0
812.0	340.0
1008.0	425.0

The primary stabilization processes are pH and alkalinity adjustment using lime or soda ash (if the pH is low), and with carbon dioxide, sulfuric acid, or hydrochloric acid if the pH is too high. Scaling can also be prevented within the treatment plan by using protective pipe linings and coatings. Various inhibitors and sequestering agents can also be used.[22] If lime is used for pH adjustment, care must be exercised that excess lime does not form clinker particles that are carried out into the water distribution system.

Some waters (including cooling waters in some industries) are always corrosive because of the presence of dissolved oxygen, carbon dioxide, and various solids, regardless of whether calcium carbonate is present. $CaCO_3$ scaling indices were never intended to be used as indices or corrosion potential of mild steel pipes, but that has occurred. LSI, RSI, and PSI are blind to the presence of chloride and sulfate content, as well as to the nature of the pipe and fitting materials and to the galvanic potential differences. For that reason, stability indices should not be used for corrosion prediction. However, the LSI, RSI, and PSI scaling potential continuums continue to be inappropriately divided by some authorities into scaling and corrosion regions.

36. DISINFECTION

Chlorination is commonly used for disinfection. Chlorine can be added as a gas or as a liquid. If it is added to the water as a gas, it is stored as a liquid, which vaporizes around $-31°F$ $(-35°C)$. Liquid chlorine is the primary form used since it is less expensive than calcium hypochlorite solid $(Ca(OCl)_2)$ and sodium hypochlorite $(NaOCl)$.

Chlorine is corrosive and toxic. Special safety and handling procedures must be followed with its use.

37. CHLORINATION CHEMISTRY

When chlorine gas dissolves in water, it forms hydrochloric acid (HCl) and hypochlorous acid (HOCl).

$$Cl_2 + H_2O \rightarrow HCl + HOCl \qquad 17.62$$

$$HCl \rightarrow H^+ + Cl^- \qquad 17.63$$

$$HOCl \rightleftharpoons H^+ + OCl^- \qquad 17.64$$

The fraction of HOCl ionized into H^+ and OCl^- depends on the pH and can be determined from Fig. 17.7.

Figure 17.7 *Ionized HOCl Fraction vs. pH*

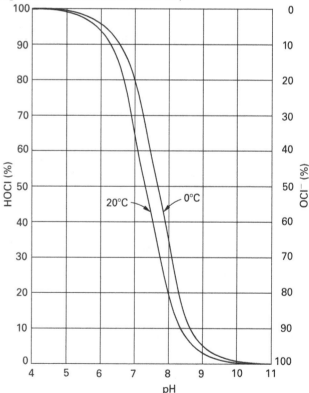

[22] A *sequestering agent* (*chelant, chelator, chelating agent*) is a compound that works by sequestering positive metallic ions in a complex negative ion.

When calcium hypochlorite solid is added to water, the ionization reaction is

$$Ca(OCl)_2 \rightarrow Ca^{2+} + 2OCl^- \qquad 17.65$$

38. CHLORINE DOSE

Figure 17.8 illustrates a breakpoint chlorination curve. Basically, the concept of *breakpoint chlorination* is to continue adding chlorine until the desired quantity of free residuals appears. This cannot occur until after the demand for combined residuals has been satisfied.

Figure 17.8 *Breakpoint Chlorination Curve*

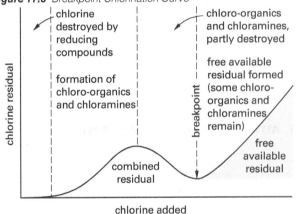

The amount of chlorine necessary for disinfection varies with the organic and inorganic material present in the water, the pH, the temperature, and the contact time. Thirty minutes of chlorine contact time is generally sufficient to deactivate giardia cysts. Satisfactory results can generally be obtained if a free chlorine residual (consisting of hypochlorous acid and hypochlorite ions) of 0.2–0.5 mg/L can be maintained throughout the distribution system. Combined residuals, if use is approved by health authorities, should be 1.0–2.0 mg/L at distant points in the distribution system.

Example 17.7

The flow rate through a treatment plant is 2 MGD (7.5 ML/day). The hypochlorite ion (OCl^-) dose is 20 mg/L as substance. The purity of calcium hypochlorite, $Ca(OCl)_2$, is 97.5% as delivered. How many pounds of calcium hypochlorite are needed to treat the water?

SI Solution

The approximate molecular weights for the components of calcium hypochlorite are

Ca:			40.1
O:	(2)(16)	=	32.0
Cl:	(2)(35.5)	=	71.0
		total	143.1

The fraction of available chlorine in the form of the hypochlorite ion is

$$G = \frac{32.0 + 71.0}{143.1}$$
$$= 0.720$$

Use the standard dose equation with adjustments for both purity and availability.

$$F_{kg/d} = \frac{D_{mg/L} Q_{ML/d}}{PG}$$
$$= \frac{\left(20 \, \frac{mg}{L}\right)\left(7.5 \, \frac{ML}{d}\right)}{(0.975)(0.720)}$$
$$= 213.7 \text{ kg/d}$$

Customary U.S. Solution

The approximate molecular weights for the components of calcium hypochlorite are

Ca:			40.1
O:	(2)(16)	=	32.0
Cl:	(2)(35.5)	=	71.0
		total	143.1

The fraction of available chlorine in the form of the hypochlorite ion is

$$\frac{32.0 + 71.0}{143.1} = 0.720$$

Use the standard dose equation, Eq. 17.19, with adjustments for both purity and availability.

$$F_{\text{lbm/day}} = \frac{D_{\text{mg/L}} Q_{\text{MGD}} \left(8.345 \ \dfrac{\text{lbm-L}}{\text{mg-MG}}\right)}{PG}$$

$$= \frac{\left(20 \ \dfrac{\text{mg}}{\text{L}}\right)(2 \ \text{MGD}) \left(8.345 \ \dfrac{\text{lbm-L}}{\text{mg-MG}}\right)}{(0.975)(0.720)}$$

$$= 475.5 \ \text{lbm/day}$$

39. CT VALUES

The EPA has defined a removal efficiency of 99.9% (3-log inactivation) for *Giardia* and a removal efficiency of 99.99% (4-log inactivation) for viruses. Breakpoint chlorination still applies, but additional factors must be considered in the design of the treatment process. These factors include the hydraulic characteristics of the chlorine contact chamber, including baffling to promote plug-flow reactor (PFR) conditions and increase contact time, as well as characteristics of the water such as pH, temperature, and turbidity.

These factors affect the *concentration × time value* (CT value) required for disinfection. The CT value is the product of C, the concentration of free chlorine, and T, the time that the chlorine is in contact with the water being disinfected. The CT value required for disinfection will vary with the microorganism being targeted as well as with the temperature and pH of the water. If the contact time and the required CT value are both known, the proper chlorine dose can be determined.

The EPA has published tables giving CT values as a function of temperature and pH for a 3-log inactivation of *Giardia* cysts by free chlorine and for a 4-log inactivation of viruses by free chlorine. The much higher CT values in the first table reflect the fact that *Giardia* cysts require considerably longer contact periods for inactivation than do viruses. [CT Values for 3-LOG Inactivation of Giardia Cysts by Free Chlorine] [CT Values for 4-LOG Inactivation of Viruses by Free Chlorine]

In determining whether the required CT value is met, the concentration and contact time may be evaluated for all components of the treatment plant downstream of the chlorination point, such as the reactor tank and pipes. However, if only the time in the reactor tank is used, any additional downstream contact time is a factor of safety.

To take hydraulic inefficiencies into account, the EPA has published baffling factors based on the ratio of $T10$ to T. These baffling factors can be used with Eq. 17.66.

EPA Baffling Factors

$$T = \frac{V \times BF}{\text{peak flow}} \qquad \qquad 17.66$$

If the reactor tank volume, V, is fixed, then Eq. 17.66 can be used to give T. If the value of T calculated with Eq. 17.66, multiplied by the free chlorine concentration, is less than the required CT value for the given pH and temperature, then the desired inactivation will not be achieved. Alternatively, the required CT value can be divided by the free chlorine concentration to give the required value of T, and then Eq. 17.66 can be solved for V. If the calculated value of V is larger than the actual reactor tank volume, then the reactor tank is too small for the desired inactivation to occur.

It may be possible to overcome the physical limitations of the reactor by applying filtration to remove pathogens ahead of disinfection. The EPA has established filtration credits for *Cryptosporidium*, *Giardia*, and viruses as a function of filtration technology. [Filtration Credits for Microbial Removal to Meet SWTR, IESWTR, and LTIESWTR]

40. ADVANCED OXIDATION PROCESSES

Alternatives to chlorination have become necessary since THMs were traced to the chlorination process. These alternatives are categorized as *advanced oxidation processes* (AOPs).

Both bromine and iodine have properties similar to chlorine and can be used for disinfection. They are seldom used because they are relatively costly and produce their own disinfection by-products.

Chlorine dioxide (ClO_2) is manufactured at the water treatment plant from chlorine and sodium chlorite. Its ionization by-products (chlorite and chlorate) and high cost have limited its use.

Ozone is used extensively throughout the world, though not in large quantities in the United States. Ozone is a more powerful disinfectant than chlorine. Ozone can be used alone or (in several developing technologies) in conjunction with hydrogen peroxide. Ozone is generated on-site by running high voltage electricity through dry air or pure oxygen. Gases (oxygen and unused ozone) developing during ozonation must be collected and destroyed in an ozone-destruct unit to ensure that no ozone escapes into the atmosphere.

Ultraviolet radiation is effective in disinfecting shallow (e.g., less than 10 cm) bodies of water. Its primary disadvantages are cost and the absence of any residual disinfection for downstream protection.

Water

41. CHLORAMINATION

Before leaving the plant, treated water may be *chloraminated* (i.e., treated with both ammonia and chlorine to form chloramines). This step ensures lasting disinfection by providing a residual level of chloramines, protecting the water against bacteria regrowth as it travels through the distribution system.

42. DECHLORINATION

In addition to its routine disinfection use, chlorine can enter water supply systems during operation and maintenance activities such as disinfection of mains, testing of hydrants, and routine flushing of distribution systems. Such flows are often discharged to wastewater sewer systems. Chlorine reacts with some organic compounds to produce THMs and is toxic to aquatic life. For these reasons, most water supply and wastewater facilities dechlorinate their discharges for regulatory compliance.

Excess chlorine can be removed with a reducing agent, referred to as a *dechlor*. Table 17.5 lists common dechlor chemicals. Dechlors reduce all forms of active chlorine (chlorine gas, hypochlorites, and chloramines) to chloride. Aeration also reduces chlorine content, as does contact with granular activated charcoal (GAC). Carbon absorption processes are more costly than chemical treatments and are used only when total dechlorination is required.

Table 17.5 Dechlorinating Chemicals Used in Water Supply and Wastewater Facilities

name	formula	stoichiometric dose[*] (mg/mg Cl_2)
ascorbic acid	$C_6H_8O_6$	2.48
calcium thiosulfate	CaS_2O_3	1.19
hydrogen peroxide	H_2O_2	0.488
sodium ascorbate	$C_6H_7NaO_6$	2.78
sodium bisulfite	$NaHSO_3$	1.46
sodium metabisulfite	$Na_2S_2O_5$	1.34
sodium sulfite	Na_2SO_3	1.78
sodium thiosulfate	$Na_2S_2O_3$	0.556
sulfur dioxide	SO_2	0.903

[*]Doses are approximate and depend on pH and assumed reaction chemistry. Theoretical values may be used for initial approximations and equipment sizing. Under the best conditions, 10% excess is required.

Equation 17.67 gives an example of a reaction equation for dechlorination using sodium sulfite.

Dechlorination of Sulfite Compounds
$$Na_2SO_3 + Cl_2 + H_2O \rightarrow Na_2SO_4 + 2HCl \quad \text{17.67}$$

Sulfur dioxide is a toxic gas and is most frequently used by large wastewater. Sodium bisulfite, sodium sulfite, and sodium thiosulfate are most frequently used by smaller water utilities. Sodium bisulfite is low in cost and has a high rate of dechlorination. Sodium sulfite tablets are easy to store and handle. Sodium thiosulfate is less hazardous and depletes oxygen less than sodium bisulfite and sodium sulfite. Ascorbic acid and sodium ascorbate are used because they do not impact DO concentrations. Monitoring is required to prevent the overapplication of chemicals that may deplete the dissolved oxygen concentration or alter the pH of receiving streams.

43. DEMINERALIZATION AND DESALINATION

Demineralization and desalination (*salt water conversion*) are required when only brackish water supplies are available.[23] These processes are carried out in distillation, electrodialysis, ion exchange, and membrane processes.

Distillation is a process whereby the raw water is vaporized, leaving the salt and minerals behind. The water vapor is reclaimed by condensation. Distillation cannot be used to economically provide large quantities of water.

Reverse osmosis is the least costly and most attractive membrane demineralization process. A thin membrane separates two solutions of different concentrations. Pore size is smaller (0.0001–0.001 μm) than with ultrafilter membranes, as salt ions are not permitted to pass through. (See Sec. 17.25.) Typical large-scale osmosis units operate at 150–500 psi (1.0–5.2 MPa).[24]

Nanofiltration is similar to ultrafiltration and reverse osmosis, with pore size (0.001 μm) and operating pressure (75–250 psig; 0.5–1.7 MPa) intermediate between the two. Nanofilters are commonly referred to as *softening membranes*.

In *electrodialysis*, positive and negative ions flow through selective membranes under the influence of an induced electrical current. Unlike pressure-driven filtration processes, however, the ions (not the water molecules) pass through the membrane. The ions removed from the water form a concentrate stream that is discarded.

[23]In the United States, Florida, Arizona, and Texas are leaders in desalinization installations. Potable water is routinely made from sea water in the Middle East.
[24]Membrane processes operate in a *fixed flux* condition. In order to keep the yield constant over time, the pressure must be constantly increased in order to compensate for the effects of fouling and compaction. Fouling by inorganic substances and biofilms is the biggest problem with membranes.

The *ion exchange* process is an excellent solution to demineralization and desalination. (See Sec. 17.33.)

44. NOMENCLATURE

A	surface area	ft^2	m^2
AI	aggressive index	–	–
b	width, length, distance	ft	m
BF	baffling factor	–	–
C	coefficient	–	–
C	concentration	mg/L	mg/L
d	diameter	ft	m
D	diameter	ft	m
f	frequency	Hz	Hz
f	removal fraction	–	–
F	feed rate	lbm/day	kg/d
g	acceleration of gravity, 32.2 (9.81)	ft/sec^2	m/s^2
g_c	gravitational constant, 32.2	lbm-ft/lbf-sec^2	n.a.
G	availability	–	–
h	head (depth) or height	ft	m
I	moment of inertia	ft^4	m^4
K_a	ionization constant	–	–
L	length	ft	m
m	mass	lbm	kg
LSI	Langelier stability index	–	–
M	alkalinity	mg/L	mg/L
n	rotational speed	rev/sec	rev/s
P	power	hp	kW
P	purity	–	–
PSI	Puckorius scaling index	–	–
Q	flow rate	ft^3/sec or MGD	m^3/s or L/s
R	distance from shaft to paddle center	ft	m
Re	Reynolds number	–	–
RSI	Ryznar stability index	–	–
SG, $S.G.$	specific gravity	–	–
T	contact time	min	min
TDS	total dissolved solids	mg/L	mg/L
TON	threshold odor number	–	–
v	velocity	ft/sec	m/s
v*	overflow rate	gal/day-ft^2	L/d·m^2
V	volume	ft^3	m^3
x	mass fraction remaining	–	–
y	vertical distance	ft	m

Symbols

γ	specific weight	lbf/ft^3	N/m^3
η	collection efficiency	–	–
θ	residence time	sec	s
μ	absolute viscosity	lbf-sec/ft^2	Pa·s
ν	kinematic viscosity	ft^2/sec	m^2/s
ρ	density	lbm/ft^3	kg/m^3

Subscripts

c	critical
D	drag
f	fluid
L	per unit length
p	particle
sp	solubility product
t	terminal settling

18 Potable Water Distribution Systems

1. STORAGE AND DISTRIBUTION

Water is stored to provide water pressure, equalize pumping rates, equalize supply and demand over periods of high consumption, provide surge relief, and furnish water during fires and other emergencies when power is disrupted. Storage may also serve as part of the treatment process, either by providing increased detention time or by blending water supplies to obtain a desired concentration.

Several methods are used to distribute water depending on terrain, economics, and other local conditions. *Gravity distribution* is used when a lake or reservoir is located significantly higher in elevation than the population.

Distribution from *pumped storage* is the most common option when gravity distribution cannot be used. Excess water is pumped during periods of low hydraulic and electrical demands (usually at night) into elevated storage. During periods of high consumption, water is drawn from the storage. With pumped storage, pumps are able to operate at a uniform rate and near their rated capacity most of the time.

Using pumps without storage to force water directly into the mains is the least desirable option. Without storage, pumps and motors will not always be able to run in their most efficient ranges since they must operate during low, average, and peak flows. In a power outage, all water supply will be lost unless a backup power source comes online quickly or water can be obtained by gravity flow.

Water is commonly stored in surface and elevated tanks. The elevation of the water surface in the tank directly determines the distribution pressure. This elevation is controlled by an *altitude valve* that operates on the differential in pressure between the height of the water and an adjustable spring-loaded pilot on the valve. Altitude valves are installed at ground level and, when properly adjusted, can maintain the water levels to within 4 in (102 mm). Tanks must be vented to the atmosphere. Otherwise, a rapid withdrawal of water will create a vacuum that could easily cause the tank to collapse inward.

The preferred location of an elevated tank is on the opposite side of the high-consumption district from the pumping station. During periods of high water use, the district will be fed from both sides, reducing the loss of head in the mains below what would occur without elevated storage.

Equalizing the pumping rate during the day ordinarily requires storage of at least 15–20% of the maximum daily use. Storage for fires and emergencies is more difficult to determine. Fire storage is essentially dictated by building ordinances and insurance (i.e., economic benefits to the public). Private on-site storage volume for large industries may be dictated by local codes and ordinances.

2. SERVICE PRESSURE

For ordinary domestic use, the minimum water pressure at the tap should be 25–40 psig (170–280 kPa). A minimum of 60 psig (400 kPa) at the fire hydrant is usually adequate, since that allows for up to a 20 psig (140 kPa) pressure drop in fire hoses. 75 psig (500 kPa) and higher is common in commercial and industrial districts. *Pressure regulators* can be installed if delivery pressure is too high.

3. WATER MANAGEMENT

Unaccounted-for water is the potable water that is produced at the water treatment plant but is not accounted for in billing. It includes known unmetered uses (such as fire fighting and hydrant flushing) and all unknown uses (e.g., from leaks, broken meters, theft, and illegal connections).

Unaccounted-for water can be controlled with proper attention to meter selection, master metering, leak detection, quality control (installation and maintenance of meters), control of system pressures, and accurate data collection.

Since delivery is under pressure, infiltration of groundwater into pipe joints is not normally an issue with distribution.

4. WATER PIPE

Several types of pipe are used in water distribution depending on the flow rate, installed location (i.e., above or below ground), depth of installation, and surface surcharge. Pipes must have adequate strength to withstand external loads from backfill, traffic, and earth movement. They must have high burst strength to withstand internal pressure, a smooth interior surface to reduce friction, corrosion resistance, and tight joints to minimize loss. Once all other requirements have been satisfied, the choice of pipe material can be made on the basis of economics.

Asbestos-cement pipe was used extensively in the past. It has a smooth inner surface and is immune to galvanic corrosion. However, it has low flexural strength. *Concrete pipe* is durable, watertight, has a smooth interior, and requires little maintenance. It is manufactured in plain and reinforced varieties. *Cast-iron pipe* (ductile and gray varieties) is strong, offers long life, and is impervious. However, it has a high initial cost and is heavy (i.e., difficult to transport and install). It may need to be coated on the exterior and interior to resist corrosion of various types. *Steel pipe* offers a smooth interior, high strength, and high ductility, but it is susceptible to corrosion inside and out. The exterior and interior may both need to be coated for protection. *Plastic pipe* (ABS and PVC) has a smooth interior and is chemically inert, corrosion resistant, and easily transported and installed. However, it has low strength.

5. PIPE MATERIALS AND SIZES

Many materials are used for pipes. The material used depends on the application. Water supply distribution, wastewater collection, and air conditioning refrigerant lines place different demands on pipe material performance. Pipe materials are chosen on the basis of tensile strength to withstand internal pressures, compressive strength to withstand external loads from backfill and traffic, smoothness, corrosion resistance, chemical inertness, cost, and other factors.

The following are characteristics of the major types of legacy and new installation commercial pipe materials that are in use.

- *asbestos cement:* immune to electrolysis and corrosion, light in weight but weak structurally; environmentally limited

- *concrete:* durable, watertight, low maintenance, smooth interior

- *copper and brass:* used primarily for water, condensate, and refrigerant lines; in some cases, easily bent by hand, good thermal conductivity

- *ductile cast iron:* long lived, strong, impervious, heavy, scour resistant, but costly

- *plastic* (PVC, CPVC, HDPE, and ABS):[1] chemically inert, resistant to corrosion, very smooth, lightweight, low cost

- *steel*: high strength, ductile, resistant to shock, very smooth interior, but susceptible to corrosion

- *vitrified clay:* resistant to corrosion, acids (e.g., hydrogen sulfide from septic sewage), scour, and erosion

The required wall thickness of a pipe is proportional to the pressure the pipe must carry. However, not all pipes operate at high pressures. Therefore, pipes and tubing may be available in different wall thicknesses (*schedules*, *series*, or *types*). Steel pipe, for example, is available in schedules 40, 80, and others.[2]

For initial estimates, the approximate schedule of steel pipe can be calculated from Eq. 18.1. p is the operating pressure in psig; S is the allowable stress in the pipe material and E is the *joint efficiency*, also known as the *joint quality factor* (typically 1.00 for seamless pipe, 0.85 for electric resistance-welded pipe, 0.80 for electric fusion-welded pipe, and 0.60 for furnace butt-welded pipe). For seamless carbon steel (A53) pipe used below 650°F (340°C), the allowable stress is approximately between 12,000 psi and 15,000 psi. So, with butt-welded joints, a value of 6500 psi is often used for the product SE.

$$\text{schedule} \approx \frac{1000p}{SE} \qquad 18.1$$

---

[1]PVC: polyvinyl chloride; CPVC: chlorinated polyvinyl chloride; HDPE: high-density polyethylene; ABS: acrylonitrile-butadiene-styrene.
[2]Other schedules of steel pipe, such as 30, 60, 120, and so on, also exist, but in limited sizes. Schedule-40 pipe roughly corresponds to the standard weight (S) designation used in the past. Schedule-80 roughly corresponds to the extra-strong (X) designation. There is no uniform replacement designation for double-extra-strong (XX) pipe.

Steel pipe is available in black (i.e., plain *black pipe*) and galvanized (inside, outside, or both) varieties. Steel pipe is manufactured in plain-carbon and stainless varieties. AISI 316 stainless steel pipe is particularly corrosion resistant.

The actual dimensions of some pipes (concrete, clay, some cast iron, etc.) coincide with their *nominal dimensions*. For example, a 12 in concrete pipe has an inside diameter of 12 in, and no further refinement is needed. However, some pipes and tubing (e.g., steel pipe, copper and brass tubing, and some cast iron) are called out by a nominal diameter that has nothing to do with the inside diameter of the pipe. For example, a 16 in schedule-40 steel pipe has an actual inside diameter of 15 in. In some cases, the nominal size does not coincide with the external diameter, either.

For calculations involving flow in steel or copper pipes, it is important to know the correct inside diameter of each pipe.[3]

PVC (polyvinyl chloride) pipe is used extensively as water and sewer pipe due to its combination of strength, ductility, and corrosion resistance. Manufactured lengths range approximately 10–13 ft (3–3.9 m) for sewer pipe and 20 ft (6 m) for water pipe, with integral gasketed joints or solvent-weld bells. Infiltration is very low (less than 50 gal/in-mile-day), even in the wettest environments. The low Manning's roughness constant (0.009 typical) allows PVC sewer pipe to be used with flatter grades or smaller diameters. PVC pipe is resistant to corrosive soils and sewerage gases and is generally resistant to abrasion from pipe-cleaning tools.

Truss pipe is a double-walled PVC or ABS pipe with radial or zigzag (diagonal) reinforcing ribs between the thin walls and with lightweight concrete filling all voids. It is used primarily for underground sewer service because of its smooth interior surface ($n = 0.009$), infiltration-resistant, impervious exterior, and resistance to bending.

Prestressed concrete pipe (PSC) or *reinforced concrete pipe* (RCP) consists of a concrete core that is compressed by a circumferential wrap of high-tensile strength wire and covered with an exterior mortar coating. *Prestressed concrete cylinder pipe* (PCCP), or *lined concrete pipe* (LCP), is constructed with a concrete core, a thin steel cylinder, prestressing wires, and an exterior mortar coating. The concrete core is the primary structural, load-bearing component. The prestressing wires induce a uniform compressive stress in the core that offsets tensile stresses in the pipe. If present, the steel cylinder acts as a water barrier between concrete layers. The mortar coating protects the prestressing wires from physical damage and external corrosion. PSC and PCCP are ideal for large-diameter pressurized service and are used primarily for water supply systems. PSC is commonly available in diameters from 16 in to 60 in (400 mm to 1500 mm), while PCCP diameters as large as 144 in (3600 mm) are available.

6. MANUFACTURED PIPE STANDARDS

There are many different standards governing pipe diameters and wall thicknesses. A pipe's nominal outside diameter is rarely sufficient to determine the internal dimensions of the pipe. A manufacturing specification and class or category are usually needed to completely specify pipe dimensions.

Cast-iron pressure pipe was formerly produced to ANSI/AWWA C106/A21.6 standards but is now obsolete. Modern cast-iron soil (i.e., sanitary) pipe is produced according to ASTM A74. Ductile iron (CI/DI) pipe is produced to ANSI/AWWA C150/A21.50 and C151/A21.51 standards. Gasketed PVC sewer pipe up to 15 in inside diameter is produced to ASTM D3034 standards. Gasketed sewer PVC pipe from 18 in to 48 in is produced to ASTM F679 standards. PVC pressure pipe for water distribution is manufactured to ANSI/AWWA C900 standards. Reinforced concrete pipe (RCP) for culvert, storm drain, and sewer applications is manufactured to ASTM/AASHTO C76/M170 standards.

7. PIPE CLASS

The term *pipe class* has several valid meanings that can be distinguished by designation and context. For plastic and metallic (i.e., steel, cast and ductile iron, cast bronze, and wrought copper) pipe and fittings, pressure class designations such as 25, 150, and 300 refer to the pressure ratings and dimensional design systems as defined in the appropriate ASME, ANSI, AWWA, and other standards. Generally, the class corresponds roughly to a maximum operating pressure category in psig.

ASME pipe classes 1, 2, and 3 refer to the maximum stress categories allowed in pipes, piping systems, components, and supports, as defined in the ASME *Boiler and Pressure Vessel Code* (BPVC), Sec. III, Div. 1, "Rules for Construction of Nuclear Facility Components."

For precast concrete pipe manufactured according to ASTM C76, the pipe classes 1, 2, 3, 4, and 5 correspond to the minimum vertical loading (D-load) capacity as determined in a *three-edge bearing test* (ASTM C497). (See Table 18.1.[4]) Each pipe has two D-load ratings: the

---

[3]It is a characteristic of standard steel pipes that the schedule number does not affect the outside diameter of the pipe. An 8 in schedule-40 pipe has the same exterior dimensions as an 8 in schedule-80 pipe. However, the interior flow area will be less for the schedule-80 pipe.
[4]The D-load rating for ASTM C14 concrete pipe is D-load = F/L in lbf/ft.

pressure that induces a crack 0.01 in (0.25 mm) wide at least 1 ft (100 mm) long ($D_{0.01}$), and the pressure that results in structural collapse (D_{ult}).[5]

$$\text{D-load} = \frac{F_{lbf}}{D_{ft}L_{ft}} \quad \text{[ASTM C76 pipe]} \qquad 18.2$$

Table 18.1 *ASTM C76 Concrete Pipe D-Load Equivalent Pipe Class**

ASTM C76 pipe class	$D_{0.01}$ load rating (lbf/ft^2)	D_{ult} load rating (lbf/ft^2)
class 1 (I)	800	1200
class 2 (II)	1000	1500
class 3 (III)	1350	2000
class 4 (IV)	2000	3000
class 5 (V)	3000	3750

(Multiply lbf/ft^2 by 0.04788 to obtain kPa.)
*As defined in ASTM C76 and AASHTO M170.

8. PAINTS, COATINGS, AND LININGS

Various materials have been used to protect steel and ductile iron pipes against rust and other forms of corrosion. *Red primer* is a shop-applied, rust-inhibiting primer applied to prevent short-term rust prior to shipment and the application of subsequent coatings. *Asphaltic coating* ("tar" coating) is applied to the exterior of underground pipes. *Bituminous coating* refers to a similar coating made from tar pitch. Both asphaltic and bituminous coatings should be completely removed or sealed with a synthetic resin prior to the pipe being finish-coated, since their oils may bleed through otherwise.

Though bituminous materials (i.e., asphaltic materials) continue to be cost effective, epoxy-based products are now extensively used. Epoxy products are delivered as a two-part formulation (a polyamide resin and liquid chemical hardener) that is mixed together prior to application. *Coal tar epoxy*, also referred to as *epoxy coal tar*, a generic name, sees frequent use in pipes exposed to high humidity, seawater, other salt solutions, and crude oil. Though suitable for coating steel penstocks of hydroelectric installations, coal tar epoxy is generally not suitable for potable water delivery systems. Though it is self-priming, appropriate surface preparation is required for adequate adhesion. Coal tar epoxy has a density range of 1.9–2.3 lbm/gal (230–280 g/L).

9. SERIES PIPE SYSTEMS

A system of pipes in series consists of two or more lengths of different-diameter pipes connected end-to-end. In the case of the series pipe from a reservoir discharging to the atmosphere shown in Fig. 18.1, the available head will be split between the velocity head and the friction loss.

$$h = h_v + h_f \qquad 18.3$$

Figure 18.1 *Series Pipe System*

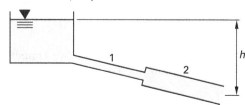

If the flow rate or velocity in any part of the system is known, the friction loss can easily be found as the sum of the friction losses in the individual sections. The solution is somewhat more simple than it first appears to be, since the velocity of all sections can be written in terms of only one velocity.

$$h_{f,t} = h_{f,1} + h_{f,2} \qquad 18.4$$

Continuity Equation

$$A_1 v_1 = A_2 v_2 \qquad 18.5$$

If neither the velocity nor the flow quantity is known, a trial-and-error solution will be required, since a friction factor must be known to calculate h_f. A good starting point is to assume fully turbulent flow.

When velocity and flow rate are both unknown, the following procedure using the Darcy friction factor can be used.[6]

step 1: Calculate the relative roughness, ϵ/D, for each section. Use the Moody diagram to determine f_a and f_b for fully turbulent flow (i.e., the horizontal portion of the curve).

step 2: Write all of the velocities in terms of one unknown velocity.

$$Q_1 = Q_2 \qquad 18.6$$

---

[5]This D-load rating for concrete pipe is analogous to the *pipe stiffness* (PS) rating (also known as *ring stiffness*) for PVC and other plastic pipe, although the units are different. The PS rating is stated in lbf/in^2 (pounds per inch of pipe length per inch of pipe diameter).

[6]If Hazen-Williams constants are given for the pipe sections, the procedure for finding the unknown velocities is similar, although considerably more difficult since v^2 and $v^{1.85}$ cannot be combined. A first approximation, however, can be obtained by replacing $v^{1.85}$ in the Hazen-Williams equation for friction loss. A trial and error method can then be used to find velocity.

$$v_2 = \left(\frac{A_1}{A_2}\right)v_1 \qquad 18.7$$

step 3: Write the total friction loss in terms of the unknown velocity.

$$h_{f,t} = \frac{f_1 L_1 v_1^2}{2 D_1 g} + \left(\frac{f_2 L_2}{2 D_2 g}\right)\left(\frac{A_1}{A_2}\right)^2 v_1^2$$

$$= \left(\frac{v_1^2}{2g}\right)\left[\frac{f_1 L_1}{D_1} + \left(\frac{f_2 L_2}{D_2}\right)\left(\frac{A_1}{A_2}\right)^2\right] \qquad 18.8$$

step 4: Solve for the unknown velocity using the Bernoulli equation between the free reservoir surface ($p = 0$, $v = 0$, $z = h$) and the discharge point ($p = 0$, if free discharge; $z = 0$). Include pipe friction, but disregard minor losses for convenience.

$$h = \frac{v_2^2}{2g} + h_{f,t}$$

$$= \left(\frac{v_1^2}{2g}\right)\left[\left(\frac{A_1}{A_2}\right)^2\left(1 + \frac{f_2 L_2}{D_2}\right) + \frac{f_1 L_1}{D_1}\right] \qquad 18.9$$

step 5: Using the value of v_1, calculate v_2. Calculate the Reynolds number, and check the values of f_1 and f_2 from step 4. Repeat steps 3 and 4 if necessary.

10. PARALLEL PIPE SYSTEMS

Adding a second pipe in parallel with a first is a standard method of increasing the capacity of a line. A *pipe loop* is a set of two pipes placed in parallel, both originating and terminating at the same junction. The two pipes are referred to as *branches* or *legs*. (See Fig. 18.2.)

Figure 18.2 *Parallel Pipe System*

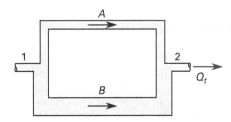

There are three principles that govern the distribution of flow between the two branches.

- The flow divides in such a manner as to make the head loss in each branch the same.

$$h_{f,A} = h_{f,B} \qquad 18.10$$

- The head loss between the junctions 1 and 2 is the same as the head loss in branches A and B.

$$h_{f,1\text{-}2} = h_{f,A} = h_{f,B} \qquad 18.11$$

- The total flow rate is the sum of the flow rates in the two branches.

$$Q_t = Q_A + Q_B \qquad 18.12$$

If the pipe diameters are known, Eq. 18.10 and Eq. 18.12 can be solved simultaneously for the branch velocities. For first estimates, it is common to neglect minor losses, the velocity head, and the variation in the friction factor, f, with velocity.

If the system has only two parallel branches, the unknown branch flows can be determined by solving Eq. 18.15 and Eq. 18.13 simultaneously. The sum found from Eq. 18.13 is equal to the total flow.

Multipath Pipeline Problems

$$(\pi D^2/4)v = (\pi D_A^2/4)v_A + (\pi D_B^2/4)v_B \qquad 18.13$$

$$Q_A + Q_B = Q_t \qquad 18.14$$

Multipath Pipeline Problems

$$h_L = f_A \frac{L_A}{D_A}\frac{v_A^2}{2g} = f_B \frac{L_B}{D_B}\frac{v_B^2}{2g} \qquad 18.15$$

However, if the system has three or more parallel branches, it is easier to use the following iterative procedure. This procedure can be used for problems (a) where the flow rate is unknown but the pressure drop between the two junctions is known, or (b) where the total flow rate is known but the pressure drop and velocity are both unknown. In both cases, the solution iteratively determines the friction coefficients, f.

step 1: Solve the expression for friction head loss, h_f (either Darcy or Hazen-Williams), for velocity in each branch. If the pressure drop is known, first convert it to friction head loss.

Head Loss Due to Flow

$$h_f = f\frac{L v^2}{D 2g} \qquad 18.16$$

Hazen-Williams Equation

$$V = k_1 C R_H^{0.63} S^{0.54}$$

$$= k_1 C R_H^{0.63}\left(\frac{h_f}{L}\right)^{0.54} \qquad 18.17$$

In Eq. 18.17, k_1 is 0.849 for SI units and 1.318 for U.S. units.

step 2: Solve for the flow rate in each branch. If they are unknown, friction factors, f, must be assumed for each branch. The fully turbulent assumption provides a good initial estimate. (The value of k' will be different for each branch.)

Continuity Equation

$$Q = A\mathrm{v} = A\sqrt{\frac{2Dgh_f}{fL}} \qquad 18.18$$
$$= k'\sqrt{h_f} \quad \text{[Darcy]}$$

step 3: Write the expression for the conservation of flow. Calculate the friction head loss from the total flow rate. For example, for a three-branch system,

$$Q_t = Q_1 + Q_2 + Q_3 = \left(k_1' + k_2' + k_3'\right)\sqrt{h_f} \quad 18.19$$

step 4: Check the assumed values of the friction factor. Repeat as necessary.

Example 18.1

3 ft^3/sec of water enter the parallel pipe network shown at junction A. All pipes are schedule-40 steel with the nominal sizes shown. Minor losses are insignificant. What is the total friction head loss between junctions A and B?

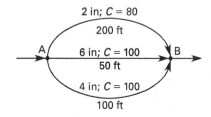

Solution

step 1: Collect the pipe dimensions from a table of pipe dimensions.

	2 in	4 in	6 in
flow area	0.0233 ft^2	0.0884 ft^2	0.2006 ft^2
diameter	0.1723 ft	0.3355 ft	0.5054 ft

Follow the procedure given in Sec. 18.10.

$$V = k_1 C R_H^{0.63} S^{0.54}$$
$$= k_1 C R_H^{0.63} \left(\frac{h_f}{L}\right)^{0.54}$$

The velocity (expressed in ft/sec) in the 2 in diameter pipe branch is

$$V_{2\,\mathrm{in}} = (1.318)(80)\left(\frac{0.1723 \text{ ft}}{4}\right)^{0.63}$$
$$\times \left(\frac{h_f}{200 \text{ ft}}\right)^{0.54}$$
$$= 0.831 h_f^{0.54}$$

The friction head loss is the same in all parallel branches. The velocities in the other two branches are

$$\mathrm{v}_{6\,\mathrm{in}} = 4.327 h_f^{0.54}$$
$$\mathrm{v}_{4\,\mathrm{in}} = 2.299 h_f^{0.54}$$

step 2: The flow rates are

Continuity Equation

$$Q = A\mathrm{v}$$
$$Q_{2\,\mathrm{in}} = (0.0233 \text{ ft}^2)0.831 h_f^{0.54}$$
$$= 0.0194 h_f^{0.54}$$
$$Q_{6\,\mathrm{in}} = (0.2006 \text{ ft}^2)4.327 h_f^{0.54}$$
$$= 0.8680 h_f^{0.54}$$
$$Q_{4\,\mathrm{in}} = (0.0884 \text{ ft}^2)2.299 h_f^{0.54}$$
$$= 0.2032 h_f^{0.54}$$

step 3: Use the total flow rate to find the friction head loss.

$$Q_t = Q_{2\,\mathrm{in}} + Q_{6\,\mathrm{in}} + Q_{4\,\mathrm{in}}$$
$$3 \frac{\text{ft}^3}{\text{sec}} = 0.0194 h_f^{0.54} + 0.8680 h_f^{0.54} + 0.2032 h_f^{0.54}$$
$$= (0.0194 + 0.8680 + 0.2032) h_f^{0.54}$$
$$h_f = 6.5 \text{ ft}$$

11. MULTIPLE RESERVOIR SYSTEMS

In the *three-reservoir problem*, there are many possible choices for the unknown quantity (pipe length, diameter, head, flow rate, etc.). In all but the simplest cases, the solution technique is by trial and error based on conservation of mass and energy. (See Fig. 18.3.)

Figure 18.3 *Three-Reservoir System*

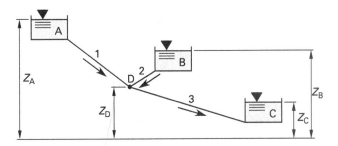

For simplification, velocity heads and minor losses are usually insignificant and can be neglected. However, the presence of a pump in any of the lines must be included in the solution procedure. This is most easily done by adding the pump head to the elevation of the reservoir feeding the pump. If the pump head is not known or depends on the flow rate, it must be determined iteratively as part of the solution procedure.

Case 1: Given all lengths, diameters, and elevations, find all flow rates.

Although an analytical solution method is possible, this type of problem is easily solved iteratively. The following procedure makes an initial estimate of a flow rate and uses it to calculate the pressure at the junction, P_D. Since this method may not converge if the initial estimate of Q_1 is significantly in error, it is helpful to use other information (e.g., normal pipe velocities) to obtain the initial estimate. An alternate procedure is simply to make several estimates of P_D and calculate the corresponding values of flow rate.

step 1: Assume a reasonable value for Q_1. Calculate the corresponding friction loss $H_{f,1}$. Use the Bernoulli equation to find the corresponding value of P_D. Disregard minor losses and velocity head.

$$v_1 = \frac{Q_1}{A_1} \qquad 18.20$$

$$Z_A = Z_D + \frac{P_D}{w} + H_{f,1} \qquad 18.21$$

step 2: Use the value of P_D to calculate the friction loss in branch 2. Use the friction loss to determine v_2. Use v_2 to determine Q_2. If flow is out of reservoir B, $H_{f,2}$ should be added. If $Z_D + (P_D/w) > Z_B$, flow will be into reservoir B; in this case, $H_{f,2}$ should be subtracted.

$$Z_B = Z_D + \frac{P_D}{w} \pm H_{f,2} \qquad 18.22$$

$$Q_2 = v_2 A_2 \qquad 18.23$$

step 3: Similarly, use the value of P_C to calculate the friction loss in branch 3. Use the friction loss to determine v_3. Use v_3 to determine Q_3.

$$Z_C = Z_D + \frac{P_D}{w} - H_{f,3} \qquad 18.24$$

$$Q_3 = v_3 A_3 \qquad 18.25$$

step 4: Check if $Q_1 \pm Q_2 = Q_3$. If it does not, repeat steps 1 through 4. After the second iteration, plot $Q_1 \pm Q_2 - Q_3$ versus Q_1. Interpolate or extrapolate the value of Q_1 that makes the difference zero.

Case 2: Given Q_1 and all lengths, diameters, and elevations except Z_C, find Z_C.

step 1: Calculate v_1.

$$v_1 = \frac{Q_1}{A_1} \qquad 18.26$$

step 2: Calculate the corresponding friction loss, $H_{f,1}$. Use the Bernoulli equation to find the corresponding value of P_D. Disregard minor losses and velocity head.

$$Z_A = Z_D + \frac{P_D}{w} + H_{f,1} \qquad 18.27$$

step 3: Use the value P_B to calculate the friction loss in branch 2, $H_{f,2}$. Use the friction loss to determine v_2. Use v_2 to determine Q_2. If flow is out of reservoir B, $H_{f,2}$ should be added. If $Z_D + (P_D/w) > Z_B$, flow will be into reservoir B; in this case, $H_{f,2}$ should be subtracted.

$$Z_B = Z_D + \frac{P_D}{w} + H_{f,2} \qquad 18.28$$

$$Q_2 = v_2 A_2 \qquad 18.29$$

step 4: From the conservation of mass, the flow rate into reservoir C is

$$Q_3 = Q_1 \pm Q_2 \qquad 18.30$$

step 5: The velocity in pipe 3 is

$$v_3 = \frac{Q_3}{A_3} \qquad 18.31$$

step 6: Calculate the friction loss in branch 3, $H_{f,3}$.

step 7: Disregarding minor losses and velocity head, the elevation of the surface in reservoir C is

$$Z_C = Z_D + \frac{P_D}{w} + H_{f,3} \qquad 18.32$$

Water

Case 3: Given Q_1, all lengths, all elevations, and all diameters except D_3, find D_3.

step 1: Repeat step 1 from case 2.

step 2: Repeat step 2 from case 2.

step 3: Repeat step 3 from case 2.

step 4: Repeat step 4 from case 2.

step 5: Calculate the friction loss in branch 3, $H_{f,3}$, from

$$Z_C = Z_D + \frac{P_D}{w} - H_{f,3} \qquad 18.33$$

step 6: Calculate D_3 from the friction loss in branch 3, $H_{f,3}$.

Case 4: Given all lengths, diameters, and elevations except Z_D, find all flow rates.

step 1: Calculate the head loss between each reservoir and junction D. Combine as many terms as possible into constant k'.

Head Loss Due to Flow

$$h_f = f\frac{Lv^2}{D2g} \qquad 18.34$$

$$\dot{V} = Av = A\sqrt{\frac{2Dgh_f}{fL}} \qquad 18.35$$

$$= k'\sqrt{h_f}$$

Hazen-Williams Equation

$$V = k_1 C R_H^{0.63}\left(\frac{h_f}{L}\right)^{0.54} \qquad 18.36$$

$$Q = AV = Ak_1 C R_H^{0.63}\left(\frac{h_f}{L}\right)^{0.54} \qquad 18.37$$

$$= k' h_f^{0.54}$$

step 2: Assume that the flow direction in all three pipes is toward junction D. Write the conservation equation for junction D.

$$Q_D = Q_1 + Q_2 + Q_3 = 0 \qquad 18.38$$

$$k_1'\sqrt{h_{f,1}} + k_2'\sqrt{h_{f,2}} + k_3'\sqrt{h_{f,3}} = 0 \qquad 18.39$$
[Darcy]

$$k_1' h_{f,1}^{0.54} + k_2' h_{f,2}^{0.54} + k_3' h_{f,3}^{0.54} = 0 \qquad 18.40$$
[Hazen-Williams]

step 3: Write the Bernoulli equation between each reservoir and junction D. Since $P_A = P_B = P_C = 0$, and $v_A = v_B = v_C = 0$, the friction loss in branch 1 is

$$h_{f,1} = Z_A - Z_D - \frac{P_D}{w} \qquad 18.41$$

However, Z_D and P_D can be combined since they are related constants in any particular situation. Define the correction, δ_D, as

$$\delta_D = Z_D + \frac{P_D}{w} \qquad 18.42$$

Then, the friction head losses in the branches are

$$h_{f,1} = Z_A - \delta_D \qquad 18.43$$

$$h_{f,2} = Z_B - \delta_D \qquad 18.44$$

$$h_{f,3} = Z_C - \delta_D \qquad 18.45$$

step 4: Assume a value for δ_D. Calculate the corresponding h_f values. Use Eq. 18.35 to find Q_1, Q_2, and Q_3. Calculate the corresponding Q value. Repeat until Q converges to zero. It is not necessary to calculate P_D or Z_D once all of the flow rates are known.

12. PIPE NETWORKS

Network flows in a *multiloop system* cannot be determined by any closed-form equation. (See Fig. 18.4.) Most real-world problems involving multiloop systems are analyzed iteratively on a computer. Computer programs are based on the *Hardy Cross method*, which can also be performed manually when there are only a few loops. In this method, flows in all of the branches are first assumed, and adjustments are made in consecutive iterations to the assumed flow.

Figure 18.4 Multiloop System

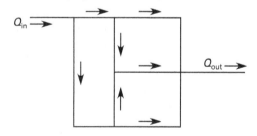

The Hardy Cross method is based on the following principles.

Principle 1: Conservation—The flows entering a junction equal the flows leaving the junction.

Principle 2: The algebraic sum of head losses around any closed loop is zero.

The friction head loss has the form $h_f = K'Q^n$, with h_f having units of feet. (k' used in Eq. 18.18 is equal to $\sqrt{1/K'}$.) The Darcy friction factor, f, is usually assumed to be the same in all parts of the network. For a Darcy head loss, the exponent is $n = 2$. For a Hazen-Williams loss, $n = 1.85$.

- For Q in ft^3/sec, L in feet, and D in feet, the friction coefficient is

$$K' = \frac{0.02517fL}{D^5} \quad \text{[Darcy]} \qquad 18.46$$

$$K' = \frac{4.727L}{D^{4.87}C^{1.85}} \quad \text{[Hazen-Williams]} \qquad 18.47$$

- For Q in gal/min, L in feet, and d in inches, the friction coefficient is

$$K' = \frac{0.03109fL}{d^5} \quad \text{[Darcy]} \qquad 18.48$$

$$K' = \frac{10.44L}{d^{4.87}C^{1.85}} \quad \text{[Hazen-Williams]} \qquad 18.49$$

- For Q in gal/min, L in feet, and D in feet, the friction coefficient is

$$K' = \frac{1.251 \times 10^{-7}fL}{D^5} \quad \text{[Darcy]} \qquad 18.50$$

$$K' = \frac{5.862 \times 10^{-5}L}{D^{4.87}C^{1.85}} \quad \text{[Hazen-Williams]} \qquad 18.51$$

- For Q in MGD (millions of gallons per day), L in feet, and D in feet, the friction coefficient is

$$K' = \frac{0.06026fL}{D^5} \quad \text{[Darcy]} \qquad 18.52$$

$$K' = \frac{10.59L}{D^{4.87}C^{1.85}} \quad \text{[Hazen-Williams]} \qquad 18.53$$

If Q_a is the assumed flow in a pipe, the true value, Q, can be calculated from the difference (correction), δ.

$$Q = Q_a + \delta \qquad 18.54$$

The friction loss term for the assumed value and its correction can be expanded as a series. Since the correction is small, higher order terms can be omitted.

$$\begin{aligned} h_f &= K'(Q_a + \delta)^n \\ &\approx K'Q_a^n + nK'\delta Q_a^{n-1} \end{aligned} \qquad 18.55$$

From Principle 2, the sum of the friction drops is zero around a loop. The correction, δ, is the same for all pipes in the loop and can be taken out of the summation. Since the loop closes on itself, all elevations can be omitted.

$$\sum h_f = \sum K'Q_a^n + n\delta \sum K'Q_a^{n-1} = 0 \qquad 18.56$$

This equation can be solved for δ.

$$\begin{aligned} \delta &= \frac{-\sum K'Q_a^n}{n\sum |K'Q_a^{n-1}|} \\ &= -\frac{\sum h_f}{n\sum \left|\dfrac{h_f}{Q_a}\right|} \end{aligned} \qquad 18.57$$

The Hardy Cross procedure is as follows.

step 1: Determine the value of n. For a Darcy head loss, the exponent is $n = 2$. For a Hazen-Williams loss, $n = 1.85$.

step 2: Arbitrarily select a positive direction (e.g., clockwise).

step 3: Label all branches and junctions in the network.

step 4: Separate the network into independent loops such that each branch is included in at least one loop.

step 5: Calculate K' for each branch in the network.

step 6: Assume consistent and reasonable flow rates and directions for each branch in the network.

step 7: Calculate the correction, δ, for each independent loop. (The numerator is the sum of head losses around the loop, taking signs into consideration.) It is not necessary for the loop to be at the same elevation everywhere. Since the loop closes on itself, all elevations can be omitted.

step 8: Apply the correction, δ, to each branch in the loop. The correction must be applied in the same sense to each branch in the loop. If clockwise has been taken as the positive direction, then δ is added to clockwise flows and subtracted from counterclockwise flows.

step 9: Repeat steps 7 and 8 until the correction is sufficiently small.

Example 18.2

A two-loop pipe network is shown. All junctions are at the same elevation. The Darcy friction factor is 0.02 for all pipes in the network. For convenience, use the nominal pipe sizes shown in the figure. Determine the flows in all branches.

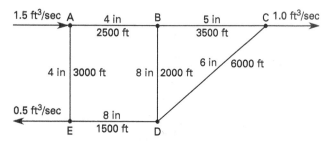

Solution

step 1: The Darcy friction factor is given, so $n = 2$.

step 2: Select clockwise as the positive direction.

step 3: Use the junction letters in the illustration.

step 4: Two independent loops are needed. Work with loops ABDE and BCD. (Loop ABCDE could also be used but would be more complex than loop BCD.)

step 5: Work with branch AB.

$$D_{AB} = \frac{4 \text{ in}}{12 \frac{\text{in}}{\text{ft}}} = 0.3333 \text{ ft}$$

Use Eq. 18.46.

$$K'_{AB} = \frac{0.02517 fL}{D^5} = \frac{(0.02517)(0.02)(2500 \text{ ft})}{(0.3333 \text{ ft})^5}$$
$$= 306.0$$

Similarly,

$$K'_{BC} = 140.5$$
$$K'_{DC} = 96.8$$
$$K'_{BD} = 7.7$$
$$K'_{ED} = 5.7$$
$$K'_{AE} = 367.4$$

step 6: Assume the direction and flow rates shown.

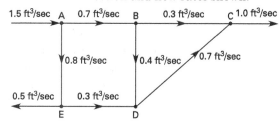

step 7: Use Eq. 18.57.

$$\delta = \frac{-\sum K'Q_a^n}{n \sum |K'Q_a^{n-1}|}$$

$$\delta_{ABDE} = \cfrac{-\left(\begin{array}{l}(306.0)\left(0.7 \ \frac{\text{ft}^3}{\text{sec}}\right)^2 + (7.7)\left(0.4 \ \frac{\text{ft}^3}{\text{sec}}\right)^2 \\ -(5.7)\left(0.3 \ \frac{\text{ft}^3}{\text{sec}}\right)^2 - (367.4)\left(0.8 \ \frac{\text{ft}^3}{\text{sec}}\right)^2\end{array}\right)}{(2)\left(\begin{array}{l}(306.0)\left(0.7 \ \frac{\text{ft}^3}{\text{sec}}\right) + (7.7)\left(0.4 \ \frac{\text{ft}^3}{\text{sec}}\right) \\ +(5.7)\left(0.3 \ \frac{\text{ft}^3}{\text{sec}}\right) + (367.4)\left(0.8 \ \frac{\text{ft}^3}{\text{sec}}\right)\end{array}\right)}$$

$$= 0.08 \text{ ft}^3/\text{sec}$$

$$\delta_{BCD} = \cfrac{-\left(\begin{array}{l}(140.5)\left(0.3 \ \frac{\text{ft}^3}{\text{sec}}\right)^2 - (96.8)\left(0.7 \ \frac{\text{ft}^3}{\text{sec}}\right)^2 \\ -(7.7)\left(0.4 \ \frac{\text{ft}^3}{\text{sec}}\right)^2\end{array}\right)}{(2)\left(\begin{array}{l}(140.5)\left(0.3 \ \frac{\text{ft}^3}{\text{sec}}\right) + (96.8)\left(0.7 \ \frac{\text{ft}^3}{\text{sec}}\right) \\ +(7.7)\left(0.4 \ \frac{\text{ft}^3}{\text{sec}}\right)\end{array}\right)}$$

$$= 0.16 \text{ ft}^3/\text{sec}$$

step 8: The corrected flows are

$$Q_{AB} = 0.7 \ \frac{\text{ft}^3}{\text{sec}} + 0.08 \ \frac{\text{ft}^3}{\text{sec}} = 0.78 \text{ ft}^3/\text{sec}$$

$$Q_{BC} = 0.3 \ \frac{\text{ft}^3}{\text{sec}} + 0.16 \ \frac{\text{ft}^3}{\text{sec}} = 0.46 \text{ ft}^3/\text{sec}$$

$$Q_{DC} = 0.7 \ \frac{\text{ft}^3}{\text{sec}} - 0.16 \ \frac{\text{ft}^3}{\text{sec}} = 0.54 \text{ ft}^3/\text{sec}$$

$$Q_{BD} = 0.4 \ \frac{ft^3}{sec} + 0.08 \ \frac{ft^3}{sec} - 0.16 \ \frac{ft^3}{sec}$$

$$= 0.32 \ ft^3/sec$$

$$Q_{ED} = 0.3 \ \frac{ft^3}{sec} - 0.08 \ \frac{ft^3}{sec} = 0.22 \ ft^3/sec$$

$$Q_{AE} = 0.8 \ \frac{ft^3}{sec} - 0.08 \ \frac{ft^3}{sec} = 0.72 \ ft^3/sec$$

13. FLOW MEASURING DEVICES

A device that measures flow can be calibrated to indicate either velocity or volumetric flow rate. There are many methods available to obtain the flow rate. Some are indirect, requiring the use of transducers and solid-state electronics, and others can be evaluated using the Bernoulli equation. Some are more appropriate for one variety of fluid than others, and some are limited to specific ranges of temperature and pressure.

Table 18.2 categorizes a few common flow measurement methods. Many other methods and variations thereof exist, particularly for specialized industries. Some of the methods listed are so basic that only a passing mention will be made of them. Others, particularly those that can be analyzed with energy and mass conservation laws, will be covered in greater detail in subsequent sections.

The utility meters used to measure gas and water usage are examples of *displacement meters*. Such devices are cyclical, fixed-volume devices with counters to record the numbers of cycles. Displacement devices are generally unpowered, drawing on only the pressure energy to overcome mechanical friction. Most configurations for positive-displacement pumps (e.g., reciprocating piston, helical screw, and nutating disk) have also been converted to measurement devices.

The venturi nozzle, orifice plate, and flow nozzle are examples of *obstruction meters*. These devices rely on a decrease in static pressure to measure the flow velocity. One disadvantage of these devices is that the pressure drop is proportional to the square of the velocity, limiting the range over which any particular device can be used.

An obstruction meter that somewhat overcomes the velocity range limitation is the *variable-area meter*, also known as a *rotameter*, illustrated in Fig. 18.5.[7] This device consists of a float (which is actually more dense than the fluid) and a transparent sight tube. With proper design, the effects of fluid density and viscosity can be minimized. The sight glass can be directly calibrated in volumetric flow rate, or the height of the float above the zero position can be used in a volumetric calculation.

Table 18.2 *Flow Measuring Devices*

I   direct (primary) measurements

        positive-displacement meters

        volume tanks

        weight and mass scales

II   indirect (secondary) measurements

        obstruction meters

          – flow nozzles

          – orifice plate meters

          – variable-area meters

          – venturi meters

        velocity probes

          – direction sensing probes

          – pitot-static meters

          – pitot tubes

          – static pressure probes

        miscellaneous methods

          – hot-wire meters

          – magnetic flow meters

          – mass flow meters

          – sonic flow meters

          – turbine and propeller meters

It is necessary to be able to measure static pressures in order to use obstruction meters and pitot-static tubes. In some cases, a *static pressure probe* is used. Figure 18.6 illustrates a simplified static pressure probe. In practice, such probes are sensitive to burrs and irregularities in the tap openings, orientation to the flow (i.e., *yaw*), and interaction with the pipe walls and other probes. A *direction-sensing probe* overcomes some of these problems.

A weather station *anemometer* used to measure wind velocity is an example of a simple *turbine meter*. Similar devices are used to measure the speed of a stream or river, in which case the name *current meter* may be used. Turbine meters are further divided into cup-type meters and propeller-type meters, depending on the orientation of the turbine axis relative to the flow direction. (The turbine axis and flow direction are parallel for propeller-type meters; they are perpendicular for cup-type meters.) Since the wheel motion is proportional to the flow velocity, the velocity is determined by counting the number of revolutions made by the wheel per unit time.

---

[7]The rotameter has its own disadvantages, however. It must be installed vertically; the fluid cannot be opaque; and it is more difficult to manufacture for use with high-temperature, high-pressure fluids.

Figure 18.5 *Variable-Area Rotameter*

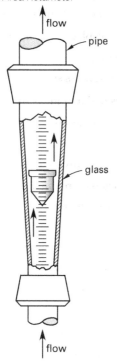

Figure 18.6 *Simple Static Pressure Probe*

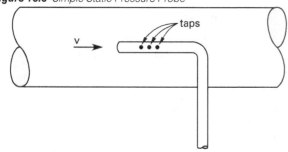

More sophisticated turbine flowmeters use a reluctance-type pickup coil to detect wheel motion. The permeability of a magnetic circuit changes each time a wheel blade passes the pole of a permanent magnet in the meter body. This change is detected to indicate velocity or flow rate.

A *hot-wire anemometer* measures velocity by determining the cooling effect of fluid (usually a gas) flowing over an electrically heated tungsten, platinum, or nickel wire. Cooling is primarily by convection; radiation and conduction are neglected. Circuitry can be used either to keep the current constant (in which case, the changing resistance or voltage is measured) or to keep the temperature constant (in which case, the changing current is measured). Additional circuitry can be used to compensate for thermal lag if the velocity changes rapidly.

Modern *magnetic flowmeters* (*magmeter, electromagnetic flowmeter*) measure fluid velocity by detecting a voltage (potential difference, electromotive force) that is generated in response to the fluid passing through a

magnetic field applied by the meter. The magnitude of the induced voltage is predicted by *Faraday's law of induction*, which states that the induced voltage is equal to the negative of the time rate of change of a magnetic field. The voltage is not affected by changes in fluid viscosity, temperature, or density. From Faraday's law, the induced voltage is proportional to the fluid velocity. For a magmeter with a dimensionless instrument constant, k, and a magnetic field strength, B, the induced voltage will be

$$V = kB\mathrm{v}D_{\mathrm{pipe}} \qquad \textit{18.58}$$

Since no parts of the meter extend into the flow, magmeters are ideal for corrosive fluids and slurries of large particles. Normally, the fluid has to be at least slightly electrically conductive, making magmeters ideal for measuring liquid metal flow. It may be necessary to dope electrically neutral fluids with precise quantities of conductive ions in order to obtain measurements by this method. Most magmeters have integral instrumentation and are direct-reading.

In an *ultrasonic flowmeter*, two electric or magnetic transducers are placed a short distance apart on the outside of the pipe. One transducer serves as a transmitter of ultrasonic waves; the other transducer is a receiver. As an ultrasonic wave travels from the transmitter to the receiver, its velocity will be increased (or decreased) by the relative motion of the fluid. The phase shift between the fluid-carried waves and the waves passing through a stationary medium can be measured and converted to fluid velocity.

14. PITOT-STATIC GAUGE

Measurements from pitot tubes are used to determine total (stagnation) energy. Piezometer tubes and wall taps are used to measure static pressure energy. The difference between the total and static energies is the kinetic energy of the flow. Figure 18.7 illustrates a comparative method of directly measuring the velocity head for an incompressible fluid.

Pitot Tube

$$\mathrm{v} = \sqrt{(2/\rho)(P_0 - P_s)} = \sqrt{2g(P_0 - P_s)/\gamma} \qquad \textit{18.59}$$

$$\frac{P_0 - P_s}{\rho} = hg \qquad \textit{18.60}$$

$$\mathrm{v} = \sqrt{2gh} \qquad \textit{18.61}$$

Figure 18.7 *Comparative Velocity Head Measurement*

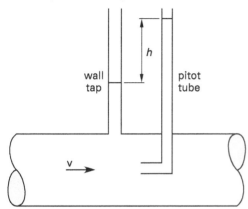

The pitot tube and static pressure tap shown in Fig. 18.7 can be combined into a *pitot-static gauge*. (See Fig. 18.8.) In a pitot-static gauge, one end of the manometer is acted upon by the static pressure (also referred to as the *transverse pressure*). The other end of the manometer experiences the total (stagnation or impact) pressure. The difference in elevations of the manometer fluid columns is the velocity head. This distance must be corrected if the density of the flowing fluid is significant.

$$\frac{v^2}{2} = \frac{P_0 - P_s}{\rho} = \frac{h(\rho_m - \rho)g}{\rho} \quad \text{[SI]} \quad 18.62(a)$$

$$\frac{v^2}{2g_c} = \frac{P_0 - P_s}{\rho} = \frac{h(\rho_m - \rho)}{\rho} \times \frac{g}{g_c} \quad \text{[U.S.]} \quad 18.62(b)$$

$$v = \sqrt{\frac{2gh(\rho_m - \rho)}{\rho}} \quad 18.63$$

Another correction, which is seldom made, is to multiply the velocity calculated from Eq. 18.63 by C_I, the *coefficient of the instrument*. Since the flow past the pitot-static tube is slightly faster than the free-fluid velocity, the static pressure measured will be slightly lower than the true value. This makes the indicated velocity slightly higher than the true value. C_I, a number close to but less than 1.0, corrects for this.

Figure 18.8 *Pitot-Static Gauge*

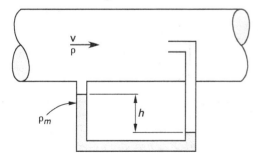

Pitot tube measurements are sensitive to the condition of the opening and errors in installation alignment. The *yaw angle* (i.e., the acute angle between the pitot tube axis and the flow streamline) should be zero.

A pitot-static tube indicates the velocity at only one point in a pipe. If the flow is laminar, and if the pitot-static tube is in the center of the pipe, v_{max} will be determined. The average velocity, however, will be only half the maximum value.

$$v = v_{max} = v_{ave} \quad \text{[turbulent]} \quad 18.64$$

$$v = v_{max} = 2v_{ave} \quad \text{[laminar]} \quad 18.65$$

Example 18.3

Water (62.4 lbm/ft³; $\rho = 1000$ kg/m³) is flowing through a pipe. A pitot-static gauge with a mercury manometer registers 3 in (0.076 m) of mercury. What is the velocity of the water in the pipe?

SI Solution

The density of mercury is $\rho = 13\,560$ kg/m³.

The velocity can be calculated directly from Eq. 18.63. [Thermal and Physical Property Tables: Selected Liquids and Solids]

$$v = \sqrt{\frac{2gh(\rho_m - \rho)}{\rho}}$$

$$= \sqrt{\frac{(2)\left(9.81 \, \frac{m}{s^2}\right)(0.076 \text{ m})}{1000 \, \frac{kg}{m^3}} \times \left(13\,560 \, \frac{kg}{m^3} - 1000 \, \frac{kg}{m^3}\right)}$$

$$= 4.33 \text{ m/s}$$

Customary U.S. Solution

The density of mercury is $\rho = 847$ lbm/ft³. [Thermal and Physical Property Tables: Selected Liquids and Solids]

From Eq. 18.63,

$$v = \sqrt{\frac{2gh(\rho_m - \rho)}{\rho}}$$

$$= \sqrt{\frac{(2)\left(32.2 \ \frac{\text{ft}}{\text{sec}^2}\right)(3 \ \text{in}) \times \left(847 \ \frac{\text{lbm}}{\text{ft}^3} - 62.4 \ \frac{\text{lbm}}{\text{ft}^3}\right)}{\left(62.4 \ \frac{\text{lbm}}{\text{ft}^3}\right)\left(12 \ \frac{\text{in}}{\text{ft}}\right)}}$$

$$= 14.23 \ \text{ft/sec}$$

15. VENTURI METER

Figure 18.9 illustrates a simple *venturi*. (Sometimes the venturi is called a *converging-diverging nozzle*.) This flow-measuring device can be inserted directly into a pipeline. Since the diameter changes are gradual, there is very little friction loss. Static pressure measurements are taken at the throat and upstream of the diameter change. These measurements are traditionally made by a manometer.

Figure 18.9 Venturi Meter

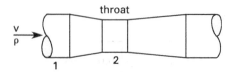

The analysis of *venturi meter performance* is relatively simple. The traditional derivation of upstream velocity starts by assuming a frictionless, incompressible, and turbulent flow. Then, the field equation is written for points 1 and 2. Equation 18.66 shows that the static pressure decreases as the velocity increases. This is known as the *venturi effect*.

Field Equation

$$\frac{P_2}{\gamma} + \frac{v_2^2}{2g} + z_2 = \frac{P_1}{\gamma} + \frac{v_1^2}{2g} + z_1 \qquad \textbf{18.66}$$

The two velocities are related by the continuity equation.

Continuity Equation

$$A_1 v_1 = A_2 v_2 \qquad \textbf{18.67}$$

Combining Eq. 18.66 and Eq. 18.67 and eliminating the unknown v_1 produces an expression for the throat

velocity. A *coefficient of velocity* is used to account for the small effect of friction. (C_v is very close to 1.0, usually 0.98 or 0.99.)

$$v_2 = \frac{C_v}{\sqrt{1 - (A_2/A_1)^2}} \sqrt{2g\left(\frac{P_1}{\gamma} + z_1 - \frac{P_2}{\gamma} - z_2\right)} \qquad \textbf{18.68}$$

Multiplying both sides by the throat area, A_2, gives an expression for the volumetric flow rate.

Venturi Meters

$$Q = \frac{C_v A_2}{\sqrt{1 - (A_2/A_1)^2}} \times \sqrt{2g\left(\frac{P_1}{\gamma} + z_1 - \frac{P_2}{\gamma} - z_2\right)} \qquad \textbf{18.69}$$

The *velocity of approach factor*, F_{va}, also known as the *meter constant*, is the reciprocal of the first term of the denominator of the first term of Eq. 18.68. The *beta ratio* can be incorporated into the formula for F_{va}.

$$\beta = \frac{D_2}{D_1} \qquad \textbf{18.70}$$

$$F_{va} = \frac{1}{\sqrt{1 - \left(\frac{A_2}{A_1}\right)^2}} = \frac{1}{\sqrt{1 - \beta^4}} \qquad \textbf{18.71}$$

If a manometer is used to measure the pressure difference directly, Eq. 18.68 can be rewritten in terms of the manometer fluid reading. (See Fig. 18.10.)

$$v_2 = C_v v_{2,\text{ideal}} = \left(\frac{C_v}{\sqrt{1 - \beta^4}}\right)\sqrt{\frac{2g(\rho_m - \rho)h}{\rho}}$$

$$= C_v F_{va}\sqrt{\frac{2g(\rho_m - \rho)h}{\rho}} \qquad \textbf{18.72}$$

Figure 18.10 Venturi Meter with Manometer

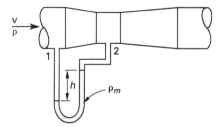

The flow rate through a venturi meter can be calculated from the throat area. There is an insignificant amount of contraction of the flow as it passes through the throat, and the *coefficient of contraction* is seldom encountered in venturi meter work. The *coefficient of discharge* ($C_d = C_c C_v$) is essentially the same as the coefficient of velocity. Values of C_d range from slightly less than 0.90 to over 0.99, depending on the Reynolds number. C_d is seldom less than 0.95 for turbulent flow. (See Fig. 18.11.)

$$\dot{V} = C_d A_2 v_{2,\text{ideal}} \qquad 18.73$$

The product $C_d F_{\text{va}}$ is known as the *coefficient of flow* or *flow coefficient*, not to be confused with the coefficient of discharge.[8] This factor is used for convenience, since it combines the losses with the meter constant.

$$C_f = C_d F_{\text{va}} = \frac{C_d}{\sqrt{1 - \beta^4}} \qquad 18.74$$

$$Q = C_f A_2 \sqrt{\frac{2g(\rho_m - \rho)h}{\rho}} \qquad 18.75$$

Figure 18.11 *Typical Venturi Meter Discharge Coefficients (long radius venturi meter)*

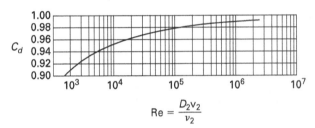

16. ORIFICE METER

The *orifice meter* (or *orifice plate*) is used more frequently than the venturi meter to measure flow rates in small pipes. It consists of a thin or sharp-edged plate with a central, round hole through which the fluid flows. Such a plate is easily clamped between two flanges in an existing pipeline. (See Fig. 18.12.)

Figure 18.12 *Orifice Meter with Differential Manometer*

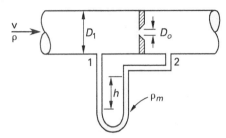

While (for small pipes) the orifice meter may consist of a thin plate without significant thickness, various types of bevels and rounded edges are also used with thicker plates. There is no significant difference in the analysis procedure between "flat plate," "sharp-edged," or "square-edged" orifice meters. Any effect that the orifice edges have is accounted for in the discharge and flow coefficient correlations. Similarly, the direction of the bevel will affect the coefficients but not the analysis method.

As with the venturi meter, pressure taps are used to obtain the static pressure upstream of the orifice plate and at the *vena contracta* (i.e., at the point of minimum pressure).[9] A differential manometer connected to the two taps conveniently indicates the difference in static pressures.

Although the orifice meter is simpler and less expensive than a venturi meter, its discharge coefficient is much less than that of a venturi meter. C_d usually ranges from 0.55 to 0.75, with values of 0.60 and 0.61 often being quoted. (The coefficient of contraction has a large effect, since $C_d = C_v C_c$.) Also, its pressure recovery is poor (i.e., there is a permanent pressure reduction), and it is susceptible to inaccuracies from wear and abrasion.[10]

The derivation of the governing equations for an orifice meter is similar to that of the venturi meter. (The obvious falsity of assuming frictionless flow through the orifice is corrected by the coefficient of discharge.) The major difference is that the coefficient of contraction is taken into consideration in writing the mass continuity equation, since the pressure is measured at the vena contracta, not the orifice.

$$A_2 = C_c A_o \qquad 18.76$$

---

[8]Some writers use the symbol K for the flow coefficient.

[9]Calibration of the orifice meter is sensitive to tap placement. Upstream taps are placed between one-half and two pipe diameters upstream from the orifice. (An upstream distance of one pipe diameter is often quoted and used.) There are three tap-placement options: flange, vena contracta, and standardized. Flange taps are used with prefabricated orifice meters that are inserted (by flange bolting) in pipes. If the location of the vena contracta is known, a tap can be placed there. However, the location of the vena contracta depends on the diameter ratio $\beta = D_o/D$ and varies from approximately 0.4 to 0.7 pipe diameters downstream. Due to the difficulty of locating the vena contracta, the standardized $1D\text{-}\frac{1}{2}D$ configuration is often used. The upstream tap is one diameter before the orifice; the downstream tap is one-half diameter after the orifice. Since approaching flow should be stable and uniform, care must be taken not to install the orifice meter less than approximately five diameters after a bend or elbow.

[10]The actual loss varies from 40% to 90% of the differential pressure. The loss depends on the diameter ratio $\beta = D_o/D_1$, and is not particularly sensitive to the Reynolds number for turbulent flow. For $\beta = 0.5$, the loss is 73% of the measured pressure difference, $P_1 - P_2$. This decreases to approximately 56% of the pressure difference when $\beta = 0.65$ and to 38%, when $\beta = 0.8$. For any diameter ratio, the pressure drop coefficient, K, in multiples of the orifice velocity head is $K = (1 - \beta^2)/C_f^2$.

Orifices

$$Q = CA_0 \sqrt{2g\left(\frac{P_1}{\gamma} + z_1 - \frac{P_2}{\gamma} - z_2\right)} \qquad \text{18.77}$$

In Eq. 18.76, C is the *coefficient of the meter*, also called the *orifice coefficient*, and is equal to

$$C = \frac{C_v C_c}{\sqrt{1 - C_c^2 (A_0/A_1)^2}} \qquad \text{18.78}$$

If flow is incompressible and the orifice meter is horizontal (so that z_1 and z_2 are equal), Eq. 18.76 can be reduced to

$$Q = CA_0 \sqrt{\frac{2}{\rho}(P_1 - P_2)} \qquad \text{18.79}$$

If a manometer is used to indicate the differential pressure $P_1 - P_2$, the velocity at the vena contracta can be calculated from Eq. 18.80.

$$v_0 = \left(\frac{C_v C_c}{1 - C_c^2 (A_0/A_1)^2}\right) \times \sqrt{2g\left(\frac{P_1}{\gamma} + z_1 - \frac{P_2}{\gamma} - z_2\right)} \qquad \text{18.80}$$

The *velocity of approach factor*, F_{va}, for an orifice meter is defined differently than for a venturi meter, since it takes into consideration the contraction of the flow. However, the velocity of approach factor is still combined with the coefficient of discharge into the flow coefficient, C_f. Figure 18.13 illustrates how the flow coefficient varies with the area ratio and the Reynolds number.

$$F_{va} = \frac{1}{\sqrt{1 - \left(\frac{C_c A_o}{A_1}\right)^2}} \qquad \text{18.81}$$

$$C_f = C_d F_{va} \qquad \text{18.82}$$

The flow rate through an orifice meter is given by Eq. 18.83.

$$\dot{V} = C_f A_o \sqrt{\frac{2g(\rho_m - \rho)h}{\rho}}$$
$$= C_f A_o \sqrt{\frac{2(P_1 - P_2)}{\rho}} \qquad \text{[SI]} \quad \text{18.83(a)}$$

$$\dot{V} = C_f A_o \sqrt{\frac{2g(\rho_m - \rho)h}{\rho}}$$
$$= C_f A_o \sqrt{\frac{2g_c(P_1 - P_2)}{\rho}} \qquad \text{[U.S.]} \quad \text{18.83(b)}$$

Figure 18.13 *Typical Flow Coefficients for Orifice Plates*

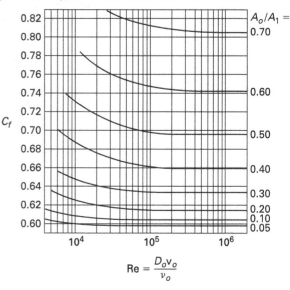

$$\text{Re} = \frac{D_o v_o}{\nu_o}$$

Example 18.4

150°F water ($\rho = 61.2$ lbm/ft³) flows in an 8 in schedule-40 steel pipe at the rate of 2.23 ft³/sec. A sharp-edged orifice with a 7 in diameter hole is placed in the line. A mercury differential manometer is used to record the pressure difference. (Mercury has a density of 848.6 lbm/ft³.) If the orifice has a flow coefficient, C_f, of 0.62, what deflection in inches of mercury is observed?

Solution

The orifice area is

$$A_o = \frac{\pi D_o^2}{4} = \frac{\pi \left(\dfrac{7 \text{ in}}{12 \dfrac{\text{in}}{\text{ft}}}\right)^2}{4} = 0.2673 \text{ ft}^2$$

Equation 18.83 is solved for h.

$$h = \frac{\dot{V}^2 \rho}{2g C_f^2 A_o^2 (\rho_m - \rho)}$$

$$= \frac{\left(2.23 \dfrac{\text{ft}^3}{\text{sec}}\right)^2 \left(61.2 \dfrac{\text{lbm}}{\text{ft}^3}\right)\left(12 \dfrac{\text{in}}{\text{ft}}\right)}{(2)\left(32.2 \dfrac{\text{ft}}{\text{sec}^2}\right)(0.62)^2}$$

$$\times (0.2673 \text{ ft}^2)^2 \left(848.6 \dfrac{\text{lbm}}{\text{ft}^3} - 61.2 \dfrac{\text{lbm}}{\text{ft}^3}\right)$$

$$= 2.62 \text{ in}$$

17. TYPES OF VALVES

Valves used for *shutoff service* (e.g., gate, plug, ball, and some butterfly valves) are used fully open or fully closed. *Gate valves* offer minimum resistance to flow. They are used in clean fluid and slurry services when valve operation is infrequent. Many turns of the hand-wheels are required to raise or lower their gates. *Plug valves* provide for tight shutoff. A 90° turn of their handles is sufficient to rotate the plugs fully open or closed. *Eccentric plug valves*, in which the plug rotates out of the fluid path when open, are among the most common wastewater valves. *Plug cock valves* have a hollow passageway in their plugs through which fluid can flow. Both eccentric plug valves and plug cock valves are referred to as "plug valves." *Ball valves* offer an unobstructed flow path and tight shutoff. They are often used with slurries and viscous fluids, as well as with cryogenic fluids. A 90° turn of their handles rotates the balls fully open or closed. *Butterfly valves* (when specially designed with appropriate seats) can be used for shutoff operation. They are particularly applicable to large flows of low-pressure (vacuum up to 200 psig (1.4 MPa)) gases or liquids, although high-performance butterfly valves can operate as high as 600 psig (4.1 MPa). Their straight-through, open-disk design results in minimal solids build-up and low pressure drops.

Other valve types (e.g., globe, needle, Y-, angle, and some butterfly valves) are more suitable for *throttling service*. *Globe valves* provide positive shutoff and precise metering on clean fluids. However, since the seat is parallel to the direction of flow and the fluid makes two right-angle turns, there is substantial resistance and pressure drop through them, as well as relatively fast erosion of the seat. Globe valves are intended for frequent operation. *Needle valves* are similar to globe valves, except that the plug is a tapered, needlelike cone. Needle valves provide accurate metering of small flows of clean fluids. Needle valves are applicable to cryogenic fluids. *Y-valves* are similar to globe valves in operation, but their seats are inclined to the direction of flow, offering more of a straight-through passage and unobstructed flow than the globe valve. *Angle valves* are essentially globe valves where the fluid makes a 90° turn. They can be used for throttling and shutoff of clean or viscous fluids and slurries. *Butterfly valves* are often used for throttling services with the same limitations and benefits as those listed for shutoff use.

Other valves are of the *check (nonreverse-flow, anti-reversal)* variety. These react automatically to changes in pressure to prevent reversals of flow. Special check valves can also prevent excess flow. Figure 18.14 illustrates *swing*, *lift*, and *angle lift check valves*, and Table 18.3 gives typical characteristics of common valve types.

18. WATER HAMMER

Water hammer in a long pipe is an increase in fluid pressure caused by a sudden velocity decrease. (See Fig. 18.15.) The sudden velocity decrease will usually be caused by a valve closing. Analysis of the water hammer phenomenon can take two approaches, depending on whether or not the pipe material is assumed to be elastic.

The *water hammer wave speed*, a, is given by Eq. 18.84. The first term on the right-hand side represents the effect of fluid compressibility, and the second term represents the effect of pipe elasticity.

$$\frac{1}{a^2} = \frac{d\rho}{dP} + \frac{\rho}{A}\frac{dA}{dP} \qquad \textit{18.84}$$

For a compressible fluid within an inelastic (rigid) pipe, $dA/dP = 0$, and the wave speed is simply the speed of sound in the fluid.

$$a = \sqrt{\frac{dP}{d\rho}} = \sqrt{\frac{E}{\rho}} \quad \begin{bmatrix} \text{inelastic pipe;} \\ \text{consistent units} \end{bmatrix} \qquad \textit{18.85}$$

If the pipe material is assumed to be inelastic (such as cast iron, ductile iron, and concrete), the time required for the water hammer shock wave to travel from the suddenly closed valve to a point of interest depends only on the velocity of sound in the fluid, a, and the distance, L, between the two points. This is also the time required to bring all of the fluid in the pipe to rest.

$$t = \frac{L}{a} \qquad \textit{18.86}$$

When the water hammer shock wave reaches the original source of water, the pressure wave will dissipate. A rarefaction wave (at the pressure of the water source) will return at velocity a to the valve. The time for the compression shock wave to travel to the source and the rarefaction wave to return to the valve is given by Eq. 18.87. This is also the length of time that the pressure is constant at the valve.

$$t = \frac{2L}{a} \qquad \textit{18.87}$$

The fluid pressure increase resulting from the shock wave is calculated by equating the kinetic energy change of the fluid with the average pressure during the compression process. The pressure increase is independent of the length of pipe. If the velocity is decreased by an amount Δv instantaneously, the increase in pressure will be

$$\Delta P = \rho a \Delta\text{v} \qquad \text{[SI]} \qquad \textit{18.88(a)}$$

$$\Delta P = \frac{\rho a \Delta\text{v}}{g_c} \qquad \text{[U.S.]} \qquad \textit{18.88(b)}$$

Figure 18.14 *Types of Valves*

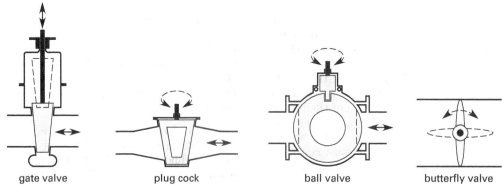

(a) valves for shutoff service

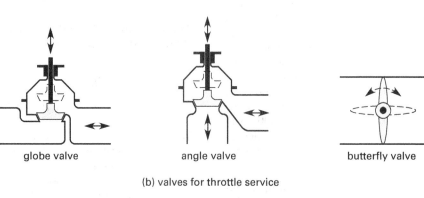

(b) valves for throttle service

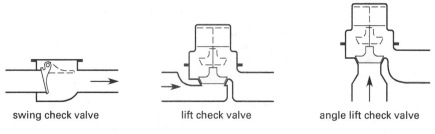

(c) valves for antireversal service

Table 18.3 *Typical Characteristics of Common Valve Types*

valve type	fluid condition	switching frequency	pressure drop (fully open)	typical control response	typical maximum pressure (atm)	typical maximum temperature (°C)
ball	clean	low	low	very poor	160	300
butterfly	clean	low	low	poor	200	400
diaphragm* (not shown)	clean to slurried	very high	low to medium	very good	16	150
gate	clean	low	low	very poor	50	400
globe	clean	high	medium to high	very good	80	300
plug	clean	low	low	very poor	160	300

(Multiply atm by 101.33 to obtain kPa.)

*Diaphragm valves use a flexible diaphragm to block the flow path. The diaphragm may be manually or pneumatically actuated. Such valves are suitable for both throttling and shutoff service.

Figure 18.15 Water Hammer

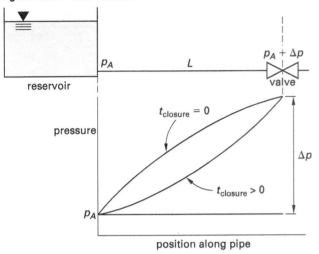

It is interesting that the pressure increase at the valve depends on Δv but not on the actual length of time it takes to close the valve, as long as the valve is closed when the wave returns to it. Therefore, there is no difference in pressure buildups at the valve for an "instantaneous closure," "rapid closure," or "sudden closure."[11] It is only necessary for the closure to occur rapidly.

Water hammer does not necessarily occur just because a return shockwave increases the pressure. When the velocity is low—less than 7 ft/sec (2.1 m/s), with 5 ft/sec (1.5 m/s) recommended—there is not enough force generated to create water hammer. The cost of using larger pipe, fittings, and valves is the disadvantage of keeping velocity low.

Having a very long pipe is equivalent to assuming an instantaneous closure. When the pipe is long, the time for the shock wave to travel round-trip is much longer than the time to close the valve. The valve will be closed when the rarefaction wave returns to the valve.

If the pipe is short, it will be difficult to close the valve before the rarefaction wave returns to the valve. With a short pipe, the pressure buildup will be less than is predicted by Eq. 18.88. (Having a short pipe is equivalent to the case of "slow closure.") The actual pressure history is complex, and no simple method exists for calculating the pressure buildup in short pipes.

Installing a *surge tank, accumulator, slow-closing valve* (e.g., a gate valve), or *pressure-relief valve* in the line will protect against water hammer damage. (See Fig. 18.16.) The surge tank (or *surge chamber*) is an open tank or reservoir. Since the water is unconfined, large pressure buildups do not occur. An accumulator is a closed tank that is partially filled with air. Since the air is much more compressible than the water, it will be compressed by the water hammer shock wave. The energy of the shock wave is dissipated when the air is compressed.

Figure 18.16 Water Hammer Protective Devices

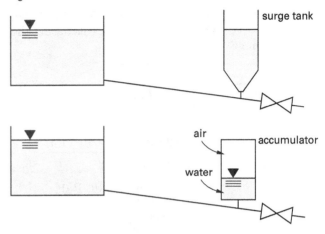

For an incompressible fluid within an elastic pipe, $d\rho/dP$ in Eq. 18.85 is zero. In that case, the wave speed is

$$a = \sqrt{\frac{A}{\rho}\frac{dP}{dA}} \quad \begin{bmatrix} \text{elastic pipe;} \\ \text{consistent units} \end{bmatrix} \quad \textbf{18.89}$$

For long elastic pipes (the common assumption for steel and plastic), the degree of pipe anchoring and longitudinal stress propagation affect the wave velocity. Longitudinal stresses resulting from circumferential stresses (i.e., Poisson's effect) must be considered by including a Poisson's effect factor, c_P. In Eq. 18.90, for pipes anchored at their upstream ends only, $c_P = 1 - 0.5\nu_{\text{pipe}}$; for pipes anchored throughout, $c_P = 1 - \nu^2_{\text{pipe}}$; for anchored pipes with expansion joints throughout, $c_P = 1$ (i.e., the commonly used *Korteweg formula*). Pipe sections joined with gasketed integral bell ends and gasketed ends satisfy the requirement for expansion joints. Equation 18.90 shows that the wave velocity can be reduced by increasing the pipe diameter.

$$a = \sqrt{\dfrac{\dfrac{E_{\text{fluid}}t_{\text{pipe}}E_{\text{pipe}}}{t_{\text{pipe}}E_{\text{pipe}} + c_P D_{\text{pipe}}E_{\text{fluid}}}}{\rho}} \quad \textbf{18.90}$$

At room temperature, the modulus of elasticity of ductile steel is approximately 2.9×10^7 lbf/in^2 (200 GPa); for ductile cast iron, 2.2–2.5×10^7 lbf/in^2 (150–170 GPa); for PVC, 3.5–4.1×10^5 lbf/in^2 (2.4–2.8 GPa); for ABS, 3.2–3.5×10^5 lbf/in^2 (2.2–2.4 GPa).

Example 18.5

Water ($\rho = 1000$ kg/m^3, $E = 2 \times 10^9$ Pa), is flowing at 4 m/s through a long length of 4 in schedule-40 steel pipe ($D_i = 0.102$ m, $t = 0.00602$ m, $E = 2 \times 10^{11}$ Pa) with expansion joints when a valve suddenly closes completely. What is the theoretical increase in pressure?

---

[11]The pressure elsewhere along the pipe, however, will be lower for slow closures than for instantaneous closures.

Solution

For pipes with expansion joints, $c_P = 1$. From Eq. 18.90, the modulus of elasticity to be used in calculating the speed of sound is

$$a = \sqrt{\dfrac{\dfrac{E_{\text{fluid}} t_{\text{pipe}} E_{\text{pipe}}}{t_{\text{pipe}} E_{\text{pipe}} + c_P D_{\text{pipe}} E_{\text{fluid}}}}{\rho}}$$

$$= \sqrt{\dfrac{\dfrac{(2 \times 10^9 \text{ Pa})(0.00602 \text{ m})(2 \times 10^{11} \text{ Pa})}{(0.00602 \text{ m})(2 \times 10^{11} \text{ Pa})}}{\begin{array}{c} + (1)(0.102 \text{ m})(2 \times 10^9 \text{ Pa}) \end{array}}}{1000 \ \dfrac{\text{kg}}{\text{m}^3}}}$$

$$= 1307.75 \text{ m/s}$$

From Eq. 18.88, the pressure increase is

$$\Delta P = \rho a \Delta \text{v} = \left(1000 \ \frac{\text{kg}}{\text{m}^3}\right)\left(1307.75 \ \frac{\text{m}}{\text{s}}\right)\left(4 \ \frac{\text{m}}{\text{s}}\right)$$

$$= 5.23 \times 10^6 \text{ Pa}$$

19. WATER DEMAND

Normal water demand is specified in gallons per capita day (gpcd) or liters per capita day (Lpcd)—the average number of gallons (liters) used by each person each day. This is referred to as *average annual daily flow* (AADF) if the average is taken over a period of a year. Residential (i.e., domestic), commercial, industrial, and public uses all contribute to normal water demand, as do waste and unavoidable loss.[12]

As Table 18.4 illustrates, an AADF of 165 gpcd (625 Lpcd) is a typical minimum for planning purposes. If large industries are present (e.g., canning, steel making, automobile production, and electronics), then their special demand requirements must be added.

Water demand varies with the time of day and season. Each community will have its own demand distribution curve. Table 18.5 gives typical multipliers, M, that might be used to estimate instantaneous demand from the average daily flow.

$$Q_{\text{instantaneous}} = M(\text{AADF}) \qquad 18.91$$

Per capita demand must be multiplied by the population to obtain the total system demand. Since a population can be expected to change in number, a supply system must be designed to handle the water demand

Table 18.4 Annual Average Water Requirements[a]

	demand	
use	gpcd	Lpcd[b]
residential	75–130	284–490
commercial and industrial	70–100	265–380
public	10–20	38–80
loss and waste	10–20	38–80
totals	165–270	625–1030

(Multiply gpcd by 3.79 to obtain Lpcd.)
[a]exclusive of fire fighting requirements
[b]liters per capita-day

Table 18.5 Typical Demand Multipliers (to be used with average annual daily flow)

period of usage	multiplier, M
maximum daily	1.5–1.8
maximum hourly	2.0–3.0
early morning	0.25–0.40
noon	1.5–2.0
winter	0.80
summer	1.30

for a reasonable amount of time into the future. Several methods can be used for estimating future demand, including mathematical, comparative, and correlative methods. Various assumptions can be made for mathematical predictions, including uniform growth rate (same as straight-line extrapolation) and constant percentage growth rate. (In the case of decreasing population, the "growth" rate would be negative.)

Economic aspects will dictate the number of years of future capacity that are installed. Excess capacities of 25% for large systems and 50% for small systems are typically built into the system initially.

20. APPLICATION: GROWTH RATES

Models of *population growth* commonly assume arithmetic or geometric growth rates over limited periods of time.[13] *Arithmetic growth rate*, also called *constant growth rate* and *linear growth rate*, is appropriate when a population increase involves limited resources or occurs at specific intervals. Means of subsistence, such as areas of farmable land by generation, and budgets and enrollments by year, for example, are commonly assumed to increase arithmetically with time. Arithmetic growth is equivalent to simple interest compounding. Given a

---

[12]"Public" use includes washing streets, flushing water and sewer mains, flushing fire hydrants, filling public fountains, and fighting fires.
[13]Another population growth model is the logistic (S-shaped) curve.

starting population, P_0, that increases every period by a constant *growth rate amount*, k, the population after t periods is

Linear Projection = Algebraic Projection

$$P_t = P_0 + k\Delta t \quad \text{[arithmetic]} \qquad 18.92$$

The *average annual growth rate* is conventionally defined as

$$r_{\text{ave},\%} = \frac{P_t - P_0}{tP_0} \times 100\% = \frac{R}{P_0} \times 100\% \qquad 18.93$$

Geometric growth (Malthusian growth) is appropriate when resources to support growth are infinite. The population changes by a fixed *growth rate fraction*, r, each period. Geometric growth is equivalent to discrete period interest compounding, and $(F/P, r\%, t)$ economic interest factors can be used. The population at time t (i.e., after t periods) is

$$P_t = P_0(1 + r)^t \equiv P_0(P/F, r\%, t) \quad \text{[geometric]} \quad 18.94$$

Geometric growth can be expressed in terms of a time constant, τ_b, associated with a specific base, b. Commonly, only three bases are used. For $b = 2$, τ is the *doubling time*, T. For $b = \frac{1}{2}$, τ is the *half-life*, $t_{1/2}$. For $b = e$, τ is the *e-folding time*, or just *growth rate*, k. Base-e growth is known as *exponential growth* or *instantaneous growth*. It is appropriate for continuous growth (not discrete time periods) and is equivalent to continuous interest compounding.

Log Growth = Exponential Growth = Geometric Growth

$$P_t = P_0 e^{k\Delta t}$$
$$= P_0(1 + r)^t$$
$$= P_0(2)^{t/T} \qquad 18.95$$
$$= P_0\left(\frac{1}{2}\right)^{t/t_{1/2}}$$

Taking the natural logarithm of Eq. 18.95 results in a *ln-linear form* that graphs as a straight line.

$$\ln P_t = \ln P_0 + k\Delta t \qquad 18.96$$

Population can also be expressed as a percentage of growth occurring over a number of periods, n, at a growth rate, k.

Percent Growth

$$P_t = P_0(1 + k)^n \qquad 18.97$$

Other methods used for projecting population are the *ratio and correlation growth* and the *decreasing-rate-of-increase growth* models. The ratio model compares population from the last census, P_1, to the projected population, P_2.

Ratio and Correlation Growth

$$\frac{P_2}{P_{2R}} = \frac{P_1}{P_{1R}} = k \qquad 18.98$$

The decreasing-rate-of-growth model uses the saturation population, S.

Decreasing-Rate-of-Increase Growth

$$P_t = P_0 + (S - P_0)(1 - e^{-k(t-t_0)}) \qquad 18.99$$

21. FIRE FIGHTING DEMAND

The amount of water that should be budgeted for municipal fire protection depends on the degree of protection desired, construction types, and the size of the population to be protected. Municipal capacity decisions are almost always made on the basis of economics driven by the insurance industry. The Insurance Services Office (ISO) rates municipalities according to multiple criteria it publishes in its Fire Suppression Rating Schedule.[14]

One of the criteria is the ability of a municipality to provide what the ISO refers to as the *needed fire flow*, NFF, at various locations around the community. The NFF methodology has been determined from measurement of historical events (i.e., past fires) over many years. The ISO awards credit points dependent on how well actual capacities (availability) matches the NFFs at these locations. Adequacy may be limited by the water works, main capacity, and hydrant distribution.

The ISO also calculates a community's *basic fire flow*, BFF, which is the fifth highest needed fire flow (not to exceed 3500 gpm (13 250 L/m) among the locations listed in the ISO batch report. BFF is used to determine the numbers of engine companies, crews, and number, nature, and size of the fire-fighting apparatus that the community needs.

Calculating the NFF is highly codified, requiring specific knowledge of construction types, floor areas, and usage (occupancy). NFF is calculated in gallons per minute and is rounded up or down to the nearest 250 gpm (16 L/s) if less than 2500 gpm (160 L/s), and up or down to the nearest 500 gpm (32 L/s) if greater than 2500 gpm (160 L/s). The minimum NFF value is 250 gpm (16 L/s), corresponding to the discharge from a single standard 1.125 in (29 mm) diameter smooth fire nozzle, and the maximum value is 12,000 gpm (760 L/s). Calculated

---

[14]Other methods exist and are in use around the world. The ISO method works well for large diameter hoses preferred in the United States and needed to achieve the high flow rates necessary with fires involving the wood-framed buildings that predominate in the United States.

values are increased by 500 gpm (32 L/s) if the building has a wood shingle roof covering that can contribute to the spread of a fire. Reductions in NFF are given for structures having sprinkler systems designed in accordance with National Fire Protection Association (NFPA) standards.

As a convenience, the NFF for 1- and 2-family dwellings not exceeding two stories in height can be read directly from Table 18.6, without the need of calculations.

Table 18.6 *Needed Fire Flow for 1- and 2-Family Dwellings*

distance between buildings (ft)	needed fire flow (gpm)
≤ 10	1500
11–30	1000
31–100	750
> 100	500

(Multiply ft by 0.3048 to obtain m.)
(Multiply gpm by 0.06309 to obtain L/s.)
Source: *Guide for Determination of Needed Fire Flow*, Chap. 7, Insurance Services Office

For other than 1- and 2-family dwellings, NFF must be calculated based on construction type, occupancy, area, and other factors. Equation 18.100 is the ISO formula for an individual, nonsprinklered building. Some of the factors are evaluated for each exposure (side) of the building. ISO provides guidance and examples of application for selecting values for the factors, but specialized knowledge and expertise is still needed to use Eq. 18.100 accurately, particularly for buildings with mixed construction classes and occupancies.

Needed Fire Flow Formula
$$NFF = (C_i)(O_i)[1.0 + (X + P)_i] \qquad \textbf{\textit{18.100}}$$

Needed Fire Flow Formula
$$C_i = 18F(A_i)^{0.5} \qquad \textbf{\textit{18.101}}$$

C_i is a factor dependent on the type of construction and is calculated with Eq. 18.101. In Eq. 18.101, F is a *construction class coefficient*. F = 1.5 for wood frame construction (class 1), 1.0 for joisted masonry (class 2), 0.8 for noncombustible construction (class 3) including masonry (class 4), and 0.6 for fire-resistive construction (classes 5 and 6). A_i is the effective area in square feet, determined as the area of the largest floor in the building plus 50% of all the other floors for classes 1, 2, 3, and 4. Other modifications apply to classes 5 and 6. Special rules apply to finding areas of multi-story, fire-resistant buildings with sprinkler systems. C is rounded up or down to the nearest 250 gpm (16 L/s) (prior to the rounding of NFF), may not be less than 500 gpm (32 L/s), is limited to a maximum of 8000 gpm (500 L/s) for

classes 1 and 2, is limited to 6000 gpm (380 L/s) for classes 3, 4, 5, and 6, and for any 1-story building of any class.

The *occupancy combustible factor*, O_i, accounts for the combustibility of a building's contents and is 1.25 for rapid burning (class C-5), 1.15 for free-burning (class C-4), 1.0 for combustible (class C-3), 0.85 for limited-combustible (class C-2), and 0.75 for noncombustible (class C-1). The exact definitions and examples of these classes are specified by the ISO.

The *exposure factor*, X, reflects the influence of an adjoining or connected building less than 100 ft (30 m) from a wall of the subject building. Its value is read from an ISO table as a function of separation, lateral exposure length, and construction class. Values vary from essentially 0 to 0.25.

The *communication factor*, P, reflects any connections (passageways) between adjacent structures. Like the exposure factor, it is read from ISO tables as a function of the length and nature of the passageway and closures. Values vary from 0 to 0.35.

The exposure and communication factors are summed for all exposures (sides) of a building. However, the total of X + P over all exposures need not exceed 0.6.

22. OTHER FORMULAS FOR FIRE FIGHTING NEEDS

Although modern methods calculate fire fighting demand based on the area and nature of construction to be protected, the earliest methods of predicting fire-fighting demand were based on population.[15] The 1911 *American Insurance Association equation* (also referred to as the *National Board of Fire Underwriters equation* after AIA's predecessor) for cities with populations less than 200,000 was used widely up until the 1980s. *P* in Eq. 18.102 is in thousands of people.

$$Q_{gpm} = 1020\sqrt{P}\left(1 - 0.01\sqrt{P}\right) \qquad \textbf{\textit{18.102}}$$

Other authorities who have developed alternative methods include Thomas (1959), Iowa State University (1967), Illinois Institutes of Technology Research Institute (1968), and Ontario Building Code (Office of the Fire Marshal, 1995).

23. DURATION OF FIRE FIGHTING FLOW

A municipality must continue to provide water to its domestic, commercial, and industrial customers while meeting its fire fighting needs. In the ISO's *Fire Suppression Rating Schedule*, in order for a municipality to get full credit, it is required that a fire system must be operable with the potable water system operating at the maximum 24 hour

---

[15]Freeman's formula (1892) and Kuchling's formula (1897) were among the earliest.

average daily rate, plus the fire demand at a minimum of 20 psig (140 kPa) for a minimum time depending on the needed fire flow: 2 hours, NFF < 3000 gpm (190 L/s); 3 hours, 3000 gpm (190 L/s) < NFF < 3500 gpm (220 L/s); and 4 hours, NFF > 3500 gpm (220 L/s).

Another approach, published[16] by the American Water Works Association, is to require the fire fighting flow to be maintained for a duration equal to $Q_{gpm}/1000$ (rounded up to the next integer) in hours.

24. FIRE HYDRANTS

Other factors considered by the ISO are the type, size, capacity, installation, and inspection program of fire hydrants. The actual separation of hydrants can be specified in building codes, local ordinances, and other published standards. However, in order to be considered for ISO credits, only hydrants within 1000 ft (300 m) of a building are considered, and this distance often is used for maximum hydrant spacing (so that no building is more than 500 ft (150 m) distant from a hydrant). Additional credit is given for hydrants within 300 ft (90 m), and this distance is often selected for hydrant spacing in residential areas. Hydrants are ordinarily located near street corners where use from four directions is possible.

25. NOMENCLATURE

a	speed of sound	ft/sec	m/s
a	water hammer wave speed	ft/sec	m/s
A	area	ft^2	m^2
A_i	effective area	ft^2	m^2
AADF	average annual daily flow	gal/day	L/d
B	magnetic field strength	T	T
c_P	Poisson's effect coefficient	–	–
C	coefficient	–	–
C	construction factor	–	–
C	Hazen-Williams coefficient	–	–
D	diameter	ft	m
E	joint efficiency	–	–
f	Darcy friction factor	–	–
F	construction class coefficient	–	–
F	force	lbf	N
g	gravitational acceleration, 32.2 (9.81)	ft/sec^2	m/s^2
g_c	gravitational constant, 32.2	lbm-ft/lbf-sec^2	n.a.
h	head or height	ft	m
k	growth rate constant	–	–
k	magmeter instrument constant	–	–
K	friction coefficient	–	–

L	length	ft	m
NFF	needed fire flow	gal/min	L/s
O	occupancy combustible factor	–	–
P	communication factor	–	–
P	population	thousands	thousands
p	pressure	lbf/ft^2	Pa
Q	flow rate	gal/min	L/s
Q	flow rate	ft^3/sec or MGD	m^3/s or L/s
r	growth rate fraction	%	%
R_H	hydraulic radius	ft	m
S	allowable stress	lbf/ft^2	Pa
S	slope of energy grade line	ft/ft	m/m
t	time	sec	s
w	specific weight	lbf/ft^3	N/m^3
v	velocity	ft/sec	m/s
V	induced voltage	V	V
X	exposure factor	–	–
z	elevation	ft	m
Z	elevation	ft	m

Symbols

β	diameter ratio	–	–
γ	specific weight	lbf/ft^3	n.a.
δ	flow rate correction	ft^3/sec	m^3/s
ϵ	specific roughness	ft	m
ν	Poisson's ratio	–	–
ρ	density	lbm/ft^3	kg/m^3
τ	time constant	hr	h

Subscripts

0	critical (yield)
b	base
c	contraction
d	discharge
f	flow or friction
i	individual
I	instrument
s	static
t	tank, theoretical, or total
v	velocity
va	velocity of approach

---

[16]*Distribution System Requirements for Fire Protection* (M31), American Water Works Association, 1992.

19 Potable Water Residuals Management

Content in blue refers to the *NCEES Handbook*.

1. RESIDUAL MATERIALS FROM WATER TREATMENT

The treatment of potable water can produce various residual materials. Depending on the processes used, these materials can consist of gases, liquids, and solids.

Gases are produced only in the minority of treatment plants that employ processes to strip volatile organics from the source water. The production of gases may occur where the source water is a contaminated groundwater such that processes of natural aeration have prevented volatilization of the chemicals before treatment.

Liquid wastes may be produced by treatment plants that employ ion exchange for softening and by those that employ membrane processes for pathogen, salt, and potentially organics removal. Ion exchange produces a relatively small amount of wastewater from media regeneration. Membrane processes produce substantial amounts of waste reject water, potentially accounting for 25% or more of the total flow.

However, the vast majority of water treatment plants in the United States employ the coagulation-precipitation-filtration sequence or combine coagulation-precipitation-filtration with precipitation softening. In these cases, the primary residual is a solid.

The solid waste that is generated from coagulation-precipitation-filtration and, where applied, precipitation softening is a dilute suspension of about 1–3% solids in 99–97% water. The treatment and management of this material involves reducing the sludge volume by removing as much of the water as feasible to allow disposal. Disposal methods vary, depending on the individual circumstances at each treatment facility. A dewatered sludge will typically contain 15–25% solids.

In most cases, the dewatered sludge does not contain a substantial amount of organic material and does not require stabilization. Depending on the quantities involved and the proximity of other treatment facilities, undeveloped land, and landfills, the final disposition of the sludge may include discharge without thickening to an on-site lagoon or to a publicly owned treatment works; land application as a soil amendment, assuming favorable climate, topographic, and hydrologic conditions; or placement in a sanitary landfill.

2. SLUDGE QUANTITIES

Sludge is the watery waste that carries off the settled floc and the water softening precipitates. The sludge volume produced is given by Eq. 19.1, in which s is the gravimetric percent solids. The gravimetric solids for coagulation sludge, when expressed as a fraction, is generally less than 0.02. For water softening sludge, it is on the order of 0.10. In Eq. 19.1, W_s can be either a specific quantity or a rate of production per unit time.

Specific Gravity for a Solids Slurry

$$V = \frac{W_s}{(s/100)\gamma S} = \frac{W_s}{[(100 - p)/100]\gamma S} \quad \textbf{19.1}$$

The most accurate way to calculate the mass of sludge is to extrapolate from jar or pilot test data. There is no absolute correlation between the mass generated and other water quality measurements. However, a few generalizations are possible.

- Each unit mass (lbm, mg/L, etc.) of alum produces 0.46 unit mass of floc.

- 100% of the reduction in suspended solids (expressed as substance) shows up as floc.

- 100% of any supplemental flocculation aids is recovered in the sludge.

Suspended solids in water may be reported in turbidity units. There is no easy way to calculate total suspended solids (TSS) in mg/L from turbidity in NTU. The ratio of TSS to NTU normally varies from 1.0–2.0, and can be as high as 10. A value of 1 or 1.5 is generally appropriate.

The rate of dry sludge production from coagulation processes can be estimated from Eq. 19.2. M is an extra factor that accounts for the use of any miscellaneous inorganic additives such as clay.

$$W_{s,\text{kg/d}} =$$

$$\frac{\left(86\,400\ \frac{\text{s}}{\text{d}}\right) Q_{\text{m}^3/\text{s}} \left(1000\ \frac{\text{L}}{\text{m}^3}\right)}{10^6\ \frac{\text{mg}}{\text{kg}}}$$
$$\times (0.46 D_{\text{alum,mg/L}} + \Delta \text{TSS}_{\text{mg/L}} \qquad \text{[SI]} \quad \textit{19.2(a)}$$
$$+ M_{\text{mg/L}})$$

$$W_{s,\text{lbm/day}} =$$

$$\left(8.345\ \frac{\text{lbm-L}}{\text{mg-MG}}\right) Q_{\text{MGD}}$$
$$\times (0.46 D_{\text{alum,mg/L}} + \Delta \text{TSS}_{\text{mg/L}} \qquad \text{[U.S.]} \quad \textit{19.2(b)}$$
$$+ M_{\text{mg/L}})$$

After a sludge dewatering process from sludge solids constant G_1 to G_2, the resulting volume of sludge will be

$$V_2 \approx V\left(\frac{G_1}{G_2}\right) \qquad \textit{19.3}$$

The volume of sludge produced per day from wastewater treatment can be calculated using the sludge production rate, M, the wet sludge density, ρ_s, and the percentage of solids in the wet sludge. This volume of sludge can be used to estimate the amount of biosolids being produced or sludge to be landfilled.

Activated Sludge

$$Q_s = \frac{M(100)}{\rho_s(\%\ \text{solids})} \qquad \textit{19.4}$$

3. QUANTITIES OF SLUDGE

The procedure for determining the volume and mass of sludge from activated sludge processing is essentially the same as for sludge from sedimentation and any other processes. Table 19.1 gives characteristics of sludge. The primary variables, other than mass and volume, are density and/or specific gravity.

$$W_{\text{wet}} = V\rho_{\text{sludge}} = V(S)\rho_{\text{water}} \qquad \textit{19.5}$$

Raw sludge is approximately 95–99% water, and the specific gravity of sludge is only slightly greater than 1.0. The actual specific gravity of sludge can be calculated from the fractional solids content, s, and specific gravity of the sludge solids, S, which is

Table 19.1 *Characteristics of Sludge*

origin of sludge	fraction solids, s	dry mass (lbm/day-person)
primary settling tank	0.06–0.08	0.12
trickling filter	0.04–0.06	0.04
mixed primary settling and trickling filter	0.05	0.16
conventional activated sludge	0.005–0.01	0.07
mixed primary and conventional activated sludge	0.02–0.03	0.19
high-rate activated sludge	0.025–0.05	0.06
mixed primary settling and high-rate activated sludge	0.05	0.18
extended aeration activated sludge	0.02	0.02
filter backwashing water	0.01–0.1	n.a.
softening sludge	0.03–0.15	n.a.

(Multiply lbm/day-person by 0.45 to obtain kg/d·person.)

approximately 2.5. (If the sludge specific gravity is already known, Eq. 19.6 can be solved for the specific gravity of the solids.) In Eq. 19.6, $(1 - s)$ is the fraction of the sludge that is water.

$$\frac{1}{S} = \frac{1-s}{1} + \frac{s}{S_s}$$
$$= \frac{1 - s_{\text{fixed}} - s_{\text{volatile}}}{1} + \frac{s_{\text{fixed}}}{S_{\text{fixed solids}}} \qquad \textit{19.6}$$
$$+ \frac{s_{\text{volatile}}}{S_{\text{volatile solids}}}$$

The specific gravity of a wet sludge can also be calculated if the weight of the water, W_w, the weight of the dry solids, W_s, and the specific gravity of the dry solids, S_s, are known or can be estimated.

Specific Gravity for a Solids Slurry

$$S = \frac{W_w + W_s}{(W_w/1.00) + (W_s/S_s)} \qquad \textit{19.7}$$

The volume of sludge can be estimated from its dried mass using Eq. 19.1.

The specific gravity of sludge is approximately 1.0, so the sludge volume can also be estimated as

$$V = \frac{W_s}{(s/100)\gamma} = \frac{W_s}{[(100 - p)/100]\gamma} \qquad \textit{19.8}$$

The dried mass of sludge solids from primary settling basins is easily determined from the decrease in solids. The decrease in suspended solids, ΔSS, due to primary settling is approximately 50% of the total incoming suspended solids.

$$W_{s,\text{kg/d}} = \frac{(\Delta\text{SS})_{\text{mg/L}}\, Q_{0,\text{m}^3/\text{d}}}{1000\ \dfrac{\text{mg}\cdot\text{m}^3}{\text{kg}\cdot\text{L}}} \qquad \text{[SI]} \quad 19.9(a)$$

$$W_{s,\text{lbm/day}} = (\Delta\text{SS})_{\text{mg/L}}\, Q_{0,\text{MGD}}$$
$$\times \left(8.345\ \frac{\text{lbm-L}}{\text{MG-mg}}\right) \qquad \text{[U.S.]} \quad 19.9(b)$$

The dried mass of solids for biological filters and secondary aeration (e.g., activated sludge) can be estimated on a macroscopic basis as a fraction of the change in BOD. In Eq. 19.10, K is the *cell yield*, also known as the *removal efficiency*, the fraction of the total influent BOD that ultimately appears as excess (i.e., settled) biological solids. The cell yield can be estimated from the food-to-microorganism ratio from Fig. 19.1. (The cell yield, K, and the *yield coefficient*, also known as the *sludge yield* or *biomass yield*, Y_{obs}, are not the same, because K is based on total incoming BOD and $Y_{\text{obs,lbm/day}}$ is based on consumed BOD. The term "cell yield" is often used for both Y_{obs} and K.)

$$W_{s,\text{kg/d}} = \frac{K S_{0,\text{mg/L}}\, Q_{0,\text{m}^3/\text{d}}}{1000\ \dfrac{\text{mg}\cdot\text{m}^3}{\text{kg}\cdot\text{L}}}$$
$$= \frac{Y_{\text{obs}}(S_0 - S)_{\text{mg/L}}\, Q_{0,\text{m}^3/\text{d}}}{1000\ \dfrac{\text{mg}\cdot\text{m}^3}{\text{kg}\cdot\text{L}}} \qquad \text{[SI]} \quad 19.10(a)$$

$$W_{s,\text{lbm/day}} = K S_{0,\text{mg/L}}\, Q_{0,\text{MGD}}$$
$$\times \left(8.345\ \frac{\text{lbm-L}}{\text{MG-mg}}\right)$$
$$= Y_{\text{obs}}(S_0 - S)_{\text{mg/L}}\, Q_{0,\text{MGD}}$$
$$\times \left(8.345\ \frac{\text{lbm-L}}{\text{MG-mg}}\right) \qquad \text{[U.S.]} \quad 19.10(b)$$

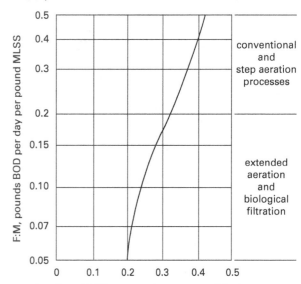

Figure 19.1 *Approximate Food: Microorganism Ratio Versus Cell Yield*, K*

*calculated assuming an effluent BOD of approximately 30 mg/L

Hammer, Mark J., *Water and Wastewater Technology*, 3rd, © 1996. Printed and electronically reproduced by permission of Pearson Education, Inc., New York, New York.

Example 19.1

A trickling filter plant processes 4 MGD of domestic wastewater with 190 mg/L BOD and 230 mg/L suspended solids. The primary sedimentation tank removes 50% of the suspended solids and 30% of the BOD. The sludge yield for the trickling filter is 0.25. The combined solids content of the sludge is 5%. What is the total sludge volume from the primary sedimentation tank and trickling filter?

Solution

The mass of solids removed in primary settling is given by Eq. 19.9.

$$W_s = (\Delta\text{SS})_{\text{mg/L}}\, Q_{0,\text{MGD}}\left(8.345\ \frac{\text{lbm-L}}{\text{MG-mg}}\right)$$
$$= (0.5)\left(230\ \frac{\text{mg}}{\text{L}}\right)(4\ \text{MGD})\left(8.345\ \frac{\text{lbm-L}}{\text{MG-mg}}\right)$$
$$= 3839\ \text{lbm/day}$$

The mass of solids removed from the trickling filter's clarifier is calculated from Eq. 19.10.

$$W_s = K\Delta S_{mg/L} Q_{0,MGD}\left(8.345 \ \frac{\text{lbm-L}}{\text{MG-mg}}\right)$$

$$= (0.25)(1-0.3)\left(190 \ \frac{\text{mg}}{\text{L}}\right)(4 \ \text{MGD})$$

$$\times \left(8.345 \ \frac{\text{lbm-L}}{\text{MG-mg}}\right)$$

$$= 1110 \ \text{lbm/day}$$

Use Eq. 19.1 to calculate the volume.

Specific Gravity for a Solids Slurry

$$V = \frac{W_s}{(s/100)\gamma S}$$

$$= \frac{3839 \ \dfrac{\text{lbm}}{\text{day}} + 1110 \ \dfrac{\text{lbm}}{\text{day}}}{\left(\dfrac{5\%}{100\%}\right)\left(62.4 \ \dfrac{\text{lbm}}{\text{ft}^3}\right)(1.0)}$$

$$= 1586 \ \text{ft}^3/\text{day}$$

4. SLUDGE THICKENING

Waste-activated sludge (WAS) has a typical solids content of 0.5–1.0%. *Thickening* of sludge is used to reduce the volume of sludge prior to digestion or dewatering. Thickening is accomplished by decreasing the liquid fraction $(1 - s)$, thus increasing the solids fraction, s. Equation 19.8 shows that the volume of wet sludge is inversely proportional to its solids content.

For dewatering, thickening to at least 4% solids (i.e., 96% moisture) is needed; for digestion, thickening to at least 5% solids is needed. Depending on the nature of the sludge, polymers may be used with all of the thickening methods. Other chemicals (e.g., lime for stabilization through pH control and potassium permanganate to react with sulfides) can also be used. However, because of cost and increased disposal problems, the trend is toward the use of fewer inorganic chemicals with new thickening or dewatering applications unless special conditions apply.

Gravity thickening occurs in circular sedimentation tanks similar to primary and secondary clarifiers. The settling process is categorized into four zones: the *clarification zone*, containing relatively clear *supernatant*; the *hindered settling zone*, where the solids move downward at essentially a constant rate; the *transition zone*, characterized by a decrease in solids settling rate; and the *compression zone*, where motion is essentially zero.

In *batch gravity thickening*, the tank is filled with thin sludge and allowed to stand. Supernatant is decanted, and the tank is topped off with more thin sludge. The operation is repeated continually or a number of times before the underflow sludge is removed. A heavy-duty (deep-truss) scraper mechanism pushes the settled solids into a hopper in the tank bottom, and the clarified effluent is removed in a peripheral weir. A doubling of solids content is usually possible with gravity thickening. Table 19.2 contains typical characteristics of gravity thickening tanks.

Table 19.2 *Typical Characteristics of Gravity Thickening Tanks*

shape	circular
minimum number of tanks*	2
overflow rate	600–800 gal/day-ft^2 (24–33 m^3/d·m^2)
maximum dry solids loading	
primary sludge	22 lbm/day-ft^2 (107 kg/d·m^2)
primary and trickling filter sludge	15 lbm/day-ft^2 (73 kg/d·m^2)
primary and modified aeration activated sludge	12 lbm/day-ft^2 (59 kg/d·m^2)
primary and conventional aeration activated sludge	8 lbm/day-ft^2 (39 mg/d·m^2)
waste-activated sludge	4 lbm/day-ft^2 (20 kg/d·m^2)
minimum detention time	6 hr
minimum sidewater depth	10 ft (3 m)
minimum freeboard	1.5 ft (0.45 m)

(Multiply gal/day-ft^2 by 0.0407 to obtain m^3/d·m^2.)
(Multiply lbm/day-ft^2 by 4.882 to obtain kg/d·m^2.)
(Multiply ft by 0.3048 to obtain m.)

*unless alternative methods of thickening are available

With *dissolved air flotation thickening* (DAFT), fine air bubbles are released into the sludge as it enters the *dissolved air flotation* (DAF) tank. The solids particles adhere to the air bubbles and float to the surface where they are skimmed away as scum. The scum has a solids content of approximately 4%. Up to 85% of the total solids may be recovered in this manner, and chemical flocculants (e.g., polymers) can increase the recovery to 97%–99%.

With *gravity belt thickening* (GBT), sludge is spread over a drainage belt. Multiple stationary plows in the path of the moving belt may be used to split, turn, and recombine the sludge so that water does not pool on top of the sludge. After thickening, the sludge cake is removed from the belt by a doctor blade.

Gravity thickening is usually best for sludges from primary and secondary settling tanks, while dissolved air flotation and centrifugal thickening are better suited for activated sludge. The current trend is toward using gravity thickening for primary sludge and flotation thickening for activated sludge, then blending the thickened sludge for further processing.

5. NOMENCLATURE

D	dose	mg/L	mg/L
G	sludge solids constant	-	-
K	cell yield	-	-
M	alkalinity	mg/L	mg/L
s	solids content	%	%
S	sludge solids	mg/L	mg/L
S	specific gravity	–	–
SS	suspended solids	mg/L	mg/L
M	sludge production rate (dry weight)	lbm/day	kg/d
p	water content	%	%
Q	flow rate	ft^3/sec or MGD	m^3/s or L/s
TSS	total suspended solids	mg/L	mg/L
V	volume	ft^3	m^3
W	weight	lbm	kg
Y_{obs}	yield coefficient	-	-

Symbols

γ	specific weight	lbf/ft^3	N/m^3
ρ	density	lbm/ft^3	kg/m^3

Subscripts

s	settling or solids
w	water

20 Sources of Pollution

Sources of pollution relevant to water resources are discussed in Chap. 5 and Chap. 14. This chapter discusses two additional sources of pollution, runoff from highways and salinity, that are often associated with urban runoff.

1. RUNOFF FROM HIGHWAYS

The *first flush* of a storm is generally considered to be the first half-inch of storm runoff or the runoff from the first 15 min of the storm. Along highways and other paved transportation corridors, the first flush contains potent pollutants such as petroleum products, asbestos fibers from brake pads, tire rubber, and fine metal dust from wearing parts. Under the National Pollutant Discharge Elimination System (NPDES), stormwater runoff in newly developed watersheds must be cleaned before it reaches existing drainage facilities, and runoff must be maintained at or below the present undeveloped runoff rate.

A good stormwater system design generally contains two separate basins or a single basin with two discrete compartments. The function of one compartment is water quality control, and the function of the other is peak runoff control.

The *water quality compartment* (WQC) should normally have sufficient volume and discharge rates to provide a minimum of one hour of detention time for 90–100% of the first flush volume. "Treatment" in the WQC consists of sedimentation of suspended solids and evaporation of volatiles. A removal goal of 75% of the suspended solids is reasonable in all but the most environmentally sensitive areas.

In environmentally sensitive areas, a filter berm of sand, a sand chamber, or a sand filter bed can further clarify the discharge from the WQC.

After the WQC becomes full from the first flush, subsequent runoff will be diverted to the *peak-discharge compartment* (PDC). This is done by designing a junction structure with an inlet for the incoming runoff and separate outlets from each compartment. If the elevation of the WQC outlet is lower than the inlet to the PDF, the first flush will be retained in the WQC.

Sediment from the WQC chamber and any filters should be removed every one to three years, or as required. The sediment must be properly handled, as it may be considered to be hazardous waste under the EPA's "mixture" and "derived-from" rules and its "contained-in" policy.

When designing chambers and filters, sizing should accommodate the first flush of a 100-year storm. The top of the berm between the WQC and PDC should include a minimum of 1 ft (0.3 m) of freeboard. In all cases, an emergency overflow weir should allow a storm greater than the design storm to discharge into the PDC or receiving water course.

Minimum chamber and berm width is approximately 8 ft (2.4 m). Optimum water depth in each chamber is approximately 2–5 ft (0.6–1.5 m). Each compartment can be sized by calculating the divided flows and staging each compartment. The outlet from the WQC should be sufficient to empty the compartment in approximately 24–28 hours after a 25-year storm.

A *filter berm* is essentially a sand layer between the WQC and the receiving chamber that filters water as it flows between the two compartments. The filter should be constructed as a layer of sand placed on geosynthetic fabric, protected with another sheet of geosynthetic fabric, and covered with coarse gravel, another geosynthetic cover, and finally a layer of medium stone. The sand, gravel, and stone layers should all have a minimum thickness of 1 ft (0.3 m). The rate of permeability is controlled by the sand size and front-of-fill material. Permeability calculations should assume that 50% of the filter fabric is clogged.

A *filter chamber* consists of a concrete structure with a removable filter pack. The filter pack consists of geosynthetic fabric wrapped around a plastic frame (core) that can be removed for backwashing and maintenance. The filter is supported on a metal screen mounted in the concrete chamber. The outlet of the chamber should be located at least 1 ft (0.3 m) behind the filter pack. The opening's size will determine the discharge rate through the filter, which should be designed as less than 2 ft/sec (60 cm/s) assuming that 50% of the filter area is clogged.

A *sand filter bed* is similar to the filter beds used for tertiary sewage treatment. The sand filter consists of a series of 4 in (100 mm) perforated PVC pipes in a gravel bed. The gravel bed is covered by geotextile fabric and 8 in (200 mm) of fine-to-medium sand. The perforated underdrains lead to the outlet channel or chamber.

2. SALINITY

Salinity can be characterized in several ways, including ionic concentrations and electrical conductivity. When salinity is characterized by relative portions of various ions, the deleterious effects are referred to as *salinity hazard*, *sodium hazard*, *boron hazard*, and *bicarbonate hazard*.

The *soluble sodium percent (percent sodium content)*, SSP, is defined by Eq. 20.1, where the concentrations are expressed in meq/L. Water quality for irrigation is roughly classified based on the SSP according to: $< 20\%$, excellent; 20–40%, good; 20–60% permissible; 60–80% doubtful; and $> 80\%$, unsuitable.

$$\text{SSP} = \frac{[\text{Na}^+] \times 100\%}{[\text{Ca}^{2+}] + [\text{Mg}^{2+}] + [\text{Na}^+] + [\text{K}^+]} \qquad 20.1$$

In waters with high concentrations of bicarbonate ions, calcium and magnesium may precipitate out as the water in the soil becomes more concentrated. As a result, the relative proportion of sodium in the form of sodium carbonate increases. The *relative proportion of sodium carbonate*, RSC, is defined by Eq. 20.2, where all of the concentrations are expressed as meq/L. The U.S. Department of Agriculture categorizes water according to its RSC as good (< 1.5 meq/L), doubtful (1.5–2.5 meq/L), and unsuitable (> 2.5 meq/L).

$$\begin{aligned} \text{RSC} = {}& \left([\text{HCO}_3^-] + [\text{CO}_3^{2-}]\right) \\ & - \left([\text{Ca}^{2+}] + [\text{Mg}^{2+}]\right) \end{aligned} \qquad 20.2$$

Percent sodium and RSC are based on chemical species, but not on reactions with the soil itself. A better measure of the sodium hazard for irrigation is the *sodium adsorption ratio*, SAR. SAR is an indicator of the amount of sodium in the water relative to calcium and magnesium. Sodium hazard is low with SAR< 9. For SAR 10–17, hazard is medium, and soil amendments (e.g., gypsum) and leaching may be required. Water with SAR 18–25 has a high sodium hazard and cannot be used continuously. An SAR in excess of 25 represents unsuitable water. The units in Eq. 20.3 are meq/L.

$$\text{SAR} = \frac{[\text{Na}^+]}{\sqrt{\dfrac{[\text{Ca}^{2+}] + [\text{Mg}^{2+}]}{2}}} \qquad 20.3$$

Salinity, like total dissolved solids, is also commonly categorized by electrical conductivity (EC) and reported in dS/m.[1,2] Water is classified according to EC as excellent (<0.25 dS/m), good (0.25–0.75 dS/m), permissible (0.75–2.0 dS/m), doubtful (2.0–3.0 dS/m), and unsuitable (>3.0 dS/m).

3. NOMENCLATURE

RSC	relative proportion of sodium carbonate	meq/L
SAR	sodium adsorption ratio	meq/L
SSP	soluble sodium percent	meq/L

---

[1] The subscripts w, iw, or e may be used (e.g., EC$_w$) to indicate water, irrigation water, or saturated soil extract.
[2] Decisiemens per meter, dS/m, is numerically the same as mS/cm. The siemens, S, is equivalent to the obsolete mho unit.

21 Watershed Management and Planning

1. PEAK RUNOFF FROM THE RATIONAL METHOD

Although total runoff volume is required for reservoir and dam design, the instantaneous peak runoff is needed to size culverts and storm drains.

The *rational formula* ("method," "equation," etc.), shown in Eq. 21.1, for peak discharge has been in widespread use in the United States since the early 1900s.[1] It is applicable to small areas (i.e., less than several hundred acres or so), but is seldom used for areas greater than 1–2 mi[2]. The intensity used in Eq. 21.1 depends on the time of concentration and the degree of protection desired (i.e., the recurrence interval).[2]

Rational Formula
$$Q = CIA \qquad \textbf{21.1}$$

Since A is in acres, Q is in ac-in/hr. However, Q is taken as ft[3]/sec since the conversion factor between these two units is 1.008.

Tables are available giving typical values of C, the runoff coefficient. [Orifices]

If more than one area contributes to the runoff, the coefficient is weighted by the areas. [Runoff Coefficients]

Accurate values of the C coefficient depend not only on the surface cover and soil type, but also on the recurrence interval, antecedent moisture content, rainfall intensity, drainage area, slope, and fraction of imperviousness. These factors have been investigated and quantified by Rossmiller (1981), who correlated these effects with the NRCS curve number. The *Schaake, Geyer, and Knapp (1967) equation* developed at Johns Hopkins University was intended for use in urban areas, correlating the impervious fraction and slope.

$$C = 0.14 + 0.65(\text{Imp}_{\text{decimal}}) + 0.05 S_{\text{percent}} \qquad \textbf{21.2}$$

The rational method assumes that rainfall occurs at a constant rate. If this is true, then the peak runoff will occur when the entire drainage area is contributing to surface runoff, which will occur at t_c. Other assumptions include (a) the recurrence interval of the peak flow is the same as for the design storm, (b) the runoff coefficient is constant, and (c) the rainfall is spatially uniform over the drainage area.

2. NRCS CURVE NUMBER

Several methods of calculating total and peak runoff have been developed over the years by the U.S. Natural Resources Conservation Service. These methods have generally been well correlated with actual experience, and the NRCS methods have become dominant in the United States.

The NRCS methods classify the land use and soil type by a single parameter called the *curve number, CN*. This method can be used for any size homogeneous watershed with a known percentage of imperviousness. If the watershed varies in soil type or in cover, it generally should be divided into regions to be analyzed separately. A composite curve number can be calculated by weighting the curve number for each region by its area. Alternatively, the runoffs from each region can be calculated separately and added.

The NRCS method of using precipitation records and an assumed distribution of rainfall to construct a synthetic storm is based on several assumptions. First, a type II storm is assumed. Type I storms, which drop most of their precipitation early, are applicable to Hawaii, Alaska, and the coastal side of the Sierra Nevada and Cascade

---

[1]In Great Britain, the rational equation is known as the *Lloyd-Davies equation*.
[2]When using intensity-duration-frequency curves to size storm sewers, culverts, and other channels, it is assumed that the frequencies and probabilities of flood damage and storms are identical. This is not generally true, but the assumption is usually made anyway.

mountains in California, Oregon, and Washington. Type II distributions are typical of the rest of the United States, Puerto Rico, and the Virgin Islands.

This method assumes that initial abstraction (depression storage, evaporation, and interception losses) is equal to 20% of the storage capacity.

$$I_a = 0.2S \quad\quad 21.3$$

For there to be any runoff at all, the gross rain must equal or exceed the initial abstraction.

$$P \geq I_a \quad\quad 21.4$$

The storage capacity must be great enough to absorb the initial abstraction plus the infiltration.

$$S \geq I_a + F \quad\quad 21.5$$

The following steps constitute the NRCS method.

step 1: Classify the soil into a *hydrologic soil group* (HSG) according to its infiltration rate. Soil is classified into HSG A (low runoff potential) through D (high runoff potential).

Group A: High infiltration rates (> 0.30 in/hr (0.76 cm/h)) even if thoroughly saturated; chiefly deep sands and gravels with good drainage and high moisture transmission. In urbanized areas, this category includes sand, loamy sand, and sandy loam.

Group B: Moderate infiltration rates if thoroughly wetted (0.15–0.30 in/hr (0.38–0.76 cm/h)), moderate rates of moisture transmission, and consisting chiefly of coarse to moderately fine textures. In urbanized areas, this category includes silty loam and loam.

Group C: Slow infiltration rates if thoroughly wetted (0.05–0.15 in/hr (0.13–0.38 cm/h)), and slow moisture transmission; soils having moderately fine to fine textures or that impede the downward movement of water. In urbanized areas, this category includes sandy clay loam.

Group D: Very slow infiltration rates (less than 0.05 in/hr (0.13 cm/h)) if thoroughly wetted, very slow water transmission, and consisting primarily of clay soils with high potential for swelling; soils with permanent high water tables; or soils with an impervious layer near the surface. In urbanized areas, this category includes clay loam, silty clay loam, sandy clay, silty clay, and clay.

(Note that as a result of urbanization, the underlying soil may be disturbed or covered by a new layer. The original classification will no longer be applicable, and the "urbanized" soil HSGs are applicable.)

step 2: Determine the preexisting soil conditions. The soil condition is classified into *antecedent runoff conditions* (ARC) I through III.[3] Generally, "average" conditions (ARC II) are assumed.

ARC I: Dry soils, prior to or after plowing or cultivation, or after periods without rain.

ARC II: Typical conditions existing before maximum annual flood.

ARC III: Saturated soil due to heavy rainfall (or light rainfall with freezing temperatures) during 5 days prior to storm.

step 3: Classify *cover type* and hydrologic condition of the soil-cover complex. For pasture, range, row crops, arid, and semi-arid lands, the NRCS method includes additional tables to classify the cover and hydrologic conditions. In order to use these tables, it is necessary to characterize the surface coverage. The condition is "good" if it is lightly grazed or has plant cover over 75% or more of its area. The condition is "fair" if plant coverage is 50–75% or not heavily grazed. The condition is "poor" if the area is heavily grazed, has no mulch, or has plant cover over less than 50% of the area.

step 4: Use Table 21.1 or Table 21.2 to determine the curve number, *CN*, corresponding to the soil classification for ARC II.

step 5: If the soil is ARC I or ARC III, convert the curve number from step 4 by using Eq. 21.6 or Eq. 21.7 and rounding up.

$$CN_{\mathrm{I}} = \frac{4.2\,CN_{\mathrm{II}}}{10 - 0.058\,CN_{\mathrm{II}}} \quad\quad 21.6$$

$$CN_{\mathrm{III}} = \frac{23\,CN_{\mathrm{II}}}{10 + 0.13\,CN_{\mathrm{II}}} \qu\quad 21.7$$

step 6: If any significant fraction of the watershed is impervious (i.e., $CN = 98$ for pavement), or if the watershed consists of areas with different curve numbers, calculate the composite curve number by weighting by the runoff areas (same as weighting by the impervious and pervious fractions). If the watershed's impervious fraction is different from the value implicit in step 3's classification, or if the impervious area is not connected directly to a storm drainage system, then the NRCS method includes direct and graphical adjustments.

step 7: Estimate the time of concentration of the watershed.[4]

step 8: Determine the *gross (total) rainfall*, *P*, from the storm. To do this, it is necessary to assume the storm length and recurrence

---

[3]The *antecedent runoff condition* (ARC) may also be referred to as the *antecedent moisture condition* (AMC).
[4]The NRCS method uses T_c, not t_c, as the symbol for time of concentration.

interval. It is a characteristic of the NRCS methods to use a 24-hr storm. Maps from the U.S. Weather Bureau can be used to read gross point rainfalls for storms with frequencies from 1 to 100 years.[5]

step 9: Multiply the gross rain point value from step 8 by a factor from Fig. 21.1 to make the gross rain representative of larger areas. This is the *areal rain*.

Figure 21.1 *Point to Areal Rain Conversion Factors*

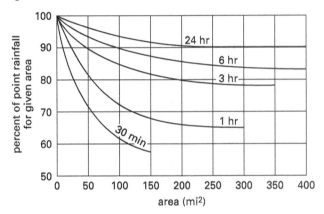

step 10: The NRCS method assumes that infiltration follows an exponential decay curve with time. Storage capacity of the soil (i.e., the potential maximum retention after runoff begins), S, is calculated from the curve number by using Eq. 21.8.

NRCS (SCS) Rainfall-Runoff

$$S = \frac{1{,}000}{CN} - 10 \qquad 21.8$$

Rearranging Eq. 21.8 gives a way to calculate the curve number when the storage capacity of the soil is known.

NRCS (SCS) Rainfall-Runoff

$$CN = \frac{1{,}000}{S + 10} \qquad 21.9$$

step 11: Calculate the total runoff (net rain, precipitation excess, etc.), Q, in inches from the areal rain. Equation 21.10 subtracts losses from interception, storm period evaporation, depression storage, and infiltration from the *gross rain* to obtain the *net rain*.

NRCS (SCS) Rainfall-Runoff

$$Q = \frac{(P - 0.2S)^2}{P + 0.8S} \qquad 21.10$$

3. NRCS GRAPHICAL PEAK DISCHARGE METHOD

Two NRCS methods are available for calculating the peak discharge. When a full hydrograph is not needed, the so-called graphical method can be used. If a hydrograph is needed, the tabular method can be used.[6] The graphical method is applicable when (a) $CN > 40$, (b) $0.1 \text{ hr} < t_c < 10 \text{ hr}$, (c) the watershed is relatively homogeneous or uniformly mixed, (d) all streams have the same time of concentration, and (e) there is no interim storage along the stream path.

Equation 21.11 is the NRCS peak discharge equation. Q_p is the peak discharge in ft^3/sec, q_u is the unit peak discharge in cubic feet per square mile per inch of runoff (csm/in), A_{mi^2} is the drainage area in square miles, Q_{in} is the runoff in inches, and F_p is a pond and swamp adjustment factor.

$$Q_p = q_u A_{\text{mi}^2} Q_{\text{in}} F_p \qquad 21.11$$

The graphical method begins with obtaining the design for the 24 hr rainfall, P, for the area. Total runoff, Q, and the curve number, CN, are determined by the same method as in Sec. 21.2. Then, the initial abstraction, I_a, is obtained from Table 21.3, and the ratio I_a/P is calculated. Next, NRCS curves are entered with t_c and the I_a/P ratio to determine the runoff in cubic feet per square mile per inch of runoff (csm/in). The appropriate pond and swamp adjustment factor is selected from the NRCS literature. ($F_p = 1$ if there are no ponds or swamps.)

4. RESERVOIR SIZING: MODIFIED RATIONAL METHOD

An effective method of preventing flooding is to store surface runoff temporarily. After the storm is over, the stored water can be gradually released. An *impounding reservoir* (*retention watershed* or *detention watershed*) is a watershed used to store excess flow from a stream or river. The stored water is released when the stream flow drops below the minimum level that is needed to meet water demand. The *impoundment depth* is the design depth. Finding the impoundment depth is equivalent to finding the design storage capacity of the reservoir.

The purpose of a *reservoir sizing* (*reservoir yield*) analysis is to determine the proper size of a reservoir or dam, or to evaluate the ability of an existing reservoir to meet water demands.

---

[5]*Rainfall Frequency Atlas of the United States for Durations from 30 Minutes to 24 Hours and Return Periods from 1 to 100 Years* (1961), U.S. Weather Bureau, Technical Paper 40.
[6]The graphical and tabular methods both rely on graphs and tables that are contained in TR-55. These graphs and tables are not included in this chapter. The NRCS tabular method is not described in this book.

Table 21.1 *Runoff Curve Numbers of Urban Areas (ARC II)*

cover description		curve numbers for hydrologic soil			
cover type and hydrologic condition	average percent impervious area	group A	group B	group C	group D
fully developed urban areas (vegetation established)					
open space (lawns, parks, golf courses, cemeteries, etc.)					
poor condition (grass cover < 50%)		68	79	86	89
fair condition (grass cover 50–75%)		49	69	79	84
good condition (grass cover > 75%)		39	61	74	80
impervious areas					
paved parking lots, roofs, driveways, etc., (excluding right-of-way)		98	98	98	98
streets and roads					
paved; curbs and storm sewers (excluding right-of-way)		98	98	98	98
paved; open ditches (including right-of-way)		83	89	92	93
gravel (including right-of-way)		76	85	89	91
dirt (including right-of-way)		72	82	87	89
western desert urban areas					
natural desert landscaping (pervious areas only)		63	77	85	88
artificial desert landscaping (impervious weed barrier, desert shrub with 1–2 in sand or gravel mulch and basin borders)		96	96	96	96
urban districts					
commercial and business	85	89	92	94	95
industrial	72	81	88	91	93
residential districts by average lot size					
$\frac{1}{8}$ acre or less (townhouses)	65	77	85	90	92
$\frac{1}{4}$ acre	38	61	75	83	87
$\frac{1}{3}$ acre	30	57	72	81	86
$\frac{1}{2}$ acre	25	54	70	80	85
1 acre	20	51	68	79	84
2 acres	12	46	65	77	82
developing urban areas					
newly graded areas (pervious areas only, no vegetation)		77	86	91	94

Reprinted from *Urban Hydrology for Small Watersheds*, Technical Release TR-55, United States Department of Agriculture, Natural Resources Conservation Service, Table 2-2a, 1986.

Table 21.2 *Runoff Curve Numbers for Cultivated Agricultural Lands (ARC II)*

cover description			curve numbers for hydrologic soil			
cover type	treatment[a]	hydrologic condition[b]	group A	group B	group C	group D
fallow	bare soil	–	77	86	91	94
	crop residue cover (CR)	poor	76	85	90	93
		good	74	83	88	90
row crops	straight row (SR)	poor	72	81	88	91
		good	67	78	85	89
	SR + CR	poor	71	80	87	90
		good	64	75	82	85
	contoured (C)	poor	70	79	84	88
		good	65	75	82	86
	C + CR	poor	69	78	83	87
		good	64	74	81	85
	contoured and terraced (C&T)	poor	66	74	80	82
		good	62	71	78	81
	C&T + CR	poor	65	73	79	81
		good	61	70	77	80
small grain	SR	poor	65	76	84	88
		good	63	75	83	87
	SR + CR	poor	64	75	83	86
		good	60	72	80	84
	C	poor	63	74	82	85
		good	61	73	81	84
	C + CR	poor	62	73	81	84
		good	60	72	80	83
	C&T	poor	61	72	79	82
		good	59	70	78	81
	C&T + CR	poor	60	71	78	81
		good	58	69	77	80
close-seeded or broadcast legumes or rotation meadow	SR	poor	66	77	85	89
		good	58	72	81	85
	C	poor	64	75	83	85
		good	55	69	78	83
	C&T	poor	63	73	80	83
		good	51	67	76	80

[a]*Crop residue cover* applies only if residue is on at least 5% of the surface throughout the year.

[b]Hydrologic condition is based on a combination of factors that affect infiltration and runoff, including (a) density and canopy of vegetative areas, (b) amount of year-round cover, (c) amount of grass or close-seeded legumes in rotations, (d) percent of residue cover on the land surface (good ≥ 20%), and (e) degree of surface roughness.

Poor: Factors impair infiltration and tend to increase runoff.

Good: Factors encourage average and better-than-average infiltration and tend to decrease runoff.

Reprinted from *Urban Hydrology for Small Watersheds*, Technical Release TR-55, United States Department of Agriculture, Natural Resources Conservation Service, Table 2-b, 1986.

Table 21.3 Initial Abstraction Versus Curve Number

curve number	I_a (in)	curve number	I_a (in)
40	3.000	70	0.857
41	2.878	71	0.817
42	2.762	72	0.778
43	2.651	73	0.740
44	2.545	74	0.703
45	2.444	75	0.667
46	2.348	76	0.632
47	2.255	77	0.597
48	2.167	78	0.564
49	2.082	79	0.532
50	2.000	80	0.500
51	1.922	81	0.469
52	1.846	82	0.439
53	1.774	83	0.410
54	1.704	84	0.381
55	1.636	85	0.353
56	1.571	86	0.326
57	1.509	87	0.299
58	1.448	88	0.273
59	1.390	89	0.247
60	1.333	90	0.222
61	1.279	91	0.198
62	1.226	92	0.174
63	1.175	93	0.151
64	1.125	94	0.128
65	1.077	95	0.105
66	1.030	96	0.083
67	0.985	97	0.062
68	0.941	98	0.041
69	0.899		

(Multiply in by 2.54 to obtain cm.)

Reprinted from *Urban Hydrology for Small Watersheds*, Technical Release TR-55, United States Department of Agriculture, Natural Resources Conservation Service, Table 4–1, 1986.

The volume of a reservoir needed to hold streamflow from a storm is simply the total area of the hydrograph. Similarly, when comparing two storms, the incremental volume needed is simply the difference in the areas of their two hydrographs.

Poertner's 1974 *modified rational method*, as illustrated in Fig. 21.2, can be used to design detention storage facilities for small areas (up to 20 ac). A trapezoidal

hydrograph is constructed with the peak flow calculated from the rational equation. The total hydrograph base (i.e., the total duration of surface runoff) is the storm duration plus the time to concentration. The durations of the rising and falling limbs are both taken as t_c.

Figure 21.2 Modified Rational Method Hydrograph

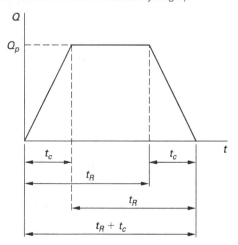

To size a detention watershed using this method, a trapezoidal hydrograph is drawn for several storm durations greater than $2t_c$ (e.g., 10, 15, 20, and 30 min). For each of the storm durations, the total rainfall is calculated as the hydrograph area. The difference between the total rainfall and the total released water (calculated as the product of the allowable release rate and the storm duration) is the required storage capacity.

Detention watershed sizing for areas larger than 20 ac should be accomplished using full hydrograph methods in combination with reservoir routing.

5. RESERVOIR SIZING: NONSEQUENTIAL DROUGHT METHOD

The *nonsequential drought method* is somewhat complex, but it has the advantage of giving an estimate of the required reservoir size, rather than merely evaluating a trial size. In the absence of synthetic drought information, it is first necessary to develop intensity-duration-frequency curves from stream flow records.

step 1: Choose a duration. Usually, the first duration used will be 7 days, although choosing 15 days will not introduce too much error.

step 2: Search the stream flow records to find the smallest flow during the duration chosen. (The first time through, for example, find the smallest discharge totaled over any 7 days.) The days do not have to be sequential.

step 3: Continue searching the discharge records to find the next smallest discharge over the number of days in the period. Continue searching and finding the smallest discharges (which gradually increase) until all of the days in the record have been used up. Do not use the same day more than once.

step 4: Give the values of smallest discharge order numbers; that is, give $M = 1$ to the smallest discharge, $M = 2$ to the next smallest, etc.

step 5: For each observation, calculate the recurrence interval as

$$F = \frac{n_y}{M} \qquad 21.12$$

n_y is the number of years of stream flow data that was searched to find the smallest discharges.

step 6: Plot the points as discharge on the y-axis versus F in years on the x-axis. Draw a reasonably continuous curve through the points.

step 7: Return to step 1 for the next duration. Repeat for all of the following durations: 7, 15, 30, 60, 120, 183, and 365 days.

A synthetic drought can be constructed for any recurrence interval. For example, in Fig. 21.3, a 5-year drought is being planned for, so the discharges V_7, V_{15}, V_{30}, ..., V_{365} are read from the appropriate curves for $F = 5$ yr.

Figure 21.3 *Sample Family of Synthetic Inflow Curves*

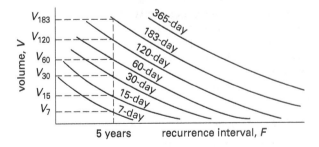

The next step is to plot the reservoir *mass diagram* (also known as a *Rippl diagram*). This is a plot of the cumulative net volume—a simultaneous plot of the cumulative

demand (known as *draft*) and cumulative inflow. The mass diagram is used to graphically determine the reservoir storage requirements (i.e., size).

As long as the slopes of the cumulative demand and inflow lines are equal, the water reserve in the reservoir will not change. When the slope of the inflow is less than the slope of the demand, the inflow cannot by itself satisfy the community's water needs, and the reservoir is drawn down to make up the difference. A peak followed by a trough is, therefore, a drought condition.

If the reservoir is to be sized so that the community will not run dry during a drought, the required capacity is the maximum separation between two parallel lines (pseudo-demand lines with slopes equal to the demand rate) drawn tangent to a peak and a subsequent trough. If the mass diagram covers enough time so that multiple droughts are present, the largest separation between peaks and subsequent troughs represents the capacity.

In order for the reservoir to supply enough water during a drought condition, the reservoir must be full prior to the start of the drought. This fact is not represented when the mass diagram is drawn, hence the need to draw a pseudo-demand line parallel to the peak.

After a drought equal to the capacity of the reservoir, the reservoir will again be empty. At the trough, however, the reservoir begins to fill up again. When the cumulative excess exceeds the reservoir capacity, the reservoir will have to "spill" (i.e., release) water. This occurs when the cumulative inflow line crosses the prior peak's pseudo-demand line, as shown in Fig. 21.4.

A *flood-control dam* is built to keep water in and must be sized so that water is not spilled. The mass diagram can still be used, but the maximum separation between troughs and subsequent peaks (not peaks followed by troughs) is the required capacity.

Figure 21.4 *Reservoir Mass Diagram (Rippl Diagram)*

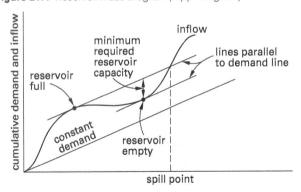

6. RESERVOIR SIZING: RESERVOIR ROUTING

Reservoir routing is the process by which the outflow hydrograph (i.e., the outflow over time) of a reservoir is determined from the inflow hydrograph (i.e., the inflow over time), the initial storage, and other characteristics of the reservoir. The simplest method is to keep track of increments in inflow, storage, and outflow period by period in a tabular simulation. This is the basis of the *storage indication method*, which is basically a bookkeeping process. The validity of this method is dependent on choosing time increments that are as small as possible.

step 1: Determine the starting storage volume, V_n. If the starting volume is zero or considerably different from the average steady-state storage, a large number of iterations will be required before the simulation reaches its steady-state results. A convergence criterion should be determined before the simulation begins.

step 2: For the next iteration, determine the inflow, discharge, evaporation, and seepage. The starting storage volume for the next iteration is found by solving Eq. 21.14.

Determination of Required Reservoir Capacity

$$\Delta S = I - O \qquad 21.13$$

$$V_{n+1} = V_n + (\text{inflow})_n - (\text{discharge})_n \\ - (\text{seepage})_n - (\text{evaporation})_n \qquad 21.14$$

Repeat step 2 as many times as necessary.

Loss due to seepage is generally very small compared to inflow and discharge, and it is often neglected. Reservoir *evaporation* can be estimated from analytical relationships or by evaluating data from *evaporation pans*. Pan data is extended to reservoir evaporation by the pan coefficient formula. In Eq. 21.15, the summation is taken over the number of days in the simulation period. Units of evaporation are typically in/day. The *pan coefficient*, p_c, is typically 0.7–0.8.

Pan Evaporation

$$E_L = p_c E_P \qquad 21.15$$

Inflow can be taken from actual past history. However, since it is unlikely that history will repeat exactly, the next method is preferred.

7. RESERVOIR ROUTING: STOCHASTIC SIMULATION

The *stochastic simulation* method of reservoir routing is the same as tabular simulation except for the method of determining the inflow. This description uses a *Monte Carlo simulation* technique that is dependent on enough historical data to establish a cumulative inflow distribution. A Monte Carlo simulation is suitable if long periods are to be simulated. If short periods are to be simulated, the simulation should be performed several times and the results averaged.

step 1: Tabulate or otherwise determine a frequency distribution of inflow quantities.

step 2: Form a cumulative distribution of inflow quantities.

step 3: Multiply the cumulative x-axis (which runs from 0 to 1) by 100 or 1000, depending on the accuracy needed.

step 4: Generate random numbers between 0 and 100 (or 0 and 1000). Use of a random number table is adequate for hand simulation.

step 5: Locate the inflow quantity corresponding to the random number from the cumulative distribution x-axis.

Example 21.1

A well-monitored stream has been observed for 50 years and has the following frequency distribution of total annual discharges.

discharge (units)	frequency (yr)	fraction of time
0 to 0.5	5	0.10
0.5 to 1.0	21	0.42
1.0 to 1.5	17	0.34
1.5 to 2.0	7	0.14

It is proposed that a dam be placed across the stream to create a reservoir with a capacity of 1.8 units. The reservoir is to support a town that will draw 1.2 units per year. Use stochastic simulation to simulate 10 yr of reservoir operation, assuming it is initially filled with 1.5 units.

Solution

step 1: The frequency distribution is given.

steps 2 and 3: Make a table to form the cumulative distribution with multiplied frequencies.

discharge	cumulative frequency	cumulative frequency × 100
0 to 0.5	0.10	10
0.5 to 1.0	0.52	52
1.0 to 1.5	0.86	86
1.5 to 2.0	1.00	100

step 4: From a random number table, choose 10 two-digit numbers. The starting point within the table is arbitrary, but the numbers must come sequentially from a row or column. Use the first row for this simulation.

78, 46, 68, 33, 26, 96, 58, 98, 87, 27

step 5: For the first year, the random number is 78. Since 78 is greater than 52 but less than 86, the inflow is in the 1.0–1.5 unit range. The midpoint of this range is taken as the inflow, which would be 1.25. The reservoir volume after the first year would be $1.5 + 1.25 - 1.20 = 1.55$. The remaining years can be similarly simulated.

year	starting volume	+ inflow	− usage	= ending volume	+ spill
1	1.5	1.25	1.2	1.55	
2	1.55	0.75	1.2	1.1	
3	1.1	1.25	1.2	1.15	
4	1.15	0.75	1.2	0.7	
5	0.7	0.75	1.2	0.25	
6	0.25	1.75	1.2	0.8	
7	0.8	1.25	1.2	0.85	
8	0.85	1.75	1.2	1.4	
9	1.4	1.75	1.2	1.95	0.15
10	1.8	0.75	1.2	1.35	

No shortages are experienced; one spill is required.

8. STORMWATER/WATERSHED MANAGEMENT MODELING

Flood routing and *channel routing* are terms used to describe the passage of a flood or runoff wave through a complex system. Due to flow times, detention, and processing, the wave front appears at different points in a system at different times. The ability to simulate flood, channel, and reservoir routing is particularly useful when evaluating competing features (e.g., treatment, routing, storage options).

The interaction of stormwater features (e.g., watersheds, sewers and storm drains, detention watersheds, treatment facilities) during and after a storm is complex. In addition to continuity considerations, the interaction is affected by hydraulic considerations (e.g., the characteristics of the flow paths) and topography (e.g., the elevation changes from point to point).

Many computer programs have been developed to predict the performance of such complex systems. These simulation programs vary considerably in complexity, degree of hydraulic detail, length of simulation interval, and duration of simulation study. The Environmental Protection Agency (EPA) Stormwater Management Model is a well-known micro-scale model, particularly well-suited to areas with a high impervious fraction. The Army Corps of Engineers' STORM model is a macro-scale model. The NRCS TR-20 program is consistent with other NRCS methodology.

9. FLOOD CONTROL CHANNEL DESIGN

For many years, the traditional method of flood control was *channelization*, converting a natural stream to a uniform channel cross section. However, this method does not always work as intended and may fail at critical moments. Some of the reasons for reduced capacities are sedimentation, increased flow resistance, and inadequate maintenance.

The design of artificial channels is often based on clear-water hydraulics without sediment. However, the capacity of such channels is greatly affected by *sedimentation*. Silting and sedimentation can double or triple the Manning roughness coefficient. Every large flood carries appreciable amounts of bed load, significantly increasing the composite channel roughness. Also, when the channel is unlined, scour and erosion can significantly change the channel cross-sectional area.

To minimize cost and right-of-way requirements, shallow supercritical flow is often intended when the channel is designed. However, supercritical flow can occur only with low bed and side roughness. When the roughness increases during a flood, the flow shifts back to deeper, slower-moving subcritical flows.

Debris carried downstream by floodwaters can catch on bridge pilings and culverts, obstructing the flow. Actual flood profiles can resemble a staircase consisting of a series of backwater pools behind obstructed bridges, rather than a uniformly sloping surface. In such situations, the most effective flood control method may be replacement of bridges or improved emergency maintenance procedures to remove debris.

Vegetation and other debris that collects during dry periods in the flow channel reduces the flow capacity. Maintenance of channels to eliminate the vegetation is often haphazard.

Flood control channel design programs frequently emphasize "creek protection" and the long-term management of natural channels. Elements of such programs include (a) excavation of a low-flow channel within the natural channel, (b) periodic intervention to keep the channel clear, (c) establishment of a wide flood plain terrace along the channel banks, incorporating wetlands, vegetation, and public-access paths, and (d) planting riparian vegetation and trees along the terrace to slow bank erosion and provide a continuous corridor for wildlife.

10. NOMENCLATURE

A	area	ft^2	m^2
A	area	ac	ac
C	constant	–	–
C	rational runoff coefficient	–	–
CN	curve number	–	–
E	evaporation	in/day	cm/d
F	adjustment factor	–	–
F	frequency of occurrence	1/yr	1/yr
F	infiltration	in	cm
I	rainfall intensity	in/hr	cm/h
I_a	initial abstraction	in	cm
Imp	imperviousness	%	%
M	order number	–	–
n_y	number of years of streamflow data	yr	yr
O	outflow volume	ft^3	m^3
p_c	pan coefficient	–	–
P	precipitation	in	cm
q	runoff	ft^3/mi^2-in	m^3/km$^2 \cdot$cm
Q	flow rate	ft^3/sec	m^3/s
S	storage capacity	in	cm
S_{percent}	slope	%	%
t	time	min	min
t_c	time to concentration	min	min
V	volume	ft^3	m^3

Subscripts

c	concentration
L	lake
n	period n
p	pan, peak, or pond
P	pan
R	rain (storm) or reservoir
u	unit

22 Source Supply and Protection

In addition to groundwater (discussed in Chap. 46 and Chap. 49) and the water held in the atmosphere, two important water sources that contribute to the water supply are

- snowpack (or snowmelt)
- rivers and lakes

1. SNOWMELT

Terms used to describe snow include

- *snowfall.* The depth of snow occurring from a single storm event.

- *snowpack.* The accumulated snowfall from multiple events.

- *density.* The volume of the melt water from a snow sample divided by the initial volume of the sample. Density varies from 0.4% for freshly fallen snow at high altitudes to 91% for compacted snow in glaciers. The assumed average value is 10% for freshly fallen snow, near 50% for aged snow.

- *water equivalent.* The depth of water that would result from melting.

- *quality.* The percentage of the total weight of a snow sample that is from ice. Most snow has a quality of 90% or more but may decrease to as low as 50% during rapid melt periods.

- *thermal quality.* The heat needed to produce a certain amount of melt from snow, divided by the amount of heat needed to produce the same amount of melt from pure ice at 32°F. At subfreezing temperatures, the thermal quality may approach 100%.

- *water content.* The amount of liquid water contained in a snow sample, with snow being composed of liquid water and ice crystals.

In some regions of the United States, the most significant form of precipitation is snowfall, and consequently the most significant source of runoff is snowmelt. Although snowfall may occur from early fall through late spring, snowmelt typically begins in the spring and may continue well into the summer. Because snowmelt presents a temporally concentrated runoff often out of sync with periods of greatest seasonal demand, large storage reservoirs are common in western states. For example, lakes where more than 90% of impounded water is from snowmelt include

- Lake Mead formed by Hoover Dam near Las Vegas. This is the largest capacity reservoir in the United States. It covers 247 mi^2, and has a capacity of 28.5 million ac-ft.

- Lake Powell formed by Glenn Canyon Dam at the Utah/Arizona border. This lake has more than 2000 mi of shoreline, covers 266 mi^2, and has a capacity of 27 million ac-ft.

Snowmelt presents unique runoff problems, especially when combined with heavy rainfall, a combination that often leads to severe flooding. For example, record snowpack and spring rains caused flooding from City Creek in downtown Salt Lake City, Utah, in 1983, a highly unusual occurrence. Rainfall does not always occur, however, and generally is significant only at higher temperatures.

Besides rainfall, other contributors to runoff from snowmelt include

- *radiation:* the influence of albedo and air temperature

- *condensation:* heat produced from condensing vapor in the snowpack

- *convection:* heat transfer from the air to the snowpack

Snowmelt occurs in several steps and may require as little as a few hours to many months to generate runoff, the final step. For runoff to occur, however, the temperature of the whole snowpack must rise to at least 32°F. At this point, water from melted snow will exceed the capacity of the snow to hold the water (the snow is *ripe*), and meltwater will percolate to the ground surface and contribute to runoff.

2. RIVERS

The water stored in rivers is analyzed through river routing. *River routing* is a method for determining hydrographs at various locations along a stream channel. This is important in flood situations where the flood propagates along the length of the channel as a wave with distinct rising and falling stages. As the wave moves farther along the channel, the peak flattens, distributing storage over a longer reach.

The basic relationship among inflow, I, and outflow, O, and storage, S, is

$$\frac{I_1 + I_2}{2} - \frac{O_1 + O_2}{2} = \frac{S_2 - S_1}{\Delta t} \qquad 22.1$$

The time difference in Eq. 22.1, Δt, is the routing period from time t_1 to time t_2. The subscripts stand for the beginning and end of the routing period.

This equation indicates that the average inflow during the routing period minus the average outflow during the routing period is equal to the storage provided by the stream channel. Channel storage can be computed from physical measurements of stream channel cross sections and stage, if sufficient gauging stations are available, but this is usually not feasible. However, where stream stage data are available from reach to reach, inflow and outflow hydrographs can define storage over successive routing periods. Because river routing calculations involve large data sets, they are best performed using spreadsheets or software specifically developed for the purpose.

Example 22.1

Using the inflow and outflow hydrographs given in the table, determine the storage profile and the peak storage (maximum volume stored) for the channel reach between cross sections A and B.

time (hr)	section A inflow (cfs)	section B outflow (cfs)
0	10	10
6	10	10
12	30	13
18	70	26
24	50	43
30	40	45
36	30	41
42	20	35
48	10	28
54	10	19
60	10	15
66	10	13
72	10	11
80	10	10

Solution

Determine the storage profile and peak storage by taking the difference between the inflow and outflow S-hydrographs. These are shown in the spreadsheet output, where SHG designates the storage hydrograph.

time (hr)	section A inflow (cfs)	section B outflow (cfs)	I-SHG (cfs)	O-SHG (cfs)	ΔSHG (cfs)	storage (10^4 cf)
0	10	10	10	10	0	0
6	10	10	20	20	0	0
12	30	13	50	33	17	37
18	70	26	120	60	60	130
24	50	43	170	103	67	145
30	40	45	210	148	62	134
36	30	41	240	189	51	110
42	20	35	260	224	36	78
48	10	28	270	252	18	39
54	10	19	280	271	9	19
60	10	15	290	286	4	8.6
66	10	13	300	299	1	2.2
72	10	11	310	310	0	0
80	10	10	320	320	0	0

The calculated hydrographs and the storage volume for the reach of river analyzed are shown in the figure. The column designated ΔSHG represents the storage profile as the wave moves through the river reach. The peak storage volume is 1,450,000 cf.

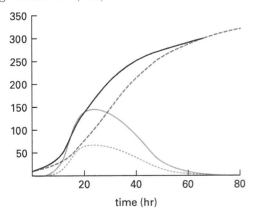

time (hr)

key

———— I-SHG (cfs)
------- O-SHG (cfs)
———— storage volume (10,000 cf)
.... storage HG (cfs)

3. RESERVOIRS

Hydrologic reservoir routing typically considers only the storage above the normal pool level (the elevation of the spillway or other discharge structure). When the reservoir is at normal pool level, storage is zero. For these reasons, storage is defined as the surcharge storage, or the storage existing above the spillway crest, and outflow is zero when storage is zero. This definition applies only to reservoir excess capacity, or the capacity generated from a rainfall event or from snowmelt. The total reservoir capacity is reduced as demand for water exceeds the flow through the reservoir, that is, as outflow exceeds inflow.

Using topographic maps, aerial photography, and other survey tools, storage can be determined for different pool elevations and correlated to outflow.

4. EROSION

In both urban and rural settings, runoff contributes to water quality by carrying pollutants to receiving rivers, streams, lakes, and reservoirs. Foremost among these pollutants are those generated by erosion and subsequent sedimentation. The eroded sediment is itself a pollutant, but the sediment can also be a carrier for other organic, mineral, and pathogenic materials. The phenomenon of wave movement described in river routing represents the period when runoff and erosion are most significant and when the sediment load in the river is highest.

Overland flow is usually laminar and is, therefore, unable to detach soil particles. If soil particles are already loosened, however, they may be transported with the runoff. Because of the requirement of detachment, sheet erosion is primarily the result of raindrop impaction. On sloped surfaces, detached soil is displaced down-slope, and when accompanied with runoff, the displaced soil particles are carried away with the water.

Raindrops have diameters that range from 0.5 mm to 6 mm and terminal velocities that range from 2 m/s to 9 m/s. The energy delivered by a raindrop as it impacts the surface is proportional to d^3v^2, where d is the diameter and v is the velocity. Consequently, the energy delivered by larger raindrops may be 10^4 times greater than that delivered by smaller raindrops. Because of the differences in energy between large and small raindrops, sheet erosion is more likely during periods of intense rainfall. At some point, the sheet flow is adequate to form into channels and turbulence develops. The turbulence creates gully and channel erosion. The eroded sediment is carried as *bed load* in the stream.

Erosion is controlled by controlling the slope and providing vegetative cover. Vegetative cover absorbs the rain drop energy and provides mechanical protection against gully erosion by holding soil in place. Vegetative cover also improves infiltration by adding organic matter to the soil; infiltration reduces runoff.

5. SEDIMENTATION

Maximum sediment production typically occurs at about 12 in of mean annual precipitation. The reason for this is that such geographic areas usually have little vegetative cover. The combination of a moderately intense rainfall and little vegetation produces comparatively high potential for sediment yield to streams. Locations with less than 12 in of mean annual rainfall typically do not experience enough regular intense precipitation to promote high sediment production. Locations with greater than 12 in of mean annual precipitation usually have adequate vegetative cover to limit erosion.

Sedimentation in reservoirs is influenced by the mass of sediment inflow, the trap efficiency of the reservoir, and the sediment particle density. The mass of sediment inflow is determined by the mechanisms of erosion discussed and represents the sediment opportunity for the reservoir. Obviously, little sedimentation will occur in a reservoir where the inflow sediment mass is small. The reservoir *trap efficiency* is the percentage of incoming sediment that is retained in the reservoir.

- Trap efficiency increases with residence time since residence time equates to particle settling time.

- Reservoir residence time is the volume divided by the flow ($t = V/Q$) and is significant to trap efficiency.

- A small reservoir with relatively high flow-through will see less sedimentation than will a larger reservoir with relatively lower flow-through.

- Trap efficiency decreases as the reservoir gradually fills with sediment.

The sediment particle density will determine the settling rate—denser particles settle more quickly than do less dense particles. Consequently, the combined factors that will produce the most sedimentation in a reservoir are

- high sediment inflow mass

- long hydraulic residence time

- high particle density

These are the conditions that exist at Lake Powell and Lake Mead, the two largest reservoirs in the United States. Both are located along the Colorado River and have drainage basins characterized by about 12 in of mean annual precipitation. Sedimentation in these and other reservoirs can seriously impact their capacities and will eventually eliminate any benefit the reservoirs have provided. For example, Lake Powell has a capacity of about 26,000,000 ac-ft with an annual sedimentation rate of about 30,000 ac-ft. Therefore, the theoretical life of Lake Powell is about

$$\frac{2.6 \times 10^7 \text{ ac-ft}}{3 \times 10^4 \text{ } \frac{\text{ac-ft}}{\text{yr}}} = 866 \text{ yr}$$

The actual life of Lake Powell will be longer than this, however, because deposited sediment will compact and because the trap efficiency will decrease over time as the reservoir fills with sediment.

6. NOMENCLATURE

Nomenclature

d	diameter	ft	m
I	inflow	ft³/sec	m³/s
O	outflow	ft³/sec	m³/s
Q	flow	ft³/sec	m³/s
S	storage	ft³	m³
t	time	hr	h
v	velocity	ft/sec	m/s
V	volume	ac-ft	m³

Topic II: Air

Air

23 Sampling and Measurement Methods

Content in blue refers to the *NCEES Handbook*.

1. INTRODUCTION

Air quality monitoring involves three different categories of measurements: emissions, ambient air, and meteorological. Emissions measurements apply to both stationary and mobile sources and are taken by sampling or monitoring at the point where the emission leaves the source. Ambient air is measured to provide a "control" against which to compare emissions monitoring results and to assess pollutant levels. Meteorological measurements include wind speed and direction, air temperature profiles, and other parameters necessary to evaluate pollutant dispersion and fate.

Monitoring may collect short-duration, single-measurement grab samples or longer-term, continuous samples. The type of sample is dictated by the pollutant being measured, the pollutant source, and the objective of the sampling event.

2. PARTICULATE MATTER

Particulate monitors are of three general types: filtering devices, impactor devices, and photometric devices.

The most common filtering device is the *high-volume* or *hi-vol sampler*. The sampler, typically operated over a period of 24 h or less, uses a blower to draw a large volume of air across a filter material. For gross measurements of total particulate, a single filter may be used. If it is necessary to classify particulate by particle size, stacked filters with openings selected to provide the desired particle size distribution are used. A hi-vol sampler is illustrated in Fig. 23.1.

Impactor (impingement) samplers rely on velocity, air flow geometry, and individual particle mass to sample and characterize particulate matter. The sampler works in a cascading fashion, with successively larger particles being collected on a membrane

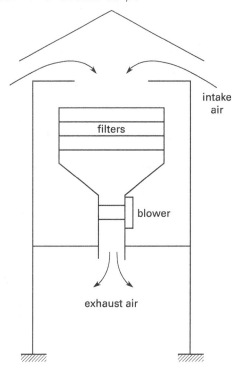

Figure 23.1 *Hi-Vol Particulate Sampler*

or film in successive stages. The air flow through each stage occurs at successively higher velocities so the higher-mass particles are unable to follow the gas flow-path and, therefore, impinge on the membrane. Impactor samplers are used for sampling particulate matter with diameters below 10 μm and are especially effective for particles with diameters less than 2.5 μm. An impactor sampler is illustrated in Fig. 23.2.

Photometric devices provide an indirect measurement of particles by correlating particulate concentration with light intensity. The air stream is passed between a light beam and detector. As concentrations of particulate increase, the light becomes more scattered, and the light intensity measured by the detector is less. Photometric devices offer the advantage of having a continuous real-time read-out, but they do not differentiate among particle sizes.

Figure 23.2 *Impactor Particulate Sampler*

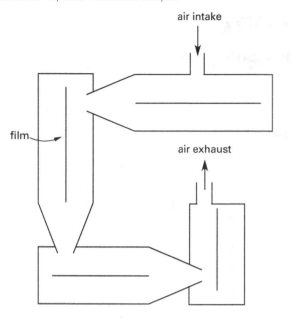

Example 23.1

Particulate sampling produced the following results for a representative 24 h period.

particle size collected, μm	less than 10	less than 2.5
clean filter mass, g	8.4910	8.7391
filter mass at 24 h, g	8.7818	8.9461
initial air flow, m³/min	1.5	1.5
final air flow, m³/min	1.46	1.47

What are the PM_{10} and $PM_{2.5}$ concentrations?

Solution

The average air flow for PM_{10} is

$$\frac{1.5 \ \dfrac{m^3}{min} + 1.46 \ \dfrac{m^3}{min}}{2} = 1.48 \ m^3/min$$

$$PM_{10} = \frac{(8.7818 \ g - 8.4910 \ g)\left(10^6 \ \dfrac{\mu g}{g}\right)}{\left(1.48 \ \dfrac{m^3}{min}\right)(24 \ h)\left(60 \ \dfrac{min}{h}\right)}$$

$$= 136 \ \mu g/m^3$$

The average air flow for $PM_{2.5}$ is

$$\frac{1.5 \ \dfrac{m^3}{min} + 1.47 \ \dfrac{m^3}{min}}{2} = 1.485 \ m^3/min$$

$$PM_{2.5} = \frac{(8.9461 \ g - 8.7391 \ g)\left(10^6 \ \dfrac{\mu g}{g}\right)}{\left(1.485 \ \dfrac{m^3}{min}\right)(24 \ h)\left(60 \ \dfrac{min}{h}\right)}$$

$$= 97 \ \mu g/m^3$$

3. GASES

Unlike particulate matter, gases do not typically require long collection periods, being well-suited to detection by monitoring instruments that can provide continuous-flow measurements with instantaneous results. Devices employed for gas monitoring are typically specific to only one gas, so multiple instruments may be required to monitor gas mixtures. Monitoring of gaseous emissions employs wet chemistry techniques; visible, infra-red, and ultraviolet light analyzers; and electrochemical analyzers. Alternatively, samples can be collected in appropriate containers and analyzed using gas chromatography. Gas chromatography is useful when there are several compounds with similar chemical structures such as N_2, O_2, CO_2, CO, and CH_4, or such as VOCs.

4. SMOKE

Smoke presents an observable, inexpensive, and unobtrusive opportunity to assess emissions from stacks. Regulatory agency staff, plant operations personnel, and other interested persons can become certified to provide quantitative measurements of smoke opacity based solely on observation.

Observations of qualified persons are compared to opacity standards represented by the *Ringelmann scale*. The Ringelmann scale uses a percent opacity from completely transparent (Ringelmann No. 0) to completely opaque (Ringelmann No. 5) and is applied to black or dark smoke. For white smoke, opacity is expressed as "percent opacity" rather than by Ringelmann number. Each Ringelmann number corresponds to 20% opacity, so 5% opacity, the accuracy with which opacity observations are typically reported, would represent $\frac{1}{4}$ Ringelmann number. The Ringelmann scale is summarized in Fig. 23.3.

Figure 23.3 *Ringelmann Scale For Smoke Opacity*

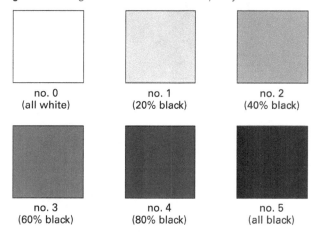

| no. 0 (all white) | no. 1 (20% black) | no. 2 (40% black) |
| no. 3 (60% black) | no. 4 (80% black) | no. 5 (all black) |

5. AIR QUALITY INDEX

The *air quality index* (AQI) is a scale used to quantify and report daily air quality to the public. The index was devised by the EPA to replace the formerly used *pollution standard index* (PSI). The AQI considers five of the six criteria pollutants identified by the NAAQS: ground-level ozone, particulates ($PM_{2.5}$ and PM_{10}), carbon monoxide, sulfur dioxide, and nitrogen dioxide. Lead is not included in the AQI. Data from these pollutants are condensed into a single number that is scaled down to a qualitative descriptor of air quality as shown in Table 23.1. As the AQI increases, public health officials may issue public health advisories and warnings and may impose restrictions on industrial and transportation activities that contribute to air pollution.

Table 23.1 *Air Quality Index (AQI) Categories*

AQI value	air quality descriptor
0–50	good
51–100	moderate
101–150	unhealthy for sensitive groups
151–200	unhealthy
201–300	very unhealthy
301–400	hazardous

Source: U.S. Environmental Protection Agency (40 CFR Part 58)

Each AQI category designates upper and lower limits for the concentration of each pollutant. These concentrates are referred to as the *breakpoints*, and are the maximum and minimum acceptable concentrations in each category. The EPA publishes a table of breakpoint values for each pollutant and AQI category. [Breakpoints for the AQI]

The AQI is the highest of the subindices, I_p, calculated for each pollutant from the following procedure.

step 1: Truncate the highest pollutant concentration, C_p, observed during the reporting period as follows.

ozone—truncate to 3 decimal places
$PM_{2.5}$—truncate to 1 decimal place
PM_{10}—truncate to an integer
CO—truncate to 1 decimal place
SO_2—truncate to an integer
NO_2—truncate to an integer

step 2: Using Eq. 23.1, find the high and low breakpoints, BP, and the high and low indices, I, that contain the truncated observed concentration, C_p.

step 3: Calculate the subindex from Eq. 23.1.

AQI Calculation

$$I_p = \frac{I_{Hi} - I_{Lo}}{BP_{Hi} - BP_{Lo}}(C_p - BP_{Lo}) + I_{Lo} \quad \textbf{23.1}$$

step 4: Round the subindex to the nearest integer.

Example 23.2

The maximum concentrations of air pollutants measured in a metropolitan area during a single day are as follows.

pollutant	duration (h)	concentration
O_3	1	0.14 ppm
CO	8	8 ppm
PM_{10}	24	190 $\mu g/m^3$
SO_2	1	0.07 ppm
NO_2	1	0.4 ppm

What is the day's AQI value and descriptor?

Solution

The concentrations for SO_2 and NO_2 are reported in ppm. In order to calculate their subindices, those concentrations must be converted to ppb.

$$C_{p,SO_2,ppb} = C_{p,SO_2,ppm}\left(1000\ \frac{ppb}{ppm}\right)$$
$$= (0.07\ ppm)\left(1000\ \frac{ppb}{ppm}\right)$$
$$= 70\ ppb$$
$$C_{p,NO_2,ppb} = C_{p,NO_2,ppm}\left(1000\ \frac{ppb}{ppm}\right)$$
$$= (0.4\ ppm)\left(1000\ \frac{ppb}{ppm}\right)$$
$$= 400\ ppb$$

Air

The 1-hour O_3 concentration is 0.14 ppm. From the EPA's table of breakpoints, this value is within the C_p range of 0.125–0.164 ppm, corresponding to the subindex range of 101–150. [Breakpoints for the AQI]

From Eq. 23.1, the O_3 subindex is

AQI Calculation

$$I_{p,O_3} = \frac{I_{Hi} - I_{Lo}}{BP_{Hi} - BP_{Lo}}(C_p - BP_{Lo}) + I_{Lo}$$

$$= \left(\frac{150 - 101}{0.164 - 0.125}\right)(0.14 - 0.125) + 101$$

$$= 120$$

Similarly,

$$I_{p,CO} = \left(\frac{100 - 51}{9.4 - 4.5}\right)(8 - 4.5) + 51 = 86$$

$$I_{p,PM_{10}} = \left(\frac{150 - 101}{254 - 155}\right)(190 - 155) + 101 = 118$$

$$I_{p,SO_2} = \left(\frac{100 - 51}{75 - 36}\right)(70 - 36) + 51 = 94$$

$$I_{p,NO_2} = \left(\frac{200 - 151}{649 - 361}\right)(400 - 361) + 151 = 158$$

The highest subindex is 158 for NO_2, so 158 is the AQI. From Table 23.1, this corresponds to the category "unhealthy."

6. NOMENCLATURE

BP	breakpoint	–	–
C	concentration	ppb	ppb
C	pollutant concentration	$\mu g/m^3$	$\mu g/m^3$
I	index value	–	–
PM	particulate matter	$\mu g/m^3$	$\mu g/m^3$

Subscripts

2.5	particle size in μg
10	particle size in μg
Hi	high
Lo	low
p	pollutant

24 Codes, Standards, Regulations, Guidelines

1. STATUTES AND REGULATIONS

Clean Air Act and Amendments

Air quality issues have been addressed by federal legislation since 1955, and many updated statutes and amendments have followed. The statute providing regulatory authority for national air quality issues is the Clean Air Act (CAA) and amendments. Federal air pollution regulations address air quality issues through a variety of enforcement and incentive mechanisms.

The National Ambient Air Quality Standards (NAAQS) attempt to implement a uniform national air pollution control program. For each pollutant included in the NAAQS, primary and secondary standards are defined. *Primary standards* are intended to protect human health, and *secondary standards* are intended to protect public welfare. The NAAQS are applied to a limited number of substances, known as the *criteria pollutants*. The NAAQS and the criteria pollutants to which they apply are presented in Table 24.1.

To ensure application of the NAAQS in each of the 50 states, the United States Environmental Protection Agency (EPA) has required the states to prepare and submit State Implementation Plans (SIPs). Each SIP must address a control strategy for each of the criteria pollutants included in the NAAQS. If a state is unable or unwilling to submit an SIP acceptable to the EPA, the EPA has the authority to develop an SIP for the state and then require the state to enforce it.

Prevention of significant deterioration (PSD) is a program that is designed to protect air quality in clean-air areas that already meet the NAAQS. To apply the PSD principle, the EPA has developed a regional classification system. Class I regions, generally including national monuments and parks and national wilderness areas, allow very little air quality deterioration and, thereby, limit economic development. Class II regions allow moderate air quality deterioration, and class III regions allow deterioration to the secondary NAAQS.

Because of geographic conditions, population density, climate, and other factors, some areas in the United States are unable to meet the NAAQS. These areas are designated as *nonattainment areas* (NA). Any new or modified sources in nonattainment areas are required to control emissions at the *lowest achievable emission rate* (LAER). Recognizing the economic problems posed by the CAA in nonattainment areas, the EPA has allowed some flexibility, through *emissions trading*, in how industry goes about meeting the compliance requirements. Emissions trading can occur by the following methods.

- *emission reduction credit:* By applying a higher level of treatment than required by regulation, businesses earn credits that can be redeemed through other CAA programs.

- *offset:* The offset allows industrial expansion by letting businesses compensate for new emissions by offsetting them with credits acquired by other businesses already existing in the region.

- *bubble:* All activities in a single plant or among a group of proximate industries can emit at various rates as long as the resulting total emission does not exceed the allowable emission for an individual source.

- *netting:* Businesses can expand without acquiring a new permit as long as the net increase in emissions is not significant. Consequently, by applying improved control technology, businesses can earn credits that can be subsequently applied to plant expansion with no net increase in emissions.

New source performance standards (NSPS) are applied to specific source categories (e.g., asphalt concrete plants, incinerators) to prevent the emergence of other air pollution problems from new construction. The NSPS are intended to reflect the best emission control measures achievable at reasonable cost, and are to be applied during initial construction of new industrial facilities.

National Emission Standards for Hazardous Air Pollutants (NESHAPs) focus on hazardous pollutants not included as criteria pollutants. *Hazardous pollutants* are compounds that pose particular, usually localized, hazards to the exposed population. On a localized level, their impact is more severe than that of the criteria pollutants. Compounds identified as hazardous pollutants are limited to asbestos, inorganic arsenic, benzene,

Table 24.1 National Ambient Air Quality Standards[a]

pollutant	time basis	primary standard[b]	secondary standard
carbon monoxide (CO)	8 h	9 ppm (10 mg/m^3)	none
	1 h	35 ppm (40 mg/m^3)	none
nitrogen dioxide (NO$_2$)	1 h	100 ppb	none
	mean annual	53 ppb (100 μg/m^3)	same
ground-level ozone (O$_3$)	8 h	0.075 ppm (147 μg/m^3)	same
sulfur dioxide (SO$_2$)	3 h	none	0.50 ppm (1300 μg/m^3)
	1 h	0.075 ppm	none
PM$_{10}$	24 h	150 μg/m^3	same
PM$_{2.5}$	annual	12 μg/m^3	15 μg/m^3
	24 h	35 μg/m^3	same
lead (Pb)	3 mo	0.15 μg/m^3	same
hydrocarbons[c]	–	–	–

[a]Typical values. The NAAQS are subject to ongoing legislation, and current values may be revised from those presented here.
[b]Equivalents in parentheses may or may not be part of the NAAQS.
[c]Hydrocarbon standards have been revoked by the EPA since hydrocarbons, as precursors of ozone, are adequately regulated by ozone standards.
Source: National Ambient Air Quality Standards (NAAQS), U.S. Environmental Protection Agency, epa.gov.

mercury, beryllium, radionuclides, and vinyl chloride. The list is short because onerous demands are placed on the EPA by the CAA for listed pollutants and because a very high level of abatement, based on risk to the exposed population, is required once a pollutant is listed. The 1990 CAA amendments listed 187 other *hazardous air pollutants* (HAPs) whose abatement is based on the *maximum achievable control technology* (MACT), a less demanding criterion than risk. Many states have adopted their own regulations to designate and control other hazardous air pollutants.

The 1990 CAA amendments placed limits on allowable emissions of sulfur oxides (SOx) and nitrogen oxides (NOx)—two categories of compounds associated with acid rain. These limits require acid deposition emission controls, which are implemented through a system of emission allowances applied to fossil fuel-fired boilers in the 48 contiguous states. Control options may include such things as use of low-sulfur fuels, flue-gas desulfurization, and replacing older facilities.

The 1977 CAA amendments authorized the EPA to regulate any substance or practice with reasonable potential to damage the stratosphere, especially with regard to ozone. In response to this regulatory authority, the EPA has closely regulated the production and sale of chlorinated fluorocarbons (CFCs) through out-and-out bans on their production and use, taxes on emissions from their use, and a permitting program.

Resource Conservation and Recovery Act (RCRA)

Regulations promulgated under the Resource Conservation and Recovery Act (RCRA) impose restrictions on emissions from hazardous waste incinerators. The regulations define *principal organic hazardous constituents* (POHCs) present in the waste feed and establish performance standards for their destruction and removal in the incineration process.

For most wastes, the performance standard for POHCs is defined by a *destruction and removal efficiency* (DRE) of 99.99% (referred to as "four nines") for each POHC in the waste feed. However, for selected wastes, primarily those associated with dioxins and furans, a DRE of 99.9999% ("six nines") is required. The DRE includes the destruction efficiency of the incinerator as well as the removal efficiency of any air pollution control equipment. The DRE is calculated from Eq. 24.1.

Incineration

$$DRE = \frac{W_{in} - W_{out}}{W_{in}} \times 100\%$$ **24.1**

The RCRA regulations also apply to hydrochloric acid (HCl) and particulate emissions. The HCl emissions are based on a mass emission rate of 1.8 kg/h, and the particulate emissions corrected for oxygen (21% oxygen by volume in the combustion air) are calculated from Eq. 24.2.

$$P_c = P_m\left(\frac{14}{21 - Y}\right)$$ **24.2**

Example 24.1

The POHC feed rate to an incinerator is 2.73 kg/h. Stack emissions monitoring shows the POHC at 0.011 kg/h, particulate at 176 mg/m^3, and O_2 at 5%. (a) What is the DRE for the POHC? (b) What is the corrected particulate matter concentration?

Solution

(a) Use Eq. 24.1.

Incineration

$$DRE = \frac{W_{in} - W_{out}}{W_{in}} \times 100\%$$

$$= \frac{2.73 \, \dfrac{kg}{h} - 0.011 \, \dfrac{kg}{h}}{2.73 \, \dfrac{kg}{h}} \times 100\%$$

$$= 99.60\%$$

(b) Use Eq. 24.2.

$$P_c = P_m \left(\frac{14}{21 - Y} \right) = \left(176 \, \frac{mg}{m^3} \right) \left(\frac{14}{21 - 5} \right)$$

$$= 154 \, mg/m^3$$

2. NOMENCLATURE

DRE	destruction and removal efficiency	%
P	particulate matter concentration	kg/m^3
W_{in}	mass feed rate of a particular POHC	kg/s
W_{out}	mass emission rate of the same POHC	kg/s
Y	oxygen concentration measured in stack gas	%

Subscripts

c	corrected
m	measured

25 Chemistry

1. TEMPERATURE: *T*

Temperature is a thermodynamic property of a substance that depends on internal energy content. Heat energy entering a substance will increase the temperature of that substance. Normally, heat energy will flow only from a hot object to a cold object. If two objects are in *thermal equilibrium* (i.e., are at the same temperature), no heat will flow between them.

If two systems are in thermal equilibrium, they must be at the same temperature. If both systems are in equilibrium with a third, then all three are at the same temperature. This concept is known as the *Zeroth Law of Thermodynamics*.

The scales most commonly used for measuring temperature are the Fahrenheit and Celsius scales.[1] The relationship between these two scales is

Temperature Conversions

$$°F = 1.8(°C) + 32 \qquad 25.1$$

$$°C = (°F - 32)/1.8 \qquad 25.2$$

The *absolute temperature scale* defines temperature independently of the properties of any particular substance. This is unlike the Celsius and Fahrenheit scales,

which are based on the freezing point of water. The absolute temperature scale should be used for all calculations.

In the customary U.S. system, the absolute scale is the *Rankine scale*.[2]

Temperature Conversions

$$°R = °F + 459.69 \qquad 25.3$$

Although the Fahrenheit and Rankine scales are numbered differently, the temperature change represented by a degree is the same. Therefore, a temperature *difference* of a given number of degrees is the same on the Fahrenheit and Rankine scales, even though the individual temperatures involved have different numerical values.

$$\Delta T_{°R} = \Delta T_{°F} \qquad 25.4$$

The absolute temperature scale in the SI system is the *Kelvin scale*.[3]

Temperature Conversions

$$K = °C + 273.15 \qquad 25.5$$

As with the Fahrenheit and Rankine scales, the temperature changes represented by one degree Celsius and one kelvin are the same. A temperature difference of a given number of degrees Celsius is equal to a difference of the same number of kelvins.

$$\Delta T_K = \Delta T_{°C} \qquad 25.6$$

The relationships between temperature differences in the customary U.S. and SI systems are independent of the freezing point of water. (See Table 25.1.)

$$\Delta T_{°C} = \tfrac{5}{9}\Delta T_{°F} \qquad 25.7$$

$$\Delta T_K = \tfrac{5}{9}\Delta T_{°R} \qquad 25.8$$

---

[1]The term *centigrade* was replaced by the term *Celsius* in 1948.
[2]Normally, three significant temperature digits (i.e., 460°) are sufficient.
[3]Normally, three significant temperature digits (i.e., 273°) are sufficient.

Table 25.1 *Temperature Scales*

	Kelvin	Celsius	Rankine	Fahrenheit
normal boiling point of water	373.15K	100.00°C	671.67°R	212.00°F
triple point of water	273.16K	0.01°C	491.69°R	32.02°F
	273.15K	0.00°C	491.67°R	32.00°F ice point
absolute zero	0K	−273.15°C	0°R	−459.67°F

2. PRESSURE: *p*

Customary U.S. pressure units are pounds per square inch (psi). Standard SI pressure units are kPa or MPa, although bars are also used in tabulations of thermodynamic data.

3. DENSITY: ρ

Customary U.S. density units in tabulations of thermodynamic data are pounds per cubic foot (lbm/ft^3). Standard SI density units are kilograms per cubic meter (kg/m^3). Density is the reciprocal of specific volume.

$$\rho = \frac{1}{v} \qquad 25.9$$

4. SPECIFIC VOLUME: *v* AND *V*

Specific volume, v, is the volume occupied by one unit mass of a substance.

State Functions (properties)

$$v = V/m \qquad 25.10$$

Customary U.S. units in tabulations of thermodynamic data are cubic feet per pound (ft^3/lbm). Standard SI specific volume units are cubic meters per kilogram (m^3/kg). *Molar specific volume*, V, with units of ft^3/lbmol (m^3/kmol), is the volume of a mole of the substance, but is seldom encountered. Specific volume is the reciprocal of density.

$$v = \frac{1}{\rho} \qquad 25.11$$

5. SPECIFIC HEAT: *c* AND *C*

An increase in internal energy is needed to cause a rise in temperature. Different substances differ in the quantity of heat needed to produce a given temperature increase. The ratio of the energy (heat), Q, added to or removed from a body, to the body's mass, m, and temperature change, ΔT, is known as the *specific heat (specific heat capacity)*, c, of the body.

Because specific heats of solids and liquids are slightly temperature dependent, the mean specific heats are used when evaluating processes covering a large temperature range.

State Functions (properties)

$$Q = mc\Delta T \qquad 25.12$$

State Functions (properties)

$$c = \frac{Q}{m\Delta T} \qquad 25.13$$

The lowercase c implies that the units are Btu/lbm-°F or J/kg·°C. Typical values of specific heat are given in Table 25.2. The *molar specific heat*, designated by the symbol C, has units of Btu/lbmol-°F or J/kmol·°C.

$$C = M \times c \qquad 25.14$$

For gases, the specific heat depends on the type of process during which the heat exchange occurs. Specific heats for constant-volume and constant-pressure processes are designated by c_v and c_p, respectively.

State Functions (properties)

$$c_v = \left(\frac{\partial u}{\partial T}\right)_v \qquad 25.15$$

State Functions (properties)

$$c_p = \left(\frac{\partial h}{\partial T}\right)_P \qquad 25.16$$

Approximate values of c_p and c_v for common gases are given in Table 25.3. c_v and c_p for solids and liquids are essentially the same. However, the designation c_p is often encountered for solids and liquids.

The law of *Dulong and Petit* predicts the approximate molar specific heat (in cal/mol·°C) at high temperatures from the atomic weight.[4] This law is valid for solid elements having atomic weights greater than 40 and for

---

[4]Dulong and Petit's law becomes valid at different temperatures for different substances, and a more specific definition of "high temperature" is impossible. For lead, the law is valid at 200K. For copper, it is not valid until above 400K.

Table 25.2 *Approximate Specific Heats of Selected Liquids and Solids**

substance	c_p Btu/lbm-°F or Btu/lbm-°R	c_p kJ/kg·°C or kJ/kg·K
aluminum, pure	0.23	0.96
aluminum, 2024-T4	0.2	0.84
ammonia	1.16	4.86
asbestos	0.20	0.84
benzene	0.41	1.72
brass, red	0.093	0.39
bronze	0.082	0.34
concrete	0.21	0.88
copper, pure	0.094	0.39
Freon-12	0.24	1.00
gasoline	0.53	2.20
glass	0.18	0.75
gold, pure	0.031	0.13
ice	0.49	2.05
iron, pure	0.11	0.46
iron, cast (4% C)	0.10	0.42
lead, pure	0.031	0.13
magnesium, pure	0.24	1.00
mercury	0.033	0.14
oil, light hydrocarbon	0.5	2.09
silver, pure	0.06	0.25
steel, 1010	0.10	0.42
steel, stainless 301	0.11	0.46
tin, pure	0.055	0.23
titanium, pure	0.13	0.54
tungsten, pure	0.032	0.13
water	1.0	4.19
wood (typical)	0.6	2.50
zinc, pure	0.088	0.37

(Multiply Btu/lbm-°F by 4.1868 to obtain kJ/kg·°C or kJ/kg·K.)

*Values in cal/g·°C are the same as Btu/lbm-°F.

most metallic elements. It is not valid at room temperature for carbon, silicon, phosphorus, and sulfur. 6.3 cal/mol·°C is known as the *Dulong and Petit value*.

$$c \times \text{AW} \approx 6.3 \pm 0.1 \qquad 25.17$$

Example 25.1

Compare the value of specific heat of pure iron calculated from Dulong and Petit's law with the value from Table 25.2.

Solution

The atomic weight of iron is 55.8. From Eq. 25.17,

$$c = \frac{6.3}{\text{AW}} = \frac{6.3 \ \dfrac{\text{cal}}{\text{mol·°C}}}{55.8 \ \dfrac{\text{g}}{\text{mol}}} = 0.11 \ \text{cal/g·°C}$$

This is the same value as is given in Table 25.2.

6. LATENT HEATS

The total energy (*total heat*, Q_t) entering a substance is the sum of the energy that changes the phase of the substance (*latent heat*, Q_l) and energy that changes the temperature of the substance (*sensible heat*, Q_s). During a phase change (solid to liquid, liquid to vapor, etc.), energy will be transferred to or from the substance without a change in temperature.[5]

$$Q_t = Q_s + Q_l \qquad 25.18$$

Examples of latent energies are the *latent heat of fusion* (i.e., change from solid to liquid), h_{sl}, *latent heat of vaporization*, h_{fg}, and *latent heat of sublimation* (i.e., direct change from solid to vapor without becoming liquid), h_{ig}.[6,7] The energy required for these latent changes to occur in water is given in Table 25.4.

Example 25.2

How much energy is required to convert 1.0 lbm (0.45 kg) of water that is originally at 75°F (24°C) and 1 atm to vapor at 212°F (100°C) and 1 atm?

SI Solution

The sensible heat required to raise the temperature of the water from 24°C to 100°C is given by Eq. 25.12.

---

[5]Changes in crystalline form are also latent changes.
[6]The subscript s (for "solid") is sometimes used in place of i (for "ice").
[7]*Sublimation* can only occur below the triple point, where it is too cold for the liquid phase to exist at all.

Table 25.3 Approximate Properties of Selected Gases

gas	chemical formula	temperature (°F)	M	customary U.S. units			SI units			k
				R ft-lbf / lbm-°R	c_p Btu / lbm-°R	c_v Btu / lbm-°R	R J / kg·K	c_p J / kg·K	c_v J / kg·K	
acetylene	C_2H_2	68	26.038	59.35	0.350	0.274	319.32	1465	1146	1.279
air		100	28.967	53.35	0.240	0.171	287.03	1005	718	1.400
ammonia	NH_3	68	17.032	90.73	0.523	0.406	488.16	2190	1702	1.287
argon	Ar	68	39.944	38.69	0.124	0.074	208.15	519	311	1.669
butane (-n)	C_4H_{10}	68	58.124	26.59	0.395	0.361	143.04	1654	1511	1.095
carbon dioxide	CO_2	100	44.011	35.11	0.207	0.162	188.92	867	678	1.279
carbon monoxide	CO	100	28.011	55.17	0.249	0.178	296.82	1043	746	1.398
chlorine	Cl_2	100	70.910	21.79	0.115	0.087	117.25	481	364	1.322
ethane	C_2H_6	68	30.070	51.39	0.386	0.320	276.50	1616	1340	1.206
ethylene	C_2H_4	68	28.054	55.08	0.400	0.329	296.37	1675	1378	1.215
Freon (R-12)*	CCl_2F_2	200	120.925	12.78	0.159	0.143	68.76	666	597	1.115
helium	He	100	4.003	386.04	1.240	0.744	2077.03	5192	3115	1.667
hydrogen	H_2	100	2.016	766.53	3.420	2.435	4124.18	14 319	10 195	1.405
hydrogen sulfide	H_2S	68	34.082	45.34	0.243	0.185	243.95	1017	773	1.315
krypton	Kr		83.800	18.44	0.059	0.035	99.22	247	148	1.671
methane	CH_4	68	16.043	96.32	0.593	0.469	518.25	2483	1965	1.264
neon	Ne	68	20.183	76.57	0.248	0.150	411.94	1038	626	1.658
nitrogen	N_2	100	28.016	55.16	0.249	0.178	296.77	1043	746	1.398
nitric oxide	NO	68	30.008	51.50	0.231	0.165	277.07	967	690	1.402
nitrous oxide	N_2O	68	44.01	35.11	0.221	0.176	188.92	925	736	1.257
octane vapor	C_8H_{18}		114.232	13.53	0.407	0.390	72.78	1704	1631	1.045
oxygen	O_2	100	32.000	48.29	0.220	0.158	259.82	921	661	1.393
propane	C_3H_8	68	44.097	35.04	0.393	0.348	188.55	1645	1457	1.129
sulfur dioxide	SO_2	100	64.066	24.12	0.149	0.118	129.78	624	494	1.263
water vapor*	H_2O	212	18.016	85.78	0.445	0.335	461.50	1863	1402	1.329
xenon	Xe		131.300	11.77	0.038	0.023	63.32	159	96	1.661

(Multiply Btu/lbm-°F by 4186.8 to obtain J/kg·K.)
(Multiply ft-lbf/lbm-°R by 5.3803 to obtain J/kg·K.)

*Values for steam and Freon are approximate and should be used only for low pressures and high temperatures.

Table 25.4 Latent Heats for Water at One Atmosphere

effect	Btu/lbm	kJ/kg	cal/g
fusion	143.4	333.5	79.7
vaporization	970.1	2256.5	539.0
sublimation	1220	2838	677.8

(Multiply Btu/lbm by 2.326 to obtain kJ/kg.)
(Multiply Btu/lbm by 5/9 to obtain cal/g.)

State Functions (properties)

$$Q_s = mc(T_2 - T_1)$$
$$= (0.45 \text{ kg})\left(4.190 \ \frac{\text{kJ}}{\text{kg·°C}}\right)(100°C - 24°C)$$
$$= 143.3 \text{ kJ}$$

From Table 25.4, the latent heat required to vaporize the water is

$$Q_l = mh_{fg} = (0.45 \text{ kg})\left(2256.5 \ \frac{\text{kJ}}{\text{kg}}\right) = 1015.4 \text{ kJ}$$

Air

The total heat required is

$$Q_t = Q_s + Q_l = 143.3 \text{ kJ} + 1015.4 \text{ kJ}$$
$$= 1158.7 \text{ kJ}$$

Customary U.S. Solution

The sensible heat required to raise the temperature of the water from 75°F to 212°F is given by Eq. 25.12.

State Functions (properties)

$$Q_s = mc(T_2 - T_1)$$
$$= (1 \text{ lbm})\left(1.0 \ \frac{\text{Btu}}{\text{lbm-°F}}\right)(212°F - 75°F)$$
$$= 137.0 \text{ Btu}$$

From Table 25.4, the latent heat required to vaporize the water is

$$Q_l = mh_{fg} = (1 \text{ lbm})\left(970.1 \ \frac{\text{Btu}}{\text{lbm}}\right) = 970.1 \text{ Btu}$$

The total heat required is

$$Q_t = Q_s + Q_l = 137.0 \text{ Btu} + 970.1 \text{ Btu}$$
$$= 1107.1 \text{ Btu}$$

7. GIBBS FUNCTION: *g* AND *G*

The *Gibbs function* is defined for a pure substance by Eq. 25.19 through Eq. 25.21.

$$g = h - Ts = u + pv - Ts \qquad 25.19$$

$$G = H - TS = U + pV - TS \qquad 25.20$$

$$G = \text{MW} \times g \qquad 25.21$$

The Gibbs function is used in investigating latent changes and chemical reactions. For a constant-temperature, constant-pressure nonflow process approaching equilibrium, the Gibbs function approaches its minimum value.

$$(dG)_{T,p} < 0 \qquad 25.22$$

Once the minimum value is obtained, the process will stop, and the Gibbs function will be constant.

$$(dG)_{T,p} = 0 \ |_{\text{equilibrium}} \qquad 25.23$$

Like enthalpy of formation, the Gibbs function, G^0, has been tabulated at the standard reference conditions of 25°C (77°F) and one atmosphere. A chemical reaction can occur spontaneously only if the change in Gibbs function is negative (i.e., the Gibbs function for the products is less than the Gibbs function for the reactants).

$$\sum_{\text{products}} nG^0 < \sum_{\text{reactants}} nG^0 \qquad 25.24$$

8. MASS, VOLUMETRIC, AND MOLE FRACTIONS

The *mass fraction*, y_i (also known as the *gravimetric fraction* and *weight fraction*), of a component i in a mixture is the ratio of the component's mass to the total mixture mass.

Ideal Gas Mixtures

$$y_i = m_i/m; \ m = \sum m_i; \sum y_i = 1 \qquad 25.25$$

The *volumetric fraction*, V_i/V, of a component i is the ratio of the component's partial volume to the overall mixture volume. It is possible to convert between mass fractions and volumetric fractions.

$$y_i = \frac{\left(\dfrac{V_i}{V}\right)M_i}{\sum\left(\dfrac{V_i}{V}\right)M_i} \qquad 25.26$$

$$\frac{V_i}{V} = \frac{\dfrac{y_i}{M_i}}{\sum\dfrac{y_i}{M_i}} \qquad 25.27$$

The *mole fraction*, x_i, of a component i is the ratio of the number of moles of substance i to the total number of moles of all substances. Mole fractions can be converted to mass fractions using Eq. 25.28.

Ideal Gas Mixtures

$$x_i = N_i/N; \ N = \sum N_i; \sum x_i = 1 \qquad 25.28$$

Mole fractions can be converted to mass fractions with the equation

Ideal Gas Mixtures

$$y_i = \frac{x_i M_i}{\sum(x_i M_i)} \qquad 25.29$$

Mass fractions can be converted to mole fractions with the equation

Ideal Gas Mixtures

$$x_i = \frac{y_i/M_i}{\sum(y_i/M_i)} \qquad 25.30$$

9. INTRODUCTION TO PSYCHROMETRICS

Atmospheric air contains small amounts of moisture and can be considered to be a mixture of two ideal gases—dry air and water vapor. All of the thermodynamic rules relating to the behavior of nonreacting gas mixtures apply to atmospheric air. From Dalton's law, for example, the total atmospheric pressure is the sum of the dry air partial pressure and the water vapor pressure.[8]

$$p = p_a + p_w \qquad 25.31$$

The study of the properties and behavior of atmospheric air is known as *psychrometrics*. Properties of atmospheric air are seldom evaluated, however, from theoretical thermodynamic principles. Rather, specialized techniques and charts have been developed for that purpose.

10. PROPERTIES OF ATMOSPHERIC AIR

At first, psychrometrics seems complicated by three different definitions of temperature. These three terms are not interchangeable.

- *dry-bulb temperature*, T_{db}: This is the equilibrium temperature that a regular thermometer measures if exposed to atmospheric air.

- *wet-bulb temperature*, T_{wb}: This is the temperature of air that has gone through an adiabatic saturation process.

- *dew-point temperature*, T_{dp}: This is the dry-bulb temperature at which water starts to condense out when moist air is cooled in a constant pressure process.

For every temperature, there is a unique vapor pressure, P_{sat}, which represents the maximum pressure the water vapor can exert. The actual vapor pressure, P_v, can be less than or equal to, but not greater than, the saturation value. The saturation pressure is found from steam tables as the pressure corresponding to the dry-bulb temperature of the atmospheric air.

$$P_v \leq P_{sat} \qquad 25.32$$

If the vapor pressure equals the saturation pressure, the air is said to be saturated.[9] *Saturated air* is a mixture of dry air and saturated water vapor. When the air is saturated, the dew-point and saturation temperatures are equal.

Psychrometrics

$$T_{dp} = T_{sat} \text{ at } P_g = P_v \qquad 25.33$$

Unsaturated air is a mixture of dry air and superheated water vapor.[10] When the air is unsaturated, the dew-point temperature will be less than the wet-bulb temperature. The *wet-bulb depression* is the difference between the dry-bulb and wet-bulb temperatures.

$$T_{dp} < T_{wb} < T_{db} \quad \text{[unsaturated]} \qquad 25.34$$

The amount of water vapor in atmospheric air is specified by three different parameters. The *humidity ratio*, ω (also known as the *specific humidity*), is the mass ratio of water vapor to dry air. If both masses are expressed in pounds (kilograms), the units of humidity ratio are lbm/lbm (kg/kg). However, since there is so little water vapor, the water vapor mass is often reported in *grains* of water. (There are 7000 grains per pound.) Accordingly, the humidity ratio will have the units of grains per pound.

Psychrometrics

$$\omega = m_v/m_a \qquad 25.35$$

Since $m = \rho V$, and since $V_v = V_a$, the humidity ratio can be written as

$$v = \frac{\rho_v}{\rho_a} \qquad 25.36$$

From the equation of state for an ideal gas, $m = pV/RT$. Since $V_v = V_a$ and $T_v = T_a$, the humidity ratio can also be written as

$$\omega = \frac{R_a P_v}{R_v P_a} = \frac{53.35 P_v}{85.78 P_a} \qquad 25.37$$

This reduces to

Psychrometrics

$$\omega = 0.622 P_v/P_a = 0.622 P_v/(P - P_v) \qquad 25.38$$

The *degree of saturation*, μ (also known as the *saturation ratio* and the *percentage humidity*), is the ratio of the actual humidity ratio to the saturated humidity ratio at the same temperature and pressure.

$$\mu = \frac{\omega}{\omega_{sat}} \qquad 25.39$$

A third index of moisture content is the *relative humidity*—the partial pressure of the water vapor divided by the saturation pressure.

Psychrometrics

$$\phi = P_v/P_g \qquad 25.40$$

---

[8]Equation 25.31 points out a problem in semantics. The term *air* means *dry air*. The term *atmosphere* refers to the combination of dry air and water vapor. It is common to refer to the atmosphere as *moist air*.

[9]Actually, the water vapor is saturated, not the air. However, this particular inconsistency in terms is characteristic of psychrometrics.

[10]As strange as it sounds, atmospheric water vapor is almost always superheated. This can be shown by drawing an isotherm passing through the vapor dome on a p-V diagram. The only place where the water vapor pressure is less than the saturation pressure is in the superheated region.

From the equation of state for an ideal gas, $\rho = P/RT$, so the relative humidity can be written as

$$\phi = \frac{\rho_v}{\rho_{sat}} \qquad 25.41$$

Combining the definitions of specific and relative humidities,

$$\phi = 1.608\omega\left(\frac{P_a}{P_g}\right) \qquad 25.42$$

11. ENERGY CONTENT OF AIR

Since moist air is a mixture of dry air and water vapor, its total enthalpy, h (i.e., energy content), takes both components into consideration. Total enthalpy is conveniently shown on the diagonal scales of the psychrometric chart, but it can also be calculated. As Eq. 25.44 indicates, the reference temperature (i.e., the temperature that corresponds to a zero enthalpy) for the enthalpy of dry air is 0°F (0°C). Steam properties correspond to a low-pressure superheated vapor at room temperature.

$$h_t = h_a + \omega h_w \qquad 25.43$$

$$h_a = c_{p,air}T \approx \left(1.005 \ \frac{kJ}{kg \cdot °C}\right)T_{°C} \qquad \text{[SI]} \quad 25.44(a)$$

$$h_a = c_{p,air}T \approx \left(0.240 \ \frac{Btu}{lbm\text{-}°F}\right)T_{°F} \qquad \text{[U.S.]} \quad 25.44(b)$$

$$h_w = c_{p,water\,vapor}T + h_{fg}$$
$$\approx \left(1.805 \ \frac{kJ}{kg \cdot °C}\right)T_{°C} + 2501 \ \frac{kJ}{kg} \qquad \text{[SI]} \quad 25.45(a)$$

$$h_w = c_{p,water\,vapor}T + h_{fg}$$
$$\approx \left(0.444 \ \frac{Btu}{lbm\text{-}°F}\right)T_{°F} + 1061 \ \frac{Btu}{lbm} \qquad \text{[U.S.]} \quad 25.45(b)$$

12. VAPOR PRESSURE

There are at least six ways of determining the partial pressure, P_v, of the water vapor in the air. The first method, derived from Eq. 25.40, is to multiply the relative humidity, ϕ, by the water's saturation pressure.

The saturation pressure, in turn, is obtained from steam tables as the pressure corresponding to the air's dry-bulb temperature.

$$P_v = \phi P_{g,db} \qquad 25.46$$

A more direct method is to read the saturation pressure (from the steam tables) corresponding to the air's dew-point temperature.

$$P_v = P_{g,dp} \qquad 25.47$$

The third method can be used if water's mole (volumetric) fraction is known.

$$P_v = x_v P = \left(\frac{V_w}{V}\right)P \qquad 25.48$$

The fourth method is to calculate the actual vapor pressure from the empirical *Carrier equation*, valid for customary U.S. units only.[11]

$$P_v = P_{g,wb} = \frac{(P - P_{g,wb})(T_{db} - T_{wb})}{2830 - 1.44\,T_{wb}} \qquad 25.49$$
$$\text{[U.S. only]}$$

The fifth method is based on the humidity ratio and is derived from Eq. 25.38.

Psychrometrics

$$P_v = \frac{P\omega}{0.622 + \omega} \qquad 25.50$$

The sixth (and easiest) method is to read the water vapor pressure from a psychrometric chart. Some, but not all, psychrometric charts have water vapor scales.

Example 25.3

Use the methods described in the previous section to determine the partial pressure of water vapor in standard atmospheric air at 60°F (15°C) dry-bulb and 50% relative humidity.

SI Solution

method 1: From the steam tables, the saturation pressure corresponding to 15°C is 1.7051 kPa. From Eq. 25.46, the partial pressure of the vapor is

$$P_v = \phi P_{g,db} = (0.5)(1.7051 \text{ kPa})$$
$$= 0.853 \text{ kPa}$$

---

[11]Equation 25.49 uses updated constants and is more accurate than the equation originally published by Carrier.

method 2: The dew-point temperature (reading straight across on the psychrometric chart) is approximately 4.5°C. Using interpolation, the saturation pressure from the steam table corresponding to 4.5°C is approximately 0.85 kPa.

method 3: The humidity ratio is 0.006 kg/kg. From Eq. 25.50,

Psychrometrics

$$
\begin{aligned}
P_v &= \frac{P\omega}{0.622 + \omega} \\[2mm]
&= \frac{(101.3 \text{ kPa})\left(0.0053 \ \dfrac{\text{kg}}{\text{kg}}\right)}{0.622 + 0.0053 \ \dfrac{\text{kg}}{\text{kg}}} \\[2mm]
&= 0.856 \text{ kPa}
\end{aligned}
$$

Customary U.S. Solution

method 1: From the steam tables, the saturation pressure corresponding to 60°F is 0.2564 lbf/in². From Eq. 25.46, the partial pressure of the vapor is

$$
\begin{aligned}
P_v &= \phi P_{g,\text{db}} = (0.50)\left(0.2564 \ \frac{\text{lbf}}{\text{in}^2}\right) \\[2mm]
&= 0.128 \text{ lbf/in}^2
\end{aligned}
$$

method 2: The dew-point temperature (reading straight across the psychrometric chart) is approximately 41°F. The saturation pressure from the steam table corresponding to 41°F is approximately 0.127 lbf/in².

method 3: Use the Carrier equation. The wet-bulb temperature of the air is approximately 50°F. From the steam tables, the saturation pressure corresponding to that temperature is 0.1780 lbf/in².

$$
\begin{aligned}
P_v &= P_{g,\text{wb}} - \frac{(P - P_{g,\text{wb}})(T_{\text{db}} - T_{\text{wb}})}{2830 - 1.44\,T_{\text{wb}}} \\[2mm]
&= 0.1780 \ \frac{\text{lbf}}{\text{in}^2} - \frac{\left(14.7 \ \dfrac{\text{lbf}}{\text{in}^2} - 0.1780 \ \dfrac{\text{lbf}}{\text{in}^2}\right)}{2830 - (1.44)(50°\text{F})} \\[2mm]
&\qquad\qquad \frac{\times (60°\text{F} - 50°\text{F})}{} \\[2mm]
&= 0.125 \text{ lbf/in}^2
\end{aligned}
$$

13. PARTIAL PRESSURE AND PARTIAL VOLUME OF GAS MIXTURES

A *gas mixture* consists of an aggregation of molecules of component gases, the molecules of any single component being distributed uniformly and moving as if they alone occupied the space. The *partial volume, V_i,* of a gas i in a mixture of nonreacting gases is the volume that gas i alone would occupy at the temperature and pressure of the mixture. (See Fig. 25.1.)

Ideal Gas Mixtures

$$
V_i = \frac{m_i R_i T}{P} \qquad \textbf{25.51}
$$

Figure 25.1 *Mixture of Ideal Gases*

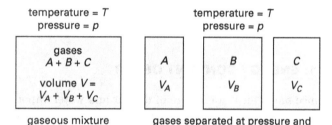

Amagat's law (also known as *Amagat-Leduc's rule*) states that the total volume of a mixture of nonreacting gases is equal to the sum of the partial volumes.

Ideal Gas Mixtures

$$
V = \sum V_i \qquad \textbf{25.52}
$$

The *partial pressure P_i,* of gas i in a mixture of nonreacting gases is the pressure gas i alone would exert in the total volume at the temperature of the mixture.

Ideal Gas Mixtures

$$
P_i = \frac{m_i R_i T}{V} \qquad \textbf{25.53}
$$

According to *Dalton's law of partial pressures,* the *total pressure* of a gas mixture is the sum of the partial pressures.

Ideal Gas Mixtures

$$
P = \sum P_i \qquad \textbf{25.54}
$$

The partial pressure can also be calculated from the mole fraction and the total pressure. However, for ideal gases, the mole fraction, the partial pressure ratio, and volumetric fraction are the same.

Ideal Gas Mixtures

$$
x_i = P_i/P = V_i/V \qquad \textbf{25.55}
$$

If the average specific gas constant, $\overline{R}$, for the gas mixture is known, it can be used with the mass fraction to calculate the partial pressure.

$$
P_i = x_i\left(\frac{R_i P}{\overline{R}}\right) \qquad \textbf{25.56}
$$

14. POLLUTANT AND GAS CHEMISTRY

Gas Laws and Chemistry

Boyle's law states that the volume of a gas varies inversely with its pressure at constant temperature. For any given gas, the product of the pressure and the volume will be constant at constant temperature. The mathematical representation of Boyle's law is

Special Cases of Closed Systems (with no change in kinetic or potential energy)

$$Pv = \text{constant} \qquad 25.57$$

Boyle's law is most useful when expressed in the following form, which is used to find the volume of a gas at a new pressure given a current pressure and volume at constant temperature.

$$P_1 v_1 = P_2 v_2 \qquad 25.58$$

Charles' law states that the volume of a gas at constant pressure varies in direct proportion to the absolute temperature. The ratio of the volume occupied by a gas and the absolute temperature remains constant for any gas. Mathematically, Charles' law is

Special Cases of Closed Systems (with no change in kinetic or potential energy)

$$T/v = \text{constant} \qquad 25.59$$

Charles' law is usually expressed as a ratio of volume and temperature under two different conditions as follows.

$$\frac{v_1}{T_1} = \frac{v_2}{T_2} \qquad 25.60$$

Dalton's law of partial pressures states that in a mixture of gases, such as air, each gas exerts pressure independently of the others. The resulting "partial pressure" of each gas is proportional to the amount (percent by volume or mole fraction) of that gas in the mixture.

Each gas exerts pressure equal to that which it would exert if it were the sole occupant of the volume available to the mixture. For example, in an atmosphere of 80% nitrogen and 20% oxygen at 1.0 atm pressure, the pressure exerted by nitrogen is 0.8 atm and by oxygen is 0.2 atm.

Henry's law states that the amount of any gas that will dissolve in a given volume of a liquid at constant temperature is directly proportional to the pressure that the gas exerts above the liquid.

Henry's Law at Constant Temperature

$$P_i = Py_i = hx_i \qquad 25.61$$

It is common to express Henry's law in terms of *Henry's law constant, h*, as

$$h = \frac{P_i}{x_i} \qquad 25.62$$

The concentration of the gas dissolved in the liquid can be taken as the mole fraction of the gas, x_i. P_i is the gas partial pressure and is usually taken as the vapor pressure, P_v, of the chemical at a known temperature. Tables are available giving values for Henry's law constant for common gases.

The *ideal gas equation* (also called the *universal gas law*) is a combination of Boyle's law and Charles' law and defines a relationship among pressure, volume, and absolute temperature for gases.

Ideal Gas Constants

$$PV = nRT \qquad 25.63$$

The ideal gas equation can be written in a variety of forms, depending on the application. Alternative forms include the following.

$$PM = \rho RT \qquad 25.64$$

$$\frac{P_1 v_1}{T_1} = \frac{P_2 v_2}{T_2} \qquad 25.65$$

The ideal gas equation can also be used to calculate the *molar volume, v/n*, of a gas.

$$\frac{v}{n} = \frac{RT}{P} \qquad 25.66$$

The molecular weight of a gas mixture is a mole- or volumetric-weighted average of the molecular weights of the component gases.

Ideal Gas Mixtures

$$M = \sum x_i M_i \qquad 25.67$$

Example 25.4

A 2.0 m^3 closed tank is filled halfway with water. The tank contains a mixture of nitrogen and oxygen gases at 20°C and 5 atm. The mixture is 72% nitrogen and 28% oxygen by volume. The Henry's law constant for oxygen at 20°C is 40 100 atm. What is the concentration of the oxygen in the water?

Solution

From Dalton's law, the partial pressure of oxygen (O_2) in the tank is $(0.28)(5 \text{ atm}) = 1.4 \text{ atm}$.

Use Eq. 25.61.

Henry's Law at Constant Temperature

$$P_i = Py_i = hx_i$$

$$x_{O_2} = \frac{P_{O_2}}{h_{O_2}}$$

$$= \frac{1.4 \text{ atm}}{40\,100 \text{ atm}}$$

$$= 3.49 \times 10^{-5}$$

x_{O_2} the mole fraction of gas in the liquid.

$$x_{O_2} = 3.5 \times 10^{-5}$$

$$= \frac{n_{gas}}{n_{gas} + n_{liquid}}$$

For 1 L of water,

$$n_{liquid} = \frac{\rho_w}{M_w}$$

$$= \frac{1000 \frac{g}{L}}{18 \frac{g}{mol}}$$

$$= 56 \text{ mol/L}$$

Solve for n_{gas}.

$$x_{O_2} = \frac{n_{gas}}{n_{gas} + n_{liquid}}$$

$$n_{gas} = \frac{x_{O_2} n_{liquid}}{1 - x}$$

$$= \frac{(3.5 \times 10^{-5})\left(56 \frac{mol}{L}\right)}{1 - 3.5 \times 10^{-5}}$$

$$= 0.00196 \text{ mol/L}$$

This can also be expressed in mass units.

$$C_{O_2} = n_{gas} M_{O_2}$$

$$= \left(0.00196 \frac{mol}{L}\right)\left(\left(32 \frac{g}{mol}\right)\left(1000 \frac{mg}{g}\right)\right)$$

$$= 64 \text{ mg/L}$$

Example 25.5

Volatile organic compound (VOC) vapors at 1.3 atm and 28°C are vented from a storage tank into the building where the tank is located. The building volume is 35 000 m³, and the allowable average VOC concentration in the building is 800 μg/m³. The temperature and pressure in the building are 18°C

and 1 atm, respectively. The molecular weight of the VOC is 165 g/mol. What volume of VOC can be vented from the tank without exceeding the allowable VOC concentration?

Solution

The maximum number of moles of VOC that can be vented into the building is

$$n = \frac{\left(800 \frac{\mu g}{m^3}\right)(35\,000 \text{ m}^3)}{\left(165 \frac{g}{mol}\right)\left(10^6 \frac{\mu g}{g}\right)} = 0.17 \text{ mol}$$

Use Eq. 25.63 to calculate the VOC volume allowed in the building, V_2.

Ideal Gas Constants

$$PV = nRT$$

$$V_2 = \frac{nRT_2}{P_2}$$

$$= \frac{(0.17 \text{ mol})\left(8.2 \times 10^{-5} \frac{atm \cdot m^3}{mol \cdot K}\right)}{1 \text{ atm}}$$

$$= 4.1 \times 10^{-3} \text{ m}^3$$

Use Eq. 25.65 to calculate the VOC volume vented that can be from the tank, V_1.

$$\frac{P_1 v_1}{T_1} = \frac{P_2 v_2}{T_2}$$

$$v_1 = \frac{P_2 T_1 v_2}{P_1 T_2}$$

$$= \frac{(1 \text{ atm})(28°C + 273.15°)}{(1.3 \text{ atm})(18°C + 273.15°)}$$

$$= 3.3 \times 10^{-3} \text{ m}^3 \quad (3.3 \text{ L})$$

Example 25.6

By volume, air is a mixture of N_2 at 78%, O_2 at 20.9%, Ar at 0.93%, CO_2 at 0.032%, and many other gases at lesser concentrations. What is the average molecular weight of air?

Solution

Use Eq. 25.67 to calculate the average molecular weight of air.

$$M_{\text{mixture}} = \sum x_i M_i$$

$$M_{\text{air}} = x_{N_2} M_{N_2} + x_{O_2} M_{O_2}$$

$$+ x_{Ar} M_{Ar} + x_{CO_2} M_{CO_2}$$

$$= (0.78)\left(28 \ \frac{g}{mol}\right)$$

$$+ (0.209)\left(32 \ \frac{g}{mol}\right)$$

$$+ (0.0093)\left(40 \ \frac{g}{mol}\right)$$

$$+ (0.00032)\left(44 \ \frac{g}{mol}\right)$$

$$= 28.9 \ \text{g/mol}$$

Transformations and Fate

Photochemical smog is produced from a variety of complex physical and chemical reactions involving nitrogen oxides, carbon monoxide, hydrocarbons, sunlight, and many other factors under favorable conditions. The primary components of photochemical smog are ozone and oxidized hydrocarbons, particularly aldehydes. Also present in photochemical smog are organic nitrates.

Nitric oxide (NO) reacts with oxygen in the air to form nitrogen dioxide (NO_2). Under favorable conditions and in the presence of sunlight, the NO_2 will photolyze back to NO and release a free oxygen atom (O). The atomic oxygen reacts with an oxygen molecule (O_2) to form ozone (O_3). Finally, the O_3 reacts with NO to form NO_2 and O_2. The cycle is shown by the following reactions.

$$2NO + O_2 \rightarrow 2NO_2 \qquad \text{25.68}$$

$$NO_2 + \text{sunlight} \rightarrow NO + O \qquad \text{25.69}$$

$$O + O_2 + N_2 \rightarrow O_3 + N_2 \qquad \text{25.70}$$

$$O_3 + NO \rightarrow NO_2 + O_2 \qquad \text{25.71}$$

This sequence represents the normal nitrogen photolytic cycle and does not contribute to ground-level ozone accumulation as long as available NO and NO_2 remain balanced. However, if conditions develop that increase the availability of NO_2 over the NO, more ozone will be produced (see Eq. 25.69 and Eq. 25.70) than is destroyed (see Eq. 25.71), and ground-level ozone concentrations will increase.

The presence of hydrocarbons in the air is an important condition affecting increased NO_2 production. The following reactions show one of many ways in which NO_2 production is increased by the presence of hydrocarbons.

$$RH + \cdot OH \rightarrow \cdot R + H_2O \qquad \text{25.72}$$

$$\cdot R + O_2 \rightarrow \cdot RO_2 \qquad \text{25.73}$$

$$\cdot RO_2 + NO \rightarrow \cdot RO + NO_2 \qquad \text{25.74}$$

$$\cdot RO + O_2 \rightarrow \cdot HO_2 + RCHO \qquad \text{25.75}$$

$$\cdot HO_2 + NO \rightarrow NO_2 + \cdot OH \qquad \text{25.76}$$

The RH in Eq. 25.72 is a hydrocarbon, $\cdot$R represents a hydrocarbon radical, and RCHO in Eq. 25.75 represents an aldehyde of a different hydrocarbon. The hydroxyl radical ($\cdot$OH) also plays an important role in the reaction and is available from reaction with atomic oxygen and water, as well as being a product of Eq. 25.76. Not only does ground-level ozone increase through the production of excess NO_2, but other constituents of photochemical smog, such as aldehydes, are also produced. Figure 25.2 illustrates these reactions.

Nitrogen and sulfur oxides are subject to transformations associated with *acid formation*. One of the primary mechanisms for NO_2 removal from the atmosphere is its photochemical conversion to nitric acid (HNO_3).

The nitric acid is subsequently removed as aerosols by dry- and wet-deposition processes. In dry deposition, the HNO_3 aerosols settle out onto surfaces. In wet deposition, the aerosols either form rain droplets or are washed out by falling rain. The reaction, involving the hydroxyl radical, is as follows.

$$NO_2 + \cdot OH \rightarrow HNO_3 \qquad \text{25.77}$$

Sulfur dioxide reacts with oxygen to form sulfur trioxide (SO_3), which subsequently reacts with water to form *sulfuric acid* (H_2SO_4). The reaction from SO_2 to H_2SO_4 is relatively slow, requiring one or more days. Since sulfuric acid dissociates in solution to sulfate (SO_4^{-2}), the resulting aerosols are known as *sulfate aerosols*. The sulfate aerosols, as well as unreacted SO_3, are removed by dry and wet deposition. Sulfur dioxide may also react directly with water in the atmosphere to form *sulfurous acid* (H_2SO_3). The reaction of SO_3 with water is rapid and SO_3 condenses at the relatively low temperature of 22°C. Simplified reactions for the formation of sulfuric acid and sulfurous acid from sulfur oxides are as follows.

For sulfuric acid formation,

$$2SO_2 + O_2 \rightarrow 2SO_3 \qquad \text{25.78}$$

$$SO_3 + H_2O \rightarrow H_2SO_4 \qquad \text{25.79}$$

For sulfurous acid formation,

$$SO_2 + H_2O \rightarrow H_2SO_3 \qquad \text{25.80}$$

Figure 25.2 *Nitrogen Photolytic Cycle and Ground Level Ozone Accumulation*

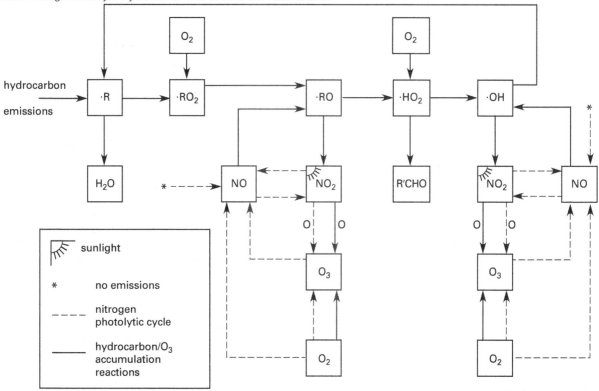

Adapted from Masters, G.M., *Introduction to Environmental Engineering and Science*, 2nd ed., © Prentice-Hall, Inc., Upper Saddle River, NJ.

Example 25.7

Assume RH in Eq. 25.72 is methane (CH_4). (a) What are Eq. 25.72 through Eq. 25.76 with the hydrocarbon, hydrocarbon radical, and aldehyde substituted for RH, $\cdot R$, and RCHO? (b) What are the chemical names of the radical and aldehyde?

Solution

(a) For Eq. 25.72 through Eq. 25.76,

$$CH_4 + \cdot OH \rightarrow \cdot CH_3 + H_2O$$
$$\cdot CH_3 + O_2 \rightarrow \cdot CH_3O_2$$
$$\cdot CH_3O_2 + NO \rightarrow \cdot CH_3O + NO_2$$
$$\cdot CH_3O + O_2 \rightarrow \cdot HO_2 + HCHO$$
$$\cdot HO_2 + NO \rightarrow NO_2 + \cdot OH$$

(b) The hydrocarbon radical, R, is methyl, and the aldehyde, RCHO, is formaldehyde (also called methyl aldehyde and methanal).

Example 25.8

During a rainstorm, 0.2 m^3 of rainwater scrubs 1 m^3 of air with an SO_2 concentration of 0.10 ppmv. The air temperature is 20°C, and the pressure is 1 atm. What is the concentration of sulfite (SO_3^{2-}) in the rainwater?

Solution

From Eq. 25.80,

$$SO_2 + H_2O \rightarrow H_2SO_3$$

One mole of SO_2 scrubbed will yield one mole of SO_3^{2-} in the rainwater.

$$M_{SO_2} = 32 \ \frac{g}{mol} + (2)\left(16 \ \frac{g}{mol}\right) = 64 \ g/mol$$

$$M_{SO_3} = 32 \ \frac{g}{mol} + (3)\left(16 \ \frac{g}{mol}\right) = 80 \ g/mol$$

Use Eq. 25.64.

$$PM = \rho RT$$

$$\rho = \frac{PM}{RT}$$

$$= \frac{(1 \text{ atm})\left(64 \ \dfrac{\text{g}}{\text{mol}}\right)}{\left(8.2 \times 10^{-5} \ \dfrac{\text{atm·m}^3}{\text{mol·K}}\right)}$$

$$\times (20°C + 273.15°)$$

$$= 2664 \text{ g/m}^3$$

$$\text{mass SO}_2 = \left(2664 \ \frac{\text{g}}{\text{m}^3}\right)\left(\frac{0.1 \text{ m}^3 \text{ SO}_2}{10^6 \text{ m}^3 \text{ air}}\right)$$

$$= 2.7 \times 10^{-4} \text{ g/m}^3 \text{ air}$$

The SO_2 concentration in rainwater is

$$\left(2.7 \times 10^{-4} \ \frac{\text{g}}{\text{m}^3 \text{ air}}\right)\left(\frac{1 \text{ m}^3 \text{ air}}{0.2 \text{ m}^3 \text{ rainwater}}\right)$$

$$= 1.3 \times 10^{-3} \text{ g/m}^3 \text{ rainwater}$$

The SO_3 concentration in rainwater is

$$\left(1.3 \times 10^{-3} \ \frac{\text{g}}{\text{m}^3 \text{ rainwater}}\right)\left(\frac{1 \text{ m}^3}{1000 \text{ L}}\right)\left(\frac{1 \text{ mol SO}_3^{2-}}{1 \text{ mol SO}_2}\right)$$

$$\times \left(80 \ \frac{\text{g}}{\text{mol SO}_3^{-2}}\right)\left(10^6 \ \frac{\mu\text{g}}{\text{g}}\right)\left(\frac{1 \text{ mol SO}_2}{64 \text{ g}}\right)$$

$$= 1.6 \ \mu\text{g/L}$$

15. NOMENCLATURE

AW	atomic weight	lbm/lbmol	kg/kmol
c	specific heat	Btu/lbm-°F	kJ/kg·°C
C	molar specific heat	Btu/lbmol-°F	J/kmol·°C
g, G	Gibbs function	Btu/lbm	kJ/kg
G	gravimetric fraction	–	–
h	enthalpy	Btu/lbm	kJ/kg
h	Henry's law constant	n.a.	Pa·m³/mol
H	molar enthalpy	Btu/lbm	kJ/kg
k	ratio of specific heats		
m	mass	lbm	kg
M	molecular weight	lbm/lbmol	kg/kmol
n, N	number of moles	–	–
p	pressure	lbf/ft³	Pa

P	pressure or partial pressure	lbf/ft³	Pa
Q	heat	Btu	kJ
R	specific gas constant	ft-lbf/lbm-°R	kJ/kg·K
$\bar{R}$	average specific gas constant	ft-lbf/lbm-°R	kJ/kg·K
°R	Rankine	R	R
s	entropy	Btu/lbm-°R	kJ/kg·K
T	temperature	°F	K
u	specific internal energy	Btu/lbmol	kJ/mol
U	molar internal energy	Btu/lbmol	kJ/mol
v	specific volume	ft³/lbm	m³/kg
V	molar specific volume	ft³/lbmol	m³/kmol
V	volume	ft³/lbmol	m³/kmol
x_i	mole fraction	–	–
y_i	mass fraction	–	–

Symbols

υ	specific volume	n.a.	m³/kg
μ	degree of saturation	–	–
ρ	mass density	lbm/ft³	kg/m³
ϕ	relative humidity	–	–
ω	humidity ratio	lbm/lbm	kg/kg

Subscripts

a	dry air or VOC
db	dry-bulb
dp	dew-point
fg	vaporization
g	gas
i	component or gas
ig	solid to gas
K	kelvin
l	latent or liquid
p, P	constant pressure
s	sensible or settling
sat	saturation
sl	solid to liquid
t	terminal or total
v	vapor or constant volume
w	water
wb	wet-bulb

Air

26 Fate and Transport

Content in blue refers to the *NCEES Handbook*.

1. DRAFT

The amount of air that flows through a furnace is determined by the furnace draft. *Draft* is the difference in static pressures that causes the flue gases to flow.[1] It is usually stated in inches of water (kPa). *Natural draft* (ND) furnaces rely on the *stack effect* (*chimney draft*) to draw off combustion gases. Air flows through the furnace and out the stack due to the pressure differential caused by reduced densities in the stack.[2]

Forced draft (FD) fans located before the furnace are used to supply air for burning. Combustion occurs under pressure, hence the descriptive term "pressure-fired unit" for such fans. FD fans are run at relatively high speeds (1200 rpm to 1800 rpm) with direct-drive motors. Two or more fans are used in parallel to provide for efficient operation at low furnace demand.

FD fans create a positive pressure (e.g., 2 in wg to 10 in wg; 0.5 kPa to 2.5 kPa). This pressure is reduced to a very small negative pressure after passing through the air heater, ducts, and windbox system. The negative pressure in the furnace serves to keep combustion gases from leaking out into the furnace and boiler areas. The pressure continues to drop as it passes through the boiler, economizer, air heater, and pollution-control equipment.

Whereas FD fans force air through the system, *induced draft* (ID) fans are used to draw combustion products through the furnace bed, stack, and pollution control system by injecting air into the stack after combustion. The term "suction units" is used with ID fans.

ID fans are located after dust collectors and precipitators (often at the base of the stack) in order to reduce the abrasive effects of fly ash. They are run at slower speeds than forced draft fans in order to reduce the abrasive effects even further. Unlike FD fans, ID fans are usually very large and powerful because they have to handle all of the combustion gases, not just combustion air.

Modern welded stacks are essentially airtight and can operate at pressures above atmospheric, often eliminating the need for ID fans.

Pure FD and ID systems are rarely used. Modern stack systems operate in a condition of *balanced draft*. Balanced draft is the term used when the static pressure is equal to atmospheric pressure. This requires the use of both ID and FD fans. In order to keep combustion products inside the combustion chamber and stack system, balanced draft systems may actually operate with a slight negative pressure.

The *draft loss* is the static pressure drop due to friction through the boiler and stack. The *available draft* is the difference between the theoretical draft (stack effect) and the draft loss. The available draft is zero in a balanced system.

$$D_{\text{available}} = D_{\text{theoretical}} - D_{\text{friction}} \qquad 26.1$$

The *net rating* or *fan boost* of the fan is the total pressure (the difference of draft losses and draft gains) supplied by the fan at maximum operating conditions. The fan power for ID and FD fans is calculated, as with any fan application, from the net rating and the flow rate. It is customary in sizing ID and FD draft fans to include increases of approximately 15% for flow rate, 30% for pressure, and 20°C (10°C) for temperature.

---

[1]The British term "draught" is synonymous with "draft."
[2]Chimneys that rely on natural draft are sometimes referred to as *gravity chimneys*.

2. STACK EFFECT

Stack effect (chimney action or natural draft) is a pressure difference caused by the difference in atmospheric air and flue gas densities. It can be relied on to draw combustion products at least partially up through the stack. The stack effect is determined with no flow of flue gas. With flow, some of the stack effect is converted to velocity head, and the remainder is used to overcome friction.

Generally, the higher the chimney, the greater the stack effect. However, the stack effect in modern plants is greatly reduced by the friction and cooling effects of economizers, air heaters, and precipitators. Fans supply the needed pressure differential. Therefore, the primary function of modern stacks is to carry the combustion products a sufficient distance upward to dilute the combustion products, not to generate draft. Modern stacks for coal-fired power plants are seldom built shorter than 200 ft (60 m) in order to meet their dispersion requirements, and most are higher than 500 ft (150 m).

The theoretical stack effect, $D_{\text{theoretical}}$, is calculated from the densities of the flue gas and atmospheric air. In Eq. 26.2, H_{stack} is the total height of the stack. The average temperature is used to determine the average density in Eq. 26.2.[3]

$$D_{\text{theoretical}} = H_{\text{stack}}(\gamma_{\text{air}} - \gamma_{\text{flue gas,ave}}) \qquad 26.2$$

If a few assumptions are made, the stack effect can be calculated from the average temperature of the flue gas (i.e., the temperature at the average stack elevation).[4] The average flue gas temperature is the temperature halfway up the stack. H_{stack} is the total height of the stack.

$$D_{\text{theoretical}} = \left(\frac{P_{\text{air}}H_{\text{stack}}g}{R_{\text{air}}}\right)\left(\frac{1}{T_{\text{air}}} - \frac{1}{T_{\text{flue gas,ave}}}\right) \quad \text{[SI]} \quad 26.3(a)$$

$$D_{\text{theoretical}} = \left(\frac{P_{\text{air}}H_{\text{stack}}g}{R_{\text{air}}g_c}\right)\left(\frac{1}{T_{\text{air}}} - \frac{1}{T_{\text{flue gas,ave}}}\right) \quad \text{[U.S.]} \quad 26.3(b)$$

The stack effect head, H_{SE}, is used to calculate the stack velocity. The coefficient of velocity, C_v, is approximately 0.30 to 0.50.

$$H_{\text{SE}} = \frac{D_{\text{theoretical}}}{\gamma_{\text{flue gas,ave}}} \qquad 26.4$$

$$v = C_v v_{\text{ideal}} = C_v\sqrt{2gh_{\text{SE}}} \qquad 26.5$$

The required stack area is

$$A_{\text{stack}} = \frac{\dot{Q}_{\text{flue gas}}}{v} \qquad 26.6$$

Example 26.1

The air surrounding an 80 ft (24 m) tall vertical stack is at 70°F (21°C) and one standard atmospheric pressure. The temperature of the flue gas at the average stack elevation is 400°F (200°C). What is the theoretical natural draft?

SI Solution

The absolute temperatures are

$$T_{\text{air}} = 21°C + 273° = 294\text{K}$$
$$T_{\text{flue gas}} = 200°C + 273° = 473\text{K}$$

Calculate the stack effect from Eq. 26.3.

$$D_{\text{theoretical}} = \left(\frac{P_{\text{air}}H_{\text{stack}}g}{R_{\text{air}}}\right)\left(\frac{1}{T_{\text{air}}} - \frac{1}{T_{\text{flue gas}}}\right)$$

$$= \left(\frac{(101.3 \text{ kPa})(24 \text{ m})\left(9.81\,\dfrac{\text{m}}{\text{s}^2}\right)}{287\,\dfrac{\text{J}}{\text{kg·K}}}\right)$$

$$\times\left(\frac{1}{294\text{K}} - \frac{1}{473\text{K}}\right)$$

$$= 0.107 \text{ kPa}$$

Customary U.S. Solution

The absolute temperatures are

$$T_{\text{air}} = 70°F + 460° = 530°R$$
$$T_{\text{flue gas}} = 400°F + 460° = 860°R$$

---

[3]The actual stack effect will be less than the value calculated with Eq. 26.2. For realistic problems, the achievable stack effect probably should be considered to be 80% of the ideal.

[4]One assumption is that the pressure in the stack is equal to atmospheric pressure. Then, the density difference will be due to only the temperature difference. Also, the flue gas is assumed to be air. This makes it unnecessary to know the flue gas composition, molecular weight, and so on.

Calculate the stack effect from Eq. 26.3.

$$D_{theoretical} = \left(\frac{P_{air}H_{stack}g}{R_{air}g_c}\right)\left(\frac{1}{T_{air}} - \frac{1}{T_{flue\,gas}}\right)$$

$$= \left[\frac{\left(14.7\,\dfrac{lbf}{in^2}\right)\left(12\,\dfrac{in}{ft}\right)^2(80\,ft)\left(32.2\,\dfrac{ft}{sec^2}\right)}{\left(53.3\,\dfrac{ft\text{-}lbf}{lbm\text{-}°R}\right)\left(32.2\,\dfrac{lbm\text{-}ft}{lbf\text{-}sec^2}\right)}\right]$$

$$\times\left(\frac{1}{530°R} - \frac{1}{860°R}\right)$$

$$= 2.3\,lbf/ft^2$$

Convert this to inches of water.

$$D_{theoretical} = \frac{P}{\gamma} = \frac{\left(2.3\,\dfrac{lbf}{ft^2}\right)\left(12\,\dfrac{in}{ft}\right)}{62.4\,\dfrac{lbf}{ft^3}}$$

$$= 0.44\,in\,wg$$

3. STACK FRICTION

Flue gas velocities of 15 ft/sec to 40 ft/sec (4.5 m/s to 12 m/s) are typical. As with most turbulent fluids, the friction pressure drop, $D_{friction}$, through stack "piping" is proportional to the velocity head.[5]

$$D_{friction} = \frac{Kv^2}{2g} \qquad\qquad 26.7$$

Values of the friction coefficient, K, for chimney fittings are similar to those used in ventilating ductwork. Values for specific pieces of equipment (burners, pressure regulators, vents, etc.) must be provided by the manufacturers.

The approximate friction loss coefficient for a circular section of ductwork or chimney of length L and diameter d is given by Eq. 26.8. Equation 26.8 assumes a Darcy friction factor of 0.0233. For other friction factors, the constant terms can be scaled up or down

proportionately. For noncircular ducts, the hydraulic diameter, d, can be replaced by four times the hydraulic radius (i.e., four times the duct area divided by the duct perimeter).

$$D_{friction,kPa} = \frac{0.119L_m\rho_{kg/m^3}Q^2_{L/s}}{d^5_{cm}} \qquad [SI] \quad 26.8(a)$$

$$D_{friction,in\,wg} = \frac{28L_{ft}\gamma_{lbf/ft^3}Q^2_{cfs}}{d^5_{in}} \qquad [U.S.] \quad 26.8(b)$$

4. ACID GAS

Acid gas generally refers to *sulfur trioxide*, SO_3, in flue gas.[6,7] *Sulfuric acid*, H_2SO_4, formed when sulfur trioxide combines with water, has a low vapor pressure and, consequently, a high boiling point. Hydrochloric (HCl), hydrofluoric (HF), and nitric (HNO_3) acids also form in smaller quantities. However, unlike sulfuric acid, they do not lower the vapor pressure of water significantly. For this reason, any sulfuric acid present will control the dew point.

Sulfur trioxide has a large affinity for water, forming a strong acid even at very low concentrations. At the elevated temperatures in a stack, sulfuric acid attacks steel, almost all plastics, cement, and mortar. Sulfuric acid can be prevented from forming by keeping the temperature of the flue gas above the dew-point temperature. This may require preheating equipment prior to start-up and postheating during shutdown.

Stack dew points have been reported by various researchers to be in the range of 225°F to 300°F. However, the actual value is dependent on the amount of SO_3 in the flue gas, and therefore, on the amount of sulfur in the fuel. The theoretical equilibrium relationships are too complex to be useful, and empirical correlations or graphical methods are used. Equation 26.9 can be used to determine the dew point based on the partial pressures, P_w and P_s, of the water vapor and sulfur trioxide, respectively, in units of atmospheres.[8,9]

$$\frac{1000}{T_{dp,K}} = 1.7842 + 0.0269\log_{10}P_w$$

$$-0.1029\log_{10}P_s \qquad\qquad 26.9$$

$$+0.0329\log_{10}P_w\log_{10}P_s$$

$$[error < \pm4K]$$

---

[5]The term *piping* is used to mean the stack passages.
[6]The term *stack gas* is used interchangeably with *flue gas.*
[7]Sulfur dioxide normally is not a source of acidity in the flue gas.
[8]This correlation was reported by F. H. Verhoff and J. T. Banchero in *Chemical Engineering Progress*, 1974, Vol. 70, p. 71.
[9]*Dalton's law* states that the partial pressure of a component is volumetrically weighted. Therefore, knowing the volumetric flue gas concentration, C_A, in ppmv is equivalent to knowing the partial pressures.

$$P_A = P\left(\frac{C_A}{10^6}\right)$$

Hydrochloric acid does not normally occur unless the fuel has a high chlorine content, as chlorinated solvents, municipal solid wastes (MSW), and refuse-derived fuels (RDF) all do. Hydrochloric acid formed during the combustion of MSW and RDF can be removed by semidry scrubbing. HCl removal efficiencies of 90–99% are common. (See also Sec. 26.5 and Sec. 26.14.)

Example 26.2

At a particular point in a stack, the flue gas has a total pressure of 30.2 in Hg. The flue gas is 8% water vapor by volume, and the sulfur trioxide concentration is 100 ppm. What is the approximate dew-point temperature at that point?

Solution

From Dalton's law, the partial pressures of water and sulfur trioxide are volumetrically weighted. 1 atm is equal to 29.92 in Hg. [Conversion Factors]

Ideal Gas Mixtures

$$x_i = P_i/P = V_i/V$$

$$P_{water} = \left(\frac{V_{water}}{V}\right)P$$

$$= (0.08)\left(\frac{30.2 \text{ in Hg}}{29.92 \frac{\text{in Hg}}{\text{atm}}}\right)$$

$$= 0.0807 \text{ atm}$$

$$P_{SO_3} = x_{SO_3}P$$

$$= \left(\frac{100 \text{ ppm}}{10^6 \text{ ppm}}\right)\left(\frac{30.2 \text{ in Hg}}{29.92 \frac{\text{in Hg}}{\text{atm}}}\right)$$

$$= 1.01 \times 10^{-4} \text{ atm}$$

Use Eq. 26.9.

$$\frac{1000}{T_{dp,K}} = 1.7842 + 0.0269\log_{10}(0.0807 \text{ atm})$$

$$- 0.1029\log_{10}(1.01 \times 10^{-4} \text{ atm})$$

$$+ 0.0329\log_{10}(0.0807 \text{ atm})$$

$$\times \log_{10}(1.01 \times 10^{-4} \text{ atm})$$

$$= 2.3097$$

$$T_{dp} = \frac{1000\text{K}}{2.3097} = 433\text{K} \quad (160°C, 320°F)$$

5. ACID RAIN

Acid rain consists of weak solutions of sulfuric, hydrochloric, and to a lesser extent, nitric acids. These acids are formed when emissions of sulfur oxides (SOx), hydrogen chloride (HCl), and nitrogen oxides (NOx) return to the ground in rain, fog, or snow, or as dry particles and gases. Acid rain affects lakes and streams, damages buildings and monuments, contributes to reduced visibility, and affects certain tree species. Acid rain may also represent a health hazard. (See also Sec. 26.4 and Sec. 26.14.)

6. CARBON DIOXIDE

Carbon dioxide, though an environmental issue, is not a hazardous material and is not regulated as a pollutant.[10] Carbon dioxide is not an environmental or human toxin, is not flammable, and is not explosive. Skin contact with solid or liquid carbon dioxide presents a freezing hazard. Other than the remote potential for causing frostbite, carbon dioxide has no long-term health effects. Its major hazard is that of asphyxiation, by excluding oxygen from the lungs.

As with other products of combustion, the fraction of carbon dioxide in a flue gas can be arbitrarily reduced by the introduction of dilution air into the flue stream. Therefore, carbon dioxide is reported on a standardized basis—typically as a dry volumetric fraction at some percentage (e.g., 3%) of oxygen.[11] The standardized value can be calculated from stoichiometric relationships. Since there is essentially no carbon dioxide in air (approximately 0.03% by volume), carbon dioxide in a flue gas has a unique source—carbon in the fuel. Knowing the theoretical flue gas composition is sufficient to calculate the standardized value. The analysis is independent of stack temperature.

For a standardized value of 3% oxygen, Table 26.1 gives the dry carbon dioxide volumetric fraction directly, based on the volumetric ratio of hydrogen to carbon in the fuel. For any other percentage of oxygen, the following procedure can be used.

step 1: Obtain the volumetric fuel composition. Gaseous fuel compositions are normally reported on a volumetric basis. Solid and liquid fuels are reported on a weight (gravimetric) basis. Convert weight basis analyses to a volumetric basis by dividing the gravimetric percentages by their respective atomic (or, molecular) weights. Combustion products are gaseous. (For gases, molar and volumetric ratios are the same.)

step 2: Write and balance the stoichiometric combustion equation using the volumetric fractions as coefficients for the fuel elements. Disregard trace emissions (NO, SO$_2$, CO, etc.) that contribute less than 1% to the flue gas volume, and disregard oxygen

---

[10]Industrial exposure is regulated by the U.S. Occupational Safety and Health Administration (OSHA).
[11]The phrases "at 3% O$_2$" and "at 3% excess air" are not equivalent. The former means that oxygen comprises 3% of the gaseous reaction products by volume. The latter means that 3% more air (3% more oxygen) is provided in reactants than is needed.

contributed by the fuel. Include a variable amount of excess air. Include nitrogen for the combustion and excess air at the ratio of 3.773 volumes of nitrogen for each volume of oxygen.

step 3: Divide the oxygen volume (i.e., the balanced reaction coefficient) by the sum of all flue gas volumes, excluding water vapor, and set this ratio equal to the standardized volumetric oxygen fraction. Solve for excess air.

step 4: Divide the carbon dioxide volume by the sum of all flue gas volumes, excluding water vapor. (The fraction can be multiplied by 10^6 to obtain the volumetric fraction in ppm, though this is seldom done for carbon dioxide.)

The carbon dioxide concentration (mass emission per dry standard volume), C, can be calculated from the carbon dioxide's molecular weight ($M = 44$) and Eq. 26.14.

Table 26.1 *Theoretical CO₂ Fraction at 3% Oxygen (dry volumetric basis)*

H/C ratio	CO_2	H/C ratio	CO_2
0	0.18	2.1	0.12723
0.1	0.17651	2.2	0.12548
0.2	0.17816	2.3	0.12378
0.3	0.16993	2.4	0.12212
0.4	0.16682	2.5	0.12050
0.5	0.16382	2.6	0.11893
0.6	0.16093	2.7	0.11740
0.7	0.15814	2.8	0.11590
0.8	0.15544	2.9	0.11445
0.9	0.15283	3.0	0.11303
1.0	0.15031	3.1	0.11165
1.1	0.14787	3.2	0.11029
1.2	0.14551	3.3	0.10898
1.3	0.14323	3.4	0.10768
1.4	0.14101	3.5	0.10643
1.5	0.13886	3.6	0.10520
1.6	0.13678	3.7	0.10400
1.7	0.13476	3.8	0.10283
1.8	0.13279	3.9	0.10168
1.9	0.13089	4.0	0.10056
2.0	0.12903		

Example 26.3

Coal has the following gravimetric composition: carbon, 86.5%; hydrogen, 11.75%; nitrogen (N_2), 0.39%; sulfur, 0.40%; ash, 0.01%; andoxygen (O_2), 0.96%. Determine the theoretical carbon dioxide concentration in parts per million on a dry volumetric (ppmvd) basis at 3% oxygen by volume.

Solution

step 1: Disregarding the combustion of sulfur to SO_2 and other elements present in small quantities,

flue gases will be products of carbon and hydrogen combustion and the excess air. The atomic weight of carbon is 12 lbm/lbmol. Since the hydrogen is present in elemental form, not as H_2 gas, the atomic weight is 1 lbm/lbmol. Consider 100 lbm of fuel. The carbon content will be $(0.865)(100 \text{ lbm}) = 86.5$ lbm. The volumetric ratios (number of moles) are

$$C: \frac{86.5 \text{ lbm}}{12 \frac{\text{lbm}}{\text{lbmol}}} = 7.208 \text{ lbmol}$$

$$H: \frac{11.75 \text{ lbm}}{1 \frac{\text{lbm}}{\text{lbmol}}} = 11.75 \text{ lbmol}$$

step 2: The unbalanced combustion reaction is

$$7.208\,C + 11.75\,H + n_1 O_2 + 3.773 n_1 N_2$$
$$\rightarrow n_2 CO_2 + n_3 H_2 O + 3.773 n_1 N_2$$

The balanced combustion reaction is

$$7.208\,C + 11.75\,H + 10.146 O_2 + 38.281 N_2$$
$$\rightarrow 7.208 CO_2 + 5.875 H_2 O + 38.281 N_2$$

Let e represent the excess air fraction. All of the excess oxygen and nitrogen will appear in the flue gas.

$$7.208\,C + 11.75\,H + (1 + e)10.146 O_2$$
$$+ (1 + e)38.281 N_2$$
$$\rightarrow 7.208 CO_2 + 5.875 H_2 O$$
$$+ (1 + e)38.281 N_2 + 10.146 e O_2$$

step 3: At 3% O_2,

$$0.03 = \frac{10.146 e}{7.208 + (1 + e)(38.281) + 10.146 e}$$
$$e \approx 0.157 \quad (15.7\% \text{ excess air})$$

The balanced combustion reaction at 3% oxygen is

$$7.208\,C + 11.75\,H + 11.739 O_2 + 44.291 N_2$$
$$\rightarrow 7.208 CO_2 + 5.875 H_2 O + 44.291 N_2$$
$$+ 1.593 O_2$$

step 4: The theoretical carbon dioxide fraction, on a dry volumetric basis at 3% oxygen, is

$$\frac{7.208}{7.208 + 44.291 + 1.593} = 0.1358 \quad (13.58\%)$$

(This is the same value as obtained from Table 26.1 for a volumetric fuel ratio of $H/C = 11.75/7.208 = 1.63$.)

Example 26.4

When the fuel described in Ex. 26.3 is burned, the carbon dioxide in the stack gas is measured on a wet, volumetric basis to be 10.4%. What is this value corrected to 3% O_2?

Solution

Since the theoretical carbon dioxide volumetric fraction can be calculated from the fuel composition, the actual carbon dioxide fraction in the stack is irrelevant. (The amount of excess air could be found from this value, however.) The theoretical carbon dioxide fraction, on a dry volumetric basis at 3% oxygen, is still 13.55%.

7. CARBON MONOXIDE

Carbon monoxide, CO, is formed during incomplete combustion of carbon in fuels. This is usually the result of an oxygen deficiency at lower temperatures. Carbon monoxide displaces oxygen in the bloodstream, so it represents an asphyxiation hazard. Carbon monoxide does not contribute to smog.

The generation of carbon monoxide can be minimized by furnace monitoring and control. For industrial sources, the American Boiler Manufacturers Association (ABMA) recommends limiting carbon monoxide to 400 ppm (corrected to 3% O_2) in oil- and gas-fired industrial boilers. This value can usually be met with reasonable ease. Local ordinances may be more limiting, however.

Most carbon monoxide released in highly populated areas comes from vehicles. Vehicular traffic may cause the CO concentration to exceed regulatory limits. For this reason, *oxygenated fuels* are required to be sold in those areas during certain parts of the year. Oxygenated gasoline has a minimum oxygen content of approximately 2.0%. Oxygen is increased in gasoline with additives such as ethanol (ethyl alcohol). Methyl tertiary butyl ether (MTBE) continues to be used as an oxygenate outside of the United States.

Minimization of carbon monoxide is compromised by efforts to minimize nitrogen oxides. Control of these pollutants is inversely related.

8. CHLOROFLUOROCARBONS

Most atmospheric oxygen is in the form of two-atom molecules, O_2. However, there is a thin layer in the stratosphere about 12 miles up where *ozone* molecules, O_3, are found in large quantities. Ozone filters out ultraviolet radiation that damages crops and causes skin cancer.

Chlorofluorocarbons (i.e., chlorinated fluorocarbons, such as Freon) contribute to the deterioration of the earth's ozone layer. Ozone in the atmosphere is depleted in a complex process involving pollutants, wind patterns, and atmospheric ice. As chlorofluorocarbon molecules rise through the atmosphere, solar energy breaks the chlorine free. The chlorine molecules attach themselves to ozone molecules, and the new structure eventually decomposes into chlorine oxide and normal oxygen, O_2. The depletion process is particularly pronounced in the Antarctic because that continent's dry, cold air is filled with ice crystals on whose surfaces the chlorine and ozone can combine. Also, the prevailing winter wind isolates and concentrates the chlorofluorocarbons.

The depletion is not limited to the Antarctic; it also occurs throughout the northern hemisphere, including virtually all of the United States.

The 1987 Montreal Protocol (conference) resulted in an international agreement to phase out worldwide production of chemical compounds that have ozone-depletion characteristics. Since 2000 (the peak of Antarctic ozone depletion), concentrations of atmospheric chlorine have decreased, and ozone layer recovery is increasing.

Over the years, the provisions of the 1987 Montreal Protocol have been modified numerous times. Substance lists have been amended, and action deadlines have been extended. In many cases, though production may have ceased in a particular country, significant recycling and stockpiling keeps chemicals in use. Voluntary compliance by some nations, particularly developing countries, is spotty or nonexistent. In the United States, provisions have been incorporated into the Clean Air Act (Title VI) and other legislation, but such provisions are subject to constant amendment.

Special allowances are made for aviation safety, national security, and fire suppression and explosion prevention if safe or effective substitutes are not available for those purposes. Excise taxes are used as interim disincentives for those who produce the compounds. Large reserves and recycling, however, probably ensure that chlorofluorocarbons and halons will be in use for many years after the deadlines have passed.

Replacements for chlorofluorocarbons (CFCs) include hydrochlorofluorocarbons (HCFCs) and hydrofluorocarbons (HFCs), both of which are environmentally more benign than CFCs, and blends of HCFCs and HFCs. (See Table 26.2.) The additional hydrogen atoms in the molecules make them less stable, allowing nearly all chlorine to dissipate in the lower atmosphere before reaching the ozone layer. The lifetime of HCFC molecules is 2 years to 25 years, compared with 100 years or longer for CFCs. The net result is that HCFCs have only 2–10% of the ozone-depletion ability of CFCs. HFCs have no chlorine and thus cannot deplete the ozone layer.

Table 26.2 *Typical Replacement Compounds for Chlorofluorocarbons*

designation	applications
HCFC 22	low- and medium-temperature refrigerant; blowing agent; propellant
HCFC 123	replacement for CFC-11; industrial chillers and applications where potential for exposure is low; somewhat toxic; blowing agent; replacement for perchloroethylene (dry cleaning fluid)
HCFC 124	industrial chillers; blowing agent
HFC 134a	replacement for CFC-12; medium-temperature refrigeration systems; centrifugal and reciprocating chillers; propellant
HCFC 141b	replacement for CFC-11 as a blowing agent; solvent
HCFC 142b	replacement for CFC-12 as a blowing agent; propellant
IPC (isopropyl chloride)	replacement for CFC-11 as a blowing agent

Most chemicals intended to replace CFCs still have chlorine, but at reduced levels. Additional studies are determining if HCFCs and HFCs accumulate in the atmosphere, how they decompose, and whether any byproducts could damage the environment.

Halon is a generic term used to refer to various liquefied, compressed gas *halomethanes* containing bromine, fluorine, and chlorine. Halons continue to be in widespread legal use for various specialty purposes known as *critical uses* (as defined by the EPA) because they are nonconducting, leave no residue upon evaporation, and are relatively safe for human exposure. For example, halons are commonly found in aircraft, ship engine compartments, military vehicles, cleanrooms, and commercial kitchen fire suppression systems. Halon 1011 (bromochloromethane, CH_2BrCl) is used in portable fire extinguishers. Halon 1211 (bromochlorodifluoromethane, $CBrClF_2$) is used when the fire extinguishing application method is liquid streaming. Halon 1301 (bromotrifluoromethane, $CBrF_3$) is a gaseous flooding agent, as is

required in aircraft suppression systems. Halons are rated for flammable liquids (class B fires) and electrical fires (class C fires), although they are also effective on common combustibles (class A fires). Halons are greenhouse gases, so at least in the United States, production of new halons has ceased. Existing inventory stockpiles, recycling, and importing are used to satisfy current demands.

9. DUST, GENERAL

Dust or *fugitive dust* is any solid particulate matter (PM) that becomes airborne, with the exception of PM emitted from the exhaust stack of a combustion process. Nonhazardous fugitive dusts are commonly generated when a material (e.g., coal) is unloaded from trucks and railcars. Dusts are also generated by manufacturing, construction, earth-moving, sand blasting, demolition, and vehicle movement.

Dusts pose three types of hazards. (a) Inhalation of airborne dust or vapors, particularly those that carry hazardous compounds, is the major concern. Even without toxic compounds, odors can be objectionable. Dusts are easily observed and can cover cars and other objects left outside. (b) Dusts can transport hazardous materials, contaminating the environment far from the original source. (c) In closed environments, even nontoxic dusts can represent an explosion hazard.

Dusts are categorized by size according to how deep they can penetrate into the respiratory system. Particle size is based on *aerodynamic equivalent diameter* (AED), the diameter of a sphere with the density of water (62.4 lbm/ft^3 (1000 kg/m^3)) that would have the same settling velocity as the particle. The distribution of dust sizes is divided into three fractions, also known as *conventions*. The *inhalable fraction* (*inhalable convention*) (< 100 μm AED; d_{50} = 100 μm) can be breathed into the nose and mouth; the *thoracic fraction* (< 25 μm AED; d_{50} = 10 μm) can enter the larger lung airways; and the *respirable fraction* (< 10 μm AED; d_{50} = 4 μm) can penetrate beyond terminal bronchioles into the gas exchange regions. (See Fig. 26.1.)

Figure 26.1 *Airborne Particle Fractions*

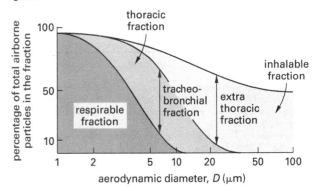

Dust emission reduction from *spot sources* (e.g., manufacturing processes such as grinders) is accomplished by *inertial separators* such as cyclone separators. Potential dust sources (e.g., truck loads and loose piles) can be covered, and dust generation can be reduced by spraying water mixed with other compounds.

There are three mechanisms of dust control by spraying. (a) In *particle capture* (as occurs in a spray curtain at a railcar unloading station), suspended particles are knocked down, wetted, and captured by liquid droplets. (b) In *bulk agglomeration* (as when a material being carried on a screw conveyor is sprayed), the moisture keeps the dust with the material being transported. (c) Spraying roads and coal piles to inhibit wind-blown dust is an example of *surface stabilization*.

Wetting agents are *surfactant* formulations added to water to improve water's ability to wet and agglomerate fine particles.[12] The resulting solution can be applied as liquid spray or as a foam.[13] *Humectant binders* (e.g., magnesium chloride and calcium chloride) and adhesive binders (e.g., waste oil) may also be added to the mixture to make the dust adhere to the contact surface if other water-based methods are ineffective.[14]

Surface stabilization of materials stored outside and exposed to wind, rain, freeze-thaw cycles, and ultraviolet radiation is enhanced by the addition of *crusting agents*.

10. FUGITIVE EMISSIONS

Equipment leaks from plant equipment are called *fugitive emissions* (FEs). Leaks from pump and valve seals are common sources, though compressors, pressure-relief devices, connectors, sampling systems, closed vents, and storage vessels are also potential sources. FEs are reduced administratively by *leak detection and repair programs* (LDARs).

Some common causes of fugitive emissions include (a) equipment in poor condition, (b) off-design pump operation, (c) inadequate seal characteristics, and (d) inadequate boiling-point margin in the seal chamber. Other pump/shaft/seal problems that can increase emissions include improper seal-face material, excessive seal-face loading and seal-face deflections, and improper pressure balance ratio.

Inadequate *boiling-point margin* (BPM) results in poor seal performance and face damage. BPM is the difference between the seal chamber temperature and the initial boiling point temperature of the pumped product at the seal chamber pressure. When seals operate close to the boiling point, seal-generated heat can cause the pumped product between the seal faces to flash and the seal to run dry. A minimum BPM of 15°F (9°C) or a 25 psig (172 kPa) pressure margin is recommended to avoid flashing. Even greater margins will result in longer seal life and reduced emissions.

11. NITROGEN OXIDES

Nitrogen oxides (NOx) are one of the primary causes of smog formation. NOx from the combustion of coal is primarily *nitric oxide* (NO) with small quantities of *nitrogen dioxide* (NO_2).[15] NO_2 can be a primary or a secondary pollutant. Although some NO_2 is emitted directly from combustion sources, it is also produced from the oxidation of nitric oxide, NO, in the atmosphere.

NOx is produced in two ways: (a) *thermal NOx* produced at high temperatures from free nitrogen and oxygen, and (b) *fuel NOx* (or *fuel-bound NOx*) formed from the decomposition/combustion of fuel-bound nitrogen.[16] When natural gas and light distillate oil are burned, almost all of the NOx produced is thermal. Residual fuel oil, however, can have a nitrogen content as high as 0.3%, and 50–60% of this can be converted to NOx. Coal has an even higher nitrogen content.[17]

Thermal NOx is usually produced in small (but significant) quantities when excess oxygen is present at the highest temperature point in a furnace, such that nitrogen (N_2) can dissociate.[18] Dissociation of N_2 and O_2 is negligible, and little or no thermal NOx is produced below approximately 3000°F (1650°C).

Formation of thermal NOx is approximately exponential with temperature. For this reason, many NOx-reduction techniques attempt to reduce the *peak flame*

---

[12]A *surface-acting agent* (*surfactant*) is a soluble compound that reduces a liquid's surface tension or reduces the interfacial tension between a liquid and a solid.

[13]Collapsible aqueous foam is an increasingly popular means of reducing the potential for explosions in secondary coal crushers.

[14]A *humectant* is a substance that absorbs or retains moisture.

[15]Other oxides are produced in insignificant amounts. These include *nitrous oxide* (N_2O), N_2O_4, N_2O_5, and NO_3, all of which are eventually oxidized to (and reported as) NO_2.

[16]Some engineers further divide the production of fuel-bound NOx into low- and high-temperature processes and declare a third fuel-related NOx-production method known as *prompt-NOx*. Prompt-NOx is the generation of the first 15–20 ppm of NOx from partial combustion of the fuel at lower temperatures.

[17]In order of increasing fuel-bound NOx production potential, common boiler fuels are: methanol, ethanol, natural gas, propane, butane, ultra-low-nitrogen fuel oil, fuel oil no. 2, fuel oil no. 6, and coal.

[18]The *mean residence time* at the high temperature points of the combustion gases is also an important parameter. At temperatures below 2500°F (1370°C), several minutes of exposure may be required to generate any significant quantities of NOx. At temperatures above 3000°F (1650°C), dissociation can occur in less than a second. At the highest temperatures—3600°F (1980°C) and above—dissociation takes less than a tenth of a second.

temperature (PFT). NOx formation can also be reduced by injecting urea or ammonia reagents directly into the furnace.[19] The relationship between NOx production and excess oxygen is complex, but NOx production appears to vary directly with the square root of the oxygen concentration.

In existing plants, retrofit NOx-reduction techniques include using fuel-rich combustion (i.e., staged air burners), recirculating flue gas, changing to a low-nitrogen fuel, reducing air preheat, installing low-NOx burners, and using overfire air.[20] Low-NOx burners using controlled flow/split flame or internal fuel staging technology are essentially direct replacement (i.e., "plug-in") units, differing from the original burners primarily in nozzle configuration. However, some fuel supply and air modifications may also be needed. Use of overfire air requires windbox modifications and separate ducts.

Lime spray-dryer scrubbers of the type found in electrical generating plants do not remove all of the NOx. Further reduction requires that the remaining NOx be destroyed. Reburn, selective catalytic reduction (SCR), and selective noncatalytic methods are required.

Scrubbing, incineration, and other end-of-pipe methods typically have been used to reduce NOx emissions from stationary sources such as gas turbine-generators, although these methods are unwieldy for the smallest units. Water or steam injection, catalytic combustion, and *selective catalytic reduction* (SCR) are particularly suited for gas-turbine and combined-cycle installations. SCR is also effective in NOx reduction in all heater and boiler applications.

The volumetric fraction (in ppm) of NOx in the flue gas is calculated from the molecular weight and mass flow rates. The molecular weight of NOx is assumed to be 46.[21] The ratio of $\dot{m}_{NOx}/\dot{m}_{flue\,gas}$, percentage of water vapor, and flue gas molecular weight are derived from flue gas analysis.

$$B_{NOx} = \left(\frac{\dot{m}_{NOx}}{\dot{m}_{flue\,gas}}\right)(10^6)\left(\frac{M_{flue\,gas}}{M_{NOx}}\right) \\ \times \left(\frac{100\%}{100\% - \%H_2O}\right)$$

26.10

Since the apparent concentration of NOx could be arbitrarily decreased without reducing the NOx production rate by simply diluting the combustion gas with excess air, it is common to correct NOx readings to 12% CO_2 by volume on a dry basis. (This is indicated by the units *ppmvd*.) Standardizing is accomplished using Eq. 26.11.

Correcting Gas Streams for Standard Conditions

$$Q_S = Q_A\left(\frac{P_A}{P_S}\right)\left(\frac{T_S}{T_A}\right)\left(\frac{1}{1-y_{H_2O(g)}}\right) \\ \times \left(\frac{\begin{array}{c}12\%\ CO_2\ \text{by volume}\\ \text{in dry gas stream}\end{array}}{\begin{array}{c}X\%\ CO_2\ \text{by volume}\\ \text{in dry gas stream}\end{array}}\right)$$

26.11

In Eq. 26.11, the subscripts A and S stand for actual and standard conditions, respectively; y is the mole fraction of water vapor in the flue gas; and $X\%$ is the actual concentration of CO_2 in the flue gas. A correction to 7% CO_2 content by volume can also be made using Eq. 26.12.

Correcting Gas Streams for Standard Conditions

$$\left(\frac{m_a}{V_{T,D,7\%O_2}}\right) = \left(\frac{m_a}{V_{T,D}}\right) \\ \times \left(\frac{(12\% - 7\%)\left(\begin{array}{c}O_2\ \text{by volume}\\ \text{in dry}\\ \text{gas stream}\end{array}\right)}{21\%\left(\begin{array}{c}O_2\ \text{by volume}\\ \text{in dry}\\ \text{gas stream}\end{array}\right) - X\%\left(\begin{array}{c}O_2\ \text{by volume}\\ \text{in dry}\\ \text{gas stream}\end{array}\right)}\right)$$

26.12

---

[19]*Urea* (NH_2CONH_2), also known as *carbamide urea*, is a water-soluble organic compound prepared from ammonia. Urea has significant biological and industrial usefulness.

[20]Air is injected into the furnace at high velocity over the combustion bed to create turbulence and to provide oxygen.

[21]46 is the molecular weight of NO_2. The predominant oxide in the flue gas is NO, and NO_2 may be only 10% to 15% of the total NOx. Furthermore, NO is measured, not NO_2. However, NO has a short half-life and is quickly oxidized to NO_2 in the atmosphere.

Some regulations specify NOx limitations in terms of pounds per hour or in terms of pounds of NOx per million Btus (MMBtu) of gross heat released. The mass emission ratio, E, in lbm/MMBtu, is calculated from the concentration in pounds per dry, standard cubic foot (dscf).

$$E_{\text{lbm/MMBtu}} = \frac{C_{\text{lbm/dscf}} V_{\text{NOx,dscf/hr}}}{q_{\text{MMBtu/hr}}}$$

$$= \frac{C_{\text{lbm/dscf}}(100) V_{\text{th,CO}_2}}{C_{m,\text{CO}_2,\%}(\text{HHV}_{\text{MMBtu/lbm}})} \qquad \textbf{26.13}$$

The relationship between the NOx concentrations, C, expressed in pounds per dry standard volume and ppm can be calculated from Eq. 26.14.[22] Conversions between ppm and pounds are made assuming NOx has a molecular weight of 46.

$$C_{\text{g/m}^3} = (4.15 \times 10^{-5}) C_{\text{ppm}} M \qquad \text{[SI]} \quad \textbf{26.14(a)}$$

$$C_{\text{lbm/ft}^3} = (2.59 \times 10^{-9}) C_{\text{ppm}} M \qquad \text{[U.S.]} \quad \textbf{26.14(b)}$$

The theoretical volume of carbon dioxide, $V_{\text{th,CO}_2}$, produced can be determined stoichiometrically. However, Table 26.3 can be used for quick estimates with an accuracy of approximately $\pm 5.9\%$.

Table 26.3 Approximate CO_2 Production for Various Fuels (with 0% excess air)

fuel	standard* ft^3/10^6 Btu	standard* m^3/10^6 cal
coal, anthracite	1980	0.222
coal, bituminous	1810	0.203
coal, lignite	1810	0.203
gas, butane	1260	0.412
gas, natural	1040	0.117
gas, propane	1200	0.135
oil	1430	0.161

*Standard conditions are 70°F (21°C) and 1 atm pressure.

Equation 26.13 is based on the theoretical volume of carbon dioxide gas produced per pound of fuel. Other approximations and correlations are based on the total dry volume of the flue gas in standard cubic feet per million Btu at 3% oxygen, $V_{t,\text{dry}}$. For quick estimates on furnaces burning natural gas, propane, and butane, V_t is approximately 10,130 ft^3/MMBtu; for fuel oil, V_t is approximately 10,680 ft^3/MMBtu.

$$E_{\text{lbm/MMBtu}} = (C_{\text{ppmv at 3\% O}_2})(V_{t,\text{dry}})$$

$$\times \left[\frac{46 \, \dfrac{\text{lbm}}{\text{lbmol}}}{\left(379.3 \, \dfrac{\text{ft}^3}{\text{lbmol}}\right)(10^6)} \right] \qquad \textbf{26.15}$$

As with all pollutants, the maximum allowable concentration or discharge of NOx is subject to continuous review and revision. Actual limits may depend on the type of geographical location, type of fuel, size of the facility, and so on. For steam or electrical (gas turbine) plants, the general target is 25 ppm to 40 ppm. The lower values apply to combustion of natural gas, and the higher values apply to combustion of distillate oil. Even lower values (down to 9 ppm to 10 ppm) are imposed in some areas.

Example 26.5

A combustion turbine produces 550,000 lbm/hr of exhaust gases. By volume, the exhaust gas contains 10% water vapor and 18% carbon dioxide (CO_2). Assume the rest of the gas is nitrogen. The flue gas temperature is 100°C at 1.2 atm pressure. What is the gas flow rate at standard conditions corrected to 12% CO_2?

Solution

The molecular weight of water (H_2O) is

$$M_{\text{H}_2\text{O}} = 2M_{\text{H}} + M_{\text{O}}$$

$$= (2)\left(1 \, \frac{\text{lbm}}{\text{lbmol}}\right) + 16 \, \frac{\text{lbm}}{\text{lbmol}}$$

$$= 18 \, \text{lbm/lbmol}$$

Nitrogen has a molecular weight of 28 lbm/lbmol. The mole fraction of water vapor is

$$y_{\text{H}_2\text{O}} = \left(\frac{m_{\text{H}_2\text{O}}}{100\%}\right)\left(\frac{M_{\text{H}_2\text{O}}}{M_{\text{N}} + M_{\text{H}_2\text{O}}}\right)$$

$$= \left(\frac{10\%}{100\%}\right)\left(\frac{18 \, \dfrac{\text{lbm}}{\text{lbmol}}}{28 \, \dfrac{\text{lbm}}{\text{lbmol}} + 18 \, \dfrac{\text{lbm}}{\text{lbmol}}}\right)$$

$$= 0.03913$$

---

[22]The constants in Eq. 26.14 are the same and can be used for any gas.

The gas flow rate at standard conditions ($T_S = 298K$, $P_S = 1$ atm) is

Correcting Gas Streams for Standard Conditions

$$Q_S = Q_A\left[\frac{P_A}{P_S}\right]\left[\frac{T_S}{T_A}\right]\left[\frac{1}{1 - y_{H_2O(g)}}\right]$$

$$\times\left(\frac{12\% \ CO_2 \ \text{by volume in dry gas stream}}{X\% \ CO_2 \ \text{by volume in dry gas stream}}\right)$$

$$= \left(550{,}000 \ \frac{lbm}{hr}\right)\left(\frac{1.2 \ atm}{1.0 \ atm}\right)\left(\frac{298K}{110°C + 273°}\right)$$

$$\times\left(\frac{1}{1 - 0.03913}\right)\left(\frac{12\%}{18\%}\right)$$

$$= 356{,}243 \ lbm/hr$$

12. SMOG

Photochemical smog (usually, just *smog*) consists of ground-level ozone and peroxyacyl nitrates (PAN). Smog is produced by the sunlight-induced reaction of ozone *precursors*, primarily nitrogen dioxide (NOx), hydrocarbons, and volatile organic compounds (VOCs). NOx and hydrocarbons are emitted by combustion sources such as automobiles, refineries, and industrial boilers. VOCs are emitted by manufacturing processes, dry cleaners, gasoline stations, print shops, painting operations, and municipal wastewater treatment plants.

13. SMOKE

Smoke results from incomplete combustion and indicates unacceptable combustion conditions. In addition to being a nuisance problem, smoke contributes to air pollution and reduced visibility. Smoke generation can be minimized by proper furnace monitoring and control.

Opacity can be measured by a variety of informal and formal methods, including transmissometers mounted on the stack. The sum of the *opacity* (the fraction of light blocked) and the *transmittance* (the fraction of light transmitted) is 1.0.

Optical density is calculated from Eq. 26.16. The *smoke spot number* (SSN) can also be used to quantify smoke levels.

Plume Opacity

$$\text{Opacity}[\%] = \left(1 - \frac{I}{I_o}\right) \times 100\% \qquad 26.16$$

Visible moisture plumes with opacities of 40% are common at large steam generators even when there are no unburned hydrocarbons emitted. High-sulfur fuels and

the presence of ammonium chloride (a by-product of some ammonia-injection processes) seem to increase formation of visible plumes. Moisture plumes from saturated gas streams can be avoided by reheating prior to discharge to the atmosphere.

14. SULFUR OXIDES

Sulfur oxides (SOx), consisting of *sulfur dioxide* (SO$_2$) and *sulfur trioxide* (SO$_3$), are the primary cause of acid rain. *Sulfurous acid* (H$_2$SO$_3$) and sulfuric acid (H$_2$SO$_4$) are produced when oxides of sulfur react with moisture in the flue gas. Both of these acids are corrosive.

$$SO_2 + H_2O \rightarrow H_2SO_3 \qquad 26.17$$

$$SO_3 + H_2O \rightarrow H_2SO_4 \qquad 26.18$$

Fuel switching (*coal substitution/blending* (CS/B)) is the burning of low-sulfur fuel. (Low-sulfur fuels are approximately 0.25–0.65% sulfur by weight, compared to high-sulfur coals with 2.4–3.5% sulfur.) However, unlike nitrogen oxides, which can be prevented during combustion, formation of sulfur oxides cannot be avoided when low-cost, high-sulfur fuels are burned.

Some air quality regulations regarding SOx production may be met by a combination of options. These options include fuel switching, flue gas scrubbing, derating, and allowance trading. The most economical blend of these options will vary from location to location.

In addition to fuel switching, available technology options for retrofitting existing coal-fired plants include wet scrubbing, dry scrubbing, sorbent injection, repowering with clean coal technology (CCT), and co-firing with natural gas.

Sulfur dioxide (like carbon dioxide and nitrogen oxides) emissions must be reported on a standardized basis. Equation 26.19 can be used to calculate the volumetric, dry basis concentration from a wet stack gas sample.

$$\frac{SO_{2,dry,at\,3\%\,O_2}}{SO_{2,wet,at\,stack\,O_2}} = \frac{CO_{2,th,dry,at\,3\%\,O_2}}{CO_{2,wet,at\,stack\,O_2}} \qquad 26.19$$

The sulfur dioxide concentration (mass per dry standard volume, C) can be calculated from sulfur dioxide's molecular weight ($M = 64.07$) and Eq. 26.14.[23]

Example 26.6

A wet stack gas analysis shows 230 ppm of SO$_2$ and 10.5% CO$_2$. Based on stoichiometric combustion, the theoretical carbon dioxide percentage at 3% oxygen should have been 13.4%. What is the SO$_2$ concentration in ppm on a dry basis, standardized to 3% oxygen?

---

[23]See Ftn. 33.

Solution

Use Eq. 26.19.

$$SO_{2,\text{dry,at}3\%O_2} = SO_{2,\text{wet,at stack}O_2}$$

$$\times \left(\frac{CO_{2,\text{th,dry,at}3\%O_2}}{CO_{2,\text{wet,at stack}O_2}} \right)$$

$$= \frac{(230 \text{ ppm})(0.134)}{0.105}$$

$$= 294 \text{ ppm}$$

15. GLOBAL ISSUES

Introduction

Pollution becomes a global issue when it crosses national boundaries. Ready examples are rivers that flow through more than one country, ocean outfalls near border cities, and air pollution carried across borders. Global pollution issues are usually politically important because pollution generated somewhere else, often by a society with different social or political objectives, is deposited at a location far removed from its source. Prominent global air pollution issues include acid rain, global warming associated with greenhouse gases, and ozone depletion.

Acid Rain

Acid rain results from deposition of nitric acid and sulfuric acid formed when nitrogen and sulfur oxides mix with atmospheric moisture. Once formed, the acids may fall to earth in rainfall or they may dry in the atmosphere and ultimately be deposited in a dry form. The acids depress the pH of rainwater and of surface water, soils, and vegetation surfaces.

Most rainfall worldwide has a natural background pH of around 5.0, which is fairly acidic. Some evidence exists to suggest that this pH has been influenced by air emissions since the early 1900s. Areas in the United States with the most significantly acidified surface waters are somewhat localized in New York and in Florida. From 14% to 23% of lakes in these regions are acidified. The close proximity of many countries of western and eastern Europe has created some significant impacts from emissions originating in less regulated countries.

One problem has been the application of stack height criteria as a means of emission control. The stack height concept is to place the pollutants high enough for them to be dispersed in the atmosphere over a wide area away from the source. This works well for local air quality control, but is not very effective for regional or global pollution control. Nevertheless, it has been a dominant air pollution control strategy in the United States and elsewhere in the world.

Example 26.7

Assume rainwater with a natural pH of 5.0 reacts with SO_3 in the air to form H_2SO_4 at 320 μg/L. What is the pH of the rainwater after reacting with SO_3?

Solution

From Eq. 26.18 and dissociation of H_2SO_4,

$$SO_3 + H_2O \rightarrow H_2SO_4 \rightarrow 2H^+ + SO_4^{2-}$$

1 mol of H_2SO_4 will yield 2 mol of H^+ in the rainwater. The molecular weight of H_2SO_4 is 98 g/mol.

At pH 5.0, the H^+ concentration of natural rainwater is 10^{-5} mol/L.

$$H^+ \text{ added} = \frac{\left(2 \frac{\text{mol } H^+}{\text{mol } H_2SO_4}\right)\left(320 \frac{\mu g}{L}\right)}{\left(98 \frac{g}{\text{mol } H_2SO_4}\right)\left(10^6 \frac{\mu\text{mol}}{\text{mol}}\right)}$$

$$= 6.5 \times 10^{-6} \text{ mol/L}$$

The final H^+ concentration of the rainwater is

$$10^{-5} \frac{\text{mol}}{L} + 6.5 \times 10^{-6} \frac{\text{mol}}{L} = 1.65 \times 10^{-5} \text{ mol/L}$$

The final pH of the rainwater is

$$pH = -\log[H^+] = -\log\left(1.65 \times 10^{-5} \frac{\text{mol}}{L}\right)$$

$$= 4.78$$

Ozone Depletion

The problem with chlorinated fluorocarbons (CFCs) in the upper atmosphere is their photolytic decomposition, which releases free chlorine atoms. The free chlorine then reacts with ozone to form oxygen. If the cycle repeats, a net decrease in ozone can occur. A simplified reaction sequence is shown below where the CFC is represented by Freon-12 (CF_2Cl_2).

$$CF_2Cl_2 + \text{light} \rightarrow CF_2 + 2Cl \qquad \textbf{26.20}$$

$$Cl + O_3 \rightarrow ClO + O_2 \qquad \textbf{26.21}$$

$$ClO + O \rightarrow Cl + O_2 \qquad \textbf{26.22}$$

International law closely regulates the production and sale of CFCs through out-and-out bans on their use and production, taxes on emissions from their use, and a permit program. A fund was also established to assist poorer countries in implementing alternatives to ozone-depleting chemicals. Research to develop alternatives to CFCs is also funded, primarily by the industries that stand to benefit.

Global Warming

The *greenhouse gases* (GHG) include carbon dioxide (CO_2), methane (CH_4), nitrous oxide (N_2O), oxygen (O_2), ozone (O_3), and water vapor (H_2O). These gases, most notably carbon dioxide, trap heat from the earth's surface and the atmosphere. Not being able to radiate into space, the trapped heat creates increased ambient temperatures. Global temperature increases of 1°C to 2°C would be very significant, causing dramatic changes in global climate and a rise in the sea level.

Several measures have been proposed to address global warming. Two primary areas of focus are reducing dependence on fossil fuels (reducing combustion sources and CO_2 production) and preventing deforestation combined with restoring deforested areas (increasing CO_2 use by increasing vegetation). These are identified as prevention and mitigation strategies and exist in conflict with economic goals, at least in the short term.

16. NOMENCLATURE

A	area	ft^2	m^2
B	volumetric fraction	–	–
C	coefficient	–	–
C	concentration	lbm/lbmol	kg/kmol
d	diameter	ft	m
D	draft	lbf/ft^3	kPa
E	mass emission ratio	lbm/MMBtu	–
F	friction coefficient	–	–
g	acceleration of gravity, 32.2 (9.81)	ft/sec^2	m/s^2
g_c	gravitational constant, 32.2	lbm-ft/lbf-sec^2	n.a.
h	head	ft	m
H	height	ft	m
K	friction coefficient	–	–
L	length	ft	m
m	mass	lbm	kg
$\dot{m}$	mass flow rate	lbm/sec	kg/s
M	molecular weight	lbm/lbmol	kg/kmol
Q	volumetric flow rate	gal/min	L/s
P	partial pressure	atm	atm
P	pressure	lbf/ft^2	kPa
R	specific gas constant	ft-lbf/lbm-°R	kJ/kg·K
T	absolute temperature	°R	K
T	temperature	°F	°C
v	velocity	ft/sec	m/s
V	volume	ft^3	m^3
x	mole fraction	–	–
X	actual concentration	%	%
y	mole fraction	–	–

Symbols

γ	specific weight	ft^3/lbm	m^3/kg
ρ	density	lbm/ft^3	kg/m^3

Subscripts

a	air
A	actual or volumetric flue gas
D	dry
m	by mass
s	sulfur trioxide
S	standard
SE	stack effect
t	total
T	temperature
th	theoretical
v	velocity
w	water

27 Atmospheric Science and Meteorology

1. METEOROLOGY

Introduction

Although the earth's atmosphere is approximately 100 mi deep, the region extending from ground surface to about 12 mi, called the *troposphere*, is where air pollution problems are encountered. Weather that affects daily life activities also occurs within the troposphere. The dispersion of air pollution in the atmosphere and subsequent human exposure is largely a function of meteorological (weather) conditions.

Wind

The earth's rotation and solar radiation induce global wind patterns. Local factors such as topography, land mass, diurnal cycles, seasonal temperature, cloud conditions, and pressure contribute to local wind patterns.

Local wind is measured as a horizontal velocity from the originating direction. Hence, a north wind originates from the north and is blowing to the south. Localized wind patterns may be represented by a *wind rose*. A variety of configurations are employed for wind roses, but they all show wind direction, magnitude, and frequency. A typical wind rose is illustrated in Fig. 27.1.

Example 27.1

Using the wind rose in Fig. 27.1, what percent of the time do winds blow from the southwest, and what is their maximum velocity?

Solution

The wind blows from the southwest about 19% of the time with a maximum velocity of between 16 mph and 30 mph.

Lapse Rate and Stability

The *lapse rate*, Γ, is the rate of air temperature change with elevation and is used to assess the stability of atmospheric conditions. The lapse rate will determine the configuration of an air pollutant plume and the dispersion pattern of the pollutants. The *dry adiabatic lapse rate*, Γ_{AD}, is the temperature change as a parcel of dry air rises or falls without exchanging heat with its surroundings. The *wet adiabatic lapse rate*, where moisture exits in the atmosphere, is slightly less than the dry adiabatic lapse rate. However, for most practical purposes, they are approximately equal, and 0.98°C per 100 m (5.4°F per 1000) can be used for either. [Selected Properties of Air]

As an air pollutant is emitted from a stack, its temperature will likely be greater than the temperature of the ambient air. As the pollutant cools, the actual lapse rate of the surrounding air (the *ambient lapse rate*) will define how the pollutant is dispersed. The four general conditions illustrated in Fig. 27.2 and described as follows can result.

- *Superadiabatic lapse rate* occurs when the surrounding air cools faster than the plume as elevation increases ($\Gamma > \Gamma_{AD}$). This represents an unstable condition that will allow pollutants to readily disperse.

- *Adiabatic lapse rate* occurs when the surrounding air and the plume cool at the same rate ($\Gamma = \Gamma_{AD}$). When this occurs, there will be no tendency for the plume to rise or fall due to temperature differences.

- *Subadiabatic lapse rate* occurs when the surrounding air cools slower than the plume as elevation increases ($\Gamma < \Gamma_{AD}$). This represents a stable condition that interferes with the dispersion of the plume.

- *Inversion* occurs when the lapse rate is inverted so that the ambient air temperature is cooler near the ground than at elevation. This represents a stable condition.

Lapse rates can vary at different elevations to produce a variety of stability conditions and plume configurations as shown in Fig. 27.3.

Example 27.2

An 80 m stack emits a plume at 32°C. The air temperature at ground level is 23°C, and the ambient lapse rate up to an elevation of 300 m is 0.81 °C/100 m. Above 300 m, the lapse rate is -0.53 °C/100 m. How high will the plume rise?

Figure 27.1 *Typical Wind Rose*

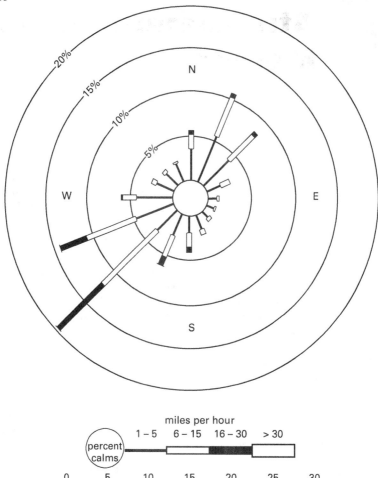

miles per hour

Source: *Meteorology and Atomic Energy* © 1955, U.S. Dept. of Commerce Weather Bureau, U.S. Atomic Energy Commission, Washington, DC.

Figure 27.2 *Lapse Rates*

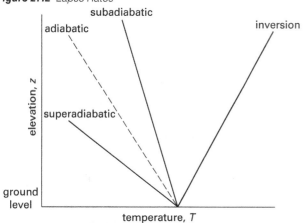

Solution

The plume cools at the dry adiabatic lapse rate of 0.98 °C/100 m as it rises and will stop rising when the plume and air temperatures are equal.

The plume temperature at 300 m is

$$32°C + \left(-0.98 \ \frac{°C}{100 \text{ m}}\right)(300 \text{ m} - 80 \text{ m}) = 29.8°C$$

The air warms at the ambient lapse rate of +0.81 °C/100 m up to an elevation of 300 m. The air temperature at 300 m is

$$23°C + \left(0.81 \ \frac{°C}{100 \text{ m}}\right)(300 \text{ m}) = 25.4°C$$

The air cools above 300 m at the ambient lapse rate of −0.53 °C/100 m.

$$29.8°C + \left(-0.98 \ \frac{°C}{100 \text{ m}}\right)z = 25.4°C + \left(-0.53 \ \frac{°C}{100 \text{ m}}\right)z$$

z is the distance above 300 m where the air and plume temperatures are equal.

$$z = \frac{29.8°C - 25.4°C}{\left(-0.53 \ \dfrac{°C}{100 \text{ m}}\right) - \left(-0.98 \ \dfrac{°C}{100 \text{ m}}\right)} = 978 \text{ m}$$

The plume will rise to 300 m + 978 m = 1278 m.

Maximum Mixing Depth

As air pollutants are emitted into the atmosphere, their dispersion is aided by mixing through convection and turbulence. The greater the volume of air that is available for mixing, the more effective dispersion will be in reducing the impact of the emission. The dry adiabatic lapse rate and the ambient lapse rate are used to define the height to which a plume will rise in the atmosphere. The two lapse rates are used to calculate the elevation at which the temperatures of the ambient air and the emitted plume are equal. This elevation defines the height of the *mixing zone*, commonly referred to as the maximum mixing depth. Figure 27.4 illustrates the *maximum mixing depth* under different conditions of atmospheric stability.

Example 27.3

Using a graphical solution, what is the maximum mixing depth for the conditions described in Ex. 27.2?

Solution

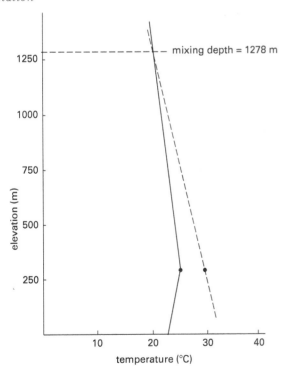

The maximum mixing depth is 1278 m.

Dispersion

Dispersion involves the horizontal and vertical spreading of a pollutant plume as it moves away from the source. This spreading is most commonly described using the *Gaussian plume* model. The physical features of the model are shown in Fig. 27.5 using a three-dimensional *xyz* coordinate system.

The model can estimate the downwind configuration of the plume and hence concentrations of emitted pollutants at specific locations of interest. Selected applications of the model are presented as follows.

- Assuming the ground is a perfect reflection, the ground-level concentration at location (x, y) downwind of the source and originating from a stack with an effective height H, with no reflections from the ground, is

  Atmospheric Dispersion Modeling (Gaussian)

  $$C = \frac{Q}{2\pi u \sigma_y \sigma_z} \exp\left[-\frac{1}{2}\frac{y^2}{\sigma_y^2}\right]$$
  $$\left[\exp\left[-\frac{1}{2}\frac{(z-H)^2}{\sigma_z^2}\right] + \exp\left[-\frac{1}{2}\frac{(z+H)^2}{\sigma_z^2}\right]\right] \quad \text{27.1}$$

The *effective stack height*, H, is equal to the stack height, h, plus an incremental height, Δh, to correct for the initial rise of the plume as it leaves the stack. However, the incremental stack height is commonly ignored, and the stack height and effective stack height are assumed to be equal. The effective stack height can be calculated from empirical relationships.

- The ground-level concentration at some distance along the plume centerline and originating from a stack with an effective height H, with no reflections from the ground, is

  $$C = \frac{Q}{\pi u \sigma_y \sigma_z} \exp\left(-\frac{1}{2}\frac{y^2}{\sigma_y^2}\right) \quad \text{27.2}$$

- The ground-level concentration at some distance along the plume centerline and originating from a ground-level source, with no reflections from the ground, is

  $$C = \frac{Q}{\pi u \sigma_y \sigma_z} \quad \text{27.3}$$

- The location of the maximum ground-level concentration along the plume centerline and originating from a stack with an effective height H, with no reflections from the ground, is

Figure 27.3 *Representative Plume Stability Configurations (adiabatic lapse rate shown as dashed line for comparison)*

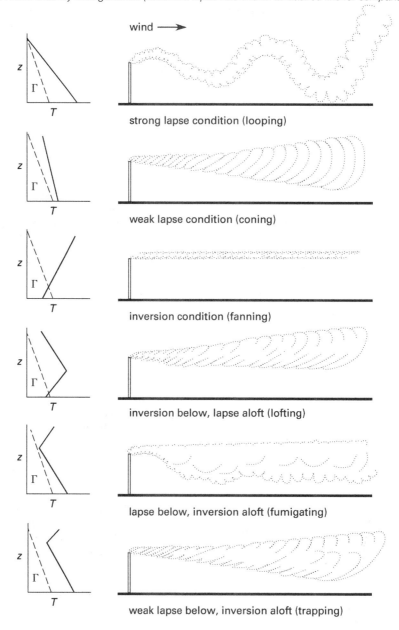

strong lapse condition (looping)

weak lapse condition (coning)

inversion condition (fanning)

inversion below, lapse aloft (lofting)

lapse below, inversion aloft (fumigating)

weak lapse below, inversion aloft (trapping)

Source: *Meteorology and Atomic Energy* © 1955, U.S. Dept. of Commerce Weather Bureau, U.S. Atomic Energy Commission, Washington, DC.

Figure 27.4 *Representative Maximum Mixing Depths*

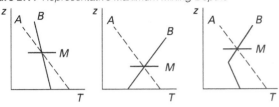

A = adiabatic lapse rate
B = ambient lapse rate
M = maximum mixing depth

Figure 27.5 *Plume Coordinates For Gaussian Dispersion Model*

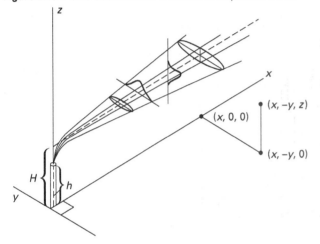

Atmospheric Dispersion Modeling (Gaussian)

$$\sigma_z = \frac{H}{\sqrt{2}} \qquad 27.4$$

The *Gaussian dispersion coefficients*, σ_y and σ_z, are taken from Fig. 27.6 and Fig. 27.7 for the atmospheric stability conditions defined in Table 27.1.

Table 27.1 *Atmospheric Stability Categories*

surface wind speed at 10 m (m/s)	day incoming solar radiation			night cloud cover	
	strong	moderate	slight	mostly overcast	mostly clear
class*	1	2	3	4	5
< 2	A	A-B	B	E	F
2–3	A-B	B	C	E	F
3–5	B	B-C	C	D	E
5–6	C	C-D	D	D	D
> 6	C	D	D	D	D

* The neutral class, D, should be assumed for overcast conditions during day or night. Class A is the most unstable, and class F is the most stable. Class B is moderately unstable, and class E is slightly stable.

Source: Wark, K. and C.F. Warner, *Air Pollution*, 2nd ed., © 1981, Harper & Row, NY.

Example 27.4

An air pollutant is emitted at 3.6 kg/s through a stack with an effective height of 225 m. The wind speed at 10 m above grade is 4 m/s, and atmospheric stability conditions are class B. (a) What is the location of the maximum ground (ground-level concentration) level concentration along the plume centerline? (b) What is the maximum ground-level concentration?

Solution

(a) Use Eq. 27.4 to find the location of maximum ground-level concentration.

Atmospheric Dispersion Modeling (Gaussian)

$$\sigma_z = \frac{H}{\sqrt{2}}$$
$$= \frac{225 \text{ m}}{\sqrt{2}}$$
$$= 159 \text{ m}$$

From Fig. 27.6 with $\sigma_z = 159$ m and using the curve for stability class B, $x = 1500$ m.

(b) From Fig. 27.7 with $x = 1500$ m and using the curve for stability class B, $\sigma_y = 200$ m.

Use Eq. 27.2 to find the maximum ground-level concentration along the plume centerline.

Atmospheric Dispersion Modeling (Gaussian)

$$C_{max} = \frac{Q}{\pi u \sigma_y \sigma_z} \exp\left[-\frac{1}{2}\frac{(H^2)}{\sigma_z^2}\right]$$

$$= \frac{\left(3.6 \frac{\text{kg}}{\text{s}}\right)\left(10^9 \frac{\mu\text{g}}{\text{kg}}\right)}{\pi\left(4 \frac{\text{m}}{\text{s}}\right)(200 \text{ m})(159 \text{ m})}$$

$$\times \exp\left[-\frac{1}{2}\frac{(225 \text{ m})^2}{(159 \text{ m})^2}\right]$$

$$= 3310 \ \mu\text{g/m}^3$$

Figure 27.6 *Vertical Gaussian Dispersion Coefficient, σ_z, for Vertical Plume Concentration at Downwind Distance, x (see Table 27.1 for definition of curves)*

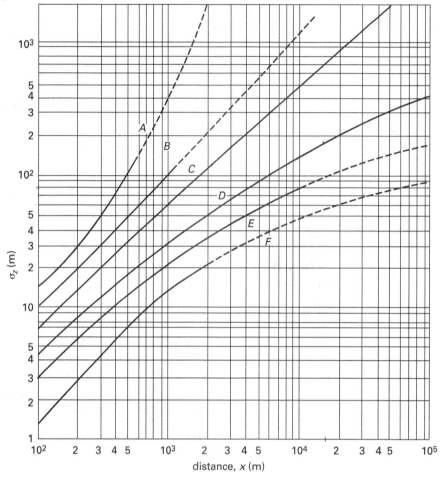

Source: Wark, K. and C.F. Warner, *Air Pollution*, 2nd ed., © 1981, Harper & Row, NY.

Figure 27.7 *Horizontal Gaussian Dispersion Coefficient, σ_y, for Horizontal Plume Concentration at Downwind Distance, x (see Table 27.1 for definition of curves)*

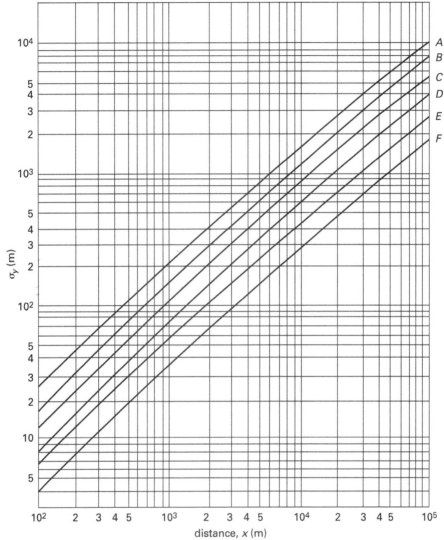

Source: Wark, K. and C.F. Warner, *Air Pollution*, 2nd ed., © 1981, Harper & Row, NY.

2. NOMENCLATURE

C	steady-state concentration at a point (x, y, z)	$\mu g/m^3$
C_{max}	maximum ground-level concentration	$\mu g/m^3$
h	stack height	m
Δh	incremental height	m
H	effective stack height	m
Q	emissions rate	$\mu g/s$
T	temperature	K
u	average wind speed at stack height	m/s
x	downwind distance along plume centerline	m
y	horizontal distance from plume	m
z	elevation	m
z	vertical distance from ground level	m

Symbols

Γ	lapse rate	$°C/m$
σ_y	Gaussian horizontal dispersion parameter	m
σ_z	Gaussian vertical dispersion parameter	m

Subscripts

AD	dry adiabatic

28 Sources of Pollution

Sources of Pollution

Content in blue refers to the *NCEES Handbook*.

1. POLLUTANT CLASSIFICATION

Introduction

Air pollutants may be classified as primary and secondary. *Primary pollutants* are those that exist in the air in the same form in which they were emitted. An example of a primary pollutant is carbon monoxide emitted directly from gasoline-powered automobiles. *Secondary pollutants* are those that are formed in the air from other emitted compounds. Ground-level ozone formed from the nitrogen dioxide photolytic cycle is an example of a common secondary pollutant. Primary and secondary pollutants are not the same as, and should not be confused with, the primary and secondary NAAQS. The major source of air pollutants in the United States and most other industrialized countries is fossil fuel combustion from both stationary and mobile sources. To a much lesser degree, pollutants may be generated from wind erosion, plant and animal bio-effluents, vapors from architectural coatings and other hydrocarbon-based materials (e.g., new asphalt pavements), industrial manufacturing, and a variety of other sources.

Particulate Matter

Suspended *particulate matter* (PM) is produced in a variety of forms. These include *dust* from manufacturing, agricultural, and construction activities, *fumes* produced by chemical processes, *mists* formed by condensed water vapor, *smoke* generated by incomplete combustion from transportation and industrial activities, and *sprays* formed by atomization of liquids. Natural wind erosion and forest fires also produce dust and smoke. The health risks associated with particulate matter exposure are exacerbated by the adsorption from the air of toxic organics onto particle surfaces.

Particulate matter may be liquid or solid with a broad size range, from 0.005 μm to 100 μm. The *particulate size* is taken as the aerodynamic diameter of an equivalent perfect sphere having the same settling velocity. Particulate matter in the size range from 0.1 μm to 10 μm presents the greatest potential impact on air quality. Particulate matter smaller than 0.1 μm tends to coagulate into larger particles, and particles larger than 10 μm settle relatively rapidly. Of particular concern is particulate matter with aerodynamic diameters less than 10 μm (PM_{10}) and less than 2.5 μm ($PM_{2.5}$) for which NAAQS are included. Particles making up $PM_{2.5}$ are called *respirable particulate matter* and present the greatest potential risk to human health because they are able to reach and remain in the lung. The sources, forms, and sizes of particulate matter are summarized in Table 28.1.

The *settling velocity* for a particle of particulate matter can be calculated from *Stokes' law*.

Stokes' Law

$$v_t = \frac{g(\rho_p - \rho_f)d^2}{18\mu} = \frac{g\rho_f(\text{S.G.} - 1)d^2}{18\mu} \qquad 28.1$$

The same parameters can also be used to calculate the *settling Reynolds number*.

Table 28.1 *Particulate Matter Types*

type	source	form	size
dust	emissions from handling and mechanical activities (e.g., grain elevators, sandblasting, construction)	solid	1 μm to 10 mm
fume	condensed vapors from chemical reactions (e.g., calcination, distillation, sublimation reactions)	solid	0.03 μm to 0.3 μm
mist	condensed vapors from chemical reactions (e.g., sulfuric acid from sulfur trioxide)	liquid	0.5 μm to 3 μm
smoke	particulate emissions from combustion activities (e.g., forest fires, coal-fired power plants)	solid	0.05 μm to 1 μm
spray	atomized liquids (e.g., sprayed coatings, pesticide application)	liquid	10 μm to 10 mm

Adapted from: Vesilind, P.A., J.J. Pierce, and R.F. Weiner, 1994, *Environmental Engineering*, 3rd ed., Butterworth-Heinemann, Newton, MA.

General Spherical

$$\mathrm{Re} = \frac{v_t \rho d}{\mu} \qquad \textbf{28.2}$$

Equation 28.1 is valid for particle diameters between about 0.1 μm and 100 μm and for settling velocities less than about 1.0 m/s, both of which are well within the range of interest in air pollution applications.

Particulate lead pollution is most pronounced in areas where automobile traffic has historically been high. Although the use of unleaded gasoline has greatly reduced airborne lead pollution from automobile emissions, residual contamination persists in soils along roadways, in parks, and in other areas where soil is exposed. This contaminated soil can act as an ongoing source of airborne lead when disturbed by vehicle traffic, construction activities, natural wind action, and other events.

An important source of airborne lead not associated with leaded gasoline use is chipping and flaking lead-based paint. When lead-based paint, used either indoors or outdoors, is sanded or when it chips or flakes, the lead-containing particles become airborne and can be inhaled or contribute to soil contamination. Lead is no longer used in paints intended for indoor applications in the United States, but it is still used in some exterior paints such as for roadway markings.

Exposure to lead-based paint presents a significant health hazard to children and is associated with behavioral changes, learning disabilities, and other health problems. Lead is included in the NAAQS.

Asbestos is a fibrous silicate material that is inert, strong, and incombustible. As *asbestos-containing materials* (ACM) age or are damaged, asbestos fibers are released into the air. Although the greatest potential exposure to asbestos occurs indoors where asbestos-containing materials were used in construction, soil containing mineral asbestos has contributed to significant asbestos exposures in some areas of the United States.

The most substantial asbestos exposure has historically been occupational. Asbestos poses health risks in relatively localized areas from industrial and construction-related activities. Asbestos is included in the national emission standards for hazardous air pollutants (NESHAP).

Example 28.1

What is the settling velocity for particles with an aerodynamic diameter of 10 μm and a density of 700 kg/m^3 in an atmosphere at 20°C and 1 atm?

Solution

From a table of air properties, the dynamic viscosity of air at 20°C is $\mu = 1.817 \times 10^{-5}$ kg/m·s, and air density is $\rho = 1.21$ kg/m^3 at 20°C.

Use Eq. 28.1.

Stokes' Law

$$
v_t = \frac{g(\rho_p - \rho_f)d^2}{18\mu}
$$

$$
= \frac{\left(9.81 \ \dfrac{\text{m}}{\text{s}^2}\right)\left(700 \ \dfrac{\text{kg}}{\text{m}^3} - 1.21 \ \dfrac{\text{kg}}{\text{m}^3}\right) \times \left(\dfrac{10 \ \mu\text{m}}{10^6 \ \dfrac{\mu\text{m}}{\text{m}}}\right)^2}{(18)\left(1.82 \times 10^{-5} \ \dfrac{\text{kg}}{\text{m·s}}\right)}
$$

$$
= 2.09 \times 10^{-3} \ \text{m/s}
$$

Gaseous Pollutants

The principal gaseous pollutants of concern include oxides of nitrogen, sulfur, and carbon; ozone; and hydrocarbons. Most gaseous pollutants are associated with combustion processes and the characteristics of the fuels burned. Potential emissions from combustion increase as combustion efficiency decreases, as available oxygen decreases, and as the presence of compounds other than carbon and hydrogen increases. The sources of gaseous pollutants and their fates in the atmosphere are summarized in Table 28.2.

Table 28.2 *Gaseous Emissions Sources*

gas	source	fate
methane and other organic gases and vapors	combustion, volatilization	reaction with ·OH
carbon monoxide and carbon dioxide	combustion	oxidation
nitrous oxide and nitrogen dioxide	combustion	oxidation, precipitation, dry deposition
sulfur dioxide and sulfur trioxide	combustion	oxidation, precipitation, dry deposition

The two nitrogen oxides (NOx) that are significant air pollutants are nitric oxide (NO) and nitrogen dioxide (NO_2). Most nitrogen oxide emissions occur from fuel release and fossil fuel combustion as relatively benign NO. However, NO can rapidly oxidize to NO_2, which is associated with respiratory ailments and which reacts with hydrocarbons in the presence of sunlight to form photochemical oxidants. Photochemical oxidants are a significant constituent of smog. Nitrogen dioxide also reacts with the hydroxyl radical (·OH) in the atmosphere to form nitric acid (HNO_3), a contributor to acid rain.

Most sulfur emissions result from the combustion of fossil fuels, especially coal. When burned, the fuels release the sulfur primarily as sulfur dioxide (SO_2) with much smaller amounts of sulfur trioxide (SO_3). These are known as *sulfur oxides* (SOx). Sulfur oxide aerosols contribute to particulate matter concentrations in the air and are a constituent of acid rain. Sulfur oxides also can significantly impact long-range visibility.

Combustion of fossil fuels under ideal conditions produces carbon dioxide (CO_2) and water. However, where conditions reduce combustion efficiency, *carbon monoxide* (CO) is formed instead of CO_2. Although CO_2 is associated with global warming, it typically is not included as an air pollutant. Carbon monoxide, however, represents as much as half the total mass of air pollutants emitted in the United States, with most of these emissions coming from motor vehicles. Carbon monoxide does not seem to present any particular environmental hazards, but it does present a potential serious threat to human health by displacing oxygen (O_2) in the blood.

Ozone (O_3) forms in the stratosphere and at ground level. Stratospheric ozone is maintained through naturally occurring processes and acts as a filter to ultraviolet light radiation from the sun. Stratospheric ozone is not an air pollutant. Ground-level ozone, on the other hand, is a secondary pollutant formed through reactions with nitrogen oxides and hydrocarbons in the presence of sunlight, making it a *photochemical oxidant*. Among the photochemical oxidants, ozone is the most common. As a secondary pollutant, emissions limitations are not applicable to ozone. Instead, emission controls for those compounds, such as nitrogen oxides and hydrocarbons, that are precursors to ozone formation are required. Ground-level ozone is a major contributor to photochemical smog and is an irritant to the mucosa and lungs.

Hydrocarbons are included as air pollutants because of their role in the formation of photochemical oxidants, particularly ground-level ozone. The hydrocarbons involved in air pollution are primarily alkanes, comprising the methane series. The methane series includes methane (CH_4), ethane (C_2H_6), propane (C_3H_8), butane (C_4H_{10}), and so on. Alkanes have the general formula C_nH_{2n+2}. If an alkane loses one hydrogen atom, it becomes a radical or alkyl (i.e., ·CH_3) that is able to readily react with other compounds. When oxidized, alkanes form compounds of the carbonyl group (*aldehydes*), where an oxygen atom is double-bonded to a terminal carbon atom. Formaldehyde (CH_2O) and acrolein (CH_2CHCOH) are common aldehyde constituents of photochemical smog that are associated with eye irritation.

Hydrocarbons include other aliphatic chemicals, in addition to alkanes, and some aromatic hydrocarbons. Chlorinated solvents such as carbon tetrachloride, tetrachloroethene, 1,1,1-trichloroethane, trichloroethene, and their degradation products; and benzene, toluene, ethylbenzene, and xylene are among the most common of these and comprise those compounds typically called *volatile organic compounds* (VOCs). Occurrence of VOCs in the air can be a major contributor to formation of photochemical smog.

Hazardous Air Pollutants

The hazardous air pollutants identified under NESHAP as HAPs comprise approximately 200 chemicals and related groups of compounds, and have been identified as potentially significant sources of health risk to humans. These pollutants occur from a variety of sources and include asbestos; compounds of antimony, arsenic, beryllium, lead, manganese, mercury, and nickel; benzene and benzene derivatives such as toluene, xylene, and phenols; halogen gases; halogenated alkanes; chloroform and related compounds; and radionuclides.

Allergens and Microorganisms

Allergens and microorganisms comprise *biological air pollutants* and include such things as algae, animal dander and hair, bacteria, excreta from animals and insects, pollens, mold spores, viruses, and other organic structures, living or dead. Exposure to biological air pollutants can be very dramatic, as in the outbreak of Legionnaires' disease at the American Legion convention in Philadelphia in 1976, or, more commonly, in the form of the symptoms of allergies and other common illnesses.

2. HYDROCARBONS

With the exception of sulfur and related compounds, most fuels are hydrocarbons. Hydrocarbons are further categorized into subfamilies such as *alkynes* (C_nH_{2n-2}, such as acetylene C_2H_2), *alkenes* (C_nH_{2n}, such as ethylene C_2H_4), and *alkanes* (C_nH_{2n+2}, such as octane C_8H_{18}). The alkynes and alkenes are referred to as *unsaturated hydrocarbons*, while the alkanes are referred to as *saturated hydrocarbons*. The alkanes are also known as the *paraffin series* and *methane series*. The alkenes are subdivided into the chain-structured *olefin series* and the ring-structured *naphthalene series*. Aromatic hydrocarbons (C_nH_{2n-6}, such as benzene C_6H_6) constitute another subfamily.

3. MOISTURE

If an ultimate analysis of a solid or liquid fuel is given, all of the oxygen is assumed to be in the form of free water.[1] The amount of hydrogen combined as free water is assumed to be one-eighth of the oxygen weight.[2] All remaining hydrogen, known as the *available hydrogen*, is assumed to be combustible. (Also, see Sec. 28.28.)

$$G_{H,combined} = \frac{G_O}{8} \qquad \text{28.3}$$

$$G_{H,available} = G_{H,total} - \frac{G_O}{8} \qquad \text{28.4}$$

Moisture in fuel is undesirable because it increases fuel weight (transportation costs) and decreases available combustion heat.[3] For coal, the *"bed" moisture level* refers to the moisture level when the coal is mined. The terms *dry* and *as fired* are often used in commercial coal specifications. The "as fired" condition corresponds to a specific moisture content when placed in the furnace. The "as fired" heating value should be used, since the moisture actually decreases the useful combustion energy. The approximate relationship between the two heating values is given by Eq. 28.5, where M is the moisture content from a proximate analysis.[4]

$$HV_{as\,fired} = HV_{dry}(1 - M) \qquad \text{28.5}$$

4. ASH AND MINERAL MATTER

Mineral matter is the noncombustible material in a fuel. *Ash* is the residue remaining after combustion. Ash may contain some combustible carbon as well as the original mineral matter. The two terms ("mineral matter" and "ash") are often used interchangeably when reporting fuel analyses.

Ash may also be categorized according to where it is recovered. Dry and wet *bottom ashes* are recovered from *ash pits*. However, as little as 10% of the total ash content may be recovered in the ash pit. *Fly ash* is carried out of the boiler by the flue gas. Fly ash can be deposited on walls and heat transfer surfaces. It will be discharged from the stack if not captured. *Economizer ash* and *air heater ash* are recovered from the devices the ashes are named after.

The finely powdered ash that covers combustion grates protects them from high temperatures.[5] If the ash has a low (i.e., below 2200°F; 1200°C) fusion temperature (melting point), it may form *clinkers* in the furnace and/or *slag* in other high-temperature areas. In extreme cases, it can adhere to the surfaces. Ashes with high melting (fusion) temperatures (i.e., above 2600°F; 1430°C) are known as *refractory ashes*. The T_{250} *temperature* is used as an index of slagging tendencies of an ash. This is the temperature at which the slag becomes molten with a viscosity of 250 poise. Slagging will be experienced when the T_{250} temperature is exceeded.

The actual melting point depends on the ash composition. Ash is primarily a mixture of silica (SiO_2), alumina (Al_2O_3), and ferric oxide (Fe_2O_3).[6] The relative proportions of each will determine the melting point, with lower melting points resulting from high amounts of ferric oxide and calcium oxide. The melting points of pure alumina and pure silica are in the 2700–2800°F (1480–1540°C) range.

---

[1]This assumes that none of the oxygen is in the form of carbonates.

[2]The value of $1/8$ follows directly from the combustion reaction of hydrogen and oxygen.

[3]A moisture content up to 5% is reported to be beneficial in some mechanically fired boilers. The moisture content contributes to lower temperatures, protecting grates from slag formation and sintering.

[4]Equation 28.5 corrects for the portion of the as-fired coal that isn't combustible, but it neglects other moisture-related losses. Section 28.28 and Sec. 28.30 contain more rigorous discussions.

[5]Some boiler manufacturers rely on the thermal protection the ash provides. For example, coal burned in cyclone boilers should have a minimum ash content of 7% to cover and protect the cyclone barrel tubes. Boiler wear and ash carryover will increase with lower ash contents.

[6]Calcium oxide (CaO), magnesium oxide ("magnesia," MgO), titanium oxide ("titania," TiO_2), ferrous oxide (FeO), and alkalies (Na_2O and K_2O) may be present in smaller amounts.

Coal ash is either of a bituminous type or lignite type. Bituminous-type ash (from midwestern and eastern coals) contains more ferric oxide than lime and magnesia. Lignite-type ash (from western coals) contains more lime and magnesia than ferric oxide.

5. SULFUR

Several forms of sulfur are present in coal and fuel oils. *Pyritic sulfur* (FeS_2) is the primary form. *Organic sulfur* is combined with hydrogen and carbon in other compounds. *Sulfate sulfur* is iron sulfate and gypsum ($CaSO_4 \cdot 2H_2O$). Sulfur in elemental, organic, and pyritic forms oxidizes to sulfur dioxide. *Sulfur trioxide* can be formed under certain conditions. Sulfur trioxide combines with water to form sulfuric acid and is a major source of boiler/stack corrosion and acid rain.

$$SO_3 + H_2O \rightarrow H_2SO_4$$

6. WOOD

Wood is not an industrial fuel, though it may be used in small quantities in developing countries. Most woods have higher heating values around 8300 Btu/lbm (19 MJ/kg), with specific values depending on the species and moisture content.

7. WASTE FUELS

Waste fuels are increasingly being used as fuels in industrial boilers and furnaces. Such fuels include digester and landfill gases, waste process gases, flammable waste liquids, and volatile organic compounds (VOCs) such as benzene, toluene, xylene, ethanol, and methane. Other waste fuels include oil shale, tar sands, green wood, seed and rice hulls, biomass refuse, peat, tire shreddings, and shingle/roofing waste.

The term *refuse-derived fuels* (RDF) is used to describe fuel produced from municipal waste. After separation (removal of glass, plastics, metals, corrugated cardboard, etc.), the waste is merely pushed into the combustion chamber. If the waste is to be burned elsewhere, it is compressed and baled.

The heating value of RDF depends on the moisture content and fraction of combustible material. For RDFs derived from typical municipal wastes, the heating value ranges from 3000 Btu/lbm to 6000 Btu/lbm (7 MJ/kg to 14 MJ/kg). Higher ranges (such as 7500–8500 Btu/lbm (17.5–19.8 MJ/kg)) can be obtained by careful selection of ingredients. Pelletized RDF (containing some coal and a limestone binder) with heating values around 8000 Btu/lbm (18.6 MJ/kg) can be used as a supplemental fuel in coal-fired units.

Scrap tires are an attractive fuel source due to their high heating values, which range from 12,000 Btu/lbm to 16,000 Btu/lbm (28 MJ/kg to 37 MJ/kg). To be compatible with existing coal-loading equipment, tires are chipped or shredded to 1 in (25 mm) size. Tires in this form are known as *tire-derived fuel* (TDF). Metal (from tire reinforcement) may or may not be present.

TDF has been shown to be capable of supplying up to 90% of a steam-generating plant's total Btu input without any deterioration in particulate emissions, pollutants, and stack opacity. In fact, compared with some low-quality coals (e.g., lignite), TDF is far superior: about 2.5 times the heating value and about 2.5 times less sulfur per Btu.

8. INCINERATION

Many toxic wastes are incinerated rather than "burned." Incineration and combustion are not the same. *Incineration* is the term used to describe a disposal process that uses combustion to render wastes ineffective (nonharmful, nontoxic, etc.). Wastes and combustible fuel are combined in a furnace, and the heat of combustion destroys the waste.[7] Wastes may themselves be combustible, though they may not be self-sustaining if the moisture content is too high.

Wastes destined for industrial incineration are usually classified by composition and heating value. The Incinerator Institute of America published broad classifications to categorize incinerator waste in 1968. Although the compositions of many wastes, particularly municipal, have changed since then, these categories remain in use. For example, biological and pathological waste is referred to as "type 4 waste." Incinerated wastes are categorized into seven types. Type 0 is *trash* (highly combustible paper and wood, with 10% or less moisture); type 1 is *rubbish* (combustible waste with up to 25% moisture); type 2 is *refuse* (a mixture of rubbish and garbage, with up to 50% moisture); type 3 is *garbage* (residential waste with up to 70% moisture); type 4 is animal solids and pathological wastes (85% moisture); type 5 is industrial process wastes in gaseous, liquid, and semiliquid form; and type 6 is industrial process wastes in solid and semisolid form requiring incineration in hearth, retort, or grate burning equipment.

9. COAL

Coal consists of volatile matter, fixed carbon, moisture, noncombustible mineral matter ("ash"), and sulfur. *Volatile matter* is driven off as a vapor when the coal is heated, and it is directly responsible for flame size. *Fixed carbon* is the combustible portion of the solid remaining after the volatile matter is driven off.

---

[7]Rotary kilns can accept waste in many forms. They are "workhorse" incinerators.

Moisture is present in the coal as free water and (for some mineral compounds) as water of hydration. Sulfur, an undesirable component, contributes to heat content.

Coals are categorized into anthracitic, bituminous, and lignitic types. *Anthracite coal* is clean, dense, and hard. It is comparatively difficult to ignite but burns uniformly and smokelessly with a short flame. *Bituminous coal* varies in composition, but generally has a higher volatile content than anthracite, starts easily, and burns freely with a long flame. Smoke and soot are possible if bituminous coal is improperly fired. *Lignite coal* is a coal of woody structure, is very high in moisture, and has a low heating value. It normally ignites slowly due to its moisture, breaks apart when burning, and burns with little smoke or soot.

Coal is burned efficiently in a particular furnace only if it is uniform in size. Screen sizes are used to grade coal, but descriptive terms can also be used.[8] *Run-of-mine coal*, ROM, is coal as mined. *Lump coal* is in the 1–6 in (25–150 mm) range. *Nut coal* is smaller, followed by even smaller *pea coal screenings*, and *fines* (dust).

10. LIQUID FUELS

Liquid fuels are lighter hydrocarbon products refined from crude petroleum oil. They include liquefied petroleum gas (LPG), gasoline, kerosene, jet fuel, diesel fuels, and heating oils. JP-4 ("jet propellant") is a 50-50 blend of kerosene and gasoline. JP-8 and Jet A are kerosene-like fuels for aircraft gas turbine engines. Important characteristics of a liquid fuel are its composition, ignition temperature, flash point,[9] viscosity, and heating value.

11. FUEL OILS

In the United States, fuel oils are categorized into grades 1 through 6 according to their viscosities.[10] Viscosity is the major factor in determining firing rate and the need for preheating for pumping or atomizing prior to burning. Grades 1 and 2 can be easily pumped at ambient temperatures. In the United States, the heaviest fuel oil used is grade 6, also known as *bunker C oil*.[11,12] Fuel oils are also classified according to their viscosities as *distillate oils* (lighter) and *residual fuel oils* (heavier).

Like coal, fuel oils contain sulfur and ash that may cause pollution, slagging on the hot end of the boiler, and corrosion in the cold end. Table 28.3 lists typical properties of common commercial fuels, while Table 28.4 lists typical properties of fuel oils.

12. GASOLINE

Gasoline is not a pure compound. It is a mixture of various hydrocarbons blended to give a desired flammability, volatility, heating value, and octane rating. There is an infinite number of blends that can be used to produce gasoline.

Gasoline's heating value depends only slightly on composition. Within a variation of $1\frac{1}{2}\%$, the heating value can be taken as 20,200 Btu/lbm (47.0 MJ/kg) for regular gasoline and as 20,300 Btu/lbm (47.2 MJ/kg) for high-octane aviation fuel.

Since gasoline is a mixture of hydrocarbons, different fractions will evaporate at different temperatures. The *volatility* is the percentage of the fuel that evaporates by a given temperature. Typical volatility specifications call for 10% or greater at 167°F (75°C), 50% at 221°F (105°C), and 90% at 275°F (135°C). Low volatility causes difficulty in starting and poor engine performance at low temperatures.

The *octane number* (ON) is a measure of *knock resistance*. It is based on comparison, performed in a standardized one-cylinder engine, with the burning of isooctane and *n*-heptane. *n*-heptane, C_7H_{16}, is rated zero and produces violent knocking. Isooctane, C_8H_{18}, is rated 100 and produces relatively knock-free operation. The percentage blend by volume of these fuels that matches the performance of the gasoline is the octane rating. The *research octane number* (RON) is a measure of the fuel's antiknock characteristics while idling; the *motor octane number* (MON) applies to high-speed, high-acceleration operations. The octane rating reported for commercial gasoline is an average of the two.

Gasolines with octanes greater than 100 (including aviation gasoline) are rated by their performance number. The *performance number* (PN) of gasoline containing antiknock compounds (e.g., tetraethyl lead, TEL, used in aviation gasoline) is related to the octane number.

$$ON = 100 + \frac{PN - 100}{3} \qquad \text{28.6}$$

---

[8]The problem with descriptive terms is that one company's "pea coal" may be as small as $\frac{1}{4}$ in (6 mm), while another's may start at $\frac{1}{2}$ in (13 mm).
[9]This is different from the *flash point* that is the temperature at which fuel oils generate enough vapor to sustain ignition in the presence of spark or flame.
[10]Grade 3 became obsolete in 1948. Grade 5 is also subdivided into light and heavy categories.
[11]120°F (48°C) is the optimum temperature for pumping no. 6 fuel oil. At that temperature, no. 6 oil has a viscosity of approximately 3000 SSU. Further heating is necessary to lower the viscosity to 150–350 SSU for atomizing.
[12]To avoid *coking* of oil, heating coils in contact with oil should not be hotter than 240°F (116°C).

Table 28.3 Typical Properties of Common Commercial Fuels

	butane	no. 1 diesel	no. 2 diesel	ethanol	gasoline	JP-4	methanol	propane
chemical formula	C_4H_{10}	–	–	C_2H_5OH	–	–	CH_3OH	C_3H_8
molecular weight	58.12	≈ 170	≈ 184	46.07	≈ 126		32.04	44.09
heating value								
higher, Btu/lbm	21,240	19,240	19,110	12,800	20,260		9838	21,646
lower, Btu/lbm	19,620	18,250	18,000	11,500	18,900	18,400	8639	19,916
lower, Btu/gal	102,400	133,332	138,110	76,152	116,485	123,400	60,050	81,855
latent heat of vaporization, Btu/lbm		115	105	361	142		511 (20°C)	147
specific gravity*	2.01	0.876	0.920	0.794	0.68–0.74	0.8017	0.793	1.55

(Multiply Btu/lbm by 2.326 to obtain kJ/kg.)

(Multiply Btu/gal by 0.2786 to obtain MJ/m^3.)

*Specific gravities of propane and butane are with respect to air.

Table 28.4 Typical Properties of Fuel Oils[a]

grade	specific gravity	heating value (MBtu/gal)[b]	heating value (GJ/m³)
1	0.805	134	37.3
2	0.850	139	38.6
4	0.903	145	40.4
5	0.933	148	41.2
6	0.965	151	41.9

(Multiply MBtu/gal by 0.2786 to obtain GJ/m^3.)

[a]Actual values will vary depending on composition.

[b]One MBtu equals one thousand Btus.

13. OXYGENATED GASOLINE

In parts of the United States, gasoline is "oxygenated" during the cold winter months. This has led to use of the term "winterized gasoline." The addition of *oxygenates* raises the combustion temperature, reducing carbon monoxide and unburned hydrocarbons.[13] Common oxygenates used in *reformulated gasoline* (RFG) include methyl tertiary-butyl ether (MTBE) and ethanol. Methanol, ethyl tertiary-butyl ether (ETBE), tertiary-amyl methyl ether (TAME), and tertiary-amyl ethyl ether (TAEE), may also be used. (See Table 28.5.) Oxygenates are added to bring the minimum oxygen level to 2–3% by weight.[14]

14. DIESEL FUEL

Properties and specifications for various grades of diesel fuel oil are similar to specifications for fuel oils. Grade 1-D ("D" for diesel) is a light distillate oil for high-speed engines in service requiring frequent speed and load changes. Grade 2-D is a distillate of lower volatility for engines in industrial and heavy mobile service. Grade 4-D is for use in medium speed engines under sustained loads.

Diesel oils are specified by a *cetane number*, which is a measure of the ignition quality (ignition delay) of a fuel. A cetane number of approximately 30 is required for satisfactory operation of low-speed diesel engines. High-speed engines, such as those used in cars, require a cetane number of 45 or more. Like the octane number for gasoline, the cetane number is determined by comparison with standard fuels. Cetane, $C_{16}H_{34}$, has a cetane number of 100. *n*-methyl-naphthalene, $C_{11}H_{10}$, has a cetane number of zero. The cetane number can be increased by use of such additives as amyl nitrate, ethyl nitrate, and ether.

A diesel fuel's *pour point* number refers to its viscosity. A fuel with a pour point of 10°F (–12°C) will flow freely above that temperature. A fuel with a high pour point will thicken in cold temperatures.

The *cloud point* refers to the temperature at which wax crystals cloud the fuel at lower temperatures. The cloud point should be 20°F (–7°C) or higher. Below that temperature, the engine will not run well.

---

[13]Oxygenation may not be successful in reducing carbon dioxide. Since the heating value of the oxygenates is lower, fuel consumption of oxygenated fuels is higher. On a per-gallon (per-liter) basis, oxygenation reduces carbon dioxide. On a per-mile (per-kilometer) basis, however, oxygenation appears to increase carbon dioxide. In any case, claims of CO_2 reduction are highly controversial, as the CO_2 footprint required to plant, harvest, dispose of decaying roots, stalks, and leaves (i.e., silage), and refine alcohol is generally ignored.

[14]Other restrictions on gasoline during the winter months intended to reduce pollution may include maximum percentages of benzene and total aromatics and limits on Reid vapor pressure, as well as specifications covering volatile organic compounds, nitric oxide (NOx), and toxins.

Table 28.5 Typical Properties of Common Oxygenates

	ethanol	MTBE[a]	TAME	ETBE	TAEE
specific gravity	0.794	0.744	0.740	0.770	0.791
octane[b]	115	110	112	105	100
heating value, MBtu/gal[c]	76.2	93.6			
Reid vapor pressure, psig[d]	18	8	15–4	3–4	2
percent oxygen by weight	34.73	18.15	15.66	15.66	13.8
volumetric percent needed to achieve gasoline					
2.7% oxygen by weight	7.8	15.1	17.2	17.2	19.4
2.0% oxygen by weight	5.6	11.0	12.4	12.7	13.0

(Multiply MBtu/gal by 0.2786 to obtain MJ/m^3.)

[a]MTBE is water soluble and does not degrade. It imparts a foul taste to water and is a possible carcinogen. It has been legislatively banned in most states, including California.
[b]Octane is equal to $(1/2)(MON + RON)$.
[c]One MBtu equals one thousand Btus.
[d]The Reid vapor pressure is the vapor pressure when heated to 100°F (38°C). This may also be referred to as the "blending vapor pressure."

15. ALCOHOL

Both methanol and ethanol can be used in internal combustion engines. *Methanol* (*methyl alcohol*) is produced from natural gas and coal, although it can also be produced from wood and organic debris. *Ethanol* (*ethyl alcohol*, *grain alcohol*) is distilled from grain, sugarcane, potatoes, and other agricultural products containing various amounts of sugars, starches, and cellulose.

Although methanol generally works as well as ethanol, only ethanol can be produced in large quantities from inexpensive agricultural products and by-products.

Alcohol is water soluble. The concentration of alcohol is measured by its *proof*, where 200 proof is pure alcohol. (180 proof is 90% alcohol and 10% water.)

Gasohol is a mixture of approximately 90% gasoline and 10% alcohol (generally ethanol).[15] Alcohol's heating value is less than gasoline's, so fuel consumption (per distance traveled) is higher with gasohol. Also, since alcohol absorbs moisture more readily than gasoline, corrosion of fuel tanks becomes problematic. In some engines, significantly higher percentages of alcohol may require such modifications as including larger carburetor jets, timing advances, heaters for preheating fuel in cold weather, tank lining to prevent rusting, and alcohol-resistant gaskets.

Mixtures of gasoline and alcohol can be designated by the first letter and the fraction of the alcohol. E10 is a mixture of 10% ethanol and 90% gasoline. M85 is a blend of 85% methanol and 15% gasoline.

Alcohol is a poor substitute for diesel fuel because alcohol's cetane number is low—from –20 to +8. Straight injection of alcohol results in poor performance and heavy knocking.

16. GASEOUS FUELS

Various gaseous fuels are used as energy sources, but most applications are limited to natural gas and *liquefied petroleum gases* (LPGs) (i.e., propane, butane, and mixtures of the two).[16,17] *Natural gas* is a mixture of methane (between 55% and 95%), higher hydrocarbons (primarily ethane), and other noncombustible gases. Typical heating values for natural gas range from 950 Btu/ft^3 to 1100 Btu/ft^3 (35 MJ/m^3 to 41 MJ/m^3).

The production of *synthetic gas* (*syngas*) through coal gasification may be applicable to large power generating plants. The cost of gasification, though justifiable to reduce sulfur and other pollutants, is too high for syngas to become a widespread substitute for natural gas.

17. IGNITION TEMPERATURE

The *ignition temperature* (*autoignition temperature*) is the minimum temperature at which combustion can be sustained. It is the temperature at which more heat is generated by the combustion reaction than is lost to the surroundings, after which combustion becomes self-sustaining. For coal, the minimum ignition temperature varies from around 800°F (425°C) for bituminous varieties to 900–1100°F (480–590°C)

---

[15]In fact, oxygenated gasoline may use more than 10% alcohol.
[16]A number of *manufactured gases* are of practical (and historical) interest in specific industries, including *coke-oven gas*, *blast-furnace gas*, *water gas*, *producer gas*, and *town gas*. However, these gases are not now in widespread use.
[17]At atmospheric pressure, propane boils at –44°F (–42°C), while butane boils at 31°F (–0.5°C).

for anthracite. For sulfur and charcoal, the ignition temperatures are approximately 470°F (240°C) and 650°F (340°C), respectively.

For gaseous fuels, the ignition temperature depends on the air-fuel ratio, temperature, pressure, and length of time the source of heat is applied. Ignition can be instantaneous or delayed, depending on the temperature. Generalizations can be made for any gas, but the generalized temperatures will be meaningless without specifying all of these factors.

18. ATMOSPHERIC AIR

It is important to make a distinction between "air" and "oxygen." Atmospheric air is a mixture of oxygen, nitrogen, and small amounts of carbon dioxide, water vapor, argon, and other inert ("rare") gases. For the purpose of combustion calculations, all constituents except oxygen are grouped with nitrogen. (See Table 28.6.) It is necessary to supply 4.32 (i.e., 1/0.2315) masses of air to obtain one mass of oxygen. Similarly, it is necessary to supply 4.773 volumes of air to obtain one volume of oxygen. The average molecular weight of air is 28.97, and its specific gas constant is 53.35 ft-lbf/lbm-°R (287.03 J/kg·K).

Table 28.6 Composition of Dry Air[a]

component	percent by weight	percent by volume
oxygen	23.15	20.95
nitrogen/inerts	76.85	79.05
ratio of nitrogen to oxygen	3.320	3.773[b]
ratio of air to oxygen	4.320	4.773

[a]Inert gases and CO_2 are included as N_2.
[b]The value is also reported by various sources as 3.76, 3.78, and 3.784.

19. COMBUSTION REACTIONS

A limited number of elements appear in combustion reactions. Carbon, hydrogen, sulfur, hydrocarbons, and oxygen are the reactants. Carbon dioxide and water vapor are the main products, with carbon monoxide, sulfur dioxide, and sulfur trioxide occurring in lesser amounts. Nitrogen and excess oxygen emerge hotter but unchanged from the stack.

Combustion reactions occur according to the normal chemical reaction principles. Balancing combustion reactions is usually easiest if carbon is balanced first, followed by hydrogen and then by oxygen. When a gaseous fuel has several combustible gases, the volumetric fuel composition can be used as coefficients in the chemical equation.

Table 28.7 lists ideal combustion reactions. These reactions do not include any nitrogen or water vapor that are present in the combustion air.

Table 28.7 Ideal Combustion Reactions

fuel	formula	reaction equation (excluding nitrogen)
carbon (to CO)	C	$2C + O_2 \rightarrow 2CO$
carbon (to CO_2)	C	$C + O_2 \rightarrow CO_2$
sulfur (to SO_2)	S	$S + O_2 \rightarrow SO_2$
sulfur (to SO_3)	S	$2S + 3O_2 \rightarrow 2SO_3$
carbon monoxide	CO	$2CO + O_2 \rightarrow 2CO_2$
methane	CH_4	$CH_4 + 2O_2 \rightarrow CO_2 + 2H_2O$
acetylene	C_2H_2	$2C_2H_2 + 5O_2 \rightarrow 4CO_2 + 2H_2O$
ethylene	C_2H_4	$C_2H_4 + 3O_2 \rightarrow 2CO_2 + 2H_2O$
ethane	C_2H_6	$2C_2H_6 + 7O_2 \rightarrow 4CO_2 + 6H_2O$
hydrogen	H_2	$2H_2 + O_2 \rightarrow 2H_2O$
hydrogen sulfide	H_2S	$2H_2S + 3O_2 \rightarrow 2H_2O + 2SO_2$
propane	C_3H_8	$C_3H_8 + 5O_2 \rightarrow 3CO_2 + 4H_2O$
n-butane	C_4H_{10}	$2C_4H_{10} + 13O_2 \rightarrow 8CO_2 + 10H_2O$
octane	C_8H_{18}	$2C_8H_{18} + 25O_2 \rightarrow 16CO_2 + 18H_2O$
olefin series	C_nH_{2n}	$2C_nH_{2n} + 3nO_2 \rightarrow 2nCO_2 + 2nH_2O$
paraffin series	C_nH_{2n+2}	$2C_nH_{2n+2} + (3n+1)O_2 \rightarrow 2nCO_2 + (2n+2)H_2O$

(Multiply oxygen volumes by 3.773 to get nitrogen volumes.)

Example 28.2

A gaseous fuel is 20% hydrogen and 80% methane by volume. What stoichiometric volume of oxygen is required to burn 120 volumes of fuel at the same conditions?

Solution

Write the unbalanced combustion reaction.

$$H_2 + CH_4 + O_2 \rightarrow CO_2 + H_2O$$

Use the volumetric analysis as coefficients of the fuel.

$$0.2H_2 + 0.8CH_4 + O_2 \rightarrow CO_2 + H_2O$$

Balance the carbons.

$$0.2H_2 + 0.8CH_4 + O_2 \rightarrow 0.8CO_2 + H_2O$$

Balance the hydrogens.

$$0.2H_2 + 0.8CH_4 + O_2 \rightarrow 0.8CO_2 + 1.8H_2O$$

Balance the oxygens.

$$0.2H_2 + 0.8CH_4 + 1.7O_2 \rightarrow 0.8CO_2 + 1.8H_2O$$

For gaseous components, the coefficients correspond to the volumes. Since one $(0.2 + 0.8)$ volume of fuel requires 1.7 volumes of oxygen, the required oxygen is

$$(1.7)(120 \text{ volumes of fuel}) = 204 \text{ volumes of oxygen}$$

20. STOICHIOMETRIC REACTIONS

Stoichiometric quantities (*ideal quantities*) are the exact quantities of reactants that are needed to complete a combustion reaction without any reactants left over. Table 28.7 contains some of the more common chemical reactions. Stoichiometric volumes and masses can always be determined from the balanced chemical reaction equation. Table 28.8 can be used to quickly determine stoichiometric amounts for some fuels.

The process of oxidizing a hydrocarbon requires air (the oxygen and nitrogen found in ambient air), and this process produces carbon dioxide, water, and unconsumed nitrogen. Equation 28.7 gives the stoichiometric reaction for the oxidation process. Depending on the specific hydrocarbon, C_xH_y, it is possible to estimate the number of moles of oxygen, b, that must be consumed to complete the reaction.

Oxidation Chemistry

$$C_xH_y + (b)O_2 + 3.76(b)N_2$$
$$\rightarrow xCO_2 + \left(\frac{y}{2}\right)H_2O + 3.76(b)N_2 \qquad 28.7$$

21. STOICHIOMETRIC AIR

Stoichiometric air (*ideal air*) is the air necessary to provide the exact amount of oxygen for complete combustion of a fuel. Stoichiometric air includes atmospheric nitrogen. For each volume of oxygen, 3.773 volumes of nitrogen pass unchanged through the reaction.[18]

Stoichiometric air can be stated in units of mass (pounds or kilograms of air) for solid and liquid fuels, and in units of volume (cubic feet or cubic meters of air) for gaseous fuels. When stated in terms of moles, the stoichiometric ratio of air to fuel is known as the *molar air-fuel ratio*, $\overline{A/F}$.

Incomplete Combustion

$$\overline{A/F} = \frac{\text{No. of moles of air}}{\text{No. of moles of fuel}} \qquad 28.8$$

When stated in terms of mass, the stoichiometric ratio of air to fuel is the *stoichiometric air-fuel ratio*, A/F.

Incomplete Combustion

$$A/F = \frac{\text{Mass of air}}{\text{Mass of fuel}} = (\overline{A/F})\left(\frac{M_{air}}{M_{fuel}}\right) \qquad 28.9$$

The ideal air-fuel ratio can be determined from the combustion reaction equation. It can also be determined by adding the oxygen and nitrogen amounts listed in Table 28.8.

When the numbers of moles of air and fuel and the molecular weight of the fuel are known, the stoichiometric air-fuel ratio is

Air to Fuel Ratio

$$AF = \frac{28.97 n_{air}}{MW_{fuel} n_{fuel}} \qquad 28.10$$

In Eq. 28.10, the number 28.97 is the molecular weight of air and has units of grams per mole.

For fuels whose ultimate analysis is known, the approximate stoichiometric air (oxygen and nitrogen) requirement in pounds of air per pound of fuel (kilograms of air per kilogram of fuel) can be quickly calculated by using Eq. 28.11. All oxygen in the fuel is assumed to be free moisture. All of the reported oxygen is assumed to be locked up in the form of water. Any free oxygen (i.e., oxygen dissolved in liquid fuels) is subtracted from the oxygen requirements.

$$A/F = 34.5\left(\frac{G_C}{3} + G_H - \frac{G_O}{8} + \frac{G_S}{8}\right) \qquad 28.11$$

[solid and liquid fuels]

---

[18]The only major change in the nitrogen gas is its increase in temperature. Dissociation of nitrogen and formation of nitrogen compounds can occur but are essentially insignificant.

Table 28.8 *Consolidated Combustion Data*[a,b,c,d]

Air

fuel	\>for 1 mole of fuel — air O₂	N₂	other products CO₂	H₂O	SO₂	\>for 1 ft³ of fuel[e] — air O₂	N₂	other products CO₂	H₂O	\>for 1 lbm of fuel — air O₂	N₂	other products CO₂	H₂O	SO₂	units of fuel
C carbon	1.0 / 379.5 / 32.0	3.773 / 1432 / 106	1.0 / 379.5 / 44.0							0.0833 / 31.63 / 2.667	0.3143 / 119.3 / 8.883	0.0833 / 31.63 / 3.667			moles / ft³ / lbm
H₂ hydrogen	0.5 / 189.8 / 16.0	1.887 / 716.1 / 53.0		1.0 / 379.5 / 18.0		0.001317 / 0.5 / 0.04216	0.004969 / 1.887 / 0.1397		0.002635 / 1.0 / 0.04747	0.248 / 94.12 / 7.936	0.9357 / 355.1 / 26.29		0.496 / 188.25 / 8.936		moles / ft³ / lbm
S sulfur	1.0 / 379.5 / 32.0	3.773 / 1432 / 106.0			1.0 / 379.5 / 64.06					0.03119 / 11.84 / 0.998	0.1177 / 44.67 / 3.306			0.03119 / 11.84 / 1.998	moles / ft³ / lbm
CO carbon monoxide	0.5 / 189.8 / 16.0	1.887 / 716.1 / 53.0	1.0 / 379.5 / 44.01			0.001317 / 0.5 / 0.04216	0.004969 / 1.887 / 0.1397	0.002635 / 1.0 / 0.1160		0.01785 / 6.774 / 0.5712	0.06735 / 25.56 / 1.892	0.03570 / 13.55 / 1.572			moles / ft³ / lbm
CH₄ methane	2.0 / 759 / 64.0	7.546 / 2864 / 212.0	1.0 / 379.5 / 44.01	2.0 / 758 / 36.03		0.00527 / 2.0 / 0.1686	0.01988 / 7.546 / 0.5586	0.002635 / 1.0 / 0.1160	0.00527 / 2.0 / 0.0949	0.1247 / 47.31 / 3.989	0.4705 / 178.5 / 13.21	0.06233 / 23.66 / 2.743	0.1247 / 47.31 / 2.246		moles / ft³ / lbm
C₂H₂ acetylene	2.5 / 948.8 / 80.0	9.433 / 3580 / 265.0	2.0 / 758 / 88.02	1.0 / 379.5 / 18.02		0.006588 / 2.5 / 0.2108	0.02486 / 9.443 / 0.6983	0.00527 / 2.0 / 0.2319	0.002635 / 1.0 / 0.04747	0.09601 / 36.44 / 3.072	0.3622 / 137.5 / 10.18	0.07681 / 29.15 / 3.380	0.03841 / 14.57 / 0.6919		moles / ft³ / lbm
C₂H₄ ethylene	3.0 / 1139 / 96.0	11.32 / 4297 / 318.0	2.0 / 758 / 88.02	2.0 / 758 / 36.03		0.007905 / 3.0 / 0.2530	0.02983 / 11.32 / 0.8380	0.00527 / 2.0 / 0.2319	0.00527 / 2.0 / 0.0949	0.1069 / 40.58 / 3.422	0.4033 / 153.1 / 11.34	0.07129 / 27.05 / 3.137	0.07129 / 27.05 / 1.284		moles / ft³ / lbm
C₂H₆ ethane	3.5 / 1328 / 112.0	13.21 / 5010 / 371.0	2.0 / 758 / 88.02	3.0 / 1139 / 54.05		0.009223 / 3.5 / 0.2951	0.03480 / 13.21 / 0.9776	0.00527 / 2.0 / 0.2319	0.007905 / 3.0 / 0.1424	0.1164 / 44.17 / 3.724	0.4392 / 166.7 / 12.34	0.06651 / 25.24 / 2.927	0.09977 / 37.86 / 1.797		moles / ft³ / lbm

(Multiply lbm/ft³ by 0.06243 to obtain kg/m³.)

[a] Rounding of molecular weights and air composition may introduce slight inconsistencies in the table values. This table is based on atomic weights with at least four significant digits, a ratio of 3.773 volumes of nitrogen per volume of oxygen, and 379.5 ft³ per mole at 1 atm and 60°F.

[b] Volumes per unit mass are at 1 atm and 60°F (16°C). To obtain volumes at other temperatures, multiply by $(T_{°F} + 460°)/520°$ or $(T_{°C} + 273°)/289°$.

[c] The volume of water applies only when the combustion products are at such high temperatures that all of the water is in vapor form.

[d] This table can be used to directly determine some SI ratios. For kg/kg ratios, the values are the same as lbm/lbm. For mixed units (e.g., ft³/lbm), conversions are required.

[e] Sulfur is not used in gaseous form.

For fuels consisting of a mixture of gases, Eq. 28.12 and the constant J_i from Table 28.9 can be used to quickly determine the stoichiometric air requirements.

$$A/F = \sum G_i J_i \quad \text{[gaseous fuels]} \qquad 28.12$$

For fuels consisting of a mixture of gases, the air-fuel ratio can also be expressed in volumes of air per volume of fuel using values of K_i from Table 28.9.

$$\text{volumetric air-fuel ratio} = \sum B_i K_i$$
$$\text{[gaseous fuels]} \qquad 28.13$$

Table 28.9 *Approximate Air-Fuel Ratio Coefficients for Components of Natural Gas**

fuel component	J (gravimetric)	K (volumetric)
acetylene, C_2H_2	13.25	11.945
butane, C_4H_{10}	15.43	31.06
carbon monoxide, CO	2.463	2.389
ethane, C_2H_6	16.06	16.723
ethylene, C_2H_4	14.76	14.33
hydrogen, H_2	34.23	2.389
hydrogen sulfide, H_2S	6.074	7.167
methane, CH_4	17.20	9.556
oxygen, O_2	-4.320	-4.773
propane, C_3H_8	15.65	23.89

*Rounding of molecular weights and air composition may introduce slight inconsistencies in the table values. This table is based on atomic weights with at least four significant digits and a ratio of 3.773 volumes of nitrogen per volume of oxygen.

Example 28.3

Use Table 28.8 to determine the theoretical volume of 90°F (32°C) air required to burn 1 volume of 60°F (16°C) carbon monoxide to carbon dioxide.

Solution

From Table 28.8, 0.5 volumes of oxygen are required to burn 1 volume of carbon monoxide to carbon dioxide. 1.887 volumes of nitrogen accompany the oxygen. The total amount of air at the temperature of the fuel is $0.5 + 1.887 = 2.387$ volumes.

This volume will expand at the higher temperature. The volume at the higher temperature is

$$v_2 = \frac{T_2 v_1}{T_1}$$
$$= \frac{(90°F + 460°)(2.387 \text{ volumes})}{60°F + 460°}$$
$$= 2.53 \text{ volumes}$$

Example 28.4

How much air is required for the ideal combustion of (a) coal with an ultimate analysis of 93.5% carbon, 2.6% hydrogen, 2.3% oxygen, 0.9% nitrogen, and 0.7% sulfur; (b) fuel oil with a gravimetric analysis of 84% carbon, 15.3% hydrogen, 0.4% nitrogen, and 0.3% sulfur; and (c) natural gas with a volumetric analysis of 86.92% methane, 7.95% ethane, 2.81% nitrogen, 2.16% propane, and 0.16% butane?

Solution

(a) Use Eq. 28.11.

$$A/F = 34.5\left(\frac{G_C}{3} + G_H - \frac{G_O}{8} + \frac{G_S}{8}\right)$$
$$= (34.5)\left(\frac{0.935}{3} + 0.026 - \frac{0.023}{8} + \frac{0.007}{8}\right)$$
$$= 11.58 \text{ lbm/lbm (kg/kg)}$$

(b) Use Eq. 28.11.

$$A/F = 34.5\left(\frac{G_C}{3} + G_H + \frac{G_S}{8}\right)$$
$$= (34.5)\left(\frac{0.84}{3} + 0.153 + \frac{0.003}{8}\right)$$
$$= 14.95 \text{ lbm/lbm (kg/kg)}$$

(c) Use Eq. 28.13 and the coefficients from Table 28.9.

$$\begin{aligned}
\frac{\text{volumetric}}{\text{air-fuel ratio}} &= \sum B_i K_i \\
&= (0.8692)\left(9.556\ \frac{\text{ft}^3}{\text{ft}^3}\right) \\
&\quad + (0.0795)\left(16.723\ \frac{\text{ft}^3}{\text{ft}^3}\right) \\
&\quad + (0.0216)\left(23.89\ \frac{\text{ft}^3}{\text{ft}^3}\right) \\
&\quad + (0.0016)\left(31.06\ \frac{\text{ft}^3}{\text{ft}^3}\right) \\
&= 10.20\ \text{ft}^3/\text{ft}^3\ (\text{m}^3/\text{m}^3)
\end{aligned}$$

22. INCOMPLETE COMBUSTION

Incomplete combustion occurs when there is insufficient oxygen to burn all of the hydrogen, carbon, and sulfur in the fuel. Without enough available oxygen, carbon burns to carbon monoxide.[19] Carbon monoxide in the flue gas indicates incomplete and inefficient combustion. Incomplete combustion is caused by cold furnaces, low combustion temperatures, poor air supply, smothering from improperly vented stacks, and insufficient mixing of air and fuel.

23. SMOKE

The amount of smoke can be used as an indicator of combustion completeness. Smoky combustion may indicate improper air-fuel ratio, insufficient draft, leaks, insufficient preheat, or misadjustment of the fuel system.

Smoke measurements are made in a variety of ways, with the standards depending on the equipment used. Photoelectric sensors in the stack are used to continuously monitor smoke. The *smoke spot number* (SSN) and ASTM smoke scale are used with continuous stack monitors. For coal-fired furnaces, the maximum desirable smoke number is SSN 4. For grade 2 fuel oil, the SSN should be less than 1; for grade 4, SSN 4; for grades 5L, 5H, and low-sulfur residual fuels, SSN 3; for grade 6, SSN 4.

The *Ringelmann scale* is a subjective method in which the smoke density is visually compared to five standardized white-black grids. Ringelmann chart no. 0 is solid white; chart no. 5 is solid black. Ringelmann chart no. 1, which is 20% black, is the preferred (and required) operating point for most power plants.

24. FLUE GAS ANALYSIS

Combustion products that pass through a furnace's exhaust system are known as *flue gases* (*stack gases*). Flue gases are almost all nitrogen.[20] Nitrogen oxides are not present in large enough amounts to be included separately in combustion reactions.

The actual composition of flue gases can be obtained in a number of ways, including by modern electronic detectors, less expensive "length-of-stain" detectors, and direct sampling with an Orsat-type apparatus.

The antiquated *Orsat apparatus* determines the volumetric percentages of CO_2, CO, O_2, and N_2 in a flue gas. The sampled flue gas passes through a series of chemical compounds. The first compound absorbs only CO_2, the next only O_2, and the third only CO. The unabsorbed gas is assumed to be N_2 and is found by subtracting the volumetric percentages of all other components from 100%. An Orsat analysis is a dry analysis; the percentage of water vapor is not usually determined. A wet volumetric analysis (needed to compute the dew-point temperature) can be derived if the volume of water vapor is added to the Orsat volumes.

The Orsat procedure is now rarely used, although the term "Orsat" may be generally used to refer to any flue gas analyzer. Modern electronic analyzers can determine free oxygen (and other gases) independently of the other gases.

25. ACTUAL AND EXCESS AIR

Complete combustion occurs when all of the fuel is burned. Under ideal conditions, the stoichiometric air (or theoretical air) would be enough to achieve complete combustion, but in actuality, more air than this is usually needed. The amount of air actually needed for complete combustion is usually expressed as a percentage of the theoretical air requirements.

Incomplete Combustion

$$\text{Percent Theoretical Air} = \frac{(A/F)_{\text{actual}}}{(A/F)_{\text{stoichiometric}}} \times 100 \quad \textbf{28.14}$$

The difference between the stoichiometric air and the actual air is called the excess air. Excess air is also usually expressed as a percentage of the theoretical air. The difference between the stoichiometric air and the actual air is called the excess air. Excess air is also usually expressed as a percentage of the theoretical air.

Incomplete Combustion

$$\text{Percent Excess Air} = \frac{(A/F)_{\text{actual}} - (A/F)_{\text{stoichiometric}}}{(A/F)_{\text{stoichiometric}}} \times 100 \quad \textbf{28.15}$$

---

[19]Toxic alcohols, ketones, and aldehydes may also be formed during incomplete combustion.
[20]This assumption is helpful in making quick determinations of the thermodynamic properties of flue gases.

Different fuel types burn more efficiently with different amounts of excess air. Coal-fired boilers need approximately 30–35% excess air, oil-based units need about 15%, and natural gas burners need about 10%.

The actual air-fuel ratio for dry, solid fuels with no unburned carbon can be estimated from the volumetric flue gas analysis and the gravimetric fractions of carbon and sulfur in the fuel.

$$A/F = \frac{M_{air,actual}}{M_{fuel}}$$

$$= \frac{3.04 B_{N_2}\left(G_C + \dfrac{G_S}{1.833}\right)}{B_{CO_2} + B_{CO}}$$ 28.16

Too much free oxygen or too little carbon dioxide in the flue gas are indicative of excess air. Because the relationship between oxygen and excess air is relatively insensitive to fuel composition, oxygen measurements are replacing standard carbon dioxide measurements in determining combustion efficiency.[21] The relationship between excess air and the volumetric fraction of oxygen in the flue gas is given in Table 28.10.

Table 28.10 *Approximate Volumetric Percentage of Oxygen in Stack Gas*

fuel*		excess air							
	0%	1%	5%	10%	20%	50%	100%	200%
fuel oils,								
no. 2–6	0	0.22	1.06	2.02	3.69	7.29	10.8	14.2
natural gas	0	0.25	1.18	2.23	4.04	7.83	11.4	14.7
propane	0	0.23	1.08	2.06	3.75	7.38	10.9	14.3

*Values for coal are only marginally lower than the values for fuel oils.

Reducing the air-fuel ratio will have several outcomes. (a) The furnace temperature will increase due to a reduction in cooling air. (b) The flue gas will decrease in quantity. (c) The heat loss will decrease. (d) The furnace efficiency will increase. (e) Pollutants will (usually) decrease.

With a properly adjusted furnace and good mixing, the flue gas will contain no carbon monoxide, and the amount of carbon dioxide will be maximized. The stoichiometric amount of carbon dioxide in the flue gas is known as the *ultimate CO₂* and is the theoretical maximum level of carbon dioxide. The air-fuel mixture should be adjusted until the maximum level of carbon dioxide is attained.

Example 28.5

Propane (C_3H_8) is burned completely with 20% excess air. What is the volumetric fraction of carbon dioxide in the flue gas?

Solution

From Table 28.7, the balanced chemical reaction equation is

$$C_3H_8 + 5O_2 \rightarrow 3CO_2 + 4H_2O$$

With 20% excess air, the oxygen volume is $(1.2)(5) = 6$.

$$C_3H_8 + 6O_2 \rightarrow 3CO_2 + 4H_2O + O_2$$

From Table 28.6, there are 3.773 volumes of nitrogen for every volume of oxygen.

$$(3.773)(6) = 22.6$$

$$C_3H_8 + 6O_2 + 22.6N_2 \rightarrow 3CO_2 + 4H_2O + O_2 + 22.6N_2$$

For gases, the coefficients can be interpreted as volumes. The volumetric fraction of carbon dioxide is

$$B_{CO_2} = \frac{3}{3 + 4 + 1 + 22.6}$$
$$= 0.0980 \quad (9.8\%)$$

26. CALCULATIONS BASED ON FLUE GAS ANALYSIS

Equation 28.17 gives the approximate percentage (by volume) of actual excess air.

$$\begin{aligned}\text{actual excess air} \\ \text{\% by volume}\end{aligned} = \frac{(B_{O_2} - 0.5 B_{CO})}{0.264 B_{N_2} - B_{O_2} + 0.5 B_{CO}} \times 100\%$$ 28.17

The ultimate CO_2 (i.e., the maximum theoretical carbon dioxide) can be determined from Eq. 28.18.

$$\begin{aligned}\text{ultimate } CO_2 \\ \text{\% by volume}\end{aligned} = \frac{B_{CO_2,actual}}{1 - 4.773 B_{O_2,actual}} \times 100\%$$ 28.18

The mass ratio of dry flue gases to solid fuel is given by Eq. 28.19.

$$\frac{\text{mass of flue gas}}{\text{mass of solid fuel}} = \frac{\begin{pmatrix}11 B_{CO_2} + 8 B_{O_2} \\ +7(B_{CO} + B_{N_2})\end{pmatrix} \times \left(G_C + \dfrac{G_S}{1.833}\right)}{3(B_{CO_2} + B_{CO})}$$ 28.19

---

[21]The relationship between excess air required and CO₂ is much more dependent on fuel type (e.g., liquid) and furnace design.

Example 28.6

A sulfur-free coal has a proximate analysis of 75% carbon. The volumetric analysis of the flue gas is 80.2% nitrogen, 12.6% carbon dioxide, 6.2% oxygen, and 1.0% carbon monoxide. Assume constant volume and constant atmospheric pressure of 1 atm, and that the molecular weight of the fuel is the same of that of carbon. Calculate the actual air-fuel ratio.

Solution

Use Eq. 28.10 and the ideal gas law.

Air to Fuel Ratio

$$AF = \frac{28.97 n_{\text{air}}}{MW_{\text{fuel}} n_{\text{fuel}}}$$

Volume and atmospheric pressure are constant, so

Ideal Gas Constants

$$PV = nRT$$
$$\frac{P}{RT} = \frac{n}{V}$$
$$P \asymp n$$

Assuming the atmospheric pressure is 1 atm, the partial pressures are equal to the percent composition of the flue gas volume.

$P_{\text{N2}} = 0.802 \text{ atm}$

$P_{\text{CO2}} = 0.126 \text{ atm}$

$P_{\text{O2}} = 0.062 \text{ atm}$

$P_{\text{CO}} = 0.010 \text{ atm}$

Assuming that the molecular weight of the fuel is the same as carbon (12 g/mol), the actual air-fuel ratio is

Air to Fuel Ratio

$$AF = \frac{28.97 n_{\text{air}}}{MW_{\text{fuel}} n_{\text{fuel}}}$$
$$= \frac{(28.97)(0.802 \text{ atm} + 0.062 \text{ atm})}{\left(12 \ \frac{\text{g}}{\text{mol}}\right)(0.126 \text{ atm} + 0.010 \text{ atm})}$$
$$= 15.3 \text{ g air/g fuel}$$

27. TEMPERATURE OF FLUE GAS

The temperature of the gas at the furnace outlet—before the gas reaches any other equipment—should be approximately 550°F (300°C). Overly low temperatures mean there is too much excess air. Overly high temperatures—above 750°F (400°C)—mean that heat is being wasted to the atmosphere and indicate other problems (ineffective heat transfer surfaces, overfiring, defective combustion chamber, etc.).

The *net stack temperature* is the difference between the stack and local environment temperatures. The net stack temperature should be as low as possible without causing corrosion of the low end.

28. HEAT OF COMBUSTION

The *heating value* of a fuel can be determined experimentally in a *bomb calorimeter*, or it can be estimated from the fuel's chemical analysis. The *higher heating value* (HHV), or *gross heating value*, of a fuel includes the heat of vaporization (condensation) of the water vapor formed from the combustion of hydrogen in the fuel. The *lower heating value* (LHV), or *net heating value*, assumes that all the products of combustion remain gaseous. The LHV is generally the value to use in calculations of thermal energy generated, since the heat of vaporization is not recovered within the furnace.

Traditionally, heating values have been reported on an HHV basis for coal-fired systems but on an LHV basis for natural gas-fired combustion turbines. There is an 11% difference between HHV and LHV thermal efficiencies for gas-fired systems and a 4% difference for coal-fired systems, approximately.

The HHV can be calculated from the LHV if the enthalpy of vaporization, Δh_v, is known at the pressure of the water vapor.[22] In Eq. 28.20, H is the hydrogen content of the combusted material

Heating Value of Waste

$$LHV = HHV - [(\Delta H_v)(9H)] \qquad \textbf{28.20}$$

As presented in Sec. 28.3, only the hydrogen that is not locked up with oxygen in the form of water is combustible. This is known as the *available hydrogen*. The correct percentage of combustible hydrogen, $G_{\text{H,available}}$, is calculated from the hydrogen and oxygen fraction. Equation 28.21 (same as Eq. 28.4) assumes that all of the oxygen is present in the form of water.

$$G_{\text{H,available}} = G_{\text{H,total}} - \frac{G_{\text{O}}}{8} \qquad \textbf{28.21}$$

---

[22]For the purpose of initial studies, the heat of vaporization is usually assumed to be 1040 Btu/lbm (2.42 MJ/kg). This corresponds to a partial pressure of approximately 1 psia (7 kPa) and a dew point of 100°F (40°C).

Dulong's formula calculates the higher heating value of coals and coke with a 2–3% accuracy range for moisture contents below approximately 10%.[23] The gravimetric or volumetric analysis percentages for each combustible element (including sulfur) are multiplied by the heating value per unit (mass or volume) from App. 28.A and summed.

$$HHV_{MJ/kg} = 32.78 G_C + 141.8 \left(G_H - \frac{G_O}{8} \right) \quad [\text{SI}] \quad \textit{28.22(a)}$$
$$+ 9.264 G_S$$

$$HHV_{Btu/lbm} = 14{,}093 G_C + 60{,}958$$
$$\times \left(G_H - \frac{G_O}{8} \right) + 3983 G_S \quad [\text{U.S.}] \quad \textit{28.22(b)}$$

The higher heating value of gasoline can be approximated from the Baumé specific gravity, °Be.

$$HHV_{gasoline,MJ/kg} = 42.61 + 0.093(°Be - 10) \quad [\text{SI}] \quad \textit{28.23(a)}$$

$$HHV_{gasoline,Btu/lbm} = 18{,}320 + 40(°Be - 10) \quad [\text{U.S.}] \quad \textit{28.23(b)}$$

The heating value of petroleum oils (including diesel fuel) can also be approximately determined from the oil's specific gravity. The values derived by using Eq. 28.24 may not exactly agree with values for specific oils because the equation does not account for refining methods and sulfur content. Equation 28.24 was originally intended for combustion at constant volume, as in a gasoline engine. However, variations in heating values for different oils are very small; therefore, Eq. 28.24 is widely used as an approximation for all types of combustion, including constant pressure combustion in industrial boilers.

$$HHV_{fuel\ oil,MJ/kg} = 51.92 - 8.792(SG)^2 \quad [\text{SI}] \quad \textit{28.24(a)}$$

$$HHV_{fuel\ oil,Btu/lbm} = 22{,}320 - 3780(SG)^2 \quad [\text{U.S.}] \quad \textit{28.24(b)}$$

Example 28.7

A coal has an ultimate analysis of 93.9% carbon, 2.1% hydrogen, 2.3% oxygen, 0.3% nitrogen, and 1.4% ash. What are its (a) higher and (b) lower heating values in Btu/lbm?

Solution

(a) The noncombustible ash, oxygen, and nitrogen do not contribute to heating value. Some of the hydrogen is in the form of water. From Eq. 28.21, the available hydrogen fraction is

$$G_{H,available} = G_{H,total} - \frac{G_O}{8}$$
$$= 2.1\% - \frac{2.3\%}{8}$$
$$= 1.8\%$$

From App. 28.A, the higher heating values of carbon and hydrogen are 14,093 Btu/lbm and 60,958 Btu/lbm, respectively. From Eq. 28.22, the total heating value per pound of coal is

$$HHV = 14{,}093 G_C + 60{,}958 G_{H,available}$$
$$= \left(14{,}093 \ \frac{Btu}{lbm} \right)(0.939)$$
$$+ \left(60{,}958 \ \frac{Btu}{lbm} \right)(0.018)$$
$$= 14{,}331 \ Btu/lbm$$

(b) From part (a), the available hydrogen is 1.8% (0.018), and the HHV is 14,331 Btu/lbm. From footnote 23, the heat of vaporization is 1040 Btu/lbm. Using Eq. 28.20, the LHV is

Heating Value of Waste

$$LHV = HHV - [(\Delta H_v)(9H)]$$
$$= 14{,}331 \ \frac{Btu}{lbm} -$$
$$\left(1040 \ \frac{Btu}{lbm} \right)(9)(0.018)$$
$$= 14{,}163 \ Btu/lbm$$

Alternatively, the lower heating value of the coal can be calculated from Eq. 28.22 by substituting the lower (net) hydrogen heating value from a table of heating values, 51,623 Btu/lbm, for the gross heating value of 60,598 Btu/lbm. This yields 14,167 Btu/lbm.

---

[23]The coefficients in Eq. 28.22 are slightly different from the coefficients originally proposed by Dulong. Equation 28.22 reflects currently accepted heating values that were unavailable when Dulong developed his formula. Equation 28.22 makes these assumptions: (1) None of the oxygen is in carbonate form. (2) There is no free oxygen. (3) The hydrogen and carbon are not combined as hydrocarbons. (4) Carbon is amorphous, not graphitic. (5) Sulfur is not in sulfate form. (6) Sulfur burns to sulfur dioxide.

29. MAXIMUM THEORETICAL COMBUSTION (FLAME) TEMPERATURE

It can be assumed that the maximum theoretical increase in flue gas temperature will occur if all of the combustion energy is absorbed adiabatically by the smallest possible quantity of combustion products. This provides a method of estimating the *maximum theoretical combustion temperature*, also sometimes called the *maximum flame temperature* or *adiabatic flame temperature*.[24]

In Eq. 28.25, the mass of the products is the sum of the fuel, oxygen, and nitrogen masses for stoichiometric combustion. The mean specific heat is a gravimetrically weighted average of the values of c_p for all combustion gases. (Since nitrogen comprises the majority of the combustion gases, the mixture's specific heat will be approximately that of nitrogen.) The heat of combustion can be found either from the lower heating value, LHV, or from a difference in air enthalpies across the furnace.

$$T_{max} = T_i + \frac{LHV}{m_{products}c_{p,mean}} \qquad 28.25$$

Due to thermal losses, incomplete combustion, and excess air, actual flame temperatures are always lower than the theoretical temperature. Most fuels produce flame temperatures in the range of 3350–3800°F (1850–2100°C).

30. COMBUSTION LOSSES

A portion of the combustion energy is lost in heating the dry flue gases (dfg).[25] This is known as *dry flue gas loss*. In Eq. 28.26, $m_{flue\,gas}$ is the mass of dry flue gas per unit mass of fuel. It can be estimated from Eq. 28.19. Although the full temperature difference is used, the specific heat should be evaluated at the average temperature of the flue gas. For quick estimates, the dry flue gas can be assumed to be pure nitrogen.

$$q_1 = m_{flue\,gas}c_p(T_{flue\,gas} - T_{incoming\,air}) \qquad 28.26$$

Heat is lost in vaporizing the water formed during the combustion of hydrogen. In Eq. 28.27, m_{vapor} is the mass of vapor per pound of fuel. G_H is the gravimetric fraction of hydrogen in the fuel. The coefficient 8.94 is essentially $8 + 1 = 9$ and converts the gravimetric mass of hydrogen to gravimetric mass of water formed. h_g is the enthalpy of superheated steam at the flue gas temperature and the partial pressure of the water vapor. h_f is the enthalpy of saturated liquid at the air's entrance temperature.

$$q_2 = m_{vapor}(h_g - h_f) = 8.94G_H(h_g - h_f) \qquad 28.27$$

Heat is lost when it is absorbed by moisture originally in the combustion air (and by free moisture in the fuel, if any). In Eq. 28.28, $m_{combustion\,air}$ is the mass of combustion air per pound of fuel. ω is the humidity ratio. h_g' is the enthalpy of superheated steam at the air's entrance temperature and partial pressure of the water vapor.

$$\begin{aligned} q_3 &= m_{atmospheric\,water\,vapor}(h_g - h_g') \\ &= \omega m_{combustion\,air}(h_g - h_g') \end{aligned} \qquad 28.28$$

When carbon monoxide appears in the flue gas, potential energy is lost in incomplete combustion. The higher heating value of carbon monoxide in combustion to CO_2 is 4347 Btu/lbm (9.72 MJ/kg).[26]

$$q_4 = \frac{2.334HHV_{CO}G_C B_{CO}}{B_{CO_2} + B_{CO}} \qquad 28.29$$

For solid fuels, energy is lost in unburned carbon in the ash. (Some carbon may be carried away in the flue gas, as well.) This is known as *combustible loss* or *unburned fuel loss*. In Eq. 28.30, m_{ash} is the mass of ash produced per pound of fuel consumed. $G_{C,ash}$ is the gravimetric fraction of carbon in the ash. The heating value of carbon is 14,093 Btu/lbm (32.8 MJ/kg).

$$q_5 = HHV_C m_{ash}G_{C,ash} \qquad 28.30$$

Energy is also lost through radiation from the exterior boiler surfaces. This can be calculated if enough information is known. The *radiation loss* is fairly insensitive to different firing rates, and once calculated it can be considered constant for different conditions.

Other conditions where energy can be lost include air leaks, poor pulverizer operation, excessive blowdown, steam leaks, missing or loose insulation, and excessive soot-blower operation. Losses due to these sources must be evaluated on a case-by-case basis. The term *manufacturer's margin* is used to describe an accumulation of various unaccounted-for losses, which can include incomplete combustion to CO, energy loss in ash, instrument errors, energy carried away by atomizing steam, sulfation and calcination reactions in fluidized bed combustion boilers, and loss due to periodic blowdown. It can be 0.25–1.5% of all energy inputs, depending on the type of combuster/boiler.

---

[24]Flame temperature is limited by the dissociation of common reaction products (CO_2, N_2, etc.). At high enough temperatures (3400–3800°F; 1880–2090°C), the endothermic dissociation process reabsorbs combustion heat and the temperature stops increasing. The temperature at which this occurs is known as the *dissociation temperature* (*maximum flame temperature*). This definition of flame temperature is not a function of heating values and flow rates.

[25]The abbreviation "dfg" for dry flue gas is peculiar to the combustion industry. It may not be recognized outside of that field.

[26]2.334 is the ratio of the molecular weight of carbon monoxide (28.01) to carbon (12), which is necessary to convert the higher heating value of carbon monoxide from a mass of CO to a mass of C. The product $2.334HHV_{CO}$ is often stated as 10,160 Btu/lbm (23.63 MJ/kg), although the actual calculated value is somewhat less. Other values such as 10,150 Btu/lbm (23.61 MJ/kg) and 10,190 Btu/lbm (23.70 MJ/kg) are encountered.

31. COMBUSTION EFFICIENCY

The *combustion efficiency* (also referred to as *boiler efficiency*, *furnace efficiency*, and *thermal efficiency*) is the overall thermal efficiency of the combustion reaction. Furnace/boilers for all fuels (e.g., coal, oil, and gas) with air heaters and economizers have 75–85% efficiency ranges, with all modern installation trending to the higher end of the range.

In Eq. 28.31, m_{steam} is the mass of steam produced per pound of fuel burned. The useful heat may also be determined from the boiler rating. One *boiler horsepower* equals approximately 33,475 Btu/hr (9.8106 kW).[27]

$$\eta = \frac{\text{useful heat extracted}}{\text{heating value}}$$
$$= \frac{m_{steam}(h_{steam} - h_{feedwater})}{\text{HHV}} \qquad \textbf{28.31}$$

Calculating the efficiency by subtracting all known losses is known as the *loss method*. Minor sources of thermal energy, such as the entering air and feedwater, are essentially disregarded.

$$\eta = \frac{\text{HHV} - q_1 - q_2 - q_3 - q_4 - q_5 - \text{radiation}}{\text{HHV}}$$
$$= \frac{\text{LHV} - q_1 - q_4 - q_5 - \text{radiation}}{\text{HHV}} \qquad \textbf{28.32}$$

Combustion efficiency can be improved by decreasing either the temperature or the volume of the flue gas or both. Since the latent heat of moisture is a loss, and since the amount of moisture generated corresponds to the hydrogen content of the fuel, a minimum efficiency loss due to moisture formation cannot be eliminated. This minimum loss is approximately 13% for natural gas, 8% for oil, and 6% for coal.

32. NOMENCLATURE

A	area	ft^2	m^2
A	moles of air	mol	mol
AF, A/F	air to fuel ratio	–	–
b	$x + (y/4)$, stoichiometric number of moles of oxygen per mole of hydrocarbon	–	–
B	volumetric fraction	–	–
°Be	Baumé specific gravity	degree	degree
c_p	specific heat at constant pressure	Btu/lbm	kJ/kg·°C
d	diameter	ft	m
F	moles of fuel	mol	mol
g	acceleration of gravity, 32.2 (9.81)	ft/sec^2	m/s^2
g_c	gravitational constant, 32.2	lbm-ft/lbf-sec^2	n.a.
G	gravimetric fraction	–	–
h	enthalpy	Btu/lbm	kJ/kg
H	hydrogen content of combusted material	–	–
ΔH_v	heat of vaporization of water, 2420	n.a.	kJ/kg
HHV, HHV	higher heating value	Btu/lbm	MJ/kg
HV	heating value	Btu/lbm	MJ/kg
J	air to fuel gravimetric ratio	–	–
K	air to fuel volume ratio	–	–
LHV, LHV	lower heating value	Btu/lbm	MJ/kg
m, M	mass	lbm	kg
M	moisture content	%	%
M	molecular weight	lbm/ lbmmol	kg/kmol
n	number of moles	–	–
ON	octane number	–	–
P	pressure	lbf/ft^2	kPa
PN	performance number	–	–
q	heat loss	Btu/lbm	kJ/kg
R	specific gas constant	ft-lbf/ lbmol-°R	kJ/kg·K
Re	Reynolds number	–	–
$S.G.$	specific gravity	–	–
t	time	sec	s
T	temperature	°F	K
v	velocity	ft/sec	m/s
v_t	terminal settling velocity	ft/sec	m/s
V	volume	ft^3	m^3

Symbols

η	efficiency	–	–
μ	absolute viscosity	lbf-sec/ft^2	Pa·s
ρ	density	lbm/ft^3	kg/m^3
ω	humidity ratio	lbm/lbm	kg/kg

Subscripts

f	filtration or fluid
g	gas (vapor)
p	particle
t	terminal
v	combusted
x	cross-sectional

---

[27]Boiler horsepower is sometimes equated with a gross heating rate of 44,633 Btu/hr (13.08 kW). However, this is the total incoming heating value assuming a standardized 75% combustion efficiency.

29 Emissions Characterization, Calculations, Inventory

1. EMISSIONS SAMPLING

There are numerous sampling procedures, as most pollutants have their own specific characteristics, nuances, and idiosyncracies.

Sampling of emissions in flue gases can either be continuous or by spot sampling using wet chemistry or whole air methods. With *wet chemistry sampling methods*, a sample is collected in a container, solid adsorbent, or liquid-impinger train. The sample is then moved from the field to the laboratory for analysis.

With *whole air sampling*, a volume of flue gas is collected in a bag-like container (e.g., a *Tedlar bag*), and the container is transported to the laboratory for analysis by gas chromatography (GC). Flame ionization detection (GC-FID) and a mass spectrophotometer (GC-MS) are both used. For some pollutants, residence time in the container is critical. Volatile organic compounds (VOCs) with short half-lives can be captured and tested in a sampling train (VOST), which allows for longer sample-holding times.

Particulates are detected and measured by EPA method 5, which, though complex, has become a de facto standard, or EPA method 17 (in-stack filtration). With method 5, the sampling device includes a series of absorbers connected in tandem. The absorption train is followed by a gas-drying tube and a vacuum pump. Several samples are usually taken. The amount of particulates is determined from the filter's weight increase. Particles in the sampling path are washed out, dried, and included in the weight. Stack gas moisture content is determined from the impinger train. Oxygen, carbon dioxide, and (by difference) nitrogen volumes are determined by analyzing a sample collected separately in a sample bag.

Tests of gas streams should be unbiased, which requires sampling to be isokinetic. The gas velocity and ratio of dry gas mass to pollutant mass are not disturbed by *isokinetic testing*.[1,2]

The number of samples taken depends on the proximity of the sampling point to upstream and downstream disturbances. The number of samples can be minimized by locating the sampling point such that flow is undisturbed by upstream and downstream changes of direction, changes in diameter, and equipment in the duct (i.e., after a "good straight run"). Sampling should be at least eight equivalent diameters downstream and two equivalent diameters upstream from disturbances in flow that would cause the flow pattern to be unsymmetrical (i.e., would cause swirling). For rectangular ducts, the equivalent diameter is

$$d_e = \frac{(2)(\text{height})(\text{width})}{\text{height} + \text{width}} \qquad 29.1$$

Since a pollutant may not be dispersed evenly across the duct, the average concentration must be determined by taking multiple samples across the flow cross section. This process is known as *traversing* the area, and the sampling points are known as *traverse points*. For circular ducts, the cross section is divided into annular areas according to a standardized procedure. A minimum of six measurements should be taken along each perpendicular diameter. 24 is the usual maximum number of traverse points. (Fig. 29.1 shows the locations for a 10-point traverse.) For rectangular ducts, the cross section is divided into a minimum of nine equal rectangular areas of the same shape and with an aspect ratio between 1:1 and 1:2. The traverse points are located at the centroids of these areas.

---

[1]With *super-isokinetic* sampling, too little pollution is indicated. This can occur if the vacuum draws a sample such that sampling velocity is higher than the bulk flow velocity. Since particles are subject to inertial forces and gases are not, the particle count will be essentially unaffected, but the metered volume will be higher, resulting in a lower apparent concentration. Conversely, if the sample is drawn too slowly (i.e., *sub-isokinetic* sampling), the metered volume would be lower, and the apparent pollution concentration will be too high.

[2]Solid and liquid substances in flue gases require isokinetic sampling. Gaseous substances do not.

Air

Figure 29.1 *Traverse Sampling Points*

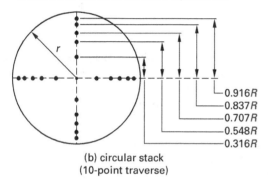

(a) rectangular stack
(measure at center of at least 9 equal areas)

0.916R
0.837R
0.707R
0.548R
0.316R

(b) circular stack
(10-point traverse)

2. EMISSIONS MONITORING SYSTEMS

Continuous emissions monitoring systems (CEMs) are used in large installations, though smaller installations may also be subject to CEM requirements. CEM not only ensures compliance with regulatory limits, but also permits the installation to participate in any allowance trading components of the regulations. CEMs are meant as compliance monitors, not merely operational indicators. The name "continuous emissions monitor system" has come to mean combined equipment for determining and recording SO_2 and NOx concentrations, volumetric flow, opacity, *diluent gases* (O_2 or CO_2), total hydrocarbons, and HCl.

CEMs consist of components that obtain the sample, analyze the sample, and acquire and record analyses in real time.[3] With *straight extractive* CEMs, a sample is drawn from the stack by a vacuum pump and sent down a heated line to an analyzer at the base of the stack. This method must keep the stack gas heated above its 250°F to 300°F (120°C to 150°C) dew point to keep the moisture from condensing. Since samples are drawn by vacuum, leaks in the tube can lead to errors.

With *dilution extraction* CEMs, the gas is diluted with clean air or other inert carrier gas and passed to the analyzer located at a lower level. Dilution ratios are generally high—between 50:1 and 300:1, although they may be as low as approximately 12:1 at the outlet of an FGD unit. Since the dew point is lowered to below −10°F (−23°C) by the dilution, samples passed to the analyzer do not need to be heated. Samples are passed under positive pressure, ensuring that outside air and other gases cannot enter the tube. Dilution extraction analyses are on a wet basis.

With in situ CEM methods, direct measurements are taken by advanced electro-optical techniques. All instrumentation is located on the stack or duct. Data is sent by wire to remote monitors and the *data acquisition/reporting system* (DAS). The term "in situ" also implies that the flue gas analysis will be "wet" (i.e., the volumetric fraction will include water vapor).

There are three common methods of determining volumetric flow rates: thermal sensing using a hotwire anemometer or thermal dispersion, differential pressure using a single-point pitot tube or multipoint annubar, and acoustic velocimetry using ultrasonic transducers.

With the advent of mass emission measurement requirements, pollutant (e.g., SO_2 and CO_2) concentration monitors and velocity monitors are being combined.

3. NOMENCLATURE

| d_e | equivalent diameter | ft | m |

---

[3]The term "continuous" is somewhat of a misnomer. Though sampling is "automatic and frequent," data are rarely collected more frequently than once every 15 minutes.

30 Pollution Treatment, Control, Minimization, Prevention

1. STANDARD AND ACTUAL FLOW RATES

Airflow through fans is typically measured in units of cubic feet per minute, ft^3/min (L/s). When the flow is at the *standard conditions* of 70°F (21°C) and 14.7 psia (101 kPa), the airflow is designated as SCFM (*standard cubic feet per minute*). The density, ρ, of standard air is 0.075 lbm/ft^3 (1.2 kg/m^3). The specific volume, v, of standard air is the reciprocal of the density of standard air, or 13.33 ft^3/lbm (0.8333 m^3/kg). Airflow at any other condition is designated as ACFM (*actual cubic feet per minute*). The two quantities are related by the density factor, K_d.[1] In Eq. 30.2, absolute temperature must be used.[2] Table 30.1 gives the ratio of p_{actual} to p_{std}.

$$ SCFM = \frac{ACFM}{K_d} \qquad 30.1 $$

$$ K_d = \frac{\rho_{std}}{\rho_{actual}} = \left(\frac{p_{std}}{p_{actual}}\right)\left(\frac{T_{actual}}{T_{std}}\right) \qquad 30.2 $$

Standard air is implicitly dry air. A correction for water vapor can be made if the relative humidity, ϕ, is known. The saturation pressure, p_{sat}, is read from a saturated steam table for the dry-bulb temperature of the air.

$$ K_d = \left(\frac{p_{std}}{p_{actual} - p_{vapor}}\right)\left(\frac{T_{actual}}{T_{std}}\right) $$
$$ = \left(\frac{p_{std}}{p_{actual} - \phi p_{sat}}\right)\left(\frac{T_{actual}}{T_{std}}\right) \qquad 30.3 $$

Table 30.1 *Pressure at Altitudes*

altitude (ft (m))	ratio of p_{actual}/p_{std}
sea level (0)	1.00
1000 (305)	0.965
2000 (610)	0.930
3000 (915)	0.896
4000 (1220)	0.864
5000 (1525)	0.832
6000 (1830)	0.801
7000 (2135)	0.772

(Multiply ft by 0.3048 to obtain m.)

Other airflow designations are ICFM (inlet cubic feet per minute), SDCFM (standard dry cubic feet per minute, the time rate of DSCF, dry standard cubic feet), MSCFD (thousand standard cubic feet per day), and MMSCFD (million standard cubic feet per day). The term ICFM is not normally used in duct design. It is used by compressor manufacturers and suppliers to specify conditions before and after filters, boosters, and

---

[1]Some sources use an *air density ratio* that is the reciprocal of the density factor, K_d, defined by Eq. 30.2. In some confusing cases, the same name (i.e., density factor) is used with the reciprocal value.
[2]The temperature correction should be based on the temperature and pressure of the air through the duct system. Though atmospheric pressure and temperature both decrease with higher altitudes, air entering any occupied space will generally be heated to normal temperatures. Therefore, the temperature correction will not generally be used unless the duct system carries air for process heating or cooling.

other equipment. If the conditions before and after the equipment are the same, then ICFM and ACFM will be identical. Otherwise, Eq. 30.4 can be used.

$$\text{ACFM} = \text{ICFM}\left(\frac{p_\text{before}}{p_\text{after}}\right)\left(\frac{T_\text{after}}{T_\text{before}}\right) \qquad 30.4$$

Example 30.1

A manufacturing application in Denver, Colorado ($p_\text{actual} = 12.2$ psia, $T_\text{actual} = 60°F$, relative humidity of 75%) requires 100 SCFM of compressed air at 125 psig. What is the ICFM in Denver?

Solution

Use Eq. 30.3 to calculate the density factor, K_d. From a saturated steam table, the saturation pressure at 60°F is 0.2564 lbf/in².

$$
\begin{aligned}
K_d &= \left(\frac{p_\text{std}}{p_\text{actual} - \phi p_\text{sat}}\right)\left(\frac{T_\text{actual}}{T_\text{std}}\right) \\[2mm]
&= \left(\frac{14.7\,\dfrac{\text{lbf}}{\text{in}^2}}{12.2\,\dfrac{\text{lbf}}{\text{in}^2} - (0.75)\left(0.2564\,\dfrac{\text{lbf}}{\text{in}^2}\right)}\right)\left(\frac{60°F + 460°}{70°F + 460°}\right) \\[2mm]
&= 1.20
\end{aligned}
$$

From Eq. 30.1, the inlet flow rate is

$$\text{ICFM} = \text{ACFM} = K_d(\text{SCFM}) = (1.20)\left(100\,\frac{\text{ft}^3}{\text{min}}\right)$$

$$= 120\ \text{ft}^3/\text{min}$$

2. ABSORPTION

Absorption processes are selected to remove gaseous air pollutants by dissolution into a liquid solvent such as water or a caustic or acid solution. Absorption is used to remove sulfur dioxide, hydrogen sulfide, chlorine, ammonia, nitrogen oxides, and hydrocarbons. Absorption air pollution control processes include various types of packed and spray towers.

In a *packed tower*, clean liquid flows from top to bottom over the tower packing media. The media typically consist of plastic or ceramic shapes designed to maximize surface area for gas-liquid contact. The contaminated gas flows countercurrently, from bottom to top, or crosscurrently, from side to side. The liquid may be recirculated, and a demister may be used to dry the effluent gas stream. Head loss through packed towers varies from 0.25 kPa to 2.0 kPa, and liquid-gas ratios are

typically 1.0–3.0 L liquid/m³ gas. Contaminant removal efficiencies exceed 90%. A typical packed tower is shown in Fig. 30.1.

Figure 30.1 *Packed Tower*

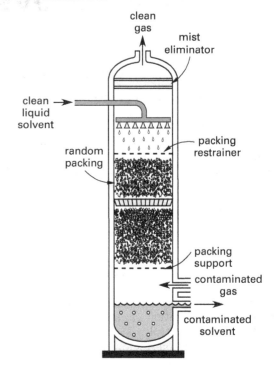

Packed tower design follows principles of fluid mechanics and equilibrium chemistry. The head loss through a packed tower with uniform media can be calculated using the following equations.

$$-\Delta P = \frac{f(1-\alpha)L\text{v}_g^2}{\alpha^3 d_p g} \qquad 30.5$$

$$f = \frac{150(1-\alpha)}{\text{Re}} + 1.75 \qquad 30.6$$

In *spray towers*, the liquid and gas usually flow countercurrently or crosscurrently. The gas moves through a liquid spray that is carried downward by gravity. As with packed towers, the liquid may be recirculated and a demister may be used to dry the effluent gas stream. Pressure drops through spray towers are somewhat lower than through packed towers, varying from 0.25 kPa to 0.5 kPa. With values between 3 L and 14 L liquid/m³ gas, liquid-gas ratios are much higher than in packed towers. Contaminant removal efficiencies are between 50% and 75%, with efficiencies near 90% reported for flue gas desulfurization. A typical spray tower is shown in Fig. 30.2(a) and Fig. 30.2(b).

Figure 30.2 *Types of Scrubbers*

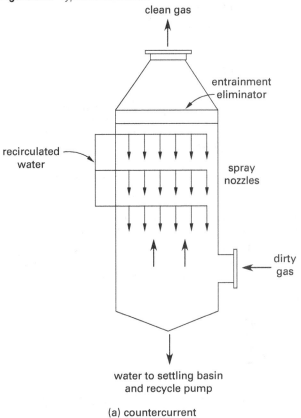

(a) countercurrent

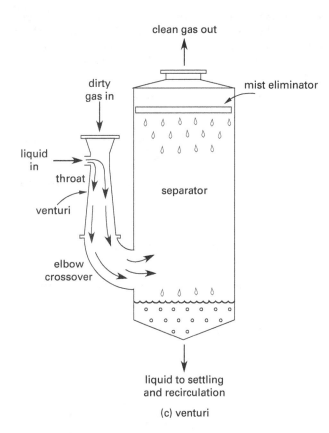

(c) venturi

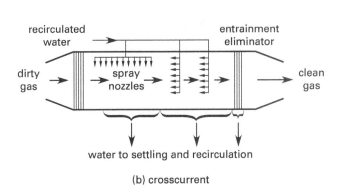

(b) crosscurrent

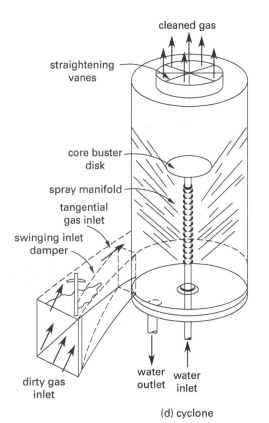

(d) cyclone

Environmental Engineering: A Design Approach, by Sincero and Sincero, © 1995. Reprinted by permission of Prentice-Hall, Inc., Upper Saddle River, NJ.

Spray tower design is similar to packed tower design. However, because water use is much higher and surface area for liquid-gas contact is much lower than in packed towers, spray towers are typically problematic to maintain and operate.

Example 30.2

A packed tower consists of a 1.5 m bed of 10 mm ceramic media. The gas velocity through the bed is 1 m/s, and the bed porosity is 0.45. The gas temperature is 39°C, and the pressure in the bed is 1 atm. What is the head loss through the bed?

Solution

From a table of air properties, $\rho_g = 1.13$ kg/m^3 and $\mu_g = 1.91 \times 10^{-5}$ kg/m·s at 39°C.

Find the Reynolds number. [Reynolds Number (Newtonian Fluid)] [General Spherical]

$$\text{Re} = \frac{v_s d_p \rho_g}{\mu_g}$$

$$= \frac{\left(1.0 \, \dfrac{\text{m}}{\text{s}}\right)(10 \text{ mm})\left(1.13 \, \dfrac{\text{kg}}{\text{m}^3}\right)}{\left(1.91 \times 10^{-5} \, \dfrac{\text{kg}}{\text{m·s}}\right)\left(1000 \, \dfrac{\text{mm}}{\text{m}}\right)}$$

$$= 592$$

Use Eq. 30.6.

$$f = \frac{150(1-\alpha)}{\text{Re}} + 1.75$$

$$= \frac{(150)(1-0.45)}{592} + 1.75$$

$$= 1.89$$

Use Eq. 30.5 and $\rho_l = 1000$ kg/m^3.

$$-\Delta P = \frac{f(1-\alpha)L v_g^2}{\alpha^3 d_p g}$$

$$= \frac{(1.89)(1-0.45)(1.5 \text{ m})\left(1.0 \, \dfrac{\text{m}}{\text{s}}\right)^2 \left(1000 \, \dfrac{\text{mm}}{\text{m}}\right)}{(10 \text{ mm})\left(9.81 \, \dfrac{\text{m}}{\text{s}^2}\right)(0.45)^3}$$

$$= 174 \text{ m of air}$$

$$-\Delta P = \frac{(174 \text{ m of air})\left(1.14 \, \dfrac{\text{kg}}{\text{m}^3}\right)\left(100 \, \dfrac{\text{cm}}{\text{m}}\right)}{1000 \, \dfrac{\text{kg}}{\text{m}^3}}$$

$$= 19.8 \text{ cm of H}_2\text{O}$$

$$-\Delta P = (19.8 \text{ cm of H}_2\text{O})\left(\frac{0.249 \text{ kPa}}{2.54 \text{ cm of H}_2\text{O}}\right)$$

$$= 1.94 \text{ kPa}$$

Example 30.3

A countercurrent packed tower uses 0.005 M sodium hydroxide (NaOH) to remove hydrochloric acid (HCl) from a gas stream at 34°C and 1 atm. The HCl concentration in the gas stream is 260 ppmv. What is the liquid-gas ratio based on the stoichiometric HCl requirement?

Solution

Use the ideal gas law to find the volume of 1 mol of HCl.

$$V = \frac{nR^*T}{P}$$

$$= \frac{(1 \text{ mol})\left(8.2 \times 10^{-5} \, \dfrac{\text{atm·m}^3}{\text{mol·K}}\right)(34°\text{C} + 273°)}{1 \text{ atm}}$$

$$= 0.025 \text{ m}^3$$

$$\text{mole weight HCl} = 1 \, \frac{\text{g}}{\text{mol}} + 35 \, \frac{\text{g}}{\text{mol}}$$

$$= 36 \text{ g/mol}$$

The HCl concentration in the gas stream is

$$\frac{(260 \text{ m}^3)\left(36 \, \dfrac{\text{g}}{\text{mol}}\right)}{(10^6 \text{ m}^3)\left(0.025 \, \dfrac{\text{m}^3}{\text{mol}}\right)} = 0.37 \text{ g/m}^3$$

The stoichiometric reaction equation is

$$\text{HCl} + \text{NaOH} \rightarrow \text{Cl}^- + \text{Na}^+ + \text{H}_2\text{O}$$

1 mol HCl requires 1 mol NaOH.

$$\frac{\left(0.37 \; \dfrac{\text{g HCl}}{\text{m}^3 \; \text{gas}}\right)\left(\dfrac{1 \; \text{L NaOH solution}}{0.005 \; \text{mol NaOH}}\right)}{\left(36 \; \dfrac{\text{g HCl}}{\text{mol HCl}}\right)\left(\dfrac{1 \; \text{mol HCl}}{1 \; \text{mol NaOH}}\right)}$$

$$= 2.1 \; \text{L NaOH solution/m}^3 \; \text{gas}$$

3. ABSORPTION, GAS (GENERAL)

Gas *absorption processes* remove a gas (the *target substance*) from a gas stream by dissolving it in a liquid solvent.[3] Absorption can be used for flue gas cleanup (FGC) to remove sulfur dioxide, hydrogen sulfide, hydrogen chloride, chlorine, ammonia, nitrogen oxides, and light hydrocarbons. Gas absorption equipment includes packed towers, spray towers and chambers, and venturi absorbers.[4]

4. ABSORPTION, GAS (SPRAY TOWERS)

In a general spray tower or spray chamber (i.e., a *wet scrubber*), liquid and gas flow countercurrently or crosscurrently. The gas moves through a liquid spray that is carried downward by gravity. A mist eliminator removes entrained liquid from the gas flow. (See Fig. 30.3.) The liquid can be recirculated. The removal efficiency is moderate.

Spray towers are characterized by low pressure drops—typically 1 in wg to 2 in wg[5] (0.25 kPa to 0.5 kPa)—and their liquid-to-gas ratios—typically 20 gal/1000 ft³ to 100 gal/1000 ft³ (3 L/m³ to 14 L/m³). For self-contained units, fan power is also low—approximately 3×10^{-4} kW/ft³ (0.01 kW/m³) of gas moved.

Flooding of spray towers is an operational difficulty where the liquid spray is carried up the column by the gas stream. Flooding occurs when the gas stream velocity approaches the *flooding velocity*. To prevent this, the tower diameter is chosen as approximately 50% to 75% of the flooding velocity.

Flue gas desulfurization (FGD) wet scrubbers can remove approximately 90% to 95% of the sulfur with efficiencies of 98% claimed by some installations. Stack effluent leaves at approximately 150°F (65°C) and is

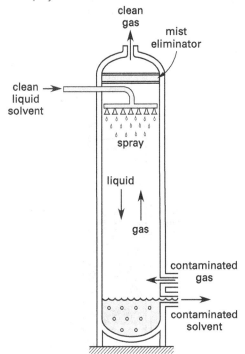

Figure 30.3 *Spray Tower Absorber*

saturated (or nearly saturated) with moisture. Dense steam plumes may be present unless the scrubbed stream is reheated.

Collected sludge waste and fly ash are removed within the scrubber by purely inertial means. After being dewatered, the sludge is landfilled.[6]

Spray towers, though simple in concept and requiring little energy to operate, are not simple to operate and maintain. Other disadvantages of spray scrubbers include high water usage, generation of wastewater or sludge, and low efficiency at removing particles smaller than 5 μm. Relative to dry scrubbing, the production of wet sludge and the requirement for a sludge-handling system is the major disadvantage of wet scrubbing.

5. ABSORPTION, GAS (IN PACKED TOWERS)

In a packed tower, clean liquid flows from top to bottom over the tower packing media, usually consisting of synthetic engineered shapes designed to maximize liquid-surface contact area. (See Fig. 30.4.) The contaminated gas flows countercurrently, from bottom to top,

---

[3]Scrubbing, gas absorption, and stripping are distinguished by their target substances, the carrier flow phase, and the directions of flow. *Scrubbing* is the removal of particulate matter from a gas flow by exposing the flow to a liquid or slurry spray. *Gas absorption* is a countercurrent operation for the removal of a target gas from a gas mixture by exposing the mixture to a liquid bath or spray. *Stripping*, also known as *gas desorption* (also a countercurrent operation), is the removal of a dissolved gas or other volatile component from liquid by exposing the liquid to air or steam. Stripping is the reverse operation of gas absorption. Packed towers can be used for both gas absorption and stripping processes, and the processes look similar. The fundamental difference is that in gas absorption processes, the target substance (i.e., the substance to be removed) is in the gas flow moving up the tower, and in stripping processes, the target substance is in the liquid flow moving down the tower.
[4]Spray and packed towers, though they are capable, are not generally used for desulfurization of furnace or incineration combustion gases, as scrubbers are better suited for this task.
[5]"inwg" and "iwg" are abbreviations for "inches water gage," also referred to as "iwc" for "inches water column" and "inches of water."
[6]A typical 500 MW plant burning high-sulfur fuel can produce as much as 10^7 ft³ (300 000 m³) of dewatered sludge per year.

although crosscurrent designs also exist. As in spray towers, the liquid can be recirculated, and a mist eliminator is used.

Figure 30.4 *Packed Bed Spray Tower*

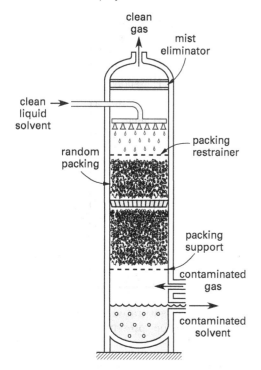

Pressure drops are in the 1 in wg to 8 in wg (0.25 kPa to 2.0 kPa) range. Typical liquid-to-gas ratios are 10 gal/1000 ft³ to 20 gal/1000 ft³ (1 L/m³ to 3 L/m³). Although the pressure drop is greater than in spray towers, the removal efficiency is much higher.

6. ADSORPTION

Adsorption occurs when the target contaminant becomes attached to the surface of the media. Typically, the media is *activated carbon* and is available in either granular (GAC) or powdered (PAC) form. The activated carbon is produced by starved air combustion of coal, coconut shells, wood, and other organic materials. Adsorption is effective in removing heavy metals such as lead and mercury, and in removing aromatic and aliphatic organic chemicals such as VOCs and dioxins. Adsorption processes may be employed to recover gases for reuse or recycling, or for destruction. Very high removal efficiencies, exceeding 99%, are possible.

The activated carbon process provides essentially constant removal until *breakthrough* occurs. At breakthrough, the effluent concentration begins to increase and continues to do so until the carbon is saturated and no pollutant removal occurs. Because optimum operation uses carbon to saturation, adsorption processes are frequently operated in a *two-in-series lead-follow mode* as shown in Fig. 30.5. The carbon is regenerated or

reactivated sometime during the interval between breakthrough and saturation. *Regeneration* is associated with contaminant recovery and is typically a chemical process. *Reactivation* is associated with contaminant destruction and is a thermal process.

Activated carbon adsorption process design is based on the X/M ratio determined from an adsorption isotherm specific to the target contaminant. The X/M ratio represents the mass of chemical adsorbed per unit mass of carbon and is influenced by the gas temperature and moisture content. Many adsorption isotherm relationships have been developed, but one of the most common is the *Freundlich equation*.

Resistivity of a Medium

$$q_e = KC_e^{1/n} \qquad \qquad 30.7$$

Head loss through the carbon bed is calculated using Eq. 30.5 and Eq. 30.6.

Example 30.4

A gas containing a VOC at 90 $\mu g/m^3$ is vented to a GAC adsorber at 1.3 m³/s. The GAC is effective in removing 99.9% of the VOC, adsorption equipment is available with 75 kg GAC capacity, and reactivation should occur at no less than 28-day intervals. The isotherm values for the VOC at the temperature and moisture conditions of the vented gas are

$$K = 210$$
$$1/n = 1.6$$
$$q_e \text{ units} = \text{mg/g}$$

What is the daily mass of GAC required to remove the VOC from the gas stream?

Solution

The final concentration is

$$C = \left(90 \ \frac{\mu g}{m^3}\right)(1 - 0.999) = 0.09 \ \mu g/m^3$$

Use Eq. 30.7.

Resistivity of a Medium

$$q_e = KC_e^{1/n}$$
$$= (210)\left(0.09 \ \frac{\mu g}{m^3}\right)^{1.6}$$
$$= 4.5 \text{ mg VOC/g GAC}$$

Figure 30.5 *Two-in-Series Activated Carbon Adsorption*

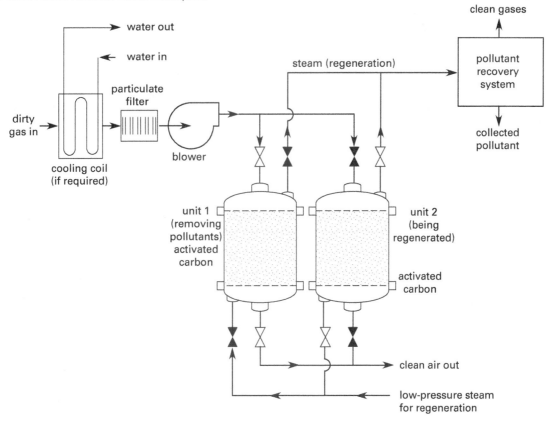

Environmental Engineering: A Design Approach by Sincero and Sincero, © 1995. Prentice-Hall, Inc., Upper SaddleRiver, NJ.

$$\dot{m} = \frac{\Delta C Q_g}{q_e}$$

$$= \frac{\left(90 \ \frac{\mu g}{m^3} - 0.09 \ \frac{\mu g}{m^3}\right)\left(1.3 \ \frac{m^3}{s}\right)\left(86\,400 \ \frac{s}{d}\right)}{\left(4.5 \ \frac{mg \ VOC}{g \ GAC}\right)\left(1000 \ \frac{\mu g}{mg}\right)\left(1000 \ \frac{g}{kg}\right)}$$

$$= 2.2 \ kg \ GAC/d$$

The maximum time between regenerations is

$$\frac{75 \ kg}{2.2 \ \frac{kg \ GAC}{d}} = 34 \ d \quad \begin{bmatrix} \text{acceptable since} \\ 34 \ d > 28 \ d \end{bmatrix}$$

7. ADSORPTION, HAZARDOUS WASTE

AC is particularly attractive for flue gas cleanup (FGC) at installations that burn spent oil, electrical cable, bio-solids (i.e., sewage sludge), waste solvents, or tires as supplemental fuels. As with liquids, flue gases can be cleaned by passing them through fixed AC. Heavy metal

dioxins are quickly adsorbed by the AC—in the first 8 in (20 cm) or so. AC injection (direct or spray) can remove 60% to 90% of the target substance present.

The spent AC creates its own waste disposal problem. For some substances (e.g., dioxins), AC can be inciner-ated. Heavy metals in AC must be removed by a wash process. Another process in use is vitrifying the AC and fly ash into a glassy, unleachable substance. *Vitrifica-tion* is a high-temperature process that turns incinerator ash into a safe, glass-like material. In some processes, heavy-metal salts are recovered separately. No gases and no hazardous wastes are formed.

8. ADVANCED FLUE GAS CLEANUP

Most air pollution control systems reduce NOx in the burner and remove SOx in the stack. *Advanced flue gas cleanup* (AFGC) methods combine processes to remove both NOx and SOx in the stack. Particulate matter is removed by an electrostatic precipitator or baghouse as is typical. Promising AFGC methods include wet scrub-bing with metal chelates such as ferrous ethylenedia-minetetraacetate (Fe(II)-EDTA) (NOx/SOx removal efficiencies of 60%/90%), adding sodium hydroxide injection to dry scrubbing operations (35% NOx removal), in-duct sorbent injection of urea (80% NOx removal), in-duct sorbent injection of sodium bicarbonate (35% NOx removal), and the NOXSO process using a fluidized bed of sodium-impregnated alumina sorbent at approximately 250°F (120°C) (70% to 90%/90% NOx/SOx removal).

9. ADVANCED OXIDATION

The term *advanced oxidation* refers to the use of ozone, hydrogen peroxide, ultraviolet radiation, and other exotic methods that produce free hydroxyl radicals (OH^-). Table 30.2 shows the relative oxidation powers of common oxidants.

Table 30.2 Relative Oxidation Powers of Common Oxidants

oxidant	oxidation power (relative to chlorine)
fluorine	2.25
hydroxyl radical (OH)	2.05
ozone (O_3)	1.52
permanganate radical (MnO_4)	1.23
chlorine dioxide (ClO_2)	1.10
hypochlorous acid (HClO)	1.10
chlorine	1.00
bromine	0.80
iodine	0.40
oxygen	0.29

10. BAGHOUSES

Baghouses consist of banks of fabric filters, usually bags, typically suspended from overhead supports. Contaminated air flows into the bottom of the baghouse and must pass through the filter fabric to escape as cleaned air through the top. The particulate is trapped at the fabric surface. Shaker and pulse jet baghouses are illustrated in Fig. 30.6.

Advantages of baghouses include high efficiency and performance that is essentially independent of flow rate, particle size, and particle (electrical) resistivity. Also, baghouses produce the lowest opacity (generally less than 10, which is virtually invisible). Disadvantages include clogging, difficult cleaning, and bag breakage.

Baghouses have a reputation for excellent particulate removal, down to 0.005 grains/ft^3 (0.18 grains/m^3) (dry), with particulate emissions of 0.01 grains/ft^3 (0.35 grains/m^3) being routine. Removal efficiencies are in excess of 99% and are often as high as 99.99% (weight basis). Fabric filters have a high efficiency for removing particular matter less than 10 μm in size. Because of this, baghouses are effective at collecting air toxics, which preferentially condense on these particles at the baghouse operating temperature—less than 300°F (150°C). Gas temperatures can be up to 500°F (260°C) with short periods up to about 550°F (288°C), depending on the filter

fabric. Most of the required operational power is to compensate for the system pressure drops due to bags, cake, and ducting.

Baghouse filter fabric has micro-sized holes but may be felted or woven, with the material depending on the nature of the flue gas and particulates. Woven fabrics are better suited for lower filtering velocities in the 1 ft/min to 2 ft/min (0.3 m/min to 0.7 m/min) range, while felted fabrics are better at 5 ft/min (1.7 m/min). The fabric is often used in tube configuration, but envelopes (i.e., flat bags) and pleated cartridges are also available.

The particulate matter collects on the outside of the bag, forming a cake-like coating. If lime has been introduced into the stream in a previous step, some of the cake will consist of unreacted lime. As the gases pass through this cake, additional neutralizing takes place.

When the pressure drop across the filter reaches a preset limit (usually 6 in wg to 8 in wg (1.5 kPa to 2 kPa)), the cake is dislodged by mechanical shaking, reverse-air cleaning, or pulse-jetting. The dislodged fly ash cake falls into collection hoppers below the bags. Fly ash is transported by pressure or vacuum conveying systems to the conditioning system (consisting of surge bins, rotary feeders, and pug mills). Figure 30.7 shows a typical baghouse.

Baghouses are characterized by their air-to-cloth ratios and pressure drop. The *air-to-cloth ratio* (A/C ratio), also known as *filter ratio*, *superficial face velocity*, and *filtering velocity*, is the ratio of the air volumetric flow rate in ft^3/min (m^3/s) to the exposed surface area in ft^2 (m^2). After canceling, the units are ft/min (m/s), hence the name *filtering velocity*. The higher the ratio, the smaller the baghouse and the higher the pressure drop. Shaker and reverse-air baghouses with woven fabrics have typical air-to-cloth ratios ranging from 1.0 ft/min to 6.0 ft/min (0.056 m/s to 0.34 m/s). Pulse-jet collectors with felted fabrics have higher ratios, ranging from 3.5 ft/min to 5.0 ft/min (0.0175 m/s to 0.025 m/s). Typical A/C ratios as a function of cleaning method are shown in Table 30.3.

Table 30.3 Air-to-Cloth (A/C) Ratios

cleaning method	A/C ratio ($m^3/m^2 \cdot s$)
shaking	0.01–0.03
reverse air	0.005–0.02
pulsed jet	0.02–0.08

Adapted from: Noll, K.E., *Fundamentals of Air Quality Systems: Design of Air Pollution Control Devices*, © 1999, American Academy of Environmental Engineers, Annapolis, MD.

Figure 30.6 *Types of Baghouses*

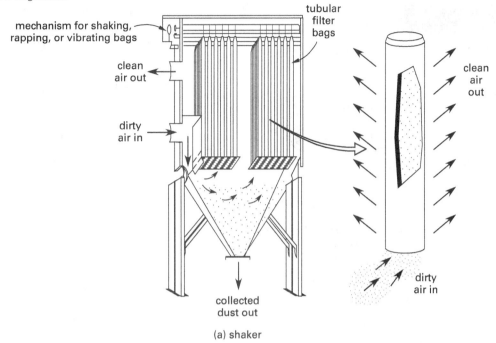

(a) shaker

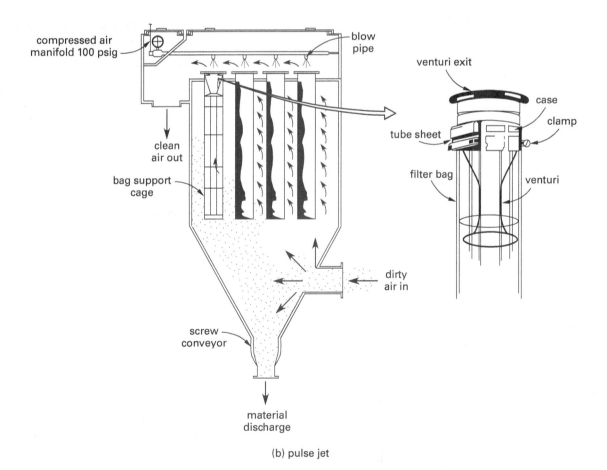

(b) pulse jet

Environmental Engineering: A Design Approach by Sincero and Sincero, © 1995. Reprinted by permission of Prentice-Hall, Inc., Upper Saddle River, NJ.

Figure 30.7 *Typical Baghouse*

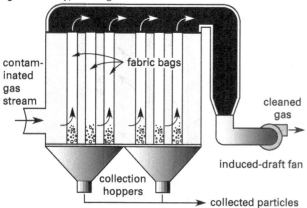

The A/C ratio is used to calculate the required filter fabric area. For this calculation, the A/C ratio is called the filtering velocity, and the following equation is applied.

$$A = \frac{Q_g}{v_f} \qquad 30.8$$

In a baghouse with N compartments, each with multiple bags, the *gross filtering velocity*, *gross air-to-cloth ratio*, and *gross filtering area*, $v_{f,N}$, refer to having all N compartments operating simultaneously. (See Eq. 30.9.) The *design filtering velocity*, *net filtering velocity*, *net air-to-cloth ratio*, and *net filtering area*, $v_{f,N-1}$, refer to one compartment taken offline for cleaning, meaning $N-1$ compartments are operating simultaneously. (See Eq. 30.10.) *Net net* refers to having $N-2$ compartments operating simultaneously.

$$v_{f,N} = \frac{Q_t}{A_{N\text{ compartments}}} = \frac{Q_t}{NA_{\text{one compartment}}} \qquad 30.9$$

$$v_{f,N-1} = \frac{Q_t}{(N-1)A_{\text{one compartment}}} \qquad 30.10$$

The number of bags is

$$n_{\text{bags}} = \frac{A_N}{A_{\text{bag}}} = \frac{A_N}{\pi dh} \qquad 30.11$$

The *areal dust density*, W, is the mass of dust cake on the filter per unit area. The areal dust density can be calculated from the incoming particle concentration (*dust loading, fabric loading*), C, filtering velocity, and collection efficiency. Typical units are lbm/ft^2 (kg/m^2).

$$W = \eta C v_f t \approx C v_f t \qquad 30.12$$

In the *filter drag model*, the pressure drop through the baghouse is calculated from the *filter drag* (*filter resistance*), S, in units of in wg-min/ft (Pa·min/m), which depends on the permeabilities of the cloth, K_1, K_0, or K_e, and the particle cake, K_2 or K_s. (K_1 is also known as the *flow resistance* of the clean fabric; K_2 is the *specific*

resistance of the cake.) Since Eq. 30.14 represents a straight line, K_e and K_s can be determined by plotting S versus W for a few test points. Maximum pressure drop is typically in the 5–20 in wg (1.2–5.0 kPa) range.

$$\Delta P = v_{\text{filtering}} S \qquad 30.13(a)$$

$$\Delta P_{\text{in wg}} = v_{f,\text{ft/min}} S_{\text{in wg-min/ft}} \qquad 30.13(b)$$

$$S = K_e + K_s W \qquad 30.14$$

Baghouses commonly have multiple compartments, and sufficient capacity must remain when one compartment is taken offline for cleaning. If there are N compartments, the time for N cycles of filtration (known as the *filtration time*) will be[7]

Fabric Filtration

$$t_f = N(t_r + t_c) - t_r \qquad 30.15$$

Filtering velocity and pressure drop depend on the number of compartments that are online (i.e., they increase when one of the compartments is taken offline). The maximum pressure drop, ΔP_{max}, occurs several times during the filtration time. The maximum pressure drop can be calculated from the maximum filter drag, which in turn, depends on the maximum areal density. The *actual filtering velocity*, $v_{f,\text{max}\Delta P}$, is the velocity at the time the maximum pressure drop occurs.[8] c in Eq. 30.19 is a scaling constant that depends on the number of compartments. (See Table 30.4.)

$$\Delta P_{\text{max}} = S_{\text{max}\Delta P} v_{f,\text{max}\Delta P} \qquad 30.16$$

$$S_{\text{max}\Delta P} = K_e + K_s W_{\text{max}\Delta P} \qquad 30.17$$

$$W_{\text{max}\Delta P} = (N-1)C(v_{f,N}t_{\text{run}} + v_{f,N-1}t_{\text{cleaning}}) \qquad 30.18$$

$$v_{f,\text{max}\Delta P} = c v_{f,N-1} \qquad 30.19$$

Table 30.4 *Values for Scaling Constant, c, Based on Number of Compartments, N*

number of compartments, N	c
3	0.87
4	0.80
5	0.76
7	0.71
10	0.67
12	0.65
15	0.64
20	0.62

The motor for the fan (blower) is sized based on the maximum pressure loss.

$$P_{\text{hp}} = \frac{Q_{t,\text{ft}^3/\text{min}}\Delta P_{\text{max,in wg}}}{6356\eta_{\text{motor}}} \quad \text{[U.S. only]} \qquad 30.20$$

---

[7]By convention, the filtration time includes only $N-1$ cleanings.
[8]The name "actual filtering velocity" is definitely confusing.

There is no single formula derived from basic principles that predicts baghouse collection efficiency, η. Baghouse designs are normally based on experience, not on fractional efficiency curves. The instantaneous collection efficiency (as well as the pressure drop) of a baghouse increases with time as pores fill with particles. The empirical rate constant, k, is derived from a curve fit of actual performance.

$$\eta = 1 - e^{-kt} \qquad 30.21$$

The minimum cleaning frequency for a desired pressure drop or the pressure drop for a specified operating period can be calculated using Eq. 30.22 in the appropriate form. K_1 and K_2 are the *fabric resistance* and *cake resistance*, respectively.

$$-\Delta P = K_1 v_f + K_2 C_p v_f^2 t \qquad 30.22$$

The pressure drop through a baghouse at a given gas flow rate is the total of the pressure drops due to the fabric, P_f, each particulate layer, P_p, and the baghouse structure, P_s. Darcy's equation for fluid flow through a porous media, fabric, and particulate layer can be used to find P_f and P_p.

Fabric Filtration

$$\Delta P = \Delta P_f + \Delta P_p + \Delta P_s \qquad 30.23$$

$$\Delta P_f = \frac{D_f \mu V}{60 K_f} \qquad 30.24$$

$$\Delta P_p = \frac{D_p \mu V}{60 K_p} \qquad 30.25$$

As contaminated air passes through a bag filter, particulate contaminants accumulate inside the filter. Equation 30.26 provides a method of estimating the depth of the layer of particulate, or dust, that accumulates in the bag filter over time t. The dust loading into the bag, L, is the mass of dust per unit volume of air. The superficial filter velocity, V, is a measurement of the air velocity through the bag. The density, ρ, is the bulk density of the particulate layer once it settles out of air.

Fabric Filtration

$$D_p = \frac{L V t}{\rho_L} \qquad 30.26$$

Example 30.5

A pulsed-jet baghouse is described by the following.

airflow rate	$8 \text{ m}^3/\text{s}$
particulate concentration	45 g/m^3
A/C ratio	$0.025 \text{ m}^3/\text{m}^2 \cdot \text{s}$
fabric resistance	$0.2 \text{ kPa} \cdot \text{s/cm}$
cake resistance	$1.8 \text{ kPa} \cdot \text{s} \cdot \text{cm/g}$
allowable pressure drop	0.65 kPa

What are the (a) required total fabric area, and (b) minimum cleaning frequency?

Solution

(a) Use Eq. 30.8.

The total fabric area is

$$A = \frac{Q_g}{v_f} = \frac{8 \ \dfrac{\text{m}^3}{\text{s}}}{0.025 \ \dfrac{\text{m}^3}{\text{m}^2 \cdot \text{s}}}$$

$$= 320 \text{ m}^2$$

(b) Use Eq. 30.22.

$$-\Delta P = K_1 v_f + K_2 C_p v_f^2 t$$

$$0.65 \text{ kPa} = \frac{\left(0.2 \ \dfrac{\text{kPa} \cdot \text{s}}{\text{cm}}\right)\left(0.025 \ \dfrac{\text{m}}{\text{s}}\right)}{\dfrac{1 \text{ m}}{100 \text{ cm}}}$$

$$+ \frac{\left(1.8 \ \dfrac{\text{kPa} \cdot \text{s} \cdot \text{cm}}{\text{g}}\right)\left(45 \ \dfrac{\text{g}}{\text{m}^3}\right)\left(0.025 \ \dfrac{\text{m}}{\text{s}}\right)^2 t}{100 \ \dfrac{\text{cm}}{\text{m}}}$$

$$t = 296 \text{ s} \quad \text{[clean every 4.9 min]}$$

11. CONDENSERS

Condensers are used for pretreatment before other emission control processes to remove VOCs when they occur as high-concentration vapors. Condensers can reduce the operating costs of subsequent emission control processes and provide a potential economic benefit from the recovery of chemical vapors. The utility and efficiency of condensers improve as VOC concentrations increase. Recovery efficiencies over 95% are possible for vapor streams containing VOCs at over 10 000 ppmv, but they fall to as low as 50% when vapor concentrations are near 500 ppmv.

The two different types of condensers are contact and noncontact. Because they allow the VOCs and the condensing liquid to become mixed and thereby create disposal problems, *contact condensers* are not as widely used as *noncontact condensers* for air pollution control. The most common noncontact condensers are shell-and-tube *surface condensers*. A noncontact surface condenser is illustrated in Fig. 30.8.

Figure 30.8 *Surface Condenser*

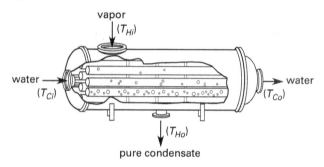

Noncontact surface condensers are designed to promote heat transfer between the VOC and the cooling liquid. The required cooling liquid flow rate, Q_l, can be calculated using Eq. 30.27 with Eq. 30.28 to estimate heat loss, q_a, from the VOC during condensation.

$$Q_l = \frac{q_a}{c_l(T_{Co} - T_{Ci})\rho_l} \qquad \text{30.27}$$

$$q_a = \dot{m}c_a(T_{Hi} - T_{Ho}) + \dot{m}h_{fg} \quad \begin{bmatrix} \text{no} \\ \text{subcooling} \end{bmatrix} \quad \text{30.28}$$

The *log mean temperature difference*, ΔT_{lm}, between the vapor and the cooling liquid can be determined from Eq. 30.29, where H designates hot liquid and C designates cold liquid for the inlet and outlet temperatures.

Log Mean Temperature Difference (LMTD)

$$\Delta T_{lm} = \frac{(T_{Ho} - T_{Ci}) - (T_{Hi} - T_{Co})}{\ln\left(\frac{T_{Ho} - T_{Ci}}{T_{Hi} - T_{Co}}\right)} \qquad \text{30.29}$$

[counterflow]

Log Mean Temperature Difference (LMTD)

$$\Delta T_{lm} = \frac{(T_{Ho} - T_{Co}) - (T_{Hi} - T_{Ci})}{\ln\left(\frac{T_{Ho} - T_{Co}}{T_{Hi} - T_{Ci}}\right)} \qquad \text{30.30}$$

[parallel flow]

The relationship among heat flow, contact area, and log mean temperature is expressed by Eq. 30.31.

Heat Exchangers

$$\dot{Q} = UAF\Delta T_{lm} \qquad \text{30.31}$$

With the heat loss and temperature difference defined, the surface area of the condenser tubes is found by Eq. 30.32.

$$A_s = \frac{q_a}{U\Delta T_m} \qquad \text{30.32}$$

Example 30.6

A countercurrent flow surface condenser is selected to remove trichloroethene (TCE) vapor from a gas stream. The mass flow rate of the TCE is 1200 kg/h, and a maximum outlet temperature of 42°C is required. The condensing liquid is water with 16°C inlet and 28°C outlet temperatures. For TCE, the latent heat of vaporization is 239 kJ/kg, the specific heat is 0.96 kJ/kg·°C, and the boiling point is 87°C at the gas stream pressure. TCE enters the surface condenser saturated.

(a) What is the log mean temperature difference?

(b) What is the needed cooling water flow rate?

Solution

(a) The TCE enters saturated, so $T_{Hi} = 87°C$. Use Eq. 30.29 to find the log mean temperature difference.

Log Mean Temperature Difference (LMTD)

$$\Delta T_{lm} = \frac{(T_{Ho} - T_{Ci}) - (T_{Hi} - T_{Co})}{\ln\left(\frac{T_{Ho} - T_{Ci}}{T_{Hi} - T_{Co}}\right)}$$

$$= \frac{(42°C - 16°C) - (87°C - 28°C)}{\ln\frac{42°C - 16°C}{87°C - 28°C}}$$

$$= 40°C$$

(b) Use Eq. 30.28 to find the heat loss during condensation.

$$q_a = \dot{m}c_a(T_{Hi} - T_{Ho}) + \dot{m}h_{fg}$$

$$= \left(\left(1200\ \frac{\text{kg}}{\text{h}}\right)\left(24\ \frac{\text{h}}{\text{d}}\right)\right)$$

$$\times \left(0.96\ \frac{\text{kJ}}{\text{kg·°C}}\right)(87°C - 42°C)$$

$$+ \left(\left(1200\ \frac{\text{kg}}{\text{h}}\right)\left(24\ \frac{\text{h}}{\text{d}}\right)\right)$$

$$\times \left(239\ \frac{\text{kJ}}{\text{kg}}\right)$$

$$= 8.1 \times 10^6\ \text{kJ/d}$$

Use Eq. 30.27 to find the needed flow rate of cooling water. For water, take the specific heat as 4.186 kJ/kg·°C and density as 1 kg/L.

$$Q_l = \frac{q_a}{c_l(T_{Co} - T_{Ci})\rho_l}$$

$$= \frac{8.1 \times 10^6 \; \dfrac{kJ}{d}}{\left(4.186 \; \dfrac{kJ}{kg \cdot °C}\right)(28°C - 16°C)\left(1 \; \dfrac{kg}{L}\right) \times \left(24 \; \dfrac{h}{d}\right)\left(60 \; \dfrac{min}{h}\right)}$$

$$= 112 \; L/min$$

12. CYCLONE SEPARATORS

Cyclone separators are used to remove particulate matter having aerodynamic diameters greater than about 15 μm. During operation, particulate matter in the incoming gas stream spirals downward at the outside and upward at the inside.[9] The particles, because of their greater mass, move toward the outside wall, where they drop into a collection bin. The clean air is exhausted out the top of the cyclone. Figure 30.9 shows a double-vortex, single cyclone separator.

Figure 30.9 Double-Vortex, Single Cyclone

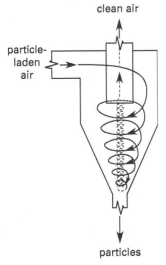

Cyclones are classified as high-throughput, conventional, or high-efficiency. *High-throughput cyclones* are designed to handle high gas volumes where lower efficiencies are acceptable. *High-efficiency cyclones* handle lower flow volumes but provide higher efficiencies. *Conventional cyclones* fall between high-throughput and high-efficiency cyclones.

Cyclone design involves selecting dimensions to provide the desired efficiency as a function of particle physical characteristics (i.e., particle size distribution and density) and terminal settling velocity. There are several different methods used to design and analyze cyclone collectors. All methods yield results that are in the same order of magnitude, but large variations in predicted performance are part of cyclone analysis. The nomenclature used in this section for the dimensions of a standard cyclone is given in Fig. 30.10. The relationships among these dimensions are different for conventional, high-efficiency, and high-throughput cyclones.

Figure 30.10 Standard Cyclone Dimensions

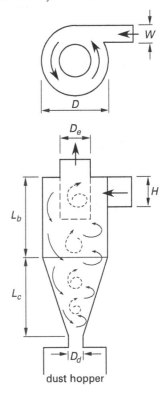

Environmental Engineering: A Design Approach by Sincero and Sincero, © 1995. Reprinted by permission of Prentice-Hall, Inc., Upper Saddle River, NJ.

The centrifugal acceleration is

$$a_c = \frac{\left(\dfrac{4Q_g}{(D - D_e)(2L_b + L_c)}\right)^2}{r_a} \qquad 30.33$$

---

[9]The number of gas revolutions varies approximately between 0.5 and 10.0, with averages of 1.5 revolutions for simple cyclones and 5 revolutions for high-efficiency cyclones.

$$r_a = r_e + 0.5(r - r_e)$$
$$= 0.5D_e + 0.25(D - D_e) \quad \text{30.34}$$

The diameter of the 100% removed particles is

$$d_p = 4\sqrt{\frac{Q_g \mu_g}{a_c \rho_p (2L_b + L_c)(D + D_e)}} \quad \text{30.35}$$

The diameter of the particle for which a cyclone has a 50% collection efficiency is

Cyclone 50% Collection Efficiency for Particle Diameter

$$d_{pc} = \left[\frac{9\mu W}{2\pi N_t V_i (\rho_p - \rho_g)}\right]^{0.5} \quad \text{30.36}$$

The terminal settling velocity of the 100% removed particles is

$$v_t = \frac{Q_g}{0.5\pi(L_b + 0.5L_c)(D + D_e)} \quad \text{30.37}$$

The *overall removal efficiency* of the cyclone uses the following equations. β is the *volume shape factor*.

$$\eta = 1 - x_o + \frac{\sum(\Delta x)v_i}{v_t} \quad \text{30.38}$$

$$v_i = \sqrt{\frac{4a_c\rho_p d_i}{3C_D\rho_g}} \quad \text{30.39}$$

$$C_D = \frac{24}{\text{Re}} \quad \text{30.40}$$

$$\text{Re} = \frac{d_i v_i \rho_g}{\mu_g} \quad \text{30.41}$$

$$d_i = 1.24\beta^{1/3}d_{\text{pi}} \quad \text{30.42}$$

The particle removal efficiency of the cyclone is

Cyclone Collection (Particle Removal) Efficiency

$$\eta = \frac{1}{1 + (d_{pc}/d_p)^2} \quad \text{30.43}$$

Collection efficiencies are not particularly high, and dusts (5 μm to 10 μm) are too fine for most cyclones. For geometrically similar cyclones, the collection efficiency varies directly with the dimensionless *separation factor*, S.

$$S = \frac{v_{\text{inlet}}^2}{rg} \quad \text{30.44}$$

Pressure drop, h, in feet (meters) of air across a cyclone can be roughly (i.e., with an accuracy of only approximately ±30%) estimated by Eq. 30.45. H and W are the height and width of the rectangular cyclone inlet duct, respectively; D_e is the gas exit duct diameter; and K is an empirical constant that varies from approximately 7.5 to 18.4.

$$h = \frac{KWH v_{\text{inlet}}^2}{2gD_e^2} \quad \text{30.45}$$

The number of effective turns the gas makes through the cyclone can be approximated as

Cyclone Effective Number of Turns Approximation

$$N_e = \frac{1}{H}\left[L_b + \frac{L_c}{2}\right] \quad \text{30.46}$$

Example 30.7

A 0.8 m diameter conventional cyclone receives a contaminated gas stream at 6 m³/s and 200°C. The particulates in the gas stream have a density of 1600 kg/m³. The density of the gas is 1.2 kg/m³, and the inlet velocity is 12 m/s. The temperature of the air is 20°C. For what particle diameter will the cyclone have a 50% collection efficiency?

Solution

The diameter of the cyclone is 0.8 m. Estimate the other dimensions from a table of cyclone dimensions. [Cyclone Ratio of Dimensions to Body Diameter]

For a conventional cyclone,

$$D = 0.8 \text{ m}$$
$$H = (0.5)(0.8 \text{ m}) = 0.4 \text{ m}$$
$$W = (0.25)(0.8 \text{ m}) = 0.2 \text{ m}$$
$$L_b = (1.75)(0.8 \text{ m}) = 1.4 \text{ m}$$
$$L_c = (2.00)(0.8 \text{ m}) = 1.6 \text{ m}$$

From Eq. 30.46, the effective number of turns made by the gas in the cyclone is

Cyclone Effective Number of Turns Approximation

$$N_e = \frac{1}{H}\left[L_b + \frac{L_c}{2}\right]$$
$$= \left(\frac{1}{0.4 \text{ m}}\right)\left(1.4 \text{ m} + \frac{1.6 \text{ m}}{2}\right)$$
$$= 5.5$$

From a table of air properties, the dynamic viscosity of 200°C air is $2.6 \times 10-5$ kg/m·s. [Thermophysical Properties of Air and Water]

The diameter of the particle for which the cyclone has a 50% collection efficiency is

Cyclone 50% Collection Efficiency for Particle Diameter

$$d_{pc} = \left[\frac{9\mu W}{2\pi N_e V_i (\rho_p - \rho_g)} \right]^{0.5}$$

$$= \left(\frac{(9)\left(2.6 \times 10^{-5} \dfrac{\text{kg}}{\text{m·s}}\right)(0.2 \text{ m})}{(2\pi)(5.5)\left(12 \dfrac{\text{m}}{\text{s}}\right) \times \left(1600 \dfrac{\text{kg}}{\text{m}^3} - 1.2042 \dfrac{\text{kg}}{\text{m}^3}\right)} \right)^{0.5}$$

$$\times \left(10^6 \frac{\mu\text{m}}{\text{m}}\right)$$

$$= 8.4 \ \mu\text{m}$$

Example 30.8

The grain size distribution for the particulate being removed by the cyclone described in Ex. 30.7 is as follows.

particle size (μm)	mass %
0–15	12
15–25	64
25–35	24

Assume a volume shape factor of 0.9 and a particulate density of 1600 kg/m³. What is the overall removal efficiency of the cyclone?

Solution

Use Eq. 30.37 and the results from Ex. 30.7.

$$v_t = \frac{Q_g}{0.5\pi(L_1 + 0.5L_2)(D + D_e)}$$

$$= \frac{6 \ \dfrac{\text{m}^3}{\text{s}}}{0.5\pi(1.6 \text{ m} + (0.5)(1.6 \text{ m}))(0.8 \text{ m} + 0.4 \text{ m})}$$

$$= 1.3 \text{ m/s}$$

Use Eq. 30.42.

$$d_i = 1.24\beta^{1/3} d_{\text{pi}} = (1.24)(0.9)^{1/3} d_{\text{pi}} = 1.2 d_{\text{pi}}$$

From a table of air properties, $\rho_g = 0.7461$ kg/m³ for air at 200°C. [Thermophysical Properties of Air and Water]

Use Eq. 30.39.

$$v_i = \sqrt{\frac{4a_c \rho_p d_i}{3 C_D \rho_g}}$$

$$= \sqrt{\frac{(4)\left(521 \dfrac{\text{m}}{\text{s}^2}\right)\left(1600 \dfrac{\text{kg}}{\text{m}^3}\right)(1.2) d_{\text{pi}}}{3 C_D \left(0.7461 \dfrac{\text{kg}}{\text{m}^3}\right)}}$$

$$= 1335 \sqrt{\frac{d_{\text{pi}}}{C_D}}$$

Use Eq. 30.41.

$$\text{Re} = \frac{d_i v_i \rho_g}{\mu_g}$$

$$= \frac{v_i (1.2) d_{\text{pi}} \left(0.748 \dfrac{\text{kg}}{\text{m}^3}\right)}{2.6 \times 10^{-5} \dfrac{\text{kg}}{\text{m·s}}}$$

$$= (3.5 \times 10^4) v_i d_{\text{pi}}$$

Use Eq. 30.40.

$$C_D = \frac{24}{\text{Re}} = \frac{24}{(3.5 \times 10^4) v_i d_{\text{pi}}}$$

$$= \frac{7.0 \times 10^{-4}}{v_i d_{\text{pi}}}$$

Substituting the result from Eq. 30.40 into the result from Eq. 30.39 gives

$$v_i = 1335 \sqrt{\frac{d_{\text{pi}}^2 v_i}{7.0 \times 10^{-4}}} = (2.5 \times 10^9) d_{\text{pi}}^2$$

particle size (μm)	mass percentage	cumulative mass fraction	d_{pi} (10^{-6} m)	v_i (m/s)
0–15	12	0.12	7.5	0.14
15–25	64	0.76	20	1.0
25–35	24	1.0	30	2.3
	100%			

In the above table, d_{pi} is taken as the average particle diameter within each particle size range (i.e., for particle size range 15 μm through 25 μm, $d_{\text{pi}} = (15 \ \mu\text{m} + 25 \ \mu\text{m})/2 = 20 \ \mu\text{m}$).

Cumulative mass fraction and incremental settling velocity can be roughly plotted.

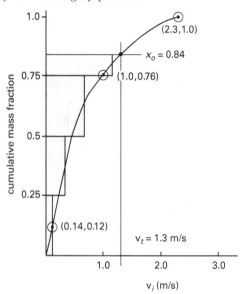

The incremental mass fraction, Δx, and the corresponding incremental settling velocity, v_i, resulting from the integration of the plot above are presented as follows.

Δx	v_i (m/s)	$\Delta x v_i$ (m/s)
0.25	0.15	0.04
0.25	0.36	0.09
0.25	0.69	0.17
0.09	1.15	0.10
		0.40

Use Eq. 30.38.

$$E = 1 - x_o + \frac{\sum \Delta x v_i}{v_t} = 1 - 0.84 + \frac{0.40 \frac{m}{s}}{1.3 \frac{m}{s}}$$

$$= 0.467 \quad (47\%)$$

13. ELECTROSTATIC PRECIPITATORS

Electrostatic precipitators (ESPs) are used to remove particulate matter from gas streams. ESPs are preferred over many other particulate removal processes because they are economical to operate, provide high efficiencies (near 99%), are dependable and predictable, and do not produce a moisture plume.

ESPs used on steam generator/electrical utility units treat gas that is approximately 280°F to 300°F (140°C to 150°C) with moisture being superheated.[10] This high temperature enhances buoyancy and plume dissipation. However, ESPs generally cannot be used with moist flows, mists, or sticky or hygroscopic particles. Scrubbers should be used in those cases. Relatively humid flows can be treated, although entrained water droplets can insulate particles, lowering their resistivities. Table 30.5 gives typical design parameters for ESPs.

Table 30.5 *Typical Electrostatic Precipitator Design Parameters*

parameter	typical range	
	U.S.	SI
efficiency	90% to 98%	
gas velocity	2 ft/sec to 4 ft/sec	0.6 m/s to 1.2 m/s
gas temperature		
standard	$\leq 700°F$	$\leq 370°C$
high-temperature	$\leq 1000°F$	$\leq 540°C$
special	$\leq 1300°F$	$\leq 700°C$
drift velocity	0.1 ft/sec to 0.7 ft/sec	0.03 m/s to 0.21 m/s
treatment/ residence time	2 sec to 10 sec	
draft pressure loss	0.1 iwg to 0.5 iwg	0.025 kPa to 0.125 kPa
plate spacing	12 in to 16 in	30 cm to 41 cm
plate height	30 ft to 50 ft	9 m to 15 m
plate length/height ratio	1.0 to 2.0	
applied voltage	30 kV to 75 kV	

Advantages of ESPs include high reliability, low maintenance, low power requirements, and low pressure drop. Disadvantages are sensitivity to particle size and resistivity, and the need to heat ESPs during start-up and shutdown to avoid corrosion from acid gas condensation.

In operation, the gas passes through a charging field created by high-voltage electrode wires called *discharge electrodes*. The negatively charged particles are attracted to positively charged collection plates known as *collectors*. The speed at which the particles migrate toward the collectors is known as the *drift velocity* or the *electric wind velocity*. Periodically, *rappers* vibrate the collectors and dislodge the particles, which drop into hoppers. Figure 30.11 illustrates several features of ESPs.

In addition to affecting drift velocity, the *specific collection area* (SCA) affects *removal (collection) efficiency*. The higher the SCA, the higher the removal efficiency, with 180 m²·s/m³ defining the upper limit. Pressure

---

[10]The flue gas, at approximately 1400°F (760°C), is cooled to this temperature range in a conditioning tower. Injected water increases the moisture content of the flue gas to approximately 25% by volume.

Figure 30.11 *Electrostatic Precipitator*

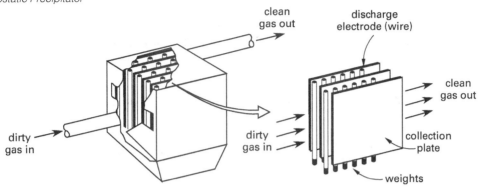

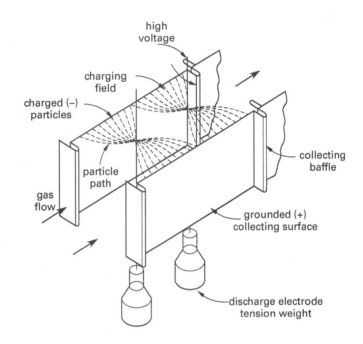

Environmental Engineering: A Design Approach by Sincero and Sincero, © 1995. Reprinted by permission of Prentice-Hall, Inc., Upper Saddle River, NJ.

drops across ESPs are low, typically on the order of 0.13 kPa. The drift velocity, w, is calculated from Eq. 30.47 and Eq. 30.48, where i_o is the permittivity constant, 8.85×10^{-12} C^2/N·m^2.

$$w = \frac{8i_o E_f^2 d_p}{24\mu_g} \qquad 30.47$$

$$E_f = \frac{V_d}{Z_e} \qquad 30.48$$

E_f is the average electric field and is expressed in units of newtons per coulomb (N/C), equivalent to volts/meter (V/m), or the applied voltage divided by the distance separating the discharge electrodes from the collectors.

The separation distance is one-half the distance between collectors. The theoretical efficiency is

Electrostatic Precipitator

$$\eta = 1 - \exp(-Aw/Q) \qquad 30.49$$

The collector area is

$$A = \frac{(-\ln(1-\eta))\,Q}{w} \qquad 30.50$$

The SCA is

$$\text{SCA} = \frac{A}{Q} \qquad 30.51$$

For a rectangular electrostatic precipitator (ESP) with interior dimensions $W \times H \times L$, the flow rate is

$$Q = v_g WH \qquad 30.52$$

The *residence time (exposure time)*, t_r, is

$$t_r = \frac{L}{v_g} \qquad 30.53$$

Collection efficiency is affected by particle *resistivity*, which is divided into three categories: low (less than 1×108 $\Omega \cdot$cm); medium, moderate, or normal (1×10^8 $\Omega \cdot$cm to 2×10^{11} $\Omega \cdot$cm); and high (more than 2×10^{11} $\Omega \cdot$cm). Particles in the medium resistivity range are collected most easily. Particles with low resistivity are easily charged, but upon contact with the charged plate, they rapidly lose their charges and are re-entrained into the gas flow. Particles with high resistivity coat the collection plate with a layer of insulation, causing *flashovers* and *back corona*, a localized electrical discharge. High resistivity particles can be brought into the moderate range by *particle conditioning*. When the gas temperature is below 350°F (177°C), adding moisture to the gas stream, reducing the temperature, or adding SO_3 and ammonia will reduce resistivity into the moderate range. When the gas temperature is above 350°F (177°C), increasing the temperature lowers particle resistivity.

Collection efficiency is also affected by the particle size. The theoretical *drift velocity (migration velocity or precipitation rate)* is proportional to the particle diameter. The theoretical drift velocity, w, of a particle in an electric field can be calculated from basic principles. In Eq. 30.54, q is the charge on the particle, which can be many times the charge on an electron, q_e (1.602×10^{-19} C); E is the electric field strength in V/m; ϵ is the particle's *dielectric constant (relative permittivity)*, approximately 1.5–2.6 for fly ash;[11] and ϵ_0 is the *permittivity of free space*, equal to 8.854×10^{-12} F/m (same as C/V, $C^2/N \cdot m^2$, and $A^2 \cdot s^4/kg \cdot m^3$). C is the *Cunningham slip factor (Cunningham correction factor)*, which accounts for the reduction in drag on particles less than 1 μm in size when gas molecules slip past them. C approaches 1.0 for particles with $d_p > 5$ μm. λ is the *mean free path length* of the gas. For air at 68°F (20°C) and 1 atm, λ is approximately 0.0665 μm. (See Eq. 30.55.)

$$w \approx \frac{qEC}{3\pi\mu_g d_p} = \frac{Nq_e EC}{3\pi\mu_g d_p} \approx \left. \frac{\epsilon\epsilon_0 E^2 d_p C}{(\epsilon+2)\mu_g} \right|_{\text{sat}} \qquad 30.54$$

$$C \approx 1 + \frac{\lambda}{d_{p,\text{Stk}}}\left(2.514 + 0.80\exp\left(\frac{-0.55 d_{p,\text{Stk}}}{\lambda}\right)\right) \qquad 30.55$$

A particle will become saturated if it remains in a strong electric field sufficiently long. The *equilibrium charge* is found from the number of electron charges on a saturated spherical particle. The charging electric field is not necessarily the same as the drift electric field.

$$N_{\text{sat,spherical}} = \frac{3\epsilon\epsilon_0 \pi E_{\text{charging}} d_p^2}{(\epsilon+2)q_e} \qquad 30.56$$

In practice, the theoretical drift velocity is seldom used. Rather, an *effective drift velocity*, w_e, is used. Although it shares the "velocity" name and units, the effective drift velocity is not an actual velocity, but is an empirical design parameter derived from pilot tests of collection efficiency.

The *Deutsch-Anderson equation* predicts the single-particle fractional *collection efficiency (removal efficiency)*, η, of an electrostatic precipitator with total collection area, A, of all plates. The exponent, y, is 1 for fly ash and for anything else in the absence of specific knowledge otherwise. *Penetration*, Pt, is the fraction of particles that pass through the ESP uncollected. Penetration is the complement of collection efficiency.

$$\eta = \frac{C_{\text{in}} - C_{\text{out}}}{C_{\text{in}}} = \frac{\dot{m}_{\text{in}} - \dot{m}_{\text{out}}}{\dot{m}_{\text{in}}}$$
$$= 1 - \exp\left(\frac{-Aw_e}{Q}\right)^y \qquad 30.57$$

$$\text{Pt} = 1 - \eta \qquad 30.58$$

The effect of flow rate on the collection efficiency is predicted by Eq. 30.59.

$$\frac{Q_1}{Q_2} = \ln(\eta_1 - \eta_2) \qquad 30.59$$

Collection efficiency can also be expressed as a function of the *residence time*. K is the ratio of collection plate area to internal volume.

$$\eta = 1 - \exp(-Kw_e t_r)$$
$$= 1 - \exp\left(-\left(\frac{A}{WHL}\right)w_e t_r\right) \qquad 30.60$$

One of the most important factors affecting the collection efficiency is the *specific collection area*, SCA, which is the ratio of the total collection surface area to the gas flow rate into the collector, usually reported in $ft^2/1000$ $ft^3 \cdot$min ($m^2/1000$ $m^3 \cdot$h). The higher the SCA, the higher the collection efficiency will be.

$$\text{SCA} = \frac{A}{Q} \qquad 30.61$$

---

[11]The dielectric constant of fly ash depends greatly on the temperature and amount of unburned carbon.

Using units common to the industry, the instantaneous collection efficiency of an ESP collector is

$$\eta = 1 - \exp(-0.06 w_{e,\text{ft/sec}} \text{SCA}_{\text{ft}^2/1000\,\text{ACFM}}) \quad \textbf{30.62}$$

The *corona power ratio*, CPR, is the power consumed by the corona in developing the electric field divided by the airflow. Generally, collection efficiency increases with increased CPR. The *corona current ratio*, CCR, is the current per unit plate area. The *power density*, P_A, is the power per unit plate area.

$$\text{CPR}_{\text{W-min/ft}^3} = \frac{P_{c,\text{W}}}{Q_{\text{ft}^3/\text{min}}} = \frac{I_{c,\text{A}} V_{\text{V}}}{Q_{\text{ft}^3/\text{min}}} \quad \textbf{30.63}$$

$$\text{CCR}_{\mu\text{A/ft}^2} = \frac{I_{c,\mu\text{A}}}{A_{\text{ft}^2}} \quad \textbf{30.64}$$

$$P_{A,\text{W/ft}^2} = \frac{P}{A} \quad \textbf{30.65}$$

Example 30.9

An ESP will treat a 260°C gas stream containing particulates with an average particle diameter of 4 μm and concentration of 6 g/m^3. The average electric field for the ESP is 350 000 N/C. The dynamic viscosity of 260°C air is 2.78×10^{-5} kg/m·s. What is the SCA required to reduce the particulate concentration to 0.08 g/m^3?

Solution

From the problem statement, $\mu_g = 2.78 \times 10^{-5}$ kg/m·s for air at 260°C. [Thermophysical Properties of Air and Water]

Use Eq. 30.47.

$$w = \frac{8 i_o E_f^2 d_p}{24 \mu_g}$$

$$= \frac{(8)\left(8.85 \times 10^{-12}\ \dfrac{\text{C}^2}{\text{N·m}^2}\right)\left(350\,000\ \dfrac{\text{N}}{\text{C}}\right)^2 (4\ \mu\text{m})}{(24)\left(2.78 \times 10^{-5}\ \dfrac{\text{kg}}{\text{m·s}}\right)\left(10^6\ \dfrac{\mu\text{m}}{\text{m}}\right)}$$

$$= 0.052 \text{ m/s}$$

The required efficiency is

$$\eta = \frac{6\ \dfrac{\text{g}}{\text{m}^3} - 0.08\ \dfrac{\text{g}}{\text{m}^3}}{6\ \dfrac{\text{g}}{\text{m}^3}} = 0.987 \quad (98.7\%)$$

Combine Eq. 30.50 and Eq. 30.51.

$$\text{SCA} = \frac{-\ln(1 - \eta)}{w} = \frac{-\ln(1 - 0.987)}{0.052\ \dfrac{\text{m}}{\text{s}}}$$

$$= 83 \text{ s/m} \quad (83 \text{ m}^2\text{·s/m}^3)$$

14. SCRUBBERS

Scrubbers for particulate control are constructed in a variety of different configurations including upward countercurrent and crosscurrent flow and incorporating various methods to contact the particulate matter with the water. Regardless of the configuration, the objective of the scrubber is to entrain the particulate matter in water droplets. The water subsequently flows from the bottom of the scrubber, the particulate is allowed to settle, and clarified water is recirculated. Compared to dry-particulate removal processes, scrubbers provide the advantages of reduced explosion risk and quenching of hot gases, but they present the disadvantage of having to settle the particulate and manage a wet sludge.

Scrubbers may be classified by the method used to contact the particles with the water droplets. General classifications are spray towers, cyclones, and venturi scrubbers. Several typical scrubber configurations are shown in Fig. 30.2, and design parameters for these are summarized in Table 30.6. The *cut diameter* in Table 30.6 represents the particle size removed at 50% efficiency.

Because particulates are removed by entrainment in water droplets, the most important factors affecting scrubber efficiency are water droplet diameter and quantity. Efficiency increases with decreasing water droplet diameter and with increasing droplet quantity. Using the Calvert model, scrubber efficiency can be calculated using Eq. 30.66.

$$\eta = (100\%)\left(1 - e^{-k(-\Delta P)}\right) \quad \textbf{30.66}$$

For countercurrent, vertical flow scrubbers, $k(-\Delta P)$ in Eq. 30.66 is determined by Eq. 30.67.

$$k(-\Delta P) = \left(\frac{3 Z_o E_f v_d}{2 d_d (v_d - v_g)}\right)\left(\frac{Q_l}{Q_g}\right) \quad \textbf{30.67}$$

For crosscurrent flow scrubbers, $k(-\Delta P)$ in Eq. 30.66 is determined by Eq. 30.68.

$$k(-\Delta P) = \left(\frac{3 Z_o E_f}{2 d_d}\right)\left(\frac{Q_l}{Q_g}\right) \quad \textbf{30.68}$$

Table 30.6 Particulate Scrubber Design Parameters

scrubber type	pressure drop (kPa)	Q_l/Q_g (m³/m³)	liquid inlet pressure (kPa)	particle cut diameter (μm)
baffle spray	0.25–0.75	$1 \times 10^{-4} - 2 \times 10^{-4}$	<100	10
spray tower	0.12–0.75	$7 \times 10^{-4} - 3 \times 10^{-3}$	65–2800	2–8
cyclone	0.35–2.5	$3 \times 10^{-4} - 2 \times 10^{-3}$	275–2800	2–3
venturi ejector	0.12–1.3	$7 \times 10^{-3} - 2 \times 10^{-2}$	100–825	1
venturi	1.3–25	$4 \times 10^{-4} - 3 \times 10^{-3}$	<5–100	0.2

Adapted from: Noll, K.E., *Fundamentals of Air Quality Systems: Design of Air Pollution Control Devices*, © 1999, American Academy of Environmental Engineers, Annapolis, MD.

The *fractional target efficiency*, E_f, in Eq. 30.67 and Eq. 30.68 is calculated from Eq. 30.69.

$$E_f = \left(\frac{\dfrac{C\rho_p d_p^2 v_p}{9\mu_g d_d}}{\dfrac{C\rho_p d_p^2 v_p}{9\mu_g d_d} + 0.7} \right)^2 \qquad 30.69$$

Values for C, the *Cunningham correction factor*, in Eq. 30.69 corresponding to liquid droplet diameters are provided in Table 30.7.

Table 30.7 Cunningham Correction Factor

particle diameter (μm)	correction factor, C^*
10	1.0
2.0	1.1
1.0	1.2
0.50	1.3
0.05	5.0
0.01	23.0

*at 1 atm and 25°C

Source: *Environmental Engineering: A Design Approach* by Sincero and Sincero, © 1995 Prentice-Hall, Inc., Upper Saddle River, NJ.

The pressure drop through a venturi scrubber must be calculated differently based on the use of SI or English units.

Venturi Scrubber

$$\Delta P = V_t^2 L \times 10^{-6} \qquad \text{[SI]} \quad 30.70(a)$$

$$\Delta P = 5 V_t^2 L \times 10^{-5} \qquad \text{[U.S.]} \quad 30.70(b)$$

The scrubber cross-sectional area can be calculated from the gas flow rate and velocity by Eq. 30.71.

$$A_x = \frac{Q_g}{v_g} \qquad 30.71$$

Example 30.10

A crosscurrent flow scrubber is used to remove particulates from a gas stream. The gas and scrubber characteristics are as follows.

gas temperature	60°C
gas velocity	32 cm/s
liquid-gas flow ratio	2×10^{-4}
scrubber contact length	4 m
liquid droplet diameter	0.025 cm
particle diameter	12 μm
particle density	2 g/cm³

What is the scrubber efficiency?

Solution

Use Eq. 30.68 and Eq. 30.69 with $\mu_g = 1.97 \times 10^{-5}$ kg/m·s at 60°C from a table of values for air viscosity and density and $C = 1.0$ from Table 30.7. Assume particle and gas velocities are equal.

To simplify, convert all units to meters and kilograms.

$$d_p = 12 \ \mu\text{m} = 1.2 \times 10^{-5} \text{ m}$$
$$d_d = 0.025 \text{ cm} = 2.5 \times 10^{-4} \text{ m}$$
$$\rho_p = 2 \text{ g/cm}^3 = 2000 \text{ kg/m}^3$$
$$v_p = 32 \text{ cm/s} = 0.32 \text{ m/s}$$

$$E_f = \left(\frac{\dfrac{C\rho_p d_p^2 \mathrm{v}_p}{9\mu_g d_d}}{\dfrac{C\rho_p d_p^2 \mathrm{v}_p}{9\mu_g d_d} + 0.7} \right)^2$$

$$= \left(\frac{\dfrac{(1.0)\left(2000 \ \dfrac{\mathrm{kg}}{\mathrm{m}^3}\right)(1.2 \times 10^{-5} \ \mathrm{m})^2 \left(0.32 \ \dfrac{\mathrm{m}}{\mathrm{s}}\right)}{(9)\left(1.97 \times 10^{-5} \ \dfrac{\mathrm{kg}}{\mathrm{m \cdot s}}\right)(2.5 \times 10^{-4} \ \mathrm{m})}}{\dfrac{(1.0)\left(2000 \ \dfrac{\mathrm{kg}}{\mathrm{m}^3}\right)(1.2 \times 10^{-5} \ \mathrm{m})^2 \left(0.32 \ \dfrac{\mathrm{m}}{\mathrm{s}}\right)}{(9)\left(1.97 \times 10^{-5} \ \dfrac{\mathrm{kg}}{\mathrm{m \cdot s}}\right)(2.5 \times 10^{-4} \ \mathrm{m})} + 0.7} \right)^2$$

$$= 0.56$$

$$k(-\Delta P) = \left(\frac{3Z_o E_f}{2d_d} \right)\left(\frac{Q_l}{Q_g} \right)$$

$$= \frac{(3)(4 \ \mathrm{m})(0.56)(2 \times 10^{-4})}{(2)(2.5 \times 10^{-4} \ \mathrm{m})}$$

$$= 2.69$$

Use Eq. 30.66.

$$\eta = (100\%)\left(1 - e^{-k(-\Delta P)}\right)$$

$$= (100\%)\left(1 - e^{-2.69}\right)$$

$$= 93\%$$

15. INCINERATION

Incineration is applied to air pollution control to destroy combustible air pollutants such as gases, vapors, and odors where the composition of these substances is limited to carbon, hydrogen, and oxygen atoms. When used for this purpose, incineration may be referred to as *afterburning*. Unlike incineration for solid waste disposal, power generation, or manufacturing uses (which are typically sources of air pollutants), incineration for air pollution control should not require subsequent emission control processes. Incineration may be divided into three classifications: direct flame, thermal, and catalytic.

Direct flame incineration is typically applied where the gas stream has sufficient energy value to maintain combustion without the need to provide a supplemental fuel source. Gases with heating values above about 3200 kJ/m³ typically satisfy this criterion. Gases with high energy values can result in temperatures exceeding 1300°C, a condition that may produce nitrogen oxides if combustion occurs with excess air. A common application of direct flame incineration is the *flare*. Flares are used to burn landfill and digester gases as well as waste refining gases.

Flares will burn without smoke when the hydrogen-carbon ratio exceeds ⅓ on a weight basis. This can be illustrated by comparing methane (CH₄) with an H:C ratio of ⅓ and acetylene (C₂H₂) with an H:C ratio of ⅟₁₂. Methane burns without smoke while acetylene burns with soot. Gases with an H:C ratio less than ⅓ can be made to burn without smoke by injecting steam into the flame zone to provide turbulence and enhance mixing between the gas and the air. Steam injection also makes for a shorter flame and hence improves stability under windy conditions.

Thermal incineration is applied to gas streams with low energy values that require supplemental fuel to sustain combustion. These gases may have heating values between about 40 kJ/m³ and 750 kJ/m³. Efficiency is based on the three Ts: time, temperature, and turbulence. Normal residence times are between 0.2 s and 0.8 s, and typical temperatures range from 500°C to 850°C. If the incinerator is deficient in any of the three Ts, carbon monoxide and intermediate oxidation products may be formed. However, if operated properly, hydrocarbon destruction efficiencies can approach 100%. Figure 30.12 presents a schematic of a thermal incinerator.

Figure 30.12 *Thermal Incinerator*

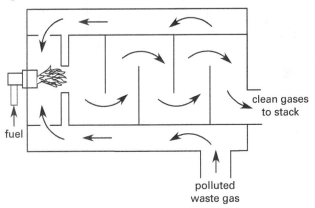

clean gases
to stack

fuel

polluted
waste gas

Catalytic incinerators rely on a catalyst material to produce oxidation at the desired temperature. Catalytic incineration can be applied to gas streams similar to those suited for thermal incineration, but because of the catalyst, residence times and temperatures can be greatly reduced. Residence times may be as much as 50 times shorter than those required for thermal incineration, and temperatures can be as much as 250°C to 300°C lower while achieving similar destruction efficiencies.

Catalytic materials include metals such as platinum, molybdenum, cobalt, and copper formed into a variety of shapes supported on inert *carriers* constructed of alumina, porcelain, and other materials. *Fouling* or *poisoning* of the catalyst is the most common operating problem. Catalytic reactivity is influenced by molecular structure in the following order for hydrocarbons: aromatics; branched alkanes; normal alkanes; alkenes; alkynes. A catalytic incinerator is illustrated in Fig. 30.13.

Figure 30.13 *Catalytic Incinerator*

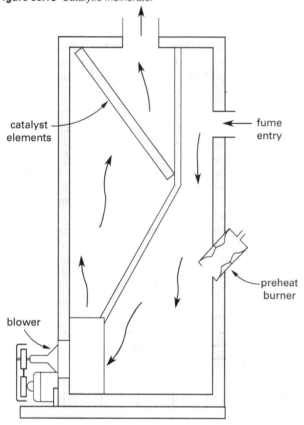

Example 30.11

Four hydrocarbons with their chemical formulas are listed as follows.

benzene	C_6H_6
cyclopropane	C_3H_6
ethene	C_2H_4
propane	C_3H_8

Will any of the listed compounds burn without smoke?

Solution

M carbon $= 12$ g/mol

M hydrogen $= 1$ g/mol

compound	carbon moles	hydrogen moles	carbon mass (g)	hydrogen mass (g)	H:C
benzene	6	6	72	6	1/12
cyclopropane	3	6	36	6	1/6
ethene	2	4	24	4	1/6
propane	3	8	36	8	1/4.5

None of the compounds will burn without smoke since none have an H:C ratio greater than $^1/_3$.

16. FLUE GAS RECIRCULATION

NOx emissions can be reduced when thermal dissociation is the primary NOx source (as it is when low-nitrogen fuels such as natural gas are burned) by recirculating a portion (15% to 25%) of the flue gas back into the furnace. This process is known as *flue gas recirculation* (FGR). The recirculated gas absorbs heat energy from the flame and lowers the peak temperature. Thermal NOx formation can be reduced by up to 50%. The recirculated gas should not be more than 600°F (315°C).

17. MOBILE SOURCE CONTROL

Introduction

The genesis of mobile source emission regulation occurred in California in 1959 when the state adopted standards to control hydrocarbons and carbon monoxide. Since then, standards have been expanded to include other pollutants as technology has developed to facilitate their control.

Mobile sources of air pollution include cars, trucks, buses, trains, planes, and so on. In the United States, there are nearly 200 million such mobile sources, compared to over 30,000 stationary sources. Mobile sources present unique problems for air pollution control and policy. They emit many of the same pollutants as stationary sources, especially hydrocarbons, carbon monoxide, and nitrogen dioxide. Also, gasoline is a significant source of benzene emissions.

Institutional Issues

One major solution to mobile source air pollution is to reduce the number of miles driven. This could be accomplished by increased mass transit access, increased bicycle use, improved pedestrian access, and so forth. However, the demand is not for additional metro lines, bike paths, and sidewalks, but for more roads and parking.

The use of the automobile is subsidized by taxes for building roads, and parking is provided free of cost to the user by those who develop commercial property, city governments, and others. Programs to facilitate bicycle use are not provided, except in a very few places. Also, it is hard to use a bicycle for transportation when people live, work, and shop in places that are widely separated. The subsidized use of the automobile has allowed people to disperse homes far and wide. It is no longer feasible to provide alternatives to automobile transportation in the United States because the dispersed population has developed a lifestyle dependent on the automobile.

One problem with automobile use is that the associated costs are not internalized. Possible ways to internalize these costs include the following.

- *higher fuel taxes:* These would be potentially very high and carry the possible disadvantage of drivers in lower pollution areas subsidizing those in higher pollution areas.

- *restricted access periods:* During rush hour, automobile traffic to higher pollution areas excludes delivery trucks and is limited to commuters and residents.

- *toll roads:* Toll fees fluctuate for various use periods and are higher for roads accessing higher pollution areas. Toll roads that cover the actual cost of building and maintaining the road are also possible deterrents to heavy vehicle use.

- *reserved lanes for buses and carpoolers:* The traffic in these lanes is lighter, which makes traveling faster. Faster travel reduces the time vehicles spend on the road and, therefore, reduces emissions.

- *parking costs:* Charge for the actual cost of providing parking.

Emissions Regulation Strategies

It is difficult to find effective control strategies because of the number of mobile sources and because those involved are individuals, not corporations. This problem is overcome by regulating mobile source emissions through activities in two categories: regulation at the point of manufacture and regulation at the point of use. A combination of these two options has been applied in the United States in the form of a preconsumer certification and enforcement program.

The certification program incorporates testing of new automobiles under controlled conditions at the manufacturers' facilities. Automobiles in each engine class are tested and can only be sold if they are able to demonstrate compliance with the certification criteria. These tests are conducted using prototype vehicles, before new engines are placed into full production.

The enforcement program applies to automobiles in production. The EPA tests a representative number of vehicles as they come off the assembly line to ensure that they comply with the performance standards established during the certification program. Only 60% of the vehicles need to pass the EPA test for the certification to remain valid. Regulations also include requirements for the manufacturer to warrant emissions control equipment, creating an incentive for the equipment to be reliable.

In some locales where mobile sources contribute significantly to nonattainment, additional restrictions may be placed on the automobile manufacturer and on the user. Stricter California emissions standards, for example, are applied to new vehicles. Also, some states require owners to have their vehicles inspected for compliance with emissions restrictions and to make repairs where noncompliance occurs.

Some locales, most notably California, have implemented regulatory and policy measures to promote the availability and sale of *low emission vehicles* (LEVs) and *zero emission vehicles* (ZEVs). In urban areas with chronic air pollution problems, LEVs and ZEVs are particularly attractive as a mechanism to improve air quality.

As mobile source operating conditions change, so do emissions. Therefore, in an effort to define a representative range of conditions for applying EPA performance standards, a standard driving cycle test is used. The *Federal Test Procedure* (FTP, also known as the *EPA 75 test* and *CVS-7.5 test*) is a constant volume sampler test conducted while a test vehicle is running on a dynamometer. It simulates cold start, idle, stop-go traffic, highway, and many other conditions modeled on a 7.5 mi course and taking 22.8 minutes. There are other federal tests, including the *Urban Dynamometer Driving Schedule* (UDDS, also known as the *LA4 test* and the *"City Test"*), which is part of the FTP and simulates light vehicle city driving, and the *Highway Fuel Economy Driving Schedule* (HWFET), which simulates highway driving conditions under 60 mph. Individual states (e.g., California and New York) and cities may have their own specialized tests.

The EPA also employs the FTP for evaluating the city part of fuel economy ratings. The highway part is determined by a simulated 10 mi, nonstop, 48 mph test. A harmonic average of the two tests, 55% for city and 45% for highway, is used to determine the average fuel economy. The average fuel economy determined in this way represents the *Corporate Average Fuel Economy* (CAFE).

Example 30.12

What would be the EPA average fuel economy for a car rated at 18 mpg city and 23 mpg highway?

Solution

CAFE applies 55% to city mileage and 45% for highway mileage. Assume 1 mi of driving. The average fuel economy is

$$\frac{1 \text{ mi}}{\dfrac{(1 \text{ mi})(0.55)}{18 \ \dfrac{\text{mi}}{\text{gal}}} + \dfrac{(1 \text{ mi})(0.45)}{23 \ \dfrac{\text{mi}}{\text{gal}}}} = 19.95 \text{ mpg}$$

Emissions Controls

A schematic of an automobile engine without air pollution control devices is presented in Fig. 30.14(a) and identifies four emission sources: (1) volatilization of hydrocarbons to the atmosphere from the carburetor or throttle body, (2) volatilization of hydrocarbons to the atmosphere from a vented gasoline tank, (3) emission of noncombusted and partially oxidized hydrocarbons from the crankcase, and (4) emission of criteria pollutants from the exhaust.

Each of the four emission sources is controlled in new and well-maintained older cars and, to varying degrees, in other types of mobile sources. The emission control devices applied to the four sources are represented by the schematic in Fig. 30.14(b) and include the following.

- a vent from the air cleaner to capture volatilized hydrocarbons. The vented hydrocarbons are adsorbed on activated carbon contained in a canister typically located in the engine compartment.

- a vent from the gasoline tank to the activated carbon canister described above to capture volatilized hydrocarbons

- a vent from the crankcase to the activated carbon canister to capture noncombusted and partially oxidized hydrocarbons from the crankcase

- a catalytic converter on the exhaust pipe to thermally oxidize criteria pollutants, primarily hydrocarbons, nitrogen oxides, and carbon monoxide

Pollution control equipment performance deteriorates with age and requires maintenance and a well-tuned engine to perform properly. Therefore, one way to remove pollutants contributed by mobile sources is to get rid of old cars, reduce the sale of used cars, and increase the sale of new cars.

Diesel engines require different approaches to air pollution control. Diesel engines operate with very lean fuel mixtures (rich in oxygen) and at relatively high temperatures. This results in inherently low hydrocarbon and carbon monoxide emissions, but creates higher nitrogen oxide emissions than do gasoline engines. The nitrogen oxides are less efficiently removed by catalytic converters on diesel exhaust systems because of the rich oxygen mixture. Furthermore, obstacles exist for controlling the

Figure 30.14 *Internal Combustion Engine*

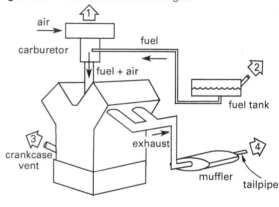

(a) emission points

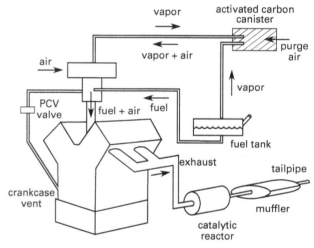

(b) control measures

carbonaceous soot emissions (designated as a probable human carcinogen) which are associated with diesel exhaust.

18. POLLUTION PREVENTION ACT

The Pollution Prevention Act (PPA) was passed by the United States Congress in 1990. The PPA requires the U.S. Environmental Protection Agency (EPA) to establish a source reduction program, provide financial assistance to states, and implement other waste minimization activities. A pollution prevention activity (also called *P2 activity* or *source reduction activity*) is described by the EPA as "any practice that reduces, eliminates, or prevents pollution at its source." Pollution prevention activities that are relevant to air pollution include

- increasing energy efficiency to reduce fuels consumption

- replacing products and materials with low-emissions alternatives

- changing formulations, such as solvents in architectural coating, to eliminate emission sources

- for stationary sources, replacing polluting fuels or energy sources with alternatives such as wind and solar energy

- for mobile sources, promoting low-emissions vehicles, including electric vehicles

- modifying processes to promote clean air practices

To encourage adoption of pollution prevention practices, the federal government and some state governments offer tax incentives, and natural gas and electrical utilities offer financial incentives through rebates and credits to institutions, industry, businesses, and individuals.

19. NOMENCLATURE

a_c	centrifugal acceleration	ft/sec²	m/s²
A	area	ft²	m²
ACFM	actual cubic feet per minute	ft³/min	n.a.
B	width	ft	m
c	scaling constant	-	-
c	specific heat	Btu/lbm-°R	kJ/kg·°C
c_p	specific heat at constant pressure	Btu/lbm-°R	kJ/kg·°C
C	concentration	n.a.	mg/L, μg/m³
C	Cunningham correction factor	–	–
CCR	corona current ratio	A/ft²	A/m²
CPR	corona power ratio	W-min/ft³	J/m²
d	diameter	ft	m
D	depth	ft	m
E	electric field strength	–	V/m
E_f	electric field	V/m	V/m
E_f	fractional target efficiency	–	–
f	friction factor	–	–
F	factor	–	–
g	acceleration of gravity, 32.2 (9.81)	ft/sec²	m/s²
g_c	gravitational constant, 32.2	lbm-ft/lbf-sec²	n.a.
h	enthalpy	Btu/lbm	kJ/kg
h	pressure drop	ft	m
H	height	ft	m
i_o	permittivity constant 8.85×10^{-12}	n.a.	C²/N·m²
ICFM	inlet cubic feet per minute	ft³/min	n.a.
k, K	empirical constant	–	–
K	adsorption capacity or factor	–	–
K	residence time	sec	s

K_d	density factor	-	-
L	length	ft	m
L	dust loading	lbm/ft³	kg/m³
L	water to gas volume ratio	n.a.	L/m³
m	mass	lbm	kg
$\dot{m}$	mass flow rate	lbm/sec	kg/s
M	molecular weight	lbm/lbmol	kg/kmol
n	number of moles	mol	mol
N	number	–	–
$1/n$	strength of adsorption	–	–
p	pressure	in wg	Pa
P	pressure	lbf/ft²	kPa
P_A	power density	W/ft²	W/m²
Pt	penetration	decimal	decimal
q	charge	n.a.	C
q	heat loss	Btu/lbm	kJ/kg
q_e	equilibrium loading	n.a.	mg/g
$\dot{Q}$	volumetric flow rate	ft³/sec	m³/s
Q	heat flow rate	ft³/sec	m³/s
R^*	universal ideal gas constant	ft-lbf/lbmol-°R	kJ/kg·K
Re	Reynolds number	–	–
S	filter drag	in wg-min/ft	Pa·min/m
S	separation factor	-	-
S	vortex finder length	ft	m
SCA	specific collection area	ft²/1000 ft³-min	m²/1000 m³·h
SCFM	standard cubic feet per minute	ft³/min	n.a.
t	time	sec, min	s, min
T	temperature	°R	K
v	velocity	ft/sec	m/s
V	velocity	ft/sec	m/s
V	volume	ft³	m³
W	areal dust density	lbm/ft²	kg/m²
w	drift velocity	ft/sec	m/sec
W	width	ft	m

Symbols

α	packing media bed porosity	–	–
β	volume shape factor	–	–
γ	specific weight	ft³/lbm	m³/kg
ϵ	dielectric constant	–	–
ϵ_0	permittivity of free space	–	F/m
η	efficiency	–	–
λ	mean free path length	–	μm
μ	dynamic viscosity	lbf-sec/ft²	kg/m·s
ρ	density	lbm/ft³	kg/m³
ϕ	relative humidity	%	%

Subscripts

1	fabric resistance
2	cake resistance
a	VOC
b	body
c	centrifugal, cleaning, or cone
d	droplet
e	effective, entry, equilibrium, equivalent, or exit
f	fabric filter, filtration, or liquid (fluid)
fg	vaporization
g	gas (vapor)
i	initial, inner, or inlet
l	liquid
L	layer
lm	log mean
p	constant pressure, particle or particulate
pc	particle collected
pi	incremental particle removed to 100%
r	residence or run
s	solid structure, or surface
sat	saturation
std	standard
t	terminal, throat or total
x	cross-sectional

Topic III: Solid and Hazardous Waste

Solid & Hazard.
Waste

31
Chemistry

Content in blue refers to the *NCEES Handbook*.

The chemistry of solid and hazardous waste follows the same chemical principles and applications as the chemistry of water (discussed in Chap. 3) and the chemistry of air (discussed in Chap. 25).

32 Fate and Transport

1. DIOXINS

Dioxins are a family of chlorinated dibenzo-*p*-dioxins (CDDs). The term *dioxin*, however, is commonly used to refer to the specific congener 2,3,7,8-tetrachlorodibenzo-*p*-dioxin (TCDD). Primary dioxin sources include herbicides containing 2,4-D, 2,4,5-trichlorophenol, and hexachlorophene. Other potential sources include incinerated municipal and industrial waste, leaded gasoline exhaust, chlorinated chemical wastes, incinerated polychlorinated biphenyls (PCBs), and any combustion in the presence of chlorine.

The exact mechanism of dioxin formation during incineration is complex but probably requires free chlorine (in the form of HCl vapor), heavy metal concentrations (often found in the ash), and a critical temperature window of 570–840°F (300–450°C). Dioxins in incinerators probably form near waste heat boilers, which operate in this temperature range.

Dioxin destruction is difficult because it is a large organic molecule with a high boiling point. Most destruction methods rely on high temperature since temperatures of 1550°F (850°C) denature the dioxins. Other methods include physical immobilization (i.e., vitrification), dehalogenation, oxidation, and catalytic cracking using catalysts such as platinum.

Dioxins liberated during the combustion of municipal solid waste (MSW) and refuse-derived fuel (RDF) can be controlled by the proper design and operation of the furnace combustion system. Once formed, they can be removed by end-of-pipe processes, including activated charcoal (AC) injection. Success has also been reported using the vanadium oxide catalyst used for NOx removal, as well as manganese oxide and tungsten oxide.

2. PCBs

Polychlorinated biphenyls (PCBs) are organic compounds (i.e., *chlorinated organics*) manufactured in oily liquid and solid forms through the late 1970s, and subsequently prohibited. PCBs are carcinogenic and can cause skin lesions and reproductive problems. PCBs build up, rather than dissipate, in the food chain, accumulating in fatty tissues. Most PCBs were used as dielectric and insulating liquids in large electrical transformers and capacitors, and in ballasts for fluorescent lights (which contain capacitors). PCBs were also used as heat transfer and hydraulic fluids, as dye carriers in carbonless copy paper, and as plasticizers in paints, adhesives, and caulking compounds.

Incineration of PCB liquids and PCB-contaminated materials (usually soil) has long been used as an effective mediation technique. Removal and landfilling of contaminated soil is expensive and regulated, but may be appropriate for quick cleanups. In addition to incineration and other thermal destruction processes, methods used to routinely remediate PCB-laden soils include biodegradation, ex situ thermal desorption and soil washing, in situ vaporization by heating or steam injection and subject vacuum extraction, and stabilization to prevent leaching. The application of quicklime (CaO) or high-calcium fly ash is now known to be ineffective.

Specialized PCB processes targeted at cleaning PCB from spent oil are also available. Final PCB concentrations are below detectable levels, and the cleaned oil can be recycled or used as fuel.

3. PESTICIDES

The term *traditional organochlorine pesticide* refers to a narrow group of persistent pesticides, including DDT, the "*drins*" (aldrin, endrin, and dieldrin), chlordane, endosulfan, hexachlorobenzene, lindane, mirex, and toxaphene. Traditional organochlorine insecticides used extensively between the 1950s and early 1970s have been widely banned because of their environmental persistence. There are, however, notable exceptions, and environmental levels of traditional organochlorine pesticides (especially DDT) are not necessarily declining throughout the world, especially in developing countries and countries with malaria. DDT, with its half-life of up to 60 years, does not always remain in the country where it is used. The semi-volatile nature of the chemicals means that at high temperatures they will tend to

Solid & Hazard. Waste

evaporate from the land, only to condense in cooler air. This *global distillation* is thought to be responsible for levels of organochlorines increasing in the Arctic.

The term *chlorinated pesticides* refers to a much wider group of insecticides, fungicides, and herbicides that contain organically bound chlorine. A major difference between traditional organochlorine and other chlorinated pesticides is that the former have been perceived to have high persistence and build up in the food chain, and the latter do not. However, even some chlorinated pesticides are persistent in the environment. There is little information available concerning the overall environmental impact of chlorinated pesticides. Far more studies exist concerning the effects of traditional organochlorines in the public domain than on chlorinated pesticides in general. Pesticides that have, for example, active organophosphate (OP), carbamate, or triazine parts of their molecules are chlorinated pesticides and may pose long-lasting environmental dangers.

In the United States, about 30–40% of pesticides are chlorinated. All the top five pesticides are chlorinated. Worldwide, half of the 10 top-selling herbicides are chlorinated (alachlor, metolachlor, 2,4-D, cyanazine, and atrazine). Four of the top 10 insecticides are chlorinated (chlorpyrifos, fenvalerate, endosulfan, and cypermethrin). Four (propiconazole, chlorothalonil, prochloraz, and triadimenol) of the 10 most popular fungicides are chlorinated.

4. PLASTICS

Plastics, of which there are six main chemical polymers as given in Table 32.1, generally do not degrade once disposed of and are considered a disposal issue. Disposal is not a problem per se, however, since plastics are lightweight, inert, and do not harm the environment when discarded.

Table 32.1 Polymers

polymer	plastic ID number	common use
polyethylene terephthalate (PET)	1	clear beverage containers
high-density polyethylene (HDPE)	2	detergent; milk bottles; oil containers; toys
polyvinyl chloride (PVC)	3	clear bottles
low-density polyethylene (LDPE)	4	grocery bags; food wrap
polypropylene (PP)	5	labels; bottles; housewares
polystyrene (PS)	6	styrofoam cups; "clam shell" food containers

A distinction is made between *biodegrading* and *recycling*. Most plastics, such as the polyethylene bags used to protect pressed shirts from the dry cleaner and to mail some magazines, are not biodegradable but are recyclable. Also, all plastics can be burned for their fuel value.

The collection and sorting problems often render low-volume plastic recycling efforts uneconomical. Complicating the drive toward recycling is the fact that many of the six different types cannot easily be distinguished visually, and they cannot be recycled successfully when intermixed. Also, some plastic products consist of layers of different polymers that cannot be separated mechanically. Some plastic products are marked with a plastic type identification number.

Sorting in low-volume applications is performed visually and manually. Commercial high-volume methods include hydrocycloning, flotation with flocculation (for all polymers), X-ray fluorescence (primarily for PVC detection), and near-infrared (NIR) spectroscopy (primarily for separating PVC, PET, PP, PE, and PS). Mass NIR spectroscopy is also promising, but has yet to be commercialized.

Unsorted plastics can be melted and reformed into some low-value products. This operation is known as *downcycling*, since each successful cycle further degrades the material. This method is suitable only for a small fraction of the overall recyclable plastic.

Other operations that can reuse the compounds found in plastic products include hydrogenation, pyrolysis, and gasification. *Hydrogenation* is the conversion of mixed plastic scrap to "syncrude" (synthetic crude oil), in a high-temperature (i.e., 750–880°F (400–470°C)), high-pressure (i.e., 2200–4400 psig (15–30 MPa)), hydrogen-rich atmosphere. Since the end product is a crude oil substitute, hydrogenation operations must be integrated into refinery or petrochemical operations.

Gasification and pyrolysis are stand-alone operations that do not require integration with a refinery. *Pyrolysis* takes place in a fluidized bed between 750°F and 1475°F (400°C and 800°C). Cracked polymer gas or other inert gas fluidizes the sand bed, which promotes good mixing and heat transfer, resulting in liquid and gaseous petroleum products.

Gasification operates at higher temperatures, 1650–3600°F (900–2000°C), and lower pressures, around 870 psig (6 MPa). The waste stream is pyrolyzed at lower temperatures before being processed by the gasifier. The gas can be used on-site to generate steam. Gasification has the added advantage of being able to treat the entire municipal solid waste stream, avoiding the need for sorting plastics.

Biodegradable plastics have focused on polymers that are derived from agricultural sources (e.g., corn, potato, tapioca, and soybean starches), rather than from petroleum. *Bioplastics* have various degrees of *biodegradation* (i.e., breaking down into carbon dioxide, water, and biomass), *disintegration* (losing their shapes and identities and becoming invisible in the compost without needing to be screened out), and *eco-toxicity* (containing no toxic material that prevents plant growth in the compost). A bioplastic that satisfies all three characteristics is a *compostable plastic*. A *biodegradable plastic* will eventually be acted on by naturally occurring bacteria or fungi, but the time required is indeterminate. Also, biodegradable plastics may leave some toxic components. A *degradable plastic* will experience a significant change in its chemical structure and properties under specific environmental conditions, although it may not be affected by bacteria (i.e., be biodegradable) or satisfy any of the requirements for a compostable plastic.

Some engineers point out that biodegrading is not even a desirable characteristic for plastics and that being nonbiodegradable is not harmful. Biodegrading converts materials (such as the paper bags often preferred over plastic bags) to water and carbon dioxide, contributing to the greenhouse effect without even receiving the energy benefit of incineration. Biodegrading of most substances also results in gases and leachates that can be more harmful to the environment than the original substance. In a landfill, biodegrading serves no useful purpose, since the space occupied by the degraded plastic does not create additional useful space (volume).

5. LANDFILL GAS

Most closed landfills are covered by a thick soil layer. As the covered refuse decays beneath the soil, the natural anaerobic biological reaction generates a low-Btu *landfill gas* (LFG) consisting of approximately 50% methane, carbon dioxide, and trace amounts of other gases. If uncontrolled, LFG migrates to the surface. If the LFG accumulates, an explosion hazard results. If it escapes, other environmental problems (including objectionable odors) occur. Therefore, synthetic and compacted clay liners and clay trenches are used to prevent gas from spreading laterally. Wells and collection pipes are used to collect and incinerate LFG in flares. However, emissions from flaring are also problematic.

Alternatives to flaring include using the LFG to produce hot water or steam for heating or electricity generation. During the 1980s, reciprocating engines and combustion turbines powered by LFG were tried. However, such engines generated relatively high emissions of their own due to impurities and composition variations in the fuel. True Rankine-cycle power plants (generally without reheat) avoid this problem, since boilers are less sensitive to impurities.

One problem with using LFG commercially is that LFG is withdrawn from landfills at less than atmospheric pressure. Conventional furnace burners need approximately 5 psig at the boiler front. Low-pressure burners requiring 2 psig are available, though expensive. Therefore, some of the plant power must be used to pressurize the LFG.

Although production is limited, LFG is produced for a long period after a landfill site is closed. Production slowly drops 3% to 5% annually to approximately 30% of its original value after about 20 to 25 years, which is considered to be the economical life of a gas-reclamation system. Approximately 40 ft^3 of gas will be produced per cubic yard (1.5 m^3 of gas per cubic meter) of landfill.

6. LEACHATES

Water and other liquids can enter landfills through disposal, precipitation, and run-on from storms. This liquid can percolate through the waste and come into contact with decomposing solid waste. When it does so, the liquid will become contaminated with dissolved and finely suspended solid matter and microbial waste. If this liquid waste, called *leachate*, then flows out of the landfill, it can contaminate surrounding soil and groundwater. Leachate becomes more concentrated as the landfill ages.

When a layer of liquid sludge is disposed of in a landfill, the consolidation of the sludge by higher layers will cause the water to be released. This released water is known as *pore-squeeze liquid*. Some water will be absorbed by the MSW and will not percolate down. The quantity of water that can be held against the pull of gravity is referred to as the *field capacity*. The potential quantity of leachate is the amount of moisture within the landfill in excess of the field capacity.

In general, the amount of leachate produced is directly related to the amount of external water entering the landfill. Theoretically, the leachate generation rate can be determined by writing a water mass balance on the landfill.

Landfills can be designed to mitigate the risk from leachates. In a lined containment landfill, leachate will percolate downward through the refuse and collect at the first landfill liner. Leachate must be removed to reduce hydraulic head on the liner and to reduce unacceptable concentrations of hazardous substances. In unlined landfills, there is nothing to prevent the leachate from flowing into the groundwater. In these cases, other means of managing the leachate can be used, including on-site treatment, collecting the leachate for disposal elsewhere, and reducing the elevation of the groundwater (known as *dewatering*) to keep it from contact with the leachate.

33 Codes, Standards, Regulations, Guidelines

1. REGULATIONS ON TOXIC SUBSTANCES

Widespread awareness of toxic substances in the environment started with the Love Canal site located in upstate New York. The site was contaminated with 22,000 tons of chemical wastes that were legally disposed in an abandoned canal in the early to mid-1940s. In 1958, some children playing near a road construction site adjacent to the canal received chemical burns, but no action was taken at that time. Later, in 1976, residents along the canal discovered chemical residue leaking into their basements. Love Canal received considerable press, was the recipient of federal disaster relief funds authorized by President Jimmy Carter, and boosted public support for environmental legislation.

With the discovery of Love Canal and several other environmental disasters in the 1970s, both state and federal statutes were enacted. On the federal level, the Toxic Substances Control Act (TSCA) was passed by Congress in 1976, and Congress enacted the Comprehensive Environmental Response, Compensation, and Liability Act (CERCLA) and amended the Solid Waste Disposal Act (SWDA) in 1980.

The new legislation brought up the issue of precisely which substances were to be considered toxic.

- *priority pollutants.* In 1977, the Clean Water Act (CWA) listed 65 pollutants that were selected for the hazards that they posed to human health and the environment and for the quantities in which they were produced and used. Through amendments, the current list of priority pollutants includes 129 chemical substances.

- *hazardous waste.* The Resource Conservation and Recovery Act (RCRA) of 1976 defined "hazardous waste" as any solid waste (discarded gas, liquid, solid material) that exhibited characteristics of ignitability, corrosivity, reactivity, or toxicity.

- *health impacts.* Health impacts from chemical exposure became a public concern and prompted the active development and application of risk assessment.

Consequently, a *toxic chemical* has become any of those listed as a priority pollutant or identified as toxic under RCRA. Risk assessment is used to evaluate the toxicity of these and other unlisted chemicals to determine whether a health impact would result from exposure.

The regulations allow some level of negotiation between the regulated industry and the permitting authority, which can lead to a balance between economic feasibility and environmental protection.

2. RESOURCE CONSERVATION AND RECOVERY ACT

There are nine subtitles in the Resource Conservation and Recovery Act (RCRA). The following are most directly relevant to practicing engineers.

- Subtitle C: Cradle to Grave. Dictates management of hazardous waste from point of generation to point of disposal, including all handling and uses in between. EPA manages this provision through the Uniform Hazardous Waste Manifest as well as through the reporting and record-keeping requirements of hazardous waste generators and transporters and hazardous waste treatment, storage, and disposal facilities.

- Subtitle D: Non-hazardous Solid Waste. Defines exemptions for Subtitle C regulations. These include hazardous waste generated by households and, under defined circumstance, those of small quantity generators. Certain resource exploration and recovery activities are also exempt. Some day-to-day activities of a municipality are included in Subtitle D. These activities are garbage, which also includes refuse (garbage and trash), nonrecycled home appliances, defined incineration residues, industrial and construction refuse, and residues from municipal wastewater and water treatment plants.

- Subtitle G: Miscellaneous. Provides whistleblower protections, allows citizen lawsuits, and requires public participation when developing regulations.

- Subtitle I: Underground Storage Tanks. Regulates tanks storing petroleum products and listed hazardous substances through a permitting and tracking program. The regulations establish standards for tank fabrication and groundwater monitoring and leak prevention and detection systems.

RCRA defines solid and hazardous waste and provides management guidelines for these materials. For example, according to RCRA, a solid waste is any discarded material whether liquid, solid, or gas (40 CFR 261.2). A hazardous waste is any solid waste that is either ignitable, toxic, corrosive, or reactive (40 CFR 261.20 to 40 CFR 261.24) or is listed in 40 CFR 261 Subpart D (Secs. 261.30 to 261.33).

The general criteria for hazardous waste are defined as follows. [Hazardous Waste Characteristics]

- characteristic of ignitability (40 CFR 261.21): has a flash point less than 60°C or 140°F

- characteristic of corrosivity (40 CFR 261.22): has a pH ≤ 2.0 or ≥ 12.5

- characteristic of reactivity (40 CFR 261.23): reacts violently with water

- characteristic of toxicity (40 CFR 261.24): is listed in Table 1 of 40 CFR 261.24 [Maximum Concentration of Contaminants for the Toxicity Characteristic]

Wastes are listed in Subpart D by specific and nonspecific source and by chemical name under the following hazard codes: ignitable waste, corrosive waste, reactive waste, toxicity characteristic waste, acute hazardous waste, and toxic waste.

The regulations describing the programs and compliance requirements for RCRA are published in 40 CFR 261. The programs authorized under RCRA are administered by the EPA Office of Solid Waste.

3. COMPREHENSIVE ENVIRONMENTAL RESPONSE, COMPENSATION, AND LIABILITY ACT

The National Oil & Hazardous Substances Pollution Contingency Plan of 1990, known as the National Contingency Plan (NCP), was originally implemented to control spills of oil and hazardous substances under the Clean Water Act (CWA). The NCP guides Superfund activities through its goal of selecting remedies that protect human health and the environment, maintaining protection over time, and minimizing untreated waste.

Comprehensive Environmental Response, Compensation, and Liability Act (CERCLA), amended in 1986 by the Superfund Amendments and Reauthorization Act (SARA), was promulgated to enforce cleanup of sites contaminated by accidental release or improper storage, handling, and disposal of hazardous substances. Because CERCLA established an initial fund to pay for cleanups where potentially responsible parties (PRPs) could not be identified or refused to act, it is commonly called Superfund. Major provisions of CERLA are to

- provide public funds for remedial actions (when no PRPs funds are available)

- provide authority for EPA to compel privately funded cleanups

- rank contaminated sites

- provide guidance for conducting remedial actions

- define emergency notification procedures for hazardous chemical releases

- require notification of chemical use, storage, and production activities

- require annual emissions reporting requirements

- support community right-to-know

Over the past few decades, Superfund has been underfunded, which has resulted in fewer enforcement actions. Consequently, most enforcement action is negotiated between EPA and the PRPs with cleanup occurring under EPA oversight. Sites subject to cleanup are evaluated and ranked using the Hazard Ranking System (HRS). This ranking populates the National Priorities List (NPL) of sites requiring cleanup.

SARA added an underground storage tank enforcement requirement to Superfund. Owners and operators of underground storage tanks must demonstrate that corrective action occurs if contamination is discovered during a tank repair, replacement, or removal.

When CERCLA is combined with RCRA, regulatory authority exists for oversight of toxic wastes regardless of whether the facility associated with their generation, treatment, storage, or disposal is in current operation or is abandoned.

The regulations describing the programs and compliance requirements for CERCLA are published in 40 CFR 300. The programs authorized through CERCLA are administered by the EPA Office of Solid Waste and Emergency Response under Superfund.

4. THIRD PARTIES

Releases of hazardous materials to surface and ground waters are regulated under the CWA, CERCLA, RCRA, and other federal and state legislation. Responses to these releases usually follow a two-part process, one under the direction of the regulatory agencies and the other as a result of citizen action.

Industry is subject to regulatory actions aimed at mitigating pollutant releases to the environment. These actions take the form of civil and criminal fines and penalties, including the use of cease-and-desist orders, to compel cleanup or other response. However, if the people responsible for activities or products that potentially pose an exposure risk do not take appropriate action to prevent the exposure of third parties to that risk, they may become liable if an exposure occurs. If it can be shown that there was intent to act irresponsibly, then criminal charges can result. If the exposure occurs without evidence of intent, then only civil liability may exist. Third parties typically employ litigation in an effort to redress wrongs and/or correct conditions that lead to exposure.

Although potentially significant, the fines and penalties imposed by the regulatory body are often inconsequential compared to those sought by third parties, especially where it is demonstrated that human health or physical impairment has resulted from the release. Consequently, it is often the threat of legal action by third parties and the associated public relations problems that provides the economic incentive for regulatory compliance.

Solid & Hazard. Waste

Risk Assessment

Risk assessment has applications in all areas of environmental engineering practice. It is applied to environmental issues related to water, air, and land, and it is used to support decision making. Regulatory standards and criteria, health policy, design standards and guidelines, and nearly every other entity, action, and authority within the field of environmental engineering is supported by some facet of risk assessment.

Risk assessment is discussed in depth in Chap. 48.

Sampling and Measurement Methods

Content in blue refers to the *NCEES Handbook*.

1. CHARACTERISTICS OF SOLID WASTE

Municipal solid waste (MSW, known for many years as "garbage") consists of the solid material discarded by a community. MSW includes food wastes, containers and packaging, residential garden wastes, other household discards, and light industrial debris. MSW does not include hazardous wastes, including paints, insecticides, lead car batteries, used crankcase oil, dead animals, raw animal wastes, and radioactive substances, which present special disposal problems.

MSW is generated at the average rate of approximately 5–8 lbm/capita-day (2.3–3.6 kg/capita·d). A value of 5 lbm/capita-day (2.3 kg/capita·d) is commonly used in design studies.

Although each community's MSW has its own characteristics, typical values are often used in calculations associated with the generation, collection, disposal, and management of MSW. Table 35.1 gives the average composition of MSW in the United States.

Table 35.1 *Average Composition of Municipal Solid Waste in the United States*

component	percentage by weight
paper and cardboard	25.9
food waste	15.1
yard waste	13.2
plastics	13.1
metal	9.1
wood	6.2
cloth	6.1
glass	4.4
rubber and leather	3.2
miscellaneous inorganic (dirt, stones, concrete bits)	1.5
other	2.0

Source: United States Environmental Protection Agency

The overall moisture content of MSW is typically approximately 20%. Tables giving typical densities, moisture content, and heating values of MSW components are widely available. [Typical Densities of As-Received Source-Separated Materials] [Typical Moisture Content of Municipal Solid Waste (MSW) Components] [Typical Heating Value of MSW Components]

Example 35.1

A city is considering alternatives for the management of the solid waste generated by its residents. The solid waste characteristics for the city are

component	% by mass	component discarded % moisture	component discarded density (kg/m^3)	component discarded energy (kJ/kg)
paper	44	6	85	16,750
garden	17	60	105	6500
food	11	70	290	4650
cardboard	9	5	50	16,300
wood	7	20	240	18,600
plastic	7	2	65	32,600
miscellaneous inert materials	5	8	480	7000

The per capita solid waste generation rate for the 82,000 residents of the city is 2.7 kg/day. (a) What is the moisture content of the bulk discarded waste? (b) What is the density of the bulk discarded waste? (c) What is the daily energy content available from the bulk discarded waste?

Solution

Calculate moisture content and density based on a 100 kg sample of the waste. The component dry mass is

$$\text{dry mass} = (\text{discarded mass})\left(\frac{1-\%\ \text{moisture}}{100\%}\right)$$

The component volume is

$$\text{volume} = \frac{\text{dry mass}}{\text{discarded density}}$$

The component energy content is

$$\text{energy} = (\text{dry mass})(\text{energy per unit mass})$$

Complete the table using these equations.

component	discarded mass (kg)	dry mass (kg)	volume (m³)	energy (kJ)
paper	44	41	0.49	686,750
garden	17	6.8	0.065	44,200
food	11	3.3	0.011	15,345
cardboard	9	8.6	0.17	140,180
wood	7	5.6	0.023	104,160
plastic	7	6.9	0.11	224,940
miscellaneous inert materials	5	4.6	0.0096	32,200
total	100	76.8	0.88	1,247,775

(a) The bulk moisture content is

$$100 \text{ kg} - 76.8 \text{ kg} = 23.2 \text{ kg}$$

For a 100 kg sample, the moisture content is 23%.

(b) The bulk discarded density is

$$\frac{100 \text{ kg}}{0.88 \text{ m}^3} = 114 \text{ kg/m}^3$$

(c) The total waste generated daily is

$$(82{,}000 \text{ people})\left(\frac{2.7 \text{ kg}}{\text{person·d}}\right) = 221{,}400 \text{ kg/d}$$

The total discarded energy content is

$$\left(1{,}247{,}775 \ \frac{\text{kJ}}{100 \text{ kg}}\right)\left(221{,}400 \ \frac{\text{kg}}{\text{d}}\right) = 2.8 \times 10^9 \text{ kJ/d}$$

36 Minimization, Reduction, Recycling

1. INTRODUCTION

Using data compiled in 2014, the EPA estimates that of all municipal solid waste (MSW) generated, 52.6% was landfilled, 34.6% was composted or recycled, and 12.8% was combusted for energy recovery. Of the MSW that was recycled, nearly half, 49.7%, was paper and paper products, with an additional 23.6% from yard clippings. Plastics, often considered to be among the most recyclable materials, accounted for only 3.5%.

These data show that recycling presents a difficult challenge. The materials recycled at the highest rates are those that are easy to separate and handle or that are inconvenient to dispose of in other ways.

Nevertheless, most municipalities in the United States have organized recycling programs. The motivation for this is not financial, as recycling typically costs more to manage than the revenue it generates. For those who embrace recycling as an activity beneficial to the environment, this is encouraging. Realizing a nearly 50% diversion from landfilling to recycling or energy recovery is a remarkable outcome given the absence of financial incentives.

As commonly used, the word *recycling* is a general description that actually involves a variety of activities —not just recycling but also reusing, reducing, converting, and minimizing—that combine to reduce the amount of material that ends up in landfills. *Reusing* indicates that the waste is still usable in its current form or needs only minor modification to serve as a raw material in some alternative application. Examples include using methane gas from wastewater treatment as a heating fuel, using spent pickle liquor from steel manufacturing as an iron source for chemical precipitation in water treatment, and refilling glass beverage bottles. Used more narrowly, *recycling* refers specifically to converting the waste to a different form for another use. For example, using plastic beverage bottles to make fleece jackets and using newspaper to make building insulation are examples of recycling.

2. RECYCLING

Recycling programs for MSW take a variety of forms. No consensus exists about which form produces the best result, but there is consensus that convenience and habit are important for promoting a favorable level of participation. In the United States, recycling programs commonly choose among alternatives for separation and collection and decide what level of service to provide to residents. Options include the following.

- Source separation in which the resident places each type of recyclable item (glass, newspapers, and so on) in a separate container; designated containers are usually provided by the municipality. Collection occurs at curbside, and each container is emptied into a corresponding bin on the collection truck.

- Source separation limited to separating recyclable materials from garbage. The resident places recyclable materials of all types into a single container. Curbside pickup is simplified as only one recycling container is collected. Further sorting occurs at a central facility.

- Source separation by the resident at a collection facility. The resident places each type of recyclable material into a designated collection bin at the facility. There is no curbside pickup.

- Nonsource separation in which both garbage and recyclables are placed in a single container for curbside pickup. All sorting occurs at a central facility.

Yard clippings, including fall leaves, make up a substantial portion of recycled materials. In some areas, there is regular curbside pickup, either in a separate container or as bulk material, bagged or unbagged. In areas with many mature deciduous trees, leaves may be picked up from windrows along the curb using a leaf-vacuum truck. Leaves and other yard clippings are converted to compost for use in landscaping. Some municipalities combine yard clippings with other wood materials (such as wood pallets and separated construction debris) that have been shredded to create a mulch that can be dyed to match local color preferences and applied as a ground cover.

Donating goods, either for resale or for distribution to people in need, is *direct recycling*. Donation centers are located to make donating convenient. Clothing and other personal and household items are donated by individuals. Restaurants donate end-of-day unsold food items, and grocery stores donate expired goods to food banks. These activities are often seen as charitable giving, but they are also an important part of recycling. Through donation programs, automobiles, electronic devices, eyeglasses, and a wide variety other goods are reused or recycled.

Beverage containers, including juice and milk containers, have evolved through design to use less material in their manufacture, so that if they are discarded, less waste results, and if they are recycled, bulk and weight are reduced. In some places, milk is sold in bags and then transferred by the buyer to a reusable pitcher for use. Packaging for food and other consumer items is often minimized to reduce the amount of packaging waste, and recyclable materials are used in the reduced packaging. Grocery bags have evolved from one-use paper to one-use plastic to reusable bags. To reduce plastic waste and promote reusable bags, some communities require a separate charge for one-use plastic bags.

Used car and truck tires cannot be compacted, so they are not easily landfilled. Tire recycling involves shredding to produce a rubber mulch that can be used for playground cover, incorporated into certain pavement applications, or combusted for energy recovery; if no alternative use is found, the mulch can landfilled.

Construction debris presents other problems for landfilling. To reduce its bulk, the debris is separated. Wood products are recycled, concrete is crushed, reinforcing steel is recovered, and the remaining rubble is reused as aggregate. In road resurfacing, the old asphalt is reduced to aggregate on site and incorporated into the new surface.

Nearly all recycling activities are low technology, involving communities of people who are motivated by environmental conservation. Higher technology activities are difficult to justify until recycled materials are valued in the marketplace, creating a financial incentive for investment.

37 Mass and Energy Balance

1. MATERIALS BALANCE

Materials balance is the accounting of all inputs to and outputs from a system, as well as any changes that accumulate inside the system. Accumulated change can be either positive or negative.

In equation form,

$$\sum \text{inputs} \pm \text{internal change} = \sum \text{outputs} \quad 37.1$$

The *materials balance* concept is illustrated in Fig. 37.1.

Figure 37.1 *Materials Balance*

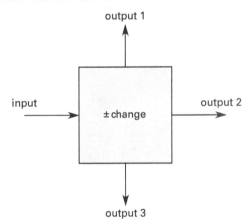

A materials balance can be based on any consistent parameter. Two parameters often used are mass and energy. A *mass balance* accounts for all mass inputs, outputs, and internal changes; an *energy balance* accounts for all energy inputs, outputs, and internal changes. Figure 37.2 through Fig. 37.5 show four inter-related materials balances that are pertinent to solid waste.

Figure 37.2 shows a goods materials balance. The input consists of goods, which may be food or consumer products. If food, the goods are literally consumed to produce metabolic energy and human digestion by-products that leave the system as wastewater and as

gases emitted to the air. Nonconsumable goods are accumulated inside the system or leave the system as solid waste.

Figure 37.2 *Goods Materials Balance*

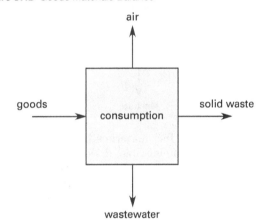

Figure 37.3 shows a wastewater materials balance. The input consists of wastewater entering a wastewater treatment plant; the wastewater contains a substrate (BOD) and very little solids. The internal change is the conversion of substrate to biomass (TSS or VSS). The outputs are air emissions, treated effluent, and the biomass sludge. The sludge enters the solid waste management stream to be either landfilled or combusted for energy recovery.

Figure 37.3 *Wastewater Materials Balance*

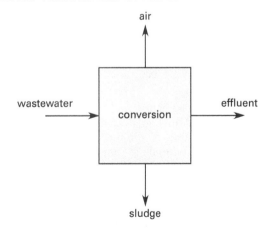

Figure 37.4 shows a landfill materials balance. The input to the landfill consists of the solid waste generated from the wastewater treatment sludge and the solid

waste discarded from the consumption of goods. Recycled materials are diverted from the landfill, and the nonrecycled goods are placed in the landfill. The landfilled solid waste decomposes to produce leachate and recoverable and nonrecoverable gases.

Figure 37.4 *Landfill Materials Balance*

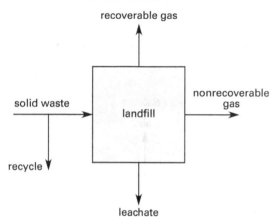

Figure 37.5 shows a fuel materials balance. The input consists of either combustible dried sludge from the wastewater treatment plant, the combustible fraction of the recoverable landfill gas, or a mixture of both. The change in the system is the heat loss from inefficiencies, and the outputs are air emissions, ash, and recoverable energy.

Figure 37.5 *Fuel Materials Balance*

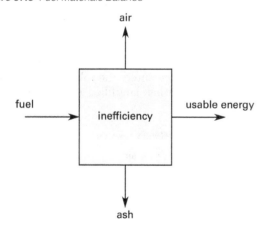

When values are assigned to each input, internal change, and output in Fig. 37.2 through Fig. 37.5, an accounting is possible of all materials as they move through each individual system and through the sequence of all four systems. Mass can be used as the parameter to track the materials through the four systems. In Fig. 37.5, if the energy content of the fuel components is known or can be determined, an energy balance will result. This sequence would then make it possible to track the goods from their original acquisition to their ultimate conversion as an energy output.

Example 37.1

A community recovers 1000 kg/d of combustible landfill gas with a heating value of 45 kJ/kg. The community's wastewater treatment plant produces 1000 kg/d of dry sludge with a heating value of 10 kJ/kg. These are incinerated in a combustion reactor with an 80% efficiency. During combustion, 5% of the incoming sludge mass is converted to ash, and 5% each of the incoming sludge and gas mass is converted to emitted gases. What is the recoverable energy?

Solution

Multiply mass by heating value to get the energy inputs.

$$Q_{in,gas} = \dot{m}_{in,gas}HV_{gas}$$
$$= \left(1000 \ \frac{kg}{d}\right)\left(45 \ \frac{kJ}{kg}\right)$$
$$= 45\,000 \ kJ/d$$
$$Q_{in,sludge} = \dot{m}_{in,sludge}HV_{sludge}$$
$$= \left(1000 \ \frac{kg}{d}\right)\left(10 \ \frac{kJ}{kg}\right)$$
$$= 10\,000 \ kJ/d$$
$$Q_{in,total} = Q_{in,gas} + Q_{in,sludge}$$
$$= 45\,000 \ \frac{kJ}{d} + 10\,000 \ \frac{kJ}{d}$$
$$= 55\,000 \ kJ/d$$

The energy outputs to the air are

$$Q_{out,gas} = (5\%) \, Q_{in,gas}$$
$$= \left(\frac{5\%}{100\%}\right)\left(45\,000 \ \frac{kJ}{d}\right)$$
$$= 2250 \ kJ/d$$
$$Q_{out,sludge} = (5\%) \, Q_{in,sludge}$$
$$= \left(\frac{5\%}{100\%}\right)\left(10\,000 \ \frac{kJ}{d}\right)$$
$$= 500 \ kJ/d$$
$$Q_{out,total} = Q_{out,gas} + Q_{out,sludge}$$
$$= 2250 \ \frac{kJ}{d} + 500 \ \frac{kJ}{d}$$
$$= 2750 \ kJ/d$$

The energy output to ash is

$$Q_{ash} = (5\%) \, Q_{in,sludge}$$

$$= \left(\frac{5\%}{100\%} \right) \left(10\,000 \, \frac{kJ}{d} \right)$$

$$= 500 \, kJ/d$$

The internal change, based on efficiency and total energy inputs, is

$$Q_{internal} = (100\% - \eta) \, Q_{in,total}$$

$$= \left(\frac{100\% - 80\%}{100\%} \right) \left(55\,000 \, \frac{kJ}{d} \right)$$

$$= 11\,000 \, kJ/d$$

Perform an energy balance and solve for the usable energy, which is all energy not lost to air, ash, or internal change.

$$\sum inputs \pm \sum internal \ change = \sum outputs$$

$$Q_{in,total} - Q_{internal} = Q_{out,total} + Q_{ash} + Q_{usable}$$

$$Q_{usable} = Q_{in,total} - Q_{internal}$$

$$- Q_{out,total} - Q_{ash}$$

$$= 55\,000 \, \frac{kJ}{d} - 11\,000 \, \frac{kJ}{d}$$

$$- 2750 \, \frac{kJ}{d} - 500 \, \frac{kJ}{d}$$

$$= 41\,750 \, kJ/d$$

2. NOMENCLATURE

HV	heating value	kJ/kg
$\dot{m}$	mass flow rate	kg/d
Q	energy flow rate	kJ/d

Symbols

η	efficiency	–

38 Hydrology, Hydrogeology, Geology

Principles of hydrology and hydrogeology are discussed in depth in Chap. 7. These principles also apply to landfill siting.

1. LANDFILL SITING

Geology is a concern where earthquake fault lines are near potential sites and where underlying formations such as karst interfere with landfill liner construction. Where these conditions are present, they represent a viable threat to the integrity of the landfill, and alternative landfill sites should be considered.

Sanitary landfills should be designed for a five-year minimum life. Sanitary landfill sites are selected on the basis of many relevant factors, including (a) economics and the availability of inexpensive land; (b) location, including ease of access, transport distance, acceptance by the population, and aesthetics; (c) availability of cover soil, if required; (d) wind direction and speed, including odor, dust, and erosion considerations; (e) flat topography; (f) dry climate and low infiltration rates; (g) location (elevation) of the water table; (h) low risk of aquifer contamination; (i) types and permeabilities of underlying strata; (j) avoidance of winter freezing; (k) future growth and capacity requirements; and (l) ultimate use.

Suitable landfill sites are becoming difficult to find, and once identified, they are subjected to a rigorous permitting process. They are also objected to by residents near the site and MSW transport corridor. This is referred to as the *NIMBY* (not in my backyard) *syndrome*. People agree that landfills are necessary, but they do not want to live near them.

2. GROUNDWATER DEWATERING

One factor in landfill siting is that the location (elevation) of the water table may require groundwater dewatering.

It may be possible to prevent or reduce contaminant migration by reducing the elevation of the groundwater table (GWT). This is accomplished by dewatering the soil with relief-type *extraction drains* (*relief drains*). The *ellipse equation*, also known as the *Donnan formula* and the *Colding equation*, used for calculating pipe spacing, L, in draining agricultural fields, can be used to determine the spacing of groundwater dewatering systems. In Eq. 38.1, K is the hydraulic conductivity with units of length/time, a is the distance between the pipe and the impervious layer barrier (a is zero if the pipe is installed on the barrier), b is the maximum allowable table height above the barrier, and Q is the *recharge rate*, also known as the *drainage coefficient*, with dimensions of length/time. The units of K and Q must be on the same time basis.

$$L = 2\sqrt{\left(\frac{K}{Q}\right)(b^2 - a^2)} = 2\sqrt{\frac{b^2 - a^2}{i}} \qquad 38.1$$

Equation 38.1 is often used because of its simplicity, but the accuracy is only approximately ±20%. To allow for possible error, the calculated spacing should be decreased by 10–20%.

For pipes above the impervious stratum, as illustrated in Fig. 38.1, the total discharge per unit length of each pipe (in ft^3/ft-sec or m^3/m·s) is given by Eq. 38.2. H is the maximum height of the water table above the pipe invert elevation. D is the average depth of flow.

$$Q_{\text{unit length}} = \frac{2\pi KHD}{L} \qquad 38.2$$

$$D = a + \frac{H}{2} \qquad 38.3$$

For pipes on the impervious stratum, the total discharge per unit length from the ends of each pipe (in ft^3/ft-sec or m^3/m·s) is

$$Q_{\text{unit length}} = \frac{4KH^2}{L} \qquad 38.4$$

Equation 38.5 gives the total discharge per pipe. If the pipe drains from both ends, the discharge per end would be half of that amount.

$$Q_{\text{pipe}} = LQ_{\text{unit length}} \qquad 38.5$$

Figure 38.1 *Geometry for Groundwater Dewatering Systems*

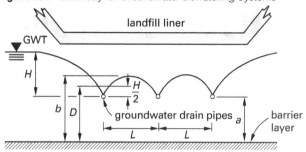

(a) pipes above impervious stratum

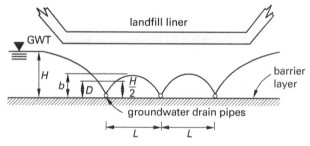

(b) pipes on impervious stratum

Example 38.1

Subsurface leachate migration from a landfill is to be mitigated by maintaining the water table that surrounds the landfill site lower than the natural level. The surrounding area consists of 15 ft of saturated, homogeneous soil over an impervious rock layer. The water table, originally at the surface, has been lowered to a depth of 9 ft. Fully pervious parallel collector drains at a depth of 12 ft are present. The hydraulic conductivity of the soil is 0.23 ft/hr. The natural water table will recharge the site in 30 days if drainage stops. What collector drain separation is required?

Solution

Referring to Fig. 38.1(a),

$$a = 15 \text{ ft} - 12 \text{ ft} = 3 \text{ ft}$$

$$b = 15 \text{ ft} - 9 \text{ ft} = 6 \text{ ft}$$

$$K = \left(0.23 \ \frac{\text{ft}}{\text{hr}}\right)\left(24 \ \frac{\text{hr}}{\text{day}}\right) = 5.52 \text{ ft/day}$$

To maintain the lowered water table level, the drainage rate must equal the recharge rate.

$$Q = \frac{9 \text{ ft}}{30 \text{ days}} = 0.3 \text{ ft/day}$$

Use Eq. 38.1 to find the drain spacing.

$$L = 2\sqrt{\left(\frac{K}{Q}\right)(b^2 - a^2)}$$

$$= 2\sqrt{\left(\frac{5.52 \ \dfrac{\text{ft}}{\text{day}}}{0.3 \ \dfrac{\text{ft}}{\text{day}}}\right)\left((6 \text{ ft})^2 - (3 \text{ ft})^2\right)}$$

$$= 44.57 \text{ ft} \quad (45 \text{ ft})$$

To allow for the possibility of error, decrease the drain separation by 10–20%, resulting in a spacing of between 36 ft and 40 ft.

3. NOMENCLATURE

a	distance	ft	m
b	distance	ft	m
D	depth of flow	ft	m
H	total hydraulic head	ft	m
i	hydraulic gradient	ft/ft	m/m
K	hydraulic conductivity	ft/sec	m/s
L	spacing or distance	ft	m
Q	recharge rate	ft³/ft-sec	m³/m·s

39 Solid Waste Storage, Collection, Transportation

1. INTRODUCTION

Solid waste properties determine the appropriate management strategy. Solid waste management typically involves at least six functional elements.

- *waste generation*: Generation rates are used to evaluate reuse and recycle feasibility and processing and disposal requirements.

- *on-site handling, storage, and processing*: The on-site preparation of the waste for reuse, recycle, or disposal. Individual households may separate recyclables from other wastes, and industry may segregate their wastes with the same purpose.

- *collection*: The physical gathering of the waste; may vary from the weekly garbage collection at homes to the collection of toxic chemical wastes from industry.

- *transfer and transport*: Consolidation of collected waste for bulk transport to a processing, recovery, or disposal facility.

- *processing and recovery*: The activities employed to recover the waste as raw materials or as energy or to prepare the waste for disposal.

- *disposal*: The ultimate fate of nonreusable and non-recyclable wastes. Incineration and landfilling are the two common disposal methods.

This chapter discusses storage, collection, and transportation.

2. STORAGE

In storing municipal solid waste (MSW), the objective is to maintain sanitary conditions. Sanitary conditions are those that limit odor, keep water out, and prevent access by insects, rodents, pets, and children. Local governments codify or otherwise define MSW storage requirements for their residents and businesses.

Containers that serve individual households should hold the discarded materials securely. The containers should be rugged and stable, their lids should fit snugly, and they should be placed in locations where they cannot be tipped by animals, rest in water, or freeze to the ground. If collection crews must lift full containers manually, a container's weight when full should not exceed 60 lbs; if mechanically lifted, the container should be compatible with the lifting device.

To prevent potential hazardous liquids from entering landfills, to prevent contaminating potentially recyclable wastes, and to protect collection crews from contact with spilled liquids, excess liquids should be absorbed with suitable materials before being placed in waste containers. Hazardous wastes should be segregated and disposed of at a hazardous waste collection facility.

Dumpsters are the most common container for MSW generated at apartment buildings, other multifamily dwellings, and commercial buildings. Dumpsters typically meet the necessary specifications for keeping waste secure, but they must be emptied frequently enough to prevent overfilling, as this can interfere with keeping the waste covered. Commercial activities may generate large volumes of bulk waste such as cardboard boxes; such waste should be segregated from other wastes and baled to prevent litter and make collection easier.

Industrial solid waste storage follows the same practices as those employed by commercial generators for similar wastes. The solid wastes from some industries, however, have characteristics that are unique to those industries and require special handling. Examples include

- scrap metals and wastes from metal fabrication

- wastes from on-site air and water pollution control process equipment

- wastes contaminated with oil, paint, solvents, and other liquids

- scrap wood, wood products, and pallets

- wastes from fruit and vegetable processing

- wastes from meat and dairy processing

Solid waste from these industries must be stored in a matter consistent with local codes, always with the purpose of protecting public health and safety and keeping waste materials from being dispersed to the environment.

3. COLLECTION

The objective of solid waste collection is to optimize routing (minimizing travel distances and collection time), equipment capacity, and crew safety and efficiency. These optimizations require rigorous analysis of collection route configurations and time/task functions. Example 39.1 and Ex. 39.2 illustrate the analyses and calculations involved.

Example 39.1

The table shows the results of a time study and route analysis for curbside residential waste collection in a planned community.

population	1200
solid waste generation rate per person	1.6 kg/day
number of residences	285
average driving time between residences	18 s
average pickup/load time at each residence	45 s
travel time from truck yard to route start	38 min
average travel time between route and landfill	65 min
time to unload at landfill	20 min
travel time from landfill to truck yard	45 min
truck compacted waste capacity	10 m³
truck compaction ratio	2.6
typical waste as-discarded density	140 kg/m³

Waste is collected once per week. How many days per week should be scheduled for one crew to collect all the waste generated by the community?

Solution

The total as-discarded waste volume that must be collected is

$$\frac{(1200 \text{ people})\left(1.6 \ \dfrac{\text{kg}}{\text{d·person}}\right)}{140 \ \dfrac{\text{kg}}{\text{m}^3}} = 13.7 \text{ m}^3/\text{d}$$

The compacted volume per week is

$$\frac{\left(13.7 \ \dfrac{\text{m}^3}{\text{d}}\right)\left(7 \ \dfrac{\text{d}}{\text{wk}}\right)}{\left(\dfrac{2.6 \text{ m}^3 \text{ as discarded}}{1 \text{ m}^3 \text{ compacted}}\right)} = 37 \text{ m}^3/\text{wk}$$

The number of truck loads that must be carried to the landfill in a one-week period is

$$\frac{37 \ \dfrac{\text{m}^3}{\text{wk}}}{10 \ \dfrac{\text{m}^3}{\text{load}}} = 3.7 \text{ loads/wk} \quad (4 \text{ loads/wk})$$

The total time available in a single 8 h day is 480 min. The total time needed is equal to the collection time plus the noncollection time. The average collection time needed for the first load is

$$\frac{\left(\dfrac{18 \text{ s} + 45 \text{ s}}{\text{stop}}\right)\left(10 \ \dfrac{\text{m}^3}{\text{load}}\right)}{\left(\dfrac{37 \text{ m}^3}{285 \text{ stops}}\right)\left(60 \ \dfrac{\text{s}}{\text{min}}\right)} = 81 \text{ min}$$

The noncollection time for the first load is 38 min from yard + 65 min to landfill + 20 min to unload + 65 min back to route = 188 min. The total time for the first load is 188 min + 81 min = 269 min.

The collection and noncollection time needed for the second load, assuming that the truck returns to the route afterward, is 81 min to load + 65 min to landfill + 20 min to unload + 65 min back to route = 231 min. The total time for the first and second loads is 269 min + 231 min = 500 min. This is greater than 480 min, so the truck must return to the yard after the second load instead of returning to the route.

The collection and noncollection time needed for the second load when the truck returns to the yard afterward is 81 min to load + 65 min to landfill + 20 min to unload + 45 min back to yard = 211 min. The total time is 269 min + 211 min = 480 min, so one truck can collect 2 loads/d. At 2 loads/d and 4 loads/wk, 2 days will be needed for the collection of waste at all residences.

Example 39.2

A population of 160,000 generates municipal solid waste at 2.3 kg/person per day with an as-discarded density of 110 kg/m³. Collection trucks have a 25 m³ compacted capacity. The trucks can compact the waste to 550 kg/m³. Truck crews of three men each work 8 h/d and 5 d/wk, with collections occurring at each stop once weekly. A truck crew can fill a truck in 145 minutes. Travel to a centrally located transfer station requires 32 min from any collection route in the city. An additional 15 min is needed for unloading at the transfer station. How many trucks are needed to meet the weekly collection schedule?

Solution

The weekly compacted volume collected is

$$\frac{(160{,}000 \text{ people}) \times \left(2.3 \, \dfrac{\text{kg}}{\text{person·d}}\right) \times \left(7 \, \dfrac{\text{d}}{\text{wk}}\right)}{550 \, \dfrac{\text{kg}}{\text{m}^3}} = 4684 \text{ m}^3/\text{wk}$$

The time available for collection per day is

$$\left(8 \, \frac{\text{h}}{\text{d}}\right)\left(60 \, \frac{\text{min}}{\text{h}}\right) = 480 \text{ min/d}$$

task	task time (min)	cumulative time (min)
leave	32	32
collect 1	145	177
return	32	209
unload	15	224
leave	32	256
collect 2	145	401
return	32	433
unload	15	448

Collecting a third load would exceed 448 min in a day, so a single truck can collect two loads in one day. The number of trucks needed is

$$\frac{\left(4684 \, \dfrac{\text{m}^3}{\text{wk}}\right)}{\left(25 \, \dfrac{\text{m}^3}{\text{truck-load}}\right)\left(5 \, \dfrac{\text{d}}{\text{wk}}\right) \times \left(\dfrac{2 \text{ loads}}{\text{d}}\right)} = 18.7 \text{ trucks} \quad (19 \text{ trucks})$$

4. TRANSPORTATION

After waste has been collected, there are in most cases just two options for transportation: the waste can be unloaded at the landfill, or it can be unloaded at a transfer station and then transported to a landfill.

Waste unloaded at the landfill directly from the collection truck will not require any additional over-the-road transportation. For a facility operating under this configuration, the landfill and the recycling or other waste processing facility will be at the same location. The only further transportation of waste that is needed is on site from the processing facility to the working face of the landfill. For this, construction dump trucks are used that are suitable for continuous use under the rough conditions present at landfill sites.

Transfer stations are used in metropolitan areas where residential and commercial activities cover large geographic areas, as well as in rural areas where the population is too small to justify a permitted landfill. In both cases, the use of a transfer station reduces inefficiencies. The transfer station may also serve as a solid waste processing or energy recovery facility, removing recyclable materials before sending the remaining waste to a landfill. Like most goods, solid waste is commonly transported by motor freight over roads and highways. For larger metropolitan areas where waste volumes justify rail transportation, the solid waste will be transported in cars designed to contain the waste. In some locations, special rail cars designed for side tipping or for rotating 180° in a tipping cradle are used. Companies contracted to transport solid waste are usually certified by the state agencies or their designated proxies.

Solid & Hazard. Waste

40 Solid Waste Treatment and Disposal

Table 40.1 *Typical Characteristics of MSW[a,b]*

component	percentage by weight
paper	40
yard waste	18
glass	8
plastic	7
steel	7
food waste	6
aluminum	2
other	12

[a]United States national average
[b]dry composition

1. MUNICIPAL SOLID WASTE

Municipal solid waste (MSW, known for many years as "garbage") consists of the solid material discarded by a community, including excess food, containers and packaging, residential garden wastes, other household discards, and light industrial debris. Hazardous wastes, including paints, insecticides, lead car batteries, used crankcase oil, dead animals, raw animal wastes, and radioactive substances present special disposal problems and are not included in MSW.

MSW is generated in the United States at the average rate of about 5 lbm/capita-day to 8 lbm/capita-day (2.3 kg/capita-day to 3.6 kg/capita-day). (Characteristics of MSW in the United States are given in Table 40.1.)

Suitable (economical and safe) landfill sites are becoming difficult to find, and once identified, are objected to by residents near the site and in the MSW transport corridor. This is referred to as the *NIMBY* (not in my backyard) *syndrome*.

2. LANDFILLS

Municipal solid waste has traditionally been disposed of in *municipal solid waste landfills* (MSWLs, previously referred to as "dumps").[1] Other wastes, including incineration ash and water treatment sludge, may be included with MSW in some landfills.

The fees charged to deposit waste are known as *tipping fees*. Tipping fees are generally higher in the eastern United States (where landfill sites are becoming scarcer). Tipping fees for resource recovery plants are approximately double those of traditional landfills.

Many early landfills were not designed with liners, not even simple compacted-soil bottom layers. Design was often by rules of thumb, such as the bottom of the landfill had to be at least 5 ft (1.5 m) above maximum elevation of groundwater and 5 ft (1.5 m) above bedrock. Percolation through the soil was believed sufficient to make the leachate bacteriologically safe. However, inorganic pollutants can be conveyed great distances, and the underlying soil itself can no longer be used as a barrier to pollution and to perform filtering and cleansing functions.

Landfills without bottom liners are known as *natural attenuation* (NA) *landfills*, since the native soil is used to reduce the concentration of leachate components. Landfills lined with clay (pure or bentonite-amended soil) or synthetic liners are known as *containment landfills*.

Landfills can be incrementally filled in a number of ways, as shown in Fig. 40.1. With the *area method*, solid waste is spread and compacted on flat ground before being covered. With the *trench method*, solid waste is placed in a trench and covered with the trench soil. In

---

[1]In 1988, 73% of the nation's MSW was landfilled, 14% was incinerated, and 13% was recycled. In 2015, 52.5% was landfilled, 12.8% was incinerated with energy recovery, 12.8% was recycled, and 8.9% was composted.

Figure 40.1 *Landfill Creation Methods*

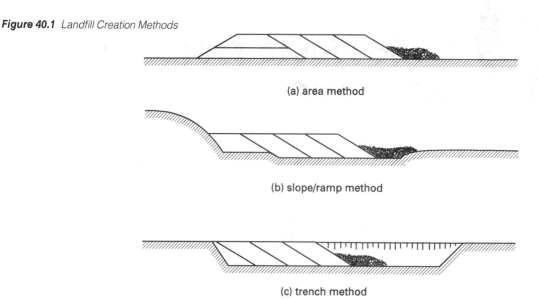

(a) area method

(b) slope/ramp method

(c) trench method

the *progressive method* (also known as the *slope/ramp method*), cover soil is taken from the front (toe) of the working face.

Landvaults are essentially above-ground piles of MSW placed on flat terrain. This design, with appropriate covers, liners, and instrumentation, has been successfully used with a variety of industrial wastes, including incineration ash and wastewater sludge. However, except when piggybacking a new landfill on top of a closed landfill, it is not widely used with MSW.

Waste is placed in layers, typically 2–3 ft thick (0.6–0.9 m), and is compacted before soil is added as a cover. Two to five passes by a tracked bulldozer are sufficient to compact the MSW to 800–1500 lbm/yd³ (470–890 kg/m³). (A density of 1000 lbm/yd³ or 590 kg/m³ is used in design studies.) The *lift* is the height of the covered layer, as shown in Fig. 40.2. When the landfill layer has reached full height, the ratio of solid waste volume to soil cover volume will be approximately between 4:1 and 3:1.

Figure 40.2 *Landfill Cells*

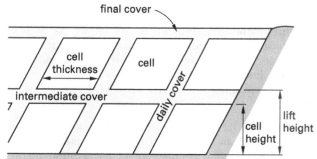

The *cell height* is typically taken as 8 ft (2.4 m) for design studies, although it can actually be much higher. The height should be chosen to minimize the cover material requirement consistent with the regulatory requirements. Cell slopes will be less than 40°, and typically 20–30°.

Each *cell* is covered by a soil layer 0.5–1.0 ft (0.15–0.3 m) in thickness. The final top and side covers should be at least 2 ft (0.6 m) thick. Daily, intermediate, and final soil covers are essential to proper landfill operation. The daily and intermediate covers prevent fly emergence, discourage rodents, reduce odors, and reduce windblown litter. In the event of ignition, the soil layers form fire stops within the landfill.

The final soil cover (cap) is intended to reduce or eliminate entering moisture. The soil cover also contains and channels landfill gas and provides a pleasing appearance and location for growing vegetation.

3. LANDFILL CAPACITY

The *compaction factor*, CF, is a multiplicative factor used to calculate the compacted volume for a particular type of waste and compaction method. The compacted volume of MSW or components within the waste is

$$V_c = (\text{CF})V_0 \qquad 40.1$$

Tables are available giving typical compaction factors for various kinds of solid waste. Compaction factors for wastes such as food, paper, plastics, and garden trimmings generally range from 0.1 to 0.5, depending on the specific kind of waste and the degree of compaction; compaction factors for wastes such as glass, dirt, ashes, and brick are generally higher. [Typical Compaction Factors for Various Solid Waste Components Placed in Landfills]

Engineers utilize the compaction factor when estimating the supply need for MSW pickup for a community. The number of garbage trucks necessary for hauling an entire community's MSW in a week depends on the capacity of each individual truck, how frequently it needs to unload, and how long it takes to pick up and unload each time. To calculate an accurate number of trucks, the average compaction factor for residential MSW is applied to the average production of MSW per household and then multiplied by the number of households in the community.

Where estimating the number of garbage trucks necessary for weekly MSW pickup, it is also critical to estimate the capacity and lifespan of a MSW landfill. Utilizing average population growth equations (Eq. 40.4 through Eq. 40.8) from the receiving community and the household MSW production, the necessary size of a landfill for a given lifespan can be determined, or if a given landfill is already being used, the estimated lifespan before a new landfill must be constructed.

The daily increase in landfill volume is predicted by using Eq. 40.2.[2] N is the population size, G is the per-capita MSW generation rate per day, and γ is the landfill overall specific weight calculated as a weighted average of soil and compacted MSW specific weights. Soils have specific weights between 70 lbf/ft^3 and 130 lbf/ft^3 (densities between 1100 kg/m^3 and 2100 kg/m^3). A value of 100 lbf/ft^3 (1600 kg/m^3) is commonly used in design studies.

$$\Delta V = \frac{NG(\text{LF})}{\gamma}$$
$$= \frac{NG(\text{LF})}{\rho g} \qquad \text{[SI]} \quad 40.2(a)$$

$$\Delta V = \frac{NG(\text{LF})}{\gamma}$$
$$= \frac{NG(\text{LF})g_c}{\rho g} \qquad \text{[U.S.]} \quad 40.2(b)$$

The *loading factor*, LF, is 1.25 for a 4:1 volumetric ratio and is calculated from Eq. 40.3 for other ratios.

$$\text{LF} = \frac{V_{\text{MSW}} + V_{\text{cover soil}}}{V_{\text{MSW}}} \qquad 40.3$$

Most large-scale sanitary landfills do not apply daily cover to deposited waste. Time, cost, and reduced capacity are typically cited as the reasons that the landfill is not covered with soil. In the absence of such cover, the loading factor has a value of 1.0.

Landfill design calculations depend on the average refuse generation per person in a community. Since a community's population does not stay uniform over time, it is necessary to calculate the population growth in the community to predict the life expectancy of the landfill's capacity. Communities' population growths are different, and growth may be linear, exponential, or a percentage per year. Landfill calculations must be based on expected growth modeled after neighboring communities or historical trends.

Where population growth is expected to be linear, the population at time t is

Linear Projection = Algebraic Projection
$$P_t = P_0 + k\Delta t \qquad 40.4$$

P_0 is the population at present or at some other time taken as time zero.

Log Growth = Exponential Growth = Geometric Growth
$$P_t = P_0 e^{k\Delta t}$$
$$\ln P_t = \ln P_0 + k\Delta t \qquad 40.5$$

Percent Growth
$$P_t = P_0(1 + k)^n \qquad 40.6$$

Ratio and Correlation Growth
$$\frac{P_2}{P_{2R}} = \frac{P_1}{P_{1R}} = k \qquad 40.7$$

Decreasing-Rate-of-Increase Growth
$$P_t = P_0 + (S - P_0)(1 - e^{-k(t - t_0)}) \qquad 40.8$$

4. ULTIMATE LANDFILL DISPOSITION

When filled to final capacity and closed, the covered landfill can be used for grassed green areas, shallow-rooted agricultural areas, and recreational areas (soccer fields, golf courses, etc.). Light construction uses are also possible, although there may be problems with gas accumulation, corrosion of pipes and foundation piles, and settlement.

Settlement will generally be uneven. Settlement in areas with high rainfall can be up to 20% of the overall landfill height. In dry areas, the settlement may be much less, 2–3%. Little information on bearing capacity of landfills is available. Some studies have suggested capacities of 500–800 lbf/ft^2 (24–38 kPa).

5. LANDFILL GAS

Once covered, the organic material within a landfill cell provides an ideal site for decomposition. Aerobic conditions prevail for approximately the first month. Gaseous products of the initial aerobic decomposition include CO_2 and H_2O. The following are typical reactions involving carbohydrates and stearic acid.

$$C_6H_{12}O_6 + 6O_2 \rightarrow 6CO_2 + 6H_2O \qquad 40.9$$

$$C_{18}H_{36}O_2 + 26O_2 \rightarrow 18CO_2 + 18H_2O \qquad 40.10$$

[2]Multiply cubic yards by 27/43,560 to get acre-feet (ac-ft).

After the first month or so, the decomposing waste will have exhausted the oxygen in the cell. Digestion continues anaerobically, producing a low-heating value *landfill gas* (LFG) consisting of approximately 50% methane (CH_4), 50% carbon dioxide, and trace amounts of other gases (e.g., CO, N_2, and H_2S). Decomposition occurs at temperatures of 100–120°F (40–50°C), but may increase to up to 160°F (70°C).

$$C_6H_{12}O_6 \rightarrow 3CO_2 + 3CH_4 \qquad 40.11$$

$$C_{18}H_{36}O_2 + 8H_2O \rightarrow 5CO_2 + 13CH_4 \qquad 40.12$$

LFG emissions through a landfill cover are monitored both for collection of usable gases, such as methane, and for pollutants that may be hazardous to a nearby community or on-site workers. Calculating the gas flux through a landfill cover requires monitoring the concentration of the target gas at the landfill surface, $C_{A,atm}$, and below the cover, $C_{A,fill}$, the porosity of the landfill, η, the diffusion coefficient of the gas, D, and the thickness of the cover, L.

Gas Flux

$$N_A = \frac{D\eta^{1/3}(C_{A_{atm}} - C_{A_{fill}})}{L} \qquad 40.13$$

LFG is essentially saturated with water vapor when formed. However, if the gas collection pipes are vertical or sloped, some of the water vapor will condense on the pipes and drain back into the landfill. Most of the remaining moisture is removed in condensate traps located along the gas collection system line.

From Dalton's law, the total gas pressure within a landfill, P, is the sum of the partial pressures of the component gases. Partial pressure is volumetrically weighted and can be found from the volumetric fraction, B, of the gas.

$$P = P_{CH_4} + P_{CO_2} + P_{H_2O} + P_{N_2} + P_{other} \qquad 40.14$$

$$P_i = B_i P \qquad 40.15$$

If uncontrolled, LFG will migrate to the surface. If the LFG accumulates, an explosion hazard results, since methane is highly explosive in concentrations of 5–15% by volume. Methane will pass through the explosive concentration as it is being diluted. Other environmental problems, including objectionable odors, can also occur. Therefore, various methods are used to prevent gas from escaping or spreading laterally.

Landfill gases can be collected by either passive or active methods. *Passive collection* uses the pressure of the gas itself as the driving force. Passive collection is applicable for sites that generate low volumes of gas and where off-site migration of gas is not expected. This generally applies to small municipal landfills—those with volumes less than approximately 50,000 yd³ (40 000 m³). Common passive control methods include (a) isolated gas

(pressure relief) vents in the landfill cover, with or without a common flare; (b) perimeter interceptor gravel trenches; (c) perimeter barrier trenches (e.g., slurry walls); (d) impermeable barriers within the landfill; and (e) sorptive barriers in the landfill.

Active gas collection draws the landfill gas out with a vacuum from *extraction wells*. Various types of horizontal and vertical wells can be used through the landfill to extract gas in vertical movement, while perimeter facilities are used to extract gases in lateral movement. Vertical wells are usually perforated PVC pipes packed in sleeves of gravel and bentonite clay. PVC pipe is resistant to the chlorine and sulfur compounds present in the landfill.

Determining the location of isolated vents is essentially heuristic—one vent per 10,000 yd³ (7500 m³) of landfill is probably sufficient. Regardless, extraction wells should be spaced with overlapping zones of influence. If the *radius of influence*, R, of each well and the desired fractional overlap, O, are known, the spacing between wells is given by Eq. 40.16. For example, if the extraction wells are placed on a square grid with spacing of $L = 1.4R$, the overlap will be 60%. For a 100% overlap, the spacing would equal the radius of influence.

$$\frac{L}{R} = 2 - O \qquad 40.16$$

In many locations, the LFG is incinerated in flares. However, emissions from flaring are problematic. Alternatives to flaring include using the LFG to produce hot water or steam for heating or electricity generation. During the 1980s, reciprocating engines and combustion turbines powered by LFG were tried. However, such engines generated relatively high emissions of their own due to impurities and composition variations in the fuel. True Rankine-cycle power plants (generally without reheat) avoid this problem, since boilers are less sensitive to impurities.

One problem with using LFG commercially is that LFG is withdrawn from landfills at less than atmospheric pressure. Conventional furnace burners need approximately 5 psig at the boiler front. Low-pressure burners that require 2 psig are available, but they are expensive. Therefore, some of the plant power must be used to pressurize the LFG in blowers.

Although production is limited, LFG is produced for a long period after a landfill site is closed. Production slowly drops 3–5% annually to approximately 30% of its original value after about 20–25 years, which is considered to be the economic life of a gas-reclamation system. The theoretical ultimate production of LFG has been estimated by other researchers as 15,000 ft³ per ton (0.45 m³/kg) of solid waste, with an estimated volumetric gas composition of 54% methane and 46% carbon dioxide. (See Table 40.2.) However, unfavorable and nonideal conditions in the landfill often reduce this yield to approximately 1000–3000 ft^3/ton (0.03–0.09 m^3/kg) or even lower.

Table 40.2 *Properties of Methane and Carbon Dioxide*

	CH$_4$	CO$_2$
color	none	none
odor	none	none
density, at STP		
(g/L)	0.717	1.977
(lbm/ft^3)	0.0447	0.123
specific gravity, at STP, ref. air	0.554	1.529
solubility (760 mm Hg, 20°C), volumes in one volume of water	0.33	0.88
solubility, qualitative	slight	moderate

(Multiply lbm/ft^3 by 16.02 to obtain g/L.)

6. LANDFILL LEACHATE

Leachates are liquid wastes containing dissolved and finely suspended solid matter and microbial waste produced in landfills. Leachate becomes more concentrated as the landfill ages. Leachate forms from liquids brought into the landfill, water run-on, and precipitation. Leachate in a natural attenuation landfill will contaminate the surrounding soil and groundwater. In a lined containment landfill, leachate will percolate downward through the refuse and collect at the first landfill liner. Leachate must be removed to reduce hydraulic head on the liner and to reduce unacceptable concentrations of hazardous substances.

When a layer of liquid sludge is disposed of in a landfill, the consolidation of the sludge by higher layers will cause the water to be released. This released water is known as *pore-squeeze liquid*.

Landfill cover holds water at a rate determined by unit area, precipitation, runoff, evapotranspiration, and percolation into compacted solid waste.

$$\text{Soil Landfill Cover Water Balance}$$
$$\Delta S_{LC} = P - R - ET - PER_{SW} \qquad 40.17$$

Some water will be absorbed by the MSW and will not percolate down. The quantity of water that can be held against the pull of gravity is referred to as the *field capacity*, FC. The potential quantity of leachate is the amount of moisture within the landfill in excess of the FC.

In general, the amount of leachate produced is directly related to the amount of external water entering the landfill. Theoretically, the leachate generation rate can be determined by writing a water mass balance on the landfill. This can be done on a preclosure and a postclosure

basis. Typical units for all the terms are units of length (e.g., "1.2 in of rain") or mass per unit volume (e.g., "a field capacity of 4 lbm/yd^3").

The preclosure leachate generation rate is

$$\begin{aligned}
&\text{preclosure leachate generation} \\
&= \text{moisture released by incoming waste,} \\
&\quad\quad \text{including pore-squeezed liquid} \\
&\quad + \text{precipitation} \\
&\quad - \text{moisture lost due to evaporation} \\
&\quad - \text{field capacity}
\end{aligned} \qquad 40.18$$

The postclosure water balance is

$$\begin{aligned}
&\text{postclosure leachate generation} \\
&= \text{precipitation} \\
&\quad - \text{surface runoff} \\
&\quad - \text{evapotranspiration} \\
&\quad - \text{moisture lost in formation} \\
&\quad\quad \text{of landfill gas and other} \\
&\quad\quad \text{chemical compounds} \\
&\quad - \text{water vapor removed along} \\
&\quad\quad \text{with landfill gas} \\
&\quad - \text{change in soil moisture storage}
\end{aligned} \qquad 40.19$$

7. LEACHATE MIGRATION FROM LANDFILLS

From Darcy's law, migration of leachate contaminants that have passed through liners into aquifers or the groundwater table is proportional to the hydraulic conductivity, K, and the hydraulic gradient, i. Hydraulic conductivities of clay liners are 1.2×10^{-7} ft/hr to 1.2×10^{-5} ft/hr (10^{-9} cm/s to 10^{-7} cm/s). However, the properties of clay liners can change considerably over time due to interactions with materials in the landfill. If the clay dries out (desiccates), it will be much more permeable. For synthetic FMLs, hydraulic conductivities are 1.2×10^{-10} ft/hr to 1.2×10^{-7} ft/hr (10^{-12} cm/s to 10^{-9} cm/s). The average permeability of high-density polyethylene is approximately 1.2×10^{-11} ft/hr (1×10^{-13} cm/s).

$$Q = KiA \qquad 40.20$$

$$i = \frac{dH}{dL} \qquad 40.21$$

8. LEACHATE RECOVERY SYSTEMS

At least two distinct leachate recovery systems are required in landfills: one within the landfill to limit the hydraulic head of leachate that has reached the top liner, and another to catch the leachate that has passed through the top liner and drainage layer and has reached the bottom liner.

Solid & Hazard. Waste

By removing leachate at the first liner, the hydrostatic pressure on the liner is reduced, minimizing the pressure gradient and hydraulic movement through the liner. A pump is used to raise the collected leachate to the surface once a predetermined level has been reached. Tracer compounds (e.g., lithium compounds or radioactive hydrogen) can be buried with the wastes to signal migration and leakage.

Leachate collection and recovery systems fail because of clogged drainage layers and pipe, crushed collection pipes due to waste load, pump failures, and faulty design.

9. LEACHATE TREATMENT

Leachate is essentially a very strong municipal wastewater, and it tends to become more concentrated as the landfill ages. Leachate from landfills contains extremely high concentrations of compounds and cannot be discharged directly into rivers or other water sources.

Leachate is treated with biological (i.e., trickling filter and activated sludge) and physical/chemical processes very similar to those used in wastewater treatment plants. For large landfills, these treatment facilities are located on the landfill site. A typical large landfill treatment facility would include an equalization tank, a primary clarifier, a first-stage activated sludge aerator and clarifier, a second-stage activated sludge aerator and clarifier, and a rapid sand filter. Additional equipment for sludge dewatering and digestion would also be required. Liquid effluent would be discharged to the municipal wastewater treatment plant.

10. LANDFILL MONITORING

Monitoring is conducted at sanitary landfills to ensure that no contaminants are released from the landfill. Monitoring is conducted in the vadose zone for gases and liquid, in the groundwater for leachate movements, and in the air for air quality monitoring.

Landfills use *monitoring wells* located outside of the covered cells and extending past the bottom of the disposal site into the aquifer. In general, individual monitoring wells should extend into each stratum upstream and downstream (based on the hydraulic gradient) of the landfill.

11. CARBON DIOXIDE IN LEACHATE

Carbon dioxide is formed during both aerobic and anaerobic decomposition. Carbon dioxide combines with water to produce *carbonic acid*, H_2CO_3. In the absence of other mitigating compounds, this will produce a slightly acidic leachate.

The concentration of carbon dioxide (in mg/L) in the leachate (assumed to be water) can be calculated by using the *absorption coefficient (solubility)*, K_s, as given in Table 40.3. The absorption coefficient for carbon dioxide at 32°F (0°C) and 1 atm is approximately $K_s = 0.88$ L/L.

$$\rho_{g/L} = \frac{PM}{RT}$$

$$= \frac{B_{CO_2} p_{t,atm}\left(44 \; \frac{g}{mol}\right)}{\left(0.08206 \; \frac{atm \cdot L}{mol \cdot K}\right) T_K} \qquad 40.22$$

Temperature Conversions

$$K = {}^{\circ}C + 273.15 \qquad 40.23$$

Table 40.3 *Typical Absorption Coefficients (L/L at 0°C and 1 atm)*

element	K_s (L/L)
hydrogen	0.017
nitrogen	0.015
oxygen	0.028
carbon monoxide	0.025
methane	0.33
carbon dioxide	0.88

At a typical internal landfill temperature of 100°F (38°C), Eq. 40.22 reduces to

$$\rho_{g/L} = 1.72 B_{CO_2} \qquad 40.24$$

The concentration of carbon dioxide in the leachate is

$$C_{CO_2,mg/L} = \rho_{g/L} K_{s,L/L}\left(1000 \; \frac{mg}{g}\right) \qquad 40.25$$

Values of solubility for carbon dioxide (or any gas) in leachate depend on the leachate composition as well as temperature and are difficult to find. Values of solubilities of gases in water, with typical units of mass of gas per mass of pure water, are commonly available, however. Solubility values with units of L/L (i.e., liters of gas per liter of water) will usually have to be calculated from published values with units of g/kg (i.e., grams of gas per kilograms of water). The total pressure within the landfill is essentially one atmosphere, and pressure of each gas is volumetrically weighted. Reasonable assumptions about the volumetric fraction (or, partial pressure ratio), B, of carbon dioxide must be made. Carbon dioxide constitutes approximately 0.033–0.039% by volume of atmospheric air, although values within a landfill will be significantly higher. (At maturity, landfill gas is 40–60% carbon dioxide, 45–60% methane, 2–5% nitrogen, and less than 1% oxygen by volume.) Solubility values are also dependent on temperature within the landfill, and solubility decreases with an increase in temperature, as Fig. 40.3 shows. Considering the heat of decomposition, an assumed landfill temperature of 100°F (38°C) is reasonable.

Figure 40.3 *Solubility of Carbon Dioxide in Water*

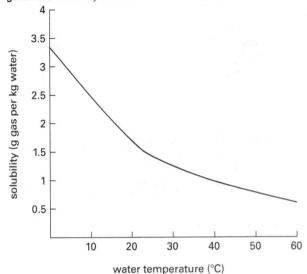

12. INCINERATION OF MUNICIPAL SOLID WASTE

Incineration of MSW results in a 90% reduction in waste disposal volume and a mass reduction of 75%.

In *no-boiler incinerators*, MSW is efficiently burned without steam generation. However, incineration at large installations is often accompanied by recycling and steam and/or electrical power generation. Incinerator/generator facilities with separation capability are known as *resource-recovery plants*. Facilities with boilers are referred to as *waste-to-steam plants*. Facilities with boilers and electrical generators are referred to as *waste-to-energy* (WTE) *facilities*. (See Fig. 40.4.) In WTE plants, the combustion heat is used to generate steam and electrical power.

Mass burning is the incineration of unprocessed MSW, typically on stoker grates or in rotary and waterwall combustors to generate steam. With mass burning, MSW is unloaded into a pit and then moved by crane to the furnace conveying mechanism. Approximately 27% of the MSW remains as ash, which consists of glass, sand, stones, aluminum, and other noncombustible materials. It and the fly ash are usually collected and disposed of in municipal landfills. (The U.S. Environmental Protection Agency (EPA) has ruled that ash from the incineration of MSW is not a hazardous waste. However, this is hotly contested and is subject to ongoing evaluation.) Capacities of typical mass burn units vary from less than 400 tons/day (360 Mg/d) to a high of 3000 tons/day (2700 Mg/d), with the majority of units processing 1000–2000 tons/day (910–1800 Mg/d).

MSW, as collected, has a heating value of approximately 4500 Btu/lbm (10 MJ/kg), though higher values have been reported. For 1000 tons/day (1000 Mg/d) of MSW incinerated, the yields are approximately 150,000–200,000 lbm/hr (75–100 Mg/h) of 650–750 psig (4.5–5.2 MPa) steam at 700–750°F (370–400°C) and 25–30 MW (27.6–33.1 MW) of gross electrical power. Approximately 10% of the electrical power is used internally, and units generating much less than approximately 10 MW (gross) may use all of their generated electrical power internally.

The *burning rate* is approximately 40–60 lbm/hr-ft^2 (200–300 kg/h·m^2) and is the fueling rate divided by the total effective grate area. Maximum heat release rates are approximately 300,000 Btu/hr-ft^2 (940 kW/m^2). The *heat release rate*, HRR, is defined as

$$\text{HRR} = \frac{(\text{fueling rate})(\text{HV})}{\text{total effective grate area}} \qquad 40.26$$

When incinerating MSW, not all the heat released can be used to generate energy. The gross heating value, or *higher heating value*, HHV, is the total heat generated, including from the production of water vapor. The net heating value, or *lower heating value*, LHV, is the leftover heat from incineration that does not include heat used in producing steam.

The HHV can be estimated using a fuel's HV. [Typical Heating Value of MSW Components]

Typical HV and moisture content of MSW are 4500 Btu/lbm and 20%, respectively. The LHV of a given quantity of MSW can be calculated as

Heating Value of Waste

$$LHV = HHV - [(\Delta H_v)(9H)] \qquad 40.27$$

The *heat of vaporization for water*, ΔH_v, and hydrogen content of the material, H, are both mass based.

By calculating the LHV, an engineer can determine how much heat a given fuel releases and therefore be used to generate energy. For example, when burning methane, each mole of methane produces two moles of water vapor.

Combustion Processes

$$CH_4 + 2\,O_2 \rightarrow CO_2 + 2\,H_2O \qquad 40.28$$

Knowing the quantity of methane fuel used allows for an estimate of water vapor produced and, by using Eq. 40.28, the LHV of methane to produce energy

13. REFUSE-DERIVED FUEL

Refuse-derived fuel (RDF) is derived from MSW. (See Table 40.4.) First-generation RDF plants use "crunch and burn" technology. In these plants, the MSW is shredded after ferrous metals are removed magnetically. First-generation RDF plants suffer from the same

Figure 40.4 *Typical Mass-Burn Waste-to-Energy Plant*

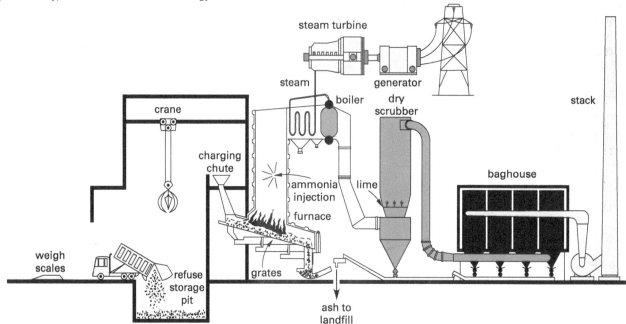

problems that have plagued early mass burn units, including ash with excessive quantities of noncombustible materials such as glass, grit and sand, and aluminum.

Table 40.4 *Typical Ultimate Analyses of MSW and RDF*

	percentage by weight	
element	MSW	RDF
carbon	26.65	31.00
water	25.30	27.14
ash	23.65	13.63
oxygen	19.61	22.72
hydrogen	3.61	4.17
chlorine	0.55	0.66
nitrogen	0.46	0.49
sulfur	0.17	0.19
total	100.00	100.00

Second-generation plants incorporate screens and air classifiers to reduce noncombustible materials and to increase the recovery of some materials. *Material recovery facilities* (MRFs) specialize in sorting out recyclables from MSW. The ash content is reduced, and the energy content of the RDF is increased. The MSW is converted into RDF pellets between 2½ in and 6 in (6.4 cm and 15 cm) in size and is introduced through feed ports above a traveling grate. Some of fuel is burned in suspension, with the rest burned on the grate. Grate speed is varied so that the fuel is completely incinerated by the time it reaches the ash rejection ports at the front of the burner.

In large (2000–3000 tons/day (1800–2700 Mg/d)), third-generation RDF plants, illustrated in Fig. 40.5, more than 95% of the original MSW combustibles are retained while reducing *mass yield* (i.e., the ratio of RDF mass to MSW mass) to below 85%.

Figure 40.5 *Typical RDF Processing*

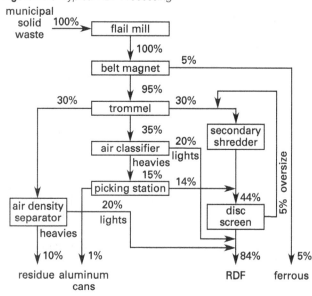

RDF has a heating value of approximately 5500–5900 Btu/lbm (12–14 MJ/kg). The moisture and ash contents of RDF are approximately 24% and 12%, respectively.

The performance of a typical third-generation facility burning RDF is similar to a mass-burn unit. For each 1000 tons/day (1000 Mg/d) of MSW collected, approximately 150,000–250,000 lbm/hr (75–125 Mg/h) of 750–850 psig

(5.2–5.9 MPa) steam at 750–825°F (400–440°C) can be generated, resulting in approximately 30–40 MW (33.1–44.1 MW) of electrical power.

Natural gas is introduced at startup to bring the furnace up to the required 1800°F (1000°C) operating temperature. Natural gas can also be used upon demand, as when a load of particularly wet RDF lowers the furnace temperature.

Coal, oil, or natural gas can be used as a back-up fuel if RDF is unavailable or as an intended part of a *co-firing* design. Co-firing installations are relatively rare, and many have discontinued burning RDF due to poor economic performance. Typical problems with co-firing are (a) upper furnace wall slagging, (b) decreased efficiencies in electrostatic precipitators, (c) increased boiler tube corrosion, (d) excessive amounts of bottoms ash, and (e) difficulties in receiving and handling RDF.

RDF is a low-sulfur fuel, but like MSW, RDF is high in ash and chlorine. Relative to coal, RDF produces less SO_2 but more hydrogen chloride. Bottom ash can also contain trace organics and lead, cadmium, and other heavy metals.

14. NOMENCLATURE

A	area	ft^2	m^2
B	volumetric fraction	-	-
C	concentration	lbmol/ft^3	g/cm^3
CF	compaction factor	-	-
d	diameter	ft	m
D	diffusion coefficient	ft^2/day	cm^2/s
ET	water lost through evapotranspiration area	in	cm
g	acceleration of gravity, 32.2 (9.81)	ft/sec^2	m/s^2
g_c	gravitational constant, 32.2	lbm-ft/lbf-sec^2	n.a.
G	MSW generation rate, per capita	lbm/day	kg/d
i	hydraulic gradient	ft/ft	m/m
H	hydrogen content of combusted material	–	–
H	total hydraulic head	ft	m
ΔH_v	heat of vaporization of water, 2420	Btu/lbm	kJ/kg
HHV	higher heating value	Btu/lbm	MJ/kg
HRR	heat release rate	Btu/hr-ft^2	W/m^2
HV	heating value	Btu/lbm	J/kg
k	growth rate constant	-	-
K	hydraulic conductivity	ft/sec	m/s
K	absolute temperature	°R	K
K_s	solubility coefficient	ft^3/ft^3	L/L
L	depth or spacing	ft	m
L	thickness	ft	cm
LF	loading factor	-	-

LHV	lower heating value	Btu/lbm	MJ/kg
M	molecular weight	lbm/lbmol	g/mol coefficient various
n	number	various	various
N	population size	-	-
N_A	gas flux	lbmol/ft^2·day	g/cm^2·s
O	fractional overlap	-	-
p	pressure	lbf/ft^2	Pa
P	precipitation	in	cm
P	population	-	-
P	pressure	lbf/ft^2	Pa
PER	amount of water percolating through unit area	in	cm
Q	flow rate	ft^3/sec	m^3/s
R	radius of influence	ft	m
R	runoff	in	cm
ΔS_{LC}	change in amount of water held in storage of landfill cover	in	cm
Δt	elapsed time	yr	yr
V	volume	ft^3	m^3

Symbols

γ	specific weight	lbf/ft^3	n.a.
η	porosity	ft^3/ft^3	cm^3/cm^3
ρ	density	lbm/ft^3	kg/m^3

Subscripts

0	initial, present, or time zero
1	at last census
1R	at last census of larger region
2	projected
2R	projected of a larger region
A,atm	compound A at surface of landfill cover
A,fill	compound A at bottom of landfill cover
c	compacted
K	kelvin
MSW	municipal solid waste
sw	solid waste
t	temperature or time

Solid & Hazard. Waste

 **Hazardous Waste Storage, Collection, Transportation**

Content in blue refers to the *NCEES Handbook*.

1. GENERAL STORAGE

Storage of *hazardous materials* (*hazmats*) is often governed by local building codes in addition to state and federal regulations. Types of construction, maximum floor areas, and building layout may all be restricted.[1]

Good engineering judgment is called for in areas not specifically governed by the building code. Engineering consideration will need to be given to the following aspects of storage facility design: spill containment provisions, chemical resistance of construction and storage materials, likelihood of and resistance to explosions, exiting, ventilation, electrical design, storage method, personnel emergency equipment, security, and spill cleanup provisions.

2. STORAGE TANKS

Underground storage tanks (USTs) have traditionally been used to store bulk chemicals and petroleum products. Fire and explosion risks are low with USTs, but subsurface pollution is common since inspection is limited. Since 1988, the U.S. Environmental Protection Agency (EPA) has required USTs to have secondary containment, corrosion protection, and leak detection. UST operators also must carry insurance in an amount sufficient to clean up a tank failure.

Because of the cost of complying with UST legislation, above-ground storage tanks (ASTs) are becoming more popular. AST strengths and weaknesses are the reverse of USTs: ASTs reduce pollution caused by leaks, but the expected damage due to fire and explosion is greatly increased. Because of this, some local ordinances prohibit all ASTs for petroleum products.[2]

The following factors should be considered when deciding between USTs and ASTs: space available, zoning ordinances, secondary containment, leak-detection equipment, operating limitations, and economics.

Most ASTs are constructed of carbon or stainless steel. These provide better structural integrity and fire resistance than fiberglass-reinforced plastic and other composite tanks. Tanks can be either field-erected or factory-fabricated (capacities greater than approximately 50,000 gal (190 kL)). Factory-fabricated ASTs are usually designed according to UL-142 (Underwriters Laboratories *Standard for Safety*), which dictates steel type, wall thickness, and characteristics of compartments, bulkheads, and fittings. Most ASTs are not pressurized, but those that are must be designed in accordance with the ASME *Boiler and Pressure Vessel Code*, Section VIII.

NFPA 30 (*Flammable and Combustible Liquids Code*, National Fire Protection Association, Quincy, MA) specifies the minimum separation distances between ASTs, other tanks, structures, and public right-of-ways. The separation is a function of tank type, size, and contents. NFPA 30 also specifies installation, spill control, venting, and testing.

ASTs must be double-walled, concrete-encased, or contained in a dike or vault to prevent leaks and spills, and they must meet fire codes. Dikes should have a capacity in excess (e.g., 110% to 125%) of the tank volume. ASTs (as do USTs) must be equipped with overfill prevention systems. Piping should be above-ground wherever possible. Reasonable protection against vandalism and hunters' bullets is also necessary.[3]

Though they are a good idea, leak-detection systems are not typically required for ASTs. Methodology for leak detection is evolving, but currently includes vacuum or pressure monitoring, electronic gauging, and optical and sniffing sensors. Double-walled tanks may also be fitted with sensors within the interstitial space.

Operationally, ASTs present special problems. In hot weather, volatile substances vaporize and represent an additional leak hazard. In cold weather, viscous content may need to be heated (often by steam tracing).

ASTs are not necessarily less expensive than USTs, but they are generally thought to be so. Additional hidden costs of regulatory compliance, secondary containment, fire protection, and land acquisition must also be considered.

Solid & Hazard. Waste

---

[1]For example, flammable materials stored in rack systems are typically limited to heights of 25 ft (8.3 m).
[2]The American Society of Petroleum Operations Engineers (ASPOE) policy statement states, "Above-ground storage of liquid hydrocarbon motor fuels is inherently less safe than underground storage. Above-ground storage of Class 1 liquids (gasoline) should be prohibited at facilities open to the public."
[3]Approximately 20% of all spills from ASTs are caused by vandalism.

3. DISPOSITION OF HAZARDOUS WASTES

When a hazardous waste is disposed of, it must be taken to a registered *treatment*, *storage*, or *disposal facility* (TSDF). The EPA's *land ban* specifically prohibits the disposal of hazardous wastes on land prior to treatment. Incineration at sea is also prohibited. Waste must be treated to specific maximum concentration limits by specific technology prior to disposal in landfills.

Once treated to specific regulated concentrations, hazardous waste residues can be disposed of by incineration, by landfilling, or, less frequently, by deep-well injection. All disposal facilities must meet detailed design and operational standards.

4. SPILLS, HAZARDOUS

Contamination by a hazardous material can occur accidentally (e.g., a spill) or intentionally (e.g., a previously used chemical-holding lagoon). Soil that has been contaminated with hazardous materials from spills or leaks from *underground storage tanks* (commonly known as *UST wastes*) is itself a hazardous waste.

The type of waste determines what laws are applicable, what permits are required, and what remediation methods are used. With contaminated soil, spilled substances can be (a) solid and nonhazardous or (b) nonhazardous liquid petroleum products (e.g., "UST nonhazardous"), and Resource Conservation and Recovery Act- (RCRA-) listed (c) hazardous substances, and (d) toxic substances.

Cleaning up a hazardous waste requires removing the waste from whatever air, soil, and water (lakes, rivers, and oceans) have been contaminated. The term *remediation* is used to mean the corrective steps taken to return the environment to its original condition. *Stabilization* refers to the act of reducing the waste concentrations to lower levels so that the waste can be transported, stored, or landfilled.

Remediation methods are classified as available or innovative. *Available methods* can be implemented immediately without being further tested. *Innovative methods* are new, unproven methods in various stages of study.

The remediation method used depends on the waste type. The two most common available methods are incineration and landfilling after stabilization.[4] Incineration can occur in rotary kilns, injection incinerators, infrared incineration, and fluidized-bed combustors. Landfilling requires the contaminated soil to be stabilized chemically or by other means prior to disposal. Technologies for general VOC-contaminated soil include vacuum extraction, bioremediation, thermal desorption, and soil washing.

---

[4]Other technologies include in situ and ex situ bioremediation, chemical treatment, in situ flushing, in situ vitrification, soil vapor extraction, soil washing, solvent extraction, and thermal desorption.

42 Hazardous Waste Treatment and Disposal

1. SUBTITLE D LANDFILLS

Since 1992, new and expanded municipal landfills in the United States have had to satisfy strict design regulations and are designated "*Subtitle D landfills*," referring to Subtitle D of the Resource Conservation and Recovery Act (RCRA). The regulations apply to any landfill designed to hold municipal solid waste, biosolids, and ash from MSW incineration.

Construction of landfills is prohibited near sensitive areas such as airports, floodplains, wetlands, earthquake zones, and geologically unstable terrain. Air quality control methods are required to control emission of dust, odors (from hydrogen sulfide and volatile organic vapors), and landfill gas. Runoff from storms must be controlled.

Subtitle D landfills must have *double-liner systems*, as shown in Fig. 42.1, and which are defined as "two or more liners and a leachate collection system above and between such liners." While states can specify greater protection, the minimum (basic) bottom layer requirements are a 30-mil flexible PVC membrane liner (FML) and at least 2 ft (0.6 m) and up to 5 ft (1.5 m) of compacted soil with a maximum hydraulic conductivity of

1.2×10^{-5} ft/hr (1×10^{-7} cm/s). If the membrane is high-density polyethylene (HDPE), the minimum thickness is 60 mils. 30-mil PVC costs less than 60-mil HDPE, but the 60-milproduct offers superior protection.

The preferred double liner consists of two FMLs separated by a drainage layer (approximately 2 ft (0.6 m) thick) of sand, gravel, or *drainage netting* and placed on low-permeability soil. This is the standard design for *Subtitle C* hazardous waste landfills. The advantage of selecting the preferred design for municipal landfills is realized in the permitting process.

The minimum thickness is 20 mils for the top FML, and the maximum hydraulic conductivity of the cover soil is 1.8×10^{-5} ft/hr (1.5×10^{-7} cm/s).

A series of wells and other instrumentation are required to detect high hydraulic heads and the accumulation of heavy metals and volatile organic compounds (VOCs) in the leachate.

2. CLAY LINERS

Liners constructed of well-compacted bentonite clays fail by fissuring (as during freeze-thaw cycles), dessication (i.e., drying out), chemical interaction, and general disruption. Permeability increases dramatically with organic fluids (hydrocarbons) and acidic or caustic leachates. Clays need to be saturated and swollen to retain their high impermeabilities. Permeabilities measured in the lab as 1.2×10^{-5} ft/hr (10^{-7} cm/s) can actually be 1.2×10^{-3} ft/hr (10^{-5} cm/s) in the field.

Although a 0.5–1.0 ft (0.15–0.30 m) thickness could theoretically contain the leachate, the probability of success is not high when typical exposure issues are considered. Clay liners with thicknesses of 4–5 ft (1.2–1.5 m) will still provide adequate protection, even when the top half has degraded.

3. FLEXIBLE MEMBRANE LINERS

Synthetic *flexible membrane liners* (FMLs, the term used by the United States EPA), also known as *synthetic membranes*, *synthetic membrane liners* (SMLs), *geosynthetics*, and *geomembranes*, are highly attractive because of their negligible permeabilities, and good

Solid & Hazard. Waste

Figure 42.1 *Subtitle D Liner and Cover Detail*

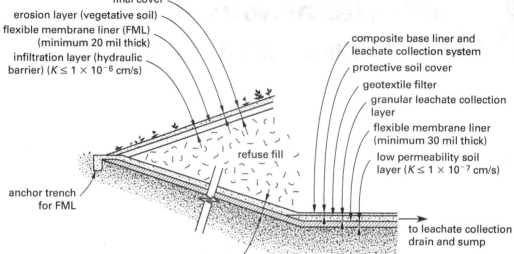

chemical resistance. The term *geocomposite liner* is used when referring to a combination of FML and one or more compacted clay layers.

FML materials include polyvinyl chloride (PVC); chlorinated polyethylene (CPE); low-density polyethylene (LDPE); linear, low-density polyethylene (LLDPE); very low density polyethylene (VLDPE); high-density polyethylene (HDPE); chlorosulfonated polyethylene, also known as Hypalon® (CSM or CSPE); ethylene propylene dienemonomer (EP or EPDM); polychloroprene, also known as neoprene (CR); isobutylene isoprene, also known as butyl (IIR); and oil-resistant polyvinyl chloride (ORPVC). LDPE covers polyethylene with densities in the range of about 56.2–58.3 lbm/ft^3 (0.900–0.935 g/cm^3), while HDPE covers the range of about 58.3–60.5 lbm/ft^3 (0.935–0.970 g/cm^3).

HDPE is preferred for municipal landfills and hazardous waste sites because of its high tensile strength, high toughness (resistance to tear and puncture), and ease of seaming. HDPE has captured the majority of the FML market.

Some FMLs can be manufactured with an internal nylon or dacron mesh, known as a *scrim*, or a spread coating on one side. The resulting membrane is referred to as a *reinforced* or *supported membrane*. An "R" is added to the name to indicate the reinforcement (e.g., "HDPE-R"). Membranes without internal scrims are *unreinforced* or *unsupported membranes*.

Geotextiles are permeable fabrics that can be used as FMLs. The transmissivity of geotextiles can be calculated using Eq. 42.1.

Geotextile Transmissivity

$$\theta = K_p t \qquad 42.1$$

Where θ is the transmissivity in cm^2/s, K_p is the in-plane hydraulic conductivity in cm/s, and t is the fabric thickness.

FMLs can be embossed or textured to increase the friction between the FML and any soil layer adjacent to it.

FMLs are sealed to the original grade and restrained against movement by deep *anchor trenches* around the periphery of the landfill. (See Fig. 42.1.)

FMLs fail primarily by puncturing, tearing, and long-term exposure to ultraviolet (UV) rays. Freezing temperatures may also weaken them. FMLs are protected from UV radiation by a soil backfill with a minimum thickness of approximately 6 in (15 cm). The cover soil should be placed promptly.

4. LANDFILL CAPS

The landfill *cap* (*top cap*) is the *final cover* placed over the landfill. Under Subtitle D regulations, caps cannot be any more permeable than the bottom liners. Therefore, only preexisting landfills can be capped with clay.

Caps fail by desiccation, penetration by roots, rodent and other animal activity, freezing and thawing cycles, erosion, settling, and breakdown of synthetic liners from sunlight exposure. Failure can also occur when the cover collapses into the disposal site due to voids opening below or excess loading (rainwater ponding) above.

Synthetic FML caps offer several advantages over clay caps—primarily lower permeability, better gas containment, faster installation, and easier testing and inspection.

In the past, some landfills had slopes as steep as 0.7:1 (horizontal:vertical). Preferred modern landfill design incorporates maximum slopes of 3:1 and occasionally as steep as 2:1 (horizontal:vertical), with horizontal tiers (benches) every so often. This is steep enough to cause concern about cap slope instability. Although clay caps can be constructed this steep, synthetic membranes have lower *interfacial friction angles* and are too smooth for use on steep slopes. (See Table 42.1.)

Slope stability analysis can be completed by using the Mohr-Coulomb failure criteria in terms of shear and normal stress on the plane of failure.

Slope Stability (Mohr-Coulomb Failure Criteria)
$$T_{ff} = c + \sigma_{ff} \times \tan\phi \qquad 42.2$$

T_{ff} is the shear stress at failure in kPa, c is the cohesion of the soil in kPa, σ is the normal stress applied at failure in kPa, and ϕ is the angle of internal friction in degrees. This represents the failure envelope.

Slippage between different layers in the cap (e.g., between the FML and the soil cover) is known as *veneer failure*. Some relief is possible with *textured synthetics*.

Table 42.1 *Typical Interfacial Friction Angles*

interface	typical friction angle
nonwoven geotextile/smooth FML	10–13°
compacted sand/smooth FML	15–25°
compacted clay/smooth FML	5–20°
nonwoven geotextile/textured FML	25–35°
compacted sand/textured FML	20–35°
compacted clay/textured FML	15–32°

A common cap system is a textured synthetic with a nonwoven geotextile. Vertical cap collapse can be prevented by incorporating a high-strength plastic *geogrid*.

5. NEUTRALIZATION

For hazardous waste that meets only the corrosivity criterion for hazardous classification, simple neutralization is an appropriate treatment technology. *Neutralization* is the process of adding an acid or a base to effect a desired pH change.

Neutralization is typically accomplished in a system containing two tanks in series. The first tank is typically mixed, and the second tank is not. Each tank has a free-fall inlet. The tank dimensions are cubic and based on a hydraulic residence time ranging from about three minutes to 30 minutes depending on reagent solubility and reaction time. The power requirements for the stirred tank can be evaluated using the velocity gradient, employed for flash mixer design.

Reagent requirements are best determined from acid or base titrations of a representative sample. The basic reagent is typically sodium hydroxide (NaOH) or lime (CaO or Ca(OH)$_2$). Lime is less expensive and less hazardous to handle, but sodium hydroxide is easier to use and reacts more quickly. The acidic reagent is most commonly sulfuric acid (H_2SO_4).

6. CHEMICAL PRECIPITATION

The *solubility product* represents the equilibrium relationship between an insoluble compound and its component ions.

Chemistry Definitions
$$A_m B_n \rightarrow m A^{n+} + n B^{m-} \qquad 42.3$$

In this reaction, $A_m B_n$ is the insoluble compound, and A^+ and B^- are the cation and anion, respectively. When this reaction is at equilibrium, $A_m B_n$ is essentially a constant, so the equilibrium equation can be written in terms of the ions only as

Chemistry Definitions
$$K_{SP} = [A^+]^m [B^-]^n \qquad 42.4$$

In Eq. 42.4, K_{SP} is the *solubility product constant*. The value of K_{SP} must be given for a specific compound.

Example 42.1

The solubility product constant for lead is 2.5×10^{-16} for the precipitation reaction.

$$Pb^{2+} + 2OH^- \rightarrow Pb(OH)_2^-$$

The concentration of lead as Pb^{2+} is 5 mg/L at pH 6.2. To what must the pH be increased for the lead to precipitate?

Solution

At 5 mg/L, the molar concentration of the lead is

$$[Pb^{2+}] = \frac{5 \frac{mg}{L}}{\left(207 \frac{g}{mol}\right)\left(1000 \frac{mg}{g}\right)}$$
$$= 2.4 \times 10^{-5} \text{ mol/L}$$

Use the solubility product constant of 2.5×10^{-16} to determine the ionic concentration of the hydroxide, $[OH^-]$.

$$[Pb^{2+}][OH^-]^2 = K_{SP}$$

$$[OH^-] = \sqrt{\frac{K_{SP}}{[Pb^{2+}]}} = \sqrt{\frac{2.5 \times 10^{-16}}{2.4 \times 10^{-5}}}$$

$$= 3.2 \times 10^{-6} \text{ mol/L}$$

The pOH is

$$pOH = -\log(3.2 \times 10^{-6}) = 5.49$$

The pH is

$$pH = 14 - pOH = 14 - 5.49 = 8.51$$

The pH must be increased to at least 8.51 to precipitate the lead.

7. SOLVENT EXTRACTION

Where two or more liquids are mixed in a waste, it may be possible to separate the liquids using *solvent extraction*. This is particularly applicable if only one of the liquids in the mixture is responsible for imparting the characteristics that make the waste hazardous.

Solvent extraction can be accomplished using a *nonaqueous phase liquid* (NAPL). An NAPL is a liquid that does not dissolve into the solvent. For example, oil is an NAPL relative to water. The NAPL is chosen so that the target liquid has a higher solubility in the NAPL than it has in the mixture or a greater affinity for the NAPL than it has for the mixture.

When the NAPL is introduced into the mixture, the target constituent will partition to the NAPL from the mixture according to its solubility. The NAPL, along the target constituent, can then be removed from the mixture by decanting or some other method. To be commercially feasible, multiple sequential extractions are often necessary.

Solvent extraction can be represented by a general equilibrium relationship as

$$\frac{C_s}{C_w} = K \qquad \textit{42.5}$$

In Eq. 42.5, K is the distribution coefficient, and C_s and C_w are the solubilities of the target chemical in the NAPL and in water, respectively. The octanol-water partition coefficient, K_{ow}, is an example of a distribution coefficient in common use.

Where multiple extractions are used, the distribution coefficient can be calculated using Eq. 42.6.

$$K = \frac{C_s}{C_w} = \frac{\dfrac{W_0 - W_1}{V_s}}{\dfrac{W_1}{V_1}} \qquad \textit{42.6}$$

In Eq. 42.6, W_0 is the mass of the substance originally present in the aqueous phase, and W_1 is the mass present in the aqueous phase after one extraction. V_w and V_s are the volumes of the water and the solvent (NAPL), respectively.

After n extractions, the mass of the substance remaining in the water can be estimated from Eq. 42.7.

$$W_n = W_0 \left(\frac{V_w}{KV_s + V_w} \right)^n \qquad \textit{42.7}$$

Equation 42.7 can also be used to estimate the volume of solvent needed to effect a desired removal.

Example 42.2

The octanol-water partition coefficient for phenol is 28.84. For a waste volume of 1000 m^3/d with an initial concentration of 1467 mg/L phenol, what daily solvent volume is needed to reach 100 mg/L after four extractions?

Solution

The original mass of phenol in the mixture is

$$W_0 = \left(1467 \ \frac{mg}{L} \right)\left(1000 \ \frac{m^3}{d} \right)\left(1000 \ \frac{g}{mg} \right)\left(1000 \ \frac{L}{m^3} \right)$$

$$= 1.467 \times 10^{12} \text{ g/d}$$

The mass of phenol present in the mixture after four extractions is

$$W_n = \left(100 \ \frac{mg}{L} \right)\left(1000 \ \frac{m^3}{d} \right)\left(1000 \ \frac{g}{mg} \right)\left(1000 \ \frac{L}{m^3} \right)$$

$$= 1.0 \times 10^{11} \text{ g/d}$$

Find the volume of solvent needed to effect a phenol reduction from 1467 mg/L to 100 mg/L after four extractions. Use Eq. 42.7, and solve for the volume of solvent.

$$W_n = W_0 \left(\frac{V_w}{KV_s + V_w} \right)^n$$

$$V_s = \left(\frac{V_w}{K} \right) \left(\left(\frac{W_0}{W_n} \right)^{1/n} - 1 \right)$$

$$= \left(\frac{1000 \, \frac{\text{m}^3}{\text{d}}}{28.84} \right) \left(\left(\frac{1.467 \times 10^{12} \, \frac{\text{g}}{\text{d}}}{1.0 \times 10^{11} \, \frac{\text{g}}{\text{d}}} \right)^{1/4} - 1 \right)$$

$$= 33.19 \, \text{m}^3/\text{d}$$

8. FLUIDIZED-BED COMBUSTORS

Fluidized-bed combustors (FBCs) are increasingly being used in steam/electric generation systems and for destruction of hazardous wastes. A *bubbling bed FBC*, as shown in Fig. 42.2, consists of four major components: (a) a windbox (plenum) that receives the fluidizing/combustion air, (b) an air distribution plate that transmits the air at 10 ft/sec to 30 ft/sec (3 m/s to 9 m/s) from the windbox to the bed and prevents the bed material from sifting into the windbox, (c) the fluid bed of inert material (usually sand or product ash), and (d) the freeboard area above the bed.

Figure 42.2 Fluidized-Bed Combustor

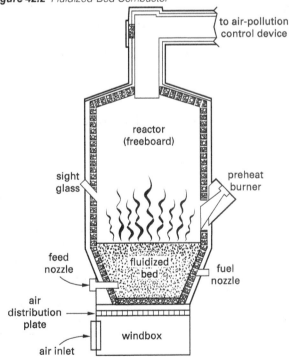

During the operation of a bubbling bed fluidized-bed combustor, the inert bed is levitated by the upcoming air, taking on many characteristics of a fluid (hence, the name FBC). The bed "boils" vigorously, offering an extremely large heat-transfer area, resulting in the thorough mixing of feed combustibles and air. Combustible material usually represents less than 1% of the bed mass, so the rest of the bed acts as a large thermal flywheel.

Combustion of volatile materials is completed in the freeboard area. Ash is generally reduced to small size so that it exits with the flue gas. In some cases (e.g., deliberate pelletization or wastes with high ash content), ash can accumulate in the bed. The fluid nature of the bed allows the ash to float on its surface, where it is removed through an overflow drain.

Most FBC systems use forced air with a single blower at the front end. If there are significant losses due to heat recovery or pollution control systems, an exhaust fan may also be used. In that case, the *null (balanced draft) point* should be in the freeboard area.

Large variations in the composition of the flue gases (known as *puffing*) is minimized by the long residence time and the large heat reservoir of the bed. Air pollution control equipment common to most boilers and incinerators is used with fluidized-bed combustors. Either wet scrubbers or baghouses can be used.

The temperature of the bed can be as high as approximately 1900°F (1040°C), though in most applications, temperatures this high are neither required nor desirable.[1] Most systems operate in the 1400°F to 1650°F (760°C to 900°C) range.

Three main options exist for reducing temperatures with overautogeneous fuels:[2] (a) water can be injected into the bed, which has the disadvantage of reducing downstream heat recovery; (b) excess air can be injected, requiring the entire system to be sized for the excess air, increasing its cost; and (c) heat-exchange coils can be placed within the bed itself, in which case, the name *fluidized-bed boiler* is applicable.

Air can be preheated to approximately 1000°F to 1250°F (540°C to 680°C) for use with subautogenous fuels, or auxiliary fuel can be used.

Modern FBC boilers for steam/electricity generation are typically of the *circulating fluidized-bed boiler design*. An important aspect of FBC boiler operation is the in-bed gas desulfurization and dechlorination that occurs when limestone and other solid reagents are injected into the combustion area. In addition, circulating fluidized-bed boilers have very low NOx emissions (i.e., less than 200 ppm).

[1]It is a common misconception that extremely high temperatures are required to destroy hazardous waste.
[2]A *subautogenous waste* has a heating value too low to sustain combustion and requires a supplemental fuel. Conversely, an *overautogeneous waste* has a heating value in excess of what is required to sustain combustion and requires temperature control.

Solid & Hazard. Waste

When combusting waste material, the amount of energy released by the process can be estimated. The higher heat value (HHV) is the maximum heat generated by the combustion. The lower heat value (LHV) is the actual heat generated by combustion after calculating the amount of heat lost to vaporize any water that is absorbing energy in or near the combustion. Equation 42.8 can be used to calculate the heating values or the change in heating value if a known quantity of water is present.

Heating Value of Waste

$$LHV = HHV - [(\Delta H_v)(9H)] \qquad 42.8$$

9. INJECTION WELLS

Properly treated and stabilized liquid and low-viscosity wastes can be injected under high pressure into appropriate strata 2000 ft to 6000 ft (600 m to 1800 m) below the surface. The wastes displace natural fluids, and the injection well is capped to maintain the pressure. Injection wells fail primarily by waste plumes through fractures, cracks, fault slips, and seepage around the well casing. Figure 42.3 shows a typical injection well installation.

Figure 42.3 *Typical Injection Well Installation*

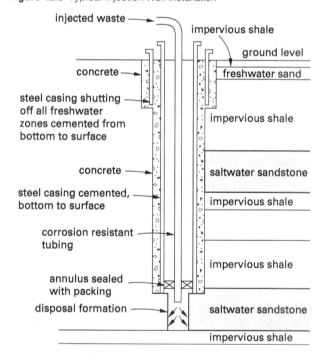

10. INCINERATORS, FLUIDIZED-BED COMBUSTION

In an FBC, combustion is efficient and excess air required is low (e.g., 25% to 50%). Destruction is essentially complete due to the long residence time (5 sec to 8 sec for gases, and even longer for solids), high turbulence, and exposure to oxygen. Combustion control is simple, and combustion is often maintained within a 15°F (8°C) band. Both overautogeneous and subautogenous feeds can be handled. Due to the thermal mass of the bed, the FBC temperature drops slowly—at the rate of about 10°F/hr (5°C/h) after shutdown. Start-up is fast and operation can be intermittent.

Relative to other hazardous waste disposal methods, NOx and metal emissions are low, and the organic content of the ash is very low (e.g., below 1%). Because of the long residence time, FBC systems do not usually require afterburners or secondary combustion chambers (SCCs). Due to the turbulent combustion process, the combustion temperature can be 200°F to 300°F (110°C to 170°C) lower than in rotary kilns. These factors translate into fuel savings, fewer or no NOx emissions, and lower metal emissions.

Most limitations of FBC systems relate to the feed. The feed must be of a form (small size and roughly regular in shape) that can be fluidized. This makes FBCs ideal for non-atomizable liquids, slurries, sludges, tars, and granular solids. It is more appropriate (and economical) to destroy atomizable liquid wastes in a boiler or special injection furnace and large bulky wastes in a rotary kiln or incinerator.

FBCs are usually inapplicable when the feed material or its ash melts below the bed temperature.[3] If the feed melts, it will agglomerate and defluidize the bed. When it is necessary to burn a feed with low melting temperature, two methods can be used. The bed can be operated as a *chemically active bed*, where operation is close to but below the ash melting point of 1350°F to 1450°F (730°C to 790°C). A small amount of controlled melting is permitted to occur, and the agglomerated ash is removed.

When a higher temperature is desirable in order to destroy hazardous materials, the melting point of feed can be increased by injecting low-cost additives (e.g., calcium hydroxide or kaolin clay). These combine with salts of alkali metals to form refractory-like materials with melting points in the 1950°F to 2350°F (1070°C to 1290°C) range.

The design of a fluidizing-bed incinerator starts with a heat balance. Energy enters the FBC during combustion of the feed and auxiliary fuel and from sensible heat contributions of the air and fuel. Some of the heat is recovered and reused. The remainder of the heat is lost through sensible heating of the combustion products, water vapor, excess air, and ash, and through radiation from the combustor vessel and associated equipment. Since these items may not all be known in advance, the following assumptions are reasonable.

- Radiation losses are approximately 5%.

- Excess air will be approximately 40%.

---

[3]The melting temperature of a eutectic mixture of two components may be lower than the individual melting temperatures.

- Sensible heat losses from the ash are minimal, particularly if the feed contains significant amounts of moisture.

- Combustion temperature will be approximately 1400°F (760°C).

- If air is preheated, the preheat temperature will be approximately 1000°F (540°C).

The fuel's *specific feed characteristic* (SFC) is the ratio of the higher (gross) heating value to the moisture content.

$$\text{SFC} = \frac{\text{total heat of combustion}}{\text{mass of water}} = \frac{m_{\text{solids}} HHV}{m_w} \quad \textbf{42.9}$$

Typical values of the specific feed characteristic range from a low of approximately 1000 Btu/lbm (2300 kJ/kg) for wastewater treatment plant biosolids, to more than 60,000 Btu/lbm (140 MJ/kg) for barks, sawdust, and RDF.[4] Making the previous assumptions and using the SFC as an indicator, the following generalizations can be made.

- Fuels with SFCs that are less than 2600 Btu/lbm (6060 kJ/kg) require a 1000°F (540°C) hot windbox, and are autogenous at that value and subautogenous below that value. The SFC drops to 2400 Btu/lbm (5600 kJ/kg) with air preheated to 1200°F (650°C). Below the subautogenous SFC, water evaporation is the controlling design factor, and auxiliary fuel is required.

- Fuels with an SFC of 4000 Btu/lbm (9300 kJ/kg) are autogenous with a cold windbox and are overautogenous above that value. Combustion is the controlling design factor for overautogenous SFCs.

Example 42.3

A wastewater treatment sludge is dewatered to 75% water by weight. The solids have a heating value of 6500 Btu/lbm (15 000 kJ/kg). The dewatered sludge enters a fluidized-bed combustor with a 1000°F (540°C) windbox at the rate of 15,000 lbm/hr (1.9 kg/s). (a) What is the specific feed characteristic? (b) Approximately what energy (in Btu/hr) must the auxiliary fuel supply?

SI Solution

(a) The dewatered sludge consists of 25% combustible solids and 75% moisture. The total mass of sludge required to contain 1.0 kg water is

$$m_{\text{sludge}} = \frac{m_w}{x_w} = \frac{1 \text{ kg}}{0.75} = 1.333 \text{ kg}$$

The mass of combustible solids per kilogram of water is

$$m_{\text{solids}} = 0.25 m_{\text{sludge}} = (0.25)(1.333 \text{ kg})$$
$$= 0.3333 \text{ kg}$$

From Eq. 42.9, the specific feed characteristic is

$$\text{SFC} = \frac{m_{\text{solids}} HHV}{m_w}$$
$$= \frac{(0.3333 \text{ kg})\left(15\,000 \frac{\text{kJ}}{\text{kg}}\right)\left(1000 \frac{\text{J}}{\text{kJ}}\right)}{1 \text{ kg}}$$
$$= 5.0 \times 10^6 \text{ J/kg}$$

(b) Making the listed assumptions, operation with a 540°C hot windbox requires an SFC of approximately 6060 kJ/kg. Therefore, the energy per kilogram of water in the fuel that an auxiliary fuel must provide is

$$(6060 \text{ kJ})\left(1000 \frac{\text{J}}{\text{kJ}}\right) - 5 \times 10^6 \text{ J} = 1.06 \times 10^6 \text{ J}$$

The energy supplied by the auxiliary fuel is

$$\dot{m}_{\text{fuel}} x_w \times \text{SFC deficit}$$
$$= \left(1.9 \frac{\text{kg}}{\text{s}}\right)(0.75)\left(1.06 \times 10^6 \frac{\text{J}}{\text{kg}}\right)$$
$$= 1.51 \times 10^6 \text{ J/s} \quad (1.5 \text{ M})$$

Customary U.S. Solution

(a) The dewatered sludge consists of 25% combustible solids and 75% moisture. The total mass of sludge required to contain 1.0 lbm water is

$$m_{\text{sludge}} = \frac{m_w}{x_w} = \frac{1 \text{ lbm}}{0.75} = 1.333 \text{ lbm}$$

The mass of combustible solids per pound of water is

$$m_{\text{solids}} = 0.25 m_{\text{sludge}} = (0.25)(1.333 \text{ lbm})$$
$$= 0.3333 \text{ lbm}$$

From Eq. 42.9, the specific feed characteristic is

$$\text{SFC} = \frac{m_{\text{solids}} HHV}{m_w} = \frac{(0.3333 \text{ lbm})\left(6500 \frac{\text{Btu}}{\text{lbm}}\right)}{1 \text{ lbm}}$$
$$= 2166 \text{ Btu/lbm}$$

---

[4] Although the units are the same, the specific feed characteristic is not the same as the heating value.

Solid & Hazard. Waste

(b) Making the listed assumptions, operation with a 1000°F hot windbox requires an autogenous SFC of approximately 2600 Btu/lbm. Therefore, the energy per pound of water in the fuel that an auxiliary fuel must provide is

$$2600 \text{ Btu} - 2166 \text{ Btu} = 434 \text{ Btu}$$

The energy supplied by the auxiliary fuel is

$$\dot{m}_{\text{fuel}} x_w \times \text{SFC deficit}$$
$$= \left(15{,}000 \ \frac{\text{lbm}}{\text{hr}}\right)(0.75)\left(434 \ \frac{\text{Btu}}{\text{lbm}}\right)$$
$$= 4.88 \times 10^6 \text{ Btu/hr}$$

11. INCINERATION, GENERAL

Most rotary kiln and liquid injection incinerators have *primary* and *secondary combustion chambers* (SCCs). Kiln temperatures are approximately 1200°F to 1400°F (650°C to 760°C) for soil incineration and up to 1700°F (930°C) for other waste types.[5]

The kiln is the primary chamber that converts solid waste to gas. The retention time in minutes can be calculated using the following equation.

Kiln Retention Time

$$t = \frac{2.28L/D}{SN}$$

42.10

L/D is the internal length-to-diameter ratio, S is the kiln rake slope in in/ft of length, and N is the rotational speed in revolutions per min.

SCC temperatures are higher—1800°F (980°C) for most hazardous wastes and 2200°F (1200°C) for liquid PCBs. The waste heat may be recovered in a boiler, but the combustion gas must be cooled prior to further processing. *Thermal ballast* can be accomplished by injecting large amounts of excess air (typical when liquid fuels are burned) or by quenching with a water spray (typical in rotary kilns).

SCCs are necessary to destroy toxics in the off-gases. SCCs are vertical units with high-swirl, vortex-type burners. These produce high *destruction removal efficiencies* (DREs) with low retention times (e.g., 0.5 sec) and moderate-to-high temperatures, even for chlorinated compounds. When soil with fine clay is incinerated, fines can build up in the SCC, causing slagging and other problems. A refractory-lined cyclone located after the primary combustion chamber can be used to reduce particle carryover to the SCC.

Prior to full operation, incinerators must be tested in a trial burn with a *principal organic hazardous constituent* (POHC) that is in or has been added to the waste. The POHC must be destroyed with a DRE of at least 99.99% by weight.

For nontoxic organics, a DRE of 95% is a common requirement. Hazardous wastes require a DRE of 99.99%. Certain hazardous wastes, including PCBs and dioxins, require a 99.9999% DRE. This is known as the *six nines rule*.

Emission limitations of some pollutants depend on the incoming concentration and the height of the stack.[6] Thus, in certain circumstances, raising the stack is the most effective method of being in compliance. This is considered justified on the basis that ground-level concentrations will be lower with higher stacks.

Rules of thumb regarding incinerator performance are as follows.

- Stoichiometric combustion requires approximately 725 lbm (330 kg) of air for each million Btu of fuel or waste burned.

- 100% excess air is required.

- Stack gas dew point is approximately 180°F (80°C).

- Water-spray-quenched flue gas is approximately 40% moisture by weight.

Table 42.2 gives performance parameters typical of incinerators.

***Table 42.2** Representative Incinerator Performance*

| | type of incinerator/use | | | |
| | rotary kiln | liquid injection | soil incinerators | |
			hazardous	non-hazardous
waste heating value				
(Btu/lbm)	15,000	20,000	0	0
(MJ/kg)	35	46	0	0
kiln temp				
(°F)	1700	–	1650	850
(°C)	925	–	900	450
SCC temp				
(°F)	1800–2200	2000–2200	1800	1400
(°C)	980–1200	1100–1200	980	750
SCC mean residence time (sec)	2	2	2	1
O$_2$ in stack gas (%)	10%	12%	9%	6%

---

[5]Temperatures higher than 1400°F (760°C) may cause incinerated soil to vitrify and clog the incinerator.
[6]Some of the metallic pollutants treated this way include antimony, arsenic, barium, beryllium, cadmium, chromium, lead, mercury, silver, and thallium.

Example 42.4

If no additional fuel is added to the incinerator, what is the mass (in lbm/hr) of water required to spray-quench combustion gases from the incineration of hazardous waste with a heating rate of 50 MBtu/hr?

Solution

Use the rules of thumb. Assume 100% excess air is required. The total dry air required is

$$m_a = \frac{(2)\left(50 \times 10^6 \, \dfrac{\text{Btu}}{\text{hr}}\right)(725 \, \text{lbm})}{10^6 \, \dfrac{\text{Btu}}{\text{hr}}}$$

$$= 72{,}500 \, \text{lbm/hr} \quad [\text{dry}]$$

Since the quenched combustion gas is 40% water by weight, it is 60% dry air by weight. The total mass of wet combustion gas produced per hour is

$$m_t = \frac{m_a}{x_a} = \frac{72{,}500 \, \dfrac{\text{lbm}}{\text{hr}}}{0.6}$$

$$= 1.21 \times 10^5 \, \text{lbm/hr}$$

The required mass of quenching water is

$$m_w = x_w m_t = (0.4)\left(1.21 \times 10^5 \, \dfrac{\text{lbm}}{\text{hr}}\right)$$

$$= 4.84 \times 10^4 \, \text{lbm/hr}$$

12. INCINERATION, HAZARDOUS WASTES

Most hazardous waste incinerators use rotary kilns. Waste in solid and paste form enters a rotating drum where it is burned at 1850°F to 2200°F (1000°C to 1200°C). (See Table 42.2 for other representative performance characteristics.) Slag is removed at the bottom, and toxic gases exit to a tall, vortex secondary combustion chamber (SCC). Gases remain in the SCC for 2 sec to 4 sec where they are completely burned at approximately 1850°F (1000°C). Liquid wastes are introduced into and destroyed by the SCC as well. Heat from off-gases may be recovered in a boiler. Typical flue gas cleaning processes include electrostatic precipitation, two-stage scrubbing, and NOx removal.

Common problems with hazardous waste incinerators include (a) inadequate combustion efficiency (easily caused by air leakage in the drum and uneven fuel loading) resulting in incomplete combustion of the primary organic hazardous component (POHC), emission of CO, NOx, and *products of incomplete combustion* (also known as *partially incinerated compounds* or PICS) and metals, (b) meeting low dioxin limits, and (c) minimizing the toxicity of slag and fly ash.

These problems are addressed by reducing air leaks in the drum and introducing air to the SCC through multiple sets of ports at specific levels. Gas is burned in the SCC in substoichiometric conditions, with the vortex ensuring adequate mixing to obtain complete combustion. Dioxin formation can be reduced by eliminating the waste-heat recovery process, since the lower temperatures present near waste-heat boilers are ideal for dioxin formation. Once formed, dioxin is removed by traditional end-of-pipe methods. Figure 42.4 shows a large-scale hazardous waste incinerator.

13. INCINERATION, INFRARED

Infrared incineration (II) is effective for reducing dioxins to undetectable levels. The basic II system consists of a waste feed conveyor, an electrical-heated primary chamber, a gas-fired afterburner, and a typical flue gas cleanup (FGC) system (i.e., scrubber, electrostatic precipitator, and/or baghouse). Electrical heating elements heat organic wastes to their combustion temperatures. Offgas is burned in a secondary combustion chamber (SCC).

14. INCINERATION, LIQUIDS

Liquid-injection incinerators can be used for atomizable liquids. Such incinerators have burners that fire directly into a refractory-lined chamber. If the liquid waste contains salts or metals, a downfired liquid-injection incinerator is used with a submerged quench to capture the molten material. A typical flue gas cleanup (FGC) system (i.e., scrubber, electrostatic precipitator, and/or baghouse) completes the system.

Incineration of organic liquid wastes usually requires little external fuel, since the wastes are overautogenous and have good heating values. The heating value is approximately 20,000 Btu/lbm (47 MJ/kg) for solvents and approximately 8000 Btu/lbm to 18,000 Btu/lbm (19 MJ/kg to 42 MJ/kg) for chlorinated compounds.

15. INCINERATION, OXYGEN-ENRICHED

Oxygen-enriched incineration is intended primarily for dioxin removal and is operationally similar to that of a rotary kiln. However, the burner includes oxidant jets. The jets aspirate furnace gases to provide more oxygen for combustion. Apparent advantages are low NOx production and increased incinerator feed rates.

Figure 42.4 *Large-Scale Hazardous Waste Incinerator*

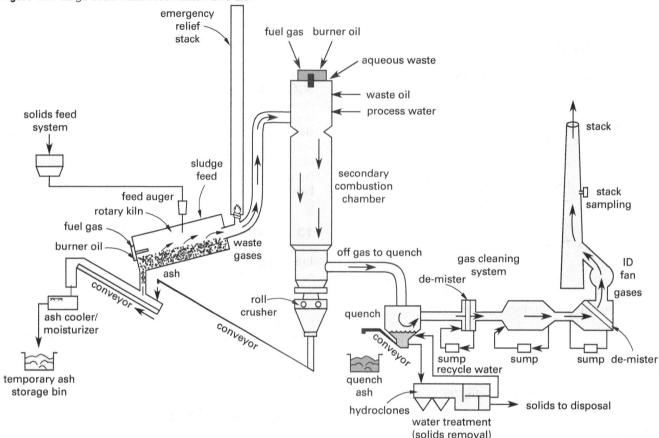

16. INCINERATION, PLASMA

A wide variety of solid, liquid, and gaseous wastes can be treated in a *plasma incinerator*. Wastes are heated by an electric arc to higher than 5000°F (2760°C), dissociating them into component atoms. Upon cooling, atoms recombine into hydrogen gas, nitrogen gas, carbon monoxide, hydrochloric acid, and particulate carbon. The ash cools to a nonleachable, vitrified matrix. Offgases pass through a normal train of cyclone, baghouse, and scrubbing operations. The process has a very high DRE. However, energy requirements are high.

17. INCINERATION, SOIL

Incineration can completely decontaminate soil. Incinerators are often thought of as being fixed facilities, as are cement kilns and special-use (e.g., Superfund) incinerators. However, mobile incinerators can be brought to sites when the soil quantities are large (e.g., 2000 tons to 100,000 tons (1800 Mg to 90,000 Mg)) and enough setup space is available.[7] (See also Sec. 42.18.)

18. INCINERATION, SOLIDS AND SLUDGES

Rotary kilns and fluidized-bed incinerators are commonly used to incinerate solids and sludges. The feed system depends on the waste's physical characteristics. Ram feeders are used for boxed or drummed solids. Bulk solids are fed via chutes or screw feeders. Sludges are fed via lances or by premixing with solids.

The constant rotation of the shell moves wastes through rotary kilns and promotes incineration. External fuel is required in rotary kilns if the heating value of the waste is below 1200 Btu/lbm (2800 kJ/mg) (i.e., is subautogenous). Additional fuel is required in the secondary chamber, as well.

Fluidized-bed incinerators work best when the waste is consistent in size and texture. An important benefit is the ability to introduce limestone and other solid reagents to the bed in order to remove HCl and SO_2.

---

[7]Modified asphalt batch processing plants can be used to incinerate soils contaminated with low-heating value, nonchlorinated hydrocarbons.

19. INCINERATION, VAPORS

Vapor incinerators, also known as *afterburners* and *flares*, convert combustible materials (gases, vapors, and particulate matter) in the stack gas to carbon dioxide and water. Afterburners can be either direct-flame or catalytic in operation.

20. LOW EXCESS-AIR BURNERS

NOx formation in gas-fired boilers can be reduced by maintaining excess air below 5%.[8] *Low excess-air burners* use a forced-draft and self-recirculating combustion chamber configuration to approximate multistaged combustion.

21. LOW-NOX BURNERS

Low (or *ultra-low*) *NOx burners* (LNB) in gas-fired applications use a combination of staged-fuel burning and internal flue gas recirculation (FGR). Recirculation within a burner is induced by either the pressure of the fuel gas or other agents (e.g., medium-pressure steam or compressed air).

22. NOMENCLATURE

A	area	ft^2	m^2
c	cohesion	–	kPa
C	concentration	various	various
C	solubility	–	–
D	diameter	in	cm
H	hydrogen content of combusted material	–	–
ΔH_v	heat of vaporization of water, 2420	n.a.	kJ/kg
HHV	higher heating value	Btu/lbm	MJ/kg
K	distribution coefficient	–	–
K_{SP}	solubility product constant		
L	length	ft	m
LHV	lower heating value	Btu/lbm	MJ/kg
m	mass	lbm	kg
$\dot{m}$	mass flow rate	lbm/hr	kg/h
n	number	–	–
N	rotational speed	rev/min	rev/min
S	slope	in/ft	cm/m
SFC	specific feed characteristic	Btu/lbm	J/kg
t	fabric thickness	–	–
T	temperature	°R	K
T_{ff}	shear stress at failure on the failure plane	–	kPa
V	volume	ft^3/day	m^3/d

x	water content	decimal	decimal
W	mass	lbm/day	g/d

Symbols

θ	transmissivity	–	cm^2/s
σ_{ff}	applied normal stress at failure on the failure plane	–	kPa
ϕ	angle of internal friction	deg	deg

Subscripts

0	initial
1	after one extraction
a	air
n	after n extractions
ow	octanol-water
p	in-plane
s	solvent
t	total
w	water

[8]Reducing excess air from 30% to 10%, for example, can reduce NOx emissions by 30%.

Topic IV: Site Assessment and Remediation

Site Assessment & Remediation

Codes, Standards, Regulations, Guidelines

1. INTRODUCTION

The two primary statutes governing site assessment and remediation are the Comprehensive Environmental Response, Compensation, and Liability Act (CERCLA) of 1980, amended in 1986 by the Superfund Amendments and Reauthorization Act (SARA), and the Resource Conservation and Recovery Act (RCRA) of 1976. This chapter discusses aspects of these statutes that are specific to site assessment and remediation.

2. SUPERFUND

Superfund is a United States federal program, established by CERCLA, to fund the cleanup of sites contaminated with hazardous substances and pollutants. The Superfund process occurs in nine steps.

step 1: Discovery

This step consists of using formal and informal mechanisms to identify a release occurrence. SARA requires the reporting of any releases. The Environmental Protection Agency (EPA) also relies on citizen complaints and whistleblowers.

step 2: Preliminary Assessment/Site Inspection

This includes regulatory agency review of site data and facilities, including surrounding land uses and groundwater uses. The preliminary assessment (PA) is usually a review of existing documentation and data specific to the site and may include a site reconnaissance. The site inspection (SI) involves an inspection of the site and may include preliminary sample collection and analysis. How the site may be reused or redeveloped may be considered.

On the basis of the PA/SI, the site is either recommended for Hazard Ranking System (HRS) scoring by the EPA or is referred to state or local agencies for action under their guidelines.

step 3: Hazard Ranking System/National Priorities Listing

The Hazard Ranking System (HRS) is a scoring process that employs a matrix worksheet where numeric values are assigned according to defined or suspected site conditions.

Factors included in the HRS scoring are:

- population at risk (how many people, how close to the site, children versus workers)
- risk to drinking water (surface or ground)
- risk of direct human contact
- risk to ecosystem
- risk to human food chain
- risk to air quality

If an HRS score greater than 28.50 results, then the site will be placed on the National Priorities List (NPL). However, responsible parties can challenge the HRS scoring.

Placement on the NPL allows Superfund financing of remedial activities and allows the EPA to compel industry to conduct cleanup under its guidelines.

step 4: Remedial Investigation/Feasibility Study

The remedial investigation (RI) is the process of investigating the site to define the contamination and site conditions. This is a complex, expensive, and time-consuming process. In many cases extensive subsurface exploration to define soil and groundwater conditions and contaminant characteristics is required. Multiple aquifer monitoring, aquifer testing, and potential conduit investigations are examples of RI-related activities.

The feasibility study (FS) is the process of identifying and evaluating potentially feasible remedial actions for a site. The potential remedial actions are based on the RI and are evaluated based on a variety of institutional and technical criteria. Public comment occurs throughout the RI/FS process through the administrative record.

Site Assessment & Remediation

step 5: Remedy Selection

This step involves preparation of a proposed plan that identifies and defends the remedial alternative recommended by those conducting the RI/FS. The proposed plan is usually developed after considerable negotiation with the regulatory agencies. The public is invited to comment on the proposed plan.

Once all parties agree on the proposed plan, the EPA issues a record of decision (ROD). The ROD summarizes the background of the site and explains the selected remedial action and why it was selected.

step 6: Remedial Design

Remedial design (RD) includes preparation of technical specifications, plans, and other documents required to implement the selected remedial action.

step 7: Remedial Action

Remedial action (RA) consists of the implementation (construction) of the selected remedial alternative.

step 8: Operation and Maintenance

This includes the ongoing operation and maintenance (O&M) of the implemented remedial action, which may continue for decades.

step 9: Deletion from the NPL

The site is removed from the NPL.

3. RESOURCE CONSERVATION AND RECOVERY ACT

Although the Resource Conservation and Recovery Act (RCRA) addresses a variety of issues pertaining to the management of solid and hazardous waste, those issues most related to soil and groundwater include:

- corrective action
- surface impoundments, underground storage tanks (USTs), above-ground storage tanks (ASTs), and waste piles
- landfills
- land treatment

Corrective Action

Corrective action (CA) is the RCRA equivalent to CERCLA. Where Superfund applies to abandoned hazardous waste facilities, RCRA corrective action applies to operating industrial facilities. Corrective action is included in Subpart F of 40 CFR 264.

The goal of CA is to use RCRA permits held by industrial facilities to compel environmental investigations and cleanups at these sites. The target sites are known as treatment, storage, and disposal (TSD) facilities and are those permitted by the EPA as hazardous waste facilities. TSD facilities include any industry that treats, stores, or disposes of hazardous waste.

As defined by RCRA, a solid waste management unit (SWMU) is "any discernable unit at which solid waste has been placed at any time irrespective of whether the unit was intended for the management of solid waste." The EPA can also look at areas of concern that do not meet the definition of SWMUs. Once the EPA enters a site, it can inspect any SWMU. The program extends to all contiguous property at a facility.

The actual CA process is largely parallel to the Superfund process. The major differences from Superfund are

- facility owners bear sole responsibility for any cleanup costs; no potentially responsible parties to share cleanup costs
- no public monies are available to assist in cleanup-related costs
- no listing process exists to allow contesting the need for an investigation; industry has no opportunity to comment on CA orders prior to their issuance

Surface Impoundments

Surface impoundments, USTs, ASTs, and waste piles are SWMUs that must be designed, operated, and maintained to minimize releases to the environment. This is also true for landfills. For these facilities, the EPA has defined design and operating requirements, action leakage rates, response actions, monitoring and inspection requirements, contingency requirements, and closure and post-closure requirements.

Land Treatment

Land treatment regulations apply to those that treat or dispose of hazardous waste in land treatment units. Such facilities are required to conduct a demonstration project for each waste being treated to prove the effectiveness of the treatment process. The EPA has defined design, operating, monitoring, and closure and postclosure requirements for these facilities.

Chemistry and Biology

1. INTRODUCTION TO ORGANIC CHEMISTRY

Organic chemistry deals with the formation and reaction of compounds of carbon, many of which are produced by living organisms. Organic compounds typically have one or more of the following characteristics.

- They are insoluble in water.[1]

- They are soluble in concentrated acids.

- They are relatively non-ionizing.

- They are unstable at high temperatures.

The method of naming organic compounds was standardized in 1930 at the International Union Chemistry meeting in Belgium. Names conforming to the established guidelines are known as *IUC names* or *IUPAC names*.[2]

2. FUNCTIONAL GROUPS

Certain combinations (groups) of atoms occur repeatedly in organic compounds and remain intact during reactions. Such combinations are called *functional groups* or *moieties*.[3] For example, the radical OH^- is known as a *hydroxyl group*. Table 44.1 contains some common functional groups. In this table and others similar to it, the symbols R and R' usually denote an attached hydrogen atom or other hydrocarbon chain of any length. They may also denote some other group of atoms.

3. FAMILIES OF ORGANIC COMPOUNDS

For convenience, organic compounds are categorized into families, or *chemical classes*. "R" is an abbreviation for the word "radical," though it is unrelated to ionic radicals. Compounds within each family have similar structures, being based on similar combinations of groups. For example, all compounds in the alcohol family have the structure [R]-OH, where [R] is any alkyl group and -OH is the hydroxyl group.

Table 44.1 Selected Functional Groups

name	standard symbol	formula	number of single bonding sites
aldehyde		CHO	1
alkyl	[R]	C_nH_{2n+1}	1
alkoxy	[RO]	$C_nH_{2n+1}O$	1
amine (amino, $n = 2$)		NH_n	$3 - n \, [n = 0, 1, 2]$
aryl (benzene ring)	[Ar]	C_6H_5	1
carbinol		COH	3
carbonyl (keto)	[CO]	CO	2
carboxyl		COOH	1
ester		COO	1
ether		O	2
halogen (halide)	[X]	Cl, Br, I, or F	1
hydroxyl		OH	1
nitrile		CN	1
nitro		NO_2	1

Families of compounds can be further subdivided into subfamilies. For example, the hydrocarbons are classified into alkanes (single carbon-carbon bond), alkenes (double carbon-carbon bond), and alkynes (triple carbon-carbon bond).[4]

Table 44.2 contains some common organic families, and Table 44.3 gives synthesis routes for various classes of organic compounds. [Important Families of Organic Compounds]

---

[1]This is especially true for hydrocarbons. However, many organic compounds containing oxygen are water soluble. The sugar family is an example of water-soluble compounds.
[2]IUPAC stands for *International Union of Pure and Applied Chemistry*.
[3]Although "moiety" is often used synonymously with "functional group," there is a subtle difference. When a functional group combines into a compound, its moiety may gain/lose an atom from/to its combinant. Therefore, moieties differ from their original functional groups by one or more atoms.
[4]Hydrocarbons with two double carbon-carbon bonds are known as *dienes*.

Table 44.2 *Families (Chemical Classes) of Organic Compounds*

family (chemical class)	structure[a]	example
organic acids		
carboxylic acids	[R]-COOH	acetic acid $((CH_3)COOH)$
fatty acids	[Ar]-COOH	benzoic acid (C_6H_5COOH)
alcohols		
aliphatic	[R]-OH	methanol (CH_3OH)
aromatic	[Ar]-[R]-OH	benzyl alcohol $(C_6H_5CH_2OH)$
aldehydes	[R]-CHO	formaldehyde (HCHO)
alkyl halides		
(haloalkanes)	[R]-[X]	chloromethane (CH_3Cl)
amides	$[R]-CO-NH_n$	β-methylbutyramide $(C_4H_9CONH_2)$
amines	$[R]_{3-n}-NH_n$	methylamine (CH_3NH_2)
	$[Ar]_{3-n}-NH_n$	aniline $(C_6H_5NH_2)$
primary amines	$n=2$	
secondary amines	$n=1$	
tertiary amines	$n=0$	
amino acids	$CH-[R]-(NH_2)COOH$	glycine $(CH_2(NH_2)COOH)$
anhydrides	[R]-CO-O-CO-[R']	acetic anhydride $(CH_3CO)_2O$
arenes (aromatics)	$ArH = C_nH_{2n-6}$	benzene (C_6H_6)
aryl halides	[AR]-[X]	fluorobenzene (C_6H_5F)
carbohydrates	$C_x(H_2O)_y$	dextrose $(C_6H_{12}O_6)$
sugars		
polysaccharides		
esters	[R]-COO-[R']	methyl acetate (CH_3COOCH_3)
ethers	[R]-O-[R]	diethyl ether $(C_2H_5OC_2H_5)$
	[Ar]-O-[R]	methyl phenyl ether $(CH_3OC_6H_5)$
	[Ar]-O-[Ar]	diphenyl ether $(C_6H_5OC_6H_5)$
glycols	$C_nH_{2n}(OH)_2$	ethylene glycol $(C_2H_4(OH)_2)$
hydrocarbons		
alkanes (single bonds)[b]	$RH = C_nH_{2n+2}$	octane (C_8H_{18})
saturated hydrocarbons		
cycloalkanes (cycloparaffins)	C_nH_{2n}	cyclohexane (C_6H_{12})
alkenes (double bonds between two carbons)[c]	C_nH_{2n}	ethylene (C_2H_4)
unsaturated hydrocarbons		
cycloalkenes	C_nH_{2n-2}	cyclohexene (C_6H_{10})
alkynes (triple bonds between two carbons)	C_nH_{2n-2}	acetylene (C_2H_2)
unsaturated hydrocarbons		
ketones	[R]-[CO]-[R]	acetone $((CH_3)_2CO)$
nitriles	[R]-CN	acetonitrile (CH_3CN)
phenols	[Ar]-OH	phenol (C_6H_5OH)

[a]See Table 44.1 for definitions of [R], [Ar], [X], and [CO].
[b]Alkanes are also known as the *paraffin series* and *methane series.*
[c]Alkenes are also known as the *olefin series.*

Table 44.3 *Synthesis Routes for Various Classes of Organic Compounds*

organic acids

    oxidation of primary alcohols

    oxidation of ketones

    oxidation of aldehydes

    hydrolysis of esters

alcohols

    oxidation of hydrocarbons

    reduction of aldehydes

    reduction of organic acids

    hydrolysis of esters

    hydrolysis of alkyl halides

    hydrolysis of alkenes (aromatic hydrocarbons)

aldehydes

    oxidation of primary and tertiary alcohols

    oxidation of esters

    reduction of organic acids

amides

    replacement of hydroxyl group in an acid with an amino group

anhydrides

    dehydration of organic acids (withdrawal of one water molecule from two acid molecules)

carbohydrates

    oxidation of alcohols

esters

    reaction of acids with alcohols (*ester alcohols*)*

    reaction of acids with phenols (*ester phenols*)

    dehydration of alcohols

    dehydration of organic acids

ethers

    dehydration of alcohol

hydrocarbons

    alkanes: reduction of alcohols and organic acids

            hydrogenation of alkenes

    alkenes: dehydration of alcohols

            dehydrogenation of alkanes

ketones

    oxidation of secondary and tertiary alcohols

    reduction of organic acids

phenols

    hydrolysis of aryl halides

*The reaction of an organic acid with an alcohol is called *esterification*.

4. SYMBOLIC REPRESENTATION

The nature and structure of organic groups and families cannot be explained fully without showing the types of bonds between the elements. Figure 44.1 illustrates the symbolic representation of some of the functional groups and families, and Fig. 44.2 illustrates reactions between organic compounds.

Figure 44.1 *Representation of Functional Groups and Families*

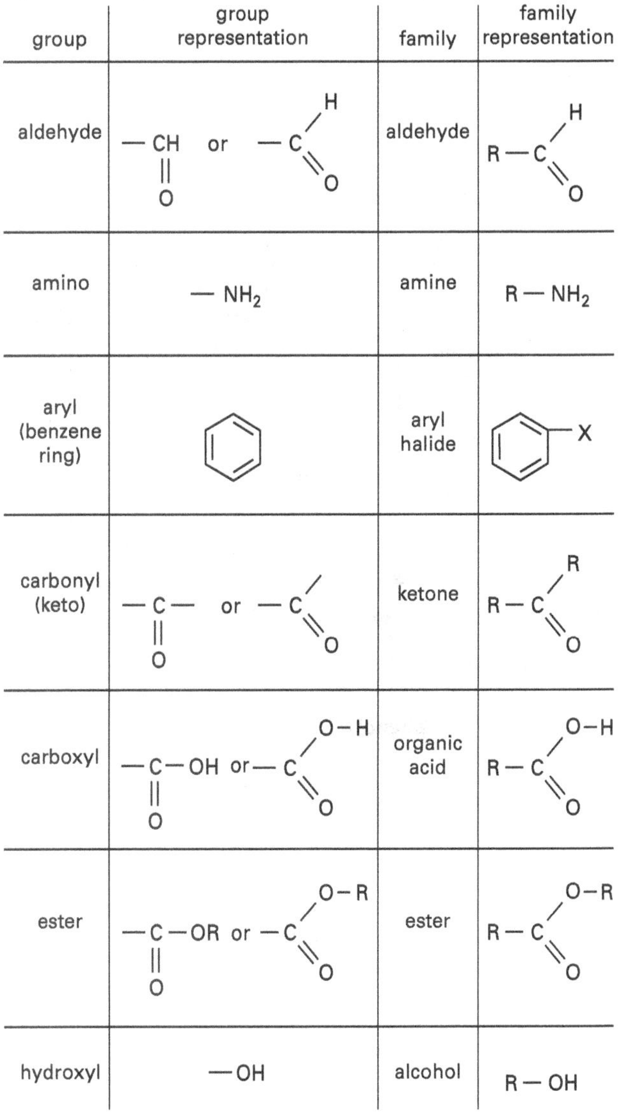

5. FORMATION OF ORGANIC COMPOUNDS

There are usually many ways of producing an organic compound. The types of reactions contained in this section deal only with the interactions between the organic families. The following processes are referred to.

- *oxidation:* replacement of a hydrogen atom with a hydroxyl group

- *reduction:* replacement of a hydroxyl group with a hydrogen atom

- *hydrolysis:* addition of one or more water molecules

- *dehydration:* removal of one or more water molecules

Figure 44.2 *Reactions Between Organic Compounds*

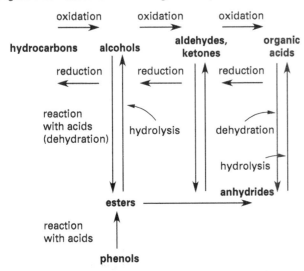

45 Hydrology and Hydrogeology

1. AQUIFERS

Underground water, also known as *subsurface water*, is contained in saturated geological formations known as *aquifers*. Aquifers are divided into two zones by the water table surface. The *vadose zone* is above the elevation of the water table. Pores in the vadose zone may be either saturated, partially saturated, or empty. The *phreatic zone* is below the elevation of the water table. Pores are always saturated in the phreatic zone.

An aquifer whose water surface is at atmospheric pressure and that can rise or fall with changes in volume is a *free aquifer*, also known as an *unconfined aquifer*. If a well is drilled into an unconfined aquifer, the water level in the well will correspond to the water table. Such a well is known as a *gravity well*.

An aquifer that is bounded on all extents is known as a *confined aquifer*. The water in confined aquifers may be under pressure. If a well is drilled into such an aquifer,

the water in the well will rise to a height corresponding to the hydrostatic pressure. The *piezometric height* of the rise is

$$H = \frac{P}{\rho g} \qquad \text{[SI]} \quad 45.1(a)$$

$$
\begin{aligned}
H &= \frac{P}{\gamma} \\
&= \frac{P}{\rho} \times \frac{g_c}{g}
\end{aligned}
\qquad \text{[U.S.]} \quad 45.1(b)
$$

If the confining pressure is high enough, the water will be expelled from the surface, and the source is known as an *artesian well*.

2. AQUIFER CHARACTERISTICS

Soil moisture content (water content), w, can be determined by oven drying a sample of soil and measuring the change in mass.[1] The water content is the ratio of the mass of water to the mass of solids, expressed as a percentage. The water content can also be determined with a *tensiometer*, which measures the vapor pressure of the moisture in the soil.

$$
\begin{aligned}
w &= \frac{m_w}{m_s} \\
&= \frac{m_t - m_s}{m_s}
\end{aligned}
\qquad 45.2
$$

The *porosity*, n, of the aquifer is the percentage of void volume to total volume.[2]

$$
\begin{aligned}
n &= \frac{V_v}{V_t} \\
&= \frac{V_t - V_s}{V_t}
\end{aligned}
\qquad 45.3
$$

---

[1]It is common in civil engineering to use the term "weight" in place of mass. For example, the *water content* would be defined as the ratio of the weight of water to the weight of solids, expressed as a percentage.

[2]The symbol θ is sometimes used for porosity.

The *void ratio, e,* is

$$
\begin{aligned}
e &= \frac{V_v}{V_s} \\
&= \frac{V_t - V_s}{V_s}
\end{aligned}
\qquad 45.4
$$

Void ratio and porosity are related.

$$
e = \frac{n}{1-n} \qquad 45.5
$$

Some pores and voids are dead ends or are too small to contribute to seepage. Only the *effective porosity, n_e,* 95–98% of the total porosity, contributes to groundwater flow.

The *hydraulic gradient, i,* is the change in hydraulic head over a particular distance. The hydraulic head at a point is determined as the piezometric head at observation wells.

$$
i = \frac{\Delta H}{L} \qquad 45.6
$$

3. PERMEABILITY

The flow of a liquid through a permeable medium is affected by both the fluid and the medium. The effects of the medium (independent of the fluid properties) are characterized by the *intrinsic permeability (specific permeability), k.* Intrinsic permeability has dimensions of length squared. (See Table 45.1.) The *darcy* has been widely accepted as the unit of intrinsic permeability. One darcy is 0.987×10^{-8} cm^2.

For studies involving the flow of water through an aquifer, effects of intrinsic permeability and the water are combined into the *hydraulic conductivity,* also known as the *coefficient of permeability* or simply the *permeability, K.* Hydraulic conductivity can be determined from a number of water-related tests.[3] It has units of volume per unit area per unit time, which is equivalent to length divided by time (i.e., units of velocity). Volume may be expressed as cubic feet and cubic meters, or gallons and liters.

Hydraulic Conductivity
$$
K = \rho g k / \mu \qquad 45.7
$$

Horizontal hydraulic conductivity represents how a liquid (i.e., water) can pass through a layer of material. The average is estimated by calculating the sum of the conductivity of each layer, K_{hm}, by the thickness of each layer, b_m, in ratio to the total thickness of all the layers the liquid is passing through, b.

Average Horizontal Conductivity (Parallel to Layering)
$$
K_h \text{ avg} = \sum_{m=1}^{n} \frac{K_{hm} b_m}{b} \qquad 45.8
$$

When a liquid is passing through multiple layers of an aquifer, from the top of the aquifer toward the bottom, the vertical hydraulic conductivity is estimated differently than for a horizontal flow of liquid in parallel with those layers. The overall vertical hydraulic conductivity, K_v avg, represents the average velocity with which the liquid will pass through the layers of the aquifer by taking a ratio of the total thickness of the layers, b, in ratio with the sum of each layer's thickness, b_m, divided by the vertical hydraulic conductivity of each layer, K_{vm}.

Overall Vertical Hydraulic Conductivity (Perpendicular to Layering)
$$
K_v \text{ avg} = \frac{b}{\sum_{m=1}^{n} \frac{b_m}{K_{vm}}} \qquad 45.9
$$

For many years in the United States, hydraulic conductivity was specified in *Meinzer units* (gallons per day per square foot). To avoid confusion related to multiple definitions and ambiguities in these definitions, hydraulic conductivity is now often specified in units of ft/day (m/d).

The coefficient of permeability is proportional to the square of the mean particle diameter.

$$
K = C D_{\text{mean}}^2 \qquad 45.10
$$

Hazen's empirical formula can be used to calculate an approximate coefficient of permeability for clean, uniform sands. D_{10} is the *effective* size in mm (i.e., the size for which 10% of the distribution is finer).

$$
K_{\text{cm/s}} \approx C D_{10,\text{mm}}^2 \quad [0.1 \text{ mm} \le D_{10,\text{mm}} \le 3.0 \text{ mm}] \qquad 45.11
$$

The coefficient C is 0.4–0.8 for very fine sand (poorly sorted) or fine sand with appreciable fines; 0.8–1.2 for medium sand (well sorted) or coarse sand (poorly sorted); and 1.2–1.5 for coarse sand (well sorted and clean).

---

[3]Permeability can be determined from constant-head permeability tests (sands), falling-head permeability tests (fine sands and silts), consolidation tests (clays), and field tests of wells (in situ gravels and sands).

Table 45.1 *Typical Permeabilities*

	k (cm^2)	k (m^2)	k (darcys)	K (cm/s)	K (gal/day-ft^2)
gravel	10^{-5} to 10^{-3}	10^{-1} to 10	10^3 to 10^5	0.5 to 50	10^4 to 10^6
gravelly sand	10^{-5}	10^{-1}	10^3	0.5	10^4
clean sand	10^{-6}	10^{-2}	10^2	0.05	10^3
sandstone	10^{-8}	10^{-3}	10	0.005	10^2
dense shale or limestone	10^{-9}	10^{-5}	10^{-1}	0.000 05	1
granite or quartzite	10^{-11}	10^{-7}	10^{-3}	0.000 000 5	10^{-2}
clay	10^{-11}	10^{-7}	10^{-3}	0.000 000 5	10^{-2}

(Multiply gal/day-ft^2 by 0.1337 to obtain ft^3/day-ft^2.)

(Multiply darcys by 0.987×10^{-8} to obtain cm^2.)

(Multiply darcys by 0.987×10^{-12} to obtain m^2.)

(Multiply cm^2 by 10^{-4} to obtain m^2.)

(Multiply ft^2 by 9.4135×10^{10} to obtain darcys.)

(Multiply m^2 by 10^4 to obtaim cm^2.)

(Multiply gal/day-ft^2 by 4.716×10^{-5} to obtain cm/s.)

4. DARCY'S LAW

Movement of groundwater through an aquifer is given by *Darcy's law*, Eq. 45.12.[4] dh/dx is the hydraulic gradient defined by Eq. 45.6. The hydraulic gradient may be specified in either ft/ft (m/m) or ft/mi (m/km), depending on the units of area used.

Darcy's Law

$$Q = -KA(dh/dx) \qquad \text{45.12}$$

The equation for the *specific discharge* (*Darcy velocity*) is the same calculation, but disregarding the area. This value is equal to the *effective velocity*.

Specific Discharge

$$q = -K(dh/dx) \qquad \text{45.13}$$

Darcy's law is applicable only when the Reynolds number is less than 1. Significant deviations have been noted when the Reynolds number is even as high as 2. In Eq. 45.14, D_{mean} is the mean grain diameter.

$$
\begin{aligned}
\text{Re} &= \frac{\rho q D_{\text{mean}}}{\mu} \\
&= \frac{q D_{\text{mean}}}{\nu}
\end{aligned}
\qquad \text{45.14}
$$

5. TRANSMISSIVITY

Transmissivity (also known as the *coefficient of transmissivity*) is an index of the rate of groundwater movement. The transmissivity of flow from a saturated aquifer of thickness b is given by Eq. 45.15. The thickness, b, of a confined aquifer is the difference in elevations of the bottom and top of the saturated formation. For permeable soil, the thickness, b, is the difference in elevations of the impermeable bottom and the water table.

Transmissivity

$$T = Kb \qquad \text{45.15}$$

Combining Eq. 45.12 and Eq. 45.15 gives

$$Q = bTi \qquad \text{45.16}$$

6. SPECIFIC YIELD, RETENTION, AND CAPACITY

The dimensionless *storage constant* (*storage coefficient*), S, of a confined aquifer is the change in aquifer water volume per unit surface area of the aquifer per unit change in head. That is, the storage constant is the amount of water that is removed from a column of the aquifer 1 ft^2 (1 m^2) in plan area when the water table drops 1 ft (1 m). For unconfined aquifers, the storage

---

[4]The negative sign in Darcy's law accounts for the fact that flow is in the direction of decreasing head. That is, the hydraulic gradient is negative in the direction of flow. The negative sign is omitted in the remainder of this chapter, or appropriate equations are rearranged to be positive.

coefficient is virtually the same as the specific yield. Various methods have been proposed for calculating the storage coefficient directly from properties of the rock and water. It can also be determined from unsteady flow analysis of wells. (See Sec. 45.12.)

The *specific yield*, S_y, is the water yielded when water-bearing material drains by gravity. It is the volume of water removed per unit area when a drawdown of one length unit is experienced. A time period may be given for different values of specific yield.

$$S_y = \frac{V_{\text{yielded}}}{V_{\text{total}}} \qquad \textit{45.17}$$

The *specific retention* S_r is the volume of water that, after being saturated, will remain in the aquifer against the pull of gravity.

$$S_r = \frac{V_{\text{retained}}}{V_{\text{total}}}$$
$$= n - S_y \qquad \textit{45.18}$$

The *specific capacity* of an aquifer is the discharge rate divided by the drawdown.

Dupuit's Formula

$$Q/D_w = \text{specific capacity} \qquad \textit{45.19}$$

7. DISCHARGE VELOCITY AND SEEPAGE VELOCITY

The *pore velocity* (*linear velocity*, *flow front velocity*, and *seepage velocity*) is given by Eq. 45.20.[5]

Specific Discharge

$$v = q/n = \frac{-K\left(\dfrac{dh}{dx}\right)}{n} \qquad \textit{45.20}$$

The gross cross-sectional area in flow depends on the aquifer dimensions. However, water can only flow through voids (pores) in the aquifer. If the gross cross-sectional area is known, it will be necessary to multiply by the porosity to reduce the area in flow. (The hydraulic conductivity is not affected.)

The *effective velocity* (also known as the *apparent velocity*, *Darcy velocity*, *Darcian velocity*, *Darcy flux*, *discharge velocity*, *specific discharge*, *superficial velocity*,

face velocity, and *approach velocity*) through a porous medium is the velocity of flow averaged over the gross aquifer cross-sectional area.

$$v_e = n v_{\text{pore}} = \frac{Q}{A} = \frac{Q}{b} = Ki \qquad \textit{45.21}$$

Contaminants introduced into an aquifer will migrate from place to place relative to the surface at the pore velocity given by Eq. 45.20. The effective velocity given in Eq. 45.21 should not be used to determine the overland time taken and distance moved by a contaminant.

8. FLOW DIRECTION

Flow direction will be from an area with a high piezometric head (as determined from observation wells) to an area of low piezometric head. Piezometric head is assumed to vary linearly between points of known head. Similarly, all points along a line joining two points with the same piezometric head can be considered to have the same head.

9. WELLS

Water in aquifers can be extracted from *gravity wells*. However, *monitor wells* may also be used to monitor the quality and quantity of water in an aquifer. *Relief wells* are used to dewater soil.

Wells may be dug, bored, driven, jetted, or drilled in a number of ways, depending on the aquifer material and the depth of the well. Wells deeper than 100 ft (30 m) are usually drilled. After construction, the well is *developed*, which includes the operations of removing any fine sand and mud. Production is *stimulated* by increasing the production rate. The fractures in the rock surrounding the well are increased in size by injecting high-pressure water or using similar operations.

Well equipment is *sterilized* by use of chlorine or other disinfectants. The strength of any chlorine solution used to disinfect well equipment should not be less than 100 ppm by weight (i.e., 100 kg of chlorine per 10^6 kg of water). Calcium hypochlorite (which contains 65% available chlorine) and sodium hypochlorite (which contains 12.5% available chlorine) are commonly used for this purpose. The mass of any chlorine-supplying compound with fractional availability A required to produce V gallons of disinfectant with concentration C is

$$m_{\text{lbm}} = (8.33 \times 10^{-6}) V_{\text{gal}} \left(\frac{C_{\text{ppm}}}{A_{\text{decimal}}} \right) \qquad \textit{45.22}$$

Figure 45.1 illustrates a typical water-supply well. Water is removed from the well through the *riser pipe* (*eductor pipe*). Water enters the well through a

---

[5]Terms used to describe pore and effective velocities are not used consistently.

perforated or slotted casing known as the *screen*. The required *open area* depends on the flow rate and is limited by the maximum permissible entrance velocity that will not lift grains larger than a certain size. Table 45.2 recommends maximum flow velocities as functions of the grain diameter. A safety factor of 1.5–2.0 is also used to account for the fact that parts of the screen may become blocked. As a general rule, the openings should also be smaller than D_{50} (i.e., smaller than 50% of the screen material particles).

Figure 45.1 *Typical Gravity Well*

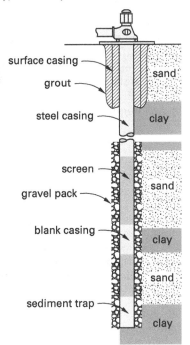

Table 45.2 *Lifting Velocity of Water*

grain diameter (mm)	maximum water velocity	
	(ft/sec)	(m/s)
0 to 0.25	0.10	0.030
0.25 to 0.50	0.22	0.066
0.50 to 1.00	0.33	0.10
1.00 to 2.00	0.56	0.17
2.00 to 4.00	2.60	0.78

(Multiply ft/sec by 0.3 to obtain m/s.)

*spherical particles with specific gravity of 2.6

Screens do not necessarily extend the entire length of the well. The required screen length can be determined from the total amount of open area required and the open area per unit length of casing. For confined aquifers, screens are usually installed in the middle 70–80% of the well. For unconfined aquifers, screens are usually installed in the lower 30–40% of the well.

It is desirable to use screen openings as large as possible to reduce entrance friction losses. Larger openings can be tolerated if the well is surrounded by a *gravel pack* to

prevent fine material from entering the well. Gravel packs are generally required in soils where the D_{90} size (i.e., the sieve size retaining 90% of the soil) is less than 0.01 in (0.25 mm) and when the well goes through layers of sand and clay.

The *octanol-water partition coefficient* estimates the amount of the contaminant found in an aqueous phase in water, and the amount in a hydrophobic or octanol phase not dissolved in water. This partition is used in situations like oil spills to estimate contaminant concentrations and determine how much contaminant will dissolve at any time in water and how much will remain undissolved at the same location.

Octanol-Water Partition Coefficient
$$K_{ow} = C_o/C_w \qquad 45.23$$

Contaminants in moisture-containing soils will have a portion of the contaminant in an aqueous phase within the moisture and a portion adhering to the solid soil. The *soil-water partition coefficient*, K_d, estimates the amount of contaminant in each phase, based on the contaminant. Since the soil-water partition coefficient is related to the organic carbon soil partition coefficient, K_{oc}, if one partition coefficient and the fraction of organic carbon present in the soil, f_{oc}, are known, the other coefficient can be calculated, as shown in Eq. 45.25.

Soil-Water Partition Coefficient $K_d = K_\rho$
$$K_d = X/C \qquad 45.24$$

$$K_d = K_{oc}f_{oc} \qquad 45.25$$

In the event of a contaminant being present in moisture-containing organic solids, the amount readily dissolved into water in an aqueous phase, C_{water}, versus the amount adhering to the organic carbon present in the soil, C_{soil}, can be estimated using the *organic carbon partition coefficient*, K_{oc}.

Organic Carbon Partition Coefficient $K_{\times}$
$$K_{oc} = C_{\text{soil}}/C_{\text{water}} \qquad 45.26$$

When a contaminant is present in groundwater, it will flow in the same direction and velocity of the aquifer if unhindered. However, contaminants are impacted by the properties of other materials found in the groundwater that the contaminant may adhere to, such as soil, resulting in a reduction in the velocity of the contaminant's spread through the aquifer. The *retardation factor*, R, represents the hindering of the contaminant from spreading at the same velocity as the groundwater.

Retardation Factor R
$$R = 1 + (\rho/\eta)K_d \qquad 45.27$$

The vadose zone penetration equation is used to estimate the depth a hydrocarbon can penetrate into the unsaturated soil depending on the volume of hydrocarbon spilled, V, the area of the spill, A, and the retention capacity of the soil the hydrocarbon was spilled on, Rv. The retention capacity changes dependent upon the type of soil and hydrocarbon spilled. [Typical Values of Rv]

Vadose Zone Penetration

$$D = \frac{RvV}{A} \qquad \textit{45.28}$$

10. DESIGN OF GRAVEL SCREENS AND POROUS FILTERS

Gravel and other porous materials may be used as filters as long as their voids are smaller than the particles to be excluded. Actually, only the largest 15% of the particles need to be filtered out, since the agglomeration of these particles will themselves create even smaller openings, and so on.

In specifying the opening sizes for screens, perforated pipe, and fabric filters, the following criteria are in widespread use as the basis for filter design in sandy gravels. D_{15}, the interpolated grain size that passes 15%, is known as the *permeability protection limit*. D_{85} is known as the *piping predicting limit*.[6,7]

$$D_{\text{opening,filter}} \leq D_{85,\text{soil}} \quad \text{[screen filters]} \qquad \textit{45.29}$$

$$[\text{filtering criterion}]^6 \; D_{15,\text{filter}} \leq 5D_{85,\text{soil}} \qquad \textit{45.30}$$
[filter beds]

$$[\text{permeability criterion}]^7 \; D_{15,\text{filter}} \leq 5D_{15,\text{soil}} \qquad \textit{45.31}$$
[filter beds]

As stated in Sec. 45.9, the screen openings should also be smaller than D_{50} (i.e., smaller than 50% of the screen material particles).

The *coefficient of uniformity*, C_u, is used to determine whether the filter particles are properly graded. Only uniform materials ($C_u < 2.5$) and well-graded materials ($2.5 < C < 6$) are suitable for use as filters.

$$C_u = \frac{D_{60}}{D_{10}} \qquad \textit{45.32}$$

11. WELL DRAWDOWN IN AQUIFERS

An aquifer with a well is shown in Fig. 45.2. Once pumping begins, the water table will be lowered in the vicinity of the well. The resulting water table

Figure 45.2 *Well Drawdown in an Unconfined Aquifer*

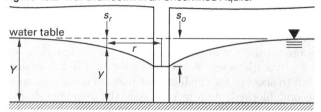

surface is referred to as a *cone of depression*. The decrease in water level at some distance r from the well is known as the *drawdown*, D_w.

If the drawdown is small with respect to the aquifer phreatic zone thickness, b, and the well completely penetrates the aquifer, the equilibrium (steady-state) well discharge is given by the *Dupuit equation*, Eq. 45.33. Equation 45.33 can only be used to determine the equilibrium flow rate a "long time" after pumping has begun.

Dupuit's Formula

$$Q = \frac{\pi k (h_2^2 - h_1^2)}{\ln\left(\dfrac{r_2}{r_1}\right)} \qquad \textit{45.33}$$

In Eq. 45.33, h_1 and h_2 are the aquifer depths at radial distances r_1 and r_2, respectively, from the well. h_1 can also be taken as the original aquifer depth, b, if r_1 is the well's *radius of influence*, the distance at which the well has no effect on the water table level.

In very thick unconfined aquifers where the drawdown is negligible compared to the aquifer thickness, or in confined aquifers where there is no cone of depression at all, $h_1 + h_2$ is essentially equal to $2b$. Then, since $h_2 - h_1 = s_1 - s_2$, $h_2^2 - h_1^2 = (h_2 - h_1)(h_2 + h_1) = 2b(D_{w,1} - D_{w,2})$. Using this equality and Eq. 45.15, Eq. 45.33 can be written as

$$Q = \frac{2\pi T(D_{w,2} - D_{w,1})}{\ln \dfrac{r_1}{r_2}} \qquad \textit{45.34}$$

For an artesian well fed by a confined aquifer of transmissivity T, the discharge is given by the *Thiem equation*.

Theim Equation

$$Q = \frac{2\pi T(h_2 - h_1)}{\ln\left(\dfrac{r_2}{r_1}\right)} \qquad \textit{45.35}$$

The rate (or amount in some period) of water that can be extracted without experiencing some undesirable result is known as the *safe yield*. Undesirable results

---

[6]Some authorities replace "5" with "9" for uniform soils.
[7]Some authorities replace "5" with "4."

include deteriorations in the water quality, large pump lifts, and infringement on the water rights of others. Extractions in excess of the safe yield are known as *overdrafts*.

Example 45.1

A 9 in (25 cm) diameter well is pumped at the rate of 50 gal/min (0.2 m³/min). The aquifer is 100 ft (30 m) thick. After some time, the well sides cave in and are replaced with an 8 in (20 cm) diameter tube. The draw-down is 6 ft (2 m). The water table recovers its original thickness 2500 ft (750 m) from the well. What will be the steady flow from the new well?

SI Solution

$$r_2 = \frac{25 \text{ cm}}{2} = 12.5 \text{ cm} \quad (0.125 \text{ m})$$

$$y_2 = 30 \text{ m} - 2 \text{ m} = 28 \text{ m}$$

Rearrange Eq. 45.33 to determine hydraulic conductivity from the given information.

Dupuit's Formula

$$k = \frac{Q \ln\left(\dfrac{r_2}{r_1}\right)}{\pi(h_2^2 - h_1^2)}$$

$$= \frac{\left(0.2 \ \dfrac{\text{m}^3}{\text{min}}\right) \ln \dfrac{750 \text{ m}}{0.125 \text{ m}}}{\pi\left((30 \text{ m})^2 - (28 \text{ m})^2\right)}$$

$$= 0.004774 \text{ m/min} \quad (0.007957 \text{ cm/s})$$

For the relined well,

$$r_2 = \frac{20 \text{ cm}}{2} = 10 \text{ cm} \quad (0.10 \text{ m})$$

Use Eq. 45.33 to solve for the new flow rate.

Dupuit's Formula

$$Q = \frac{\pi k(h_2^2 - h_1^2)}{\ln\left(\dfrac{r_2}{r_1}\right)}$$

$$= \pi\left(0.004774 \ \frac{\text{m}}{\text{min}}\right)\left(\frac{(30 \text{ m})^2 - (28 \text{ m})^2}{\ln \dfrac{750 \text{ m}}{0.10 \text{ m}}}\right)$$

$$= 0.195 \text{ m}^3/\text{min}$$

Customary U.S. Solution

$$r_2 = \frac{9 \text{ in}}{(2)\left(12 \ \dfrac{\text{in}}{\text{ft}}\right)} = 0.375 \text{ ft}$$

$$y_2 = 100 \text{ ft} - 6 \text{ ft} = 94 \text{ ft}$$

$$Q = \left(50 \ \frac{\text{gal}}{\text{min}}\right)\left(0.002228 \ \frac{\text{ft}^3\text{-min}}{\text{sec-gal}}\right)$$

$$= 0.1114 \text{ ft}^3/\text{sec}$$

Rearrange Eq. 45.33 to find the hydraulic conductivity.

Dupuit's Formula

$$k = \frac{Q \ln\left(\dfrac{r_2}{r_1}\right)}{\pi(h_2^2 - h_1^2)}$$

$$= \frac{\left(50 \ \dfrac{\text{gal}}{\text{min}}\right)\left(1440 \ \dfrac{\text{min}}{\text{day}}\right) \ln \dfrac{2500 \text{ ft}}{0.375 \text{ ft}}}{\pi\left((100 \text{ ft})^2 - (94 \text{ ft})^2\right)}$$

$$= 173.4 \text{ gal/day-ft}^2$$

For the relined well,

$$r_2 = \frac{8 \text{ in}}{(2)\left(12 \ \dfrac{\text{in}}{\text{ft}}\right)} = 0.333 \text{ ft}$$

Use Eq. 45.33 to solve for the new flow rate.

Dupuit's Formula

$$Q = \frac{\pi k(h_2^2 - h_1^2)}{\ln\left(\dfrac{r_2}{r_1}\right)}$$

$$= \frac{\pi\left(173.4 \ \dfrac{\text{gal}}{\text{day-ft}^2}\right)\left((100 \text{ ft})^2 - (94 \text{ ft})^2\right)}{\ln \dfrac{2500 \text{ ft}}{0.333 \text{ ft}}}$$

$$= 71,057 \text{ gal/day}$$

12. UNSTEADY FLOW

When pumping first begins, the removed water also comes from the aquifer above the equilibrium cone of depression. Therefore, Eq. 45.34 cannot be used, and a nonequilibrium analysis is required.

Table 45.3 Well Function W(u) for Various Values of u

u	1.0	2.0	3.0	4.0	5.0	6.0	7.0	8.0	9.0
$\times 1$	0.219	0.049	0.013	0.0038	0.0011	0.00036	0.00012	0.000038	0.000012
$\times 10^{-1}$	1.82	1.22	0.91	0.70	0.56	0.45	0.37	0.31	0.26
$\times 10^{-2}$	4.04	3.35	2.96	2.68	2.47	2.30	2.15	2.03	1.92
$\times 10^{-3}$	6.33	5.64	5.23	4.95	4.73	4.54	4.39	4.26	4.14
$\times 10^{-4}$	8.63	7.94	7.53	7.25	7.02	6.84	6.69	6.55	6.44
$\times 10^{-5}$	10.94	10.24	9.84	9.55	9.33	9.14	8.99	8.86	8.74
$\times 10^{-6}$	13.24	12.55	12.14	11.85	11.63	11.45	11.29	11.16	11.04
$\times 10^{-7}$	15.54	14.85	14.44	14.15	13.93	13.75	13.60	13.46	13.34
$\times 10^{-8}$	17.84	17.15	16.74	16.46	16.23	16.05	15.90	15.76	15.65
$\times 10^{-9}$	20.15	19.45	19.05	18.76	18.54	18.35	18.20	18.07	17.95
$\times 10^{-10}$	22.45	21.76	21.35	21.06	20.84	20.66	20.50	20.37	20.25
$\times 10^{-11}$	24.75	24.06	23.65	23.36	23.14	22.96	22.81	22.67	22.55
$\times 10^{-12}$	27.05	26.36	25.96	25.67	25.44	25.26	25.11	24.97	24.86
$\times 10^{-13}$	29.36	28.66	28.26	27.97	27.75	27.56	27.41	27.28	27.16
$\times 10^{-14}$	31.66	30.97	30.56	30.27	30.05	29.87	29.71	29.58	29.46
$\times 10^{-15}$	33.96	33.27	32.86	32.58	32.35	32.17	32.02	31.88	31.76

Reprinted from L. K. Wenzel, "Methods for Determining Permeability of Water Bearing Materials with Special Reference to Discharging Well Methods," U.S. Geological Survey, 1942, Water-Supply Paper 887.

Nonequilibrium solutions to well problems have been formulated in terms of dimensionless numbers. For small drawdowns compared with the initial thickness of the aquifer, the *Theis equation* for the drawdown at a distance r from the well and after pumping for time t is

$$D_{w,r,t} = \left(\frac{Q}{4\pi Kb}\right) W(u)$$

$$= \left(\frac{Q}{4\pi T}\right) W(u) \quad \begin{bmatrix} \text{consistent} \\ \text{units} \end{bmatrix}$$

45.36

$W(u)$ is a dimensionless *well function*. (See Table 45.3.) Though it is possible to obtain $W(u)$ from u, extracting u from $W(u)$ is more difficult. The relationship between u and $W(u)$ is often given in tabular or graphical forms. In Eq. 45.38, S is the aquifer *storage constant*.

$$W(u) = -0.577216 - \ln u + u - \frac{u^2}{(2)(2!)}$$

$$+ \frac{u^3}{(3)(3!)} - \frac{u^4}{(4)(4!)} + \cdots$$

45.37

$$u = \frac{r^2 S}{4Kbt} = \frac{r^2 S}{4Tt} \quad \begin{bmatrix} \text{consistent} \\ \text{units} \end{bmatrix}$$

45.38

Accordingly, for any two different times in the pumping cycle,

$$D_{w,1} - D_{w,2} = y_2 - y_1$$

$$= \left(\frac{Q}{4\pi Kb}\right)(W(u_1) - W(u_2))$$

45.39

If $u < 0.01$, then *Jacob's equation* can be used.

$$D_{w,r,t} = \left(\frac{Q}{4\pi T}\right) \ln \frac{2.25\,Tt}{r^2 S}$$

45.40

13. PUMPING POWER

Various types of pumps are used in wells. Problems with excessive suction lift are avoided by the use of submersible pumps.

Pumping power can be determined from hydraulic (water) power equations. The total head is the sum of static lift, velocity head, drawdown, pipe friction, and minor entrance losses from the casing, strainer, and screen. The Hazen-Williams equation is commonly used with a coefficient of $C = 100$ to determine the pipe friction.

14. FLOW NETS

Groundwater seepage is from locations of high hydraulic head to locations of lower hydraulic head. Relatively complex two-dimensional problems may be evaluated using a graphical technique that shows the decrease in hydraulic head along the flow path. The resulting graphic representation of pressure and flow path is called a *flow net*.

The flow net concept as discussed here is limited to cases where the flow is steady, two-dimensional, incompressible, and through a homogeneous medium, and where the liquid has a constant viscosity. This is the ideal case of groundwater seepage.

Flow nets are constructed from streamlines and equipotential lines. *Streamlines* (*flow lines*) show the path taken by the seepage. *Equipotential lines* are contour lines of constant driving (differential) hydraulic head. (This head does not include static head, which varies with depth.)

The object of a graphical flow net solution is to construct a network of flow paths (outlined by the streamlines) and equal pressure drops (bordered by equipotential lines). No fluid flows across streamlines, and a constant amount of fluid flows between any two streamlines.

Flow nets are constructed according to the following rules.

Rule 1: Streamlines enter and leave pervious surfaces perpendicular to those surfaces.

Rule 2: Streamlines approach the line of seepage (above which there is no hydrostatic pressure) asymptotically to (i.e., parallel but gradually approaching) that surface.

Rule 3: Streamlines are parallel to but cannot touch impervious surfaces that are streamlines.

Rule 4: Streamlines are parallel to the flow direction.

Rule 5: Equipotential lines are drawn perpendicular to streamlines such that the resulting cells are approximately square and the intersections are 90° angles. Theoretically, it should be possible to draw a perfect circle within each cell that touches all four boundaries, even though the cell is not actually square.

Rule 6: Equipotential lines enter and leave impervious surfaces perpendicular to those surfaces.

Many flow nets with differing degrees of detail can be drawn, and all will be more or less correct. Generally, three to five streamlines are sufficient for initial graphical evaluations. The size of the cells is determined by the number of intersecting streamlines and equipotential lines. As long as the rules are followed, the ratio of stream flow channels to equipotential drops will be approximately constant regardless of whether the grid is coarse or fine.

Figure 45.3 shows flow nets for several common cases. A careful study of the flow nets will help to clarify the rules and conventions previously listed.

15. SEEPAGE FROM FLOW NETS

Once a flow net is drawn, it can be used to calculate the seepage. First, the number of flow channels, N_f, between the streamlines is counted. Then, the number of equipotential drops, N_p, between equipotential lines is counted. The total hydraulic head, H, is determined as a function of the water surface levels.

$$Q = KH\left(\frac{N_f}{N_p}\right) \quad \text{[per unit width]} \qquad 45.41$$

$$H = H_1 - H_2 \qquad 45.42$$

Figure 45.3 *Typical Flow Nets*

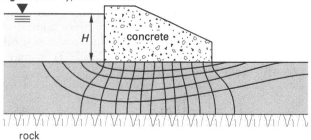

rock

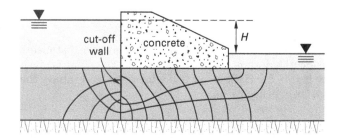

cut-off wall   concrete

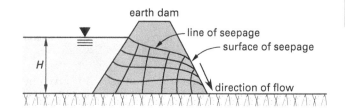

earth dam — line of seepage — surface of seepage — direction of flow

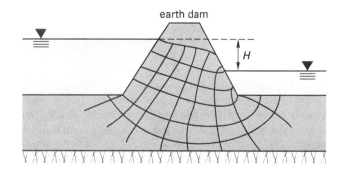

earth dam

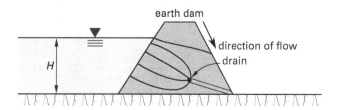

earth dam — direction of flow — drain

Site Assessment & Remediation

16. HYDROSTATIC PRESSURE ALONG FLOW PATH

The hydrostatic pressure at equipotential drop j (counting the last line on the downstream side as zero) in a flow net with a total of N_p equipotential drops is given by Eq. 45.43. If the point being investigated is along the bottom of a dam or other structure, the hydrostatic pressure is referred to as *uplift pressure*. Since uplift due to velocity is negligible, all uplift is due to the neutral pore pressure. Therefore, *neutral pressure* is synonymous with uplift pressure.

$$P_u = \left(\frac{j}{N_p}\right) H g \rho_w \qquad \text{[SI]} \qquad 45.43(a)$$

$$P_u = \left(\frac{j}{N_p}\right) H \gamma_w \qquad \text{[U.S.]} \qquad 45.43(b)$$

If the point being investigated is located a distance z below the datum, then the pressure at the point is given by Eq. 45.44. If the point being investigated is above the datum, z is negative.

$$P_u = \left(\left(\frac{j}{N_p}\right) H + z\right) g \rho_w \qquad \text{[SI]} \qquad 45.44(a)$$

$$P_u = \left(\left(\frac{j}{N_p}\right) H + z\right) \gamma_w \qquad \text{[U.S.]} \qquad 45.44(b)$$

The actual *uplift force* on a surface of area A is given by Eq. 45.45. N is the *neutral stress coefficient*, defined as the fraction of the surface that is exposed to the hydrostatic pressure. For soil, $N \approx 1$. However, for fractured rock, concrete, porous limestone, or sandstone, it varies from 0.5 to 1.0. For impervious materials such as marble or granite, it can be as low as 0.1.

$$U = N P_u A \qquad 45.45$$

As Table 45.1 indicates, some clay layers are impervious enough to prevent the passing of water. However, the hydrostatic uplift pressure below a clay layer can still be transmitted through the clay layer. The transfer geometry for thick layers is complex, making estimates of the amount of uplift force difficult. Upward forces on an impervious layer due to hydrostatic pressure below is known as *heave*. A factor of safety, FS, against heave of at least 1.5 is desirable.

$$(\text{FS})_{\text{heave}} = \frac{\text{downward pressure}}{\text{uplift pressure}} \qquad 45.46$$

17. INFILTRATION

Aquifers can be recharged (refilled) in a number of natural and artificial ways. The *Horton equation* gives a lower bound on the *infiltration capacity* (in inches (centimeters) of rainfall per unit time), which is the capacity of an aquifer to absorb water from a saturated source above as a function of time. When the rainfall supply exceeds the infiltration capacity, infiltration decreases exponentially over time. In Eq. 45.47, f_c is the final or equilibrium capacity, f_0 is the initial infiltration capacity, and f_t is the infiltration capacity at time t. k is an infiltration decay constant with dimensions of 1/hr determined from a *double-ring infiltration test*.

$$f_t = f_c + (f_0 - f_c) e^{-kt} \qquad 45.47$$

The cumulative infiltration will not correspond to the increase in water table elevation due to the effects of porosity. However, the increase in water table elevation can be determined from a knowledge of the aquifer properties.

Infiltration rates for intermediate, silty soils are approximately 50% of those for sand. Clay infiltration rates are approximately 10% of those for sand. The actual infiltration rate at any time is equal (in the Horton model) to the smaller of the rainfall intensity, $i(t)$, and the instantaneous capacity.

$$f = i \qquad [i \leq f_t] \qquad 45.48$$

$$f = f_t \qquad [i > f_t] \qquad 45.49$$

The cumulative infiltration over time is

$$F_t = f_c t + \left(\frac{f_0 - f_c}{k}\right)(1 - e^{-kt}) \qquad 45.50$$

The average infiltration rate is found by dividing Eq. 45.50 by the duration of infiltration.

$$\bar{f} = \frac{F_t}{t_{\text{infiltration}}} \qquad 45.51$$

The depth of water stored in the layer, S, is equal to the total depth of infiltration, F_t, minus the depth of water that has leaked out the bottom of the layer. Except in very permeable soils (such as sand or gravel), water leaks out the bottom of the layer slowly. In that case, the water storage is

$$S = F_t - f_c t \qquad 45.52$$

The maximum depth of water that the layer can store is

$$S_{\text{max}} = \frac{f_0 - f_c}{k} \qquad 45.53$$

Horton's equation assumes that the rate of precipitation, I, is greater than the infiltration rate, f_{in}, throughout the storm period. The overland flow, $Q_{overland}$, experienced when the rainfall rate is greater than the infiltration rate is

$$Q_{overland} = I - f_{in} \qquad \text{45.54}$$

The rate at which water is stored in the layer is equal to the infiltration rate minus the leakage outflow rate through the bottom of the layer.

$$\frac{dS}{dt} = f_{in} - f_{out} \qquad \text{45.55}$$

Equation 45.55 is easily solved for constant precipitation rates greater than the infiltration rate. If the precipitation rate is ever lower than the infiltration rate, the layer will lose water to lower levels, and Horton's theory must be modified. Rather than Eq. 45.55, finite difference techniques are used to calculate the infiltration rate and storage incrementally. The finite difference infiltration equation is given by Eq. 45.56. Of course, the infiltration can never exceed the rate of precipitation, no matter what the capacity.

$$S(t + \Delta t) = S(t) + \Delta t (f_{in} - f_c) \qquad \text{45.56}$$

Example 45.2

A soil has the following characteristics.

> initial infiltration capacity, $f_0 = 6.8$ cm/h
>
> equilibrium infiltration capacity, $f_c = 0.3$ cm/h
>
> infiltration decay constant, $k = 0.75$ 1/h

Calculate the (a) infiltration rate and (b) the total depth of infiltrated water after 3 h of hard rain. (c) What is the maximum depth of water that the layer can store?

Solution

(a) Use Eq. 45.47 to calculate the infiltration rate.

$$\begin{aligned}
f_t &= f_c + (f_0 - f_c) e^{-kt} \\
&= 0.3 \; \frac{\text{cm}}{\text{h}} + \left(6.8 \; \frac{\text{cm}}{\text{h}} - 0.3 \; \frac{\text{cm}}{\text{h}} \right) \\
&\quad \times e^{-\left(0.75 \frac{1}{\text{h}}\right)(3 \text{ h})} \\
&= 0.985 \; \text{cm/h}
\end{aligned}$$

(b) Use Eq. 45.50 to calculate the total depth of infiltrated water.

$$\begin{aligned}
F(t) &= f_c t + \left(\frac{f_0 - f_c}{k} \right)(1 - e^{-kt}) \\
&= \left(0.3 \; \frac{\text{cm}}{\text{h}} \right)(3 \text{ h}) \\
&\quad + \left(\frac{6.8 \; \frac{\text{cm}}{\text{h}} - 0.3 \; \frac{\text{cm}}{\text{h}}}{0.75 \; \frac{1}{\text{h}}} \right)\left(1 - e^{-\left(0.75 \frac{1}{\text{h}}\right)(3 \text{ h})} \right) \\
&= 8.65 \; \text{cm}
\end{aligned}$$

(c) The maximum storage is calculated from Eq. 45.53.

$$\begin{aligned}
S_{max} &= \frac{f_0 - f_c}{k} = \frac{6.8 \; \frac{\text{cm}}{\text{h}} - 0.3 \; \frac{\text{cm}}{\text{h}}}{0.75 \; \frac{1}{\text{h}}} \\
&= 8.67 \; \text{cm}
\end{aligned}$$

18. NOMENCLATURE

A	cross-sectional area of flow	ft^2	m^2
A	availability	decimal	decimal
b	thickness	ft	m
b_m	thickness of the mth layer	ft	m
C	coefficient	n.a.	mm
C	concentration	ppm	ppm
C	concentration	ppb	μg/L
C	constant	various	various
C_{soil}	concentration of chemical in soil	–	μg/kg, ppb
C_{water}	concentration of chemical in aqueous phase	–	mg/L, μg/L
D	diameter (grain size)	ft	m
D_w	well drawdown	ft	m
e	void ratio	–	–
f	fraction	–	–
f_c	final or equilibrium capacity	in/hr	cm/h
f_{in}	infiltration rate	in/hr	cm/h
f_t	infiltration capacity at time t	in/hr	cm/h
f_0	initial infiltration capacity	in/hr	cm/h
F_t	total depth of infiltration	in	cm
F	cumulative infiltration	in	cm
FS	factor of safety	–	–
g	acceleration of gravity, 32.2 (9.81)	ft/sec^2	m/s^2
g_c	gravitational constant, 32.2	lbm-ft/ lbf-sec^2	n.a.
h	hydraulic head or height	ft	m

H	height or total hydraulic head	ft	m
i	hydraulic gradient	ft/ft	m/m
$i(t), I$	rainfall intensity	in/hr	cm/h
j	equipotential index	–	–
k	infiltration decay constant	1/hr	1/h
k	intrinsic permeability	ft^2	m^2
K	coefficient of permeability	–	–
K	hydraulic conductivity	ft/sec	m/s
K_d	distribution or soil-water partition coefficient	–	–
K_h avg	average horizontal hydraulic conductivity	ft/day	m/d
K_{hm}	horizontal hydraulic conductivity of the mth layer	ft/day	m/d
K_v avg	average vertical hydraulic conductivity	ft/day	m/d
K_{vm}	vertical hydraulic conductivity of the mth layer	ft/day	m/d
L	length	ft	m
m	mass	lbm	kg
n	porosity	–	–
N	neutral stress coefficient	–	–
N	quantity (number of)	–	–
P	pressure	lbf/ft^2	Pa
q	specific discharge	ft^3/ft^2-sec	m^3/m^2·s
Q	discharge rate or flow quantity	ft^3/sec	m^3/s
r	radial distance from well	ft	m
R	retardation factor	–	–
Re	Reynolds number	–	–
Rv	retention capacity and viscosity constant	–	–
s	drawdown	ft	m
S	storage constant	–	–
S_r	specific retention	–	–
S_y	specific yield	–	–
t	time	sec	s
T	transmissivity	ft^3/day-ft	m^3/d·m
u	well function argument	–	–
U	uplift force	lbf	N
v, v	velocity	ft/sec	m/s
V	volume	ft^3	m^3
w	moisture content	–	–
$W(u)$	well function	–	–
X	concentration of chemical in soil	ppb	μg/kg
y	aquifer thickness after drawdown	ft	m

Symbols

γ	specific weight	lbf/ft^3	n.a.
η	porosity	–	–
μ	absolute viscosity	lbf-sec/ft^2	Pa·s
μ	dynamic viscosity	n.a.	kg/(m·s)
ν	kinematic viscosity	ft^2/sec	m^2/s
ρ	density	lbm/ft^3	kg/m^3

Subscripts

0	initial
10	effective size for which 10% is finer
c	equilibrium
d	soil-water
e	effective or equilibrium
f	flow or flow channels
o	octanol or at well
oc	organic carbon
ow	octanol-water
p	equipotential drops
r	radius
s	solid
t	time or total
u	uniformity or uplift
v	void
w	water

Site Assessment & Remediation

Content in blue refers to the *NCEES Handbook*.

1. GROUNDWATER CLASSIFICATION

Groundwaters in the United States are characterized over a wide range of natural water quality, typically based on total dissolved solids (TDS). TDS levels may range from less than 100 mg/L (very high quality) to more than 10 000 mg/L (very salty). Given the diversity of groundwater quality, the EPA applies the following classification to groundwater resources.

- Class I: Special Groundwaters—groundwater resources with a high beneficial use (drinking water) and located in a potentially vulnerable setting

- Class II: Current and Potential Sources of Drinking Water and Having Other Beneficial Uses—all groundwaters that are neither Class I nor Class III

- Class III: Groundwaters Not Considered Potential Sources of Drinking Water and of Limited Beneficial Use—usually limited to waters with a TDS greater than 10 000 mg/L

2. SOLUTE EQUILIBRIUM CONCENTRATION

Chemicals may dissolve in solution with water to varying degrees. The water solubility of a chemical is commonly taken as the mass of that chemical that will dissolve in a given volume of water, expressed in milligrams per liter. Water solubility depends on temperature; groundwater is typically about 6°C to 8°C.

For some chemicals, water solubility is very high. Ethanol, for example, is infinitely soluble in water. For a large number of chemicals, however, solubility in water is limited by some upper value.

For organic chemicals, whenever the mass of the liquid chemical released into the groundwater exceeds the water solubility, the chemical will partition between the dissolved and nondissolved phases. The chemical in the dissolved phase is called the *solute*, and the chemical in the nondissolved phase is called *nonaqueous phase liquid*, or NAPL.

Motor vehicle fuels and chemicals formulated for many other uses are complex mixtures of a variety of organic chemicals. When these materials are released into the groundwater, the individual chemical constituents will partition into solute and NAPL according to their proportions in the mixture and their water solubility.

In such mixtures, the dissolved fraction of each individual chemical in the mixture can be estimated by

$$C = xS \qquad 46.1$$

In Eq. 46.1, C is the equilibrium concentration of the chemical in the mixture, x is the mole fraction, and S is the water solubility.

One chemical that is widely used as an additive in gasoline presents a particular problem because of its solubility characteristics. Methyl *tert*-butyl ether (MTBE) is used to enhance gasoline combustion and reduce exhaust emissions. MTBE is less readily biodegradable than many other gasoline constituents, and it has a solubility in water of 40 000 mg/L to 52 000 mg/L. Consequently, when gasoline containing MTBE mixes with water, the MTBE will be present in solution with the water at a much greater concentration than the other gasoline constituents.

Example 46.1

A chemical mixture released to the groundwater contains the following mix of chemicals.

chemical	% of mixture by mass	solubility in water (mg/L)	molecular weight (mg/mmol)
trichloroethene (TCE)	83	1100	131.5
tetrachloroethene (PCE)	8	150	166
trans-1,2-dichloroethene (*t*-1,2-DCE)	5	600	97
cis-1,2-dichloroethene (*c*-1,2-DCE)	4	800	97

Site Assessment & Remediation

Calculate the number of moles, the mole fraction, and the equilibrium concentration for each chemical.

Solution

For ease of calculation, assume 100 mg of the mixture consisting of 83 mg TCE, 8 mg PCE, 5 mg *t*-1,2-DCE, and 4 mg *c*-1,2-DCE. Using the equation for molecular weight, solve for the number of moles of each chemical.

Ideal Gas Mixtures

$$M = m/N$$

$$N = \frac{m}{M}$$

chemical	% of mixture by mass	molecular weight, M (mg/mmol)	number of moles, N (mmol)
TCE	83	131.5	0.63
PCE	8	166	0.048
t-1,2-DCE	5	97	0.052
c-1,2-DCE	4	97	0.041
			0.771

Calculate the mole fraction of each chemical.

Ideal Gas Mixtures

$$x_i = N_i / N$$

$$N = \sum N_i$$

$$x_i = N_i / \sum N_i$$

chemical	number of moles, N (mmol)	mole fraction, x
TCE	0.63	0.817
PCE	0.048	0.062
t-1,2-DCE	0.052	0.067
c-1,2-DCE	0.041	0.053
	0.771	1.0

Calculate the equilibrium concentration for each chemical.

$$C = xS$$

chemical	mole fraction, x	solubility in water, S (mg/L)	equilibrium concentration, C (mg/L)
TCE	0.817	1100	900
PCE	0.062	150	9.4
t-1,2-DCE	0.067	600	40
c-1,2-DCE	0.053	800	43

3. DISTRIBUTION COEFFICIENT

The distribution coefficient, K_d, is a measure of the affinity of a given chemical for soils that contain organic matter. Consequently, K_d is specific to site soil conditions. It is used to determine the velocity of the solute relative to the velocity of the groundwater.

K_d is the ratio of a chemical's concentration in soil, X, to its concentration in water, C.

Soil-Water Partition Coefficient $K_d = K_p$

$$K_d = X / C \qquad \textbf{46.2}$$

The *organic carbon partition coefficient*, K_{oc}, is the ratio of a chemical's concentration in the organic carbon component of a soil, C_{soil}, to its concentration in water, C_{water}.

Organic Carbon Partition Coefficient K_{oc}

$$K_{oc} = C_{\text{soil}} / C_{\text{water}} \qquad \textbf{46.3}$$

K_{oc} multiplied by the fraction of organic carbon in the soil, f_{oc}, is equal to K_d.

Soil-Water Partition Coefficient $K_d = K_p$

$$K_d = K_{oc} f_{oc} \qquad \textbf{46.4}$$

The *octanol-water partition coefficient*, K_{ow}, is the ratio of a chemical's concentration in the octanol phase of a two-phase octanol-water system, C_o, to its concentration in the aqueous phase, C_w.

Octanol-Water Partition Coefficient

$$K_{ow} = C_o / C_w \qquad \textbf{46.5}$$

Tables of values are available for the water solubility, organic carbon partition coefficient, and octanol-water partition coefficient of various chemicals. [Water Solubility, Vapor Pressure, Henry's Law Constant, K_{oc}, and K_{ow} Data for Selected Chemicals]

K_d can be determined in the laboratory using bench-scale batch adsorption isotherm tests with the solute and the media. This is the best way to measure K_d because it is specific to the solute and the media, and it applies to inorganic and organic chemicals. K_d is the slope of the isotherm plot.

Example 46.2

Calculate the distribution coefficient for trichloroethene (TCE) in groundwater with soil having an organic carbon content of 210 mg/kg.

Solution

Express the organic carbon content as a decimal fraction.

$$f_{oc} = \left(210 \ \frac{mg}{kg}\right)\left(10^{-6} \ \frac{kg}{mg}\right) = 0.00021$$

From a table of values for the organic carbon partition coefficient, the value of K_{oc} for TCE is 152 mL/g. [Water Solubility, Vapor Pressure, Henry's Law Constant, K_{oc}, and K_{ow} Data for Selected Chemicals]

Calculate the distribution coefficient.

Soil-Water Partition Coefficient $K_d = K_p$

$$K_d = K_{oc} f_{oc}$$

$$= \left(\left(152 \ \frac{mL}{g}\right)\left(1 \ \frac{cm^3}{mL}\right)\right)(0.00021)$$

$$= 0.032 \ cm^3/g$$

4. GENERAL CHARACTERISTICS OF NAPL

A *nonaqueous phase liquid* (NAPL) is any material that can exist in a separate liquid phase from groundwater. If a material is released in quantities greater than the material's solubility in water, it will exist as an NAPL. Oil is the obvious example, but compounds such as paint thinner can be nonaqueous phase at higher concentrations.

If the NAPL has a specific gravity greater than 1.0 (that is, a density greater than water), it is called a *dense nonaqueous phase liquid*, or DNAPL. If the NAPL has a specific gravity less than 1.0, it is called a *light nonaqueous phase liquid*, or LNAPL.

For example:

- Acetone has infinite solubility, so it will never exist as an NAPL.

- Benzene has a specific gravity of 0.88 and a solubility of 1750 mg/L. At concentrations greater than 1750 mg/L, it will exist as an LNAPL.

- Carbon tetrachloride has a specific gravity of 1.59 and a solubility of 757 mg/L. At concentrations greater than 757 mg/L, it will exist as a DNAPL.

Gasoline, diesel fuel, lubricating oils, and other fuel oils are mixtures of compounds that may contain both DNAPLs and LNAPLs. However, the LNAPLs occur as a greater percent of the mixture, and the result is that fuel and lubricating oils almost always exist as LNAPL.

5. CHEMICAL ANALYSIS AND FINGERPRINTING

The occurrence and concentration of organic chemicals commonly associated with groundwater contamination are determined using analytical instruments such as gas chromatographs (GC) and mass spectrophotometers (MS) coupled with GCs. Samples are identified by the purge time required to pass through the instrument, creating a chromatograph, or compared with the mass spectrum of a known chemical.

Using chromatographs created for mixtures of known compounds (gasoline, diesel fuel, jet fuel, or BP regular gasoline, Shell premium gasoline, etc.), a comparison against a groundwater sample containing an unknown contaminant can identify the type of contaminant and/ or its origin. Comparisons of this type can also be used to evaluate the age of the release. This procedure is called *fingerprinting*. For example, assume the gas stations for two different fuel suppliers were located on opposite corners of an intersection and that gasoline was discovered in groundwater wells nearby. By collecting a sample of the gasoline from each station and comparing the resulting chromatographs to the chromatographs resulting from an analysis of the groundwater sample, a match between one or the other of the gasoline chromatographs with the groundwater sample chromatograph will identify the source.

Fingerprinting can also be used to help determine how long ago a chemical mixture (gasoline, diesel fuel) was released to groundwater. Comparing two chromatographs of samples collected from the same source and widely spaced time intervals, or of the chemical mixture and a groundwater sample contaminated with the chemical mixture, will show the change in relative concentrations of chemical components with increased evaporation. Changes in the chromatograph from evaporation and other physical, chemical, and biological reactions result in weathering. Fresh samples will show higher and more distinct peaks in the chromatograph than will weathered samples, although the peaks will occur at the same purge times.

6. NOMENCLATURE

C	concentration in water	mg/L	mg/L
C	constant	various	various
C_{soil}	concentration of chemical in soil	μg/kg	μg/kg
C_{water}	concentration of chemical in aqueous phase	mg/L	mg/L
f_{oc}	fraction of organic carbon in the soil	decimal	decimal
K	coefficient	–	–
m	mass	lbm	kg
M	molecular weight	lbm/lbmol	kg/kmol
N	number	–	–
S	water solubility	mg/L	mg/L
x	mole fraction	–	–
X	concentration of chemical in soil	mg/L	mg/L

Subscripts

d	distribution coefficient
oc	organic carbon partition
ow	octonal-water partition
o	octanol phase
w	aqueous phase

47 Site Assessment and Characterization

1. SOLUTES IN GRANULAR MEDIA

Solute transport, or *mass transport*, occurs when solutes are transported by the bulk movement of water. This transport is called *advection*.

If the solute is nonreactive (that is, if there is no physical, chemical, or biological reaction within the water or between the solute and the water), it will move at the average bulk velocity of the groundwater in which it is dissolved. The groundwater average bulk velocity, called the *rate of advection*, can be found using Darcy's law.

$$\text{Darcy's Law}$$
$$Q = -KA(dh/dx) \qquad \textbf{47.1}$$

Figure 47.1 shows the plan view of an idealized groundwater flow net associated with an identified contamination source. The flow line is shown propagating from the source, representing the confined path of solute flow. Under conditions of ideal flow, as Fig. 47.1 shows, the solute will follow the flow line without deviation (with no spreading of the solute away from the flow line).

Figure 47.1 Ideal Solute Flow Path

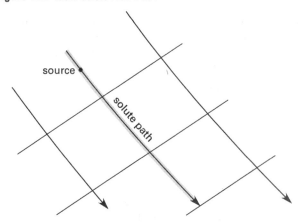

Under actual conditions, however, the solute spreads out as it moves away from the source. This spreading may occur in all three dimensions, but it is most pronounced longitudinally, with the significance of vertical spreading being influenced by solute density. This spreading phenomenon is called *hydrodynamic dispersion*. It is a combination of *mechanical dispersion* (commonly referred to simply as *dispersion*) and *diffusion*.

Dispersion and diffusion cause the dilution of the solute. In more permeable materials, dispersion is far more significant than diffusion; diffusion becomes more important in less permeable soils and may occur even where groundwater flow is zero.

Dispersion is caused by mixing within the pore spaces and by differential flow in flow channels. Assuming that the source is continuous, the highest concentrations of solute will occur near the flow line propagating from the source. Concentrations of the solute will decrease toward the lateral and longitudinal boundaries. The effects of dispersion are illustrated in Fig. 47.2.

Figure 47.2 Effect of Dispersion on Plume Propagation

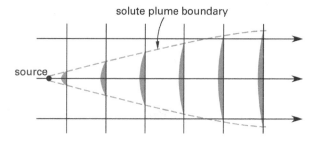

Diffusion accounts for the molecular spreading of the solute across the boundary between relatively contaminated water and clean water. The fluid does not have to be moving for diffusion to occur.

Diffusion is most pronounced in its impact at the boundary. It tends to flatten the concentration gradient across the boundary, which increases the time between breakthrough and saturation at the boundary. The solute front, therefore, creeps. The overall result is a plume that disperses as it moves away from the source, with a diffusion of the concentration front across the plume boundary. The plume configuration is represented by isoconcentration contour lines in Fig. 47.3.

Figure 47.3 *Effect of Diffusion on Plume Propagation*

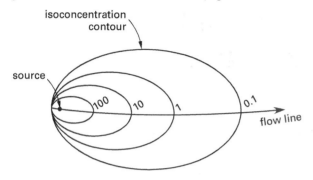

In soils with lower permeabilities, such as those containing significant amounts of silts and clays, diffusion may dominate contaminant transport away from the source, especially where the groundwater flow is small.

Fractured Media Flow

Although equally important as porous media flow, flow through fractured rock media is much more complex. The phenomena of dispersion and diffusion are present in fractured media flow, and they produce similar effects on contaminant distribution when considered on a large scale. The contaminant will diffuse into the porous rock, will diffuse at the boundary between the contaminant and fresh water in the fracture, and will disperse as the contaminant moves through multiple fractures.

Fractured rock presents additional problems when dead-end fractures are considered. When contaminated groundwater enters these fractures, it becomes essentially nonmobile, and long periods are allowed for diffusion of contaminant into the porous rock. These dead-end fractures can later serve as contaminant sources as cleanup actions progress or when flow conditions change.

2. VERTICAL AND HORIZONTAL HYDRAULIC CONDUCTIVITY

The overall vertical hydraulic conductivity, K_v avg, of layered strata is calculated from the individual vertical hydraulic conductivities of the layers using Eq. 47.2. b is the total aquifer thickness; K_{vm} and b_m are the vertical hydraulic conductivity and thickness, respectively, of the mth layer.

Overall Vertical Hydraulic Conductivity (Perpendicular to Layering)

$$K_v \, \text{avg} = \frac{b}{\sum\limits_{m=1}^{n} \dfrac{b_m}{K_{vm}}} \qquad \textit{47.2}$$

The average horizontal hydraulic conductivity, K_h avg, of layered strata is similarly calculated using Eq. 47.3. K_{hm} is the vertical hydraulic conductivity of the mth layer.

Average Horizontal Conductivity (Parallel to Layering)

$$K_h \, \text{avg} = \sum_{m=1}^{n} \frac{K_{hm} b_m}{b} \qquad \textit{47.3}$$

Example 47.1

Boring logs show interbedded soil layers with the characteristics shown. What is the overall vertical hydraulic conductivity for the aquifer?

layer	thickness (cm)	vertical hydraulic conductivity (cm/s)
1	70	0.017
2	109	0.20
3	88	0.70
4	46	0.028

Solution

Use Eq. 47.2 to calculate the overall vertical hydraulic conductivity.

Overall Vertical Hydraulic Conductivity (Perpendicular to Layering)

$$K_v \, \text{avg} = \frac{b}{\sum\limits_{m=1}^{n} \dfrac{b_m}{K_{vm}}}$$

$$= \frac{70 \text{ cm} + 109 \text{ cm} + 88 \text{ cm} + 46 \text{ cm}}{\dfrac{70 \text{ cm}}{0.017 \, \frac{\text{cm}}{\text{s}}} + \dfrac{109 \text{ cm}}{0.20 \, \frac{\text{cm}}{\text{s}}} + \dfrac{88 \text{ cm}}{0.70 \, \frac{\text{cm}}{\text{s}}} + \dfrac{46 \text{ cm}}{0.028 \, \frac{\text{cm}}{\text{s}}}}$$

$$= 0.041 \text{ cm/s}$$

3. VERTICAL DISTRIBUTION

Vertical distribution (assuming homogeneous media) is influenced by the density of the solute. When the density is less than or only slightly greater than that of water, the solute will tend to remain near the groundwater surface. As the solute's density increases to values greater than water, the degree of vertical migration will increase. Where varying permeability conditions are encountered (heterogeneous media), the solute may be directed upward or downward.

The conditions affecting vertical distribution of the contaminant are particularly relevant where layered sequences of multiple aquifers are encountered. Figure 47.4, Fig. 47.5, and Fig. 47.6 show some possible scenarios.

The groundwater flow between aquifers can be calculated for a unit area of aquifer by applying a variation of Darcy's law.

$$Q = K_v A \frac{dh}{dx} \qquad \textit{47.4}$$

Figure 47.4 *Little or No Vertical Distribution*

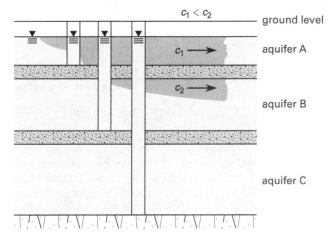

static conditions (no pumping)
solute density ≈ water density
competent aquitard

Figure 47.5 *Moderate Vertical Distribution*

static conditions
solute density > water density
leaky aquitard

Figure 47.6 *High Vertical Distribution*

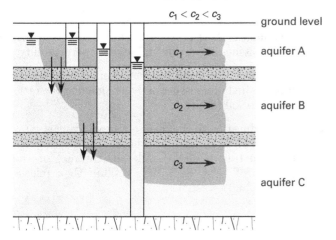

pumped conditions
solute density > water density
leaky aquitard

Example 47.2

Three layered aquifers and the aquitards between them are shown. The vertical hydraulic conductivity, K_v, for each layer is as shown.

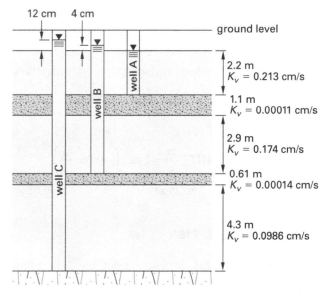

(not to scale)

Assume a unit distance, L, between wells of 1 m. What is the unit groundwater flow and direction between aquifers B and C?

Solution

Use Eq. 47.1 to calculate the overall vertical hydraulic conductivity.

Overall Vertical Hydraulic Conductivity (Perpendicular to Layering)

$$K_p \, \mathrm{avg} = \frac{b}{\displaystyle\sum_{m=1}^{n} \frac{b_m}{K_{cm}}}$$

$$= \frac{\left(\dfrac{290 \ \mathrm{cm} + 61 \ \mathrm{cm} + 430 \ \mathrm{cm}}{100 \ \dfrac{\mathrm{cm}}{\mathrm{m}}}\right)}{\dfrac{290 \ \mathrm{cm}}{0.174 \ \dfrac{\mathrm{cm}}{\mathrm{s}}} + \dfrac{61 \ \mathrm{cm}}{0.00014 \ \dfrac{\mathrm{cm}}{\mathrm{s}}} + \dfrac{430 \ \mathrm{cm}}{0.0986 \ \dfrac{\mathrm{cm}}{\mathrm{s}}}}$$

$$= 0.0000177 \ \mathrm{m/s}$$

Calculate the hydraulic gradient.

$$\frac{dh}{dx} = \frac{\Delta h_{\mathrm{B}} - \Delta h_{\mathrm{C}}}{L} = \frac{\left(\dfrac{4 \text{ cm} - 12 \text{ cm}}{100 \ \dfrac{\text{cm}}{\text{m}}} \right)}{1 \text{ m}}$$
$$= -0.08 \text{ m/m}$$

Use Darcy's law to calculate the unit groundwater flow rate.

Darcy's Law

$$Q = -KA(dh/dx)$$
$$= \left(-0.0000177 \ \frac{\text{m}}{\text{s}} \right)(1 \text{ m}^2)\left(-0.08 \ \frac{\text{m}}{\text{m}} \right)$$
$$= 1.4 \times 10^{-6} \text{ m}^3/\text{s·m}^2$$

4. HETEROGENEITIES

When isoconcentration lines are drawn on a site plan, the inherent assumption is that the contamination exists in convenient patterns with smoothed curves and even distribution. In reality, the contaminant transport is significantly influenced by soil heterogeneities. The flow patterns are influenced by macroscopic media differences such as interbedded lenses of varying sizes, as well as by microscopic media differences brought about by slight variations in clay and silt content.

Microscopic influences are difficult to predict and require careful investigation to identify. The primary characteristic of microscopic influences is *fingering*, which results in a jagged horizontal and vertical boundary and creates fingers of high-velocity flow channels. Figure 47.7 illustrates interbedding from macroscopic influences.

Figure 47.7 *Interbedding from Macroscopic Influences*

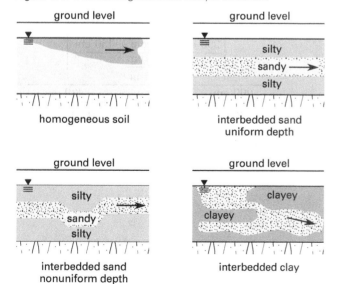

5. NAPL PLUME PROPAGATION

On being released into the subsurface, a nonaqueous phase liquid (NAPL) will flow under the influence of gravity and pressure forces, and the NAPL will flow downward through the unsaturated zone as a distinct liquid. This will be true both for a light NAPL (LNAPL) and for a dense NAPL (DNAPL).

Any pore water will be displaced by the migrating NAPL. The degree to which the pore space is filled with the NAPL is called the *saturation ratio*, r_{sat}. For a completely filled pore, the saturation ratio is 1.0, and for a completely empty pore, the saturation ratio is 0.0. For a saturation ratio of 1.0, the fraction of the soil volume occupied by NAPL is equal to the soil porosity, n_e. The soil volume occupied by the NAPL is

$$V_{\mathrm{NAPL}} = r_{\mathrm{sat}} n_e \qquad \textbf{47.5}$$

As vertical migration progresses, lateral spreading will occur because of capillary forces and interbedding of sand, silts, and clays within the media. With continuing vertical migration, the NAPL is left behind as residual liquid (*residual saturation*) trapped in the pore spaces. The ability of the media to affect residual saturation is known as *retention capacity*. The retention capacity influences the depth of NAPL penetration, D, into the vadose zone.

Vadose Zone Penetration

$$D = \frac{R v V}{A} \qquad \textbf{47.6}$$

If the NAPL contains a volatile fraction, some of the NAPL may volatilize and form a vapor plume. The boundaries of the vapor plume may be far greater than those of the NAPL plume. If a sufficiently large volume of NAPL is released, the NAPL will eventually encounter the capillary fringe. What occurs at the capillary fringe depends on whether the NAPL is dense or light. However, residual saturation will result within the unsaturated zone, including the capillary fringe, and may act as an ongoing source of dissolved contaminant to the groundwater for some time.

Factors affecting NAPL migration in the subsurface are

- the volume of NAPL released (large versus small), including issues of wetting and relative permeability

- the area over which infiltration occurs (broad versus confined), including issues of dispersion and the ratio of liquid to soil (dilution)

- the time duration of the release (chronic versus catastrophic), which determines whether the limiting factor is the rate of NAPL release (chronic) or the capacity of the soil to retain the release (catastrophic)

Site Assessment & Remediation

K	hydraulic conductivity	ft/sec	m/s
L	length	ft	m
Q	discharge or discharge rate	ft³/sec	m³/s
r_{sat}	saturation ratio	decimal	decimal
Rv	retention capacity and viscosity constant	–	–
V	volume	ft³	m³
x	horizontal distance	ft	m

- properties of the NAPL (viscosity, density, solubility, volatility), including issues of migration rate, migration direction (horizontal and vertical), interaction with the water table, formation of a dissolved (solute) plume, and formation of a vapor plume

- properties of the media (moisture content, porosity, permeability, particle shape, interbedding), including issues of soil retention capacity, ultimate boundaries of NAPL plume, wetting (pre-wetting), and migration rate

- subsurface flow conditions (gradient, water table fluctuations), including issues of migration rate (gradient establishes slope) and volume of contacted soil (water table fluctuation contacts more soil)

A fluctuating water table influences the distribution of the LNAPL by contacting it with an increased soil volume as the water table moves up and down. When a monitoring well is constructed into an aquifer where the LNAPL is present above the water table, direct measurements of the water table elevation are not possible. At some location within the LNAPL plume, the LNAPL will exist above the residual saturation and will, therefore, be mobile. As the LNAPL flows into the monitoring well to the water table, the weight of the LNAPL will depress the surface of the water in the monitoring well.

This creates two conditions that can lead to false conclusions if not understood. First, because the NAPL in the well depresses the water table, the true elevation of the water table will be higher than the measured value. Second, because the LNAPL has drained into the open well through the screen, the LNAPL thickness measured in a monitoring well will be greater than the actual thickness of the mobile LNAPL in the vadose zone.

Subscripts

avg	average
m	mth layer
h	horizontal
NAPL	nonaqueous phase liquid
v	vertical

6. NOMENCLATURE

A	area or cross-sectional area of flow	ft²	m²
b	thickness	ft	m
D	maximum depth of penetration	ft	m
h	height or hydraulic head	ft	m
n_e	porosity	decimal	decimal
K	hydraulic conductivity	ft/sec	m/s
K_h avg	average horizontal hydraulic conductivity	ft/day	m/d
K_{hm}	horizontal hydraulic conductivity of the mth layer	ft/day	m/d
K_v avg	average vertical hydraulic conductivity	ft/day	m/d
K_{vm}	vertical hydraulic conductivity of the mth layer	ft/day	m/d

48 Risk Assessment

1. INTRODUCTION

The study of risk involves two aspects, risk analysis and risk management, that are related but distinctly different. *Risk analysis*, also known as *risk assessment*, is a *technical* evaluation of the overall probability of an undesirable consequence of exposure to a hazard. *Risk management* is a *political* (or management) decision regarding the selection of actions to be taken in response to exposure to a risk. While an environmental engineer may make the key decisions involved in risk analysis, upper-level managers, politicians, or attorneys may be the key decision-makers in risk management decisions, which typically involve choosing between social, economic, and technical alternatives or impacts. This section will cover only risk analysis.

Risk analysis can be applied to many aspects of the human-built and natural environments. Our everyday lives involve numerous risk analysis evaluations that we take for granted; often we are unaware of the mental processes involved. Without much effort, we quickly analyze whether we can make it through an intersection, or whether another cigarette or doughnut is an acceptable risk to our health. These risks are known as *assumed risks* because we "assume," or take, them and accept the consequences, even if we are not fully aware of what those consequences are.

Assumed risks are collectively known as the *background risk*, meaning the sum of the risks to which we are exposed excluding the risk of additional activities being evaluated. *Incremental risk* is the risk associated with the additional activities being evaluated, and *total risk* is the sum of the background risk and the incremental risk. Incremental risk is also known as *imposed risk* because the risk associated with the activity may be involuntarily imposed on the exposed population.

Environmental engineers readily understand that judgmental factors are involved in making risk management decisions. Policy and remediation alternatives must be weighed with social, economic, and political factors to come up with a solution. It is important to realize also that there is always a level of uncertainty in the risk analysis process. However, to the extent possible, the risk analysis process attempts to use facts to evaluate the environmental and human health effects of exposure to hazardous materials and situations.

Quantitative risk analysis determines a quantitative, numerical estimate of risk associated with an activity. (This is contrasted with a *qualitative analysis* whereby the potential risk may be identified but its magnitude is not estimated.) Quantitative risk analysis is typically described as a four-step process:

step 1: hazard identification

step 2: toxicity assessment

step 3: exposure assessment

step 4: risk characterization

The risk analysis process involves the application of scientific knowledge and engineering principles in the areas of toxicology, chemistry, and fate and transport modeling. The four-step process is applied by the EPA and state regulatory agencies to the remediation of hazardous waste sites in order to determine appropriate levels of cleanup and to select alternative remedies.

2. HAZARD IDENTIFICATION

Hazard identification involves identifying substances and determining that they are indeed health hazards. A hazard is not the same as a risk. *Hazard* refers to the capability of the substance to cause an adverse effect on humans, aquatic life, or any other organism. The hazard posed by the substance depends on its toxicity, mobility, and persistence, and on the measures taken to prevent its release to the environment (e.g., noncorrosive containers, spill containment). By contrast, *risk* is the probability of the adverse effect occurring. Exposure, or potential exposure, to the hazard is required for a risk to be present.

Although an environmental engineer will most commonly deal with hazard identification and risk analysis in situations involving the remediation of hazardous waste sites, another important situation occurs in the design and operation of industrial processes that involve hazardous substances. Process hazard analysis should be a part of the design of industrial processes involving hazardous substances. The Occupational Safety and Health Administration (OSHA) has identified the procedure for performing a process hazard analysis.

For a typical risk analysis of chemicals encountered, the first step is to identify the chemicals that are present at the site and the media that are affected. Samples are collected to determine the contaminant levels in the air, groundwater, surface water, soils, and sediments associated with the site. The concentration and distribution of the chemicals present in each sample are determined. Moreover, the ways by which the chemicals can move off the site to expose the receptor population need to be known in order to perform the subsequent fate and transport analysis.

A large number of chemicals may be identified at a site, so an initial screening is necessary to reduce the number of chemicals that must be evaluated to a manageable number (perhaps 10 to 20). The objective of the initial screening is to identify the chemicals that will account for 99% of the risk for both carcinogens and noncarcinogens. "99% of the risk" does not mean 99% of the chemicals; only a few chemicals might account for this level of risk.

The initial screening procedure involves six steps.

step 1: Sort the chemicals by medium, and then into carcinogens and noncarcinogens.

step 2: Calculate the mean and maximum concentrations for each chemical.

step 3: Determine the slope factors (SF) for carcinogens and the reference doses (RfD) for noncarcinogens. (See Table 48.1.)

step 4: Determine the toxicity scores for each chemical by medium.

step 5: Rank the chemicals by toxicity scores for each medium.

step 6: Select the chemicals comprising 99% of the total toxicity score for each medium.

Beyond initial screening based on toxicity score, the chemicals present should also be analyzed for persistence, mobility, and other factors that may increase or decrease the hazard. Chemicals that are highly persistent and readily mobile by the pathways present at the site should be retained for further analysis.

In step 4, the toxicity scores are determined in the initial screening as given in Eq. 48.1 and Eq. 48.2. Strictly, *toxicity score*, TS, contains units, though it is generally reported unitless.

For noncarcinogens,

$$TS_n = \frac{C_{max}}{RfD} \qquad 48.1$$

For carcinogens,

$$TS_c = C_{max}(SF) \qquad 48.2$$

3. TOXICITY ASSESSMENT

The purpose of toxicity assessment is to quantify the hazard of the substances. Although identified as a separate step of risk analysis, in practice at least part of the toxicity assessment must be done concurrently with hazard identification because the reference doses (noncarcinogens) and the slope factors (carcinogens) are used in the initial screening.

In most hazardous waste site remediation projects, EPA toxicity indices are used. The toxicity data are available in the Integrated Risk Information Systems (IRIS) database. Where data are missing or ambiguous, the regulatory agency (state or EPA) will usually provide toxicity indices.

The toxicity indices are different for noncarcinogens and carcinogens. Noncarcinogens are assigned thresholds below which the toxicant dose does not result in an observable adverse health effect. The threshold, known as the *no observable adverse effect level* (NOAEL), is used to calculate a *reference dose* (RfD). The RfD is calculated by dividing the NOAEL by uncertainty factors of several orders of magnitude. The IRIS database contains oral and inhalation RfDs for more than 500 chemicals.

The RfD is the lifetime dose that a healthy person could be exposed to daily (*chronic daily intake*, CDI) without an appreciable risk. The RfD is expressed in mg of chemical per kg body mass per day (mg/kg·d). The *hazard index* is used to analyze the risk for noncarcinogens. It is the ratio of the actual dose to the RfD, which if greater than 1.0 represents the possibility of an adverse health effect from the given level of exposure.

Carcinogens are believed not to have thresholds, and any exposure is believed to cause a *response* (mutation of the human DNA). Since there is no level of carcinogens that can be considered safe for continued human exposure, the EPA method extrapolates the dose-response curve from high doses to very low exposures in order to specify what are known as *slope factors* (SFs). (Slope factors are also known as *potency factors* and *carcinogen potency factors*.) The SF is the probability of cancer produced by lifetime exposure to 1.0 mg/kg·d of the carcinogen. The SF is selected at the 95% upper confidence limit of the dose-response curve. The SF is the inverse of dose and is expressed as $(mg/kg \cdot d)^{-1}$. The risk (probability) is the product of the SF and the dose.

Extracts from the EPA's IRIS database are given in Table 48.1. The EPA's carcinogenicity classification system is given in Table 48.2.

Table 48.1 Typical Toxicological Data from IRIS Database*

chemical	class	RfD (oral) (mg/kg·d)	RfD (inhalation) (mg/kg·d)	SF (oral) (mg/kg·d)$^{-1}$	SF (inhalation) (mg/kg·d)$^{-1}$
benzene	A	–	–	0.0290	0.0290
1,2-dichloroethane	B2	–	–	0.091	0.091
1,2-dibromoethane (EDB)	B2	–	–	85	0.77
1,3-dichloropropene	B2	0.0003	0.00572	no data	no data
acrylamide	B2	0.0002	–	4.50	4.50
bromodichloromethane	B2	0.002	no data	0.062	no data
chloroform	B2	0.01	–	0.0061	0.081
dieldrin	B2	0.00005	no data	16	16
chlorobenzene	D	0.02	pending	no data	no data
endrin	D	0.0003	no data	no data	no data
nitrobenzene	D	0.0005	–	no data	no data
pentachlorobenzene	D	0.0008	no data	no data	no data
pyrene	D	0.003	no data	no data	no data
toluene	D	0.2	1.40	–	–

*Source: Environmental Protection Agency

Table 48.2 EPA Carcinogenicity Classification System*

group	description
A	human carcinogen
B1 or B2	probable human carcinogen:
	B1 indicates that human data are available.
	B2 indicates sufficient evidence in animals but inadequate or no evidence in humans.
C	possible human carcinogen
D	not classifiable as to human carcinogenicity
E	evidence of noncarcinogenicity for humans

*Source: Environmental Protection Agency

Example 48.1

The project engineer for cleanup of a hazardous waste site in eastern Oregon has obtained data on chemicals present at the site and their potential pathways of exposure to residents in the area. She needs to perform an initial screening to determine the chemicals or surrogate chemicals that will be used in the exposure assessment. The project engineer notes that she will need to develop a slope factor for 1,3-dichloropropene for the oral route since no data is given in the IRIS database. She reviews available toxicity studies and, with the help of a toxicologist, determines that a slope factor of 8×10^{-3} (mg/kg·d)$^{-1}$ would be appropriate. Which of the chemicals noted below should she select for subsequent analysis for the groundwater route?

chemical	groundwater mean (mg/L)	groundwater max (mg/L)	air mean (mg/m^3)	air max (mg/m^3)
carcinogens				
1,2-dichloroethane	0.12	0.92	61	103
1,2-dibromoethane (EDB)	0.034	0.096	4.3	7.9
1,3-dichloropropene	9.0×10^{-4}	9.9×10^{-4}	5.8	9.9
dieldrin	0.4	0.44	ND	ND
noncarcinogens				
chlorobenzene	60	73	0.33	4.6
endrin	6.1	8.1	0.041	0.093
pyrene	9	48	ND	ND
toluene	0.8	4.3	0.71	0.98

Solution

Steps 1 and 2 of the screening procedure appear in the problem statement.

step 3: Determine the SFs for carcinogens and the RfDs for noncarcinogens.

The reference doses and slope factors are determined from the EPA IRIS database, extracts of which are given in Table 48.1. The problem statement gives a slope factor of 8×10^{-3} (mg/kg·d)$^{-1}$.

step 4: Determine the toxicity scores for each chemical by medium.

step 5: Rank the chemicals by toxicity scores by medium.

In this case, only the groundwater route is being considered.

For noncarcinogens, $TS_n = C_{max}/RfD$.

For carcinogens, $TS_c = C_{max}(SF)$.

The calculations are shown in the following table for steps 4 and 5.

chemical	noncarcinogens			
	C_{max} (mg/L)	RfD, oral (mg/kg·d)	TS_n (kg·d/L)	rank
endrin	8.1	3×10^{-4}	27 000	1
pyrene	48	3×10^{-3}	16 000	2
dieldrin	0.44	5×10^{-5}	8800	3
chlorobenzene	73	0.02	3650	4
toluene	4.3	0.2	21.5	5
1,3-dichloro-propene	9.9×10^{-4}	3×10^{-4}	3.3	6
		total	55 474.8	

chemical	carcinogens			
	C_{max} (mg/L)	SF, oral $(mg/kg·d)^{-1}$	TS_c	rank
1,2-dibromo-ethane (EDB)	0.096	85	8.16	1
dieldrin	0.44	16	7.04	2
1,2-dichloro-ethane	0.92	9.1×10^{-2}	0.08372	3
1,3-dichloro-propene	9.9×10^{-4}	8×10^{-3}	7.92×10^{-6}	4
		total	15.2837	

Note that dieldrin and 1,3-dichloropropene show both carcinogenic and noncarcinogenic effects, so they must be analyzed for both effects.

step 6: Select the chemicals comprising 99% of the total toxicity score by medium.

The chemicals ranked 1 through 4 comprise 99% of the total toxicity score for noncarcinogens.

The correct answer for noncarcinogens is endrin, pyrene, dieldrin, and chlorobenzene.

The chemicals ranked 1 and 2 comprise 99% of the total toxicity score for carcinogens.

The correct answer for carcinogens is 1,2-dibromoethane (EDB) and dieldrin.

4. EXPOSURE ASSESSMENT

Exposure assessment is the process of tracking a substance from its source to the potential receptors. This is by no means a trivial undertaking and can quickly become very complicated due to the multiplicities of chemicals, release mechanisms, exposure pathways, fate and transport mechanisms, and receptors, among other things. The general approach is to formulate conservative but realistic scenarios of the mechanisms and processes involved, and then to quantitatively evaluate the intake (exposure) of the receptor. Factors to consider for hazardous waste sites often include workers at industrial sites or other nearby activities, trespassers, residential use, recreational use, and construction of remediation elements. The necessary components of the scenarios are described below.

Sources

The potential sources of chemical releases must first be identified. In some cases, this may be very obvious, such as a pile of drums. In other instances, more evaluation is required, such as to identify buried wastes, to determine if a lagoon has an impermeable liner, or to delineate contaminated soil areas. The concentrations and total masses of the chemicals in these sources must be determined or estimated.

Release Mechanisms

The release mechanisms must be identified. Some may be obvious, such as visible leakage from corroded drums. Others will require more investigation to determine, such as volatilization from open pools of chemicals, leakage from the bottom of a lagoon, or spillage from tanks or ponds. Contaminated soil particles can be windblown from the site, and waste may be burned, releasing chemicals with the combustion products. Leachate may be generated from precipitation onto waste piles or burial areas. Stormwater may wash contaminated soil from a site.

Transport

All of the relevant *transport mechanisms* associated with the release mechanisms must be identified in the scenario. Transport mechanisms may include air transport of volatilized chemicals, groundwater transport of leakage, surface water transport of leachate and contaminated storm water, and many others.

Transfer

Transfer means the exchange of contaminants from one transport mechanism to another. For example, leachate may flow into surface water that subsequently percolates into the soil and reaches groundwater. Fugitive dust might be blown from the site and settle in surface water. Chemicals may volatilize from surface water or groundwater, or be adsorbed onto the soils or sediments through which they pass. Plants that are used as food for humans or animals can take up chemicals. Precipitation may cause chemicals in soil to dissolve and be transported in runoff. Chemicals may be washed out of the air by precipitation, and airborne particles may settle onto lawns and gardens.

Transformation

Transformation occurs when a chemical is transformed into different substances that may be more or less toxic. While many organic compounds biodegrade to less toxic constituents, others (such as vinyl chloride) may change to more potent products. Photochemical degradation and oxidation by atmospheric ozone are additional examples of transformation.

Exposure Point

The *exposure point* is the location where the receptor and the chemical make contact or have the potential to do so. Exposure points include the original site, a water supply well for a community or residence, planting areas with soil contaminated from airborne particles, and surface water where contaminated fish are caught, to name a few. Identification of the receptor population needs to be done concurrently with identification of the exposure point.

Receptor Population

The *receptor population* is the group of humans that could potentially be exposed at the exposure points. Both existing and future populations need to be identified. Receptors at high risk, such as the elderly and children, need to be identified. The remediation site workers must also be included in the receptor population.

Exposure Point Concentrations

Major effort may be required to calculate the concentrations of the selected chemicals at the exposure points. All pathways and contaminants must be evaluated for present and future conditions as hypothesized in the scenarios. The support of an experienced modeler is essential in performing fate and transport modeling.

Receptor Doses

The last part of an exposure assessment is to calculate the receptor doses. This involves determining the quantity of the medium to which the receptor is exposed and multiplying it by the exposure point concentration. If the receptor is exposed to the same contaminant from more than one pathway, the total dose for that contaminant is the sum of the dose by each pathway. Although *chronic exposure* is the usual case, *acute exposure*, such as from a spill to surface water, will need to be evaluated also.

The *receptor dose* is the total mass of contaminant taken into the body divided by the body mass of the receptor and averaged over the days on which the risk factor is

based. For noncarcinogens, the averaging time is the same as the exposure period, while for carcinogens the averaging time is 70 yr to correspond to lifetime risk.

The chronic daily intake (CDI) can be calculated with the following equations.

ingestion in drinking water:

$$CDI = \frac{(CW)(IR)(EF)(ED)}{(BW)(AT)} \qquad \text{Exposure} \atop 48.3$$

ingestion while swimming:

$$CDI = \frac{(CW)(CR)(ET)(EF)(ED)}{(BW)(AT)} \qquad \text{Exposure} \atop 48.4$$

ingestion of chemicals in soil:

$$CDI = \frac{(CS)(IR)(CF)(FI)(EF)(ED)}{(BW)(AT)} \qquad \text{Exposure} \atop 48.5$$

inhalation of airborne chemicals:

$$CDI = \frac{(CA)(IR)(ET)(EF)(ED)}{(BW)(AT)} \qquad \text{Exposure} \atop 48.6$$

ingestion of contaminated fruits, vegetables, fish, and shellfish:

$$CDI = \frac{(CF)(IR)(FI)(EF)(ED)}{(BW)(AT)} \qquad \text{Exposure} \atop 48.7$$

Example 48.2

For the situation given in Ex. 48.1, the project engineer and modeler have developed five scenarios for the release, transport, transfer, transformation, and exposure point concentrations for potential receptors. They have determined the exposure point concentrations for two of the selected contaminants, as follows.

exposure point	receptor	pathway	chemical	concentration
residential water supply well	adult	groundwater, oral	dieldrin	0.36 mg/L
			chlorobenzene	36 mg/L
air in residence	adult	volatilization of groundwater, inhalation	dieldrin	no data available
			chlorobenzene	2.4 mg/m^3

The engineer and modeler estimate that natural attenuation will reduce the concentrations of the indicated chemicals to nondetectable levels after 20 years and estimate the exposure duration to be 350 days per year. Estimate ingestion/inhalation rate (IR) factors are 2 L/d adult water consumption and 20 m^3/d adult air consumption. What are the CDIs for each toxicant?

Solution

For water supply well exposure,

IR = 2 L/d adult water consumption

EF = 350 d/yr

ED = 75 yr for carcinogens

ED = 20 yr for noncarcinogens

BW = 70 kg

AT = ED × 365 d/yr

For dieldrin, a carcinogen, the well exposure is

Exposure

$$CDI = \frac{(CW)(IR)(EF)(ED)}{(BW)(AT)}$$

$$= \frac{\left(0.36 \; \frac{mg}{L}\right)\left(2 \; \frac{L}{d}\right)\left(350 \; \frac{d}{yr}\right)(75 \; yr)}{(70 \; kg)\left((75 \; yr)\left(365 \; \frac{d}{yr}\right)\right)}$$

$$= 9.863 \times 10^{-3} \; mg/kg{\cdot}d \quad (0.010 \; mg/kg{\cdot}d)$$

For chlorobenzene, a noncarcinogen, the well exposure is

Exposure

$$CDI = \frac{(CW)(IR)(EF)(ED)}{(BW)(AT)}$$

$$= \frac{\left(36 \; \frac{mg}{L}\right)\left(2 \; \frac{L}{d}\right)\left(350 \; \frac{d}{yr}\right)(20 \; yr)}{(70 \; kg)\left((20 \; yr)\left(365 \; \frac{d}{yr}\right)\right)}$$

$$= 0.9863 \; mg/kg{\cdot}d$$

For air exposure,

IR = 20 m^3/d adult air consumption

ET = 24 hr/d = 1

EF = 350 d/yr

ED = 75 yr for carcinogens

ED = 20 yr for noncarcinogens

BW = 70 kg

AT = ED × 365 d/yr

For dieldrin, there is no data available for the concentration in air, so the CDI for air exposure is unknown.

For chlorobenzene, the air exposure is

Exposure

$$CDI = \frac{(CA)(IR)(ET)(EF)(ED)}{(BW)(AT)}$$

$$= \frac{\left(2.4 \; \frac{mg}{m^3}\right)\left(20 \; \frac{m^3}{d}\right)(1)\left(350 \; \frac{d}{yr}\right)(20 \; yr)}{(70 \; kg)\left((20 \; yr)\left(365 \; \frac{d}{yr}\right)\right)}$$

$$= 0.6575 \; mg/kg{\cdot}d$$

The CDIs are

	CDI (mg/kg·d)	
chemical	water well	air
dieldrin	0.010	no data
chlorobenzene	0.9863	0.6575

5. RISK CHARACTERIZATION

The final step in the risk analysis process is to quantify the total noncarcinogenic and carcinogenic risks by combining the estimated exposures with the calculated potencies.

For noncarcinogens, the risk is characterized by the *hazard index*, which is the ratio of the dose to the reference dose. An acceptable exposure occurs when the total exposures by all pathways and exposure routes for each chemical and for the sum of the hazard indices for all chemicals are less than 1.0.

Noncarcinogens

$$HI = CDI_{noncarcinogen}/RfD \qquad \text{48.8}$$

For carcinogens, the risks for each exposure scenario are summed for each receptor group by exposure route. The receptor groups should be differentiated between the group experiencing average exposure and the group experiencing maximum exposure. Both the mean and maximum exposure point concentrations should be used to characterize the range of risk. The *carcinogen risk* is calculated as the product of the chronic daily intake and the slope factor.

Carcinogens

$$Risk = dose \times toxicity = CDI \times CSF \qquad \text{48.9}$$

Table for Example 48.3

pathway	exposure point	exposure route	chemical	CDI (mg/kg·d)	RfD (mg/kg·d)	CSF ((mg/kg·d)$^{-1}$)
groundwater	water well	oral	chlorobenzene	1.028		
			pyrene	0.036		
			dieldrin	0.003		
groundwater	volatilization at tap	inhalation	chlorobenzene	0.686		
			pyrene	0.044		
			dieldrin	ND		
fugitive dust	air at residence	inhalation	chlorobenzene	0.045		
			pyrene	0.033		
			dieldrin	0.003		
surface water	garden at residence	oral	chlorobenzene	0.006		
			pyrene	0.005		
			dieldrin	0.009		
total		oral	chlorobenzene	1.034	0.02	–
			pyrene	0.041	0.003	–
			dieldrin	0.012	–	16
total		inhalation	chlorobenzene	0.731	0.002	–
			pyrene	0.077	0.0003	–
			dieldrin	0.003	–	16

Example 48.3

The project engineer is now ready to characterize the risks for the situation given in Ex. 48.1. She is evaluating one of the scenarios for exposure of an adult in a residence near the site. The pathways, exposure routes, exposure points, chemicals, reference doses, and slope factors are given in *Table for Example 48.3*. Characterize the risk to this receptor.

Solution

For noncarcinogens, use Eq. 48.8.

Noncarcinogens
$$HI = CDI_{\text{noncarcinogen}}/RfD$$

chemical	CDI	RfD	HI
chlorobenzene			
oral	1.034	0.02	51.7
inhalation	0.731	0.002	365.5
pyrene			
oral	0.041	0.003	13.7
inhalation	0.077	0.0003	256.7
total HI			
oral			65.4
inhalation			622.2

The risk due to noncarcinogens is unacceptable because the hazard index is much greater than 1.0 for individual and combined exposures.

For carcinogens, use Eq. 48.9.

Carcinogens
$$\text{Risk} = \text{dose} \times \text{toxicity} = CDI \times CSF$$

chemical	CDI	CSF	risk
dieldrin			
oral	0.012	16	0.192
inhalation	0.003	16	0.048

The risk due to carcinogens is also unacceptable because the risk for dieldrin is outside the range of the normally acceptable EPA risk of 10^{-4} to 10^{-6}.

6. NOMENCLATURE

AT	averaging time	d
BW	body mass	kg
C	concentration	mg/L, mg/m^3
CDI, CDI	chronic daily intake	mg/kg·d
CF	conversion factor for soil, 10^{-6}	1 L/1000 cm^3, 10^{-6} kg/mg
CF	concentration in food	mg/g
CR	contact rate	L/h
CS	chemical concentration in soil	mg/kg
CSF	cancer slope factor	$(\text{mg/kg·d})^{-1}$
CW	chemical concentration in water	mg/L
ED	exposure duration	yr
EF	exposure frequency	d/yr, events/yr
ET	exposure time	h/d, h/event
FI	fraction ingested	–
HI	hazard index	–
IR	ingestion rate	L/d, mg soil/d, kg/meal
IR	inhalation rate	m^3/h
RfD, RfD	chronic reference dose	mg/kg·d
SF	slope factor	$(\text{mg/kg·d})^{-1}$
TS	toxicity score	kg·d/L

Subscripts

c	carcinogens
n	noncarcinogens
max	maximum

Site Assessment & Remediation

Fate and Transport

1. INTRODUCTION

The movement of soluble organic pollutants (solutes) in groundwater assumes they are nonreactive. That is, the solute moves with the groundwater, unimpeded by chemical interactions between the solute and the soil. This is true for some solutes, but for the majority of solutes, interaction with the soil is significant. The factors affecting solute transport are represented by the following.

- *advection:* The gross movement of the solute with the water through the soil. This has a dilution effect as it acts to spread the solute over an ever-increasing distance away from the source.

- *dispersion:* The mixing of solute with the groundwater as it travels through soil pores and channels. This has a dilution affect as it acts to spread the solute vertically, longitudinally, and laterally over an ever-increasing area.

- *sorption:* The capture of solute on soil media surfaces that creates a slowing of the solute velocity relative to that of the groundwater.

- *reaction:* The transformation of solute to other products with dissimilar transport characteristics. This may increase or decrease solute velocity.

Advection and dispersion are discussed in Chap. 47. This chapter addresses phenomena influencing sorption and reaction.

2. SORPTION AND SOLUTE TRANSPORT

Sorption is primarily the attraction and trapping of solute on the surfaces of soil media. It may occur because of ionic interactions in mineral-rich clayey soils or by molecular entrapment in small fissures in organic soils.

The phenomenon of sorption in a soil-groundwater system is called *retardation* because the sorption tends to slow the solute velocity. Retardation is the velocity of a specific solute relative to the velocity of the groundwater in a given soil-groundwater system. The amount of retardation is measured by the *retardation factor, R.* Dividing the velocity of the groundwater by the retardation factor gives the solute velocity.

$$v_s = \frac{v_w}{R} \qquad 49.1$$

The retardation factor will always have a value of one or greater. If $R = 1$, then the solute is not retarded and moves at the same velocity as the groundwater, v_w.

The resulting value of the solute velocity, v_s, is taken to represent the velocity of the solute front. This is taken to be at the center of mass of the chemical plume, which occurs downgradient of the source at the point where the contaminant concentration, C, is one-half the contaminant concentration at the source, C_o.

$$\frac{C}{C_o} = 0.5 \qquad 49.2$$

Figure 49.1 illustrates the influence of retardation on contaminant transport. Pore volumes in the figure may be understood as the number of times that groundwater is transported through a representative pore space.

Figure 49.1 Influence of Retardation on Solute

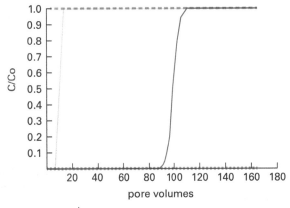

The retardation factor can be defined in terms of the aquifer soil characteristics, as shown in Eq. 49.3. ρ is the bulk density of the soil, and η is the porosity of the aquifer.

Retardation Factor R

$$R = 1 + (\rho/\eta)K_d \qquad 49.3$$

The distribution coefficient, K_d, is specific to the combination of solute and soil; its impact on solute transport varies with the chemical of interest and the soil characteristics. A nonreactive (i.e., conservative) contaminant will be transported with the bulk movement of the groundwater. Under these conditions, no retardation occurs (the solute controls in this case), and $R \doteq 1$. For inorganic contaminants (which often occur as ions), K_d is influenced by the clay content of the soil. Higher clay content results in a higher K_d, because the surface area and surface charge of the soil particles will be greater. Preferential retardation occurs in this sequence: divalent cations, then monovalent cations, and then anions. For inorganics in clean sands, K_d is close to zero, and R is close to one. For organics, K_d is influenced by the organic carbon content of the soil and by the solubility of the chemical in water. Low-solubility chemicals in organic soils are essentially immobile; inversely, for high-solubility chemicals in clean sands R is close to one.

3. TRANSPORT OF NONAQUEOUS PHASE LIQUIDS

The flow velocity of a nonaqueous phase liquid (NAPL) can be determined by dividing the specific discharge by the average effective porosity of the soil to get the actual velocity. *Specific discharge*, also called *Darcy velocity*, is

Specific Discharge

$$q = -K(dh/dx) \qquad 49.4$$

Dividing both sides by the average effective porosity, n, gives the flow velocity, v.

Specific Discharge

$$v = q/n = \frac{-K\left(\dfrac{dh}{dx}\right)}{n} \qquad 49.5$$

Hydraulic conductivity, K, can be further defined in terms of intrinsic permeability, k. In Eq. 49.6, μ is the dynamic viscosity of the fluid.

Hydraulic Conductivity

$$K = \rho g k/\mu \qquad 49.6$$

When expressed in terms of intrinsic permeability, the hydraulic conductivity for an NAPL can be calculated by using the NAPL's density and dynamic viscosity. It

is necessary, however, to account for temperature in this calculation. Intrinsic permeability, k, is independent of the fluid and so is not influenced by temperature, but temperature does influence hydraulic conductivity, K, by affecting dynamic viscosity, μ, and density, ρ. In calculating K, then, select values of μ and ρ for the temperature of interest. The viscosity and specific gravity of selected fuel oils is given in Table 49.1.

Table 49.1 *Viscosity and Specific Gravity of Lubricants and Fuel Oils*

| | | kinematic viscosity (cSt) | |
oil	specific gravity	at 5°C	at 15°C
SAE 10	0.88–0.935	330–550	132–198
SAE 30	0.88–0.935	1980–3300	660–990
SAE 90	0.88–0.935	3000	1210
gasoline	0.68–0.74	1.0	1.0
jet fuel	0.74–0.85	2.56	2.56
kerosene	0.78–0.82	4.74	4.11
no. 2 fuel	0.82–0.95	12.8	7.40
no. 2 diesel	0.82–0.95	20.2	12.4
no. 3 fuel	0.82–0.95	17.8	12.4
navy no. 1	0.989	880	230
olive	0.912–0.918	330	121

Example 49.1

No. 2 fuel oil is released into a soil-groundwater system. The soil is gravel and coarse sand with a hydraulic conductivity of 4.9 m/d, an effective porosity of 0.3, and a gradient of 0.005. The ambient pore space air temperature is 5°C. Determine the migration velocity of the oil.

Solution

Use Eq. 49.5 to compute the groundwater flow velocity.

Specific Discharge

$$
\begin{aligned}
v_w &= \frac{-K\left(\dfrac{dh}{dx}\right)}{n} \\
&= \frac{\left(-4.9\ \dfrac{\text{m}}{\text{d}}\right)(-0.005)}{0.3} \\
&= 0.082\ \text{m/d}
\end{aligned}
$$

From a table of water properties, find the density and dynamic viscosity of water at 5°C. [Properties of Water (SI Metric Units)]

$$\rho_w = 1000.0\ \text{kg/m}^3$$
$$\mu_w = 0.001518\ \text{Pa·s} = 0.001518\ \text{kg/m·s}$$

Use Eq. 49.6 and solve for the soil permeability, using the values of density and dynamic viscosity for 5°C water.

Hydraulic Conductivity

$$K = \rho g k / \mu$$

$$
\begin{aligned}
k &= \frac{K \mu_w}{\rho_w g} \\
&= \frac{\left(\dfrac{4.9 \, \frac{m}{d}}{86\,400 \, \frac{s}{d}} \right) \left(0.001518 \, \frac{kg}{m \cdot s} \right)}{\left(1000 \, \frac{kg}{m^3} \right) \left(9.81 \, \frac{m}{s^2} \right)} \\
&= 8.776 \times 10^{-12} \, m^2
\end{aligned}
$$

From Table 49.1, a midrange value for the specific gravity of no. 2 fuel oil is about 0.9, and the oil's kinematic viscosity, ν_o, is 12.8 cSt. The density and absolute viscosity of the oil are

$$
\begin{aligned}
\rho_o &= (SG)\rho_w = (0.90)\left(1000 \, \frac{kg}{m^3} \right) \\
&= 900 \, kg/m^3
\end{aligned}
$$

$$
\begin{aligned}
\mu_o &= \nu_o \rho_o = \left(\frac{12.8 \, cSt}{10^6 \, \dfrac{cSt}{\dfrac{m^2}{s}}} \right) \left(900 \, \frac{kg}{m^3} \right) \\
&= 0.01152 \, kg/m \cdot s
\end{aligned}
$$

Calculate the hydraulic conductivity of the oil.

$$
\begin{aligned}
K_o &= \frac{\rho_o g k}{\mu_o} \\
&= \frac{\left(900 \, \frac{kg}{m^3} \right) \left(9.81 \, \frac{m}{s^2} \right) (8.776 \times 10^{-12} \, m^2) \left(86\,400 \, \frac{s}{d} \right)}{0.01152 \, \frac{kg}{m \cdot s}} \\
&= 0.5811 \, m/d
\end{aligned}
$$

Compute fuel oil flow velocity at 5°C assuming the gradient and porosity are the same for water and fuel oil.

$$
\begin{aligned}
v_o &= \frac{-K \left(\dfrac{dh}{dx} \right)}{n} = \frac{\left(-0.5811 \, \frac{m}{d} \right)(-0.005)}{0.3} \\
&= 0.0097 \, m/d
\end{aligned}
$$

4. ORGANIC CHEMICAL DEGRADATION

Essentially all contaminants, whether organic or inorganic and whether present as solute or NAPL, undergo some degree of chemical, physical, and/or biological transformation in the subsurface environment. This transformation is almost always a degradation of the chemical structure—for example, through a loss of halogens and cleavage of carbon-carbon bonds.

Transformation influences the characteristics of the contamination and consequently the behavior and impacts of the contaminant in the subsurface environment. The following are examples of transformation.

- Fuel oils are known to weather. *Weathering* is the process of losing volatile fractions. These more volatile fractions are usually simpler in structure, making them more susceptible to biological action as well. A fuel oil such as gasoline, for example, is composed of many different chemicals.

  Early in the weathering process, the soil volume influenced by the contamination may increase through vapor transport. As weathering progresses, the contamination may become dominant in heavier hydrocarbon fractions that are less mobile and do not contribute to solute plumes. Further weathering may reduce the longer chain hydrocarbons to more mobile compounds and resultant generation of a latent plume.

- Metals and other inorganic contaminants undergo reactions with mineral groups in the soil and their dissolved components in the groundwater.

  Many oxidation and reduction reactions involving inorganic contaminants are possible, with the usual consequence being some form of fixing the contaminant to reduce its mobility. Cationic metals combine to anionic materials to form insoluble or immobile complexes. Many metals are more soluble at low pH, and as waste containing soluble metals in a low pH solution is introduced into the soil or groundwater, carbonate species may act to increase the pH and reduce the metal solubility.

- Organic solvents, a common source of groundwater contamination, undergo degradation to reduce halogenation and to remove methyl groups. The products of the degradation of these organic solvents are sometimes more mobile, volatile, or toxic than the parent compounds.

Chemical Degradation Pathways

The transformation and fate of a contaminant in the soil and groundwater system are controlled by a variety of factors that apply to both solute and NAPL.

- *biological degradation.* Under proper conditions, all organic compounds are biodegradable, and inorganic compounds are involved with and influenced by biological reactions. Indigenous organisms exist in soils to depths of several hundred feet. When exposed to a contaminant, they have an opportunity to adapt to use the contaminant as a food source. Where contamination is chronic and may exist for several years before discovered or remediated, the organisms may become uniquely adapted to use the chemical as food.

- *physical/chemical degradation.* Organic compounds in water experience a phenomenon known as *hydrolysis*, in which one atom of a molecule is replaced by a hydrogen ion from the water. Hydrolysis can be significant in dehalogenation of chlorinated organics when conditions are favorable. The presence of adequate free oxygen (DO) will also create conditions where metals may be oxidized to form less soluble products.

- *volatilization.* Volatilization is limited to organic contaminants in relatively granular soils. Volatilization may be enhanced by a fluctuating water table. As volatile components are lost as vapor to soil pores, the remaining contaminants may become enriched in less mobile constituents.

- *dissolution and precipitation.* These are opposite phenomena that are usually associated with inorganic contaminants. Reactions with mineral constituents in the soil and groundwater can shift the equilibrium of some compounds between their soluble and insoluble form. The significance of these factors is also influenced by the mixing of the contaminant with the groundwater and exposure of the contaminants to unreacted constituents in the soil-groundwater system.

- *physical and chemical fixation.* Adsorption and complex formation are two examples of physical and chemical fixation. Retardation illustrates the impact of adsorption on contaminant mobility and is a function of organic matter in the soil. Charged clays also retard contaminant movement where ionic interactions can occur. Some inorganic contaminants will react chemically with mineral constituents in the soil and groundwater to form immobile complexes.

All these mechanisms can be enhanced through engineered activities.

In addition to the physical and chemical characteristics of the contaminants (viscosity, volatility, adsorbability, reactivity, biodegradability), the soil characteristics play a significant role in the fate and mobility of both inorganic and organic contaminants.

- *clays.* Clays are composed of charged particles that will increase ionic interactions with inorganics resulting in low mobility mineral complexes. The clay can also act as an adsorption media, and its tight structure allows low permeability to most fluids. Some organic solvents, however, displace water and may

act as drying agents to increase the permeability of clays.

- *sands.* Sands increase the opportunity for volatilization of organic contaminants. They also create a greater opportunity for mixing of inorganic contaminants with mineral constituents that may lead to formation of insoluble substances. Sands are likely to allow increased DO, compared to clays, which would increase aerobic biodegradation. Because of their loose structure, sands allow higher mobility of soluble contaminants.

- *soil origins.* The mineral origin of the soil will influence its chemical characteristics. Soils originating from limestone formations, for example, will exhibit characteristics of high alkalinity and contain materials likely to react with inorganic contaminants.

Example 49.2

The degradation sequence for perchloroethene to dichloroethene isomers is as follows.

$$\text{perchloroethene } (CCl_2{=}CCl_2)$$
$$\Downarrow$$
$$\text{trichloroethene } (CHCl{=}CCl_2)$$
$$\Downarrow$$
$$\text{dichloroethene isomers}$$
$$(CH_2{=}CCl_2, \textit{ cis-} \text{ and } \textit{trans}{-}CHCl{=}CHCl)$$

Values for the octanol-water partition coefficient, Henry's constant, and solubility for selected chlorinated organic solvents are summarized in the table.

compound	octanol-water partition coefficient, K_{ow}	Henry's law constant (atm×m³/ mol)	solubility (mg/L)
chloroethane	35	0.0085	5740
chloroform	93	0.0038	9300
1,1-dichloroethene	135	0.021	2730
cis-1,2-dichloroethene	30	0.0037	3500
trans-1,2-dichloroethene	62	0.38	6300
methylene chloride	0.95	0.0032	16 700
perchloroethene	398	0.018	150
trichloroethene	240	0.010	1080
vinyl chloride	24	2.8	1100

(a) What compound would follow the dichloroethene isomers in the degradation sequence for perchloroethene?

(b) Which compound in the table is likely to be most mobile in a soil-groundwater system?

(c) Which compound in the table is most likely to partition to the vapor phase?

(d) Which compound in the table is most likely to exist as a nonaqueous phase liquid?

Solution

(a) The degradation sequence from the dichloroethene isomers would be to a monochloroethene. Therefore, each dichloroethene will degrade to vinyl chloride $(CH_2 = CHCl)$.

(b) Mobility in a soil-groundwater system generally increases with decreasing values of the octanol-water partition coefficient. The compound in the table with the lowest octanol-water partition coefficient is methylene chloride.

(c) The potential to partition to the vapor phase, or volatility, increases with increasing values of Henry's constant. The compound in the table with the greatest Henry's constant is vinyl chloride.

(d) As solubility decreases, the potential for a compound to exist as a nonaqueous phase liquid increases. The compound in the table with the lowest solubility is perchloroethene.

5. NOMENCLATURE

C	concentration	mg/L	mg/L
h	height or hydraulic head	ft	m
g	gravitational acceleration, 32.2 (9.81)	ft/sec^2	m/s^2
k	intrinsic permeability	ft^2	m^2
K	hydraulic conductivity	ft/sec	m/s
K_d	distribution coefficient	–	–
n	porosity	–	–
q	specific discharge	ft/sec	m/s
R	retardation factor	–	–
SG	specific gravity	–	–
v	velocity	ft/sec	m/s
x	horizontal distance	ft	m

Symbols

ρ	density	lbm/ft^3	kg/m^3
η	porosity	–	–
μ	dynamic viscosity	lbf-sec/ft^2	kg/(m·s)
ν	kinematic viscosity	ft^2/sec	m^2/s

Subscripts

o	oil or source (origin)
s	solute
w	water

Site Assessment & Remediation

50 Remediation Alternative Identification

1. REMEDIATION STRATEGIES

Remediation strategies require substantial lead time to develop and implement because of the need to understand the site and contamination characteristics. Whether following Comprehensive Environmental Response, Compensation, and Liability Act (CERCLA) or Resource Conservation and Recovery Act (RCRA) guidelines or complying with a state or local cleanup order, the process generally proceeds according to four broadly defined steps.

step 1: Identify the contaminants and their distribution. This step involves site reconnaissance, historical review, and the collection of soil and groundwater samples, with the objective of gaining a thorough understanding at reasonable cost. As Fig. 50.1 illustrates, after a point, there is little advantage to be gained from additional investigation because small additional gains in understanding come at increasingly high additional costs.

Figure 50.1 *Overall Remedial Action Cost Compared to Discovery Costs*

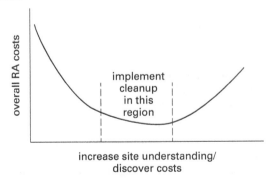

step 2: Identify contaminant transport and transformations. This involves defining site geology and hydrogeology and characterizing the contaminating chemicals.

step 3: Evaluate human health and environmental effects. The objective is to identify and quantify risks and to define treatment and cleanup criteria. Answer the question "How clean is clean?"

step 4: Develop a remedial program to meet the conditions defined in steps 1 through 3.

2. AQUIFER REMEDIATION STRATEGIES

Groundwater remediation strategies can be grouped into four general categories.

- no direct cleanup action (institutional remedy)
- in situ physical, chemical, and biological methods
- physical and hydraulic containment
- extraction and surface treatment

The selection of one or more of these four strategies depends largely on the contaminant transformation and fate. Selection factors include

- contaminant transformations and mobility
- physical and chemical characteristics of the soil
- aquifer setting (gradient, potential conduits, potential beneficial uses)
- local political and regulatory setting
- availability of alternative water resources

No Direct Cleanup Action

Where the combination of soil and groundwater conditions and contaminant characteristics are favorable, it may be possible to justify that no direct cleanup action of a contaminated aquifer will occur.

"No direct cleanup" does not mean that the responsible party can simply walk away from the site. What it does mean is that direct action to remediate the contamination will not occur (contaminated groundwater will not be treated, contained, or extracted), but the following activities will be implemented at the site.

- Ongoing monitoring of the groundwater will be required to ensure that conditions remain static. If mobility of the contaminant is detected, the no-action alternative may have to be abandoned.

Ongoing monitoring will likely involve regular collection of groundwater samples and analysis for the contaminants and degradation products. Groundwater elevations will be measured as part of regular sampling, and groundwater contour maps will be updated. Each monitoring event will result in a summary report being prepared and submitted to the responsible regulatory agency. The summary reports are part of the public record.

- There will be limitations on current uses and certain kinds of future development of the site. For example, activities involving deep excavations where dewatering would be required may be restricted, as would any kind of pumping well. Agricultural uses involving heavy irrigation that would significantly influence the groundwater gradient or water table elevation may be restricted. Deed restrictions would be placed on the property to convey any restricted land uses to subsequent owners. Because of these restrictions, it may not be economically desirable to apply the no-action alternative at some sites that meet the technical requirements.

Under some conditions involving complicated soil-groundwater systems and contaminants that are exceptionally difficult to remediate, or where excessive costs may be incurred, a technically or economically feasible alternative to no direct action may not exist.

Sites that are most likely candidates for no direct cleanup action are those with the following two characteristics.

- The exposure potential to workers or residents of the community is limited. In evaluating this, consideration must be made of population distribution, land uses in the vicinity, contaminant transport by air and water routes, likelihood of site surface soil disturbances, and other conditions.

- Contaminant distribution is relatively confined. This implies either that the release was relatively small or that the mobility of the contaminant is low. In either case, this condition allows the plume to be monitored, and if mobility is detected or other conditions change, control over the plume may be implemented.

One significant advantage of no direct cleanup action is that exposure from cleanup activities to residents of the area or to workers is eliminated. The possibility that remedial equipment may be in place for several decades may also present an incentive for selecting no direct cleanup. No direct cleanup action requires skilled public relations people to gain community acceptance.

Regardless of the remedial alternative ultimately selected for the aquifer, no direct cleanup is usually the alternative with the lowest cost and is used as a basis for comparing other alternatives.

In Situ Methods

In situ methods for aquifer restoration can be placed into one of two general groups: physical and chemical methods and biological methods.

Physical and Chemical Methods (Neutralization/Detoxification)

Neutralization/detoxification involves the injection of reagents into the plume to induce reactions that will change the contaminant characteristics. If groundwater is relatively shallow, a cutoff wall backfilled with reacting material such as crushed limestone or activated carbon may also be effective.

For inorganic contaminants, desired reactions reduce or oxidize the compound to an insoluble form and then allow the demobilized contaminant to precipitate. Any oxidizing or reducing reagent employed must be completely reactive in that no undesired residual remains. For example, hydrogen peroxide is a suitable oxidizing reagent because it decomposes to free oxygen and water, products that are not contaminants themselves. However, chlorine compounds are not suitable for in situ oxidation because trihalomethane and other unwanted byproducts may result.

Although potentially applicable to either organic or inorganic contaminants, inorganic contaminants present the best possibility for success when employing in situ neutralization and detoxification methods. The advantage with inorganic contaminants is better control of byproduct formation and their fate.

Physical and Chemical Methods (Fixation/Immobilization)

Physical and chemical methods involve adding reactive materials to the contaminated aquifer to physically bind the contaminants in some kind of solid matrix. Reactive materials include portland cement, lime, fly ash, organic polymers, silicate compounds, or combinations of these. The reactive materials are placed by direct injection into the aquifer, by cutoff walls, by drilled caissons, and by other methods. As opposed to containment by placing the materials as a barrier to migration, fixation/immobilization methods act directly on the contaminants. These methods require treatability studies prior to implementation to ensure the effectiveness of the proposed techniques.

Fixation and immobilization methods are best suited for relatively localized contamination where concentrations of contaminants in the groundwater are high. A potential drawback is that some contaminated materials may require removal and other treatment and disposal to accommodate the treatment materials.

The processes involving neutralization/detoxification and fixation/immobilization are frequently referred to collectively as *solidification/stabilization*.

Biological Methods

In situ biological treatment of contaminated groundwater relies on concepts of attached growth typically associated with municipal wastewater treatment. The soil grains act as the attached growth medium, and the contaminated groundwater is the wastewater. If conditions for biological activity are—or can be made to be—favorable, the soil bacteria will degrade the contaminants.

The important factors controlling in situ bioremediation are the following.

- *oxygen source.* Depending on the availability of free oxygen, the indigenous bacterial population may have adapted to use the contaminants as a food source under either aerobic or anaerobic conditions. If the bacteria are efficient anaerobically, then the anaerobic environment should be maintained. However, for several common groundwater contaminants, aerobic processes proceed more rapidly than do anaerobic processes (if they occur at all). Consequently, an evaluation of remedial performance under both aerobic and anaerobic conditions is required. If the indigenous bacteria are aerobic, free oxygen (DO) may be deficient and limit degradation rates.

- *nutrients.* Nitrogen and phosphorus are usually deficient and will need to be added for optimum degradation to occur.

- *bacteria.* Bacteria naturally exist in soil to depths of several hundred feet. When exposed to a contaminant, bacteria may adapt to use it as a substrate. However, the indigenous bacteria may not be the most efficient organism at degrading the waste. The indigenous bacterial population may be augmented or replaced with bacteria better suited to utilize the waste as a food source.

Where additional DO, nutrients, or bacteria are required, two general methods may be employed to introduce these into the aquifer.

- *direct injection.* An injection well is placed within the contaminant plume at a location that will allow dispersion of the DO, nutrients, or bacteria. Well placement will be different for DO than for nutrients or bacteria. The DO can be in the form of gas (usually air, but pure oxygen is available) that is sparged into the aquifer, or it may be in a liquid form as hydrogen peroxide. Wells for oxygen addition may be relatively densely located. Nutrients are generally limited to nitrogen and phosphorus, but other trace compounds may also be required. Ammonia and nitrate compounds are common for nitrogen, and phosphates are common for phosphorus. Bacteria may be purchased from firms that prepare special cultures to target specific contaminants. Although they will be more expensive than bacteria taken from the local POTW, they may or may not be more effective.

- *extraction/injection.* The groundwater is extracted and aerated in a reactor where nutrient and bacteria are added. This enriched groundwater is then injected to mix with the aquifer and effect treatment.

Physical and Hydraulic Containment

When treatment is infeasible for either technical or economic reasons, the preferred response may be to contain the plume within a defined area. Containment may take two general forms: physical containment and hydraulic containment. Both forms require the use of monitoring wells to ensure that containment is effective. Either method may be used with other remediation schemes.

- *physical containment.* This involves constructing a physical barrier around the entire plume or a portion of it. The barrier may consist of sheet piles driven through the depth of the aquifer into an underlying confining layer, or it may consist of a trench that has been excavated and backfilled with an impermeable material. Groundwater extraction may be required to equalize hydraulic pressure within the contained area.

- *hydraulic containment.* This involves placing a series of wells (either injection or extraction) to create a mound or trough in the groundwater surface. Hydraulic containment requires maintenance and does not represent the same long-term solution as does physical containment.

Extraction and Surface Treatment

Extraction and surface treatment is commonly called *pump and treat*, which defines what occurs in this remediation scheme. One or more extraction wells are constructed at a down gradient location to capture the plume. It is essentially the same as hydraulic containment, but upgradient injection wells may or may not be used. The extracted groundwater is subjected to treatment through typical unit processes (physical, chemical, or biological), and the treated water is discharged to the sanitary sewer or to surface water.

3. SOIL AND UNSATURATED ZONE REMEDIATION STRATEGIES

Like groundwater remediation strategies, soil remediation strategies may include

- no direct cleanup action (institutional remedy)

- physical, chemical, and biological methods for in situ and excavated soils

- physical containment

Physical, Chemical, and Biological Methods

This may occur for soils in situ or for excavated materials. Common remedial activities include the following.

- *encapsulation.* Containment by excluding surface water infiltration and preventing soil dispersion is commonly applied to soil as a physical in situ remedial measure. In this case, the contaminant may already be physically or chemically bound within the soil matrix, and the purpose, therefore, is to prevent conditions that would mobilize the contaminants. Encapsulation involves placing a surface barrier over the impacted area to prevent infiltration, surface erosion from wind and rain, and other surface disturbances from mechanical activities (i.e., vehicular traffic). The surface barrier may consist of asphalt overlying clay or a geotextile membrane.

- *excavation and disposal.* Simply excavating the contaminated soil and hauling it to a landfill for disposal is usually an expensive, high-risk, and environmentally least desirable alternative. However, it is fast and may be appropriate in cases where cleanup must proceed immediately and be completed in a short time period. Excavation and disposal are discouraged by regulatory agencies because those actions consume scarce landfill space and merely transfer the problem from one site to another. Also, some contaminants are banned from landfill disposal.

- *excavation and treatment.* The soil is excavated and then either treated on site or transported to an off-site facility for treatment. There are a wide variety of processes employed for treatment of excavated soil. Where nonchlorinated petroleum hydrocarbons are the contaminant, the soil may be treated by incineration in an asphalt plant or dedicated incinerator. The burned soil is then discarded as clean fill. Chlorinated petroleum hydrocarbons are not typically treated by incineration, as when incinerated they may form dioxins.

Where the contaminants are volatile, the soil may be placed on the ground to air-dry or allow the contaminants to volatilize to the atmosphere. Frequent turning of the soil, and windy warm weather, can help this method.

The soil may be subjected to washing by employing processes similar to heap leaching of metal ore. The soil is mixed with a liquid so that the contaminant is released into the liquid (liquid extraction). Soil washing processes require additional treatment facilities for the liquid wastes produced.

Soils may also be excavated and placed on the ground for land farming or biological treatment. This is common with petroleum hydrocarbon contaminated soils and is effective. Exposure to sunlight and air, adequate nutrients, warm weather, and moisture are important for effective treatment.

- *in situ methods.* For in situ methods, the purpose may be either to immobilize the contaminants, to degrade them, or to affect their mobilization so they may be removed from the soil. Methods of soil washing, stabilization/fixation, and bioremediation, as described earlier, may also be applied as in situ treatment methods.

In situ soil venting may be employed where the contaminant is a volatile chemical and exists in a relatively permeable soil or in fractured rock. Soil venting involves placing several wells in the unsaturated zone and inducing a vacuum at each well. As airflow is induced through the soil to the wells, the chemicals are volatilized and removed from soil pore spaces. Because this usually requires treatment of collected vapors, the wells are generally operated at negative (vacuum) rather than positive (forced draft) pressure.

4. SELECTION CRITERIA

Selection criteria may be evaluated using a weighted average analysis where each factor influencing a remedial alternative is identified and a value is assigned to its relative importance. Each alternative, then, is evaluated on how well it meets the criteria. Criteria of economics, technical feasibility, human health risk, environmental justice, environmental protection, and public acceptance are examples of selection criteria.

Example 50.1 illustrates the application of weighted average analysis. The example uses a weighting factor (WF) for criteria scaled from 1 as more important to 7 as less important, and it uses a rating (R) for how well each alternative meets the criteria scaled from 1 as more satisfactory to 7 as less satisfactory. The weighted rating (WR) is the product of WF and R. Using this scaling scheme, the alternative with the lowest relative rating is the preferred choice. Other weighting and rating schemes can be used to meet individual needs.

Example 50.1

Four remedial action alternatives are being evaluated. The characteristics of the alternatives are summarized in the table.

			alternatives			
			A	B	C	D
category	criterion	WF	R	R	R	R
economics	capital costs	2	1	3	2	3
	O&M costs	3	4	2	3	1
feasibility	mature technology	5	2	5	4	7
	pilot testing	7	6	3	2	5
	lead time	4	3	4	3	2
health risk	pathways	2	3	7	2	1
	dose	2	3	4	3	3
justice	balanced impacts	4	3	4	2	3
environmental protection	surface water	5	3	2	3	2
	groundwater	3	2	2	1	2
community	population	1	4	6	5	3
	land uses	4	3	5	2	2

Which site best meets the desired criteria?

Solution

Multiply each rating (R) by its weighting factor (WF) to get the weighted rating (WR). For each alternative, add the WRs for all the criteria. Divide each total score by the lowest of the total scores to get relative ratings.

			alternatives			
			A	B	C	D
category	criterion	WF	WR	WR	WR	WR
economics	capital costs	2	2	6	4	6
	O&M costs	3	12	6	9	3
feasibility	mature technology	5	10	25	20	35
	pilot testing	7	42	21	14	35
	lead time	4	12	16	12	8
health risk	pathways	2	6	14	4	2
	dose	2	6	8	6	6
justice	balanced impacts	4	12	16	8	12
environmental protection	surface water	5	15	10	15	10
	groundwater	3	6	6	3	6
community	population	1	4	6	5	3
	land uses	4	12	20	8	8
	total score		139	154	108	134
	relative rating		1.29	1.43	1.00	1.24

Alternative C has the lowest weighted average and relative rating and, therefore, best satisfies the specified criteria.

Site Assessment & Remediation

Remediation Technologies and Management

1. INTRODUCTION

This chapter discusses (in alphabetical order) the methods and equipment that can be used to reduce or eliminate pollution. Legislation often requires use of the *best available control technology* (BACT—also known as the *best available technology*—BAT) and the *maximum achievable control technology* (MACT), *lowest achievable emission rate* (LAER), and *reasonably available control technology* (RACT) in the design of pollution-prevention systems.

2. ADSORPTION (LIQUID PHASE)

Adsorption is a surface phenomenon associated with the high surface area of the medium, usually granular activated carbon (GAC). As the target contaminants contact the GAC, removal occurs at high efficiencies. Removal is propagated as a wave front through the GAC bed until trace concentrations of the target contaminant are detected in the effluent. When these trace concentrations are first detected, *breakthrough* occurs. As operation continues for an additional time beyond breakthrough, the concentration of the target contaminant continues to increase until it reaches the influent level. This condition defines *saturation* or *exhaustion* of the GAC bed.

To maximize the full capacity of the media between breakthrough and saturation, adsorption vessels are usually configured in a two-in-series, lead-follow, operating mode. This allows the lead vessel to be operated until the bed is completely saturated because the following vessel will capture the target chemical lost from the lead vessel during that interval between breakthrough and saturation.

Isotherms

Adsorption process design utilizes adsorption isotherms either available from published sources for pure chemical solutions or developed using bench scale tests for chemical mixtures. They provide an approximation of the media use rate for pure solutions. As more and more complex mixtures of chemicals are encountered, the published isotherms provide a less and less reliable estimate of the media use rate or time to breakthrough. Isotherms developed for specific chemical mixtures are more reliable than published isotherms but require laboratory studies to develop the data for their construction. Although design based on isotherms should be considered preliminary, they provide the advantage of being able to analyze adsorption processes without having to conduct laboratory studies. Isotherms are valid only over the range of concentrations used in the isotherm development, and caution should be used when extrapolating to values away from this range.

Adsorption isotherms, either for pure solution or for chemical mixtures, are commonly based on either the Freundlich or Langmuir model. The Freundlich model is more common. The Freundlich isotherm equation is

Resistivity of a Medium

$$q_e = KC_e^{1/n} \qquad \textit{51.1}$$

To solve the Freundlich equation directly for the required adsorbent dose to affect a desired level of treatment, the following form is used.

$$\frac{C_0 - C_e}{M} = KC_e^{1/n} \qquad \textit{51.2}$$

Here, C_0 is the initial concentration of the chemical in solution in units of milligrams per liter. In this form, K has units of milligrams per gram, M has units of grams per liter, and C_0 and C_e have units of milligrams per liter. Values for K and $1/n$ for selected chemicals are provided in Table 51.1.

Table 51.1 Freundlich Isotherm Constants

compound	K (mg/g)	$1/n$
hexachlorobutadiene	360	0.63
anethole	300	0.42
phenyl mercuric acetate	270	0.44
p-nonylphenol	250	0.37
acridine yellow	230	0.12
benzidine dihydrochloride	220	0.37
n-butylphthalate	220	0.45
n-nitrosodiphenylamine	220	0.37
dimethylphenylcarbinol	210	0.33
bromoform	200	0.83
β-naphthol	100	0.26
acridine orange	180	0.29
α-naphthol	180	0.31
α-naphthylamine	160	0.34
2,4,6-trichlorophenol	155	0.40
pentachlorophenol	150	0.42
p-nitroaniline	140	0.27
l-chloro-2-nitrobenzene	130	0.46
1,2-dichlorobenzene	129	0.43
benzothiazole	120	0.27
diphenylamine	120	0.31
guanine	120	0.40
styrene	120	0.56
dimethyl phthalate	97	0.41
chlorobenzene	93	0.98
hydroquinone	90	0.25
p-xylene	85	0.16
acetophenone	74	0.44
1,2,3,4-tetrahydronaphthalene	74	0.81
adenine	71	0.38
nitrobenzene	68	0.43
dibromochloromethane	63	0.93
ethylbenzene	53	0.79
tetrachloroethene	51	0.56
trichloroethene	28	0.62
toluene	26	0.44
phenol	21	0.54
carbon tetrachloride	11	0.83
1,1,2-trichloroethane	5.8	0.60
1,1-dichloroethylene	4.9	0.54
trans-1,2-dichlorethane	3.1	0.51
1,1,1-trichloroethane	2.5	0.34
1,1-dichloroethane	1.8	0.53
methylene chloride	1.3	1.16
benzene	1.0	1.6

Adsorption system design using isotherms is based on evaluating the media use rate and satisfying criteria for empty bed contact time (typically EBCT > 10 min) and hydraulic loading rate (typically HLR > 2 gal/min-ft²). To perform these calculations, GAC bed area and bed volume need to be obtained from manufacturer specifications. Some typical specifications are listed in Table 51.2.

Table 51.2 Typical Manufacturer Vessel Specifications

GAC capacity (lbm)	GAC bed volume (ft³)	bed diameter (ft)
2000	72	3
10,000	357	7.5
20,000	720	10

The typical density of GAC is 27.5 lbm/ft³.

Example 51.1

An extraction well can pump 25 gal/min of water from an aquifer contaminated with 1,1,1-trichloroethane (1,1,1-TCA) at 900 μg/L and trichloroethene (TCE) at 400 μg/L. Using the vessel specifications in Table 51.2, design a GAC adsorption system to treat the water to satisfy maximum contaminant level (MCL) standards.

Solution

The maximum contaminant level (MCL) for 1,1,1-trichloroethane (1,1,1-TCA) is 0.2 mg/L, and the MCL for trichloroethene (TCE) is 0.005 mg/L. [Regulated Drinking Water Contaminants]

These standards will be satisfied when a two-in-series lead-follow adsorber configuration is used and when the lead vessel is operated until breakthrough occurs. Because neither adsorption isotherm column test results nor a bench scale isotherm for the groundwater containing the mixture of both chemicals is available, use published isotherms constants with the Freundlich isotherm equation for each chemical individually. To do this, determine which contaminant will control design.

From Eq. 51.2 and Table 51.1, the isotherm equations for 1,1,1-TCA and TCE are

$$\frac{C_0 - C_e}{M} = KC_e^{1/n}$$

$$= \left(2.5 \ \frac{\text{mg}}{\text{g}}\right)C_e^{0.34} \quad \text{[for 1,1,1-TCA]}$$

$$= \left(28 \ \frac{\text{mg}}{\text{g}}\right)C_e^{0.62} \quad \text{[for TCE]}$$

Calculate the mass of carbon required for 1,1,1-TCA.

$$\frac{C_0 - C_e}{M} = KC_e^{1/n}$$

$$M = \frac{C_0 - C_e}{KC_e^{1/n}}$$

$$= \frac{0.9 \frac{mg}{L} - 0.2 \frac{mg}{L}}{\left(2.5 \frac{mg}{g}\right)\left(0.2 \frac{mg}{L}\right)^{0.34}}$$

$$= 0.49 \frac{g\ GAC}{L\ water\ treated}$$

For TCE, the mass of carbon required is

$$M = \frac{C_0 - C_e}{KC_e^{1/n}}$$

$$= \frac{0.4 \frac{mg}{L} - 0.005 \frac{mg}{L}}{\left(28 \frac{mg}{g}\right)\left(0.005 \frac{mg}{L}\right)^{0.62}}$$

$$= 0.38 \frac{g\ GAC}{L\ water\ treated}$$

The mass is greater for 1,1,1-TCA, so 1,1,1-TCA will control design. Because TCE is present, its concentration must be accounted for, but how this is done is left primarily to the judgment of the engineer. For example, a safety factor could be applied to increase the carbon required, the carbon required for each chemical could be added together, or the overall concentration of contaminants in the water could be adjusted to let them be "as 1,1,1-TCA." One way this might be done is

$$900 \frac{\mu g}{L} + \left(400 \frac{\mu g}{L}\right)\left(\frac{0.49}{0.38}\right)$$

$$= 1416 \frac{\mu g}{L} \text{ as 1,1,1-TCA}$$

$$\frac{1.416 \frac{mg}{L} - 0.2 \frac{mg}{L}}{\left(2.5 \frac{mg}{g}\right)\left(0.2 \frac{mg}{L}\right)^{0.34}}$$

$$= 0.84 \frac{g\ GAC}{L\ water\ treated}$$

This will provide a conservative estimate of GAC use. By adding GAC required for each chemical, the use rate would be 0.87 mg of GAC per liter of water treated. Using 0.84 g GAC per liter would provide a safety factor of 0.84/0.87 or nearly 1.0. This seems reasonable and is supported by adding together the GAC required for each chemical.

The daily GAC mass required to remove 1,1,1-TCA and TCE is

$$\left(\left(25 \frac{gal}{min}\right)\left(3.785 \frac{L}{gal}\right)\left(1440 \frac{min}{day}\right)\right)$$

$$\times \left(\left(0.84 \frac{g\ GAC}{L}\right)\left(10^{-3} \frac{kg}{g}\right)\right.$$

$$\left. \times \left(2.204 \frac{lbm}{kg}\right)\right)$$

$$= 252 \text{ lbm GAC/day}$$

For two standard sizes of adsorption vessels, the number of days to saturation would be

$$\frac{10{,}000 \text{ lbm GAC}}{252 \frac{\text{lbm GAC}}{\text{day}}} = 40 \text{ days}$$

$$\frac{20{,}000 \text{ lbm GAC}}{252 \frac{\text{lbm GAC}}{\text{day}}} = 80 \text{ days}$$

An operating period before change-out of more than about 30 days may lead to problems from calcium carbonate precipitation due to water hardness. Therefore, choose the 10,000 lbm vessel operated in a two-in-series, lead-follow mode. From Table 51.2, a 10,000 lbm vessel typically has a 357 ft³ bed volume with a 7.5 ft diameter. Using these specifications, calculate the empty bed contact time and the hydraulic loading rate.

$$EBCT = \frac{(357 \text{ ft}^3)\left(7.5 \frac{gal}{\text{ft}^3}\right)}{25 \frac{gal}{min}}$$

$$= 107 \text{ min} \quad [> 10 \text{ min, so OK}]$$

$$HLR = \frac{25 \frac{gal}{min}}{(7.5 \text{ ft})^2\left(\frac{\pi}{4}\right)}$$

$$= 0.57 \text{ gal/min-ft}^2 \quad \begin{bmatrix} < 2 \text{ gal/min-ft}^2\text{, so} \\ \text{not OK; recirculate} \end{bmatrix}$$

Try a recirculation ratio of 4.0 for HLR_{recirc}. Check the recirculated flow rate and empty bed contact time.

$$HLR_{recirc} = (4.0)\left(0.57 \frac{gal}{min\text{-}ft^2}\right)$$

$$= 2.28 \text{ gal/min-ft}^2$$

$$[> 2 \text{ gal/min-ft}^2\text{, so OK}]$$

$$\text{EBCT}_{\text{recirc}} = \frac{(357 \text{ ft}^3)\left(7.5 \dfrac{\text{gal}}{\text{ft}^3}\right)}{\left(25 \dfrac{\text{gal}}{\text{min}}\right)(4)}$$

$$= 27 \text{ min} \quad [>10 \text{ min, so OK}]$$

3. ADSORPTION (SOLVENT RECOVERY)

AC for solvent recovery is used in a cyclic process where it is alternately exposed to the target substance and then regenerated by the removal of the target substance. For solvent recovery to be effective, the VOC inlet concentration should be at least 700 ppm. Regeneration of the AC is accomplished by heating, usually by passing low-pressure (e.g., 5 psig) steam over the AC to raise the temperature above the solvent-capture temperature. Figure 51.1 shows a stripping-AC process with recovery.

Figure 51.1 Stripping-AC Process with Recovery

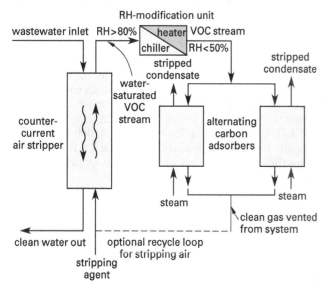

There are three main processes for solvent recovery. In a traditional *open-loop recovery system*, a countercurrent air stripper separates VOC from the incoming stream. The resulting VOC air/vapor stream passes through an AC bed. Periodically, the air/vapor stream is switched to an alternate bed, and the first bed is regenerated by passing steam through it. The VOC is recovered from the steam. The VOC-free air stream is freely discharged. Operation of a *closed-loop recovery system* is the same as an open loop system, except that the VOC-free air stream is returned to the air stripper.

For VOC-laden gas flows from 5000 SCFM to 100,000 SCFM (2.3 kL/s to 47 kL/s), a third type of solvent recovery system involves a *rotor concentrator*. VOCs are continuously adsorbed onto a multilayer, corrugated wheel whose honeycomb structure has been coated with powdered AC. The wheel area is divided into three zones: one for adsorption, one for desorption

(i.e., regeneration), and one for cooling. Each of the zones is isolated from the other by tight sealing. The wheel rotates at low speed, continuously exposing new portions of the streams to each of the three zones. Though the equipment is expensive, operational efficiencies are high—95% to 98%.

4. AIR STRIPPING

Air strippers are primarily used to remove volatile organic compounds (VOCs) or other target substances from water. In operation, contaminated water enters a stripping tower at the top, and fresh air enters at the bottom. The effectiveness of the process depends on the volatility of the compound, its temperature and concentration, and the liquid-air contact area. However, removal efficiencies of 80% to 90% (and above) are common for VOCs. Figure 51.2 shows a typical air stripping operation.

Figure 51.2 Schematic of Air Stripping Operation

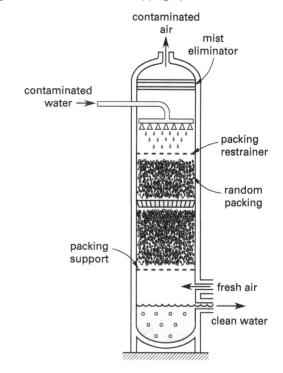

There are three types of stripping towers—*packed towers* filled with synthetic *packing media* (polypropylene balls, rings, saddles, etc.) like those shown in Fig. 51.3, *spray towers*, and (less frequently for VOC removal) *tray towers* with horizontal trays arranged in layers vertically. *Redistribution rings* (*wall wipers*) prevent channeling down the inside of the tower. (*Channeling* is the flow of liquid through a few narrow paths rather than an over-the-bed packing.) The stripping air is generated by a small blower at the column base. A mist eliminator at the top eliminates entrained water from the air.

As the contaminated water passes over packing media in a packed tower, the target substance leaves the liquid and enters the air where the concentration is lower. The mole fraction of the target substance in the water, x, decreases; the mole fraction of the target substance in the air, y, increases.

The discharged air, known as *off-gas*, is discharged to a process that destroys or recovers the target substance. (Since the quantities are small, recovery is rarer.) Destruction of the target substance can be accomplished by flaring, carbon absorption, and incineration. Since flaring is dangerous and carbon absorption creates a secondary waste if the GAC is not regenerated, incineration is often preferred.

Figure 51.3 *Packing Media Types*

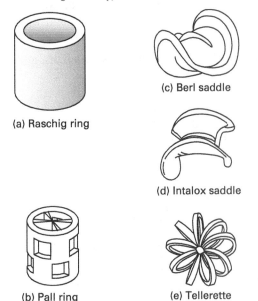

(a) Raschig ring

(b) Pall ring

(c) Berl saddle

(d) Intalox saddle

(e) Tellerette

Organic chemicals possess varying properties of volatility as measured using Henry's law constant, H.

Convection: Overall Coefficients

$$p_A^* = HC_{AL} \qquad 51.3$$

In Eq. 51.3, p_A^* is the partial pressure of the contaminant, which is assumed to be equal to the contaminant's vapor pressure. C_{AL} is the contaminant's concentration in water, which is assumed to be equal to the contaminant's water solubility. Rearranging Eq. 51.3 gives

$$H = \frac{p_A^*}{C_{AL}} \qquad 51.4$$

Henry's law constant can be dimensionless or have units of pressure or atm·m^3/mol. For air stripper design, units of atm·m^3/mol are used. A compound is typically considered to be volatile if its Henry's law constant is greater than 0.001 atm·m^3/mol at 20°C. As Henry's law constant increases, the likelihood of a compound to partition from the water into the atmosphere increases. For these units, Henry's law constant is computed by

$$H = \frac{p_A^* M}{C_{AL} RT} \qquad 51.5$$

In Eq. 51.5, p_A^* is in units of atmospheres, and C_{AL} is in units of milligrams per liter or grams per cubic meter.

The air-water ratio is another parameter used in stripper design. The value of the air-water ratio can be calculated using Eq. 51.6.

$$\frac{V_a}{V_w} = \frac{\left(\dfrac{C_0}{C} - 1\right)RT}{H} \qquad 51.6$$

In Eq. 51.6, the volume of air, V_a, and the volume of water, V_w, can be in any units as long as they are the same. C_0 and C are the initial and final concentrations in water, and can also be in any units as long as they are the same.

The air-water ratio can be approximated for air stripping design by dividing the stripping factor, S, by Henry's law constant. Values for S range from 1 to 5, with typical values of 3 to 4. Selecting a higher value for S produces a more conservative design. When used in air-water ratio calculations, Henry's law constant is usually dimensionless.

$$\frac{V_a}{V_w} = \frac{S}{H} \qquad 51.7$$

In designing air strippers to volatilize organics from solution, another parameter commonly used is the *overall mass transfer rate coefficient*, $K_L a$ (commonly called the *mass transfer coefficient*). $K_L a$ has units of time^{-1}. The value of $K_L a$ is specific to the physical parameters of the testing apparatus. For example, different values of $K_L a$ must be determined for different types of packing media used in air stripping process equipment.

Equation 51.8 through Eq. 51.13 are also used in air stripper design.

$$Q_a = Q\left(\frac{V_a}{V_w}\right) \qquad 51.8$$

$$\mathrm{HLR}_m = \mathrm{HLR}\rho_w \qquad 51.9$$

$$\mathrm{NTU} = \frac{S}{S-1} \ln\left(\frac{\left(\dfrac{C_0}{C}\right)(S-1)+1}{S}\right) \qquad 51.10$$

$$\text{HTU} = \frac{\text{HLR}_m}{(K_L a)\rho_w} \qquad 51.11$$

$$\text{packing height} = (\text{NTU})(\text{HTU}) \qquad 51.12$$

$$\text{column area} = \frac{Q}{\text{HLR}} \qquad 51.13$$

Example 51.2

Water at 20°C contains trichloroethene (TCE) at a concentration of 200 μg/L. The TCE is to be stripped from the water to reach a concentration of 5 μg/L, the maximum contaminant level. Determine the air-water ratio needed.

Solution

From a table of values for Henry's law constant, for TCE, H is 9.10×10^{-3} atm·m^3/mol. [Water Solubility, Vapor Pressure, Henry's Law Constant, K_{oc}, and K_{ow} Data for Selected Chemicals]

From a table of values, the universal gas constant is 82.05 atm·cm^3/mol·K. [Ideal Gas Constants]

$$R = \frac{82.05 \ \dfrac{\text{atm·cm}^3}{\text{mol·K}}}{\left(100 \ \dfrac{\text{cm}}{\text{m}}\right)^3}$$

$$= 0.000082 \ \text{atm·m}^3/\text{mol·K}$$

Use Eq. 51.6 to find the air-water ratio needed.

$$\frac{V_a}{V_w} = \frac{\left(\dfrac{C_0}{C} - 1\right)RT}{H}$$

$$= \frac{\left(\dfrac{200 \ \dfrac{\mu g}{L}}{5 \ \dfrac{\mu g}{L}} - 1\right)}{9.10 \times 10^{-3} \ \dfrac{\text{atm·m}^3}{\text{mol}}}$$
$$\times \left(0.000082 \ \dfrac{\text{atm·m}^3}{\text{mol·K}}\right)$$
$$\times (293\text{K})$$

$$= 103$$

Example 51.3

Contaminated groundwater is pumped at 375 gal/min and contains chloroform (CHCl$_3$) at a concentration of 2000 μg/L. The chloroform concentration must be reduced to 20 μg/L. Air stripping is used to remove the chloroform from the water. For chloroform, molecular weight is 120 g/mol, vapor pressure is 0.26 atm at 25°C, and solubility in water is 9300 mg/L at 25°C.

Bench tests are used to determine a mass transfer coefficient of 0.019 s^{-1} using an appropriate packing material. The desired hydraulic loading rate is 30 gal/min-ft^2, and the stripping factor, S, is 3.0. The air and water are at 25°C, and the atmospheric pressure is 1.0 atm.

Determine the airflow rate, packing height, and column diameter.

Solution

From a table of values, the universal gas constant is 82.05 atm·cm^3/mol·K. [Ideal Gas Constants]

$$R = \frac{82.05 \ \dfrac{\text{atm·cm}^3}{\text{mol·K}}}{\left(100 \ \dfrac{\text{cm}}{\text{m}}\right)^3}$$

$$= 0.000082 \ \text{atm·m}^3/\text{mol·K}$$

Use Eq. 51.5 to calculate the dimensionless Henry's law constant.

$$H = \frac{p_A^* M}{C_{AL} R T}$$

$$= \frac{(0.26 \ \text{atm})\left(120 \ \dfrac{g}{\text{mol}}\right)}{\left(\dfrac{\left(9300 \ \dfrac{\text{mg}}{L}\right)\left(1000 \ \dfrac{\text{m}^3}{L}\right)}{1000 \ \dfrac{\text{mg}}{g}}\right)}$$
$$\times \left(0.000082 \ \dfrac{\text{atm·m}^3}{\text{mol·K}}\right)$$
$$\times (25°\text{C} + 273°)$$

$$= 0.14$$

From Eq. 51.7, the air-water ratio is

$$\frac{V_a}{V_w} = \frac{S}{H} = \frac{3.0}{0.14} = 21.4 \quad (22)$$

From Eq. 51.8, the airflow rate is

$$Q_a = Q\left(\frac{V_a}{V_w}\right)$$

$$= \left(\frac{375\ \dfrac{\text{gal}}{\text{min}}}{60\ \dfrac{\text{sec}}{\text{min}}}\right)(22)\left(3.785\ \frac{\text{L}}{\text{gal}}\right)$$

$$= 520\ \text{L/s}$$

From Eq. 51.10, the number of transfer units is

$$\text{NTU} = \frac{S}{S-1}\ln\left(\frac{\left(\dfrac{C_0}{C}\right)(S-1)+1}{S}\right)$$

$$= \frac{3}{3-1}\ln\left(\frac{\left(\dfrac{2000\ \dfrac{\mu\text{g}}{\text{L}}}{20\ \dfrac{\mu\text{g}}{\text{L}}}\right)(3-1)+1}{3}\right)$$

$$= 6.3$$

From Eq. 51.9, the mass hydraulic loading rate is

$$\text{HLR}_m = \text{HLR}\rho_w$$

$$= \left(\frac{\left(30\ \dfrac{\text{gal}}{\text{min-ft}^2}\right)\left(3.785\ \dfrac{\text{L}}{\text{gal}}\right)}{\left(60\ \dfrac{\text{s}}{\text{min}}\right)\left(1000\ \dfrac{\text{L}}{\text{m}^3}\right)}\right)$$

$$\times\left(1000\ \frac{\text{kg}}{\text{m}^3}\right)$$

$$= 20\ \text{kg/m}^2\text{·s}$$

From Eq. 51.11, the height of each transfer unit is

$$\text{HTU} = \frac{\text{HLR}_m}{(K_L a)\rho_w}$$

$$= \frac{20\ \dfrac{\text{kg}}{\text{m}^2\text{·s}}}{(0.019\ \text{s}^{-1})\left(1000\ \dfrac{\text{kg}}{\text{m}^3}\right)}$$

$$= 1.05\ \text{m}$$

From Eq. 51.12, the packing height is

$$(\text{NTU})(\text{HTU}) = (1.05\ \text{m})(6.3)$$

$$= 6.6\ \text{m}$$

From Eq. 51.13, the column area is

$$\frac{Q}{\text{HLR}} = \frac{375\ \dfrac{\text{gal}}{\text{min}}}{30\ \dfrac{\text{gal}}{\text{min-ft}^2}} = 12.5\ \text{ft}^2$$

Find the column diameter.

$$A = \frac{\pi d^2}{4}$$

$$d = \sqrt{\frac{4A}{\pi}}$$

$$= \frac{\sqrt{\dfrac{(4)(12.5\ \text{ft}^2)}{\pi}}}{3.28\ \dfrac{\text{ft}}{\text{m}}}$$

$$= 1.2\ \text{m}$$

5. BIOFILTRATION

The term *biofiltration* refers to the use of composting and soil beds. A *biofilter* is a bed of soil or compost through which runs a distribution system of perforated pipe. Contaminated air or liquid flows through the pipes and into the bed. Volatile organic compounds (VOCs) are oxidized to CO_2 by microorganisms. Volatile inorganic compounds (VICs) are oxidized to acids (e.g., HNO_3 and H_2SO_4) and salts.

Biofilters require no fuel or chemicals when processing VOCs, and the operational lifetime is essentially infinite. For VICs, the lifetime depends on the soil's capacity to neutralize acids. Though reaction times are long and absorption capacities are low (hence the large areas required), the oxidation continually regenerates (rather than depletes) the treatment capacity. Once operational, biofiltration is probably the least expensive method of eliminating VOCs and VICs.

The *removal efficiency* of a biofilter is given by Eq. 51.14. k is an empirical *reaction rate constant*, and t is the *bed residence time* of the carrier fluid (water or air) in the bed. The reaction rate constant depends on the temperature but is primarily a function of the biodegradability of the target substance. Biofiltration typically removes 80% to 99% of volatile organic and inorganic compounds (VOCs and VICs).

$$\eta = 1 - \frac{C_{\text{out}}}{C_{\text{in}}} = 1 - e^{-kt} \qquad \text{51.14}$$

6. BIOREACTION

Bioreactors (*reactor tanks*) are open or closed tanks containing dozens or hundreds of slowly rotating disks covered with a biological film of microorganisms (i.e., colonies). Closed tanks can be used to maintain anaerobic or other design atmosphere conditions. For example, methanotrophic bacteria are useful in breaking down chlorinated hydrocarbons (e.g., trichlorethylene (TCE), dichloroethylene (DCE), and vinyl chloride (VC)) that would otherwise be considered nonbiodegradable. Methanotrophic bacteria derive their food from methane gas that is added to the bioreactor. An enzyme, known as MMO, secreted by the bacteria breaks down the chlorinated hydrocarbons.

7. BIOREMEDIATION

Bioremediation encompasses the methods of biofiltration, bioreaction, bioreclamation, activated sludge, trickle filtration, fixed-film biological treatment, landfilling, and injection wells (for in situ treatment of soils and groundwater). Bioremediation relies on microorganisms in a moist, oxygen-rich environment to oxidize solid, liquid, or gaseous organic compounds, producing carbon dioxide and water. Bioremediation is effective for removing volatile organic compounds (VOCs) and easy-to-degrade organic compounds such as BTEXs (benzene, toluene, ethylbenzene, and xylene).[1] Wood-preserving wastes, such as creosote and other polynuclear aromatic hydrocarbons (PAHs), can also be treated.

Bioremediation can be carried out in open tanks, packed columns, beds of porous synthetic materials, composting piles, or soil. The effectiveness of bioremediation depends on the nature of the process, the time and physical space available, and the degradability of the substance. Table 51.3 categorizes gases according to their general degradabilities.

Table 51.3 *Degradability of Volatile Organic and Inorganic Gases*

rapidly degradable VOCs	rapidly reactive VICs	slowly degradable VOCs	very slowly degradable VOCs
alcohols	H_2S	hydrocarbons[a]	halogenated
aldehydes	NOx (but	phenols	hydrocarbons[b]
ketones	not N_2O)	methylene	polyaromatic
ethers	SO_2	chloride	hydrocarbons
esters	HCl		CS_2
organic acids	NH_3		
amines	PH_3		
thiols	SiH_4		
other molecules	HF		
containing			
O, N, or S			
functional			
groups			

[a]Aliphatics degrade faster than aromatics, such as xylene, toluene, benzene, and styrene.
[b]These include trichloroethylene, trichloroethane, carbon tetrachloride, and pentachlorophenol.

Though slow, limited by microorganisms with the specific affinity for the chemicals present, and susceptible to compounds that are toxic to the microorganisms, bioremediation has the advantage of destroying substances rather than merely concentrating them. Bioremediation is less effective when a variety of different compounds are present simultaneously.

8. BIOVENTING

Bioventing is the treatment of contaminated soil in a large plastic-covered tank. Clean air, water, and nutrients are continuously supplied to the tank while off-gases are suctioned off. The off-gas is cleaned with activated carbon (AC) adsorption or with thermal or catalytic oxidation prior to discharge. Bioventing has been used successfully to remove volatile hydrocarbon compounds (e.g., gasoline and BTEX compounds) from soil.

9. CARBON ADSORPTION

Adsorption uses high surface-area activated carbon to remove organic contaminants. Adsorption can use *granular activated carbon* (GAC) in column or fluidized-bed reactors or *powdered activated carbon* (PAC) in complete-mix reactors. Activated carbon is relatively nonspecific, and it will remove a wide variety of refractory organics as well as some inorganic contaminants. It should generally be considered for organic contaminants that are nonpolar, have low solubility, or have high molecular weights.

The most common problems associated with columns are breakthrough, excessive headloss due to plugging, and premature exhaustion. *Breakthrough* occurs when the carbon becomes saturated with the target

---

[1]BTEXs are common ingredients in gasoline.

compound and can hold no more. *Plugging* occurs when biological growth blocks the spaces between carbon particles. *Premature exhaustion* occurs when large and high molecular-weight molecules block the internal pores in the carbon particles. These latter two problems can be prevented by locating the carbon columns downstream of filtration.

10. SOIL WASHING

Soil washing is effective in removing heavy metals, wood preserving wastes (PAHs), and BTEX compounds from contaminated soil. Soil washing is a two-step process. In the first step, soil is mixed with water to dissolve the contaminants. Additives are used as required to improve solubility. In the second step, additional water is used to flush the soil and to separate the fine soil from coarser particles. (Semi-volatile materials concentrate in the fines.) Metals are extracted by adding chelating agents to the wash water.[2] The contaminated wash water is subsequently treated.

11. SORBENT INJECTION

Sorbent injection (FSI) involves injecting a limestone slurry directly into the upper parts of the combustion chamber (or into a thermally favorable location downstream) to reduce SOx. Heat calcines the limestone into reactive lime. Fast drying prevents wet particles from building up in the duct. Lime particles are captured in a scrubber with or without an electrostatic precipitator (ESP). With only an ESP, the SOx removal efficiency is approximately 50%.

12. SPARGING

Sparging is the process of using air injection wells to bubble air through groundwater. The air pushes volatile contaminants into the overlying soil above the aquifer where they can be captured by vacuum extraction.

13. STEAM STRIPPING

Steam stripping is more effective than air stripping for removing semi- and non-volatile compounds, such as diesel fuel, oil, and other organic compounds with boiling points up to approximately 400°F (200°C). Operation of a steam stripper is similar to that of an air stripper, except that steam is used in place of the air. Steam strippers can be operated at or below atmospheric pressure. Higher vacuums will remove greater amounts of the compound.

14. THERMAL DESORPTION

Thermal desorption is primarily used to remove volatile organic compounds (VOCs) from contaminated soil. The soil is heated directly or indirectly to approximately 1000°F (540°C) to evaporate the volatiles. This method differs from incineration in that the released gases are not burned but are captured in a subsequent process step (e.g., activated carbon filtration).

15. VACUUM EXTRACTION

Vacuum extraction is used to remove many types of volatile organic compounds (VOCs) from soil. The VOCs are pulled from the soil through a well dug in the contaminated area. Air is withdrawn from the well, vacuuming volatile substances with it.

16. VAPOR CONDENSING

Some vapors can be removed simply by cooling them to below their dew points. Traditional contact (open) and surface (closed) condensers can be used.

17. VITRIFICATION

Vitrification melts and forms slag and ash wastes into glass-like pellets. Heavy metals and toxic compounds cannot leach out, and the pellets can be disposed of in hazardous waste landfills. Vitrification can occur in the incineration furnace or in a stand-alone process. Stand-alone vitrification occurs in an electrically heated vessel where the temperature is maintained at 2200°F to 2370°F (1200°C to 1300°C) for up to 20 hr or so. Since the electric heating is nonturbulent, flue gas cleaning systems are not needed.

18. NOMENCLATURE

A	area	ft^2	m^2
C	concentration	mg/L	mg/L
C_{AL}	contaminant concentration in water	mg/L	mg/L
d	diameter	ft	m
EBCT	empty bed contact time	min	min
H	Henry's law constant	atm·m^3/mol	atm·m^3/mol
HLR	hydraulic loading rate	gal/ft^2·min	L/m^2·min
HLR$_m$	mass hydraulic loading rate	lbm/ft^2-sec	kg/m^2·s
HTU	height of a transfer unit	ft	m
k	reaction rate constant	1/sec	1/s

Site Assessment & Remediation

---

[2]A *chelate* is a ring-like molecular structure formed by unshared electrons of neighboring atoms. A *chelating agent* (*chelant*) is an organic compound in which atoms form bonds with metals in solution. By combining with metal ions, chelates control the damaging effects of trace metal contamination. Ethylenediaminetetraacetic acid (EDTA) types are the leading industrial chelants.

K	adsorption capacity at unit concentration	mg/g	mg/g
$K_L a$	mass transfer coefficient	time^{-1}	time^{-1}
M	mass of medium	g/L	g/L
M	molecular weight	lbm/lbmol	g/mol
$1/n$	strength of adsorption	–	–
NTU	number of transfer units	–	–
p_A^*	contaminant partial pressure	atm	Pa
q_e	equilibrium loading on the medium, equal to X/M	mg/g	mg/g
Q	water flow rate	gal/min	L/s
Q_a	airflow rate	ft^3/sec	L/s
R	universal gas constant 82.05	atm·ft^3/ lbmol·°R	atm·m^3/ mol·K
S	stripping factor	–	–
t	time	sec	s
T	temperature	°R	K
V	volume	ft^3	m^3
x	fraction by weight	–	–
X	mass of chemical adsorbed	mg	mg

Symbols

η	efficiency	%	%
ρ	density	lbm/ft^3	kg/m^3

Subscripts

0	initial
a	air
e	equilibrium
recirc	recirculations
w	water

Topic V: Environmental Health and Safety

Environmental
Health & Safety

52

Health and Safety

1. INTRODUCTION

The environmental engineer will most likely encounter health and safety concerns while working with remediation of hazardous waste sites. Although health and safety are always a concern in the workplace, especially around industrial process equipment or construction, remediation of hazardous waste sites involves the purposeful exposure of workers to known and unknown hazardous substances. This chapter will focus primarily on health and safety related to work at hazardous waste sites.

Hazardous waste site remediation work can expose workers to many types of physical, chemical, and biological hazards. Such hazards include

- heat and cold stress

- noise

- ionizing radiation

- cumulative trauma disorders

- biological hazards

- electrical hazards

- physical safety hazards

- fire and explosion

- oxygen deficiency

- chemical exposure

The principal focus of this chapter is on chemical exposure; fire and explosion from gases and vapors along with oxygen deficiency are also covered briefly.

This chapter describes chemical hazard assessment, health and safety plans for hazardous chemicals, air monitoring for chemical gases and vapors, and personal protective equipment.

2. HAZARD ASSESSMENT

From an OSHA (legal) standpoint, a *health hazard* is a chemical, mixture of chemicals, or pathogen for which there is statistically significant evidence based on at least one study conducted in accordance with established scientific principles that acute or chronic health effects may occur in exposed employees. Health hazards include chemicals that are carcinogens, toxic or highly toxic agents, reproductive toxins, irritants, corrosives, sensitizers, hepatotoxins (affecting the liver), nephrotoxins (affecting the kidneys), neurotoxins (affecting the nervous system), and agents that act on the hematopoietic system (the production of blood cells), along with agents that damage the lungs, skin, eyes, or mucous membranes.

One of the first tasks associated with remediation at a hazardous waste site is assessment of the hazards that must be dealt with so that the necessary precautions (such as the use of personal protective equipment) can be taken. In addition, an assessment must be made to identify conditions that may be immediately dangerous to life or health (IDLH). Chemicals pose three primary types of hazards: exposure to chemicals, fire or explosion, and oxygen deficiency.

Chemical Exposure

Chemical exposure can be acute or chronic. *Acute exposure* is exposure to a high dose over a short period of time (hours or a few days) with symptoms appearing during or within a short time (a few days to two weeks) after exposure. *Chronic exposure* is exposure to a low dose over a long period (months or years) with the onset of symptoms not observable for years or even decades after exposure. Because a toxic chemical may be colorless and odorless (or may desensitize a person's sense of smell (e.g., hydrogen sulfide)), hazard assessment should not depend on the ability of the worker to sense the chemical or feel discomfort.

The primary exposure route for chemicals is inhalation. Toxic chemicals can readily enter the bloodstream through the lungs, a fact that makes this exposure route extremely significant to workers at hazardous waste remediation sites. Another important exposure route is through contact with the skin (dermal) or the eyes. Chemicals can directly injure the skin or the eyes or can pass through the skin or eyes into the bloodstream to attack target organs or tissues anywhere in the body. Exposure via ingestion, although unlikely, can result

Environmental Health & Safety

from chewing gum, smoking, drinking, or eating at a hazardous waste site. Exposure through injection can occur through puncture wounds sustained while working with materials at the site.

Chemical hazard assessment involves determining the concentrations of chemicals present and their associated health hazards. There are four primary sources of information about chemical hazards. These sources are

- *Integrated Risk Information System* (IRIS): an EPA database

- *Threshold Limit Values for Chemical Substances and Physical Agents, Biological Exposure Indices*: published annually by the American Conference of Governmental Industrial Hygienists (ACGIH)

- *NIOSH Pocket Guide to Chemical Hazards*: published by the National Institute for Occupational Safety and Health (NIOSH)

- 29 CFR 1910.1000, Subpart Z, *Toxic and Hazardous Substances*: workplace exposure limits established by the Occupational Safety and Health Administration

Because workers performing remediation at a hazardous waste site will need to meet the requirements of the Occupational Safety and Health Act, the environmental engineer should first review the workplace exposure limits of 29 CFR 1910.1000, Subpart Z, *Toxic and Hazardous Substances*, in assessing the chemical exposure hazard. These exposure limits establish the legal air contaminant levels to which workers can safely be exposed over an 8 h workday without personal protection.

The other three sources of data on air contaminants should also be reviewed because they may be more restrictive. The OSHA standards may be more or less restrictive than the NIOSH-, ACGIH-, or EPA-recommended air contaminant levels, so the environmental engineer should understand the differences in order to be able to make a conservative (protective) decision on acceptable exposure levels.

Fire or Explosion

The causes of fires and explosions at hazardous waste sites include chemical reactions, ignition of flammable or explosive chemicals, ignition of materials due to oxygen enrichment, disturbance of shock-sensitive materials, and the sudden release of pressure of contained chemicals. Common causes of fire or explosion at hazardous waste sites are from handling containers or mixing chemicals or exposing chemicals (usually as vapors) to sparks from equipment.

Oxygen Deficiency

The term *oxygen deficiency* refers to a concentration of oxygen below which atmosphere-supplying respiratory protection must be provided. This condition exists in atmospheres where the percentage of oxygen by volume is less than 19.5%.

Oxygen deficiency can result from the displacement of normal air, which contains 21% oxygen by volume at sea level, with other gases present at the hazardous waste site. Oxygen deficiency is an especially important hazard in enclosed spaces, such as tanks or buildings, or in low-lying areas, such as in ravines, ditches, or depressions at landfills. Although the main physiological effects of oxygen deprivation become apparent at 16% oxygen, to account for differences in individuals and measurement errors, air with 19.5% or less oxygen is considered the level at which personal protection is required. Physiological symptoms of oxygen deficiency include impaired attention and judgment and increased breathing and heart rate.

Example 52.1

Workers are remediating an old warehouse that contains a variety of chemicals. The warehouse is made of concrete and has little ventilation. Approximately 7 kg of a combustible carbon material was accidentally burned in the warehouse a few days ago. Assume that 60% of the carbon formed carbon dioxide and 40% formed carbon monoxide. The warehouse contains a maximum of 2 kmol (i.e., 2000 mol or 2 kg·mol) of oxygen in air. Determine whether the air in the warehouse is oxygen deficient.

Solution

Air is approximately 21% O_2 and 79% N_2 by volume. The combustion occurs as follows.

$$C + O_2 \longrightarrow CO_2$$

$$2C + O_2 \longrightarrow 2CO$$

For CO_2, the number of moles of O_2 consumed per mole of carbon is

$$(0.6)\left(\frac{7 \text{ kg C}}{12 \dfrac{\text{kg C}}{\text{kmol C}}}\right)\left(1 \ \frac{\text{kmol O}_2}{\text{kmol C}}\right)$$
$$= 0.35 \text{ kmol O}_2 \text{ consumed}$$

Environmental Health & Safety

For CO, the number of moles of O_2 consumed per mole of carbon is

$$(0.4)\left(\frac{7 \text{ kg C}}{12 \frac{\text{kg}}{\text{kmol C}}}\right)\left(\frac{1 \text{ kmol O}_2}{2 \text{ kmol C}}\right)$$
$$= 0.12 \text{ kmol O}_2 \text{ consumed}$$

The total O_2 consumed is

$$0.35 \text{ kmol} + 0.12 \text{ kmol} = 0.47 \text{ kmol O}_2$$

Calculate the concentration of each gas component in the air.

$$(0.35 \text{ kmol O}_2)\left(1 \frac{\text{kmol CO}_2}{\text{kmol O}_2}\right) = 0.35 \text{ kmol CO}_2$$

$$(0.12 \text{ kmol O}_2)\left(2 \frac{\text{kmol CO}}{\text{kmol O}_2}\right) = 0.23 \text{ kmol CO}$$

$$2 \text{ kmol O}_2 - 0.47 \text{ kmol O}_2 \text{ consumed}$$
$$= 1.53 \text{ kmol O}_2 \text{ remaining}$$

$$(2 \text{ kmol O}_2)\left(\frac{0.79 \text{ kmol N}_2}{0.21 \text{ kmol O}_2}\right) = 7.52 \text{ kmol N}_2$$

The total moles of gases is

$$0.35 \text{ kmol CO}_2 + 0.23 \text{ kmol CO}$$
$$+ 1.53 \text{ kmol O}_2 + 7.52 \text{ kmol N}_2$$
$$= 9.64 \text{ kmol}$$

The mole fraction of O_2 is

$$\frac{1.53 \text{ kmol O}_2}{9.64 \text{ kmol total}} = 0.159 \quad (15.9\%)$$

Because the mole fraction of O_2 is less than 19.5%, the atmosphere in the warehouse is oxygen deficient.

Identification of IDLH Conditions

The phrase "immediately dangerous to life or health" (IDLH) refers to an atmospheric concentration of any toxic, corrosive, or asphyxiant substance that poses an immediate threat to life or would interfere with an individual's ability to escape from a dangerous atmosphere. It is extremely important that conditions that may be immediately dangerous to life or health are systematically identified in the initial hazard assessment. There are several indicators of conditions that may be IDLH.

Tanks that must be entered by remediation personnel are prime candidates for IDLH conditions because vapors may be present that are toxic or explosive or that create an oxygen-deficient environment. Even tanks that are open at the top can present an IDLH hazard because some gases are heavier than air and can accumulate in the bottom of the tank. Buildings and trenches that must be entered present similar types of IDLH hazards.

Flammable or explosive conditions are also IDLH hazards. The presence of gas-generating materials, such as from incompatible mixtures or as suggested by bulging drums, should be regarded as potential for flammable or explosive conditions. Instrument readings can also indicate the presence of such conditions.

The presence of extremely toxic materials, such as cyanide or phosgene, also presents potential IDLH conditions. Depending on concentration, many other chemicals can present IDLH hazards. The National Institute for Occupational Safety and Health (NIOSH) *Pocket Guide to Chemical Hazards* gives the IDLH concentration for many chemicals and also gives flammable and explosive concentrations.

Visible vapor clouds should raise concerns about IDLH conditions, as should the presence of dead vegetation or dead animals.

Site Control Zones

Hazardous waste remediation and emergency response sites are typically divided into zones for control of personnel entry and personal protective equipment requirements, as well as for designation of the activities that can be conducted in each zone. The EPA nomenclature for the various zones and perimeters is shown in Fig. 52.1 along with other common terms for the same items.

Figure 52.1 Site Control Zone Nomenclature

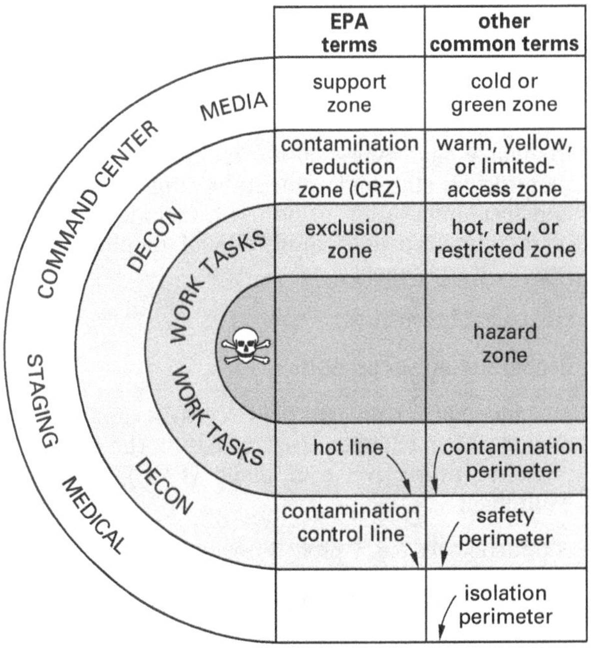

3. HEALTH AND SAFETY PLANS

Often, an environmental engineer will be in charge of remediation at a hazardous waste site and will be assisted by a health and safety officer. The health and safety officer will normally prepare a health and safety *plan* and be responsible for conducting the health and safety *program*. (That is, the "program" is the implementation of the plan.) However, the environmental engineer needs to understand the elements comprising the health and safety plan and program so he or she can ensure that work tasks are conducted accordingly. The environmental engineer must also be able to identify changing conditions that require revisions or exceptions to a plan or program.

Employers are required to develop and implement a written health and safety plan for their employees involved in hazardous waste operations. The program must be designed to identify, evaluate, and control health and safety hazards, and provide for emergency response for hazardous waste operations. The written health and safety plan must be made available to any contractor or subcontractor who will be involved with the hazardous waste operation; to employees or union representatives; to OSHA personnel; and to personnel of federal, state, or local agencies with regulatory authority over the site.

The health and safety plan must be in writing and address the health and safety hazards of each phase of site operation and include the requirements and procedures for employee protection. The plan must include at least the following.

- a health and safety risk or hazard analysis for each site task and operation found in the work plan

- employee training assignments

- personal protective equipment to be used by workers for each of the site tasks and operations being conducted

- medical surveillance requirements

- frequency and types of air monitoring, personnel monitoring, and environmental sampling techniques and instrumentation to be used, including methods of maintenance and calibration of monitoring and sampling equipment

- site control measures

- decontamination procedures

- an emergency response plan for safe and effective responses to emergencies, including the necessary personal protective equipment (PPE) and other equipment

- confined space entry procedures

- a spill-containment program

- requirements for pre-entry briefings

The site-specific health and safety plan must provide for pre-entry briefings prior to initiating any site activity, and whenever needed to ensure that workers understand the plan and that the plan is being followed.

The health and safety supervisor is responsible for conducting inspections to determine the effectiveness of the site health and safety plan and to ensure that any deficiencies in the plan are corrected.

4. AIR MONITORING

Ambient air monitoring should include monitoring for toxic substances, combustible gases, inorganic and organic particulates and vapors, oxygen deficiency, and specific chemicals suspected to be present. Air monitoring is necessary both to assess air contaminant hazards at the site for which personal protection may be required and to monitor ongoing exposure to air contaminants so the medical monitoring requirements can be determined. Air monitoring will provide information for preparing a health and safety plan that will define the location where specific levels of personal protection are needed and the degree of hazards present.

Direct-Reading Instruments

Direct-reading air-monitoring instruments are commonly used at hazardous waste remediation sites to determine the level of personal protection needed, identify the types of instruments needed to further characterize the air contaminants, and provide information needed to prepare the health and safety plan. Also, these types of instruments can detect conditions that are immediately dangerous to life and health (IDLH).

Direct-reading instruments are designed for, and are limited to, measuring high concentrations of air contaminants. They are unreliable at detecting low concentrations. Also, though designed to detect air contaminants relative to specific chemicals, these instruments detect other chemicals, thereby giving false readings.

OSHA has established a database of chemical sampling information. The database provides general information about chemicals, including exposure limits, health factors, and monitoring requirements and methods. Similar information can also be obtained from the NIOSH *Manual of Analytical Methods* and from the OSHA *Analytical Methods*.

Limit of Detection

Published sampling procedures give minimum sampling requirements. Small samples may not contain enough of the contaminant to produce a signal distinguishable from the instrument's background noise.

Typically, the *signal-to-noise ratio* ranges from 3:1 to 5:1. The chemical mass that produces a 3:1 signal-to-noise ratio is known as the *limit of detection* (LOD). The LOD is defined as the lowest concentration that can be determined to be statistically different from a blank. For example, if the LOD of a contaminant is 20 ng, the laboratory would report a finding of less than 20 ng as ND or "not detected." Although the actual mass and concentration of the contaminant cannot be determined from such a result, it is possible to use the LOD mass and the sample volume to compute an estimated concentration, which can be useful in the absence of better data. If the LOD concentration is less than 25% of the threshold limit value (TLV) at a specified mass (and preferably less than 10% of the TLV), the exposure can be considered acceptable. If the LOD concentration is more than 25% of the TLV, it would be good practice to resurvey the exposure area.

Limit of Quantification

The *limit of quantification* (LOQ) is the minimum mass of a contaminant above which the precision of a reported result is better than a specified level. *Precision* is the reproducibility of replicate analyses of a sample with the same mass or concentration. Precision refers to the size of deviations from the mean observation. Numerically, precision is the *coefficient of variation* (CV), which is calculated by dividing the estimated standard deviation by the mean of a series of measurements. The ratio of the standard deviation to the mean is a decimal and when multiplied by 100 can be reported as a percent precision.

Normal and Standard Conditions

After air monitoring is performed, the results are compared with one or more criteria. In industrial hygiene air monitoring, the criteria are based on a defined set of conditions called *normal* conditions, or *normal temperature and pressure* (NTP). This set of conditions differs from the set of conditions used in basic science and engineering called *standard* conditions, or *standard temperature and pressure* (STP). NTP may be called "standard" conditions in the context of industrial hygiene or air monitoring, but note the difference between NTP conditions and STP conditions listed in Table 52.1.

Table 52.1 *Properties of Atmospheric Air*

parameter	NTP	STP
temperature	25°C = 298.15K	0°C = 273.15K
pressure	1 atm, 760 mm Hg, 101 325 Pa	1 atm, 760 mm Hg, 101 325 Pa
dry air molecular weight	28.96 g/mol	28.96 g/mol

Often it is necessary to convert the volume of gases and vapors (or air) from field conditions to normal conditions and vice versa. The conversions are based on the ideal gas law. The ideal gas equation can be written at two different environmental conditions, standard condition (S) and field condition (F).

$$V_F = V_S\left(\frac{P_S}{P_F}\right)\left(\frac{T_F}{T_S}\right) \qquad \text{52.1}$$

$$V_S = V_F\left(\frac{P_F}{P_S}\right)\left(\frac{T_S}{T_F}\right) \qquad \text{52.2}$$

Concentration Conversions

For typical industrial hygiene air monitoring, gases and vapors are presumed to behave as ideal gases. The *ideal gas equation* can be used to calculate the density of a presumed ideal gas.

Ideal Gas Constants

$$PV = nRT \qquad \text{52.3}$$

The *number of moles* is the mass divided by the molecular weight. By rearranging the molecular weight, [Periodic Table of Elements]

$$N = \frac{m}{M} \qquad \text{52.4}$$

Density is the mass of the gas or vapor divided by its volume.

$$\rho = \frac{m}{V} \qquad \text{52.5}$$

Concentration is the mass of a substance A divided by the volume of substance B.

$$C_{m/V,\text{A}} = \frac{m_\text{A}}{V_\text{B}} \qquad \text{52.6}$$

Concentration can also be expressed volumetrically as the volume of substance A to 10^6 volumes of substance A plus B, or parts per million (ppm or ppmv).

$$C_{\text{ppm},\text{A}} = \left(\frac{V_\text{A}}{V_\text{A} + V_\text{B}}\right)(10^6) \qquad \text{52.7}$$

However, in virtually all industrial hygiene or air sampling of gas and vapor monitoring, the volume of substance A (the gas or vapor contaminant) is small compared to the volume of substance B (typically air), so Eq. 52.7 can be simplified to eliminate the volume of substance A in the denominator. This assumption results in an error of only about 1% at 10 000 ppm and much less at the lower concentrations typically encountered. The simplified equation is

$$C_{\text{ppm,A}} \approx \left(\frac{V_A}{V_B}\right)(10^6) \qquad 52.8$$

The density can be found by combining Eq. 52.3 through Eq. 52.5.

$$\rho_A = \frac{m_A}{V_A} = \frac{P(M_A)}{R^*T} \qquad 52.9$$

Equation 52.8 and Eq. 52.9 can be combined to provide a way to convert between concentrations as ppm and mass per volume.

$$C_{\text{ppmv,A}} = \left(\frac{C_{\text{kg/m}^3,A}R^*T}{(M_A)P}\right)(10^6) \qquad 52.10$$

$$C_{\text{kg/m}^3} = \left(\frac{C_{\text{ppmv,A}}(M_A)P}{R^*T}\right)(10^{-6}) \qquad 52.11$$

In the case of air contaminants, a volumetric concentration expressed in ppmv may be converted to a mass-per-unit volume concentration (e.g., mg/m^3), which is still considered to be a volumetric concentration. In some cases, concentrations calculated as ratios of two masses will also be reported in parts per million. These gravimetric ratios will be distinguished from volumetric ratios by the designation "ppmw" (i.e., parts per million by weight).

Example 52.2

Workers at a hazardous waste site remediation project are expected to encounter toluene ($C_6H_5CH_3$) vapors, so the project manager decides to perform a preliminary survey to collect and analyze air samples to determine the level of personal protection needed. The lab provides the following data on toluene from NIOSH Method No. 1500.

conditions	NTP
ppm conversion	1 ppm = 3.77 mg/m^3 at NTP
range studied	548–2190 mg/m^3 or 1.13–4.51 mg/sample
ACGIH TLV-TWA	188 mg/m^3
estimated LOD	53 μg
air sampling flow rate	≤ 0.2 L/min
air sampling media	100 mg/50 mg charcoal
molecular weight	92 g/mol

A preliminary sample analyzed by the lab resulted in "not detected" (ND) as the mass of toluene measured. The volume collected was 1.8 L, and the field temperature and pressure during sample collection were 32°C and 720 mm Hg, respectively.

(a) Was the TLV-TWA exceeded in the preliminary sample?

(b) If the concentration of toluene is 30 ppm at field conditions, what sample air pump flow rate at field conditions is needed to collect the minimum sample mass in 1 h? Is this flow rate acceptable?

(c) What is the maximum sample volume that should be collected at field conditions if the concentration of toluene is twice the TLV?

Solution

(a) Correct the sample volume to NTP. [Unit and Conversion Factors]

At NTP, $T_S = 25°C + 273° = 298\text{K}$. [Temperature Conversions]

$$P_S = (760 \text{ mm Hg})\left(\frac{1 \text{ atm}}{760 \text{ mm Hg}}\right) = 1 \text{ atm}$$

At field conditions, $T_F = 32°C + 273° = 305\text{K}$. [Temperature Conversions]

$$P_F = (720 \text{ mm Hg})\left(\frac{1 \text{ atm}}{760 \text{ mm Hg}}\right) = 0.947 \text{ atm}$$
$$V_F = 1.8 \text{ L}$$

Use Eq. 52.2. At NTP,

$$V_S = V_F\left(\frac{P_F}{P_S}\right)\left(\frac{T_S}{T_F}\right) = (1.8 \text{ L})\left(\frac{720 \text{ mm Hg}}{760 \text{ mm Hg}}\right)\left(\frac{298\text{K}}{305\text{K}}\right)$$
$$= 1.67 \text{ L}$$

The maximum value of the reported concentration of toluene is ND, which is 53 μg. Calculate the LOD concentration at NTP.

From Eq. 52.6,

$$C_{\text{LOD}} = \frac{m}{V_S}$$
$$= \left(\frac{53 \ \mu\text{g}}{1.67 \text{ L}}\right)\left(\frac{1 \text{ g}}{10^6 \ \mu\text{g}}\right)\left(10^3 \ \frac{\text{mg}}{\text{g}}\right)\left(10^3 \ \frac{\text{L}}{\text{m}^3}\right)$$
$$= 31.8 \text{ mg/m}^3$$

The percent of TLV is

$$\frac{31.8 \ \frac{\text{mg}}{\text{m}^3}}{188 \ \frac{\text{mg}}{\text{m}^3}} = 0.17 \quad (17\%)$$

The TLV is not exceeded. The preliminary sample result was 17% of the TLV at NTP. The preliminary result could be considered acceptable exposure, but additional monitoring could also be appropriate.

(b) From Eq. 52.9,

$$\rho_{\text{toluene}} = \frac{m_{\text{toluene}}}{V_{\text{toluene}}} = \frac{P_F(M_{\text{toluene}})}{R^*T}$$

$$= \frac{(0.947 \ \text{atm})\left(92 \ \frac{\text{g}}{\text{mol}}\right)}{\left(0.08205 \ \frac{\text{L}\cdot\text{atm}}{\text{mol}\cdot\text{K}}\right)(305\text{K})}$$

$$= 3.48 \ \text{g/L}$$

Find the mass concentration of toluene at field conditions.

From Eq. 52.8,

$$C_{\text{ppmv,A}} \approx \left(\frac{V_A}{V_B}\right)(10^6)$$

$$\frac{V_A}{V_B} \approx \frac{C_{\text{ppmv,A}}}{10^6}$$

$$\frac{V_{\text{toluene}}}{V_{\text{air}}} \approx \frac{C_{\text{ppmv,toluene}}}{10^6} = \frac{30}{10^6}$$

Use Eq. 52.6.

$$C_{m/V,A} = \frac{m_A}{V_B} = \left(\frac{m_A}{V_A}\right)\left(\frac{V_A}{V_B}\right)$$

$$C_{m/V,\text{toluene}} = \left(\frac{m_{\text{toluene}}}{V_{\text{toluene}}}\right)\left(\frac{V_{\text{toluene}}}{V_{\text{air}}}\right)$$

$$= \left(3.48 \ \frac{\text{g}}{\text{L}}\right)\left(\frac{30}{10^6}\right)$$

$$\times \left(10^3 \ \frac{\text{mg}}{\text{g}}\right)\left(10^3 \ \frac{\text{L}}{\text{m}^3}\right)$$

$$= 104 \ \text{mg/m}^3 \quad [\text{field conditions}]$$

Find the flow rate at field conditions to collect the minimum mass.

$$m_{\text{sample}} = Q_F C_F t = 1.13 \ \text{mg}$$

$$Q_F = \frac{1.13 \ \text{mg}}{C_F t}$$

$$= \frac{(1.13 \ \text{mg})\left(10^3 \ \frac{\text{L}}{\text{m}^3}\right)}{\left(104 \ \frac{\text{mg}}{\text{m}^3}\right)(60 \ \text{min})}$$

$$= 0.180 \ \text{L/min} \quad [\text{field conditions}]$$

Convert the maximum allowable flow rate at NTP to field conditions and compare with the calculated flow rate needed.

From Eq. 52.1 and the given air sampling flow rate from NIOSH Method No. 1500,

$$Q_F = Q_S\left(\frac{P_S}{P_F}\right)\left(\frac{T_F}{T_S}\right)$$

$$= \left(0.2 \ \frac{\text{L}}{\text{min}}\right)\left(\frac{760 \ \text{mm Hg}}{720 \ \text{mm Hg}}\right)\left(\frac{305\text{K}}{298\text{K}}\right)$$

$$= 0.216 \ \text{L/min}$$

Since the 0.180 L/min calculated is less than the 0.216 L/min allowable (field), the flow rate is acceptable.

(c) Find the mass concentration of toluene at field conditions. Use the ACGIH TLV-TWA.

$$C_{m/V} = 2 \times \text{TLV} = (2)\left(188 \ \frac{\text{mg}}{\text{m}^3}\right)$$

$$= 376 \ \text{mg/m}^3 \quad [\text{at NTP}]$$

$$C_{m/V} = \left(376 \ \frac{\text{mg}}{\text{m}^3}\right)\left(\frac{720 \ \text{mm Hg}}{760 \ \text{mm Hg}}\right)\left(\frac{298\text{K}}{305\text{K}}\right)$$

$$= 348 \ \text{mg/m}^3 \quad [\text{field conditions}]$$

Find the volume to collect the mass required. From the data supplied by NIOSH Method No. 1500, the maximum mass per sample is $m_{\text{max}} = 4.51 \ \text{mg}$.

From Eq. 52.6,

$$V = \frac{m}{C_{m/V}} = \frac{(4.51 \ \text{mg})\left(10^3 \ \frac{\text{L}}{\text{m}^3}\right)}{348 \ \frac{\text{mg}}{\text{m}^3}} = 13.0 \ \text{L}$$

Environmental Health & Safety

Dynamic Mixing of Gases and Vapors (Dilution)

When performing air monitoring, it is sometimes necessary to prepare continuously flowing mixtures of dilute concentrations of a contaminant in air. Usually a contaminant in high concentrations is diluted to low concentrations after mixing with a flow of clean air. In such situations, the phrase "clean air" means that the contaminant is absent. The principle of the conservation of mass and the continuity equation are used to determine the mixture volumes.

When a contaminant is diluted by mixing with air, the volume increases and the concentration decreases; however, the total mass of contaminant remains constant. The mass of the contaminant entering the mixing point equals the mass of the contaminant leaving the mixing point (conservation of mass). Also, total flow in a system is equal to the sum of the individual flows (continuity equation). A dynamic mixing system is illustrated in Fig. 52.2.

Figure 52.2 *Dynamic Mixing*

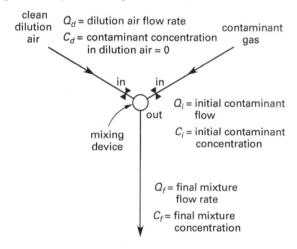

The conservation of mass equation is as follows.

$$Q_d C_d + Q_i C_i = Q_f C_f \qquad 52.12$$

Because clean air is used for dilution, $C_d = 0$. Therefore,

$$Q_f C_f = Q_i C_i \qquad 52.13$$

The continuity equation is as follows.

$$Q_f = Q_i + Q_d \qquad 52.14$$

Example 52.3

A continuous flow of a mixture of nitrogen dioxide (NO_2) and clean air is to be prepared for equipment testing. The desired final mixed concentration is 100 ppm NO_2 with a flow rate of 25 cm^3/min at the field conditions. The system is to operate at 710 mm Hg and 35°C. The NO_2 is available from a cylinder producing a concentration of 945 mg/m^3 at 25°C and 760 mm Hg. What dilution and NO_2 flow rates are required?

Solution

Find the temperature at field conditions. [Unit and Conversion Factors]

Temperature Conversions

$$T_F = 35°C + 273° = 308K$$

$$P_F = (710 \text{ mm Hg})\left(\frac{1 \text{ atm}}{760 \text{ mm Hg}}\right) = 0.934 \text{ atm}$$

Correct the NO_2 concentration to field conditions. From Eq. 52.1,

$$
\begin{aligned}
V_F &= V_S\left(\frac{P_S}{P_F}\right)\left(\frac{T_F}{T_S}\right) \\
&= (1 \text{ m}^3)\left(\frac{760 \text{ mm Hg}}{710 \text{ mm Hg}}\right)\left(\frac{308K}{25°C + 273°}\right) \\
&= 1.11 \text{ m}^3
\end{aligned}
$$

$$
\begin{aligned}
C_{m/V_F, NO_2} &= C_{m/V_S, NO_2}\left(\frac{V_S}{V_F}\right) \\
&= \left(945 \ \frac{\text{mg}}{\text{m}^3}\right)\left(\frac{1 \text{ m}^3}{1.11 \text{ m}^3}\right) \\
&= 854 \text{ mg/m}^3
\end{aligned}
$$

Convert the NO_2 concentration to ppm at field conditions. Use a periodic table to find the molecular weight of NO_2. [Periodic Table of Elements]

Ideal Gas Mixtures

$$
\begin{aligned}
M_{NO_2} &= (1)\left(14 \ \frac{\text{g}}{\text{mol}}\right) + (2)\left(16 \ \frac{\text{g}}{\text{mol}}\right) \\
&= 46 \text{ g/mol}
\end{aligned}
$$

From Eq. 52.10,

$$C_{\text{ppm,NO}_2} = \left(\frac{C_{m/V,\text{NO}_2} R^* T}{(M_{\text{NO}_2})P}\right)(10^6)$$

$$= \frac{\left(854 \frac{\text{mg}}{\text{m}^3}\right)\left(0.08205 \frac{\text{L·atm}}{\text{mol·K}}\right)}{\left(46 \frac{\text{g}}{\text{mol}}\right)(0.934 \text{ atm})}$$
$$\times (308\text{K})(10^6)$$
$$\times \left(10^3 \frac{\text{L}}{\text{m}^3}\right)\left(10^3 \frac{\text{mg}}{\text{g}}\right)$$

$$= 500 \text{ ppm} \quad \text{[field conditions]}$$

From the conservation of mass equation (see Eq. 52.13),

$$Q_i = \frac{Q_f C_f}{C_i} = \frac{\left(25 \frac{\text{cm}^3}{\text{min}}\right)(100 \text{ ppm})}{500 \text{ ppm}}$$

$$= 5 \text{ cm}^3/\text{min} \quad \text{[field conditions]}$$

From the continuity equation (see Eq. 52.14),

$$Q_d = Q_f - Q_i = 25 \frac{\text{cm}^3}{\text{min}} - 5 \frac{\text{cm}^3}{\text{min}}$$

$$= 20 \text{ cm}^3/\text{min} \quad \text{[field conditions]}$$

Determining Concentrations Using Partial Pressures

Raoult's law is used to determine the concentration of the components of a mixture in a closed space. The law applies to the components of a mixture containing two or more volatile components where the mixed vapor phase is in equilibrium with the solution. The solution mole fraction, x^*, is used in these calculations, rather than the vapor phase mole fractions.

$$P_{\text{PVP},i} = x_i^* P_{\text{VP},i} \qquad \text{52.15}$$

$$P_{\text{TVP,mixture}} = \sum x_i^* P_{\text{VP},i} \qquad \text{52.16}$$

The concentration of any gaseous component can be found from the ratio of the partial pressure to its mole fraction in the entire gaseous mixture. This relationship provides a way to determine concentration (volume-

based) of an individual component in a gaseous mixture. P_{total} is the atmospheric pressure, which includes the pressures of the air as well as all vapors.

$$C_{\text{ppm},i} = \left(\frac{P_{\text{PVP},i}}{P_{\text{total}}}\right)(10^6)$$
$$= \left(\frac{x_i P_{\text{VP},i}}{P_{\text{total}}}\right)(10^6) \qquad \text{52.17}$$

Example 52.4

At 20°C, the vapor pressures of pure ethanol (CH_3CH_2OH) and pure water (H_2O) are 44 mm Hg and 17.54 mm Hg, respectively. 90-proof whiskey contains 45% ethanol and 55% water by mass. What would the partial vapor pressures of ethanol and water be above a glass containing 400 g of such whiskey in a closed room at 20°C and 750 mm Hg?

Solution

Find the molecular weights of ethanol and water. [Periodic Table of Elements]

Ideal Gas Mixtures

$$M_{\text{ethanol}} = (6)\left(1 \frac{\text{g}}{\text{mol}}\right) + (2)\left(12 \frac{\text{g}}{\text{mol}}\right)$$
$$+ (1)\left(16 \frac{\text{g}}{\text{mol}}\right)$$
$$= 46 \text{ g/mol}$$

$$M_{\text{water}} = (2)\left(1 \frac{\text{g}}{\text{mol}}\right) + (1)\left(16 \frac{\text{g}}{\text{mol}}\right)$$
$$= 18 \text{ g/mol}$$

The masses of ethanol and water are

$$m_{\text{ethanol}} = (0.45)(400 \text{ g}) = 180 \text{ g}$$
$$m_{\text{water}} = (0.55)(400 \text{ g}) = 220 \text{ g}$$

Use Eq. 52.4 to determine the number of moles of ethanol and water. [Periodic Table of Elements]

Ideal Gas Mixtures

$$N_{\text{ethanol}} = \frac{m_{\text{ethanol}}}{M_{\text{ethanol}}} = \frac{180 \text{ g}}{46 \frac{\text{g}}{\text{mol}}} = 3.91 \text{ mol}$$

$$N_{\text{water}} = \frac{m_{\text{water}}}{M_{\text{water}}} = \frac{220 \text{ g}}{18 \frac{\text{g}}{\text{mol}}} = 12.2 \text{ mol}$$

Environmental Health & Safety

The total number of moles is

Ideal Gas Mixtures

$$N_{total} = N_{ethanol} + N_{water}$$
$$= 3.91 \text{ mol} + 12.2 \text{ mol}$$
$$= 16.1 \text{ mol}$$

The mole fraction of ethanol is

Ideal Gas Mixtures

$$x_{ethanol} = \frac{N_{ethanol}}{N_{total}} = \frac{3.91 \text{ mol}}{16.1 \text{ mol}} = 0.24$$

The mole fraction of water is

Ideal Gas Mixtures

$$x_{water} = \frac{N_{water}}{N_{total}} = \frac{12.2 \text{ mol}}{16.1 \text{ mol}} = 0.76$$

The partial vapor pressure can be determined from Eq. 52.15.

$$P_{PVP,ethanol} = x_{ethanol} P_{VP,ethanol}$$
$$= (0.24)(44 \text{ mm Hg})$$
$$= 10.7 \text{ mm Hg}$$
$$P_{PVP,water} = x_{water} P_{VP,water}$$
$$= (0.76)(17.54 \text{ mm Hg})$$
$$= 13.3 \text{ mm Hg}$$

Use Eq. 52.17 to solve for the ethanol and water concentrations.

$$C_{ppm,ethanol} = \left(\frac{P_{PVP,ethanol}}{P_{total}}\right)(10^6)$$
$$= \left(\frac{10.7 \text{ mm Hg}}{750 \text{ mm Hg}}\right)(10^6)$$
$$= 14\,200 \text{ ppm}$$
$$C_{ppm,water} = \left(\frac{P_{PVP,water}}{P_{total}}\right)(10^6)$$
$$= \left(\frac{13.3 \text{ mm Hg}}{750 \text{ mm Hg}}\right)(10^6)$$
$$= 17\,700 \text{ ppm}$$

Exposure Assessment

The purpose of air sampling and analysis is to determine worker exposure to airborne contaminants in relation to standards and guides developed by regulatory and technical service organizations. The time-weighted average calculation method is the primary method used in assessing worker exposure.

Time-Weighted Average Exposure Method

The *time-weighted average* (TWA) method apportions the *measured actual concentrations* (MACs) from air sampling to the times over which the exposures occurred. This method averages the high and low exposures experienced based on the associated time period of each exposure and gives a more accurate representative measure of exposure than averaging measurements alone. The TWA is calculated by multiplying each concentration by the exposure time interval and dividing the sum by the total time period.

Time-Weighted Average (TWA)

$$\text{TWA} = \frac{\sum_{i=1}^{n} c_i t_i}{\sum_{i=1}^{n} t_i} \qquad 52.18$$

The *averaging time* (*averaging period*) is either 8 h or 10 h, depending on the exposure standard that is being used for the exposure assessment. The standards in use in the United States are given in Table 52.2.

The OSHA and NIOSH standards typically specify allowable air sampling flow rates for the various chemicals. The flow rates are established to prevent loss of sample mass due to breakthrough or collection of a sample mass lower than the limit of detection (LOD) for the method. These standards also indicate total volumes that should be collected during the sampling period.

Example 52.5

A worker is exposed to toluene (C_7H_8) during an 8 h shift as follows. The measured actual concentration (MAC) data are reported at NTP.

duration (h)	measured actual concentration (MAC) (mg/m³)
1	300
3	400
3	200
1	0

Compare the TWA exposure to the TLV-TWA of 188 mg/m³ and the PEL-TWA of 200 ppm.

Solution

Find the temperature and pressure at NTP.

Temperature Conversions

$$T = 25°C + 273° = 298K$$
$$p = 1 \text{ atm}$$

Table 52.2 Exposure Assessment Standards

organization	standard	averaging period
American Conference of Governmental Industrial Hygienists (ACGIH)	threshold limit value—time-weighted average (TLV-TWA)	480 min (8 h) per day, 40 h per week
American Conference of Governmental Industrial Hygienists (ACGIH)	threshold limit value—short-term exposure limit (TLV-STEL)	15 min TWA; TLV-TWA must not be exceeded for the day
American Conference of Governmental Industrial Hygienists (ACGIH)	threshold limit value—ceiling (TLV-C)	instantaneous reading; use 15 min TWA if no instantaneous reading available
Occupational Safety and Health Administration (OSHA)	permissible exposure limit—time-weighted average (PEL-TWA)	480 min (8 h) per day, 40 h per week
Occupational Safety and Health Administration (OSHA)	permissible exposure limit—short-term limit (PEL-STEL)	15 min TWA unless another exposure period is given; PEL-TWA must not be exceeded
Occupational Safety and Health Administration (OSHA)	permissible exposure limit—ceiling (PEL-C)	instantaneous reading; use 15 min TWA if no instantaneous reading available
National Institute of Occupational Safety and Health (NIOSH)	recommended exposure limit—time-weighted average (REL-TWA)	up to 10 h per day, 40 h per week
National Institute of Occupational Safety and Health (NIOSH)	recommended exposure limit—short-term exposure limit (REL-STEL)	15 min TWA
National Institute of Occupational Safety and Health (NIOSH)	recommended exposure limit—ceiling (REL-C)	instantaneous reading
American Industrial Hygiene Association (AIHA)	workplace environmental exposure level—time-weighted average (WEEL-TWA)	up to 8 h per day, 40 h per week
American Industrial Hygiene Association (AIHA)	workplace environmental exposure level—short-term time-weighted average (WEEL-short term TWA)	averaging time is specified for each chemical
American Industrial Hygiene Association (AIHA)	workplace environmental exposure level—ceiling (WEEL-C)	instantaneous reading

Find the TWA of the MAC values reported.

Use Eq. 52.18.

Time-Weighted Average (TWA)

$$\text{TWA} = \frac{\sum c_{m/V,i} t_i}{\sum t_i}$$

$$= \frac{\left(300 \ \frac{\text{mg}}{\text{m}^3}\right)(1 \ \text{h}) + \left(400 \ \frac{\text{mg}}{\text{m}^3}\right)(3 \ \text{h}) + \left(200 \ \frac{\text{mg}}{\text{m}^3}\right)(3 \ \text{h}) + \left(0 \ \frac{\text{mg}}{\text{m}^3}\right)(1 \ \text{h})}{1 \ \text{h} + 3 \ \text{h} + 3 \ \text{h} + 1 \ \text{h}}$$

$$= 262.5 \ \text{mg/m}^3$$

Convert the PEL-TWA to mass concentration.

Find the molecular weight of toluene. [Periodic Table of Elements]

Ideal Gas Mixtures

$$M_{\text{toluene}} = (8)\left(1 \ \frac{\text{g}}{\text{mol}}\right) + (7)\left(12 \ \frac{\text{g}}{\text{mol}}\right) = 92 \ \text{g/mol}$$

From Eq. 52.11,

$$C_{m/V,\text{toluene}} = \left(\frac{C_{\text{ppm,toluene}} M_{\text{toluene}} P}{R^* T}\right)(10^{-6})$$

$$= \left|\frac{\begin{array}{c}(200\ \text{ppm})\left(92\ \dfrac{\text{g}}{\text{mol}}\right)(1\ \text{atm})\\[4pt] \times \left(10^3\ \dfrac{\text{mg}}{\text{g}}\right)\left(10^3\ \dfrac{\text{L}}{\text{m}^3}\right)\end{array}}{\left(0.08205\ \dfrac{\text{L·atm}}{\text{mol·K}}\right)(298\text{K})}\right|(10^{-6})$$

$$= 753\ \text{mg/m}^3$$

The MAC-TWA for toluene ($262.5\ \text{mg/m}^3$) exceeds the TLV-TWA ($188\ \text{mg/m}^3$) but not the PEL-TWA ($753\ \text{mg/m}^3$).

Short-Term Exposure Method

When measuring short-term and ceiling exposures, it is sometimes not possible to get an instantaneous reading to compare to ceiling levels. In these cases, the analyst must estimate the MAC peak from TWA 15 min measurements. The ceiling can be estimated by assuming that the MAC rises from the pre-short term value to a peak and then declines to the post-short term value in the form of a triangle. The duration of the peak can be assumed to be 5 min (or some other rational duration) out of the 15 min short-term sampling period. The peak concentration can then be estimated by assuming that the area of the triangle equals the short-term MAC times 15 min. This procedure is illustrated in the following example.

Example 52.6

Workers are exposed to chemical A ($M_A = 78$ g/mol) for a 60 min period as follows.

period	sample no.	MAC
pre-peak: 30 min	1	$100\ \text{mg/m}^3$
peak: 15 min	2	$160\ \text{mg/m}^3$
post-peak: 30 min	3	$120\ \text{mg/m}^3$

The short-term exposure limits for chemical A are as follows.

TLV-STEL	TLV-C	PEL-STEL	PEL-C
50 ppm	100 ppm	$239.3\ \text{mg/m}^3$	$478.6\ \text{mg/m}^3$

Have the short-term exposure limits been exceeded?

Solution

Assume the peak occurs for 5 min out of the 15 min short-term exposure period. Draw a diagram of exposure and estimate the ceiling exposure.

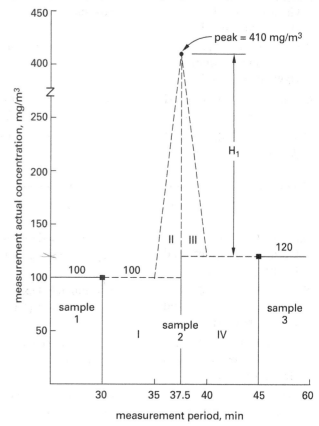

The average exposure during the short-term exposure period is 160 mg/m³. The concentration multiplied by time for the short-term period is

$$Ct = \left(160\ \frac{\text{mg}}{\text{m}^3}\right)(15\ \text{min}) = 2400\ \text{mg·min/m}^3$$

Assume the peak occurs as shown in the illustration over a 5 min period. The sum of areas I through IV must equal the Ct of 2400 mg·min/m³.

$$A_I = \left(100\ \frac{\text{mg}}{\text{m}^3}\right)(7.5\ \text{min}) = 750\ \text{mg·min/m}^3$$

$$A_{II} = \left(\frac{1}{2}\right)\left(H_1 + 20\ \frac{\text{mg}}{\text{m}^3}\right)(2.5\ \text{min})$$

$$A_{III} = \left(\frac{1}{2}\right)(H_1)(2.5\ \text{min})$$

$$A_{IV} = \left(120\ \frac{\text{mg}}{\text{m}^3}\right)(7.5\ \text{min}) = 900\ \text{mg·min/m}^3$$

$$Ct = A_{\text{I}} + A_{\text{II}} + A_{\text{III}} + A_{\text{IV}}$$

$$2400\ \frac{\text{mg} \cdot \text{min}}{\text{m}^3} = 750\ \frac{\text{mg} \cdot \text{min}}{\text{m}^3} + 25\ \frac{\text{mg} \cdot \text{min}}{\text{m}^3}$$
$$+ \left(\frac{2.5\ \text{min}}{2}\right) H_1 + \left(\frac{2.5\ \text{min}}{2}\right) H_1$$
$$+ 900\ \frac{\text{mg} \cdot \text{min}}{\text{m}^3}$$

$$H_1 = 290\ \text{mg/m}^3$$

The peak is

$$120\ \frac{\text{mg}}{\text{m}^3} + 290\ \frac{\text{mg}}{\text{m}^3} = 410\ \text{mg/m}^3$$

Find the temperature and pressure at NTP.

Temperature Conversions

$$T = 25°\text{C} + 273° = 298\text{K}$$

$$P = (760\ \text{mm Hg})\left(\frac{1\ \text{atm}}{760\ \text{mm Hg}}\right) = 1\ \text{atm}$$

Use Eq. 52.10.

$$C_{\text{ppm,A}} = \left(\frac{C_{m/V,\text{A}} R^* T}{(M_\text{A}) P}\right)(10^6)$$

$$= \left(\frac{\left(410\ \dfrac{\text{mg}}{\text{m}^3}\right)\left(0.08205\ \dfrac{\text{L} \cdot \text{atm}}{\text{mol} \cdot \text{K}}\right)(298\text{K})}{\left(78\ \dfrac{\text{g}}{\text{mol}}\right)(1\ \text{atm})\left(10^3\ \dfrac{\text{L}}{\text{m}^3}\right)\left(10^3\ \dfrac{\text{mg}}{\text{g}}\right)}\right)(10^6)$$

$$= 129\ \text{ppm}$$

Similarly, a concentration of 160 mg/m^3 is 50.2 ppm. Compare the results.

	TLV-STEL (ppm)	TLV-C (ppm)	PEL-STEL (mg/m³)	PEL-C (mg/m³)
standard	50	100	239.3	478.6
MAC/estimated	50.2	129	160	410
compliance	no	no	yes	yes

Upper and Lower Confidence Limits

It is sometimes necessary to know how close a calculated or measured result is to the standard when *sampling and analytical errors* (SAEs) are considered. It is assumed that the measured value may be higher or lower than the true value by plus or minus the SAE. Usually a 95% confidence level is used in calculating the SAE. The SAE is calculated as follows.

$$\text{SAE} = (1.645)(\text{CV}_{\text{total}}) \qquad 52.19$$

The constant 1.645 relates to the 95% confidence level for a normal distribution. With a 99% confidence level, this factor would be 2.329.

The coefficient of variation including sampling and analytical errors, CV$_{\text{total}}$, is found from the following.

$$\text{CV}_{\text{total}} = \frac{s}{\bar{x}} \qquad 52.20$$

The CV and the SAE are used to calculate the upper and lower confidence limits, UCL and LCL, as follows.

$$\text{UCL}\ (95\%) = \frac{\bar{x}}{\text{PEL}} + \text{SAE} \qquad 52.21$$

$$\text{LCL}\ (95\%) = \frac{\bar{x}}{\text{PEL}} - \text{SAE} \qquad 52.22$$

The term "$\bar{x}$/PEL" is often referred to as the *standardized concentration*. When the LCL is less than or equal to 1.0, occupants are being underexposed, and OSHA regulations are not being violated. When the LCL is greater than 1.0, an OSHA violation exists. If the LCL is less than or equal to 1.0 and the UCL is greater than 1.0, a possible overexposure may have occurred.

Example 52.7

Workers were exposed to solvent X49 for an 8 h period. The laboratory reported a TWA concentration of 90 ppm of X49 in samples collected for analysis. The PEL for X49 is 102 ppm. The total coefficient of variation including sampling and analytical errors, CV$_{\text{total}}$, is 30%. Was the PEL exceeded?

Solution

Find the SAE. From Eq. 52.19,

$$\text{SAE} = (1.645)(\text{CV}_{\text{total}}) = (1.645)(0.30)$$
$$= 0.49$$

Find the standardized concentration.

$$\frac{\bar{x}}{\text{PEL}} = \frac{90\ \text{ppm}}{102\ \text{ppm}} = 0.88$$

Find the UCL and LCL. From Eq. 52.21,

$$\text{UCL}\ (95\%) = \frac{\bar{x}}{\text{PEL}} + \text{SAE} = 0.88 + 0.49$$
$$= 1.38$$

From Eq. 52.22,

$$\text{LCL}\ (95\%) = \frac{\bar{x}}{\text{PEL}} - \text{SAE} = 0.88 - 0.49$$
$$= 0.39$$

Environmental Health & Safety

Although the 90 ppm did not exceed the PEL of 102 ppm, the 95% UCL did exceed 1.0, so it's 95% certain that a violation has occurred.

5. PERSONAL PROTECTIVE EQUIPMENT

OSHA Requirements

The Occupational Safety and Health Administration has promulgated regulations covering *personal protective equipment* (PPE) for worker respiratory protection as well as regulations for other workplace hazards. The regulations include standards and guidelines for respiratory protection and are found in 29 CFR 1910, Subpart I, *Personal Protective Equipment*. The following standards are relevant.

- 1910.132—General Requirements

- 1910.134—Respiratory Protection

- 1910.134, Appendix A—Fit Testing Procedures (Mandatory)

- 1910.134, Appendix B-1—User Seal Check Procedures (Mandatory)

- 1910.134, Appendix B-2—Respiratory Cleaning Procedures (Mandatory)

- 1910.134, Appendix C—OSHA Respirator Medical Evaluation Questionnaire (Mandatory)

- 1910.134, Appendix D—Information for Employees Using Respirators When Not Required Under Standard (Mandatory)

- 1910 Subpart I, Appendix B—Nonmandatory Compliance Guidelines for Hazard Assessment and Personal Protective Equipment Selection

The first four OSHA standards for respiratory personal protection describe the general requirements for personal protective equipment in the workplace, the standards for the use of respiratory protective equipment, fit testing procedures for respirators, and procedures for the user to check the respirator seal.

The standards for respiratory protection are legal requirements that should be implemented at hazardous waste remediation sites as well as at other workplace locations. Guidelines for application of the OSHA standards for personal protection are given in the *OSHA Technical Manual*.

EPA Levels of Protection

The levels of protection developed by the EPA for work at hazardous waste sites are described in Table 52.3. These levels are also accepted by OSHA for hazardous waste site entry with the additional requirement that each PPE ensemble must be tailored to the specific situation to provide the most appropriate level of protection.

6. NOMENCLATURE

c, C	concentration	ppm, kg/m^3
CV	coefficient of variation including sampling error	–
LCL	lower confidence limit	–
m	mass	kg
M	molecular weight	g/mol
n, N	number of moles	–
p, P	pressure	atm
PEL	permissible exposure limit	ppm, mg/m^3
Q	flow rate	m^3/s
R, R^*	universal gas constant	L·atm/mol·K
s	sample standard deviation	–
SAE	sampling and analytical errors	–
t	time	s
T	temperature	K
TLV	threshhold limit value	ppm, mg/m^3
UCL	upper confidence limit	–
V	volume	L, m^3
x^*	solution phase mole fraction	–
$\bar{x}$	concentration	ppm

Symbols

ρ	density	mg/m^3

Subscripts

d	dilution
f	final condition
F	field
i	initial condition or ith interval
LOD	limit of detection
PVP	partial vapor
S	standard
TVP	total vapor
VP	vapor

Table 52.3 *EPA Levels of Protection*

aspect	level A	level B
required	vapor protective suit that meets NFPA 1991; pressure-demand, full-face SCBA; inner chemical-resistant gloves; chemical-resistant safety boots; two-way radio communication	liquid splash-protective suit that meets NFPA 1992; pressure-demand, full-facepiece SCBA; inner chemical-resistant gloves; chemical-resistant safety boots; two-way radio communications; hard hat
optional	cooling system; outer gloves; hard hat	cooling system; outer gloves
protection provided	highest available level of respiratory, skin, and eye protection from solid, liquid, and gaseous chemicals	provides same level of respiratory protection as level A, but less skin protection; liquid splash protection but no protection against chemical vapors or gases
used when	chemicals have been identified and have high level of hazards to respiratory system, skin, and eyes; substances are present with known or suspected skin toxicity or carcinogenicity; operations must be conducted in confined or poorly ventilated areas	chemicals have been identified but do not require a high level of skin protection; initial site surveys are required until higher levels of hazards are identified; primary hazards associated with site entry are from liquid and not vapor contact
limitations	protective clothing must resist permeation by the chemical or mixtures present; en-semble items must allow integration without loss of performance	protective clothing items must resist penetration by the chemicals or mixtures present; ensemble items must allow integration without loss of performance

aspect	level C	level D
required	support function protective garment that meets NFPA 1993; full-facepiece, air-purifying, canister-equipped respirator; chemical-resistant gloves and safety boots; two-way communications system; hard hat	coveralls; safety boots/shoes; safety glasses or chemical splash goggles
optional	face-shield; escape SCBA	gloves; escape SCBA; face-shield
protection provided	same level of skin protection as level B, but a lower level of respiratory protection; liquid splash protection but no protection against chemical vapors or gases	no respiratory protection; minimal skin protection
used when	contact with site chemicals will not affect the skin; air contaminants have been iden-tified and concentrations measured; a can-ister is available that can remove the contaminant; the site and its hazards have been completely characterized	atmosphere contains no known hazard; work functions preclude splashes, immersion, potential for inhalation, or direct contact with hazardous chemicals
limitations	protective clothing items must resist pene-tration by the chemical or mixtures present; chemical airborne concentration must be less than IDLH levels; atmosphere must contain at least 19.5% oxygen; not acceptable for chemical emergency response	this level should not be worn in the hot zone; the atmosphere must contain at least 19.5% oxygen; not acceptable for chemical emergency response

Environmental Health & Safety

53 Security, Emergency Plans, Incident Response

1. TYPES OF EMERGENCIES

Incident Types

There are seven major types of incidents that may require emergency response.

- highway accidents
- railway accidents
- ship and barge accidents
- aircraft accidents
- plant accidents
- pipeline accidents
- container failures

Highway Accidents

Highway accidents are by far the most common type of emergency incident to which an environmental engineer may offer expertise. Hazardous materials are carried by trucks every day over highways and streets. These materials are transported through residential as well as commercial and rural areas, and they often pose a risk to humans and the natural environment. A highway accident may involve injuries and multiple vehicles, so the additional complication of dealing with hazardous materials presents a challenging problem for emergency response personnel. The effects resulting from spills of toxic materials into drains or waterways or the release of toxic gases into the air may be catastrophic. Highway accidents may also result in long-term chronic effects from contamination of soil and water that cannot be cleaned up completely after the accident.

The cause of most highway accidents is human error. However, accidents also occur due to weather, mechanical failure, roadway obstructions, and roadbed defects.

Railway Accidents

Railway accidents are less common than highway accidents, but the potential effects are greater due to the larger volumes of hazardous materials that may be involved in the incident. For example, a typical railroad tank car has a capacity of approximately 130 m^3, while a typical highway tanker has a capacity of approximately 30 m^3 to 40 m^3. Moreover, rail accidents may occur at locations with difficult access compared to highway accidents.

Although track deterioration and failure are common causes of rail accidents, other causes include collision with other trains or vehicles at crossings, terrorist acts, and signal or mechanical failures.

Ship and Barge Accidents

Ship and barge accidents can have catastrophic effects on waterways. Containment and clean-up of ship and barge accidents are very difficult and frequently are insufficient to prevent both short-term catastrophic effects and long-term chronic effects on aquatic life, wildlife, and humans. Releases to waterways can result in an accumulation of pollutants in sediments that is difficult to remediate and may result in decades of chronic effects.

Aircraft Accidents

Fortunately, aircraft accidents are rare compared to other transportation accidents, and they seldom involve large quantities of hazardous cargoes. Aircraft accidents may occur anywhere and can involve releases of petroleum products from the aircraft (as well as release of cargo).

Plant Accidents

Accidents at plants or facilities are not uncommon. They often involve serious health and safety threats to humans as well as to the natural environment. Plant accidents are usually associated with manufacturing operations but may also occur from the use, transportation, or storage of both hazardous and nonhazardous materials. Causes of plant accidents include equipment

Environmental Health & Safety

failure, human error (such as accidental mixing of reactive products), product or tank overflow, physical damage to containers, and exposure to fire, water, or heat.

Besides the concern for the safety of employees at a plant during an accident, there is also concern about hazardous material or pollutant releases to the natural environment (air and water) and effects on humans in the vicinity of the plant site (so-called "downwinders"). In many industrial plants, discharges to the water or emissions to the air are a routine part of operation. During an incident, these routine pathways to the environment may become routes for hazardous materials or additional pollutants to escape.

Pipeline Accidents

Pipeline accidents are not common due to the care taken to design pipelines with generous safety factors, and the fact that pipelines are, for the most part, buried and are not exposed to potential hazards. When they occur, however, pipeline accidents can be quite dramatic and are often catastrophic. Pipelines carry a wide variety of products such as natural gas, water, petroleum, and wastewater. Some gases and liquids transported by pipelines pose more serious hazards than others. Accidents involving natural gas pipelines may involve explosions and fire, which are often a serious threat to human safety even when the effects on the natural environment may be minimal. On the other hand, failures of wastewater pipelines may not pose a serious threat to the safety of human life, but could have a catastrophic effect on aquatic life.

Container Failures

Container failures can occur in conjunction with highway, railroad, and plant accidents. They can also occur independently of a transportation accident, such as at a storage facility.

Although mishandling of a container is a common cause of failure, other causes include thermal, mechanical, and chemical failures. *Thermal failures* occur when the container is subjected to heat or flame. In a fire, a pressure relief valve may vent or blow out due to pressure buildup, causing a "blowtorch" effect. (See Fig. 53.1.) *Mechanical failures* occur due to overfilling, a defect in the container or pressure relief valve, a gasket or connection failure, or damage from other events. *Chemical failures* result when products cause corrosion.

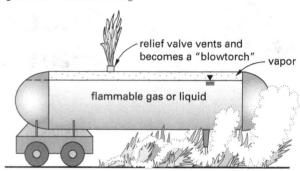

Figure 53.1 *Pressure Venting*

Emergency Types

Flammable and Combustible Liquid Emergencies

The majority of U.S. hazardous material responses involve flammable and combustible liquids. Flammable and combustible liquids are handled by truck, train, barge, and pipeline and are of great concern to incident responders. (The principal distinction between flammable and combustible liquids is that *flammable liquids* give off vapors at temperatures below 37.8°C that may be easily ignited. *Combustible liquids* also give off vapors at temperatures below 37.8°C, but not enough to present an ignition hazard. However, combustible liquids produce large amounts of heat when they are ignited.)

The U.S. Department of Transportation (USDOT) defines a flammable liquid as "a liquid having a flash point of not more than 60.5°C, or any material in a liquid phase with a flash point at or above 37.8°C that is intentionally heated and offered for transportation or is transported at its flash point in a bulk packaging. . ." (49 CFR 173.120(a)). The USDOT definition excludes liquefied compressed gases, compressed gases in solution, cryogenic liquids, and mixtures with flash points above 60.5°C.

USDOT defines a combustible liquid as "any liquid that does not meet the definition of any other hazard class. . . and (1) has a flash point above 60.5°C and below 93°C, or (2) is a flammable liquid with a flash point at or above 37.8°C" (49 CFR 173.120(b)).

Gas Emergencies

Emergencies involving gases are one of the most common types that emergency responders must handle. Gases are widely used throughout the nation and are shipped in several forms: compressed, liquefied, and cryogenic. USDOT has established specific definitions for gases shipped on the nation's transportation network.

Flammable gas is defined as "any material which is a gas at 20°C or less and 101.3 kPa of pressure which (1) is ignitable at 101.3 kPa when in a mixture of 13% or less by volume with air, or (2) has a flammable range (at

Environmental
Health & Safety

101.3 kPa) with air of at least 12% regardless of the lower limit" (49 CFR 173.115(a)). Flammable gases are assigned USDOT Hazard Class 2.1.

Compressed gas is defined as "any material or mixture which (1) exerts on the packaging a gage pressure of 200 kPa or greater at 20°C, and (2) does not meet the definition of a flammable gas or a gas poisonous by inhalation. This category includes compressed gas, liquefied gas, pressurized cryogenic gas, and compressed gas in solution" (49 CFR 173.115(b)). Compressed gases are assigned USDOT Hazard Class 2.2.

Gas that is *poisonous* by inhalation is defined as "a material which is a gas at 20°C or less and a pressure of 101.3 kPa and which (1) is known to be so toxic to humans as to pose a hazard to health during transportation, or (2) in the absence of adequate data on human toxicity, is presumed to be toxic to humans based on tests conducted on laboratory animals" (49 CFR 173.115(c)). Gases that are poisonous by inhalation are assigned USDOT Hazard Class 2.3.

Liquefied compressed gas is defined as "a gas which, when packaged under pressure for transport, is partially liquid at temperatures above −50°C" (49 CFR 173.115 (e)).

A *cryogenic liquid* is defined as "a refrigerated liquefied gas having a boiling point colder than −90°C at 101.3 kPa" (49 CFR 173.115(g)).

2. ROLE OF THE ENVIRONMENTAL ENGINEER

Potential Roles

The environmental engineer may have a wide variety of roles relative to emergency management. An environmental engineer employed by a public agency may function as a key member of a team having direct responsibility for response and mitigation or cleanup of an emergency, such as an oil spill from a ship or tanker collision or grounding. An environmental engineer employed by a public agency may also have a more limited role in the emergency and may function as an advisor to utility services, facilities, or environmental regulatory agencies that may be affected by the emergency.

An environmental engineer may also be employed by an emergency response contractor or as a consultant to a potentially affected service provider, such as a water supply utility. The environmental engineer may be asked to evaluate the risk to human health and the environment from the release of hazardous substances and a wide variety of other pollutants in order to advise emergency response personnel and to identify mitigation or remedial measures. Unless prepared by appropriate training and experience, the environmental engineer should not function as a hazmat responder or as a

member of an emergency service team (Emergency Medical Services (EMS) providers, police officers, and firefighters). However, the environmental engineer may be called upon to provide advice to hazardous waste responders about potential effects on, or protection of, the natural environment, human health, or water supplies.

Working with Hazardous Materials Responders

There are typically four categories of hazardous materials responders.

- first responders
- technicians
- specialists
- incident (on-scene) commander

The *first responders* may include police, firefighters, public works personnel, industrial spill teams, and EMS personnel. The first responders initially recognize the problem, identify the hazards involved, and notify more qualified responders. They also work to contain spills and try to minimize harm.

Technicians may include hazardous materials response teams, industrial crews, and emergency response teams. Their role is to control the spill and stop the release.

Specialists may include teams with special skills and expertise, state and local regulators, and industrial experts. The specialists provide support to the first responders through consultation and advice about the effects of courses of action, possible mitigation measures, and technical characteristics of the hazardous materials. The environmental engineer may be considered a specialist and may be called upon to fulfill this role.

The *incident commander* is also known as the *on-scene commander*. Incident commanders may be fire chiefs, battalion fire chiefs, sheriffs and deputies, state troopers, or emergency response team supervisors. The incident commander is responsible for the overall safety of personnel on the scene and for directing the proper containment, cleanup, mitigation, and disposal of the hazardous materials.

3. LEGAL REQUIREMENTS

As hazardous waste remediation and emergency response operations have matured over the last two or three decades, so have the legal regulations and requirements. The Occupational Safety and Health Administration (OSHA) has developed explicit standards that apply to workers involved with emergency response and hazardous waste site operations. These standards provide legal obligations incumbent on employers to protect the health and safety of workers who perform such

Environmental Health & Safety

activities. The standards are given in 29 CFR 1910.119 along with four appendices (1910.119 Appendices A through D).

While 29 CFR 1910.120 contains provisions for worker safety during hazardous clean-up, emergency response, and corrective actions, 29 CFR 1910.119 contains requirements for preventing releases of toxic, reactive, flammable, and explosive chemicals. This section and its four appendices are applicable, among other criteria, to processes with accumulations of chemicals at or above the specified *threshold quantities* (TQs) listed in 19 CFR 1910.119 Appendix A (and presented in App. 53.A). Appendix 53.A, which is from 29 CFR 1910.119, does not relate to the protection of emergency response and clean-up personnel.

4. PROPERTIES OF GASES AND LIQUIDS RELEVANT TO EMERGENCY RESPONSE

Vapor Pressure

When liquids evaporate into a confined space, such as in the top of a tank or railroad tank car, the *vapor pressure* increases until the maximum vapor pressure is reached for the temperature of the space. Exposure of a tank to fire can increase the vapor pressure above atmospheric pressure, at which point the liquid will boil. The greater the mass of vapor and the higher the vapor pressure, the greater the potential for ignition or explosion. Pressure relief valves, designed to release some of the accumulated vapors, may be damaged in an emergency incident and not function properly. A relief valve may also function as a blowtorch if the escaping vapors are ignited.

Raoult's law states that the vapor pressure of a solution at a particular temperature is equal to the mole fraction of the solvent (in the liquid phase) multiplied by the vapor pressure of the pure solvent at the same temperature. For an ideal vapor (gas), the *partial pressure ratio*, *mole fraction*, and *volumetric fraction* are equivalent for a given temperature and pressure. As a result, Eq. 53.1 can be used to determine flammable limits.

$$x_v = \frac{p_{vp,pure}x_l}{p} \qquad 53.1$$

Vapor Specific Gravity

Vapor specific gravity (sometimes incorrectly referred to as "vapor density") is the specific gravity of a gas or vapor referred to as the density of air. (The specific gravity of air is 1.0.) A gas or vapor with a vapor specific gravity less than 1.0 is lighter than air and will rise upward to the top of a confining space, such as a tank or building. A gas or vapor with a vapor specific gravity greater than 1.0 is heavier than air and will sink to the bottom of a confining space. Table 53.1 gives

approximate specific gravities of some common vapors. Emergency responders must consider the vapor specific gravity of the gas or vapor involved in an incident.

Table 53.1 *Approximate Specific Gravities of Vapors*

gas	vapor specific gravity	conditions (at 1 atm, 101.325 kPa)
acetylene	0.906	0°C
air	1.000	21°C
ammonia (anhydrous)	0.597	0°C
carbon dioxide	1.522	21°C
chlorine	2.482	0°C
ethylene	0.978	0°C
hydrogen	0.0695	0°C
hydrogen sulfide	1.2	15°C
methane	0.5549	16°C
nitrogen	0.967	21°C
oxygen	1.105	21°C
sulfur dioxide	2.262	21°C
vinyl chloride	2.15	15°C

Source: Onguard, Inc. 1996 (Appendix D)

Liquid Specific Gravity

The *specific gravity* of a liquid is the ratio of the density of the liquid to the density of pure water. Since most flammable liquids have specific gravities less than 1.0, they will normally accumulate and float on the surface of a body of water.

Boiling Point

The *boiling point* of a liquid is the temperature at which a liquid rapidly becomes a vapor. Below its boiling point, a liquid changes to a vapor slowly through *evaporation*. A liquid with a low boiling point is more dangerous because it will produce vapors sooner than a liquid with a high boiling point.

When a solid and a liquid are in contact and the surface temperature of the solid, T_s, is greater than the saturation temperature of the liquid, T_{sat}, heat travels from the solid to the liquid. The heat flux is

Boiling
$$q'' = h(T_s - T_{sat}) = h\Delta T_e \qquad 53.2$$

Expansion Ratio

The *expansion ratio* is defined as the ratio of the volume of gas produced when liquid is vaporized to the original volume of liquid. Examples of expansion ratios for several liquids are given in Table 53.2.

Table 53.2 Approximate Expansion Ratios

liquid	boiling point (°C)	expansion ratio*
helium	−269	745
methane	−161	625
nitrogen	−196	696
oxygen	−183	860

*volume expansion ratio from liquid at boiling point and 1 atm to gas at 21°C and 1 atm

Critical Point

The *critical point* refers to the condition at which a substance can exist in both gas and liquid states simultaneously. When a liquid is at its *critical temperature*, further heating will cause part or all of it to vaporize instantly. *Critical pressure* is the pressure required to liquefy a gas. These concepts are important in emergency response situations because liquefied gas above its critical temperature will instantly convert to a gas, which may cause a container to fail catastrophically.

Flash Point

The *flash point* is defined as the minimum temperature at which a liquid will ignite if an ignition source is present. Below the flash point, a liquid will not produce enough vapor for combustion (i.e., the mixture will be too lean).

Fire Point

The *fire point* is the minimum temperature at which a liquid will produce enough vapor for sustained combustion. This differs from the flash point, at which a vapor flash can occur but combustion will not be sustained. The fire point is usually only one or two degrees Celsius above the flash point.

Flammable Limits of Gases and Vapors

The concentrations of a substance in air below which there is insufficient fuel to burn or explode (i.e., the mixture is too lean) are known as the *lower flammable limit* (LFL) and *lower explosion limit* (LEL), respectively. The concentrations above which there is too much fuel to burn or explode (i.e., the mixture is too rich) are known as the *upper flammable limit* (UFL) and *upper explosion limit* (UEL), respectively. The LFL and the UFL are typically expressed at normal temperature and pressure (NTP) of 25°C and 1 atm, but this should be verified when using data from a reference.

The *flammable range* refers to the range of volumetric percentages of a gas or vapor in air that will form a flammable mixture. Gases and vapors with low LFLs are inherently more dangerous than those with higher LFLs. Vapors with a wide flammable range also are of greater concern than those with a narrow flammable range. Increased pressure or temperature will lower the LFL and raise the UFL, thereby widening the flammability range. Decreased pressure or temperature will narrow the range between the LFL and the UFL. Flammability limits of some typical flammable products are shown in Fig. 53.2.

The LFL of a flammable gas can be estimated by dividing the vapor pressure of the gas at its flash point by the pressure at one atmosphere (760 mm Hg, 101.325 kPa, or 14.7 psia).

$$\text{LFL} = \frac{P_{\text{vapor}}}{P_{\text{atm}}} \quad\quad 53.3$$

Figure 53.2 Flammable Limits

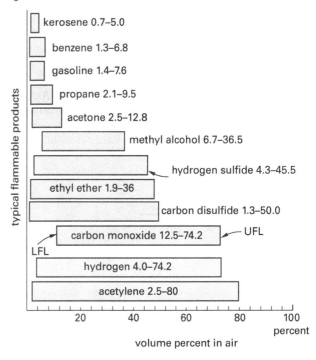

For a flammable gas, the flash point may be in the cryogenic range. In that case, a gas is classified as flammable if the LFL is less than or equal to 13% by volume at NTP or if the flammability range is more than 12% at NTP. Note that most flammable gases are lighter than air, but flammable vapors are often heavier than air.

The LFL and UFL of mixtures of vapors can be determined from Eq. 53.4 and Eq. 53.5.

Predicting Lower Flammable Limits of Mixtures of Flammable Gases (Le Chatelier's Rule)

$$\text{LFL}_m = \frac{100}{\sum_{i=1}^{n}(C_{fi}/\text{LFL}_i)} \quad \textbf{53.4}$$

Predicting Lower Flammable Limits of Mixtures of Flammable Gases (Le Chatelier's Rule)

$$\text{UFL}_m = \frac{100}{\sum_{i=1}^{n}(C_{fi}/\text{UFL}_i)} \quad \textbf{53.5}$$

Example 53.1

Workers are clearing a large warehouse containing drums and cylinders that are believed to contain flammable or explosive liquids and gases. While they are working in one of the rooms, the control valve on a cylinder marked "Warning—Hydrogen Gas—Contents Under Pressure" is broken. The room is 10 m by 20 m by 10 m high. The estimated pressure in the cylinder is 20 MPa (gage), and the gas has an estimated volume of 3000 L. Is the resulting mixture flammable?

Solution

Find the volume of hydrogen (H_2) in the room. Assume the H_2 mixes thoroughly in the room almost instantaneously and the cylinder is at the same temperature as the air in the room. Assume the pressure in the room remains at 1 atm.

The volume of the cylinder is

$$v_1 = \frac{3000 \text{ L}}{1000 \ \dfrac{\text{L}}{\text{m}^3}} = 3 \text{ m}^3$$

The pressure in the cylinder in kilopascals is

$$P_1 = (20 \text{ MPa})\left(1000 \ \frac{\text{kPa}}{\text{MPa}}\right)$$
$$+ (1 \text{ atm})\left(101.325 \ \frac{\text{kPa}}{\text{atm}}\right)$$
$$= 20\,101 \text{ kPa}$$

The pressure in the room in kilopascals is

$$P_2 = (1 \text{ atm})\left(101.325 \ \frac{\text{kPa}}{\text{atm}}\right)$$
$$= 101.3 \text{ kPa}$$

Use Boyle's law to find the volume of the gas when it leaves the cylinder.

Special Cases of Closed Systems (with no change in kinetic or potential energy)

$$Pv = \text{constant}$$
$$P_1 v_1 = P_2 v_2$$
$$v_2 = \frac{P_1 v_1}{P_2} = \frac{(20\,101 \text{ kPa})(3 \text{ m}^3)}{101.3 \text{ kPa}}$$
$$= 595.3 \text{ m}^3$$

Find the volume of air in the room.

$$v_{\text{air}} = (10 \text{ m})(20 \text{ m})(10 \text{ m})$$
$$= 2000 \text{ m}^3$$

Find the percentage by volume of hydrogen in the room air.

$$B_{H_2} = \frac{v_2}{v_{\text{total}}} = \frac{595.3 \text{ m}^3}{2000 \text{ m}^3 + 595.3 \text{ m}^3}$$
$$= 0.2294 \quad (23\%)$$

Determine the flammability limits of hydrogen gas from Fig. 53.2.

$$\text{LFL} = 4.0\%$$
$$\text{UFL} = 74.2\%$$

The room contents are flammable because the concentration of hydrogen gas is between the LFL and the UFL.

Example 53.2

A mixture contains three gases.

gas	actual volume	LFL % volume	UFL % volume
methane	1	5	15
propane	4	2.1	9.5
ethyl alcohol	2	3.3	19.0

What are the lower and upper flammability limits of the mixture?

Solution

Determine the percentage by volume of each gas in the mixture.

$$C_{f,\text{methane}} = \frac{v_{\text{methane}}}{v_{\text{total}}} = \frac{1}{1+4+2} \times 100\%$$
$$= 14.3\%$$

$$C_{f,\text{propane}} = \frac{v_{\text{propane}}}{v_{\text{total}}} = \frac{4}{1+4+2} \times 100\%$$
$$= 57.1\%$$

$$C_{f,\text{ethyl alcohol}} = \frac{v_{\text{ethyl alcohol}}}{v_{\text{total}}} = \frac{2}{1+4+2} \times 100\%$$
$$= 28.6\%$$

From Eq. 53.4, the lower flammability limit of the mixture is

Predicting Lower Flammable Limits of Mixtures of Flammable Gases (Le Chatelier's Rule)

$$\text{LFL}_m = \frac{100}{\sum\limits_{i=1}^{n}(C_{fi}/\text{LFL}_i)}$$

$$= \frac{100\%}{\dfrac{14.3\%}{5\%} + \dfrac{57.1\%}{2.1\%} + \dfrac{28.6\%}{3.3\%}}$$

$$= 2.58\%$$

From Eq. 53.4, the upper flammability limit of the mixture is

Predicting Lower Flammable Limits of Mixtures of Flammable Gases (Le Chatelier's Rule)

$$\text{UFL}_m = \frac{100\%}{\sum\limits_{i=1}^{n}(C_{fi}/\text{LFL}_i)}$$

$$= \frac{100\%}{\dfrac{14.3\%}{15\%} + \dfrac{57.1\%}{9.5\%} + \dfrac{28.6\%}{19.0\%}}$$

$$= 11.81\%$$

Autoignition

The *autoignition* temperature is the minimum temperature at which a substance (liquid, gas, or vapor) will spontaneously initiate self-sustained combustion without the presence of an ignition source. Autoignition can result from an adiabatic (without heat transfer) compression of a gas. In emergency response situations, such compression may occur from damage to or collapse of tanks, tanker trucks, or structures. The gas temperature resulting from adiabatic compression can be found from Eq. 53.6.

$$\frac{T_2}{T_1} = \left(\frac{P_2}{P_1}\right)^{(k-1)/k} \quad \text{[adiabatic]} \qquad 53.6$$

k is the ratio of specific heats.

$$k = \frac{c_p}{c_v} \qquad 53.7$$

For air at room temperature and atmospheric pressure, c_p is approximately 1005 J/kg·K, and c_v is approximately 718 J/kg·K. k is approximately 1.4.

Example 53.3

A large tank contains ethane gas (C_2H_6, $c_p = 1616$ J/kg·K, $c_v = 1340$ J/kg·K) at 2 atm and 100°C. The autoignition temperature of ethane is 472°C. During an emergency incident, the gas in the tank mixes with air and enters its flammable range. If the tank is compressed due to structural failure until the internal pressure in the tank rises to 10 MPa, will an explosion occur?

Solution

The ratio of specific heats is

$$k = \frac{c_p}{c_v} = \frac{1616 \, \dfrac{\text{J}}{\text{kg·K}}}{1340 \, \dfrac{\text{J}}{\text{kg·K}}}$$

$$= 1.206$$

$$T_1 = 100°C + 273°$$
$$= 373\text{K}$$

$$P_1 = (2 \text{ atm})\left(\frac{101.325 \text{ kPa}}{1 \text{ atm}}\right)$$
$$= 202.65 \text{ kPa}$$

$$P_2 = (10 \text{ MPa})\left(\frac{10^3 \text{ kPa}}{1 \text{ MPa}}\right)$$
$$= 10 \times 10^3 \text{ kPa}$$

Environmental Health & Safety

Use Eq. 53.6.

$$T_2 = T_1\left(\frac{P_2}{P_1}\right)^{(k-1)/k}$$

$$= (373\text{K})\left(\frac{10 \times 10^3 \text{ kPa}}{202.65 \text{ kPa}}\right)^{(1.206-1)/1.206}$$

$$= 726\text{K} - 273°$$

$$= 453°\text{C}$$

Since $T_2 = 453°$C is less than the autoignition temperature of ethane ($472°$C), combustion and explosion will not occur.

Explosive Energy Release

Pressurized containers are often encountered in emergency response situations. A sudden, catastrophic release of compressed gas can release tremendous amounts of energy, causing destruction of property and injury or death to emergency responders or others in the vicinity. The energy released from an adiabatic failure of a pressurized tank or cylinder can be calculated from the following equation.

$$W = \left(\frac{P_1 v_1}{k-1}\right)\left[1 - \left(\frac{P_2}{P_1}\right)^{(k-1)/k}\right] \quad \textit{53.8}$$

Example 53.4

Workers at a remediation site uncovered an old tank of compressed nitrogen gas (N_2) with an indicated volume of 20 m³ and pressure of 1500 kPa (absolute). For nitrogen, $c_p = 1043$ J/kg·K, and $c_v = 746$ J/kg·K. If the tank fails catastrophically and its contents are released under adiabatic conditions, what is the energy release in equivalent mass of TNT? Assume TNT releases 4.7 MJ/kg TNT.

Solution

$$P_2 = (1 \text{ atm})\left(\frac{101.325 \text{ kPa}}{1 \text{ atm}}\right)$$

$$= 101.325 \text{ kPa}$$

Use Eq. 53.7.

$$k = \frac{c_p}{c_v} = \frac{1043 \dfrac{\text{J}}{\text{kg·K}}}{746 \dfrac{\text{J}}{\text{kg·K}}}$$

$$= 1.398$$

$$v_1 = 20 \text{ m}^3$$

Use Eq. 53.8.

$$W = \left(\frac{P_1 v_1}{k-1}\right)\left[1 - \left(\frac{P_2}{P_1}\right)^{(k-1)/k}\right]$$

$$= \left(\frac{(1500 \text{ kPa})(20 \text{ m}^3)}{1.398 - 1}\right)$$

$$\times \left[1 - \left(\frac{101.325 \text{ kPa}}{1500 \text{ kPa}}\right)^{(1.398-1)/1.398}\right]$$

$$= 40.4 \times 10^3 \text{ kJ} \quad (40.4 \text{ MJ})$$

The TNT equivalent is

$$m = \frac{40.4 \text{ MJ}}{4.7 \dfrac{\text{MJ}}{\text{kg}}} = 8.60 \text{ kg}$$

Solubility

The solubility of liquids and gases is an important consideration in emergency response situations. Gases that are soluble in water may produce a product that has dangerous properties or is a hazard or pollutant. An example is the solution of hydrogen chloride gas in water—hydrochloric acid, HCl.

BLEVEs

A *BLEVE* is a *boiling liquid expanding vapor explosion*. A BLEVE can occur when a liquid (including water) in a closed container (i.e., tank, drum) is heated well above its boiling point at normal atmospheric pressure. A BLEVE explosion can cause catastrophic loss of life and property.

5. ASSESSMENT OF CHRONIC EFFECTS FROM EMERGENCY INCIDENTS

Releases to Air

Releases to air during emergency incidents are usually rapid and may result in acute effects on hazardous materials responders and the community. Such releases include smoke and vapors containing toxic pollutants, vapor clouds, and toxic or hazardous gases.

The intake of a chemical from long-term exposure to inhalation of vapors can be calculated from Eq. 53.9.

$$\text{CDI} = \frac{\overset{\text{Exposure}}{(\text{CA})(\text{IR})(\text{ET})(\text{EF})(\text{ED})}}{(\text{BW})(\text{AT})} \quad \textit{53.9}$$

In evaluating chronic effects, the *exposure frequency*, EF, is the number of days per year that the affected population is exposed to vapors from the incident, and the *averaging time*, AT, is the expected time the affected population would be exposed (for noncarcinogens). The same procedure is applied to chronic exposure over several years as for acute exposure.

Releases to Water

Water quality emergencies involve release of petroleum products or toxic or hazardous chemicals to the natural environment, such as to drainage ways or directly to waterbodies. The concern of the environmental engineer may be the potential effects on human health from contamination of drinking water supplies or the potential effects from pollutants toxic to aquatic life.

The chronic effects from drinking water sources can be evaluated by determining the instream concentration of the pollutant at the point of use. One approach is to ensure that the instream concentration of the pollutant is at an acceptable level prior to treatment. This approach is appropriate when it is risky to rely upon treatment processes for long-term protection of human health after an emergency response incident. Decay of the pollutant may be accounted for as a part of the mitigation approach. Dilution can also be used, at least in part, to reduce the pollutant concentration at the point of use. The safest approach is to ensure that discharges leaving the control of the responsible party at the incident site are within acceptable limits so that any downstream exposure will not cause unsafe chronic exposure.

The evaluation procedure for chronic effects starts with determining the *chronic daily intake* (CDI) of the pollutant through the oral route from drinking water and through other applicable routes, such as vapors released from the water. The intake can then be compared to the *reference dose* (RfD) for noncarcinogens by computing the *hazard ratio* (HR), CDI/RfD. If the hazard ratio exceeds 1.0, the exposure is unacceptable, and further cleanup efforts will be needed. For carcinogens, the *risk* (R) can be determined by the product of the CDI and the *slope factor* (SF) or the *potency factor* (PF). The SF and the PF both refer to the risk associated with 1 mg/kg·d daily intake. Judgment will be involved in determining an *acceptable risk* for the specific conditions. Acceptable risk is usually set at 1×10^{-4} to 1×10^{-6}, which is a range of risk levels typically used at U.S. Environmental Protection Agency cleanup projects.

6. NOMENCLATURE

AT	averaging time (same as ED for noncarcinogens, 70 yr for carcinogens)	d
B	volumetric fraction	decimal
BW	body mass	kg
c	specific heat	J/kg·K
C	concentration	ppm, kg/m^3
C_{fi}	volume percent of gas in fuel gas	%
CA	contaminant concentration in air	mg/m^3
CDI	chronic daily intake (also known as intake)	mg/kg·d
ED	exposure duration	yr
EF	exposure frequency	d/yr
ET	exposure time	h/d or h/event
h	heat transfer coefficient	W/m^2·K
IR	inhalation rate	m^3/h
k	specific heat ratio	–
LCL	lower confidence limit	–
LFL	lower flammable limit	–
m	mass	kg
p, P	pressure	atm
q'	heat flux	W/m^2
T	temperature	K
UFL	upper flammability limit	–
v	volume	L, m^3
W	energy release from adiabatic compression of gas or vapor	kJ
x	mole fraction	–

Subscripts

atm	atmosphere
e	excess
f	final condition
i	initial condition or gas
l	liquid
m	mixture
p	constant pressure
s	solid
sat	saturation
v	constant volume, vapor, or in vapor phase
vp	vapor

54 Codes, Standards, Regulations, Guidelines

1. LEGAL (OSHA) STANDARDS FOR WORKER PROTECTION

While the Environmental Protection Agency (EPA) provides exposure limitations and risk factors for environmental cleanup projects, the American Conference of Governmental Industrial Hygienists (ACGIH) provides recommendations to industrial hygienists about workplace exposure, and the National Institute for Occupational Safety and Health (NIOSH) provides research and recommendations for workplace exposure limits (recommended exposure limits), it is the *Occupational Safety and Health Administration* (OSHA) that sets the legally enforceable workplace exposure limits. OSHA standards are given in the 29 Code Federal Regulations (CFR). Although OSHA deals with many other aspects of workplace safety, of key importance relative to toxicology are the regulations on air contaminants given in 29 CFR 1910.00, Subpart Z, "Toxic and Hazardous Substances."

Subpart Z includes three tables, Z-1, Z-2, and Z-3. The current standards should always be reviewed when performing any type of air contaminant evaluation because changes to the exposure limits occur from time to time.

Table Z-1 gives *permissible exposure limits* (PELs) for air contaminants based on an 8 hr TWA and identifies substances for which protection from skin exposure is required. Substances with ceiling limits are also identified. Table Z-2 gives limits for air contaminants on an 8 hr TWA basis with acceptable ceiling concentrations and acceptable maximum peak concentrations and durations for exposure allowed above the ceiling concentrations. Table Z-3 gives exposure limits for mineral dusts.

The calculation procedure for use of the PELs is the same as for ACGIH threshold limit values (TLVs). The procedure for mixtures is also the same.

2. RADIATION EXPOSURE REGULATIONS

Part 20 of Title 10 of the Code of Federal Regulations (10 CFR 20) contains the U.S. Nuclear Regulatory Commission's radiation exposure limits.

Occupational Dose to Adults

The occupational dose to adults (10 CFR 20.1201) is the more limiting of

- a total effective dose equivalent, $H_{E,\text{tot}}$, of 5 rems (0.05 Sv) in one year

- the sum of the deep-dose equivalent, H_D, and the committed dose equivalent, $H_{T,50}$, to any individual organ or tissue, other than the lens of the eye, equal to 50 rems (0.5 Sv) in one year

The annual limits to the lens of the eye, to the skin, and to the *extremities* (e.g., hand, elbow, arm below the elbow, foot, knee, and leg below the knee) are

- a lens dose equivalent, H_L, of 15 rems (0.15 Sv) in one year

- a shallow-dose equivalent, H_S, of 50 rems (0.50 Sv) to the skin or to any extremity in one year

Occupational Dose to Minors

The annual occupational dose limit for minors is 10% of the annual dose limits for adult workers (10 CFR 20.1207).

Embryo/Fetus

The term *embryo/fetus* in the regulations means the developing human organism from conception until the time of birth. A *declared pregnant woman* is a woman who has voluntarily informed her employer, in writing, of her pregnancy and the estimated date of conception. The total effective dose equivalent, $H_{E,\text{tot}}$, limit for the embryo/fetus during the entire pregnancy, due to the occupational exposure of a declared pregnant woman, is 0.5 rem (5 mSv) (10 CFR 20.1208). Additionally, substantial variation from a uniform monthly exposure rate to a declared pregnant woman must be avoided. The dose equivalent to the embryo/fetus is the sum of

- the deep-dose equivalent to the declared pregnant woman

- the dose equivalent to the embryo/fetus resulting from radionuclides in the embryo/fetus and radionuclides in the declared pregnant woman

Individual Members of the Public

Public dose means the dose received by a member of the public from exposure to radiation or radioactive material, not including occupational dose or doses received from background radiation, from any medical exposure the individual has received, from exposure to individuals administered radioactive material, from voluntary participation in medical research programs, or from disposal of radioactive material into sanitary sewage. The total effective dose equivalent, H_E, limit for individual members of the public is 0.1 rem (1 mSv) in a year (10 CFR 20.1301).

As Low As Reasonably Achievable (ALARA) Regulation

As low as reasonably achievable (ALARA) means making every reasonable effort to maintain exposures to radiation as far below the dose limits as is practical consistent with the purpose for which the activity is undertaken. The USNRC's regulations in 10 CFR 20 require that employees use, to the extent that is practical, procedures and engineering controls based upon sound radiation protection principles to limit occupational doses (and doses to members of the public) to those that are as low as is reasonably achievable. In other words, it is possible to violate the ALARA regulation even if no exposure exceeds the regulatory limits if a reasonable way to reduce radiation exposure is not taken advantage of. The ALARA concept must be applied both to individual doses and to collective doses. *Collective dose* is the sum of all doses received by persons and is reported in person·rem or person·Sv.

Example 54.1

Ten people are exposed to a 2 R/hr X-ray field for 3 hr. What is the collective dose equivalent?

Solution

$$(10 \text{ persons})\left(2 \ \frac{\text{R}}{\text{h}}\right)(3 \text{ h}) = 60 \text{ person·R}$$

For radiation protection purposes, 1 R $\approx$ 1 rem.

60 person·R $\approx$ 60 person·rem = 0.60 person·Sv

The average dose equivalent is

$$\frac{60 \text{ person·rem}}{10 \text{ persons}} = 6.0 \text{ rem/person}$$

$$= 0.060 \text{ person·Sv}$$

3. NOMENCLATURE

H_D	deep-dose equivalent	Sv
H_E	total effective dose equivalent	rems, Sv
$H_{E,\text{tot}}$	total effective dose equivalent	rems, Sv
H_L	lens dose equivalent	Sv
H_S	shallow-dose equivalent	Sv
$H_{T,50}$	committed dose equivalent	Sv

55 Industrial Hygiene

1. INTRODUCTION

Industrial hygiene is the art and science of identifying, evaluating, and controlling environmental factors (including stress)· that may cause sickness, health impairment, or discomfort among workers or citizens of the community. Industrial hygiene involves the recognition of health hazards associated with work operations and processes, evaluation and measurement of the magnitudes of hazards, and determining applicable control methods. Occupational health hazards specifically involve illness or impairment for which a worker may be compensated under a worker protection program.

The fundamental law governing worker protection in the United States is the 1970 federal Occupational Safety and Health Act. It requires employers to provide a workplace that is free from hazards by complying with specified safety and health standards. Employees must also comply with standards that apply to their own conduct. The federal regulatory agency responsible for administering the Occupational Safety and Health Act is the Occupational Safety and Health Administration (OSHA). OSHA sets standards, investigates violations of the standards, performs inspections of plants and other facilities, investigates complaints, and takes enforcement action against violators. OSHA also funds state programs, which are permitted if they are at least as stringent as the federal program.

The 1970 act also established the National Institute for Occupational Safety and Health (NIOSH). NIOSH is responsible for safety and health research and makes recommendations for regulations. The recommendations are known as *recommended exposure limits* (RELs). Among other activities, NIOSH also publishes health and safety criteria and health hazard alerts, and is responsible for testing and certifying respiratory protective equipment.

2. HAZARD IDENTIFICATION

Overview of Hazards

There are four basic types of hazards with which industrial hygiene is concerned: chemical hazards, physical hazards, ergonomic hazards, and biological hazards. *Chemical hazards* result from airborne chemicals such as gases, vapors, and particulates in harmful concentrations. Besides inhalation, chemical hazards may affect workers by absorption through the skin. *Physical hazards* include radiation, noise, vibration, and excessive heat or cold. *Ergonomic hazards* include work procedures and arrangements that require motions that result in biomechanical stress and injury. *Biological hazards* include exposure to biological organisms that may lead to illness. Any of these types of hazards may occur at high intensity, resulting in acute or immediate effects, or they may occur at low intensity, resulting in long-term or chronic effects. Industrial hygiene is focused on control or elimination of these hazards in the workplace.

In respect to chemical hazards, the terms "toxicity" and "hazard" are not synonymous. *Toxicity* is the capacity of the chemical to produce harm when it has reached a sufficient concentration at a particular site in the body. *Hazard* refers to the probability that this concentration will occur.

Hazard Communication

Two important preventative measures required by OSHA are *safety data sheets* (SDSs) and labeling of containers of hazardous materials. A third important OSHA requirement is that all covered employers must provide the necessary information and training to affected workers. The OSHA Hazard Communication Standard is given in Title 29 of the *Code of Federal Regulations* (CFR) Part 1910.1200. Other OSHA requirements are also given in Title 29.

Safety Data Sheets

An SDS provides key information about a chemical or substance so that users or emergency responders can determine safe use procedures and necessary emergency response actions. An SDS provides information on the identification of the material and its manufacturer, identification of hazardous components and their characteristics, physical and chemical characteristics of the ingredients, fire and explosion hazard data, reactivity

Environmental Health & Safety

data, health hazard data, precautions for safe handling and use, and recommended control measures for use of the material. The information on SDSs is essential for dealing with hazardous chemicals and should be complete; however, it may be necessary to contact the manufacturer if information is lacking.

Container Labeling

Labels are required on hazardous material containers. Labels should provide essential information for the safe use and storage of hazardous materials. Failure to provide adequate labeling of hazardous material containers is a common violation of OSHA standards.

Worker Information and Training

The OSHA standard requires that employers provide workers with information about the potential health hazards from exposure to hazardous chemicals that they use in the workplace. It also requires employers to provide adequate training to workers on how to safely handle and use hazardous materials. The standard requires training in use of the hazardous materials in normal operations and actions to take during emergencies.

3. GASES, VAPORS, AND SOLVENTS

Exposure Factors for Gases and Vapors

The most frequently encountered hazard in the workplace is exposure to gases and vapors from solvents and chemicals. Several factors define the exposure potential for gases and vapors. The most important are how a material is used and what engineering or personal protective controls exist. If the inhalation route of entry is controlled, dermal contact may still be a major route of exposure.

Vapor pressure of a substance is related to temperature. Vapor pressure affects the concentration of the substance in vapor form above the liquid and is dependent upon the temperature and the properties of the substance. Processes that operate at lower temperatures are inherently less hazardous than processes that operate at higher temperatures because fewer chemicals have boiling points and vapor pressures in the operating temperature range.

The *concentration* of a chemical in a liquid (a solution) affects the potential hazard because the vapor pressure varies directly with the mole fraction (concentration) of the solute according to Raoult's law. Thus, solutions with a lower concentration will have lower potentials for volatilization of the hazardous fraction.

Reactivity affects the hazard potential because the products may be volatile or nonvolatile depending on the properties of the combining substances.

Two terms are used to quantify the concentration of a gas in air that a worker can be safely exposed to. These are *threshold limit value* (TLV) and *permissible exposure limit* (PEL). The TLV is a concentration in air that nearly all workers can be exposed to daily without adverse effects. The PEL is a regulatory exposure limit for workers; OSHA publishes PELs as standards. A PEL may be more restrictive than the corresponding TLV. TLVs are updated annually by the American Conference of Governmental Industrial Hygienists (ACGIH). PELs are updated less frequently. Both should be reviewed when evaluating potential hazards from gases and vapors.

Solvents

Solvents are widely used throughout industry for many purposes, and their safe use is an important industrial hygiene concern. It is essential that accurate SDS information be provided to employees on the physical properties and the toxicological effects of exposure to solvents.

Solvent vapors can be inhaled or may be absorbed through the skin. Vapors that are inhaled are absorbed through the lungs into the bloodstream and are then deposited in tissue with a high content of fat and lipids. The central nervous system, liver, and bone marrow are the key organs of the body that are affected. The toxicity of a solvent is greatly affected by the conditions of use.

Water as a Solvent

Aqueous solutions have low vapor pressures at ambient temperatures, so the potential hazard through inhalation is low. Salts in an aqueous solution are essentially not volatile.

Organic Solvents

Organic solvents, and all organic compounds in general, can be classified by their molecular structure. The classification is by the number of carbon atoms in the basic skeletal chain, bonding, and molecular arrangement. Organic solvents can be classified as aliphatic, cyclic, aromatic, halogenated hydrocarbon, ketone, ester, alcohol, and ether. Each class has characteristic molecular structures, properties, and health effects. Since the names of organic solvents can be similar or misleading, the environmental engineer should always refer to the label, the MSDS, or a laboratory before assessing the hazard content of an organic solvent.

Gases and Flammable or Combustible Liquids

Hazardous gases fall into four main types: cryogenic liquids, simple asphyxiants, chemical asphyxiants, and all other gases whose hazards depend on their properties.

Cryogenic liquids can vaporize rapidly, producing a cold gas that is more dense than air and displacing oxygen in confined spaces. After reaching thermal equilibrium, gas from a cryogenic liquid may spread out in the available space. Gas from liquid nitrogen, for example, can cause oxygen to condense out of the atmosphere, creating an explosion hazard.

Simple asphyxiants, which include helium, neon, nitrogen, hydrogen, and methane, can dilute or displace oxygen. *Chemical asphyxiants*, which include carbon monoxide, hydrogen cyanide, and hydrogen sulfide, can pass into blood cells and tissue and interfere with blood-carrying oxygen.

In addition to the hazards from gases and vapors, some liquids pose hazards because their vapors are flammable or combustible when mixed with air in certain percentages. The term *flammable* refers to the ability of an ignition source to propagate a flame throughout the vapor-air mixture and have a closed-cup flash point below 37.8°C (100°F) and a vapor pressure not exceeding 272 atm at 37.8°C (100°F). The phrase *closed-cup flash point* refers to a method of testing for flash points of liquids. The term *combustible* refers to liquids with flash points above 37.8°C (100°F).

For each airborne flammable substance, there are minimum and maximum concentrations in air between which flame propagation will occur. The lower concentration in air is known as the *lower flammable limit* (LFL) or *lower explosive limit* (LEL). The upper limit is known as the *upper flammable limit* (UFL) or *upper explosive limit* (UEL). Below the LEL, there is not enough fuel to propagate a flame. Above the UEL, there is not enough air to propagate a flame. The lower the LEL, the greater the hazard from a flammable liquid. For many common liquids and gases, the LEL is a few percent, and the UEL is 6–12%. If a concentration in air is less than the PEL or the TLV, the concentration will be less than the LEL. The occupational safety requirements for handling and using flammable and combustible liquids are given in Subpart H of 29 CFR 1910.106.

For liquids that are vaporized in a ventilated area, the concentration can be calculated. T is the absolute ambient temperature, k is the nonideal mixing factor, Q_V is the ventilation rate, and P is the absolute ambient pressure.

Concentration (C_{ppm}) of Vaporized Liquid in Ventilated Space

$$C_{ppm} = [Q_m R_g T \times 10^6 / (k Q_V P M)] \qquad 55.1$$

See Table 55.1 for a representative listing of combustible materials and their LELs and UELs.

Evaluation and Control of Hazards

The toxicological effects from aqueous solutions include dermatitis, throat irritation, and bronchitis. The hazard is less with aqueous solutions because their vapor pressures are usually low. On the other hand, organic compounds represent a greater hazard because their vapor pressures are usually higher and their toxicities greater. The effects from organic solvents include central nervous system disorders, narcosis, and death.

Vapor-Hazard Ratio

One indicator of hazards from vapors and gases from solvents is the *vapor-hazard ratio*, which is the equilibrium vapor pressure in ppm at 25°C (77°F) divided by the TLV in ppm. The higher the ratio, the greater the hazard. The vapor-hazard ratio accounts for the volatility of a solvent as well as its toxicity. To assess the overall hazard, the vapor-hazard ratio should be evaluated in conjunction with the TLV, ignition temperature, flash point, toxicological information, and degree of exposure. The vapor-hazard ratio accounts for the volatility of a solvent as well as its toxicity.

The best control method is not to use a solvent that is hazardous. Sometimes a process can be redesigned to eliminate the use of a solvent. The following evaluation steps are recommended.

- Use water or an aqueous solution when possible.

- Use a *safety solvent* if it is not possible to use water. Safety solvents have inhibitors and high flash points.

- Use a different process when possible to avoid use of a hazardous solvent.

- Provide a properly designed ventilation system if toxic solvents must be used.

- Never use highly toxic or highly flammable solvents (benzene, carbon tetrachloride, gasoline).

Ventilation

The most effective way to prevent inhalation of vapors from solvents is to provide closed systems or adequate local exhaust ventilation. If limitations exist on the use of closed systems or local exhaust ventilation, then workers should be provided with personal protective equipment.

Personal Protective Equipment

Respirators provide emergency and backup protection but are unreliable as a primary source of protection from hazardous vapors. Face masks can leak or become contaminated around the edges, reduce the efficiency of the worker, and increase the lack of oxygen in oxygen-deficient areas. Other drawbacks are the need to have the respirator properly fitted to the worker and the need for the worker to be trained in its proper use. Additionally, the worker may feel a false sense of security while wearing a respirator.

Besides inhalation, dermal contact is an important concern when working with hazardous solvents. Mechanical equipment should be provided to keep the worker isolated from contact with the solvent. However, since some contact may occur even with mechanical equipment in use, protective clothing should be provided. Protective clothing

Table 55.1 *Representative Hazardous Concentrations in Air*

	combustibles						toxics				
material	TLV/ TWA (ppm)	LFL (%/vol)	UFL (%/vol)	IDLH (ppm)	specific gravity (air = 1.0)	material	TLV/ TWA (ppm)	IDLH (ppm)	LFL (ppm)	LFL (%/vol)	speci gravi (air = 3
acetone	750	2.5	12.8	2500	2.0	acetone	750	2500	25 000	2.5	2.0
acetylene	-A-	2.5	100.0	-A-	0.9	ammonia	25	300	160 000	16.0	0.6
ammonia	25	15.0	28.0	300	0.6	benzene	1.0	-C-	12 000	1.2	2.6
benzene	1.0	1.2	7.8	500	2.6	butane	800	-U-	16 000	1.6	2.0
butane	800	1.6	8.4	-U-	2.0	n-butyl acetate	150	1700	17 000	1.7	4.0
n-butyl acetate	150	1.7	7.6	1700	4.0	carbon dioxide	5000	40 000	N/C	N/C	1.5
diborane	0.1	0.8	88.0	15	1.0	carbon monoxide	25	1200	125 000	12.5	1.0
ethane	-A-	3.0	12.5	-A-	1.0	chlorine	0.5	10	N/C	N/C	2.5
ethanol	1000	3.3	19.0	-U-	1.6	ethylene oxide	1	-C-	30 000	3.0	1.5
ethyl acetate	400	2.0	11.5	2000	3.0	ethyl ether	400	19 000	19 000	1.9	2.6
ethyl ether	400	1.9	36.0	1900	2.6	gasoline	300	-U-	14 000	1.4	3–4.0
ethylene oxide	1	3.0	100.0	-C-	1.5	heptane	400	750	10 500	1.05	3.5
gasoline	300	1.4	7.6	-U-	3–4.0	hexane	50	1100	11 000	1.0	3.0
heptane	400	1.05	6.7	750	3.5	hydrogen cyanide	10	50	56 000	5.6	0.9
hexane	50	1.1	7.5	1100	3.0	hydrogen sulfide	10	100	40 000	4.0	1.2
hydrogen	-A-	4.0	75.0	-A-	0.1	isopropyl alcohol	400	2000	20 000	2.0	2.1
isopropyl alcohol	400	2.0	12.0	2000	2.1	methyl acetate	200	3100	31 000	3.1	2.6
methane	-A-	5.0	15.0	-A-	0.6	methanol	200	6000	60 000	6.0	1.1
methanol	200	6.0	36.0	6000	1.1	methyl chloride	50	2000	81 000	8.1	1.8
methyl ethyl ketone	200	1.4	11.4	3000	2.5	methyl ethyl ketone	200	3000	14 000	1.4	2.5
pentane	600	1.5	7.8	15 000	2.5	methyl methacrylate	100	1000	17 000	1.7	3.5
propane	1000	2.1	9.5	2100	1.6	nitric oxide	25	100	N/C	N/C	1.0
propylene oxide	20	2.3	36.0	400	2.0	nitrogen dioxide	3	20	N/C	N/C	1.6
styrene	50	0.9	6.8	700	3.6	pentane	600	15 000	15 000	1.5	2.5
toluene	50	1.1	7.1	500	3.1	n-propyl acetate	200	1700	17 000	1.7	3.5
turpentine	100	0.8	-U-	800	4.7	styrene	50	700	9000	0.9	3.6
vinyl acetate	10	2.6	13.4	-U-	3.0	sulfur dioxide	2	100	N/C	N/C	2.2
vinyl chloride	1.0	3.6	33.0	-C-	2.2	1,1,1-trichloroethane	350	700	75 000	7.5	4.6
xylene	100	0.9	6.7	900	3.7	toluene	50	500	11 000	1.1	3.2
						trichloroethylene	50	1000	80 000	8.0	4.5
						turpentine	100	800	8000	0.8	4.7
						vinyl chloride	1.0	-C-	36 000	3.6	2.2
						xylene	100	900	9000	0.9	3.7

Key: A, asphyxiant; C, carcinogen; U, data not available; N/C, noncombustible

*subject to change without notice

includes aprons, face shields, goggles, and gloves. The manufacturer's recommendations should be followed for use of all protective clothing and equipment.

One common problem with protective clothing is incorrect selection or misuse of gloves. The time for particular solvents to penetrate gloves that are commonly thought of as "protective" is surprisingly short. Both the permeability and the abrasion resistance of gloves must be considered in their selection and use. For example, methyl chloride will permeate a neoprene glove in less than 15 minutes. The manufacturer should provide the *breakthrough time* and the *permeation rates* for the glove

being evaluated. The breakthrough time and permeation rate are dependent on the specific chemical and the composition and thickness of the glove.

Protective eyewear should be provided where the risk of splashing of chemicals is present. Of course, mechanical equipment, barriers, guards, and other engineering measures should be provided as the first line of defense. For chemical splash protection, unvented chemical goggles, indirect-vented chemical goggles, or indirect-vented eye-cup goggles should be used. A face shield may also be needed. Direct-vented goggles and normal eyeglasses should not be used, and contact lenses should not be worn.

OSHA incident reports requires the incident rate within an organization to be calculated. In Eq. 55.2, N is the number of injuries, illnesses, and fatalities, and T is the total hours worked by all employees during the time-frame desired. Equation 55.2 can be used for injury rate, illness rate, fatality rate, lost workday cases rate, number of lost workdays rate, specific hazard rate, and lost workday injuries rate.

$$IR = N \times 200{,}000 \div T \qquad \text{\small Incidence Rates} \atop \textbf{\small 55.2}$$

Example 55.1

A rubber safety glove has an 8 h breakthrough time for ethanol based on a steady-state permeation rate of 12 μg/cm^2·min. The glove has been in continuous use for two 8 h shifts by the same worker who has 300 cm^2 of skin per hand. How many grams of ethanol have reached both hands of the worker at the end of the second shift?

Solution

The exposure duration is

$$\text{total time} - \text{breakthrough time} = 16 \text{ h} - 8 \text{ h}$$
$$= 8 \text{ h}$$

The total exposure is

$$\begin{pmatrix} \text{no. of} \\ \text{hands} \end{pmatrix}\begin{pmatrix} \dfrac{\text{skin area}}{\text{hand}} \end{pmatrix}\begin{pmatrix} \text{steady-state} \\ \text{permeation rate} \end{pmatrix}\begin{pmatrix} \text{exposure} \\ \text{duration} \end{pmatrix}$$

$$= (2)(300 \text{ cm}^2)\left(12 \ \dfrac{\mu\text{g}}{\text{cm}^2 \cdot \text{min}}\right)$$

$$\times (8 \text{ h})\left(60 \ \dfrac{\text{min}}{\text{h}}\right)\left(\dfrac{1 \text{ g}}{10^6 \ \mu\text{g}}\right)$$

$$= 3.456 \text{ g}$$

The ethanol exposure at the end of the second shift is 3.456 g.

4. PARTICULATES

Particulates include dusts, fumes, fibers, and mists. *Dusts* have a wide range of sizes and usually result from a mechanical process such as grinding. *Fumes* are extremely small particles, less than 1 μm in diameter, and result from combustion and other processes. *Fibers* are thin and long particulates, with asbestos being a prime example. *Mists* are suspended liquids that float in air, such as from the atomization of cutting oil. All of these types of particulate can pose an inhalation hazard if they reach the lungs.

With one known exception, particles larger than approximately 5 μm cannot reach the alveoli or inner recesses of the lungs before being trapped and expelled from the body through the digestive system or from the mouth and nose. Protection is afforded by the presence of mucus and cilia in the nasal passages, throat, larynx, trachea, and bronchi. The exception is asbestos fibers, which can reach the alveoli even though fibers may be larger. Particles smaller than 5 μm are considered respirable dusts and pose an exposure hazard when present in the breathing zone.

The body's reaction to particulates depends primarily on the type of particulate. Lung diseases result from the accumulation of particulates in the lungs, which restrict the ability of the lungs to transfer oxygen and carbon dioxide. This can cause the heart to overwork, leading to damage. Lung disease includes *fibrosis* (scar tissue formation), *bronchitis* (inflammation of the bronchi and an overproduction of mucus), *asthma* (constriction of the bronchial tubes), and cancer. Systemic reactions occur when the blood absorbs inorganic toxic particulates such as lead, mercury, and organic compounds. *Metal fume fever* results from breathing fine fumes of zinc, copper, and other metals. Allergic reactions occur from inhalation or dermal contact with organic particulates such as flour, grains, and chemicals. Bacterial and fungal infections result from the inhalation of particulates containing live organisms such as anthrax from wool particulates. Irritation of the nose and throat results from exposure to acid, alkali, or other irritating dusts or mists. Damage to internal tissues can result from inhalation of radioactive particulates.

There are four factors that affect the health risk from exposure to particulates: the types of particulate, the length of exposure, the concentration of particulates in the breathing zone, and the size of particulates in the breathing zone.

The type of particulate can determine the type of health effect that may result from the exposure. Both organic and inorganic dusts can produce allergic effects, dermatitis, and systemic toxic effects. Particulates that contain free silica can produce pneumoconiosis from chronic exposure. *Pneumoconiosis* is lung disease caused by fibrosis from exposure to both organic and inorganic particulates. Other particulates can cause systemic toxicity to the kidneys, blood, and central nervous system. Asbestos fibers can cause lung scarring and cancer.

The critical duration of exposure varies with the type of particulate. Metal fumes can cause metal fume fever with just a few hours of exposure. Toxic metal particulates can cause toxic effects from exposure of a few days to several months. Pneumoconioses may take several years to become disabling.

The concentration of particulates in the breathing zone is the primary factor in determining the health risk from

Environmental Health & Safety

particulates. The American Conference of Governmental Industrial Hygienists (ACGIH) has established TLVs that should not be exceeded. OSHA establishes PELs for safe exposure levels in the workplace.

The fourth exposure factor is the size of the particulates. Particles larger than 5 μm will normally be filtered out through the upper respiratory system before reaching the alveoli of the lungs.

Silica

Silica (SiO_2) has several associated health hazards. The crystalline form of free silica (quartz) deposited in the lungs causes the growth of fibrous tissue around the deposit. The fibrous tissue reduces the amount of normal lung tissue, thereby reducing the ability of the lungs to transfer oxygen. When the heart tries to pump more blood to compensate, heart strain and permanent damage or death may result. This condition is known as *silicosis*. Mycobacterial infection occurs in about 25% of silicosis cases.

Silica-containing dust is generated from a variety of occupations, including rock mining, sandblasting, stone cutting, and foundry work. Nonspecific defense mechanisms, such as mucus in the nasal passages, entrap and remove some of the particles. However, the particles from 0.5 μm to 3 μm in diameter may reach the alveoli where they can cause damage. Smokers exposed to silica dust have a significantly increased chance of developing lung cancer.

The OSHA PEL for dusts containing crystalline silica (quartz) is based on a formula that uses the percent of silica present in the sample. The PEL for crystalline silica can be found from 29 CFR 1910.1000, Table Z-3, as follows.

$$\text{PEL}_{\text{mg/m}^3} = \frac{10}{\%\text{SiO}_2 + 2} \qquad 55.3$$

The constant 2 is designed to limit the PEL to 5 mg/m^3 when the percent of free silica ($\%\text{SiO}_2$) is low (less than 1%). The time-weighted average (TWA) of the TLV for free silica is 0.1 mg/m^3. This is consistent with the PEL formula for 100% free silica, which yields 0.1 mg/m^3.

The TLVs for other forms of free silica are given in Table 55.2.

Table 55.2 *Typical TLVs for Various Forms of Free Silica*

form of free silica	TLV (mg/m^3)
quartz	0.1
cristobalite (diatomite)	0.05
tridymite	0.05
fused silica dust	0.1
tripoli and silica flour	0.1

Source: 29 CFR 1910.1000

Workers are crushing and grading rock for a drainfield in a subsurface disposal system. Samples of the air in the breathing zone of the workers show the following.

time	dust concentration (mg/m^3)	percent SiO$_2$
0800	1.5	5.8
1000	1.1	6.3
1300	2.1	4.2
1500	1.6	3.9
1700	1.8	4.1

Has worker exposure exceeded the OSHA PEL for quartz?

Solution

For mixtures containing free silica, use Eq. 55.3.

$$\text{PEL} = \frac{10 \, \dfrac{\text{mg}}{\text{m}^3}}{\%\text{SiO}_2 + 2}$$

percent SiO$_2$	PEL (mg/m^3)
5.8	1.28
6.3	1.20
4.2	1.61
3.9	1.69
4.1	1.64

The readings at 0800, 1300, and 1700 hours exceeded the PEL.

Asbestos

Asbestos is generically described as a naturally occurring, fibrous, hydrated mineral silicate. Inhalation of short asbestos fibers can cause *asbestosis*, a kind of pneumoconiosis, as a nonmalignant scarring of the lungs. *Bronchogenic carcinoma* is a malignancy (cancer) of the lining of the lung's air passages. *Mesothelioma* is a diffuse malignancy of the lining of the chest cavity or the lining of the abdomen.

Asbestos mining, construction activities, and working in shipyards are possible exposure activities that may involve the environmental engineer. The asbestos must be in a form that allows it to be airborne in order for inhalation to occur. Moreover, inhalation does not necessarily mean that lung damage will occur.

The onset of illness seems to be correlated with length and diameter of inhaled asbestos fibers. Fibers 2 μm in length cause asbestosis. Mesothelioma is associated with fibers 5 μm long. Fibers longer than 10 μm produce lung cancer. Fiber diameters greater than 3 μm are more likely to cause asbestosis or lung cancer, while fibers 3 μm or less in diameter are associated with mesothelioma.

There is a considerable latency period for asbestos-related illness. It may be 20 years after exposure to respirable asbestos before lung cancer is clinically evident, and it may be 30 years before peak incidence is noted.

OSHA has established comprehensive regulations relative to exposure to asbestos. The key documents are

- 29 CFR 1910.1001—regulations for general industry

- 29 CFR 1926.1101—regulations for the construction industry

- 29 CFR 1915.1001—regulations for the shipyard industry

- 40 CFR 763—Asbestos-Containing Materials in Schools Rule

The regulations should be reviewed for the requirements associated with the particular workplace activity (e.g., construction—29 CFR 1926.1101). The principal requirement is that no worker can be exposed to airborne asbestos fibers in excess of the PEL.

1. 0.1 fiber/cm^3 of air, 8 h TWA (PEL)

2. 1.0 fiber/cm^3 of air, 30 min short-term exposure limit (STEL)

The OSHA regulations for protection from exposure to asbestos are extensive. They require an employer to perform a negative exposure assessment in many cases. Monitoring must be performed by a competent person who is capable of identifying asbestos hazards and selecting control strategies and who has the authority to make corrective changes. The regulations also specify when medical surveillance is required, when personal protection must be provided, and the engineering controls and work practices that must be implemented.

Lead

The body does not use lead for any metabolic purpose, so any exposure to lead is undesirable. Lead dust and fumes can pose a severe hazard. Acute large doses of lead can cause systemic poisoning or seizures. Chronic exposure can damage the blood-forming bone marrow and the urinary, reproductive, and nervous systems. Lead is probably a human carcinogen, although whether it is causative or facilitative is subject to research.

OSHA has developed several regulations pertaining to exposure to lead in the workplace.

- 29 CFR 1910.1025—general exposure to lead in industry

- 29 CFR 1926.62—lead in construction standard

The general industry exposure standard for lead is a PEL of 50 μg/m^3, 8 h TWA (29 CFR 1910.1025(c)(1)).

Beryllium

Inhalation of metallic beryllium, beryllium oxide, or soluble beryllium compounds can lead to *chronic beryllium disease* (*berylliosis*). Ingestion and dermal contact do not pose a documented hazard, so maintaining beryllium dusts and fumes below the TLV in the breathing zone is a critical protection measure. As with asbestosis, it may be 20 years before the effects of exposure to beryllium dust are detectable.

Chronic beryllium disease is characterized by granulomas on the lungs, skin, and other organs. The disease can result in lung and heart dysfunction and enlargement of certain organs. Beryllium has been classified as a suspected human carcinogen.

The OSHA standard for exposure to beryllium is 2.0 μg/m^3, 8 h TWA with a ceiling concentration of 5.0 μg/m^3. The standard allows a 30 min exposure of up to 25 μg/m^3 during an 8 h shift.

Coal Dust

Coal dust can cause chronic bronchitis, silicosis, and *coal worker's pneumoconiosis*, also known as *black lung disease.* In coal mines, the Mine Safety and Health Administration (MSHA) regulates exposure to coal dust. The MSHA standard is 2 mg/m^3, 8 h TWA, of respirable coal dust. In other settings, OSHA regulates exposure to coal dust. The OSHA PEL is 2.4 mg/m^3 if less than 5% free silica is present; otherwise, the limit is the same as for crystalline silica.

Welding Fumes

Exposure to welding fumes can cause a disease known as *metal fume fever.* This disease results from inhalation of extremely fine oxide particles that have been freshly formed as fume. Zinc oxide fume is the most common source, but magnesium oxide, copper oxide, and other metallic oxides can also cause metal fume fever. Metal fume fever is of short duration, with symptoms including fever and shaking chills appearing 4–12 hours after exposure.

Since a wide variety of welding processes and alloys may be involved, the nature of the process and the system employed need to be analyzed with the fumes in order to determine the hazard and controls. The specific TLVs should be evaluated for the specific constituent involved. Fumes may come from welding aluminum, titanium, steel, ferrous alloys, and stainless steel. Fumes may contain hexavalent chromium, ozone, carbon monoxide, iron, manganese, silicon, nickel, and fluoride.

Radioactive Dusts

Radioactive dusts can cause toxicity in addition to the effects from ionizing radiation. Inhalation of radioactive dust can result in deposition of the radionuclide in the body, which may enter the bloodstream and affect individual organs.

Control measures should be instituted to prevent workers from inhaling radioactive dust, either by restricting access or by providing appropriate personal protection such as respirators. Engineering controls to capture radioactive dust are an absolute necessity to minimize worker exposure.

Biological Particulates

A wide variety of biological organisms can be inhaled as particulates causing respiratory diseases and allergies. Examples include dust that contains anthrax spores from the wool or bones of infected animals, and fungi spores from grain and other agricultural produce.

Control of Particulates

Ventilation

Ventilation is the most effective method for controlling particulates. Enclosed processes should be used wherever possible. Equipment can be enclosed so that only the feed and discharge openings are open. With adequate pressure, enclosed equipment can be nearly as effective as closed processes. Large automated equipment can sometimes be placed in separate enclosures. Workers would have to wear personal protection to enter the enclosures. Local exhaust ventilation with hooded enclosures can be very effective at controlling particulate emissions into general work areas. Where complete enclosure and local exhaust methods are not sufficient, *dilution ventilation* will be necessary to control particulates in the work area. In some instances, the work process can be changed from a dry to a wet process to reduce particulate generation.

Personal Protection

Respirators are an effective means of controlling worker exposure to particulates that remain in the work area after engineering controls have been applied or when access to dusty areas is intermittent. Respirators may also be used to provide additional protection or comfort to workers in areas where local or general ventilation is effective. The NIOSH guidelines for selection of respirators should be followed to ensure that the respirator will be effective at removing the specific particulate to which

the workers are exposed. The OSHA respiratory protection regulations in 29 CFR 1910.134 should also be followed.

Besides respirators, workers may also need protective clothing such as suits, eye protection, gloves, hard hats, boots, and so on. 29 CFR 1910.132 provides OSHA general requirements for protective equipment.

5. HEAT AND COLD STRESS

Thermal Stress

Heat and cold, or *thermal*, stress involves three zones of consideration relative to industrial hygiene. In the middle is the *comfort zone*, where workers feel comfortable in the work environment. On either side of the comfort zone is a *discomfort zone*, where workers feel uncomfortable with the heat or cold, but a health risk is not present. Outside of each discomfort zone is a *health risk zone*, where there is a significant risk of health disorders due to heat or cold. Industrial hygiene is primarily concerned with controlling worker exposure in the health risk zone.

The analysis of thermal stress involves taking a *heat balance* of the human body with the objective of determining whether the net heat storage is positive, negative, or zero. A simplified form of the heat balance is

$$Q_S = Q_M + Q_R + Q_C + Q_E \qquad \text{55.4}$$

If the storage, Q_S, is zero, heat gain is balanced by heat loss, and the body is in equilibrium. If Q_S is positive, the body is gaining heat; and if Q_S is negative, the body is losing heat.

The heat balance is affected by environmental and climatic conditions, work demands, and clothing. The metabolic rate, Q_M, is more significant for heat stress than for cold stress when compared with radiation and convection. The metabolic rate can affect heat gain by one to two orders of magnitude compared to radiation and convection, but it affects heat loss to about the same extent as radiation and convection.

Clothing affects the thermal balance through insulation, permeability, and ventilation. *Insulation* provides resistance to heat flow by radiation, convection, and conduction. *Permeability* affects the movement of water vapor and the amount of evaporative cooling. *Ventilation* influences evaporative and convective cooling.

Heat Stress

Heat stress can increase body temperature, heart rate, and sweating, which together constitute *heat strain*. As body temperature increases, blood circulation transports heat from the body core to the skin, where the blood is cooled and returned to the core. At the skin surface, heat is lost by convection and radiation at a rate that depends on the difference in temperature between the skin and the environment. Heat is also lost by evaporation of sweat, when the body attempts to increase sweating to the point where the heat storage rate is zero. Heart rate increases because more blood must be pumped between the body core and the skin to increase the heat loss. Sweat rate (and the total volume of sweat) increases to increase evaporative cooling.

The most serious heat disorder is *heatstroke*, because it involves a high risk of death or permanent damage. Fortunately, heatstroke is rare. Of lesser severity, *heat exhaustion* is the most commonly observed heat disorder for which treatment is sought. *Dehydration* is usually not noticed or reported, but without restoration of water loss, dehydration leads to heat exhaustion. The symptoms of these key heat stress disorders are as follows.

- heatstroke: chills, restlessness, irritability

- heat exhaustion: fatigue, weakness, blurred vision, dizziness, headache

- dehydration: no early symptoms, fatigue or weakness, headache, dry mouth

Appropriate first aid and medical attention should be sought when any heat stress disorder is recognized.

Evaluation of Heat Stress

NIOSH has established a threshold limit value (TLV) for prolonged exposure of 38°C for the acceptable body core temperature. (The *core temperature* is commonly assumed to equal the *oral temperature* plus 0.5°C.) Also, the core temperature can be 39°C for short periods with adequate recovery so long as the time-weighted average (TWA) is no more than 38°C. Moreover, the average heart rate over a day should not exceed 110 beats per minute (bpm). "Prolonged exposure" is assumed to be an 8 h work shift with breaks every 2 h.

NIOSH has established additional heat stress TLVs that can be used to ensure that the core temperature will not exceed safe levels. The TLVs are a function of the metabolic rate and an index of environmental heat. Three equations can be used to determine the ceiling limit, the recommended exposure limit, and the recommended alert limit.

The *ceiling limit* (CL) is given by

$$\text{CL} = 70.0°\text{C} - 13.5 \log Q_{M,\text{watts}} \qquad 55.5$$

NIOSH recommends that the CL not be exceeded for more than 15 min during the workday.

The NIOSH *recommended exposure limit* (REL) for acclimatized workers is given by

$$\text{REL} = 56.7°\text{C} - 11.5 \log Q_{M,\text{watts}} \qquad 55.6$$

The REL is based on summer-weight work clothes and is based on 8 h workdays with breaks every 2 h for acclimatized workers.

The *recommended alert level* (RAL) for unacclimatized workers is given by

$$\text{RAL} = 60.0°\text{C} - 14.1 \log Q_{M,\text{watts}} \qquad 55.7$$

Heat stress on unacclimatized workers should not be allowed to exceed the RAL.

To use the formulas for CL, REL, or RAL, the time-weighted average of the metabolic rate must be estimated. The metabolic rate for a given task can be estimated from Table 55.3.

Table 55.3 *Metabolic Rates for a Task*

task	average metabolic rate (W)
basal metabolism, B	70
posture metabolism, P	
sitting	20
standing	40
walking	170
activity metabolism, A	
light hand	30
heavy hand	65
light one arm	70
heavy one arm	120
light both arms	105
heavy both arms	175
light whole body	245
moderate whole body	350
heavy whole body	490
very heavy whole body	630
climbing metabolism, V	$56 V_v^*$

$^*V_v$ = rate of vertical ascent in m/min
Source: Adapted from American Conference of Government Industrial Hygienists, *1999 TLVs and BEIs, Threshold Limit Values for Chemical Substances and Physical Agents, Biological Exposure Indices*, 1999.

Using the metabolic rates for a given task and the following equation, the time-weighted average of the metabolism can be determined for an 8 h workday. B, P, A, and V are determined from Table 55.3. A time-weighted average of the metabolic rate (TWA-M) is then calculated from the duration of the metabolism for each task.

$$Q_{M,\text{task}} = B + P + A + V \qquad 55.8$$

After the REL or RAL is determined, the *wet-bulb globe temperature* (WBGT) must be determined for the working conditions. The WBGT affects heat removal through perspiration and is determined from Eq. 55.9 or Eq. 55.10 for indoor and outdoor (direct sunlight) exposure. Equation 55.9 is used to find the average WBGT.

$$\text{WBGT} = 0.7\text{NWB} + 0.3\text{GT} \qquad \begin{array}{c}\text{Heat Stress}\\ \textbf{55.9}\end{array}$$

$$\text{WBGT} = 0.7\text{NWB} + 0.2\text{GT} + 0.1\text{DB} \qquad \begin{array}{c}\text{Heat Stress}\\ \textbf{55.10}\end{array}$$

$$\begin{array}{l}\underset{\text{WBGT}}{\text{Average}} = (\text{WBGT}_1)(t_1) + (\text{WBGT}_2)(t_2) \\ \qquad + \cdots + (\text{WBGT}_n)(t_n) \\ \qquad /[(t_1) + (t_2) + \cdots (t_n)]\end{array} \qquad \begin{array}{c}\text{Heat Stress}\\ \textbf{55.11}\end{array}$$

For varying conditions, the TWA-WBGT must be determined. Also, the TWA-WBGT must be adjusted for other than summer work clothes by adding an adjustment factor to the calculated WBGT. The adjustment factors are given in Table 55.4.

Table 55.4 *WBGT Clothing Adjustment Factors*

clothing type	adjustment factor (°C)
summer work clothes	0
coveralls	+2
winter uniform	+4
vapor transmitting suit	+6
vapor barrier suit	+8
full-body encapsulating unit	+11

Source: Adapted from American Conference of Government Industrial Hygienists, *1999 TLVs and BEIs, Threshold Limit Values for Chemical Substances and Physical Agents, Biological Exposure Indices*, 1999.

Another aspect of heat stress evaluation is determining how long a worker can safely work in a particular environment. One approach to this problem is to establish conditions for recovery, such as sitting in a cool area in work clothes, and then to determine the environmental heat stress factors associated with the recovery period. The duration of the recovery period can then be changed until the REL or RAL is no longer exceeded. This analysis can also be performed for a 1 h period for a single set of activities and environmental conditions in order to get a ratio of recovery to work time for those conditions.

Other methods to estimate time-limited exposure to heat stress involve comparison of the TWA-WBGT with exposure limit charts to get the allowable exposure time, graphical solutions, and sophisticated heat balance analyses. Additionally, the International Organization for Standardization (ISO) has published a method for determining the allowable exposure time based on a standard perspiration rate.

Besides evaluating the exposure to heat stress through formulas based on environmental conditions, certain physiological functions can be measured directly to confirm that heat stress is within allowable limits. These functions are core temperature, heart rate, and sweating.

The instantaneous core temperature should never exceed 39°C, and the TWA core temperature should not exceed 38°C. The average heart rate should not exceed 110 beats per minute (bpm). The recovery heart rate after 1 min should be less than 110 bpm. Alternatively, the heart rate at 3 min should be less than 90 bpm, or the 1 min rate minus the 3 min rate should be at least 10 bpm.

Perspiration mass is the initial body mass plus the mass of food or drink minus the mass of excretions minus the final body mass. The perspiration volume in liters is the perspiration mass in kg × 1 L/kg. Perspiration volume should not exceed 5 L, and perspiration rate should not exceed 1 L/h (measured over a 2 h to 4 h period).

Example 55.3

A crew at a toxic spill site along a railroad in Arizona must conduct the cleanup during daylight hours and within one 8 h workday. The anticipated heat exposure levels for the crew are as follows.

activity	duration (h)	clothing
1 sitting, writing	1	coveralls
2 walking	2	coveralls
3 carrying heavy tools	2	coveralls
4 digging soil by hand (moderate, whole body)	1	vapor transmitting
5 filling sample containers (standing, both arms, light)	1	vapor transmitting
6 operating excavating equipment (sitting, both arms, light)	1	encapsulated

All workers are acclimatized to the local environment, and nominal rest breaks are provided in a cool trailer. It is a sunny day, and the environmental heat factors are

$$\text{DB} = 22°\text{C}$$
$$\text{GT} = 25°\text{C}$$
$$\text{NWB} = 18°\text{C}$$
$$\text{wind speed} = 2 \text{ m/s}$$

Will the crew exceed the NIOSH recommended exposure level (REL) for this work?

Solution

Calculate the TWA-WBGT that will be experienced by the crew while working in direct sunlight.

From Eq. 55.10,

Heat Stress

$$WBGT = 0.7NWB + 0.2GT + 0.1DB$$

$$= (0.7)(18°C) + (0.2)(25°C) + (0.1)(22°C)$$

$$= 19.8°C$$

activity	adjustment factor (°C)	duration (h)	TWA adjustment factor (°C)
1	+2	1	2
2	+2	2	4
3	+2	2	4
4	+6	1	6
5	+6	1	6
6	+11	1	11
total		8	33
TWA			4.1

$$TWA\text{-}WBGT = 19.8°C + 4.1°C$$
$$= 23.9°C$$

Calculate the TWA metabolic rate (TWA-M) of the crew.

Use Eq. 55.8.

$$Q_{M,\text{task}} = B + P + A + V$$

From Table 55.3,

activity	B	P	A	V	$Q_{M,\text{task}}$	duration	TWA-M
1	70	20	30	0	120	1	120
2	70	170	0	0	240	2	480
3	70	170	65	0	305	2	610
4	70	40	350	0	460	1	460
5	70	40	105	0	215	1	215
6	70	20	105	0	195	1	195
total						8	2080
TWA							260

Calculate the REL using Eq. 55.6.

$$REL = 56.7°C - 11.5 \log Q_M$$
$$= 56.7°C - 11.5 \log 260$$
$$= 28.9°C$$

Since the TWA-WBGT of 23.9°C is less than the REL of 28.9°C, the crew will not exceed the recommended exposure limit.

Control of Heat Stress

Controls that are applicable to any heat stress situation are known as *general controls*. General controls include worker training, heat stress hygiene, and medical monitoring.

Worker training will help the worker understand heat stress and recognize heat stress disorders. The training can also show the worker first aid for heat stress and appropriate hygiene to prevent heat stress problems.

Heat stress hygiene comprises actions taken by workers to prevent heat stress problems. These include fluid replacement, self-determination of heat stress effects, and acclimation to heat stress exposure. Lifestyle, general health, and diet also affect a worker's response to heat stress.

Medical monitoring by a qualified physician of individuals who are at extraordinary risk of heat-related disorders is also a general control measure.

Specific controls are controls that are put in place for a particular job. They include engineering controls, administrative controls, and personal protection. *Engineering controls* include changing the physical work demands to reduce the metabolic heat gain, reducing external heat gain from the air or surfaces, and enhancing external heat loss by increasing sweat evaporation and decreasing air temperature. *Administrative controls* include scheduling the work to allow worker acclimatization to occur, leveling work activity to reduce peak metabolic activity, and sharing or scheduling work so the heat exposure of individual workers is reduced. *Personal protection* includes using systems to circulate air or water through tubes or channels around the body, wearing ice garments, and wearing reflective clothing.

Cold Stress

The body reacts to cold stress by reducing blood circulation to the skin to insulate itself. The body also shivers to increase metabolism. These mechanisms are ineffective against long-term extreme cold stress, so humans react by increasing clothing for more insulation, increasing body activity to increase metabolic heat gain, and finding a warmer location.

There are two main hazards from cold stress: hypothermia and tissue damage. *Hypothermia* depresses the central nervous system, causing sluggishness and slurred speech, and progresses to disorientation and unconsciousness. To avoid hypothermia, the minimum core body temperature must be above 36°C for prolonged exposure and above 35°C for occasional exposure of short duration. Fatigue can increase the risk of severe hypothermia. Tissue damage (e.g., *frostbite* or *frostnip*) can occur when contact is made with objects or air that cause the skin temperature to fall below −1°C (the temperature required for skin to freeze).

First aid should be applied and medical attention should be sought whenever cold stress problems are recognized.

Environmental Health & Safety

The *equivalent chill temperature* (ECT), also known as the *wind chill index*, is calculated by determining the heat loss under combinations of air temperature and wind speed and then computing an equivalent air temperature with no air motion giving the same rate of heat loss. When workplace temperatures fall below 16°C, monitoring should be performed. Below −1°C, the dry-bulb temperature and air speed should be measured and recorded every 4 h. When the air speed exceeds 2 m/s, the ECT should be determined. When the ECT falls below −7°C, the ECT should be recorded.

Worker training, cold stress hygiene, and medical surveillance can control cold stress. Engineering controls, administrative controls, and personal protection measures can also be used to control cold stress.

6. ERGONOMICS

Ergonomics is the study of human characteristics to determine how a work environment should be designed to make work activities safe and efficient. It includes both physiological and psychological effects on the worker, as well as health, safety, and productivity aspects.

An environmental engineer should have a general knowledge of ergonomics because he or she may assume a position of responsibility for a project, a program, or a corporation that includes responsibility for the health and safety of employees or contractors. On large projects or programs, an environmental engineer will probably be assisted by an industrial hygienist or a health and safety specialist who will have specific responsibility to identify and report ergonomic hazards along with other hazards on the project or in the workplace. On small projects, however, an environmental engineer may not have such expertise readily available and needs to be able to identify situations where expert advice is needed. This section describes situations or conditions that imply a need for further ergonomic evaluation by an industrial hygienist. It also focuses on projects (such as hazardous waste site remediation or treatment plant operation) rather than on traditional manufacturing processes.

Work-Rest Cycles

An environmental engineer should be aware of the need for workers to be given rest periods appropriate for the metabolic energy used on the tasks. Excessively heavy work should be broken by frequent short rest periods to reduce cumulative fatigue. The percentage of time a worker should rest can be estimated by the following equation.

$$t_{\text{rest}} = \frac{Q_{M,\text{max}} - Q_M}{Q_{M,\text{rest}} - Q_M} \times 100\% \qquad \textit{55.12}$$

The rate of metabolic energy expended in typical activities for a 70 kg male worker is given in Table 55.5.

Table 55.5 *Typical Total Metabolic Energy Expenditure Rate of Various Activities*

activity	typical total metabolic energy expenditure rate (W)
resting, prone	105
resting, seated	115
standing, at ease	130
light bench work	120
medium bench work	190
driving	200
walking, casual	230
walking, hard	410
pushing wheelbarrow	410
shoveling	450
rock drilling	500
climbing stairs	700
heavy digging	800

Source: Adapted from Plog, Barbara A., Jill Niland, and Patricia J. Quinlan, *Fundamentals of Industrial Hygiene*, 4th ed., National Safety Council, Itasca, Illinois, 1996.

The maximum energy a worker can produce is approximately 1200 W, about half of which goes to maintain the body and the other half to actual work. The heaviest work a fit young man can sustain is approximately 580 W, but the general population can sustain only about 465–495 W. Women, on average, can sustain about 80% of these values.

The procedure for estimating rest time is illustrated in Ex. 55.4.

Besides metabolic energy, heart rate can be used to evaluate work in terms of light, medium, heavy, very heavy, and extremely heavy work, as shown in Table 55.6.

Table 55.6 *Metabolic Classification of Work*

class of work	total energy expenditure (W)	heart rate (beats/min)
light	175	90 or less
medium	350	100
heavy	525	120
very heavy	700	140
extremely heavy	1050	160 or more

Source: Adapted from Plog, Barbara A., Jill Niland, and Patricia J. Quinlan, *Fundamentals of Industrial Hygiene*, 4th ed., National Safety Council, Itasca, Illinois, 1996.

The direct correlation between heart rate and metabolic energy allows work to be evaluated through the relatively simple measurement of pulse. During light, medium, and heavy work, the metabolic and other physiological functions of the body can attain a steady-state condition for the duration of the work. During very heavy work when the heart rate is 140 bpm or more, the oxygen deficit and the buildup of metabolic by-products increase, which requires the worker to take rest periods or to cease work. At 160 bpm, frequent rest periods are required; this level of activity usually cannot be sustained for a full workday by even the most capable of workers.

Example 55.4

Workers at a hazardous waste remediation site must dig hardpan from around several tanks in an area not accessible to excavating equipment. The male workers are in poor physical condition, so the maximum metabolic energy the project manager believes is appropriate for sustained work on this project is 300 W. She estimates that the workers will need to spend 2 h pushing a wheelbarrow, 2 h shoveling, and 4 h digging with pickaxes to complete the project. The workers can sit when they rest. How much time should the project manager allocate to rest?

Solution

Calculate the TWA of the metabolic energy for the project.

activity	metabolic energy, Q_M (W)	duration (h)	TWA-M
pushing wheelbarrow	410	2	820
shoveling	450	2	900
digging	800	4	3200
total		8	4920
TWA-M			615

The TWA metabolic energy, Q_M, to be sustained for the duration of the project is 615 W. From Table 55.5, resting metabolism while sitting is 115 W.

Calculate the resting time required. Use Eq. 55.12.

$$t_{\text{rest}} = \frac{Q_{M,\text{max}} - Q_M}{Q_{M,\text{rest}} - Q_M} \times 100\% = \frac{300 \text{ W} - 615 \text{ W}}{115 \text{ W} - 615 \text{ W}} \times 100\%$$
$$= 63\%$$

The workers should be given 38 min rest out of every hour (63%).

Manual Handling of Loads

On many projects, loads must be handled manually. Improper handling is the most common cause of injury and of the most severe injuries in the workplace.

Heavy loads can strain the body, particularly the lower back. Even light or small objects can cause risk of injury to the body if they are handled in a way that requires strain-inducing stretching, reaching, or lifting.

Guidelines for Manual Material Handling

Safe manual material handling depends both on the design of the task and on the employee following safe practices. Table 55.7 provides guidelines for both the task design and the work practices.

Table 55.7 *Guidelines for Task Design and Work Practices for Manual Material Handling*

Task Design Guidelines

> Move materials predominately horizontally by pushing and pulling. Avoid lifting, lowering, and bending.
>
> Lift and lower materials between knuckle height and shoulder height.
>
> Lift and lower close to and in front of the body. Avoid bending forward or twisting the body.
>
> Make the material light, compact, and safe to grasp whenever possible. Provide good handles.
>
> Make sure the material does not have sharp edges, corners, or pinch points.
>
> If material is in bins, make sure the material can be easily removed.

Work Practices Guidelines

> Train workers in safe and efficient procedures.
>
> Select workers capable of performing the work.
>
> Encourage workers to be in good physical shape.
>
> Place material conveniently within reach.
>
> Use handling aids where appropriate.
>
> Keep the workspace clear.
>
> Get a good grip on the load.
>
> Test the weight before trying to move the load.
>
> Get the load close to the body.
>
> Place the feet close to the load.
>
> Stand in a stable position with the feet pointing in direction of movement.
>
> Lift mostly by straightening the legs.
>
> Do not twist the back or bend sideways.
>
> Get help when needed.
>
> Do not lift or lower with the arms extended.
>
> Do not continue heaving when the load is too heavy.

Environmental Health & Safety

Load Limits for Manual Material Handling

NIOSH has established a recommended weight limit (RWL) for manually lifting and lowering loads (*NIOSH Guide to Manual Lifting*). The maximum load under the most favorable circumstances is 51 lbm (23 kg). This limit is the maximum load that 90% of American workers, male or female, physically fit, and accustomed to physical labor, can lift and lower without injury. The maximum load of 51 lbm (23 kg) is reduced when the most favorable conditions are not achieved.

NIOSH provides an equation to reduce the maximum load based on six less-than-favorable conditions, as follows.

Recommended Weight Limit (RWL)

$$RWL = 51(10/H)(1 - 0.0075|V-30|)$$
$$\times (0.82 + 1.8/D) \qquad 55.13$$
$$\times (1 - 0.0032A)(FM)(CM)$$

Equation 55.13 must be calculated for both the start and the end of the lift. H is the horizontal distance between the ankles and the center of the load. V is the height of the hands above the floor. D is the vertical distance that the hands must travel between the start and end of the lift. All distances must be in inches.

A is the asymmetry angle measuring how far the object is displaced from the front of the worker's body. FM is the frequency multiplier, a dimensionless number that takes into account the number of lifts per minute, the value of V, and the duration of continuous lifting. Values for FM can be found in tables published by NIOSH. [Frequency Multiplier Table]

CM is the coupling multiplier, a dimensionless number that measures the ease and comfort with which the load can be grasped. The CM varies from 0.90 for a poor coupling (for example, an object with an off-center load or sharp edges) to 1.00 for a good coupling (for example, an object with handles that can be gripped comfortably). Values for CM can be found in tables published by NIOSH. [Coupling Multiplier (CM) Table (Function of Coupling of Hands to Load)]

Equation 55.13 is not dimensionally consistent, and units are disregarded. The equation gives a recommended weight limit in pounds.

Use of the NIOSH equation for calculating the recommended weight limit is illustrated in Ex. 55.5.

Example 55.5

A worker in a laboratory will need to lift 60 soil samples from the floor to a workbench in 1 hr. The tops of the sample containers have handles that are optimally designed. Not all the samples will be placed on the table directly in front of the worker; some will be put off to the side, and the average angular offset from a straight-ahead ending location is 20°. The worker's position is shown here.

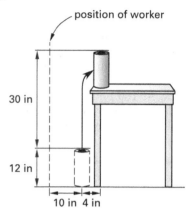

What is the NIOSH recommended weight limit for each container of soil?

Solution

From the illustration, the horizontal distance from the worker and the center of the load, H, is 10 in at the start of the lift and 14 in at the end of the lift. The height of the worker's hands, V, is 12 in at the start of the lift and 42 in at the end of the lift. The vertical distance that the hands must travel, D, is 30 in. The average asymmetry angle is 20°.

From a NIOSH frequency multiplier table, the frequency multiplier, FM, corresponding to a frequency of 1/min for a duration of 1 hr is 0.94. [Frequency Multiplier Table]

From a NIOSH coupling multiplier table, a container with optimally designed handles has good coupling, and its coupling multiplier, CM, is 1.00. [Coupling Multiplier (CM) Table (Function of Coupling of Hands to Load)]

From Eq. 55.13, the recommended weight limit for the start of the lift is

Recommended Weight Limit (RWL)

$$RWL = 51(10/H)(1 - 0.0075|V-30|)$$
$$\times (0.82 + 1.8/D)(1 - 0.0032A)$$
$$\times (FM)(CM)$$
$$= (51)\left(\frac{10}{10}\right)(1 - 0.0075|12 - 30|)$$
$$\times \left(0.82 + \frac{1.8}{30}\right)(1 - (0.0032)(20))$$
$$\times (0.94)(1.00)$$
$$= 34.16 \text{ lbf}$$

From Eq. 55.13, the recommended weight limit for the end of the lift is

Recommended Weight Limit (RWL)

$$RWL = 51(10/H)(1 - 0.0075|V - 30|)$$
$$\times (0.82 + 1.8/D)(1 - 0.0032A)$$
$$\times (FM)(CM)$$

$$= (51)\left(\frac{10}{14}\right)(1 - 0.0075|42 - 30|)$$
$$\times \left(0.82 + \frac{1.8}{30}\right)(1 - (0.0032)(20))$$
$$\times (0.94)(1.00)$$
$$= 25.67 \text{ lbf}$$

The recommended weight limit is the lower of the two values, or 25.67 lbf.

Cumulative Trauma Disorders

Cumulative trauma disorders (CTDs) can occur in almost any work situation. CTDs result from repeated stresses that are not excessive individually but, over time, cause disorders, injuries, and the inability to perform a job. High repetitiveness, or continuous use of the same body part, results in fatigue followed by cumulative muscle strain. These cumulative injuries are usually incurred by tendons, tendon sheaths, and soft tissue. Moreover, the cumulative injuries can result in damage to nerves and restricted blood flow. CTDs are common in the hand, wrist, forearm, shoulder, neck, and back. Bone and the spinal vertebrae may also be damaged.

The manifestations of CTDs on soft tissues include stretched and strained muscles, rough or torn tendons, inflammation of tendon sheaths, irritation and inflammation of bursa, and stretched (sprained) ligaments. Nerves can be affected by pressure from tendons or other soft tissue, resulting in loss of muscle control, numbness, tingling, or pain, and loss of response of nerves that control automatic functions such as body temperature and sweating. Blood vessels may be compressed, resulting in restricted blood flow and impaired control of tissues (muscles) dependent on that blood supply. Vibration, such as from operating vibrating tools, can cause the arteries in the fingers and hands to close down, resulting in numbness, tingling, and eventually loss of sensation and control.

Industrial hygienists have defined *high repetitiveness* as a cycle time of less than 30 s, or more than 50% of a cycle time spent performing the same fundamental motion. If the work activity requires the muscles to remain contracted at about 15–20% of their maximum capability, circulation can be restricted, which also contributes to CTDs. Also, severe deviation of the wrists, forearms, and other body parts can contribute to CTDs.

7. NOMENCLATURE

A	activity metabolism	W
A	asymmetry angle for lifting	deg
B	basal metabolism	W
C	concentration	ppm, mg/m^3
CL	ceiling heat limit	°C
CM	coupling multiplier for lifting	–
D	vertical travel distance of hands between start and end of lift	in
DB	dry-bulb temperature	°C
FM	frequency multiplier for lifting	–
GT	globe temperature	°C
H	horizontal multiplier for lifting	in
IR	total injury/illness incidence rate	–
k	non-ideal mixing factor	–
LFL	lower flammability limit	–
M	molecular weight	kg/kmol
N	number of injuries	–
NWB	natural wet-bulb temperature	°C
P	absolute ambient pressure	Pa
P	posture metabolism	W
PEL	permissible exposure limit	mg/m^3
Q	heat flow	W
Q_m	mass flow rate	kg/s
Q_V	ventilation rate	L/s·person, L/s·m^2
R_g	ideal gas constant, 8.314	kJ/kmol·K
RAL	recommended heat alert limit	°C
REL	recommended heat exposure limit	°C
RWL	recommended weight limit	kg
t	rest time, percent of period	%
t	time	s
T	temperature	°C, K
T	total hours	h
TLV	threshold limit value	–
UFL	upper flammability limit	–
V	velocity metabolism	W
V_v	rate of vertical ascent	m/min
V	vertical distance of hands from floor	in
WBGT	wet-bulb globe temperature	°C

Subscripts

C	convection
E	evaporation
M	metabolic
R	radiation
S	storage

Environmental Health & Safety

56 Exposure Assessments: Chemical and Biological

1. INTRODUCTION

Sixteenth century physician Paracelsus is known as the father of toxicology. He articulated the following four toxicological principles while studying the illnesses and diseases of workers exposed to toxic substances.

- Experiments are needed to identify and understand responses to chemicals.

- The curative and toxic properties of chemicals are different.

- The healing and harmful effects of chemicals can be differentiated based on dose.

- The properties and effects of specific chemicals can be identified.

From these principles grew the science of toxicology, which is presently defined as the study of adverse effects of chemicals on living organisms. Over the nearly five centuries since Paracelsus suggested his principles of toxicology, the effects of ionizing radiation, biological agents, and naturally occurring toxins have been added to the science and study of toxicology.

This chapter examines the pathways of human exposure to chemicals and other toxicants, the effects on workers of exposure to toxicants, dose-response relationships, methods for determining safe human doses, and standards for worker protection.

2. EXPOSURE PATHWAYS

Before a toxicant can do damage to a "target" tissue or organ, it must first enter the body. "Target" means the tissue or organ where the toxicant exerts its effects. There are three major exposure pathways: dermal absorption, inhalation, and ingestion (see Fig. 56.1). The eyes are an additional exposure pathway because they are particularly vulnerable to damage in the workplace.

Dermal Absorption

A common exposure pathway for toxicants is through the skin. The skin is composed of the epidermis, the dermis, and the subcutaneous layer. The *epidermis* is the upper layer, which is composed of several layers of flattened and scale-like cells. These cells do not contain blood vessels; they obtain their nutrients from the underlying dermis. The cells of the epidermis migrate to the surface, die, and leave behind a protein called *keratin*. Keratin is the most insoluble of all proteins, and, together with the scale-like cells, provides extreme resistance to substances and environmental conditions. Beneath the epidermis lies the *dermis*, which contains blood vessels, connective tissue, hair follicles, sweat glands, and other glands. The dermis supplies the nutrients for itself and for the epidermis. The innermost layer is called the *subcutaneous fatty tissue*, which provides a cushion for the skin and connection to the underlying tissue.

The condition of the skin, and the chemical nature of any toxic substance it contacts, affect whether and at what rate the skin absorbs that substance. The epidermis is impermeable to many gases, water, and chemicals. However, if the epidermis is damaged by cuts and abrasions, or is broken down by repeated exposure to soaps, detergents, or organic solvents, toxic substances can readily penetrate and enter the bloodstream. Chemical burns, such as those from acids, can also destroy the protection afforded by the epidermis, allowing toxicants to enter the bloodstream. Inorganic chemicals (and organic chemicals that are dissolved in water) are not readily absorbed through healthy skin. However, many organic solvents are lipid- (fat-) soluble and can easily penetrate skin cells and enter the body. After a toxicant has penetrated the skin and entered the bloodstream, the blood can transport it to target organs in the body.

Inhalation

Because the lungs are intended to transfer oxygen and carbon dioxide to and from the blood, they are a "super-highway" for toxicants to enter the body through the same process of molecular absorption. The respiratory tract consists of the nasal cavity, pharynx, larynx, trachea, primary bronchi, bronchioles, and alveoli. The first line of defense for the lungs consists of the mucus membranes found in the nasal cavity, the pharynx, the larynx, the trachea, and the bronchi. Particles are

Figure 56.1 *Exposure Routes for Chemical Agents*

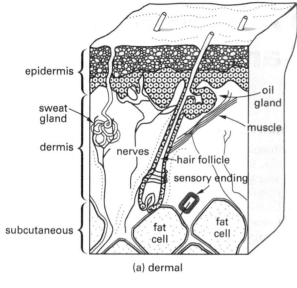

(a) dermal

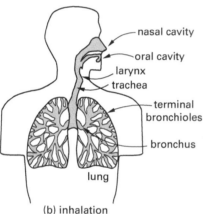

(b) inhalation

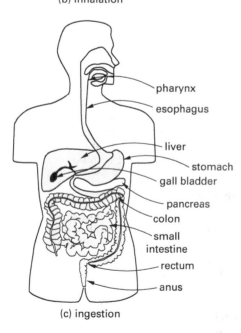

(c) ingestion

trapped in the mucus linings of the respiratory tract and are moved by the action of small hairs, called *cilia*, to the throat and then into the digestive system or are expectorated through the nose or mouth. Molecular transfer of oxygen and carbon dioxide between the bloodstream and the lungs occurs in the millions of air sacs, known as *alveoli*. Toxicants that reach the alveoli can be transferred to the blood through the respiratory system.

Toxicants that reach the alveoli will not be transferred to the blood at the same rate; various factors can increase or decrease the transfer rate of one toxicant relative to another. As in all gas transfer, the rate of diffusion depends on the ratio of the concentration of the toxicant in the air to its concentration in the blood. When the concentration of the toxicant is the same in the air and the blood, equilibrium exists and diffusion ceases. The solubility of a toxicant in blood will affect its absorption in the bloodstream. Toxicants with a high solubility in blood, such as chloroform, will be rapidly absorbed into the bloodstream, while other less-soluble toxicants will diffuse more slowly. Large molecule toxicants will transfer more slowly across the cell membrane in the alveoli than toxicants of smaller-size molecules. Both the respiration rate and the duration of exposure will affect the mass of toxicant transferred to the bloodstream over a given period. Finally, lungs that have been damaged through fibrosis or another type of disease will not be able to function as efficiently as those in a healthy individual, which will affect the transfer of a toxicant to the bloodstream.

Ingestion

Toxicants can enter the bloodstream through the digestive tract, which consists of the mouth, pharynx, esophagus, stomach, small intestine, large intestine, and rectum. The small intestine is the principal site in the digestive tract where the beneficial nutrients from food, and toxics from contaminated substances, are absorbed. Within the small intestine, millions of *villi* (projections) provide a huge surface area to absorb substances into the bloodstream. The large intestine is small, with much less surface area, and completes the digestion process by absorbing primarily water and inorganic salts.

Toxic substances will be absorbed in the intestines at various rates depending on the specific toxicant, its molecular size, and its degree of lipid (fat) solubility. Small molecular size and high lipid solubility facilitate diffusion of toxicants in the digestive tract.

The Eye

The eye is a pathway for toxicants to enter the body, through exposure to splashed chemicals, mists, particulates, and other workplace hazards. Transparent tissue in the front of the eye is known as the *cornea* and is the most likely eye tissue to come in contact with toxic substances. Toxicants can damage the cornea beyond its ability to repair itself by replacement of the outer cells from lower, basal cells. Under the cornea are the *anterior chamber*, the *lens*, and the *posterior chamber*. The transparency of the cornea and the lens is critical for the eye to function properly. Permanent damage can occur from scar tissue and death of the cells of the cornea.

Site of Effect

The exposure pathways described are routes of entry of toxicants into the body. However, the effects of such exposure may be either localized or systemic. If *localized*, the immediate area of entry is affected. If *systemic*, the toxicant's effects will occur at other sites in the body and may affect an organ or the central nervous system, causing damage to one or more functional systems. An example of a localized effect would be acid burns on the skin at the site of the exposure. An example of a systemic effect would be a toxicant that enters the bloodstream and is transported to a target organ where its effects occur.

For systemic effects to occur, the rate of accumulation of a toxicant must exceed the body's ability to *excrete* (eliminate) it or to *biotransform* it (transform it to less harmful substances). A toxicant can be eliminated from the body through the *kidneys*, which are the primary organs for eliminating toxicants from the body. The kidneys biotransform a toxicant into a water-soluble form and then eliminate it though the urine. The *liver* is also an important organ for eliminating a toxicant from the body, first by biotransformation, then by excretion into the bile where it is eliminated through the small intestine as feces.

A toxicant may also be stored in tissues for long periods before an effect occurs. Toxic substances stored in tissue (primarily in fat but also in bones, the liver, and the kidneys) may exert no effect for many years, or at all within the affected person's life. DDT, a pesticide, for instance, can be stored in body fat for many years and not exert any adverse effect on the body.

When toxic substances cannot be eliminated fast enough to keep up with the exposure, or when they cannot be stored without adverse effect in fatty tissue, then the target organ is overwhelmed and systemic effects occur. The liver, kidneys, and central nervous system are target organs that are commonly affected systemically.

3. EFFECTS OF EXPOSURE TO TOXICANTS

After toxicants are absorbed into the body through one or more of the pathways, a wide variety of effects on the human body is possible. When the toxic agents concentrate in target tissue or organs, the agents may interfere with the normal functioning of enzymes and cells or may cause genetic mutations.

Pulmonary Toxicity

Pulmonary toxicity refers to adverse effects on the respiratory system from toxic agents.

- Various toxicants, particulates, and gases and vapors can irritate or damage the nasal passages and the nerve cells responsible for smell. Excessive mucus flow, constriction of airflow, and a decrease in the effectiveness of the mucus system to clear toxicants may occur.

- *Nasal cancer* may result from prolonged exposure to chromate, nickel, isopropyl alcohol, hexavalent chromium, formaldehyde, nitrosamines, acrolein, actaldehyde, and vinyl chloride.

- Nitrogen dioxide, sulfur dioxide, ozone, and cigarette smoke may damage the cilia, which move particulates to the mouth for removal from the body. These toxicants, plus ammonia, may also lead to the proliferation of cells and the development of cancer.

- Chlorine, hexavalent chromium, ammonia, sulfur dioxide, and cigarette smoke may cause excessive mucus secretion, known as *bronchitis*.

- Ammonia, chlorine, hydrogen fluoride, nickel, and oxides of nitrogen, ozone, and perchloroethylene can cause *pulmonary edema*, the excessive accumulation of fluid in the alveoli of the lungs. This can interfere with transfer of oxygen and carbon dioxide and may develop into pneumonia.

- Silica, coal dust, asbestos, beryllium, cadmium, and the pesticide paraquat can result in an increased amount of connective tissue around the alveoli. This is known as *fibrosis*. Asbestos fibers can also cause lung cancer.

- Crystalline silica can cause increased deposition of connective tissue around the alveoli, known as *silicosis*.

- Cigarette smoke, ozone, nitrogen dioxide, and cadmium can cause enlargement and joining of the alveoli, causing the lungs to lose the ability to expand and contract, known as *emphysema*.

- Benzo[*a*]pyrene, nickel, chromium, arsenic, beryllium, and cigarette smoke can cause lung cancer.

Cardiotoxicity

Cardiotoxicity refers to the effects of toxic agents on the heart. The heart pumps blood through the aorta to the arteries and capillaries where oxygen and nutrients are delivered to tissues. Waste products are pumped from the capillaries to the lungs, kidney, and liver for removal from the body. The heart rate is controlled by the

Environmental Health & Safety

autonomic nervous system, and contractions are caused by electrochemical activity in the heart muscle.

- Toxicants can interfere with the normal contractions of the heart, which decreases its ability to transport oxygen and nutrients. The heart rate may be changed, and the strength of contractions may be diminished.

- Certain metals can affect the contractions of the heart and can interfere with cell metabolism.

- Organic solvents can alter the heart rate and the strength of contractions. Solvents act by suppressing the central nervous system, which affects the portion of the brain responsible for regulating the heart rate.

- Carbon monoxide can result in a decrease in the oxygen supply, causing improper functioning of the nervous system controlling the heart rate.

Hematoxicity

Hematoxicity refers to damage to the body's blood supply, which includes red blood cells, white blood cells, platelets, and plasma. The *red blood cells* transport oxygen to the body's cells and carbon dioxide to the lungs. *White blood cells* perform a variety of functions associated with the immune system. *Platelets* are important in blood clotting. *Plasma* is the noncellular portion of blood and contains proteins, nutrients, gases, and waste products.

- Benzene, lead, methylene chloride, nitrobenzene, naphthalene, and insecticides are capable of red blood cell destruction and can cause a decrease in the oxygen-carrying capacity of the blood. The resulting anemia can affect normal nerve cell functioning and control of the heart rate, and can cause shortness of breath, pale skin, and fatigue.

- Carbon tetrachloride, pesticides, benzene, and ionizing radiation can affect the ability of the bone marrow to produce red blood cells.

- Mercury, cadmium, and other toxicants can affect the ability of the kidneys to stimulate the bone marrow to produce more red blood cells when needed to counteract low oxygen levels in the blood.

- Some chemicals, including carbon monoxide, can interfere with the blood's capacity to carry oxygen, resulting in lowered blood pressure, dizziness, fainting, increased heart rate, muscular weakness, nausea, and, after prolonged exposure, death.

- Hydrogen cyanide and hydrogen sulfide can stimulate cells in the aorta, causing increased heart and respiratory rate. At high concentrations, death can result from respiratory failure.

- Benzene, carbon tetrachloride, and trinitrotoluene can suppress stem cell production and the

production of white blood cells. This can affect the clotting mechanism and the immune system.

- Benzene can cause high levels of white blood cells, a condition known as *leukemia.*

Hepatoxicity

Hepatoxicity refers to adverse effects on the liver that impede its ability to function properly. The liver converts carbohydrates, fats, and proteins to maintain the proper levels of glucose in the blood and converts excess protein and carbohydrates to fat. It also converts excess amino acids to ammonia and urea, which are removed in the kidneys. The liver also provides storage of vitamins and beneficial metals, as well as carbohydrates, fats, and proteins. Red blood cells that have degenerated are removed by the liver. Substances needed for other metabolic processes are provided by the liver. Finally, the liver detoxifies metabolically produced substances and toxicants that enter the body.

- Hexavalent chromium and arsenic cause cell damage in the liver.

- Carbon tetrachloride and alcohol can cause damage and death of liver cells, a condition known as *cirrhosis of the liver.*

- Chemicals or viruses can cause inflammation of the liver, known as *hepatitis.* Cell death and enlargement of the liver can occur.

Nephrotoxicity

Nephrotoxicity refers to adverse effects on the kidneys. The kidneys excrete ammonia as urea to rid the body of metabolic wastes. They maintain blood pH by exchanging hydrogen ions for sodium ions, and maintain the ion and water balance by excreting excess ions or water as needed. They also secrete hormones needed to regulate blood pressure. Like the liver, the kidneys function to detoxify substances.

- Heavy metals—primarily lead, mercury, and cadmium—cause impaired cell function and cell death. These metals can be stored in the kidneys, interfering with the functioning of enzymes in the kidneys.

- Chloroform and other organic substances can cause cell dysfunction, cell death, and cancer.

- Ethylene glycol can cause renal failure from obstruction of the normal flow of liquid through the kidneys.

Neurotoxicity

Neurotoxicity refers to toxic effects on the nervous system, which consists of the central nervous system (CNS) and the peripheral nervous system (PNS). The *central nervous system* includes the brain and the spinal cord, while the *peripheral nervous system* includes the remaining nerves, which are distinguished as sensory and motor nerves.

The basic functional unit of the nervous system is the *neuron*, which has the job of transmitting stimuli to and from the CNS. *Axons* are fiber-like groups of cells wrapped in a membrane sheath that carry the impulses within the neuron. The nerves function through electrochemical changes that occur in response to *neurotransmitters*, which are chemical substances released to initiate a response in a nerve, muscle, or gland. The exchange of sodium and potassium ions across a nerve membrane produces a voltage potential that stimulates the receptor. Enzymes cause the neurotransmitter to be rapidly broken down so the receptor will not be constantly stimulated.

Neurotoxic effects fall into two basic types: *destruction* of nerve cells and *interference* with neurotransmission.

- Methylmercury causes neurotoxicity in both high and low doses. In adults, damage from methylmercury includes loss of muscle control and coordination, a condition known as *ataxia*. In children, it can cause mental retardation and paralysis.

- Aluminum has been associated with Alzheimer's disease, but its effects on nerve cell functioning are not clear.

- Manganese can result in damage to the CNS. Its effects include irritability, difficulty in walking, and speech abnormalities. Prolonged exposure to manganese can cause symptoms similar to Parkinson's disease, including tremors and spastic muscle contractions.

- Methanol can cause nerve damage leading to blindness and death.

- Carbon monoxide and hydrogen cyanide can disrupt nerve cell metabolism and lead to death.

- Acrylamide, carbon disulfide, hexacarbons, and organophosphate pesticides can cause loss of peripheral sensations and can impair skeletal muscle functioning.

- Hexachlorophene is an antibacterial agent that interferes with neurotransmission and can cause blurred vision, muscle weakness, confusion, seizures, coma, and death.

- Lead affects the motor nerves.

- DDT interferes with neurotransmission, which can lead to tremors, irritability, hypersensitivity, dizziness, and convulsions.

- Methylene dichloride, carbon tetrachloride, and chloroform depress the CNS and can lead to dizziness, loss of muscle control, unconsciousness, and respiratory or cardiac arrest.

- Organophosphate pesticides and carbamate insecticides can cause muscle twitching, weakness, paralysis, and respiratory arrest.

Immunotoxicity

Immunotoxicity refers to toxic effects on the immune system, which includes the lymph system, blood cells, and antibodies in the blood.

The *lymph system* includes lymph fluid from the blood that passes through lymph nodules where microorganisms and cellular wastes are filtered out. The *spleen* also serves to destroy microorganisms as blood passes through it, and the *thymus* produces cells that protect the body from microorganisms. The cells of the immune system are produced in the *bone marrow*, and have the capabilities to destroy bacteria- and virus-infected cells. These cells are known as *natural killer cells*. *Antibodies* that are specific for antigens are also produced in the cells of the immune system.

- Halogenated aromatic hydrocarbons (HAHs) include polychlorinated biphenyls (PCBs), polybrominated biphenyls (PBBs), polychlorinated dibenzo-*p*-dioxins (TCDDs), and dibenzofurans. HAHs can cause atrophy of the thymus, decreased antibody production, and tumor promotion. TCDDs can affect stem cell division in the bone marrow.

- Polycyclic aromatic hydrocarbons (PAHs) are carcinogens due (possibly) to their ability to suppress the body's immune system. Examples of PAHs are benzo[*a*]pyrene (BaP) and 7,12-dimethylbenz[*a*]anthracene (DMBA).

- Heavy metals (e.g., lead, arsenic, mercury, and cadmium) can suppress the immune system. Lead, arsenic, and cadmium cause decreased levels of antibodies, resulting in susceptibility to infection. Mercury can decrease antibody response and cause an autoimmune response whereby the immune system attacks normal tissue, primarily in the kidneys, which is a major storage site for mercury.

- Organophosphate pesticides can decrease production of antibodies and killer cells, leading to greater susceptibility to infection.

- Some organic solvents affect antibody production and the formation of natural killer cells.

Reproductive Toxicity

The effect of toxicants on the male or female reproductive system is known as *reproductive toxicity*. For the male reproductive system, toxicants primarily affect the division of sperm cells and the development of healthy sperm. For the female reproductive system, toxicants can affect the endocrine system, the brain, and the reproductive tract.

Environmental Health & Safety

The male reproductive system can be affected by the following toxicants.

- the fungicide 1,2-dibromo-3-chloropropane

- the fumigant ethylene dibromide

- glycol ethers used in paint

- lead

- ethanol

- acetaldehyde

- phthalate esters

- methylene chloride

- the soil fumigant dibromochloropropane

- cadmium

- pesticides (e.g., DDT, DDE, methoxychlor, toxaphene, chlordecone, and endosulfan)

The female reproductive system can be affected by the following toxicants.

- heavy metals (e.g., inorganic mercury, lead, manganese, tin, and cadmium)

- styrene and polystyrene

- polycyclic aromatic hydrocarbons

- cigarette smoke

- pesticides

Toxic Effects on the Eye

There are a wide variety of toxic substances that can cause damage to the eye through contact with the cornea. Acids and bases can change the pH of the eye tissue and cause the cells to break down or disintegrate through coagulation of the proteins in the eye. Lime, metals, solvents, arsenicals, organic mercury, organic toxic chemicals, methanol, smog, dust, and allergens can cause damage to the eye by causing interference with the neurological functions, irritation or cloudiness of the cornea, damage to blood vessels, scratches on the cornea, formation of cataracts in the lens, and decreased vision, among other effects. Caustics, hydroxides of light metals, can cause greater damage to the eye than acids because acids precipitate a protein barrier that prevents further penetration into the tissue. Caustics, however, will continue to penetrate the eye until they are removed.

4. DOSE-RESPONSE RELATIONSHIPS

Dose-response relationships can be used to relate the response of an organism to increasing dose levels of toxicants. They show the level of death or other effect at specific doses of a toxic substance.

Studies involving animals are known as *toxicological studies*, whereas studies involving humans are called *epidemiological studies*. The results from both types of studies are sometimes used together to establish dose-response relationships and to establish safe human doses.

Dose-Response Curves

An objective of toxicity tests is to establish the *dose-response curve*, as illustrated in Fig. 56.2.

Figure 56.2 *Dose-Response Relationships*

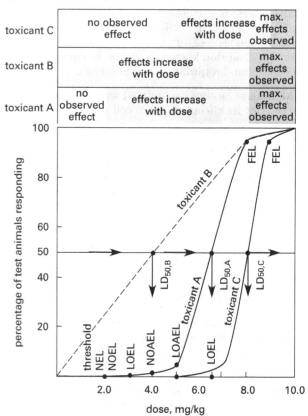

Several important features of a dose-response curve are illustrated and described as follows for toxicant A of Fig. 56.2.

- *response:* The ordinate is typically the percent of the test animals that respond at the given dose.

- *dose:* The abscissa is typically the dose applied to the test animal, such as milligram of toxicant per kilogram of body mass. Other measures of dose are possible.

- *no observed effect:* The range of the curve below which no effect is observed is the range of no observed effect. The upper end of the range is known as the *threshold*, the *no effect level* (NEL), or the *no observed effect level* (NOEL). The terms all mean the dose below which no measurable effect can be observed.

- *lowest observed effect:* The dose where minor effects first can be measured, but the effects are not directly related to the response being measured, is known as the *lowest observed effect level* (LOEL).

- *no observed adverse effect:* The dose where effects related to the response being measured first can be measured is known as the *no observed adverse effect level* (NOAEL). However, at this level, the effects observed at the higher doses are not observed. The terms all mean the dose below which no measurable effect can be observed.

- *lowest observed adverse effect:* The dose where effects related to the response being measured first can be measured, and are the same effects as the effects observed at the higher doses, is known as the *lowest observed adverse effect level* (LOAEL).

- *frank effect:* The *frank effect level* (FEL) dose marks the point where maximum effects are observed with little increase in effect for increasing dose.

For some toxicants, primarily those believed to be carcinogens, there is no apparent threshold. Any dose is considered to have an effect even though such effect may be unmeasurable at low doses. Such toxicants have no safe exposure level. This is illustrated as toxicant B in Fig. 56.2. Lead is an example of a toxicant with no threshold dose.

Toxicants A and C in Fig. 56.2 have similarly shaped dose-response curves, which means that they show the same relative effects at both high and low doses (i.e., toxicant A is more toxic than toxicant C at both high and low doses). If the dose-response curves for toxicants A and C were to cross, then one toxicant would be more toxic at low doses, and the other would be more toxic at high doses. Toxicant A is more toxic than toxicant C because a lower dose of toxicant A results in a pronounced response.

Acute and Chronic Toxicity Tests

To establish a dose-response relationship, toxicity tests are necessary. Both acute and chronic toxicity tests can be used for this purpose.

Acute toxicity tests are conducted for a relatively short duration, typically 24 hours to 14 days, with death of the test animal typically being the "response." Test animals typically include rats, mice, rabbits, and guinea pigs. The acute test must include a *control group* that is subjected to identical environmental conditions as the test animals except that it is not exposed to the

toxicant. The response of the control group enables the responses at low toxicant doses to be differentiated from environmental or other factors that may affect the test animals.

The results of acute toxicity tests are typically expressed as the *lethal dose* or *lethal concentration*—the concentration of toxicant at which a specified percentage of the test animals died. The lethal dose is expressed as the mass of toxicant per unit mass of test animal. Thus, LD_{50} means the dose in milligrams of toxicant per kilogram of body mass at which 50% of the test animals died.

For acute tests involving inhalation as the exposure pathway, the concentration (in parts per million) of the toxicant in air is used. If the toxicant is in particulate form, the concentration in milligrams of toxic particles per cubic meter of air is used. For example, LC_{50} means the concentration of the toxicant in air at which 50% of the test animals died.

The lethal concentration is associated with the inhalation pathway, while the lethal dose may be associated with either the ingestion or skin absorption pathway. However, this convention is not always followed, and the LD_{50} sometimes refers to the inhalation pathway. (Usually the pathway is noted with the lethal dose.)

Chronic toxicity tests can also be performed to determine the long-term dose-response relationship of a toxicant. Chronic tests may be conducted for 30 days or more, for years, or for the life of the test animal. Although death is often the end point in acute toxicity tests, in chronic toxicity tests the administered doses are selected such that most of the animals survive for the duration of the study. Chronic tests monitor diet, perform urinalysis, observe changes in blood composition and chemistry, and perform gross and microscopic examination of major tissues and organs.

5. SAFE HUMAN DOSE

EPA Methods

Several approaches exist for selecting a safe human dose from the data obtained from toxicological and epidemiological studies. The approach most likely to be encountered by the environmental engineer is the one recommended by the U.S. Environmental Protection Agency (EPA).

First, the differences between carcinogens and noncarcinogens (referred to by the EPA as *systemic toxicants*) are compared. Chemicals are classified as carcinogens if they can, or are believed to, produce tumors after exposure. *Carcinogens* generally do not exhibit thresholds of response at low doses of exposure, and any exposure is assumed to have an associated risk. The EPA model used to evaluate carcinogenic risk assumes no threshold and a linear response to any amount of exposure.

Environmental Health & Safety

Noncarcinogens (i.e., systemic toxicants) are chemicals that do not produce tumors (or gene mutations) but instead adversely interfere with the functions of enzymes in the body, thereby causing abnormal metabolic responses. Noncarcinogens have a dose threshold below which no adverse health response can be measured. Some substances can produce both carcinogenic and noncarcinogenic responses.

Noncarcinogens

The threshold below which adverse health effects in humans are not measurable or observable is defined by the EPA as the *reference dose* (RfD). The reference dose is used by the EPA and often results in more restrictive intake values than the values published by other organizations. The reference dose is the safe daily intake that is believed not to cause adverse health effects. The reference dose relates to the ingestion and dermal contact pathways and is route specific. For gases and vapors (exposure by the inhalation pathway), the threshold may be defined as the *reference concentration* (RfC). Sometimes the term "reference dose" is also used for exposure by the inhalation pathway.

The reference dose is set for the most sensitive response when more than one response, or endpoint, is measured. For example, a chemical may affect the central nervous system, the cardiovascular system, or other particular organs. The most sensitive of these responses is the basis for selecting the reference dose.

The reference dose (or reference concentration) is determined by dividing the NOAEL by the total *uncertainty factor* (UF). Uncertainty factors are used (with noncancer risk assessments only) to give a margin of safety when the procedure used to determine the NOAEL contains some degree of uncertainty, such as when data from tests on animals are extrapolated to determine the NOAEL for humans. All applicable uncertainty factors are multiplied together to give the total uncertainty factor. Some typical uncertainty factors used by the EPA and the World Health Organization are given in Table 56.1.

Reference Dose

$$RfD = \frac{NOAEL}{UF} \qquad 56.1$$

After an RfD has been identified for one or more toxicants, the safe human dose and the hazard ratio can be determined in order to assess whether exposures indicate an unacceptable hazard. The safe human dose (SHD) is the amount that a human of a given weight, W, can safely be exposed to in a day.

Reference Dose

$$SHD = RfD \times W = \frac{NOAEL \times W}{UF} \qquad 56.2$$

The hazard ratio (HR) is the *estimated exposure dose* (EED) for a toxicant divided by the reference dose for that toxicant. When there is exposure to more than one toxicant, the hazard ratio is calculated for each of the toxicants from all routes of exposure. If the sum of the hazard ratios exceeds 1.0, the risk is unacceptable. The hazard ratio should be considered a preliminary assessment.

$$HR = \frac{EED}{RfD} \qquad 56.3$$

Example 56.1

A chronic oral toxicity study is conducted to determine the effects of copper cyanide (CuCN) on rats. The quality of the data is judged by the toxicologist to be "average." The following results are obtained from the study.

NOEL	adverse effect observed at all doses
LOEL	adverse effect observed at all doses
NOAEL	adverse effect observed at all doses
LOAEL	25 mg/kg·d
FEL	100 mg/kg·d

In the absence of other, more reliable published information, what should be used as the RfD for human exposure from drinking water containing CuCN?

Solution

From Table 56.1, uncertainty factors 10A, 10H, and 10L apply. Therefore,

$$UF = (10)(10)(10) = 1000$$

From Eq. 56.1,

Reference Dose

$$RfD = \frac{NOAEL}{UF} = \frac{25 \; \dfrac{mg}{kg \cdot d}}{1000}$$

$$= 2.5 \times 10^{-2} \; mg/kg \cdot d \quad (25 \; \mu g/kg \cdot d)$$

Example 56.2

A community water supply well is found to be contaminated with copper cyanide (CuCN), methanol (CH$_3$OH), and potassium cyanide (KCN). The concentrations in the water and the EPA reference doses are as follows.

Table 56.1 *Typical EPA Factors Used to Calculate Reference Dose*

designation	description	uncertainty factor (UF)
10A, interspecies variability	used when extrapolating from results of long-term studies on experimental animals when results from human exposure studies are inadequate or unavailable	10
10H, intraspecies variability	used when extrapolating from valid experimental results in studies using prolonged exposure to average healthy humans	10
10L	used when deriving a reference dose (RfD) from LOAEL values rather than NOAEL values	10
10S	used when extrapolating from subchronic results on experimental animals when long-term human results are inadequate or absent	10

toxicant	exposure (μg/L)	*RfD* (oral) (μg/kg·d)
CuCN	40	5
CH$_3$OH	1000	500
KCN	600	50

Is a 70 kg person who drinks 2 L of water daily exceeding the safe human dose for these noncarcinogens?

Solution

Determine the estimated exposure dose for each of the toxicants. For copper cyanide,

$$EED_{CuCN} = \frac{E_{CuCN}(CR)}{m} = \frac{\left(40\,\dfrac{\mu g}{L}\right)\left(2\,\dfrac{L}{d}\right)}{70\text{ kg}}$$

$$= 1.14\ \mu\text{g/kg·d}$$

From Eq. 56.3,

$$HR_{CuCN} = \frac{EED_{CuCN}}{RfD_{CuCN}} = \frac{1.14\,\dfrac{\mu g}{kg\cdot d}}{5\,\dfrac{\mu g}{kg\cdot d}}$$

$$= 0.229$$

Similarly, for the other toxicants,

toxicant	exposure, E (μg/L)	estimated exposure dose, *EED* (μg/kg·d)	*RfD* (μg/kg·d)	hazard ratio, *HR*
CuCN	40	1.14	5	0.229
CH$_3$OH	1000	28.57	500	0.057
KCN	600	17.14	50	0.343
			total	0.629

Because the sum of the hazard ratios is less than 1.0, the safe human dose is not exceeded.

Carcinogens

The distinguishing feature of cancer is the uncontrolled growth of cells into masses of tissue called *tumors.* Tumors may be *benign*, in which the mass of cells remains localized, or *malignant*, in which the tumors spread through the bloodstream to other sites within the body. This latter process is known as *metastasis* and determines whether the disease is characterized as cancer. The term *neoplasm* (new and abnormal tissue) is also used to describe tumors.

Cancer occurs in three stages: initiation, promotion, and progression. During the *initiation* stage, a cell mutates and the DNA is not repaired by the body's normal DNA repair mechanisms. During the *promotion* stage, the mutated cells increase in number and undergo differentiation to create new genes. During *progression*, the cancer cells invade adjacent tissue and move through the bloodstream to other sites in the body. It is believed that continued exposure to the agent that initiated genetic mutation is necessary for progression to continue. Many mutations are believed to be required for the progression of cancer cells to occur at remote sites in the body.

Direct Human Exposure

The EPA's classification system for carcinogenicity is based on a consensus of expert opinion called *weight of evidence.*

The EPA maintains a database of toxicological information known as the Integrated Risk Information System (IRIS). The IRIS data include chemical names, chemical abstract service registry numbers (CASRN), reference doses for systemic toxicants, carcinogen potency factors (CPF) for carcinogens, and the carcinogenicity group classification, which is shown in Table 56.2.

Environmental Health & Safety

Table 56.2 *EPA Carcinogenicity Classification System*

group	description
A	human carcinogen
B1 or B2	probable human carcinogen
	B1 indicates that human data are available. B2 indicates sufficient evidence in animals and inadequate or no evidence in humans.
C	possible human carcinogen
D	not classifiable as to human carcinogenicity
E	evidence of noncarcinogenicity for humans

The dose-response for carcinogens differs substantially from that of noncarcinogens. For carcinogens it is believed that any dose can cause a response (mutation of DNA).

Since there are no levels (i.e., no thresholds) of carcinogens that could be considered safe for continued human exposure, a judgment must be made as to the acceptable level of exposure, which is typically chosen to be an excess lifetime cancer risk of 1×10^{-6} (0.0001%). *Excess lifetime cancer risk* refers to the incidence of cancers developed in the exposed animals minus the incidence in the unexposed control animals. For whole populations exposed to carcinogens, the number of total excess cancers, EC, is the product of the lifetime probability of excess cancer and the total exposed population, EP.

$$EC = (EP)Risk \qquad 56.4$$

Under the EPA approach, the *cancer slope factor* (CSF) is the slope of the dose-response curve at very low exposures. The CSF (also called the *cancer potency factor*, or CPF) has units of $(mg/kg \cdot d)^{-1}$. The CSF is pathway (route) specific. The CSF is obtained by extrapolation from the high doses typically used in toxicological studies. (See Table 56.3.)

The CSF is the probability of risk produced by lifetime exposure to 1.0 mg/kg·d of the known or potential human carcinogen. The CSF can be multiplied by the long-term daily intake (*chronic daily intake*, CDI) to obtain the lifetime probability of risk for daily doses other than 1.0 mg/kg·d. For less than lifetime exposure, the *exposure duration* must be used to calculate the total intake, which must be divided by the *averaging duration* of 75 years for carcinogens. For noncarcinogens, the averaging duration is the same as the exposure duration.

Carcinogens

$$Risk = dose \times toxicity = CDI \times CSF \qquad 56.5$$

For dermal contact, CDI is expressed as the adsorbed dose (AD). The EPA has developed equations for calculating CDI and AD exposure, the equation to be used depending on the exposure pathway. [Reference Dose: Exposure]

Table 56.3 *EPA Standard Values for Intake Calculations*

parameter	standard value
average body weight, female adult	65.4 kg
average body weight, male adult	78 kg
average body weight, child	
6–11 months	9 kg
1–5 years	16 kg
6–12 years	33 kg
amount of water ingested, adult	2.3 L/d
amount of water ingested, child	1.5 L/d
amount of air breathed, female adult	11.3 m³/d
amount of air breathed, male adult	15.2 m³/d
amount of air breathed, child (3–5 years)	8.3 m³/d
amount of fish consumed, adult	6 g/d
exposure duration, lifetime	75 yr

Example 56.3

A village with a stable population of 50,000 has a water supply that has been contaminated with benzene (C_6H_6) from a leaking underground storage tank. The leak occurred during the 20 yr before it was removed. The estimated average concentration of benzene during this period of leaking was 50 µg/L. It is expected that it will take another 10 yr before the benzene will be below detectable levels. The average estimated concentration of benzene during this period is 20 µg/L. The cancer slope factor for benzene by the oral route is 2.0×10^{-2} $(mg/kg \cdot d)^{-1}$. From Table 56.3, use a 75 yr lifespan, 72 kg adults, 16 kg (1–5 years) children, 2.3 L/d adult water consumption, 1.5 L/d child water consumption, and 10% children in the village. If the acceptable risks of additional cancer deaths due to benzene in the water are 1 adult and 0.5 child, can the water supply be used for the next 10 yr, or should the village abandon the supply?

Solution

step 1: Find the chronic daily intake (CDI) for adults and children for each time period.

Given a total lifespan (AT) of 75 yr and an exposure factor (EF) of 1, use Table 56.3 and the EPA equation for ingestion in drinking water.

Exposure

$$CDI = \frac{(CW)(IR)(EF)(ED)}{(BW)(AT)}$$

	BW (kg)	IR (L/d)	ED (yr)	CW (mg/L)	CDI (mg/kg·d)
adults	72	2.3	10	20×10^{-3}	9.00×10^{-5}
			20	50×10^{-3}	4.50×10^{-4}
children	16	1.5	10	20×10^{-3}	6.41×10^{-4}
			20	50×10^{-3}	3.20×10^{-3}

step 2: Find the probable risk of additional cancers, R, for adults and children for each time period.

Given a slope factor for orally ingested benzene of 2.0×10^{-2} $(mg/kg \cdot d)^{-1}$, use Eq. 56.5.

Carcinogens

$$Risk = CDI \times CSF$$

Equation 56.5 yields the following probabilities of excess cancer risk.

	ED (yr)	CDI (mg/kg·d)	risk
adults	10	9.00×10^{-5}	1.80×10^{-6}
	20	4.50×10^{-4}	9.00×10^{-6}
children	10	6.41×10^{-4}	1.28×10^{-5}
	20	3.20×10^{-3}	6.40×10^{-5}

step 3: Find the total excess cancers for adults and children for each time period.

From Eq. 56.4,

$$EC = (EP)Risk$$

	EP	ED (yr)	risk	EC
adults	45,000	10	1.80×10^{-6}	0.0812
		20	9.00×10^{-6}	0.404
children	5000	10	1.28×10^{-5}	0.0641
		20	6.40×10^{-5}	0.320

The total adult excess cancer risk is

$$EC_{a,\text{total}} = EC_{a,10} + EC_{a,20}$$
$$= 0.0812 + 0.404$$
$$= 0.485$$

Similarly, the total children excess cancer risk is

$$EC_{c,\text{total}} = EC_{c,10} + EC_{c,20}$$
$$= 0.0641 + 0.320$$
$$= 0.384$$

The total excess cancer risks for adults and children are less than 1.0 and 0.5, respectively, so the water supply can be used.

Bioconcentration Factors

Besides setting factors for direct human exposure to toxicants through water ingestion, inhalation, and skin contact, the EPA also has developed *bioconcentration factors* (BCF) (also referred to as *steady-state BCF*) so that the human intake from consumption of fish and other foods can be determined. Bioconcentration factors have been developed for many toxicants and provide a relationship between the toxicant concentration in the tissue of the organism (e.g., fish) and the concentration in the medium (e.g., water). The BCF is the ratio of the concentration in the organism to the concentration in the medium. Not all chemicals or other substances will bioaccumulate, and the BCF pertains to a specific organism, such as fish.

Bioconcentration Factor BCF

$$BCF = C_{org} / C$$
56.6

Selected bioconcentration factors (BCF) for selected chemicals in fish are given in Table 56.4. The substances are arranged in descending order of BCFs to illustrate the substances that have a high potential to bioaccumulate in fish. These substances are of great importance when the oral pathway is present in a particular situation.

Table 56.4 *Typical Bioconcentration Factors for Fish*[*]

substance	BCF (L/kg)
polychlorinated biphenyls	100 000
4,4′ DDT	54 000
DDE	51 000
heptachlor	15 700
chlordane	14 000
toxaphene	13 100
mercury	5500
2,3,7,8 tetrachlorodibenzo-*p*-dioxin (TCDD)	5000
dieldrin	4760
copper	200
cadmium	81
lead	49
zinc	47
arsenic	44
tetrachloroethylene	31
aldrin	28
carbon tetrachloride	19
chromium	16
chlorobenzene	10
benzene	5.2
chloroform	3.75
vinyl chloride	1.17
antimony	1

[*]For illustrative purposes only. Subject to change without notice. Local regulations may be more restrictive than federal.

Environmental Health & Safety

The BCF factors can be applied to determine the total dose to humans who ingest fish from water contaminated with toxicants that bioaccumulate. This dose would be added to the dose received from drinking the contaminated water.

Example 56.4

The town of Central has a water supply from the Middle River that is discovered to have been contaminated with 0.03 μg/L of heptachlor (carcinogen class B2) for 5 yr. In addition to getting their drinking water from this supply, people in the town have continually enjoyed the large trout from the Middle River and consume twice the EPA standard factor for fish consumption. The slope factor for heptachlor is 4.5 $(\text{mg/kg·d})^{-1}$. What is the risk of excess cancer over the lifetime of an adult if the heptachlor contamination is removed this year?

Solution

Determine the standard factors. From Table 56.3 and Table 56.4,

$$\text{IR}_{\text{fish}} = \left(6.0 \ \frac{\text{g}}{\text{d}}\right)(2) = 12 \ \text{g/d}$$
$$\text{IR}_{\text{water}} = 2.3 \ \text{L/d}$$
$$\text{ED} = 5 \ \text{yr}$$
$$\text{BW} = 75 \ \text{kg}$$
$$\text{BCF} = 15\,700 \ \text{L/kg}$$
$$\text{CW} = 0.03 \ \mu\text{g/L}$$
$$\text{SF} = 4.5 \ (\text{mg/kg·d})^{-1}$$

Using the EPA equation for ingestion in drinking water, calculate the chronic daily intake, CDI, for ingestion of heptachlor in drinking water. For carcinogens, the averaging time, AT, is considered to be a lifetime, or 75 yr, converted to days.

Exposure

$$\text{CDI}_{\text{water}} = \frac{(\text{CW})(\text{IR})(\text{EF})(\text{ED})}{(\text{BW})(\text{AT})}$$

$$= \frac{\left(\left(0.03 \times 10^{-6} \ \frac{\text{g}}{\text{L}}\right)\left(1000 \ \frac{\text{mg}}{\text{g}}\right)\right) \times \left(2.3 \ \frac{\text{L}}{\text{d}}\right)\left(365 \ \frac{\text{d}}{\text{yr}}\right)(5 \ \text{yr})}{(72 \ \text{kg})\left((75 \ \text{yr})\left(365 \ \frac{\text{d}}{\text{yr}}\right)\right)}$$

$$= 6.388 \times 10^{-8} \ \text{mg/kg·d}$$

Calculate the concentration of heptachlor in the fish.

Bioconcentration Factor BCF

$$BCF = C_{\text{org}}/C$$
$$C_{\text{fish}} = (BCF)\,C_{\text{heptachlor}}$$

$$= \left(15\,700 \ \frac{\text{L}}{\text{kg}}\right)\left(\frac{0.03 \ \frac{\mu\text{g}}{\text{L}}}{1000 \ \frac{\mu\text{g}}{\text{mg}}}\right)$$

$$= 0.471 \ \text{mg/kg}$$

Calculate the CDI for ingestion of heptachlor in contaminated fish. CF is the concentration in food measured in milligrams per kilogram.

Exposure

$$\text{CDI}_{\text{fish}} = \frac{(\text{CF})(\text{IR})(\text{FI})(\text{EF})(\text{ED})}{(\text{BW})(\text{AT})}$$

$$= \frac{\left(0.471 \ \frac{\text{mg}}{\text{kg}}\right)\left(\frac{12 \ \frac{\text{g}}{\text{d}}}{1000 \ \frac{\text{g}}{\text{kg}}}\right) \times (1)\left(365 \ \frac{\text{d}}{\text{yr}}\right)(5 \ \text{yr})}{(72 \ \text{kg})\left((75 \ \text{yr})\left(365 \ \frac{\text{d}}{\text{yr}}\right)\right)}$$

$$= 5.233 \times 10^{-6} \ \text{mg/kg·d}$$

The total CDI is

$$\text{CDI}_{\text{total}} = \text{CDI}_{\text{water}} + \text{CDI}_{\text{fish}}$$
$$= 6.388 \times 10^{-8} \ \frac{\text{mg}}{\text{kg·d}} + 5.233 \times 10^{-6} \ \frac{\text{mg}}{\text{kg·d}}$$
$$= 5.30 \times 10^{-6} \ \text{mg/kg d}$$

From Eq. 56.5, the excess cancer risk is

Carcinogens

$$Risk = CDI_{\text{total}} \times CSF_{\text{heptachlor}}$$

$$= \left(4.5 \ \frac{\text{kg·d}}{\text{mg}}\right)\left(5.30 \times 10^{-6} \ \frac{\text{mg}}{\text{kg·d}}\right)$$

$$= 23.8 \times 10^{-6}$$

The total risk of excess cancer is 23.8×10^{-6}.

ACGIH Methods

The American Conference of Governmental Industrial Hygienists (ACGIH) uses methods for determining the safe human dose that are somewhat different from the EPA methods previously described.

Threshold Limit Values

The ACGIH method uses predetermined *threshold limit values* (TLV) for both noncarcinogens and carcinogens. The TLVs are the concentrations in air that workers could be repeatedly exposed to on a daily basis without adverse health effects. The term TLV-TWA means the maximum time-weighted average concentration that all workers may be exposed to during an 8 hr day and 40 hr week. The TLV-TWA is for the inhalation route of exposure.

The time-weighted average can be calculated with Eq. 56.7. For each interval i, t_i is the interval's duration, and c_i is the concentration during the interval.

Time-Weighted Average (TWA)

$$\text{TWA} = \frac{\sum_{i=1}^{n} c_i t_i}{\sum_{i=1}^{n} t_i} \qquad 56.7$$

ACGIH also determines *short-term exposure limits* (TLV-STEL) for airborne toxicants, which are the recommended concentrations workers may be exposed to for short periods during the workday without suffering certain adverse health effects (e.g., irritation, chronic tissue damage, and narcosis). The TLV-STEL is the TWA concentration in air that should not be exceeded for more than 15 min of the workday. The TLV-STEL should not occur more than four times daily, and there should be at least 60 min between successive STEL exposures. In such cases, the excursions may exceed three times the TLV-TWA for no more than a total of 30 min during the workday, but shall not exceed five times the TLV-TWA under any circumstances. In all cases, the 8 hr TLV-TWA may not be exceeded. Short-term exposure limits have not been established by ACGIH for some toxicants.

ACGIH also publishes *ceiling threshold limit values* (TLV-C) that should not be exceeded at any time during the workday. If instantaneous sampling is infeasible, the sampling period for the TLV-C can be up to 15 min in duration. Also, the TLV-TWA should not be exceeded.

For mixtures of substances, the *equivalent exposure* over 8 hr is the sum of the individual exposures.

$$E = \frac{1}{8} \sum_{i=1}^{n} C_i T_i \qquad 56.8$$

The *hazard ratio* is the concentration of the contaminant divided by the exposure limit of the contaminant. For mixtures of substances, the total hazard ratio is the sum of the individual hazard ratios and must not exceed unity. This is known as the *law of additive effects*. For this law to apply, the effects from the individual substances in the mixture must act on the same organ. If the effects do not act on the same organ, then each of the individual hazard ratios must not exceed unity. The equivalent exposure of a mixture of gases is

$$E_m = \sum_{i=1}^{n} \frac{C_i}{L_i} \qquad 56.9$$

The TLV values published by ACGIH also include a "skin" designation (i.e., classification) for toxicants for which the dermal exposure route should also be considered. This notation means that ACGIH recognizes that the potential exposure through the skin, particularly the mucus membranes, of gases and vapors and the direct contact of the skin to liquids may be significant for the chemical. ACGIH then recommends biological monitoring to determine the relative exposure by the dermal route compared to the total dose.

Example 56.5

A worker is exposed to contaminants according to the following schedule.

substance	duration and concentration (ppm)				peak
	2 h	1 h	1 h	4 h	
acetone	1200	900	800	900	–
n-butyl alcohol	150	100	50	50	–
toluene	200	250	250	100	450, 10 min

The TWA-TLVs for these substances are as follows.

substance	TWA-TLV (ppm)	ceiling (ppm)	allowable peak (ppm)	duration of peak (min)
acetone	1000	–	–	–
n-butyl alcohol	100	–	–	–
toluene	200	300	500	10

Assume that the target organs are the same for acetone and *n*-butyl alcohol but different for toluene. Has the worker been exposed in excess of the allowable limitations?

Solution

Calculate the equivalent exposure of each contaminant. Use Eq. 56.8.

$$E = \frac{1}{8} \sum_{i=1}^{n} C_i T_i$$

For acetone,

$$E_{\text{acetone}} = \left(\frac{1}{8 \text{ h}} \right) \left(\begin{array}{l} (1200 \text{ ppm})(2 \text{ h}) + (900 \text{ ppm})(1 \text{ h}) \\ \quad + (800 \text{ ppm})(1 \text{ h}) + (900 \text{ ppm})(4 \text{ h}) \end{array} \right)$$
$$= 962.5 \text{ ppm}$$

For n-butyl alcohol,

$$E_{n\text{-butyl}} = \left(\frac{1}{8 \text{ h}} \right) \left(\begin{array}{l} (150 \text{ ppm})(2 \text{ h}) + (100 \text{ ppm})(1 \text{ h}) \\ \quad + (50 \text{ ppm})(1 \text{ h}) + (50 \text{ ppm})(4 \text{ h}) \end{array} \right)$$
$$= 81.3 \text{ ppm}$$

For toluene,

$$E_{\text{toluene}} = \left(\frac{1}{8 \text{ h}} \right) \left(\begin{array}{l} (200 \text{ ppm})(2 \text{ h}) + (250 \text{ ppm})(1 \text{ h}) \\ \quad + (250 \text{ ppm})(1 \text{ h}) + (100 \text{ ppm})(4 \text{ h}) \end{array} \right)$$
$$= 162.5 \text{ ppm}$$

Check the mixture, using the law of additive effects for the components, acetone and n-butyl, that affect the same target organs.

Use Eq. 56.9.

$$E_m = \sum_{i=1}^{n} \frac{C_i}{L_i} = \frac{C_{\text{acetone}}}{L_{\text{acetone}}} + \frac{C_{n\text{-butyl}}}{L_{n\text{-butyl}}}$$
$$= \frac{962.5 \text{ ppm}}{1000 \text{ ppm}} + \frac{81.3 \text{ ppm}}{100 \text{ ppm}}$$
$$= 0.963 + 0.813 = 1.775 > 1.0$$

Check the individual components.

$$\text{acetone} = 0.963 < 1.0$$
$$n\text{-butyl alcohol} = 0.813 < 1.0$$
$$\text{toluene} = \frac{162.5 \text{ ppm}}{200 \text{ ppm}} = 0.813 < 1.0$$

Check the ceiling and peak.

Toluene is the only contaminant with exposure limitations for a ceiling and a peak.

$$\text{maximum} = 250 \text{ ppm} < 300 \text{ ppm ceiling}$$
$$\text{peak} = 450 \text{ ppm}, 10 \text{ min} < 500 \text{ ppm}, 10 \text{ min}$$

The worker is not in compliance due to the mixture of acetone and n-butyl alcohol exceeding the hazard ratio of 1.0.

Biological Exposure Indices

Biological exposure indices (BEIs) are published by ACGIH for biological monitoring. BEIs are reference values for evaluation of the total exposure of a worker to chemicals in the workplace. The BEIs are the levels of contaminants that would be observed in specimens collected from a healthy worker who was exposed to chemicals to the same extent as a worker with inhalation exposure to the TLV. The BEIs are based on a normal workday of 8 hr and a workweek of 5 days. ACGIH does not provide safe exposure values for dermal or ingestion pathways except as referenced in the BEIs.

The BEIs require collection of urine, exhaled air, and blood specimens from exposed workers. The assistance of medical personnel is required to collect the specimens, and a qualified industrial hygienist is needed to collect data and interpret the results.

This medical approach provides the total body burden by accounting for substances that are absorbed through the skin, the pulmonary system, and the gastrointestinal tract. The urine analysis accounts for the level of metabolites of toxic agents in the urine. The blood analysis measures the concentration of the toxic agent present in the blood. Analysis of exhaled air is used to determine the rate at which fat-soluble gases and vapors that are not metabolized are cleared from the body.

Carcinogenicity Classification

ACGIH and other organizations do not use the same classification system as the EPA system described previously. ACGIH's carcinogenicity categories are given in Table 56.5.

Table 56.5 *ACGIH Carcinogenicity Categories*

category	description
A1	confirmed human carcinogen (sufficient weight of evidence from epidemiologic studies)
A2	suspected human carcinogen (limited evidence from human data but sufficient evidence from animal data)
A3	confirmed animal carcinogen with unknown relevance to humans (limited animal data not confirmed by human data)
A4	not classifiable as a human carcinogen (lack of sufficient human and animal data)
A5	not suspected as a human carcinogen (sufficient human or animal data)

ACGIH recommendations are contained in *Threshold Limit Values for Chemical Substances and Physical Agents, Biological Exposure Indices (TLVs and BEIs)*, published annually by the American Conference of Governmental Industrial Hygienists. This publication contains threshold limit values for chemical substances and physical agents in the workplace, along with biological exposure indices. Samples of the data provided in this publication for benzene are given in Table 56.6. A qualified industrial hygienist should be consulted regarding the interpretation and application of TLVs.

Table 56.6 *Sample TLV Data Provided by ACGIH**

TLV section	Benzene [71-43-2]
TWA	0.5 ppm/mg/m^3
STEL/C	2.5 ppm/mg/m^3
notations	skin; A1: BEI
molecular weight	78.11
TLV basis—critical effect(s)	leukemia
BEI section	
determinant	s-phenylmercapturic acid in urine
sampling time	end of shift
BEI	25 μg/g creatine
notation	B
notation section	
A	refers to Appendix A— carcinogenicity
BEI	identifies substances for which there are BEIs (see BEI section)
B	refers to Appendix B— substances of variable composition

*Adapted from *2012 TLVs and BEIs*, ACGIH, Cincinnati, OH

NIOSH Methods

The National Institute for Occupational Safety and Health (NIOSH) was established by the Occupational Safety and Health Act of 1970. NIOSH is part of the Centers for Disease Control and Prevention (CDC) and is the only federal institute responsible for conducting research and making recommendations for the prevention of work-related illnesses and injuries. The Institute's responsibilities include

- investigating hazardous working conditions as requested by employers or workers
- evaluating hazards ranging from chemicals to machinery
- creating and disseminating methods for preventing disease, injury, and disability

- conducting research and providing recommendations for protecting workers
- providing education and training to persons preparing for or actively working in the field of occupational safety and health

NIOSH maintains and publishes important databases and references on chemical hazards and worker protection, including the following.

- Immediately Dangerous to Life and Health (IDLH) Concentrations
- International Chemical Safety Cards
- *NIOSH Manual of Analytical Methods* (NMAM)
- *NIOSH Pocket Guide to Chemical Hazards* (NPG)
- Recommendations for Chemical Protective Clothing
- Specific Medical Tests Published for OSHA Regulated Substances
- Toxicologic Review of Selected Chemicals
- Certified Equipment List

The information given in the IDLH concentration database applies to exposure to airborne contaminants when that exposure is likely either to cause death or immediate or delayed permanent adverse health effects or to prevent escape from such an environment. The purpose of establishing an IDLH concentration is to ensure that workers can escape from a given contaminated environment in the event of failure of the respiratory protection equipment.

The NIOSH chemical *safety cards* database contains chemical information integrated from various national and international organizations. A sample of an International Chemical Safety Card is given in Fig. 56.3.

The *NIOSH Manual of Analytical Methods* (NMAM) is a collection of methods for sampling and analysis of contaminants in workplace air and in the blood and urine of workers who are occupationally exposed. These methods have been developed specifically to have adequate sensitivity to detect the lowest concentrations as regulated by OSHA and recommended by NIOSH, and to have sufficient range to measure concentrations exceeding safe levels of exposure. NMAM also includes chapters on quality assurance, strategies of sampling airborne substances, method development, and discussions of some portable direct-reading instrumentation.

Probably the most important NIOSH toxicological publication is the *NIOSH Pocket Guide to Chemical Hazards*. The *Pocket Guide* is intended as a source of general industrial hygiene information for workers, employers, and occupational health professionals. The *Pocket Guide* presents key information and data in

Environmental Health & Safety

Figure 56.3 *Sample International Chemical Safety Card*

International Chemical Safety Cards

BENZENE **ICSC: 0015**

National Institute for Occupational Safety and Health — NIOSH

BENZENE
Cyclohexatriene
Benzol
C_6H_6
Molecular mass: 78.1

CAS # 71-43-2
RTECS # CY1400000
ICSC # 0015
UN # 1114
EC # 601-020-00-8

TYPES OF HAZARD/ EXPOSURE	ACUTE HAZARDS/ SYMPTOMS	PREVENTION	FIRST AID/ FIRE FIGHTING
FIRE	Highly flammable.	NO open flames, NO sparks, and NO smoking.	Powder, AFFF, foam, carbon dioxide.
EXPLOSION	Vapour/air mixtures are explosive. Risk of fire and explosion: see chemical dangers.	Closed system, ventilation, explosion-proof electrical equipment and lighting. Do NOT use compressed air for filling, discharging, or handling. Use non-sparking handtools.	In case of fire: keep drums, etc., cool by spraying with water.
EXPOSURE		AVOID ALL CONTACT!	
• INHALATION	Dizziness. Drowsiness. Headache. Nausea. Shortness of breath. Convulsions. Unconsciousness.	Ventilation, local exhaust, or breathing protection.	Fresh air, rest. Refer for medical attention.
• SKIN	MAY BE ABSORBED! Dry skin (further see Inhalation).	Protective gloves. Protective clothing.	Remove contaminated clothes. Rinse skin with plenty of water or shower. Refer for medical attention.
• EYES		face shield, or eye protection in combination with breathing protection.	First rinse with plenty of water for several minutes (remove contact lenses if easily possible), then take to a doctor.
• INGESTION	Abdominal pain. Sore throat. Vomiting (further see Inhalation).	Do not eat, drink, or smoke during work.	Rinse mouth. Do NOT induce vomiting. Refer for medical attention.

SPILLAGE DISPOSAL	STORAGE	PACKAGING & LABELLING
Collect leaking and spilled liquid in sealable containers as far as possible. Absorb remaining liquid in sand or inert absorbent and remove to safe place. Do NOT wash away into sewer (extra personal protection: complete protective clothing including self-contained breathing apparatus).	Fireproof. Separated from food and feedstuffs, oxidants and halogens.	Do not transport with food and feedstuffs. F symbol T symbol R: 45-11-48/23/24/25 S: 53-45 UN Hazard Class: 3 UN Packing Group: II

SEE IMPORTANT INFORMATION ON BACK

ICSC: 0015

IMPORTANT DATA

PHYSICAL STATE; APPEARANCE:
COLOURLESS LIQUID, WITH CHARACTERISTIC ODOUR.

PHYSICAL DANGERS:
The vapour is heavier than air and may travel along the ground; distant ignition possible.

CHEMICAL DANGERS:
Reacts violently with oxidants and halogens causing fire and explosion hazard.

OCCUPATIONAL EXPOSURE LIMITS (OELs):
TLV: 10 ppm, 32 mg/m^3 (as TWA) A2 (ACGIH 1991-1992).
OSHA PEL: 1910.1028 TWA 1 ppm ST 5 ppm See Appendix F
NIOSH REL: Ca TWA 0.1 ppm ST 1 ppm See Appendix A
NIOSH IDLH: Potential occupational carcinogen 500 ppm

ROUTES OF EXPOSURE:
The substance can be absorbed into the body by inhalation and through the skin.

INHALATION RISK:
A harmful contamination of the air can be reached rather quickly on evaporation of this substance at 20°C; on spraying or dispersion, however, much faster.

EFFECTS OF SHORT-TERM EXPOSURE:
The substance irritates the skin and the respiratory tract. Swallowing the liquid may cause aspiration into the lungs with the risk of chemical pneumonitis. The substance may cause effects on the central nervous system. Exposure far above the occupational exposure limit may result in unconsciousness.

EFFECTS OF LONG-TERM OR REPEATED EXPOSURE:
The liquid defats the skin. The substance may have effects on the blood forming organs, liver and immune system. This substance is carcinogenic to humans.

PHYSICAL PROPERTIES

Boiling point: 80°C
Melting point: 6°C
Relative density (water = 1): 0.9
Solubility in water, g/100 ml at 25°C: 0.18
Vapour pressure, kPa at 20°C: 10
Relative vapour density (air = 1): 2.7
Relative density of the vapour/air-mixture at 20°C (air = 1): 1.2
Flash point: -11°C (c.c.)°C
Auto-ignition temperature: about 500°C
Explosive limits, vol% in air: 1.2-8.0
Octanol/water partition coefficient as log Pow: 2.13

ENVIRONMENTAL DATA

NOTES

Use of alcoholic beverages enhances the harmful effect. Depending on the degree of exposure, periodic medical examination is indicated. The odour warning when the exposure limit value is exceeded is insufficient.

Transport Emergency Card: TEC (R)-7
NFPA Code: H2; F3; R0;

ADDITIONAL INFORMATION

ICSC: 0015 **BENZENE**

IMPORTANT LEGAL NOTICE: Neither NIOSH, the CEC or the IPCS nor any person acting on behalf of NIOSH, the CEC or the IPCS is responsible for the use which might be made of this information. This card contains the collective views of the IPCS Peer Review Committee and may not reflect in all cases all the detailed requirements included in national legislation on the subject. The user should verify compliance of the cards with the relevant legislation in the country of use. The only modifications made to produce the U.S. version is inclusion of the OSHA PELs, NIOSH RELs and IDLH values.

Source: International Chemical Safety Cards (ICSC), Centers for Disease Control and Prevention and National Institute for Occupational Safety and Health.

abbreviated tabular form for 677 chemicals or substance groupings (e.g., manganese compounds, tellurium compounds, inorganic tin compounds) that are found in the work environment. The industrial hygiene information in the *Pocket Guide* is intended to help users recognize and control occupational chemical hazards. The chemicals or substances include all substances for which the National Institute for Occupational Safety and Health (NIOSH) has recommended exposure limits (RELs) and those with permissible exposure limits (PELs) as found in the Occupational Safety and Health Administration (OSHA) *General Industry Air Contaminants Standard* (29 CFR 1910.1000). A sample of the information is given in Fig. 56.4.

The NIOSH recommended exposure limits (RELs) given in the *Pocket Guide* are time-weighted average (TWA) concentrations for up to a 10 hr workday during a 40 hr workweek. A short-term exposure limit (STEL) is a 15 min TWA exposure that should not be exceeded at any time during a workday. A ceiling REL should not be exceeded at any time. The "skin" designation means there is a potential for dermal absorption, so skin exposure should be prevented as necessary through the use of good work practices and gloves, coveralls, goggles, and other appropriate equipment.

The NIOSH database "Recommendations for Chemical Protective Clothing, A Companion to the NIOSH Pocket Guide to Chemical Hazards" provides recommendations for *chemical protective clothing* (CPC) to prevent direct skin contact and contamination. The database also includes recommendations to prevent physical injury to the unprotected skin from thermal hazards, such as rapidly evaporating liquified gases.

6. BIOLOGICAL HAZARDS

Since at least the mid-nineteenth century, sanitary engineers and public health engineers have dealt with biological organisms in many different situations. More recently, environmental engineers, who embrace a broader range of concerns and responsibilities for the natural and human environment than did their predecessors, need to be aware of biological hazards that may be encountered both directly with and incidentally to projects. This section describes biological hazards that an environmental engineer needs to be familiar with in order to recognize when to obtain the advice of an industrial hygienist or biosafety specialist.

Biological Agents

Approximately 200 biological agents are known to produce infectious, allergenic, toxic, and carcinogenic reactions in workers. These agents and their reactions are as follows.

- Microorganisms (viruses, bacteria, fungi) and the toxins they produce cause infection and allergic reactions.

- Arthropod (crustaceans), arachnid (spiders, scorpions, mites, and ticks), and insect bites and stings cause skin inflammation, systemic intoxication, transmission of infectious agents, and allergic reactions.

- Allergens and toxins from plants cause dermatitis from skin contact, rhinitis (inflammation of the nasal mucus membranes), and asthma from inhalation.

- Protein allergens (urine, feces, hair, saliva, and dander) from vertebrate animals cause allergic reactions.

Also posing potential biohazards are lower plants other than fungi (e.g., lichens, liverworts, and ferns) and invertebrate animals other than arthropods (e.g., parasites, flatworms, and roundworms).

Microorganisms may be divided into prokaryotes and eukaryotes. *Prokaryotes* are organisms having DNA that is not physically separated from its cytoplasm (cell plasma that does not include the nucleus). They are small, simple, one-celled structures, less than 5 μm in diameter, with a primitive nuclear area consisting of one chromosome. Reproduction is normally by binary fission in which the parent cell divides into two daughter cells. All bacteria, both single-celled and multicellular, are prokaryotes, as are blue-green algae.

Eukaryotes are organisms having a nucleus that is separated from the cytoplasm by a membrane. Eukaryotes are larger cells (greater than 20 μm) than prokaryotes, with a more complex structure, and each cell contains a distinct membrane-bound nucleus with many chromosomes. They may be single-celled or multicellular, reproduction may be asexual or sexual, and complex life cycles may exist. This class of microorganisms includes fungi, algae (except blue-green), and protozoa.

Since prokaryotes and eukaryotes have all of the enzymes and biological elements to produce metabolic energy, they are considered organisms.

In contrast, a *virus* does not contain all of the elements needed to reproduce or sustain itself and must depend on its host for these functions. Viruses are nucleic acid molecules enclosed in a protein coat. A virus is inert outside of a host cell and must invade the host cell and use its enzymes and other elements for the virus's own reproduction. Viruses can infect very small organisms such as bacteria, as well as humans and animals. Viruses are 20 μm to 300 μm in diameter.

Smaller than the viruses by an order of magnitude are *prions*, small proteinaceous infectious particles. Prions have properties similar to viruses and cause degenerative diseases in humans and animals.

Infection

The invasion of the body by pathogenic microorganisms and the reaction of the body to them and to the toxins they produce is called an *infection*. Infection may be

Figure 56.4 *Sample of NIOSH Pocket Guide to Chemical Hazards*

NIOSH Pocket Guide to Chemical Hazards

Benzene		CAS 71-43-2
C_6H_6		RTECS CY1400000
Synonyms & Trade Names Benzol, Phenyl hydride		**DOT ID & Guide** 1114 130

Exposure Limits	NIOSH REL: Ca TWA 0.1 ppm ST 1 ppm See Appendix A		
	OSHA PEL: [1910.1028] TWA 1 ppm ST 5 ppm See Appendix F		
IDLH Ca [500 ppm] See: 71432		**Conversion** 1 ppm = 3.19 mg/m^3	

Physical Description
Colorless to light-yellow liquid with an aromatic odor. [Note: A solid below 42°F.]

MW: 78.1	BP: 176°F	FRZ: 42°F	Sol: 0.07%
VP: 75 mmHg	IP: 9.24 eV		Sp.Gr: 0.88
Fl.P: 12°F	UEL: 7.8%	LEL: 1.2%	

Class IB Flammable Liquid: Fl.P. below 73°F and BP at or above 100°F.

Incompatibilities & Reactivities
Strong oxidizers, many fluorides & perchlorates, nitric acid

Measurement Method
Charcoal tube; CS$_2$; Gas chromatography/Flame ionization detection; IV [#1500, Hydrocarbons] [Also #3700, #1501] See: NMAM INDEX

Personal Protection & Sanitation	**First Aid** (See procedures)
Skin: Prevent skin contact	Eye: Irrigate immediately
Eyes: Prevent eye contact	Skin: Soap wash immediately
Wash skin: When contaminated	Breathing: Respiratory support
Remove: When wet (flammable)	Swallow: Medical attention immediately
Change: N.R.	
Provide: Eyewash, Quick drench	

Respirator Recommendations NIOSH
At concentrations above the NIOSH REL, or where there is no REL, at any detectable concentration: (APF = 10,000) Any self-contained breathing apparatus that has a full facepiece and is operated in a pressure-demand or other positive-pressure mode/(APF = 10,000) Any supplied-air respirator that has a full facepiece and is operated in a pressure-demand or other positive-pressure mode in combination with an auxiliary self-contained positive-pressure breathing apparatus
Escape: (APF = 50) Any air-purifying, full-facepiece respirator (gas mask) with a chin-style, front- or back-mounted organic vapor canister/Any appropriate escape-type, self-contained breathing apparatus

Exposure Routes inhalation, skin absorption, ingestion, skin and/or eye contact

Symptoms irritation eyes, skin, nose, respiratory system; giddiness; headache, nausea, staggered gait; fatigue, anorexia, lassitude (weakness, exhaustion); dermatitis; bone marrow depressant/depression; [Potential occupational carcinogen]

Target Organs Eyes, skin, respiratory system, blood, central nervous system, bone marrow

Cancer Site [leukemia]

See also: INTRODUCTION   See ICSC CARD: 0015   See MEDICAL TESTS: 0022

Source: *NIOSH Pocket Guide to Chemical Hazards*, DHHS (NIOSH) Publication No. 2005-149, National Institute for Occupational Safety and Health.

endogenous where microorganisms that are normally present in the body (*indigenous*) at a particular site (such as *E. coli* in the intestinal tract) reach another site (such as the urinary tract), causing infection there.

Infections from microorganisms not normally found on the body are called *exogenous* infections. The mechanisms of entry include inhalation, indirect or direct contact, penetration (e.g., mosquito bites), and ingestion. Sometimes infectious agents can be transmitted from coworkers by inhalation, such as with measles or tuberculosis. Workers who do not show signs of the disease but are able to transmit the infection are known as *carriers*. The effect of the infection is dependent upon the virulence of the infectious agent, the route of infection, and the relative immunity of the worker.

The most common routes of exposure to infectious agents are through cuts, punctures and bites (insect and animal), abrasions of the skin, inhalation of aerosols generated by accidents or work practices, contact between mucous membranes or contaminated material, and ingestion. In laboratory and medical settings, transmission of blood-borne pathogens can occur through handling of blood products and human tissue.

Biohazardous Workplaces and Activities

Although environmental engineers have long been concerned with waterborne diseases and their prevention in the design and operation of water supply and wastewater systems, pathogens may also be encountered in the workplace through air or direct contact.

Microbiology and Public Health Laboratories

Workers in laboratories handling infectious agents have a potential risk of infection from accidental exposure or work practices. The majority of laboratory-acquired infections occur in research laboratories. Research workers often handle concentrated preparations of infectious agents, and some test procedures require complex manipulation of the samples. Also, these workers often handle infected animals that can expose workers to infection from bites or scratches. Laboratory-acquired infections have included brucellosis, Q fever, hepatitis, typhoid fever, tularemia, tuberculosis, human immunodeficiency virus (HIV), hantavirus (Korean hemorrhagic fever), herpes B virus, and Ebola-related filovirus.

Environmental engineers should use special care when dealing with projects involving laboratories that handle infectious microorganisms. Such situations require the support of a qualified industrial hygienist, and in some cases a biosafety hygienist.

Health Care Facilities

Health care facilities—such as hospitals, medical offices, blood banks, and outpatient clinics—present numerous opportunities for exposure to a wide variety of hazardous and toxic substances, as well as to infectious agents.

Hospitals usually have a doctor and an infectious disease control nurse or microbiologist to manage and oversee infection control functions. An environmental engineer may become involved with projects involving ventilation systems, containment systems, and handling of infectious wastes at health care facilities. Infectious wastes include blood, blood products, and tissues, and can cause infection from blood-borne pathogens.

Biotechnology Facilities

Biotechnology is one of the newest technologies and involves a much greater scope and complexity than the historical use of microorganisms in the chemical and pharmaceutical industries. This technology now deals with DNA manipulation and the development of products for medicine, industry, and agriculture. The microorganisms used by the biotechnology industry often are genetically engineered plant and animal cells. Allergies can be a major health issue.

Animal Facilities

Workers exposed to animals are at risk for animal-related allergies and infectious agents. Occupations include agricultural workers, veterinarians, workers in zoos and museums, taxidermists, and workers in animal-product processing plants.

Zoonotic diseases (diseases that affect both humans and animals) are the most common diseases reported by laboratory workers. Infection can occur from bites and scratches, puncture with contaminated needles, aerosols from animal respiration or excretions, and contact with infected tissue. Work-acquired infections from nonhuman primates are common.

Some of the diseases of concern in animal facilities include Q fever, hantavirus, Ebola and Marburg viruses, and simian immunodeficiency viruses. An environmental engineer could encounter potential infectious agents while dealing with solid and liquid wastes from animal processing facilities and animal agricultural activities.

Agriculture

Agricultural workers are exposed to infectious microorganisms through inhalation of aerosols, contact with broken skin or mucus membranes, and inoculation from injuries. Farmers and horticultural workers may be exposed to fungal diseases. Food and grain handlers may be exposed to parasitic diseases. Workers who process animal products may acquire bacterial skin diseases such as anthrax from contaminated hides, tularemia from skinning infected animals, and erysipelas from contaminated fish, shellfish, meat, or poultry. Infected turkeys, geese, and ducks can expose poultry workers to *psittacosis*, a bacterial infection. Workers handling grain may be exposed to *mycotoxins* from fungi and *endotoxins* from bacteria.

Environmental engineers or project workers may potentially be exposed to any number of these diseases while working on projects involving the treatment and disposal of solid and liquid wastes, such as through waste treatment processes and land application of wastewater. Landfills and composting facilities may also handle wastes containing contaminated agricultural and animal wastes or by-products.

Utility Workers

Workers maintaining water systems may be exposed to *Legionella pneumophila* (Legionnaires' disease). Sewage collection and treatment workers may be exposed to enteric bacteria, hepatitis A virus, infectious bacteria, parasitic protozoa (*Giardia*), and allergenic fungi.

Solid waste handling and disposal facility workers may be exposed to blood-borne pathogens from infectious wastes.

Wood-Processing Facilities

Wood-processing workers may be exposed to bacterial endotoxins and allergenic fungi. Environmental engineers may be involved with disposal of wastewater or solid wastes that contain these microorganisms.

Mining

Miners may be exposed to zoonotic bacteria, mycobacteria, fungi, and untreated runoff water and wastewater.

Forestry

Forestry workers may be exposed to zoonotic diseases (rabies virus, Russian spring fever virus, Rocky Mountain spotted fever, Lyme disease, and tularemia) transmitted by ticks and fungi. Environmental engineers may be involved with forestry activities associated with wetlands protection, land application of wastewater, landfills, and other projects where such infectious agents are present.

Blood-Borne Pathogens

Environmental engineers should have a general knowledge of blood-borne pathogens and be able to recognize when expert advice should be obtained on a project involving exposure to infectious wastes.

The risk from hepatitis B and human immunodeficiency virus (HIV) in health care and laboratory situations led OSHA to publish standards for occupational exposure to blood-borne pathogens. Some blood-borne pathogens are summarized as follows.

Human Immunodeficiency Virus (HIV)

HIV is the blood-borne virus that causes acquired immunodeficiency syndrome (AIDS). Contact with infected blood or other body fluids can transmit HIV. Transmission may occur from unprotected sexual intercourse, sharing of infected needles, accidental puncture wounds from contaminated needles or sharp objects, or transfusion with contaminated blood.

Symptoms of HIV include swelling of lymph nodes, pneumonia, intermittent fever, intestinal infections, weight loss, and tuberculosis. Death typically occurs from severe infection causing respiratory failure due to pneumonia.

Hepatitis

The hepatitis virus affects the liver. Symptoms of infection include jaundice, cirrhosis and liver failure, and liver cancer.

Hepatitis A can be contracted through contaminated food or water or by direct contact with body fluids such as blood or saliva. Hepatitis B, known as *serum hepatitis*, may be transmitted through contact with infected blood or other body fluids and through blood transfusions. Hepatitis B is the most significant occupational infector of health care and laboratory workers. Hepatitis C is similar to hepatitis B, but can also be transmitted by shared needles, accidental puncture wounds, blood transfusions, and unprotected sex.

Hepatitis D requires one of the other hepatitis viruses to replicate. Individuals with chronic hepatitis D often develop cirrhosis of the liver. Chronic hepatitis may be present in carriers.

Syphilis

The bacterium responsible for the transmission of syphilis is *Treponema pallidum* (subspecies *pallidum*). Syphilis is most often transmitted through sexual contact, but may also be transmitted from mother to fetus. Symptoms begin with an open sore in the genital region, followed by a rash on the palms and soles of the feet, fever, headache, and loss of appetite. The final stage produces degeneration of the aorta of the heart and affects the central nervous system, producing psychosis. There may be years between stages of the disease.

Toxoplasmosis

Toxoplasmosis is caused by a parasitic organism called *Toxoplasma gondii*, which may be transmitted by ingestion of contaminated meat, across the placenta, and through blood transfusions and organ transplants. The organism can be shed in the fecal matter of infected pets. Infection of pregnant women is of great concern with this disease, which can include blindness, mental retardation, jaundice, anemia, and central nervous system disorders.

Rocky Mountain Spotted Fever

Ticks infected with the pathogen *Rickettsia rickettsii* pass this disease from pets and other animals to humans. Symptoms and effects include headache, rash, fever, chills, nausea, vomiting, cardiac arrhythmia, and kidney dysfunction. Death may occur from renal failure and shock.

Bacteremia

Bacteremia is the presence of bacteria in the bloodstream, whether associated with active disease or not. The bacteria may be introduced through contamination during transfusions, through surgical procedures that allow bacteria to be released from sequestered sites, and from routine medical procedures. Fever is the most common symptom, and severe cases can lead to death.

Bacteria- and Virus-Derived Toxins

Some toxins are derived from bacteria and viruses. The effects of these toxins vary from mild illness to debilitating illness or death.

Botulism

The organism *Clostridium botulinum* produces the toxin that is responsible for botulism. There are four types of botulism: food-borne, infant, adult enteric (intestinal), and wound. *Food botulism* is associated with poorly preserved foods and is the most widely recognized form. *Infant botulism* can occur in the second month after birth when the bacteria colonize the intestinal tract and produce the toxin. *Adult enteric botulism* is similar to infant botulism. *Wound botulism* occurs when the spores enter a wound through contaminated soil or needles. The toxin is absorbed in the bloodstream and blocks the release of a neurotransmitter. Severe cases can result in respiratory paralysis and death.

Lyme Disease

Lyme disease is transmitted to humans through bites of ticks infected with *Borrelia burgdorferi*. Lyme disease follows three stages of development. Stage I involves transmission of the organism throughout the body through the bloodstream. Symptoms include fever, chills, fatigue, headache, pneumonia, hepatitis, and meningitis. Stages II and III result in neurological symptoms, meningitis, and encephalitis, and they produce an irregular heartbeat. Stage III results in arthritis of the large joints.

Tetanus

Tetanus occurs from infection by the bacterium *Clostridium tetani*, which produces two exotoxins, tetanolysin and tetanospasmin. These toxins affect the red blood cells and the nervous system. Symptoms include muscle spasms and a constant state of muscle contraction. Routine immunizations prevent the disease.

Toxic Shock Syndrome

Toxic shock syndrome (TSS) is caused by the bacterium *Staphylococcus aureus*, which produces a pyrogenic toxin. Symptoms include sore throat, fever, fatigue, headache, vomiting, diarrhea, low blood pressure, and rash. Shock or kidney failure can occur, which can be lethal. TSS may occur in children and men as well as in menstruating females.

Ebola (African Hemorrhagic Fever)

The Ebola and Marburg viruses produce an acute hemorrhagic fever in humans. Symptoms include headache, progressive fever, sore throat, and diarrhea. Later, hepatitis and bleeding from the mucus membranes and internal organs occur. Blood does not coagulate and shock may develop; death usually occurs by the ninth day.

Hantavirus

The hantavirus is found in rodents and shrews of the southwest and is spread by contact with their excreta. Symptoms during the early phase of the disease include chills, fever, headache, backache, weakness, and general malaise. Later, gastrointestinal bleeding and hypotension occur, followed by reduced and then increased amounts of urine being produced. Death may result without proper treatment.

Tuberculosis

Tuberculosis (TB) is a bacterial disease from *Mycobacterium tuberculosis*. Humans are the primary source of infection. TB affects a third of the world's population outside the United States. A drug-resistant strain is a serious problem worldwide, including in the United States. The risk of contracting active TB is increased among HIV-infected individuals. The usual route of transmission is by inhalation of infectious droplets suspended in air from coughing, sneezing, singing, or talking by an infected person. Symptoms include fatigue, fever, weight loss, hoarseness, cough, and blood in sputum.

Legionnaires' Disease

Legionnaires' disease (legionellosis) is a type of pneumonia caused by inhaling the bacteria *Legionella pneumophilia*. Symptoms include fever, cough, headache, muscle aches, and abdominal pain. People usually recover in a few weeks and suffer no long-term consequences. *Legionellae* are common in nature and are associated with heat-transfer systems, warm-temperature water, and stagnant water. Sources of exposure include sprays from cooling towers or evaporative condensers and fine mists from showers and humidifiers. Proper design and operation of ventilation, humidification, and water-cooled heat-transfer equipment and other water systems equipment can reduce the risk. Good system maintenance includes regular cleaning and disinfection.

Environmental Health & Safety

7. NOMENCLATURE

AT	averaging time	yr
BCF	bioconcentration factor	mg/L, mg/kg
BW	body mass (weight)	kg
c	concentration	ppm
C	concentration	ppm, mg/m^3
CDI	chronic daily intake	mg/kg·d
CF	concentration in food	mg/kg
CPF	carcinogen potency factor (same as cancer slope factor)	$(mg/kg \cdot d)^{-1}$
CR	contact rate	d^{-1}
CSF	cancer slope factor (same as carcinogen potency factor)	$(mg/kg \cdot d)^{-1}$
CW	chemical concentration in water	mg/L
E	exposure	$\mu g/L$
E	time-weighted average exposure	ppm, mg/m^3
E_m	equivalent exposure of mixture	–
EC	excess cancers	–
ED	exposure duration	yr
EED	estimated exposure dose	mg/kg·d
EF	exposure factor	–
EP	exposed population	–
FI	fraction ingested	–
HR	hazard ratio	–
IR	ingestion rate	L/d
L	exposure limit of particular contaminants	mg/m^3
LC_{50}	lethal concentration	mg/m^3
LD_{50}	lethal dose	mg/kg
LOAEL	lowest observed adverse effect level	mg/kg·d
LOEL	lowest observed effect level	mg/kg·d
$NOAEL$	no observed adverse effect level	mg/kg·d
NOEL	no observed effect level	mg/kg·d
$Risk$	risk; probability of excess cancer	–
RfD	reference dose	mg/kg·d
SHD	safe human dose	mg/d
SF	slope factor (same as carcinogen potency factor)	$(mg/kg \cdot d)^{-1}$
t	time	s
T	time of exposure	s
TWA	time-weighted average	ppm
UF	uncertainty factor	–
W	weight of adult male	kg

Subscripts

I	individual or interval
org	organism

57 Exposure Assessments: Sound and Radiation

1. SOUND AND NOISE

Characteristics of Sound

Sound is pressure variation in air, water, or some other medium that the human ear can detect. *Noise* is unwanted, unpleasant, or painful sound. The *frequency* of sound is the number of pressure variations per second, measured in cycles per second, or hertz (Hz). The frequency range of human audible sound is approximately 20–20 000 Hz.

Sound produces a sensory response in the brain called *hearing*, but excessive pressure can cause pain in the ears. Sound is produced by vibration of a source that causes longitudinal vibration of the medium, which is usually air, but could also be water or solids. Sound waves are elastic waves that can occur in media with both elasticity and mass.

Another characteristic of sound is the *wavelength*, denoted as λ, which is the distance between two analogous and successive points of the sound wave. Wavelength is important because sound will bend around objects with dimensions less than the sound wavelength but will be reflected or scattered by objects larger than the sound wavelength. Thus, barriers will be effective for a short wavelength sound.

Velocity is the rate at which analogous successive sound pressure points pass a point. Velocity is called the *speed of sound* and is equal to the product of the wavelength and the frequency.

$$a = f\lambda \qquad \textit{57.1}$$

The speed of sound is dependent upon the medium, as illustrated in Table 57.1.

Table 57.1 *Speed of Sound in Various Media*

medium	speed of sound (m/s)	condition
air	330	1 atm, 0°C
water	1490	1 atm, 20°C
aluminum	4990	1 atm
steel	5150	1 atm

The speed of sound in an ideal gas can be determined from the following formula.

$$a = \sqrt{\frac{kRT}{M}} \qquad \textit{57.2}$$

For air, k is approximately 1.40.

The characteristics of a sound wave are illustrated in Fig. 57.1.

Sound Pressure

Sound pressure measures the intensity of sound and is the variation in atmospheric pressure caused by the disturbance of the air by a vibrating object. (See Fig. 57.1.) Positive pressures are known as *compressions*, and negative pressures are *rarefactions*. Because the sound wave produces both positive and negative pressure, the mean value would be zero, which is not a meaningful measurement. Therefore, the root-mean-square (rms) pressure is used.

Figure 57.1 *Characteristics of a Sound Wave*

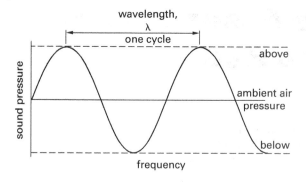

$$P_{\text{rms}} = \sqrt{\frac{\sum P_i^2}{n}} \qquad 57.3$$

The rms pressure eliminates the negative pressures by taking the square root of the average of the sum of the squares of the sound pressure over each instant of time. Sound pressure is typically measured in micropascals (μPa) (10^{-6} N/m^2). Because humans can perceive absolute sound pressure in a very wide range of 20–200 million μPa (which is also the range of the *threshold of hearing* to the *threshold of pain*), it is more convenient to use a relative measure called the *bel*. The *decibel* (0.1 bel), abbreviated dB, is the common measure and is the minimum sound difference humans can perceive.

Sound pressure level, SPL, is measured in decibels relative to a reference level, P_0, of 20 μPa, the threshold of hearing at a reference frequency of 1000 Hz, as follows.

$$\begin{aligned} \text{SPL} &= 10 \log_{10}\left(\frac{P}{P_0}\right)^2 \\ &= 20 \log_{10}\left(\frac{P}{P_0}\right) \end{aligned} \qquad 57.4$$

Sound Power

Sound power, W, is the absorbed or transmitted sound energy per unit time (in watts). The sound power level, L_W, is measured in decibels relative to a reference level, W_0, of 10^{-12} W.

$$L_W = 10 \log_{10}\left(\frac{W}{W_0}\right) \qquad 57.5$$

Sound Intensity

Sound intensity is an areal function of the sound power of a source.

$$I = \frac{W}{4\pi r^2} \qquad 57.6$$

This relationship can be visualized as the sound power on a unit area on the surface of a sphere of radius r from the source (a point source of sound). (The area of the surface of the sphere is $4\pi r^2$.) Sound intensity cannot be measured directly.

Loudness

Loudness is primarily determined by sound pressure but is also affected by frequency because the human ear is more sensitive to high-frequency sounds than low-frequency sounds. The upper sound limit that humans can perceive is between 16 000 Hz and 20 000 Hz. Because loudness is related to both sound pressure and frequency, a sound pressure-weighting scheme has been developed. This scheme uses sound level meters that have weighting characteristics, called weighting networks A, B, and C. The A network approximates equal loudness curves at low sound pressure levels. The B network approximates medium sound pressure levels, and the C network approximates high levels. The A network is widely used because it relates well to industrial noises and disturbances to the community. OSHA has adopted the A network as the preferred unit of measurement. Measurements using the A-weighted scale are designated dBA. *Slow response* refers to the operating mode of the instrument, in which the indicator will respond more slowly for easier reading.

Noise

Noise is vibration conducted through solids, liquids, or gases. Noise can cause psychological and physiological damage and can interfere with workers' communication, thereby affecting safety. Exposure to excessive noise for a sufficient time can result in hearing loss. Threshold limit value (TLV) criteria have been developed to protect against hearing loss in the speech-frequency range. In 29 CFR 1910.95, OSHA has established the permissible exposure levels (PELs) and duration of exposure to noise to prevent harm to workers. Generally, the TLVs for noise are more restrictive than the PELs. Employers are required by OSHA to implement a hearing conservation program when sound levels exceed 85 dBA on an 8 h time-weighted average (TWA) basis. The permissible noise exposures are given in Table 57.2.

When the daily noise exposure is composed of two or more periods of noise exposure at different levels, their combined effect should be considered, rather than the individual effect of each. If the sum of the fractions $C_1/T_1 + C_2/T_2 + \cdots + C_n/T_n$ exceeds unity, then the mixed exposure should be considered to exceed the limit value. C_n indicates the total time of exposure at a specified noise level, and T_n indicates the total time of exposure permitted at that level.

Table 57.2 Permissible Noise Exposures

duration per day (h)	sound level (dBA slow response)*
32	80
16	85
8	90
6	92
4	95
3	97
2	100
1.5	102
1	105
0.5	110
0.25 or less	115

*"Slow response" refers to the metering circuit being set on slow mode to reduce rapid, hard-to-read needle excursions. Compliance with OSHA regulations requires measurement set on slow response mode.
Source: 29 CFR 1910.95, Table G-16

Hearing Loss

Hearing loss can be caused by sudden intense noise over only a few exposures. This type of loss is known as *acoustic trauma*. Hearing loss can also be caused by exposure over a long duration (months or years) to hazardous noise levels. This type is known as *noise-induced hearing loss*. The permanence and nature of the injury depends on the type of hearing loss.

The main risk factors associated with hearing loss are the intensity of the noise (sound pressure level), the type of noise (frequency), daily exposure time (hours per day), and the total work duration (years of exposure). These are known as *noise exposure factors*. Generally, exposure to sound levels above 115 dBA is considered hazardous, and exposure to levels below 70–75 dBA is considered safe from risk of permanent hearing loss. Also, noise with predominant frequencies above 500 Hz is considered to have a greater potential to cause hearing loss than lower-frequency sounds.

Sound Measurements

Sound measurements are typically made with a sound level meter, which can be set to the A, B, or C network. Meters typically have a range of 40–140 dB (referenced to 20 μPa) and can be set for slow or fast response. In fast mode, the meter responds quickly to changing noise levels. In slow mode, the meter will be easier to read in a factory setting and is the mode required by OSHA for compliance with its requirements.

Other instruments used in noise measurements are octave-band analyzers and noise dosimeters. *Octave-band analyzers* can determine the frequency distribution of sound pressure. *Noise dosimeters* record the total noise energy to which a worker has been exposed during a shift.

Classes of Noise Exposure

Noise exposure can be classified as continuous noise, intermittent noise, and impact noise. *Continuous noise* is broadband noise of a nearly constant sound pressure level and frequency to which a worker is exposed 8 hours daily and 40 hours weekly. *Intermittent noise* involves exposure to a specific broadband sound pressure level several times a day. *Impact noise* is a sharp burst of short duration sound.

OSHA has established permissible noise exposures for a specific duration each workday, known as *permissible exposure levels* (PELs). The PELs are based on continuous 8 hour exposure at a sound pressure level of 90 dBA, which is established as 100%. For other exposure durations, OSHA has established relationships between the sound level and the exposure time. Every 5 dBA increase in noise cuts the allowable exposure time in half. Sound pressure levels below 90 dBA are not considered hazardous and do not have to be determined.

When workers are exposed to different noise levels during the day, the *noise dose*, D, must be calculated as follows.

Noise Pollution

$$D = 100\% \times \sum \frac{C_i}{T_i} \qquad \textbf{57.7}$$

$$T_i = \frac{8 \text{ hr}}{2^{(L_i - 90)/5}} \qquad \textbf{57.8}$$

C_i is the time in hours spent at each sound pressure level (SPL), and T_i is the time permitted at that SPL, taken from Table 57.2 or calculated from Eq. 57.8. If D equals or exceeds 100%, the noise dose exceeds the OSHA standard.

For intermittent noise, the time characteristics of the noise must also be determined. Both short-term and long-term exposure must be measured. A dosimeter is typically used for intermittent noises.

For impact noises, workers should not be exposed to peaks of more than 140 dBA under any circumstances. The threshold limit value for impulse noise should not exceed the values provided in Table 57.3.

Table 57.3 Typical Threshold Limit Values for Impact Noise

peak sound level (dB)	maximum number of daily impacts
140	100
130	1000
120	10 000

Environmental Health & Safety

Example 57.1

A machine shop worker is exposed to the following noise levels during the workday. Is the exposure excessive?

$$80 \text{ dBA for } 2.5 \text{ h}$$
$$92 \text{ dBA for } 2.0 \text{ h}$$
$$95 \text{ dBA for } 3.0 \text{ h}$$
$$115 \text{ dBA for } 0.5 \text{ h}$$

Solution

From Table 57.2, determine the permissible noise exposure time at each noise level. The noise dose exposure for the worker is given by Eq. 57.7.

Noise Pollution

$$D = 100\% \times \sum \frac{C_i}{T_i}$$

$$= 100\% \times \left(\frac{2.5 \text{ h}}{32 \text{ h}} + \frac{2.0 \text{ h}}{6 \text{ h}} + \frac{3.0 \text{ h}}{4 \text{ h}} + \frac{0.5 \text{ h}}{0.25 \text{ h}} \right)$$

$$= 316\%$$

Because the D of 316% is greater than 100%, the exposure is excessive.

Noise Control

A *hearing conservation program* should include noise measurements, noise control measures, hearing protection, audiometric testing of workers, and information and training programs. Employees are required to properly use the protective equipment provided by employers.

After the noise exposure is compared with acceptable noise levels, the degree of noise reduction needed can be determined. Noise reduction measures can comprise the following three basic methods applied in order.

1. changing the process or equipment

2. limiting the exposure

3. using hearing protection

Noise reduction involves looking at the source of noise generation, the path of noise from the source to the worker, and protection of the worker. Reducing noise generation may require modifying or replacing equipment or reengineering the manufacturing process. The noise path can be altered by enclosing the noise source, increasing the distance from the worker to the source, or placing sound barriers or sound absorbing materials along the noise path. Worker protection can include placing an enclosure around the worker and/or providing ear protection.

The decrease in sound pressure level or sound power level due to insertion of an enclosure or wall between a noise source and a receiver (the *insertion loss*) can be found from the following equation. The decrease in sound pressure level is the same as the decrease in sound power level.

$$\text{IL} = \text{SPL}_1 - \text{SPL}_2 = L_{W,1} - L_{W,2}$$

$$= 10 \log \left(\frac{L_{W,1}}{L_{W,2}} \right)$$

$$= 20 \log \left(\frac{P_1}{P_2} \right)$$

57.9

Noise control measures can include engineering controls, administrative controls, and personal hearing protection. Engineering controls are the most reliable and effective approach to noise control. Engineering controls involve modifying a machine to operate with lower noise generation, for example, by reducing vibrating components or substituting quieter components or functions. Engineering controls also include reducing noise along the noise path, for example, through the use of sound barriers. Noise generation should be considered early in the machine or manufacturing design process and should be a part of all procurement decisions.

Administrative controls include changing the exposure of workers to high noise levels by modifying work schedules or locations so as to reduce workers' exposure times. Administrative controls include any administrative decision that limits a worker's exposure to noise.

Personal hearing protection is the final noise control measure, to be implemented only after engineering controls are implemented. Protective devices do not reduce the noise hazard and may not be totally effective, so engineering controls are preferred over hearing protection. Protective devices include helmets, earplugs, canal caps, and earmuffs. Earplugs may be used with helmets to increase the level of noise reduction.

An important characteristic of personal hearing protection is the *noise reduction rating* (NRR). The NRR is established by the Environmental Protection Agency (EPA) and must be printed on the package of a device. The NRR can be used to determine whether a device provides sufficient hearing protection.

Audiometry

Audiometry is the measurement of hearing acuity. It is used to assess a worker's hearing ability by measuring the individual's threshold sound pressure level at various frequencies (250–6000 Hz). The threshold audiogram can be used to create a baseline of hearing ability and to determine changes over time and identify changes resulting from noise control measures. Baseline and annual hearing tests are required where workers are exposed to more than a *time-weighted average* (TWA) over 8 hours of 85 dBA. The average change from the baseline is used to measure the degree of hearing impairment.

2. SOUND METER MEASUREMENTS

Rather than attempt to describe sound in terms of subjective loudness, simple *sound meters (noise meters)* measure noise using various response curves. (A *response curve* is basically a set of weighting values that can be applied to each frequency detected.) Meter response curves have been standardized worldwide, and most industrial-quality meters respond accurately to three curves, designated as the A-, B-, and C-scales. The slow-response A-scale approximates human hearing the most accurately, so it is used the most frequently.[1,2]

When sound measurements are reported, a sound "level" always means a logarithmic value expressed in decibels. Sound level units are reported along with the scale used. For example, "15 dBA" (also written as "dB(A)" and "dB-A") indicates that the A-scale was used.

Some meters often also provide a fourth, unweighted reading using the basic frequency response of the microphone and meter circuitry. This reading is different for different meters. The terms *unweighted response, flat response, 20 kHz response,* and *linear response* all refer to this case.

Peak levels from impulse sources are measured with special equipment (i.e., an impact meter or oscilloscope) that captures and holds the maximum instantaneous noise level.

3. SOUND PRESSURE LEVEL

Sound propagates as pressure fluctuates. The actual pressure at a particular point is the *sound pressure level*, SPL. Sound pressure level is the quantity that is actually measured in sound meters.

Since the human ear responds to variations in air pressure over a range of more than a million to one, a logarithmic power ratio scale is used to measure sound in decibels. The minimum perceptible pressure amplitude (i.e., the approximate *threshold of hearing*) has been standardized as 20 μPa, and this value is used as a reference pressure.[3] The acoustic energy is proportional to the square of the sound pressure. For an rms (root-mean-squared or "effective") sound pressure, P, in pascals, the sound pressure level is

Noise Pollution

$$\text{SPL (dB)} = 10\log_{10}(P^2/P_0^2) \qquad 57.10$$

Usually, the sound pressure level is sampled at several single frequencies over the audio spectrum. These samples are referred to as "narrow band" or "octave band" measurements. The individual measurements can be combined into a *broad band* value by weighting each by standardized values.[4]

4. SOUND POWER LEVEL

The *sound power level* is used to measure the total acoustic power emitted by a source of sound. Sound power is, therefore, independent of the environment. Sound power is not measured by sound meters; it must be calculated. A standard reference level of 1×10^{-12} W is used.[5] The literature uses L_W and PWL as symbols for sound power level. For a source of sound emitting W watts, the sound power level in decibels is

$$L_W = 10\log_{10}\frac{W}{W_{\text{ref}}} = 10\log_{10}\frac{W}{1 \times 10^{-12}\text{ W}} \qquad 57.11$$

5. CONVERTING SOUND PRESSURE INTO SOUND POWER

Sound pressure level and sound power level are both reported in decibels. However, there is no simple conversion between them because sound pressure level is affected by directivity and separation distance from the source, while sound power level is not. However, the two scales have been established such that a decibel change in one quantity will produce the same decibel change in the other quantity. Therefore, a decibel decrease or increase in sound power will produce the same decibel decrease or increase in sound pressure.

The sound pressure at a distance r from a continuous noise source with known sound power, directivity, and room constant, R, in a confined space large enough to have both direct and reverberant sound, is

$$\text{SPL}_{\text{dBA}} = 0.2 + L_W + 10\log_{10}\left(\frac{Q}{4\pi r^2} + \frac{4}{R}\right) \quad \text{[SI]} \quad 57.12(a)$$

$$\text{SPL}_{\text{dBA}} = 10.5 + L_W + 10\log_{10}\left(\frac{Q}{4\pi r^2} + \frac{4}{R}\right) \quad \text{[U.S.]} \quad 57.12(b)$$

Directivity, Q, depends on the orientation of the source and the physical location of the source in relationship to walls and other solid surfaces. Directivity is 1 for an isotropic source radiating into free space, 2 for a source on the surface of an infinite flat plane, 4 for a source at the

---

[1]The A-curve discriminates against lower frequencies, the C-curve is nearly flat, and the B-curve falls in between.
[2]Refer to ANSI S1.4 for sound level meters.
[3]At one time, this was thought to be the threshold of hearing for an average young person at 1000 Hz.
[4]Weighting and combining are done automatically by the sound meter.
[5]Many years ago, a reference of 10^{-13} W was used. There may still be some references to this value in older literature.

intersection of two perpendicular planes, and 8 for a source in the corner of three mutually perpendicular reflecting planes.

Example 57.2

The combined sound pressure level from four identical machines at a point equally distant from each is 100 dBA. Three machines are shut down. What is the new sound pressure level?

Solution

When three of the four machines are shut down, the total radiated sound power will be 25% of the original condition. The reduction in sound power level will be

$$L_{W,1} - L_{W,2} = 10 \log_{10} \frac{W_1}{W_2} = 10 \log_{10} 4$$
$$= 6.02 \text{ dB} \quad \text{[use 6 dB]}$$

The change in sound pressure level is the same as the change in sound power level. Therefore, the new sound pressure level will be

$$\text{SPL} = 100 \text{ dBA} - 6 \text{ dB} = 94 \text{ dBA}$$

6. FREQUENCY CONTENT OF NOISE

Detailed knowledge of the frequency content of noise is required for most noise control problems. *Frequency content* is described by dividing the sound spectrum into a series of frequency ranges called *bands*. Frequency bands are established by dividing the frequency range of interest into bands of either equal bandwidths or equal percentages.[6] The sound level in each frequency band can be reported or plotted against the midpoint of the frequency band.

7. OCTAVE BAND ANALYSIS

Percentage bandwidths and center frequencies have been standardized worldwide. The most common (and widest) standard bandwidth is 50%, which means that the ratio between the frequencies at the two ends of the band is 1:2. For example, the band whose center frequency is 1000 Hz spans the range from 707 Hz to 1414 Hz. Since musical pitch frequency ratios of 1:2 are called *octaves*, the name *octave band* has been adopted.

Center frequencies for octave bands have been standardized at 63 Hz, 125 Hz, 250 Hz, 500 Hz, 1000 Hz, 2000 Hz, 4000 Hz, and 8000 Hz.[7,8] These bands are numbered 1, 2, 3, 4, 5, 6, 7, and 8, respectively. Octave bands can also be subdivided for better frequency detail. The most common subdivision is the *one-third octave band*.

8. COMBINING MULTIPLE SOURCES

Multiple sound sources combine to produce more sound. Sound pressures from multiple correlated sources combine linearly, while sound pressures combine as sums of squares. However, multiple sound pressure levels or sound power levels expressed in decibels cannot be directly added. Two 90 dB sources do not produce a 180 dB source. For multiple sound sources, the combined power is given by Eq. 57.13. This is known as an *unweighted sum* because the individual readings are used without modification.

Noise Pollution

$$\text{SPL}_{\text{total}} = 10 \log_{10} \sum 10^{\text{SPL}/10} \qquad \textbf{57.13}$$

Using Eq. 57.13 to add or subtract the sound from two or more sources is not the same as determining the *change* in sound level. As described in Sec. 57.9, Sec. 57.10, and Sec. 57.11, the relationship between the old, new, and change in sound levels is that of a simple linear combination (i.e., addition or subtraction).

Equation 57.13 illustrates an important noise reduction principle: The noisiest source must be identified and treated before significant overall noise reduction can be achieved. Sources with sound levels more than a few decibels lower than the noisiest source make a minor contribution to the total sound level.

Example 57.3

What is the unweighted combined sound pressure from two machines, one with a sound pressure of 89 dB and the other with a sound pressure of 94 dB?

Solution

Use Eq. 57.13.

Noise Pollution

$$\text{SPL}_{\text{total}} = 10 \log_{10} \sum 10^{\text{SPL}/10}$$
$$= 10 \log_{10} \left(10^{89 \, \text{dB}/10} + 10^{94 \, \text{dB}/10} \right)$$
$$= 95.2 \text{ dB}$$

---

[6]Equal bandwidths are generally used only when the frequency range of interest is small. There is little standardization of narrow bandwidths, primarily because narrow band analysis is used more for detailed investigation of noise sources than for reporting absolute levels.
[7]A now-obsolete series of octave bands was used until the late 1950s.
[8]Octave bands centered at 31.5 Hz and 16,000 Hz are also occasionally encountered. However, these two end bands are usually omitted, as they are at the extreme limits of most peoples' hearing abilities. For example, the A-weighted correction for 31.5 Hz is −39.2 dB. This is such a large reduction that the contribution from this band is negligible except for all but the most powerful or pure-tone sources.

9. NOISE REDUCTION

Received sound can be reduced by modifying either the sound source, the transmission path, or the area occupied by the source. The process of reducing the noise is known as *noise reduction*, *sound abatement*, and *attenuation*. The amount of noise reduction, NR, is a decrease in sound pressure level due to such modifications.

$$\text{NR} = \text{SPL}_1 - \text{SPL}_2 = L_{W,1} - L_{W,2} \qquad 57.14$$

When adding noise reduction material to a room, the new surface absorption (sabin area) should be 3 to 10 times the original surface absorption. In a room with well-distributed absorption surfaces and at large distances from a single noise source, or for the case of many small noise sources distributed in the room, the noise reduction is

$$\text{NR} = 10 \log_{10} \frac{\sum S_1}{\sum S_2} = 10 \log_{10} \frac{R_1(1 - \overline{\alpha}_1)}{R_2(1 - \overline{\alpha}_2)} \qquad 57.15$$

10. INSERTION LOSS

Insertion loss, IL, is the term used to describe the amount of noise reduction based on sound pressure levels at a particular place after the noise source has been isolated. The source is usually isolated by covering it with an enclosure or by constructing a wall between it and occupants. (The wall or enclosure has been "inserted" into the sound transmission path.)

$$\text{IL} = \text{SPL}_1 - \text{SPL}_2 = 20 \log_{10} \frac{P_1}{P_2} \qquad 57.16$$

Insertion loss takes into account structure-borne sound and flanking and diffracted sound, which can bend around the edges of a barrier. Including these factors produces a better estimate of the noise reduction than transmission loss.

11. TRANSMISSION LOSS

The *transmission loss*, TL, (also known as the *sound reduction index*, SRI) is based on sound power levels and is the actual portion of the sound energy lost when sound passes through a wall or barrier. The fraction of the incident energy transmitted through a barrier is designated as the *transmission coefficient*, τ.

$$\text{TL} = L_{W,1} - L_{W,2} = 10 \log_{10} \frac{W_1}{W_2} = 10 \log_{10} \frac{1}{\tau} \qquad 57.17$$

Unlike insertion loss, which is determined from measurements of actual sound pressure levels, with and without a sound barrier, transmission loss is solely related to a material's ability to reduce the flow of sound energy passing through it. Basically, transmission loss is a function of the materials used, while insertion loss is affected by materials, installation, and leaks. According to the *mass law* (i.e., a doubling of the mass of the wall will result in a 6 dB reduction in transmitted sound), transmission loss is greatly dependent on the mass per unit area, ρ_S, (also known as *surface density*) of the barrier. Transmission loss can be predicted at normal incidence for a simple, homogeneous panel (e.g., a concrete block wall or single layer of sheetrock on vertical wood studs) at a specific frequency, f, from the mathematical expression of the mass law, Eq. 57.18. Predicting transmission loss of walls receiving frequencies near their resonant frequencies, at other than normal incidence, and of more complex walls, including double panels with airspace and/or insulation between, requires more sophisticated expressions.

$$\text{TL}_{\text{dB}} = 20 \log_{10}(f_{\text{Hz}} \rho_{S,\text{panel,kg/m}^2}) - 47 \qquad \text{[SI]} \qquad 57.18(a)$$

$$\text{TL}_{\text{dB}} = 20 \log_{10}\left(4.88 f_{\text{Hz}} \rho_{S,\text{panel,lbm/ft}^2}\right) - 47 \quad \text{[U.S.]} \qquad 57.18(b)$$

12. ATTENUATION

Attenuation is the amount by which sound is decreased as it travels from the source to the receiver. This general term is used interchangeably with noise reduction, insertion loss, and transmission loss. An *attenuator* is any device (e.g., a muffler) that can be inserted into the sound path to reduce the noise received.

Point source attenuation is calculated with Eq. 57.19.

Noise Pollution

$$\Delta\text{SPL}(\text{dB}) = 10 \log_{10}(r_1/r_2)^2 \qquad 57.19$$

13. TIME-WEIGHTED AVERAGE

In the United States, the parameter used in the Occupational Safety and Health Act (OSHA) regulations to assess a worker's exposure to noise is the *time-weighted average*, TWA. TWA represents a worker's noise exposure normalized to an eight-hour day, taking into account the average levels of noise and the time spent in each area. Determining the TWA starts with calculating the noise dose, D, using Eq. 57.7. Since they are different representations of the same level, the *OSHA action levels* are based on either TWA or dose percent. These action levels are 85 dB (or 50% dose) and 90 dB (or 100% dose). Preventative steps must be taken if the daily noise dose is more than 1.0.

Noise Pollution

$$\text{TWA} \approx 90 + 16.61\left(\log_{10} \frac{D}{100}\right) \qquad 57.20$$

Environmental Health & Safety

Unlike a traditional sound meter, which measures the instantaneous sound level, a *noise dosimeter* (*dose badge*) worn by an employee monitors and integrates the sound level over the entire work shift in order to determine the dose, D, time-weighted average, TWA, average sound level, SPL, and other parameters directly.

14. ALLOWABLE EXPOSURE LIMITS

OSHA sets maximum limits on daily sound exposure. (See Table 57.2.) The "all-day" eight-hour noise level limit in the United States is 90 dBA. This is higher than the maximum level permitted in other countries.[9] In the United States, employees may not be exposed to steady sound levels above 115 dBA, regardless of the duration. Impact sound levels are limited to 140 dBA.

In the United States, hearing protection, educational programs, periodic examinations, and other actions are required for workers whose eight-hour exposure is more than 85 dBA or whose noise dose exceeds 50% of the action levels.

15. RADIATION

Radiation can be either nonionizing or ionizing. *Nonionizing radiation* includes electric fields, magnetic fields, electromagnetic radiation, radio frequency and microwave radiation, and optical radiation and lasers.

Dealing with *ionizing radiation* from nuclear sources requires special skills and knowledge. A specialist in health physics should be consulted whenever ionizing radiation is encountered.

Nuclear Radiation

Nuclear radiation is a term that applies to all forms of radiation energy that originate in the nuclei of radioactive atoms. Nuclear radiation includes alpha particles, beta particles, neutrons, X-rays, and gamma rays. The common property of all nuclear radiation is an ability to be absorbed by and transfer energy to the absorbing body.

The preferred unit of ionizing radiation, given in the National Council on Radiation Protection's (NCRP's) *Recommended Limitations for Exposure to Ionizing Radiation*, is the mSv. Sv is the symbol for sievert, which is the SI unit of absorbed dose times the *weighting factor* (previously known as the *quality factor*) of the radiation as compared to gamma radiation. The absorbed dose is measured in grays (Gy). The gray is equal to 1 J of absorbed energy per kilogram of matter. A summary of ionizing radiation units is given in Table 57.4.

Table 57.4 Units for Measuring Ionizing Radiation

property	SI
energy absorbed	gray (Gy)
	1 J/kg
	1 Gy = 100 rad (obsolete)
biological effect	sievert (Sv)
	Gy × quality factor
	1 Sv = 100 rem (obsolete)

Alpha Particles

Alpha particles consist of two protons and two neutrons, with an atomic mass of four. Alpha particles combine with electrons from the absorbed material and become helium atoms. Alpha particles have a positive charge of two units and react electrically with human tissue. Because of their large mass, they can travel only about 10 cm in air and are stopped by the outer layer of the skin. Alpha-emitters are considered to be only internal radiation hazards, which requires alpha particles to be ingested by eating or breathing. They affect the bones, kidney, liver, lungs, and spleen.

Beta Particles

Beta particles are electrically charged particles ejected from the nuclei of radioactive atoms during disintegration. They have a negative charge of one unit and the mass of an electron. High-energy beta particles can penetrate in human tissue to a depth of 20 mm to 130 mm and travel up to 9 m in air. Skin burns can result from an extremely high dose of low-energy beta radiation, and some high-energy beta sources can penetrate deep into the body, but beta-emitters are primarily internal radiation hazards, which would require them to be ingested. Beta particles are more hazardous than alpha particles because they can penetrate deeper into tissue. High-energy beta radiation can produce a secondary radiation called *bremsstrahlung*. These are X-rays produced when electrons (i.e., beta particles) pass near the nuclei of other atoms. Bremsstrahlung radiation is proportional to the energy of the beta particle and the atomic number of the adjacent nucleus. Materials with low atomic numbers (e.g., plexiglass) are preferred shielding materials.

Neutrons

Neutron particles have no electrical charge and are released upon disintegration of certain radioactive materials. Their range in air and in human tissue depends on

---

[9]The 2003 *European Union Noise Directive* (2003/10/EC) became effective in 2006. It sets the maximum "all day" eight-hour noise level exposure limit in EU countries at 87 dBA. The first action level, to train personnel and make hearing protection available, occurs at 80 dBA. The second action level, to implement a noise reduction program and require usage of hearing protection, occurs at 85 dBA.

their kinetic energy, but the average depth of penetration in human tissue is 60 mm. Neutrons lose velocity when they are absorbed or deflected by the nuclei with which they collide. However, the nuclei are left with higher energy that is later released as protons, gamma rays, beta particles, or alpha particles. It is these secondary emissions from neutrons that produce damage in tissue.

X-Rays

X-rays are produced by electron bombardment of target materials and are highly penetrating electromagnetic radiation. X-rays have a valuable scientific and commercial use in producing shadow pictures of objects. The energy of an X-ray is inversely proportional to its wavelength. X-rays of short wavelength are called *hard*, and they can penetrate several centimeters of steel. Long wavelength X-rays are called *soft*, and they are less penetrating. The power of X-rays and gamma rays to penetrate matter is called *quality*. *Intensity* is the energy flux density.

Gamma Rays

Gamma rays, or gamma radiation, are a class of electromagnetic photons (radiation) emitted from the nuclei of radioactive atoms. They are highly penetrating and are an external radiation hazard. Gamma rays are emitted spontaneously from radioactive materials, and the energy emitted is specific to the radionuclide. Gamma rays present an internal exposure problem because of their deep penetrating ability.

Isotopes

An isotope of an element contains the same number of protons as the element but a different number of neutrons. The notation for element X that has atomic number Z and mass number A is: $_{Z}^{A}X$ or X-A. For example, nickel-63, Ni-63, $_{28}^{63}Ni$, or ^{63}Ni are recognized as being equivalent. The atomic number, Z, is determined by the number of protons in the nucleus. Hydrogen has an atomic number of 1; all atoms with only one proton in their nuclei are hydrogen. However, hydrogen can have zero neutrons (H-1, $_{1}^{1}H$), one neutron (H-2, $_{1}^{2}H$, deuterium), or two neutrons (H-3, $_{1}^{3}H$, tritium).

Not all isotopes of an element are radioactive. Some are stable and do not decay. For example, hydrogen-3 is radioactive, but hydrogen-2 and hydrogen-1 are not radioactive. All elements having an atomic number greater than 80 have radioactive isotopes, and all isotopes of elements with an atomic number above 83 are radioactive.

Until a radioisotope undergoes a decay event, no radiation is emitted. An isotope may emit one or more different types of ionizing radiation, depending on the mode of radioactive decay.

Radioactive Decay

The activity, A, of a radioactive substance is its *decay rate*, the rate of disintegration of radioactive atoms. The SI unit of activity is the *becquerel* (Bq), equivalent to 1/s (i.e., one disintegration per second). The non-SI curie (Ci) unit, equivalent to 3.7×10^{10} disintegrations per second, was previously used and may still be encountered, particularly in medical research.

Activity is not necessarily proportional to physical size or any other physical dimension. A radioactive source the size of a thin wire can contain more activity than a 55 gallon drum full of a radioactive liquid.

Radioactive decay is measured in terms of *half-life*, the time to lose half of the activity of the original material. (See Table 57.5.) Decay activity can be calculated as follows.

$$N = N_0 e^{-0.693 t / \tau} \qquad \textbf{57.21}$$

$$N = N_0 (0.5)^{t/\tau} \qquad \textbf{57.22}$$

$$A = A_0 e^{-\lambda t} \qquad \textbf{57.23}$$

The half-life can be calculated from the rate constant.

$$\tau = \frac{0.693}{\lambda} \qquad \textbf{57.24}$$

Table 57.5 Approximate Half-Lives of Selected Radioisotopes

radioisotope	half-life, τ
Au-198	2.696 d
C-11	20.38 min
Co-57	270.9 d
Cr-51	27.704 d
Cs-137	30.0 yr
Cu-64	12.701 h
Fe-55	2.7 yr
H-3	12.35 yr
I-125	60.14 d
I-131	8.04 d
K-40	1.277×10^9 yr
O-15	122.24 s
P-32	14.29 d
Pd-109	13.427 h
Po-210	138.38 d
Rb-86	18.66 d
Rn-222	3.8235 d
S-35	87.44 d
Th-230	7.7×10^4 yr
U-235	703.8×10^6 yr

Source: Kocher, D.C., *Radioactive Decay Data Tables*, DOE-TICS-11026; USDOE; Washington, DC, 1981.

Environmental Health & Safety

The daughter product consists of the remaining nuclei after radioactive decay of the parent nuclei. λ_1 and λ_2 are the decay constants of the parent and daughter nuclei, respectively, and N_{10} is the initial activity of the parent nuclei.

Daughter Product Activity

$$N_2 = \frac{\lambda_1 N_{10}}{\lambda_2 - \lambda_1}(e^{-\lambda_1 t} - e^{-\lambda_2 t}) \qquad 57.25$$

The daughter activity will be zero when t equals zero or infinity. The maximum activity occurs when the parent and daughter are equal (or when the parent has one daughter). The time of maximum activity can be calculated using Eq. 57.26.

Half-Life (Radioactive Decay): Daughter Product Maximum Activity Time

$$t' = \frac{\ln\lambda_2 - \ln\lambda_1}{\lambda_2 - \lambda_1} \qquad 57.26$$

Example 57.4

(a) How long will it take a 1000 becquerel (Bq) radionuclide to decay to 300 Bq if the decay constant is $0.090 \ \text{d}^{-1}$? (b) What is the half-life?

Solution

(a) Use Eq. 57.21.

$$N = N_0 e^{-0.693 t / \tau}$$

$$\frac{N}{N_0} = e^{-0.693 t / \tau}$$

$$\ln\frac{N}{N_0} = \frac{-0.693 t}{\tau} = \frac{-0.693 t}{\dfrac{0.693}{\lambda}} = -t\lambda$$

$$t = \frac{\ln\dfrac{N}{N_0}}{-\lambda} = \frac{\ln\dfrac{300 \ \text{Bq}}{1000 \ \text{Bq}}}{-0.090 \ \dfrac{1}{\text{d}}} = 13.4 \ \text{d}$$

(b) Find the half-life. From Eq. 57.24,

Half-Life (Radioactive Decay)

$$\tau = \frac{0.693}{\lambda}$$

$$= \frac{0.693}{0.090 \ \text{d}^{-1}}$$

$$= 7.7 \ \text{d}$$

Example 57.5

What radioactivity would remain from 2 Bq cobalt-60, which has a half-life of 5.24 yr, after a 20 yr period?

Solution

Use Eq. 57.21.

Half-Life (Radioactive Decay)

$$N = N_0 e^{-0.693 t / \tau}$$

$$= (2 \ \text{Bq})e^{(-0.693)(20 \, \text{yr})/5.24 \, \text{yr}}$$

$$= 0.143 \ \text{Bq}$$

Example 57.6

What is the activity of an I-131 ($\tau = 8.04$ days) source now if the activity was 3.7 TBq 11 days ago?

Solution

From Eq. 57.24,

Half-Life (Radioactive Decay)

$$\tau = \frac{0.693}{\lambda}$$

$$\lambda = \frac{0.693}{\tau} = \frac{0.693}{8.04 \ \text{d}}$$

$$= 0.0862 \ \text{d}^{-1}$$

Half-Life (Radioactive Decay)

From Eq. 57.23,

$$A = A_0 e^{-\lambda t}$$

$$= (3.7 \ \text{TBq})e^{-(0.0862 \ \text{d}^{-1})(11 \ \text{d})}$$

$$= 1.43 \ \text{TBq}$$

Radiation Effects on Humans

Ionizing radiation transfers energy to human tissue when it passes through the body. *Dose* refers to the amount of radiation that a body absorbs when exposed to ionizing radiation. The effects on the body from external radiation are quite different from the effects from internal radiation. Internal radiation is spread throughout the body to tissues and organs according to the chemical properties of the radiation. The effects of internal radiation depend on the energy and the residence time within the body. The principal effect of radiation on the body is destruction of or damage to cells. Damage may affect reproduction of cells or cause mutation of cells.

The effects of ionizing radiation on individuals include skin, lung, and other cancers; bone damage; cataracts; and a shortening of life. Effects on the population as a whole include possible damage to human reproductive elements, thereby affecting the genes of future generations.

Exposure

Exposure is a measure of X-ray and gamma radiation based on its ability to ionize the molecules in air. The term "exposure" does not apply to any absorbing material except air. In the past, the *roentgen* (R) was used to measure the exposure. One roentgen produces 1 statcoulomb of charge in 1 cm^3 of air at standard temperature and pressure (STP).

The roentgen is no longer used in any of the regulations relating to occupational or public exposure to radiation. It remains useful in several other fields, such as medical diagnosis and therapy, radiation sterilization, and shielding design. The SI unit of exposure is the coulomb per kilogram (C/kg). A roentgen is equivalent to a dose of 2.58×10^{-4} C/kg.

Biological damage caused by radiation passing through a volume of a target is roughly proportional to the amount of radiation energy absorbed by the target. *Absorbed dose*, D_T, is the total energy deposited by ionizing radiation per unit mass of irradiated material. In the case of radiation's biological effects, the absorbing material of interest is human tissue.

The units of absorbed dose are the rad and the gray. The *rad* is an absorbed dose of 100 ergs/gram or 0.01 J/kg (0.01 Gy). The *gray* (Gy) is the SI unit for an absorbed dose of 1 J/kg (100 rad).

Dose Equivalent and Weighting Factor

Even when depositing the same amount of energy per mass (i.e., the same absorbed dose), different types of radiation cause different amounts of biological damage in living tissue. To reflect this difference, a quantity called the *dose equivalent*, H_T, is used to indicate biological effects on a specific tissue type.

To quantify the sensitivity of a biological tissue to different types of radiation (or energies in the case of neutrons), a *weighting factor*, w_R, is used. The weighting factor reflects the ability of a particular type of radiation to cause biological damage, relative to gamma radiation.

The rem and sievert (Sv) are used to measure dose equivalent in radiation protection applications. Given the relationship 1 Gy = 100 rad, the relationship between rem and Sv is 1 Sv = 100 rem. The dose equivalent, H_T, is equal to the absorbed tissue dose, D_T, multiplied by the weighting factor, w_R.

$$H_T = D_T w_R \qquad 57.27$$

Consistent units should be maintained in Eq. 57.27. If the absorbed dose is given in rads, the dose equivalent is in rems, while if the absorbed dose is given in grays, the dose equivalent is in sieverts.

As seen in Table 57.6, the weighting factor for beta particles is 1 Sv/Gy (or rem/rad), and the weighting factor for alpha particles is 20 Sv/Gy (or rem/rad). Since the weighting factor for alphas is higher than for betas, alphas do more biological damage than betas for the same amount of energy deposited per unit mass of tissue (i.e., for the same dose).

Example 57.7

A person is exposed to 5 rad of alpha radiation, 1 rad of beta radiation, and 1 rad of neutrons of unknown energy. What is the dose equivalent for this person?

Solution

From Eq. 57.27 and Table 57.6,

$$
\begin{aligned}
H_T &= \sum D_T w_R \\
&= (5 \text{ rad})\left(20 \ \frac{\text{rem}}{\text{rad}}\right) + (1 \text{ rad})\left(1 \ \frac{\text{rem}}{\text{rad}}\right) \\
&\quad + (1 \text{ rad})\left(10 \ \frac{\text{rem}}{\text{rad}}\right) \\
&= 111 \text{ rem} \quad (1.11 \text{ Sv})
\end{aligned}
$$

Permissible Exposure Levels

To protect humans from radiation effects, maximum permissible levels of exposure have been developed based on the maximum radiation dose that can be received with little risk of later development of adverse effects. The National Council on Radiation Protection and Measurement (NCRP) has published maximum permissible levels of external and internal radiation. The Nuclear Regulatory Commission (NRC) has established permissible doses, levels, and concentrations in 10 CFR 20. The International Commission on Radiological Protection (ICRP) has also published guides for maximum external exposure. For occupational exposure, the NCRP recommendations are given in Table 57.7.

Table 57.7 NCRP-Recommended Limitations for Exposure to Ionizing Radiation

occupational exposure	mSv
effective dose limit—annual	50
effective dose limit—cumulative	$10 \times$ age
equivalent dose annual limits for tissues and organs—lens of eye	150
equivalent dose annual limits for tissues and organs—skin, hands, and feet	500

Source: Adapted from NCRP Report No. 116, *Limitation of Exposure to Ionizing Radiation*, 1993.

Safety Factors

An environmental engineer should be aware of the basic safety factors for limiting dose. These factors are time, distance, and shielding.

Table 57.6 *Weighting Factors and Absorbed Dose Equivalencies*

type of radiation	weighting factor, w_R (Sv/Gy or rem/rad)	absorbed dose equal to a unit dose equivalent (Gy/Sv or rad/rem)
X-, gamma, or beta radiation	1	1
alpha particles, multiple-charged particles, fission fragments, and heavy particles of unknown charge	20	0.05
neutrons of unknown energy	10	0.1
high-energy protons	10	0.1*

*The absorbed dose in rad equal to 1 rem, or the absorbed dose in Gy equal to 1 Sv.

Source: Table 1004(b).1, 10 CFR 20.1004, *Units of radiation dose*, U.S. Nuclear Regulatory Commission, Washington, DC.

The dose received is directly related to the time exposed, so reducing the time of exposure will reduce the dose. An individual's time of exposure can also be limited by spreading the exposure time among more workers.

Distance is another safety factor that can be changed to reduce the dose. The intensity of external radiation decreases as the inverse of the square of the distance. By increasing the distance to a source from 2 m to 20 m, for example, the exposure would be reduced to 1% (2 m/20 m)^2.

Shielding involves placing a mass of material between a source and workers. The objective is to use a high-density material that will act as a barrier to X-ray and gamma-ray radiation. Lead and concrete are often used, with lead being the more effective material because of its greater density. For neutrons, different material is needed than for X-rays and gamma rays because neutrons produce secondary radiation from collisions with nuclei. Neutron shielding requires a light nucleus material. Typically water or graphite is used.

The shielding properties of materials are often compared using the *half-value thickness*, which is the thickness of the material required to reduce the radiation to half of the incident value. The half-value properties vary with the radiation source.

An environmental engineer should keep the following suggestions in mind when radiation is encountered on a project site.

- Obtain the advice of a qualified health physicist or industrial hygienist when dealing with radioactive materials.

- Treat radioactive materials as hazardous to human health.

- Be aware that the properties and hazards associated with internal and external radiation are different.

- Ensure that only properly calibrated instruments are used by trained personnel to monitor radiation.

- Apply the safety factors of time, distance, and shielding to keep exposure at safe levels.

- Ensure that exposure guidelines are achieved and that workers receive the minimum exposure possible in any situation.

16. BIOLOGICAL EFFECTS

Background Radiation

Background radiation includes natural radiation sources and human-made radiation sources. *Natural radiation sources* include cosmic sources and *naturally occurring radioactive material* (NORM). Approximately 340 nuclides occur in nature, and 70 of these, such as radon, are radioactive. *Human-made radiation sources* include fallout from nuclear explosive devices, some consumer products, and nuclear accidents such as the one at Chernobyl.

Together, natural and human-made background radiation expose each person to an average of 360 mrem (3.6 mSv) each year.

Acute Versus Chronic Effects

The duration of a radiation exposure is an important factor in determining how the human body reacts. Exposure durations are categorized as acute or chronic. *Acute exposure* is exposure to a large amount of radiation in a short period of time. In an occupational setting, the exposure period is typically less than eight hours. Table 57.8 gives a summary of the clinical effects of acute ionizing radiation doses. *Chronic exposure* is long-term, low-level exposure. Background radiation is an example of chronic radiation exposure.

Stochastic Versus Nonstochastic Effects

The term *stochastic* means involving chance or probability. Radiation increases the probability of stochastic events, such as cancer, occurring.

Table 57.8 *Summary of Clinical Effects of Acute Ionizing Radiation Doses*

	dose equivalent (rem)					
	0–100	100–200	200–600	600–1000	1000–5000	over 5000
incidence of vomiting	none	100 rem=5% 200 rem=50%	300 rem= 100%	100%		
time of onset	–	3 hr	2 hr	1 hr	30 min	
principal affected organs	none	hematopoietic tissue			gastrointestinal tract	central nervous system
characteristic signs	none	moderate leukopenia	severe leukopenia; purpura; hemorrhage; infection; epilation above 300 rem		diarrhea; fever; disturbance of electrolyte balance	convulsions; tremor; ataxia; lethargy
critical period postexposure	–	–	4 to 6 wk		5 to 14 days	1 to 48 hr
prognosis	excellent	excellent	good	guarded	hopeless	
convalescent period	none	several weeks	1 to 12 months	long	–	–
incidence of death	none	none	0 to 80%	80% to 100%	90% to 100%	
death occurs within	–	–	2 months		2 weeks	2 days
cause of death	–	–	hemorrhage; infection		circulatory collapse	respiratory failure; brain edema

Glasstone, S., *The Effects of Nuclear Weapons*, USAEC, Washington, DC, 1962.

Stochastic effects are biological effects that occur randomly and for which the probability of the effect occurring, not its severity, is assumed to be a linear function of dose without threshold. Hereditary effects and cancer incidence are examples of stochastic effects.

Nonstochastic (also called *deterministic*) *effects* are biological effects for which the severity varies with the dose and a threshold is believed to exist. Skin reddening and radiation-induced cataract formation are examples of nonstochastic effects.

Somatic, Genetic, and Fetal Effects

Exposure to ionizing radiation can cause three types of biological effects: somatic effects, genetic effects, and fetal effects.

Somatic effects are experienced directly by the irradiated individual and are either prompt or delayed, depending upon the period of time before the effects manifest in the exposed individual. These effects include damage to body tissues and organs that can impair their ability to function normally. The symptoms exhibited during the course of the Chernobyl accident were prompt somatic effects. Delayed somatic effects can be long-term, 20 years to 30 years. The main delayed somatic effect of ionizing radiation is an increase in the probability of developing different types of cancers. For example, without any artificial exposure, of 10,000 people, 2500 may exhibit some form of cancer during their lifetime. If 10,000 people are each irradiated with 1 rem of whole-body radiation, it is estimated that the radiation may cause three additional cases of cancer in the group. Of 10,000 people, 1640 may succumb to some form of cancer. If 10,000 people are each irradiated with 1 rem of whole-body radiation, it is estimated that the radiation may cause one additional cancer death in the group.

Genetic effects may be passed on to future generations. Inherited characteristics in human reproduction are controlled by genes in the reproductive cells. Radiation can alter genes and produce mutations that could eventually result in anomalies in offspring. Because these mutations are generally recessive, several generations may pass before the effects become apparent. The current incidence of all types of genetic disorders and traits that cause some type of serious handicap at some time during an individual's lifetime is about 1000 incidents per 10,000 live births. If each 30 year generation receives 1 rem of whole-body irradiation, the radiation may cause an additional 5 to 75 genetic disorders in the first generation, or 60 to 1000 disorders at genetic equilibrium about four generations later.

Environmental Health & Safety

Fetal effects are those effects that result from the exposure of a fetus or embryo to penetrating radiation. A number of studies have indicated that embryos or fetuses are more sensitive to radiation than adults, particularly during the first three months after conception, when a woman may not know that she is pregnant. The main concerns during this period of the pregnancy may be developmental abnormalities during the growth of the fetus. As the pregnancy progresses, the sensitivity of the fetus to radiation decreases. The main concern later in the pregnancy is an increase in the risk of leukemia in the first 10 years of the child's life. The current incidence of all types of fetal effects is about 700 incidents per 10,000 live births. This includes effects due to measles, alcohol, and drugs. If 10,000 embryos were to receive 1 rem of whole-body irradiation before birth, it is estimated that the radiation may affect 10 additional children.

17. RADIATION DOSE CALCULATIONS

Radiation exposure can come from sources inside as well as outside of the body. To quantify both internal and external exposures, the *total effective dose equivalent* (TEDE) is used. The TEDE is the sum of the deep-dose equivalent, H_D, for external exposures and the committed effective dose equivalent, $H_{E,50}$, for internal exposures over the next 50 years (assumed remaining lifespan).

$$\frac{\text{total radiation}}{\text{exposure}} = \frac{\text{external}}{\text{exposure}} + \frac{\text{internal}}{\text{exposure}} \quad \textbf{57.28}$$
$$\text{TEDE} = H_D + H_{E,50}$$

The *external dose* is the portion of the dose equivalent received from radiation sources outside the whole body. For radiation protection purposes, "whole body" means head, trunk (including male gonads), arms above the elbow, and legs above the knee.

The *deep-dose equivalent*, H_D, is the external whole-body exposure: the portion of the dose equivalent at a tissue depth of 1 cm (1 g/cm^2).

The *shallow-dose equivalent*, H_S, is the external exposure of the skin or an extremity and is taken as the dose equivalent at a tissue depth of 0.007 cm (0.007 g/cm^2) averaged over an area of 1 cm^2.

The *lens dose equivalent* or *eye dose equivalent*, H_L, is the external exposure of the lens of the eye, taken as the dose equivalent at a tissue depth of 0.3 cm (0.3 g/cm^2).

Beta Doses

For beta emitters, the dose equivalent is a function of the activity of the radioactive source, the distance from the source, the energy of the beta particles, and the number of beta particles per decay (the "yield" or "abundance"). The energy of the beta particles and the

number of beta particles per decay are incorporated into a unit called the *electron dose rate factor*, DF. (See Table 57.9.)

$$H_T = X_b t = \text{DF}\left(\frac{A}{A_s}\right)t \quad \textbf{57.29}$$

Example 57.8

What is the dose equivalent rate at a depth of 0.007 cm for a 3.7×10^7 Bq paste of phosphorus-32 uniformly deposited on 1 in^2 of skin?

Solution

$$A_s = (1 \text{ in}^2)\left(2.54 \frac{\text{cm}}{\text{in}}\right)^2 = 6.45 \text{ cm}^2$$

From Table 57.9 for P-32 at 0.007 cm,

$$\text{DF} = 2.1 \times 10^{-2} \text{ Sv·cm}^2/\text{Bq·yr}$$

Use Eq. 57.29.

$$\begin{aligned}
X_b &= \text{DF}\left(\frac{A}{A_s}\right) \\
&= \left(2.1 \times 10^{-2} \frac{\text{Sv·cm}^2}{\text{Bq·yr}}\right)\left(\frac{3.7 \times 10^7 \text{ Bq}}{6.45 \text{ cm}^2}\right) \\
&= 1.2 \times 10^5 \text{ Sv/yr}
\end{aligned}$$

Gamma Doses

For photon emitters, the dose equivalent is a function of the activity of the radioactive source, the distance from the source, the energy of the photons, and the number of photons per decay. The energy of the photons and the number of photons per decay are incorporated into a unit called the *specific gamma ray dose constant*, Γ. (See Table 57.10.)

$$H_T = X_g t = \Gamma A t \quad \text{[at 100 cm]} \quad \textbf{57.30}$$

Example 57.9

What is the whole-body dose equivalent rate 100 cm from an unshielded 1.59×10^5 MBq cobalt-57 source?

Solution

From Table 57.10, for Co-57,

$$\Gamma = 4.087 \times 10^{-5} \text{ (mSv/h)/MBq} \quad \text{[at 100 cm]}$$

Table 57.9 *Electron Dose Rate Factors in Skin from Selected Radionuclides Deposited on the Body Surface*

dose rate factors (Sv/yr per Bq/cm^2) versus depth in tissue[a,b]

nuclide	0.004 cm	0.008 cm	0.040 cm	0.007 cm
^{3}H	0.0	0.0	0.0	0.0
^{14}C	7.9×10^{-3}	2.1×10^{-3}	0.0	2.9×10^{-3}
^{32}P	2.4×10^{-2}	2.0×10^{-2}	1.1×10^{-2}	2.1×10^{-2}
^{51}Cr	0.0	0.0	0.0	0.0
^{54}Mn	0.0	0.0	0.0	0.0
^{55}Fe	0.0	0.0	0.0	0.0
^{58}Co	3.6×10^{-3}	2.6×10^{-3}	4.4×10^{-4}	2.8×10^{-3}
^{60}Co	1.6×10^{-2}	8.7×10^{-3}	2.5×10^{-4}	9.9×10^{-3}
^{86}Rb	2.3×10^{-2}	1.9×10^{-2}	1.0×10^{-2}	2.0×10^{-2}
^{89}Sr	2.3×10^{-3}	1.9×10^{-2}	9.8×10^{-3}	2.0×10^{-2}
^{90}Sr	2.1×10^{-2}	1.5×10^{-2}	3.4×10^{-3}	1.6×10^{-2}
^{90}Y	2.4×10^{-2}	2.0×10^{-2}	1.2×10^{-2}	2.1×10^{-2}
^{91}Y	2.3×10^{-2}	1.9×10^{-2}	9.9×10^{-3}	2.0×10^{-2}
^{95}Zr	1.7×10^{-2}	1.0×10^{-2}	7.4×10^{-4}	1.2×10^{-2}
^{95}Nb	6.4×10^{-3}	1.7×10^{-3}	1.8×10^{-5}	2.3×10^{-3}
^{99}Mo	2.3×10^{-2}	1.8×10^{-2}	7.1×10^{-3}	1.9×10^{-2}
^{99m}Tc	2.9×10^{-3}	1.8×10^{-3}	0.0	2.1×10^{-3}
^{103}Ru	1.1×10^{-2}	4.8×10^{-3}	2.1×10^{-4}	5.8×10^{-3}
^{106}Ru	0.0	0.0	0.0	0.0
^{105}Rh	1.8×10^{-2}	1.2×10^{-2}	2.1×10^{-3}	1.3×10^{-2}
^{127}Sb	2.2×10^{-2}	1.7×10^{-2}	5.9×10^{-3}	1.8×10^{-2}
^{127}Te	2.1×10^{-2}	1.5×10^{-2}	4.0×10^{-3}	1.6×10^{-2}
^{127m}Te	1.6×10^{-2}	2.5×10^{-3}	9.4×10^{-5}	4.7×10^{-3}
^{129}Te	2.3×10^{-2}	1.9×10^{-2}	9.1×10^{-3}	2.0×10^{-2}
^{129m}Te	2.3×10^{-2}	1.1×10^{-2}	3.5×10^{-3}	1.3×10^{-2}
^{131}Te	2.8×10^{-2}	2.2×10^{-2}	1.0×10^{-2}	2.3×10^{-2}
^{132}Te	1.3×10^{-2}	5.9×10^{-3}	4.7×10^{-5}	7.0×10^{-3}
^{131}I	2.1×10^{-2}	1.4×10^{-2}	3.0×10^{-3}	1.5×10^{-2}
^{132}I	2.3×10^{-2}	1.8×10^{-2}	8.2×10^{-3}	1.9×10^{-2}
^{135}I	2.2×10^{-2}	1.7×10^{-2}	6.5×10^{-3}	1.8×10^{-2}
^{134}Cs	1.6×10^{-2}	1.1×10^{-2}	2.7×10^{-3}	1.2×10^{-2}
^{136}Cs	2.0×10^{-2}	1.2×10^{-2}	5.9×10^{-4}	1.3×10^{-2}
^{137}Cs	2.0×10^{-2}	1.3×10^{-2}	2.3×10^{-3}	1.4×10^{-2}
^{137m}Ba	2.4×10^{-3}	2.0×10^{-3}	1.2×10^{-3}	2.1×10^{-3}
^{140}Ba	2.2×10^{-2}	1.6×10^{-2}	5.0×10^{-3}	1.7×10^{-2}
^{140}La	2.4×10^{-2}	1.9×10^{-2}	9.2×10^{-3}	2.0×10^{-2}
^{141}Ce	2.5×10^{-2}	1.5×10^{-2}	1.6×10^{-3}	1.7×10^{-2}
^{143}Ce	2.4×10^{-2}	1.8×10^{-2}	7.7×10^{-3}	1.9×10^{-2}
^{144}Ce	1.5×10^{-2}	7.6×10^{-3}	1.7×10^{-4}	8.9×10^{-3}
^{143}Pr	2.2×10^{-2}	1.7×10^{-2}	6.2×10^{-3}	1.8×10^{-2}
^{147}Nd	2.3×10^{-2}	1.5×10^{-2}	4.2×10^{-3}	1.7×10^{-2}
^{239}Np	3.6×10^{-2}	2.0×10^{-2}	1.2×10^{-3}	2.3×10^{-2}
^{238}Pu	0.0	0.0	0.0	0.0
^{239}Pu	3.8×10^{-6}	0.0	0.0	0.0
^{241}Pu	0.0	0.0	0.0	0.0
^{241}Am	4.8×10^{-4}	1.1×10^{-5}	0.0	2.2×10^{-5}
^{242}Cm	0.0	0.0	0.0	0.0
^{244}Cm	0.0	0.0	0.0	0.0[c]

[a]The first three depths are values recommended by Whitton (1973) for various parts of the body surface; the last depth is the average value over the body surface recommended in ICRP Publication 26 1977.

[b]Depths in g/cm^2 are equivalent to depths in cm for water.

[c]Source: Kocher, D.C. and Eckerman, K.F., "Electron Dose-Rate Conversion Factors for External Exposure of the Skin for Uniformly Deposited Activity on the Body Surface," Health Physics, 53:135-141; 1987.

Environmental Health & Safety

Use Eq. 57.30.

$$X_g = \Gamma A = \left(4.087 \times 10^{-5}\ \frac{\text{mSv/h}}{\text{MBq}}\right)(1.59 \times 10^5\ \text{MBq})$$

$$= 6.5\ \text{mSv/h} \quad [\text{at } 100\ \text{cm}]$$

Table 57.10 Specific Gamma Ray Dose Constants (whole body; point sources at 100 cm distance)*

radioisotope	Γ $\left(\dfrac{\text{mSv/h}}{\text{MBq}}\right)$
Be-7	9.292×10^{-6}
C-11	1.908×10^{-4}
O-15	1.911×10^{-4}
F-18	1.851×10^{-4}
Na-22	3.590×10^{-4}
K-40	2.197×10^{-5}
Cr-51	6.320×10^{-6}
Fe-59	1.787×10^{-4}
Co-57	4.087×10^{-5}
Co-60	3.697×10^{-4}
Rb-86	1.458×10^{-5}
Pd-109	1.290×10^{-7}
I-125	7.432×10^{-5}
I-131	7.640×10^{-5}
Cs-137	1.017×10^{-4}
Au-198	7.881×10^{-5}
Po-210	1.422×10^{-9}
Rn-222	7.280×10^{-8}
Th-230	1.861×10^{-5}
U-235	9.159×10^{-5}

*Use the inverse-square law to calculate dose equivalents at other distances.
Adapted from Unger, L.M. and Trubey, D.K., *Specific Gamma-Ray Dose Constants for Nuclides Important to Dosimetry and Radiological Assessment.* ORNL/RSIC-45; Oak Ridge National Laboratory; Oak Ridge, TN; 1981, as reprinted 1982.

Neutron Doses

For neutrons, the *fluence*, f, can be used to calculate the neutron dose equivalent. 1 rem (0.01 Sv) of neutron radiation of unknown energies may be assumed to be caused by a total fluence of 25 million neutrons per square centimeter hitting the body. If the approximate energy distribution of the neutrons is known, Table 57.11 can be used to calculate the tissue dose equivalent.

$$H_T = D_T w_R = X_n t$$

$$= \frac{f}{\substack{\text{fluence per unit} \\ \text{dose equivalent}}} \qquad 57.31$$

Table 57.11 Neutron Weighting Factors and Absorbed Dose Equivalencies (whole body)

neutron energy (MeV)	weighting factor[a], w_R	fluence per unit dose equivalent[b], f/H_T (neutrons/cm²·rem)
2.5×10^{-8} (thermal)	2	980×10^6
1×10^{-7}	2	980×10^6
1×10^{-6}	2	810×10^6
1×10^{-5}	2	810×10^6
1×10^{-4}	2	840×10^6
1×10^{-3}	2	980×10^6
1×10^{-2}	2.5	1010×10^6
1×10^{-1}	7.5	170×10^6
5×10^{-1}	11	39×10^6
1	11	27×10^6
2.5	9	29×10^6
5	8	23×10^6
7	7	24×10^6
10	6.5	24×10^6
14	7.5	17×10^6
20	8	16×10^6
40	7	14×10^6
60	5.5	16×10^6
1×10^2	4	20×10^6
2×10^2	3.5	19×10^6
3×10^2	3.5	16×10^6
4×10^2	3.5	14×10^6

[a] Value of weighting factor, w_R, at the point where the dose equivalent is maximum in a 30 cm diameter cylinder tissue-equivalent phantom
[b] Monoenergetic neutrons incident normally on a 30 cm diameter cylinder tissue-equivalent phantom
Source: Table 1004(b).2, 10 CFR 20.1004, *Units of Radiation Dose*, U.S. Nuclear Regulatory Commission, Washington, DC.

Example 57.10

What is the dose equivalent, H_T, in rems from an unshielded 5 MeV neutron source with a fluence, f, of 10^9 neutrons/cm²?

Solution

From Table 57.11, for 5 MeV neutrons,

$$\frac{f}{H_T} = 23 \times 10^6 \text{ neutrons/cm}^2 \cdot \text{rem}$$

$$H_T = \frac{f}{\text{fluence per unit}}$$
$$\text{dose equivalent}$$

$$= \frac{10^9 \dfrac{\text{neutrons}}{\text{cm}^2}}{23 \times 10^6 \dfrac{\text{neutrons}}{\text{cm}^2 \cdot \text{rem}}}$$

$$= 43 \text{ rem} \quad (0.43 \text{ Sv})$$

18. INTERNAL EXPOSURES

The *internal* (or *committed*) *dose* is that portion of the dose equivalent received from radioactive material taken into the body. The longer a radioactive material stays in the body, the larger the radiation dose received.

The *committed dose equivalent*, $H_{T,50}$, is the dose equivalent to organs or tissues that will be received from an intake of radioactive material by an individual during the 50-year period following the intake.

The *committed effective dose equivalent*, $H_{E,50}$, is the sum of the products of (1) the weighting factors applicable to each of the body organs or tissues that are irradiated, and (2) the $H_{T,50}$ to these organs or tissues.

The *tissue weighting factor*, w_T, for an organ or tissue is the proportion of the risk of stochastic effects resulting from irradiation of that organ or tissue to the total risk of stochastic effects when the whole body is irradiated uniformly. (See Table 57.12.)

$$H_{E,50} = \sum w_T H_{T,50} \qquad 57.32$$

Table 57.12 *Organ Dose Tissue Weighting Factors*

organ or tissue	w_T
gonads	0.25
breast	0.15
red bone marrow	0.12
lung	0.12
thyroid	0.03
bone surface	0.03
remainder[a]	0.30
whole body[b]	1.00

[a] 0.30 results from 0.06 for each of 5 "remainder" organs (excluding the skin and the lens of the eye) that receive the highest doses.
[b] For the purpose of weighting the external whole-body dose (for adding it to the internal dose), a single weighting factor, $w_T = 1.0$, has been specified.
Source: 10 CFR 20.1003, *Definitions*, U.S. Nuclear Regulatory Commission, Washington, DC.

The human body cannot distinguish between stable isotopes and radioactive isotopes of the same element. This means that radioactive isotopes inside the body will experience the same absorption and removal rates, and occupy the same deposition sites, that stable isotopes would. Clearance of internal radioisotopes involves two processes: (1) radioactive decay and (2) normal biological removal. In performing clearance calculations, both radiological and biological clearances are assumed to follow exponential laws.

The *radiological clearance rate* is

$$q_{\text{rad},t} = q_0 e^{-\lambda_{\text{rad}} t} \qquad 57.33$$

The *biological clearance rate* is

$$q_{\text{bio},t} = q_0 e^{-\lambda_{\text{bio}} t} \qquad 57.34$$

The *effective clearance rate* due to the combined effects of biological and physical clearance is

$$q_t = q_0 e^{-\lambda_{\text{rad}} t} e^{-\lambda_{\text{bio}} t}$$
$$= q_0 e^{-(\lambda_{\text{rad}} + \lambda_{\text{bio}}) t} \qquad 57.35$$
$$= q_0 e^{-\lambda_{\text{eff}} t}$$

The *effective half-life*, τ_e, is the time it takes for a radioactive material in the body to decrease from the original activity to one-half of that activity, considering both the biological and physical clearance. The effective half-life is related to the biological and radiological half-lives by

Effective Half-Life

$$\frac{1}{\tau_e} = \frac{1}{\tau_r} + \frac{1}{\tau_b} \qquad 57.36$$

Example 57.11

What is the effective half-life of Mn-54 ($\tau_{\text{rad}} = 312.5$ days), which is part of a molecule that is cleared from the body with τ_{bio} of 25 days?

Solution

Use Eq. 57.36.

Effective Half-Life

$$\frac{1}{\tau_e} = \frac{1}{\tau_r} + \frac{1}{\tau_b}$$

$$= \frac{1}{312.5 \text{ days}} + \frac{1}{25 \text{ days}}$$

$$= 0.0432 \text{ day}$$

$$\tau_e = 23.15 \text{ days}$$

Environmental Health & Safety

Calculating Internal Dose

The *annual limit on intake* (ALI) and the *derived air concentration* (DAC) can be used to calculate internal doses for intakes of radioactive material.

ALI is the derived limit for the amount of radioactive material that can be taken into the body of an adult worker by inhalation or ingestion in a year. ALI is the smaller value of intake of a given radionuclide in a year that would result in either a committed effective dose equivalent of 5 rems (0.05 Sv) or a committed dose equivalent of 50 rems (0.5 Sv) in any individual organ or tissue.

DAC is the concentration of a given radionuclide in air that, if breathed for a working year of 2000 hours under conditions of light work (i.e., inhalation rate 1.2 m^3 of air per hour), would result in an intake of one ALI. The DAC relates to one of two modes of exposure: either external submersion or the internal committed dose equivalents resulting from inhalation of radioactive materials. DACs based upon submersion are for immersion in a semi-infinite cloud of uniform concentration and apply to each radionuclide separately.

A "DAC-hour" is the product of the actual concentration of radioactive material in air (expressed as a fraction or multiple of the DAC for each radionuclide) and the time of exposure for that radionuclide, in hours. Two thousand DAC-hours equal one ALI and produce a committed effective dose equivalent of 5 rems (0.05 Sv).

The DAC is a limit intended to control chronic occupational exposures. The relationship between the DAC and the ALI is given by

$$\text{DAC} = \frac{\text{ALI}}{(2000 \text{ h})\left(60 \ \dfrac{\text{min}}{\text{h}}\right)\left(2 \times 10^4 \ \dfrac{\text{mL}}{\text{min}}\right)} \quad \textit{57.37}$$

$$= \frac{\text{ALI}}{2.4 \times 10^9 \text{ mL}}$$

2×10^4 mL/min is the standard breathing rate under conditions of light work.

Inhalation Class

Class (*lung class* or *inhalation class*) as used in Table 57.13 is a classification scheme for inhaled material categorized according to the rate of clearance from the pulmonary region of the lung. Materials are classified as D, W, or Y as follows.

class D (days) = less than 10 days

class W (weeks) = from 10 to 100 days

class Y (years) = greater than 100 days

Occupational Values

The ALIs in Table 57.13, part 1 are the annual intakes of a given radionuclide that would result in either

- a committed effective dose equivalent, $H_{E,50}$, of 5 rem (stochastic ALI)

- a committed dose equivalent, $H_{T,50}$, of 50 rem to an organ or tissue (nonstochastic ALI)

When the ALI is defined by the stochastic dose limit, it is the only ALI value listed in Table 57.13. When the committed dose equivalent, $H_{T,50}$, of a particular organ would reach 50 rem before the committed effective dose equivalent, $H_{E,50}$, would reach 5 rem, the ALI is limited by the nonstochastic dose limit to an organ. In that case, the organ or tissue to which the nonstochastic (committed dose equivalent, $H_{T,50} \le 50$ rem) limit applies is listed, and the ALI for the stochastic (committed effective dose equivalent, $H_{E,50} \le 5$ rem) limit is shown in parentheses.

Effluents, Air, and Water

The concentration values given in columns 1 and 2 of part 2 of Table 57.13 are equivalent to the radionuclide concentrations that, if inhaled or ingested continuously over a year, would produce a total effective dose equivalent, $H_{E,\text{tot}}$, of 0.05 rem (50 mrem or 0.5 mSv).

Sewer Disposal

The sewer disposal concentrations are calculated by taking the most restrictive occupational stochastic oral ingestion ALI and dividing by (a) 7.3×10^5 mL (the standard per-capita annual water intake by a "reference man") and (b) a factor of 10. The factor of 10 reduces the 5 rem produced by intake of one ALI to the 0.5 rem committed effective dose equivalent, $H_{E,50}$, regulatory limit for a member of the public.

Example 57.12

A worker inhaled 8 μCi of iodine-131. Calculate (a) the committed dose equivalent, $H_{T,50}$, to the thyroid and (b) the committed effective dose equivalent, $H_{E,50}$.

Solution

(a) From Table 57.13, part 1, column 2, for I-131, the thyroid gland has a nonstochastic ALI of 5×10^1 μCi. A nonstochastic ALI produces 50 rem committed dose equivalent, $H_{T,50}$, in the specified organ (thyroid).

$$50 \text{ rem} = 5 \times 10^1 \ \mu\text{Ci}$$

For the thyroid, the committed dose equivalent produced by 8 μCi is

$$H_{T,50} = 8 \ \mu\text{Ci}$$

Table 57.13 *Representative Annual Limits on Intake (ALIs) and Derived Air Concentrations (DACs) of Radionuclides for Occupational Exposure, Effluent Concentrations, and Concentrations for Release to Sewerage*

| atomic no. | radionuclide | class | part 1 occupational values | | | part 2 effluent concentrations | | part 3 releases to sewers |
			col. 1 oral inges-tion ALI (μCi)	col. 2 inhalation ALI (μCi)	col. 3 inhalation DAC (μCi/mL)	col. 1 air (μCi/mL)	col. 2 water (μCi/mL)	monthly average concentration (μCi/mL)
1	hydrogen-3	water, DAC includes absorption	8×10^4	8×10^4	2×10^{-5}	1×10^{-7}	1×10^{-3}	1×10^{-2}
6	carbon-11	monoxide	–	1×10^6	5×10^{-4}	2×10^{-6}	–	–
		dioxide	–	6×10^5	3×10^{-4}	9×10^{-7}	–	–
		compounds	4×10^5	4×10^5	2×10^{-4}	6×10^{-7}	6×10^{-3}	6×10^{-2}
6	carbon-14	monoxide	–	2×10^6	7×10^{-4}	2×10^{-6}	–	–
		dioxide	–	2×10^5	9×10^{-5}	3×10^{-7}	–	–
		compounds	2×10^3	2×10^3	1×10^{-6}	3×10^{-9}	3×10^{-5}	3×10^{-4}
15	phosphorus-32	D, all compounds except phosphates given for W	6×10^2	9×10^2	4×10^{-7}	1×10^{-9}	9×10^{-6}	9×10^{-5}
		W, phosphates of Zn^{2+}, S^{3+}, Mg^{2+}, Fe^{3+}, Bi^{3+}, and lanthanides	–	4×10^2	2×10^{-7}	5×10^{-10}	–	–
28	nickel-63	D, see Ni-56	9×10^3	2×10^3	7×10^{-7}	2×10^{-9}	1×10^{-4}	1×10^{-3}
		W, see Ni-56	–	3×10^3	1×10^{-6}	4×10^{-9}	–	–
		vapor	–	8×10^2	3×10^{-7}	1×10^{-9}	–	–
20	calcium-45	W, all compounds	2×10^3	8×10^2	4×10^{-7}	1×10^{-9}	2×10^{-5}	2×10^{-4}
53	iodine-125	D, all compounds	4×10^1 thyroid	6×10^1 thyroid	3×10^{-8}	–	–	–
			(1×10^2)	(2×10^2)	–	3×10^{-10}	2×10^{-6}	2×10^{-5}
53	iodine-131	D, all compounds	3×10^1 thyroid	5×10^1 thyroid	2×10^{-8}	–	–	–
			(9×10^1)	(2×10^2)	–	2×10^{-10}	1×10^{-6}	1×10^{-5}
95	americium-241	W, all compounds	8×10^{-1} bone surface	6×10^{-3} bone surface	3×10^{-12}	–	–	–
			(1×10^0)	(1×10^{-2})	–	2×10^{-14}	2×10^{-8}	2×10^{-7}

Source: Appendix B to 10 CFR 20, Index of Radioisotopes, U.S. Nuclear Regulatory Commission, Washington, DC.

Therefore,

$$\frac{H_{T,50}}{8\ \mu\text{Ci}} = \frac{50\ \text{rem}}{5 \times 10^1\ \mu\text{Ci}}$$

$$H_{T,50} = \frac{(8\ \mu\text{Ci})(50\ \text{rem})}{5 \times 10^1\ \mu\text{Ci}} = 8\ \text{rem} \quad (0.08\ \text{Sv})$$

(b) From Table 57.13, part 1, column 2, for I-131, the stochastic ALI is 2×10^2 μCi. This would produce a 5 rem committed effective dose equivalent, $H_{E,50}$.

$$5\ \text{rem} = 2 \times 10^2\ \mu\text{Ci}$$

The committed effective dose equivalent produced by 8 μCi is

$$H_{E,50} = 8\ \mu\text{Ci}$$

Environmental Health & Safety

Therefore,

$$\frac{H_{E,50}}{8 \ \mu\text{Ci}} = \frac{5 \text{ rem}}{2 \times 10^2 \ \mu\text{Ci}}$$

$$H_{E,50} = \frac{(8 \ \mu\text{Ci})(5 \text{ rem})}{2 \times 10^2 \ \mu\text{Ci}} = 0.2 \text{ rem} \quad (0.002 \text{ Sv})$$

Example 57.13

What is the total effective dose equivalent, $H_{E,\text{tot}}$, to a person who inhaled iodine-131 at an average concentration of $9 \times 10^{-12} \ \mu\text{Ci/mL}$ for one year?

Solution

From Table 57.13, part 2, column 1, for I-131, $2 \times 10^{-10} \ \mu\text{Ci/mL}$ produces a 0.05 rem total effective dose equivalent, $H_{E,\text{tot}}$.

$$0.05 \text{ rem} = 2 \times 10^{-10} \ \mu\text{Ci/mL}$$

The total effective dose equivalent produced by $9 \times 10^{-12} \ \mu\text{Ci/mL}$ is

$$H_{E,\text{tot}} = 9 \times 10^{-12} \ \mu\text{Ci/mL}$$

Therefore,

$$\frac{H_{E,\text{tot}}}{9 \times 10^{-12} \ \frac{\mu\text{Ci}}{\text{mL}}} = \frac{0.05 \text{ rem}}{2 \times 10^{-10} \ \frac{\mu\text{Ci}}{\text{mL}}}$$

$$H_{E,\text{tot}} = \frac{\left(9 \times 10^{-12} \ \frac{\mu\text{Ci}}{\text{mL}}\right)(0.05 \text{ rem})}{2 \times 10^{-10} \ \frac{\mu\text{Ci}}{\text{mL}}}$$

$$= 2.3 \times 10^{-3} \text{ rem} \quad (2.3 \times 10^{-5} \text{ Sv})$$

Example 57.14

What is the committed effective dose equivalent, $H_{E,50}$, to a person whose drinking water was supplied by a facility containing Ni-63 at an average monthly concentration of $5 \times 10^{-5} \ \mu\text{Ci/mL}$?

Solution

From Table 57.13, part 3 for Ni-63, $1 \times 10^{-3} \ \mu\text{Ci/mL}$ produces 0.5 rem committed effective dose equivalent, $H_{E,50}$, so

$$0.5 \text{ rem} = 1 \times 10^{-3} \ \mu\text{Ci/mL}$$

The committed effective dose equivalent produced by $5 \times 10^{-5} \ \mu\text{Ci/mL}$ is

$$H_{E,50} = 5 \times 10^{-5} \ \mu\text{Ci/mL}$$

Therefore,

$$\frac{H_{E,50}}{5 \times 10^{-5} \ \frac{\mu\text{Ci}}{\text{mL}}} = \frac{0.5 \text{ rem}}{1 \times 10^{-3} \ \frac{\mu\text{Ci}}{\text{mL}}}$$

$$H_{E,50} = \frac{\left(5 \times 10^{-5} \ \frac{\mu\text{Ci}}{\text{mL}}\right)(0.5 \text{ rem})}{1 \times 10^{-3} \ \frac{\mu\text{Ci}}{\text{mL}}}$$

$$= 2.5 \times 10^{-2} \text{ rem} \quad (2.5 \times 10^{-4} \text{ Sv})$$

19. CONTROLLING EXTERNAL RADIATION EXPOSURE

External radiation exposure is a problem primarily with mid- to high-energy beta emitters and gamma and X-ray emitters. The three most effective ways to reduce the exposure from external radiation hazards are (1) reduced exposure time, (2) increased distance, and (3) thicker shielding.

Time

The exposure received is directly proportional to the amount of time spent near a radiation source. That is, if the time spent near a radiation source is doubled, the radiation dose is doubled.

Distance

The greater the distance from a radiation source, the less radiation received. For sources that can be treated as a *point source* (a source that is a distance at least three times the longest dimension of the source), the radiation dose received is inversely proportional to the square of the distance of separation. This is referred to as the *inverse-square law*. The intensity of a radiation field decreases with the square of the distance from the source, so if the distance from a source is doubled, the intensity of the radiation field will decrease to one-fourth the initial exposure.

Inverse Square Law

$$\frac{I_1}{I_2} = \frac{(R_2)^2}{(R_1)^2} \qquad 57.38$$

Not all sources follow the inverse-square law.

- *Point sources:* Radiation intensity decreases with distance approximately as $1/R^2$ (doubling the distance quarters the dose rate).

- *Line sources:* Radiation intensity decreases with distance approximately as $1/R$ (doubling the distance cuts the radiation intensity to about $1/2$).

- *Disk and cylindrical sources:* Radiation intensity decreases with distance between $1/R$ and $1/R^2$ (doubling the distance reduces the radiation intensity to between $1/2$ and $1/4$).

Example 57.15

For the scenario in Ex. 57.9, what is the dose equivalent rate 20 cm from the same source?

Solution

From Eq. 57.38,

Inverse Square Law

$$\frac{I_1}{I_2} = \frac{(R_2)^2}{(R_1)^2}$$

$$I_2 = I_1\left(\frac{R_1}{R_2}\right)^2 = \left(6.5 \ \frac{\text{mSv}}{\text{h}}\right)\left(\frac{1 \text{ m}}{0.2 \text{ m}}\right)^2$$

$$= 1.6 \times 10^2 \ \text{mSv/h}$$

Shielding

Any material can act as a radiation shield, and, in general, the denser the material, the better shield it makes. In most cases, the closer the shielding to the source, the better economically, since less shielding material will be required. Selecting the proper shielding material depends on the type of radiation involved.

- Proper shielding for low-energy gamma emitters, such as I-125, requires only thin sheets of lead foil.

- For medium-energy gamma emitters, such as Co-57, about $1/4$ in (6.4 mm) of lead is needed.

- High-energy gamma emitters such as cobalt-60, sodium-22, manganese-54, chromium-51, and iodine-131 require lead bricks to effectively attenuate the gamma radiation.

- Proper shielding of high-energy beta emitters can be achieved with about 1 cm of plexiglass.

- Shielding is not required for low-energy beta emitters, such as sulfur-35 or carbon-14, as these beta particles have very limited ranges in air.

- Shielding is not required for alpha particles, though alpha emitters, such as americium-241, usually also emit gamma radiation.

The resulting intensity of radiation can be calculated using Eq. 57.39 and Eq. 57.40. I is the intensity, μ is the attenuation coefficient, X is the thickness, and B is the buildup.

Shielding

$$I = I_0 e^{-\mu X} \qquad \textbf{57.39}$$

$$I = I_0 B e^{-\mu X} \qquad \textbf{57.40}$$

20. NOMENCLATURE

a	speed of sound	m/s
A_s	surface area	m^2
ALI	annual limit on intake	μCi
B	buildup	-
C_i	time spent at specified SPL	hr
C_n	time of noise exposure at specified level	s
D	noise dose	%
D_T	absorbed dose	(Sv/yr)/(Bq/cm^2) or rad
DAC	derived air concentration	μCi/mL
DF	dose factor	(Sv/yr)/(Bq/cm^2)
f	fluence	neutrons/cm^2
f	frequency of sound	Hz
$H_{E,50}$	committed effective dose equivalent	Sv
$H_{E,\text{tot}}$	total effective dose equivalent	Sv
H_T	dose equivalent	Sv
$H_{T,50}$	committed equivalent	Sv
H_D	deep-dose equivalent	Sv
H_L	lens dose equivalent	Sv
H_S	shallow-dose equivalent	Sv
I	intensity	various
I	radiation source intensity	C/kg
I	sound intensity	W/m^2
IL	insertion loss	dB
k	constant	-
L	level	dB
M	molecular weight	g/mol
n	number of moles	–
N	number of atoms	–
N_{10}	initial activity of parent nuclei	curies
NR	noise reduction	dB
P	pressure	Pa
q	radioactivity in an organ or the body	Bq
Q	directivity	-
Q	flow rate	W
r	distance from sound source	m
R	acoustic resistance	N·s/m^5
R	distance from source	m
R	room constant	m^2
R	specific gas constant	kJ/kg·K

S	surface absorption	sabins
SPL	sound pressure level	dB
SPL (dB)	sound pressure level in decibels	dB
T	reference duration	h
t, T	time	s
t_n	total time of exposure permitted at specified level	s
T	temperature	°C, K
T_i	time permitted at specified SPL	hr
TEDE	total effective dose equivalent	rem, Sv
TL	transmission loss	dB
TWA	time-weighted average	dB
w_R	weighting factor	Sv/Gy or rem/rad
w_T	tissue weighting factor	–
W	sound power	W
X	dose equivalent rate	Sv/s
X	thickness	cm

Symbols

Γ	gamma constant	(mSv/h)/mBq
λ	decay constant for radionuclides	time^{-1}
λ	wavelength	m
μ	attenuation coefficient	cm^2/g
τ	half-life	time
τ	transmission coefficient	–
ρ	density	kg/m^3

Subscripts

0	initial condition or reference
b	beta or biological
bio	biological
e	effective
eff	effective
g	gamma
i	interval
n	neutron
r	radioactive
rad	radiological
ref	reference
rms	root-mean-square
S	surface per unit area
tot	total
T	tissue
W	sound power

58

Indoor Air Quality

1. INDOOR AIR

Introduction

Pollution of indoor air is a concern for at least three reasons.

- The average person in the United States spends about 87% of his/her time indoors.

- Multiple air pollution sources emitting a variety of pollutants are located indoors.

- Interior spaces tend to concentrate pollutants since dispersion into the atmosphere is limited by the finite volume of the room.

Table 58.1 lists the major indoor air pollutants and their sources. Of these, environmental tobacco smoke (ETS) and radon are the most significant because of their relative toxicity, frequency of occurrence, and potential to exist at elevated concentrations. Of somewhat lesser significance are formaldehyde and asbestos, followed by the other listed pollutants.

Environmental Tobacco Smoke

Tobacco smoke may be characterized as either mainstream or sidestream. *Mainstream* tobacco smoke is that taken directly into the lungs by the smoker. *Sidestream* smoke is what is commonly referred to as secondhand tobacco smoke. Sidestream smoke and exhaled mainstream smoke are the sources of ETS. Indoor levels of the criteria pollutants CO and PM2.5 attributable to tobacco smoke may be as much as one to two orders of magnitude greater than in similar spaces where smoking does not occur. Environmental tobacco smoke is also the source of a variety of other chemicals including formaldehyde and other aldehydes, acetone and other ketones, benzene, phenol, and other aromatics, organic acids, amines, methyl chloride, and nicotine, and many more.

The size of particulate matter associated with ETS is from about 0.1 μm to 0.5 μm, making it particularly prone to deposition in the lungs. Radon exposure is increased by ETS since it provides a mechanism for transport into the lungs.

Radon

Radon, a noble gas, results from the radioactive decay of radium which is found naturally in soils in many regions of the United States. The half-life of radon is a relatively short 3.8 days. However, the decay products, or progeny, of radon have much shorter half-lives of no more than several minutes. Although radon is often described as an alpha source, Fig. 58.1 shows that three forms of radiation are emitted during the decay lifetime.

Table 58.1 Indoor Air Pollutant Sources and Concerns

source	pollutants of concern
combustion	criteria pollutants
tobacco smoke	environmental tobacco smoke (ETS)
woodburning stoves	carbon monoxide
kerosene heaters	nitrogen dioxide
gas stoves	hydrocarbons
automobiles in attached garages	
building materials and furnishings	formaldehyde
particleboard	VOCs
paints	hydrocarbons
adhesives	
caulks	
carpeting	
consumer products	VOCs
pesticides	hydrocarbons
cleaning materials	particulates
waxes, polishes	
cosmetics, personal products	
hobby materials	
people and pets	particulate bioeffluents
outdoor air	criteria pollutants

Environmental
Health & Safety

Figure 58.1 *Radon Decay Products*

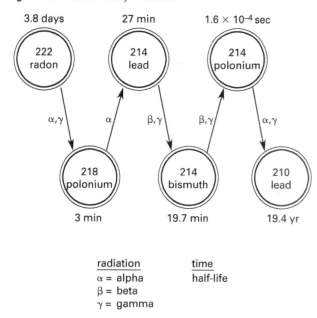

radiation
α = alpha
β = beta
γ = gamma

time
half-life

The radon decay products are electrically charged and, consequently, adsorb to airborne particulates which can be inhaled and deposited in the lungs. The primary risk from radon exposure is lung cancer.

The primary source of radon in indoor air is soil, contributing up to 90% of all exposures. Masonry and other building products account for some of the remaining 10% and are not the primary source as is sometimes reported. The radon gas enters buildings through cracks and other breaches in foundations, floors, and other parts of the structure proximate to the soil. Negative pressures caused by temperature gradients between indoor and outdoor air, wind, and fluctuating barometric pressure draw the radon into buildings. If buildings are drafty, a chimney effect may act to distribute radon gas in upper floors as well.

Radon concentrations are measured in pCi/L. The EPA guideline for action to mitigate indoor radon exposure is 4 pCi/L. Test kits to measure radon concentrations in indoor air are easy to use, inexpensive (typically costing less than $25), and available from retail outlets and online.

Radon mitigation techniques are implemented with the purpose of reducing indoor air concentrations. Examples of mitigation techniques include the following.

- improving ventilation in basement and crawlspace areas by passive measures such as simply opening windows or installing vent openings, or by more active measures

- sealing cracks and other breaches and creating physical barriers to prevent entry

- sealing drafty windows and doors to reduce air currents that draw radon from lower to upper floors

- constructing subslab and crawlspace vents to create pressure barriers that prevent entry and exhaust the gas to the outside atmosphere

Formaldehyde

The sources of formaldehyde in indoor air are mostly associated with building materials, especially plywood, particle board, and fiberboard, which contain urea formaldehyde resins. When exposed to moisture from humid air or other sources, the urea formaldehyde hydrolyzes to release formaldehyde vapor to the indoor air. Other, far less significant sources are furnishings and fabrics such as carpets, draperies, and clothing.

Symptoms of low level exposure to formaldehyde include headaches, sinus congestion, depression, and other ailments. The National Institute for Occupational Safety and Health (NIOSH) has set a recommended action level for formaldehyde exposure at 0.1 ppm.

Asbestos

Asbestos has been used in many building materials such as floor tiles, textured ceilings, heating pipe insulation, and structural fire protection insulation. When any of these asbestos-containing materials (ACMs) are disturbed or deteriorate, asbestos fibers may become airborne and eventually find their way into the lungs. Consequently, mitigation measures may be directed at simply protecting the ACM from activities that would disturb it. However, public pressure may require removal of asbestos as the only acceptable mitigation measure.

Nonoccupational exposure to asbestos has been linked to lung cancer and *mesothelioma. Asbestosis* is typically associated with occupational exposure. One outcome of the potential health risks associated with exposure to asbestos in indoor air was the Asbestos Hazard Emergency Response Act (AHERA) of 1986, specifically passed by the United States Congress to address asbestos in schools.

2. MASS BALANCE

Indoor air pollution can be modeled using a *mass balance*. For the mass balance, it is assumed that pollutants can

- enter an occupied space from sources outside the space

- be emitted from sources within the space

- be degraded within the space

- escape from the space to the outside

Equation 58.1 shows the mass balance in its most basic form.

$$\sum \text{inputs} \pm \sum \text{internal changes} = \sum \text{outputs} \quad \textbf{58.1}$$

Inputs consist of pollutants entering from outside. Internal changes consist of pollutants from indoor sources. Outputs consist of pollutants degraded and pollutants escaping to the outside. This is shown by Eq. 58.2.

$$Q_o C_o + I = Q_e C_e + kVC_e \quad \textbf{58.2}$$

Equation 58.2 assumes steady state. Q_o and Q_e are, respectively, the rates of airflow entering and leaving the space, and C_o and C_e are the concentrations of pollutants in the incoming and outgoing air. I is the rate at which pollutants are emitted from sources within the space, and k is the rate at which the pollutants degrade. Values of k for some common indoor air pollutants are given in Table 58.2.

Table 58.2 *Pollutant Reaction Rate Values for Common Indoor Air Pollutants*

indoor air pollutant	$k\,(\text{h}^{-1})$
carbon monoxide (CO)	0
formaldehyde (HCHO)	0.35
nitrogen monoxide (NO)	0
nitrogen oxides (NOx)	0.12
particulate (diameter $< 0.5\ \mu\text{m}$)	0.50
radon	0.0075

Example 58.1

An occupied space has a volume of 300 m³. Airflows into and out of the space are equal at 0.8 m³/s. Assume 100% of the outside concentration enters the space with airflow from outside. The nitrogen dioxide concentration inside the space is 90 μg/m³, and the concentration outside is 29 μg/m³. What is the emission rate of nitrogen dioxide from sources inside the space?

Solution

From Table 58.2, the pollutant reaction rate, k, for nitrogen oxides is 0.12 h^{-1}. Use Eq. 58.2 to take a mass balance for the space, and solve for the emission rate from sources inside the space, I.

$$Q_o C_o + I = Q_e C_e + kVC_e$$
$$I = Q_e C_e + kVC_e - Q_o C_o$$

$$= \left(0.8\ \frac{\text{m}^3}{\text{s}}\right)\left(90\ \frac{\mu\text{g}}{\text{m}^3}\right) + \left(\frac{0.12\ \text{h}^{-1}}{3600\ \dfrac{\text{s}}{\text{h}}}\right)(300\ \text{m}^3)$$

$$\times \left(90\ \frac{\mu\text{g}}{\text{m}^3}\right) - \left(0.8\ \frac{\text{m}^3}{\text{s}}\right)\left(29\ \frac{\mu\text{g}}{\text{m}^3}\right)$$

$$= 49.7\ \mu\text{g/s}$$

3. RADON

Radon gas is a radioactive gas produced from the natural decay of radium within the rocks beneath a building. Radon accumulates in unventilated areas (e.g., basements), in stagnant water, and in air pockets formed when the ground settles beneath building slabs. Radon also can be brought into the home by radon-saturated well water used in baths and showers. The EPA's action level of 4 pCi/L for radon in air is contested by many as being too high.

Radon mitigation methods include (a) pressurizing to prevent the infiltration of radon, (b) installing depressurization systems to intercept radon below grade and vent it safely, (c) removing radon-producing soil, and (d) abandoning radon-producing sites.

4. NOMENCLATURE

C_e	pollutant concentration in air escaping to outside	μg/m³
C_o	pollutant concentration in air entering from outside	μg/m³
I	pollutants emitted from sources inside a space	μg/s
k	pollutant reaction rate	h^{-1}
Q_e	airflow escaping to outside	m³/s
Q_o	airflow entering from outside	m³/s
V	volume of space	m³

Environmental Health & Safety

Topic VI: Associated Engineering Principles

Assoc. Eng.
Principles

59

Statistics

1. SET THEORY

A *set* (usually designated by a capital letter) is a population or collection of individual items known as *elements* or *members*. The *null set*, $\varnothing$, is empty (i.e., contains no members). If A and B are two sets, A is a *subset* of B if every member in A is also in B. A is a *proper subset* of B if B consists of more than the elements in A. These relationships are denoted as follows.

$$A \subseteq B \quad \text{[subset]}$$

$$A \subset B \quad \text{[proper subset]}$$

The *universal set*, U, is one from which other sets draw their members. If A is a subset of U, then A' (also designated as A^{-1}, $\tilde{A}$, $-A$, and $\bar{A}$) is the *complement* of A and consists of all elements in U that are not in A. This is illustrated by the *Venn diagram* in Fig. 59.1(a).

The *union of two sets*, denoted by $A \cup B$ and shown in Fig. 59.1(b), is the set of all elements that are in either A or B or both. The *intersection of two sets*, denoted by $A \cap B$ and shown in Fig. 59.1(c), is the set of all elements that belong to both A and B. If $A \cap B = \varnothing$, A and B are said to be *disjoint sets*.

Figure 59.1 *Venn Diagrams*

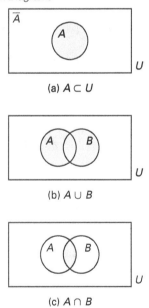

(a) $A \subset U$

(b) $A \cup B$

(c) $A \cap B$

If A, B, and C are subsets of the universal set, the following laws apply.

- *identity laws*

$$A \cup \varnothing = A \qquad 59.1$$

$$A \cup U = U \qquad 59.2$$

$$A \cap \varnothing = \varnothing \qquad 59.3$$

$$A \cap U = A \qquad 59.4$$

- *idempotent laws*

$$A \cup A = A \qquad 59.5$$

$$A \cap A = A \qquad 59.6$$

- *complement laws*

$$A \cup A' = U \qquad 59.7$$

$$(A')' = A \qquad 59.8$$

$$A \cap A' = \varnothing \qquad 59.9$$

$$U' = \varnothing \qquad 59.10$$

- *commutative laws*

$$A \cup B = B \cup A \qquad 59.11$$

$$A \cap B = B \cap A \qquad 59.12$$

- *associative laws*

$$(A \cup B) \cup C = A \cup (B \cup C) \qquad 59.13$$

$$(A \cap B) \cap C = A \cap (B \cap C) \qquad 59.14$$

- *distributive laws*

$$A \cup (B \cap C) = (A \cup B) \cap (A \cup C) \qquad 59.15$$

$$A \cap (B \cup C) = (A \cap B) \cup (A \cap C) \qquad 59.16$$

- *de Morgan's laws*

$$(A \cup B)' = A' \cap B' \qquad 59.17$$

$$(A \cap B)' = A' \cup B' \qquad 59.18$$

2. COMBINATIONS OF ELEMENTS

There are a finite number of ways in which n elements can be combined into distinctly different groups of r items. For example, suppose a farmer has a hen, a rooster, a duck, and a cage that holds only two birds. The possible *combinations* of three birds taken two at a time are (hen, rooster), (hen, duck), and (rooster, duck). The birds in the cage will not remain stationary, and the combination (rooster, hen) is not distinctly different from (hen, rooster). That is, the groups are not *order conscious*.

The number of combinations of n items taken r at a time is written $C(n, r)$, C_r^n, $_nC_r$, or $\binom{n}{r}$ (pronounced "n choose r") and given by Eq. 59.19. It is sometimes referred to as the *binomial coefficient*.

Permutations and Combinations

$$C(n, r) = \frac{P(n, r)}{r!} = \frac{n!}{[r!(n - r)!]} \qquad 59.19$$

Example 59.1

Six people are on a sinking yacht. There are four life jackets. How many combinations of survivors are there?

Solution

The groups are not order conscious. From Eq. 59.19,

Permutations and Combinations

$$C(n, r) = \frac{n!}{[r!(n - r)!]}$$

$$C(6, 4) = \frac{6!}{[4!(6 - 4)!]}$$

$$= \frac{6 \times 5 \times 4 \times 3 \times 2 \times 1}{(4 \times 3 \times 2 \times 1)(2 \times 1)}$$

$$= 15$$

3. PERMUTATIONS

An order-conscious subset of r items taken from a set of n items is the *permutation* $P(n, r)$, also written P_r^n and $_nP_r$. The permutation is order conscious because the arrangement of two items (say a_i and b_i) as a_ib_i is different from the arrangement b_ia_i. The number of permutations of r items from a set of n items is

Permutations and Combinations

$$P(n, r) = \frac{n!}{(n - r)!} \qquad 59.20$$

If groups of the entire set of n items are being enumerated, the number of permutations of n items taken n at a time is

$$P(n, n) = \frac{n!}{(n - n)!} = \frac{n!}{0!} = n! \qquad 59.21$$

If some items in the group of n items are identical, so that there are k distinct types of items, with n_1 items of type 1, n_2 items of type 2, and so on to n_k items of type k (with $n_1 + n_2 + ... + n_k$ totaling n), then the number of distinct permutations of n items taken n at a time is

Permutations and Combinations

$$P(n; \ n_1, \ n_2, \ \cdots, \ n_k) = \frac{n!}{n_1! n_2! \cdots n_k!}$$

A *ring permutation* is a special case of n items taken n at a time. There is no identifiable beginning or end, and the number of permutations is divided by n.

$$P_{\text{ring}}(n, n) = \frac{P(n, n)}{n} = (n - 1)! \qquad 59.22$$

Example 59.2

A pianist knows four pieces but will have enough stage time to play only three of them. Pieces played in a different order constitute a different program. How many different programs can be arranged?

Solution

The groups are order conscious. From Eq. 59.20,

$$P(4,3) = \frac{n!}{(n-r)!} = \frac{4!}{(4-3)!} = \frac{4 \times 3 \times 2 \times 1}{1} = 24$$

Example 59.3

Seven diplomats from different countries enter a circular room. The only furnishings are seven chairs arranged around a circular table. How many ways are there of arranging the diplomats?

Solution

All seven diplomats must be seated, so the groups are permutations of seven objects taken seven at a time. Since there is no head chair, the groups are ring permutations. From Eq. 59.22,

$$P_{\text{ring}}(7,7) = (7-1)! = 6 \times 5 \times 4 \times 3 \times 2 \times 1$$
$$= 720$$

4. PROBABILITY THEORY

The act of conducting an experiment (trial) or taking a measurement is known as *sampling*. *Probability theory* determines the relative likelihood that a particular event will occur. An *event*, e, is one of the possible outcomes of the *trial*. Taken together, all of the possible events constitute a finite *sample space*, $E = [e_1, e_2, \ldots, e_n]$. The trial is drawn from the *population* or *universe*. Populations can be finite or infinite in size.

Events can be numerical or nonnumerical, discrete or continuous, and dependent or independent. An example of a nonnumerical event is getting tails on a coin toss. The number from a roll of a die is a discrete numerical event. The measured diameter of a bolt produced from an automatic screw machine is a numerical event. Since the diameter can (within reasonable limits) take on any value, its measured value is a continuous numerical event.

An event is *independent* if its outcome is unaffected by previous outcomes (i.e., previous runs of the experiment) and *dependent* otherwise. Whether or not an event is independent depends on the population size and how the sampling is conducted. Sampling (a trial) from an infinite population is implicitly independent. When the population is finite, *sampling with replacement* produces independent events, while *sampling without replacement* changes the population and produces dependent events.

The terms *success* and *failure* are loosely used in probability theory to designate obtaining and not obtaining, respectively, the tested-for condition. "Failure" is not the same as a *null event* (i.e., one that has a zero probability of occurrence).

The *probability* of event e_1 occurring is designated as $P(e_1)$ and is calculated as the ratio of the total number of ways the event can occur to the total number of outcomes in the sample space.

Example 59.4

There are 380 students in a rural school—200 girls and 180 boys. One student is chosen at random and is checked for gender and height. (a) Define and categorize the population. (b) Define and categorize the sample space. (c) Define the trials. (d) Define and categorize the events. (e) In determining the probability that the student chosen is a boy, define success and failure. (f) What is the probability that the student is a boy?

Solution

(a) The population consists of 380 students and is finite.

(b) In determining the gender of the student, the sample space consists of the two outcomes $E = [\text{girl}, \text{boy}]$. This sample space is nonnumerical and discrete. In determining the height, the sample space consists of a range of values and is numerical and continuous.

(c) The trial is the actual sampling (i.e., the determination of gender and height).

(d) The events are the outcomes of the trials (i.e., the gender and height of the student). These events are independent if each student returns to the population prior to the random selection of the next student; otherwise, the events are dependent.

(e) The event is a success if the student is a boy and is a failure otherwise.

(f) From the definition of probability,

$$P(\text{boy}) = \frac{\text{no. of boys}}{\text{no. of students}} = \frac{180}{380} = 0.47$$

5. JOINT PROBABILITY

Joint probability rules specify the probability of a combination of events. If n mutually exclusive events from the set X have probabilities $P(X_i)$, the probability of any one of these events occurring in a given trial is the sum of the individual probabilities.

$$P(A + B + \cdots + X) = P(A) + P(B) + \cdots + P(X) \quad \text{59.23}$$

When given two independent sets of events, A and B, Eq. 59.24 will give the probability that events A and B will both occur.

$$P(A, B) = P(A)P(B) \qquad 59.24$$

When given two independent sets of events, A and B, Eq. 59.25 will give the probability that either event A or B will occur.

Laws of Probability: Property 2. Law of Total Probability
$$P(A + B) = P(A) + P(B) - P(A, B) \qquad 59.25$$

Example 59.5

A bowl contains five white balls, two red balls, and three green balls. What is the probability of getting either a white ball or a red ball in one draw from the bowl?

Solution

Since the two possible events are mutually exclusive and come from the same sample space, Eq. 59.23 can be used.

$$P(\text{white} + \text{red}) = P(\text{white}) + P(\text{red})$$
$$= \frac{5}{10} + \frac{2}{10}$$
$$= 7/10$$

Example 59.6

One bowl contains five white balls, two red balls, and three green balls. Another bowl contains three yellow balls and seven black balls. What is the probability of getting a red ball from the first bowl and a yellow ball from the second bowl in one draw from each bowl?

Solution

Equation 59.24 can be used because the two events are independent.

$$P(\text{red and yellow}) = P(\text{red})P(\text{yellow})$$
$$= \left(\frac{2}{10}\right)\left(\frac{3}{10}\right)$$
$$= 6/10$$

6. COMPLEMENTARY PROBABILITIES

The probability of an event occurring is equal to one minus the probability of the event not occurring. This is known as *complementary probability*.

$$P(e_i) = 1 - P(\text{not } e_i) \qquad 59.26$$

Equation 59.26 can be used to simplify some probability calculations. Specifically, calculation of the probability of numerical events being "greater than" or "less than" or quantities being "at least" a certain number can often be simplified by calculating the probability of the complementary event.

Example 59.7

A fair coin is tossed five times.[1] What is the probability of getting at least one tail?

Solution

The probability of getting at least one tail in five tosses could be calculated as

$$P(\text{at least 1 tail}) = P(1 \text{ tail}) + P(2 \text{ tails})$$
$$+ P(3 \text{ tails}) + P(4 \text{ tails})$$
$$+ P(5 \text{ tails})$$

However, it is easier to calculate the complementary probability of getting no tails (i.e., getting all heads).

From Eq. 59.25 and Eq. 59.26 (for calculating the probability of getting no tails in five successive tosses),

$$P(\text{at least 1 tail}) = 1 - P(0 \text{ tails})$$
$$= 1 - (0.5)^5$$
$$= 0.96875$$

7. CONDITIONAL PROBABILITY

Given two dependent sets of events, A and B, the probability that event A_j will occur given the fact that the dependent event B_j has already occurred is written as $P(A_j | B_j)$ and given by *Bayes' theorem*, Eq. 59.27.

Bayes' Theorem
$$P(B_j | A) = \frac{P(B_j)P(A|B_j)}{\sum_{i=1}^{n} P(A|B_i)P(B_i)} \qquad 59.27$$

---

[1] It makes no difference whether one coin is tossed five times or five coins are each tossed once.

8. PROBABILITY DENSITY FUNCTIONS

A *density function* is a nonnegative function whose integral taken over the entire range of the independent variable is unity. A *probability density function* is a mathematical formula that gives the probability of a discrete numerical event occurring. A *discrete numerical event* is an occurrence that can be described (usually) by an integer. For example, 27 cars passing through a bridge toll booth in an hour is a discrete numerical event. Fig. 59.2 shows a graph of a typical probability density function.

A probability density function, $f(x)$, gives the probability that discrete event x will occur. That is, $P(x) = f(x)$. Important discrete probability density functions are the binomial, hypergeometric, and Poisson distributions.

Figure 59.2 Probability Density Function

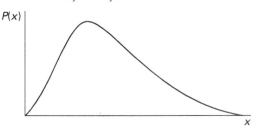

If a random event X may have any of the discrete values x_1, x_2, ..., x_n, then the probability of one of these values occurring is called a *discrete probability*. The *probability mass function* of the discrete random variable X is defined as

Probability Functions, Distributions, and Expected Values

$$f(x_k) = P(X = x_k), \quad k = 1, 2, ..., n \qquad 59.28$$

The *cumulative distribution function* of the discrete random variable X is defined as

Cumulative Distribution Functions

$$F(x_m) = \sum_{k=1}^{m} P(x_k) = P(X \leq x_m), \quad m = 1, 2, ..., n \quad 59.29$$

9. BINOMIAL DISTRIBUTION

The *binomial probability density function* (*binomial distribution*) is used when all outcomes can be categorized as either successes or failures. The probability of success in a single trial is p, and the probability of failure is the complement, $q = 1 - p$. The population is assumed to be infinite in size so that sampling does not change the values of p and q. (The binomial distribution can also be used with finite populations when sampling with replacement.)

Equation 59.30 gives the probability of x successes in n independent *successive trials*. The quantity $\binom{n}{x}$ is the *binomial coefficient*, identical to the number of combinations of n items taken x at a time.

Binomial Distribution

$$P_n(x) = C(n, x)p^x q^{n-x} = \frac{n!}{x!\,(n-x)!}p^x q^{n-x} \quad 59.30$$

$$\binom{n}{x} = \frac{n!}{(n-x)!\,x!} \qquad 59.31$$

Equation 59.30 is a discrete distribution, taking on values only for discrete integer values up to n. The mean, μ, and variance, σ^2, of this distribution are

$$\mu = np \qquad 59.32$$

Binomial Distribution

$$\sigma^2 = npq \qquad 59.33$$

Example 59.8

Five percent of a large batch of high-strength steel bolts purchased for bridge construction are defective. (a) If seven bolts are randomly sampled, what is the probability that exactly three will be defective? (b) What is the probability that two or more bolts will be defective?

Solution

(a) The bolts are either defective or not, so the binomial distribution can be used.

$$p = 0.05 \quad [\text{success} = \text{defective}]$$

$$q = 1 - 0.05 = 0.95 \quad [\text{failure} = \text{not defective}]$$

From Eq. 59.30,

Binomial Distribution

$$P_n(x) = C(n, x)p^x q^{n-x}$$

$$P_7(3) = \binom{7}{3}p^3 q^{7-3}$$

$$= \frac{7 \times 6 \times 5 \times 4 \times 3 \times 2 \times 1}{(3 \times 2 \times 1)(4 \times 3 \times 2 \times 1)}$$

$$\times (0.05)^3 (0.95)^4$$

$$= 0.00356$$

(b) The probability that two or more bolts will be defective could be calculated as

$$P(x \geq 2) = P(2) + P(3) + P(4) + P(5) + P(6) + P(7)$$

This method would require six probability calculations. It is easier to use the complement of the desired probability.

$$P_n(x) = C(n, x)p^x q^{n-x}$$

$$P_7(0) = \binom{7}{0}p^0 q^{7-0}$$

$$= \frac{7 \times 6 \times 5 \times 4 \times 3 \times 2 \times 1}{(1)(7 \times 6 \times 5 \times 4 \times 3 \times 2 \times 1)}$$

$$\times (0.05)^0 (0.95)^7$$

$$= 0.6983$$

$$P_7(1) = \binom{7}{1}p^1 q^{7-1}$$

$$= \frac{7 \times 6 \times 5 \times 4 \times 3 \times 2 \times 1}{(1)(6 \times 5 \times 4 \times 3 \times 2 \times 1)}$$

$$\times (0.05)^1 (0.95)^6$$

$$= 0.2573$$

$$P_7(x \geq 2) = 1 - P_7(x < 2)$$
$$= 1 - (P_7(0) + P_7(1))$$
$$1 - (0.6983 + 0.2573)$$
$$= 0.0444$$

10. HYPERGEOMETRIC DISTRIBUTION

Probabilities associated with sampling from a finite population without replacement are calculated from the *hypergeometric distribution*. If a population of finite size M contains K items with a given characteristic (e.g., red color, defective construction), then the probability of finding x items with that characteristic in a sample of n items is

$$P(x) = f(x) = \frac{\binom{K}{x}\binom{M-K}{n-x}}{\binom{M}{n}} \quad \text{[for } x \leq n\text{]} \quad 59.34$$

11. MULTIPLE HYPERGEOMETRIC DISTRIBUTION

Sampling without replacement from finite populations containing several different types of items is handled by the *multiple hypergeometric distribution*. If a population of finite size M contains K_i items of type i (such that

$\Sigma K_i = M$), the probability of finding x_1 items of type 1, x_2 items of type 2, and so on, in a sample size of n (such that $\Sigma x_i = n$) is

$$P(x_1, x_2, x_3, \ldots) = \frac{\binom{K_1}{x_1}\binom{K_2}{x_2}\binom{K_3}{x_3}\cdots}{\binom{M}{n}} \quad 59.35$$

12. POISSON DISTRIBUTION

Certain discrete events occur relatively infrequently but at a relatively regular rate. The probability of such an event occurring is given by the *Poisson distribution*. Suppose an event occurs, on the average, λ times per period. The probability that the event will occur x times per period is

$$P(x) = f(x) = \frac{e^{-\lambda}\lambda^x}{x!} \quad [\lambda > 0] \quad 59.36$$

λ is both the mean and the variance of the Poisson distribution.

$$\mu = \lambda \quad 59.37$$
$$\sigma^2 = \lambda \quad 59.38$$

Example 59.9

The number of customers arriving at a hamburger stand in the next period is a Poisson distribution having a mean of eight. What is the probability that exactly six customers will arrive in the next period?

Solution

$\lambda = 8$ and $x = 6$. From Eq. 59.36,

$$P(6) = \frac{e^{-\lambda}\lambda^x}{x!} = \frac{e^{-8}(8)^6}{6!} = 0.122$$

13. CONTINUOUS DISTRIBUTION FUNCTIONS

Most numerical events are *continuously distributed* and are not constrained to discrete or integer values. For example, the resistance of a 10% 1 Ω resistor may be any value between 0.9 Ω and 1.1 Ω. The probability of an exact numerical event is zero for continuously distributed variables. That is, there is no chance that a numerical event will be *exactly* x.[2] It is possible to

---

[2]It is important to understand the rationale behind this statement. Since the variable can take on any value and has an infinite number of significant digits, we can infinitely continue to increase the precision of the value. For example, the probability is zero that a resistance will be exactly 1 Ω because the resistance is really 1.03 or 1.0260008 or 1.02600080005, and so on.

determine only the probability that a numerical event will be less than x, greater than x, or between the values of x_1 and x_2, but not exactly equal to x.

Since an expression, $f(x)$, for a probability density function cannot always be written, it is more common to specify the *continuous distribution function*, $F(x_0)$, which gives the probability of numerical event x_0 or less occurring, as illustrated in Fig. 59.3.

$$P(X < x_0) = F(x_0) = \int_0^{x_0} f(x)\,dx \qquad 59.39$$

$$f(x) = \frac{dF(x)}{dx} \qquad 59.40$$

Figure 59.3 *Continuous Distribution Function*

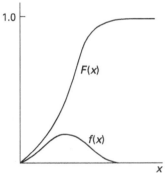

14. EXPONENTIAL DISTRIBUTION

The continuous *exponential distribution* is given by its probability density and continuous distribution functions.

$$f(x) = \lambda e^{-\lambda x} \qquad 59.41$$

$$P(X < x) = F(x) = 1 - e^{-\lambda x} \qquad 59.42$$

The mean and variance of the exponential distribution are

$$\mu = \frac{1}{\lambda} \qquad 59.43$$

$$\sigma^2 = \frac{1}{\lambda^2} \qquad 59.44$$

15. NORMAL DISTRIBUTION

The *normal distribution* (*Gaussian distribution*) is a symmetrical distribution commonly referred to as the *bell-shaped curve*, which represents the distribution of outcomes of many experiments, processes, and phenomena.

(See Fig. 59.4.) The probability density function for the normal distribution with mean μ and variance σ^2 is

Normal Distribution (Gaussian Distribution)

$$f(x) = \frac{1}{\sigma\sqrt{2\pi}}\, e^{-\frac{1}{2}\left(\frac{x-\mu}{\sigma}\right)^2} \qquad 59.45$$

A *standardized distribution*, or *unit normal distribution*, is a distribution for which the mean value, μ, is zero, and the standard deviation, σ, is one. For a standardized distribution, Eq. 59.45 reduces to

Normal Distribution (Gaussian Distribution)

$$f(x) = \frac{1}{\sqrt{2\pi}}\, e^{-x^2/2} \qquad 59.46$$

Figure 59.4 *Normal Distribution*

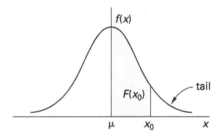

Since $f(x)$ is difficult to integrate, Eq. 59.45 is seldom used directly, and a *standard normal table* (also called a *unit normal table*) is used instead. The standard normal table is based on a standardized distribution with a mean of zero and a standard deviation of one. The range of values from an experiment or phenomenon will not generally correspond to the standard normal table, but any distribution can be *normalized*, or converted to a standardized distribution, by changing each value, x, to a *standard normal value*, Z, using Eq. 59.47. In Eq. 59.47, μ and σ are the mean and standard deviation, respectively, of the distribution from which x comes. For all practical purposes, all normal distributions are completely bounded by $\mu \pm 3\sigma$.

Normal Distribution (Gaussian Distribution)

$$Z = \frac{x - \mu}{\sigma} \qquad 59.47$$

Numbers in the standard normal table are the probabilities of the normalized x being between zero and Z and represent the areas under the curve up to point Z. When x is less than μ, Z will be negative. However, the curve is symmetrical, so the table value corresponding to positive Z can be used. The probability of x being greater than Z is the complement of the table value. The curve area past point Z is known as the *tail of the curve*.

Example 59.10

The mass, m, of a particular hand-laid fiberglass part is normally distributed with a mean of 66 kg and a standard deviation of 5 kg. (a) What percentage of the parts will have a mass less than 72 kg? (b) What percentage of the parts will have a mass in excess of 72 kg? (c) What percentage of the parts will have a mass between 61 kg and 72 kg?

Solution

(a) The 72 kg value must be normalized, so use Eq. 59.47. The standard normal variable is

Normal Distribution (Gaussian Distribution)

$$Z = \frac{x - \mu}{\sigma} = \frac{72\,\text{kg} - 66\,\text{kg}}{5\,\text{kg}} = 1.2$$

Reading from a unit normal table, the area under the normal curve to the left of $x = 1.2$ is 0.8849. This represents the probability of the mass being less than 72 kg (i.e., Z being less than 1.2).

$$P(m < 72\,\text{kg}) = P(Z < 1.2) = 0.5 + 0.3849 = 0.8849$$

(b) The probability of the mass exceeding 72 kg is the area under the tail past point z.

$$P(m > 72\,\text{kg}) = P(Z > 1.2) = 1 - 0.8849 = 0.1151$$

(c) The standard normal value corresponding to 61 kg is

Normal Distribution (Gaussian Distribution)

$$Z = \frac{x - \mu}{\sigma} = \frac{61\,\text{kg} - 66\,\text{kg}}{5\,\text{kg}} = -1$$

It is easiest to calculate the areas on the left and right sides of the mean separately and add them together. A normal distribution is symmetrical, so the area under the curve between $x = -1$ and $x = 0$ is the same as the area between $x = 0$ and $x = 1$. From the unit normal table, take the value for the area under the curve to the left of $x = 1$, and subtract the area to the left of the mean, or 0.5.

$$P(-1 \le x \le 0) = P(x \le 1) - P(x \le 0)$$
$$= 0.8413 - 0.5$$
$$= 0.3413$$

Similarly, the area under the curve between $x = 0$ and $x = 1.2$ is

$$P(0 \le x \le 1.2) = P(x \le 1.2) - P(x \le 0)$$
$$= 0.8849 - 0.5$$
$$= 0.3849$$

The total of the two areas is

$$P(-1 \le x \le 1.2) = P(-1 \le x \le 0)$$
$$+(0 \le x \le 1.2)$$
$$= 0.3413 + 0.3849$$
$$= 0.7262$$

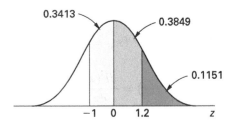

16. STUDENT'S *t*-DISTRIBUTION

In many cases requiring statistical analysis, including setting of confidence limits and hypothesis testing, the true population standard deviation, σ, and mean, μ, are not known. In such cases, the sample standard deviation, s, is used as an estimate for the population standard deviation, and the sample mean, $\bar{x}$, is used to estimate the population mean, μ. To account for the additional uncertainty of not knowing the population parameters exactly, a *t-distribution* is used rather than the normal distribution. The *t*-distribution essentially relaxes or expands the confidence intervals to account for these additional uncertainties.

The *t*-distribution is actually a family of distributions. They are similar in shape (symmetrical and bell-shaped) to the normal distribution, although wider, and flatter in the tails. *t*-distributions are more likely to result in (or, accept) values located farther from the mean than the normal distribution. The specific shape is dependent on the sample size, n. The smaller the sample size, the wider and flatter the distribution tails. The shape of the distribution approaches the standard normal curve as the sample size increases. Generally, the two distributions have the same shape for $n > 50$.

An important parameter needed to define a *t*-distribution is the *degrees of freedom*, ν. The degrees of freedom is usually 1 less than the sample size.

t-Distribution

$$\nu = n - 1 \qquad 59.48$$

Student's t-distribution is a standardized *t*-distribution, centered on zero, just like the standard normal variable, Z. With a normal distribution, the population standard deviation would be multiplied by $Z = 1.96$ to get a 95% confidence interval. When using the *t*-distribution, the sample standard deviation, s, is multiplied by a number, t, coming from the *t*-distribution.

t-Distribution

$$t = \frac{\bar{x} - \mu}{s/\sqrt{n}} \qquad 59.49$$

That number is designated as $t_{\alpha,\nu}$ or $t_{\alpha,n-1}$ for a one-tail test/confidence interval, and would be designated as $t_{\alpha/2,\nu}$ or $t_{\alpha/2,n-1}$ for a two-tail test/confidence interval. For example, for a two-tail confidence interval with confidence level $C = 1 - \alpha$, the upper and lower confidence limits are

Normal Distribution (Gaussian Distribution): Confidence Limits

$$UCL95 = \overline{x} + t_{n-1,0.05}(sd/\sqrt{n}) \qquad \textbf{59.50}$$

$$LCL95 = \overline{x} - t_{n-1,0.05}(sd/\sqrt{n})$$

17. CHI-SQUARED DISTRIBUTION

The *chi-squared* (*chi square*, χ^2) *distribution* is a distribution of the sum of squared standard normal deviates, Z_i. It has numerous useful applications. The distribution's only parameter, its *degrees of freedom*, ν, is the number of standard normal deviates being summed. Chi-squared distributions are positively skewed, but the skewness decreases and the distribution approaches a normal distribution as degrees of freedom increases. The mean and variance of a chi-squared distribution are related to its degrees of freedom.

$$\mu = \nu \qquad \textbf{59.51}$$

$$\sigma^2 = 2\nu \qquad \textbf{59.52}$$

After taking n samples from a standard normal distribution, the *chi-squared statistic* is defined as

χ^2-Distribution

$$\chi^2 = Z_1^2 + Z_2^2 + \ldots + Z_n^2 \qquad \textbf{59.53}$$

The sum of chi-squared variables is itself a chi-squared variable. For example, after taking three measurements from a standard normal distribution, squaring each term, and adding all three squared terms, the sum is a chi-squared variable. A chi-squared distribution with three degrees of freedom can be used to determine the probability of that summation exceeding another number. This characteristic makes the chi-squared distribution useful in determining whether or not a population's variance has shifted, or whether two populations have the same variance. The distribution is also extremely useful in categorical hypothesis testing.

18. LOG-NORMAL DISTRIBUTION

With a *log-normal* (*lognormal* or *Galton's*) *distribution*, the logarithm of the independent variable, $\ln(x)$ or $\log(x)$, is normally distributed, not x. Log-normal distributions are rare in engineering; they are more common in social, political, financial, biological, and environmental applications.[3] The log-normal distribution may be applicable whenever a normal-appearing distribution is skewed to the left, or when the independent variable is a function (e.g., product) of

multiple positive independent variables. Depending on its parameters (i.e., skewness), a log-normal distribution can take on different shapes, as shown in Fig. 59.5.

The symbols μ and σ are used to represent the *parameter values* of the transformed (logarithmitized) distribution; μ is the *location parameter* (*log mean*), the mean of the transformed values; and σ is the *scale parameter* (*log standard deviation*), whose unbiased estimator is the sample standard deviation, s, of the transformed values. These parameter values are used to calculate the standard normal variable, z, in order to determine event probabilities (i.e., areas under the normal curve). The *expected value*, $E(x)$, and standard deviation, σ_x, of the nontransformed distribution, x, are

$$E(x) = e^{\mu + \frac{1}{2}\sigma^2} \qquad \textbf{59.54}$$

$$\sigma_x = \sqrt{(e^{\sigma^2} - 1)e^{2\mu + \sigma^2}} \qquad \textbf{59.55}$$

$$Z = \frac{x_0 - \mu}{\sigma} \qquad \textbf{59.56}$$

Figure 59.5 *Log-Normal Probability Density Function*

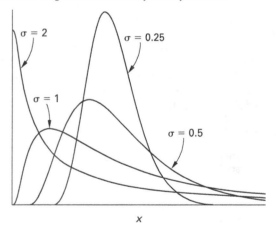

19. ERROR FUNCTION

The *error function*, erf(x), and its complement, the *complementary error function*, erfc(x), are defined by Eq. 59.57 and Eq. 59.58. The functions can be used to determine the probable error of a measurement, but they also appear in many engineering formulas. Values are seldom calculated from Eq. 59.57 or Eq. 59.58. Rather, approximation and tabulations are used.

$$\text{erf}(x_0) = \frac{2}{\sqrt{\pi}} \int_0^{x_0} e^{-x^2} dx \qquad \textbf{59.57}$$

$$\text{erfc}(x_0) = 1 - \text{erf}(x_0) \qquad \textbf{59.58}$$

[3]Log-normal is a popular reliability distribution. In electrical engineering, some transmission losses are modeled as log-normal. Some geologic mineral concentrations follow the log-normal pattern.

The error function can be used to calculate areas under the normal curve. Combining Eq. 59.46, Eq. 59.47, and Eq. 59.57,

$$\frac{1}{\sqrt{2\pi}}\int_0^z e^{-u^2/2}\,du = \tfrac{1}{2}\text{erf}\!\left(\frac{Z}{\sqrt{2}}\right) \qquad 59.59$$

The error function has the following properties.

$$\begin{aligned}
\text{erf}(0) &= 0 \\
\text{erf}(+\infty) &= 1 \\
\text{erf}(-\infty) &= -1 \\
\text{erf}(-x_0) &= -\text{erf}(x_0)
\end{aligned}$$

20. APPLICATION: RELIABILITY

Introduction

Reliability, $R(t)$, is the probability that an item will continue to operate satisfactorily up to time t. The *bathtub distribution*, Fig. 59.6, is often used to model the probability of failure of an item (or, the number of failures from a large population of items) as a function of time. Items initially fail at a high rate, a phenomenon known as *infant mortality*. For the majority of the operating time, known as the *steady-state operation*, the failure rate is constant (i.e., is due to random causes). After a long period of time, the items begin to deteriorate and the failure rate increases. (No mathematical distribution describes all three of these phases simultaneously.)

Figure 59.6 *Bathtub Reliability Curve*

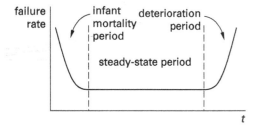

The *hazard function*, $Z(t)$, represents the *conditional probability of failure*—the probability of failure in the next time interval, given that no failure has occurred thus far.

$$Z(t) = \frac{f(t)}{R(t)} = \frac{\dfrac{dF(t)}{dt}}{1 - F(t)} \qquad 59.60$$

A *proof test* is a comprehensive validation of an item that checks 100% of all failure mechanisms. It tests for faults and degradation so that the item can be certified as being in "as new" condition. The *proof test interval* is the time after initial installation at which an item must be either proof tested or replaced. If the policy is to replace rather than test an item, the item's lifetime is equal to the proof test interval and is known as *mission time*.

Exponential Reliability

Steady-state reliability is often described by the *negative exponential distribution*. This assumption is appropriate whenever an item fails only by random causes and does not experience deterioration during its life. The parameter λ is related to the *mean time to failure* (MTTF) of the item.[4]

$$R(t) = e^{-\lambda t} = e^{-t/\text{MTTF}} \qquad 59.61$$

$$\lambda = \frac{1}{\text{MTTF}} \qquad 59.62$$

Equation 59.61 and the exponential continuous distribution function, Eq. 59.42, are complementary.

$$R(t) = 1 - F(t) = 1 - (1 - e^{-\lambda t}) = e^{-\lambda t} \qquad 59.63$$

The hazard function for the negative exponential distribution is

$$Z(t) = \lambda \qquad 59.64$$

Therefore, the hazard function for exponential reliability is constant and does not depend on t (i.e., on the age of the item). In other words, the expected future life of an item is independent of the previous history (length of operation). This lack of memory is consistent with the assumption that only random causes contribute to failure during steady-state operations. And since random causes are unlikely discrete events, their probability of occurrence can be represented by a Poisson distribution with mean λ. That is, the probability of having x failures in any given period is

$$P(x) = \frac{e^{-\lambda}\lambda^x}{x!} \qquad 59.65$$

Serial System Reliability

In the analysis of system reliability, the binary variable X_i is defined as 1 if item i operates satisfactorily and 0 if otherwise. Similarly, the binary variable Φ is 1 only if the entire system operates satisfactorily. Therefore, Φ will depend on a *performance function* containing the X_i.

A *serial system* is one for which all items must operate correctly for the system to operate. Each item has its own reliability, R_i. For a serial system of n items, the performance function is

$$\Phi = X_1 X_2 X_3 \cdots X_n = \min(X_i) \qquad 59.66$$

---

[4]The term "mean time *between* failures" is improper. However, the term *mean time before failure* (MTBF) is acceptable.

The probability of a serial system operating correctly is

$$P(\Phi = 1) = R_{\text{serial system}} = R_1 R_2 R_3 \cdots R_n \quad 59.67$$

Parallel System Reliability

A *parallel system* with n items will fail only if all n items fail. Such a system is said to be *redundant* to the nth degree. Using redundancy, a highly reliable system can be produced from components with relatively low individual reliabilities.

The performance function of a redundant system is

$$\begin{aligned}\Phi &= 1 - (1 - X_1)(1 - X_2)(1 - X_3) \cdots (1 - X_n) \\ &= \max(X_i)\end{aligned} \quad 59.68$$

The reliability of the parallel system is

$$\begin{aligned}R &= P(\Phi = 1) \\ &= 1 - (1 - R_1)(1 - R_2)(1 - R_3) \cdots (1 - R_n)\end{aligned} \quad 59.69$$

With a fully redundant, parallel k-out-of-n system of n independent, identical items, any k of which maintain system functionality, the ratio of redundant MTTF to single-item MTTF ($1/\lambda$) is given by Table 59.1.

Table 59.1 *MTTF Multipliers for k-out-of-n Systems*

k			n		
	1	2	3	4	5
1	1	3/2	11/6	25/12	137/60
2		1/2	5/6	13/12	77/60
3			1/3	7/12	47/60
4				1/4	9/20
5					1/5

Example 59.11

The reliability of an item is exponentially distributed with mean time to failure (MTTF) of 1000 hr. What is the probability that the item will not have failed before 1200 hr of operation?

Solution

The probability of not having failed before time t is the reliability. From Eq. 59.62 and Eq. 59.63,

$$\begin{aligned}\lambda &= \frac{1}{\text{MTTF}} = \frac{1}{1000 \text{ hr}} \\ &= 0.001 \text{ hr}^{-1}\end{aligned}$$

$$\begin{aligned}R(1200) &= e^{-\lambda t} = e^{(-0.001 \text{ hr}^{-1})(1200 \text{ hr})} \\ &= 0.3\end{aligned}$$

Example 59.12

What are the reliabilities of the following systems?

(a)

(b)

Solution

(a) This is a serial system. From Eq. 59.67,

$$\begin{aligned}R &= R_1 R_2 R_3 R_4 = (0.93)(0.98)(0.91)(0.87) \\ &= 0.72\end{aligned}$$

(b) This is a parallel system. From Eq. 59.69,

$$\begin{aligned}R &= 1 - (1 - R_1)(1 - R_2)(1 - R_3) \\ &= 1 - (1 - 0.76)(1 - 0.52)(1 - 0.39) \\ &= 0.93\end{aligned}$$

21. ANALYSIS OF EXPERIMENTAL DATA

Experiments can take on many forms. An experiment might consist of measuring the mass of one cubic foot of concrete or measuring the speed of a car on a roadway. Generally, such experiments are performed more than once to increase the precision and accuracy of the results.

Both systematic and random variations in the process being measured will cause the observations to vary, and the experiment would not be expected to yield the same result each time it was performed. Eventually, a collection of experimental outcomes (observations) will be available for analysis.

The *frequency distribution* is a systematic method for ordering the observations from small to large, according to some convenient numerical characteristic. The *step interval* should be chosen so that the data are presented in a meaningful manner. If there are too many intervals, many of them will have zero frequencies; if there are too few intervals, the frequency distribution will have little value. Generally, 10 to 15 intervals are used.

Once the frequency distribution is complete, it can be represented graphically as a *histogram*. The procedure in drawing a histogram is to mark off the interval limits (also known as *class limits*) on a number line and then draw contiguous bars with lengths that are proportional to the frequencies in the intervals and that are centered on the midpoints of their respective intervals. The continuous nature of the data can be depicted by a *frequency polygon*. The number or percentage of observations that occur, up to and including some value, can be shown in a *cumulative frequency table*.

Example 59.13

The number of cars that travel through an intersection between 12 noon and 1 p.m. is measured for 30 consecutive working days. The results of the 30 observations are

79, 66, 72, 70, 68, 66, 68, 76, 73, 71, 74, 70, 71, 69, 67, 74, 70, 68, 69, 64, 75, 70, 68, 69, 64, 69, 62, 63, 63, 61

(a) What are the frequency and cumulative distributions? (Use a distribution interval of two cars per hour.) (b) Draw the histogram. (Use a cell size of two cars per hour.) (c) Draw the frequency polygon. (d) Graph the cumulative frequency distribution.

Solution

(a) Tabulate the frequency, cumulative frequency, and cumulative percent distributions.

cars per hour	frequency	cumulative frequency	cumulative percent
60–61	1	1	3
62–63	3	4	13
64–65	2	6	20
66–67	3	9	30
68–69	8	17	57
70–71	6	23	77
72–73	2	25	83
74–75	3	28	93
76–77	1	29	97
78–79	1	30	100

(b) Draw the histogram.

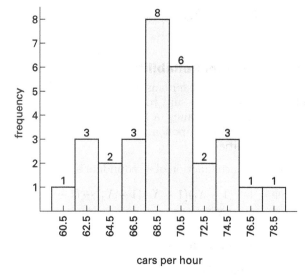

cars per hour

(c) Draw the frequency polygon.

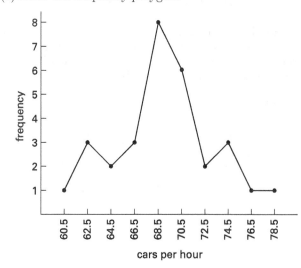

cars per hour

(d) Graph the cumulative frequency distribution.

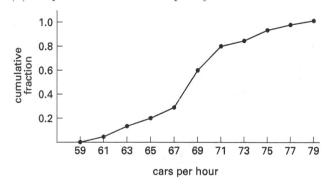

cars per hour

22. MEASURES OF EXPERIMENTAL ADEQUACY

An experiment is said to be *accurate* if it is unaffected by experimental error. In this case, *error* is not synonymous with *mistake*, but rather includes all variations not within the experimenter's control.

For example, suppose a gun is aimed at a point on a target and five shots are fired. The mean distance from the point of impact to the sight in point is a measure of the alignment accuracy between the barrel and sights. The difference between the actual value and the experimental value is known as *bias*.

Precision is not synonymous with accuracy. Precision is concerned with the repeatability of the experimental results. If an experiment is repeated with identical results, the experiment is said to be precise.

The average distance of each impact from the centroid of the impact group is a measure of the precision of the experiment. It is possible to have a highly precise experiment with a large bias.

Most of the techniques applied to experiments in order to improve the accuracy (i.e., reduce bias) of the experimental results (e.g., repeating the experiment, refining the experimental methods, or reducing variability) actually increase the precision.

Sometimes the word *reliability* is used with regard to the precision of an experiment. In this case, a "reliable estimate" is used in the same sense as a "precise estimate."

Stability and *insensitivity* are synonymous terms. A stable experiment will be insensitive to minor changes in the experimental parameters. For example, suppose the centroid of a bullet group is 2.1 in from the target point at 65°F and 2.3 in away at 80°F. The sensitivity of the experiment to temperature change would be

$$\text{sensitivity} = \frac{\Delta x}{\Delta T} = \frac{2.3 \text{ in} - 2.1 \text{ in}}{80°F - 65°F} = 0.0133 \text{ in/°F}$$

23. MEASURES OF CENTRAL TENDENCY

It is often unnecessary to present the experimental data in their entirety, either in tabular or graphical form. In such cases, the data and distribution can be represented by various parameters. One type of parameter is a measure of *central tendency*. Mode, median, and mean are measures of central tendency.

The *mode* is the observed value that occurs most frequently. The mode may vary greatly between series of observations. Therefore, its main use is as a quick measure of the central value since little or no computation is required to find it. Beyond this, the usefulness of the mode is limited.

The *median* is the point in the distribution that partitions the total set of observations into two parts containing equal numbers of observations. It is not influenced by the extremity of scores on either side of the distribution. The median is found by counting up (from either end of the frequency distribution) until half of the observations have been accounted for.

For even numbers of observations, the median is estimated as some value (i.e., the average) between the two center observations.

Similar in concept to the median are *percentiles* (*percentile ranks*), *quartiles*, and *deciles*. The median could also have been called the *50th percentile* observation. Similarly, the 80th percentile would be the observed value (e.g., the number of cars per hour) for which the cumulative frequency was 80%. The quartile and decile points on the distribution divide the observations or distribution into segments of 25% and 10%, respectively.

The *arithmetic mean* is the arithmetic average of the observations. The sample mean, $\bar{X}$, can be used as an *unbiased estimator* of the population mean, μ. The *mean* may be found without ordering the data (as was necessary to find the mode and median). The mean can be found from the following formula.

Dispersion, Mean, Median, and Mode Values

$$\bar{X} = (1/n)(X_1 + X_2 + \cdots + X_n) = (1/n)\sum_{i=1}^{n} X_i \quad \textbf{59.70}$$

The weighted arithmetic mean is defined as

Dispersion, Mean, Median, and Mode Values

$$\bar{X}_w = \frac{\sum w_i X_i}{\sum w_i} \quad \textbf{59.71}$$

The *geometric mean* is used occasionally when it is necessary to average ratios. The geometric mean is calculated as

$$\text{geometric mean} = \sqrt[n]{X_1 X_2 X_3 \cdots X_n} \quad [X_i > 0] \quad \textbf{59.72}$$

The *harmonic mean* is defined as

$$\text{harmonic mean} = \frac{n}{\dfrac{1}{X_1} + \dfrac{1}{X_2} + \cdots + \dfrac{1}{X_n}} \quad \textbf{59.73}$$

The *root-mean-squared* (*rms*) *value* of a series of observations is defined as

Dispersion, Mean, Median, and Mode Values

$$\text{sample root-mean-square value} = \sqrt{(1/n)\sum X_i^2} \quad \textbf{59.74}$$

Assoc. Eng. Principles

The *ratio of exceedance* of a single measurement, X_i, from the mean (or, any other value), μ, is

$$\text{ER} = \left| \frac{X_i - \mu}{\mu} \right| \qquad 59.75$$

The exceedance of a single measurement from the mean can be expressed in standard deviations. For a normal distribution, this exceedance is usually given the variable z and is referred to as the *standard normal value*, *variable*, *variate*, or *deviate*.

$$Z = \left| \frac{X_i - \mu}{\sigma} \right| \qquad 59.76$$

Example 59.14

Find the mode, median, and arithmetic mean of the distribution represented by the data given in Ex. 59.13.

Solution

First, resequence the observations in increasing order.

61, 62, 63, 63, 64, 64, 66, 66, 67, 68, 68, 68, 68, 69, 69, 69, 69, 70, 70, 70, 70, 71, 71, 72, 73, 74, 74, 75, 76, 79

The mode is the interval 68–69, since this interval has the highest frequency. If 68.5 is taken as the interval center, then 68.5 would be the mode.

The 15th and 16th observations are both 69, so the median is

$$\frac{69 + 69}{2} = 69$$

The mean can be found from the raw data or from the grouped data using the interval center as the assumed observation value. Using the raw data,

$$\overline{X} = (1/n) \sum_{i=1}^{n} X_i = \frac{2069}{30} = 68.97$$

24. MEASURES OF DISPERSION

The simplest statistical parameter that describes the variability in observed data is the *range*. The range is found by subtracting the smallest value from the largest. Since the range is influenced by extreme (low probability) observations, its use as a measure of variability is limited.

The *population standard deviation* is a better estimate of variability because it considers every observation. That is, in Eq. 59.77, N is the total population size, not the sample size, n.

Dispersion, Mean, Median, and Mode Values

$$\sigma_{\text{population}} = \sqrt{(1/N) \sum (X_i - \mu)^2} \qquad 59.77$$

The standard deviation of a sample (particularly a small sample) is a biased (i.e., not a good) estimator of the population standard deviation. An *unbiased estimator* of the population standard deviation is the *sample standard deviation*, s, also known as the *standard error* of a sample.[5]

Dispersion, Mean, Median, and Mode Values

$$s = \sqrt{1/(n-1) \sum_{i=1}^{n} (X_i - \overline{X})^2} \qquad 59.78$$

If the sample standard deviation, s, is known, the standard deviation of the sample, σ_{sample}, can be calculated.

$$\sigma_{\text{sample}} = s \sqrt{\frac{n-1}{n}} \qquad 59.79$$

The *variance* is the square of the standard deviation. Since there are two standard deviations, there are two variances. The *variance of the sample* is σ^2, and the *sample variance* is s^2.

Dispersion, Mean, Median, and Mode Values

$$s^2 = [1/(n-1)] \sum_{i=1}^{n} (X_i - \overline{X})^2 \qquad 59.80$$

The *relative dispersion* is defined as a measure of dispersion divided by a measure of central tendency. The *coefficient of variation*, *CV*, is a relative dispersion calculated from the sample standard deviation and the mean.

Dispersion, Mean, Median, and Mode Values

$$CV = s/\overline{X} \qquad 59.81$$

Example 59.15

The rainfall totals (in inches) last month for six sampled locations in a state were 4, 6, 7, 9, 9, and 13. Calculate (a) the sample range, (b) the sample standard deviation and (c) the sample variance.

Solution

(a) The sample range is $15 - 4 = 11$.

---

[5]There is a subtle yet significant difference between *standard deviation of the sample,* σ (obtained from Eq. 59.77 for a finite sample drawn from a larger population), and the *sample standard deviation,* s (obtained from Eq. 59.78). While σ can be calculated, it has no significance or use as an estimator. It is true that the difference between σ and s approaches zero when the sample size, n, is large, but this convergence does nothing to legitimize the use of σ as an estimator of the true standard deviation. (Some people say "large" is 30, others say 50 or 100.)

(b) Use Eq. 59.70 to find the mean of the six values.

Dispersion, Mean, Median, and Mode Values

$$\overline{X} = (1 / n)(X_1 + X_2 + \cdots + X_n)$$

$$= \left(\frac{1}{6}\right)(4 + 6 + 7 + 9 + 9 + 13)$$

$$= 8$$

Square the deviation of each data point from the mean, then total the results.

$$\left(X_1 - \overline{X}\right)^2 = (4 - 8)^2 = 16$$

$$\left(X_2 - \overline{X}\right)^2 = (6 - 8)^2 = 4$$

$$\left(X_3 - \overline{X}\right)^2 = (7 - 8)^2 = 1$$

$$\left(X_4 - \overline{X}\right)^2 = (9 - 8)^2 = 1$$

$$\left(X_5 - \overline{X}\right)^2 = (9 - 8)^2 = 1$$

$$\left(X_6 - \overline{X}\right)^2 = (13 - 8)^2 = 25$$

$$\sum_{i=1}^{n}\left(X_i - \overline{X}\right)^2 = 16 + 4 + 1 + 1 + 1 + 25 = 48$$

Use Eq. 59.78 to calculate the sample standard deviation.

Dispersion, Mean, Median, and Mode Values

$$s = \sqrt{[1 / (n - 1)]\sum_{i=1}^{n}\left(X_i - \overline{X}\right)^2}$$

$$= \sqrt{\left(\frac{1}{6 - 1}\right)(48)}$$

$$= 3.098$$

(c) Use Eq. 59.80 to calculate the sample variance.

Dispersion, Mean, Median, and Mode Values

$$s^2 = [1 / (n - 1)]\sum_{i=1}^{n}\left(X_i - \overline{X}\right)^2$$

$$= \left(\frac{1}{6 - 1}\right)(48)$$

$$= 9.6$$

25. SKEWNESS

Skewness is a measure of a distribution's lack of symmetry. Distributions that are pushed to the left have negative skewness, while distributions pushed to the right have positive skewness. Various formulas are used to calculate skewness. *Pearson's skewness*, sk, is a simple normalized difference between the mean and mode. (See Eq. 59.82.) Since the mode is poorly represented when sampling from a distribution, the difference is estimated

as three times the deviation from the mean. *Fisher's skewness*, γ_1, and more modern methods are based on the *third moment about the mean*. (See Eq. 59.83.) Normal distributions have zero skewness, although zero skewness is not a sufficient requirement for determining normally distributed data. Square root, log, and reciprocal transformations of a variable can reduce skewness.

$$\text{sk} = \frac{\mu - \text{mode}}{\sigma} \approx \frac{3(\overline{X} - \text{median})}{s} \qquad 59.82$$

$$\gamma_1 = \frac{\displaystyle\sum_{i=1}^{N}\left(X_i - \mu\right)^3}{N\sigma^3} \approx \frac{n\displaystyle\sum_{i=1}^{n}\left(X_i - \overline{X}\right)^3}{(n - 1)(n - 2)s^3} \qquad 59.83$$

26. KURTOSIS

While skewness refers to the symmetry of the distribution about the mean, *kurtosis* refers to the contribution of the tails to the distribution, or alternatively, how flat the peak is. A *mesokurtic distribution* ($\beta_2 = 3$) is statistically normal. Compared to a normal curve, a *leptokurtic distribution* ($\beta_2 < 3$) is "fat in the tails," longer-tailed, and more sharp-peaked, while a *platykurtic distribution* ($\beta_2 > 3$) is "thin in the tails," shorter-tailed, and more flat-peaked. Different methods are used to calculate kurtosis. The most common, *Fisher's kurtosis*, β_2, is defined as the *fourth standardized moment (fourth moment of the mean)*.

$$\beta_2 = \frac{\displaystyle\sum_{i=1}^{N}(X_i - \mu)^4}{N\sigma^4} \qquad 59.84$$

Some statistical analyses, including those in Microsoft Excel, calculate a related statistic, *excess kurtosis (kurtosis excess or Pearson's kurtosis)*, by subtracting 3 from the kurtosis. A normal distribution has zero excess kurtosis; a peaked distribution has positive excess kurtosis; and a flat distribution has negative excess kurtosis.

$$\gamma_2 = \beta_2 - 3$$

$$\approx \frac{n(n + 1)\displaystyle\sum_{i=1}^{n}(X_i - \overline{X})^4}{(n - 1)(n - 2)(n - 3)s^4} - \frac{3(n - 1)^2}{(n - 2)(n - 3)} \qquad 59.85$$

27. CENTRAL LIMIT THEOREM

Measuring a sample of n items from a population with mean μ and standard deviation σ is the general concept of an experiment. The sample mean, $\overline{X}$, is one of the parameters that can be derived from the experiment. This experiment can be repeated k times, yielding a set of averages $(\overline{X}_1, \overline{X}_2, \ldots, \overline{X}_k)$. The k numbers in the set themselves represent samples from distributions of averages. The average of averages, $\mu_{\overline{y}}$, and sample standard deviation of averages, $s_{\overline{x}}$ (known as the *standard error of the mean*), can be calculated.

The *central limit theorem* characterizes the distribution of the sample averages. The theorem can be stated in several ways, but the essential elements are the following points.

1. The averages, $\overline{X}_i$, are normally distributed variables, even if the original data from which they are calculated are not normally distributed.

2. The grand average, $\mu_{\overline{y}}$ (i.e., the average of the averages), approaches and is an unbiased estimator of μ.

Normal Distribution (Gaussian Distribution): The Central Limit Theorem

$$\mu_{\overline{y}} = \mu \qquad \textit{59.86}$$

The standard deviation of the original distribution, σ, is much larger than the standard error of the mean.

Normal Distribution (Gaussian Distribution): The Central Limit Theorem

$$\sigma_{\overline{y}} = \frac{\sigma}{\sqrt{n}} \qquad \textit{59.87}$$

28. CONFIDENCE LEVEL

The results of experiments are seldom correct 100% of the time. Recognizing this, researchers accept a certain probability of being wrong. In order to minimize this probability, experiments are repeated several times. The number of repetitions depends on the desired level of confidence in the results.

If the results have a 5% probability of being wrong, the *confidence level*, C, is 95% that the results are correct, in which case the results are said to be *significant*. If the results have only a 1% probability of being wrong, the confidence level is 99%, and the results are said to be *highly significant*. Other confidence levels (90%, 99.5%, etc.) are used as appropriate.

The complement of the confidence level is α, referred to as *alpha*. α is the *significance level* and may be given as a decimal value or percentage. Alpha is also known as *alpha risk* and *producer risk*, as well as the probability of a type I error. A *type I error*, also known as a *false positive error*, occurs when the null hypothesis is incorrectly rejected, and an action occurs that is not actually required. For a random sample of manufactured products, the null hypothesis would be that the distribution of sample measurements is not different from the historical distribution of those measurements. If the null hypothesis is rejected, all of the products (not just the

sample) will be rejected, and the producer will have to absorb the expense. It is not uncommon to use 5% as the producer risk in noncritical business processes.

$$\alpha = 100\% - C \qquad \textit{59.88}$$

β (*beta* or *beta risk*) is the *consumer risk*, the probability of a type II error. A *type II error* occurs when the null hypothesis (e.g., "everything is fine") is incorrectly accepted, and no action occurs when action is actually required. If a batch of defective products is accepted, the products are distributed to consumers who then suffer the consequences. Generally, smaller values of α coincide with larger values of β because requiring overwhelming evidence to reject the null increases the chances of a type II error. β can be minimized while holding α constant by increasing sample sizes. The *power of the test* is the probability of rejecting the null hypothesis when it is false.

$$\text{power of the test} = 1 - \beta \qquad \textit{59.89}$$

29. NULL AND ALTERNATIVE HYPOTHESES

All statistical conclusions involve constructing two mutually exclusive hypotheses, termed the *null hypothesis* (written as H_0) and *alternative hypothesis* (written as H_1). Together, the hypotheses describe all possible outcomes of a statistical analysis. The purpose of the analysis is to determine which hypothesis to accept and which to reject.

Usually, when an improvement is made to a program, treatment, or process, the change is expected to make a difference. The null hypothesis is so named because it refers to a case of "no difference" or "no effect." Typical null hypotheses are:

H_0: There has been no change in the process.
H_0: The two distributions are the same.
H_0: The change has had no effect.
H_0: The process is in control and has not changed.
H_0: Everything is fine.

The alternative hypothesis is that there has been an effect, and there is a difference. The null and alternative hypotheses are mutually exclusive.

30. APPLICATION: CONFIDENCE LIMITS

As a consequence of the central limit theorem, sample means of n items taken from a normal distribution with mean μ and standard deviation σ will be normally distributed with mean μ and variance σ^2/n. The probability that any given average, $\overline{X}$, exceeds some value, L, is given by Eq. 59.90.

$$P(\overline{X} > L) = P\left(Z > \left|\frac{L-\mu}{\frac{\sigma}{\sqrt{n}}}\right|\right) \qquad 59.90$$

L is the *confidence limit* for the confidence level $1 - P(\overline{X} > L)$ (normally expressed as a percent). Values of Z are read directly from the standard normal table. As an example, $Z = 1.645$ for a 95% confidence level since only 5% of the curve is above that value of Z in the upper tail. This is known as a *one-tail confidence limit* because all of the probability is given to one side of the variation. Similar values are given in Table 59.2.

Table 59.2 *Values of Z for Various Confidence Levels*

confidence level, C	one-tail limit, Z	two-tail limit, Z
90%	1.28	1.645
95%	1.645	1.96
97.5%	1.96	2.17
99%	2.33	2.575
99.5%	2.575	2.81
99.75%	2.81	3.00

With *two-tail confidence limits*, the probability is split between the two sides of variation. There will be upper and lower confidence limits, UCL and LCL, respectively.

$$P(\text{LCL} < \overline{X} < \text{UCL}) = P\left(\left|\frac{\text{LCL}-\mu}{\frac{\sigma}{\sqrt{n}}}\right| < Z < \left|\frac{\text{UCL}-\mu}{\frac{\sigma}{\sqrt{n}}}\right|\right) \qquad 59.91$$

31. APPLICATION: BASIC HYPOTHESIS TESTING

A *hypothesis test* is a procedure that answers the question, "Did these data come from [a particular type of] distribution?" There are many types of tests, depending on the distribution and parameter being evaluated. The simplest hypothesis test determines whether an average value obtained from n repetitions of an experiment could have come from a population with known mean, μ, and standard deviation, σ. A practical application of this question is whether a manufacturing process has changed from what it used to be or should be. Of course,

the answer (i.e., "yes" or "no") cannot be given with absolute certainty—there will be a confidence level associated with the answer.

The following procedure is used to determine whether the average of n measurements can be assumed (with a given confidence level) to have come from a known population.

step 1: Assume random sampling from a normal population.

step 2: Choose the desired confidence level, C.

step 3: Decide on a one-tail or two-tail test. If the hypothesis being tested is that the average has or has not *increased* or *decreased*, choose a one-tail test. If the hypothesis being tested is that the average has or has not *changed*, choose a two-tail test.

step 4: Use Table 59.2 or the standard normal table to determine the Z-value corresponding to the confidence level and number of tails.

step 5: Calculate the actual standard normal variable, Z'.

$$Z' = \left|\frac{\overline{x}-\mu}{\frac{\sigma}{\sqrt{n}}}\right| \qquad 59.92$$

step 6: If $Z' \geq Z$, the average can be assumed (with confidence level C) to have come from a different distribution.

Example 59.16

When it is operating properly, a cement plant has a daily production rate that is normally distributed with a mean of 880 tons/day and a standard deviation of 21 tons/day. During an analysis period, the output is measured on 50 consecutive days, and the mean output is found to be 871 tons/day. With a 95% confidence level, determine whether the plant is operating properly.

Solution

step 1: The production rate samples are known to be normally distributed.

step 2: $C = 95\%$ is given.

step 3: Since a specific direction in the variation is not given (i.e., the example does not ask whether the average has decreased), use a two-tail hypothesis test.

step 4: The population mean and standard deviation are known. The standard normal distribution may be used. From Table 59.2, $Z = 1.96$.

step 5: From Eq. 59.92,

$$Z' = \left| \frac{\bar{x} - \mu}{\frac{\sigma}{\sqrt{n}}} \right| = \left| \frac{871 - 880}{\frac{21}{\sqrt{50}}} \right| = 3.03$$

Since $3.03 > 1.96$, the distributions are not the same. There is at least a 95% probability that the plant is not operating correctly.

32. APPLICATION: STATISTICAL PROCESS CONTROL

All manufacturing processes contain variation due to random and nonrandom causes. Random variation cannot be eliminated. *Statistical process control* (SPC) is the act of monitoring and adjusting the performance of a process to detect and eliminate nonrandom variation.

Statistical process control is based on taking regular (hourly, daily, etc.) samples of n items and calculating the mean, $\bar{X}$, and range, R, of the sample. To simplify the calculations, the range is used as a measure of the dispersion. These two parameters are graphed on their respective X-bar and R *control charts*, as shown in Fig. 59.7.[6] Confidence limits are drawn at $\pm 3\sigma / \sqrt{n}$. From a statistical standpoint, the control chart tests a hypothesis each time a point is plotted. When a point falls outside these limits, there is a 99.75% probability that the process is out of control. Until a point exceeds the control limits, no action is taken.[7]

Figure 59.7 *Typical Statistical Process Control Charts*

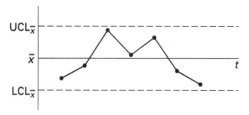

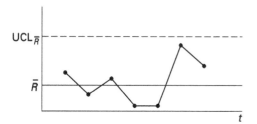

33. LINEAR REGRESSION

If it is necessary to draw a straight line ($y = mx + b$) through n data points $(x_1, y_1), (x_2, y_2), \ldots, (x_n, y_n)$, the following method based on the *method of least squares* can be used.

step 1: Calculate the following nine quantities.

$$\sum x_i \quad \sum x_i^2 \quad \left(\sum x_i\right)^2 \quad \bar{x} = \frac{\sum x_i}{n} \quad \sum x_i y_i$$

$$\sum y_i \quad \sum y_i^2 \quad \left(\sum y_i\right)^2 \quad \bar{y} = \frac{\sum y_i}{n}$$

step 2: Calculate the slope, m, of the line.

$$m = \frac{n \sum x_i y_i - \sum x_i \sum y_i}{n \sum x_i^2 - \left(\sum x_i\right)^2} \qquad 59.93$$

step 3: Calculate the y-intercept, b.

$$b = \bar{y} - m\bar{x} \qquad 59.94$$

step 4: To determine the goodness of fit, calculate the *correlation coefficient*, r.

$$r = \frac{n \sum x_i y_i - \sum x_i \sum y_i}{\sqrt{\left(n \sum x_i^2 - \left(\sum x_i\right)^2\right) \times \left(n \sum y_i^2 - \left(\sum y_i\right)^2\right)}} \qquad 59.95$$

If m is positive, r will be positive; if m is negative, r will be negative. As a general rule, if the absolute value of r exceeds 0.85, the fit is good; otherwise, the fit is poor. r equals 1.0 if the fit is a perfect straight line.

A low value of r does not eliminate the possibility of a nonlinear relationship existing between x and y. It is possible that the data describe a parabolic, logarithmic, or other nonlinear relationship. (Usually this will be apparent if the data are graphed.) It may be necessary to convert one or both variables to new variables by taking squares, square roots, cubes, or logarithms, to name a few of the possibilities, in order to obtain a linear relationship. The apparent shape of the line through the data will give a clue to the type of variable transformation that is required. The curves in Fig. 59.8 may be used as guides to some of the simpler variable transformations.

---

[6]Other charts (e.g., the *sigma chart, p-chart,* and *c-chart*) are less common but are used as required.
[7]Other indications that a correction may be required are seven measurements on one side of the average and seven consecutively increasing measurements. Rules such as these detect shifts and trends.

Figure 59.8 *Nonlinear Data Curves*

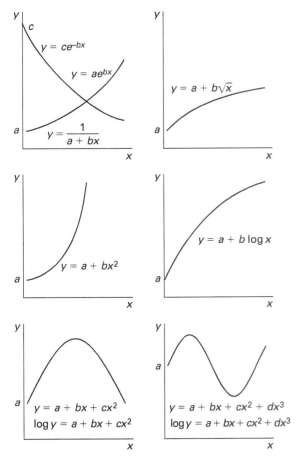

Figure 59.9 *Common Regression Difficulties*

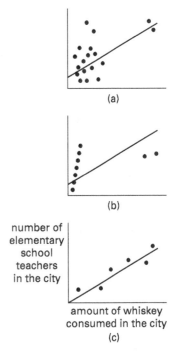

Figure 59.9 illustrates several common problems encountered in trying to fit and evaluate curves from experimental data. Fig. 59.9(a) shows a graph of clustered data with several extreme points. There will be moderate correlation due to the weighting of the extreme points, although there is little actual correlation at low values of the variables. The extreme data should be excluded, or the range should be extended by obtaining more data.

Figure 59.9(b) shows that good correlation exists in general, but extreme points are missed, and the overall correlation is moderate. If the results within the small linear range can be used, the extreme points should be excluded. Otherwise, additional data points are needed, and curvilinear relationships should be investigated.

Figure 59.9(c) illustrates the problem of drawing conclusions of cause and effect. There may be a predictable relationship between variables, but that does not imply a cause and effect relationship. In the case shown, both variables are functions of a third variable, the city population. But there is no direct relationship between the plotted variables.

Example 59.17

An experiment is performed in which the dependent variable, y, is measured against the independent variable, x. The results are as follows.

x	y
1.2	0.602
4.7	5.107
8.3	6.984
20.9	10.031

(a) What is the least squares straight line equation that best represents this data? (b) What is the correlation coefficient?

Solution

(a) Calculate the following quantities.

$$\sum x_i = 35.1$$
$$\sum y_i = 22.72$$
$$\sum x_i^2 = 529.23$$
$$\sum y_i^2 = 175.84$$
$$\left(\sum x_i\right)^2 = 1232.01$$
$$\left(\sum y_i\right)^2 = 516.38$$
$$\bar{x} = 8.775$$
$$\bar{y} = 5.681$$
$$\sum x_i y_i = 292.34$$
$$n = 4$$

From Eq. 59.93, the slope is

$$m = \frac{n\sum x_i y_i - \sum x_i \sum y_i}{n\sum x_i^2 - \left(\sum x_i\right)^2} = \frac{(4)(292.34) - (35.1)(22.72)}{(4)(529.23) - (35.1)^2}$$
$$= 0.42$$

From Eq. 59.94, the y-intercept is

$$b = \bar{y} - m\bar{x} = 5.681 - (0.42)(8.775)$$
$$= 2.0$$

The equation of the line is

$$y = 0.42x + 2.0$$

(b) From Eq. 59.95, the correlation coefficient is

$$r = \frac{n\sum x_i y_i - \sum x_i \sum y_i}{\sqrt{\left(n\sum x_i^2 - \left(\sum x_i\right)^2\right)\left(n\sum y_i^2 - \left(\sum y_i\right)^2\right)}}$$
$$= \frac{(4)(292.34) - (35.1)(22.72)}{\sqrt{\begin{array}{l}\left((4)(529.23) - 1232.01\right)\\ \quad \times \left((4)(175.84) - 516.38\right)\end{array}}}$$
$$= 0.914$$

Example 59.18

Repeat Ex. 59.17 assuming the relationship between the variables is nonlinear.

Solution

The first step is to graph the data. Since the graph has the appearance of the fourth case in Fig. 59.8, it can be assumed that the relationship between the variables has the form of $y = a + b\log x$. Therefore, the variable change $z = \log x$ is made, resulting in the following set of data.

z	y
0.0792	0.602
0.672	5.107
0.919	6.984
1.32	10.031

If the regression analysis is performed on this set of data, the resulting equation and correlation coefficient are

$$y = 7.599z + 0.000247$$

$$r = 0.999$$

This is a very good fit. The relationship between the variable x and y is approximately

$$y = 7.599\log x + 0.000247$$

34. NOMENCLATURE

b	y-intercept	–
C	confidence level	%
CV	coefficient of variation	–
ER	ratio of exceedance	–
L	confidence limit	–
$LCL95$	95% lower confidence limit	–
m	mass	kg
m	slope	–
MTTF	mean time to failure	sec
n	number of items in a set or number of trials	–
N	population size	–
p	probability of success	–
q	probability of failure	–
r	correlation coefficient	–
r	number of items taken from a set	–
R	reliability	–
s	sample standard derivation	–
sd	standard deviation	–
sk	Pearson's skewness	–
t	critical value of a t-distribution	–
t	time	sec
T	temperature	°F
$UCL95$	95% upper confidence limit	–
w	weight (in a weighted mean)	–
x	distance	in
x	number of successes in a series of n trials	–
$\bar{x}, \bar{X}$	sample mean	–
Z	standard normal value	–
Z'	actual standard normal value	–

Symbols

α	significance level	%
β	consumer risk	–
β_2	Fisher's kurtosis	–
γ_1	Fisher's skewness	–
γ_2	excess kurtosis	–
λ	expected number of occurrences	–
μ	population mean	–
ν	number of degrees of freedom	–
Φ	binary variable	–
σ	population standard deviation	–
σ^2	population variance	–

Assoc. Eng. Principles

60 Sustainability

1. INTRODUCTION

Sustainability is the underlying objective of nearly all environmental engineering activities. Wastewater is treated to recover the water for use downstream and to remove pollutants that would otherwise enter the environment. Wastewater sludge is processed for methane and carbon dioxide recovery and stabilized for use as a soil amendment. Air quality is maintained by air pollution control processes and devices on stationary and mobile sources and is a motivating force behind alternative energy development and application. Solid waste has been a focus for source reduction and materials recovery for several decades, with most municipalities, institutions, and commercial and industrial enterprises in the United States operating materials reduction, reuse, and recycling programs.

2. SUSTAINABLE WATER RESOURCES MANAGEMENT

Water presents the best opportunity for environmental engineers to engage in sustainable resource management. When not confronted with flood or drought, few people notice the day-to-day use and reuse of water in the water-wastewater treatment sequence, yet the water-wastewater treatment sequence for a surface water source represents a sustainable water cycle of use and reuse. Water is withdrawn from a river for treatment to potable water quality standards and then distributed to homes and business. These users consume some portion of the water, primarily through landscape irrigation, and discharge the remainder to sewers. The sewers convey the collected wastewater to a publicly owned treatment works (POTW), where it is treated to remove pollutants that are harmful to the environment or create a risk to public health. The water is then discharged back to the river, where the next community downstream withdraws it for their uses and begins the sequence again. All major and most minor waterways in the United States experience this sequence of withdrawal and discharge hundreds of times as the water flows to the oceans.

One area of concern, however, is overuse of a limited water supply. A water supply can be depleted by extracting water from an aquifer at rates higher than recharge. Aquifer depletion is an issue in the Southwest and southern Midwest. Water loss to evaporation and diversion to agriculture also depletes the resource. This is particularly evident for the Colorado River and its drainage.

The water cycle includes surface water (freshwater lakes, rivers, and reservoirs and saltwater oceans), groundwater, and atmospheric water. Water is also stored in plants and is transpired through plants from the ground to the atmosphere. However, although the overall water supply is stable, the uneven distributions of population and water supply create problems of drought and floods. The water supply is not stable in locations like metropolitan Phoenix, Arizona, and other southwestern and southern Midwest cities. In many of these places, available surface water is used, and water stored in aquifers is being consumed more quickly than it can be replenished.

This water use is unsustainable. Eventually, measures such as no-water landscaping; restricted agriculture irrigation; adoption of low volume irrigation methods; reduced flow fixtures in homes, hotels, institutions, businesses, and industry; restrictions on vehicle washing and other aesthetic uses; limits on high water use in mining and industrial activity; and a host of other measures will be required. Population and business growth may also become restricted as demand for water exceeds the available resource.

The problem is somewhat reversed in the Pacific Northwest, Great Lakes region, New England, and much of the rest of the country outside of the Southwest and southern Midwest. Water supply in these locations is more than adequate to meet demand, but excess rainfall or sudden melting of snowpack creates problems of flooding. Much of water resources engineering is involved with flood mitigation.

3. WATER LAW

Water law reflects the differences of water availability across the United States. In most western states where water is relatively limited, prior appropriation rights are common. A *prior appropriation right* gives precedence to whomever was the first to put the water to beneficial use. These rights can be bought and sold, or they may convey when property is bought and sold. In states with more generous rainfall, such as eastern states, the most common form of water law is riparian. *Riparian rights* allow anyone whose property includes or abuts a river, lake, reservoir, or other waterway to use its water. Riparian rights convey with the land.

61

Economics: Cash Flow, Interest, Decision-Making

1. INTRODUCTION

In its simplest form, an *engineering economic analysis* is a study of the desirability of making an investment.[1] The decision-making principles in this chapter can be applied by individuals as well as by companies. The nature of the spending opportunity or industry is not important. Farming equipment, personal investments, and multimillion dollar factory improvements can all be evaluated using the same principles.

Similarly, the applicable principles are insensitive to the monetary units. Although *dollars* are used in this chapter, it is equally convenient to use pounds, yen, or euros.

Finally, this chapter may give the impression that investment alternatives must be evaluated on a year-by-year basis. Actually, the *effective period* can be defined as a day, month, century, or any other convenient period of time.

2. MULTIPLICITY OF SOLUTION METHODS

Most economic conclusions can be reached in more than one manner. There are usually several different analyses that will eventually result in identical answers.[2] Other than the pursuit of elegant solutions in a timely manner, there is no reason to favor one procedural method over another.[3]

3. PRECISION AND SIGNIFICANT DIGITS

The full potential of electronic calculators will never be realized in engineering economic analyses. Considering that calculations are based on estimates of far-future cash flows and that unrealistic assumptions (no inflation, identical cost structures of replacement assets, etc.) are routinely made, it makes little sense to carry cents along in calculations.

The calculations in this chapter have been designed to illustrate and review the principles presented. Because of this, greater precision than is normally necessary in everyday problems may be used. Though used, such precision is not warranted.

Unless there is some compelling reason to strive for greater precision, the following rules are presented for

---

[1]This subject is also known as *engineering economics* and *engineering economy*. There is very little, if any, true economics in this subject.
[2]Because of round-off errors, particularly when factors are taken from tables, these different calculations will produce slightly different numerical results (e.g., $49.49 versus $49.50). However, this type of divergence is well known and accepted in engineering economic analysis.
[3]This does not imply that approximate methods, simplifications, and rules of thumb are acceptable.

use in reporting final answers to engineering economic analysis problems.

- Omit fractional parts of the dollar (i.e., cents).

- Report and record a number to a maximum of four significant digits unless the first digit of that number is 1, in which case, a maximum of five significant digits should be written. For example,

$49	not	$49.43
$93,450	not	$93,453
$1,289,700	not	$1,289,673

4. NONQUANTIFIABLE FACTORS

An engineering economic analysis is a quantitative analysis. Some factors cannot be introduced as numbers into the calculations. Such factors are known as *non-quantitative factors*, *judgment factors*, and *irreducible factors*. Typical nonquantifiable factors are

- preferences

- political ramifications

- urgency

- goodwill

- prestige

- utility

- corporate strategy

- environmental effects

- health and safety rules

- reliability

- political risks

Since these factors are not included in the calculations, the policy is to disregard the issues entirely. Of course, the factors should be discussed in a final report. The factors are particularly useful in breaking ties between competing alternatives that are economically equivalent.

5. YEAR-END AND OTHER CONVENTIONS

Except in short-term transactions, it is simpler to assume that all receipts and disbursements (cash flows) take place at the end of the year in which they occur.[4] This is known as the *year-end convention*. The exceptions to the year-end convention are initial project cost (purchase cost), trade-in allowance, and other cash flows that are associated with the inception of the project at $t = 0$.

On the surface, such a convention appears grossly inappropriate since repair expenses, interest payments, corporate taxes, and so on seldom coincide with the end of

a year. However, the convention greatly simplifies engineering economic analysis problems, and it is justifiable on the basis that the increased precision associated with a more rigorous analysis is not warranted (due to the numerous other simplifying assumptions and estimates initially made in the problem).

There are various established procedures, known as *rules* or *conventions*, imposed by the Internal Revenue Service on U.S. taxpayers. An example is the *half-year rule*, which permits only half of the first-year depreciation to be taken in the first year of an asset's life when certain methods of depreciation are used. These rules are subject to constantly changing legislation and are not covered in this book. The implementation of such rules is outside the scope of engineering practice and is best left to accounting professionals.

6. CASH FLOW DIAGRAMS

Although they are not always necessary in simple problems (and they are often unwieldy in very complex problems), *cash flow diagrams* can be drawn to help visualize and simplify problems having diverse receipts and disbursements.

The following conventions are used to standardize cash flow diagrams.

- The horizontal (time) axis is marked off in equal increments, one per period, up to the duration (or *horizon*) of the project.

- Two or more transfers in the same period are placed end to end, and these may be combined.

- Expenses incurred before $t = 0$ are called *sunk costs*. Sunk costs are not relevant to the problem unless they have tax consequences in an after-tax analysis.

- *Receipts* are represented by arrows directed upward. *Disbursements* are represented by arrows directed downward. The arrow length is proportional to the magnitude of the cash flow.

Example 61.1

A mechanical device will cost $20,000 when purchased. Maintenance will cost $1000 each year. The device will generate revenues of $5000 each year for five years, after which the salvage value is expected to be $7000. Draw and simplify the cash flow diagram.

---

[4]A *short-term transaction* typically has a lifetime of five years or less and has payments or compounding that are more frequent than once per year.

Solution

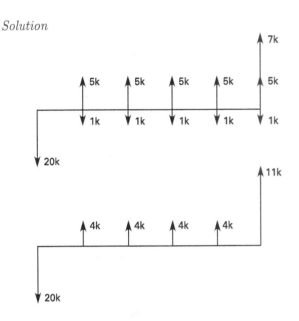

7. TYPES OF CASH FLOWS

To evaluate a real-world project, it is necessary to present the project's cash flows in terms of standard cash flows that can be handled by engineering economic analysis techniques. The standard cash flows are single payment cash flow, uniform series cash flow, gradient series cash flow, and the infrequently encountered exponential gradient series cash flow. (See Fig. 61.1.)

Figure 61.1 *Standard Cash Flows*

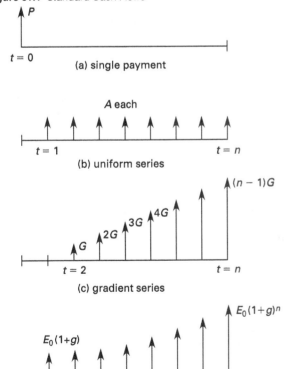

A *single payment cash flow* can occur at the beginning of the time line (designated as $t = 0$), at the end of the time line (designated as $t = n$), or at any time in between. The symbol P (for *present*) is typically used to represent a single payment at $t = 0$, and the symbol F (for *future*) to represent a single payment at any other time.

The *uniform series cash flow* consists of a series of equal transactions starting at $t = 1$ and ending at $t = n$. The symbol A (for *annual*) is typically given to the magnitude of each individual cash flow.[5]

The *gradient series cash flow* starts with a cash flow (typically given the symbol G) at $t = 2$ and increases by G each year until $t = n$, at which time the final cash flow is $(n - 1)G$.

An *exponential gradient series cash flow* is based on a phantom value (typically given the symbol E_0) at $t = 0$ and grows or decays exponentially according to the following relationship.[6]

$$\text{amount at time } t = E_t = E_0(1 + g)^t$$
$$[t = 1, 2, 3, \ldots, n]$$

61.1

In Eq. 61.1, g is the *exponential growth rate*, which can be either positive or negative. Exponential gradient cash flows are rarely seen in economic justification projects assigned to engineers.[7]

8. TYPICAL PROBLEM TYPES

There is a wide variety of problem types that, collectively, are considered to be engineering economic analysis problems.

By far, the majority of engineering economic analysis problems are *alternative comparisons*. In these problems, two or more mutually exclusive investments compete for limited funds. A variation of this is a *replacement/retirement analysis*, which is repeated each year to determine if an existing asset should be replaced. Finding the percentage return on an investment is a *rate of return problem*, one of the alternative comparison solution methods.

Investigating interest and principal amounts in loan payments is a *loan repayment problem*. An *economic life analysis* will determine when an asset should be retired. In addition, there are miscellaneous problems involving economic order quantity, learning curves, break-even points, product costs, and so on.

9. IMPLICIT ASSUMPTIONS

Several assumptions are implicitly made when solving engineering economic analysis problems. Some of these assumptions are made with the knowledge that they are or will be poor approximations of what really will happen. The assumptions are made, regardless, for the benefit of obtaining a solution.

The most common assumptions are the following.

- The year-end convention is applicable.

- There is no inflation now, nor will there be any during the lifetime of the project.

- Unless otherwise specifically called for, a before-tax analysis is needed.

- The effective interest rate used in the problem will be constant during the lifetime of the project.

- Nonquantifiable factors can be disregarded.

- Funds invested in a project are available and are not urgently needed elsewhere.

- Excess funds continue to earn interest at the effective rate used in the analysis.

This last assumption, like most of the assumptions listed, is almost never specifically mentioned in the body of a solution. However, it is a key assumption when comparing two alternatives that have different initial costs.

For example, suppose two investments, one costing $10,000 and the other costing $8000, are to be compared at 10%. It is obvious that $10,000 in funds is available, otherwise the costlier investment would not be under consideration. If the smaller investment is chosen, what is done with the remaining $2000? The last assumption yields the answer: The $2000 is "put to work" in some investment earning (in this case) 10%.

10. EQUIVALENCE

Industrial decision makers using engineering economic analysis are concerned with the magnitude and timing of a project's cash flow as well as with the total profitability of that project. In this situation, a method is required to compare projects involving receipts and disbursements occurring at different times.

By way of illustration, consider $100 placed in a bank account that pays 5% effective annual interest at the end of each year. After the first year, the account will have grown to $105. After the second year, the account will have grown to $110.25.

Assume that you will have no need for money during the next two years, and any money received will immediately go into your 5% bank account. Then, which of the following options would be more desirable?

As illustrated, none of the options is superior under the assumptions given. If the first option is chosen, you will immediately place $100 into a 5% account, and in two years the account will have grown to $110.25. In fact, the account will contain $110.25 at the end of two years regardless of the option chosen. Therefore, these alternatives are said to be *equivalent*.

option A: $100 now

option B: $105 to be delivered in one year

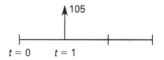

option C: $110.25 to be delivered in two years

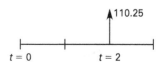

Equivalence may or may not be the case, depending on the interest rate, so an alternative that is acceptable to one decision maker may be unacceptable to another. The interest rate that is used in actual calculations is known as the *effective interest rate*.[8] If compounding is once a year, it is known as the *effective annual interest rate*. However, effective quarterly, monthly, daily, and so on, interest rates are also used.

The fact that $100 today grows to $105 in one year (at 5% annual interest) is an example of what is known as the *time value of money* principle. This principle simply articulates what is obvious: Funds placed in a secure investment will increase to an equivalent future amount. The procedure for determining the present investment from the equivalent future amount is known as *discounting*.

---

[5]The cash flows do not begin at $t = 0$. This is an important concept with all of the series cash flows. This convention has been established to accommodate the timing of annual maintenance (and similar) cash flows for which the year-end convention is applicable.
[6]By convention, for an exponential cash flow series: The first cash flow, E_0, is at $t = 1$, as in the uniform annual series. However, the first cash flow is $E_0(1 + g)$. The cash flow of E_0 at $t = 0$ is absent (i.e., is a *phantom cash flow*).
[7]For one of the few discussions on exponential cash flow, see *Capital Budgeting*, Robert V. Oakford, The Ronald Press Company, New York, 1970.
[8]The adjective *effective* distinguishes this interest rate from other interest rates (e.g., nominal interest rates) that are not meant to be used directly in calculating equivalent amounts.

11. SINGLE-PAYMENT EQUIVALENCE

The equivalence of any present amount, P, at $t = 0$, to any future amount, F, at $t = n$, is called the *future worth* and can be calculated from Eq. 61.2.

$$F = P(1 + i)^n \qquad \text{Economics} \quad 61.2$$

The factor $(1 + i)^n$ is known as the *single payment compound amount factor.* Tables are available giving the values of this and other cash flow factors for common values of i and n. [Factor Table]

The equivalence of any future amount to any present amount is called the *present worth* and can be calculated from Eq. 61.3.

$$P = F(1 + i)^{-n} = \frac{F}{(1 + i)^n} \qquad \text{Economics} \quad 61.3$$

The factor $(1 + i)^{-n}$ is known as the *single payment present worth factor.*[9]

The interest rate used in Eq. 61.2 and Eq. 61.3 must be the effective rate per period. Also, the basis of the rate (annually, monthly, etc.) must agree with the type of period used to count n. Therefore, it would be incorrect to use an effective annual interest rate if n was the number of compounding periods in months.

Example 61.2

How much should you put into a 10% (effective annual rate) savings account in order to have $10,000 in five years?

Solution

This problem could also be stated: What is the equivalent present worth of $10,000 five years from now if money is worth 10% per year?

$$P = F(1 + i)^{-n} \qquad \text{Economics}$$
$$= (\$10,000)(1 + 0.10)^{-5}$$
$$= \$6209$$

The factor 0.6209 would usually be obtained from the tables.

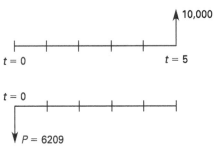

12. STANDARD CASH FLOW FACTORS AND SYMBOLS

Equation 61.2 and Eq. 61.3 may give the impression that solving engineering economic analysis problems involves a lot of calculator use, and, in particular, a lot of exponentiation. Such calculations may be necessary from time to time, but most problems are simplified by the use of tabulated values of the factors.

Rather than actually writing the formula for the compound amount factor (which converts a present amount to a future amount), it is common convention to substitute the standard functional notation of $(F/P, i\%, n)$. Therefore, the future value in n periods of a present amount would be symbolically written as

$$F = P(F/P, i\%, n) \qquad 61.4$$

Similarly, the present worth factor has a functional notation of $(P/F, i\%, n)$. Therefore, the present worth of a future amount n periods in the future would be symbolically written as

$$P = F(P/F, i\%, n) \qquad 61.5$$

Values of these *cash flow (discounting) factors* can be looked up in tables. There is often initial confusion about whether the (F/P) or (P/F) column should be used in a particular problem. There are several ways of remembering what the functional notations mean.

One method of remembering which factor should be used is to think of the factors as conditional probabilities. The conditional probability of event **A** given that event **B** has occurred is written as $p\{\mathbf{A}|\mathbf{B}\}$, where the given event comes after the vertical bar. In the standard notational form of discounting factors, the given amount is similarly placed after the slash. What you want comes before the slash. (F/P) would be a factor to find F given P.

---

[9]The *present worth* is also called the *present value* and *net present value*. These terms are used interchangeably and no significance should be attached to the terms *value*, *worth*, and *net*.

Another method of remembering the notation is to interpret the factors algebraically. The (F/P) factor could be thought of as the fraction F/P. Algebraically, Eq. 61.4 would be

$$F = P\left(\frac{F}{P}\right) \qquad 61.6$$

This algebraic approach is actually more than an interpretation. The numerical values of the discounting factors are consistent with this algebraic manipulation. The (F/A) factor could be calculated as $(F/P) \times (P/A)$. This consistent relationship can be used to calculate other factors that are often omitted from tables but might be occasionally needed, such as (F/G) or (G/P). For instance, the annual cash flow that would be equivalent to a uniform gradient may be found from

$$A = G(P/G, i\%, n)(A/P, i\%, n) \qquad 61.7$$

Formulas for the compounding and discounting factors are contained in Table 61.1. Normally, it will not be necessary to calculate factors from the formulas. A table of cash flow equivalent factors is adequate for solving most problems. [Factor Table]

Example 61.3

What factor will convert a gradient cash flow ending at $t = 8$ to a future value at $t = 8$? (That is, what is the $(F/G, i\%, n)$ factor?) The effective annual interest rate is 10%.

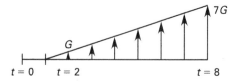

Solution

method 1:

From Table 61.1, the $(F/G, 10\%, 8)$ factor is

$$(F/G, 10\%, 8) = \frac{(1+i)^n - 1}{i^2}$$

$$= \frac{(1+0.10)^8 - 1}{(0.10)^2} - \frac{8}{0.10}$$

$$= 34.3589$$

method 2:

The values of (P/G) and (F/P) taken from a table of cash flow equivalent factors can be used to calculate the needed factor. [Factor Table]

$$(F/G, 10\%, 8) = (P/G, 10\%, 8)(F/P, 10\%, 8)$$
$$= (16.0287)(2.1436)$$
$$= 34.3591$$

The (F/G) factor could also have been calculated as the product of the (A/G) and (F/A) factors.

$F = 34.3591 G$

$t = 0$ $t = 8$

13. CALCULATING UNIFORM SERIES EQUIVALENCE

A cash flow that repeats each year for n years without change in amount is known as an *annual amount* and is given the symbol A. As an example, a piece of equipment may require annual maintenance, and the maintenance cost will be an annual amount. Although the equivalent value for each of the n annual amounts could be calculated and then summed, it is more expedient to use one of the uniform series factors. For example, it is possible to convert from an annual amount to an equivalent future amount by use of the (F/A) factor.

$$F = A(F/A, i\%, n) \qquad 61.8$$

A *sinking fund* is a fund or account into which annual deposits of A are made in order to accumulate an amount F at $t = n$ in the future. Since the annual deposit needed to achieve this is calculated as $A = F(A/F, i\%, n)$, the (A/F) factor is known as the *sinking fund factor*. An *annuity* is a series of equal payments (A) made over a period of time.[10] Usually, it is necessary to "buy into" an investment (a bond, an insurance policy, etc.) in order to ensure the annuity. In the simplest case of an annuity that starts at the end of the first year and continues for n years, the purchase price, P, that is equivalent to the value of the annuity is

$$P = A(P/A, i\%, n) \qquad 61.9$$

The present worth of an *infinite (perpetual) series* of annual amounts is known as a *capitalized cost*. There is no $(P/A, i\%, \infty)$ factor in the tables, but the capitalized cost can be calculated simply as

Capitalized Costs

$$P = \frac{A}{i} \quad [i \text{ in decimal form}] \qquad 61.10$$

Alternatives with different lives will generally be compared by way of *equivalent uniform annual cost* (EUAC). An EUAC is the annual amount that is equivalent to all of the cash flows in the alternative. The

---

[10]An annuity may also consist of a lump sum payment made at some future time. However, this interpretation is not considered in this chapter.

Table 61.1 *Discount Factors for Discrete Compounding*

factor name	converts	symbol	formula
single payment compound amount	P to F	$(F/P, i\%, n)$	$(1+i)^n$
single payment present worth	F to P	$(P/F, i\%, n)$	$(1+i)^{-n}$
uniform series sinking fund	F to A	$(A/F, i\%, n)$	$\dfrac{i}{(1+i)^n - 1}$
capital recovery	P to A	$(A/P, i\%, n)$	$\dfrac{i(1+i)^n}{(1+i)^n - 1}$
uniform series compound amount	A to F	$(F/A, i\%, n)$	$\dfrac{(1+i)^n - 1}{i}$
uniform series present worth	A to P	$(P/A, i\%, n)$	$\dfrac{(1+i)^n - 1}{i(1+i)^n}$
uniform gradient present worth	G to P	$(P/G, i\%, n)$	$\dfrac{(1+i)^n - 1}{i^2(1+i)^n} - \dfrac{n}{i(1+i)^n}$
uniform gradient future worth	G to F	$(F/G, i\%, n)$	$\dfrac{(1+i)^n - 1}{i^2} - \dfrac{n}{i}$
uniform gradient uniform series	G to A	$(A/G, i\%, n)$	$\dfrac{1}{i} - \dfrac{n}{(1+i)^n - 1}$

EUAC differs in sign from all of the other cash flows. Costs and expenses expressed as EUACs, which would normally be considered negative, are actually positive. The term *cost* in the designation EUAC serves to make clear the meaning of a positive number.

Example 61.4

Maintenance costs for a machine are $250 each year. The interest rate is 8%. What is the present worth of these maintenance costs over a 12-year period?

Solution

From a table of cash flow equivalent factors, the factor $(P/A, 8\%, 12)$ is equal to 7.5361. [Factor Table]

$$P = A(P/A, 8\%, 12) = (-\$250)(7.5361)$$
$$= -\$1884$$

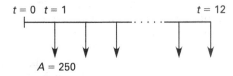

$A = 250$

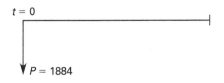

$P = 1884$

---

14. FINDING PAST VALUES

From time to time, it will be necessary to determine an amount in the past equivalent to some current (or future) amount. For example, you might have to calculate the original investment made 15 years ago given a current annuity payment.

Such problems are solved by placing the $t = 0$ point at the time of the original investment, and then calculating the past amount as a P value. For example, the original investment, P, can be extracted from the annuity, A, by using the standard cash flow factors.

$$P = A(P/A, i\%, n) \qquad \textit{61.11}$$

The choice of $t = 0$ is flexible. As a general rule, the $t = 0$ point should be selected for convenience in solving a problem.

Example 61.5

You currently pay $250 per month to lease your office phone equipment. You have three years (36 months) left on the five-year (60-month) lease. The effective interest rate per month is 1%. What would have been an equivalent purchase price two years ago?

Solution

The solution of this example is not affected by the fact that investigation is being performed in the middle of the horizon. This is a simple calculation of present worth.

$$P = A(P/A, 1\%, 60)$$
$$= (-\$250)(44.9550)$$
$$= -\$11,239$$

15. TIMES TO DOUBLE AND TRIPLE AN INVESTMENT

If an investment doubles in value (in n compounding periods and with $i\%$ effective interest), the ratio of current value to past investment will be 2.

$$F/P = (1+i)^n = 2 \qquad 61.12$$

Similarly, the ratio of current value to past investment will be 3 if an investment triples in value. This can be written as

$$F/P = (1+i)^n = 3 \qquad 61.13$$

It is a simple matter to extract the number of periods, n, from Eq. 61.12 and Eq. 61.13 to determine the *doubling time* and *tripling time*, respectively. For example, the doubling time is

$$n = \frac{\log 2}{\log(1+i)} \qquad 61.14$$

When a quick estimate of the doubling time is needed, the *rule of 72* can be used. The doubling time is approximately $72/i$.

The tripling time is

$$n = \frac{\log 3}{\log(1+i)} \qquad 61.15$$

Equation 61.14 and Eq. 61.15 form the basis of Table 61.2.

16. VARIED AND NONSTANDARD CASH FLOWS

Gradient Cash Flow

A common situation involves a uniformly increasing cash flow. If the cash flow has the proper form, its present worth can be determined by using the *uniform gradient factor*, $(P/G, i\%, n)$. The uniform gradient factor finds the present worth of a uniformly increasing cash flow that starts in year two (not in year one).

Table 61.2 Doubling and Tripling Times for Various Interest Rates

interest rate (%)	doubling time (periods)	tripling time (periods)
1	69.7	110.4
2	35.0	55.5
3	23.4	37.2
4	17.7	28.0
5	14.2	22.5
6	11.9	18.9
7	10.2	16.2
8	9.01	14.3
9	8.04	12.7
10	7.27	11.5
11	6.64	10.5
12	6.12	9.69
13	5.67	8.99
14	5.29	8.38
15	4.96	7.86
16	4.67	7.40
17	4.41	7.00
18	4.19	6.64
19	3.98	6.32
20	3.80	6.03

There are three common difficulties associated with the form of the uniform gradient. The first difficulty is that the initial cash flow occurs at $t = 2$. This convention recognizes that annual costs, if they increase uniformly, begin with some value at $t = 1$ (due to the year-end convention) but do not begin to increase until $t = 2$. The tabulated values of (P/G) have been calculated to find the present worth of only the increasing part of the annual expense. The present worth of the base expense incurred at $t = 1$ must be found separately with the (P/A) factor.

The second difficulty is that, even though the factor $(P/G, i\%, n)$ is used, there are only $n - 1$ actual cash flows. It is clear that n must be interpreted as the *period number* in which the last gradient cash flow occurs, not the number of gradient cash flows.

Finally, the sign convention used with gradient cash flows may seem confusing. If an expense increases each year (as in Ex. 61.6), the gradient will be negative, since it is an expense. If a revenue increases each year, the gradient will be positive. (See Fig. 61.2.) In most cases, the sign of the gradient depends on whether the cash flow is an expense or a revenue.[11]

---

[11]This is not a universal rule. It is possible to have a uniformly decreasing revenue as in Fig. 61.2(c). In this case, the gradient would be negative.

Figure 61.2 *Positive and Negative Gradient Cash Flows*

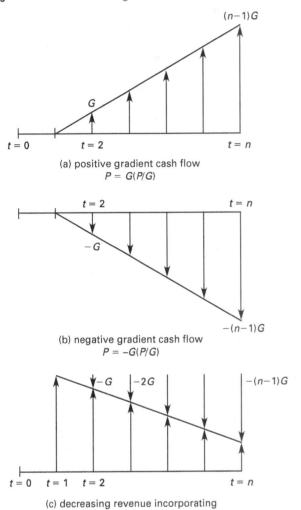

(a) positive gradient cash flow
$P = G(P/G)$

(b) negative gradient cash flow
$P = -G(P/G)$

(c) decreasing revenue incorporating
a negative gradient
$P = A(P/A) - G(P/G)$

Example 61.6

Maintenance on an old machine is $100 this year but is expected to increase by $25 each year thereafter. What is the present worth of five years of the costs of maintenance? Use an interest rate of 10%.

Solution

In this problem, the cash flow must be broken down into parts. (The five-year gradient factor is used even though there are only four nonzero gradient cash flows.) [Factor Table]

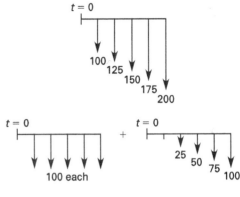

$$P = A(P/A, 10\%, 5) + G(P/G, 10\%, 5)$$
$$= (-\$100)(3.7908) - (\$25)(6.8618)$$
$$= -\$551$$

Stepped Cash Flows

Stepped cash flows are easily handled by the technique of *superposition of cash flows*. This technique is used in Ex. 61.7.

Example 61.7

An investment costing $1000 returns $100 for the first five years and $200 for the following five years. How would the present worth of this investment be calculated?

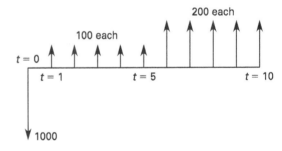

Solution

Using the principle of superposition, the revenue cash flow can be thought of as $200 each year from $t = 1$ to $t = 10$, with a negative revenue of $100 from $t = 1$ to $t = 5$. Superimposed, these two cash flows make up the actual performance cash flow.

$$P = -\$1000 + (\$200)(P/A, i\%, 10) - (\$100)(P/A, i\%, 5)$$

Missing and Extra Parts of Standard Cash Flows

A missing or extra part of a standard cash flow can also be handled by superposition. For example, suppose an annual expense is incurred each year for ten years, except in the ninth year. (The cash flow is illustrated in Fig. 61.3.) The present worth could be calculated as a subtractive process.

$$P = A(P/A, i\%, 10) - A(P/F, i\%, 9) \qquad 61.16$$

Figure 61.3 *Cash Flow with a Missing Part*

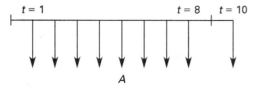

Alternatively, the present worth could be calculated as an additive process.

$$P = A(P/A, i\%, 8) + A(P/F, i\%, 10) \qquad 61.17$$

Delayed and Premature Cash Flows

There are cases when a cash flow matches a standard cash flow exactly, except that the cash flow is delayed or starts sooner than it should. Often, such cash flows can be handled with superposition. At other times, it may be more convenient to shift the time axis. This shift is known as the *projection method*. Example 61.8 demonstrates the projection method.

Example 61.8

An expense of $75 is incurred starting at $t = 3$ and continues until $t = 9$. There are no expenses or receipts until $t = 3$. Use the projection method to determine the present worth of this stream of expenses.

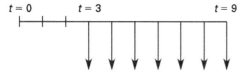

Solution

First, determine a cash flow at $t = 2$ that is equivalent to the entire expense stream. If $t = 0$ was where $t = 2$ actually is, the present worth of the expense stream would be

$$P' = (-\$75)(P/A, i\%, 7)$$

P' is a cash flow at $t = 2$. It is now simple to find the present worth (at $t = 0$) of this future amount.

$$P = P'(P/F, i\%, 2) = (-\$75)(P/A, i\%, 7)(P/F, i\%, 2)$$

Cash Flows at Beginnings of Years: The Christmas Club Problem

This type of problem is characterized by a stream of equal payments (or expenses) starting at $t = 0$ and ending at $t = n - 1$, (See Fig. 61.4.) It differs from the standard annual cash flow in the existence of a cash flow at $t = 0$ and the absence of a cash flow at $t = n$. This problem gets its name from the service provided by some savings institutions whereby money is automatically deposited each week or month (starting immediately, when the savings plan is opened) in order to accumulate money to purchase Christmas presents at the end of the year.

Figure 61.4 *Cash Flows at Beginnings of Years*

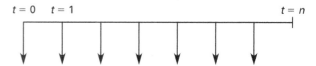

It may seem that the present worth of the savings stream can be determined by directly applying the (P/A) factor. However, this is not the case, since the Christmas Club cash flow and the standard annual cash flow differ. The Christmas Club problem is easily handled by superposition, as demonstrated by Ex. 61.9.

Example 61.9

How much can you expect to accumulate by $t = 10$ for a child's college education if you deposit $300 at the beginning of each year for a total of ten payments?

Solution

The first payment is made at $t = 0$, and there is no payment at $t = 10$. The future worth of the first payment is calculated with the (F/P) factor. The absence of the payment at $t = 10$ is handled by superposition. This "correction" is not multiplied by a factor.

$$F = (\$300)(F/P, i\%, 10) + (\$300)(F/A, i\%, 10) - \$300$$
$$= (\$300)(F/A, i\%, 11) - \$300$$

17. THE MEANING OF PRESENT WORTH AND *i*

If $100 is invested in a 5% bank account (using annual compounding), you can remove $105 one year from now; if this investment is made, you will receive a *return on investment* (ROI) of $5. The cash flow diagram, (see Fig. 61.5) and the present worth of the two transactions are

$$P = -\$100 + (\$105)(P/F, 5\%, 1)$$
$$= -\$100 + (\$105)(0.9524)$$
$$= 0$$

Figure 61.5 *Cash Flow Diagram*

The present worth is zero even though you will receive a $5 return on your investment.

However, if you are offered $120 for the use of $100 over a one-year period, the cash flow diagram, (see Fig. 61.6.) and present worth (at 5%) would be

$$P = -\$100 + (\$120)(P/F, 5\%, 1)$$
$$= -\$100 + (\$120)(0.9524)$$
$$= \$14.29$$

Figure 61.6 *Cash Flow Diagram*

Therefore, the present worth of an alternative is seen to be equal to the equivalent value at $t = 0$ of the increase in return above that which you would be able to earn in an investment offering $i\%$ per period. In the previous case, $14.29 is the present worth of ($20 – $5), the difference in the two ROIs.

The present worth is also the amount that you would have to be given to dissuade you from making an investment, since placing the initial investment amount along with the present worth into a bank account earning $i\%$ will yield the same eventual return on investment. Relating this to the previous paragraphs, you could be dissuaded from investing $100 in an alternative that would return $120 in one year by a $t = 0$ payment of $14.29. Clearly, ($100 + $14.29) invested at $t = 0$ will also yield $120 in one year at 5%.

Income-producing alternatives with negative present worths are undesirable, and alternatives with positive present worths are desirable because they increase the average earning power of invested capital. (In some cases, such as municipal and public works projects, the present worths of all alternatives are negative, in which case, the least negative alternative is best.)

The selection of the interest rate is difficult in engineering economics problems. Usually, it is taken as the average rate of return that an individual or business

organization has realized in past investments. Alternatively, the interest rate may be associated with a particular level of risk. Usually, i for individuals is the interest rate that can be earned in relatively *risk-free investments*.

18. SIMPLE AND COMPOUND INTEREST

If $100 is invested at 5%, it will grow to $105 in one year. During the second year, 5% interest continues to be accrued, but on $105, not on $100. This is the principle of *compound interest*: The interest accrues interest.[12]

If only the original principal accrues interest, the interest is said to be *simple interest*. Simple interest is rarely encountered in long-term engineering economic analyses, but the concept may be incorporated into short-term transactions.

19. EXTRACTING THE INTEREST RATE: RATE OF RETURN

An intuitive definition of the *rate of return* (ROR) is the effective annual interest rate at which an investment accrues income. That is, the rate of return of a project is the interest rate that would yield identical profits if all money were invested at that rate. Although this definition is correct, it does not provide a method of determining the rate of return.

It was previously seen that the present worth of a $100 investment invested at 5% is zero when $i = 5\%$ is used to determine equivalence. Therefore, a working definition of rate of return would be the effective annual interest rate that makes the present worth of the investment zero. Alternatively, rate of return could be defined as the effective annual interest rate that will discount all cash flows to a total present worth equal to the required initial investment.

It is tempting, but impractical, to determine a rate of return analytically. It is simply too difficult to extract the interest rate from the equivalence equation. For example, consider a $100 investment that pays back $75 at the end of each of the first two years. The present worth equivalence equation (set equal to zero in order to determine the rate of return) is

$$P = 0 = -\$100 + (\$75)(1 + i)^{-1}$$
$$+ (\$75)(1 + i)^{-2}$$

61.18

Solving Eq. 61.18 requires finding the roots of a quadratic equation. In general, for an investment or project spanning n years, the roots of an nth-order polynomial would have to be found. It should be obvious that an

---

[12]This assumes, of course, that the interest remains in the account. If the interest is removed and spent, only the remaining funds accumulate interest.

analytical solution would be essentially impossible for more complex cash flows. (The rate of return in this example is 31.87%.)

If the rate of return is needed, it can be found from a trial-and-error solution. To find the rate of return of an investment, proceed as follows.

step 1: Set up the problem as if to calculate the present worth.

step 2: Arbitrarily select a reasonable value for i. Calculate the present worth.

step 3: Choose another value of i (not too close to the original value), and again solve for the present worth.

step 4: Interpolate or extrapolate the value of i that gives a zero present worth.

step 5: For increased accuracy, repeat steps 2 and 3 with two more values that straddle the value found in step 4.

A common, although incorrect, method of calculating the rate of return involves dividing the annual receipts or returns by the initial investment. (See Sec. 62.28.) However, this technique ignores such items as salvage, depreciation, taxes, and the time value of money. This technique also is inadequate when the annual returns vary.

It is possible that more than one interest rate will satisfy the zero present worth criteria. This confusing situation occurs whenever there is more than one change in sign in the investment's cash flow.[13] Table 61.3 indicates the numbers of possible interest rates as a function of the number of sign reversals in the investment's cash flow.

Table 61.3 *Multiplicity of Rates of Return*

number of sign reversals	number of distinct rates of return
0	0
1	0 or 1
2	0, 1, or 2
3	0, 1, 2, or 3
4	0, 1, 2, 3, or 4
m	0, 1, 2, 3, ..., $m-1$, m

Difficulties associated with interpreting the meaning of multiple rates of return can be handled with the concepts of external investment and external rate of return. An *external investment* is an investment that is distinct from the investment being evaluated (which becomes known as the internal investment). The *external rate of*

return, which is the rate of return earned by the external investment, does not need to be the same as the rate earned by the internal investment.

Generally, the multiple rates of return indicate that the analysis must proceed as though money will be invested outside of the project. The mechanics of how this is done are not covered here.

Example 61.10

What is the rate of return on invested capital if $1000 is invested now with $500 being returned in year 4 and $1000 being returned in year 8?

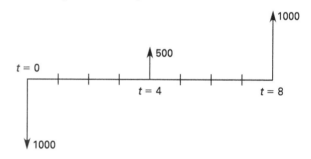

Solution

First, set up the problem as a present worth calculation. Try $i = 5\%$.

Factor Table

$$P = -\$1000 + (\$500)(P/F, 5\%, 4)$$
$$+ (\$1000)(P/F, 5\%, 8)$$
$$= -\$1000 + (\$500)(0.8227) + (\$1000)(0.6768)$$
$$= \$88$$

Next, try a larger value of i to reduce the present worth. If $i = 10\%$,

Factor Table

$$P = -\$1000 + (\$500)(P/F, 10\%, 4)$$
$$+ (\$1000)(P/F, 10\%, 8)$$
$$= -\$1000 + (\$500)(0.6830) + (\$1000)(0.4665)$$
$$= -\$192$$

Using simple interpolation, the rate of return is

$$\text{ROR} = 5\% + \left(\frac{\$88}{\$88 + \$192}\right)(10\% - 5\%)$$
$$= 6.57\%$$

---

[13]There will always be at least one change of sign in the cash flow of a legitimate investment. (This excludes municipal and other tax-supported functions.) At $t = 0$, an investment is made (a negative cash flow). Hopefully, the investment will begin to return money (a positive cash flow) at $t = 1$ or shortly thereafter. Although it is possible to conceive of an investment in which all of the cash flows were negative, such an investment would probably be classified as a *hobby*.

A second iteration between 6% and 7% yields 6.39%.

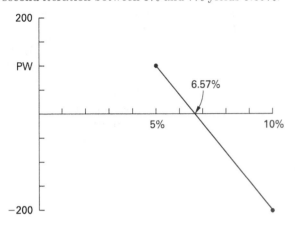

Example 61.11

A biomedical company is developing a new drug. A venture capital firm gives the company $25 million initially and $55 million more at the end of the first year. The drug patent will be sold at the end of year 5 to the highest bidder, and the biomedical company will receive $80 million. (The venture capital firm will receive everything in excess of $80 million.) The firm invests unused money in short-term commercial paper earning 10% effective interest per year through its bank. In the meantime, the biomedical company incurs development expenses of $50 million annually for the first three years. The drug is to be evaluated by a government agency and there will be neither expenses nor revenues during the fourth year. What is the biomedical company's rate of return on this investment?

Solution

Normally, the rate of return is determined by setting up a present worth problem and varying the interest rate until the present worth is zero. Writing the cash flows, though, shows that there are two reversals of sign: one at $t = 2$ (positive to negative) and the other at $t = 5$ (negative to positive). Therefore, there could be two interest rates that produce a zero present worth. (In fact, there actually are two interest rates: 10.7% and 41.4%.)

time	cash flow (millions)
0	+25
1	+55 − 50 = +5
2	−50
3	−50
4	0
5	+80

However, this problem can be reduced to one with only one sign reversal in the cash flow series. The initial $25 million is invested in commercial paper (an *external*

investment having nothing to do with the drug development process) during the first year at 10%. The accumulation of interest and principal after one year is

$$(25)(1 + 0.10) = 27.5$$

This 27.5 is combined with the 5 (the money remaining after all expenses are paid at $t = 1$) and invested externally, again at 10%. The accumulation of interest and principal after one year (i.e., at $t = 2$) is

$$(27.5 + 5)(1 + 0.10) = 35.75$$

This 35.75 is combined with the development cost paid at $t = 2$.

The cash flow for the development project (the internal investment) is

time	cash flow (millions)
0	0
1	0
2	35.75 − 50 = −14.25
3	−50
4	0
5	+80

There is only one sign reversal in the cash flow series. The *internal rate of return* on this development project is found by the traditional method to be 10.3%. This is different from the rate the company can earn from investing externally in commercial paper.

20. RATE OF RETURN VERSUS RETURN ON INVESTMENT

Rate of return (ROR) is an effective annual interest rate, typically stated in percent per year. *Return on investment* (ROI) is a dollar amount. *Rate of return* and *return on investment* are not synonymous.

Return on investment can be calculated in two different ways. The accounting method is to subtract the total of all investment costs from the total of all net profits (i.e., revenues less expenses). The time value of money is not considered.

In engineering economic analysis, the return on investment is calculated from equivalent values. Specifically, the present worth (at $t = 0$) of all investment costs is subtracted from the future worth (at $t = n$) of all net profits.

When there are only two cash flows, a single investment amount and a single payback, the two definitions of return on investment yield the same numerical value. When there are more than two cash flows, the returns on investment will be different depending on which definition is used.

21. MINIMUM ATTRACTIVE RATE OF RETURN

A company may not know what effective interest rate, i, to use in engineering economic analysis. In such a case, the company can establish a minimum level of economic performance that it would like to realize on all investments. This criterion is known as the *minimum attractive rate of return* (MARR). Unlike the effective interest rate, i, the minimum attractive rate of return is not used in numerical calculations.[14] It is used only in comparisons with the rate of return.

Once a rate of return for an investment is known, it can be compared to the minimum attractive rate of return. To be a viable alternative, the rate of return must be greater than the minimum attractive rate of return.

The advantage of using comparisons to the minimum attractive rate of return is that an effective interest rate, i, never needs to be known. The minimum attractive rate of return becomes the correct interest rate for use in present worth and equivalent uniform annual cost calculations.

22. TYPICAL ALTERNATIVE-COMPARISON PROBLEM FORMATS

With the exception of some investment and rate of return problems, the typical problem involving engineering economics will have the following characteristics.

- An interest rate will be given.

- Two or more alternatives will be competing for funding.

- Each alternative will have its own cash flows.

- It will be necessary to select the best alternative.

23. DURATIONS OF INVESTMENTS

Because they are handled differently, short-term investments and short-lived assets need to be distinguished from investments and assets that constitute an infinitely lived project. Short-term investments are easily identified: a drill press that is needed for three years or a temporary factory building that is being constructed to last five years.

Investments with perpetual cash flows are also (usually) easily identified: maintenance on a large flood control dam and revenues from a long-span toll bridge. Furthermore, some items with finite lives can expect renewal on a repeated basis.[15] For example, a major freeway with a pavement life of 20 years is unlikely to be abandoned; it will be resurfaced or replaced every 20 years.

Actually, if an investment's finite lifespan is long enough, it can be considered an infinite investment because money 50 or more years from now has little impact on current decisions. The $(P/F, 10\%, 50)$ factor, for example, is 0.0085. Therefore, one dollar at $t = 50$ has an equivalent present worth of less than one penny. Since these far-future cash flows are eclipsed by present cash flows, long-term investments can be considered finite or infinite without significant impact on the calculations.

24. CHOICE OF ALTERNATIVES: COMPARING ONE ALTERNATIVE WITH ANOTHER ALTERNATIVE

Several methods exist for selecting a superior alternative from among a group of proposals. Each method has its own merits and applications.

Present Worth Method

When two or more alternatives are capable of performing the same functions, the superior alternative will have the largest present worth. The *present worth method* is restricted to evaluating alternatives that are mutually exclusive and that have the same lives. This method is suitable for ranking the desirability of alternatives.

Example 61.12

Investment A costs $10,000 today and pays back $11,500 two years from now. Investment B costs $8000 today and pays back $4500 each year for two years. If an interest rate of 5% is used, which alternative is superior?

Solution

$$P(\text{A}) = -\$10{,}000 + (\$11{,}500)(P/F, 5\%, 2)$$
$$= -\$10{,}000 + (\$11{,}500)(0.9070)$$
$$= \$431$$
$$P(\text{B}) = -\$8000 + (\$4500)(P/A, 5\%, 2)$$
$$= -\$8000 + (\$4500)(1.8594)$$
$$= \$367$$

Alternative A is superior and should be chosen.

Capitalized Cost Method

The present worth of a project with an infinite life is known as the *capitalized cost* or *life-cycle cost*. Capitalized cost is the amount of money at $t = 0$ needed to

---

[14]Not everyone adheres to this rule. Some people use "minimum attractive rate of return" and "effective interest rate" interchangeably.
[15]The term *renewal* can be interpreted to mean replacement or repair.

perpetually support the project on the earned interest only. Capitalized cost is a positive number when expenses exceed income.

In comparing two alternatives, each of which is infinitely lived, the superior alternative will have the lowest capitalized cost.

Normally, it would be difficult to work with an infinite stream of cash flows since most economics tables do not list factors for periods in excess of 100 years. However, the (A/P) discounting factor approaches the interest rate as n becomes large. Since the (P/A) and (A/P) factors are reciprocals of each other, it is possible to divide an infinite series of equal cash flows by the interest rate in order to calculate the present worth of the infinite series. This is the basis of Eq. 61.19.

$$\text{capitalized cost} = \text{initial cost} + \frac{\text{annual costs}}{i} \quad \textit{61.19}$$

Equation 61.19 can be used when the annual costs are equal in every year. If the operating and maintenance costs occur irregularly instead of annually, or if the costs vary from year to year, it will be necessary to somehow determine a cash flow of equal annual amounts (EAA) that is equivalent to the stream of original costs.

The equal annual amount may be calculated in the usual manner by first finding the present worth of all the actual costs and then multiplying the present worth by the interest rate (the (A/P) factor for an infinite series). However, it is not even necessary to convert the present worth to an equal annual amount since Eq. 61.20 will convert the equal amount back to the present worth.

$$\begin{aligned}\text{capitalized cost} &= \text{initial cost} + \frac{\text{EAA}}{i} \\ &= \text{initial cost} + \frac{\text{present worth}}{\text{of all expenses}} \quad \textit{61.20}\end{aligned}$$

Example 61.13

What is the capitalized cost of a public works project that will cost $25,000,000 now and will require $2,000,000 in maintenance annually? The effective annual interest rate is 12%.

Solution

Worked in millions of dollars, from Eq. 61.19, the capitalized cost is

$$\begin{aligned}\text{capitalized cost} &= 25 + (2)(P/A, 12\%, \infty) \\ &= 25 + \frac{2}{0.12} \\ &= 41.67\end{aligned}$$

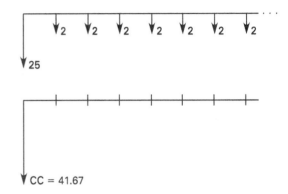

Annual Cost Method

Alternatives that accomplish the same purpose but that have unequal lives must be compared by the *annual cost method*.[16] The annual cost method assumes that each alternative will be replaced by an identical twin at the end of its useful life (infinite renewal). This method, which may also be used to rank alternatives according to their desirability, is also called the *annual return method* or *capital recovery method*.

Restrictions are that the alternatives must be mutually exclusive and repeatedly renewed up to the duration of the longest-lived alternative. The calculated annual cost is known as the *equivalent uniform annual cost* (EUAC) or just *equivalent annual cost*. Cost is a positive number when expenses exceed income.

Example 61.14

Which of the following alternatives is superior over a 30-year period if the interest rate is 7%?

	alternative A	alternative B
type	brick	wood
life	30 years	10 years
initial cost	$1800	$450
maintenance	$5/year	$20/year

---

[16]Of course, the annual cost method can be used to determine the superiority of assets with identical lives as well.

Solution

$$EUAC(A) = (\$1800)(A/P, 7\%, 30) + \$5$$
$$= (\$1800)(0.0806) + \$5$$
$$= \$150$$
$$EUAC(B) = (\$450)(A/P, 7\%, 10) + \$20$$
$$= (\$450)(0.1424) + \$20$$
$$= \$84$$

Alternative B is superior since its annual cost of operation is the lowest. It is assumed that three wood facilities, each with a life of 10 years and a cost of $450, will be built to span the 30-year period.

25. CHOICE OF ALTERNATIVES: COMPARING AN ALTERNATIVE WITH A STANDARD

With specific economic performance criteria, it is possible to qualify an investment as acceptable or unacceptable without having to compare it with another investment. Two such performance criteria are the benefit-cost ratio and the minimum attractive rate of return.

Benefit-Cost Ratio Method

The *benefit-cost ratio method* is often used in municipal project evaluations where benefits and costs accrue to different segments of the community. With this method, the present worth of all benefits (irrespective of the beneficiaries) is divided by the present worth of all costs. The project is considered acceptable if the ratio equals or exceeds 1.0, that is, if $B/C \geq 1.0$. [Benefit-Cost Analysis]

When the benefit-cost ratio method is used, disbursements by the initiators or sponsors are *costs*. Disbursements by the users of the project are known as *disbenefits*. It is often difficult to determine whether a cash flow is a cost or a disbenefit (whether to add it to the denominator or subtract it from the numerator of the benefit-cost ratio calculation).

Regardless of where the cash flow is placed, an acceptable project will always have a benefit-cost ratio greater than or equal to 1.0, although the actual numerical result will depend on the placement. For this reason, the benefit-cost ratio method should not be used to rank competing projects.

The benefit-cost ratio method of comparing alternatives is used extensively in transportation engineering where the ratio is often (but not necessarily) written in terms of annual benefits and annual costs instead of present

worths. Another characteristic of highway benefit-cost ratios is that the route (road, highway, etc.) is usually already in place and that various alternative upgrades are being considered. There will be existing benefits and costs associated with the current route. Therefore, the *change* (usually an increase) in benefits and costs is used to calculate the benefit-cost ratio.[17]

$$B/C = \frac{\Delta \text{ user benefits}}{\Delta \text{ investment cost} + \Delta \text{ maintenance} - \Delta \text{ residual value}} \quad 61.21$$

The change in *residual value* (*terminal value*) appears in the denominator as a negative item. An increase in the residual value would decrease the denominator.

Example 61.15

By building a bridge over a ravine, a state department of transportation can shorten the time it takes to drive through a mountainous area. Estimates of costs and benefits (due to decreased travel time, fewer accidents, reduced gas usage, etc.) have been prepared. Should the bridge be built? Use the benefit-cost ratio method of comparison.

	millions
initial cost	40
capitalized cost of perpetual annual maintenance	12
capitalized value of annual user benefits	49
residual value	0

Solution

If Eq. 61.21 is used, the benefit-cost ratio is

$$B/C = \frac{49}{40 + 12 - 0} = 0.942$$

Since the benefit-cost ratio is less than 1.00, the bridge should not be built.

If the maintenance costs are placed in the numerator (per Ftn. 16), the benefit-cost ratio value will be different, but the conclusion will not change.

$$B/C_{\text{alternate method}} = \frac{49 - 12}{40} = 0.925$$

---

[17]This discussion of highway benefit-cost ratios is not meant to imply that everyone agrees with Eq. 61.21. In *Economic Analysis for Highways* (International Textbook Company, Scranton, PA, 1969), author Robley Winfrey took a strong stand on one aspect of the benefits versus disbenefits issue: highway maintenance. According to Winfrey, regular highway maintenance costs should be placed in the numerator as a subtraction from the user benefits. Some have called this mandate the *Winfrey method*.

Rate of Return Method

The minimum attractive rate of return (MARR) has already been introduced as a standard of performance against which an investment's actual *rate of return* (ROR) is compared. If the rate of return is equal to or exceeds the minimum attractive rate of return, the investment is qualified. This is the basis for the *rate of return method* of alternative selection.

Finding the rate of return can be a long, iterative process. Usually, the actual numerical value of rate of return is not needed; it is sufficient to know whether or not the rate of return exceeds the minimum attractive rate of return. This *comparative analysis* can be accomplished without calculating the rate of return simply by finding the present worth of the investment using the minimum attractive rate of return as the effective interest rate (i.e., $i = \text{MARR}$). If the present worth is zero or positive, the investment is qualified. If the present worth is negative, the rate of return is less than the minimum attractive rate of return.

26. RANKING MUTUALLY EXCLUSIVE MULTIPLE PROJECTS

Ranking of multiple investment alternatives is required when there is sufficient funding for more than one investment. Since the best investments should be selected first, it is necessary to place all investments into an ordered list.

Ranking is relatively easy if the present worths, future worths, capitalized costs, or equivalent uniform annual costs have been calculated for all the investments. The highest ranked investment will be the one with the largest present or future worth, or the smallest capitalized or annual cost. Present worth, future worth, capitalized cost, and equivalent uniform annual cost can all be used to rank multiple investment alternatives.

However, neither rates of return nor benefit-cost ratios should be used to rank multiple investment alternatives. Specifically, if two alternatives both have rates of return exceeding the minimum acceptable rate of return, it is not sufficient to select the alternative with the highest rate of return.

An *incremental analysis*, also known as a *rate of return on added investment study*, should be performed if rate of return is used to select between investments. An incremental analysis starts by ranking the alternatives in order of increasing initial investment. Then, the cash flows for the investment with the lower initial cost are subtracted from the cash flows for the higher-priced alternative on a year-by-year basis. This produces, in effect, a third alternative representing the costs and benefits of the added investment. The added expense of the higher-priced investment is not warranted unless the rate of return of this third alternative exceeds the

minimum attractive rate of return as well. The choice criterion is to select the alternative with the higher initial investment if the incremental rate of return exceeds the minimum attractive rate of return.

An incremental analysis is also required if ranking is to be done by the benefit-cost ratio method. The incremental analysis is accomplished by calculating the ratio of differences in benefits to differences in costs for each possible pair of alternatives. If the ratio exceeds 1.0, alternative 2 is superior to alternative 1. Otherwise, alternative 1 is superior.[18]

$$\frac{B_2 - B_1}{C_2 - C_1} \geq 1 \quad \text{[alternative 2 superior]} \qquad 61.22$$

27. ALTERNATIVES WITH DIFFERENT LIVES

Comparison of two alternatives is relatively simple when both alternatives have the same life. For example, a problem might be stated: "Which would you rather have: car A with a life of three years, or car B with a life of five years?"

However, care must be taken to understand what is going on when the two alternatives have different lives. If car A has a life of three years and car B has a life of five years, what happens at $t = 3$ if the five-year car is chosen? If a car is needed for five years, what happens at $t = 3$ if the three-year car is chosen?

In this type of situation, it is necessary to distinguish between the length of the need (the *analysis horizon*) and the lives of the alternatives or assets intended to meet that need. The lives do not have to be the same as the horizon.

Finite Horizon with Incomplete Asset Lives

If an asset with a five-year life is chosen for a three-year need, the disposition of the asset at $t = 3$ must be known in order to evaluate the alternative. If the asset is sold at $t = 3$, the salvage value is entered into the analysis (at $t = 3$) and the alternative is evaluated as a three-year investment. The fact that the asset is sold when it has some useful life remaining does not affect the analysis horizon.

Similarly, if a three-year asset is chosen for a five-year need, something about how the need is satisfied during the last two years must be known. Perhaps a rental asset will be used. Or, perhaps the function will be "farmed out" to an outside firm. In any case, the costs of satisfying the need during the last two years enter the analysis, and the alternative is evaluated as a five-year investment.

---

[18]It goes without saying that the benefit-cost ratios for all investment alternatives by themselves must also be equal to or greater than 1.0.

If both alternatives are "converted" to the same life, any of the alternative selection criteria (present worth method, annual cost method, etc.) can be used to determine which alternative is superior.

Finite Horizon with Integer Multiple Asset Lives

It is common to have a long-term horizon (need) that must be met with short-lived assets. In special instances, the horizon will be an integer number of asset lives. For example, a company may be making a 12-year transportation plan and may be evaluating two cars: one with a three-year life, and another with a four-year life.

In this example, four of the first car or three of the second car are needed to reach the end of the 12-year horizon.

If the horizon is an integer number of asset lives, any of the alternative selection criteria can be used to determine which is superior. If the present worth method is used, all alternatives must be evaluated over the entire horizon. (In this example, the present worth of 12 years of car purchases and use must be determined for both alternatives.)

If the equivalent uniform annual cost method is used, it may be possible to base the calculation of annual cost on one lifespan of each alternative only. It may not be necessary to incorporate all of the cash flows into the analysis. (In the running example, the annual cost over three years would be determined for the first car; the annual cost over four years would be determined for the second car.) This simplification is justified if the subsequent asset replacements (renewals) have the same cost and cash flow structure as the original asset. This assumption is typically made implicitly when the annual cost method of comparison is used.

Infinite Horizon

If the need horizon is infinite, it is not necessary to impose the restriction that asset lives of alternatives be integer multiples of the horizon. The superior alternative will be replaced (renewed) whenever it is necessary to do so, forever.

Infinite horizon problems are almost always solved with either the annual cost or capitalized cost method. It is common to (implicitly) assume that the cost and cash flow structure of the asset replacements (renewals) are the same as the original asset.

28. OPPORTUNITY COSTS

An *opportunity cost* is an imaginary cost representing what will not be received if a particular strategy is rejected. It is what you will lose if you do or do not do something. As an example, consider a growing company with an existing operational computer system. If the company trades in its existing computer as part of an upgrade plan, it will receive a *trade-in allowance*. (In other problems, a *salvage value* may be involved.)

If one of the alternatives being evaluated is not to upgrade the computer system at all, the trade-in allowance (or, salvage value in other problems) will not be realized. The amount of the trade-in allowance is an opportunity cost that must be included in the problem analysis.

Similarly, if one of the alternatives being evaluated is to wait one year before upgrading the computer, the *difference in trade-in allowances* is an opportunity cost that must be included in the problem analysis.

29. NOMENCLATURE

A	annual amount or annuity	$
A/F	sinking fund factor	-
B	benefit or present worth of all benefits	$
B/C	cost-benefit ratio	-
C	cost or present worth of all costs	$
E_0	initial amount of an exponentially growing cash flow	$
EAA	equivalent annual amount	$
EUAC	equivalent uniform annual cost	$
F	forecasted quantity	various
F	future payment or future worth	$
g	exponential growth rate	decimal
G	gradient series cash flow or uniform gradient amount	$
i	effective interest rate	decimal
i	number of years	-
$i\%$	effective interest rate	%
n	number of compounding periods or years in life of asset	−
P	present payment or present worth or purchase price	$
ROI	return on investment	$
ROR	rate of return	decimal per unit time
t	time	years (typical)

62 Economics: Capitalization, Depreciation, Accounting

1. REPLACEMENT STUDIES

An investigation into the retirement of an existing process or piece of equipment is known as a *replacement study.* Replacement studies are similar in most respects to other alternative comparison problems: An interest rate is given, two alternatives exist, and one of the previously mentioned methods of comparing alternatives is used to choose the superior alternative. Usually, the annual cost method is used on a year-by-year basis.

In replacement studies, the existing process or piece of equipment is known as the *defender*. The new process or piece of equipment being considered for purchase is known as the *challenger*.

2. TREATMENT OF SALVAGE VALUE IN REPLACEMENT STUDIES

Since most defenders still have some market value when they are retired, the problem of what to do with the salvage arises. It seems logical to use the salvage value of the defender to reduce the initial purchase cost of the challenger. This is consistent with what would actually happen if the defender were to be retired.

By convention, however, the defender's salvage value is subtracted from the defender's present value. This does not seem logical, but it is done to keep all costs and benefits related to the defender with the defender. In this case, the salvage value is treated as an opportunity cost that would be incurred if the defender is not retired.

If the defender and the challenger have the same lives and a present worth study is used to choose the superior alternative, the placement of the salvage value will have no effect on the net difference between present worths for the challenger and defender. Although the values of the two present worths will be different depending on the placement, the difference in present worths will be the same.

If the defender and the challenger have different lives, an annual cost comparison must be made. Since the salvage value would be "spread over" a different number of years depending on its placement, it is important to abide by the conventions listed in this section.

There are a number of ways to handle salvage value in retirement studies. The best way is to calculate the cost of keeping the defender one more year. In addition to the usual operating and maintenance costs, that cost includes an opportunity interest cost incurred by not selling the defender, and also a drop in the salvage value if the defender is kept for one additional year. Specifically,

$$\underset{\text{(defender)}}{\text{EUAC}} = \text{next year's maintenance costs}$$
$$+\ i(\text{current salvage value}) \quad\quad \textit{62.1}$$
$$+\ \text{current salvage}$$
$$-\ \text{next year's salvage}$$

It is important in retirement studies not to double count the salvage value. That is, it would be incorrect to add the salvage value to the defender and at the same time subtract it from the challenger.

Equation 62.1 contains the difference in salvage value between two consecutive years. This calculation shows that the defender/challenger decision must be made on a year-by-year basis. One application of Eq. 62.1 will not usually answer the question of whether the defender should remain in service indefinitely. The calculation must be repeatedly made as long as there is a drop in salvage value from one year to the next.

3. ECONOMIC LIFE: RETIREMENT AT MINIMUM COST

As an asset grows older, its operating and maintenance costs typically increase. Eventually, the cost to keep the asset in operation becomes prohibitive, and the asset is retired or replaced. However, it is not always obvious when an asset should be retired or replaced.

As the asset's maintenance cost is increasing each year, the amortized cost of its initial purchase is decreasing. It is the sum of these two costs that should be evaluated to determine the point at which the asset should be retired or replaced. Since an asset's initial purchase price is likely to be high, the amortized cost will be the controlling factor in those years when the maintenance costs are low. Therefore, the EUAC of the asset will decrease in the initial part of its life.

However, as the asset grows older, the change in its amortized cost decreases while maintenance cost increases. Eventually, the sum of the two costs reaches a minimum and then starts to increase. The age of the asset at the minimum cost point is known as the *economic life* of the asset. The economic life generally is less than the length of need and the technological lifetime of the asset (see Fig. 62.1.)

Figure 62.1 *EUAC Versus Age at Retirement*

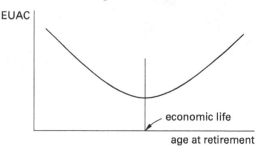

The determination of an asset's economic life is illustrated by Ex. 62.1.

Example 62.1

Buses in a municipal transit system have the characteristics listed. In order to minimize its annual operating expenses, when should the city replace its buses if money can be borrowed at 8%?

initial cost of bus: $120,000

year	maintenance cost	salvage value
1	35,000	60,000
2	38,000	55,000
3	43,000	45,000
4	50,000	25,000
5	65,000	15,000

Solution

The annual maintenance is different each year. Each maintenance cost must be spread over the life of the bus. This is done by first finding the present worth and then amortizing the maintenance costs. Find the annual cost if a bus is kept for one year and then sold. [Factor Table]

$$\begin{aligned}
\text{EUAC}(1) &= (\$120,000)(A/P, 8\%, 1)\\
&\quad + (\$35,000)(A/F, 8\%, 1)\\
&\quad - (\$60,000)(A/F, 8\%, 1)\\
&= (\$120,000)(1.0800) + (\$35,000)(1.000)\\
&\quad - (\$60,000)(1.000)\\
&= \$104,600
\end{aligned}$$

If a bus is kept for two years and then sold, the annual cost will be

$$
\begin{aligned}
\text{EUAC}(2) = \big(&\$120{,}000 + (\$35{,}000)(P/F, 8\%, 1)\big) \\
&\times (A/P, 8\%, 2) \\
&+ (\$38{,}000 - \$55{,}000)(A/F, 8\%, 2) \\
= \big(&\$120{,}000 + (\$35{,}000)(0.9259)\big)(0.5608) \\
&+ (\$38{,}000 - \$55{,}000)(0.4808) \\
= \ &\$77{,}296
\end{aligned}
$$

If a bus is kept for three years and then sold, the annual cost will be

$$
\begin{aligned}
\text{EUAC}(3) = \big(&\$120{,}000 + (\$35{,}000)(P/F, 8\%, 1) \\
&+ (\$38{,}000)(P/F, 8\%, 2)\big)(A/P, 8\%, 3) \\
&+ (\$43{,}000 - \$45{,}000)(A/F, 8\%, 3) \\
= \big(&\$120{,}000 + (\$35{,}000)(0.9259) \\
&+ (\$38{,}000)(0.8573)\big)(0.3880) \\
&- (\$2000)(0.3080) \\
= \ &\$71{,}158
\end{aligned}
$$

This process is continued until the annual cost begins to increase. In this example, EUAC(4) is $71,700. Therefore, the buses should be retired after three years.

4. LIFE-CYCLE COST

The *life-cycle cost* of an alternative is the equivalent value (at $t = 0$) of the alternative's cash flow over the alternative's lifespan. Since the present worth is evaluated using an effective interest rate of i (which would be the interest rate used for all engineering economic analyses), the life-cycle cost is the same as the alternative's present worth. If the alternative has an infinite horizon, the life-cycle cost and capitalized cost will be identical.

5. CAPITALIZED ASSETS VERSUS EXPENSES

High expenses reduce profit, which in turn reduces income tax. It seems logical to label each and every expenditure, even an asset purchase, as an expense. As an alternative to this *expensing the asset*, it may be decided to capitalize the asset. *Capitalizing the asset* means that the cost of the asset is divided into equal or unequal parts, and only one of these parts is taken as an expense each year. Expensing is clearly the more desirable alternative, since the after-tax profit is increased early in the asset's life.

There are long-standing accounting conventions as to what can be expensed and what must be capitalized.[1] Some companies capitalize everything—regardless of cost—with expected lifetimes greater than one year. Most companies, however, expense items whose purchase costs are below a cutoff value. A cutoff value in the range of $250–500, depending on the size of the company, is chosen as the maximum purchase cost of an expensed asset. Assets costing more than this are capitalized.

It is not necessary for a large corporation to keep track of every lamp, desk, and chair for which the purchase price is greater than the cutoff value. Such assets, all of which have the same lives and have been purchased in the same year, can be placed into groups or *asset classes*. A group cost, equal to the sum total of the purchase costs of all items in the group, is capitalized as though the group was an identifiable and distinct asset itself.

6. PURPOSE OF DEPRECIATION

Depreciation is an *artificial expense* that spreads the purchase price of an asset or other property over a number of years.[2] Depreciating an asset is an example of capitalization, as previously defined. The inclusion of depreciation in engineering economic analysis problems will increase the after-tax present worth (profitability) of an asset. The larger the depreciation, the greater will be the profitability. Therefore, individuals and companies eligible to utilize depreciation want to maximize and accelerate the depreciation available to them.

Although the entire property purchase price is eventually recognized as an expense, the net recovery from the expense stream never equals the original cost of the asset. That is, depreciation cannot realistically be thought of as a fund (an annuity or sinking fund) that accumulates capital to purchase a replacement at the end of the asset's life. The primary reason for this is that the depreciation expense is reduced significantly by the impact of income taxes, as will be seen in later sections.

---

[1]For example, purchased vehicles must be capitalized; payments for leased vehicles can be expensed. Repainting a building with paint that will last five years is an expense, but the replacement cost of a leaking roof must be capitalized.

[2]In the United States, the tax regulations of Internal Revenue Service (IRS) allow depreciation on almost all forms of *business property* except land. The following types of property are distinguished: *real* (e.g., buildings used for business), *residential* (e.g., buildings used as rental property), and *personal* (e.g., equipment used for business). Personal property does *not* include items for personal use (such as a personal residence), despite its name. *Tangible personal property* is distinguished from *intangible property* (goodwill, copyrights, patents, trademarks, franchises, agreements not to compete, etc.).

7. DEPRECIATION BASIS OF AN ASSET

The *depreciation basis* of an asset is the part of the asset's purchase price that is spread over the *depreciation period*, also known as the *service life*.[3] Usually, the depreciation basis and the purchase price are not the same.

A common depreciation basis is the difference between the purchase price and the expected salvage value at the end of the depreciation period. That is,

$$\text{depreciation basis} = C - S_n \qquad \text{62.2}$$

There are several methods of calculating the year-by-year depreciation of an asset. Equation 62.2 is not universally compatible with all depreciation methods. Some methods do not consider the salvage value. This is known as an *unadjusted basis*. When the depreciation method is known, the depreciation basis can be rigorously defined.[4]

8. DEPRECIATION METHODS

Generally, tax regulations do not allow the cost of an asset to be treated as a deductible expense in the year of purchase. Rather, portions of the depreciation basis must be allocated to each of the n years of the asset's depreciation period. The amount that is allocated each year is called the *depreciation*.

Various methods exist for calculating an asset's depreciation each year.[5] Although the depreciation calculations may be considered independently (for the purpose of determining book value or as an academic exercise), it is important to recognize that depreciation has no effect on engineering economic analyses unless income taxes are also considered.

Straight-Line Method

With the *straight-line* (SL) *method*, depreciation is the same each year. The depreciation basis ($C - S_n$) is allocated uniformly to all of the n years in the depreciation period. Each year, the depreciation will be

Depreciation: Straight Line

$$D_j = \frac{C - S_n}{n} \qquad \text{62.3}$$

Constant Percentage Method

The *constant percentage method*[6] is similar to the straight-line method in that the depreciation is the same each year. If the fraction of the basis used as depreciation is $1/n$, there is no difference between the constant percentage and straight-line methods. The two methods differ only in what information is available. (With the straight-line method, the life is known. With the constant percentage method, the depreciation fraction is known.)

Each year, the depreciation will be

$$\begin{aligned} D &= (\text{depreciation fraction})(\text{depreciation basis}) \\ &= (\text{depreciation fraction})(C - S_n) \end{aligned} \qquad \text{62.4}$$

Sum-of-the-Years' Digits Method

In *sum-of-the-years' digits* (SOYD) depreciation, the digits from 1 to n inclusive are summed. The total, T, can also be calculated from

$$T = \frac{1}{2}n(n+1) \qquad \text{62.5}$$

The depreciation in year j can be found from Eq. 62.6. Notice that the depreciation in year j, D_j, decreases by a constant amount each year.

$$D_j = \frac{(C - S_n)(n - j + 1)}{T} \qquad \text{62.6}$$

Double Declining Balance Method[7]

Double declining balance[8] (DDB) depreciation is independent of salvage value. Furthermore, the book value never stops decreasing, although the depreciation decreases in magnitude. Usually, any book value in excess of the salvage value is written off in the last year of the asset's depreciation period. Unlike any of the other depreciation methods, double declining balance depends on accumulated depreciation.

$$D_{\text{first year}} = \frac{2C}{n} \qquad \text{62.7}$$

$$D_j = \frac{2\left(C - \sum_{m=1}^{j-1} D_m\right)}{n} \qquad \text{62.8}$$

Calculating the depreciation in the middle of an asset's life appears particularly difficult with double declining balance, since all previous years' depreciation amounts seem to be required. It appears that the depreciation in the sixth year, for example, cannot be calculated unless the values of depreciation for the first five years are calculated. However, this is not true.

Depreciation in the middle of an asset's life can be found from the following equations. (d is known as the *depreciation rate*.)

$$d = \frac{2}{n} \quad \begin{bmatrix} \text{double declining} \\ \text{balance} \end{bmatrix} \qquad 62.9$$

$$D_j = dC(1 - d)^{j-1} \qquad 62.10$$

Statutory Depreciation Systems

In the United States, property placed into service in 1981 and thereafter must use the *Accelerated Cost Recovery System* (ACRS), and after 1986, the *Modified Accelerated Cost Recovery System* (MACRS) or other statutory method. Other methods (straight-line, declining balance, etc.) cannot be used except in special cases.

Property placed into service in 1980 or before must continue to be depreciated according to the method originally chosen (e.g., straight-line, declining balance, or sum-of-the-years' digits). ACRS and MACRS cannot be used.

Under ACRS and MACRS, the cost recovery amount in the jth year of an asset's cost recovery period is calculated by multiplying the initial cost by a factor.

Modified Accelerated Cost Recovery System (MACRS)

$$D_j = (\text{factor})\,C \qquad 62.11$$

The initial cost used is not reduced by the asset's salvage value for ACRS and MACRS calculations. The factor used depends on the asset's cost recovery period. (See Table 62.1.) Such factors are subject to continuing legislation changes. Current tax publications should be consulted before using this method.

Table 62.1 *Representative MACRS Depreciation Factors*[*]
[Taxation]

year, j	depreciation rate for recovery period, n			
	3 years	5 years	7 years	10 years
1	33.33%	20.00%	14.29%	10.00%
2	44.45%	32.00%	24.49%	18.00%
3	14.81%	19.20%	17.49%	14.40%
4	7.41%	11.52%	12.49%	11.52%
5		11.52%	8.93%	9.22%
6		5.76%	8.92%	7.37%
7			8.93%	6.55%
8			4.46%	6.55%
9				6.56%
10				6.55%
11				3.28%

*Values are for the "half-year" convention. This table gives typical values only. Since these factors are subject to continuing revision, they should not be used without consulting an accounting professional.

Production or Service Output Method

If an asset has been purchased for a specific task and that task is associated with a specific lifetime amount of output or production, the depreciation may be calculated by the fraction of total production produced during the year. Under the *units of production* method, the depreciation is not expected to be the same each year.

$$D_j = (C - S_n)\left(\frac{\text{actual output in year } j}{\text{estimated lifetime output}} \right) \quad 62.12$$

Sinking Fund Method

The *sinking fund method* is seldom used in industry because the initial depreciation is low. The formula for sinking fund depreciation (which increases each year) is

$$D_j = (C - S_n)(A/F, i\%, n)(F/P, i\%, j-1) \quad 62.13$$

Example 62.2

An asset is purchased for $9000. Its estimated economic life is 10 years, after which it will be sold for $200. Find the depreciation in the first three years using straight-line, double declining balance, and sum-of-the-years' digits depreciation methods.

Solution

$$\text{SL: } D = \frac{\$9000 - \$200}{10} = \$880 \text{ each year}$$

$$\text{DDB: } D_1 = \frac{(2)(\$9000)}{10} = \$1800 \text{ in year 1}$$

$$D_2 = \frac{(2)(\$9000 - \$1800)}{10} = \$1440 \text{ in year 2}$$

$$D_3 = \frac{(2)(\$9000 - \$3240)}{10} = \$1152 \text{ in year 3}$$

$$\text{SOYD: } T = \left(\frac{1}{2}\right)(10)(11) = 55$$

$$D_1 = \left(\frac{10}{55}\right)(\$9000 - \$200) = \$1600 \text{ in year 1}$$

$$D_2 = \left(\frac{9}{55}\right)(\$8800) = \$1440 \text{ in year 2}$$

$$D_3 = \left(\frac{8}{55}\right)(\$8800) = \$1280 \text{ in year 3}$$

9. ACCELERATED DEPRECIATION METHODS

An *accelerated depreciation method* is one that calculates a depreciation amount greater than a straight-line amount. Double declining balance and sum-of-the-years' digits methods are accelerated methods. The ACRS and MACRS methods are explicitly accelerated methods. Straight-line and sinking fund methods are not accelerated methods.

Assoc. Eng. Principles

Use of an accelerated depreciation method may result in unexpected tax consequences when the depreciated asset or property is disposed of. Professional tax advice should be obtained in this area.

10. BONUS DEPRECIATION

Bonus depreciation is a special, one-time, depreciation authorized by legislation for specific types of equipment, to be taken in the first year, in addition to the standard depreciation normally available for the equipment. Bonus depreciation is usually enacted in order to stimulate investment in or economic recovery of specific industries (e.g., aircraft manufacturing).

11. BOOK VALUE

The difference between original purchase price and accumulated depreciation is known as *book value*.[9] At the end of each year, the book value (which is initially equal to the purchase price) is reduced by the depreciation in that year.

It is important to properly synchronize depreciation calculations. It is difficult to answer the question, "What is the book value in the fifth year?" unless the timing of the book value change is mutually agreed upon. It is better to be specific about an inquiry by identifying when the book value change occurs. For example, the following question is unambiguous: "What is the book value at the end of year 5, after subtracting depreciation in the fifth year?" or "What is the book value after five years?"

Unfortunately, this type of care is seldom taken in book value inquiries, and it is up to the respondent to exercise reasonable care in distinguishing between beginning-of-year book value and end-of-year book value. To be consistent, the book value equations in this chapter have been written in such a way that the year subscript, j, has the same meaning in book value and depreciation calculations. That is, BV_5 means the book value at the end of the fifth year, after five years of depreciation, including D_5, have been subtracted from the original purchase price.

There can be a great difference between the book value of an asset and the *market value* of that asset. There is no legal requirement for the two values to coincide, and no intent for book value to be a reasonable measure of market value.[10] Therefore, it is apparent that book value is merely an accounting convention with little

practical use. Even when a depreciated asset is disposed of, the book value is used to determine the consequences of disposal, not the price the asset should bring at sale.

The calculation of book value is relatively easy, even for the case of the declining balance depreciation method.

For the straight-line depreciation method, the book value at the end of the jth year, after the jth depreciation deduction has been made, is

$$\text{BV}_j = C - \frac{j(C - S_n)}{n} = C - jD \qquad \textbf{62.14}$$

For the sum-of-the-years' digits method, the book value is

$$\text{BV}_j = (C - S_n)\left(1 - \frac{j(2n + 1 - j)}{n(n + 1)}\right) + S_n \qquad \textbf{62.15}$$

For the declining balance method, including double declining balance, the book value is

$$\text{BV}_j = C(1 - d)^j \qquad \textbf{62.16}$$

For the sinking fund method, the book value is calculated directly as

$$\text{BV}_j = C - (C - S_n)(A/F, i\%, n)(F/P, i\%, j) \qquad \textbf{62.17}$$

Of course, the book value at the end of year j can always be calculated for any method by successive subtractions (i.e., subtraction of the accumulated depreciation), as Eq. 62.18 illustrates.

Book Value

$$BV = \text{initial cost} - \sum D_j \qquad \textbf{62.18}$$

Figure 62.2 illustrates the book value of a hypothetical asset depreciated using several depreciation methods. Notice that the double declining balance method initially produces the fastest write-off, while the sinking fund method produces the slowest write-off. Also, the book value does not automatically equal the salvage value at the end of an asset's depreciation period with the double declining balance method.[11]

---

[9]The balance sheet of a corporation usually has two asset accounts: the *equipment account* and the *accumulated depreciation account*. There is no book value account on this financial statement, other than the implicit value obtained from subtracting the accumulated depreciation account from the equipment account. The book values of various assets, as well as their original purchase cost, date of purchase, salvage value, and so on, and accumulated depreciation appear on detail sheets or other peripheral records for each asset.

[10]Common examples of assets with great divergences of book and market values are buildings (rental houses, apartment complexes, factories, etc.) and company luxury automobiles (Porsches, Mercedes, etc.) during periods of inflation. Book values decrease, but actual values increase.

[11]This means that the straight-line method of depreciation may result in a lower book value at some point in the depreciation period than if double declining balance is used. A *cut-over* from double declining balance to straight line may be permitted in certain cases. Finding the *cut-over point*, however, is usually done by comparing book values determined by both methods. The analytical method is complicated.

Figure 62.2 *Book Value with Different Depreciation Methods*

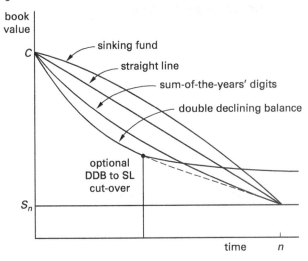

Example 62.3

For the asset described in Ex. 62.2, calculate the book value at the end of the first three years if sum-of-the-years' digits depreciation is used. The book value at the beginning of year 1 is $9000.

Solution

From Eq. 62.18,

$$BV_1 = \$9000 - \$1600 = \$7400$$
$$BV_2 = \$7400 - \$1440 = \$5960$$
$$BV_3 = \$5960 - \$1280 = \$4680$$

12. AMORTIZATION

Amortization and depreciation are similar in that they both divide up the cost basis or value of an asset. In fact, in certain cases, the term "amortization" may be used in place of the term "depreciation." However, depreciation is a specific form of amortization.

Amortization spreads the cost basis or value of an asset over some base. The base can be time, units of production, number of customers, and so on. The asset can be tangible (e.g., a delivery truck or building) or intangible (e.g., goodwill or a patent).

If the asset is tangible, if the base is time, and if the length of time is consistent with accounting standards and taxation guidelines, then the term "depreciation" is appropriate. However, if the asset is intangible, if the base is some other variable, or if some length of time other than the customary period is used, then the term "amortization" is more appropriate.[12]

Example 62.4

A company purchases complete and exclusive patent rights to an invention for $1,200,000. It is estimated that once commercially produced, the invention will have a specific but limited market of 1200 units. For the purpose of allocating the patent right cost to production cost, what is the amortization rate in dollars per unit?

Solution

The patent should be amortized at the rate of

$$\frac{\$1,200,000}{1200 \text{ units}} = \$1000 \text{ per unit}$$

13. DEPLETION

Depletion is another artificial deductible operating expense, designed to compensate mining organizations for decreasing mineral reserves. Since original and remaining quantities of minerals are seldom known accurately, the *depletion allowance* is calculated as a fixed percentage of the organization's gross income. These percentages are usually in the 10–20% range and apply to such mineral deposits as oil, natural gas, coal, uranium, and most metal ores.

14. BASIC INCOME TAX CONSIDERATIONS

The issue of income taxes is often overlooked in academic engineering economic analysis exercises. Such a position is justifiable when an organization (e.g., a non-profit school, a church, or the government) pays no income taxes. However, if an individual or organization is subject to income taxes, the income taxes must be included in an economic analysis of investment alternatives.

Assume that an organization pays a fraction, f, of its profits to the federal government as income taxes. If the organization also pays a fraction s of its profits as state income taxes and if state taxes paid are recognized by the federal government as tax-deductible expenses, then the composite tax rate is

$$t = s + f - sf \qquad \text{62.19}$$

The basic principles used to incorporate taxation into engineering economic analyses are the following.

- Initial purchase expenditures are unaffected by income taxes.

- Salvage revenues are unaffected by income taxes.

---

[12]From time to time, the U.S. Congress has allowed certain types of facilities (e.g., emergency, grain storage, and pollution control) to be written off more rapidly than would otherwise be permitted in order to encourage investment in such facilities. The term "amortization" has been used with such write-off periods.

- Deductible expenses, such as operating costs, maintenance costs, and interest payments, are reduced by the fraction t (i.e., multiplied by the quantity $(1 - t)$).

- Revenues are reduced by the fraction t (i.e., multiplied by the quantity $(1 - t)$).

- Since tax regulations allow the depreciation in any year to be handled as if it were an actual operating expense, and since operating expenses are deductible from the income base prior to taxation, the after-tax profits will be increased. If D is the depreciation, the net result to the after-tax cash flow will be the addition of tD. Depreciation is multiplied by t and added to the appropriate year's cash flow, increasing that year's present worth.

For simplicity, most engineering economics practice problems involving income taxes specify a single income tax rate. In practice, however, federal and most state tax rates depend on the income level. Each range of incomes and its associated tax rate are known as *income bracket* and *tax bracket*, respectively. For example, the state income tax rate might be 4% for incomes up to and including $30,000, and 5% for incomes above $30,000. The income tax for a taxpaying entity with an income of $50,000 would have to be calculated in two parts.

$$\text{tax} = (0.04)(\$30{,}000) + (0.05)(\$50{,}000 - \$30{,}000)$$
$$= \$2200$$

Income taxes and depreciation have no bearing on municipal or governmental projects since municipalities, states, and the U.S. government pay no taxes.

Example 62.5

A corporation that pays 53% of its profit in income taxes invests $10,000 in an asset that will produce a $3000 annual revenue for eight years. If the annual expenses are $700, salvage after eight years is $500, and 9% interest is used, what is the after-tax present worth? Disregard depreciation.

Solution

Calculate the after-tax present worth. [Factor Table]

$$P = -\$10{,}000 + (\$3000)(P/A, 9\%, 8)(1 - 0.53)$$
$$- (\$700)(P/A, 9\%, 8)(1 - 0.53)$$
$$+ (\$500)(P/F, 9\%, 8)$$
$$= -\$10{,}000 + (\$3000)(5.5348)(0.47)$$
$$- (\$700)(5.5348)(0.47) + (\$500)(0.5019)$$
$$= -\$3766$$

15. TAXATION AT THE TIMES OF ASSET PURCHASE AND SALE

There are numerous rules and conventions that governmental tax codes and the accounting profession impose on organizations. Engineers are not normally expected to be aware of most of the rules and conventions, but occasionally it may be necessary to incorporate their effects into an engineering economic analysis.

Tax Credit

A *tax credit* (also known as an *investment tax credit* or *investment credit*) is a one-time credit against income taxes.[13] Therefore, it is added to the after-tax present worth as a last step in an engineering economic analysis. Such tax credits may be allowed by the government from time to time for equipment purchases, employment of various classes of workers, rehabilitation of historic landmarks, and so on.

A tax credit is usually calculated as a fraction of the initial purchase price or cost of an asset or activity.

$$TC = \text{fraction} \times \text{initial cost} \qquad \textit{62.20}$$

When the tax credit is applicable, the fraction used is subject to legislation. A professional tax expert or accountant should be consulted prior to applying the tax credit concept to engineering economic analysis problems.

Since the investment tax credit reduces the buyer's tax liability, a tax credit should be included only in after-tax engineering economic analyses. The credit is assumed to be received at the end of the year.

Gain or Loss on the Sale of a Depreciated Asset

If an asset that has been depreciated over a number of prior years is sold for more than its current book value, the difference between the book value and selling price is taxable income in the year of the sale. Alternatively, if the asset is sold for less than its current book value, the difference between the selling price and book value is an expense in the year of the sale.

Example 62.6

One year, a company makes a $5000 investment in a historic building. The investment is not depreciable, but it does qualify for a one-time 20% tax credit. In that same year, revenue is $45,000 and expenses (exclusive of the $5000 investment) are $25,000. The company pays a total of 53% in income taxes. What is the after-tax present worth of this year's activities if the company's interest rate for investment is 10%?

---

[13]Strictly, *tax credit* is the more general term, and applies to a credit for doing anything creditable. An *investment tax credit* requires an investment in something (usually real property or equipment).

Solution

The tax credit is

$$TC = (0.20)(\$5000) = \$1000$$

This tax credit is assumed to be received at the end of the year. The after-tax present worth is [Factor Table]

$$
\begin{aligned}
P &= -\$5000 + (\$45{,}000 - \$25{,}000)(1 - 0.53) \\
&\quad \times (P/F, 10\%, 1) + (\$1000)(P/F, 10\%, 1) \\
&= -\$5000 + (\$20{,}000)(0.47)(0.9091) \\
&\quad + (\$1000)(0.9091) \\
&= \$4455
\end{aligned}
$$

16. DEPRECIATION RECOVERY

The economic effect of depreciation is to reduce the income tax in year j by tD_j. The present worth of the asset is also affected: The present worth is increased by $tD_j(P/F, i\%, j)$. The after-tax present worth of all depreciation effects over the depreciation period of the asset is called the *depreciation recovery* (DR).[14]

$$DR = t\sum_{j=1}^{n} D_j(P/F, i\%, j) \qquad 62.21$$

There are multiple ways depreciation can be calculated, as summarized in Table 62.2. *Straight-line* (SL) *depreciation recovery* from an asset is easily calculated, since the depreciation is the same each year. Assuming the asset has a constant depreciation of D and depreciation period of n years, the depreciation recovery is

$$DR = tD(P/A, i\%, n) \qquad 62.22$$

Depreciation: Straight Line

$$D_j = \frac{C - S_n}{n} \qquad 62.23$$

Sum-of-the-years' digits (SOYD) *depreciation recovery* is also relatively easily calculated, since the depreciation decreases uniformly each year.

$$
\begin{aligned}
DR &= \left(\frac{t(C - S_n)}{T}\right) \\
&\quad \times \big(n(P/A, i\%, n) - (P/G, i\%, n)\big)
\end{aligned} \qquad 62.24
$$

Finding *declining balance depreciation recovery* is more involved. There are three difficulties. The first (the apparent need to calculate all previous depreciations in order to determine the subsequent depreciation) has already been addressed by Eq. 62.10.

The second difficulty is that there is no way to ensure (that is, to force) the book value to be S_n at $t = n$. Therefore, it is common to write off the remaining book value (down to S_n) at $t = n$ in one lump sum. This assumes $BV_n \geq S_n$.

The third difficulty is that of finding the present worth of an *exponentially decreasing cash flow*. Although the proof is omitted here, such exponential cash flows can be handled with the *exponential gradient factor*, (P/EG).[15]

$$(P/EG, z - 1, n) = \frac{z^n - 1}{z^n(z - 1)} \qquad 62.25$$

$$z = \frac{1 + i}{1 - d} \qquad 62.26$$

Then, as long as $BV_n > S_n$, the declining balance depreciation recovery is

$$DR = tC\left(\frac{d}{1 - d}\right)(P/EG, z - 1, n) \qquad 62.27$$

Example 62.7

For the asset described in Ex. 62.2, calculate the after-tax depreciation recovery with straight-line and sum-of-the-years' digits depreciation methods. Use 6% interest with 48% income taxes.

Solution

Using SL, the depreciation recovery is

$$
\begin{aligned}
DR &= (0.48)(\$880)(P/A, 6\%, 10) \\
&= (0.48)(\$880)(7.3601) \\
&= \$3109
\end{aligned}
$$

Using SOYD, the depreciation series can be thought of as a constant $1600 term with a negative $160 gradient.

$$
\begin{aligned}
DR &= (0.48)(\$1600)(P/A, 6\%, 10) \\
&\quad - (0.48)(\$160)(P/G, 6\%, 10) \\
&= (0.48)(\$1600)(7.3601) \\
&\quad - (0.48)(\$160)(29.6023) \\
&= \$3379
\end{aligned}
$$

The ten-year (P/G) factor is used even though there are only nine years in which the gradient reduces the initial $1600 amount.

Assoc. Eng. Principles

---

[14]Since the depreciation benefit is reduced by taxation, depreciation cannot be thought of as an annuity to fund a replacement asset.
[15]The (P/A) columns in a table of cash flow equivalent factors can be used for (P/EG) as long as the interest rate is assumed to be $z - 1$.

Table 62.2 *Depreciation Calculation Summary [Depreciation: Straight Line]*

method	depreciation basis	depreciation in year j, D_j	book value after jth depreciation, BV_j	present worth of after-tax depreciation recovery, (DR)	supplementary formulas
straight-line (SL)	$C - S_n$	$\dfrac{C - S_n}{n}$ (constant)	$C - jD$	$tD(P/A, i\%, n)$	
constant percentage	$C - S_n$	$\dfrac{\text{fraction} \times (C - S_n)}{\text{(constant)}}$	$C - jD$	$tD(P/A, i\%, n)$	
sum-of-the-years' digits (SOYD)	$C - S_n$	$\dfrac{(C - S_n) \times (n - j + 1)}{T}$	$(C - S_n) \times \left[1 - \dfrac{j(2n+1-j)}{n(n+1)}\right] + S_n$	$\dfrac{t(C-S_n)}{T} \times \left(n(P/A, i\%, n) - (P/G, i\%, n) \right)$	$T = \frac{1}{2}n(n+1)$
double declining balance (DDB)	C	$dC(1-d)^{j-1}$	$C(1-d)^j$	$tC\left(\dfrac{d}{1-d}\right) \times (P/EG, z - 1, n)$	$d = \dfrac{2}{n};\ z = \dfrac{1+i}{1-d}$ $(P/EG, z - 1, n) = \dfrac{z^n - 1}{z^n(z - 1)}$
sinking fund (SF)	$C - S_n$	$(C - S_n) \times (A/F, i\%, n) \times (F/P, i\%, j-1)$	$C - (C - S_n) \times (A/F, i\%, n) \times (F/A, i\%, j)$	$\dfrac{t(C-S_n)(A/F, i\%, n)}{1 + i}$	
accelerated cost recovery system (ACRS/ MACRS)	C	$C \times \text{factor}$	$C - \displaystyle\sum_{m=1}^{j} D_m$	$t\displaystyle\sum_{j=1}^{n} D_j(P/F, i\%, j)$	
units of production or service output	$C - S_n$	$(C - S_n) \times \left(\dfrac{\text{actual output in year } j}{\text{lifetime output}}\right)$	$C - \displaystyle\sum_{m=1}^{j} D_m$	$t\displaystyle\sum_{j=1}^{n} D_j(P/F, i\%, j)$	

Example 62.8

What is the after-tax present worth of the asset described in Ex. 62.5 if straight-line, sum-of-the-years' digits, and double declining balance depreciation methods are used?

Solution

Using SL, the depreciation recovery is

$$\text{DR} = (0.53)\left(\frac{\$10{,}000 - \$500}{8}\right)(P/A, 9\%, 8)$$

$$= (0.53)\left(\frac{\$9500}{8}\right)(5.5348)$$

$$= \$3483$$

Using SOYD, the depreciation recovery is calculated as follows.

$$T = \left(\tfrac{1}{2}\right)(8)(9) = 36$$

$$\text{depreciation base} = \$10{,}000 - \$500 = \$9500$$

$$D_1 = \left(\tfrac{8}{36}\right)(\$9500) = \$2111$$

$$G = \left(\tfrac{1}{36}\right)(\$9500) = \$264$$

$$\text{DR} = (0.53)\big((\$2111)(P/A, 9\%, 8)$$
$$\quad - (\$264)(P/G, 9\%, 8)\big)$$
$$= (0.53)\big((\$2111)(5.5348)$$
$$\quad - (\$264)(16.8877)\big)$$
$$= \$3830$$

Using DDB, the depreciation recovery is calculated as follows.[16]

$$d = \frac{2}{8} = 0.25$$

$$z = \frac{1+0.09}{1-0.25} = 1.4533$$

$$(P/EG, z-1, n) = \frac{(1.4533)^8 - 1}{(1.4533)^8(0.4533)} = 2.095$$

From Eq. 62.27,

$$DR = (0.53)\left(\frac{(0.25)(\$10{,}000)}{0.75}\right)(2.095)$$

$$= \$3701$$

The after-tax present worth, neglecting depreciation, was previously found to be $-\$3766$.

The after-tax present worths, including depreciation recovery, are

$$SL: P = -\$3766 + \$3483 = -\$283$$
$$SOYD: P = -\$3766 + \$3830 = \quad \$64$$
$$DDB: P = -\$3766 + \$3701 = -\ \$65$$

17. OTHER INTEREST RATES

The *effective interest rate per period*, i (also called *yield* by banks), is the only interest rate that should be used in equivalence equations. The interest rates given in tables of cash flow equivalent factors are implicitly all effective interest rates. Usually, the period will be one year, hence the name *effective annual interest rate*. However, there are other interest rates in use as well.

The term *nominal interest rate*, r (*rate per annum*), is encountered when compounding is more than once per year. The nominal rate does not include the effect of compounding and is not the same as the effective rate. And, since the effective interest rate can be calculated from the nominal rate only if the number of compounding periods per year is known, nominal rates cannot be compared unless the method of compounding is specified. The only practical use for a nominal rate per year is for calculating the effective rate per period.

18. RATE AND PERIOD CHANGES

If there are m compounding periods during the year (two for semiannual compounding, four for quarterly compounding, twelve for monthly compounding, etc.) and the nominal rate is r, the *effective rate per compounding period* is

$$\phi = \frac{r}{m} \qquad \qquad 62.28$$

The effective annual rate, i_e, can be calculated by using Eq. 62.29.

Non-Annual Compounding

$$i_e = \left(1 + \frac{r}{m}\right)^m - 1 \qquad \qquad 62.29$$

Sometimes, only the effective rate per period (e.g., per month) is known. However, that will be a simple problem since compounding for m periods at an effective rate per period is not affected by the definition or length of the period.

The following rules may be used to determine which interest rate is given.

- Unless specifically qualified, the interest rate given is an annual rate.

- If the compounding is annual, the rate given is the effective rate. If compounding is other than annual, the rate given is the nominal rate.

The effective annual interest rate determined on a *daily compounding basis* will not be significantly different than if *continuous compounding* is assumed.[17] In the case of continuous (or daily) compounding, the discounting factors can be calculated directly from the nominal interest rate and number of years, without having to find the effective interest rate per period. Table 62.3 can be used to determine the discount factors for continuous compounding.

Table 62.3 Discount Factors for Continuous Compounding (n is the number of years)

symbol	formula
$(F/P, r\%, n)$	e^{rn}
$(P/F, r\%, n)$	e^{-rn}
$(A/F, r\%, n)$	$\dfrac{e^r - 1}{e^{rn} - 1}$
$(F/A, r\%, n)$	$\dfrac{e^{rn} - 1}{e^r - 1}$
$(A/P, r\%, n)$	$\dfrac{e^r - 1}{1 - e^{-rn}}$
$(P/A, r\%, n)$	$\dfrac{1 - e^{-rn}}{e^r - 1}$

Example 62.9

A savings and loan offers a nominal rate of 5.25% compounded daily over 365 days in a year. What is the effective annual rate?

---

[16]This method should start by checking that the book value at the end of the depreciation period is greater than the salvage value. In this example, such is the case. However, the step is not shown.
[17]The number of *banking days in a year* (250, 360, etc.) must be specifically known.

Solution

method 1: Use Eq. 62.29.

Non-Annual Compounding

$$r = 0.0525, m = 365$$

$$i_e = \left(1 + \frac{r}{m}\right)^m - 1 = \left(1 + \frac{0.0525}{365}\right)^{365} - 1 = 0.0539$$

method 2: Assume daily compounding is the same as continuous compounding.

$$i = (F/P, r\%, 1) - 1$$
$$= e^{0.0525} - 1$$
$$= 0.0539$$

Example 62.10

A real estate investment trust pays $7,000,000 for an apartment complex with 100 units. The trust expects to sell the complex in 10 years for $15,000,000. In the meantime, it expects to receive an average rent of $900 per month from each apartment. Operating expenses are expected to be $200 per month per occupied apartment. A 95% occupancy rate is predicted. In similar investments, the trust has realized a 12% effective annual return on its investment. Compare to those past investments the expected present worth of this investment when calculated assuming (a) annual compounding (i.e., the year-end convention), and (b) monthly compounding. Disregard taxes, depreciation, and all other factors.

Solution

(a) The net annual income will be

$$(0.95)(100 \text{ units})\left(\frac{\$900}{\text{unit-mo}} - \frac{\$200}{\text{unit-mo}}\right)\left(12 \frac{\text{mo}}{\text{yr}}\right)$$
$$= \$798,000/\text{yr}$$

The present worth of ten years of operation is

$$P = -\$7,000,000 + (\$798,000)(P/A, 12\%, 10)$$
$$\quad + (\$15,000,000)(P/F, 12\%, 10)$$
$$= -\$7,000,000 + (\$798,000)(5.6502)$$
$$\quad + (\$15,000,000)(0.3220)$$
$$= \$2,344,000$$

(b) The net monthly income is

$$(0.95)(100 \text{ units})\left(\frac{\$900}{\text{unit-mo}} - \frac{\$200}{\text{unit-mo}}\right)$$
$$= \$66,500/\text{mo}$$

Rearrange Eq. 62.29 to calculate the effective monthly rate, r/m, from the effective annual rate, $i_e = 12\%$, and the number of compounding periods per year, $m = 12$.

Non-Annual Compounding

$$i_e = \left(1 + \frac{r}{m}\right)^m - 1$$

$$\frac{r}{m} = (1 + i_e)^{1/m} - 1$$
$$= (1 + 0.12)^{1/12} - 1$$
$$= 0.009489 \quad (0.9489\%)$$

The number of compounding periods in 10 years is

$$n = (10 \text{ yr})\left(12 \frac{\text{mo}}{\text{yr}}\right) = 120 \text{ mo}$$

The present worth of 120 months of operation is

$$P = -\$7,000,000 + (\$66,500)(P/A, 0.9489\%, 120)$$
$$\quad + (\$15,000,000)(P/F, 0.9489\%, 120)$$

Since table values for 0.9489% discounting factors are not available, use the formulas for the uniform series present worth factor and the single payment present worth factor. [Economics]

$$(P/A, 0.9489\%, 120) = \frac{(1 + i)^n - 1}{i(1 + i)^n}$$
$$= \frac{(1 + 0.009489)^{120} - 1}{(0.009489)(1 + 0.009489)^{120}}$$
$$= 59.638$$
$$(P/F, 0.9489\%, 120) = (1 + i)^{-n} = (1 + 0.009489)^{-120}$$
$$= 0.3220$$

The present worth over 120 monthly compounding periods is

$$P = -\$7,000,000 + (\$66,500)(64.261)$$
$$\quad + (\$15,000,000)(0.3220)$$
$$= \$1,796,000$$

Assoc. Eng. Principles

19. BONDS

A *bond* is a method of long-term financing commonly used by governments, states, municipalities, and very large corporations.[18] The bond represents a contract to pay the bondholder specific amounts of money at specific times. The holder purchases the bond in exchange for specific payments of interest and principal. Typical municipal bonds call for quarterly or semiannual interest payments and a payment of the *face value of the bond* on the *date of maturity* (end of the bond period).[19] Due to the practice of discounting in the bond market, a bond's face value and its purchase price generally will not coincide.

In the past, a bondholder had to submit a coupon or ticket in order to receive an interim interest payment. This has given rise to the term *coupon rate*, which is the nominal annual interest rate on which the interest payments are made. Coupon books are seldom used with modern bonds, but the term survives. The coupon rate determines the magnitude of the semiannual (or otherwise) interest payments during the life of the bond. The bondholder's own effective interest rate should be used for economic decisions about the bond.

Actual *bond yield* is the bondholder's actual rate of return of the bond, considering the purchase price, interest payments, and face value payment (or, value realized if the bond is sold before it matures). By convention, bond yield is calculated as a nominal rate (rate per annum), not an effective rate per year. The bond yield should be determined by finding the effective rate of return per payment period (e.g., per semiannual interest payment) as a conventional rate of return problem. Then, the nominal rate can be found by multiplying the effective rate per period by the number of payments per year, as in Eq. 62.29.

Example 62.11

What is the maximum amount an investor should pay for a 25-year bond with a $20,000 face value and 8% coupon rate (interest only paid semiannually)? The bond will be kept to maturity. The investor's effective annual interest rate for economic decisions is 10%.

Solution

For this problem, take the compounding period to be six months. Then, there are 50 compounding periods. Since 8% is a nominal rate, the effective bond rate per period is calculated from Eq. 62.28 as

$$\phi_{\text{bond}} = \frac{r}{m} = \frac{8\%}{2} = 4\%$$

The bond payment received semiannually is

$$(0.04)(\$20{,}000) = \$800$$

10% is the investor's effective rate per year. Rearrange Eq. 62.29 to calculate the effective analysis rate per period.

Non-Annual Compounding

$$i_e = \left(1 + \frac{r}{m}\right)^{m} - 1$$

$$\frac{r}{m} = (1 + i_e)^{1/m} - 1 = (1 + 0.10)^{1/2} - 1$$

$$= 0.04881 \quad (4.88\%)$$

The maximum amount that the investor should be willing to pay is the present worth of the investment.

$$P = (\$800)(P/A, 4.88\%, 50)$$
$$+ (\$20{,}000)(P/F, 4.88\%, 50)$$

Calculate the needed cash flow factors using the formulas for the uniform series present worth factor and the single payment present worth factor. [Economics]

$$(P/A, 4.88\%, 50) = \frac{(1 + i)^n - 1}{i(1 + i)^n}$$

$$= \frac{(1 + 0.0488)^{50} - 1}{(0.0488)(1.0488)^{50}}$$

$$= 18.600$$

$$(P/F, 4.88\%, 50) = \frac{1}{(1 + i)^n}$$

$$= \frac{1}{(1 + 0.0488)^{50}}$$

$$= 0.09233$$

Then, the present worth is

$$P = (\$800)(18.600) + (\$20{,}000)(0.09233)$$
$$= \$16{,}727$$

20. PROBABILISTIC PROBLEMS

If an alternative's cash flows are specified by an implicit or explicit probability distribution rather than being known exactly, the problem is *probabilistic*.

Probabilistic problems typically possess the following characteristics.

---

[18]In the past, 30-year bonds were typical. Shorter term 10-year, 15-year, 20-year, and 25-year bonds are also commonly issued.
[19]A *fully amortized bond* pays back interest and principal throughout the life of the bond. There is no balloon payment.

- There is a chance of loss that must be minimized (or, rarely, a chance of gain that must be maximized) by selection of one of the alternatives.

- There are multiple alternatives. Each alternative offers a different degree of protection from the loss. Usually, the alternatives with the greatest protection will be the most expensive.

- The magnitude of loss or gain is independent of the alternative selected.

Probabilistic problems are typically solved using annual costs and expected values. An *expected value* is similar to an *average value* since it is calculated as the mean of the given probability distribution. If cost 1 has a probability of occurrence, p_1, cost 2 has a probability of occurrence, p_2, and so on, the expected value is

$$\mathcal{E}\{\text{cost}\} = p_1(\text{cost 1}) + p_2(\text{cost 2}) + \cdots \quad 62.30$$

Example 62.12

Flood damage in any year is given according to the following table. What is the present worth of flood damage for a 10-year period? Use 6% as the effective annual interest rate.

damage	probability
0	0.75
$10,000	0.20
$20,000	0.04
$30,000	0.01

Solution

The expected value of flood damage in any given year is

$$\mathcal{E}\{\text{damage}\} = (0)(0.75) + (\$10{,}000)(0.20)$$
$$+ (\$20{,}000)(0.04) + (\$30{,}000)(0.01)$$
$$= \$3100$$

Calculate the present worth of 10 years of expected flood damage. [Factor Table]

$$\text{present worth} = (\$3100)(P/A, 6\%, 10)$$
$$= (\$3100)(7.3601)$$
$$= \$22{,}816$$

Example 62.13

A dam is being considered on a river that periodically overflows and causes $600,000 damage. The damage is essentially the same each time the river causes flooding. The project horizon is 40 years. A 10% interest rate is being used.

Three different designs are available, each with different costs and storage capacities.

design alternative	cost	maximum capacity
A	$500,000	1 unit
B	$625,000	1.5 units
C	$900,000	2.0 units

The National Weather Service has provided a statistical analysis of annual rainfall runoff from the watershed draining into the river.

units annual rainfall	probability
0	0.10
0.1–0.5	0.60
0.6–1.0	0.15
1.1–1.5	0.10
1.6–2.0	0.04
2.1 or more	0.01

Which design alternative would you choose assuming the dam is essentially empty at the start of each rainfall season?

Solution

The sum of the construction cost and the expected damage should be minimized. If alternative A is chosen, it will have a capacity of 1 unit. Its capacity will be exceeded (causing $600,000 damage) when the annual rainfall exceeds 1 unit. Find the expected value of the annual cost of alternative A. [Factor Table]

$$\mathcal{E}\{\text{EUAC(A)}\} = (\$500{,}000)(A/P, 10\%, 40)$$
$$+ (\$600{,}000)(0.10 + 0.04 + 0.01)$$
$$= (\$500{,}000)(0.1023) + (\$600{,}000)(0.15)$$
$$= \$141{,}150$$

Similarly,

$$\mathcal{E}\{\text{EUAC(B)}\} = (\$625{,}000)(A/P, 10\%, 40)$$
$$+ (\$600{,}000)(0.04 + 0.01)$$
$$= (\$625{,}000)(0.1023) + (\$600{,}000)(0.05)$$
$$= \$93{,}938$$

$$\mathcal{E}\{\text{EUAC(C)}\} = (\$900{,}000)(A/P, 10\%, 40)$$
$$+ (\$600{,}000)(0.01)$$
$$= (\$900{,}000)(0.1023) + (\$600{,}000)(0.01)$$
$$= \$98{,}070$$

Alternative B should be chosen.

21. WEIGHTED COSTS

The reliability of preliminary cost estimates can be increased by considering as much historical data as possible. For example, the cost of finishing a concrete slab when the contractor has not yet been selected should be estimated from actual recent costs from as many local jobs as is practical. Most jobs are not directly comparable, however, because they differ in size or in some other characteristic. A *weighted cost* (*weighted average cost*) is a cost that has been averaged over some rational basis. The weighted cost is calculated by weighting the individual cost elements, C_i, by their respective weights, w_i. Respective weights are usually *relative fractions* (*relative importance*) based on other characteristics, such as length, area, number of units, frequency of occurrence, points, etc.

$$C_{\text{weighted}} = w_1 C_1 + w_2 C_2 + \cdots + w_N C_N \qquad \text{62.31}$$

$$w_j = \frac{A_j}{\sum\limits_{i=1}^{N} A_i} \qquad \text{[weighting by area]} \qquad \text{62.32}$$

It is implicit in calculating weighted average costs that the costs with the largest weights are more important to the calculation. For example, costs from a job of 10,000 ft^2 are considered to be twice as reliable (important, relevant, etc.) as costs from a 5000 ft^2 job. This assumption must be carefully considered.

Determining a weighted average from Eq. 62.31 disregards the fixed and variable natures of costs. In effect, fixed costs are allocated over entire jobs, increasing the apparent variable costs. For that reason, it is important to include in the calculation only costs of similar cases. This requires common sense segregation of the initial cost data. For example, it is probably not appropriate to include very large (e.g., supermarket and mall) concrete finishing job costs in the calculation of a weighted average cost used to estimate residential slab finishing costs.

Example 62.14

Calculate a weighted average cost per unit length of concrete wall to be used for estimating future job costs. Data from three similar walls are available.

    wall 1: length, 50 ft; actual cost, $9000

    wall 2: length, 90 ft; actual cost, $15,000

    wall 3: length, 65 ft; actual cost, $11,000

Solution

Consider wall 1. The cost per foot of wall is $9000/50 ft. According to Eq. 62.32, accuracy of cost data is proportional to wall length (i.e., cost per foot from a wall of length $2X$ is twice as reliable as cost per foot from a wall

of length X). This cost per foot value is given a weight of 50 ft/(50 ft + 90 ft + 65 ft). So, the weighted cost per foot for wall 1 is

$$w_1 C_{\text{ft},1} = \frac{(50\ \text{ft})\left(\dfrac{\$9000}{50\ \text{ft}}\right)}{50\ \text{ft} + 90\ \text{ft} + 65\ \text{ft}} = \frac{\$9000}{50\ \text{ft} + 90\ \text{ft} + 65\ \text{ft}}$$

Similarly, the weighted cost per foot for wall 2 is $w_2 C_{\text{ft},2} = \$15,000/(50\ \text{ft} + 90\ \text{ft} + 65\ \text{ft})$, and the weighted cost per foot for wall 3 is $w_3 C_{\text{ft},3} = \$11,000/(50\ \text{ft} + 90\ \text{ft} + 65\ \text{ft})$.

From Eq. 62.31, the total weighted cost per foot is

$$\begin{aligned} C_{\text{ft}} &= w_1 C_{\text{ft},1} + w_2 C_{\text{ft},2} + w_3 C_{\text{ft},3} \\ &= \frac{\$9000 + \$15,000 + \$11,000}{50\ \text{ft} + 90\ \text{ft} + 65\ \text{ft}} \\ &= \$170.73\ \text{per foot} \end{aligned}$$

22. FIXED AND VARIABLE COSTS

The distinction between fixed and variable costs depends on how these costs vary when an independent variable changes. For example, factory or machine production is frequently the independent variable. However, it could just as easily be vehicle miles driven, hours of operation, or quantity (mass, volume, etc.). Examples of fixed and variable costs are given in Table 62.4.

If a cost is a function of the independent variable, the cost is said to be a *variable cost*. The change in cost per unit variable change (i.e., what is usually called the *slope*) is known as the *incremental cost*. Material and labor costs are examples of variable costs. They increase in proportion to the number of product units manufactured.

If a cost is not a function of the independent variable, the cost is said to be a *fixed cost*. Rent and lease payments are typical fixed costs. These costs will be incurred regardless of production levels.

Some costs have both fixed and variable components, as Fig. 62.3 illustrates. The fixed portion can be determined by calculating the cost at zero production.

An additional category of cost is the *semivariable cost*. This type of cost increases stepwise. Semivariable cost structures are typical of situations where *excess capacity* exists. For example, supervisory cost is a stepwise function of the number of production shifts. Also, labor cost for truck drivers is a stepwise function of weight (volume) transported. As long as a truck has room left (i.e., excess capacity), no additional driver is needed. As soon as the truck is filled, labor cost will increase.

Table 62.4 *Summary of Fixed and Variable Costs*

fixed costs
  rent
  property taxes
  interest on loans
  insurance
  janitorial service expense
  tooling expense
  setup, cleanup, and tear-down expenses
  depreciation expense
  marketing and selling costs
  cost of utilities
  general burden and overhead expense
variable costs
  direct material costs
  direct labor costs
  cost of miscellaneous supplies
  payroll benefit costs
  income taxes
  supervision costs

Figure 62.3 *Fixed and Variable Costs*

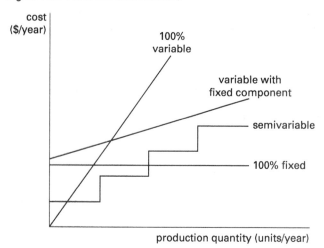

23. ACCOUNTING COSTS AND EXPENSE TERMS

The accounting profession has developed special terms for certain groups of costs. When annual costs are incurred due to the functioning of a piece of equipment, they are known as *operating and maintenance* (O&M) *costs.* The annual costs associated with operating a business (other than the costs directly attributable to production) are known as *general, selling, and administrative* (GS&A) *expenses.*

Figure 62.4 *Costs and Expenses Combined*

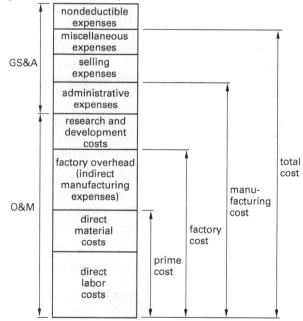

Direct labor costs are costs incurred in the factory, such as assembly, machining, and painting labor costs. *Direct material costs* are the costs of all materials that go into production.[20] Typically, both direct labor and direct material costs are given on a per-unit or per-item basis. The sum of the direct labor and direct material costs is known as the *prime cost.*

There are certain additional expenses incurred in the factory, such as the costs of factory supervision, stock-picking, quality control, factory utilities, and miscellaneous supplies (cleaning fluids, assembly lubricants, routing tags, etc.) that are not incorporated into the final product. Such costs are known as *indirect manufacturing expenses* (IME) or *indirect material and labor costs.*[21] The sum of the per-unit indirect manufacturing expense and prime cost is known as the *factory cost.*

Research and development (R&D) *costs* and *administrative expenses* are added to the factory cost to give the *manufacturing cost* of the product.

Additional costs are incurred in marketing the product. Such costs are known as *selling expenses* or *marketing expenses.* The sum of the selling expenses and manufacturing cost is the *total cost* of the product. Figure 62.4 illustrates these terms.[22]

The distinctions among the various forms of cost (particularly with overhead costs) are not standardized. Each company must develop a classification system to deal with the various cost factors in a consistent manner. There are also other terms in use (e.g., *raw materials, operating supplies, general plant overhead*), but

---

[20]There may be problems with pricing the material when it is purchased from an outside vendor and the stock on hand derives from several shipments purchased at different prices.
[21]The *indirect material and labor costs* usually exclude costs incurred in the office area.
[22]*Total cost* does not include income taxes.

these terms must be interpreted within the framework of each company's classification system. Table 62.5 is typical of such classification systems.

24. ACCOUNTING PRINCIPLES

Basic Bookkeeping

An accounting or *bookkeeping system* is used to record historical financial transactions. The resultant records are used for product costing, satisfaction of statutory requirements, reporting of profit for income tax purposes, and general company management.

Bookkeeping consists of two main steps: recording the transactions, followed by categorization of the transactions.[23] The transactions (receipts and disbursements) are recorded in a *journal* (*book of original entry*) to complete the first step. Such a journal is organized in a simple chronological and sequential manner. The transactions are then categorized (into interest income, advertising expense, etc.) and posted (i.e., entered or written) into the appropriate *ledger account*.[24]

The ledger accounts together constitute the *general ledger* or *ledger*. All ledger accounts can be classified into one of three types: *asset accounts*, *liability accounts*, and *owners' equity accounts*. Strictly speaking, income and expense accounts, kept in a separate journal, are included within the classification of owners' equity accounts.

Together, the journal and ledger are known simply as "the books" of the company, regardless of whether bound volumes of pages are actually involved.

Balancing the Books

In a business environment, *balancing the books* means more than reconciling the checkbook and bank statements. All accounting entries must be posted in such a way as to maintain the equality of the *basic accounting equation*,

$$\text{assets} = \text{liability} + \text{owner's equity} \qquad 62.33$$

In a *double-entry bookkeeping system*, the equality is maintained within the ledger system by entering each transaction into two balancing ledger accounts. For example, paying a utility bill would decrease the cash account (an asset account) and decrease the utility expense account (a liability account) by the same amount.

Transactions are either *debits* or *credits*, depending on their sign. Increases in asset accounts are debits; decreases are credits. For liability and equity accounts, the opposite is true: Increases are credits, and decreases are debits.[25]

Cash and Accrual Systems[26]

The simplest form of bookkeeping is based on the *cash system*. The only transactions that are entered into the journal are those that represent cash receipts and disbursements. In effect, a checkbook register or bank deposit book could serve as the journal.

During a given period (e.g., month or quarter), expense liabilities may be incurred even though the payments for those expenses have not been made. For example, an invoice (bill) may have been received but not paid. Under the *accrual system*, the obligation is posted into the appropriate expense account before it is paid.[27] Analogous to expenses, under the accrual system, income will be claimed before payment is received. Specifically, a sales transaction can be recorded as income when the customer's order is received, when the outgoing invoice is generated, or when the merchandise is shipped.

Financial Statements

Each period, two types of corporate financial statements are typically generated: the *balance sheet* and *profit and loss* (P&L) *statements worth*.[28] The profit and loss statement, also known as a *statement of income and retained earnings*, is a summary of sources of *income* or *revenue* (interest, sales, fees charged, etc.) and *expenses* (utilities, advertising, repairs, etc.) for the period. The expenses are subtracted from the revenues to give a *net income* (generally, before taxes).[29] Figure 62.5 illustrates a simple profit and loss statement.

[23]These two steps are not to be confused with the *double-entry bookkeeping method*.

[24]The two-step process is more typical of a *manual bookkeeping system* than a computerized *general ledger system*. However, even most computerized systems produce reports in journal entry order, as well as account summaries.

[25]There is a difference in sign between asset and liability accounts. An increase in an expense account is actually a decrease. The accounting profession, apparently, is comfortable with the common confusion that exists between debits and credits.

[26]There is also a distinction made between cash flows that are known and those that are expected. It is a *standard accounting principle* to record losses in full, at the time they are recognized, even before their occurrence. In the construction industry, for example, losses are recognized in full and projected to the end of a project as soon as they are foreseeable. Profits, on the other hand, are recognized only as they are realized (typically, as a percentage of project completion). The difference between cash and accrual systems is a matter of *bookkeeping*. The difference between loss and profit recognition is a matter of *accounting convention*. Engineers seldom need to be concerned with the accounting tradition.

[27]The expense for an item or service might be accrued even *before* the invoice is received. It might be recorded when the purchase order for the item or service is generated, or when the item or service is received.

[28]Other types of financial statements (*statements of changes in financial position*, *cost of sales statements*, inventory and asset reports, etc.) also will be generated, depending on the needs of the company.

[29]Financial statements also can be prepared with percentages (of total assets and net revenue) instead of dollars, in which case they are known as *common size financial statements*.

Table 62.5 Typical Classification of Expenses

direct labor expenses

  machining and forming

  assembly

  finishing

  inspection

  testing

direct material expenses

  items purchased from other vendors

  manufactured assemblies

factory overhead expenses (indirect manufacturing expenses)

  supervision

  benefits (e.g., pension, medical insurance, vacations)

  wages overhead (e.g., unemployment comp., SS, & disability taxes)

  stock-picking

  quality control and inspection

  expediting

  rework

  maintenance

  miscellaneous supplies (e.g., routing tags, assembly lubricants, cleaning fluids, wiping cloths, janitorial supplies)

  packaging (materials and labor)

  factory utilities

  laboratory

  depreciation on factory equipment

research and development expenses

  engineering (labor)

  patents

  testing

  prototypes (material and labor)

  drafting

  O&M of R&D facility

administrative expenses

  corporate officers

  accounting

  secretarial/clerical/reception

  security (protection)

  medical (nurse)

  employment (personnel)

  reproduction

  data processing

  production control

  depreciation on nonfactory equipment

  office supplies

  office utilities

  O&M of offices

selling expenses

  marketing (labor)

  advertising

  transportation (if not paid by customer)

  outside sales force (labor and expenses)

  demonstration units

  commissions

  technical service and support

  order processing

  branch office expenses

miscellaneous expenses

  insurance

  property taxes

  interest on loans

nondeductible expenses

  federal income taxes

  fines and penalties

Figure 62.5 Simplified Profit and Loss Statement

revenue		
interest	2000	
sales	237,000	
returns	(23,000)	
net revenue		216,000
expenses		
salaries	149,000	
utilities	6000	
advertising	28,000	
insurance	4000	
supplies	1000	
net expenses		188,000
period net income		28,000
beginning retained earnings		63,000
net year-to-date earnings		91,000

The *balance sheet* presents the *basic accounting equation* in tabular form. The balance sheet lists the major categories of assets and outstanding liabilities. The difference between asset values and liabilities is the *equity*, as defined in Eq. 62.33. This equity represents what would be left over after satisfying all debts by liquidating the company.

There are several terms that appear regularly on balance sheets (see Fig. 62.6).

- *current assets:* cash and other assets that can be converted quickly into cash, such as accounts receivable, notes receivable, and merchandise (inventory). Also known as *liquid assets.*

- *fixed assets:* relatively permanent assets used in the operation of the business and relatively difficult to convert into cash. Examples are land, buildings, and equipment. Also known as *nonliquid assets.*

- *current liabilities:* liabilities due within a short period of time (e.g., within one year) and typically paid out of current assets. Examples are accounts payable, notes payable, and other accrued liabilities.

- *long-term liabilities:* obligations that are not totally payable within a short period of time (e.g., within one year).

Figure 62.6 *Simplified Balance Sheet*

ASSETS

current assets

cash	14,000	
accounts receivable	36,000	
notes receivable	20,000	
inventory	89,000	
prepaid expenses	3000	
total current assets		162,000

plant, property, and equipment

land and buildings	217,000	
motor vehicles	31,000	
equipment	94,000	
accumulated depreciation	(52,000)	
total fixed assets		290,000
total assets		452,000

LIABILITIES AND OWNERS' EQUITY

current liabilities

accounts payable	66,000	
accrued income taxes	17,000	
accrued expenses	8000	
total current liabilities		91,000

long-term debt

notes payable	117,000	
mortgage	23,000	
total long-term debt		140,000

owners' and stockholders' equity

stock	130,000	
retained earnings	91,000	
total owners' equity		221,000
total liabilities and owners' equity		452,000

Analysis of Financial Statements

Financial statements are evaluated by management, lenders, stockholders, potential investors, and many other groups for the purpose of determining the *health of the company*. The health can be measured in terms of *liquidity* (ability to convert assets to cash quickly), *solvency* (ability to meet debts as they become due), and *relative risk* (of which one measure is *leverage*—the portion of total capital contributed by owners).

The analysis of financial statements involves several common ratios, usually expressed as percentages. The following are some frequently encountered ratios.

- *current ratio:* an index of short-term paying ability.

$$\text{current ratio} = \frac{\text{current assets}}{\text{current liabilities}} \quad 62.34$$

- *quick* (or *acid-test*) *ratio:* a more stringent measure of short-term debt-paying ability. The *quick assets* are defined to be current assets minus inventories and prepaid expenses.

$$\text{quick ratio} = \frac{\text{quick assets}}{\text{current liabilities}} \quad 62.35$$

- *receivable turnover:* a measure of the average speed with which accounts receivable are collected.

$$\text{receivable turnover} = \frac{\text{net credit sales}}{\text{average net receivables}} \quad 62.36$$

- *average age of receivables:* number of days, on the average, in which receivables are collected.

$$\text{average age of receivables} = \frac{365}{\text{receivable turnover}} \quad 62.37$$

- *inventory turnover:* a measure of the speed with which inventory is sold, on the average.

$$\text{inventory turnover} = \frac{\text{cost of goods sold}}{\text{average cost of inventory on hand}} \quad 62.38$$

- *days supply of inventory on hand:* number of days, on the average, that the current inventory would last.

$$\text{days supply of inventory on hand} = \frac{365}{\text{inventory turnover}} \quad 62.39$$

- *book value per share of common stock:* number of dollars represented by the balance sheet owners' equity for each share of common stock outstanding.

$$\text{book value per share of common stock} = \frac{\text{common shareholders' equity}}{\text{number of outstanding shares}} \quad 62.40$$

- *gross margin:* gross profit as a percentage of sales. (Gross profit is sales less cost of goods sold.)

$$\text{gross margin} = \frac{\text{gross profit}}{\text{net sales}} \quad 62.41$$

- *profit margin ratio:* percentage of each dollar of sales that is net income.

$$\text{profit margin} = \frac{\text{net income before taxes}}{\text{net sales}} \quad 62.42$$

Assoc. Eng.
Principles

- *return on investment ratio:* shows the percent return on owners' investment.

$$\text{return on investment} = \frac{\text{net income}}{\text{owners' equity}} \qquad 62.43$$

- *price-earnings ratio:* indication of relationship between earnings and market price per share of common stock, useful in comparisons between alternative investments.

$$\text{price-earnings} = \frac{\text{market price per share}}{\text{earnings per share}} \qquad 62.44$$

25. COST ACCOUNTING

Cost accounting is the system that determines the cost of manufactured products. Cost accounting is called *job cost accounting* if costs are accumulated by part number or contract. It is called *process cost accounting* if costs are accumulated by departments or manufacturing processes.

Cost accounting is dependent on historical and recorded data. The unit product cost is determined from actual expenses and numbers of units produced. Allowances (i.e., budgets) for future costs are based on these historical figures. Any deviation from historical figures is called a *variance*. Where adequate records are available, variances can be divided into *labor variance* and *material variance*.

When determining a unit product cost, the direct material and direct labor costs are generally clear-cut and easily determined. Furthermore, these costs are 100% variable costs. However, the indirect cost per unit of product is not as easily determined. Indirect costs (*burden, overhead,* etc.) can be fixed or semivariable costs. The amount of indirect cost allocated to a unit will depend on the unknown future overhead expense as well as the unknown future production (*vehicle size*).

A typical method of allocating indirect costs to a product is as follows.

step 1: Estimate the total expected indirect (and overhead) costs for the upcoming year.

step 2: Determine the most appropriate vehicle (basis) for allocating the overhead to production. Usually, this vehicle is either the number of units expected to be produced or the number of direct hours expected to be worked in the upcoming year.

step 3: Estimate the quantity or size of the overhead vehicle.

step 4: Divide expected overhead costs by the expected overhead vehicle to obtain the unit overhead.

step 5: Regardless of the true size of the overhead vehicle during the upcoming year, one unit of overhead cost is allocated per unit of overhead vehicle.

Once the prime cost has been determined and the indirect cost calculated based on projections, the two are combined into a *standard factory cost* or *standard cost*, which remains in effect until the next budgeting period (usually a year).

During the subsequent manufacturing year, the standard cost of a product is not generally changed merely because it is found that an error in projected indirect costs or production quantity (vehicle size) has been made. The allocation of indirect costs to a product is assumed to be independent of errors in forecasts. Rather, the difference between the expected and actual expenses, known as the *burden (overhead) variance*, experienced during the year is posted to one or more *variance accounts.*

Burden (overhead) variance is caused by errors in forecasting both the actual indirect expense for the upcoming year and the overhead vehicle size. In the former case, the variance is called *burden budget variance*; in the latter, it is called *burden capacity variance.*

Example 62.15

A company expects to produce 8000 items in the coming year. The current material cost is $4.54 each. Sixteen minutes of direct labor are required per unit. Workers are paid $7.50 per hour. 2133 direct labor hours are forecasted for the product. Miscellaneous overhead costs are estimated at $45,000.

Find the per-unit (a) expected direct material cost, (b) direct labor cost, (c) prime cost, (d) burden as a function of production and direct labor, and (e) total cost.

Solution

(a) The direct material cost was given as $4.54.

(b) The direct labor cost is

$$\left(\frac{16 \text{ min}}{60 \frac{\text{min}}{\text{hr}}} \right)\left(\frac{\$7.50}{\text{hr}} \right) = \$2.00$$

(c) The prime cost is

$$\$4.54 + \$2.00 = \$6.54$$

(d) If the burden vehicle is production, the burden rate is $45,000/8000 = $5.63 per item.

If the burden vehicle is direct labor hours, the burden rate is $45,000/2133 = $21.10 per hour.

(e) If the burden vehicle is production, the total cost is

$$\$4.54 + \$2.00 + \$5.63 = \$12.17$$

If the burden vehicle is direct labor hours, the total cost is

$$\$4.54 + \$2.00 + \left(\dfrac{16 \text{ min}}{60 \ \dfrac{\text{min}}{\text{hr}}}\right)\left(\dfrac{\$21.10}{\text{hr}}\right) = \$12.17$$

Example 62.16

The actual performance of the company in Ex. 62.15 is given by the following figures.

actual production: 7560

actual overhead costs: $47,000

What are the burden budget variance and the burden capacity variance?

Solution

The burden capacity variance is

$$\$45,000 - (7560)(\$5.63) = \$2437$$

The burden budget variance is

$$\$47,000 - \$45,000 = \$2000$$

The overall burden variance is

$$\$47,000 - (7560)(\$5.63) = \$4437$$

The sum of the burden capacity and burden budget variances should equal the overall burden variance.

$$\$2437 + \$2000 = \$4437$$

26. COST OF GOODS SOLD

Cost of goods sold (COGS) is an accounting term that represents an inventory account adjustment.[30] Cost of goods sold is the difference between the starting and ending inventory valuations. That is,

$$\begin{aligned} \text{COGS} &= \text{starting inventory valuation} \\ &\quad - \text{ending inventory valuation} \end{aligned} \qquad 62.45$$

Cost of goods sold is subtracted from *gross profit* to determine the *net profit* of a company. Despite the fact that cost of goods sold can be a significant element in the profit equation, the inventory adjustment may not be made each accounting period (e.g., each month) due to the difficulty in obtaining an accurate inventory valuation.

With a *perpetual inventory system*, a company automatically maintains up-to-date inventory records, either through an efficient stocking and stock-releasing system or through a *point of sale* (POS) *system* integrated with the inventory records. If a company only counts its inventory (i.e., takes a *physical inventory*) at regular intervals (e.g., once a year), it is said to be operating on a *periodic inventory system*.

Inventory accounting is a source of many difficulties. The inventory value is calculated by multiplying the quantity on hand by the standard cost. In the case of completed items actually assembled or manufactured at the company, this standard cost usually is the manufacturing cost, although factory cost also can be used. In the case of purchased items, the standard cost will be the cost per item charged by the supplying vendor. In some cases, delivery and transportation costs will be included in this standard cost.

It is not unusual for the elements in an item's inventory to come from more than one vendor, or from one vendor in more than one order. Inventory valuation is more difficult if the price paid is different for these different purchases. There are four methods of determining the cost of elements in inventory. Any of these methods can be used (if applicable), but the method must be used consistently from year to year. The four methods are as follows.

- *specific identification method:* Each element can be uniquely associated with a cost. Inventory elements with serial numbers fit into this costing scheme. Stock, production, and sales records must include the serial number.

- *average cost method:* The standard cost of an item is the average of (recent or all) purchase costs for that item.

- *first-in, first-out* (FIFO) *method:* This method keeps track of how many of each item were purchased each time and the number remaining out of each purchase, as well as the price paid at each purchase. The inventory system assumes that the oldest elements are issued first.[31] Inventory value is a weighted average dependent on the number of elements from each purchase remaining. Items issued no longer contribute to the inventory value.

---

[30]The cost of goods sold inventory adjustment is posted to the COGS *expense account.*

[31]If all elements in an item's inventory are identical, and if all shipments of that item are agglomerated, there will be no way to guarantee that the oldest element in inventory is issued first. But, unless *spoilage* is a problem, it really does not matter.

• *last-in, first-out* (LIFO) *method:* This method keeps track of how many of each item were purchased each time and the number remaining out of each purchase, as well as the price paid at each purchase.[32] The inventory value is a weighted average dependent on the number of elements from each purchase remaining. Items issued no longer contribute to the inventory value.

27. BREAK-EVEN ANALYSIS

Break-even analysis is a method of determining when the value of one alternative becomes equal to the value of another. A common application is that of determining when costs exactly equal revenue. If the manufactured quantity is less than the break-even quantity, a loss is incurred. If the manufactured quantity is greater than the break-even quantity, a profit is made. (See Fig. 62.7.)

Assuming no change in the inventory, the *break-even point* can be found by setting costs equal to revenue ($C = R$).

$$C = f + aQ \qquad 62.46$$

$$R = pQ \qquad 62.47$$

$$Q^* = \frac{f}{p - a} \qquad 62.48$$

Figure 62.7 *Break-Even Quality*

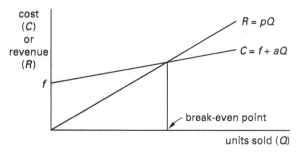

An alternative form of the break-even problem is to find the number of units per period for which two alternatives have the same total costs. Fixed costs are to be spread over a period longer than one year using the equivalent uniform annual cost (EUAC) concept. One of the alternatives will have a lower cost if production is less than the break-even point. The other will have a lower cost for production greater than the break-even point.

Example 62.17

Two plans are available for a company to obtain automobiles for its sales representatives. How many miles must the cars be driven each year for the two plans to have the same costs? Use an interest rate of 10%. (Use the year-end convention for all costs.)

plan A: Lease the cars and pay $1.15 per mile with gas, oil, and insurance included in the price.

plan B: Purchase the cars for $20,000. Each car has an economic life of three years, after which it can be sold for $4800. Gas and oil cost $0.60 per mile. Insurance is $1400 per car.

Solution

Let x be the number of miles driven per year. Calculate the EUAC per car for both alternatives. [Factor Table]

$$\text{EUAC(A)} = \$1.15x$$

$$\begin{aligned}
\text{EUAC(B)} &= \$0.60x + \$1400 + (\$20{,}000)(A/P, 10\%, 3) \\
&\quad - (\$4800)(A/F, 10\%, 3) \\
&= \$0.60x + \$1400 + (\$20{,}000)(0.4021) \\
&\quad - (\$4800)(0.3021) \\
&= \$0.60x + \$10{,}892
\end{aligned}$$

Setting EUAC(A) and EUAC(B) equal and solving for x yields 19,804 miles per year as the break-even point.

28. PAYBACK PERIOD

The *payback period* is defined as the length of time, usually in years, for the cumulative net annual profit to equal the initial investment. It is tempting to introduce equivalence into payback period calculations, but by convention, this is generally not done.[33]

$$\text{payback period} = \frac{\text{initial investment}}{\text{net annual profit}} \qquad 62.49$$

Example 62.18

A ski resort installs two new ski lifts at a total cost of $1,800,000. The resort expects the annual gross revenue to increase by $500,000 while it incurs an annual expense of $50,000 for lift operation and maintenance. What is the payback period?

---

[32]See previous footnote.

[33]Equivalence (i.e., interest and compounding) generally is not considered when calculating the "payback period." However, if it is desirable to include equivalence, then the term *payback period* should not be used. Other terms, such as *cost recovery period* or *life of an equivalent investment*, should be used. Unfortunately, this convention is not always followed in practice.

Assoc. Eng. Principles

Solution

From Eq. 62.49,

$$\text{payback period} = \frac{\$1,800,000}{\dfrac{\$500,000}{yr} - \dfrac{\$50,000}{yr}} = 4 \text{ yr}$$

29. MANAGEMENT GOALS

Depending on many factors (market position, age of the company, age of the industry, perceived marketing and sales windows, etc.), a company may select one of many production and marketing strategic goals. Three such strategic goals are

- maximization of product demand

- minimization of cost

- maximization of profit

Such goals require knowledge of how the dependent variable (e.g., demand quantity or quantity sold) varies as a function of the independent variable (e.g., price). Unfortunately, these three goals are not usually satisfied simultaneously. For example, minimization of product cost may require a large production run to realize economies of scale, while the actual demand is too small to take advantage of such economies of scale.

If sufficient data are available to plot the independent and dependent variables, it may be possible to optimize the dependent variable graphically. (See Fig. 62.8.) Of course, if the relationship between independent and dependent variables is known algebraically, the dependent variable can be optimized by taking derivatives or by use of other numerical methods.

30. INFLATION

It is important to perform economic studies in terms of *constant value dollars*. One method of converting all cash flows to constant value dollars is to divide the flows by some annual *economic indicator* or price index.

If indicators are not available, cash flows can be adjusted by assuming that inflation is constant at a decimal rate, f, per year. Then, all cash flows can be converted to $t = 0$ dollars by dividing by $(1 + f)^n$, where n is the year of the cash flow.

An alternative is to replace the effective annual interest rate i, with a value corrected for inflation. This corrected value, d, is

Inflation

$$d = i + f + (i \times f) \qquad \textbf{62.50}$$

Figure 62.8 *Graphs of Management Goal Functions*

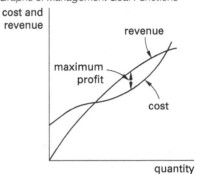

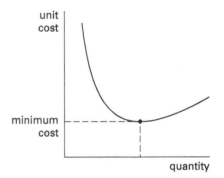

This method has the advantage of simplifying the calculations. However, tables of cash flow equivalent factors are not available for nonintegral values of d. The needed factors must be calculated from their formulas.

Example 62.19

What is the uninflated present worth of a $2000 future value in two years if the average inflation rate is 6% and i is 10%?

Solution

$$P = \frac{F}{(1 + i)^n (1 + f)^n}$$
$$= \frac{\$2000}{(1 + 0.10)^2 (1 + 0.06)^2}$$
$$= \$1471$$

Example 62.20

Repeat Ex. 62.19 using Eq. 62.50.

Solution

Use Eq. 62.50 to calculate the interest rate corrected for inflation.

Inflation

$$d = i + f + (i \times f)$$
$$= 0.10 + 0.06 + (0.10)(0.06)$$
$$= 0.166$$

Assoc. Eng.
Principles

Use the formula for the single payment compound amount factor, using the corrected interest rate.

Economics

$$P = F(P/F, \ i\%, \ n) = F(1+i)^{-n}$$
$$= (\$2000)(1+0.166)^{-2}$$
$$= \$1471$$

31. CONSUMER LOANS

Many different arrangements can be made between a borrower and a lender. With the advent of creative financing concepts, it often seems that there are as many variations of loans as there are loans made. Nevertheless, there are several traditional types of transactions. Real estate or investment texts, or a financial consultant, should be consulted for more complex problems.

Simple Interest

Interest due does not compound with a *simple interest loan*. The interest due is merely proportional to the length of time that the principal is outstanding. Because of this, simple interest loans are seldom made for long periods (e.g., more than one year). (For loans less than one year, it is commonly assumed that a year consists of 12 months of 30 days each.)

A \$12,000 simple interest loan is taken out at 16% per annum interest rate. The loan matures in two years with no intermediate payments. How much will be due at the end of the second year?

Solution

The interest each year is

$$PI = (0.16)(\$12,000) = \$1920$$

The total amount due in two years is

$$PT = \$12,000 + (2)(\$1920) = \$15,840$$

\$4000 is borrowed for 75 days at 16% per annum simple interest. How much will be due at the end of 75 days?

Solution

$$\text{amount due} = \$4000 + (0.16)\left(\dfrac{\dfrac{75 \text{ days}}{360 \ \dfrac{\text{days}}{\text{bank yr}}}}\right)(\$4000)$$
$$= \$4133$$

Loans with Constant Amount Paid Toward Principal

With this loan type, the payment is not the same each period. The amount paid toward the principal is constant, but the interest varies from period to period. (See Fig. 62.9.) The equations that govern this type of loan are

$$\text{BAL}_j = \text{LV} - j(\text{PP}) \qquad \textbf{62.51}$$

$$\text{PI}_j = \phi(\text{BAL})_{j-1} \qquad \textbf{62.52}$$

$$\text{PT}_j = \text{PP} + \text{PI}_j \qquad \textbf{62.53}$$

$$\text{PP} = \frac{\text{LV}}{N} \qquad \textbf{62.54}$$

$$N = \frac{\text{LV}}{\text{PP}} \qquad \textbf{62.55}$$

$$\text{LV} = (\text{PP} + \text{PI}_1)(P/A, \phi, N) \\ - \text{PI}_N(P/G, \phi, N) \qquad \textbf{62.56}$$

$$1 = \left(\frac{1}{N} + \phi\right)(P/A, \phi, N) \\ - \left(\frac{\phi}{N}\right)(P/G, \phi, N) \qquad \textbf{62.57}$$

Figure 62.9 *Loan with Constant Amount Paid Toward Principal*

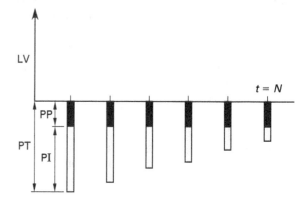

Example 62.23

A \$12,000 six-year loan is taken from a bank that charges 15% effective annual interest. Payments toward the principal are uniform, and repayments are made at the end of each year. Tabulate the interest, total payments, and the balance remaining after each payment is made.

Solution

The amount of each principal payment is

$$PP = \frac{LV}{N} = \frac{\$12,000}{6} = \$2000$$

At the end of the first year (before the first payment is made), the principal balance is \$12,000 (i.e., $BAL_0 = \$12,000$). From Eq. 62.52, the interest payment is

$$PI_1 = \phi(BAL)_0 = (0.15)(\$12,000) = \$1800$$

The total first payment is

$$PT_1 = PP + PI = \$2000 + \$1800$$
$$= \$3800$$

The following table is similarly constructed.

j	BAL_j	PP_j	PI_j	PT_j
	(in dollars)			
0	12,000	–	–	–
1	10,000	2000	1800	3800
2	8000	2000	1500	3500
3	6000	2000	1200	3200
4	4000	2000	900	2900
5	2000	2000	600	2600
6	0	2000	300	2300

Direct Reduction Loans

This is the typical "interest paid on unpaid balance" loan. The amount of the periodic payment is constant, but the amounts paid toward the principal and interest both vary. (See Fig. 62.10.)

$$BAL_{j-1} = PT\left(\frac{1-(1+\phi)^{j-1-N}}{\phi}\right) \qquad 62.58$$

$$PI_j = \phi(BAL)_{j-1} \qquad 62.59$$

$$PP_j = PT - PI_j \qquad 62.60$$

$$BAL_j = BAL_{j-1} - PP_j \qquad 62.61$$

$$N = \frac{-\ln\left(1-\dfrac{\phi(LV)}{PT}\right)}{\ln(1+\phi)} \qquad 62.62$$

Equation 62.62 calculates the number of payments necessary to pay off a loan. This equation can be solved with effort for the total periodic payment (PT) or the initial value of the loan (LV). It is easier, however, to use the $(A/P, i\%, n)$ factor to find the payment and loan value.

$$PT = LV(A/P, \phi\%, N) \qquad 62.63$$

If the loan is repaid in yearly installments, then i is the effective annual rate. If the loan is paid off monthly, then i should be replaced by the effective rate per month (ϕ from Eq. 62.29). For monthly payments, N is the number of months in the loan period.

Figure 62.10 Direct Reduction Loan

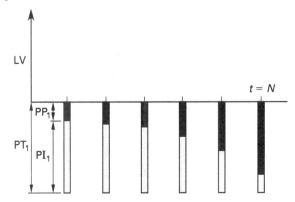

Example 62.24

A \$45,000 loan is financed at 9.25% per annum. The monthly payment is \$385. What are the amounts paid toward interest and principal in the 14th period? What is the remaining principal balance after the 14th payment has been made?

Solution

The effective rate per month is

$$\phi = \frac{r}{m} = \frac{0.0925}{12}$$
$$= 0.0077083\ldots \quad \text{[say 0.007708]}$$

$$N = \frac{-\ln\left(1 - \frac{\phi(\mathrm{LV})}{\mathrm{PT}}\right)}{\ln(1+\phi)}$$

$$= \frac{-\ln\left(1 - \frac{(0.007708)(45{,}000)}{385}\right)}{\ln(1 + 0.007708)}$$

$$= 301$$

$$\mathrm{BAL}_{14-1} = \mathrm{PT}\left(\frac{1 - (1+\phi)^{14-1-N}}{\phi}\right)$$

$$= (\$385)\left(\frac{1 - (1 + 0.007708)^{14-1-301}}{0.007708}\right)$$

$$= \$44{,}476.39$$

$$\mathrm{PI}_{14} = \phi(\mathrm{BAL})_{14-1}$$

$$= (0.007708)(\$44{,}476.39)$$

$$= \$342.82$$

$$\mathrm{PP}_{14} = \mathrm{PT} - \mathrm{PI}_{14} = \$385 - \$342.82 = \$42.18$$

Therefore, using Eq. 62.61, the remaining principal balance is

$$\mathrm{BAL}_{14} = \mathrm{BAL}_{14-1} - \mathrm{PP}_{14}$$

$$= \$44{,}476.39 - \$42.18$$

$$= \$44{,}434.21$$

Direct Reduction Loans with Balloon Payments

This type of loan has a constant periodic payment, but the duration of the loan is insufficient to completely pay back the principal (i.e., the loan is not fully amortized). Therefore, all remaining unpaid principal must be paid back in a lump sum when the loan matures. This large payment is known as a *balloon payment*.[34] (See Fig. 62.11.)

Equation 62.58 through Eq. 62.62 also can be used with this type of loan. The remaining balance after the last payment is the balloon payment. This balloon payment must be repaid along with the last regular payment calculated.

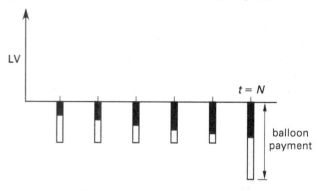

Figure 62.11 *Direct Reduction Loan with Balloon Payment*

32. FORECASTING

There are many types of forecasting models, although most are variations of the basic types.[35] All models produce a *forecast*, F_{t+1}, of some quantity (*demand* is used in this section) in the next period based on actual measurements, D_j, in current and prior periods. All of the models also try to provide *smoothing* (or *damping*) of extreme data points.

Forecasts by Moving Averages

The method of *moving average forecasting* weights all previous demand data points equally and provides some smoothing of extreme data points. The amount of smoothing increases as the number of data points, n, increases.

$$F_{t+1} = \frac{1}{n} \sum_{m=t+1-n}^{t} D_m \qquad \textit{62.64}$$

Forecasts by Exponentially Weighted Averages

With *exponentially weighted forecasts*, the more current (most recent) data points receive more weight. This method uses a *weighting factor*, α, also known as a *smoothing coefficient*, which typically varies between 0.01 and 0.30. An initial forecast is needed to start the method. Forecasts immediately following are sensitive to the accuracy of this first forecast. It is common to choose $F_0 = D_1$ to get started.

$$F_{t+1} = \alpha D_t + (1-\alpha)F_t \qquad \textit{62.65}$$

---

[34]The term *balloon payment* may include the final interest payment as well. Generally, the problem statement will indicate whether the balloon payment is inclusive or exclusive of the regular payment made at the end of the loan period.
[35]For example, forecasting models that take into consideration steady (linear), cyclical, annual, and seasonal trends are typically variations of the exponentially weighted model. A truly different forecasting tool, however, is *Monte Carlo simulation*.

33. LEARNING CURVES

The more products that are made, the more efficient the operation becomes due to experience gained. Therefore, direct labor costs decrease.[36] Usually, a *learning curve* is specified by the decrease in cost each time the cumulative quantity produced doubles. If there is a 20% decrease per doubling, the curve is said to be an 80% learning curve (i.e., the *learning curve rate*, R, is 80%).

Then, the time to produce the nth item is

$$T_n = T_1 n^{-b} \qquad \text{62.66}$$

The total time to produce units from quantity n_1 to n_2 inclusive is approximately given by Eq. 62.67. T_1 is a constant, the time for item 1, and does not correspond to n unless $n_1 = 1$.

$$\int_{n_1}^{n_2} T_n dn$$
$$\approx \left(\frac{T_1}{1-b} \right) \left(\left(n_2 + \frac{1}{2} \right)^{1-b} - \left(n_1 - \frac{1}{2} \right)^{1-b} \right) \qquad \text{62.67}$$

The *average time per unit* over the production from n_1 to n_2 is the above total time from Eq. 62.67 divided by the quantity produced, $(n_2 - n_1 + 1)$.

$$T_{\text{ave}} = \frac{\int_{n_1}^{n_2} T_n \, dn}{n_2 - n_1 + 1} \qquad \text{62.68}$$

Table 62.6 lists representative values of the *learning curve constant*, b. For learning curve rates not listed in the table, Eq. 62.69 can be used to find b.

$$b = \frac{-\log_{10} R}{\log_{10}(2)} = \frac{-\log_{10} R}{0.301} \qquad \text{62.69}$$

Table 62.6 *Learning Curve Constants*

learning curve rate, R	b
0.70 (70%)	0.515
0.75 (75%)	0.415
0.80 (80%)	0.322
0.85 (85%)	0.234
0.90 (90%)	0.152
0.95 (95%)	0.074

Example 62.25

A 70% learning curve is used with an item whose first production time is 1.47 hr. (a) How long will it take to produce the 11th item? (b) How long will it take to produce the 11th through 27th items?

Solution

(a) From Eq. 62.66,

$$T_{11} = T_1 n^{-b} = (1.47 \text{ hr})(11)^{-0.515}$$
$$= 0.428 \text{ hr}$$

(b) The time to produce the 11th item through 27th item is given by Eq. 62.67.

$$T \approx \left(\frac{T_1}{1-b} \right) \left(\left(n_{27} + \frac{1}{2} \right)^{1-b} - \left(n_{11} - \frac{1}{2} \right)^{1-b} \right)$$
$$T \approx \left(\frac{1.47 \text{ hr}}{1-0.515} \right) \left((27.5)^{1-0.515} - (10.5)^{1-0.515} \right)$$
$$= 5.643 \text{ hr}$$

34. ECONOMIC ORDER QUANTITY

The *economic order quantity* (EOQ) is the order quantity that minimizes the inventory costs per unit time. Although there are many different EOQ models, the simplest is based on the following assumptions.

- Reordering is instantaneous. The time between order placement and receipt is zero. (See Fig. 62.12.)

- Shortages are not allowed.

- Demand for the inventory item is deterministic (i.e., is not a random variable).

- Demand is constant with respect to time.

- An order is placed when the inventory is zero.

Figure 62.12 *Inventory with Instantaneous Reorder*

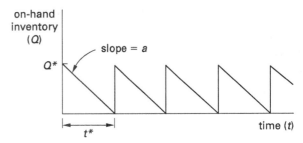

---

[36]Learning curve reductions apply only to direct labor costs. They are not applied to indirect labor or direct material costs.

If the original quantity on hand is Q, the stock will be depleted at

$$t^* = \frac{Q}{a} \qquad 62.70$$

The total inventory storage cost between t_0 and t^* is

$$H = \tfrac{1}{2}Qht^* = \frac{Q^2 h}{2a} \qquad 62.71$$

The total inventory and ordering cost per unit time is

$$C_t = \frac{aK}{Q} + \frac{hQ}{2} \qquad 62.72$$

C_t can be minimized with respect to Q. The economic order quantity and time between orders are

$$Q^* = \sqrt{\frac{2aK}{h}} \qquad 62.73$$

$$t^* = \frac{Q^*}{a} \qquad 62.74$$

35. SENSITIVITY ANALYSIS

Data analysis and forecasts in economic studies require estimates of costs that will occur in the future. There are always uncertainties about these costs. However, these uncertainties are insufficient reason not to make the best possible estimates of the costs. Nevertheless, a decision between alternatives often can be made more confidently if it is known whether or not the conclusion is sensitive to moderate changes in data forecasts. Sensitivity analysis provides this extra dimension to an economic analysis.

The sensitivity of a decision is determined by inserting a range of estimates for critical cash flows and other parameters. If radical changes can be made to a cash flow without changing the decision, the decision is said to be *insensitive* to uncertainties regarding that cash flow. However, if a small change in the estimate of a cash flow will alter the decision, that decision is said to be very *sensitive* to changes in the estimate. If the decision is sensitive only for a limited range of cash flow values, the term *variable sensitivity* is used. Figure 62.13 illustrates these terms.

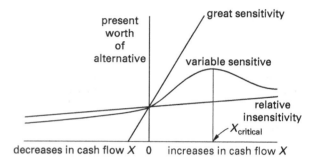

Figure 62.13 *Types of Sensitivity*

An established semantic tradition distinguishes between risk analysis and uncertainty analysis. *Risk analysis* addresses variables that have a known or estimated probability distribution. In this regard, statistics and probability theory can be used to determine the probability of a cash flow varying between given limits. On the other hand, *uncertainty analysis* is concerned with situations in which there is not enough information to determine the probability or frequency distribution for the variables involved.

As a first step, sensitivity analysis should be applied one at a time to the dominant factors. Dominant cost factors are those that have the most significant impact on the present value of the alternative.[37] If warranted, additional investigation can be used to determine the sensitivity to several cash flows varying simultaneously. Significant judgment is needed, however, to successfully determine the proper combinations of cash flows to vary. It is common to plot the dependency of the present value on the cash flow being varied in a two-dimensional graph. Simple linear interpolation is used (within reason) to determine the critical value of the cash flow being varied.

36. VALUE ENGINEERING

The *value* of an investment is defined as the ratio of its return (performance or utility) to its cost (effort or investment). The basic object of *value engineering* (VE, also referred to as *value analysis*) is to obtain the maximum per-unit value.[38]

Value engineering concepts often are used to reduce the cost of mass-produced manufactured products. This is done by eliminating unnecessary, redundant, or superfluous features, by redesigning the product for a less expensive manufacturing method, and by including features for easier assembly without sacrificing utility and function.[39] However, the concepts are equally applicable to one-time investments, such as buildings, chemical

---

[37]In particular, engineering economic analysis problems are sensitive to the choice of effective interest rate, i, and to accuracy in cash flows at or near the beginning of the horizon. The problems will be less sensitive to accuracy in far-future cash flows, such as salvage value and subsequent generation replacement costs.

[38]Value analysis, the methodology that has become today's value engineering, was developed in the early 1950s by Lawrence D. Miles, an analyst at General Electric.

[39]Some people say that value engineering is the act of going over the plans and taking out everything that is interesting.

processing plants, and space vehicles. In particular, value engineering has become an important element in all federally funded work.[40]

Typical examples of large-scale value engineering work are using stock-sized bearings and motors (instead of custom manufactured units), replacing rectangular concrete columns with round columns (which are easier to form), and substituting custom buildings with prefabricated structures.

Value engineering is usually a team effort. And, while the original designers may be on the team, usually outside consultants are utilized. The cost of value engineering is usually returned many times over through reduced construction and life-cycle costs.

37. NOMENCLATURE

a	constant depletion rate (items/unit time)	decimal
a	*incremental cost* to produce one additional item (also called *marginal cost* or *differential cost*)	$
A	annual amount	$
b	learning curve constant	–
BAL_j	balance after the jth payment	$
BV	book value	$
BV_j	book value at end of the jth year	$
C	cost or present worth of all costs	$
C	total cost	$
C_i	individual cost elements per area	$/ft
COGS	cost of goods sold	$
d	declining balance depreciation rate	decimal
d	inflation adjusted interest rate per interest period	decimal
D	demand	various
D	depreciation	$
DR	present worth of after-tax depreciation recovery	$
e	constant inflation rate	decimal
$\mathscr{E}$	expected value	various
EUAC	equivalent uniform annual cost	$
f	federal income tax fraction	decimal
f	fixed cost that does not vary with production	$
f	general inflation rate per interest period	decimal
F	forecasted quantity	various
F	future worth	$
G	uniform gradient amount	$
h	inventory storage cost	$/item-unit time

H	total inventory storage cost between orders	$
i	effective interest rate	decimal
i	interest rate per interest period	decimal
i_e	effective annual interest	decimal
j	number of years	–
j	payment or period number	–
K	fixed cost of placing an order	$
LV	principal total value loaned (cost minus down payment)	$
m	number of compounding periods per year	–
n	number of compounding periods or years in life of asset	–
n	total number of items produced	–
n, N	number	–
N	total number of payments to pay off the loan	–
p	incremental value (price)	$
p	probability	decimal
P	present worth	$
PI_j	jth interest payment	–
PP_j	jth principal payment	–
PT_j	jth total payment	–
Q	order quantity (original quantity on hand)	–
Q	quantity sold	–
Q^*	quantity at break-even point	–
r	nominal rate per year (rate per annum)	decimal per unit time
R	total revenue	$
R	decimal learning curve rate (2^{-b})	decimal
s	state income tax fraction	decimal
S_n	expected salvage value in year n	$
t	composite tax rate	decimal
t	time	years (typical)
t^*	time at depletion	time
T	a quantity equal to $\frac{1}{2}n(n+1)$	–
T_1	time or cost for the first item	time or $
T_n	time or cost for the nth item	time or $
TC	tax credit	$
w_i	weight of individual cost elements by area	$
w_j	weighting by area	$
z	a quantity equal to $\frac{1+i}{1-d}$	decimal

[40]U.S. Government Office of Management and Budget Circular A-131 outlines value engineering for federally funded construction projects.

Assoc. Eng. Principles

Symbols

α	smoothing coefficient for forecasts	–
ϕ	effective rate per period	decimal

Subscripts

0	initial
j	at time j
n	at time n
N	number
t	at time t

63 Project Management

1. PROJECT MANAGEMENT

Project management is the coordination of the entire process of completing a job, from its inception to final move-in and post-occupancy follow-up. In many cases, project management is the responsibility of one person. Large projects can be managed with *partnering*. With this method, the various stakeholders of a project, such as the architect, owner, contractor, engineer, vendors, and others are brought into the decision making process. Partnering can produce much closer communication on a project and shared responsibilities. However, the day-to-day management of a project may be difficult with so many people involved. A clear line of communications and delegation of responsibility should be established and agreed to before the project begins.

Many project managers follow the procedures outlined in *A Guide to the Project Management Body of Knowledge* (PMBOK Guide), published by the Project Management Institute. The PMBOK Guide is an internationally recognized standard (IEEE Std 1490) that defines the fundamentals of project management as they apply to a wide range of projects, including construction, engineering, software, and many other industries. The PMBOK Guide is process-based, meaning it describes projects as being the outcome of multiple processes. Processes overlap and interact throughout the various phases of a project. Each process occurs within one of five *process groups*, which are related as shown in Fig. 63.1. The five PMBOK process groups are (1) initiating, (2) planning, (3) controlling and monitoring, (4) executing, and (5) closing.

Figure 63.1 *PMBOK Process Groups*

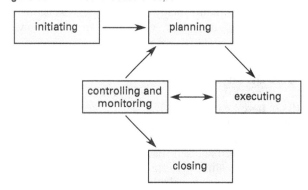

The PMBOK Guide identifies 10 project management knowledge areas that are typical of nearly all projects. The 10 knowledge areas and their respective processes are summarized as follows.

1. *integration:* develop project charter, project scope statement, and management plan; direct and manage execution; monitor and control work; integrate change control; close project

2. *scope:* plan, define, create work breakdown structure (WBS), verify, and control scope

3. *schedule:* define, sequence, estimate resources, and estimate duration of activities; develop and control schedule

4. *cost:* estimate, budget, and control costs

5. *quality:* plan; perform quality assurance and quality control

6. *resources:* plan; acquire, develop, and manage project team

7. *communications:* plan; distribute project information; report on performance

8. *risk:* identify risks; plan risk management and response; perform qualitative and quantitative risk analysis; monitor and control

9. *procurement:* plan purchases, acquisitions, and contracts; request seller responses; select sellers; administer and close contracts

10. *stakeholders:* identify people and organizations affected by project; analyze expectations; engage stakeholders in project decisions and execution

Each of the processes also falls into one of the five basic process groups, creating a matrix so that every process is related to one knowledge area and one process group. Additionally, processes are described in terms of inputs (documents, plans, designs, etc.), tools and techniques (mechanisms applied to inputs), and outputs (documents, products, etc.).

Establishing a budget (knowledge area 4) and scheduling design and construction (knowledge area 3) are two of the most important parts of project management because they influence many of the design decisions to follow and can determine whether a project is even feasible.

2. BUDGETING

Budgets may be established in several ways. For speculative or for-profit projects, the owner or developer works out a pro forma statement listing the expected income of the project and the expected costs to build it. An estimated selling price of the developed project or rent per square foot is calculated and balanced against all the various costs, one of which is the construction price. In order to make the project economically feasible, there will be a limit on the building costs. This becomes the budget within which the work must be completed.

Budgets for municipal and other public projects are often established through public funding or legislation. In these cases, the construction budget is often fixed without the architect's or engineer's involvement, and the project must be designed and built for the fixed amount. Unfortunately, when public officials estimate the cost to build a project, they sometimes neglect to include all aspects of development, such as professional fees, furnishings, and other items.

Budgets may also be based on the proposed project specifics. This is the most realistic and accurate way to establish a preliminary budget because it is based on the actual characteristics of the project. However, this method assumes funding will be available for the project as designed.

There are four basic variables in developing any construction budget: quantity, quality, available funds, and time. There is always a balance among these four variables, and changing one or more affects the others. For instance, if an owner needs a certain area (quantity), needs the project built at a certain time, and has a fixed amount of money to spend, then the quality of construction will have to be adjusted to meet the other constraints. If time, quality, and the total budget are fixed, then the area must be adjusted. In some cases, *value engineering* can be performed during which individual systems and materials are reviewed to see if the same function can be accomplished in a less expensive way.

Fee Projections

A *fee projection* is one of the earliest and most important tasks that a project manager must complete. A fee projection takes the total fee the designer will receive for the project and allocates it to the schedule and staff members who will work on the project, after deducting amounts for profit, overhead, and other expenses that will not be used for professional time.

Ideally, fee projections should be developed from a careful projection of the scope of work, its associated costs (direct personnel expenses, indirect expenses, and overhead), consultant fees, reimbursable expenses, and profit desired. These should be determined as a basis for setting the final fee agreement with the client. If this is done correctly, there should be enough money to complete the project within the allotted time.

There are many methods for estimating and allocating fees. Figure 63.2 shows a simple manual form that combines time scheduling with fee projections. In this example, the total working fee, that is, the fee available to pay people to do the job after subtracting for profit, consultants, and other expenses, is listed in the upper right corner of the chart. The various phases or work tasks needed to complete the job are listed in the left column, and the time periods (most commonly in weeks) are listed across the top of the chart.

The project manager estimates the percentage of the total amount of work or fee that he or she thinks each phase will require. This estimate is based on experience and any common rules of thumb the design or construction office may use. The percentages are placed in the third column on the right and multiplied by the total working fee to get the allotted fee for each phase (the figure in the second column on the right). This allotted fee is then divided among the number of time periods in the schedule and placed in the individual columns under each time period.

If phases or tasks overlap (as they do in Fig. 63.2), sum the fees in each period and place this total dollar amount at the bottom of the chart. The total dollar amount can then be divided by an average billing rate for the people working on the project to determine an approximate budgeted number of hours that the office can afford to spend on the project each week and still make a profit. Of course, if the number of weekly hours exceeds about 40, then more than one person will be needed to do the work.

By monitoring time sheets, the project manager can compare the actual hours (or fees) expended against the budgeted time (or fees) and take corrective action if actual time exceeds budgeted time.

Quality planning involves determining with the client what the expectations are concerning design, cost, and other aspects of the project. Quality does not simply mean high-cost finishes, but rather the requirements of the client based on his or her needs. These needs should

Figure 63.2 *Fee Projection Chart*

Project: Mini-mall						Project No.: 9274					Date: 10/14/2017		
Completed by: JBL						Project Manager: JBL					Total Fee: $26,400		

Phase or Task	Period	1	2	3	4	5	6	7	8	9	% of total fee	fee allocation by phase or task	person-hrs est.
	Date	11/16-22	11/23	11/30	12/7	12/14	12/21	12/28	1/4	1/11			
SD—design		1320	1320								10	2640	
SD—presentation			1320								5	1320	
DD—arch. work				1980	1980						15	3960	
DD—consultant coord.				530	790						5	1320	
DD—approvals					1320						5	1320	
CD—plans/elevs.						1056	1056	1056	1056	1056	20	5280	
CD—details								2640	2640		20	5280	
CD—consultant coord.						440		440	440		5	1320	
CD—specs.									1320	1320	10	2640	
CD—material sel.						660	660				5	1320	
budgeted fees/period		1320	2640	2510	4090	2156	1716	4136	5456	2376	100%	$26,400	
person-weeks or hours		53 / 1.3	106 / 2.6	100 / 2.5	164 / 4	108 / 2.7	86 / 2.2	207 / 5	273 / 6.8	119 / 3			
staff assigned		JLK	JLK AST JBC	JLK AST EMW-(1/2)	JLK AST JBC EMW	JLK AST EMW	JLK AST	JLK AST EMW ——→	JLK SBS BFD	JLK AST EMW			
actual fees expended													

be clearly defined in the programming phase of a project and written down and approved by the client before design work begins.

Cost Estimating

Estimators compile and analyze data on all of the factors that can influence costs, such as materials, labor, location, duration of the project, and special machinery requirements. The methods for estimating costs can differ greatly by industry. On a construction project, for example, the estimating process begins with the decision to submit a bid. After reviewing various preliminary drawings and specifications, the estimator visits the site of the proposed project. The estimator needs to gather information on access to the site; the availability of electricity, water, and other services; and surface topography and drainage. The estimator usually records this information in a signed report that is included in the final project estimate.

After the site visit, the estimator determines the quantity of materials and labor the firm will need to furnish. This process, called the quantity survey or "takeoff," involves completing standard estimating forms, filling in dimensions, numbers of units, and other information. Table 63.1 is a small part of a larger takeoff report illustrating the degree of detail needed to estimate the project cost. A cost estimator working for a general contractor, for example, uses a construction project's plans and specifications to estimate the materials dimensions and count the quantities of all items associated with the project.

Though the quantity takeoff process can be done manually using a printout, a red pen, and a clicker, it can also be done with a digitizer that enables the user to take measurements from paper bid documents, or with an integrated *takeoff viewer* program that interprets electronic bid documents. In any case, the objective is to generate a set of takeoff elements (counts, measurements, and other conditions that affect cost) that is used to establish cost estimates. Table 63.1 is an example of a typical quantity take-off report for lumber needed for a construction project.

Although subcontractors estimate their costs as part of their own bidding process, the general contractor's cost estimator often analyzes bids made by subcontractors. Also during the takeoff process, the estimator must make decisions concerning equipment needs, the sequence of operations, the size of the crew required, and physical constraints at the site. Allowances for wasted materials, inclement weather, shipping delays, and other factors that may increase costs also must be incorporated in the estimate. After completing the quantity surveys, the estimator prepares a cost summary for the entire project, including the costs of labor, equipment, materials, subcontracts, overhead, taxes, insurance, markup, and any other costs that may affect

Assoc. Eng. Principles

Table 63.1 *Partial Lumber and Hardware Take-Off Report*

size	description	usage	pieces	total length
foundation framing				
2×4	DF STD/BTR	stud	38	8
2×6	DF #2/BTR	stud	107	8
2×6	PTDF	mudsill	–	RL
2×6	DF #2/BTR	bracing	–	RL
2×4	DF STD/BTR	blocking	–	RL
$11^7/_8''$	TJI/250	floor joist	11	18
$11^7/_8''$	TJI/250	floor joist	12	10
2×4	DF STD/BTR	plate	–	RL
2×6	DF #2/BTR	plate	–	RL
4×4	PTDF	posts	7	4
4×6	PTDF	posts	23	4
6×6	PTDF	posts	1	4
2×12	DF #2/BTR	rim	–	RL
$^{15}/_{32}''$	CDX plywood	subfloor	12	4×8
pre-cut doors				
$3^1/_2'' \times 7^1/_4''$	LVL	header	1	100″
$3^1/_2'' \times 7^1/_4''$	LVL	header	1	130″
$3^1/_2'' \times 7^1/_4''$	LVL	header	10	24″
exterior sheathing and shear wall				
$^1/_2''$	CDX plywood	shear wall	100	4×8
$^1/_2''$	CDX plywood	exterior sheathing	89	4×8
roof sheathing				
$^1/_2''$	CDX plywood	roof sheathing	67	4×8
building A & B hardware				
Simpson HD64	6 pieces each			
Simpson HD22	23 pieces each			

Source: Sample Lumber Take-Off Report by Modern Estimating Services, LLC.

the project. The chief estimator then prepares the bid proposal for submission to the owner. Construction cost estimators also may be employed by the project's architect or owner to estimate costs or to track actual costs relative to bid specifications as the project develops.

Estimators often specialize in large construction companies employing more than one estimator. For example, one may estimate only electrical work and another may concentrate on excavation, concrete, and forms.

Computers play an integral role in cost estimation because estimating often involves numerous mathematical calculations requiring access to various historical databases. For example, to undertake a parametric analysis (a process used to estimate costs per unit based on square footage or other specific requirements of a project), cost estimators use a computer database containing information on the costs and conditions of many other similar projects. Although computers cannot be used for the entire estimating process, they can relieve estimators of much of the drudgery associated with routine, repetitive, and time-consuming calculations.

Cost Influences

There are many variables that affect project cost. Construction cost is only one part of the total project development budget. Other factors include such things as site acquisition, site development, fees, and financing. Table 63.2 lists most of the items commonly found in a project budget and a typical range of values based on construction cost. Not all of these are part of every development, but they illustrate the things that must be considered.

Building cost is the money required to construct the building, including structure, exterior cladding, finishes, and electrical and mechanical systems. *Site development costs* are usually a separate item. They include such things as parking, drives, fences, landscaping, exterior lighting, and sprinkler systems. If the development is large and affects the surrounding area, a developer may be required to upgrade roads, extend utility lines, and do other major off-site work as a condition of getting approval from public agencies.

Movable equipment and furnishings include furniture, accessories, window coverings, and major equipment necessary to put the facility into operation. These are often listed as separate line items because the funding for them may come out of a separate budget and because they may be supplied under separate contracts.

Professional services are architectural and engineering fees as well as costs for such things as topographic surveys, soil tests, special consultants, appraisals and legal fees, and the like. Inspection and testing involve money required for special on-site, full-time inspection (if required), and testing of such things as concrete, steel, window walls, and roofing.

Because construction takes a great deal of time, a factor for inflation should be included. Generally, the present budget estimate is escalated to a time in the future at the expected midpoint of construction. Although it is impossible to predict the future, by using past cost indexes and inflation rates and applying an estimate to the expected condition of the construction, the architect can usually make an educated guess.

A *contingency cost* should also be added to account for unforeseen changes by the client and other conditions that add to the cost. For an early project budget, the percentage of the contingency should be higher than contingencies applied to later budgets, because there are more unknowns. Normally, 5% to 10% should be included.

Financing includes not only the long-term interest paid on permanent financing but also the immediate costs of loan origination fees, construction loan interest, and other administrative costs. On long-term loans, the cost of financing can easily exceed all of the original building and development costs. In many cases, long-term interest, called *debt service*, is not included in the project budget because it is an ongoing cost to the owner, as are maintenance costs.

Finally, many clients include moving costs in the development budget. For large companies and other types of clients, the money required to physically relocate, including changing stationery, installing telephones, and the like, can be a substantial amount.

Methods of Budgeting

The costs described in the previous section and shown in Table 63.2 represent a type of budget done during programming or even prior to programming to test the feasibility of a project. The numbers are preliminary, often based on sketchy information. For example, the building cost may simply be an estimated cost per square foot multiplied by the number of gross square feet needed. The square footage cost may be derived from similar buildings in the area, from experience, or from commercially available cost books.

Budgeting, however, is an ongoing activity. At each stage of the design process, there should be a revised budget reflecting the decisions made to that time. As shown in the example, pre-design budgets are usually based only on area, but other units can also be used. For example, many companies have rules of thumb for making estimates based on cost per hospital bed, cost per student, cost per hotel room, or similar functional units.

After the pre-programming budget, the architect usually begins to concentrate on the building and site development costs. At this stage an average cost per square foot may still be used, or the building may be divided into several functional parts and different square footage prices may be assigned to each part. A school, for example, may be classified into classroom space, laboratory space, shop space, office space, and gymnasium space, each having a different cost per square foot. This type of division can be developed concurrently with the programming of the space requirements.

During schematic design, when more is known about the space requirements and general configuration of the building and site, *system budgeting* is based on major subsystems. Historical cost information on each type of subsystem can be applied to the design. At this point it is easier to see where the money is being used in the building. Design decisions can then be based on studies of alternative systems. A typical subsystem budget is shown in Table 63.3.

Table 63.2 *Project Budget Line Items*

	line item		example
A	site acquisition		$1,100,000
B	building costs	area times cost per ft^2	(assume) $6,800,000
C	site development	10% to 20% of B	(15%) $1,020,000
D	total construction cost	B + C	$7,820,000
E	movable equipment	5% to 10% of B	(5%) $340,000
F	furnishings		$200,000
G	total construction and furnishings	D + E + F	$8,360,000
H	professional services	5% to 10% of D	(7%) $547,400
I	inspection and testing		$15,000
J	escalation estimate	2% to 20% of G per year	(10%) $836,000
K	contingency	5% to 10% of G	(8%) $668,800
L	financing costs		$250,000
M	moving expenses		(assume) $90,000
N	total project budget	G + H through M	$11,867,200

Table 63.3 *System Cost Budget of Office Buildings*

	average cost	
subsystem	($/ft^2)	(% of total)
foundations	3.96	5.2
floors on grade	3.08	4.0
superstructure	16.51	21.7
roofing	0.18	0.2
exterior walls	9.63	12.6
partitions	5.19	6.8
wall finishes	3.70	4.8
floor finishes	3.78	5.0
ceiling finishes	2.79	3.7
conveying systems	6.45	8.5
specialties	0.70	0.9
fixed equipment	2.74	3.6
HVAC	9.21	12.1
plumbing	3.61	4.6
electrical	4.68	6.1
	76.21	100.0

Values for low-, average-, and high-quality construction for different building types can be obtained from cost databases and published estimating manuals and applied to the structure being budgeted. The dollar amounts included in system cost budgets usually include markup for contractor's overhead and profit and other construction administrative costs.

During the later stages of schematic design and early stages of construction documents, more detailed estimates are made. The procedure most often used is the *parameter method*, which involves an expanded itemization of construction quantities and assignment of unit costs to these quantities. For example, instead of using one number for floor finishes, the cost is broken down into carpeting, vinyl tile, wood strip flooring, unfinished concrete, and so forth. Using an estimated cost per square foot, the cost of each type of flooring can be estimated based on the area. With *parametric budgeting*, it is possible to evaluate the cost implications of each building component and to make decisions concerning both quantity and quality in order to meet the original budget estimate. If floor finishes are over budget, the architect and the client can review the parameter estimate and decide, for example, that some wood flooring must be replaced with less expensive carpeting. Similar decisions can be made concerning any of the parameters in the budget.

Another way to compare and evaluate alternative construction components is with *matrix costing*. With this technique, a matrix is drawn showing, along one side, the various alternatives and, along the other side, the individual elements that combine to produce the total cost of the alternatives. For example, in evaluating alternatives for workstations, all of the factors that would comprise the final cost could be compared. These factors might include the cost of custom-built versus pre-manufactured workstations, task lighting that could be planned with custom-built units versus higher-wattage ambient lighting, and so on.

Parameter line items are based on commonly used units that relate to the construction element under study. For instance, a gypsum board partition would have an assigned cost per square foot of complete partition of a particular construction type rather than separate costs for metal studs, gypsum board, screws, and finishing. There would be different costs for single-layer gypsum board partitions, 1-hour rated walls, 2-hour rated walls, and other partition types.

Overhead and Profit

Two additional components of construction cost are the contractor's overhead and profit. Overhead can be further divided into general overhead and project overhead. *General overhead* is the cost to run a contracting business, and involves office rent, secretarial help, heat, and other recurring costs. *Project overhead* is the money it takes to complete a particular job, not including labor, materials, or equipment. Temporary offices, project telephones, sanitary facilities, trash removal, insurance, permits, and temporary utilities are examples of project overhead. The total overhead costs, including both general and project expenses, can range from about 10% to 20% of the total costs for labor, materials, and equipment.

Profit is the last item a contractor adds onto an estimate and is listed as a percentage of the total of labor, materials, equipment, and overhead. This is one of the most highly variable parts of a budget. Profit depends on the type of project, its size, the amount of risk involved, how much money the contractor wants to make, the general market conditions, and, of course, whether or not the job is being bid.

During extremely difficult economic conditions, a contractor may cut the profit margin to almost nothing simply to get the job and keep his or her workforce employed. If the contract is being negotiated with only one contractor, the profit percentage will be much higher. In most cases, however, profit will range from 5% to 20% of the total cost of the job. Overall, overhead and profit can total about 15% to 40% of construction cost.

Cost Information

One of the most difficult aspects of developing project budgets is obtaining current, reliable prices for the kinds of construction units being used. There is no shortage of commercially produced cost books that are published yearly. These books list costs in different ways; some are very detailed, giving the cost for labor and materials for individual construction items, while others list parameter costs and subsystem costs. The detailed price listings are of little use to architects because they are too specific and make comparison of alternate systems difficult.

There are also computerized cost estimating services that only require the architect or engineer to provide general information about the project, location, size, major materials, and so forth. The computer service then applies its current price database to the information and produces a cost budget. Many architects and engineers also work closely with general contractors to develop a realistic budget.

Commercially available cost information, however, is the average of many past construction projects from around the country. Local variations and particular conditions may affect the value of their use on a specific project.

Two conditions that must be accounted for in developing any project budget are geographical location and inflation. These variables can be adjusted by using cost indexes that are published in a variety of sources, including the major architectural and construction trade magazines. Using a base year as index 1000, for example, for selected cities around the country, new indexes are developed each year that reflect the increase in costs (both material and labor) that year.

The indexes can be used to apply costs from one part of the country to another and to escalate past costs to the expected midpoint of construction of the project being budgeted.

Example 63.1

The cost index in your city is 1257 and the cost index for another city in which you are designing a building is 1308. If the expected construction cost is $1,250,000 based on prices for your city, what will be the expected cost in the other region?

Solution

$$\frac{\text{cost A}}{\text{index A}} = \frac{\text{cost B}}{\text{index B}}$$

$$\text{cost in}_{\text{other region}} = \left(\text{index in}_{\text{other region}}\right)\left(\frac{\text{your city cost}}{\text{your city index}}\right)$$

$$= (1308)\left(\frac{\$1,250,000}{1257}\right)$$

$$= \$1,300,716$$

Life-Cycle Cost Analysis

Life-cycle cost analysis (LCC) is a method for determining the total cost of a building or building component or system over a specific period of time. It takes into account the initial cost of the element or system under consideration as well as the cost of financing, operation, maintenance, and disposal. Any residual value of the components is subtracted from the other costs. The costs are estimated over a length of time called the *study period*. The duration of the study period varies with the needs of the client and the useful life of the material or system. For example, investors in a building may be interested in comparing various alternate materials over the expected investment time frame, while a city government may be interested in a longer time frame representing the expected life of the building. All future costs are discounted back to a common time, usually the base date, to account for the time value of money. The *discount rate* is used to convert future costs to their equivalent present values.

Using life-cycle cost analysis allows two or more alternatives to be evaluated and their total costs to be compared. This is especially useful when evaluating energy conservation measures where one design alternative may have a higher initial cost than another, but a lower overall cost because of energy savings. Some of the specific costs involved in an LCC of a building element include the following.

- initial costs, which include the cost of acquiring and installing the element

- operational costs for electricity, water, and other utilities

- maintenance costs for the element over the length of the study period, including any repair costs

- replacement costs, if any, during the length of the study period

- finance costs required during the length of the study period

- taxes, if any, for initial costs and operating costs

The *residual value* is the remaining value of the element at the end of the study period based on resale value, salvage value, value in place, or scrap value. All of the costs listed are estimated, discounted to their present value, and added together. Any residual value is discounted to its present value and then subtracted from the total to get the final life-cycle cost of the element.

A life-cycle cost analysis is not the same as a *life-cycle assessment* (LCA). An LCA analyzes the environmental impact of a product or building system over the entire life of the product or system.

3. SCHEDULING

There are two primary elements that affect a project's schedule: design sequencing and construction sequencing. The architect has control over design and the production of contract documents, while the contractor has control over construction. The entire project should be scheduled for the best course of action to meet the client's goals. For example, if the client must move by a certain date and normal design and construction sequences make this impossible, the engineer, architect, or contractor may recommend a fast-track schedule or some other approach to meet the deadline.

Design Sequencing

Prior to creating a design, the architect needs to gather information about a client's specific goals and objectives, as well as analyzing any additional factors that may influence a project's design. (This gathering of information is known as *programming*.)

Once this preliminary information has been gathered, the design process may begin. The design process normally consists of several clearly defined phases, each of which must be substantially finished and approved by the client before the next phase may begin. These phases are outlined in the "Owner-Architect Agreement," which is published by the American Institute of Architects (AIA), as well as in other AIA documents. The architectural profession commonly refers to the phases as follows.

1. *schematic design phase:* develops the general layout of the project through schematic design drawings, along with any preliminary alternate studies for materials and building systems

2. *design development phase:* refines and further develops any decisions made during the schematic design phase; preliminary or outline specifications are written and a detailed project budget is created

3. *construction documents phase:* final working drawings, as well as the project manual and any bidding or contract documents, are solidified

4. *bidding or negotiation phase:* bids from several contractors are obtained and analyzed; negotiations with a contractor begin and a contractor is selected

5. *construction phase:* see the following section, "Construction Sequencing"

The time required for each of these five phases is highly variable and depends on the following factors.

- *size and complexity of the project:* A 500,000 ft² (46 450 m²) hospital will take much longer to design than a 30,000 ft² (2787 m²) office building.

- *number of people working on the project:* Although adding more people to the job can shorten the schedule, there is a point of diminishing returns. Having too many people only creates a management and coordination problem, and for some phases only a few people are needed, even for very large jobs.

- *abilities and design methodology of the project team:* Younger, less-experienced designers will usually need more time to do the same amount of work than would more senior staff members.

- *type of client, client decision-making, and approval processes of the client:* Large corporations or public agencies are likely to have a multilayered decision-making and approval process. Getting necessary information or approval for one phase from a large client may take weeks or even months, while a small, single-authority client might make the same decision in a matter of days.

The construction schedule may be established by the contractor or construction manager, or it may be estimated by the architect during the programming phase so that the client has some idea of the total time required from project conception to move-in.

Many variables can affect construction time. Most can be controlled in one way or another, but others, like weather, are independent of anyone's control. Beyond the obvious variables of size and complexity, the following is a partial list of some of the more common variables.

- management ability of the contractor to coordinate the work of direct employees with that of any subcontractors
- material delivery times
- quality and completeness of the architect's drawings and specifications
- weather
- labor availability and labor disputes
- new construction or remodeling (remodeling generally takes more time and coordination than for new buildings of equal areas)
- site conditions (construction sites or those with subsurface problems usually take more time to build on)
- characteristics of the architect (some professionals are more diligent than others in performing their duties during construction)
- lender approvals
- agency and governmental approvals

Construction Sequencing

Construction sequencing involves creating and following a work schedule that balances the timing and sequencing of land disturbance activities (e.g., earthwork) and the installation of *erosion and sedimentation control* (ESC) measures. The objective of construction sequencing is to reduce on-site erosion and off-site sedimentation that might affect the water quality of nearby water bodies.

The project manager should confirm that the general construction schedule and the construction sequencing schedule are compatible. Key construction activities and associated ESC measures are listed in Table 63.4.

Time/Cost Trade-Off

A project's completion time and its cost are intricately related. Though some costs are not directly related to the time a project takes, many costs are. This is the essence of the time/cost trade-off: The cost increases as the project time is decreased and vice versa. A project

Table 63.4 Construction Activities and ESC Measures

construction activity	ESC measures
designate site access	Stabilize exposed areas with gravel and/or temporary vegetation. Immediately apply stabilization to areas exposed throughout site development.
protect runoff outlets and conveyance systems	Install principal sediment traps, fences, and basins prior to grading; stabilize stream banks and install storm drains, channels, etc.
land clearing	Mark trees and buffer areas for preservation.
site grading	Install additional ESC measures as needed during grading.
site stabilization	Install temporary and permanent seeding, mulching, sodding, riprap, etc.
building construction and utilities installation	Install additional ESC measures as needed during construction.
landscaping and final site stabilization*	Remove all temporary control measures; install topsoil, trees and shrubs, permanent seeding, mulching, sodding, riprap, etc.; stabilize all open areas, including borrow and spoil areas.

*This is the last construction phase.

manager's roles include understanding the time/cost relationship, optimizing a project's pace for minimal cost, and predicting the impact of a schedule change on project cost.

The costs associated with a project can be classified as direct costs or indirect costs. The project cost is the sum of the direct and indirect costs.

Direct costs, also known as *variable costs*, *operating costs*, *prime costs*, and *on costs*, are costs that vary directly with the level of output (e.g., labor, fuel, power, and the cost of raw material). Generally, direct costs increase as a project's completion time is decreased, since more resources need to be allocated to increase the pace.

Indirect costs, also known as *fixed costs*, are costs that are not directly related to a particular function or product. Indirect costs include taxes, administration, personnel, and security costs. Such costs tend to be relatively steady over the life of the project and decrease as the project duration decreases.

The time required to complete a project is determined by the critical path, so to compress (or "crash") a project schedule (accelerate the project activities in order to complete the project sooner), a project manager must focus on critical path activities.

Assoc. Eng. Principles

A procedure for determining the optimal project time, or time/cost trade-off, is to determine the normal completion time and direct cost for each critical path activity and compare it to its respective "crash time" and direct cost. The *crash time* is the shortest time in which an activity can be completed. If a new critical path emerges, consider this in subsequent time reductions. In this way, one can step through the critical path activities and calculate the total direct project cost versus the project time. (To minimize the cost, those activities that are not on the critical path can be extended without increasing the project completion time.) The indirect, direct, and total project costs can then be calculated for different project durations. The optimal duration is the one with the lowest cost. This model assumes that the normal cost for an activity is lower than the crash cost, the time and cost are linearly related, and the resources needed to shorten an activity are available. If these assumptions are not true, then the model would need to be adapted. Other cost considerations include incentive payments, marketing initiatives, and the like.

Fast Tracking

Besides efficient scheduling, construction time can be compressed with *fast-track scheduling*. This method overlaps the design and construction phases of a project. Ordering of long-lead materials and equipment can occur, and work on the site and foundations can begin before all the details of the building are completely worked out. With fast-track scheduling, separate contracts are established so that each major system can be bid and awarded by itself to avoid delaying other construction.

Although the fast-track method requires close coordination between the architect, contractor, subcontractors, owner, and others, it makes it possible to construct a high-quality building in 10% to 30% less time than with a conventional construction contract.

Schedule Management

Several methods are used to schedule and monitor projects. The most common and easiest is the *bar chart* or *Gantt chart*, such as Fig. 63.3. The various activities of the schedule are listed along the vertical axis. Each activity is given a starting and finishing date, and overlaps are indicated by drawing the bars for each activity so that they overlap. Bar charts are simple to make and understand and are suitable for small to midsize projects. However, they cannot show all the sequences and dependencies of one activity on another.

Critical path techniques are used to graphically represent the multiple relationships between stages in a complicated project. The graphical network shows the *precedence relationships* between the various activities. The graphical network can be used to control and monitor the progress, cost, and resources of a project. A critical path technique will also identify the most critical activities in the project.

Critical path techniques use *directed graphs* to represent a project. These graphs are made up of *arcs* (arrows) and *nodes* (junctions). The placement of the arcs and nodes completely specifies the precedences of the project. Durations and precedences are usually given in a *precedence table* (matrix).

4. RESOURCE LEVELING

Resource leveling is used to address *overallocation* (i.e., situations that demand more resources than are available). Usually, people and equipment are the limited resources, although project funding may also be limited if it becomes available in stages. Two common ways are used to level resources: (1) Tasks can be delayed (either by postponing their start dates or extending their completion dates) until resources become available. (2) Tasks can be split so that the parts are completed when planned and the remainders are completed when resources becomes available. The methods used depend on the limitations of the project, including budget, resource availability, finish date, and the amount of flexibility available for scheduling tasks. If resource leveling is used with tasks on a project's critical path, the project's completion date will inevitably be extended.

Because of a shortage of reusable forms, a geotechnical contractor decides to build a 250 ft long concrete wall in 25 ft segments. Each segment requires the following crews and times.

- set forms: three carpenters; two days

- pour concrete: three carpenters; one day

- strip forms: four carpenters; one day

The contractor has budgeted for only three carpenters. Compared to the ideal schedule, how many additional days will it take to construct the wall if the number of carpenters is leveled to three?

Solution

Having only three carpenters affects the form stripping operation, but it does not affect the form setting and concrete pouring operations. If the contractor had four carpenters, stripping the forms from each 25 ft segment would take 1 day; stripping forms for the entire 250 ft wall would take (10 segments)(1 day/segment) = 10 days.

With only three carpenters, form stripping productivity is reduced to $\frac{3}{4}$ of the four-carpenter rate. Each form stripping operation takes 4/3 days, and the entire wall takes (10 segments)(4/3 days/segment) = 13.3 days. The increase in time is 13.3 days − 10 days = 3.3 days.

Figure 63.3 *Gantt Chart*

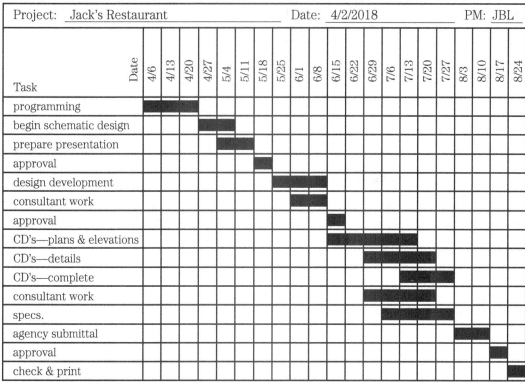

Task	Date	4/6	4/13	4/20	4/27	5/4	5/11	5/18	5/25	6/1	6/8	6/15	6/22	6/29	7/6	7/13	7/20	7/27	8/3	8/10	8/17	8/24
programming		■	■	■																		
begin schematic design					■																	
prepare presentation						■																
approval								■														
design development									■													
consultant work									■													
approval											■											
CD's—plans & elevations											■	■										
CD's—details													■	■								
CD's—complete															■	■						
consultant work																■						
specs.															■	■						
agency submittal																		■	■			
approval																				■		
check & print																					■	

Project: Jack's Restaurant   Date: 4/2/2018   PM: JBL

5. ACTIVITY-ON-NODE NETWORKS

The *critical path method* (CPM) is one of several critical path techniques that uses a directed graph to describe the precedence of project activities. CPM requires that all activity durations be specified by single values. That is, CPM is a *deterministic method* that does not intrinsically support activity durations that are distributed as random variables.

Another characteristic of CPM is that each activity (task) is traditionally represented by a node (junction), hence the name *activity-on-node network*. Each node is typically drawn on the graph as a square box and labeled with a capital letter, although these are not absolute or universally observed conventions. Each activity can be thought of as a continuum of work, each with its own implicit "start" and "finish" *events*. For example, the activity "grub building site" starts with the event of a bulldozer arriving at the native site, followed by several days of bulldozing, and ending with the bulldozer leaving the cleaned site. An activity, including its start and finish events, occurs completely within its box (node).

Each activity in a CPM diagram is connected by arcs (connecting arrows, lines, etc.) The arcs merely show precedence and dependencies. Events are not represented on the graph, other than, perhaps, as the heads of tails of the arcs. Nothing happens along the arcs, and the arcs have zero durations. Because of this, arcs are not labeled. (See Fig. 63.4.)

For convenience, when a project starts or ends with multiple simultaneous activities, *dummy nodes* with zero durations may be used to specify the start and/or finish of the entire project. Dummy nodes do not add time to the project. They are included for convenience. Since all CPM arcs have zero durations, the arcs connecting dummy nodes are not different from arcs connecting other activities.

Figure 63.4 *Activity-on-Node Network*

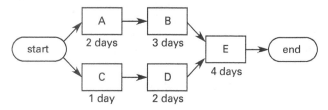

A CPM graph depicts the activities required to complete a project and the sequence in which the activities must be completed. No activity can begin until all of the activities with arcs leading into it have been completed. The duration and various dates are associated with each node, and these dates can be written on the CPM graph for convenience. These dates include the earliest possible start date, ES; the latest start date, LS; the earliest finish date, EF; and, the latest finish date, LF. The ES, EF, LS, and LF dates are calculated from the durations and the activity interdependencies.

After the ES, EF, LS, and LF dates have been determined for each activity, they can be used to identify the critical path. The *critical path* is the sequence of activities (the *critical activities*) that must all be started and finished exactly on time in order to not delay the project. Delaying the starting time of any activity on the critical path, or increasing its duration, will delay the entire project. The critical path is the longest path through the network. If it is desired that the project be completed sooner than expected, then one or more activities in the critical path must be shortened. The critical path is generally identified on the network with heavier (thicker) arcs.

Activities not on the critical path are known as *noncritical activities*. Other paths through the network will require less time than the critical path, and hence, will have inherent delays. The noncritical activities can begin or finish earlier or later (within limits) without affecting the overall schedule. The amount of time that an activity can be delayed without affecting the overall schedule is known as the *float* (*float time* or *slack time*). Float can be calculated in two ways, with identical results. The first way uses Eq. 63.1. The second way is described in the following section, "Solving a CPM Problem."

Nomenclature

$$\text{Float} = \text{LS} - \text{ES or LF} - \text{EF} \qquad 63.1$$

Float is zero along the critical path. This fact can be used to identify the activities along the critical path from their respective ES, EF, LS, and LF dates.

It is essential to maintain a distinction between time, date, length, and duration. Like the timeline for engineering economics problems, all projects start at time = 0, but this corresponds to starting on day = 1. That is, work starts on the first day. Work ends on the last day of the project, but this end rarely corresponds to midnight. So, if a project has a critical path length (duration-to-completion) of 15 days and starts on (at the beginning of) May 1, it will finish on (at the end of) May 15, not May 16.

6. SOLVING A CPM PROBLEM

As previously described, the solution to a critical path method problem reveals the earliest and latest times that an activity can be started and finished, and it also identifies the critical path and generates the float for each activity.

As an alternative to using Eq. 63.1, the following procedure may be used to solve a CPM problem. To facilitate

the solution, each node should be replaced by a square that has been quartered. The compartments have the meanings indicated by the key.

ES	EF
LS	LF

key

ES: Earliest Start

EF: Earliest Finish

LS: Latest Start

LF: Latest Finish

step 1: Place the project start time or date in the ES and EF positions of the start activity. The start time is zero for relative calculations.

step 2: Consider any unmarked activity, all of whose predecessors have been marked in the EF and ES positions. (Go to step 4 if there are none.) Mark in its ES position the largest number marked in the EF position of those predecessors.

step 3: Add the activity time to the ES time and write this in the EF box. Go to step 2.

step 4: Place the value of the latest finish date in the LS and LF boxes of the finish mode.

step 5: Consider unmarked predecessors whose successors have all been marked. Their LF is the smallest LS of the successors. Go to step 7 if there are no unmarked predecessors.

step 6: The LS for the new node is LF minus its activity time. Go to step 5.

step 7: The float for each node is LS − ES and LF − EF.

step 8: The critical path encompasses nodes for which the float equals LS − ES from the start node. There may be more than one critical path.

Example 63.3

Using the precedence table given, construct the precedence matrix and draw an activity-on-node network.

activity	duration (days)	predecessors
A, start	0	–
B	7	A
C	6	A
D	3	B
E	9	B, C
F	1	D, E
G	4	C
H, finish	0	F, G

7. ACTIVITY-ON-ARC NETWORKS

Another variety of deterministic critical path techniques represents project activities as arcs. With *activity-on-arc networks* (also known as *activity-on-branch networks*), the continuum of work occupies the arcs, while the nodes represent instantaneous starting and ending events. The arcs have durations, while the nodes do not. Nothing happens on a node, as it represents an instant in time only. As shown in Fig. 63.5, each node is typically drawn on the graph as a circle and labeled with a number, although this format is not universally adhered to. Since the activities are described by the same precedence table as would be used with an activity-on-node graph, the arcs are labeled with the activity capital letter identifiers or the activity descriptions. The duration of the activity may be written adjacent to its arc. (This is the reason for identifying activities with letters, so that any number appearing with an arc can be interpreted as a duration.)

Although the concepts of ES, EF, LS, and LF dates, critical path, and float are equally applicable to activity-on-arc and activity-on-node networks, the two methods cannot be combined within a single project graph. Calculations for activity-on-arc networks may seem less intuitive, and there are other possible complications.

The activity-on-arc method is complicated by the frequent requirement for *dummy activities* and nodes to maintain precedence. Consider the following part of a precedence table.

activity	predecessors
L	–
M	–
N	L, M
P	M

Activity P depends on the completion of only M. Figure 63.6(a) is an activity-on-arc representation of this precedence. However, N depends on the completion of both L and M. It would be incorrect to draw the network as Fig. 63.6(b) since the activity N appears twice. To represent the project, the dummy activity X must be used, as shown in Fig. 63.6(c).

If two activities have the same starting and ending events, a *dummy node* is required to give one activity a uniquely identifiable completion event. This is illustrated in Fig. 63.7(b).

The solution method for an activity-on-arc problem is essentially the same as for the activity-on-node problem, requiring forward and reverse passes to determine earliest and latest dates.

Solution

The precedence matrix is

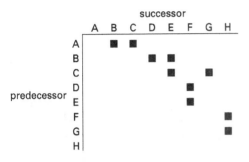

The activity-on-node network is

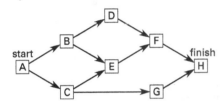

Example 63.4

Complete the network for the previous example and find the critical path. Assume the desired completion duration is in 19 days.

Solution

The critical path is shown with darker lines.

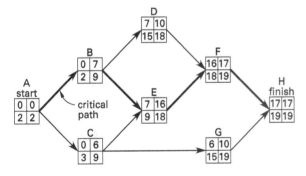

Figure 63.5 Activity-on-Arc Network

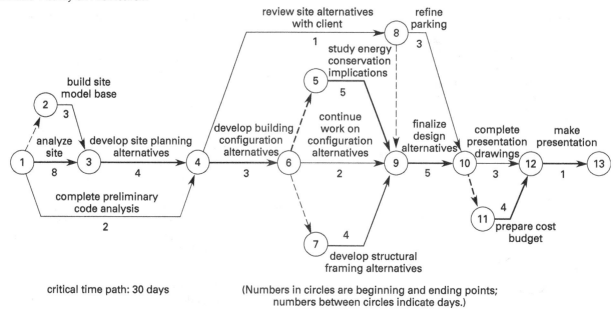

critical time path: 30 days

(Numbers in circles are beginning and ending points;
numbers between circles indicate days.)

Example 63.5

Represent the project in Ex. 63.3 as an activity-on-arc network.

Solution

event	event description
0	start project
1	finish B, start D
2	finish C, start G
3	finish B and C, start E
4	finish D and E, start F
5	finish F and G

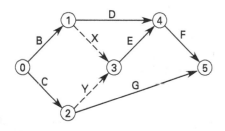

Figure 63.6 Activity-on-Arc Network with Predecessors

(a)

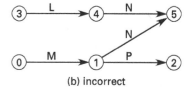

(b) incorrect

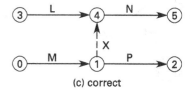

(c) correct

Figure 63.7 Use of a Dummy Node

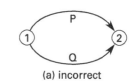

(a) incorrect

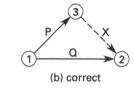

(b) correct

8. STOCHASTIC CRITICAL PATH MODELS

Stochastic models differ from deterministic models only in the way in which the activity durations are found. Whereas durations are known explicitly for the deterministic model, the time for a stochastic activity is distributed as a random variable.

This stochastic nature complicates the problem greatly since the actual distribution is often unknown. Such problems are solved as a deterministic model using the mean of an assumed duration distribution as the activity duration.

The most common stochastic critical path model is PERT, which stands for *program evaluation and review technique*. PERT diagrams have traditionally been drawn as activity-on-arc networks. In PERT, all duration variables are assumed to come from a *beta distribution*, with mean and standard deviation given by Eq. 63.2 and Eq. 63.3, respectively.

$$u_{ij} = \frac{a_{ij} + 4b_{ij} + c_{ij}}{6} \qquad \text{Pert} \quad 63.2$$

$$\sigma_{ij} = \frac{c_{ij} - a_{ij}}{6} \qquad \text{Pert} \quad 63.3$$

The project *completion time* for large projects is assumed to be normally distributed with mean equal to the critical path length and with overall variance equal to the sum of the variances along the critical path.

The probability that a project duration will exceed some length, x, can be found from Eq. 63.4. Z is the standard normal variable.

$$P(\text{duration} > x) = P(t > Z) \qquad 63.4$$

$$Z = \frac{x - \mu}{\sigma} \qquad \text{Normal Distribution (Gaussian Distribution)} \quad 63.5$$

Example 63.6

The mean times and variances for activities along a PERT critical path are given. What is the probability that the project's completion time will be (a) less than 14 days, (b) more than 14 days, (c) more than 23 days, and (d) between 14 and 23 days?

activity mean time (days)	activity standard deviation (days)
9	1.3
4	0.5
7	2.6

Solution

The most likely completion time is the sum of the mean activity times.

$$\mu_{\text{critical path}} = 9 \text{ days} + 4 \text{ days} + 7 \text{ days}$$
$$= 20 \text{ days}$$

The variance of the project's completion times is the sum of the variances along the critical path. Variance, σ^2, is the square of the standard deviation, σ.

$$\sigma^2 = \sum_{(i,j) \in CP} \sigma_{ij}^2 \qquad \text{Pert}$$
$$= (1.3 \text{ days})^2 + (0.5 \text{ days})^2$$
$$+ (2.6 \text{ days})^2$$
$$= 8.7 \text{ days}^2$$
$$\sigma = \sqrt{8.7 \text{ days}^2}$$
$$= 2.95 \text{ days} \quad [\text{use 3 days}]$$

The standard normal variable corresponding to 14 days is given by Eq. 63.5.

$$Z = \frac{x - \mu}{\sigma} \qquad \text{Normal Distribution (Gaussian Distribution)}$$
$$= \frac{14 \text{ days} - 20 \text{ days}}{3 \text{ days}}$$
$$= -2.0$$

From a unit normal distribution table, the area under the standard normal curve is 0.4772 for $0.0 < Z < -2.0$. Since the normal curve is symmetrical, the negative sign is irrelevant in determining the area.

The standard normal variable corresponding to 23 days is given by Eq. 63.5.

$$Z = \frac{x - \mu}{\sigma} \qquad \text{Normal Distribution (Gaussian Distribution)}$$
$$= \frac{23 \text{ days} - 20 \text{ days}}{3 \text{ days}}$$
$$= 1.0$$

Assoc. Eng. Principles

The area under the standard normal curve is 0.3413 for $0.0 < Z < 1.0$. From Eq. 63.4,

(a) $P(\text{duration} < 14) = P(Z < -2.0) = 0.5 - 0.4772$
$$= 0.0228 \quad (2.28\%)$$

(b) $P(\text{duration} > 14) = P(Z > -2.0) = 0.4772 + 0.5$
$$= 0.9772 \quad (97.72\%)$$

(c) $P(\text{duration} > 23) = P(Z > 1.0) = 0.5 - 0.3413$
$$= 0.1587 \quad (15.87\%)$$

(d) $P(14 < \text{duration} < 23) = P(-2.0 < Z < 1.0)$
$$= 0.4772 + 0.3413$$
$$= 0.8185 \quad (81.85\%)$$

9. MONITORING

Monitoring is keeping track of the progress of the job to see if the planned aspects of time, fee, and quality are being accomplished. The original fee projections can be monitored by comparing weekly time sheets with the original estimate. This can be done manually or with project management software. A manual method is shown in Fig. 63.8, which uses the same example project estimated in Fig. 63.2.

In Fig. 63.8, the budgeted weekly costs are placed in the table under the appropriate time-period column and phase-of-work row. The actual costs expended are written next to them. At the bottom of the chart, a cumulative graph is plotted that shows the actual money expended against the budgeted fees. The cumulative ratio of percentage completion to cost can also be plotted.

Monitoring quality is more difficult. At regular times during a project, the project manager, designers, and office principals should review the progress of the job to determine if the original project goals are being met and if the job is being produced according to the client's and design firm's expectations. The work in progress can also be reviewed to see whether it is technically correct and if all the contractual obligations are being met.

10. COORDINATING

During the project, the project manager must constantly coordinate the various people involved: the architect's staff, the consultants, the client, the building code officials, firm management, and, of course, the construction contractors. This may be done on a weekly, or even daily, basis to make sure the schedule is being maintained and the necessary work is getting done.

The coordination can be done by using checklists, holding weekly project meetings to discuss issues and assign work, and exchanging drawings or project files among the consultants.

11. DOCUMENTATION

Everything that is done on a project must be documented in writing. This documentation provides a record in case legal problems develop and serves as a project history to use for future jobs. Documentation is also a vital part of communication. An email or written memo is more accurate, communicates more clearly, and is more difficult to forget than a simple phone call, for example.

Most design firms have standard forms or project management software for documents such as transmittals, job observation reports, time sheets, and the like. Such software makes it easy to record the necessary information. In addition, all meetings should be documented with meeting notes. Phone call logs (listing date, time, participants, and discussion topics), emails, personal daily logs, and formal communications like letters and memos should also be generated and preserved to serve as documentation.

Two types of documents, *change orders* due to unexpected conditions or changes to the plans after bidding, and *as-built construction documents* to record what was actually installed (as opposed to what was shown in the original construction documents) are particularly important.

12. EARNED VALUE METHOD

The *earned value method* (EVM), also known as *earned value management*, is a project management technique that correlates actual *project value* (PV) with *earned value* (EV).[1] *Value* is generally defined in dollars, but it may also be defined in hours. Close monitoring of earned value makes it possible to forecast cost and schedule overruns early in a project. In its simplest form, the method monitors the project plan, actual work performed, expenses, and cumulative value to see if the project is on track. Earned value shows how much of the budget and time should have been spent, with regard to the amount of work actually completed. The earned value method differs from typical budget versus expenses models by requiring the cost of work in progress to be quantified. Because of its complexity, the method is best implemented in its entirety in very large projects.

A *work breakdown structure* (WBS) is at the core of the method. A WBS is a hierarchical structure used to organize tasks for reporting schedules and tracking

---

[1]The earned value method may also be referred to as *performance measurement, management by objectives, budgeted cost of work performed,* and *cost/schedule control systems.*

Figure 63.8 *Project Monitoring Chart*

Project: Mini-mall		time												
Phase/People/Departments		1	2	3	4	5	6	7	8	9	10	11	12	total
schematic design	budgeted	1320	2640											
	actual	2000	2900											
design development	budgeted			2510	4090									
	actual			3200										
construction docs.	budgeted					2156	1716	4136	5456	2376				
	actual													
	budgeted													
	actual													
	budgeted													
	actual													
	budgeted													
	actual													
	budgeted													
	actual													
total (cumulative)	budgeted	1320	3960	6470	10,560	12,716	14,432	18,568	24,024	26,400				
	actual	2000	4900	8100										

At beginning of job, plot budgeted total dollars (or hours) on graph. Plot actual expended dollars (or hours) as job progresses. Also plot estimated percentage complete as job progresses.

Budgeted − − − −
Actual ———

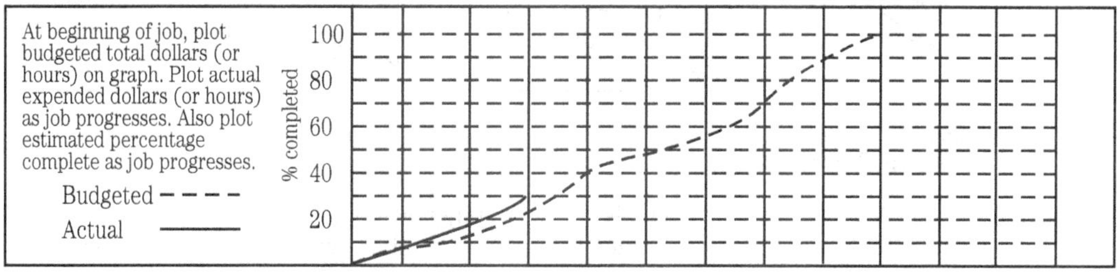

costs. For monitoring, the WBS is subsequently broken down into manageable *work packages*—small sets of activities at lower and lowest levels of the WBS that collectively constitute the overall project scope. Each work package has a relatively short completion time and can be divided into a series of milestones whose status can be objectively measured. Each work package has start and finish dates and a budget value.

The earned value method uses specific terminology for otherwise common project management and accounting principles. There are three primary measures of project performance: BCWS, ACWP, and BCWP. The *budgeted cost of work scheduled* (BCWS) is a spending plan for the project as a function of schedule and performance. For any specified time period, the *cumulative planned expenditures* is the total amount budgeted for the project up to that point. With EVM, the spending plan serves as a performance baseline for making predictions about cost and schedule variance and estimates of completion.

The *actual cost of work performed* (ACWP) is the actual spending as a function of time or performance. It is the cumulative actual expenditures on the project viewed at regular intervals within the project duration. The *budgeted cost of work performed* (BCWP) is the actual earned value based on the technical accomplishment. BCWP is the cumulative budgeted value of the work actually completed. It may be calculated as the sum of the values budgeted for the work packages actually completed, or it may be calculated by multiplying the fraction of work completed by the planned cost of the project.

These primary measures are used to derive secondary measures that give different views of the project's current and future health. Figure 63.9 illustrates how some of these measures can be presented graphically.

Cost variance (CV) is the difference between the budgeted and actual costs of the work completed.

CPM Precedence Relationships: Variances

$$CV = BCWP - ACWP \qquad 63.6$$

Schedule variance (SV) is the difference between the value of the work accomplished in a given period and the value of the work planned. Schedule variance is a measure of how much a project is ahead of or behind schedule, but in terms of value (e.g., dollars), not time.

CPM Precedence Relationships: Variances

$$SV = BCWP - BCWS \qquad 63.7$$

Cost performance index (CPI) is a measure of cost efficiency, comparing the actual cost expended with the earned value. A CPI of one or greater suggests efficiency.

Indices

$$CPI = \frac{BCWP}{ACWP} \qquad 63.8$$

Figure 63.9 *Earned Value Management Measures*

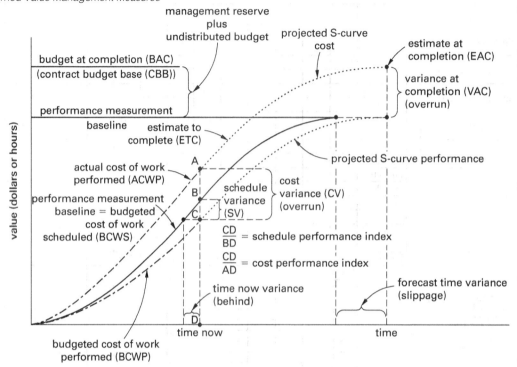

Schedule performance index (SPI) is a measure of schedule effectiveness, comparing the earned value of the work done with the earned value of the work scheduled to be done. An SPI of one or greater suggests that work is on or ahead of schedule.

Indices

$$SPI = \frac{BCWP}{BCWS} \qquad \text{63.9}$$

Budget at completion (BAC), also known as *performance measurement baseline*, is the total of all costs in the original budget.

Estimate to complete (ETC) is an estimate of the total cost of the work needed to complete the project.

Forecasting

$$ETC = \frac{BAC - BCWP}{CPI} \qquad \text{63.10}$$

Estimate at completion (EAC) is an estimate of what the total cost of work will be when the project is completed. The EAC is the sum of the money already spent (the ACWP) and the estimated cost to complete the project (the ETC). Unlike the BAC, which is determined at the start of the project, the EAC is continually revised as the project progresses.

Forecasting

$$EAC = (ACWP + ETC) \qquad \text{63.11}$$

Variance at completion (VAC) is the difference between the original budgeted cost and the projected final cost.

$$VAC = BAC - EAC \qquad \text{63.12}$$

Example 63.7

A project was baselined at $10,000 per week for eight weeks. The project was supposed to be completed at eight weeks but has already taken 11 weeks; however, the cost per week has not changed. One additional week is estimated to complete the project. Determine the

(a) budgeted cost of work scheduled

(b) actual cost of work performed

(c) budgeted cost of work performed

(d) cost variance

(e) schedule variance

(f) cost performance index.

(g) schedule performance index

(h) budget at completion

(i) estimate to complete

(j) estimate at completion

(k) variance at completion

Solution

(a) The budgeted cost of work scheduled is

$$\text{BCWS} = (8\ \text{wk})\left(10{,}000\ \frac{\$}{\text{wk}}\right) = \$80{,}000$$

(b) The actual cost of work performed is

$$\text{ACWP} = (11\ \text{wk})\left(10{,}000\ \frac{\$}{\text{wk}}\right) = \$110{,}000$$

(c) The budgeted cost of work performed is

$$\begin{aligned}\text{BCWP} = \text{EV} &= (\text{fraction completed})(\text{BCWS}) \\ &= (0.917)(\$80{,}000) \\ &= \$73{,}360\end{aligned}$$

(d) The cost variance is

CPM Precedence Relationships: Variances

$$\begin{aligned}\text{CV} = \text{BCWP} - \text{ACWP} &= \$73{,}360 - \$110{,}000 \\ &= -\$36{,}640\end{aligned}$$

The CV is negative, indicating a cost overrun.

(e) The schedule variance, SV, is

CPM Precedence Relationships: Variances

$$\begin{aligned}\text{SV} = \text{BCWP} - \text{BCWS} &= \$73{,}360 - \$80{,}000 \\ &= -\$6{,}640\end{aligned}$$

(f) The cost performance index is

Indices

$$\begin{aligned}\text{CPI} = \frac{\text{BCWP}}{\text{ACWP}} &= \frac{\$73{,}360}{\$110{,}000} \\ &= 0.6667\end{aligned}$$

The CPI is less than 1.0, which indicates that the project is over budget.

(g) The schedule performance index is

$$\begin{aligned}\text{SPI} = \frac{\text{BCWP}}{\text{BCWS}} &= \frac{\$73{,}360}{\$80{,}000} \\ &= 0.9167\end{aligned}$$

(h) The budget at completion is

$$\text{BAC} = (8\ \text{wk})\left(10{,}000\ \frac{\$}{\text{wk}}\right) = \$80{,}000$$

(i) The estimate to complete is

$$\text{ETC} = (1\ \text{wk})\left(10{,}000\ \frac{\$}{\text{wk}}\right) = \$10{,}000$$

(j) The estimate at completion, EAC, is

Forecasting

$$\begin{aligned}\text{EAC} = (\text{ACWP} + \text{ETC}) &= \$110{,}000 + \$10{,}000 \\ &= \$120{,}000\end{aligned}$$

(k) The variance at completion is

$$\begin{aligned}\text{VAC} = \text{BAC} - \text{EAC} &= \$80{,}000 - \$120{,}000 \\ &= -\$40{,}000\end{aligned}$$

The VAC is negative, indicating a cost overrun.

13. NOMENCLATURE

a	optimistic duration of activity	d
b	most likely duration of activity	d
c	pessimistic duration of activity	d
d	duration of activity	d
ACWP	actual cost of work performed	$
BAC	budget at completion	$
BCWP	budgeted cost of work performed	$
BCWS	budgeted cost of work scheduled	$
CPI	cost performance index	-
CV	cost variance	$
EAC	estimate at completion	$
EF	earliest finish	date
ES	earliest start	date
ETC	estimate to complete	$
EV	earned value	$
LF	latest finish	date
LS	latest start	date
P	probability	-
SPI	schedule performance index	-
SV	schedule variance	$
t	time	time
T	duration of project	d
VAC	variance at completion	$
x	length of time	time
z	standard normal variable	-
Z	normal distribution	-

Symbols

μ	mean	various
σ	standard deviation	various

Subscripts

CP	critical path
i	beginning of period
j	end of period

Assoc. Eng. Principles

64

Mass and Energy Balance

1. INTRODUCTION

The concepts of *mass balance* and *energy balance* are common to many environmental engineering practices encompassing water, air, solid waste, and their associated applications. A discussion of mass balance applied to wastewater treatment, wetland hydrology, and stream discharge, with example problems, is presented here as a representative sample.

2. MASS BALANCE IN GROWTH KINETICS AND ACTIVATED SLUDGE

Secondary wastewater treatment effects the removal of dissolved and colloidal organics by

- biological oxidation to carbon dioxide and water

- conversion to biomass that can be settled by gravity

- adsorption/absorption of refractories to settleable biomass

For these mechanisms to occur, secondary wastewater treatment will always consist of a bioreactor followed by a clarifier. The design of the bioreactor relies on the relationship between biomass growth and substrate utilization and employs concepts of mass balance.

Organism growth results in biomass formation. Biomass is represented by X_A and may be considered to represent the organism population present to consume the waste. Volatile suspended solids (VSS) are the means of measuring X_A.

The *substrate* (or food) is the waste constituent available to the organisms to sustain growth. The substrate is represented by S_e. Biochemical oxygen demand (BOD) is used to measure S_e.

Biomass growth and substrate utilization are related as illustrated by the *growth curve* in Fig. 64.1.

Figure 64.1 *Biomass Growth Curve*

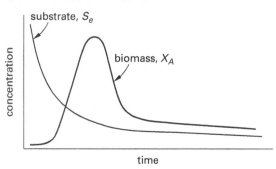

The four segments of the growth curve are as follows.

- *lag phase:* The organism is not adapted to use the substrate as a food source. The lag phase represents the time required for the organism to adapt to use the substrate.

- *log-growth phase:* Substrate is abundant, and the organisms have adapted to use it readily as a food source. Biomass growth occurs at a logarithmic rate.

- *stationary phase:* Stable growth occurs. The rate of organism growth is balanced by the rate of organism death. Organism population is balanced by substrate availability.

- *endogenous phase:* Death and endogenous respiration (starvation) dominate. Substrate is limited so that organisms are required to utilize their own cell tissue to survive.

Biomass growth can be expressed by Eq. 64.1.

$$\frac{dX_A}{dt} = kX_A \qquad 64.1$$

In Eq. 64.1, dX/dt is the biomass growth rate. The growth rate constant, k, can be defined using an empirical relationship known as the *Monod equation*.

$$k = \frac{k_o S_e}{K_s + S_e} \qquad 64.2$$

In Eq. 64.2, k_o is the maximum growth rate constant. K_s is the half-saturation constant, defined as the concentration of S_e when $k = k_o/2$.

For simplification, express dX/dt as r_x. Since r_x requires the conversion of substrate to biomass, as biomass is created, substrate is consumed. However, not all the substrate consumed goes to making biomass; some is used for energy to keep the organism alive. The fraction of substrate that goes to make biomass is defined as the yield coefficient, Y.

Define the rate of substrate use as $-dS/dt$ (negative because it is being consumed), and simplify it to r_s. Relate r_x to r_s using Y.

$$r_x = Y(-r_s) = -Yr_s \qquad 64.3$$

Combining Eq. 64.3 with the Monod equation, Eq. 64.2, produces Eq. 64.4 and Eq. 64.5.

$$r_x = kX_A = \frac{X_A k_o S_e}{K_s + S_e} \qquad 64.4$$

$$r_s = \frac{-kX_A}{Y} = \frac{-X_A k_o S_e}{Y(K_s + S_e)} \qquad 64.5$$

Accounting for the loss of organisms due to death, the biomass is depleted as a function of a constant, the endogenous decay coefficient, k_d, which has units of $1/t$.

$$r_{x,\text{endogenous}} = -k_d X_A \qquad 64.6$$

The net biomass growth is expressed by

$$r_{x,\text{net}} = \frac{X_A k_o S_e}{K_s + S_e} - k_d X_A \qquad 64.7$$

This equation represents the net biomass growth occurring in the bioreactor under steady state conditions.

The bioreactor with its associated clarifier is represented by the mass balance schematic in Fig. 64.2, labeled to show all inputs, outputs, and changes to the system.

Figure 64.2 *Mass Balance Schematic of Activated Sludge*

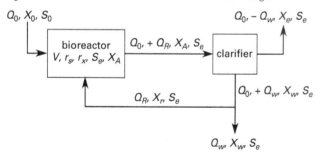

With the schematic labeled, a mass balance can be performed around the system. As X_r and X_w are equal, for consistency only X_w will be used in the mass balance. The mass balance is performed separately for biomass and for substrate.

For biomass,

$$\text{biomass in} + \text{biomass growth} = \text{biomass out} \qquad 64.8$$

$$Q_0 X_0 + V r_{x,\text{net}} = (Q_0 - Q_w)X_e + Q_w X_w \qquad 64.9$$

$$Q_0 X_0 + V\left(\frac{X_A k_o S_e}{K_s + S_e} - k_d X_A\right) \qquad 64.10$$
$$= (Q_0 - Q_w)X_e + Q_w X_w$$

For substrate,

$$\begin{array}{c}\text{substrate in} + \text{substrate growth} \\ = \text{substrate out}\end{array} \qquad 64.11$$

$$Q_0 S_0 + V r_s = (Q_0 - Q_w)S_e + Q_w S_e \qquad 64.12$$

$$Q_0 S_0 + V\left(\frac{X_A k_o S_e}{Y(K_s + S_e)}\right) = (Q_0 - Q_w)S_e + Q_w S_e \qquad 64.13$$

The biomass and substrate mass balance equations can be simplified with the following assumptions.

- The concentrations of X_0 and X_e are negligible compared to X_A and X_w. All X_0 and X_e terms go to zero.

- The bioreactor is complete mix, so that S_0 immediately becomes S_e as it enters the system.

- All biomass growth and substrate use reactions occur only in the bioreactor.

- The complete volume of the bioreactor is represented by V.

With these assumptions, the biomass equation can be rewritten as

$$\frac{k_o S_e}{K_s + S_e} = \frac{Q_w X_w}{V X_A} + k_d \qquad 64.14$$

The substrate equation can be rewritten as

$$\frac{k_o S_e}{K_s + S_e} = \frac{Q_0 Y(S_0 - S_e)}{V X_A} \qquad 64.15$$

The terms on the left side of Eq. 64.14 and Eq. 64.15 are equal. Combining terms on the right sides of the equations gives

$$\frac{Q_w X_w}{V X_A} + k_d = \frac{Q_0 Y(S_0 - S_e)}{V X_A} \qquad 64.16$$

The hydraulic residence time, θ, is

$$\theta = V/Q \qquad 64.17$$

The solids residence time, θ_c, is

Activated Sludge

$$\theta_c = \frac{V(X_A)}{Q_w X_w + Q_e X_e} \qquad 64.18$$

As above, assume X_e is negligible. Equation 64.18 then becomes

$$\theta_c = \frac{VX_A}{Q_w X_w} \qquad 64.19$$

If θ is substituted for V/Q_0 and θ_c is substituted for $VX_A/Q_w X_w$, Eq. 64.16 becomes

$$\frac{1}{\theta_c} + k_d = \frac{Y(S_0 - S_e)}{\theta X_A} \qquad 64.20$$

Rearranging gives

$$\frac{1}{\theta_c} = \frac{Y(S_0 - S_e)}{\theta X_A} - k_d \qquad 64.21$$

Equation 64.21 is the design equation for complete-mix activated sludge, one of the most common methods of wastewater treatment employed in the world for municipal wastewater and biodegradable industrial wastewaters. Solving for X_A gives another form of the activated sludge design equation.

Activated Sludge

$$X_A = \frac{\theta_c Y(S_0 - S_e)}{\theta(1 + k_d \theta_c)} \qquad 64.22$$

3. MASS BALANCE IN WETLANDS HYDROLOGY

Hydrology is the most important factor in determining the composition, richness, and productivity of wetlands. Changes in hydrology can affect chemical and physical properties of the wetlands system, such as nutrient availability, degree of substrate anoxia, soil salinity, sedimentation, and pH. Incoming waters are the principal source of nutrients; water outflows remove decomposition products and wastes. Changes in hydrology can result in rapid and considerable changes in wetlands vegetation and animal species, as well as in productivity of wetlands.

Hydroperiod

A key characteristic of a wetland is its *hydroperiod*, which is the seasonal pattern of the rise and fall of its surface and subsurface water levels. The hydroperiod distinctly characterizes each type of wetland. If the hydroperiod is reasonably consistent from year to year, the wetland will be stable, and its productivity and value will be predictable.

The hydroperiod can be determined from the change in volume over a particular time period, a calculation known as a *water balance*, also known as a *water budget*. A water balance records all inflows and outflows of water during the particular time period.

Flood duration is the amount of time that a wetland has standing water. (The term *flood duration* does not apply to wetlands that are permanently flooded.) *Flood frequency* is the average number of times that a wetland is flooded over a given period (i.e., one year). Both flood duration and flood frequency are used to characterize periodically flooded wetlands.

Hydroperiods are qualitatively described by the U.S. Fish and Wildlife Service in Table 64.1.

Illustrative hydroperiods for several types of wetlands are shown in Fig. 64.3. A cypress dome (see Fig. 64.3 (a)) may have water above the land surface during the wet summer season as well as during the dry periods in the late summer and early spring. A coastal marsh (see Fig. 64.3(b)) may have a hydroperiod composed of semidiurnal flooding and dewatering superimposed on semimonthly spring and ebb tides. A bottomland forest (see Fig. 64.3(c)) may have rapid changes due to seasonal flooding from local precipitation and thawing conditions. A tropical floodplain (see Fig. 64.3(d)) may exhibit large seasonal changes in hydroperiod due to the flooding of rivers.

The hydroperiod for a particular site can vary from year to year as well as seasonally and monthly or daily. Water depths can vary over periods of decades or just a few years. The seasonal fluctuations of water levels occurring with riverine wetlands is known as *pulsing*. Pulsing nourishes the wetland by providing nutrients and removing waste products. Pulse-fed wetlands are often the most productive and provide valuable support for adjacent ecosystems. A characteristic common to most wetlands is a seasonally fluctuating water level.

Water Budget

The *water budget*, or *water balance*, is calculated from the sum of the net precipitation, P_n, surface water inflows, S_i, and outflows, S_o, groundwater inflows, G_i, and outflows, G_o, evapotranspiration, ET, and tidal inflow, $+T$, or outflow, $-T$.

$$\frac{\Delta V}{\Delta t} = P_n + S_i + G_i - \text{ET} - S_o - G_o \pm T \qquad 64.23$$

The terms in Eq. 64.23 can have units of either depth per unit of time for a specific area (mm/y) or volume per unit of time applicable to the whole area (m^3/d).

Assoc. Eng. Principles

Table 64.1 Qualitative Descriptions of Hydroperiods of Wetlands (U.S. Fish and Wildlife Service)

hydroperiod	water regime	characteristic
tidal	subtidal	land surface is permanently flooded by tidal water
tidal	irregularly exposed	land surface is exposed by tides less often than daily
tidal	regularly flooded	tidal water alternately floods and exposes the land surface at least once daily
tidal	irregularly flooded	tidal water floods the land surface less often than daily
nontidal	permanently flooded	water covers the land surface throughout the year in all years
nontidal	intermittently exposed	surface water is present throughout the year except in years of extreme drought
nontidal	semipermanently flooded	surface water persists throughout the growing season in most years, or water table is very near the land surface
nontidal	seasonally flooded	surface water is present for extended periods especially early in the growing season, but is absent by the end of the season in most years; when surface water is absent, water table is near the land surface
nontidal	saturated	land is saturated to the surface for extended periods during the growing season, but surface water is seldom present
nontidal	temporarily flooded	surface water is present for brief periods during the growing season, but the water table is usually below the soil surface for most of the season
nontidal	intermittently flooded	land surface is usually exposed, but surface water is present for variable periods without detectable seasonal periodicity; weeks, months, or years may occur between inundation events
nontidal	artificially flooded	the amount and duration of flooding is controlled by pumps, siphons, dikes, or dams

The *average depth* at a particular time is the volume of the water stored divided by the area of the wetland.

$$d_{\text{ave}} = \frac{V}{A} \qquad \textit{64.24}$$

Over an annual cycle, the net change in storage of water in wetlands may be close to zero. Precipitation will vary with the climate of the area, and may have distinct wet and dry seasons. The surface inflows and outflows usually vary seasonally with precipitation, runoff, and flooding. Groundwater inflows and outflows vary less than surface inflows and outflows and may be absent altogether. Evapotranspiration is typically seasonal, with high rates in the summer and low rates in the winter. For wetlands that are tidally influenced, flooding will vary with elevation and will exhibit one or two tidal cycles daily.

Example 64.1

A hypothetical water budget for the Okefenokee Swamp in Georgia is given.

season	precipitation (mm)	surface +ground-water inflow (mm)	ET (mm)	surface outflow (mm)	ground-water outflow (mm)
spring	4000	1300	1300	3200	100
summer	2000	800	4000	1000	100
fall	2000	600	3000	800	100
winter	5100	1200	1000	2300	100
annual total	13 100	3900	9300	7300	400

The Okefenokee Swamp is 1770 km² in total area. Assume the average water depth is 8 m at the start of the spring season, precipitation is net, and the swamp is not affected by tides. What will be the average water depth at the end of spring?

Figure 64.3 *Illustrative Hydroperiods*

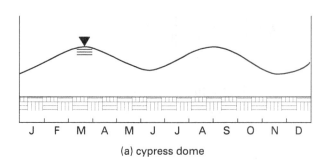

(a) cypress dome

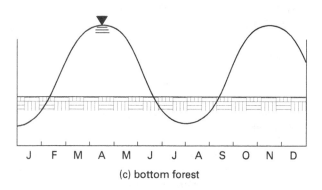

(c) bottom forest

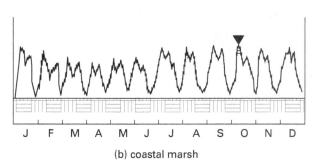

(b) coastal marsh

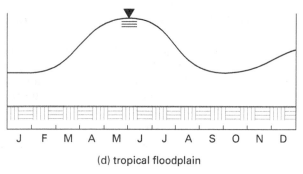

(d) tropical floodplain

Solution

For the spring season,

$$P_n = P = 4000 \text{ mm} \quad (4 \text{ m})$$
$$S_i + G_i = 1300 \text{ mm} \quad (1.30 \text{ m})$$
$$\text{ET} = 1300 \text{ mm} \quad (1.30 \text{ m})$$
$$S_o = 3200 \text{ mm} \quad (3.20 \text{ m})$$
$$G_o = 100 \text{ mm} \quad (0.10 \text{ m})$$
$$T = 0 \text{ mm}$$

The average water volume at the beginning of the spring is calculated from Eq. 64.24.

$$d_{\text{ave},1} = \frac{V_1}{A}$$

$$V_1 = A d_{\text{ave},1} = \frac{(1770 \text{ km}^2)(8 \text{ m})}{1000 \dfrac{\text{m}}{\text{km}}}$$

$$= 14.2 \text{ km}^3$$

The water balance is given by Eq. 64.23.

$$\frac{\Delta V}{\Delta t} = P_n + S_i + G_i - \text{ET} - S_o - G_o \pm T$$

For the spring season, the change in swamp water volume is

$$\Delta V = \frac{\begin{array}{c}(4 \text{ m} + 1.30 \text{ m} - 1.30 \text{ m} - 3.20 \text{ m} - 0.10 \text{ m}) \\ \times (1770 \text{ km}^2)\end{array}}{1000 \dfrac{\text{m}}{\text{km}}}$$

$$= 1.24 \text{ km}^3$$
$$V_2 = V_1 + \Delta V = 14.2 \text{ km}^3 + 1.24 \text{ km}^3$$
$$= 15.4 \text{ km}^3$$

The average water depth at the end of the spring season is given by Eq. 64.24.

$$d_{\text{ave},2} = \frac{V_2}{A} = \frac{(15.4 \text{ km}^3)\left(1000 \dfrac{\text{m}}{\text{km}}\right)}{1770 \text{ km}^2}$$
$$= 8.70 \text{ m}$$

Alternatively, since the change in the water depth per unit area for the spring season is 700 mm (0.70 m), the final average water depth is

$$d_{\text{ave},2} = d_{\text{ave},1} + \Delta d = 8 \text{ m} + 0.70 \text{ m}$$
$$= 8.70 \text{ m}$$

4. MASS BALANCE AND STREAM DISCHARGE

Although dilution is not acceptable as a treatment mechanism, its natural occurrence does provide a very substantial influence on assimilative capacity. Dilution can be assessed using a mass balance around the discharge point, taking the discharge and the upstream flow as inputs and the downstream flow as the output. In addition to dilution, the dissolved oxygen (DO) concentration in a stream segment will be influenced by microbial activity and oxygen transfer at the air-water interface. The mass balance is represented by Eq. 64.25.

$$DO_I + DO_w + DO_a + DO_b + DO_o = 0 \quad \text{64.25}$$

In Eq. 64.25, DO_I stands for the DO mass flow into the stream segment, DO_w for the DO mass in the wastewater discharged to the stream segment, DO_a for the DO mass added from the atmosphere, DO_b for the DO mass consumed by biological activity, and DO_o for the DO mass flow leaving the stream segment

---

Example 64.2

Before it was replaced by a new facility, an old wastewater treatment plant (WWTP) discharged raw sewage during periods of high rainfall. Typical discharge flow was 3.0 MGD with a DO concentration of 1.2 mg/L. During these periods, the flow in the river receiving the raw sewage was 15,000 cfs with a DO concentration of 8.9 mg/L. Assume influences on the DO concentration from reaeration and microbial activity are negligible. Was the capacity of the river sufficient to assimilate the raw sewage discharge with respect to the DO concentration?

Solution

Begin by preparing a mass balance diagram showing inputs and outputs.

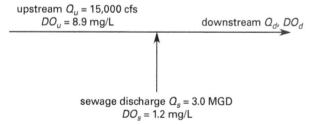

Convert the flow rates to liters per second.

$$Q_u = \left(15,000 \ \frac{ft^3}{sec}\right)\left(28.32 \ \frac{L}{ft^3}\right)$$
$$= 424,800 \ L/sec$$

$$Q_s = \frac{\left(3.0 \times 10^6 \ \frac{gal}{day}\right)\left(3.785 \ \frac{L}{gal}\right)}{86,400 \ \frac{sec}{day}}$$
$$= 131.4 \ L/sec$$

$$Q_d = Q_u + Q_s = 424,800 \ \frac{L}{sec} + 131.4 \ \frac{L}{sec}$$
$$= 424,931 \ L/sec$$

Take a mass balance around the discharge point.

$$DO_I + DO_w + DO_a - DO_b - DO_o = 0$$

From the problem statement, assume that DO_a and DO_b are negligible.

$$DO_I + DO_w - DO_o = 0$$
$$Q_u DO_u + Q_s DO_s - Q_d DO_d = 0$$

Solve for the unknown term, DO_d.

$$DO_d = \frac{Q_u DO_u + Q_s DO_s}{Q_d}$$

$$= \frac{\left(424,800 \ \frac{L}{sec}\right)\left(8.9 \ \frac{mg}{L}\right) + \left(131.4 \ \frac{L}{sec}\right)\left(1.2 \ \frac{mg}{L}\right)}{424,931 \ \frac{L}{sec}}$$

$$= 8.898 \ mg/L$$

The impact of the discharge on the river DO was negligible.

5. NOMENCLATURE

A	area	ft^3	m^3
d	depth of flow	ft	m
DO	dissolved oxygen concentration	mg/L	mg/L
ET	evapotranspiration	mm/y	m^3/s
G	groundwater flow	mm/y	m^3/s
k	growth rate constant	day^{-1}	d^{-1}
k_d	endogenous decay coefficient	day^{-1}	d^{-1}
k_o	maximum growth rate constant	day^{-1}	d^{-1}
K_s	half-saturation constant	mg/L-day	mg/L·d
P	precipitation	mm/y	m^3/s
Q	flow rate	ft^3/sec	m^3/s
r_s	rate of substrate use, $-dS/dt$	mg/L-day	mg/L·d
r_x	biomass growth rate, dX/dt	mg/L-day	mg/L·d
S	surface flow	mm/y	m^3/s
S_e	substrate concentration	mg/L	mg/L
t	time	day^{-1}	day^{-1}
T	tidal flow	mm/y	m^3/s
V	volume	mL	mL
X_A	biomass concentration	mg/L	mg/L
X_e	effluent suspended solids concentration	mg/L	mg/L
X_r	recycled sludge suspended solids concentration	mg/L	mg/L
X_w	waste sludge suspended solids concentration	mg/L	mg/L
Y	yield coefficient	–	–

Symbols

θ	hydraulic residence time	day^{-1}	d^{-1}
θ_c	solids residence time	day^{-1}	d^{-1}

Subscripts

0	initial
a	atmosphere
b	biological activity
d	downstream
e	effluent
I	input
o	output
R	recycle
s	sewage or substrate
u	upstream
w	waste sludge or wastewater

Assoc. Eng. Principles

65 Data Management

Content in blue refers to the *NCEES Handbook*.

Data management involves using big data and modeling to manage and predict outcomes that inform decision-making. Asset management involves tracking assets to optimize performance and minimize costs over the asset design life. They combine to optimize resources to maintain a desired level of performance.

Asset management programs of interest to environmental engineers typically emphasize planning for maintenance and replacement of infrastructure such as wastewater collection and treatment, water treatment and distribution, and solid waste facilities. Industry may apply asset management to control pollution of water, air, and land resources. A central element of this planning is to ensure that maintenance and replacement occurs before failure of physical components and processes and that funds are available to implement plans on schedule.

The Environmental Protection Agency (EPA) has identified benefits of asset management applied to water systems. These can be applied with equal benefit to other public and private facilities. The potential benefits are

- prolong asset life

- improve decision-making for asset rehabilitation, repair, and replacement

- sustainably meet consumer demands

- set rates on transparent operational data and operational and financial planning

- budget for critical activities for sustained performance

- meet service expectations and regulatory requirements

- improve emergency response to system failures

- improve asset security and safety

- optimize overall costs for capital and operations and maintenance expenditures

Asset management includes an inventory of all assets and developing an understanding of the characteristics of each asset. Asset characteristics that will focus an asset management program toward achieving potential benefits include the asset current condition and remaining service life, the asset current and future service requirements, and the criticality of the asset to the overall sustained operation of the system. In larger systems or where interconnectivity among the assets of a municipality and industrial and commercial enterprises operating within the municipality who receive water, wastewater, and solid waste services, asset management may incorporate data visualization and mapping technologies such as geographic information system (GIS) mapping, light detection and ranging (LIDAR), and the global positioning system (GPS).

The EPA has adopted guidelines administered through the Office of Enforcement and Compliance that provides a methodology for municipalities to implement asset management strategies. The guidelines called Capacity, Management, Operations, and Maintenance (CMOM) apply to sanitary wastewater collection systems with the stated objectives of improved operation, management, and maintenance; investigating segments of collection systems that are operating over design capacity; and responding to sanitary sewer overflow events. The municipality selects performance goals and develops activities required to meet them. The goals and implemented activates are reported to the EPA but are intended to provide useful planning for the municipality in the form of a working document.

The International Organization for Standardization (ISO) has also developed asset management guidance in its ISO 55000 series standard. The ISO 55000 series contains three standards:

- ISO 55000: Asset Management: Overview, Principles and Terminology

- ISO 55001: Asset Management: Management Systems—Requirements

- ISO 55002: Asset Management: Management Systems—Guidelines for the Application of ISO 55001

The Governmental Accounting Standards Board has published Statement 34, which establishes financial reporting standards for state and local governments, including school districts and public utilities. The reporting requirements include basic financial statements with analysis that contribute to the overall scope of an asset management program.

Topic VII: Support Material

Appendices

APPENDIX 1.A
Viscosity Index Chart: 0–100 VI

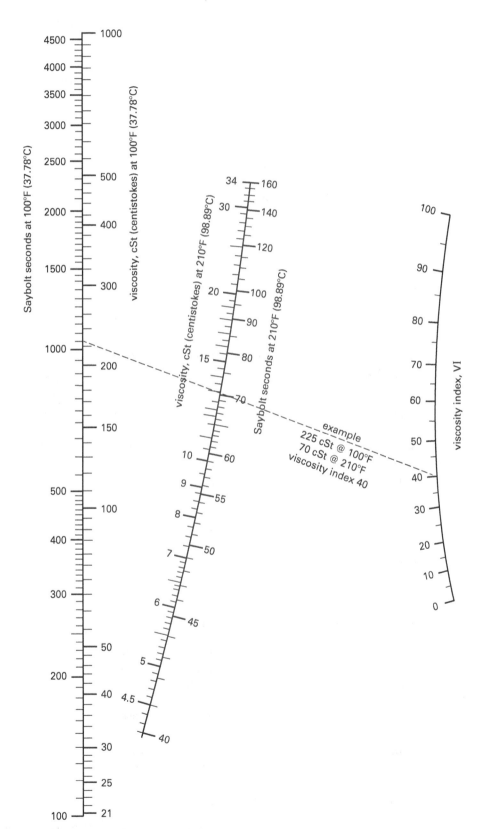

Based on correlations presented in ASTM D2270.

Support Material

APPENDIX 2.A
Dimensions of Welded and Seamless Steel Pipe[a,b]
(selected sizes)[c] (customary U.S. units)

nominal diameter (in)	schedule	outside diameter (in)	wall thickness (in)	internal diameter (in)	internal area (in²)	internal diameter (ft)	internal area (ft²)
$\frac{1}{8}$	40 (S)	0.405	0.068	0.269	0.0568	0.0224	0.00039
	80 (X)		0.095	0.215	0.0363	0.0179	0.00025
$\frac{1}{4}$	40 (S)	0.540	0.088	0.364	0.1041	0.0303	0.00072
	80 (X)		0.119	0.302	0.0716	0.0252	0.00050
$\frac{3}{8}$	40 (S)	0.675	0.091	0.493	0.1909	0.0411	0.00133
	80 (X)		0.126	0.423	0.1405	0.0353	0.00098
$\frac{1}{2}$	40 (S)	0.840	0.109	0.622	0.3039	0.0518	0.00211
	80 (X)		0.147	0.546	0.2341	0.0455	0.00163
	160		0.187	0.466	0.1706	0.0388	0.00118
	(XX)		0.294	0.252	0.0499	0.0210	0.00035
$\frac{3}{4}$	40 (S)	1.050	0.113	0.824	0.5333	0.0687	0.00370
	80 (X)		0.154	0.742	0.4324	0.0618	0.00300
	160		0.218	0.614	0.2961	0.0512	0.00206
	(XX)		0.308	0.434	0.1479	0.0362	0.00103
1	40 (S)	1.315	0.133	1.049	0.8643	0.0874	0.00600
	80 (X)		0.179	0.957	0.7193	0.0798	0.00500
	160		0.250	0.815	0.5217	0.0679	0.00362
	(XX)		0.358	0.599	0.2818	0.0499	0.00196
$1\frac{1}{4}$	40 (S)	1.660	0.140	1.380	1.496	0.1150	0.01039
	80 (X)		0.191	1.278	1.283	0.1065	0.00890
	160		0.250	1.160	1.057	0.0967	0.00734
	(XX)		0.382	0.896	0.6305	0.0747	0.00438
$1\frac{1}{2}$	40 (S)	1.900	0.145	1.610	2.036	0.1342	0.01414
	80 (X)		0.200	1.500	1.767	0.1250	0.01227
	160		0.281	1.338	1.406	0.1115	0.00976
	(XX)		0.400	1.100	0.9503	0.0917	0.00660
2	40 (S)	2.375	0.154	2.067	3.356	0.1723	0.02330
	80 (X)		0.218	1.939	2.953	0.1616	0.02051
	160		0.343	1.689	2.240	0.1408	0.01556
	(XX)		0.436	1.503	1.774	0.1253	0.01232
$2\frac{1}{2}$	40 (S)	2.875	0.203	2.469	4.788	0.2058	0.03325
	80 (X)		0.276	2.323	4.238	0.1936	0.02943
	160		0.375	2.125	3.547	0.1771	0.02463

APPENDIX 2.A *(continued)*
Dimensions of Welded and Seamless Steel Pipe[a,b]
(selected sizes)[c] (customary U.S. units)

nominal diameter (in)	schedule	outside diameter (in)	wall thickness (in)	internal diameter (in)	internal area (in²)	internal diameter (ft)	internal area (ft²)
	(XX)		0.552	1.771	2.464	0.1476	0.01711
3	40 (S)	3.500	0.216	3.068	7.393	0.2557	0.05134
	80 (X)		0.300	2.900	6.605	0.2417	0.04587
	160		0.437	2.626	5.416	0.2188	0.03761
	(XX)		0.600	2.300	4.155	0.1917	0.02885
$3\frac{1}{2}$	40 (S)	4.000	0.226	3.548	9.887	0.2957	0.06866
	80 (X)		0.318	3.364	8.888	0.2803	0.06172
	(XX)		0.636	2.728	5.845	0.2273	0.04059
4	40 (S)	4.500	0.237	4.026	12.73	0.3355	0.08841
	80 (X)		0.337	3.826	11.50	0.3188	0.07984
	120		0.437	3.626	10.33	0.3022	0.07171
	160		0.531	3.438	9.283	0.2865	0.06447
	(XX)		0.674	3.152	7.803	0.2627	0.05419
5	40 (S)	5.563	0.258	5.047	20.01	0.4206	0.1389
	80 (X)		0.375	4.813	18.19	0.4011	0.1263
	120		0.500	4.563	16.35	0.3803	0.1136
	160		0.625	4.313	14.61	0.3594	0.1015
	(XX)		0.750	4.063	12.97	0.3386	0.09004
6	40 (S)	6.625	0.280	6.065	28.89	0.5054	0.2006
	80 (X)		0.432	5.761	26.07	0.4801	0.1810
	120		0.562	5.501	23.77	0.4584	0.1650
	160		0.718	5.189	21.15	0.4324	0.1469
	(XX)		0.864	4.897	18.83	0.4081	0.1308
8	20	8.625	0.250	8.125	51.85	0.6771	0.3601
	30		0.277	8.071	51.16	0.6726	0.3553
	40 (S)		0.322	7.981	50.03	0.6651	0.3474
	60		0.406	7.813	47.94	0.6511	0.3329
	80 (X)		0.500	7.625	45.66	0.6354	0.3171
	100		0.593	7.439	43.46	0.6199	0.3018
	120		0.718	7.189	40.59	0.5990	0.2819
	140		0.812	7.001	38.50	0.5834	0.2673
	(XX)		0.875	6.875	37.12	0.5729	0.2578
	160		0.906	6.813	36.46	0.5678	0.2532
10	20	10.75	0.250	10.250	82.52	0.85417	0.5730
	30		0.307	10.136	80.69	0.84467	0.5604
	40 (S)		0.365	10.020	78.85	0.83500	0.5476
	60 (X)		0.500	9.750	74.66	0.8125	0.5185
	80		0.593	9.564	71.84	0.7970	0.4989
	100		0.718	9.314	68.13	0.7762	0.4732
	120		0.843	9.064	64.53	0.7553	0.4481

APPENDIX 2.A *(continued)*
Dimensions of Welded and Seamless Steel Pipe[a,b]
(selected sizes)[c] (customary U.S. units)

nominal diameter (in)	schedule	outside diameter (in)	wall thickness (in)	internal diameter (in)	internal area (in²)	internal diameter (ft)	internal area (ft²)
10 *(continued)*	140 (XX)		1.000	8.750	60.13	0.7292	0.4176
	160		1.125	8.500	56.75	0.7083	0.3941
12	20	12.75	0.250	12.250	117.86	1.0208	0.8185
	30		0.330	12.090	114.80	1.0075	0.7972
	(S)		0.375	12.000	113.10	1.0000	0.7854
	40		0.406	11.938	111.93	0.99483	0.7773
	(X)		0.500	11.750	108.43	0.97917	0.7530
	60		0.562	11.626	106.16	0.96883	0.7372
	80		0.687	11.376	101.64	0.94800	0.7058
	100		0.843	11.064	96.14	0.92200	0.6677
12	120 (XX)		1.000	10.750	90.76	0.89583	0.6303
	140		1.125	10.500	86.59	0.87500	0.6013
	160		1.312	10.126	80.53	0.84383	0.5592
14 O.D.	10	14.00	0.250	13.500	143.14	1.1250	0.9940
	20		0.312	13.376	140.52	1.1147	0.9758
	30 (S)		0.375	13.250	137.89	1.1042	0.9575
	40		0.437	13.126	135.32	1.0938	0.9397
	(X)		0.500	13.000	132.67	1.0833	0.9213
	60		0.593	12.814	128.96	1.0679	0.8956
	80		0.750	12.500	122.72	1.0417	0.8522
	100		0.937	12.126	115.48	1.0105	0.8020
	120		1.093	11.814	109.62	0.98450	0.7612
	140		1.250	11.500	103.87	0.95833	0.7213
	160		1.406	11.188	98.31	0.93233	0.6827
16 O.D.	10	16.00	0.250	15.500	188.69	1.2917	1.3104
	20		0.312	15.376	185.69	1.2813	1.2895
	30 (S)		0.375	15.250	182.65	1.2708	1.2684
	40 (X)		0.500	15.000	176.72	1.2500	1.2272
	60		0.656	14.688	169.44	1.2240	1.1767
	80		0.843	14.314	160.92	1.1928	1.1175
	100		1.031	13.938	152.58	1.1615	1.0596
	120		1.218	13.564	144.50	1.1303	1.0035
	140		1.437	13.126	135.32	1.0938	0.9397
	160		1.593	12.814	128.96	1.0678	0.8956
18 O.D.	10	18.00	0.250	17.500	240.53	1.4583	1.6703
	20		0.312	17.376	237.13	1.4480	1.6467
	(S)		0.375	17.250	233.71	1.4375	1.6230
	30		0.437	17.126	230.36	1.4272	1.5997
	(X)		0.500	17.000	226.98	1.4167	1.5762
	40		0.562	16.876	223.68	1.4063	1.5533
	60		0.750	16.500	213.83	1.3750	1.4849
	80		0.937	16.126	204.24	1.3438	1.4183
	100		1.156	15.688	193.30	1.3073	1.3423
	120		1.375	15.250	182.65	1.2708	1.2684
	140		1.562	14.876	173.81	1.2397	1.2070
	160		1.781	14.438	163.72	1.2032	1.1370

APPENDIX 2.A *(continued)*
Dimensions of Welded and Seamless Steel Pipe[a,b]
(selected sizes)[c] (customary U.S. units)

nominal diameter (in)	schedule	outside diameter (in)	wall thickness (in)	internal diameter (in)	internal area (in²)	internal diameter (ft)	internal area (ft²)
20 O.D.	10	20.00	0.250	19.500	298.65	1.6250	2.0739
	20 (S)		0.375	19.250	291.04	1.6042	2.0211
	30 (X)		0.500	19.000	283.53	1.5833	1.9689
	40		0.593	18.814	278.00	1.5678	1.9306
	60		0.812	18.376	265.21	1.5313	1.8417
	80		1.031	17.938	252.72	1.4948	1.7550
	100		1.281	17.438	238.83	1.4532	1.6585
	120		1.500	17.000	226.98	1.4167	1.5762
	140		1.750	16.500	213.83	1.3750	1.4849
	160		1.968	16.064	202.67	1.3387	1.4075
24 O.D.	10	24.00	0.250	23.500	433.74	1.9583	3.0121
	20 (S)		0.375	23.250	424.56	1.9375	2.9483
	(X)		0.500	23.000	415.48	1.9167	2.8852
	30		0.562	22.876	411.01	1.9063	2.8542
	40		0.687	22.626	402.07	1.8855	2.7922
	60		0.968	22.060	382.20	1.8383	2.6542
	80		1.218	21.564	365.21	1.7970	2.5362
	100		1.531	20.938	344.32	1.7448	2.3911
	120		1.812	20.376	326.92	1.6980	2.2645
	140		2.062	19.876	310.28	1.6563	2.1547
	160		2.343	19.310	292.87	1.6092	2.0337
30 O.D.	10	30.00	0.312	29.376	677.76	2.4480	4.7067
	(S)		0.375	29.250	671.62	2.4375	4.6640
	20 (X)		0.500	29.000	660.52	2.4167	4.5869
	30		0.625	28.750	649.18	2.3958	4.5082

(Multiply in by 25.4 to obtain mm.)

(Multiply in² by 645 to obtain mm².)

[a]Designations are per ANSI B36.10.

[b]The "S" wall thickness was formerly designated as "standard weight." Standard weight and schedule-40 are the same for all diameters through 10 in. For diameters between 12 in and 24 in, standard weight pipe has a wall thickness of 0.375 in. The "X" (or "XS") wall thickness was formerly designated as "extra strong." Extra strong weight and schedule-80 are the same for all diameters through 8 in. For diameters between 10 in and 24 in, extra strong weight pipe has a wall thickness of 0.50 in. The "XX" (or "XXS") wall thickness was formerly designed as "double extra strong." Double extra strong weight pipe does not have a corresponding schedule number.

[c]Pipe sizes and weights in most common usage are listed. Other weights and sizes exist.

Support Material

APPENDIX 2.B
Darcy Friction Factors (turbulent flow)

Reynolds no.	relative roughness, ϵ/D								
	0.00000	0.000001	0.0000015	0.00001	0.00002	0.00004	0.00005	0.00006	0.00008
2×10^3	0.0495	0.0495	0.0495	0.0495	0.0495	0.0495	0.0495	0.0495	0.0495
2.5×10^3	0.0461	0.0461	0.0461	0.0461	0.0461	0.0461	0.0461	0.0461	0.0461
3×10^3	0.0435	0.0435	0.0435	0.0435	0.0435	0.0436	0.0436	0.0436	0.0436
4×10^3	0.0399	0.0399	0.0399	0.0399	0.0399	0.0399	0.0400	0.0400	0.0400
5×10^3	0.0374	0.0374	0.0374	0.0374	0.0374	0.0374	0.0374	0.0375	0.0375
6×10^3	0.0355	0.0355	0.0355	0.0355	0.0355	0.0356	0.0356	0.0356	0.0356
7×10^3	0.0340	0.0340	0.0340	0.0340	0.0340	0.0341	0.0341	0.0341	0.0341
8×10^3	0.0328	0.0328	0.0328	0.0328	0.0328	0.0328	0.0329	0.0329	0.0329
9×10^3	0.0318	0.0318	0.0318	0.0318	0.0318	0.0318	0.0318	0.0319	0.0319
1×10^4	0.0309	0.0309	0.0309	0.0309	0.0309	0.0309	0.0310	0.0310	0.0310
1.5×10^4	0.0278	0.0278	0.0278	0.0278	0.0278	0.0279	0.0279	0.0279	0.0280
2×10^4	0.0259	0.0259	0.0259	0.0259	0.0259	0.0260	0.0260	0.0260	0.0261
2.5×10^4	0.0245	0.0245	0.0245	0.0245	0.0246	0.0246	0.0246	0.0247	0.0247
3×10^4	0.0235	0.0235	0.0235	0.0235	0.0235	0.0236	0.0236	0.0236	0.0237
4×10^4	0.0220	0.0220	0.0220	0.0220	0.0220	0.0221	0.0221	0.0222	0.0222
5×10^4	0.0209	0.0209	0.0209	0.0209	0.0210	0.0210	0.0211	0.0211	0.0212
6×10^4	0.0201	0.0201	0.0201	0.0201	0.0201	0.0202	0.0203	0.0203	0.0204
7×10^4	0.0194	0.0194	0.0194	0.0194	0.0195	0.0196	0.0196	0.0197	0.0197
8×10^4	0.0189	0.0189	0.0189	0.0189	0.0190	0.0190	0.0191	0.0191	0.0192
9×10^4	0.0184	0.0184	0.0184	0.0184	0.0185	0.0186	0.0186	0.0187	0.0188
1×10^5	0.0180	0.0180	0.0180	0.0180	0.0181	0.0182	0.0183	0.0183	0.0184
1.5×10^5	0.0166	0.0166	0.0166	0.0166	0.0167	0.0168	0.0169	0.0170	0.0171
2×10^5	0.0156	0.0156	0.0156	0.0157	0.0158	0.0160	0.0160	0.0161	0.0163
2.5×10^5	0.0150	0.0150	0.0150	0.0151	0.0152	0.0153	0.0154	0.0155	0.0157
3×10^5	0.0145	0.0145	0.0145	0.0146	0.0147	0.0149	0.0150	0.0151	0.0153
4×10^5	0.0137	0.0137	0.0137	0.0138	0.0140	0.0142	0.0143	0.0144	0.0146
5×10^5	0.0132	0.0132	0.0132	0.0133	0.0134	0.0137	0.0138	0.0140	0.0142
6×10^5	0.0127	0.0128	0.0128	0.0129	0.0131	0.0133	0.0135	0.0136	0.0139
7×10^5	0.0124	0.0124	0.0124	0.0126	0.0127	0.0131	0.0132	0.0134	0.0136
8×10^5	0.0121	0.0121	0.0121	0.0123	0.0125	0.0128	0.0130	0.0131	0.0134
9×10^5	0.0119	0.0119	0.0119	0.0121	0.0123	0.0126	0.0128	0.0130	0.0133
1×10^6	0.0116	0.0117	0.0117	0.0119	0.0121	0.0125	0.0126	0.0128	0.0131
1.5×10^6	0.0109	0.0109	0.0109	0.0112	0.0114	0.0119	0.0121	0.0123	0.0127
2×10^6	0.0104	0.0104	0.0104	0.0107	0.0110	0.0116	0.0118	0.0120	0.0124
2.5×10^6	0.0100	0.0100	0.0101	0.0104	0.0108	0.0113	0.0116	0.0118	0.0123
3×10^6	0.0097	0.0098	0.0098	0.0102	0.0105	0.0112	0.0115	0.0117	0.0122
4×10^6	0.0093	0.0094	0.0094	0.0098	0.0103	0.0110	0.0113	0.0115	0.0120
5×10^6	0.0090	0.0091	0.0091	0.0096	0.0101	0.0108	0.0111	0.0114	0.0119
6×10^6	0.0087	0.0088	0.0089	0.0094	0.0099	0.0107	0.0110	0.0113	0.0118
7×10^6	0.0085	0.0086	0.0087	0.0093	0.0098	0.0106	0.0110	0.0113	0.0118

APPENDIX 2.B *(continued)*
Darcy Friction Factors (turbulent flow)

Reynolds no.	relative roughness, ϵ/D								
	0.00000	0.000001	0.0000015	0.00001	0.00002	0.00004	0.00005	0.00006	0.00008
8×10^6	0.0084	0.0085	0.0085	0.0092	0.0097	0.0106	0.0109	0.0112	0.0118
9×10^6	0.0082	0.0083	0.0084	0.0091	0.0097	0.0105	0.0109	0.0112	0.0117
1×10^7	0.0081	0.0082	0.0083	0.0090	0.0096	0.0105	0.0109	0.0112	0.0117
1.5×10^7	0.0076	0.0078	0.0079	0.0087	0.0094	0.0104	0.0108	0.0111	0.0116
2×10^7	0.0073	0.0075	0.0076	0.0086	0.0093	0.0103	0.0107	0.0110	0.0116
2.5×10^7	0.0071	0.0073	0.0074	0.0085	0.0093	0.0103	0.0107	0.0110	0.0116
3×10^7	0.0069	0.0072	0.0073	0.0084	0.0092	0.0103	0.0107	0.0110	0.0116
4×10^7	0.0067	0.0070	0.0071	0.0084	0.0092	0.0102	0.0106	0.0110	0.0115
5×10^7	0.0065	0.0068	0.0070	0.0083	0.0092	0.0102	0.0106	0.0110	0.0115

Reynolds no.	relative roughness, ϵ/D								
	0.0001	0.00015	0.00020	0.00025	0.00030	0.00035	0.0004	0.0006	0.0008
2×10^3	0.0495	0.0496	0.0496	0.0496	0.0497	0.0497	0.0498	0.0499	0.0501
2.5×10^3	0.0461	0.0462	0.0462	0.0463	0.0463	0.0463	0.0464	0.0466	0.0467
3×10^3	0.0436	0.0437	0.0437	0.0437	0.0438	0.0438	0.0439	0.0441	0.0442
4×10^3	0.0400	0.0401	0.0401	0.0402	0.0402	0.0403	0.0403	0.0405	0.0407
5×10^3	0.0375	0.0376	0.0376	0.0377	0.0377	0.0378	0.0378	0.0381	0.0383
6×10^3	0.0356	0.0357	0.0357	0.0358	0.0359	0.0359	0.0360	0.0362	0.0365
7×10^3	0.0341	0.0342	0.0343	0.0343	0.0344	0.0345	0.0345	0.0348	0.0350
8×10^3	0.0329	0.0330	0.0331	0.0331	0.0332	0.0333	0.0333	0.0336	0.0339
9×10^3	0.0319	0.0320	0.0321	0.0321	0.0322	0.0323	0.0323	0.0326	0.0329
1×10^4	0.0310	0.0311	0.0312	0.0313	0.0313	0.0314	0.0315	0.0318	0.0321
1.5×10^4	0.0280	0.0281	0.0282	0.0283	0.0284	0.0285	0.0285	0.0289	0.0293
2×10^4	0.0261	0.0262	0.0263	0.0264	0.0265	0.0266	0.0267	0.0272	0.0276
2.5×10^4	0.0248	0.0249	0.0250	0.0251	0.0252	0.0254	0.0255	0.0259	0.0264
3×10^4	0.0238	0.0239	0.0240	0.0241	0.0243	0.0244	0.0245	0.0250	0.0255
4×10^4	0.0223	0.0224	0.0226	0.0227	0.0229	0.0230	0.0232	0.0237	0.0243
5×10^4	0.0212	0.0214	0.0216	0.0218	0.0219	0.0221	0.0223	0.0229	0.0235
6×10^4	0.0205	0.0207	0.0208	0.0210	0.0212	0.0214	0.0216	0.0222	0.0229
7×10^4	0.0198	0.0200	0.0202	0.0204	0.0206	0.0208	0.0210	0.0217	0.0224
8×10^4	0.0193	0.0195	0.0198	0.0200	0.0202	0.0204	0.0206	0.0213	0.0220
9×10^4	0.0189	0.0191	0.0194	0.0196	0.0198	0.0200	0.0202	0.0210	0.0217
1×10^5	0.0185	0.0188	0.0190	0.0192	0.0195	0.0197	0.0199	0.0207	0.0215
1.5×10^5	0.0172	0.0175	0.0178	0.0181	0.0184	0.0186	0.0189	0.0198	0.0207
2×10^5	0.0164	0.0168	0.0171	0.0174	0.0177	0.0180	0.0183	0.0193	0.0202
2.5×10^5	0.0158	0.0162	0.0166	0.0170	0.0173	0.0176	0.0179	0.0190	0.0199
3×10^5	0.0154	0.0159	0.0163	0.0166	0.0170	0.0173	0.0176	0.0188	0.0197
4×10^5	0.0148	0.0153	0.0158	0.0162	0.0166	0.0169	0.0172	0.0184	0.0195
5×10^5	0.0144	0.0150	0.0154	0.0159	0.0163	0.0167	0.0170	0.0183	0.0193
6×10^5	0.0141	0.0147	0.0152	0.0157	0.0161	0.0165	0.0168	0.0181	0.0192
7×10^5	0.0139	0.0145	0.0150	0.0155	0.0159	0.0163	0.0167	0.0180	0.0191

APPENDIX 2.B *(continued)*
Darcy Friction Factors (turbulent flow)

Reynolds no.	relative roughness, ϵ/D								
	0.0001	0.00015	0.00020	0.00025	0.00030	0.00035	0.0004	0.0006	0.0008
8×10^5	0.0137	0.0143	0.0149	0.0154	0.0158	0.0162	0.0166	0.0180	0.0191
9×10^5	0.0136	0.0142	0.0148	0.0153	0.0157	0.0162	0.0165	0.0179	0.0190
1×10^6	0.0134	0.0141	0.0147	0.0152	0.0157	0.0161	0.0165	0.0178	0.0190
1.5×10^6	0.0130	0.0138	0.0144	0.0149	0.0154	0.0159	0.0163	0.0177	0.0189
2×10^6	0.0128	0.0136	0.0142	0.0148	0.0153	0.0158	0.0162	0.0176	0.0188
2.5×10^6	0.0127	0.0135	0.0141	0.0147	0.0152	0.0157	0.0161	0.0176	0.0188
3×10^6	0.0126	0.0134	0.0141	0.0147	0.0152	0.0157	0.0161	0.0176	0.0187
4×10^6	0.0124	0.0133	0.0140	0.0146	0.0151	0.0156	0.0161	0.0175	0.0187
5×10^6	0.0123	0.0132	0.0139	0.0146	0.0151	0.0156	0.0160	0.0175	0.0187
6×10^6	0.0123	0.0132	0.0139	0.0145	0.0151	0.0156	0.0160	0.0175	0.0187
7×10^6	0.0122	0.0132	0.0139	0.0145	0.0151	0.0155	0.0160	0.0175	0.0187
8×10^6	0.0122	0.0131	0.0139	0.0145	0.0150	0.0155	0.0160	0.0175	0.0187
9×10^6	0.0122	0.0131	0.0139	0.0145	0.0150	0.0155	0.0160	0.0175	0.0187
1×10^7	0.0122	0.0131	0.0138	0.0145	0.0150	0.0155	0.0160	0.0175	0.0186
1.5×10^7	0.0121	0.0131	0.0138	0.0144	0.0150	0.0155	0.0159	0.0174	0.0186
2×10^7	0.0121	0.0130	0.0138	0.0144	0.0150	0.0155	0.0159	0.0174	0.0186
2.5×10^7	0.0121	0.0130	0.0138	0.0144	0.0150	0.0155	0.0159	0.0174	0.0186
3×10^7	0.0120	0.0130	0.0138	0.0144	0.0150	0.0155	0.0159	0.0174	0.0186
4×10^7	0.0120	0.0130	0.0138	0.0144	0.0150	0.0155	0.0159	0.0174	0.0186
5×10^7	0.0120	0.0130	0.0138	0.0144	0.0150	0.0155	0.0159	0.0174	0.0186

Reynolds no.	relative roughness, ϵ/D								
	0.001	0.0015	0.002	0.0025	0.003	0.0035	0.004	0.006	0.008
2×10^3	0.0502	0.0506	0.0510	0.0513	0.0517	0.0521	0.0525	0.0539	0.0554
2.5×10^3	0.0469	0.0473	0.0477	0.0481	0.0485	0.0489	0.0493	0.0509	0.0524
3×10^3	0.0444	0.0449	0.0453	0.0457	0.0462	0.0466	0.0470	0.0487	0.0503
4×10^3	0.0409	0.0414	0.0419	0.0424	0.0429	0.0433	0.0438	0.0456	0.0474
5×10^3	0.0385	0.0390	0.0396	0.0401	0.0406	0.0411	0.0416	0.0436	0.0455
6×10^3	0.0367	0.0373	0.0378	0.0384	0.0390	0.0395	0.0400	0.0421	0.0441
7×10^3	0.0353	0.0359	0.0365	0.0371	0.0377	0.0383	0.0388	0.0410	0.0430
8×10^3	0.0341	0.0348	0.0354	0.0361	0.0367	0.0373	0.0379	0.0401	0.0422
9×10^3	0.0332	0.0339	0.0345	0.0352	0.0358	0.0365	0.0371	0.0394	0.0416
1×10^4	0.0324	0.0331	0.0338	0.0345	0.0351	0.0358	0.0364	0.0388	0.0410
1.5×10^4	0.0296	0.0305	0.0313	0.0320	0.0328	0.0335	0.0342	0.0369	0.0393
2×10^4	0.0279	0.0289	0.0298	0.0306	0.0315	0.0323	0.0330	0.0358	0.0384
2.5×10^4	0.0268	0.0278	0.0288	0.0297	0.0306	0.0314	0.0322	0.0352	0.0378
3×10^4	0.0260	0.0271	0.0281	0.0291	0.0300	0.0308	0.0317	0.0347	0.0374
4×10^4	0.0248	0.0260	0.0271	0.0282	0.0291	0.0301	0.0309	0.0341	0.0369
5×10^4	0.0240	0.0253	0.0265	0.0276	0.0286	0.0296	0.0305	0.0337	0.0365
6×10^4	0.0235	0.0248	0.0261	0.0272	0.0283	0.0292	0.0302	0.0335	0.0363
7×10^4	0.0230	0.0245	0.0257	0.0269	0.0280	0.0290	0.0299	0.0333	0.0362

APPENDIX 2.B *(continued)*
Darcy Friction Factors (turbulent flow)

Reynolds no.	relative roughness, ϵ/D								
	0.001	0.0015	0.002	0.0025	0.003	0.0035	0.004	0.006	0.008
8×10^4	0.0227	0.0242	0.0255	0.0267	0.0278	0.0288	0.0298	0.0331	0.0361
9×10^4	0.0224	0.0239	0.0253	0.0265	0.0276	0.0286	0.0296	0.0330	0.0360
1×10^5	0.0222	0.0237	0.0251	0.0263	0.0275	0.0285	0.0295	0.0329	0.0359
1.5×10^5	0.0214	0.0231	0.0246	0.0259	0.0271	0.0281	0.0292	0.0327	0.0357
2×10^5	0.0210	0.0228	0.0243	0.0256	0.0268	0.0279	0.0290	0.0325	0.0355
2.5×10^5	0.0208	0.0226	0.0241	0.0255	0.0267	0.0278	0.0289	0.0325	0.0355
3×10^5	0.0206	0.0225	0.0240	0.0254	0.0266	0.0277	0.0288	0.0324	0.0354
4×10^5	0.0204	0.0223	0.0239	0.0253	0.0265	0.0276	0.0287	0.0323	0.0354
5×10^5	0.0202	0.0222	0.0238	0.0252	0.0264	0.0276	0.0286	0.0323	0.0353
6×10^5	0.0201	0.0221	0.0237	0.0251	0.0264	0.0275	0.0286	0.0323	0.0353
7×10^5	0.0201	0.0221	0.0237	0.0251	0.0264	0.0275	0.0286	0.0322	0.0353
8×10^5	0.0200	0.0220	0.0237	0.0251	0.0263	0.0275	0.0286	0.0322	0.0353
9×10^5	0.0200	0.0220	0.0236	0.0251	0.0263	0.0275	0.0285	0.0322	0.0353
1×10^6	0.0199	0.0220	0.0236	0.0250	0.0263	0.0275	0.0285	0.0322	0.0353
1.5×10^6	0.0198	0.0219	0.0235	0.0250	0.0263	0.0274	0.0285	0.0322	0.0352
2×10^6	0.0198	0.0218	0.0235	0.0250	0.0262	0.0274	0.0285	0.0322	0.0352
2.5×10^6	0.0198	0.0218	0.0235	0.0249	0.0262	0.0274	0.0285	0.0322	0.0352
3×10^6	0.0197	0.0218	0.0235	0.0249	0.0262	0.0274	0.0285	0.0321	0.0352
4×10^6	0.0197	0.0218	0.0235	0.0249	0.0262	0.0274	0.0284	0.0321	0.0352
5×10^6	0.0197	0.0218	0.0235	0.0249	0.0262	0.0274	0.0284	0.0321	0.0352
6×10^6	0.0197	0.0218	0.0235	0.0249	0.0262	0.0274	0.0284	0.0321	0.0352
7×10^6	0.0197	0.0218	0.0234	0.0249	0.0262	0.0274	0.0284	0.0321	0.0352
8×10^6	0.0197	0.0218	0.0234	0.0249	0.0262	0.0274	0.0284	0.0321	0.0352
9×10^6	0.0197	0.0218	0.0234	0.0249	0.0262	0.0274	0.0284	0.0321	0.0352
1×10^7	0.0197	0.0218	0.0234	0.0249	0.0262	0.0273	0.0284	0.0321	0.0352
1.5×10^7	0.0197	0.0217	0.0234	0.0249	0.0262	0.0273	0.0284	0.0321	0.0352
2×10^7	0.0197	0.0217	0.0234	0.0249	0.0262	0.0273	0.0284	0.0321	0.0352
2.5×10^7	0.0196	0.0217	0.0234	0.0249	0.0262	0.0273	0.0284	0.0321	0.0352
3×10^7	0.0196	0.0217	0.0234	0.0249	0.0262	0.0273	0.0284	0.0321	0.0352
4×10^7	0.0196	0.0217	0.0234	0.0249	0.0262	0.0273	0.0284	0.0321	0.0352
5×10^7	0.0196	0.0217	0.0234	0.0249	0.0262	0.0273	0.0284	0.0321	0.0352

Support Material

APPENDIX 2.B *(continued)*
Darcy Friction Factors (turbulent flow)

relative roughness, ϵ/D

Reynolds no.	0.01	0.015	0.02	0.025	0.03	0.035	0.04	0.045	0.05
2×10^3	0.0568	0.0602	0.0635	0.0668	0.0699	0.0730	0.0760	0.0790	0.0819
2.5×10^3	0.0539	0.0576	0.0610	0.0644	0.0677	0.0709	0.0740	0.0770	0.0800
3×10^3	0.0519	0.0557	0.0593	0.0628	0.0661	0.0694	0.0725	0.0756	0.0787
4×10^3	0.0491	0.0531	0.0570	0.0606	0.0641	0.0674	0.0707	0.0739	0.0770
5×10^3	0.0473	0.0515	0.0555	0.0592	0.0628	0.0662	0.0696	0.0728	0.0759
6×10^3	0.0460	0.0504	0.0544	0.0583	0.0619	0.0654	0.0688	0.0721	0.0752
7×10^3	0.0450	0.0495	0.0537	0.0576	0.0613	0.0648	0.0682	0.0715	0.0747
8×10^3	0.0442	0.0489	0.0531	0.0571	0.0608	0.0644	0.0678	0.0711	0.0743
9×10^3	0.0436	0.0484	0.0526	0.0566	0.0604	0.0640	0.0675	0.0708	0.0740
1×10^4	0.0431	0.0479	0.0523	0.0563	0.0601	0.0637	0.0672	0.0705	0.0738
1.5×10^4	0.0415	0.0466	0.0511	0.0553	0.0592	0.0628	0.0664	0.0698	0.0731
2×10^4	0.0407	0.0459	0.0505	0.0547	0.0587	0.0624	0.0660	0.0694	0.0727
2.5×10^4	0.0402	0.0455	0.0502	0.0544	0.0584	0.0621	0.0657	0.0691	0.0725
3×10^4	0.0398	0.0452	0.0499	0.0542	0.0582	0.0619	0.0655	0.0690	0.0723
4×10^4	0.0394	0.0448	0.0496	0.0539	0.0579	0.0617	0.0653	0.0688	0.0721
5×10^4	0.0391	0.0446	0.0494	0.0538	0.0578	0.0616	0.0652	0.0687	0.0720
6×10^4	0.0389	0.0445	0.0493	0.0536	0.0577	0.0615	0.0651	0.0686	0.0719
7×10^4	0.0388	0.0443	0.0492	0.0536	0.0576	0.0614	0.0650	0.0685	0.0719
8×10^4	0.0387	0.0443	0.0491	0.0535	0.0576	0.0614	0.0650	0.0685	0.0718
9×10^4	0.0386	0.0442	0.0491	0.0535	0.0575	0.0613	0.0650	0.0684	0.0718
1×10^5	0.0385	0.0442	0.0490	0.0534	0.0575	0.0613	0.0649	0.0684	0.0718
1.5×10^5	0.0383	0.0440	0.0489	0.0533	0.0574	0.0612	0.0648	0.0683	0.0717
2×10^5	0.0382	0.0439	0.0488	0.0532	0.0573	0.0612	0.0648	0.0683	0.0717
2.5×10^5	0.0381	0.0439	0.0488	0.0532	0.0573	0.0611	0.0648	0.0683	0.0716
3×10^5	0.0381	0.0438	0.0488	0.0532	0.0573	0.0611	0.0648	0.0683	0.0716
4×10^5	0.0381	0.0438	0.0487	0.0532	0.0573	0.0611	0.0647	0.0682	0.0716
5×10^5	0.0380	0.0438	0.0487	0.0531	0.0572	0.0611	0.0647	0.0682	0.0716
6×10^5	0.0380	0.0438	0.0487	0.0531	0.0572	0.0611	0.0647	0.0682	0.0716
7×10^5	0.0380	0.0438	0.0487	0.0531	0.0572	0.0611	0.0647	0.0682	0.0716
8×10^5	0.0380	0.0437	0.0487	0.0531	0.0572	0.0611	0.0647	0.0682	0.0716
9×10^5	0.0380	0.0437	0.0487	0.0531	0.0572	0.0610	0.0647	0.0682	0.0716
1×10^6	0.0380	0.0437	0.0487	0.0531	0.0572	0.0610	0.0647	0.0682	0.0716
1.5×10^6	0.0379	0.0437	0.0487	0.0531	0.0572	0.0610	0.0647	0.0682	0.0716
2×10^6	0.0379	0.0437	0.0487	0.0531	0.0572	0.0610	0.0647	0.0682	0.0716
2.5×10^6	0.0379	0.0437	0.0487	0.0531	0.0572	0.0610	0.0647	0.0682	0.0716
3×10^6	0.0379	0.0437	0.0487	0.0531	0.0572	0.0610	0.0647	0.0682	0.0716
4×10^6	0.0379	0.0437	0.0486	0.0531	0.0572	0.0610	0.0647	0.0682	0.0716
5×10^6	0.0379	0.0437	0.0486	0.0531	0.0572	0.0610	0.0647	0.0682	0.0716
6×10^6	0.0379	0.0437	0.0486	0.0531	0.0572	0.0610	0.0647	0.0682	0.0716
7×10^6	0.0379	0.0437	0.0486	0.0531	0.0572	0.0610	0.0647	0.0682	0.0716

APPENDIX 2.B *(continued)*
Darcy Friction Factors (turbulent flow)

Reynolds no.	relative roughness, ϵ/D								
	0.01	0.015	0.02	0.025	0.03	0.035	0.04	0.045	0.05
8×10^6	0.0379	0.0437	0.0486	0.0531	0.0572	0.0610	0.0647	0.0682	0.0716
9×10^6	0.0379	0.0437	0.0486	0.0531	0.0572	0.0610	0.0647	0.0682	0.0716
1×10^7	0.0379	0.0437	0.0486	0.0531	0.0572	0.0610	0.0647	0.0682	0.0716
1.5×10^7	0.0379	0.0437	0.0486	0.0531	0.0572	0.0610	0.0647	0.0682	0.0716
2×10^7	0.0379	0.0437	0.0486	0.0531	0.0572	0.0610	0.0647	0.0682	0.0716
2.5×10^7	0.0379	0.0437	0.0486	0.0531	0.0572	0.0610	0.0647	0.0682	0.0716
3×10^7	0.0379	0.0437	0.0486	0.0531	0.0572	0.0610	0.0647	0.0682	0.0716
4×10^7	0.0379	0.0437	0.0486	0.0531	0.0572	0.0610	0.0647	0.0682	0.0716
5×10^7	0.0379	0.0437	0.0486	0.0531	0.0572	0.0610	0.0647	0.0682	0.0716

Support Material

APPENDIX 3.A
Periodic Table of the Elements
(referred to carbon-12)

The Periodic Table of Elements (Long Form)

The number of electrons in filled shells is shown in the column at the extreme left; the remaining electrons for each element are shown immediately below the symbol for each element. Atomic numbers are enclosed in brackets. Atomic weights (rounded, based on carbon-12) are shown above the symbols. Atomic weight values in parentheses are those of the isotopes of longest half-life for certain radioactive elements whose atomic weights cannot be precisely quoted without knowledge of origin of the element.

metals — transition metals — nonmetals

periods	I A	II A	III B	IV B	V B	VI B	VII B		VIII		I B	II B	III A	IV A	V A	VI A	VII A	0
1 / 0	1.00794 H[1] 1																	4.00260 He[2] 2
2 / 2	6.941 Li[3] 1	9.01218 Be[4] 2											10.811 B[5] 3	12.0107 C[6] 4	14.0067 N[7] 5	15.9994 O[8] 6	18.9984 F[9] 7	20.1797 Ne[10] 8
3 / 2,8	22.9898 Na[11] 1	24.3050 Mg[12] 2											26.9815 Al[13] 3	28.0855 Si[14] 4	30.9738 P[15] 5	32.065 S[16] 6	35.453 Cl[17] 7	39.948 Ar[18] 8
4 / 2,8	39.0983 K[19] 8,1	40.078 Ca[20] 8,2	44.9559 Sc[21] 9,2	47.867 Ti[22] 10,2	50.9415 V[23] 11,2	51.9961 Cr[24] 13,1	54.9380 Mn[25] 13,2	55.845 Fe[26] 14,2	58.9332 Co[27] 15,2	58.6934 Ni[28] 16,2	63.546 Cu[29] 18,1	65.38 Zn[30] 18,2	69.723 Ga[31] 18,3	72.64 Ge[32] 18,4	74.9216 As[33] 18,5	78.96 Se[34] 18,6	79.904 Br[35] 18,7	83.798 Kr[36] 18,8
5 / 2,8,18	85.4678 Rb[37] 8,1	87.62 Sr[38] 8,2	88.9059 Y[39] 9,2	91.224 Zr[40] 10,2	92.9064 Nb[41] 12,1	95.96 Mo[42] 13,1	(98) Tc[43] 14,1	101.07 Ru[44] 15,1	102.906 Rh[45] 16,1	106.42 Pd[46] 18	107.868 Ag[47] 18,1	112.411 Cd[48] 18,2	114.818 In[49] 18,3	118.710 Sn[50] 18,4	121.760 Sb[51] 18,5	127.60 Te[52] 18,6	126.904 I[53] 18,7	131.293 Xe[54] 18,8
6 / 2,8,18	132.905 Cs[55] 18,8,1	137.327 Ba[56] 18,8,2	* (57-71)	178.49 Hf[72] 32,10,2	180.948 Ta[73] 32,11,2	183.84 W[74] 32,12,2	186.207 Re[75] 32,13,2	190.23 Os[76] 32,14,2	192.217 Ir[77] 32,15,2	195.084 Pt[78] 32,17,1	196.967 Au[79] 32,18,1	200.59 Hg[80] 32,18,2	204.383 Tl[81] 32,18,3	207.2 Pb[82] 32,18,4	208.980 Bi[83] 32,18,5	(209) Po[84] 32,18,6	(210) At[85] 32,18,7	(222) Rn[86] 32,18,8
7 / 2,8,18,32	(223) Fr[87] 18,8,1	(226) Ra[88] 18,8,2	† (89-103)	(265) Rf[104] 32,10,2	(268) Db[105] 32,11,2	(271) Sg[106] 32,12,2	(272) Bh[107] 32,13,2	(270) Hs[108] 32,14,2	(276) Mt[109] 32,15,2	(281) Ds[110] 32,17,2	(280) Rg[111] 32,18,1	(285) Cn[112] 32,18,2						

*lanthanide series	138.905 La[57] 18,9,2	140.116 Ce[58] 20,8,2	140.908 Pr[59] 21,8,2	144.242 Nd[60] 22,8,2	(145) Pm[61] 23,8,2	150.36 Sm[62] 24,8,2	151.964 Eu[63] 25,8,2	157.25 Gd[64] 25,9,2	158.925 Tb[65] 27,8,2	162.500 Dy[66] 28,8,2	164.930 Ho[67] 29,8,2	167.259 Er[68] 30,8,2	168.934 Tm[69] 31,8,2	173.054 Yb[70] 32,8,2	174.967 Lu[71] 32,9,2
†actinide series	(227) Ac[89] 18,9,2	232.038 Th[90] 18,10,2	231.036 Pa[91] 20,9,2	238.029 U[92] 21,9,2	(237) Np[93] 23,8,2	(244) Pu[94] 24,8,2	(243) Am[95] 25,8,2	(247) Cm[96] 25,9,2	(247) Bk[97] 26,9,2	(251) Cf[98] 28,8,2	(252) Es[99] 29,8,2	(257) Fm[100] 30,8,2	(258) Md[101] 31,8,2	(259) No[102] 32,8,2	(262) Lr[103] 32,9,2

Support Material

APPENDIX 15.A
Critical Depths in Circular Channels

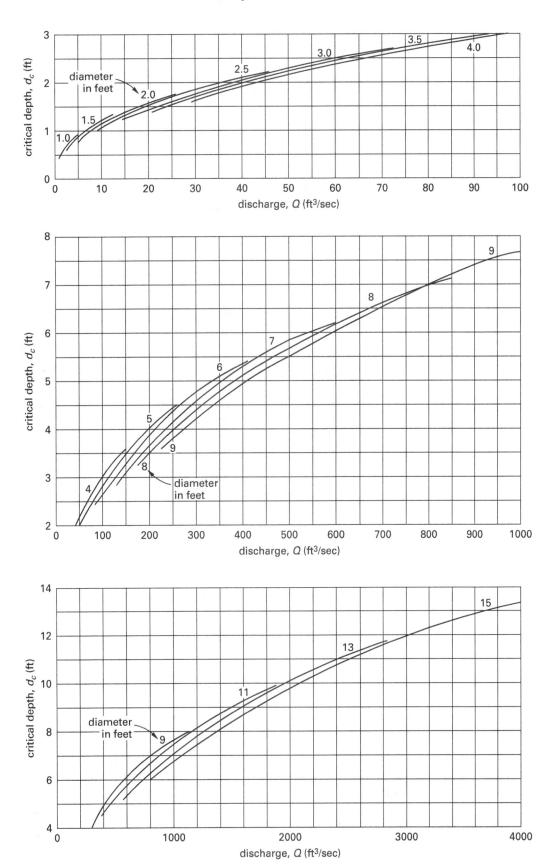

APPENDIX 15.B
Conveyance Factor, K
Symmetrical Rectangular,[a] Trapezoidal, and V-Notch[b] Open Channels

(use for determining Q or d when b is known)

(customary U.S. units[b,c])

$$K' \text{ in } Q = K'\left(\frac{1}{n}\right)b^{8/3}\sqrt{S}$$

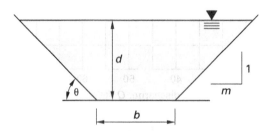

	m and θ									
	0.0	0.25	0.5	0.75	1.0	1.5	2.0	2.5	3.0	4.0
$x = d/b$	90°	76.0°	63.4°	53.1°	45.0°	33.7°	26.6°	21.8°	18.4°	14.0°
0.01	0.00068	0.00068	0.00069	0.00069	0.00069	0.00069	0.00069	0.00069	0.00070	0.00070
0.02	0.00213	0.00215	0.00216	0.00217	0.00218	0.00220	0.00221	0.00222	0.00223	0.00225
0.03	0.00414	0.00419	0.00423	0.00426	0.00428	0.00433	0.00436	0.00439	0.00443	0.00449
0.04	0.00660	0.00670	0.00679	0.00685	0.00691	0.00700	0.00708	0.00716	0.00723	0.00736
0.05	0.00946	0.00964	0.00979	0.00991	0.01002	0.01019	0.01033	0.01047	0.01060	0.01086
0.06	0.0127	0.0130	0.0132	0.0134	0.0136	0.0138	0.0141	0.0148	0.0145	0.0150
0.07	0.0162	0.0166	0.0170	0.0173	0.0175	0.0180	0.0183	0.0187	0.0190	0.0197
0.08	0.0200	0.0206	0.0211	0.0215	0.0219	0.0225	0.0231	0.0236	0.0240	0.0250
0.09	0.0241	0.0249	0.0256	0.0262	0.0267	0.0275	0.0282	0.0289	0.0296	0.0310
0.10	0.0284	0.0294	0.0304	0.0311	0.0318	0.0329	0.0339	0.0348	0.0358	0.0376
0.11	0.0329	0.0343	0.0354	0.0364	0.0373	0.0387	0.0400	0.0413	0.0424	0.0448
0.12	0.0376	0.0393	0.0408	0.0420	0.0431	0.0450	0.0466	0.0482	0.0497	0.0527
0.13	0.0425	0.0446	0.0464	0.0480	0.0493	0.0516	0.0537	0.0556	0.0575	0.0613
0.14	0.0476	0.0502	0.0524	0.0542	0.0559	0.0587	0.0612	0.0636	0.0659	0.0706
0.15	0.0528	0.0559	0.0585	0.0608	0.0627	0.0662	0.0692	0.0721	0.0749	0.0805
0.16	0.0582	0.0619	0.0650	0.0676	0.0700	0.0740	0.0777	0.0811	0.0845	0.0912
0.17	0.0638	0.0680	0.0716	0.0748	0.0775	0.0823	0.0866	0.0907	0.0947	0.1026
0.18	0.0695	0.0744	0.0786	0.0822	0.0854	0.0910	0.0960	0.1008	0.1055	0.1148
0.19	0.0753	0.0809	0.0857	0.0899	0.0936	0.1001	0.1059	0.1115	0.1169	0.1277
0.20	0.0812	0.0876	0.0931	0.0979	0.1021	0.1096	0.1163	0.1227	0.1290	0.1414
0.22	0.0934	0.1015	0.109	0.115	0.120	0.130	0.139	0.147	0.155	0.171
0.24	0.1061	0.1161	0.125	0.133	0.140	0.152	0.163	0.173	0.184	0.204
0.26	0.119	0.131	0.142	0.152	0.160	0.175	0.189	0.202	0.215	0.241
0.28	0.132	0.147	0.160	0.172	0.182	0.201	0.217	0.234	0.249	0.281
0.30	0.146	0.163	0.179	0.193	0.205	0.228	0.248	0.267	0.287	0.324
0.32	0.160	0.180	0.199	0.215	0.230	0.256	0.281	0.304	0.327	0.371
0.34	0.174	0.198	0.219	0.238	0.256	0.287	0.316	0.343	0.370	0.423
0.36	0.189	0.216	0.241	0.263	0.283	0.319	0.353	0.385	0.416	0.478
0.38	0.203	0.234	0.263	0.288	0.312	0.353	0.392	0.429	0.465	0.537
0.40	0.218	0.253	0.286	0.315	0.341	0.389	0.434	0.476	0.518	0.600
0.42	0.233	0.273	0.309	0.342	0.373	0.427	0.478	0.526	0.574	0.668
0.44	0.248	0.293	0.334	0.371	0.405	0.467	0.525	0.580	0.633	0.740
0.46	0.264	0.313	0.359	0.401	0.439	0.509	0.574	0.636	0.696	0.816
0.48	0.279	0.334	0.385	0.432	0.474	0.553	0.625	0.695	0.763	0.897
0.50	0.295	0.355	0.412	0.463	0.511	0.598	0.679	0.757	0.833	0.983
0.55	0.335	0.410	0.482	0.548	0.609	0.722	0.826	0.926	1.025	1.22

APPENDIX 15.B *(continued)*
Conveyance Factor, K
Symmetrical Rectangular,[a] Trapezoidal, and V-Notch[b] Open Channels

(use for determining Q or d when b is known)
(customary U.S. units[b,c])

$$K' \text{ in } Q = K'\left(\frac{1}{n}\right)b^{8/3}\sqrt{S}$$

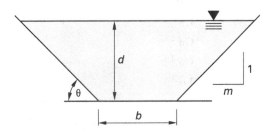

					m and θ					
	0.0	0.25	0.5	0.75	1.0	1.5	2.0	2.5	3.0	4.0
$x = d/b$	90°	76.0°	63.4°	53.1°	45.0°	33.7°	26.6°	21.8°	18.4°	14.0°
0.60	0.375	0.468	0.557	0.640	0.717	0.858	0.990	1.117	1.24	1.49
0.70	0.457	0.592	0.722	0.844	0.959	1.17	1.37	1.56	1.75	2.12
0.80	0.542	0.725	0.906	1.078	1.24	1.54	1.83	2.10	2.37	2.90
0.90	0.628	0.869	1.11	1.34	1.56	1.98	2.36	2.74	3.11	3.83
1.00	0.714	1.022	1.33	1.64	1.93	2.47	2.99	3.48	3.97	4.93
1.20	0.891	1.36	1.85	2.33	2.79	3.67	4.51	5.32	6.11	7.67
1.40	1.07	1.74	2.45	3.16	3.85	5.17	6.42	7.64	8.84	11.2
1.60	1.25	2.16	3.14	4.14	5.12	6.99	8.78	10.52	12.2	15.6
1.80	1.43	2.62	3.93	5.28	6.60	9.15	11.6	14.0	16.3	20.9
2.00	1.61	3.12	4.82	6.58	8.32	11.7	14.9	18.1	21.2	27.3
2.25	1.84	3.81	6.09	8.46	10.8	15.4	19.8	24.1	28.4	36.7

[a]For rectangular channels, use the 0.0 (90°, vertical sides) column.

[b]Q = flow rate, ft³/sec; d = depth of flow, ft; b = bottom width of channel, ft; S = geometric slope, ft/ft; n = Manning's roughness constant.

[c]For SI units (i.e., Q in m³/s, and d and b in m), divide each table value by 1.486.

APPENDIX 16.A
Water Chemistry CaCO$_3$ Equivalents

cations	formula	ionic weight	equivalent weight	substance to CaCO$_3$ factor
aluminum	Al^{+3}	27.0	9.0	5.56
ammonium	NH_4^+	18.0	18.0	2.78
calcium	Ca^{+2}	40.1	20.0	2.50
cupric copper	Cu^{+2}	63.6	31.8	1.57
cuprous copper	Cu^{+3}	63.6	21.2	2.36
ferric iron	Fe^{+3}	55.8	18.6	2.69
ferrous iron	Fe^{+2}	55.8	27.9	1.79
hydrogen	H^+	1.0	1.0	50.00
manganese	Mn^{+2}	54.9	27.5	1.82
magnesium	Mg^{+2}	24.3	12.2	4.10
potassium	K^+	39.1	39.1	1.28
sodium	Na^+	23.0	23.0	2.18

anions	formula	ionic weight	equivalent weight	substance to CaCO$_3$ factor
bicarbonate	HCO_3^-	61.0	61.0	0.82
carbonate	CO_3^{-2}	60.0	30.0	1.67
chloride	Cl^-	35.5	35.5	1.41
fluoride	F^-	19.0	19.0	2.66
hydroxide	OH^-	17.0	17.0	2.94
nitrate	NO_3^-	62.0	62.0	0.81
phosphate (tribasic)	PO_4^{-3}	95.0	31.7	1.58
phosphate (dibasic)	HPO_4^{-2}	96.0	48.0	1.04
phosphate (monobasic)	$H_2PO_4^-$	97.0	97.0	0.52
sulfate	SO_4^{-2}	96.1	48.0	1.04
sulfite	SO_3^{-2}	80.1	40.0	1.25

compounds	formula	molecular weight	equivalent weight	substance to CaCO$_3$ factor
aluminum hydroxide	$Al(OH)_3$	78.0	26.0	1.92
aluminum sulfate	$Al_2(SO_4)_3$	342.1	57.0	0.88
aluminum sulfate	$Al_2(SO_4)_3 \cdot 18H_2O$	666.1	111.0	0.45
alumina	Al_2O_3	102.0	17.0	2.94
sodium aluminate	$Na_2Al_2O_4$	164.0	27.3	1.83
calcium bicarbonate	$Ca(HCO_3)_2$	162.1	81.1	0.62
calcium carbonate	$CaCO_3$	100.1	50.1	1.00
calcium chloride	$CaCl_2$	111.0	55.5	0.90
calcium hydroxide (pure)	$Ca(OH)_2$	74.1	37.1	1.35
calcium hydroxide (90%)	$Ca(OH)_2$	–	41.1	1.22
calcium oxide (lime)	CaO	56.1	28.0	1.79
calcium sulfate (anhydrous)	$CaSO_4$	136.2	68.1	0.74
calcium sulfate (gypsum)	$CaSO_4 \cdot 2H_2O$	172.2	86.1	0.58

APPENDIX 16.A *(continued)*
Water Chemistry CaCO₃ Equivalents

compounds	formula	molecular weight	equivalent weight	substance to CaCO$_3$ factor
calcium phosphate	$Ca_3(PO_4)_2$	310.3	51.7	0.97
disodium phosphate	$Na_2HPO_4 \cdot 12H_2O$	358.2	119.4	0.42
disodium phosphate (anhydrous)	Na_2HPO_4	142.0	47.3	1.06
ferric oxide	Fe_2O_3	159.6	26.6	1.88
iron oxide (magnetic)	Fe_3O_4	321.4	–	–
ferrous sulfate (copperas)	$FeSO_4 \cdot 7H_2O$	278.0	139.0	0.36
magnesium oxide	MgO	40.3	20.2	2.48
magnesium bicarbonate	$Mg(HCO_3)_2$	146.3	73.2	0.68
magnesium carbonate	$MgCO_3$	84.3	42.2	1.19
magnesium chloride	$MgCl_2$	95.2	47.6	1.05
magnesium hydroxide	$Mg(OH)_2$	58.3	29.2	1.71
magnesium phosphate	$Mg_3(PO_4)_2$	263.0	43.8	1.14
magnesium sulfate	$MgSO_4$	120.4	60.2	0.83
monosodium phosphate	$NaH_2PO_4 \cdot H_2O$	138.1	46.0	1.09
monosodium phosphate (anhydrous)	NaH_2PO_4	120.1	40.0	1.25
metaphosphate	$NaPO_3$	102.0	34.0	1.47
silica	SiO_2	60.1	30.0	1.67
sodium bicarbonate	$NaHCO_3$	84.0	84.0	0.60
sodium carbonate	Na_2CO_3	106.0	53.0	0.94
sodium chloride	$NaCl$	58.5	58.5	0.85
sodium hydroxide	$NaOH$	40.0	40.0	1.25
sodium nitrate	$NaNO_3$	85.0	85.0	0.59
sodium sulfate	Na_2SO_4	142.0	71.0	0.70
sodium sulfite	Na_2SO_3	126.1	63.0	0.79
tetrasodium EDTA	$(CH_2)_2N_2(CH_2COONa)_4$	380.2	95.1	0.53
trisodium phosphate	$Na_3PO_4 \cdot 12H_2O$	380.2	126.7	0.40
trisodium phosphate (anhydrous)	Na_3PO_4	164.0	54.7	0.91
trisodium NTA	$(CH_2)_3N(COONa)_3$	257.1	85.7	0.58

gases	formula	molecular weight	equivalent weight	substance to CaCO$_3$ factor
ammonia	NH_3	17	17	2.94
carbon dioxide	CO_2	44	22	2.27
hydrogen	H_2	2	1	50.00
hydrogen sulfide	H_2S	34	17	2.94
oxygen	O_2	32	8	6.25

Support Material

APPENDIX 16.A *(continued)*
Water Chemistry $CaCO_3$ Equivalents

acids	formula	molecular weight	equivalent weight	substance to $CaCO_3$ factor
carbonic	H_2CO_3	62.0	31.0	1.61
hydrochloric	HCl	36.5	36.5	1.37
phosphoric	H_3PO_4	98.0	32.7	1.53
sulfuric	H_2SO_4	98.1	49.1	1.02

(Multiply the concentration (in mg/L) of the substance by the corresponding factors to obtain the equivalent concentration in mg/L as $CaCO_3$. For example, 70 mg/L of Mg^{++} would be $(70 \text{ mg/L})(4.1) = 287$ mg/L as $CaCO_3$.)

APPENDIX 28.A
Heats of Combustion for Common Compounds[a]

				heat of combustion			
				Btu/ft^3		Btu/lbm	
		molecular	specific volume	gross	net	gross	net
substance	formula	weight	(ft^3/lbm)	(high)	(low)	(high)	(low)
carbon	C	12.01				14,093	14,093
carbon dioxide	CO_2	44.01	8.548				
carbon monoxide	CO	28.01	13.506	322	322	4347	4347
hydrogen	H_2	2.016	187.723	325	275	60,958	51,623
nitrogen	N_2	28.016	13.443				
oxygen	O_2	32.000	11.819				
paraffin series (alkanes)							
methane	CH_4	16.041	23.565	1013	913	23,879	21,520
ethane	C_2H_6	30.067	12.455	1792	1641	22,320	20,432
propane	C_3H_8	44.092	8.365	2590	2385	21,661	19,944
n-butane	C_4H_{10}	58.118	6.321	3370	3113	21,308	19,680
isobutane	C_4H_{10}	58.118	6.321	3363	3105	21,257	19,629
n-pentane	C_5H_{12}	72.144	5.252	4016	3709	21,091	19,517
isopentane	C_5H_{12}	72.144	5.252	4008	3716	21,052	19,478
neopentane	C_5H_{12}	72.144	5.252	3993	3693	20,970	19,396
n-hexane	C_6H_{14}	86.169	4.398	4762	4412	20,940	19,403
olefin series (alkenes and alkynes)							
ethylene	C_2H_4	28.051	13.412	1614	1513	21,644	20,295
propylene	C_3H_6	42.077	9.007	2336	2186	21,041	19,691
n-butene	C_4H_8	56.102	6.756	3084	2885	20,840	19,496
isobutene	C_4H_8	56.102	6.756	3068	2869	20,730	19,382
n-pentene	C_5H_{10}	70.128	5.400	3836	3586	20,712	19,363
aromatic series							
benzene	C_6H_6	78.107	4.852	3751	3601	18,210	17,480
toluene	C_7H_8	92.132	4.113	4484	4284	18,440	17,620
xylene	C_8H_{10}	106.158	3.567	5230	4980	18,650	17,760
miscellaneous fuels							
acetylene	C_2H_2	26.036	14.344	1499	1448	21,500	20,776
air		28.967	13.063				
ammonia	NH_3	17.031	21.914	441	365	9668	8001
digester gas[b]	–	25.8	18.3	658	593	15,521	13,988
ethyl alcohol	C_2H_5OH	46.067	8.221	1600	1451	13,161	11,929
hydrogen sulfide	H_2S	34.076	10.979	647	596	7100	6545
iso-octane	C_8H_{18}	114.2	0.0232[c]	106	98.9	20,590	19,160
methyl alcohol	CH_3OH	32.041	11.820	868	768	10,259	9078
naphthalene	$C_{10}H_8$	128.162	2.955	5854	5654	17,298	16,708
sulfur	S	32.06				3983	3983
sulfur dioxide	SO_2	64.06	5.770				
water vapor	H_2O	18.016	21.017				

(Multiply Btu/lbm by 2.326 to obtain kJ/kg.)
(Multiply Btu/ft^3 by 37.25 to obtain kJ/m^3.)
[a]Gas volumes listed are at 60°F (16°C) and 1 atm.
[b]Digester gas from wastewater treatment plants is approximately 65% methane and 35% carbon dioxide by volume. Use composite properties of these two gases.
[c]liquid form; stoichiometric mixture

Support Material

APPENDIX 53.A
Highly Hazardous Chemicals, Toxics and Reactives
(Standards—29 CFR 1910.119, App. A)

chemical name	CAS[a]	TQ[b]
Acetaldehyde	75-07-0	2500
Acrolein (2-Propenal)	107-02-8	150
Acrylyl Chloride	814-68-6	250
Allyl Chloride	107-05-1	1000
Allylamine	107-11-9	1000
Alkylaluminums	Varies	5000
Ammonia, Anhydrous	7664-41-7	10000
Ammonia solutions (greater than 44% ammonia by weight)	7664-41-7	15000
Ammonium Perchlorate	7790-98-9	7500
Ammonium Permanganate	7787-36-2	7500
Arsine (also called Arsenic Hydride)	7784-42-1	100
Bis(Chloromethyl) Ether	542-88-1	100
Boron Trichloride	10294-34-5	2500
Boron Trifluoride	7637-07-2	250
Bromine	7726-95-6	1500
Bromine Chloride	13863-41-7	1500
Bromine Pentafluoride	7789-30-2	2500
Bromine Trifluoride	7787-71-5	15000
3-Bromopropyne (also called Propargyl Bromide)	106-96-7	100
Butyl Hydroperoxide (Tertiary)	75-91-2	5000
Butyl Perbenzoate (Tertiary)	614-45-9	7500
Carbonyl Chloride (see Phosgene)	75-44-5	100
Carbonyl Fluoride	353-50-4	2500
Cellulose Nitrate (concentration greater than 12.6% nitrogen)	9004-70-0	2500
Chlorine	7782-50-5	1500
Chlorine Dioxide	10049-04-4	1000
Chlorine Pentafluoride	13637-63-3	1000
Chlorine Trifluoride	7790-91-2	1000
Chlorodiethylaluminum (also called Diethylaluminum Chloride)	96-10-6	5000
1-Chloro-2,4-Dinitrobenzene	97-00-7	5000
Chloromethyl Methyl Ether	107-30-2	500
Chloropicrin	76-06-2	500
Chloropicrin and Methyl Bromide mixture	None	1500
Chloropicrin and Methyl Chloride mixture	None	1500
Commune Hydroperoxide	80-15-9	5000
Cyanogen	460-19-5	2500
Cyanogen Chloride	506-77-4	500
Cyanuric Fluoride	675-14-9	100
Diacetyl Peroxide (concentration greater than 70%)	110-22-5	5000
Diazomethane	334-88-3	500
Dibenzoyl Peroxide	94-36-0	7500
Diborane	19287-45-7	100
Dibutyl Peroxide (Tertiary)	110-05-4	5000
Dichloro Acetylene	7572-29-4	250
Dichlorosilane	4109-96-0	2500
Diethylzinc	557-20-0	10000
Diisopropyl Peroxydicarbonate	105-64-6	7500
Dilauroyl Peroxide	105-74-8	7500
Dimethyldichlorosilane	75-78-5	1000
Dimethylhydrazine, 1,1-	57-14-7	1000
Dimethylamine, Anhydrous	124-40-3	2500
2,4-Dinitroaniline	97-02-9	5000
Ethyl Methyl Ketone Peroxide (also Methyl Ethyl Ketone Peroxide; concentration greater than 60%)	1338-23-4	5000

APPENDIX 53.A (continued)
Highly Hazardous Chemicals, Toxics and Reactives
(Standards—29 CFR 1910.119, App. A)

chemical name	CAS[a]	[b]
Ethyl Nitrite	109-95-5	5000
Ethylamine	75-04-7	7500
Ethylene Fluorohydrin	371-62-0	100
Ethylene Oxide	75-21-8	5000
Ethyleneimine	151-56-4	1000
Fluorine	7782-41-4	1000
Formaldehyde (Formalin)	50-00-0	1000
Furan	110-00-9	500
Hexafluoroacetone	684-16-2	5000
Hydrochloric Acid, Anhydrous	7647-01-0	5000
Hydrofluoric Acid, Anhydrous	7664-39-3	1000
Hydrogen Bromide	10035-10-6	5000
Hydrogen Chloride	7647-01-0	5000
Hydrogen Cyanide, Anhydrous	74-90-8	1000
Hydrogen Fluoride	7664-39-3	1000
Hydrogen Peroxide (52% by weight or greater)	7722-84-1	7500
Hydrogen Selenide	7783-07-5	150
Hydrogen Sulfide	7783-06-4	1500
Hydroxylamine	7803-49-8	2500
Iron, Pentacarbonyl	13463-40-6	250
Isopropylamine	75-31-0	5000
Ketene	463-51-4	100
Methacrylaldehyde	78-85-3	1000
Methacryloyl Chloride	920-46-7	150
Methacryloyloxyethyl Isocyanate	30674-80-7	100
Methyl Acrylonitrile	126-98-7	250
Methylamine, Anhydrous	74-89-5	1000
Methyl Bromide	74-83-9	2500
Methyl Chloride	74-87-3	15000
Methyl Chloroformate	79-22-1	500
Methyl Ethyl Ketone Peroxide (concentration greater than 60%)	1338-23-4	5000
Methyl Fluoroacetate	453-18-9	100
Methyl Fluorosulfate	421-20-5	100
Methyl Hydrazine	60-34-4	100
Methyl Iodide	74-88-4	7500
Methyl Isocyanate	624-83-9	250
Methyl Mercaptan	74-93-1	5000
Methyl Vinyl Ketone	79-84-4	100
Methyltrichlorosilane	75-79-6	500
Nickel Carbonly (Nickel Tetracarbonyl)	13463-39-3	150
Nitric Acid (94.5% by weight or greater)	7697-37-2	500
Nitric Oxide	10102-43-9	250
Nitroaniline (para Nitroaniline)	100-01-6	5000
Nitromethane	75-52-5	2500
Nitrogen Dioxide	10102-44-0	250
Nitrogen Oxides (NO; NO(2); N2O4; N2O3)	10102-44-0	250
Nitrogen Tetroxide (also called Nitrogen Peroxide)	10544-72-6	250
Nitrogen Trifluoride	7783-54-2	5000
Nitrogen Trioxide	10544-73-7	250
Oleum (65% to 80% by weight; also called Fuming Sulfuric Acid)	8014-94-7	1000
Osmium Tetroxide	20816-12-0	100
Oxygen Difluoride (Fluorine Monoxide)	7783-41-7	100
Ozone	10028-15-6	100
Pentaborane	19624-22-7	100

APPENDIX 53.A *(continued)*
Highly Hazardous Chemicals, Toxics and Reactives
(Standards—29 CFR 1910.119, App. A)

chemical name	CAS[a]	[b]
Peracetic Acid (concentration greater than 60% Acetic Acid; also called Peroxyacetic Acid)	79-21-0	1000
Perchloric Acid (concentration greater than 60% by weight)	7601-90-3	5000
Perchloromethyl Mercaptan	594-42-3	150
Perchloryl Fluoride	7616-94-6	5000
Peroxyacetic Acid (concentration greater than 60% Acetic Acid; also called Peracetic Acid)	79-21-0	1000
Phosgene (also called Carbonyl Chloride)	75-44-5	100
Phosphine (Hydrogen Phosphide)	7803-51-2	100
Phosphorus Oxychloride (also called Phosphoryl Chloride)	10025-87-3	1000
Phosphorus Trichloride	7719-12-2	1000
Phosphoryl Chloride (also called Phosphorus Oxychloride)	10025-87-3	1000
Propargyl Bromide	106-96-7	100
Propyl Nitrate	627-3-4	2500
Sarin	107-44-8	100
Selenium Hexafluoride	7783-79-1	1000
Stibine (Antimony Hydride)	7803-52-3	500
Sulfur Dioxide (liquid)	7446-09-5	1000
Sulfur Pentafluoride	5714-22-7	250
Sulfur Tetrafluoride	7783-60-0	250
Sulfur Trioxide (also called Sulfuric Anhydride)	7446-11-9	1000
Sulfuric Anhydride (also called Sulfur Trioxide)	7446-11-9	1000
Tellurium Hexafluoride	7783-80-4	250
Tetrafluoroethylene	116-14-3	5000
Tetrafluorohydrazine	10036-47-2	5000
Tetramethyl Lead	75-74-1	1000
Thionyl Chloride	7719-09-7	250
Trichloro (chloromethyl) Silane	1558-25-4	100
Trichloro (dichlorophenyl) Silane	27137-85-5	2500
Trichlorosilane	10025-78-2	5000
Trifluorochloroethylene	79-38-9	10000
Trimethyoxysilane	2487-90-3	1500

[a]Chemical Abstract Service Number
[b]Threshold Quantity in Pounds (amount necessary to be covered by this standard).

APPENDIX 59.A
Values of the Error Function and Complementary Error Function
(for positive values of x)

x	erf(x)	erfc(x)	x	erf(x)	erfc(x)	x	erf(x)	erfc(x)	x	erf(x)	erfc(x)	x	erf(x)	erfc(x)
0	0.0000	1.0000												
0.01	0.0113	0.9887	0.51	0.5292	0.4708	1.01	0.8468	0.1532	1.51	0.9673	0.0327	2.01	0.9955	0.0045
0.02	0.0226	0.9774	0.52	0.5379	0.4621	1.02	0.8508	0.1492	1.52	0.9684	0.0316	2.02	0.9957	0.0043
0.03	0.0338	0.9662	0.53	0.5465	0.4535	1.03	0.8548	0.1452	1.53	0.9695	0.0305	2.03	0.9959	0.0041
0.04	0.0451	0.9549	0.54	0.5549	0.4451	1.04	0.8586	0.1414	1.54	0.9706	0.0294	2.04	0.9961	0.0039
0.05	0.0564	0.9436	0.55	0.5633	0.4367	1.05	0.8624	0.1376	1.55	0.9716	0.0284	2.05	0.9963	0.0037
0.06	0.0676	0.9324	0.56	0.5716	0.4284	1.06	0.8661	0.1339	1.56	0.9726	0.0274	2.06	0.9964	0.0036
0.07	0.0789	0.9211	0.57	0.5798	0.4202	1.07	0.8698	0.1302	1.57	0.9736	0.0264	2.07	0.9966	0.0034
0.08	0.0901	0.9099	0.58	0.5879	0.4121	1.08	0.8733	0.1267	1.58	0.9745	0.0255	2.08	0.9967	0.0033
0.09	0.1013	0.8987	0.59	0.5959	0.4041	1.09	0.8768	0.1232	1.59	0.9755	0.0245	2.09	0.9969	0.0031
0.1	0.1125	0.8875	0.6	0.6039	0.3961	1.1	0.8802	0.1198	1.6	0.9763	0.0237	2.1	0.9970	0.0030
0.11	0.1236	0.8764	0.61	0.6117	0.3883	1.11	0.8835	0.1165	1.61	0.9772	0.0228	2.11	0.9972	0.0028
0.12	0.1348	0.8652	0.62	0.6194	0.3806	1.12	0.8868	0.1132	1.62	0.9780	0.0220	2.12	0.9973	0.0027
0.13	0.1459	0.8541	0.63	0.6270	0.3730	1.13	0.8900	0.1100	1.63	0.9788	0.0212	2.13	0.9974	0.0026
0.14	0.1569	0.8431	0.64	0.6346	0.3654	1.14	0.8931	0.1069	1.64	0.9796	0.0204	2.14	0.9975	0.0025
0.15	0.1680	0.8320	0.65	0.6420	0.3580	1.15	0.8961	0.1039	1.65	0.9804	0.0196	2.15	0.9976	0.0024
0.16	0.1790	0.8210	0.66	0.6494	0.3506	1.16	0.8991	0.1009	1.66	0.9811	0.0189	2.16	0.9977	0.0023
0.17	0.1900	0.8100	0.67	0.6566	0.3434	1.17	0.9020	0.0980	1.67	0.9818	0.0182	2.17	0.9979	0.0021
0.18	0.2009	0.7991	0.68	0.6638	0.3362	1.18	0.9048	0.0952	1.68	0.9825	0.0175	2.18	0.9980	0.0020
0.19	0.2118	0.7882	0.69	0.6708	0.3292	1.19	0.9076	0.0924	1.69	0.9832	0.0168	2.19	0.9980	0.0020
0.2	0.2227	0.7773	0.7	0.6778	0.3222	1.2	0.9103	0.0897	1.7	0.9838	0.0162	2.2	0.9981	0.0019
0.21	0.2335	0.7665	0.71	0.6847	0.3153	1.21	0.9130	0.0870	1.71	0.9844	0.0156	2.21	0.9982	0.0018
0.22	0.2443	0.7557	0.72	0.6914	0.3086	1.22	0.9155	0.0845	1.72	0.9850	0.0150	2.22	0.9983	0.0017
0.23	0.2550	0.7450	0.73	0.6981	0.3019	1.23	0.9181	0.0819	1.73	0.9856	0.0144	2.23	0.9984	0.0016
0.24	0.2657	0.7343	0.74	0.7047	0.2953	1.24	0.9205	0.0795	1.74	0.9861	0.0139	2.24	0.9985	0.0015
0.25	0.2763	0.7237	0.75	0.7112	0.2888	1.25	0.9229	0.0771	1.75	0.9867	0.0133	2.25	0.9985	0.0015
0.26	0.2869	0.7131	0.76	0.7175	0.2825	1.26	0.9252	0.0748	1.76	0.9872	0.0128	2.26	0.9986	0.0014
0.27	0.2974	0.7026	0.77	0.7238	0.2762	1.27	0.9275	0.0725	1.77	0.9877	0.0123	2.27	0.9987	0.0013
0.28	0.3079	0.6921	0.78	0.7300	0.2700	1.28	0.9297	0.0703	1.78	0.9882	0.0118	2.28	0.9987	0.0013
0.29	0.3183	0.6817	0.79	0.7361	0.2639	1.29	0.9319	0.0681	1.79	0.9886	0.0114	2.29	0.9988	0.0012
0.3	0.3286	0.6714	0.8	0.7421	0.2579	1.3	0.9340	0.0660	1.8	0.9891	0.0109	2.3	0.9989	0.0011
0.31	0.3389	0.6611	0.81	0.7480	0.2520	1.31	0.9361	0.0639	1.81	0.9895	0.0105	2.31	0.9989	0.0011
0.32	0.3491	0.6509	0.82	0.7538	0.2462	1.32	0.9381	0.0619	1.82	0.9899	0.0101	2.32	0.9990	0.0010
0.33	0.3593	0.6407	0.83	0.7595	0.2405	1.33	0.9400	0.0600	1.83	0.9903	0.0097	2.33	0.9990	0.0010
0.34	0.3694	0.6306	0.84	0.7651	0.2349	1.34	0.9419	0.0581	1.84	0.9907	0.0093	2.34	0.9991	0.0009
0.35	0.3794	0.6206	0.85	0.7707	0.2293	1.35	0.9438	0.0562	1.85	0.9911	0.0089	2.35	0.9991	0.0009
0.36	0.3893	0.6107	0.86	0.7761	0.2239	1.36	0.9456	0.0544	1.86	0.9915	0.0085	2.36	0.9992	0.0008
0.37	0.3992	0.6008	0.87	0.7814	0.2186	1.37	0.9473	0.0527	1.87	0.9918	0.0082	2.37	0.9992	0.0008
0.38	0.4090	0.5910	0.88	0.7867	0.2133	1.38	0.9490	0.0510	1.88	0.9922	0.0078	2.38	0.9992	0.0008
0.39	0.4187	0.5813	0.89	0.7918	0.2082	1.39	0.9507	0.0493	1.89	0.9925	0.0075	2.39	0.9993	0.0007
0.4	0.4284	0.5716	0.9	0.7969	0.2031	1.4	0.9523	0.0477	1.9	0.9928	0.0072	2.4	0.9993	0.0007
0.41	0.4380	0.5620	0.91	0.8019	0.1981	1.41	0.9539	0.0461	1.91	0.9931	0.0069	2.41	0.9993	0.0007
0.42	0.4475	0.5525	0.92	0.8068	0.1932	1.42	0.9554	0.0446	1.92	0.9934	0.0066	2.42	0.9994	0.0006
0.43	0.4569	0.5431	0.93	0.8116	0.1884	1.43	0.9569	0.0431	1.93	0.9937	0.0063	2.43	0.9994	0.0006
0.44	0.4662	0.5338	0.94	0.8163	0.1837	1.44	0.9583	0.0417	1.94	0.9939	0.0061	2.44	0.9994	0.0006
0.45	0.4755	0.5245	0.95	0.8209	0.1791	1.45	0.9597	0.0403	1.95	0.9942	0.0058	2.45	0.9995	0.0005
0.46	0.4847	0.5153	0.96	0.8254	0.1746	1.46	0.9611	0.0389	1.96	0.9944	0.0056	2.46	0.9995	0.0005
0.47	0.4937	0.5063	0.97	0.8299	0.1701	1.47	0.9624	0.0376	1.97	0.9947	0.0053	2.47	0.9995	0.0005
0.48	0.5027	0.4973	0.98	0.8342	0.1658	1.48	0.9637	0.0363	1.98	0.9949	0.0051	2.48	0.9995	0.0005
0.49	0.5117	0.4883	0.99	0.8385	0.1615	1.49	0.9649	0.0351	1.99	0.9951	0.0049	2.49	0.9996	0.0004
0.5	0.5205	0.4795	1	0.8427	0.1573	1.5	0.9661	0.0339	2	0.9953	0.0047	2.5	0.9996	0.0004

Index

INDEX - G

INDEX - O

INDEX - P

INDEX - S